This is the first volume of a two volume set that provides a modern account of basic Banach algebra theory including all known results on general Banach *-algebras. This account emphasizes the role of *-algebra structure and explores the algebraic results which underlie the theory of Banach algebras and *-algebras. Both volumes contain previously unpublished results.

This first volume is an independent, self-contained reference on Banach algebra theory. Each topic is treated in the maximum interesting generality within the framework of some class of complex algebras rather than topological algebras.

In both volumes proofs are presented in complete detail at a level accessible to graduate students. In addition, the books contain a wealth of historical comments, background material, examples, particularly in noncommutative harmonic analysis, and an extensive bibliography. Together these books will become the standard reference for the general theory of *-algebras.

ENCYCLOPEDIA OF MATHEMATICS AND ITS APPLICATIONS

EDITED BY G.-C. ROTA

Volume 49

Banach Algebras and The General Theory of *-Algebras

ENCYCLOPEDIA OF MATHEMATICS AND ITS APPLICATIONS

Banach Algebras and
The General Theory of *-Algebras
Volume I: Algebras and Banach Algebras

THEODORE W. PALMER

Department of Mathematics
University of Oregon

CAMBRIDGE
UNIVERSITY PRESS

CAMBRIDGE UNIVERSITY PRESS
Cambridge, New York, Melbourne, Madrid, Cape Town, Singapore,
São Paulo, Delhi, Dubai, Tokyo

Cambridge University Press
The Edinburgh Building, Cambridge CB2 8RU, UK

Published in the United States of America by Cambridge University Press, New York

www.cambridge.org
Information on this title: www.cambridge.org/9780521124102

First published 1994
This digitally printed version 2009

A catalogue record for this publication is available from the British Library

Library of Congress Cataloguing in Publication data
Palmer, Theodore, W.
Banach algebras and the general theory of *-algebras / Theodore W. Palmer.
p.   cm.—(Encyclopedia of mathematics and its applications ; v. 49)
Includes bibliographical references.
Contents: v. 1. Algebras and Banach algebras.
ISBN 0-521-36637-2
1. Banach algebras.   I. Title.   II. Series.
QA326.P35   1993
512′.55—dc20
93-10615
CIP

ISBN 978-0-521-36637-3 Hardback
ISBN 978-0-521-12410-2 Paperback

PREFACE

This volume provides a gentle introduction to most of the main areas of research on general Banach algebras. It also serves the more specific purpose of providing the background for Volume II which will deal more intensively with *-algebras (*i.e.* algebras with fixed involutions, normally denoted by *). The focus is on the algebraic, and sometimes the geometric, underpinnings of the analytic theory. The subject is rich with aesthetic appeal, and many topics are pursued just as far as I found them attractive. References are given to more thorough expositions when they are available or to original sources. I have tried to make the book readable for beginning graduate students. Towards this end, I sometimes include a bit of undergraduate level material when it may not have been absorbed by such readers. There are also generous comments and historical remarks. They are all intended to serve a pedagogic purpose. I have tried to document the original source of most ideas, but sometimes I have failed. I apologize to those thus slighted. The knowledgeable reader will also find numerous previously unpublished results and technical improvements.

Readers should note the Symbol Index at the end of the volume. I have chosen notation carefully and used it consistently throughout the work. For instance, $\mathcal{A}$ always represents an algebra and $\mathcal{A}$ with a subscript always represents a subset of that algebra. Each entry in the bibliography displays the numbers of the sections in this volume to which it is related. A few of these entries, primarily those recording recent papers, are not actually referred to in the text but have been included to record the names of current research workers. When it is convenient to state several parallel cases in a single definition or result, the various options are enclosed in angle brackets $\langle \rangle$ and separated by an ordinary slash $/$. Internal references are given by the familiar device of tripartite numbers separated by periods where the $\langle$ first $/$ second $/$ third $\rangle$ number refers to the $\langle$ chapter $/$ section $/$ subsection or statement $\rangle$. Additional notation and conventions are introduced at the beginning of Chapter 1.

In 1970 I began writing a book on *-algebras. Repeatedly I discovered that existing references did not cover the background material in sufficient detail or from the viewpoint I needed. Thus this first volume began life as

v

a series of appendices. Chapter 2 on the spectrum and spectral algebras is a direct descendent of the first of these appendices which attained the status of a complete independent exposition. When the appendices became as long as the main text, I realized that they needed to come first and could easily be expanded to provide a relatively complete introduction to Banach algebra theory. I learned this theory from the ground-breaking book of Charles E. Rickart [1960] and those familiar with his book will see the strong influence his organization of the subject still has on my own.

By 1978 I had written a relatively complete manuscript. Unfortunately it was never quite finished for publication, and I devoted the decade of the 1980's mainly to administrative work. During that whole period, I tried to keep current with work on general Banach algebras, and I continually revised sections and incorporated really striking new results which obviously belonged. However, there was no time to complete the manuscript for publication. A daunting pile of typed pages was the result in those pre-computer days. Finally it became nearly impossible to trace down and change all the cross references when new material was added.

In December 1987 Robert S. Doran (Texas Christian University) asked me how my book was progressing and whether I had a publisher. Wholly involved with dean's work at the time, I replied that I did not see how I would ever finish it without a coauthor. Within a few days he expressed willingness to revise the book as a coauthor and I immediately accepted. Bob quickly arranged a contract with Cambridge University Press, and I began to withdraw from further administrative commitments. In December 1988 Bob sent me a preprint of Thomas J. Ransford's beautiful proof of Barry Johnson's uniqueness of norm theorem. Within a few hours I had used Ransford's method to give a new proof of the fundamental theorem of spectral semi-norms (Theorem 2.3.6). Since I had known for years that an easy early proof of this result was a key to a smooth exposition of many of my ideas, I decided then and there to leave the dean's office and work to complete this book. With Bob's help it seemed a task of two or three years at most. Unfortunately, during the 21 months it took me to free myself of all administrative commitments, Bob was drawn more and more into administration himself. In March 1991 he had to drop his role as coauthor. This book would have greatly benefitted had he been able to continue. Besides securing a contract for publication, Bob worked with his colleague David Addis to develop TEX macros for the book and arranged to have his wife Shirley Doran prepare a preliminary TEX version of my old manuscript. Without all their contributions, I could not have TEXed the whole book. Bob also made numerous suggestions on style. Many readers will thank him for convincing me to give up the use of the Fraktur alphabet. (In my heart, I still believe that 𝔄 is a typical Banach algebra.)

Many other people have helped with completion of this book. All the words and mistakes are my own, but most of the commas were contributed by my wife Laramie and by Kenneth A. Ross, both of whom have proof-read essentially the whole work. Richard M. Koch has repeatedly come to my rescue when some computer or T$_E$Xnicality defeated me. Laramie has helped with the book from the beginning; Ken and Dick were also essential. John Duncan (University of Arkansas), Robert B. Burckel (Kansas State University) and Michael J. Meyer (Georgia State University) read most of the manuscript, sometimes in earlier versions, and made valuable sugges-tions. Barry E. Johnson (University of Newcastle) and H. Garth Dales (Leeds University) helped on more limited portions. Numerous colleagues have provided preprints or valuable information. Beginning in 1970 several generations of graduate students at the University of Oregon have seen pre-liminary versions of the book. They have pointed out obscurities or even errors and in the later stages have contributed to the proofreading efforts. I cannot list them all but here are a few: Abdullah H. Al Moajil, Robert Bekes, Michael Boardman, Sean Bradley, Jon M. Clauss, David Collett, Dan Hendrick, Thomas W. Judson, Michael Leen, Chung Lin, Jorge M. López, Michael Ottinger, William L. Paschke, Paul L. Patterson III, John Phillips, James Rowell, Richard C. Vrem and Sheng L. Wu. To all those mentioned above by name or category, I extend heart-felt appreciation.

With deep filial respect, this volume is dedicated to:

Ernest Jesse Palmer
(April 8, 1875 to February 25, 1962)

and

Elizabeth McDougall Palmer
(March 14, 1902 to April 25, 1972).

CONTENTS

Tentative Contents of Volume II

1

Introduction to Normed Algebras; Examples

Introduction

This chapter begins by stating some basic conventions, definitions and notation that will be used throughout the work. Additional standard notations will be introduced from time to time, as needed. The reader should consult the index of notation for reference. Many of the ideas presented in the first section will be familiar to some readers. They are mentioned for the sake of review and to fix our notation. Also, of course, some standard concepts are defined in slightly different ways by different authors, and we wish to make clear our own conventions. The chapter concludes with a number of examples discussed in some depth. We urge readers to acquaint themselves with these since an abstract theory, such as that presented in this work, lacks substance without knowledge of examples.

The first section deals primarily with basic elementary results on normed, semi-normed, or topological linear spaces and algebras. Such topics as ideals, homomorphisms, quotient norms, etc. are discussed, and the role of semi-norms in locally convex topological linear spaces is quickly surveyed. The unitization of an algebra and an important convention about it are also introduced.

In order to enliven the section, we have introduced several interesting or slightly unusual topics which need relatively few prerequisites. Some basic facts about commutant subsets of algebras and maximal commutative subalgebras are presented. For any submultiplicative semi-norm σ on an algebra $\mathcal{A}$, we present the important properties of the non-negative real-valued function on $\mathcal{A}$ defined by

$$\lim_{n \to \infty} \sigma(a^n)^{1/n} \qquad \forall\, a \in \mathcal{A}.$$

Section 1.2 deals with the double centralizer algebra $\mathcal{D}(\mathcal{A})$ of an algebra $\mathcal{A}$. We regard $\mathcal{D}(\mathcal{A})$ as a more natural unitization of a non-unital algebra $\mathcal{A}$. It also allows the classification of extensions of well-behaved Banach algebras under mild hypotheses. This interesting theory is also presented.

Section 1.3 discusses a number of ways in which algebras and Banach algebras can be combined to make new algebras. It introduces direct sums, direct products, subdirect products, both projective and injective limits, ultraproducts and ultrapowers.

1

Section 1.4 is devoted to the Arens product. This interesting, fundamental and elementary construction, which provides a product on the double dual Banach space of any Banach algebra, is surveyed in some detail and explored more thoroughly in some of the examples which follow.

The remaining six sections present examples. Some of these will probably be familiar but most are discussed in an elementary fashion to show how the ideas arise naturally and to make them accessible even to the beginner. In many years of teaching this material, we have often noted students who become facile with the theory without knowing a reasonable stock of examples. Thus we urge that these sections be read in detail. Algebras of functions are dealt with very briefly since they belong more to the subject matter of Chapter 3. Matrix algebras, operator algebras and group algebras are presented in more detail.

A number of simple examples, which we might have included here are algebras with involutions. Since the second volume of this book will be devoted to a much deeper study of involutions, we omit most of these examples here. For this reason, most algebras of operators on Hilbert space are omitted or slighted even though they rank among the simplest examples, as will be seen when they are presented in Volume II.

1.1 Norms and Semi-norms on Algebras

Sets, Functions and Notation

If $\mathcal{D}$ and $\mathcal{S}$ are sets, we write $\mathcal{D}\backslash\mathcal{S}$ for the *difference set* $\{a \in \mathcal{D} : a \notin \mathcal{S}\}$. If f is any function with domain $\mathcal{D}$, and $\mathcal{S}$ is any set, $f|_{\mathcal{S}}$ denotes the *restriction of f* to the domain $\mathcal{D}\cap\mathcal{S}$. In neither case do we insist that $\mathcal{S}$ be a subset of $\mathcal{D}$. We sometimes use the notation $f^{\leftarrow}$ for the relation which is the *inverse* of a function f, particularly when $f^{\leftarrow}$ is not a function. (However if f is a function in an algebra $\mathcal{A}$ of linear functions under composition, we always use f^{-1} for the inverse when it is a function in $\mathcal{A}$. Conversely, if f is a function with values in a group in which multiplicative notation is used or in the invertible elements in some ring, and if f belongs to a group or ring of functions in which multiplication is defined pointwise, then f^{-1} will always represent the function defined by $f^{-1}(x) = f(x)^{-1}$ for each x in the domain of f.) If $\mathcal{F}$ is a set of functions, each with domain including $\mathcal{X}$, we write $\mathcal{F}(\mathcal{X})$ for the set $\{f(x) : f \in \mathcal{F}, x \in \mathcal{X}\}$.

In any topological space we denote the *boundary, closure* and *interior* of a subset $\mathcal{S}$ by $\partial\mathcal{S}$, $\mathcal{S}^{-}$, and $\mathcal{S}^{\circ}$, respectively. The *support* supp$f$ of a real- or complex-valued function f defined on a topological space Ω is the closure of the set where f is non-zero.

We will use $\mathbb{C}$, $\mathbb{D}$, $\mathbb{T}$, $\mathbb{R}$, $\mathbb{R}_{+}$, $\mathbb{R}_{+}^{\bullet}$, $\mathbb{Z}$, $\mathbb{N}$, $\mathbb{N}^0$ and $\emptyset$ to denote, respectively, the set of *complex numbers,* the *unit disc* $\{\lambda \in \mathbb{C} : |\lambda| \leq 1\}$, the 1-*torus*

$\{\lambda \in \mathbb{C} : |\lambda| = 1\}$, the set of *real numbers*, the set of *non-negative real numbers*, the set of *positive real numbers*, the set of all *integers*, the set $\{1, 2, 3, \ldots\}$ of *natural numbers* or *positive integers*, the set $\{0, 1, 2, \ldots\}$ of *non-negative integers* and the *empty set*. We endow each of these sets with all its usual structure so that, for instance, $\mathbb{R}$ is an ordered normed field. *Open* and *closed intervals* are denoted by $]\cdot, \cdot[$ and $[\cdot, \cdot]$, respectively. The *supremum* of a set in $\mathbb{R}$ is ∞ if the set is unbounded and $-\infty$ if the set is empty. Similar conventions hold for the *infimum*. The *complex conjugate* of a complex number λ will be denoted by λ^*. We frequently use the *Kronecker delta*

$$\delta_{ij} = \begin{cases} 1 & \text{if } i = j \\ 0 & \text{if } i \neq j. \end{cases}$$

The indices i and j may be any type of mathematical object.

It is often convenient to state several definitions or results in a parallel fashion. We do so by listing the various choices involved in order, enclosed in angular brackets and separated by slashes: $\langle \ldots / \ldots / \ldots \rangle$. References within this work are given by a three part number made up of the chapter number, section number and result or subsection number. References to other books and articles are given by mentioning the author's name in the text and then giving the year of publication enclosed in square brackets. Full details are located in the bibliography.

Algebras and Subalgebras

1.1.1 Definition An *algebra* $\mathcal{A}$ over a field $\mathbb{F}$ is a ring which is also a linear space over $\mathbb{F}$ under the same addition and satisfies

$$(\lambda a)b = \lambda(ab) = a(\lambda b) \qquad \forall \, \lambda \in \mathbb{F}; \, a, b \in \mathcal{A}.$$

N. B. Throughout this work all linear spaces or algebras will have the **complex** field as their field of scalars unless the contrary is explicitly stated. Occasionally $\mathbb{F}$ will denote either the real or the complex field: $\mathbb{R}$ or $\mathbb{C}$.

An algebra is said to be *unital* if it has a multiplicative identity (*i.e.*, an element 1 satisfying $1a = a1 = a$ for all $a \in \mathcal{A}$).

A *subalgebra* is a subring which is also a linear subspace (*i.e.* it is a linear subspace which contains the product of any of its elements). A subalgebra of a unital algebra which contains the multiplicative identity element of the larger algebra is called a *unital subalgebra*.

We will always state definitions and results for algebras even when the theory for arbitrary rings is no more complicated. When feasible, we will try to state definitions, propositions and arguments so that they also apply to algebras with the real numbers as scalar field but this is not always possible. Note that a subalgebra may be a unital algebra (because it contains its own identity element) without being a unital subalgebra (because the larger algebra is either nonunital or has a different identity element).

An element a in an algebra $\mathcal{A}$ is called an $\langle$ *idempotent* / *nilpotent* $\rangle$ if it satisfies $\langle$ $a^2 = a$ / $a^n = 0$ for some $n \in \mathbb{N}$ $\rangle$. An idempotent is said to be *proper* if it is not a multiplicative identity element for the algebra to which it belongs. A non-zero proper idempotent is said to be *nontrivial*. Finally, two idempotents e and f are *orthogonal* if they satisfy $ef = fe = 0$.

An element a in an algebra $\mathcal{A}$ is called a $\langle$ *left* / *right* $\rangle$ *divisor of zero* if there is some non-zero $b \in \mathcal{A}$ satisfying $\langle$ $ab = 0$ / $ba = 0$ $\rangle$. An element which is either a left or right divisor of zero is called a *divisor of zero* and an element which is both is called a *two-sided divisor of zero* and a *joint divisor of zero* if the same element b can be used on both sides.

Historical Remarks on Algebras

The term "algebra", was first applied, in the sense used in this work, by Benjamin Peirce [1870]. His interesting paper, which was published posthumously in 1881, was in part a philosophical discussion intended to establish the modern 20th century view that "Mathematics is the science which draws necessary conclusions." As an example, he defined "linear associative algebras", by axioms and derived a number of consequences. For another third of a century most mathematicians studied algebras as "systems of hypercomplex numbers". This term denoted a linear space with a given distinguished basis and a multiplication table for the basis elements giving each product as a specified linear combination of basis elements. Shortly after the turn of the century, under the influence of Leonard Eugene Dickson, the term "algebra", and the axiomatic definition and viewpoint were generally adopted by mathematicians. For instance, at Dickson's suggestion the term "algebra", (but not the axiomatic definition) is used in the text of Joseph H. Maclagan Wedderburn's classic paper [1907] establishing his decomposition theorems. Dickson used both the term "algebra", and the axiomatic definition in his influential book [1927]. See also Karen Hunger Parshall [1985].

Linear Spans, Convex Hulls, Products and Sums of Subsets

When $\mathcal{S}$ is a subset of a linear space $\mathcal{X}$ we write span($\mathcal{S}$) for the linear span of $\mathcal{S}$ and co($\mathcal{S}$) for the convex hull of $\mathcal{S}$. If $\mathcal{X}$ is the linear span of a linearly independent set, we call the set a *Hamel basis* for $\mathcal{X}$. A subset $\mathcal{S}$ of a linear space is said to be *balanced* if λx belongs to $\mathcal{S}$ for all complex λ satisfying $|\lambda| \leq 1$ and all $x \in \mathcal{S}$. The *balanced convex hull* of a subset $\mathcal{S}$ of $\mathcal{X}$, denoted by ba($\mathcal{S}$), is the set

$$\left\{ \sum_{j=1}^{n} \lambda_j x_j : n \in \mathbb{N}; \ \lambda_j \in \mathbb{C}, \ x_j \in \mathcal{S} \text{ for } j = 1, 2, \ldots, n; \ \sum_{j=1}^{n} |\lambda_j| \leq 1 \right\}.$$

Let $\mathcal{A}$ be an algebra. If $\mathcal{S}$ and $\mathcal{T}$ are linear subspaces of $\mathcal{A}$, we will denote the set span$\{ab : a \in \mathcal{S}, \ b \in \mathcal{T}\}$ by $\mathcal{S}\mathcal{T}$. Also if a is an element of $\mathcal{A}$ and $\mathcal{S}$ and

$\mathcal{T}$ are linear subspaces of $\mathcal{A}$, we denote the set span$\{bac : b \in \mathcal{S}, c \in \mathcal{T}\}$ by $\mathcal{S}a\mathcal{T}$. Similar variants are also used including expressions like $(1 - a)\mathcal{A} = \{b - ab : b \in \mathcal{A}\}$ and $(\lambda - a)\mathcal{A} = \{\lambda b - ab : b \in \mathcal{A}\}$ even in nonunital algebras. In general, the algebras we study will be noncommutative. Hence it is important to agree that the symbol $\Pi_{j=1}^{n}a_j$ means $a_1 a_2 \cdots a_n$. For any integer n strictly greater than one and any linear subspace $\mathcal{S}$, we denote the set span$\{\Pi_{j=1}^{n}a_j : a_j \in \mathcal{S}\}$ by $\mathcal{S}^n$. If $\mathcal{S}$ and $\mathcal{T}$ are any subsets of $\mathcal{A}$, we denote the set $\{a+b : a \in \mathcal{S}, b \in \mathcal{T}\}$ by $\mathcal{S}+\mathcal{T}$. If $\mathcal{S} = \{a\}$ is a singleton, we replace $\mathcal{S} + \mathcal{T}$ by $a + \mathcal{T}$. (These notations were introduced into the theory of algebras by Wedderburn [1907] following similar notation used in group theory by George Frobenius.)

The multiplication of linear subspaces just defined satisfies the associative law and the distributive law relative to addition. The addition of subsets satisfies the commutative and associative laws. We will use these properties without further comment. (The reader should note the contrast with the notation $\mathcal{F}(\mathcal{X}) = \{f(x) : f \in \mathcal{F}; x \in \mathcal{X}\}$, where $\mathcal{F}$ is a set of functions with domains including $\mathcal{X}$. In this case no linear span is taken even if $\mathcal{F}$ and $\mathcal{X}$ are both linear spaces.) The *kernel* of a linear map $\varphi \colon \mathcal{X} \to \mathcal{Y}$ is the linear subspace $\ker(\varphi) = \{x \in \mathcal{X} : \varphi(x) = 0\}$,

Ideals, Homomorphisms and Quotients of Algebras

We use the word *ideal* to mean a two-sided ideal. Thus a *left ideal* or *right ideal* is not a special kind of ideal. In an algebra, a one-sided ideal or an ideal is a linear subspace since it is a subalgebra. A *homomorphism* between algebras is a ring homomorphism which is also a linear map. Thus it is a linear map φ satisfying

$$\varphi(ab) = \varphi(a)\varphi(b) \qquad \forall \, a, b \in \mathcal{A}$$

if $\mathcal{A}$ is the domain of φ. Its *kernel* is $\{a \in \mathcal{A} : \varphi(a) = 0\}$. If $\mathcal{A}$ and $\mathcal{B}$ are algebras, the set of homomorphisms of $\mathcal{A}$ into $\mathcal{B}$ will be denoted by $\mathrm{Hom}(\mathcal{A}, \mathcal{B})$. As usual, a bijective homomorphism of algebras is called an *isomorphism* and an *automorphism* if it maps $\mathcal{A}$ onto itself. If $\mathcal{A}$ is a unital algebra, the simplest automorphisms are the *inner automorphisms* $a \mapsto b^{-1}ab$ where b is invertible and b^{-1} is its inverse. A linear map or homomorphism $\varphi \colon \mathcal{A} \to \mathcal{B}$ is said to be *injective* if its kernel is $\{0\}$, and *surjective* if each element of $\mathcal{B}$ is the image under φ of some element of $\mathcal{A}$.

A linear map $\varphi \colon \mathcal{A} \to \mathcal{B}$ which is anti-multiplicative (*i.e.*, $\varphi(ab) = \varphi(b)\varphi(a)$ for all $a, b \in \mathcal{A}$) is called an *anti-homomorphism*. "Isomorphic", "anti-isomorphism", "anti-isomorphic", etc. are defined as usual. If $\varphi \colon \mathcal{A} \to \mathcal{B}$ is an algebra homomorphism (or anti-homomorphism), then its kernel is an ideal of $\mathcal{A}$. Conversely, if $\mathcal{I}$ is an ideal of an algebra $\mathcal{A}$, then the set $\mathcal{A}/\mathcal{I} = \{a + \mathcal{I} : a \in \mathcal{A}\}$ of cosets has (in the usual way) the structure of an algebra such that the natural map $\varphi \colon \mathcal{A} \to \mathcal{A}/\mathcal{I}$ is a homomorphism with kernel $\mathcal{I}$ (*i.e.*, $(a+\mathcal{I})+(b+\mathcal{I}) = a+b+\mathcal{I}$, $\lambda(a+\mathcal{I}) = \lambda a + \mathcal{I}$, $(a+\mathcal{I})(b+\mathcal{I}) = (ab+\mathcal{I})$) and $\varphi(a) = a + \mathcal{I}$.

If $\mathcal{I}, \mathcal{K}$ and $\mathcal{L}$ are ideals in an algebra $\mathcal{A}$ and $\mathcal{L}$ is included in $\mathcal{I}$, we expect the reader to be familiar with the natural isomorphisms between $\mathcal{I}/(\mathcal{I} \cap \mathcal{K})$ and $(\mathcal{I} + \mathcal{K})/\mathcal{K}$: $a + \mathcal{I} \cap \mathcal{K} \longleftrightarrow a + \mathcal{K}$; and between $\mathcal{A}/\mathcal{I}$ and $(\mathcal{A}/\mathcal{L})/(\mathcal{I}/\mathcal{L})$: $a + \mathcal{I} \longleftrightarrow a + \mathcal{L} + (\mathcal{I}/\mathcal{L})$. They are both easily checked.

Historical Remark on Ideals and Quotients

We wish to comment on the significance of the word "ideal.", In attempting to prove Fermat's last theorem, both Augustin-Louis Cauchy and Ernst Eduard Kummer were led to consider, at least implicitly, algebraic number fields, *i.e.*, finite algebraic extensions of the rational field. At first they each made the mistake of implicitly assuming that the fundamental theorem of arithmetic (the uniqueness of factorization into primes) held. When Peter Gustav Lejeune Dirichlet pointed out the error, Kummer in *ca.* 1845 invented what he called ideal numbers in order to restore unique factorization. Richard Dedekind replaced Kummer's ideal numbers with the special subrings which we now use, and he quite naturally called them ideals as we still do. In the theory of algebras, ideals were considered for several decades under the name "invariant sub-complexes", *cf.* Theodor Molien [1893], Élie Cartan [1898], and Wedderburn [1907]. The term "ideal", was used in the context of algebras by Dickson in his influential book [1927].

The quotient algebra of an algebra modulo an ideal was introduced by Molien [1893]. In the case of certain special commutative algebras, the idea was accepted much earlier and can be traced back to Karl Friedrich Gauss' theory of congruences.

Hereditary Subalgebras and Normal Subalgebras

The two classes of subalgebras mentioned in the heading have some similarity to ideals in that they are defined by requiring that certain products of sets have special properties. A subalgebra $\mathcal{B}$ of an algebra $\mathcal{A}$ is said to be *hereditary* if it satisfies $\mathcal{B}\mathcal{A}\mathcal{B} \subseteq \mathcal{B}$. Obviously the intersection of a left ideal and right ideal has this property. The concept arose to generalize this situation to closed *-subalgebras of C*-algebras which were generated by their positive parts. The term seems to have been first used by Francois Combes in [1969]. This followed the recognition by Reese T. Prosser [1963] that these algebras have the form $\mathcal{L} \cap \mathcal{L}^*$ where $\mathcal{L}$ is a closed left ideal, and a series of papers by Gert K. Pedersen, beginning with [1966], which had the goal of understanding the non-closed case.

The concept of normal subalgebras was introduced by Marc A. Rieffel [1979a], [1979b]. It is motivated by the following consideration which relates to Section 1.9 below. Suppose G is a locally compact group and H is an open subgroup. Then $L^1(H)$ can be regarded as a subalgebra of $L^1(G)$ simply by extending functions on H to be zero on the rest of G. The problem then is to distinguish the subalgebras arising from normal subgroups from those arising from non-normal subgroups. Suppose H is normal. If $\mathcal{I}$ is an ideal of $L^1(G)$, then $\mathcal{K} = \mathcal{I} \cap L^1(H)$ would satisfy $L^1(G)\mathcal{K} = \mathcal{K}L^1(G)$, and this would only be true in general when H is a normal subgroup. Thus Rieffel defined a subalgebra of an algebra to be *normal* if the above invariance property (*i.e.*, for $\mathcal{B}$ a subalgebra of $\mathcal{A}$ and $\mathcal{I}$ an ideal of $\mathcal{A}$, $\mathcal{A}(\mathcal{I} \cap \mathcal{B}) = (\mathcal{I} \cap \mathcal{B})\mathcal{A})$ holds for a suitable class of ideals. In the cited paper he was able to use this concept to generalize certain aspects of the theory of induced representations from normal subgroups to the setting of normal subalgebras of algebras. We mention this concept here in part because it has not yet been exploited despite its obvious potential. With further refinements, possibly involving double centralizer algebras, it may play an important role.

The Reverse Algebra

If $\mathcal{A}$ is an algebra, the *reverse* of $\mathcal{A}$ is the algebra $\mathcal{A}^R$ with the same underlying linear space as $\mathcal{A}$ but with multiplication defined by setting ab in $\mathcal{A}^R$ equal to ba in $\mathcal{A}$. Thus the identity map of $\mathcal{A}$ onto $\mathcal{A}^R$ is an anti-isomorphism, and an algebra $\mathcal{B}$ is anti-isomorphic to $\mathcal{A}$ if and only if it is isomorphic to $\mathcal{A}^R$.

Semi-topological Linear Spaces and Algebras

Much of this work deals with the interplay between algebraic and topological or metric properties. We now introduce the first of these hybrid concepts. *N. B.* Throughout this work, a set $\mathcal{U}$ is called a *neighborhood* of a point x if x is in the interior of $\mathcal{U}$. Thus a neighborhood need not be open.

1.1.2 Definition   A *semi-topological linear space* $\mathcal{X}$ is a linear space together with a topology such that the maps

$$\mathbb{C} \times \mathcal{X} \to \mathcal{X} \text{ defined by } (\lambda, x) \mapsto \lambda x \tag{1}$$

$$\mathcal{X} \times \mathcal{X} \to \mathcal{X} \text{ defined by } (x, y) \mapsto x + y \tag{2}$$

are continuous. A *semi-topological algebra* $\mathcal{A}$ is an algebra together with a topology such that the linear space of $\mathcal{A}$ is a semi-topological linear space and the maps of $\mathcal{A}$ into $\mathcal{A}$ defined by

$$a \mapsto ab \text{ and } a \mapsto ba \tag{3}$$

are both continuous for each fixed $b \in \mathcal{A}$. A *topological linear space* or a *topological algebra* is, respectively, a semi-topological linear space or a semi-topological algebra in which the topology is Hausdorff.

Some authors require that the multiplication in a topological algebra $\mathcal{A}$ be jointly continuous from $\mathcal{A} \times \mathcal{A}$ to $\mathcal{A}$. In this terminology, a semi-topological algebra is often a Hausdorff topological algebra with singly rather than jointly continuous multiplication.

Separation Properties

Recall that for a semi-topological linear space the T_0, T_1, Hausdorff (T_2) and regular Hausdorff (T_3) separation axioms are equivalent. To see this, consider a neighborhood $\mathcal{U}$ of zero in a topological linear space $\mathcal{X}$. Subtraction is continuous since addition and multiplication by -1 are. Thus there is a neighborhood $\mathcal{V}$ of zero satisfying $\mathcal{V} - \mathcal{V} \subseteq \mathcal{U}$, and hence $\mathcal{V}^- \subseteq \mathcal{U}$. Since translation is a homeomorphism, this shows that for any neighborhood $\mathcal{U}$ of a point in $\mathcal{X}$ there is a neighborhood $\mathcal{V}$ satisfying $\mathcal{V}^- \subseteq \mathcal{U}$. This result shows that T_0 implies T_3. In fact a topological linear space is completely regular since it is uniformizable (*cf. e.g.*, Kelley [1955]). We will say more on this subject when we discuss topological groups. In the next chapter, Section 2.9 will also give more information on topological algebras.

Quotients of Semi-topological Algebras

If $\mathcal{I}$ is an ideal in a semi-topological algebra $\mathcal{A}$, then $\mathcal{A}/\mathcal{I}$ can be furnished with the quotient topology. We remind the reader that the *quotient topology* defined by the map $\varphi \colon \mathcal{A} \to \mathcal{A}/\mathcal{I}$ is the topology on $\mathcal{A}/\mathcal{I}$ in which a set $\mathcal{U} \subseteq \mathcal{A}/\mathcal{I}$ is open if and only if $\varphi^{\leftarrow}(\mathcal{U})$ is open in $\mathcal{A}$.

1.1.3 Proposition *Let $\mathcal{A}$ be a semi-topological algebra and let $\mathcal{I}$ be an ideal in $\mathcal{A}$.*

(a) $\mathcal{A}/\mathcal{I}$ is a semi-topological algebra under the quotient topology. It is a topological algebra if and only if $\mathcal{I}$ is closed.

(b) If $\mathcal{I}$ is the kernel of a continuous homomorphism into a topological algebra, then $\mathcal{I}$ is closed.

Proof (a): In order to check that $\mathcal{A}/\mathcal{I}$ is a semi-topological algebra we must check the continuity of the maps $(a + \mathcal{I}, b + \mathcal{I}) \mapsto a + b + \mathcal{I}$, $(\lambda, a + \mathcal{I}) \mapsto \lambda a + \mathcal{I}$, $a + \mathcal{I} \mapsto ab + \mathcal{I}$ and $a + \mathcal{I} \mapsto ba + \mathcal{I}$. We show only the first since the others are similar. Suppose $\mathcal{U}$ is an open neighborhood of $a + b + \mathcal{I}$ in $\mathcal{A}/\mathcal{I}$. By definition of the quotient topology, this means that the preimage $\mathcal{U}'$ of $\mathcal{U}$ under the natural homomorphism $\varphi \colon \mathcal{A} \to \mathcal{A}/\mathcal{I}$ is open. Thus there are open neighborhoods $\mathcal{V}$ and $\mathcal{W}$ of a and b, respectively, satisfying $\mathcal{V} + \mathcal{W} \subseteq \mathcal{U}'$. Since $\mathcal{U}'$ is a union of cosets of $\mathcal{I}$, we have $(\mathcal{V} + \mathcal{I}) + (\mathcal{W} + \mathcal{I}) \subseteq \mathcal{U}'$. However, $\mathcal{V} + \mathcal{I}$ (and similarly $\mathcal{W} + \mathcal{I}$) is open since it is the union of the open sets $\mathcal{V} + a$. Hence $\varphi(\mathcal{V})$ and $\varphi(\mathcal{W})$ are open neighborhoods of $a + \mathcal{I}$ and $b + \mathcal{I}$, respectively, and they satisfy $\varphi(\mathcal{V}) + \varphi(\mathcal{W}) \subseteq \mathcal{U}$.

If $\mathcal{I}$ is closed, $\mathcal{A}/\mathcal{I}$ is T_1 and hence a topological algebra. Conversely if $\mathcal{A}/\mathcal{I}$ is T_1, then $\mathcal{I}$ is closed.

(b): If $\varphi \colon \mathcal{A} \to \mathcal{B}$ is continuous and $\mathcal{B}$ is a topological algebra, then the kernel $\mathcal{I} = \varphi^{\leftarrow}(\{0\})$ is closed by the definition of continuity. □

If $\mathcal{A}$ and $\mathcal{B}$ are semi-topological algebras, denote the set of continuous algebra homomorphisms $\varphi \colon \mathcal{A} \to \mathcal{B}$ by $\mathrm{CHom}(\mathcal{A}, \mathcal{B})$. Any $\varphi \in \mathrm{CHom}(\mathcal{A}, \mathcal{B})$ can be factored

$$\mathcal{A} \to \mathcal{A}/\mathcal{I} \to \varphi(\mathcal{A}) \to \mathcal{B} \tag{4}$$

where $\mathcal{I}$ is the kernel of φ. If $\mathcal{B}$ is a topological algebra, then $\mathcal{I}$ is closed so $\mathcal{A}/\mathcal{I}$ and $\varphi(\mathcal{A})$ are topological algebras.

It is easy to see that the closure of a subalgebra, left ideal, right ideal or ideal in a semi-topological algebra is a subset of the same type. A subset S of an algebra $\mathcal{A}$ is called *commutative* if any two of its elements commute (*i.e.*, if a, $b \in S$ implies $ab = ba$). It is easy to see that the closure of a commutative subset in a topological algebra is commutative.

Commutants

John von Neumann [1929] introduced both the notation and an appreciation of the importance of commutants. Commutants are sometimes called centralizers, but this word also has other meanings.

1.1.4 Definition If S is a subset of an algebra $\mathcal{A}$, then the *commutant* of S is the subset S' defined by

$$S' = \{a \in \mathcal{A} : ab = ba \quad \forall\, b \in S\}.$$

We denote the double commutant $(S')'$ by S'' and similarly for higher order commutants. The set $\mathcal{A}'$ is called the *center* (German *Zentrum*) of $\mathcal{A}$ and is denoted by $\mathcal{A}_Z$.

1.1.5 Proposition *The commutant S' of a subset S of an algebra $\mathcal{A}$ satisfies:*
(a) *S' is a subalgebra of $\mathcal{A}$.*
(b) *$S \subseteq S''$.*
(c) *$S \subseteq S'$ if and only if S is commutative.*
(d) *$S_1 \subseteq S_2$ implies $S_2' \subseteq S_1'$.*
(e) *$S' = S'''$.*
(f) *A subset C of $\mathcal{A}$ is a maximal (under inclusion order) commutative subset if and only if it satisfies $C = C'$. Hence maximal commutative subsets are subalgebras.*
(g) *Every commutative subset C is included in some maximal commutative subalgebra.*

Proof Most of these results are immediate consequences of the definition. Result (e) comes from applying (d) to (b), and (f) can be seen by noting that $\mathcal{C} \cup \{c\}$ is commutative if $\mathcal{C}$ is and c belongs to $\mathcal{C}'$. Obviously (g) depends on an application of Zorn's lemma.   □

We now note the topological consequences of these results. A variant of result (b) was first systematically exploited by Paul Civin and Bertram Yood [1959].

1.1.6 Proposition *Let $\mathcal{A}$ be a topological algebra. For every $\mathcal{S} \subseteq \mathcal{A}$, $\mathcal{S}'$ is closed. Hence the center of an algebra and every maximal commutative subalgebra are closed.*

Proof For each $b \in \mathcal{S}$, the set $\{a \in \mathcal{A} : ab - ba = 0\}$ is closed. The intersection of all these sets for $b \in \mathcal{S}$ is just $\mathcal{S}'$. The center is $\mathcal{A}'$ and (f) of the last proposition shows $\mathcal{C} = \mathcal{C}'$.   □

For another interesting property of maximal commutative subalgebras refer to Proposition 2.5.3.

Semi-norms

Nearly all of the topological or semi-topological algebras and linear spaces in this work will have topologies defined by a semi-norm or a family of semi-norms. We introduce this simple but important concept now.

1.1.7 Definition A *semi-norm* σ on a linear space $\mathcal{X}$ is a map $\sigma : \mathcal{X} \to \mathbb{R}_+$ satisfying:

$$\text{(a)} \qquad \sigma(x + y) \;\leq\; \sigma(x) + \sigma(y) \quad \text{(subadditivity)} \tag{5}$$
$$\text{(b)} \qquad \sigma(\lambda x) \;=\; |\lambda|\sigma(x) \quad \text{(absolute homogeneity)} \tag{6}$$

for all $x, y \in \mathcal{X}$ and $\lambda \in \mathbb{C}$. A semi-norm σ on an algebra $\mathcal{A}$ is called an *algebra semi-norm* if it also satisfies:

$$\text{(c)} \qquad \sigma(ab) \;\leq\; \sigma(a)\sigma(b) \quad \text{(submultiplicativity)} \tag{7}$$

for all $a, b \in \mathcal{A}$. An algebra semi-norm σ on an algebra $\mathcal{A}$ is called *nontrivial* unless $\mathcal{A}$ has a multiplicative identity at which σ is zero. For any semi-norm σ on a linear space $\mathcal{X}$, define $\mathcal{X}_\sigma$ by

$$\mathcal{X}_\sigma = \{x \in \mathcal{X} : \sigma(x) = 0\}. \tag{8}$$

A semi-norm σ on a linear space $\mathcal{X}$ is called a *norm* if $\mathcal{X}_\sigma = \{0\}$, (*i.e.*, when $\sigma(x) = 0$ implies $x = 0$). A linear space together with a semi-norm or norm is called a *semi-normed* or *normed linear space*. An algebra together with an algebra semi-norm or algebra norm is called a *semi-normed* or *normed algebra*. A normed linear space or normed algebra which is complete in its norm is called a *Banach space* or *Banach algebra*, respectively. A normed linear space is said to be *separable* if it has a countable dense subset. Finally, a map φ between semi-normed linear spaces $(\mathcal{X},\ \sigma)$ and $(\mathcal{Y},\ \tau)$ is said to be $\langle$ *isometric / contractive* $\rangle$ if it satisfies

$$\langle\ \tau(\varphi(x)) = \sigma(x)\ /\ \tau(\varphi(x)) \le \sigma(x)\ \rangle \qquad \forall\ x \in \mathcal{X}.$$

An isometric linear map is called an *isometry*.

The above definition of a nontrivial algebra semi-norm is certainly not what one might expect at first, since it calls the identically zero semi-norm on a non-unital algebra nontrivial. However, we will see in several places (*e.g.*, Theorem 2.2.1) that this definition is very convenient. H. Garth Dales [1981b] gives examples of normable and non-normable algebras.

Historical Remarks on Banach Algebras

Before discussing several aspects of this definition which need further comment, we discuss the early history of Banach algebras. Both Arne Beurling (*cf. e.g.*, [1938]) and Norbert Wiener [1932] raised questions and used some techniques which we would now place in the theory of Banach algebras. Furthermore, John von Neumann alone [1929], [1936a], [1940], [1943] and with Francis J. Murray [1935], [1937], [1943] began the study of von Neumann algebras (they called them rings of operators) which are among the most important examples of Banach algebras and are now often treated as such.

Curiously, those first attempting to abstract von Neumann algebras did not recognize the importance of the norm. Stourton William Peile Steen began his abstract study of algebras of operators with the commutative case [1936], [1937] in which he correctly noted the importance of order properties for the set of hermitian elements. As he began to generalize to not necessarily commutative algebras [1938] and then explicitly to von Neumann algebras [1939], he continued to use the order properties which were no longer so well suited. Finally, in the last paper of the series [1940], he introduced the norm (the referee of the paper supplied a reference to Nagumo [1936], see below) and actually defined what were later called H^*-algebras by their re-discoverer and developer Warren Ambrose [1945], who apparently was not aware of Steen's work.

Banach algebras, under the name linear metric rings, seem to have been first defined explicitly by Mitio Nagumo [1936]. He remarks in the introduction to his paper that he intends to treat the analytic rather than algebraic theory (for instance "the ideal theory") of his rings. His main result, which is still of considerable interest, will be presented as part of Theorem 2.1.12 below. Nagumo's terminology was immediately adopted by Kôsaku Yosida who wrote two papers [1936a], [1936b] related to Nagumo's results. These three papers were widely noted. They are cited by Stanisław Mazur in his note [1938] stating the Gelfand–Mazur theorem (Corollary 2.2.3 below). This paper in turn led to the two interesting papers of Meier Eidelheit

[1940a], [1940b] which explicitly begin to develop a general theory of linear metric rings. Eidelheit's paper [1940a] contains the first example of a result of the type treated in Chapter 6 below (*cf.* 1.7.15 below).

Aristotle D. Michal and Robert S. Martin [1934] studied the expansion of analytic functions in an awkwardly, but correctly, defined Banach algebra with a trace function as part of its basic structure, but this promising start does not seem to have borne fruit. Pessach Hebroni's concept of normed rings (with no linear structure) [1938] provides another adumbration of Banach algebra theory. However, it was not until Israel Moiseevič Gelfand [1939a], [1939b], [1939c] and his collaborators (Vitalli Arsenievič Ditkin [1939], Georgi Evgenyevič Šilov [1939], Gelfand and Andrei Nikolaevič Kolmogoroff [1939]) took up the theory that it became an important part of the mathematical literature. It has often been remarked that Gelfand's elementary proof [1941d] of Wiener's theorem [1932] (that the Fourier series of the inverse of a function with absolutely convergent Fourier series converges absolutely) showed the mathematical community the power of Banach algebra theory.

Gelfand's papers of 1939 were merely announcements without details. By the time his complete paper [1941a] appeared, World War II had disrupted communications and mathematical research so that further developments did not come so fast as they might have otherwise. Nevertheless, by 1941 the following additional papers dealing with Banach algebras had been published by Russian authors: Gelfand [1941b], [1941c]; Gelfand and Šilov [1941]; Iakov Isaevič Khourguine and N. Tschetinine [1940]; Mark Grigorevič Kreĭn [1940a], [1940b], [1941a], [1941b], [1941c]; Šilov [1940a], [1940b]; Vitold L. Šmulian [1940]. Furthermore, in 1941 John Williams Calkin published his important study [1941] of ideals in $\mathcal{B}(\mathcal{H})$, where $\mathcal{H}$ is a Hilbert space, and Irving E. Segal [1941] published an abstract of his Yale thesis [1940] on group algebras. Unfortunately the details of Segal's work were not published until [1947a], [1947b] because of the war, and then in a considerably different form. By 1942, Banach algebra theory had developed beyond infancy. See also Antonie F. Monna [1973].

Just as the Soviet Union was the site of the first flowering of research on Banach algebras, it was also the place where the first monographs/textbooks appeared. In 1940 Gelfand, Raĭkov and Šilov wrote an expanded account of a number of their early papers which was first published in truncated form in [1946]. It was translated into English in 1957. This was greatly expanded in [1960] and the enlarged version was translated into both German and English in 1964. This work is primarily addressed to commutative Banach algebras, but contains a substantial treatment of Banach *-algebras also. Richard F. Arens had noted a nontrivial error in MR 22 #5907 which went uncorrected. See Arens' review of the German edition for details, Eva Kallin [1963] for a counterexample, and Arens [1961] for a correction.

David Abramovič Raĭkov [1945] treats commutative harmonic analysis from the viewpoint of Banach algebras. This book was translated into German in 1954. In [1947a] Šilov gave a 118 page treatment of completely regular commutative Banach algebras. Formal power series are also studied in some detail (cf. 3.4.18 below). A 19-page English summary included complete statements of most results.

Mark Aronovič Naĭmark [1948] is a comprehensive treatment of the theory of *-algebras as it was then known. In the same year Einar Hille included a substantial treatment of Banach algebra theory in his book on semigroups [1948]. Chapters V and XXII provided the best introductory treatment in English up to that time. The revision, published jointly with Ralph S. Phillips [1957], expanded this treatment in Chapters IV and IX.

Lynn H. Loomis [1953] presented the material from a course initiated at Harvard under George W. Mackey and continued by Loomis. It was sharply focused on commutative harmonic analysis from a Banach algebra viewpoint and might be considered a successor to both André Weil [1940] and Raĭkov [1945]. A small slip about the uniqueness of the extensions of left ideals in a nonunital algebra to the unitization was repeated in Naĭmark [1956]. See MR 22-#1825 for a correction. Gelfand and Naĭmark [1948] covers the theory of *-algebras.

The first full length textbook covering all of Banach algebra theory was Naĭmark's important monograph [1956]. Starting with a detailed treatment of basic functional analysis, the book went up to the frontiers of research in the decomposition theory of von Neumann algebras. It was translated into both German and English in 1959. Unfortunately the final chapter was based on some work by Tomita which claimed to remove separability restrictions from the decomposition theory. It was soon discovered that this claim was overly optimistic. (See MR 29 #5120 for an actual counterexample.) Thus a second English edition was issued in 1964 (MR 34 #4928) in which Naĭmark returned to von Neumann's original treatment of decomposition theory. This version was republished in 1970 without essential change (MR 50 #8075). A somewhat more thoroughly updated version appeared in Russian in 1968 (MR 50 #8076). Finally, this version was translated into English in 1972 with further improvements.

This record shows that there were substantial book length treatments of Banach algebra theory before Charles E. Rickart [1960] appeared. Nevertheless we consider that Rickart's book defined the classic scope of the general theory of Banach algebras. His treatment was more algebraic and thus somewhat less oriented towards analysis than previous works. In particular it placed the theory much more solidly into the context of other algebraic investigations. This has proved fruitful. After 1960, most books on functional analysis devoted substantial space to Banach algebra theory, and many more specialized books appeared. We have not discussed early books on von Neumann algebras nor the more recent flowering of books on C*-algebras, but we should mention that Marshall Harvey Stone's [1932] book prepared American readers for highly algebraic excursions into analysis.

The Topology Defined by a Semi-norm

A semi-normed linear space $(\mathcal{X}, \sigma)$ is a semi-topological linear space with basis $\{N^\sigma_\varepsilon(x) : \varepsilon \in \mathbb{R}^*_+, x \in \mathcal{X}\}$ for its topology, where

$$N^\sigma_\varepsilon(x) = \{y \in \mathcal{X} : \sigma(x - y) < \varepsilon\}. \tag{9}$$

This topology defined by σ is called the *σ-topology*. Topological statements about a semi-normed space will always refer to this topology unless some other topology is explicitly mentioned. Two semi-norms on the same space are called *equivalent* if they define the same topology. A semi-norm σ_1 is said to $\langle$ *majorize* / *dominate* $\rangle$ a semi-norm σ_2 on a linear space $\mathcal{X}$ if $\langle$ $\sigma_2(x) \le \sigma_1(x)$ holds for all $x \in \mathcal{X}$ / there is a constant k satisfying

$$\sigma_2(x) \le k\sigma_1(x) \qquad \forall\, x \in \mathcal{X} \,\rangle. \tag{10}$$

Two semi-norms are $\langle$ equal / equivalent $\rangle$ if and only if each $\langle$ majorizes / dominates $\rangle$ the other.

Whenever $(\mathcal{X}, \sigma)$ is a semi-normed linear space, we will denote the *closed unit ball* by

$$\mathcal{X}_1 = \{x \in \mathcal{X} : \sigma(x) \le 1\}. \tag{11}$$

The *open unit ball* is then $\mathcal{X}_1^\circ$ and the *unit sphere* is $\partial\mathcal{X}_1$. For $t > 0$ we write $t\mathcal{X}_1$ for the closed ball of radius t. The set $\{x \in \mathcal{X} : \sigma(x) = 0\}$, previously denoted by $\mathcal{X}_\sigma$, is the closure of $\{0\}$ in the σ-topology. Clearly $(\mathcal{X}, \sigma)$ is a topological linear space exactly when σ is a norm. (Norms will frequently be denoted by $\|\cdot\|$.)

If $(\mathcal{A}, \sigma)$ is a semi-normed algebra, then it is a semi-topological algebra in which multiplication is jointly continuous in the σ-topology. It is a topological algebra if and only if σ is a norm. Clearly $\mathcal{A}_\sigma$ is an ideal.

Algebra Semi-norms

1.1.8 Proposition *Let $(\mathcal{A}, \sigma)$ be a semi-normed algebra and let $\mathcal{I}$ be an ideal. Then the quotient topology on $\mathcal{A}/\mathcal{I}$ is given by a semi-norm σ' defined by*

$$\sigma'(a + \mathcal{I}) = \inf\{\sigma(b) : b \in a + \mathcal{I}\} \qquad \forall\, a \in \mathcal{A} \tag{12}$$

and σ' is an algebra semi-norm. Furthermore, σ' is a norm if and only if $\mathcal{I}$ is closed.

Proof Check it.   □

We will call σ' the *quotient semi-norm*. Directly, or from Proposition 1.1.3, it is obvious that σ' is a norm if and only if $\mathcal{I}$ is σ-closed. Quotient norms for Banach algebras modulo closed ideals were used by both Calkin [1941] and Gelfand [1941a].

We assume the reader is familiar with the completion of a normed linear space or normed algebra. We normally avoid completions of semi-normed spaces. When it is necessary to use the *completion* of $(\mathcal{A}, \sigma)$, it is just the completion of $\mathcal{A}/\mathcal{A}_\sigma$ in the norm σ' introduced above.

The insistence that an algebra norm be submultiplicative is not really as restrictive as it may seem. If multiplication in some algebra $\mathcal{A}$ is jointly continuous with respect to some linear space semi-norm σ, then the absolute homogeneity of σ shows that there is an equivalent algebra semi-norm on $\mathcal{A}$. We incorporate this observation in the following more interesting result, which is due to Gelfand [1941a]. See also Ma. José Mijangos [1982], Ali Ülger [1990b] and Subhash J. Bhatt and D. J. Karia [1992].

1.1.9 Proposition *Let $\mathcal{A}$ be an algebra and let $(\mathcal{A}, \sigma)$ be a Banach space which is a topological algebra in the topology defined by σ. Then there is an algebra norm $\| \cdot \|$ on $\mathcal{A}$ equivalent to σ such that $(\mathcal{A}, \| \cdot \|)$ is a Banach algebra. Moreover if S is some bounded multiplicative semigroup in $\mathcal{A}$ (i.e., a bounded subset closed under multiplication), then the algebra norm can be chosen so that each element of S has norm at most 1. For this last result, completeness is not required if the norm is already assumed to be submultiplicative.*

Proof The uniform boundedness theorem shows that multiplication is actually jointly continuous, *i.e.*, continuous as a map from $\mathcal{A} \times \mathcal{A} \to \mathcal{A}$. Hence, using the absolute homogeneity of σ, there is a constant M satisfying $\sigma(ab) \leq M\sigma(a)\sigma(b)$ for all $a, b \in \mathcal{A}$. We may assume $M \geq 1$. Define a norm by

$$\|a\|' = \sup\{\sigma(\lambda a + ab) : |\lambda| + \sigma(b) \leq 1;\ \lambda \in \mathbb{C};\ b \in \mathcal{A}\} \qquad \forall\, a \in \mathcal{A} \quad (13)$$

and note $\|a\|' \leq M\sigma(a)$ since $\sigma(\lambda a + ab) \leq |\lambda|\sigma(a) + M\sigma(a)\sigma(b)$. By choosing $\lambda + b = 1 + 0$, we see $\|a\|' \geq \sigma(a)$. Hence $\| \cdot \|'$ is at least an equivalent linear space norm. However, any $a, c \in \mathcal{A}$ satisfy

$$\begin{aligned}
\|ac\|' &\leq \sup\{\sigma(\lambda ac + acb) : \sigma(\lambda c + cb) \leq \|c\|';\ \lambda \in \mathbb{C};\ b \in \mathcal{A}\} \\
&\leq \|c\|' \sup\{\sigma(ad) : \sigma(d) \leq 1;\ d \in \mathcal{A}\} \leq \|a\|'\,\|c\|'
\end{aligned}$$

so that $\| \cdot \|'$ is an equivalent algebra norm.

Now suppose S is a bounded multiplicative semigroup in the original norm (or topology). It is still bounded in the equivalent algebra norm, so we will assume that $\| \cdot \|'$ is submultiplicative and S is bounded by N with $N \geq 1$. Define a new linear space norm by

$$\||a\|| = \sup\{\|ca\|',\ \|a\|' : c \in S\} \qquad \forall\, a \in \mathcal{A}.$$

It is easy to check that this is a linear space norm satisfying

$$\|a\|' \le \||a\|| \le N\|a\|' \quad \text{and} \quad \||ca\|| \le \||a\|| \qquad \forall\, a \in \mathcal{A};\ c \in \mathcal{S}.$$

Now define the desired algebra norm by

$$\|a\| = \sup\{\||\lambda a + ab\|| \,:\, |\lambda| + \||b\|| \le 1;\ \lambda \in \mathbb{C};\ b \in \mathcal{A}\} \qquad \forall\, a \in \mathcal{A}.$$

By the same argument as before, we check that this is an equivalent algebra norm in which each element of $\mathcal{S}$ has norm at most 1. □

The Limit σ^∞ for an Algebra Semi-norm σ

The following interesting and elementary properties of a submultiplicative semi-norm are extremely useful. They were discovered by Gelfand [1941a]. For related results see Theorems 2.2.2 and 2.2.5 below.

1.1.10 Theorem *Let σ be an algebra semi-norm on an algebra $\mathcal{A}$. For each $a \in \mathcal{A}$, define $\sigma^\infty(a)$ by*

$$\sigma^\infty(a) = \inf\{\sigma(a^n)^{1/n} : n \in \mathbb{N}\}. \tag{14}$$

For all a, $b \in \mathcal{A}$, we have:
(a) $\sigma^\infty(a) = \lim_{n\to\infty} \sigma(a^n)^{1/n}$.
(b) $0 \le \sigma^\infty(a) \le \sigma(a)$.
(c) $\sigma^\infty(\lambda a) = |\lambda|\sigma^\infty(a) \qquad \forall\, \lambda \in \mathbb{C}$.
(d) $\sigma^\infty(ab) = \sigma^\infty(ba)$.
(e) $\sigma^\infty(a^n) = \sigma^\infty(a)^n \qquad \forall\, n \in \mathbb{N}$.
(f) *If $\mathcal{A}$ has a multiplicative identity 1, then $\sigma^\infty(1) = 1$ holds unless σ is trivial.*
(g) *If a and b commute, they satisfy:*

$$\sigma^\infty(a + b) \le \sigma^\infty(a) + \sigma^\infty(b) \qquad \sigma^\infty(ab) \le \sigma^\infty(a)\sigma^\infty(b).$$

Hence σ^∞ is an algebra semi-norm on any commutative subalgebra of $\mathcal{A}$.
(h) *Any sequence $\{a_n\}_{n\in\mathbb{N}} \subseteq \mathcal{A}$ which converges to a satisfies:*

$$\limsup_{n\to\infty} \sigma^\infty(a_n) \le \sigma^\infty(a).$$

(i) *Finally, σ satisfies*

$$\sigma^\infty(a) = \inf\{\|a\| : \|\cdot\| \text{ is an algebra norm on } \mathcal{A} \text{ equivalent to } \sigma\},$$

and if $\mathcal{A}$ is unital we may restrict the norms $\|\cdot\|$ in the above infimum to satisfy $\|1\| = 1$.

Proof Note that (b), (c), and (f) are immediate from equation (14). Thus in proving (a), we may assume $\sigma^\infty(a) \le \sigma(a) < 1$. Choose $\varepsilon > 0$ satisfying

$\varepsilon < 1 - \sigma^\infty(a)$. Find $m \in \mathbb{N}$ satisfying $\sigma(a^m)^{1/m} < \sigma^\infty(a) + \varepsilon/2$. Find $r \in \mathbb{N}$ satisfying $\sigma^\infty(a) + (\varepsilon/2) < (\sigma^\infty(a) + \varepsilon)^{(r+1)/r}$. Now for any $n \geq mr$, we can write $n = mp + q$ with $p \geq r$, $0 \leq q < m$. This implies

$$
\begin{aligned}
\sigma^\infty(a) &\leq \sigma(a^n)^{1/n} \leq \sigma(a^{mp})^{1/n}\sigma(a^q)^{1/n} \\
&\leq (\sigma(a^m)^{1/m})^{mp/n}\sigma(a)^{q/n} < (\sigma^\infty(a) + (\varepsilon/2))^{mp/n} \\
&< (\sigma^\infty(a) + \varepsilon)^{mp(r+1)/nr} < \sigma^\infty(a) + \varepsilon,
\end{aligned}
$$

since $mp(r+1) > mpr + qr = nr$. It follows that $\sigma^\infty(a) = \lim \sigma(a^n)^{1/n}$.

From (a) it is easy to prove (d) (consider $(ab)^n = a(ba)^{n-1}b$), (e) and the second inequality of (g). We now prove the first inequality of (g).

Let a, $b \in \mathcal{A}$ commute and choose $s > \sigma^\infty(a)$ and $t > \sigma^\infty(b)$. Write a_0 and b_0 for $s^{-1}a$ and $t^{-1}b$. Since $\sigma(a_0{}^n)^{1/n}$ and $\sigma(b_0{}^n)^{1/n}$ are both strictly less than 1 for sufficiently large n, there is a finite number M satisfying $\sup\{\sigma(a_0^n), \sigma(b_0^n) : n \in \mathbb{N}\} \leq M$. For any $n \in \mathbb{N}$, we get

$$
\begin{aligned}
\sigma((a+b)^n) &= \sigma\left(\sum_{k=0}^n \binom{n}{k} a^{n-k}b^k\right) \\
&\leq \sum_{k=0}^n \binom{n}{k} s^{n-k}t^k \, \sigma(a_0{}^{n-k})\sigma(b_0{}^k) \\
&\leq \sum_{k=0}^n \binom{n}{k} s^{n-k}t^k M^2 = (s+t)^n M^2.
\end{aligned}
$$

Thus taking the limit as n increases, we have

$$
\sigma^\infty(a+b) = \lim \sigma((a+b)^n)^{1/n} \leq s + t.
$$

This completes the proof of (g), since $s > \sigma^\infty(a)$ and $t > \sigma^\infty(b)$ were arbitrary.

(h): As in the proof of (a), we may assume $\sigma(a) < 1$ and $\sigma(a_n) < 1$ for all n. Let $\varepsilon > 0$ be given. The definition shows that it is enough to find n_0 and m in $\mathbb{N}$ so that

$$
\sigma(a_n{}^m)^{1/m} < \sigma^\infty(a) + \varepsilon/2 \qquad \forall \, n \geq n_0. \tag{15}
$$

Choose m to satisfy $\sigma(a^m)^{1/m} < \sigma^\infty(a) + \varepsilon/4$. The inequality

$$
\begin{aligned}
\sigma(a_n{}^m - a^m) &= \sigma\left(\sum_{j=0}^{m-1} a_n{}^j (a_n - a)a^{m-j-1}\right) \\
&\leq m \max\{\sigma(a_n), \sigma(a)\}^{m-1}\sigma(a_n - a) < m\sigma(a_n - a)
\end{aligned}
$$

implies the existence of an n_0 so that

$$\sigma(a_n{}^m - a^m) < m(\sigma^\infty(a) + \varepsilon/4)^{m-1}\varepsilon/4$$

for all $n \geq n_0$. The binomial expansion for $((\sigma^\infty(a) + \varepsilon/4) + \varepsilon/4)^m$ shows

$$
\begin{aligned}
\sigma(a_n{}^m) &= \sigma(a^m + (a_n{}^m - a^m)) \leq \sigma(a^m) + \sigma(a_n{}^m - a^m) \\
&< (\sigma^\infty(a) + \varepsilon/4)^m + m(\sigma^\infty(a) + \varepsilon/4)^{m-1}\varepsilon/4 \\
&< (\sigma^\infty(a) + \varepsilon/2)^m
\end{aligned}
$$

for all $n \geq n_0$. Taking the mth root gives inequality (15).

(i): If $\sigma^\infty(a)$ is strictly less than 1, $\{a^n : n \in \mathbb{N}\}$ is a bounded multiplicative semigroup. Hence by the last remark of Proposition 1.1.9, we may choose an equivalent algebra norm satisfying $\|a^n\| \leq 1$ for all $n \in \mathbb{N}$. If $\mathcal{A}$ is unital, we may include 1 in the semigroup. □

An element a in a semi-normed algebra $(\mathcal{A}, \sigma)$ is said to be *topologically nilpotent* if it satisfies $\sigma^\infty(a) = 0$. (The term *quasi-nilpotent* is sometimes used instead of topologically nilpotent, but we prefer to retain the prefix "quasi-" for another concept introduced in the next chapter.) In many semi-normed algebras, this happens if and only if a is not invertible but $\lambda 1 - a$ is invertible in the unitization which we are about to discuss for all non-zero $\lambda \in \mathbb{C}$. See Theorem 2.2.5 below.

Unital Algebras and Unitization

We have already called an algebra *unital* if it contains a multiplicative identity element. The multiplicative identity element in a unital algebra will usually be denoted by 1. It is easy to see that 1 in a normed unital algebra must satisfy $\|1\| \geq 1$. Note that the term unital is used by some authors (*e.g.*, Bonsall and Duncan [1973b]) to signify that a normed algebra contains an identity element of norm one. We do not make this restriction. A map φ between unital algebras is called *unital* if it satisfies $\varphi(1) = 1$.

Many of the algebras that we will consider are not unital. On the other hand, many constructions and arguments are much easier to formulate and understand when an identity element is present. For this reason it is desirable to have a procedure for enlarging an algebra so that the enlarged algebra contains an identity element. There are several ways of doing this. The simplest way, which was already known to Benjamin Peirce [1870], is also the least natural but it suffices for most arguments. Let $\mathcal{A}$ be an algebra. Consider the linear space $\mathbb{C} \oplus \mathcal{A}$ with multiplication defined by

$$(\lambda \oplus a)(\mu \oplus b) = \lambda\mu \oplus (\lambda b + \mu a + ab) \qquad \forall \, \lambda, \mu \in \mathbb{C}; \; a, b \in \mathcal{A}. \tag{16}$$

Clearly the map $a \mapsto 0 \oplus a$ embeds $\mathcal{A}$ isomorphically into this algebra. Furthermore $1 \oplus 0$ is an identity element. We can partially alleviate the awkwardness of this construction by the following device, which was already used by Segal [1941].

1.1.11 Definition Let $\mathcal{A}$ be an algebra. If $\mathcal{A}$ is not unital, denote the algebra described above by $\mathcal{A}^1$. If $\mathcal{A}$ is unital, denote the algebra $\mathcal{A}$ itself by $\mathcal{A}^1$. We will call $\mathcal{A}^1$ the *unitization* of $\mathcal{A}$.

The advantages of this convention on the meaning of $\mathcal{A}^1$ will begin to become apparent when we discuss the spectrum. From now on we will always consider $\mathcal{A}$ as a subset of $\mathcal{A}^1$ under the embedding $a \mapsto 0 \oplus a$. We will also suppress the direct sum notation $\oplus$ and will usually denote the multiplicative identity of $\mathcal{A}^1$ by 1 whether or not $\mathcal{A}$ is unital. Furthermore, for any $\lambda \in \mathbb{C}$, the element $\lambda 1$ of $\mathcal{A}^1$ will be denoted simply by λ.

Even when calculating in an algebra $\mathcal{A}$, which may not be unital, it is frequently convenient to use notation which must be interpreted in $\mathcal{A}^1$. For instance, the expression $a(\lambda + b)(\mu + c)$ represents an element in $\mathcal{A}$. We use it since it is easier to understand than $\lambda\mu a + \lambda ac + \mu ab + abc$. Similarly we will routinely denote the smallest ideal of an algebra $\mathcal{A}$ containing an element a of $\mathcal{A}$ by $\mathcal{A}^1 a \mathcal{A}^1$. We remark that if $\mathcal{I}$ is an ideal of $\mathcal{A}$, then it is also an ideal of $\mathcal{A}^1$ since $\mathcal{A}^1 \mathcal{I} \mathcal{A}^1 \subseteq \mathcal{I}$ holds.

If $(\mathcal{A}, \sigma)$ is a nonunital semi-normed algebra, there are several ways of extending the semi-norm σ to a semi-norm σ^1 on $\mathcal{A}^1$. Since different extensions are useful under different circumstances, we make no permanent choice. One method of extending σ which always works, but is frequently unnatural, is to define

$$\sigma^1(\lambda + a) = |\lambda| + \sigma(a) \qquad \forall \, \lambda + a \in \mathcal{A}^1. \tag{17}$$

We restrict attention to algebra norms in the rest of this discussion.

1.1.12 Definition Let $\| \cdot \|$ be an algebra norm on an algebra $\mathcal{A}$.
  (a) If $\| \cdot \|$ satisfies $\|a\| = \sup\{\|ab\|, \|ba\| : b \in \mathcal{A}_1\}$ for all $a \in \mathcal{A}$, then it is called a *regular norm*.
  (b) If $\mathcal{A}$ is nonunital, define the *unitization* $\| \cdot \|_1$ of $\| \cdot \|$ on $\mathcal{A}^1$ by

$$\|\lambda + a\|_1 = |\lambda| + \|a\| \qquad \forall \, \lambda + a \in \mathcal{A}^1.$$

  (c) If $\mathcal{A}$ is nonunital and $\| \cdot \|$ is regular, define the *regular unitization* $\| \cdot \|_R$ of $\| \cdot \|$ on $\mathcal{A}^1$ by

$$\|\lambda + a\|_R = \sup\{\|\lambda b + ab\|, \|\lambda b + ba\| : b \in \mathcal{A}_1\} \qquad \forall \, \lambda + a \in \mathcal{A}^1.$$

A number of authors have defined an analogue of $\| \cdot \|_R$ by $\|\lambda + a\|_R' = \sup\{\|\lambda b + ab\| : b \in \mathcal{A}_1\}$. However, if there is a right identity j in $\mathcal{A}$ (but no left identity), then the equation $\| -1 + j\|_R' = 0$ shows that $\| \cdot \|_R'$ is not a norm on $\mathcal{A}^1$. Our definition avoids this problem.

1.1.13 Proposition *Let $\| \cdot \|$ be an algebra norm on an algebra $\mathcal{A}$.*

(a) *If $\mathcal{A}$ is unital, then $\| \cdot \|$ is regular if and only if it satisfies $\|1\| = 1$.*

(b) *If $\mathcal{A}$ is nonunital, $\| \cdot \|_1$ is a regular norm on $\mathcal{A}^1$ which extends $\| \cdot \|$ on $\mathcal{A}$, and any regular norm $\||| \cdot \|||$ on $\mathcal{A}^1$ which extends $\| \cdot \|$ on $\mathcal{A}$ satisfies:*

$$\||| \lambda + a \||| \le \| \lambda + a \|_1 \qquad \forall \, \lambda + a \in \mathcal{A}^1.$$

(c) *If $\mathcal{A}$ is nonunital and $\| \cdot \|$ is regular, $\| \cdot \|_R$ is a regular norm on $\mathcal{A}^1$ which extends $\| \cdot \|$ on $\mathcal{A}$, and any regular norm $\||| \cdot \|||$ on $\mathcal{A}^1$ which extends $\| \cdot \|$ on $\mathcal{A}$ satisfies:*

$$\| \lambda + a \|_R \le \||| \lambda + a \||| \le \| \lambda + a \|_1 \le (1 + 2e) \| \lambda + a \|_R \qquad \forall \, \lambda + a \in \mathcal{A}^1.$$

Proof (a) and (b): Obvious.

(c): Any $\lambda + a \in \mathcal{A}^1$ and $b \in \mathcal{A}_1$ satisfy $\| \lambda b + ab \| = \||| \lambda b + ab \||| \le \||| \lambda + a \||| \, \||| b \||| \le \||| \lambda + a \|||$ and the similar inequality with b on the left. This gives the first inequality and the second has already been noted.

In the final inequality, e is the base of the natural logarithms. Since the proof is easy with the numerical range techniques of Section 2.6, we postpone it until just after Theorem 2.6.4.   □

Locally Convex Spaces

Besides semi-normed linear spaces, there is another class of semi-topological linear spaces that we will frequently meet. These are the locally convex semi-topological linear spaces. We will review their theory briefly and informally. For proofs and more details, the reader should consult a book on functional analysis or one of the many books on topological linear spaces, *e.g.*, Kelley and Namioka *et al.* [1963], Robertson and Robertson [1964], Köthe [1971] and Schaefer [1971].

1.1.14 Definition A *locally convex semi-topological linear space* is a semi-topological linear space $\mathcal{X}$ that satisfies one of the following two equivalent properties:

(a) There is a neighborhood base $\mathcal{U}$ at zero consisting of balanced convex subsets.

(b) There is a collection P of semi-norms on $\mathcal{X}$ such that the topology of $\mathcal{X}$ is the weakest topology in which each of these semi-norms is continuous.

Given condition (a), condition (b) follows by constructing the *Minkowski functional* σ_U of each U contained in $\mathcal{U}$, where σ_U is defined by

$$\sigma_U(x) = \inf\{t \in \mathbb{R}_+ : x \in tU\}, \qquad \forall \, x \in \mathcal{X}.$$

Then P may be chosen as $\{\sigma_U : U \in \mathcal{U}\}$. Conversely, if the family P of semi-norms is given, one can take $\mathcal{U}$ to be the set of all subsets

$$U_{F,n} = \{x \in \mathcal{X} : \sigma(x) < n^{-1}, \quad \sigma \in F\},$$

where F ranges over finite subsets of P and n ranges over $\mathbb{N}$. These statements require some checking, which we omit. A standard form of the Hahn–Banach theorem applied to the second definition shows that there are enough continuous linear functionals on $\mathcal{X}$ to separate the cosets of the subspace which is the closure of zero. This fact is more familiar in the case of locally convex topological linear spaces (*i.e.*, locally convex semitopological linear spaces satisfying the Hausdorff separation axiom), where it asserts that there are enough continuous linear functionals to separate points. Finally, we mention that a locally convex topological linear space is metrizable if and only if the set P in condition (b) can be chosen as a countable set. In this case, the metric may be chosen as

$$d(x,y) = \sum_{n=1}^{\infty} 2^{-n} \min\{1, \sigma_n(x-y)\} \qquad \forall \ x, y \in \mathcal{X}, \qquad (18)$$

where $\{\sigma_n : n \in \mathbb{N}\}$ is an enumeration of P. This metric is translation invariant, but it is not a norm since it is not absolutely homogeneous. There are interesting and important cases in which the topology it defines is not given by any norm.

This work is sharply focused on normed algebras and Banach algebras. We tend to explore generalizations to abstract (complex) algebras while omitting many generalizations to more general topological algebras. However, the brief Section 2.9 in the next chapter will give a survey of a few facts about more general topological algebras which are needed.

The Dual Space

1.1.15 Definition Let $\mathcal{X}$ be a linear space. A linear map $\omega \colon \mathcal{X} \to \mathbb{C}$ is called a *linear functional* on $\mathcal{X}$. The linear space of all linear functionals on $\mathcal{X}$ (under pointwise operations) is denoted by $\mathcal{X}^\dagger$ and is called the *algebraic dual* of $\mathcal{X}$. If $\mathcal{X}$ and $\mathcal{Y}$ are linear spaces and $T \colon \mathcal{X} \to \mathcal{Y}$ is a linear map, then the *algebraic dual of* T, denoted by $T^\dagger \colon \mathcal{Y}^\dagger \to \mathcal{X}^\dagger$, is defined by $T^\dagger(\omega) = \omega \circ T$ for all $\omega \in \mathcal{Y}^\dagger$.

Let $\mathcal{X}$ be a normed linear space. If $\omega \in \mathcal{X}^\dagger$ is continuous, define the *dual norm* of ω by

$$\|\omega\| = \sup\{|\omega(x)| : x \in \mathcal{X}_1\}.$$

The linear subspace of $\mathcal{X}^\dagger$ consisting of all continuous linear functionals on $\mathcal{X}$ with the dual norm is denoted by $\mathcal{X}^*$ and called the *Banach space dual* of $\mathcal{X}$. If T is a continuous linear map between normed linear spaces $\mathcal{X}$ and $\mathcal{Y}$, then the restriction of the algebraic dual map $T^\dagger \colon \mathcal{Y}^\dagger \to \mathcal{X}^\dagger$ to $\mathcal{Y}^*$ will be denoted by $T^* \colon \mathcal{Y}^* \to \mathcal{X}^*$ and called the *dual map* again. (We shall reserve the term "adjoint map", which is often used as a synonym for the dual map, for the case of the Hilbert space adjoint to be defined later.)

We denote $(\mathcal{X}^*)^*$ by $\mathcal{X}^{**}$ and call it the *double dual* of $\mathcal{X}$. The *natural map* κ or $\kappa_{\mathcal{X}}$ of $\mathcal{X}$ into its double dual is defined by evaluation:

$$\kappa(x)(\omega) = \omega(x) \qquad \forall\, x \in \mathcal{X};\, \omega \in \mathcal{X}^*.$$

The weakest topology on $\mathcal{X}$ relative to which each element of $\mathcal{X}^*$ is continuous will be called the *weak topology* or the $\mathcal{X}^*$-*topology*. The weakest topology on $\mathcal{X}^*$ relative to which each element of $\kappa(\mathcal{X})$ is continuous will be called the *weak*-topology* or the $\mathcal{X}$-*topology*.

Note that the Banach space dual $\mathcal{X}^*$ is always a Banach space, whether or not $\mathcal{X}$ is complete. The natural map κ is always an isometric linear injection. It is also easy to see that the weak and weak* topologies are locally convex.

The Algebra of Linear Maps

The linear space (under pointwise linear operations) of all linear maps of one linear space $\mathcal{X}$ into another $\mathcal{Y}$ (normed or not) will be denoted by $\mathcal{L}(\mathcal{X}, \mathcal{Y})$. If $\mathcal{Y}$ equals $\mathcal{X}$, we simplify this to $\mathcal{L}(\mathcal{X})$. Then $\mathcal{L}(\mathcal{X})$ is an algebra when multiplication is composition of maps. Unless the contrary is specifically stated, all subalgebras of $\mathcal{L}(\mathcal{X})$ will be given these same algebraic operations. We state the next definition formally because of its importance in what follows.

1.1.16 Definition For any normed linear spaces $\mathcal{X}$ and $\mathcal{Y}$ the subspace of $\mathcal{L}(\mathcal{X}, \mathcal{Y})$ consisting of continuous linear maps will be denoted by $\mathcal{B}(\mathcal{X}, \mathcal{Y})$ and considered as a normed linear space under the *operator norm*

$$\|T\| = \sup\{\|Tx\| : x \in \mathcal{X}_1\} \qquad \forall\, T \in \mathcal{B}(\mathcal{X}). \tag{19}$$

Whenever the linear spaces are obvious from context, we will simply call elements of $\mathcal{B}(\mathcal{X}, \mathcal{Y})$ *operators*.

Topological statements about $\mathcal{B}(\mathcal{X}, \mathcal{Y})$ always refer to the norm topology unless another topology is specifically mentioned. The *strong operator topology* on $\mathcal{B}(\mathcal{X}, \mathcal{Y})$ is the topology of pointwise convergence with the norm topology on $\mathcal{Y}$. The *weak operator topology* on $\mathcal{B}(\mathcal{X}, \mathcal{Y})$ is the topology of pointwise convergence when $\mathcal{Y}$ carries its weak topology.

When $\mathcal{Y}$ equals $\mathcal{X}$, we write $\mathcal{B}(\mathcal{X})$ for $\mathcal{B}(\mathcal{X}, \mathcal{Y})$ and consider it as a normed algebra under composition as multiplication.

Many readers will already know that $\mathcal{B}(\mathcal{X})$ is a Banach algebra if $\mathcal{X}$ is a Banach space, and that $\mathcal{B}(\mathcal{X})$ is a topological algebra under both the strong operator topology and the weak operator topology. In any case, these statements are easily verified. Observe that $\mathcal{B}(\mathcal{X}, \mathcal{Y})$ is complete if $\mathcal{Y}$ is complete (whether or not $\mathcal{X}$ is complete).

The Ideal of Finite-rank Operators

1.1.17 Definition Let W, $\mathcal{X}$, $\mathcal{Y}$ and $\mathcal{Z}$ be normed linear spaces. For elements $y \in \mathcal{Y}$ and $\omega \in \mathcal{X}^*$, we define an operator $y \otimes \omega \in \mathcal{B}(\mathcal{X}, \mathcal{Y})$ by

$$(y \otimes \omega)(z) = \omega(z)y \qquad \forall \, z \in \mathcal{X}. \tag{20}$$

We denote the linear span of all such operators by $\mathcal{B}_F(\mathcal{X}, \mathcal{Y})$ and call operators in this space *finite-rank operators*. We denote $\mathcal{B}_F(\mathcal{X}, \mathcal{X})$ by $\mathcal{B}_F(\mathcal{X})$.

The tensor product notation $(\otimes)$ will be explained in Section 1.10. For arbitrary $S \in \mathcal{B}(W, \mathcal{X})$, $T \in \mathcal{B}(\mathcal{Y}, \mathcal{Z})$, $y \in \mathcal{Y}$, $\omega \in \mathcal{X}^*$, $z \in \mathcal{Z}$ and $\tau \in \mathcal{Y}^*$ it is easy to check the formulas

$$
\begin{aligned}
(z \otimes \tau)(y \otimes \omega) &= \tau(y)z \otimes \omega \\
T(y \otimes \omega) &= T(y) \otimes \omega \\
(y \otimes \omega)S &= y \otimes S^*(\omega).
\end{aligned}
\tag{21}
$$

where $z \otimes \tau$ belongs to $\mathcal{B}_F(\mathcal{Y}, \mathcal{Z})$, $z \otimes \omega$ and $T(y) \otimes \omega$ belong to $\mathcal{B}_F(\mathcal{X}, \mathcal{Z})$ and $y \otimes S^*(\omega)$ belongs to $\mathcal{B}_F(W, \mathcal{Y})$. Also the dual map of $y \otimes \omega \in \mathcal{B}_F(\mathcal{X}, \mathcal{Y})$ is given by

$$(y \otimes \omega)^* = \omega \otimes \kappa(y) \in \mathcal{B}_F(\mathcal{Y}^*, \mathcal{X}^*).$$

Now suppose W, $\mathcal{Y}$ and $\mathcal{Z}$ all equal $\mathcal{X}$ in the above formulas. Then these formulas show that $\mathcal{B}_F(\mathcal{X})$ is an ideal of $\mathcal{B}(\mathcal{X})$. We prove more.

1.1.18 Proposition *Let $\mathcal{X}$ be a normed linear space. The set of operators in $\mathcal{B}(\mathcal{X})$ with finite-dimensional range is the set*

$$\mathcal{B}_F(\mathcal{X}) = \{\sum_{j=1}^{n} x_j \otimes \omega_j : n \in \mathbb{N}; \, x_j \in \mathcal{X}; \, \omega_j \in \mathcal{X}^*\} \tag{22}$$

of finite-rank operators. It is an ideal of $\mathcal{B}(\mathcal{X})$ which is included in every non-zero ideal.

Proof Clearly every operator in $\mathcal{B}_F(\mathcal{X})$ has finite-dimensional range. Conversely, if $T \in \mathcal{B}(\mathcal{X})$ has finite-dimensional range, let $\{x_1, x_2, \ldots, x_n\}$ be a basis for its range. Use the Hahn–Banach theorem to extend linear functionals on this range space to elements $\{\omega_1, \omega_2, \ldots, \omega_n\}$ of $\mathcal{X}^*$ satisfying $\omega_j(x_k) = \delta_{jk}$. Then

$$T = \sum_{j=1}^{n} x_j \otimes T^*(\omega_j)$$

is the desired expansion.

Suppose $\mathcal{I}$ is a non-zero ideal of $\mathcal{B}(\mathcal{X})$. Then there is a non-zero operator $T \in \mathcal{I}$ and a vector $z \in \mathcal{X}$ satisfying $Tz \neq 0$. By the Hahn–Banach theorem there is a continuous linear functional $\tau \in \mathcal{X}^*$ satisfying $\tau(Tz) = 1$. Hence for any $x \in \mathcal{X}$ and any $\omega \in \mathcal{X}^*$ we can write $x \otimes \omega$ as

$$x \otimes \omega = (x \otimes \tau)(Tz \otimes \omega) = (x \otimes \tau)T(z \otimes \omega)$$

which proves that $x \otimes \omega$ belongs to the ideal $\mathcal{I}$.   □

This proposition explains why we call $\mathcal{B}_F(\mathcal{X})$ the *ideal of finite-rank operators*. Note that the equations in (21) show that $\mathcal{B}_F(\mathcal{X}, \mathcal{Y})$ is a type of generalized ideal in the complex of spaces $\mathcal{B}(\mathcal{X}, \mathcal{Y})$ as $\mathcal{X}$ and $\mathcal{Y}$ vary. The ideal $\mathcal{B}_K(\mathcal{X})$ of compact operators will be defined and discussed in §1.7.7.

1.2 Double Centralizers and Extensions

A double centralizer (sometimes called a double multiplier) of an algebra $\mathcal{A}$ is a pair of maps of $\mathcal{A}$ into itself. The set of double centralizers of $\mathcal{A}$ forms a unital algebra $\mathcal{D}(\mathcal{A})$ under natural operations. We will begin by introducing the regular representation which defines a homomorphism of the original algebra onto an ideal in $\mathcal{D}(\mathcal{A})$. The mapping is surjective if and only if $\mathcal{A}$ is unital and, in this case, it is actually an isomorphism. In the situations of most interest, this homomorphism is injective and one often thinks of the original algebra $\mathcal{A}$ as a subalgebra of the double centralizer algebra $\mathcal{D}(\mathcal{A})$. In many circumstances this embedding of $\mathcal{A}$ into $\mathcal{D}(\mathcal{A})$ is a more natural way of adding an identity to a nonunital algebra. (After Proposition 1.2.6, we will mention three classes of examples which are given in more detail later in this chapter.) Moreover, $\mathcal{D}(\mathcal{A})$ is an example of an extension of the algebra $\mathcal{A}$: that is, there is a short exact sequence

$$\{0\} \to \mathcal{A} \to \mathcal{D}(\mathcal{A}) \to \mathcal{D}(\mathcal{A})/\mathcal{A} \to \{0\}.$$

It turns out that this short exact sequence is universal for the algebra $\mathcal{A}$ in the sense that all short exact sequences

$$\{0\} \to \mathcal{A} \to \mathcal{B} \to \mathcal{C} \to \{0\}$$

can be constructed from this one and a naturally defined homomorphism of $\mathcal{C}$ into $\mathcal{D}(\mathcal{A})/\mathcal{A}$. When we think of $\mathcal{B}$ in such a short exact sequence as being constructed from the "simpler" algebras $\mathcal{A}$ and $\mathcal{C}$, we call it an extension. Thus, this is a satisfying way of constructing and classifying extensions of $\mathcal{A}$. We explore all these ideas in this section starting with the regular representation.

The Regular Representations

1.2.1 Definition Let $\mathcal{A}$ be an algebra. For each $a \in \mathcal{A}$, let L_a and R_a be the linear maps in $\mathcal{L}(\mathcal{A})$ defined by

$$L_a(b) = ab \quad \text{and} \quad R_a(b) = ba \qquad \forall\, b \in \mathcal{A}. \tag{1}$$

We call the map L (*i.e.*, $a \mapsto L_a$) the *left regular representation* of $\mathcal{A}$ and the map R (*i.e.*, $a \mapsto R_a$) the *right regular representation* of $\mathcal{A}$. The kernel $\mathcal{A}_{LA} = \{a \in \mathcal{A} : ab = 0 \ \forall \ b \in \mathcal{A}\}$ of L is called the *left annihilator* of $\mathcal{A}$ and the kernel $\mathcal{A}_{RA} = \{a \in \mathcal{A} : ba = 0 \ \forall \ b \in \mathcal{A}\}$ of R is called the *right annihilator* of $\mathcal{A}$. The intersection $\mathcal{A}_A$ of the left and right annihilators of $\mathcal{A}$ is called the *annihilator* of $\mathcal{A}$. The *extended* left and right regular representations are the maps L^1 and R^1 of $\mathcal{A}$ into $L(\mathcal{A}^1)$ defined by

$$
\begin{aligned}
L_a^1(\lambda + b) &= \lambda a + ab \\
R_a^1(\lambda + b) &= \lambda a + ba
\end{aligned}
\qquad \forall \ a \in \mathcal{A}, \ \lambda + b \in \mathcal{A}^1. \tag{2}
$$

Note that the left and right regular and extended regular representations are bounded if the original algebra is normed. In that case, we give them their operator norms, using the norm

$$
\|\lambda + a\| = |\lambda| + \|a\| \qquad \forall \ \lambda \in \mathbb{C}; \ a \in \mathcal{A} \tag{3}
$$

for $\mathcal{A}^1$ if $\mathcal{A}$ is nonunital. Then it is easy to see

$$
\|a\| = \|L_a^1\| = \|R_a^1\| \qquad \forall \ a \in \mathcal{A}. \tag{4}
$$

In particular, L^1 and R^1 are injective. When $\mathcal{A}$ is noncommutative, the right regular and extended right regular representations are anti-representations rather than representations in the general terminology introduced later, but the present usage is well established. It is of some interest that the extended left regular representation was used by Charles Saunders Peirce [1881] to prove that every finite-dimensional algebra could be represented by matrices. See Thomas Hawkins [1972] for remarks on the context in which this proof was given.

Note that the annihilator ideals satisfy

$$
(\mathcal{A}_{LA})^2 = (\mathcal{A}_{RA})^2 = (\mathcal{A}_A)^2 = \{0\} \tag{5}
$$

and hence the annihilator ideals themselves are $\{0\}$ if $\mathcal{A}$ is semiprime (Definition 4.4.1 below) and *a fortiori* if $\mathcal{A}$ is semisimple (Definition 4.3.1 below). Of course, they vanish if $\mathcal{A}$ is unital or even approximately unital (Definition 5.1.1 below). Following Wang [1961], Ronald Larsen [1971] says a Banach algebra $\mathcal{A}$ is *without order* if $\mathcal{A}_{LA} = \{0\}$ or $\mathcal{A}_{RA} = \{0\}$. This ungainly terminology is more common in commutative algebras.

Centralizers

When the annihilator ideal is zero, there is another more natural way of embedding $\mathcal{A}$ into a unital algebra. The following construction was first explicitly studied by Gerhard P. Hochschild [1947]. It was introduced into analysis by Barry E. Johnson [1964a], [1964b], [1966], who was apparently unaware of Hochschild's work and abstracted the idea from the work of James G. Wendel [1952] and Jeffery D. Weston [1960] (*cf.* our remarks below on Ju-kwei Wang's [1961] study of multipliers). See also Sigurdur Helgason [1956], John Dauns [1969] and Dauns and Karl H. Hofmann [1968], [1969].

1.2.2 Definition   Let $\mathcal{A}$ be an algebra. A $\langle$ *left centralizer / right centralizer* $\rangle$ of $\mathcal{A}$ is an element $\langle$ L / R $\rangle \in \mathcal{L}(\mathcal{A})$ satisfying

$$\langle L(ab) = L(a)b \ / \ R(ab) = aR(b) \ \rangle \qquad \forall \, a, \, b \in \mathcal{A}. \tag{6}$$

A *double centralizer* of $\mathcal{A}$ is a pair (L, R) where L is a left centralizer, R is a right centralizer and together they satisfy

$$aL(b) = R(a)b \qquad \forall \, a, \, b \in \mathcal{A}. \tag{7}$$

The *algebra* $\mathcal{D}(\mathcal{A})$ *of double centralizers* of $\mathcal{A}$ is the set of double centralizers with pointwise linear operations and with multiplication defined by

$$(L_1, R_1)(L_2, R_2) = (L_1 L_2, R_2 R_1) \tag{8}$$

for all (L_1, R_1) and (L_2, R_2) in $\mathcal{D}(\mathcal{A})$.

If $\mathcal{A}$ is a normed algebra, then the set $\mathcal{D}_B(\mathcal{A}) = \{(L, R) \in \mathcal{D}(\mathcal{A}) : L, \, R \in \mathcal{B}(\mathcal{A})\}$ is called the *algebra of bounded double centralizers* and is given the norm

$$\|(L, R)\| = \max\{\|L\|, \|R\|\} \qquad \forall \, (L, R) \in \mathcal{D}_B(\mathcal{A}), \tag{9}$$

where $\|L\|$ and $\|R\|$ are the operator norms of L and R, respectively.

A $\langle$ left / right $\rangle$ centralizer is simply a map in the commutant $\langle$ $(R_\mathcal{A})'$ / $(L_\mathcal{A})'$ $\rangle$ of the $\langle$ right / left $\rangle$ regular representation in $\mathcal{L}(\mathcal{A})$. As mentioned above, this definition of double centralizers is due to Hochschild [1947]. Johnson [1964a] considered any pair (L, R) of maps in $\mathcal{L}(\mathcal{A})$ which satisfies equation (7). We will show in Theorem 1.2.4 below that such a pair is a double centralizer in our terminology if both the one-sided annihilator ideals are zero. Therefore, our definition agrees with Johnson's definition in this case, which is the most frequently considered case anyway. (Note that the identities $(\mathcal{A}_{LA})^2 = (\mathcal{A}_{RA})^2 = \{0\}$ imply that any semisimple or even any semiprime algebra (as defined in Chapter 4 below) satisfies this condition.) When $\mathcal{A}_{LA}$ and $\mathcal{A}_{RA}$ are not both zero (particularly if their intersection $\mathcal{A}_\mathcal{A}$ is zero), the definition given here is more convenient.

It is easy to check that $\mathcal{D}(\mathcal{A})$ is an algebra under the stated operations and that $\mathcal{D}_B(\mathcal{A})$ is a subalgebra when it is defined. Clearly (I, I) is an identity element for $\mathcal{D}(\mathcal{A})$, where I is the identity map in $\mathcal{L}(\mathcal{A})$. Moreover the map $a \mapsto (L_a, R_a)$ is a homomorphism (called the *regular homomorphism*) of $\mathcal{A}$ into $\mathcal{D}(\mathcal{A})$ which maps into $\mathcal{D}_B(\mathcal{A})$ when $\mathcal{A}$ is normed. The kernel of the regular homomorphism is obviously the annihilator ideal $\mathcal{A}_{LA} \cap \mathcal{A}_{RA} = \mathcal{A}_A$ of $\mathcal{A}$. If $\mathcal{A}$ is unital, then for all $a \in \mathcal{A}$ and all $(L, R) \in \mathcal{D}(\mathcal{A})$ we have $L(a) = 1L(a) = R(1)a = L_{R(1)}(a)$ and $R(a) = R(a)1 = aL(1) = R_{L(1)}(a)$. Note also that $L(1) = 1L(1) = R(1)1 = R(1)$. Thus the regular homomorphism is surjective if and only if $\mathcal{A}$ is unital.

If $\mathcal{A}$ is a normed algebra, it is simple to check that $\mathcal{D}_B(\mathcal{A})$ is a normed algebra under the norm given in the definition. Of course the identity element of $\mathcal{D}_B(\mathcal{A})$ satisfies $\|1\| = \|(I, I)\| = 1$.

For all a, $b \in \mathcal{A}$ and $(L, R) \in \mathcal{D}(\mathcal{A})$, we have

$$
\begin{array}{rcccccl}
L_a L(b) &=& aL(b) &=& R(a)b &=& L_{R(a)}b, \\
RR_a(b) &=& R(ba) &=& bR(a) &=& R_{R(a)}b, \\
LL_a(b) &=& L(ab) &=& L(a)b &=& L_{L(a)}b, \quad \text{and} \\
R_a R(b) &=& R(b)a &=& bL(a) &=& R_{L(a)}b.
\end{array}
\tag{10}
$$

Therefore the image of the regular homomorphism is an ideal in $\mathcal{D}(\mathcal{A})$ and we have

$$
\begin{array}{rcl}
(L, R)(L_a, R_a) &=& (L_{L(a)}, R_{L(a)}) \\
&& \qquad\qquad\qquad \forall\, (L, R) \in \mathcal{D}(\mathcal{A}); \; a \in \mathcal{A}. \tag{11} \\
(L_a, R_a)(L, R) &=& (L_{R(a)}, R_{R(a)})
\end{array}
$$

If the annihilator ideal is zero and $\mathcal{A}$ is considered as embedded in $\mathcal{D}(\mathcal{A})$ (*via* the regular homomorphism), then these relations can be written as

$$
(L, R)a = L(a), \quad a(L, R) = R(a) \qquad \forall\, (L, R) \in \mathcal{D}(\mathcal{A}); \; a \in \mathcal{A}. \tag{12}
$$

We will now state some of these results formally and add a few more.

1.2.3 Proposition *Let $\mathcal{A}$ be an algebra.*

(a) *$\mathcal{D}(\mathcal{A})$ is a unital algebra.*

(b) *The regular homomorphism of $\mathcal{A}$ into $\mathcal{D}(\mathcal{A})$ has the annihilator ideal $\mathcal{A}_A$ as kernel and its range is an ideal. It is surjective if and only if $\mathcal{A}$ is unital, in which case it is an isomorphism.*

(c) *If either $\mathcal{A}^2 = \mathcal{A}$ or the annihilator ideal $\mathcal{A}_A$ is zero, then the following conditions on $(L, R) \in \mathcal{D}(\mathcal{A})$ are equivalent:*

     (c_1) *(L, R) belongs to the center of $\mathcal{D}(\mathcal{A})$.*

     (c_2) *$L = R$.*

     (c_3) *L and R are both left and right centralizers.*

(d) *If the annihilator ideal is zero, then $\mathcal{A}$ is commutative if and only if $\mathcal{D}(\mathcal{A})$ is commutative.*

Proof It only remains to show (c) and (d). Under the hypothesis of (d) the regular homomorphism is injective, so $\mathcal{A}$ is commutative if $\mathcal{D}(\mathcal{A})$ is commutative. The implication (c_3) $\Rightarrow$ (c_1) shows the converse. Hence it remains only to prove the equivalence of the conditions in (c).

First we will show, without any restriction on $\mathcal{A}$, that either (c_1) or (c_2) implies (c_3). The second of these implications is immediate from the definitions. If (c_1) holds, then any $a, b \in \mathcal{A}$ satisfy $L(ab) = LL_a(b) = L_aL(b) = aL(b)$ and $R(ab) = RR_b(a) = R_bR(a) = R(a)b$. Hence (c_1) implies (c_3).

Next we will show that (c_3) implies (c_1) and (c_2) under either of the hypotheses of (c). Suppose (c_3) holds. Then any $a, b \in \mathcal{A}$ and any left centralizer L' satisfy $LL'(ab) = L(L'(a)b) = L'(a)L(b) = L'(aL(b)) = L'L(ab)$. A similar calculation for right centralizers shows that (c_3) implies (c_1) when $\mathcal{A}$ satisfies $\mathcal{A}^2 = \mathcal{A}$. Also any $a, b \in \mathcal{A}$ and any double centralizer (L', R') satisfy the two relations

$$bLL'(a) = L(bL'(a)) = L(R'(b)a) = R'(b)L(a) = bL'L(a)$$

$$LL'(a)b = L'(a)L(b) = L'(aL(b)) = L'(L(a)b) = L'L(a)b.$$

A similar calculation for $R'R$ shows that (c_3) implies (c_1) when $\mathcal{A}$ satisfies $\mathcal{A}_A = \{0\}$.

If $\mathcal{A}_A = \{0\}$ and (c_3) are both true, then the identities $L(a)b = L(ab) = aL(b) = R(a)b$ and $bL(a) = R(b)a = R(ba) = bR(a)$ for all $a, b \in \mathcal{A}$ imply (c_2). If $\mathcal{A}^2 = \mathcal{A}$ and (c_3) both hold, then the identity $L(ab) = aL(b) = R(a)b = R(ab)$ for all $a, b \in \mathcal{A}$ implies (c_2).   □

A linear operator $T = L = R \in \mathcal{L}(\mathcal{A})$ satisfying condition (c) in the above theorem is an example of a multiplier. For any algebra $\mathcal{A}$, we define $T \in \mathcal{L}(\mathcal{A})$ to be a *multiplier* if it satisfies

$$T(a)b = aT(b) \qquad \forall\, a, b \in \mathcal{A}.$$

If the $\langle$ left / right $\rangle$ annihilator ideal of $\mathcal{A}$ is zero, then the calculation $\langle cT(ab) = T(c)ab = cT(a)b$ / $T(ab)c = abT(c) = aT(b)c \rangle$ for all $c \in \mathcal{A}$ shows that T is a $\langle$ left / right $\rangle$ centralizer. Hence Proposition 1.2.3(c) shows that when the annihilator ideal is zero, the set of multipliers is simply the center of the double centralizer algebra. Multipliers were defined by Helgason [1956] and first systematically studied on commutative Banach algebras (with annihilator ideal $\{0\}$) by Ju-kwei Wang in his thesis [1961]. They are still most often considered in this context. The standard reference for their theory is Ronald Larsen [1971]. However, we are aware of a number of papers which purport to use multipliers on noncommutative algebras but apply properties which hold only in the commutative case. Hence we shall not use the term "multiplier" again.

The second statement in the next theorem is our first example of automatic continuity, in that a purely algebraic condition on a map implies that it is continuous. Both results are due to Johnson [1964a]. A related result is given in Proposition 5.2.6.

1.2.4 Theorem *Let $\mathcal{A}$ be an algebra satisfying $\mathcal{A}_{LA} = \mathcal{A}_{RA} = \{0\}$.*

(a) If L and R are arbitrary (not necessarily linear) maps of $\mathcal{A}$ into $\mathcal{A}$ satisfying the double centralizer condition (7)

$$aL(b) = R(a)b \qquad \forall\, a,\, b \in \mathcal{A},$$

then (L, R) is a double centralizer.

(b) If $\mathcal{A}$ is a Banach algebra and (L, R) is a double centralizer, then L and R are bounded linear maps, so $\mathcal{D}_B(\mathcal{A})$ equals $\mathcal{D}(\mathcal{A})$. In this case, $\mathcal{D}(\mathcal{A}) = \mathcal{D}_B(\mathcal{A})$ is a Banach algebra under its norm and the regular homomorphism is contractive.

Proof (a): For any a, b, $c \in \mathcal{A}$ and any λ, $\mu \in \mathbb{C}$, we get $cL(\lambda a + \mu b) = R(c)(\lambda a + \mu b) = \lambda R(c)a + \mu R(c)b = c(\lambda L(a) + \mu L(b))$, and $cL(ab) = R(c)ab = cL(a)b$. Since $c \in \mathcal{A}$ is arbitrary and the right annihilator of $\mathcal{A}$ is zero, we conclude that L is a left centralizer. Similarly, R is a right centralizer.

(b): Suppose $\mathcal{A}$ is a Banach algebra, $\{a_n\}_{n\in\mathbb{N}}$ is a sequence in $\mathcal{A}$ converging to zero and the sequence $\{L(a_n)\}_{n\in\mathbb{N}}$ also converges to some element b in $\mathcal{A}$. Then any $c \in \mathcal{A}$ satisfies

$$cb = \lim_{n\to\infty} cL(a_n) = \lim_{n\to\infty} R(c)a_n = 0.$$

Hence, $\mathcal{A}_{RA} = \{0\}$ implies $b = 0$. Thus L is closed and, by the closed graph theorem, it is continuous. Similarly R is continuous.

It is easy to check that $\mathcal{D}_B(\mathcal{A})$ is a Banach algebra since the limit in $\mathcal{B}(\mathcal{A})$ of a sequence of left or right centralizers has the same form. Clearly the regular homomorphism is contractive. $\square$

If $\mathcal{A}$ is an algebra satisfying $\mathcal{A}_A = \{0\}$, then $\mathcal{D}(\mathcal{A})$ is a unital algebra which contains an ideal isomorphic to $\mathcal{A}$ under the regular homomorphism. If $\mathcal{A}$ is a Banach algebra satisfying $\mathcal{A}_{LA} = \mathcal{A}_{RA} = \{0\}$, the last theorem shows that $\mathcal{D}(\mathcal{A}) = \mathcal{D}_B(\mathcal{A})$ is a unital Banach algebra which contains an ideal continuously isomorphic to $\mathcal{A}$ under the regular homomorphism. Hence, under suitable restrictions on the annihilator ideals, the construction of $\mathcal{D}(\mathcal{A})$ provides another way to embed an algebra in a unital algebra and a Banach algebra in a unital Banach algebra. (Examples are mentioned after Proposition 1.2.6.) Note that $\mathcal{A}$ equals $\mathcal{D}(\mathcal{A})$ if and only if $\mathcal{A}$ is unital. For technical reasons, it is sometimes desirable to replace $\mathcal{D}(\mathcal{A})$ by the possibly smaller unital subalgebra $\mathcal{D}(\mathcal{A})_Z + \mathcal{A}$. For a discussion of this, see Dauns [1969] and Dauns and Hofmann [1968], [1969].

Left and Right Idealizers

1.2.5 Definition Let $\mathcal{A}$ be a subalgebra of an algebra $\mathcal{B}$. The *left idealizer* of $\mathcal{A}$ in $\mathcal{B}$ is the set $\{b \in \mathcal{B} : b\mathcal{A} \subseteq \mathcal{A}\}$. The *right idealizer* is defined similarly. The *idealizer* $\mathcal{B}_{\mathcal{A}}$ of $\mathcal{A}$ in $\mathcal{B}$ is the intersection of the left and right idealizer.

Clearly the ⟨ left / right ⟩ idealizer is the largest subalgebra of $\mathcal{B}$ in which $\mathcal{A}$ is a ⟨ left / right ⟩ ideal. Similarly the idealizer is the largest subalgebra in which $\mathcal{A}$ is an ideal. If b belongs to the left idealizer of $\mathcal{A}$ in $\mathcal{B}$, then L_b maps $\mathcal{A}$ into $\mathcal{A}$ and its restriction to $\mathcal{A}$ is a left centralizer. The same remarks hold for the right idealizer and right regular representation. Hence if b belongs to the idealizer of $\mathcal{A}$ in $\mathcal{B}$, then $(L_b|\mathcal{A}, R_b|\mathcal{A})$ is a double centralizer. This is the most natural way in which double centralizers arise. See the proof of the following result.

1.2.6 Proposition *Let $\mathcal{A}$ be a subalgebra of an algebra $\mathcal{B}$ and let $\mathcal{B}_\mathcal{A}$ be the idealizer of $\mathcal{A}$ in $\mathcal{B}$. Then there is a natural homomorphism θ of $\mathcal{B}_\mathcal{A}$ into $\mathcal{D}(\mathcal{A})$ defined by:*

$$\theta(b) = (L_b|\mathcal{A}, R_b|\mathcal{A}) \qquad \forall\, b \in \mathcal{B}_\mathcal{A}. \tag{13}$$

The homomorphism θ extends the regular homomorphism of $\mathcal{A}$ into $\mathcal{D}(\mathcal{A})$. It is the only such homomorphism if the annihilator of $\mathcal{A}$ is $\{0\}$. The kernel of θ is $\{b \in \mathcal{B} : b\mathcal{A} = \mathcal{A}b = \{0\}\}$.

If $\mathcal{A}$ is an ideal in $\mathcal{B}$, then θ maps $\mathcal{B}$ into $\mathcal{D}(\mathcal{A})$. If $\mathcal{B}$ is a normed algebra, then θ is a contractive map into $\mathcal{D}_\mathcal{B}(\mathcal{A})$.

Proof The description of the kernel of θ and the last sentence are obvious. Thus we only need to check the uniqueness assertion. Let $\theta\colon \mathcal{B}_\mathcal{A} \to \mathcal{D}(\mathcal{A})$ be any homomorphism which extends the regular homomorphism. Let $a \in \mathcal{A}$ and $b \in \mathcal{B}_\mathcal{A}$ be arbitrary and denote $\theta(b)$ by (L, R). Then formula (10) gives $(L_{ab}, R_{ab}) = \theta(ab) = \theta(a)\theta(b) = (L_a, R_a)(L, R) = (L_{R(a)}, R_{R(a)})$. Since $\mathcal{A}$ satisfies $\mathcal{A}_\mathcal{A} = \{0\}$, this implies $R(a) = ab$. A similar argument gives $L(a) = ba$. Hence we conclude that $\theta(b) = (L, R) = (L_b|\mathcal{A}, R_b|\mathcal{A})$ (where we use the left and right regular representations of $\mathcal{B}$), proving the uniqueness. □

Examples of Double Centralizer Algebras

Let Ω be a locally compact topological space. In §1.5.1 we will more formally introduce the following two important commutative Banach algebras of bounded continuous complex-valued functions on Ω. Both are Banach algebras under pointwise multiplication and the supremum norm:

$$\|f\|_\infty = \sup\{|f(\omega)| \;:\; \omega \in \Omega\}.$$

Let $C(\Omega)$ be the algebra of all bounded continuous functions and let $C_0(\Omega)$ be the algebra of continuous functions vanishing at infinity. Clearly $C_0(\Omega)$ is a closed ideal in $C(\Omega)$. Hence by Proposition 1.2.6 there is a natural homomorphism of $C(\Omega)$ into $\mathcal{D}(C_0(\Omega))$. In §1.5.1 we show that this is an isometric algebra isomorphism onto $\mathcal{D}(C_0(\Omega))$.

Similarly, in §1.7.1 we will show that the double centralizer algebra of the ideal $\mathcal{B}_K(\mathcal{X})$ of compact operators on any Banach space $\mathcal{X}$ can be identified with the algebra $\mathcal{B}(\mathcal{X})$ of all bounded linear operators on $\mathcal{X}$. In fact, we show that this result extends to a general class of ideals in $\mathcal{B}(\mathcal{X})$.

Finally in §1.9.13 we will show that for any locally compact group G the double centralizer algebra of $L^1(G)$ can be identified with the measure algebra $M(G)$. We have already noted that the use of this identification by Wendel [1952] provided an important stimulus to the study of double centralizer algebras in analysis.

Automorphisms

The proof of the next result is straightforward. When derivations are introduced in Chapter 6, a similar result will be proved there.

1.2.7 Proposition *Let $\mathcal{A}$ be an algebra in which the annihilator ideal is zero and let α be an automorphism of $\mathcal{A}$. Then we may define an automorphism $\overline{\alpha}$ of $\mathcal{D}(\mathcal{A})$ by*

$$\overline{\alpha}(L, R) = (\alpha \circ L \circ \alpha^{-1}, \alpha \circ R \circ \alpha^{-1})$$

which extends α when $\mathcal{A}$ is viewed as a subalgebra of $\mathcal{D}(\mathcal{A})$. Then $\alpha \to \overline{\alpha}$ defines an isomorphism of $Aut(\mathcal{A})$ onto the subgroup of $Aut(\mathcal{D}(\mathcal{A}))$ consisting of those automorphisms for which $\mathcal{A}$ is invariant as a subset of $\mathcal{D}(\mathcal{A})$. If $\mathcal{A}$ is normed, we may replace the word "automorphisms", by "homeomorphic automorphisms", everywhere. If α is an isometry, so is $\overline{\alpha}$.

Topological and Geometric Categories of Normed Algebras

There are two categories of normed algebras (or normed linear spaces) in common use. Since confusion between them sometimes occurs, it is worthwhile to point out the distinction clearly here. We will only describe the case of normed algebras, since the case of normed linear spaces is exactly similar.

In the more common and important category, the morphisms are continuous algebra homomorphisms. That is, $\varphi : \mathcal{A} \to \mathcal{B}$ is a morphism if it is an algebra homomorphism and there is some real number k satisfying $\|\varphi(a)\| \leq k\|a\|$ for all $a \in \mathcal{A}$. The isomorphisms are homeomorphic algebra isomorphisms. Thus the objects are not really normed algebras but rather topological algebras with a topology defined by some algebra norm. The objects may also be identified with equivalence classes of normed algebras, where two normed algebras are equivalent if they are isomorphic as algebras and the algebra isomorphism is a homeomorphism. (Equivalently, the algebra isomorphism $\varphi : \mathcal{A} \to \mathcal{A}'$ satisfies $k\|a\| \leq \|\varphi(a)\| \leq k'\|a\|$ for some non-zero constants k and k' and all $a \in \mathcal{A}$.) We follow the usual custom in such cases of denoting an equivalence class of normed algebras by any single element in the equivalence class. This is the *topological category of normed algebras*.

The second category has contractive algebra homomorphisms as its morphisms (*i.e.*, algebra homomorphisms $\varphi\colon \mathcal{A} \to \mathcal{B}$ satisfying $\|\varphi(a)\| \leq \|a\|$ for all $a \in \mathcal{A}$). Hence, an isomorphism is an isometric algebra isomorphism. In this category an algebra with two unequal but equivalent algebra norms corresponds to two non-isomorphic objects. Thus, in particular, a finite-dimensional algebra, which is a single object in the topological category, is the underlying space of infinitely many non-isomorphic objects in this second category. We will call this the *geometric category of normed algebras.* If we restrict the objects in these two categories to be Banach algebras, we get the *topological* and *geometric categories of Banach algebras.* The terms "dominate" and "majorize" for comparison of two norms which we introduced above, correspond to these two categories.

By simply deleting all reference to multiplicative structure, we get the *topological* and *geometric categories of normed linear spaces* and the *topological* and *geometric categories of Banach spaces.*

Many arguments and constructions in this work may be carried out in either category. We will seldom point this out explicitly. Instead we will carry out the construction in whichever category seems best.

From time to time, the differences in the properties of these two categories of Banach algebras do become significant. The following is a good example. Example 4.8.11, which is due to Robert C. Busby [1971a], shows that the short five lemma of category theory fails in the geometric category of Banach algebras. Hence the theory of extensions is more complicated in that category. (In the geometric category, the bijection of Theorem 1.2.11 has a (well identified) subset of $\mathrm{Ext}(\mathcal{C}, \mathcal{A})$ as its range. For details, see the original paper by Busby [1971a]. In the terminology of that article, our construction of the double centralizer algebra shows that a Banach algebra with zero right and left annihilator ideals is faithful, and the various arguments using the open mapping theorem establish the algebraic properties of the topological category of Banach algebras.)

Classification of Extensions

We now turn to the use of the double centralizer algebra in the classification of extensions. With any class of mathematical objects, one wishes to analyze the complicated examples in terms of the simple examples. To do this, one needs to know how to build up complicated examples from simple ones. There are many such constructions for topological algebras—direct sums, tensor products, direct and inverse limits, etc. However, we will investigate here the fundamental case of extensions. One is given two algebras $\mathcal{A}$ and $\mathcal{C}$ and asks how to classify the algebras $\mathcal{B}$ in which $\mathcal{A}$ can be embedded as an ideal in such a way that $\mathcal{B}/\mathcal{A}$ is isomorphic to $\mathcal{C}$. For algebras, the theory was partially worked out by Hochschild [1947]. The Banach algebra case turns out to be no harder. We follow the treatment of Busby [1971a] which was originally limited to C*-algebras [1967], [1968]. See the first reference for further details. (The reader should be aware that in some contexts $\mathcal{B}$ is called an extension of $\mathcal{A}$ whenever $\mathcal{A}$ is a subalgebra of $\mathcal{B}$. See Example 3.4.17 for information on this.)

An algebra is *simple* (in the technical sense) if it has no ideals except $\{0\}$ and itself. A topological algebra is *topologically simple* if it has no *closed* ideals except $\{0\}$ and itself. Many algebras with which we will deal can be built up in a finite number of steps by successive extensions from topologically simple algebras. We need some technical definitions.

1.2.8 Definition   A *short exact sequence of Banach algebras* is a triple of Banach algebras $\mathcal{A}, \mathcal{B}$ and $\mathcal{C}$ together with a pair of continuous homomorphisms $\varphi \colon \mathcal{A} \to \mathcal{B}$ and $\psi \colon \mathcal{B} \to \mathcal{C}$ such that φ is injective, its image $\varphi(\mathcal{A})$ equals $\ker(\psi)$, and ψ is surjective. This short exact sequence is denoted by

$$0 \;\longrightarrow\; \mathcal{A} \;\xrightarrow{\;\varphi\;}\; \mathcal{B} \;\xrightarrow{\;\psi\;}\; \mathcal{C} \;\longrightarrow\; 0 \qquad\qquad (14)$$

The terminal arrow $\langle\, 0 \to \mathcal{A} \,/\, \mathcal{C} \to 0 \,\rangle$ encodes the assumption that $\langle\, \varphi$ is injective $/\ \psi$ is surjective $\rangle$. The open mapping theorem shows that φ is a homeomorphism of $\mathcal{A}$ onto its range $\ker(\psi)$ and that ψ is open. Hence in particular, $\mathcal{C}$ is homeomorphically isomorphic to the quotient $\mathcal{B}/\varphi(\mathcal{A})$ with its quotient norm.

We now change our viewpoint slightly and thus give a new name for essentially the same object.

1.2.9 Definition   Let $\mathcal{A}$ and $\mathcal{C}$ be Banach algebras. An *extension of $\mathcal{A}$ by $\mathcal{C}$* is a short exact sequence

$$0 \;\longrightarrow\; \mathcal{A} \;\xrightarrow{\;\varphi\;}\; \mathcal{B} \;\xrightarrow{\;\psi\;}\; \mathcal{C} \;\longrightarrow\; 0 \qquad\qquad (15)$$

of Banach algebras. Two extensions

$$0 \;\longrightarrow\; \mathcal{A} \;\xrightarrow{\;\varphi\;}\; \mathcal{B} \;\xrightarrow{\;\psi\;}\; \mathcal{C} \;\longrightarrow\; 0 \qquad\qquad (16)$$

$$0 \;\longrightarrow\; \mathcal{A} \;\xrightarrow{\;\varphi'\;}\; \mathcal{B}' \;\xrightarrow{\;\psi'\;}\; \mathcal{C} \;\longrightarrow\; 0$$

of $\mathcal{A}$ by $\mathcal{C}$ are said to be *equivalent* if there is a homeomorphic isomorphism $\theta \colon \mathcal{B} \to \mathcal{B}'$ such that the diagram

$$
\begin{array}{ccccccccc}
0 & \longrightarrow & \mathcal{A} & \xrightarrow{\;\varphi\;} & \mathcal{B} & \xrightarrow{\;\psi\;} & \mathcal{C} & \longrightarrow & 0 \\
 & & \downarrow{\scriptstyle I_{\mathcal{A}}} & & \downarrow{\scriptstyle \theta} & & \downarrow{\scriptstyle I_{\mathcal{C}}} & & \\
0 & \longrightarrow & \mathcal{A} & \xrightarrow{\;\varphi'\;} & \mathcal{B}' & \xrightarrow{\;\psi'\;} & \mathcal{C} & \longrightarrow & 0
\end{array}
\qquad (17)
$$

commutes, where $I_\mathcal{A}$ and $I_\mathcal{C}$ are the identity maps on $\mathcal{A}$ and $\mathcal{C}$, respectively. The set of equivalence classes of extensions of $\mathcal{A}$ by $\mathcal{C}$ is denoted by $\mathrm{Ext}(\mathcal{C},\mathcal{A})$. The extension (15) is called a *semidirect product* if there is a continuous homomorphism $\chi\colon\mathcal{C}\to\mathcal{B}$ with $\psi\circ\chi$ the identity map on $\mathcal{C}$. A short exact sequence is said to be *split* if it represents a semidirect product. A semidirect product is called a *direct product* if $\chi(\mathcal{C})$ is the kernel of a continuous surjective homomorphism π onto $\mathcal{A}$.

A simple but interesting example of a split extension is given in 5.3. When the maps φ and ψ in (15) are obvious, one often speaks informally of $\mathcal{B}$ as the extension of $\mathcal{A}$ by $\mathcal{C}$. However, some authors reverse $\mathcal{A}$ and $\mathcal{C}$ in this phrase.

If θ is any continuous homomorphism which makes the diagram in the definition of equivalence commute, it is in fact a homeomorphic isomorphism. To see this, suppose b belongs to $\ker(\theta)$. Then we find successively $\psi(b) = \psi'\circ\theta(b) = 0$, $b = \varphi(a)$ for some $a\in\mathcal{A}$, $\varphi'(a) = \theta\circ\varphi(a)$, $a = 0$, $b = 0$. Similarly any $b'\in\mathcal{B}'$ satisfies $\psi'(b') = \psi(b) = \psi'\circ\theta(b)$ for some $b\in\mathcal{B}$, $b' - \theta(b) = \varphi'(a)$ for some $a\in\mathcal{A}$, $b' = \theta(b) + \varphi'(a) = \theta(b + \varphi(a))$. The open mapping theorem completes the proof. Hence, to prove equivalence, it is enough to produce a continuous homomorphism which makes (17) commute. (*N. B.* This is the argument which fails in the geometric category of Banach algebras.)

In a direct product, any element b of $\mathcal{B}$ can be identified with the pair $(\pi(b),\ \psi(b))$ and thus $\mathcal{B}$ is isomorphic to the Cartesian product, $\mathcal{A}\times\mathcal{C}$ with pointwise operations. This Cartesian product algebra can be given the ℓ^∞-norm

$$\|(a,c)\| = \max\{\|a\|, \|c\|\} \qquad \forall\, a\in\mathcal{A};\ c\in\mathcal{C}$$

which makes the isomorphism into a homeomorphic isomorphism. The images of $\mathcal{A}$ and $\mathcal{C}$ in $\mathcal{B}$ are both closed ideals, and another way to describe a direct product is to say that it is the sum of two closed ideals with intersection $\{0\}$.

In a semidirect product, each element b is the sum of $b - \chi\circ\psi(b)$ and $\chi\circ\psi(b)$, where the first summand is in the image of φ since it is in the kernel of ψ. Thus again, $\mathcal{B}$ can be identified with the Cartesian product $\mathcal{A}\times\mathcal{C}$ with the ℓ^∞-norm, but this time only the linear operations are pointwise. The image of $\mathcal{A}$ is a closed ideal, but the image of $\mathcal{C}$ is only a closed subalgebra in general. (Both of these constructions remain correct in the geometric category of Banach algebras.)

1.2.10 Theorem *Let*

$$0 \ \longrightarrow\ \mathcal{A} \ \xrightarrow{\ \varphi\ }\ \mathcal{B} \ \xrightarrow{\ \psi\ }\ \mathcal{C} \ \longrightarrow\ 0 \tag{18}$$

be an extension of $\mathcal{A}$ by $\mathcal{C}$ and let $\tau: \mathcal{C}' \to \mathcal{C}$ be a continuous homomorphism. Then there is an extension

$$0 \longrightarrow \mathcal{A} \xrightarrow{\varphi'} \mathcal{B}' \xrightarrow{\psi'} \mathcal{C}' \longrightarrow 0 \qquad (19)$$

of $\mathcal{A}$ by $\mathcal{C}'$ and a continuous homomorphism $\theta: \mathcal{B}' \to \mathcal{B}$ such that the diagram

$$
\begin{array}{ccccccccc}
0 & \longrightarrow & \mathcal{A} & \xrightarrow{\varphi} & \mathcal{B} & \xrightarrow{\psi} & \mathcal{C} & \longrightarrow & 0 \\
 & & \uparrow{\scriptstyle I_A} & & \uparrow{\scriptstyle \theta} & & \uparrow{\scriptstyle \tau} & & \\
0 & \longrightarrow & \mathcal{A} & \xrightarrow{\varphi'} & \mathcal{B}' & \xrightarrow{\psi'} & \mathcal{C}' & \longrightarrow & 0
\end{array}
\qquad (20)
$$

is commutative. This extension of $\mathcal{A}$ by $\mathcal{C}'$ is unique up to equivalence.

Proof Let $\mathcal{B} \times \mathcal{C}'$ be the Banach algebra with $\mathcal{B} \times \mathcal{C}'$ as its underlying set, with pointwise algebraic operations and with norm given by $\|(b,c')\| = \max\{\|b\|, \|c'\|\}$. Let $\mathcal{B}'$ be the closed subalgebra $\{(b,c') \in \mathcal{B} \times \mathcal{C}' : \psi(b) = \tau(c')\}$. Define the maps φ', θ, and ψ' by $\varphi'(a) = (\varphi(a), 0)$, $\theta(b,c') = b$ and $\psi'(b,c') = c'$ for all $a \in \mathcal{A}$ and all $(b,c') \in \mathcal{B}'$. This clearly makes (20) commute. It is obvious that ψ' is contractive and surjective (since ψ is surjective) and $\psi' \circ \varphi' = 0$. Furthermore, φ' is continuous since θ restricted to its range is an isometry and φ is continuous. Hence it only remains to show the inclusion $\ker(\psi') \subseteq \varphi'(\mathcal{A})$. However, $\ker(\psi') = \{(b,0) \in \mathcal{B}'\} = \{(b,0) : \psi(b) = 0\} = \{(\varphi(a),0) : a \in \mathcal{A}\} = \varphi'(\mathcal{A})$.

Suppose the diagram

$$
\begin{array}{ccccccccc}
0 & \longrightarrow & \mathcal{A} & \xrightarrow{\varphi} & \mathcal{B} & \xrightarrow{\psi} & \mathcal{C} & \longrightarrow & 0 \\
 & & \uparrow{\scriptstyle I_A} & & \uparrow{\scriptstyle \theta''} & & \uparrow{\scriptstyle \tau} & & \\
0 & \longrightarrow & \mathcal{A} & \xrightarrow{\varphi''} & \mathcal{B}'' & \xrightarrow{\psi''} & \mathcal{C}' & \longrightarrow & 0
\end{array}
\qquad (21)
$$

also commutes. Define $\theta': \mathcal{B}'' \to \mathcal{B}'$ by $\theta'(b'') = (\theta''(b''), \psi''(b''))$, so that it is a continuous homomorphism which makes the analogue of diagram (17) commute. The argument preceding this theorem shows that θ' is a homeomorphic isomorphism, so that the two extensions are equivalent. □

Recall that after Proposition 1.1.3 we defined $\mathrm{CHom}(\mathcal{A}, \mathcal{B})$ to be the set of continuous homomorphisms from a normed algebra $\mathcal{A}$ to another $\mathcal{B}$.

1.2.11 Theorem *Let $\mathcal{A}$ and $\mathcal{C}$ be Banach algebras with the first satisfying $\mathcal{A}_{RA} = \mathcal{A}_{LA} = \{0\}$. Let $0 \longrightarrow \mathcal{A} \xrightarrow{\rho} \mathcal{D}(\mathcal{A}) \xrightarrow{\sigma} \mathcal{D}(\mathcal{A})/\mathcal{A} \longrightarrow 0$ be the short exact sequence in which ρ is the regular homomorphism and σ is the natural map. The construction of Theorem 1.2.10 applied to this short exact sequence gives a bijection of $\mathrm{CHom}(\mathcal{C}, \mathcal{D}(\mathcal{A})/\mathcal{A})$ onto $\mathrm{Ext}(\mathcal{C}, \mathcal{A})$. Furthermore, the extension corresponding to $\tau \in \mathrm{CHom}(\mathcal{C}, \mathcal{D}(\mathcal{A})/\mathcal{A})$ is a semidirect product if and only if there exists a map $\omega \in \mathrm{CHom}(\mathcal{C}, \mathcal{D}(\mathcal{A}))$ satisfying*

$\tau = \sigma \circ \omega$ *and this extension is a direct product if and only if τ is the zero map.*

Proof The remarks following Theorem 1.2.4 show that $\mathcal{D}(\mathcal{A}) = \mathcal{D}_B(\mathcal{A})$ is a Banach algebra. For any $\tau \in \mathrm{CHom}(\mathcal{C}, \mathcal{D}(\mathcal{A})/\mathcal{A})$, the construction of Theorem 1.2.10 gives the following commutative diagram.

$$
\begin{array}{ccccccccc}
0 & \longrightarrow & \mathcal{A} & \xrightarrow{\ \rho\ } & \mathcal{D}(\mathcal{A}) & \xrightarrow{\ \sigma\ } & \mathcal{D}(\mathcal{A})/\mathcal{A} & \longrightarrow & 0 \\
 & & \big\uparrow{\scriptstyle I_{\mathcal{A}}} & & \big\uparrow{\scriptstyle \theta} & & \big\uparrow{\scriptstyle \tau} & & \\
0 & \longrightarrow & \mathcal{A} & \xrightarrow{\ \varphi\ } & B & \xrightarrow{\ \psi\ } & \mathcal{C} & \longrightarrow & 0
\end{array}
\tag{22}
$$

where the lower line is an extension of $\mathcal{A}$ by $\mathcal{C}$, which is unique up to equivalence. This establishes a map of $\mathrm{CHom}(\mathcal{C}, \mathcal{D}(\mathcal{A})/\mathcal{A})$ into $\mathrm{Ext}(\mathcal{C}, \mathcal{A})$. We must show that the map is both injective and surjective.

We will prove injectivity first. Suppose that $0 \longrightarrow \mathcal{A} \xrightarrow{\varphi} B \xrightarrow{\psi} \mathcal{C} \longrightarrow 0$ and $0 \longrightarrow \mathcal{A} \xrightarrow{\varphi'} B' \xrightarrow{\psi'} \mathcal{C} \longrightarrow 0$ are two extensions of $\mathcal{A}$ by $\mathcal{C}$ constructed from $\tau \colon \mathcal{C} \to \mathcal{D}(\mathcal{A})/\mathcal{A}$ and $\tau' \colon \mathcal{C} \to \mathcal{D}(\mathcal{A})/\mathcal{A}$, respectively. Let $\theta \colon B \to \mathcal{D}(\mathcal{A})$ and $\theta' \colon B' \to \mathcal{D}(\mathcal{A})$ be the maps associated with this construction, which satisfy $\theta \circ \varphi = \rho = \theta' \circ \varphi'$. If these extensions are equivalent, there is a homeomorphic isomorphism $\theta'' \colon B \to B'$ which makes the following diagram commute.

$$
\begin{array}{ccccccccc}
0 & \longrightarrow & \mathcal{A} & \xrightarrow{\ \varphi\ } & B & \xrightarrow{\ \psi\ } & \mathcal{C} & \longrightarrow & 0 \\
 & & \big\downarrow{\scriptstyle I_{\mathcal{A}}} & & \big\downarrow{\scriptstyle \theta''} & & \big\downarrow{\scriptstyle I_{\mathcal{C}}} & & \\
0 & \longrightarrow & \mathcal{A} & \xrightarrow{\ \varphi'\ } & B' & \xrightarrow{\ \psi'\ } & \mathcal{C} & \longrightarrow & 0
\end{array}
$$

We have $\theta \circ \varphi = \rho = \theta' \circ \varphi' = (\theta' \circ \theta'') \circ \varphi$. Therefore, the uniqueness assertion of Proposition 1.2.6 implies $\theta = \theta' \circ \theta''$. Since any element of $\mathcal{C}$ can be written as $\psi(b)$ for some $b \in B$, the calculation

$$
\begin{aligned}
\tau(\psi(b)) &= \sigma \circ \theta(b) = \sigma \circ \theta' \circ \theta''(b) \\
&= \tau' \circ \psi' \circ \theta''(b) = \tau'(\psi(b)) \qquad \forall\, b \in B
\end{aligned}
$$

proves $\tau = \tau'$. Hence the map is injective.

Now suppose an extension

$$
0 \longrightarrow \mathcal{A} \xrightarrow{\ \varphi\ } B \xrightarrow{\ \psi\ } \mathcal{C} \longrightarrow 0
$$

of $\mathcal{A}$ by $\mathcal{C}$ is given. Proposition 1.2.6 shows that there is a continuous homomorphism $\theta \colon B \to \mathcal{D}(\mathcal{A})$ satisfying $\rho = \theta \circ \varphi$. Define $\tau \in \mathrm{CHom}(\mathcal{C}, \mathcal{D}(\mathcal{A})/\mathcal{A})$ by $\tau(\psi(b)) = \sigma \circ \theta(b)$ for all $\psi(b) \in \mathcal{C}$. First we check that τ is well defined. If $\psi(b') = \psi(b)$, then there is some $a \in \mathcal{A}$ satisfying $b' = b + \varphi(a)$, so $\sigma(\theta(b')) = \sigma(\theta(b) + \rho(a)) = \sigma(\theta(b))$ as desired. The remark following

Definition 1.2.8, that $\mathcal{C}$ is homeomorphically isomorphic to $\mathcal{B}/\varphi(\mathcal{A})$ with its quotient norm, shows that τ is continuous.

Now the diagram in (22) satisfies all the conditions of diagram (20) and thus, by uniqueness, $0 \longrightarrow \mathcal{A} \overset{\varphi}{\longrightarrow} \mathcal{B} \overset{\psi}{\longrightarrow} \mathcal{C} \longrightarrow 0$ is the extension defined by τ up to equivalence. Hence the map is surjective.

Finally we prove the results about semidirect and direct products. First, suppose τ can be written as $\tau = \sigma \circ \omega$ for $\omega \in \mathrm{CHom}(\mathcal{C}, \mathcal{D}(\mathcal{A}))$. Then it is easy to check that $\chi \colon \mathcal{C} \to \mathcal{B} \subseteq \mathcal{D}(\mathcal{A}) \times \mathcal{C}$ defined by $\chi(c) = (\omega(c), c)$ shows that the extension splits (*i.e.*, the extension is semidirect). Conversely, if the splitting map $\chi \colon \mathcal{C} \to \mathcal{B}$ is given, we may define ω to be $\theta \circ \chi$. The statement about when the extension is a direct product is even easier to check. $\qquad\qquad\qquad\qquad\qquad\qquad\qquad\qquad\qquad\qquad\qquad\qquad \Box$

We have discussed extensions of Banach algebras because this is the case in which we are most interested. However the reader can easily strip away all the norm and continuity assumptions from the above discussion and obtain an improvement of Hochschild's original theory [1947] for extensions of algebras. Hochschild also considers the case in which the annihilator ideal is not necessarily zero.

1.3   Sums, Products and Limits

In this section we introduce a number of ways in which new algebras or Banach algebras can be constructed from a collection of algebras. No deep results are proved. The last section mentioned the direct product of two algebras as a special case of an extension. When more than a finite number of algebras are involved, direct products and direct sums are two very different generalizations of this simple case.

Direct Products

Let $\mathcal{A}$ and $\mathcal{C}$ be Banach algebras and give the Cartesian product $\mathcal{A} \times \mathcal{C}$ pointwise algebra operations. Define a norm $\| \cdot \|$ by

$$\|(a, c)\| = \max\{\|a\|, \|c\|\} \qquad \forall\, (a, c) \in \mathcal{A} \times \mathcal{C}.$$

This direct product satisfies a universal mapping property. If $\{\mathcal{A}^\alpha : \alpha \in A\}$ is a family of objects in a category, then an object $\mathcal{A}$ together with maps (called projections) $\pi_\alpha \colon \mathcal{A} \to \mathcal{A}^\alpha$ for each $\alpha \in A$ is called a *product* of $\{\mathcal{A}^\alpha : \alpha \in A\}$ if, whenever another object $\mathcal{A}'$ and another collection of maps $\pi'_\alpha \colon \mathcal{A}' \to \mathcal{A}^\alpha$ for each $\alpha \in A$ is given, there exists a unique map $\varphi \colon \mathcal{A}' \to \mathcal{A}$ such that $\pi'_\alpha = \pi_\alpha \circ \varphi$ for all $\alpha \in A$. This definition determines the product up to isomorphism. It is easy to see that the direct product introduced above with projection $\pi_1(a, c) = a$ and $\pi_2(a, c) = c$ satisfies this definition in both the topological and the geometric category of Banach algebras. More generally, if $\{\mathcal{A}^\alpha : \alpha \in A\}$ is any finite set of Banach algebras, the Cartesian product $\times_{\alpha \in A}\mathcal{A}^\alpha$ can be made into a Banach algebra by coordinatewise algebraic operations and the ℓ^∞-norm:

$$\|(a_\alpha)_{\alpha \in A}\| = \max\{\|a_\alpha\| : \alpha \in A\} \qquad \forall \, (a_\alpha)_{\alpha \in A} \in \times_{\alpha \in A}\mathcal{A}^\alpha.$$

It is again clear that this construction with the obvious definition of projections gives a product in either category of Banach algebras.

In the topological category of Banach algebras this construction satisfies another universal mapping property, so long as $\{\mathcal{A}^\alpha : \alpha \in A\}$ is finite. Let $\{\mathcal{A}^\alpha : \alpha \in A\}$ be a collection of objects in a category. An object $\mathcal{A}$ together with maps (called injections) $\eta_\alpha \colon \mathcal{A}^\alpha \to \mathcal{A}$ is called a *coproduct* of $\{\mathcal{A}^\alpha : \alpha \in A\}$ if, whenever another object $\mathcal{A}'$ and another collection of maps $\eta'_\alpha \colon \mathcal{A}^\alpha \to \mathcal{A}'$ for each $\alpha \in A$ is given, there exists a unique map $\varphi \colon \mathcal{A} \to \mathcal{A}'$ satisfying $\eta'_\alpha = \varphi \circ \eta_\alpha$ for each $\alpha \in A$. This definition determines a coproduct up to isomorphism. It is again easy to check that the construction described above, together with the obvious injection maps, defines a coproduct in the topological category of Banach algebras. To get a coproduct in the geometric category, we merely need to change the norm, so long as $\{\mathcal{A}^\alpha : \alpha \in A\}$ is finite. The norm we need is the ℓ^1-*norm*:

$$\|(a_\alpha)_{\alpha \in A}\|_1 = \sum_{\alpha \in A} \|a_\alpha\| \qquad \forall \, (a_\alpha)_{\alpha \in A} \in \times_{\alpha \in A}\mathcal{A}^\alpha.$$

When we come to infinite families $\{\mathcal{A}^\alpha : \alpha \in A\}$ of algebras, normed algebras or Banach algebras, the construction of the product and coproduct diverges more decisively, and becomes dependent on whether algebras, normed algebras or Banach algebras are involved.

Throughout this work we always interpret the *Cartesian product* $\times_{\alpha \in A}\mathcal{A}^\alpha$ of a family of sets as the collection of all functions $a \colon A \to \cup_{\alpha \in A}\mathcal{A}^\alpha$ satisfying $a(\alpha) \in \mathcal{A}^\alpha$ for each $\alpha \in A$. An element of the Cartesian product is called a *cross section* when considered as a function.

The *direct product* $\Pi_{\alpha \in A}\mathcal{A}^\alpha$ of a family $\{\mathcal{A}^\alpha : \alpha \in A\}$ of algebras is simply the Cartesian product $\times_{\alpha \in A}\mathcal{A}^\alpha$ made into an algebra by coordinatewise algebraic operations.

1.3.1 Definition Let $\{\mathcal{A}^\alpha : \alpha \in A\}$ be a family of normed algebras or Banach algebras. The ℓ^∞-*direct product* $\Pi^\infty_{\alpha \in A}\mathcal{A}^\alpha$ of $\{\mathcal{A}^\alpha : \alpha \in A\}$ is the subset of the Cartesian product $\times_{\alpha \in A}\mathcal{A}^\alpha$ consisting of all cross sections a such that

$$\|a\|_\infty = \sup\{\|a(\alpha)\| : \alpha \in A\} \tag{1}$$

is finite. The subset is made into a normed algebra or Banach algebra by coordinatewise algebraic operations and the ℓ^∞-norm defined by equation (1).

It is easy to see that the above definitions subsume those given earlier and define products in all of the relevant categories when the projections are defined in the obvious way.

Direct Sums

Let $\{\mathcal{A}^\alpha : \alpha \in A\}$ be a collection of algebras or normed algebras. The *direct sum* $\oplus_{\alpha \in A} \mathcal{A}^\alpha$ of $\{\mathcal{A}^\alpha : \alpha \in A\}$ is the subset of the Cartesian product $\times_{\alpha \in A} \mathcal{A}^\alpha$ consisting of all cross sections which are zero except at a finite number of elements of A. (The finite set of α with $a(\alpha)$ non-zero varies with $a \in \oplus_{\alpha \in A} \mathcal{A}^\alpha$.) This set is made into an algebra by coordinatewise algebraic operations. If the summands $\mathcal{A}^\alpha$ are normed algebras, then the direct sum is a normed algebra with the ℓ^1-*norm* defined by

$$\|a\|_1 = \sum_{\alpha \in A} \|a(\alpha)\|.$$

The direct sum just described is frequently called the *algebraic direct sum*.

In order to define the direct sum in the categories of Banach algebras, and for many other purposes in this work, we need the notion of an unordered sum of numbers. Let A be an arbitrary index set and for each $\alpha \in A$, let a_α be a complex number. The *unordered sum* $\sum_{\alpha \in A} a_\alpha$ of this arbitrary set of numbers is defined as follows. Let $\mathcal{F}$ be the collection of all finite subsets of A. When $\mathcal{F}$ is ordered by inclusion, it is a directed set. For each $F \in \mathcal{F}$ define $a_F = \sum_{\alpha \in F} a_\alpha$. Then $\{a_F\}_{F \in \mathcal{F}}$ is a net. If this net has a limit, then this limit is the unordered sum. If the net has no limit, then the sum does not exist. If it should happen that each number a_α $(\alpha \in A)$ is non-negative, then it is easy to see that $\sup\{a_F : F \in \mathcal{F}\}$ is the limit of the net (and hence the unordered sum) if and only if it is finite. Hence for non-negative series we define $\sup\{a_F : F \in \mathcal{F}\}$ to be the unordered sum, whether or not it is finite.

1.3.2 Definition   Let $\{\mathcal{A}^\alpha : \alpha \in A\}$ be a collection of Banach algebras. The *Banach algebra direct sum* $\oplus_{\alpha \in A}^{\ell^1} \mathcal{A}^\alpha$ of $\{\mathcal{A}^\alpha : \alpha \in A\}$ is the subset of the Cartesian product $\times_{\alpha \in A} \mathcal{A}^\alpha$ consisting of all the cross sections a such that the unordered sum

$$\|a\|_1 = \sum_{\alpha \in A} \|a(\alpha)\| \tag{2}$$

is finite. This set is made into a Banach algebra by coordinatewise linear operations and the ℓ^1-norm defined by equation (2).

The definitions of direct sums of algebras, normed algebras, and Banach algebras just given subsume the previous definitions and give coproducts in each of the relevant categories with respect to the obvious injections. Suppose $\{\mathcal{A}^\alpha : \alpha \in A\}$ is a collection of Banach algebras. Then its Banach algebra direct sum $\oplus_{\alpha \in A}^{\ell^1} \mathcal{A}^\alpha$ is just the completion of its normed algebraic direct sum $\oplus_{\alpha \in A} \mathcal{A}^\alpha$.

Internal Direct Sums

The direct sums we have just discussed are sometimes called *external direct sums*. There are corresponding notions of *internal direct sums*. Let $\mathcal{A}$ be an algebra, let A be an index set, and let $\{\mathcal{I}^\alpha : \alpha \in A\}$ be a family of ideals of $\mathcal{A}$. Then the *sum* $\sum_{\alpha \in A} \mathcal{I}^\alpha$ of this family is the set of finite sums of elements from $\cup_{\alpha \in A} \mathcal{I}^\alpha$. The algebra $\mathcal{A}$ is called the *internal direct sum* of the family $\{\mathcal{I}^\alpha : \alpha \in A\}$ if $\mathcal{A}$ is the sum of the family and satisfies

$$\mathcal{I}^\alpha \cap \sum_{\beta \in A \setminus \{\alpha\}} \mathcal{I}^\beta = \{0\} \qquad \forall \, \alpha \in A.$$

When these conditions hold, we write $\mathcal{A} = \oplus_{\alpha \in A} \mathcal{I}^\alpha$. It is clear that the (external) algebraic direct sum $\oplus_{\alpha \in A} \mathcal{A}^\alpha$ is the internal direct sum of the ideals $\mathcal{I}^\alpha = \{a \in \oplus_{\alpha \in A} \mathcal{A}^\alpha : a(\beta) = 0 \text{ for all } \beta \neq \alpha\}$. Conversely, if $\mathcal{A}$ is the internal direct sum of $\{\mathcal{I}^\alpha : \alpha \in A\}$, then $\mathcal{A}$ is naturally isomorphic to the external direct sum $\oplus_{\alpha \in A} \mathcal{I}^\alpha$.

If $\mathcal{A}$ is a normed algebra, we say it is the *internal direct sum* of the family $\{\mathcal{I}^\alpha : \alpha \in A\}$ if each $\mathcal{I}^\alpha$ is a closed ideal, $\mathcal{A} = \sum_{\alpha \in A} \mathcal{I}^\alpha$ and

$$\mathcal{I}^\alpha \cap \overline{\sum_{\beta \in A \setminus \{\alpha\}} \mathcal{I}^\beta} = \{0\} \qquad \forall \, \alpha \in A, \tag{3}$$

where the bar denotes closure. Again it is easy to check that the external direct sum $\oplus_{\alpha \in A} \mathcal{A}^\alpha$ of a family $\{\mathcal{A}^\alpha : \alpha \in A\}$ of normed algebras is the internal direct sum of $\{\mathcal{I}^\alpha : \alpha \in A\}$, where each $\mathcal{I}^\alpha$ is defined by $\mathcal{I}^\alpha = \{a \in \oplus_{\alpha \in A} \mathcal{A}^\alpha : a(\beta) = 0 \text{ for all } \beta \neq \alpha\}$. Suppose $\mathcal{A}$ is the internal direct sum of $\{\mathcal{I}^\alpha : \alpha \in A\}$. The definition of a coproduct guarantees that there is a map (in either category) of the external direct sum onto $\mathcal{A}$. The map is an algebra isomorphism but not necessarily an isomorphism in either category of normed algebras.

In either category of Banach algebras we say that $\mathcal{A}$ is the *internal direct sum* of $\{\mathcal{I}^\alpha : \alpha \in A\}$ if each $\mathcal{I}^\alpha$ is a closed ideal, $\sum_{\alpha \in A} \mathcal{I}^\alpha$ is dense in $\mathcal{A}$ and $\mathcal{I}^\alpha \cap \overline{\sum_{\beta \in A \setminus \{\alpha\}} \mathcal{I}^\beta}$ is zero for each $\alpha \in A$. Clearly an external direct sum is an internal direct sum as before. If $\mathcal{A}$ is the internal direct sum of $\{\mathcal{I}^\alpha : \alpha \in A\}$, then the definition of a coproduct shows that there is a map (in either category) of the external direct sum $\oplus_{\alpha \in A}^{\ell^1} \mathcal{I}^\alpha$ onto a dense subset of $\mathcal{A}$ (which includes $\sum_{\alpha \in A} \mathcal{I}^\alpha$), but this map need not be a homeomorphism nor onto $\mathcal{A}$.

Subdirect Products

Finally, we discuss what are usually called subdirect sums of algebras and normed algebras, but what are more correctly called subdirect products. These arise naturally in many contexts as will be shown in Chapter 4 and elsewhere. It should be noted at the outset that representation of an algebra as a subdirect product furnishes relatively little information about its structure unless additional information, such as that discussed in Chapter 6, is available. Subdirect products were first considered by H. Prüfer [1925] for commutative rings and by Gottfried Köthe [1930] for noncommutative rings. For further information and history, consult Neal H. McCoy [1947]. The normed case is an obvious extension of the non-normed case. Subdirect products are simply algebras, normed algebras or Banach algebras together with a suitable embedding as a subalgebra, or normed subalgebra of a direct product. Direct sums are examples of subdirect products when provided with the obvious embedding into the direct product.

1.3.3 Definition Let A be an index set and let $\{\mathcal{A}^\alpha : \alpha \in A\}$ be a family of algebras. A *subdirect product* of $\{\mathcal{A}^\alpha : \alpha \in A\}$ is a subalgebra $\mathcal{A}$ of $\Pi_{\alpha \in A}\mathcal{A}^\alpha$ such that for each $\alpha \in A$, $\pi_\alpha : \mathcal{A} \to \mathcal{A}^\alpha$ is surjective. The *subdirect product* in either category of normed algebras or Banach algebras is simply a normed or complete normed subalgebra $\mathcal{A}$ of $\Pi_{\alpha \in A}\mathcal{A}^\alpha$ defined in the appropriate category with the same restriction that each $\pi_\alpha : \mathcal{A} \to \mathcal{A}^\alpha$ should be surjective.

Let $\mathcal{A}$ be a subdirect product in any of these categories and, for each $\beta \in A$, define $\mathcal{I}_\beta$ to be $\{a \in \mathcal{A} \subseteq \times_{\alpha \in A}\mathcal{A}^\alpha : a(\beta) = 0\}$. Clearly, $\mathcal{I}_\alpha$ is an ideal which is closed in the normed cases. Furthermore, each π_α can be identified with the natural map $\mathcal{A} \to \mathcal{A}/\mathcal{I}_\alpha \simeq \mathcal{A}^\alpha$. Finally, $\cap_{\alpha \in A}\mathcal{I}_\alpha = \{0\}$ holds. Conversely, if $\{\mathcal{I}_\alpha : \alpha \in A\}$ is any collection of ideals (closed in the normed case) with $\cap_{\alpha \in A}\mathcal{I}_\alpha = \{0\}$, then $\mathcal{A}$ is isomorphic to a subdirect product of $\{\mathcal{A}/\mathcal{I}_\alpha : \alpha \in A\}$. This is the way in which subdirect products most commonly arise.

All of the foregoing discussion of direct products, direct sums, internal direct sums and subdirect products applies to linear spaces, normed linear spaces and Banach spaces. Simply dropping all references to multiplicative structure, (and thus replacing ideals by linear subspaces) makes this clear. Later we will need to deal with direct sums in the category of Hilbert spaces. The discussion of that case is postponed until needed.

Inductive or Direct Limits

We begin by defining inductive limits which can be considered as a massive extension of direct sums. We will then give one example which is important in the definition of K-theory. The example is often called the algebra of matrices of arbitrary size.

1.3.4 Definition Let $(A, \leq)$ be a directed set, and for each $\alpha \in A$ let $\mathcal{A}_\alpha$ be an algebra. For each $\alpha, \beta \in A$ with $\alpha \leq \beta$, let $\varphi_{\beta\alpha} : \mathcal{A}_\alpha \to \mathcal{A}_\beta$ be a homomorphism; these will be called *connecting homomorphisms*. For each

$\alpha \leq \beta \leq \gamma \in A$, let these homomorphisms satisfy:

$$\varphi_{\gamma\alpha} = \varphi_{\gamma\beta} \circ \varphi_{\beta\alpha}. \tag{4}$$

The above set of data will be called an *inductive* (or *direct*) *system of algebras*. If, in addition, each $\mathcal{A}_\alpha$ is a normed algebra and each φ_α is continuous, the inductive system will be called an *inductive system of normed algebras*. Finally if all these conditions hold and, for each $\alpha \in A$, the limit superior

$$\limsup_{\beta \in A} \|\varphi_{\beta\alpha}\| \tag{5}$$

is finite, then the inductive system will be called a *normed inductive system of normed algebras*. (When all the $\mathcal{A}_\alpha$ are Banach algebras, the uniform boundedness principle shows that this last condition is satisfied if and only if $\limsup_{\beta \in A} \|\varphi_{\beta\alpha}(a)\|$ is bounded for each $\alpha \in A$ and each $a \in \mathcal{A}_\alpha$).

The *normed inductive limit* (or *normed direct limit*) of such a normed inductive system of normed algebras $(\mathcal{A}_\alpha, \varphi_{\beta\alpha} : \alpha, \beta \in A)$ is the algebra $\mathcal{A}$, denoted by $\mathcal{A} = \varinjlim_\alpha \mathcal{A}_\alpha$ and the continuous homomorphisms $\varphi_\alpha : \mathcal{A}_\alpha \to \mathcal{A}$, defined as follows. Let $\tilde{A}$ be the disjoint union of all the algebras $\mathcal{A}_\alpha$ for $\alpha \in A$. Consider the equivalence relation $\sim$ defined on this set by $a_\alpha \sim b_\beta$ if and only if there exists a $\gamma \in A$ with $\alpha \leq \gamma$, $\beta \leq \gamma$ and $\varphi_{\gamma\alpha}(a_\alpha) = \varphi_{\gamma\beta}(b_\beta)$. Let $\tilde{\tilde{A}}$ be the quotient of $\tilde{A}$ by this equivalence relation. It is easy to check that, for any finite set of elements in $\tilde{\tilde{A}}$, there is an index $\gamma \in A$ large enough so that all of the equivalence classes have elements from $\mathcal{A}_\gamma$. Hence $\tilde{\tilde{A}}$ has the structure of an algebra if we merely retain the algebra operations from each of the $\mathcal{A}_\alpha$ which are consistent by the inductive nature of the homomorphisms. Finally, it is obvious that $\limsup_{\beta \in A} \|\varphi_{\beta\alpha}(a)\|$ defines an algebra semi-norm on $\tilde{\tilde{A}}$. Let $(A, \|\cdot\|)$ be the normed algebra obtained by dividing out the ideal on which this semi-norm vanishes. For each $\alpha \in A$ and each $a \in \mathcal{A}_\alpha$, define $\varphi_\alpha(a)$ to be the equivalence class in $\mathcal{A}$ of a.

The *inductive limit* (or *algebraic inductive limit*) of an inductive system of algebras is defined by simply omitting the last quotient by the ideal on which the semi-norm vanishes.

We leave it to the reader to show that the normed inductive limit satisfies $\varphi_\alpha = \varphi_\beta \circ \varphi_{\beta\alpha}$ for all $\alpha \leq \beta$ and is the unique normed algebra such that any other normed algebra $\mathcal{B}$ and system of continuous homomorphisms $\{\Psi_\alpha : \alpha \in A\}$ satisfying $\Psi_\alpha = \Psi_\beta \circ \varphi_{\beta\alpha}$ for all $\alpha \leq \beta$ factors through $\mathcal{A}$ in the obvious way. (A similar universal property holds for algebraic inductive limits.) Furthermore, it is clear that if each of the homomorphisms in the inductive system is injective and the limit superior semi-norm is a norm (*e.g.*, if all the homomorphisms are isometries), then the inductive limit is essentially just the increasing union of the normed algebras $\mathcal{A}_\alpha$. Even for general inductive systems, the inductive limit is essentially an increasing union of the $\varphi_\alpha(\mathcal{A}_\alpha)$.

1.3.5 The Algebra $M_\infty(\mathcal{A})$ of Matrices of Arbitrarily Large Size over an Algebra $\mathcal{A}$

This is an important general example of a normed inductive limit of algebras. (In this case and others where the connecting homomorphisms are isometries, the normed inductive limit and algebraic inductive limit coincide.) Let $\mathcal{A}$ be an algebra or a normed algebra and let $M_n(\mathcal{A})$ be the algebra of $n \times n$ matrices over $\mathcal{A}$ described in Definition 1.6.9. If $\mathcal{A}$ is a normed algebra or a Banach algebra, then $M_n(\mathcal{A})$ is a normed algebra or a Banach algebra under various natural norms as noted before Proposition 1.6.10. Construct an inductive system with the positive integers $\mathbb{N}$ as index set, the algebra $M_n(\mathcal{A})$ corresponding to $n \in \mathbb{N}$ and with the connecting maps defined by sending each smaller matrix into a larger one by putting it in the upper left-hand corner and completing the matrix with zeroes. Clearly this is an inductive limit, and each homomorphism is an isometry under any of the norms mentioned above if $\mathcal{A}$ is a normed algebra or a Banach algebra. The limit is essentially the union of matrix algebras of arbitrary size over $\mathcal{A}$, so that any algebraic operation with a finite number of elements can always be carried out in a specific size matrix algebra over $\mathcal{A}$, but the size may increase as needed in any series of computations. Operations in this algebra bring out some of the more subtle structure of the original algebra $\mathcal{A}$. It is used in most definitions of K-theory.

Of course we could complete $M_\infty(\mathcal{A})$ or any other normed algebra. It is worth noting that if $\mathcal{A}$ is a C*-algebra, then the completion of $M_\infty(\mathcal{A})$ has an interesting interpretation. First, $M_\infty(\mathcal{A})$ itself is isometrically isomorphic to a dense subalgebra of the tensor product of $\mathcal{A}$ with the Banach algebra of all compact operators on separable Hilbert space. This tensor product has a natural norm. The appropriate norm for the inductive limit (relative to which this isomorphism is an isometry) comes from applying a certain operator norm to the individual matrix algebras, which extends the C*-norm of the original algebra. Then the completions of the algebra of arbitrarily large matrices and of the tensor product are naturally isometrically isomorphic. The importance of this tensor product was recognized before the natural inductive limit construction of $M_\infty(\mathcal{A})$ was appreciated. (See Section 1.10 and Volume II.)

Projective or Inverse Limits

We again begin with a definition of this concept which has about the same relationship to direct products as the inductive limit has to direct sums. In some sense it is the dual construction to that of inductive limits. In the case of inductive limits, we were interested in limits in the category of normed algebras, but here we are primarily interested in the category of topological algebras. We will again give one example.

1.3.6 Definition Let $(A, \leq)$ be a directed set, and for each $\alpha \in A$ let $\mathcal{A}_\alpha$ be an algebra. For each $\alpha, \beta \in A$ with $\alpha \leq \beta$, let $\varphi_{\beta\alpha} \colon \mathcal{A}_\beta \to \mathcal{A}_\alpha$ be a homomorphism. For each $\alpha \leq \beta \leq \gamma \in A$, let these homomorphisms satisfy:

$$\varphi_{\gamma\alpha} = \varphi_{\beta\alpha} \circ \varphi_{\gamma\beta}. \tag{6}$$

The above set of data will be called a *projective* (or *inverse*) *system of algebras*. If in addition each $\mathcal{A}_\alpha$ is a normed algebra and each φ_α is continuous, then the projective system will be called a *normed projective system of normed algebras*. The *projective limit* of such a projective system $(\mathcal{A}_\alpha, \varphi_{\beta\alpha} : \alpha, \beta \in A)$ is the algebra $\mathcal{A}$, denoted by $\varprojlim_\alpha \mathcal{A}_\alpha$ and the homomorphisms $\varphi_\alpha \colon \mathcal{A} \to \mathcal{A}_\alpha$ satisfying $\varphi_{\beta\alpha} \circ \varphi_\beta = \varphi_\alpha$ defined as follows. Let

$$\mathcal{A} = \{a \in \prod_{\alpha \in A} \mathcal{A}_\alpha : \varphi_{\beta\alpha}(a_\beta) = a_\alpha \quad \text{for all } \beta \geq a\}, \tag{7}$$

and endow this collection of coherent cross sections with the obvious pointwise operations. For each α, the map φ_α is just the projection map onto the α factor. In the case of a projective system of normed algebras, this same algebraic projective limit can be given the weakest topology making all of the projection maps φ_α continuous. This topology, which is just the relativized topology from the direct product, is seldom a norm topology.

It easy to check that these projective limits satisfy the universal property that if $\mathcal{B}$ is another algebra (or topological algebra) with a family of (continuous) homomorphisms $\Psi_\alpha \colon \mathcal{B} \to \mathcal{A}_\alpha$ satisfying $\varphi_{\beta\alpha} \circ \Psi_\beta = \Psi_\alpha$, then there is a unique map $\Psi \colon \mathcal{B} \to \mathcal{A}$ satisfying $\varphi_\alpha \circ \Psi$ for all $\alpha \in \mathcal{A}$.

1.3.7 A Projective Limit Algebra This example should help make the concept more comprehensible. Consider the directed set of all compact subsets of the real line ordered by inclusion. For each compact subset K, let $\mathcal{A}_K$ be the Banach algebra $C(K)$ of all continuous complex-valued functions on K with the supremum norm $\| \cdot \|_\infty$. Then φ_{LK} is just the restriction map when L includes K. As an algebra, the projective limit is just the collection of all continuous (not necessarily bounded) complex-valued functions on $\mathbb{R}$ since any coherent family must certainly come from such a function. (Or define one, depending on one's viewpoint.) However, the topology, when this is considered as a projective limit of topological algebras, is the topology of uniform convergence on each compact subset. This easy example exhibits one surprising phenomenon. Although the directed set is very complicated, it has a cofinal subset with the order structure of the positive integers: $\{[-n, n] : n \in \mathbb{N}\}$. In general, projective limits are much better behaved when such a cofinal subset is available.

Ultraproducts and Ultrapowers

These concepts have proved valuable in Banach space theory, but are only beginning to be used in Banach algebra theory. We will provide a brief introduction. Details can be found in Brailey Sims [1982] and Stefan Heinrich [1980].

Let I be an infinite index set. A *filter* of subsets of I is a nonempty family $\mathcal{F}$ of nonempty subsets of I satisfying:

(1) If $A, B \in \mathcal{F}$, then $A \cap B \in \mathcal{F}$, and
(2) If $A \in \mathcal{F}$ and $A \subseteq B \subseteq I$, then $B \in \mathcal{F}$.

(Some authors allow a filter of subsets to contain the empty set $\emptyset$. Then the collection of all subsets of I is the only filter of subsets containing $\emptyset$. With this convention, our filters are simply called *proper filters*.) Probably the most familiar filter is the family of all neighborhoods of a point in a topological space. Given any nonempty subset S of I, the family of all subsets of I which include S is a filter denoted by $\mathcal{F}_S$. Still another filter on I is the *Fréchet filter* of all the subsets of I which have finite complement in I.

A filter of subsets is called an *ultrafilter* if it is maximal (under inclusion) among filters. Every filter is included in some ultrafilter. It is easy to see that a filter of subsets is an ultrafilter if and only if for each subset A of I exactly one of A and $I \setminus A$ belongs to the filter. Another easy argument shows that this is equivalent to:

If $A \cup B$ belongs to $\mathcal{F}$, then either $A \in \mathcal{F}$ or $B \in \mathcal{F}$ holds.

Hence for any ultrafilter $\mathcal{F}$, $\bigcap \{A : A \in \mathcal{F}\}$ is either a singleton set or empty. Note that $\mathcal{F}_S$ introduced above is an ultrafilter if and only if $S = \{p\}$ is a singleton set. In this case we write $\mathcal{F}_p$ instead of $\mathcal{F}_S$. The only ultrafilters with $\bigcap \{A : A \in \mathcal{F}\}$ nonempty are those of the form $\mathcal{F}_p$. We call these *fixed (or trivial) ultrafilters* and all others *free ultrafilters*. Hence an ultrafilter is fixed if and only if it contains a finite subset, and it is free if and only if $\bigcap \{A : A \in \mathcal{F}\} = \emptyset$. Every filter of subsets is included in an ultrafilter. We will require that our ultrafilters be *countably incomplete*. This means that the ultrafilter is not closed under countable intersections. It is equivalent to requiring that the intersection of the sets belonging to some countable subset of the ultrafilter be the empty set. It certainly implies that the ultrafilter is free. From now on $\mathcal{F}$ will be a particular countably incomplete ultrafilter on I.

For each $i \in I$, let $\mathcal{A}^i$ be a $\langle$ Banach algebra / Banach space $\rangle$. Our construction begins with the ℓ^∞-direct product $\Pi_{i \in I}^\infty \mathcal{A}^i$ of $\{\mathcal{A}^i : i \in I\}$ of Definition 1.3.1. The case in which all the $\mathcal{A}^i$ equal a fixed $\mathcal{A}$ is important. For any $a \in \Pi_{i \in I}^\infty \mathcal{A}^i$, we can define $\|a\|$ to be the unique non-negative number such that for every $\varepsilon > 0$ the set $\{i \in I : \|a\| - \varepsilon < \|a(i)\| < \|a\| + \varepsilon\}$ is in $\mathcal{F}$. We will take the existence, uniqueness and other properties of $\| \cdot \|$

for granted. Let $\mathcal{B}$ be the quotient of $\Pi_{i \in I}^{\infty} \mathcal{A}^i$ modulo the subset of elements with norm 0. Then $\mathcal{B}$ has the structure of a ⟨ Banach algebra / Banach space ⟩.

1.3.8 Definition   The ⟨ Banach algebra / Banach space ⟩ $\mathcal{B}$ defined above is called the *ultraproduct with respect to $\mathcal{F}$ of the family* $\{\mathcal{A}^i\}_{i \in I}$. If all $\mathcal{A}^i = \mathcal{A}$, then it is called the *ultrapower of $\mathcal{A}$ with respect to $\mathcal{F}$*.

It is not easy to give an illuminating example, but the reader may look at the theory of ultraprime ideals introduced in Definition 4.1.11. The paper by Michael Cowling and Gero Fendler [1984] has interesting applications to representations on Banach spaces (*cf.* Chapter 4 below). See also the proceedings of a conference edited by Nigel J. Cutland [1988].

1.4   Arens Multiplication

Introduction

Richard Arens defined two products on the double dual $\mathcal{A}^{**}$ of any normed algebra $\mathcal{A}$ [1951a], [1951b]. The first systematic study of these products was carried out by Paul Civin and Bertram Yood [1961]. A useful compilation and simplification of results up to that date was published by John Duncan and Seyed Ali Reza Hosseiniun [1979]. (See also Harald Hanche-Olsen [1980] for a different approach to a related problem.)

Each product makes $\mathcal{A}^{**}$ into a Banach algebra and the canonical injection $\kappa: \mathcal{A} \to \mathcal{A}^{**}$ is a homomorphism for both products. (This canonical injection is just the evaluation map: $\kappa(a)(\omega) = \omega(a)$ for any $a \in \mathcal{A}$ and $\omega \in \mathcal{A}^*$.) We shall denote the two products by juxtaposition and by a dot, respectively, (*i.e.*, we use fg and $f \cdot g$ for the two products of f and g in $\mathcal{A}^{**}$).

In most cases the two products do not agree and the double dual algebra with either product is too large, too complicated and too badly behaved to be useful without some simplification. However, there are special cases (notably the C*-algebras which will be considered in the second volume of this work) in which this algebra is easy to comprehend and exploit. In other cases, quotients of the algebra modulo various natural ideals are easier to exploit. In all cases that have been studied in detail, the structure is intimately connected with that of the original algebra and has interesting, if sometimes complicated, relationships to significant mathematical objects.

In this section we explore identities and one-sided identities in the Arens products, connections with double centralizers, the case in which the two products agree and a few simple examples. Two representations which also stem from Arens' very general construction are defined.

Although the two Arens products are usually called the first and the second Arens product, they stand on a completely equal footing since their definition simply depends on choosing left or right first. The first product is continuous in the weak* topology in its first (left) variable for any choice of the second (right) variable in $\mathcal{A}^{**}$; it is also continuous in the same topology in its second variable for any choice of the first variable in the canonical image of $\mathcal{A}$ in $\mathcal{A}^{**}$. (That is, $f \mapsto fg$ is continuous in $f \in \mathcal{A}^{**}$ for any $g \in \mathcal{A}^{**}$, and $g \mapsto \kappa(a)g$ is continuous in g for any $a \in \mathcal{A}$.) The second product enjoys the reflected continuity properties. (That is, $g \mapsto f \cdot g$ is continuous in $g \in \mathcal{A}^{**}$ for any $f \in \mathcal{A}^{**}$, and $f \mapsto f \cdot \kappa(a)$ is continuous in f for any $a \in \mathcal{A}$.) Basic properties of the weak* topology guarantee that any element $f \in \mathcal{A}^{**}$ is the limit of a net in $\kappa(\mathcal{A})$ within the ball of radius $\|f\|$ (cf. Dunford and Schwartz [1958] V.4.2). These facts allow the calculation of both products, but the reader should be warned that this is a subject on which an unusual number of false results have been published, and several of the errors stem from attempting to use continuity results in proofs. The entirely algebraic original construction is less subject to erroneous interpretation. We now give that definition, which guarantees the existence of products with the continuity properties just described.

Definitions and Basic Properties

1.4.1 Definition   Let $\mathcal{A}$ be a normed algebra, let $\kappa: \mathcal{A} \to \mathcal{A}^{**}$ be the natural injection and let L and R be the left and right regular representations of $\mathcal{A}$. For any $a \in \mathcal{A}$ and $\omega \in \mathcal{A}^*$, define elements ω_a and $_a\omega$ of $\mathcal{A}^*$ by

$$\omega_a = (L_a)^*(\omega) \quad \text{and} \quad _a\omega = (R_a)^*(\omega). \tag{1}$$

For any $\omega \in \mathcal{A}^*$ and $f \in \mathcal{A}^{**}$, define elements $_f\omega$ and ω_f of $\mathcal{A}^*$ by

$$_f\omega(a) = f(\omega_a) \quad \text{and} \quad \omega_f(a) = f(_a\omega) \tag{2}$$

for all $a \in \mathcal{A}$. Finally for any $f, g \in \mathcal{A}^{**}$ and $\omega \in \mathcal{A}^*$, define elements fg and $f \cdot g$ of $\mathcal{A}^{**}$ by

$$fg(\omega) = f(_g\omega) \quad \text{and} \quad f \cdot g(\omega) = g(\omega_f). \tag{3}$$

These two products in $\mathcal{A}^{**}$ are called the *first* and *second Arens product*, respectively. When the two products coincide, the algebra $\mathcal{A}$ is said to be *Arens regular*.

We gather some elementary properties of these two products.

1.4.2 Theorem   *The elements ω_a and $_a\omega$ belong to $\mathcal{A}^*$ and satisfy:*

$$\omega_a(b) = \omega(ab) \qquad _a\omega(b) = \omega(ba) \tag{4}$$

$$\|\omega_a\| \le \|\omega\| \, \|a\| \qquad \|_a\omega\| \le \|\omega\| \, \|a\| \tag{5}$$

$$(\omega_a)_b = \omega_{ab} \quad (_a\omega)_b = {}_a(\omega_b) \quad {}_a(_b\omega) = {}_{ab}\omega \tag{6}$$

for all $\omega \in \mathcal{A}^*$ and $a, b \in \mathcal{A}$. The elements $_f\omega$ and ω_f belong to $\mathcal{A}^*$ and satisfy

$$\|_f\omega\| \le \|\omega\|\|f\| \quad\quad \|\omega_f\| \le \|\omega\|\|f\| \tag{7}$$

$$(_f\omega)_a = {}_f(\omega_a) \quad\quad {}_a(\omega_f) = (_a\omega)_f \tag{8}$$

$$\kappa(a)\omega = {}_a\omega \quad\quad \omega_{\kappa(a)} = \omega_a \tag{9}$$

for all $a \in \mathcal{A}$, $\omega \in \mathcal{A}^*$ and $f \in \mathcal{A}^{**}$. Finally the two products satisfy

$$\|fg\| \le \|f\| \, \|g\| \quad\quad \|f \cdot g\| \le \|f\| \, \|g\| \tag{10}$$

$$_{fg}\omega = {}_f(_g\omega) \quad\quad \omega_{f \cdot g} = (\omega_f)_g \tag{11}$$

$$\kappa(a)f = \kappa(a) \cdot f = (L_a)^{**}(f) \quad f\kappa(a) = f \cdot \kappa(a) = (R_a)^{**}(f) \tag{12}$$

$$\kappa(a)\kappa(b) = \kappa(a) \cdot \kappa(b) = \kappa(ab) \tag{13}$$

for all $a, b \in \mathcal{A}$, $\omega \in \mathcal{A}^*$, $f, g \in \mathcal{A}^{**}$.

Each product makes $\mathcal{A}^{**}$ into a Banach algebra, and κ is an injective homomorphism from $\mathcal{A}$ to $\mathcal{A}^{**}$ with respect to either Arens product. The two products agree whenever one of the factors is in $\kappa(\mathcal{A})$. Furthermore, the map $\langle\, \overline{L}: \mathcal{A}^{**} \to \mathcal{B}(\mathcal{A}^*) \,/\, \overline{R}: \mathcal{A}^{**} \to \mathcal{B}(\mathcal{A}^*) \,\rangle$ defined by $\langle\, \overline{L}_f(\omega) = {}_f\omega \,/\, \overline{R}_f(\omega) = \omega_f \,\rangle$ is a $\langle$ homomorphism / anti-homomorphism $\rangle$ with respect to the $\langle$ first / second $\rangle$ Arens product.

The following maps are continuous in the weak* topology (i.e., the $\mathcal{A}^*$-topology) on $\mathcal{A}^{**}$:

$$f \mapsto fg \quad g \mapsto f \cdot g \tag{14}$$

$$g \mapsto \kappa(a)g \quad f \mapsto f \cdot \kappa(a) \tag{15}$$

for all $a \in \mathcal{A}$ and $f, g \in \mathcal{A}^{**}$.

If $\varphi: \mathcal{A} \to \mathcal{B}$ is a continuous $\langle$ homomorphism / anti-homomorphism $\rangle$, then $\varphi^{**}: \mathcal{A}^{**} \to \mathcal{B}^{**}$ is a continuous $\langle$ homomorphism / anti-homomorphism $\rangle$ with respect to $\langle$ either Arens product / the opposite Arens products $\rangle$ on $\mathcal{A}^{**}$ and $\mathcal{B}^{**}$.

Proof All these results are easy to obtain from step by step calculations starting with the three step definition.   □

We will show in Proposition 2.5.3(e) that $\kappa(\mathcal{A})$ is a spectral subalgebra of $\mathcal{A}^{**}$ with respect to either Arens product and in 4.1.15 that various kinds of ideals in $\mathcal{A}$ and $\mathcal{A}^{**}$ are closely related. In Proposition 8.3.7 and Theorem 8.7.14 we will give more details on the relationship between minimal ideals in $\mathcal{A}$ and $\mathcal{A}^{**}$.

Fereidoun Ghahramani, Anthony To-Ming Lau [1988] and Ghahramani, Lau and Viktor Losert [1990] show that, for locally compact groups G and

H, any isometric isomorphism $\varphi \colon L^1(G)^{**} \to L^1(H)^{**}$ sends $L^1(G)$ onto $L^1(H)$ and thus comes from a homeomorphic isomorphism of G onto H.

Michael Grosser and Losert [1984] consider a natural quotient of $\mathcal{A}^{**}$ which has an interesting interpretation in the case of group algebras $L^1(G)$ of locally compact abelian groups. For compact groups, Nilgun Isik, John Pym and Ali Ülger [1987] give somewhat complicated convolution formulas for $L^1(G)^{**}$ considered as a measure algebra on a semigroup which is the product of the original group and a rather degenerate semigroup constructed from it. See also Robert S. Doran and Wayne Tiller [1983], D. J. Parsons [1984] and Lau [1979], [1981], [1983], [1986], [1987], Lau and Losert [1988], Lau and Pym [1990].

Bruno Iochum and G. Loupias [1989] use an ultraproduct construction to study the Arens product. S. L. Gulick [1966a] and [1966b] and Nicholas J. Young in [1976] extended the definition of Arens products from normed algebras to more general algebras. The latter paper considers locally convex algebras in which the product of any two bounded sets is bounded. See also Nilgün Arikan [1981], [1982], [1983] and Pym and Ülger [1989].

Identities and One-sided Identities for $\mathcal{A}^{**}$

If $\mathcal{A}$ has an identity element 1, it is obvious that $\kappa(1)$ is an identity element for $\mathcal{A}^{**}$ with either Arens product. However, it is common for $\mathcal{A}^{**}$ to have an identity, or at least a one-sided identity, even when $\mathcal{A}$ does not. Left and right one-sided identities behave quite differently with respect to the two Arens products. It is comparatively easy for there to be a ⟨ right / left ⟩ identity for the ⟨ first / second ⟩ Arens product and e is such an element if and only if it satisfies

$$\begin{array}{ll} \langle\, e(\omega_a) = \omega(a) \,/\, e(_a\omega) = \omega(a) \,\rangle & \text{or equivalently} \\ \langle\, _e\omega = \omega \,/\, \omega_e = \omega \,\rangle & \end{array} \qquad (16)$$

for all $a \in \mathcal{A}$ and $\omega \in \mathcal{A}^*$. A ⟨ left / right ⟩ identity for the ⟨ first / second ⟩ Arens product must also be a ⟨ left / right ⟩ identity for the ⟨ second / first ⟩ Arens product and is less common. This implication follows from the fact that an element $e \in \mathcal{A}^{**}$ is a ⟨ left / right ⟩ identity for the ⟨ first / second ⟩ Arens product if and only if it satisfies

$$\langle\, e(_f\omega) = f(\omega) \,/\, e(\omega_f) = f(\omega) \,\rangle \qquad \forall\, f \in \mathcal{A}^{**}\ \omega \in \mathcal{A}^*, \qquad (17)$$

and these conditions imply (16) (in reverse order) by equation (9).

For reasons to be considered in Chapter 5 below, it is fairly common, and quite interesting, for $\mathcal{A}^{**}$ to have a right identity for the first Arens product which is also a left identity for the second. We call such an element a *mixed identity* for $\mathcal{A}^{**}$. Let us denote one by $e \in \mathcal{A}^{**}$. This condition is equivalent to assuming $_e\omega = \omega = \omega_e$ or, equivalently, that $e(\omega_a) = \omega(a) = e(_a\omega)$ for all $\omega \in \mathcal{A}^*$ and $a \in \mathcal{A}$. There is no reason to expect that a mixed identity is unique. In fact, any $e' \in \mathcal{A}^{**}$ such that e and

e' agree on $(L_A)^*(A^*) + (R_A)^*(A^*)$ is also a mixed identity. Anticipating the terminology and results of Chapter 5, we note that A^{**} has a mixed identity if and only if A has a bounded two-sided approximate identity. In this situation we say that A is *approximately unital*. Similar results involving one-sided approximate identities hold for a right identity for the first Arens product or a left identity for the second product individually even if they are not mixed identities. If A^{**} has a ⟨ left / right ⟩ identity for either product, then equations (16) and (4) show that the annihilator ideal ⟨ A_{RA} / A_{LA} ⟩ ⟩ is zero.

The situation on uniqueness of mixed identities changes if A^{**} has a (two-sided) identity element for either one of the Arens products. In that case any mixed identity must equal this unique (two-sided) identity since ⟨ $e = ee' = e'$ / $e = e'e = e'$ ⟩ if e is an identity and e' is a ⟨ right / left ⟩ identity.

When A^{**} is unital or when it just has a mixed identity, there is another way to consider the Arens product which is helpful in some situations. In slightly less generality, it was first published by Duncan and Hosseiniun [1979]. Recall that Theorem 1.4.2 already showed the existence of a ⟨ homomorphism $\overline{L}: A^{**} \to B(A^*)$ / anti-homomorphism $\overline{R}: A^{**} \to B(A^*)$ ⟩ defined by ⟨ $\overline{L}_f(\omega) = {}_f\omega$ / $\overline{R}_f(\omega) = \omega_f$ ⟩. In fact we have already derived in equations (1) and (8), but not yet stated explicitly, that the range of ⟨ $\overline{L}$ / $\overline{R}$ ⟩ is included in the commutant ⟨ $((L_A)^*)'$ / $((R_A)^*)'$ ⟩. When A^{**} has a one- or two-sided identity we get the following.

1.4.3 Proposition *Let A be a normed algebra. Then ⟨ $\overline{L}$ / $\overline{R}$ ⟩ has the whole commutant ⟨ $((L_A)^*)'$ / $((R_A)^*)'$ ⟩ as its range if and only if A^{**} has a ⟨ right / left ⟩ identity for the ⟨ first / second ⟩ Arens product. Furthermore, ⟨ $\overline{L}$ / $\overline{R}$ ⟩ is a homeomorphic ⟨ isomorphism / anti-isomorphism ⟩ of A^{**} with the ⟨ first / second ⟩ Arens product onto ⟨ $((L_A)^*)'$ / $((R_A)^*)'$ ⟩ if and only if there is a two-sided identity for the ⟨ first / second ⟩ Arens product. Finally, ⟨ $\overline{L}$ / $\overline{R}$ ⟩ is an isometry if and only if the two-sided identity for the ⟨ first / second ⟩ Arens product has norm one.*

Proof We work only with $\overline{L}$ since the case of $\overline{R}$ is similar. Suppose $\overline{L}$ has all of the commutant as its range. Then there is some $e \in A^{**}$ satisfying $\overline{L}_e = I \in B(A^*)$. This means ${}_e\omega = \omega$, from which it follows that e is a right identity for the first Arens product. If $\overline{L}$ is an isomorphism, then e must be a two-sided identity and if $\overline{L}$ is an isometry, $\|e\|$ must be one.

Conversely, suppose $e \in A^{**}$ is a right identity for the first product. Let S be arbitrary in the commutant $((L_A)^*)'$. Define $f_S \in A^{**}$ by $f_S(\omega) = e(S(\omega))$ for all $\omega \in A^*$. Then equation (16) shows $f_S(\omega_a) = e(S(\omega_a)) = e(S \circ (L_a)^*(\omega)) = e((L_a)^* \circ S(\omega)) = e(S(\omega)_a) = S(\omega)(a)$, so $\overline{L}$ maps f_S onto S. Thus $\overline{L}$ is surjective. Now suppose e is a two-sided identity for the first product. Since ${}_f\omega$ is just $\overline{L}_f(\omega)$, equation (17) shows that $\overline{L}$ is injective.

Finally if e is a two-sided identity element of norm one, then the map we have just constructed from the commutant to $\mathcal{A}^{**}$ is contractive. Since $\overline{L}$ is obviously always contractive, we conclude that it is an isometry. □

This result shows that when $\mathcal{A}^{**}$ is unital for one of the Arens products its Arens multiplication may be defined in terms of the multiplication in $((L_\mathcal{A})^*)'$ or $((R_\mathcal{A})^*)'$ as a subset of $\mathcal{B}(\mathcal{A}^*)$ induced by the appropriate map $\overline{L}$ or $\overline{R}$. The existence of an identity in $\mathcal{A}^{**}$ can be described in terms of the existence of a particularly strong kind of approximate identity in $\mathcal{A}$, but we will not discuss this here since approximate identities are not introduced until Chapter 5.

It is of considerable interest that if either f is a right idealizer of $\kappa(\mathcal{A})$ in $\mathcal{A}^{**}$ or g is left idealizer of $\kappa(\mathcal{A})$ in $\mathcal{A}^{**}$, then $\overline{L}_f$ and $\overline{R}_g$ commute in $\mathcal{B}(\mathcal{A}^*)$. It is easy to see that these two operators commute exactly when the mixed associative law $g \cdot (hf) = (g \cdot h)f$ holds for all $h \in \mathcal{A}^{**}$. Hence they commute for all $f, g \in \mathcal{A}^{**}$ if $\mathcal{A}$ is Arens regular or $\kappa(\mathcal{A})$ is a one-sided ideal in $\mathcal{A}^{**}$ with respect to one, and hence both, Arens products.

Relationship to the Double Centralizer Algebra

It should not be surprising that there is an intimate connection between the Arens products on $\mathcal{A}^{**}$ and the double centralizer algebra $\mathcal{D}(\mathcal{A})$. We will define various maps between subalgebras of $\mathcal{A}^{**}$, $\mathcal{D}(\mathcal{A})$ and $\mathcal{D}(\mathcal{A}^{**})$ under minimal hypotheses on $\mathcal{A}$ beginning with the following two easy propositions. Some of these ideas are related to László Máté [1967] and Pak Ken Wong [1985].

1.4.4 Proposition *Let (L, R) be a double centralizer of a Banach algebra $\mathcal{A}$ in which the annihilator $\langle$ ideal $\mathcal{A}_A$ is / ideals $\mathcal{A}_{LA}$ and $\mathcal{A}_{RA}$ are $\rangle$ $\{0\}$. Then*

$$(L, R) \mapsto (L^{**}, R^{**}) \tag{18}$$

is an isometric injective isomorphism of $\langle$ $\mathcal{D}_B(\mathcal{A})$ / $\mathcal{D}(\mathcal{A}) = \mathcal{D}_B(\mathcal{A})$ $\rangle$ into $\mathcal{D}_B(\mathcal{A}^{**})$ when $\mathcal{A}^{**}$ is endowed with either Arens multiplication. Note that L^{**} and R^{**} leave $\kappa(\mathcal{A})$ invariant as a subset of $\mathcal{A}^{**}$.

Proof Under the second alternative hypothesis, all double centralizers are bounded by Theorem 1.2.4. (In that case the same theorem shows that it is enough to consider the third and sixth lines below.) One easily checks the following results for all $a \in \mathcal{A}$, $\omega \in \mathcal{A}^*$, and $g \in \mathcal{A}^{**}$:

$$
\begin{array}{rclrcl}
L^*(\omega)_a &=& \omega_{L(a)} & {}_aL^*(\omega) &=& L^*({}_a\omega) \\
R^*(\omega)_a &=& R^*(\omega_a) & {}_aR^*(\omega) &=& {}_{R(a)}\omega \\
L^*(\omega_a) &=& \omega_{R(a)} & {}_{L(a)}\omega &=& R^*({}_a\omega) \\
{}_gL^*(\omega) &=& L^*({}_g\omega) & L^*(\omega)_g &=& \omega_{L^{**}(g)} \\
{}_gR^*(\omega) &=& {}_{R^{**}(g)}\omega & R^*(\omega)_g &=& R^*(\omega_g) \\
R^*({}_g\omega) &=& {}_{L^{**}(g)}\omega & L^*(\omega_g) &=& \omega_{R^{**}(g)}.
\end{array}
$$

These equations show that $\langle\ L^{**}\ /\ R^{**}\ /\ (L^{**},\ R^{**})\ \rangle$ is a bounded $\langle$ left / right / double $\rangle$ centralizer whenever $\langle\ L\ /\ R\ /\ (L, R)\ \rangle$ has the same property, no matter which of the Arens products we use. Since $(S \circ T)^{**}$ equals $S^{**} \circ T^{**}$ and $\|S\|$ equals $\|S^{**}\|$, the maps $L \mapsto L^{**}$ and $R \mapsto R^{**}$ are homomorphisms and isometries. (This proof could also be based on looking at L^{**} and R^{**} as extensions by continuity of L and R, respectively.) $\qquad\square$

1.4.5 Corollary *If $\mathcal{A}^{**}$ has a mixed identity e and (L, R) is a double centralizer of $\mathcal{A}$, then they satisfy the following identities:*

$$R^{**}(f) = fR^{**}(e) = fL^{**}(e) \quad and$$
$$L^{**}(f) = L^{**}(e) \cdot f = R^{**}(e) \cdot f \tag{19}$$

for all $f \in \mathcal{A}^{**}$. Thus the map $(L, R) \mapsto (L^{**}, R^{**})$ of $\mathcal{D}(\mathcal{A})$ into $\mathcal{D}(\mathcal{A}^{**})$ has range in (the image of) $\mathcal{A}^{**}$ for one of the Arens products if and only if there is a two-sided identity $e \in \mathcal{A}^{**}$ with respect to this product. In that case, $(L, R) \mapsto L^{**}(e) = R^{**}(e)$ is a homeomorphic isomorphism of $\mathcal{D}(\mathcal{A}) = \mathcal{D}_B(\mathcal{A})$ into $\mathcal{A}^{**}$, which is an isometry if and only if the identity has norm one.

Proof As noted already, the existence of a mixed identity ensures that $\mathcal{A}_{LA}$ and $\mathcal{A}_{RA}$ are zero so the stronger form of the last proposition is available. Equations (19) now follow by applying the basic properties of double centralizers to (L^{**}, R^{**}) and the equations $fe = f$ or $e \cdot f = f$, respectively. In particular, (I^{**}, I^{**}) is a norm one, two-sided identity for $\mathcal{D}(\mathcal{A}^{**})$ in whichever Arens product is being used. It can be identified with e in $\mathcal{A}^{**}$ if and only if e is an identity for $\mathcal{A}^{**}$ in that Arens product. (Recall that, if any algebra like $\mathcal{A}^{**}$ has an identity of norm strictly greater than one, then the embedding of $\mathcal{A}^{**}$ into $\mathcal{D}(\mathcal{A}^{**})$ is an isomorphism but not an isometry.) $\qquad\square$

When $\mathcal{A}^{**}$ is not unital, the situation becomes more complicated, and in particular it may be necessary to consider quotient algebras. We begin to study this situation by noting that the idealizer $\mathcal{A}_{\mathcal{A}}^{**}$ of $\kappa(\mathcal{A})$ in $\mathcal{A}^{**}$ does not depend on which Arens product we use, since the two products agree when one factor is in $\kappa(\mathcal{A})$:

$$\mathcal{A}_{\mathcal{A}}^{**} = \{f \in \mathcal{A}^{**} : f\kappa(\mathcal{A}) = f \cdot \kappa(\mathcal{A}) \subseteq \kappa(\mathcal{A})$$
$$\text{and } \kappa(\mathcal{A})f = \kappa(\mathcal{A}) \cdot f \subseteq \kappa(\mathcal{A})\}. \tag{20}$$

Proposition 1.2.3 describes the obvious homomorphism of this set into $\mathcal{D}(\kappa(\mathcal{A}))$, and we may consider this as a map into $\mathcal{D}(\mathcal{A})$. Technically we can define this as

$$f \mapsto \tilde{\theta}(f) = (\tilde{L}_f, \tilde{R}_f) \quad \text{where} \tag{21}$$
$$\tilde{L}_f = \kappa^{-1} \circ L_f \circ \kappa \quad \text{and} \quad \tilde{R}_f = \kappa^{-1} \circ R_f \circ \kappa$$

with $\langle\, L_f\, /\, R_f\, \rangle$ being the $\langle$ left / right $\rangle$ representation of f in $\mathcal{A}^{**}$ with either of its Arens products. Now it is easy to check

$$(\tilde{L}_f)^* = \overline{R}_f \in ((R_\mathcal{A})^*)' \quad \text{and} \quad (\tilde{R}_f)^* = \overline{L}_f \in ((L_\mathcal{A})^*)' \qquad (22)$$

in terms of the notation used in Theorem 1.4.2 and Proposition 1.4.3. In the next proposition, we examine the second conjugates of the actions $\tilde{L}_f$ and $\tilde{R}_f$. For any subset $\mathcal{S}$ of $\mathcal{A}^*$, we use $\mathcal{S}^\perp$ to denote the subset of $\mathcal{A}^{**}$ consisting of those functionals which vanish on $\mathcal{S}$.

1.4.6 Proposition *The subset $\langle\, (L_\mathcal{A})^*(\mathcal{A}^*)^\perp\, /\, (R_\mathcal{A})^*(\mathcal{A}^*)^\perp\, \rangle$ is an ideal in $\mathcal{A}^{**}$ with respect to the $\langle$ first / second $\rangle$ Arens product which is the kernel of the maps $\langle\, \overline{L}$ and $\tilde{R}\, /\, \overline{R}$ and $\tilde{L}\, \rangle$. In fact, any element f from this ideal satisfies*

$$\langle\, \mathcal{A}^{**}f = 0\, /\, f \cdot \mathcal{A}^{**} = 0\, \rangle. \qquad (23)$$

The subset $\mathcal{A}_N^{**} = (R_\mathcal{A})^*(\mathcal{A}^*)^\perp \cap (L_\mathcal{A})^*(\mathcal{A}^*)^\perp = \{(R_\mathcal{A})^*(\mathcal{A}^*) + (L_\mathcal{A})^*(\mathcal{A}^*)\}^\perp$ is an ideal in $\mathcal{A}_\mathcal{A}^{**}$ and is the kernel of the homomorphism $\tilde{\theta} : \mathcal{A}_\mathcal{A}^{**} \to \mathcal{D}(\mathcal{A})$, defined above as

$$\tilde{\theta}(f) = (\tilde{L}_f, \tilde{R}_f) \qquad \forall\, f \in \mathcal{A}_\mathcal{A}^{**}. \qquad (24)$$

The two Arens products agree on $\mathcal{A}_\mathcal{A}^{**}/\mathcal{A}_N^{**}$ and the map induced by $\tilde{\theta}$ is an isomorphism onto $\mathcal{D}(\mathcal{A})$. In particular, $\mathcal{A}_N^{**}$ is zero if $\mathcal{A}^{**}$ has an identity for either Arens product.

When the map $\tilde{\theta}\colon \mathcal{A}_\mathcal{A}^{**} \to \mathcal{D}(\mathcal{A})$ is followed by the embedding $(L, R) \mapsto (L^{**}, R^{**})$, the image of $f \in \mathcal{A}_\mathcal{A}^{**}$ is a double centralizer in either Arens product which agrees as a pair of maps with $(L_f^{(2)}, R_f^{(1)})$ where the superscript numerals indicate the regular representation with respect to the second and first Arens multiplication, respectively.

Proof We consider only the second case (for variety). For $f \in (R_\mathcal{A})^*(\mathcal{A}^*)^\perp$, we have $f(_a\omega) = 0$ for all $a \in \mathcal{A}$ and $\omega \in \mathcal{A}^*$; this implies $\omega_f = 0$ from which the identification as kernels and equation (23) follow directly. (Note that $\langle\, \tilde{L}\, /\, \tilde{R}\, \rangle$ can be defined on the appropriate one-sided idealizer set which includes the annihilator set $\langle\, (R_\mathcal{A})^*(\mathcal{A}^*)^\perp\, /\, (L_\mathcal{A})^*(\mathcal{A}^*)^\perp\, \rangle$.) Since $\overline{R}$ is an anti-homomorphism with respect to the second Arens product defined on all of $\mathcal{A}$, the set is a (two-sided) ideal even though (23) only shows that it is a right ideal. We can show directly that $g \cdot f$ is in this ideal for any $g \in \mathcal{A}^{**}$. We see $g \cdot f(_a\omega) = f((_a\omega)_g) = f(_a(\omega_g)) = 0$ if $f \in (R_\mathcal{A})^*(\mathcal{A}^*)^\perp$.

Suppose $f \in \mathcal{A}^{**}$, $g \in \mathcal{A}_\mathcal{A}^{**}$ and $g\kappa(a) = \kappa(b)$. Then

$$fg(_a\omega) = f(_b\omega) = g\kappa(a)(\omega_f) = f \cdot g(_a\omega).$$

The case of ω_a is similar so the two products do agree on the quotient. Now consider the last statement of the proposition. We again prove the second case. For $f \in \mathcal{A}^{**}$, $g \in \mathcal{A}_{\mathcal{A}}^{**}$ and $\omega \in \mathcal{A}^*$, we get $(\tilde{R}_g)^{**}(f)(\omega) = f((\tilde{R}_g)^*(\omega)) = f(_g\omega) = fg(\omega) = R_g^{(1)}(f)(\omega)$.

Note that this proof has established more than stated in the proposition in several cases.   □

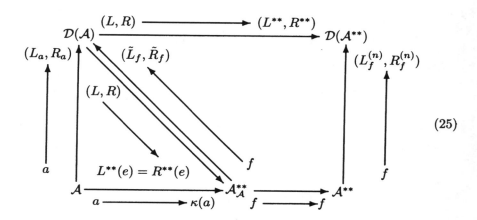

$$(25)$$

The last three propositions can best be visualized in terms of the commutative diagram above, which represents the case in which the nth Arens product has an identity e either for $n = 1$ or 2. (All the maps are defined and interesting without this restriction, but the diagram is no longer commutative in general.) Each map is a homeomorphic isomorphism onto its range and is isometric if e has norm one.

Thus we see that if at least one of the Arens products has an identity, then a substantial part of $\mathcal{A}^{**}$ behaves very well indeed: just like the double centralizer algebra. Consider approximately unital algebras for a moment. If such an algebra is Arens regular, then it has only one mixed identity which is in fact a (two-sided) identity for both Arens products. Hence, in the context of approximately unital normed algebras, $\mathcal{A}^{**}$ being unital for one of the Arens products is a weak version of $\mathcal{A}$ being Arens regular.

Arens Semiregularity

A still weaker version of Arens regularity has been introduced by Michael Grosser [1984] and studied there and subsequently by the same author [1987] and by others. An approximately unital Banach algebra $\mathcal{A}$ is said to be *semiregular* if it satisfies $R^{**}(e) = L^{**}(e)$ for all mixed identities e. (The idea behind the definition is that an algebra is semiregular if and only if the two embeddings of double centralizers, based on using their left or their right centralizers, agree for any mixed identity.) Corollary 1.4.5 shows this is true whenever $\mathcal{A}^{**}$ has an identity for at least one of the Arens products. The above remarks show that Arens regular algebras are semiregular. The latter result and the easy fact that commutative approximately unital Banach algebras are semiregular were established in the first cited paper. The case of commutative algebras shows that semiregularity is strictly weaker than the possession of an identity for at least one Arens product.

Losert and Harald Rindler [1984] showed that, for a locally compact group G, the algebra $L^1(G)$ is semiregular if and only if it is either discrete or commutative. This contrasts with the case of regularity where $L^1(G)$ is regular if and only if G is finite, a result proved by Young [1973a] after Civin and Yood [1961] handled the abelian case. Similarly, the algebra $\mathcal{B}_A(\mathcal{X})$ of *approximable operators* (*i.e.*, those uniformly approximable by finite-rank operators (Definition 1.1.17 above)) on a Banach space $\mathcal{X}$ is regular if and only if $\mathcal{X}$ is reflexive (Young [1976]). Much weaker conditions are known for semiregularity (Grosser [1987]). We shall treat these results in a preliminary way later in this section and in more detail in Section 1.7 after introducing more of the relevant concepts. However, two special cases completely solved in the last reference are particularly interesting. Consider the algebra $\mathcal{B}_A(C(\Omega))$ of operators on the Banach space of all continuous complex-valued functions on a compact Hausdorff space Ω. This algebra of operators is semiregular if and only if Ω is scattered, where scattered signifies that every nonempty subset contains an isolated point. Let μ be a finite measure and let $L^1(\mu)$ be the usual space of (equivalence classes) of absolutely μ-integrable functions. Then $\mathcal{B}_A(L^1(\mu))$ is semiregular only when $L^1(\mu)$ is finite-dimensional (in which case, of course, the algebra is Arens regular).

Before discussing Arens regularity, we introduce three simple examples.

Examples of Arens Multiplication for Commutative Algebras

For a commutative algebra $\mathcal{A}$, Definition 1.4.1 immediately implies $_a\omega = \omega_a$ and $_f\omega = \omega_f$, hence $fg = g \cdot f$ (note reversal of factors) for all $a \in \mathcal{A}$, $\omega \in \mathcal{A}^*$ and $f, g \in \mathcal{A}^{**}$. From this it easily follows that a commutative algebra is Arens regular if and only if one, hence both, Arens products are commutative. In the following examples, this observation will be useful. For a fuller discussion of this and many other results on the Arens products for general commutative Banach algebras, see Theorem 3.1.12.

1.4.7 Arens Multiplication in $(c_0)^{**} \simeq \ell^\infty$ This is perhaps the simplest example in a nonreflexive space and shows that the Arens multiplication is often the natural multiplication on the double dual. (Of course, in a Banach algebra which is reflexive as a Banach space so that $\mathcal{A}$ and $\mathcal{A}^{**}$ are canonically isomorphic under κ, both the Arens products agree with the original multiplication.) As usual, let c_0 be the collection of all sequences converging to zero with elementwise algebra operations and the supremum norm:

$$\|a\|_\infty = \sup\{|a_n| : n \in \mathbb{N}\}.$$

We identify ℓ^1 (the space of all sequences for which the norm,

$$\|b\|_1 = \sum_{n=1}^{\infty} |b_n|,$$

is finite, with elementwise linear operations) with the dual of c_0 as follows:

$$\omega: \ell^1 \to (c_0)^* \quad \text{is defined by} \quad \omega_b(a) = \sum_{n=1}^{\infty} a_n b_n$$

for all $a \in c_0$ and $b \in \ell^1$. Then we can identify ℓ^∞ (the space of all bounded sequences with the same supremum norm as c_0 and elementwise linear operations again) with the dual of $\ell^1 \simeq (c_0)^*$ by:

$$f: \ell^\infty \to (\ell^1)^* \quad \text{where} \quad f_a(b) = \sum_{n=1}^{\infty} a_n b_n$$

for all $a \in \ell^\infty$ and $b \in \ell^1$. With these identifications (which the reader can easily check are isometric linear isomorphisms), the canonical injection of c_0 into its double dual identified as ℓ^∞ is just the usual inclusion as spaces of sequences. Thus we will suppress the symbol for this injection.

Now the following results are immediate:

$$a(\omega_b) = (\omega_b)_a = \omega_{ab} \quad \text{where} \quad (ab)_n = a_n b_n \qquad \forall\, a \in c_0;\ b \in \ell^1;$$

$$f(\omega_b) = (\omega_b)_f = \omega_{fb} \quad \text{where} \quad (fb)_n = f_n b_n \qquad \forall\, f \in \ell^\infty;\ b \in \ell^1;$$

$$(fg)_n = (f \cdot g)_n = f_n g_n \qquad \forall\, f, g \in \ell^\infty.$$

Thus, in this case, the two products agree and just give the elementwise multiplication with which we would certainly have endowed ℓ^∞ in the first place.

This example is unusual. Most algebras are not Arens regular and generally the double dual is quite complicated. The fact is that c_0 is an example of a commutative C*-algebra. Commutative C*-algebras will be introduced in Chapter 3, particularly Theorem 3.2.12. Not necessarily commutative C*-algebras are briefly introduced in §1.7.17 and will be studied at more length in the second volume of this work. There we will show that any C*-algebra is Arens regular and that its double dual has a very natural interpretation as a universal W*-algebra (or von Neumann algebra) "completion", of the original algebra.

Theorem 3.2.12 shows that every unital commutative C*-algebra has the form $C(\Omega)$ for some compact space Ω and vice versa. Hence it is obvious that $c_0^{**} \simeq \ell^\infty$ is $C(\Omega)$ for what turns out to be a rather complicated extremely disconnected connected space. For generalizations, see Samuel Kaplan [1985].

1.4.8   Arens Multiplication on $(\ell^1)^{**}$ when ℓ^1 has Pointwise Multiplication   There are two possible fairly natural multiplications on ℓ^1: pointwise and convolution multiplication. We will examine the Arens multiplication in both cases. Although both multiplications are commutative on ℓ^1, only the first is Arens regular. This is particularly interesting since, as we will note below, certain Banach space properties force Arens regularity. This example shows that no Banach space property can characterize Arens regularity.

To make these two examples more interesting, we wish to exhibit a reasonably concrete interpretation of the double dual space of ℓ^1. It turns out that there are two concrete representations, each of which has some interest. But before we even describe these concrete representations of $(\ell^1)^{**} \simeq (c_0)^{***}$, we note a special property of any triple dual Banach space $\mathcal{X}^{***}$: it has a canonical contractive projection π onto $\mathcal{X}^*$. Any linear subspace which is the range of a bounded projection π is complemented by the closed linear space $(I-\pi)(\mathcal{X}^{***})$. In this case this complementary subspace is precisely $\kappa(\mathcal{X})^\perp$, as the reader may easily check. The construction of this projection will be given in Section 1.7.

The first concrete realization of $(\ell^\infty)^* \simeq (\ell^1)^{**} \simeq (c_0)^{***}$ we mention is that given by Dunford and Schwartz ([1958] IV.8.16) as the space $ba(\mathbb{N})$ of all complex, bounded, finitely additive set functions where the duality is implemented by simply integrating the sequence in ℓ^∞ by the set function in $ba(\mathbb{N})$. In this case the two complementary subspaces defined in the last paragraph are simply:

$\pi(ba(\mathbb{N}))$ is the space of countably additive measures on $\mathbb{N}$ which can obviously be identified with ℓ^1 by the Radon–Nikodym theorem; and

$(I - \pi)(ba(\mathbb{N}))$ is the space of finitely additive measures which vanish on all finite sets.

The second representation of the dual of ℓ^∞ comes from noting that ℓ^∞ is a unital, commutative C*-algebra. Hence, by what is essentially a corollary of Theorem 3.2.12 described in §3.2.15, ℓ^∞ can be identified with the Banach space $C(\beta\mathbb{N})$ of all continuous complex-valued functions on a compact Hausdorff space $\beta\mathbb{N}$, called the Stone–Čech compactification of $\mathbb{N}$. There is a continuous injection of $\mathbb{N}$ onto a dense subspace of $\beta\mathbb{N}$, and the function in $C(\beta\mathbb{N})$ corresponding to $f \in \ell^\infty$ may be considered as just the extension of f by continuity to all of $\beta\mathbb{N}$. Then the Riesz representation theorem identifies $(\ell^\infty)^* \simeq (C(\beta\mathbb{N}))^*$ with the Banach space $M(\beta\mathbb{N})$ of all complex regular Borel measures on $\beta\mathbb{N}$. This time the pair of complementary spaces consists of the subspace of measures supported on $\mathbb{N} \subseteq \beta\mathbb{N}$ (naturally identified with ℓ^1, again by the Radon–Nikodym theorem) and the space of measures supported on the complementary space $\beta\mathbb{N} \setminus \mathbb{N}$.

We now consider the Arens products arising from elementwise multiplication on ℓ^1 following Civin and Yood [1961], Theorem 1.4.2. In describing the situation, we use several concepts which will not be introduced until later in this work, and some of the ideas come from our discussion in Chapter 3 of the Arens multiplication in commutative algebras 3.1.12. To begin with, the situation is similar to that of the last example. We identify ℓ^∞ with the dual of $\ell^1 \simeq (c_0)^*$ by:

$$\omega: \ell^\infty \to (\ell^1)^* \quad \text{where} \quad \omega_a(b) = \sum_{n=1}^\infty a_n b_n$$

for all $a \in \ell^\infty$ and $b \in \ell^1$. We will use $\langle\ \kappa_0\ /\ \kappa_1\ \rangle$ for the natural map of $\langle\ c_0\ /\ \ell^1\ \rangle$ into its double dual space. We immediately see

$$_b(\omega_a) = (\omega_a)_b = \omega_{ab} \quad \text{where} \quad (ab)_n = a_n b_n$$

for all $a \in \ell^\infty$ and $b \in \ell^1$. However, this shows that these linear functionals are all in $\kappa_0(c_0)$, from which it follows that all Arens products with at least one factor from the complementary subspace to $\kappa_1(\ell^1)$ in the triple dual space decomposition must be zero. Call this complementary subspace $\mathcal{R} = \kappa_0(c_0)^\perp$. Thus the direct sum decomposition $(\ell^1)^{**} \simeq \kappa_1(\ell^1) \oplus \mathcal{R}$ is an algebra direct sum, multiplication on $(\ell^1)^{**}$ is commutative for either Arens product and thus the two products agree. Moreover, it is easy to see that any non-zero homomorphism of ℓ^1 (and hence of $(\ell^1)^{**}$ under either interpretation of this as a space of measures) into $\mathbb{C}$ is given by evaluation at a positive integer n. All these evaluations belong to the dual space of ℓ^1 since they are obviously all contractions. Their closed linear span is just the image of c_0 in ℓ^∞ under the embedding $\kappa_0: c_0 \to \ell^\infty$. Thus the subspace of elements of $(\ell^1)^{**}$ which vanish on all these homomorphisms is just $\mathcal{R}$. Hence $\mathcal{R}$ is the Gelfand radical and the Jacobson radical of $(\ell^1)^{**}$, and ℓ^1 (or equivalently $\kappa_1(\ell^1)$) is Jacobson semisimple. Thus, with pointwise multiplication on ℓ^1, the Arens multiplication on its double dual is extremely simple but badly behaved in that it has an enormous annihilator ideal.

1.4.9 Arens Multiplication on $(\ell^1)^{**} \simeq M(\beta\mathbb{N})$ when ℓ^1 has Convolution Multiplication In this example, it will be tidier if we consider the entries in the sequences in ℓ^1 to be indexed starting with zero. Obviously this makes no difference to the Banach space structure. The convolution product is then defined by

$$(b * d)_n = \sum_{k=0}^n b_{n-k} d_k \qquad \forall\, n \in \mathbb{N}^0;\ b, d \in \ell^1$$

which is easily seen to be commutative. Let $\delta_n \in \ell^1$ be the sequence which is zero except at n where it is one (so $(\delta_n)_m = \delta_{nm}$, the Kronecker delta). Then δ_0 is a multiplicative identity and in general $\delta_n * \delta_m = \delta_{n+m}$ so that ℓ^1 with this product may be interpreted as the semigroup algebra of $\mathbb{N}^0$ with addition for its semigroup operation. This time an easy calculation shows

$$_b(\omega_a) = (\omega_a)_b = \omega_{a \circ b} \quad \text{where} \quad (a \circ b)_n = \sum_{m=0}^{\infty} a_{n+m} b_m$$

for all $n \in \mathbb{N}^0$, $a \in \ell^\infty$ and $b \in \ell^1$. However, the formula for the next step in the definition would not help us to see that this algebra is not Arens regular. Instead we look at a very concrete construction using property (f) of Theorem 1.4.11 below.

Consider two sequences $\{b_n\}$ and $\{d_n\}$ in ℓ^1 where $b_0 = \delta_0$ and, for $n \geq 1$, $b_n = \delta_{2^{2n-1}}$ and $d_n = \delta_{2^{2n}}$. These are chosen so that simple inequalities show that all $b_n * d_m$ are distinct if their indices are different. Hence the characteristic function of $\{2^{2n-1} + 2^{2m} : m < n\}$ in ℓ^∞ gives 0 for one of the limits in equation (28) of Theorem 1.4.11(f) and 1 for the other limit. Therefore this algebra is not Arens regular. The idea for this example comes from Young [1973b] where he proves the following surprising and comprehensive theorem. Semigroup algebras are defined in §1.9.15 below.

1.4.10 Theorem *The following are equivalent for any locally compact topological semigroup S in which multiplication is at least singly continuous:*

(a) *The measure algebra $M(S)$ (with the topology of S) is Arens regular.*

(b) *The algebra $\ell^1(S)$ is Arens regular.*

(c) *There do not exist sequences $\{u_n\}$ and $\{v_m\}$ in S such that the sets $\{u_n v_m : m > n\}$ and $\{u_n v_m : m < n\}$ are disjoint.*

(d) *The semigroup operation can be extended to the Stone–Čech compactification βS of S as a discrete space.*

When these conditions hold, $\ell^1(S)^{**}$ with its Arens product can be naturally identified with $M(\beta S)$ equipped with the convolution product based on the extended semigroup operation.

We refer the reader to the original paper for a proof, although one can be based on our next result and is suggested by the last example.

Arens Regularity

Recall that an algebra is called Arens regular if the two Arens products agree on its double dual. This topic has attracted most of the attention devoted to Arens multiplication, at least until recently. As we will note, Arens regularity is uncommon. Duncan and Hosseiniun [1979] first formulated the following comprehensive theorem this way, using ideas crystallized by Young [1976] from Pym [1965], and J. Hennefeld [1968].

For any $\omega \in \mathcal{A}^*$ define

$$\langle \Lambda_\omega : \mathcal{A} \to \mathcal{A}^* \ / \ \mathrm{P}_\omega : \mathcal{A} \to \mathcal{A}^* \rangle \quad \text{by} \quad \langle \Lambda_\omega(a) = \omega_a \ / \ \mathrm{P}_\omega(a) = {}_a\omega \rangle \quad (26)$$

for all $a \in \mathcal{A}$. These maps satisfy

$$\langle \Lambda_\omega^*(f) = {}_f\omega \ / \ \mathrm{P}_\omega^*(f) = \omega_f \rangle \quad \text{and} \quad (27)$$

$$\langle \Lambda_\omega^{**}(f)(g) = fg(\omega) \ / \ \mathrm{P}_\omega^{**}(g)(f) = f \cdot g(\omega) \rangle$$

for all $f, g \in \mathcal{A}^{**}$. The linear functional ω on an algebra $\mathcal{A}$ is called *weakly almost periodic* if either (and hence both) Λ_ω and P_ω are weakly compact operators. This means explicitly that the weak closure of the sets $\{\omega_a : a \in \mathcal{A}_1\}$ and $\{{}_a\omega : a \in \mathcal{A}_1\}$ have compact closure in the weak topology (or $\mathcal{A}^{**}$ topology) on $\mathcal{A}^*$. A well known result (*cf.* Section 1.7 below or Dunford and Schwartz [1958] VI.4.2) states that a linear map is weakly compact if and only if its double dual is included in the canonical image of the original target space of the linear map. We will use this criterion below.

1.4.11 Theorem *The following are equivalent for a Banach algebra $\mathcal{A}$:*
(a) *$\mathcal{A}$ is Arens regular.*
(b) *For each $f \in \mathcal{A}^{**}$ the map $g \mapsto fg$ is weak* continuous.*
(c) *For each $g \in \mathcal{A}^{**}$ the map $f \mapsto f \cdot g$ is weak* continuous.*
(d) *For each $\omega \in \mathcal{A}^*$ the map Λ_ω is weakly compact.*
(e) *For each $\omega \in \mathcal{A}^*$ the map P_ω is weakly compact.*
(f) *Any two bounded sequences $\{a_n\}$ and $\{b_n\}$ in $\mathcal{A}$ and $\omega \in \mathcal{A}^*$ satisfy*

$$\lim_n \lim_m \omega(a_n b_m) = \lim_m \lim_n \omega(a_n b_m). \quad (28)$$

Proof The equivalence of (a), (b) and (c) follow from the weak* continuity properties of the Arens products contained in Theorem 1.4.2 and the density of $\kappa(\mathcal{A})$ in $\mathcal{A}^{**}$. The second result in equation (27) shows that if $\mathcal{A}$ is Arens regular, then it satisfies (d) and (e). Conversely, if either of the maps in (d) and (e) is weakly compact, then its dual is continuous from the weak* topology on $\mathcal{A}^{**}$ to the weak topology (*op. cit.* VI.4.7) on $\mathcal{A}^*$ which, by the first part of equation (27), implies (b) or (c). Finally, equation (28) is Alexander Grothendieck's criterion [1954] for weak compactness of either of the sets $\{\omega_a : a \in \mathcal{A}_1\}$ or $\{{}_a\omega : a \in \mathcal{A}_1\}$. $\qquad \square$

The following corollary can be based on criterion (f) of the last theorem, but the results were first noted by Civin and Yood [1961] where they were derived from the last sentence of Theorem 1.4.2 above.

1.4.12 Corollary *A subalgebra or quotient algebra of an Arens regular algebra is Arens regular.*

Arens [1951b] gave the first example of a non Arens regular commutative algebra. Civin and Yood [1961] proved many interesting results chiefly relating to group algebras. In particular, they showed that for a locally compact abelian group G, $L^1(G)$ is never regular unless G is finite. This was extended to arbitrary locally compact groups by Young [1973a]. Ian G. Craw and Young [1974] show that no locally compact group G has a weight function with Arens regular Beurling algebra (cf. 1.9.15 below) unless G is discrete and countable. Despite these negative results, Sheila A. McKilligan and Alan J. White [1972] proved interesting properties of quotient algebras for group algebras as well as more general Banach algebras. See also Civin [1961], [1962a], [1962b], Oshobi and Pym [1981], [1984], [1987] and Oshobi [1986] (see review).

We will postpone until Volume II the interesting proof that for a Hilbert space $\mathcal{H}$, $\mathcal{B}(\mathcal{H})$ and therefore all its closed subalgebras are Arens regular. Since the Banach algebra $C_0(\Omega)$ of continuous complex-valued functions vanishing at infinity on any locally compact space Ω is a closed subalgebra of some $\mathcal{B}(\mathcal{H})$ (cf. Section 1.5, §1.7.17 and Theorem 3.2.12), any uniform algebra (3.1.8) or quotient of a uniform algebra is Arens regular. The norm closure $\mathcal{B}_A(\mathcal{X})$ is Arens regular for a Banach space $\mathcal{X}$ if and only if $\mathcal{X}$ is reflexive, but little positive is known about the Arens regularity of $\mathcal{B}(\mathcal{X})$. We mentioned additional results on Arens regularity in our discussion of semiregularity above.

Our examples have shown that two different multiplications on the same Banach space can differ with respect to Arens regularity. Nevertheless, sufficiently strong conditions on the Banach space structure, short of reflexivity, do imply regularity. (See Gilles Godefroy and Iochum [1988].) Similar results have been obtained by Ülger [1987] and [1990a] and Ángel Rodríguez-Palacios [1987], who consider a kind of super regularity for Banach spaces. The Banach space of any C*-algebra satisfies this condition.

Pym and Ülger [1989] consider the Arens regularity of inductive limit algebras and Ülger [1988a] deals with projective tensor product algebras. Ülger [1990c] shows that the Banach algebra $C(\Omega, \mathcal{A})$ of all continuous functions from a compact Hausdorff space Ω to a Banach algebra $\mathcal{A}$ (with supremum norm) is Arens regular if and only if $\mathcal{A}$ is Arens regular.

Duncan and Hosseiniun [1979], following McKilligan and White [1972], establish a variant of Theorem 1.4.11 in which the maps Λ_ω and P_ω are required to be compact. This is equivalent to either Arens product being jointly continuous in each variable in the bounded weak* topology described in Dunford and Schwartz [1958] V.5.3.

Before the last theorem, weakly almost periodic elements of the dual space $\mathcal{A}^*$ of an algebra $\mathcal{A}$ were defined. A linear functional ω on an algebra $\mathcal{A}$ is called *almost periodic* if both Λ_ω and P_ω are compact operators. This means explicitly that the the sets $\{\omega_a : a \in \mathcal{A}_1\}$ and $\{_a\omega : a \in \mathcal{A}_1\}$ have compact closure in the norm topology on $\mathcal{A}^*$. If $\mathcal{A}$ is the usual group algebra $L^1(G)$ of a locally compact group G, then $\mathcal{A}^*$ can be identified with $L^\infty(G)$. This example motivates the name assigned to these functionals. When G is the real line under addition, the concept agrees with classical terminology. The sets of weakly almost periodic and almost periodic functionals on $\mathcal{A}$ will be denoted by $\mathcal{A}^*_{WAP}$ and $\mathcal{A}^*_{AP}$, respectively. The book by Robert B. Burckel [1970] gives a useful compendium up to the time of its publication. See also Andrew Tonge [1980] and Jesús Gil de Lamadrid and Loren N. Argabright [1990].

$\kappa(\mathcal{A})$ as an Ideal in $\mathcal{A}^{**}$

The ideas used in Theorem 1.4.11 allow us to say when $\kappa(\mathcal{A})$ is a one- or two-sided ideal in $\mathcal{A}^{**}$. So many papers have been published on this interesting question that the following simple straightforward characterization due to Seiji Watanabe [1974] is particularly satisfying. See also Michael Grosser [1979b], Ghahramani [1984].

1.4.13 Proposition *The following are equivalent for a normed algebra $\mathcal{A}$:*
(a) $\kappa(\mathcal{A})$ *is a* ⟨ *left / right* ⟩ *ideal in* $\mathcal{A}^{**}$.
(b) ⟨ R_a / L_a ⟩ *is weakly compact for each* $a \in \mathcal{A}$.
(c) ⟨ $(R_a)^*$ / $(L_a)^*$ ⟩ *is weakly compact for each* $a \in \mathcal{A}$.
Hence, $\kappa(\mathcal{A})$ is a two-sided ideal if and only if both R_a and L_a are weakly compact for all $a \in \mathcal{A}$.

Proof Gantmacher's theorem (*cf.* Dunford and Schwartz [1958] VI.4.8) shows the equivalence of the last two conditions. Condition (b) is equivalent to (R_a^{**}, L_a^{**}) having range in $\kappa(\mathcal{A})$ (*op. cit.* VI.4.2). However, as we have noted earlier, $(R_a)^{**}(f) = f\kappa(a)$ and $(L_a)^{**}(f) = \kappa(a)f$, proving the equivalence to (a).   □

Arens Representations

In his original papers, Arens' construction was extremely general. In particular, the situation he described applies not only to algebras but to certain representations as well. Representations will be studied systematically in Chapter 4, but here we will define a *continuous* ⟨ *representation / anti-representation* ⟩ T of a Banach algebra $\mathcal{A}$ on a Banach space $\mathcal{X}$ to be a continuous ⟨ homomorphism / anti-homomorphism ⟩ $T: \mathcal{A} \to \mathcal{B}(\mathcal{X})$ with the image of an element $a \in \mathcal{A}$ under T denoted by T_a. The Arens representations arising from a representation T are two representations $T^1, T^2: \mathcal{A}^{**} \to \mathcal{B}(\mathcal{X}^{**})$ where T^n is a representation with respect to the nth Arens product for $n = 1, 2$. These representations extend T in the sense that the following diagram is commutative where ** denotes the map which assigns to any operator S its double dual S^{**}. (The map on top is a homomorphism (or representation) with respect to the nth Arens product, as noted above.)

$$
\begin{array}{ccc}
\mathcal{A}^{**} & \xrightarrow{T^n} & \mathcal{B}(\mathcal{X}^{**}) \\
\uparrow{\scriptstyle \kappa} & & \uparrow{\scriptstyle **} \\
\mathcal{A} & \xrightarrow{T} & \mathcal{B}(\mathcal{X})
\end{array}
\tag{29}
$$

Although the construction we are about to give is derived from Arens' original papers, it was hardly utilized until quite recently. The present treatment is based on Palmer [1985] (*cf.* Grosser [1987]). The basic definition is similar to that of the Arens products.

1.4.14 Definition Let $T: \mathcal{A} \to \mathcal{B}(\mathcal{X})$ be a continuous representation of a normed algebra $\mathcal{A}$ on a normed linear space $\mathcal{X}$. For any $a \in \mathcal{A}$, $x \in \mathcal{X}$ and $\tau \in \mathcal{X}^*$, define elements $\tau_a \in \mathcal{X}^*$ and $_x\tau \in \mathcal{A}^*$ by

$$
\tau_a = (T_a)^*(\tau) \quad \text{and} \quad _x\tau(a) = \tau(T_a(x)) = \tau_a(x).
\tag{30}
$$

For any $a \in \mathcal{A}$, $x \in \mathcal{X}$, $\tau \in \mathcal{X}^*$, $f \in \mathcal{A}^{**}$ and $F \in \mathcal{X}^{**}$, define elements $_F\tau \in \mathcal{A}^*$ and $\tau_f \in \mathcal{X}^*$ by

$$
_F\tau(a) = F(\tau_a) \quad \text{and} \quad \tau_f(x) = f(_x\tau).
\tag{31}
$$

Finally, for any $f \in \mathcal{A}^{**}$, $\tau \in \mathcal{X}^*$ and $F \in \mathcal{X}^{**}$, define elements T_f^1 and T_f^2 of $\mathcal{B}(\mathcal{X}^{**})$ by

$$
T_f^1(F)(\tau) = f(_F\tau) \quad \text{and} \quad T_f^2(F)(\tau) = F(\tau_f).
\tag{32}
$$

The two maps T^1 and T^2 of $\mathcal{A}^{**}$ into $\mathcal{B}(\mathcal{X}^{**})$ are called the *first and second Arens representations*, respectively.

We again gather some elementary properties of these two representations.

1.4.15 Theorem *Let $T: \mathcal{A} \to \mathcal{B}(\mathcal{X})$ be a continuous representation. The elements τ_a and $_x\tau$ belong to $\mathcal{X}^*$ and $\mathcal{A}^*$, respectively, and satisfy:*

$$
\|\tau_a\| \le \|T\| \, \|\tau\| \, \|a\| \qquad \|_x\tau\| \le \|T\| \, \|\tau\| \, \|x\|
\tag{33}
$$

$$
(\tau_a)_b = \tau_{ab} \qquad _b(_x\tau)(a) = \tau_{ab}(x)
\tag{34}
$$

for all $a, b \in \mathcal{A}$, $x \in \mathcal{X}$ and $\tau \in \mathcal{X}^*$. The elements $_F\tau$ and τ_f belong to $\mathcal{A}^*$ and $\mathcal{X}^*$, respectively, and satisfy:

$$
\|_F\tau\| \le \|T\| \, \|\tau\| \, \|F\| \qquad \|\tau_f\| \le \|T\| \, \|\tau\| \, \|f\|
\tag{35}
$$

$$
(_F\tau)_a = {_F(\tau_a)} \quad _{\kappa(x)}\tau = {_x\tau} \quad _x(\tau_f) = (_x\tau)_f \quad \tau_{\kappa(a)} = \tau_a \quad \tau_{f \cdot g} = (\tau_f)_g
\tag{36}
$$

for all $a \in \mathcal{A}$, $x \in \mathcal{X}$, $\tau \in \mathcal{X}^*$, $f, g \in \mathcal{A}^{**}$ and $F \in \mathcal{X}^{**}$. Finally, the two maps T^1 and T^2 of $\mathcal{A}^{**}$ into $B(\mathcal{X}^{**})$ are continuous representations with respect to the first and second Arens product, respectively. Also T^1 and T^2 have bound $\|T\|$ and each makes diagram (29) commutative.

Proof All these results are again easy to obtain from step by step calculations starting with the three step definition. The only intermediate equation we have not displayed above is $T^1_{g(F)}\tau = g(_F\tau)$ for all $\tau \in \mathcal{X}^*$, $g \in \mathcal{A}^{**}$ and $F \in \mathcal{X}^{**}$.   □

Similar constructions are possible for anti-representations. Section 1.7 gives an explicit construction for the first Arens representation of the algebra of approximable operators. Since the construction involves ideas and notation which will not be introduced until that section, we postpone it.

1.5 Algebras of Functions

The next six sections, beginning with this one, are simply intended to give the reader a stock of examples to which the theory presented in this work can be applied. The discussion in all these sections shows how naturally Banach algebra concepts arise in examples. At least at the beginning of each section, the exposition has been kept elementary.

The present section is extremely brief because the topic belongs more naturally to the subject matter of Chapter 3, which studies commutative algebras. In that chapter, we will show that a commutative Banach algebra, and more generally a commutative spectral algebra (*cf.* Section 2.4 for the definition), can be represented as an algebra of bounded continuous functions with pointwise multiplication. In this section we briefly introduce such algebras.

Let Ω be a set and let $\mathcal{C}$ be a collection of complex-valued functions on Ω. If this collection of functions is closed under addition, multiplication and scalar multiplication, then it is an algebra under pointwise operations. Let us agree that, when we call any such collection of functions an algebra, we mean that the algebraic operations are pointwise unless we clearly specify some other operations. If the functions in $\mathcal{C}$ are all bounded, then the *supremum norm* or *uniform norm* defined by

$$\|f\|_\infty = \sup\{|f(\omega)| : \omega \in \Omega\} \quad f \in \mathcal{C}$$

is an algebra norm on $\mathcal{C}$. Convergence with respect to this norm is uniform convergence. Thus if $\mathcal{C}$ should be closed under uniform convergence, then it would be complete under the supremum norm. Hence $\mathcal{C}$ would be a Banach algebra.

Let ω be a point in Ω. Then evaluation at ω defines a homomorphism $\gamma_\omega : \mathcal{C} \to \mathbb{C}$ by $\gamma_\omega(f) = f(\omega)$ for all $f \in \mathcal{C}$. Unless all the functions in $\mathcal{C}$

vanish at ω, γ_ω will have all of $\mathbb{C}$ as its range. Hence $\ker(\gamma_\omega)$ will be an ideal of $\mathcal{C}$ which is maximal among proper ideals. (To see this, note that if $\mathcal{I}$ were an ideal properly between $\ker(\gamma_\omega)$ and $\mathcal{C}$, then $\gamma_\omega(\mathcal{I})$ would be a non-zero proper ideal of $\mathbb{C}$.) Results in the next two chapters (particularly Corollary 2.2.3, Proposition 2.4.12 and the first portion of Section 3.1) will show that in many circumstances, every ideal which is maximal among proper ideals of $\mathcal{C}$ will arise in this way. In particular, Theorem 3.2.12 and §3.2.15 show that the algebras of continuous functions considered below have this property.

Consider the algebra $B(\Omega)$ of all bounded functions on Ω. This is obviously uniformly closed so the remarks above show that it is a Banach algebra under the supremum norm. There are various uniformly closed subalgebras of $B(\Omega)$ which we will consider in this work including certain subalgebras of measurable functions. However, the most important examples arise when Ω is a topological space and we consider certain classes of bounded continuous functions on Ω.

1.5.1　Algebras of Continuous Functions　Let Ω be a topological space. We will define two algebras of bounded continuous complex-valued functions on Ω, each of which is a commutative Banach algebra under the supremum norm. Let $C(\Omega)$ be the algebra of all bounded continuous complex-valued functions on Ω, and let $C_0(\Omega)$ be the algebra of all continuous complex-valued functions on Ω which vanish at infinity. (A function f on Ω is said to *vanish at infinity* if for every constant $\varepsilon > 0$ there is a compact subset K of Ω satisfying $|f(\omega)| < \varepsilon$ for all $\omega \in \Omega \setminus K$.) It is easy to check that each of these algebras is a Banach algebra under the supremum norm.

We usually assume that the space Ω is *completely regular* and Hausdorff. (Some authors include Hausdorff as part of the definition of completely regular, *e.g.*, Gillman and Jerison [1960].) This condition can be defined by requiring that singletons are closed sets (*i.e.*, Ω is T_1) and that, for any closed subset $K \subseteq \Omega$ and any point $\omega \in \Omega \setminus K$, there is a continuous function $f: \Omega \to [0, 1]$ satisfying

$$f(K) = \{0\}, \quad f(\omega) = 1.$$

It is easy to see that the concept of complete regularity is not changed if we merely assume that f belongs to $C(\Omega)$, instead of requiring that it have range in $[0, 1]$. In §3.2.15, we show that for any topological space Ω there is a completely regular Hausdorff topological quotient space $\tilde{\Omega}$ such that the natural map of $C(\tilde{\Omega})$ onto $C(\Omega)$ is an isometric isomorphism. Thus the assumption that Ω is completely regular is no loss of generality from the viewpoint of Banach algebra theory.

Recall that Urysohn's lemma shows that a normal space (which we assume Hausdorff) is completely regular. It is easy to show that a compact space is normal, that a subspace of completely regular space is completely regular and that a locally compact space (which is a subspace of its one-point compactification) is therefore completely regular. In §3.2.15 we also show that $C_0(\Omega)$ is naturally isometrically isomorphic to $C_0(\tilde{\Omega})$, where $\tilde{\Omega}$ is a locally compact quotient of Ω. Hence we will usually assume that Ω is locally compact when discussing $C_0(\Omega)$. We remark that the algebra $C_{00}(\Omega)$ of continuous complex-valued functions on Ω with compact support is a dense ideal of $C_0(\Omega)$ when Ω is locally compact.

The Stone–Weierstrass theorem gives a useful criterion for a subalgebra of $C_0(\Omega)$ to be dense. In particular, it establishes the last remark. Since this theorem is frequently stated in less generality than we will want, we take this opportunity to state the result in the form we will need. If Ω is locally compact, then a subalgebra of $C_0(\Omega)$ is dense if it is closed under complex conjugation, separates the points of Ω and does not vanish identically at any point of Ω. The last two requirements simply mean that, for any two distinct points ω and ω' of Ω, there are functions f and g in the subalgebra satisfying $f(\omega) \neq f(\omega')$ and $g(\omega) \neq 0$. See Robert B. Burckel [1972], [1984], Thomas J. Ransford [1984b], Jaroslav Zemánek [1978] for recent extensions and proofs.

The commutative Banach algebras $C(\Omega)$ and $C_0(\Omega)$ are characterized as Banach algebras by Theorem 3.2.12. Note that $C(\Omega)$ is always unital and that $C_0(\Omega)$ is unital if and only if Ω is compact, in which case the two algebras coincide. Furthermore $C_0(\Omega)$ is an ideal in $C(\Omega)$. We will now show that extension of the regular homomorphism described in Proposition 1.2.6 naturally identifies $C(\Omega)$ with the double centralizer algebra of $C_0(\Omega)$.

Theorem *Let Ω be a locally compact Hausdorff space. Then the natural homomorphism of $C(\Omega)$ into the double centralizer algebra $\mathcal{D}(C_0(\Omega))$ of $C_0(\Omega)$ is an isometric isomophism onto $\mathcal{D}(C_0(\Omega))$.*

Proof To simplify notation, we write $\mathcal{C}$, $\mathcal{D}$ and $\mathcal{A}$ for $C(\Omega)$, $\mathcal{D}(C_0(\Omega))$ and $C_0(\Omega)$, respectively. Proposition 1.2.3(c) and (d) show that $\mathcal{D}$ is commutative and that any double centralizer $(L, R) \in \mathcal{D}$ satisfies $L = R$ so that L and R are both left and right centralizers. Hence we may identify $\mathcal{D}$ with the Banach algebra of left centralizers on $\mathcal{A}$ under composition as product and the operator norm. With this identification the natural map $\theta : \mathcal{C} \to \mathcal{D}$ has the form

$$\theta(f)(g) = fg \qquad \forall \, f \in \mathcal{C}; \, g \in \mathcal{A}.$$

We will now construct an inverse for this map. Choose $L \in \mathcal{D}$. Let g and h be two functions in $\mathcal{A}$ which are non-zero at $\omega \in \Omega$. Then the equation $L(g)h = gL(h)$ shows that $L(g)(\omega)/g(\omega)$ is independent of g so long as $g(\omega) \neq 0$. Hence we can define $f_L : \Omega \to \mathbb{C}$ by $f_L(\omega) = L(g)(\omega)/g(\omega)$ for any g which is non-zero at ω. Since the same g may be used in some neighborhood of ω, we see that f_L is continuous. It is bounded by $\|L\|$, so

f_L belongs to $\mathcal{C}$. Clearly the map $L \mapsto f_L$ is an inverse for θ. Both maps are contractive, so they are isometric isomorphisms between $\mathcal{C}$ and $\mathcal{D}$. $\quad\square$

For later use we also need the following result on $C_0(\Omega)$.

Theorem *Let Ω be a locally compact Hausdorff space and let $\mathcal{I}$ be an ideal of $C_0(\Omega)$. Either there is some $\omega \in \Omega$ on which every $g \in \mathcal{I}$ vanishes or $\mathcal{I}$ includes $C_{00}(\Omega)$, so $\mathcal{I}$ is dense in $C_0(\Omega)$.*

If $\mathcal{I}$ is the kernel of a non-zero homomorphism $\gamma : C_0(\Omega) \to \mathbb{C}$, then the former case holds and $\mathcal{I} = \{f \in C_0(\Omega) : f(\omega) = 0\}$ so $\gamma(f) = f(\omega)$ for all $f \in C_0(\Omega)$

Proof Suppose there is no $\omega \in \Omega$ on which every $g \in \mathcal{I}$ vanishes. Choose a non-zero $h \in C_{00}(\Omega)$. For each $\omega \in \text{supp}(h)$, choose $g_\omega \in \mathcal{I}$ such that $g_\omega(\omega) = 1$. If g_ω^* is defined by $g_\omega^*(\tau) = g_\omega(\tau)^*$ for all $\tau \in \Omega$, then $g_\omega^* g_\omega = |g_\omega|^2$ belongs to $\mathcal{I}$ and is non-negative-valued. By the compactness of $\text{supp}(h)$, we can find $g_1, g_2, \ldots, g_n \in \mathcal{I}$ non-negative valued and such that $\tilde{g} = \sum_{j=1}^n g_j \in \mathcal{I}$ satisfies $1 \le \tilde{g}(\omega)$ for $\omega \in \text{supp}(h)$. Next choose $k \in C_{00}(\Omega)$ so that $k(\omega) = 1$ for all $\omega \in \text{supp}(h)$ and $k(\omega) = 0$ if $g(\omega) < 1/2$. Then kg^{-1} is defined and belongs to $C_{00}(\Omega)$, so $h = (kg^{-1})gh$ belongs to $\mathcal{I}$.

Suppose $\mathcal{I} = \ker(\gamma)$. The former case holds since γ is non-zero. Hence there is some $\omega \in \Omega$ on which every $g \in \ker(\gamma)$ vanishes. The range of a homomorphism is a subalgebra, so γ maps onto $\mathbb{C}$. Choose $h \in C_0(\Omega)$ so that $h(\omega) = 1$. Then for all $f \in C_0(\Omega)$, γ vanishes on $(f - \gamma(f))h$, so $(f(\omega) - \gamma(f))1 = 0$. $\quad\square$

If K is a subset of Ω, then it is *clopen* (= closed and open) if and only if its characteristic function belongs to $C(\Omega)$. Furthermore, an element f of $C(\Omega)$ is an *idempotent* (*i.e.*, $f^2 = f$) if and only if it is a characteristic function. Hence, Ω is disconnected if and only if $C(\Omega)$ contains idempotents different from 0 and 1. Similarly, $C_0(\Omega)$ contains idempotents different from 0 if and only if Ω has at least one nonempty compact clopen subset. A deep generalization of these results, due to Šilov [1953], is given in Corollary 3.4.13 below.

A simple application of the Stone–Weierstrass theorem to the remarks of the last paragraph shows that, for a compact Hausdorff space Ω, $C(\Omega)$ is generated as a Banach algebra by its set of idempotents if and only if Ω is totally disconnected. Many other connections are known between the algebra structures of $C(\Omega)$ and $C_0(\Omega)$ and the topological structure of Ω. Some will be given in Chapter 3.

The Gelfand theory presented in Chapter 3 shows that for any commutative Banach algebra $\mathcal{C}$ there is a canonical locally compact space $\Gamma_\mathcal{C}$ and a canonical homomorphism $\theta : \mathcal{C} \to C_0(\Gamma_\mathcal{C})$. This space $\Gamma_\mathcal{C}$ is compact if and only if $\mathcal{C}$ is unital in which case $C_0(\Gamma_\mathcal{C}) = C(\Gamma_\mathcal{C})$. In many interesting cases, the canonical homomorphism θ is injective. Theorem 3.2.12 characterizes those algebras for which the canonical homomorphism is an isometric isomorphism.

1.5.2 The Disc Algebra   The above remarks justify the claim made earlier that $C_0(\Omega)$ and $C(\Omega)$ are universal commutative algebras in some sense. However, they are far from typical. We now present a simple but somewhat more general example.

Let $\mathbb{D}$ be the unit disc, $\mathbb{D} = \{\lambda \in \mathbb{C} : |\lambda| \leq 1\}$, and denote its interior by $\mathbb{D}^\circ = \{\lambda \in \mathbb{C} : |\lambda| < 1\}$ as usual. The disc algebra $A(\mathbb{D})$ is the algebra of continuous complex-valued functions on $\mathbb{D}$ which are analytic (=holomorphic) on its interior. These functions are bounded since $\mathbb{D}$ is compact. The algebra is closed in $C(\mathbb{D})$ and therefore complete in the supremum norm since the uniform limit of analytic functions is analytic. (To see this, one may use Morera's theorem: a continuous function on $\mathbb{D}^\circ$ is analytic if its integral around any circle in $\mathbb{D}^\circ$ vanishes.) Hence $A(\mathbb{D})$ is a commutative unital Banach algebra. The maximum modulus principle shows that the map which restricts a function in $A(\mathbb{D})$ to the boundary $\mathbb{T}$ of $\mathbb{D}$ is actually an isometric isomorphism onto its range. This range is a proper subset of $C(\mathbb{T})$ since the continuous function which takes each $\zeta \in \mathbb{T}$ to its complex conjugate certainly does not belong to it. These facts will be proved together with many other interesting properties of the disc algebra in §3.2.13.

1.5.3 The Algebra $C^1(\mathbb{T})$ as the Quotient of a Uniform Algebra   The algebra $C^1(\mathbb{T})$ is just the linear space of functions on $\mathbb{T}$ which are continuously differentiable in the sense of differentiation along $\mathbb{T}$. (An easy definition of this differentiation is given at the beginning of Section 1.8 if the reader wishes to see it.) This is a Banach algebra under pointwise multiplication and the norm

$$\|f\|_{C^1} = \|f\|_\infty + \|f'\|_\infty \qquad \forall\, f \in C^1(\mathbb{T}).$$

(To check completeness, write f as an indefinite integral of f'.)

The following construction is due to Sten Kaijser [1976]. Consider the collection of all continuous functions $f : \mathbb{T} \times \mathbb{D} \to \mathbb{C}$. This is a Banach algebra under the supremum norm and pointwise product. We will consider a closed subalgebra $\mathcal{A}$ defined by two conditions. First, for each $\zeta \in \mathbb{T}$ the function $\lambda \mapsto f(\zeta, \lambda)$ should belong to the disc algebra. Any such function clearly has a series expansion $f(\zeta, \lambda) = \sum_{n=0}^{\infty} f_n(\zeta)\lambda^n$ in terms of continuous functions f_n on $\mathbb{T}$. Note that the first two terms in the expansion of the product fg have coefficient functions $(fg)_0 = f_0 g_0$ and $(fg)_1 = f_0 g_1 + f_1 g_0$, respectively. The second condition for a function f to be a member of $\mathcal{A}$ is that it should satisfy

$$f_0(e^{it}) = f_0(1) - i\int_0^t f_1(e^{is})e^{is}\,ds.$$

This condition obviously defines a subspace closed in the supremum norm. It can be restated as requiring that f_0 be differentiable and satisfy $f_0' = f_1$, in which form we see that it defines a subalgebra. Hence $\mathcal{A}$ is a Banach algebra of functions under pointwise multiplication and the supremum (or uniform) norm. Such algebras are called *uniform algebras*. Note that the previous two examples discussed uniform algebras also.

We wish to write $C^1(\mathbb{T})$ as a quotient of $\mathcal{A}$. The ideal $\mathcal{I}$ involved is defined to be the collection of functions f in $\mathcal{A}$ with expansions satisfying $f_0 = f_1 \equiv 0$. This also gives a nontrivial example of a split extension in the terminology of Definition 1.2.9 since we may embed $C^1(\mathbb{T})$ into $\mathcal{A}$ by an isomorphism χ:

$$\chi(f)(\zeta, \lambda) = f(\zeta) + f'(\zeta)\lambda.$$

This embedding is isometric by the maximum modulus principle. It is obvious that the quotient $\mathcal{A}/\mathcal{I}$ can be identified with $C^1(\mathbb{T})$ and that χ is a right inverse for the natural quotient map.

1.6   Matrix Algebras

For each $n \in \mathbb{N}$, the set M_n of all $n \times n$ (complex) matrices is an algebra under the usual linear operations and matrix multiplication. The expression $(\lambda_{ij})_{n \times n}$ stands for the matrix

$$\begin{pmatrix} \lambda_{11} & \lambda_{12} & \cdots & \lambda_{1n} \\ \lambda_{21} & \lambda_{22} & \cdots & \lambda_{2n} \\ \vdots & \vdots & \ddots & \vdots \\ \lambda_{n1} & \lambda_{n2} & \cdots & \lambda_{nn} \end{pmatrix}.$$

The product of $(\lambda_{ij})_{n \times n}$ and $(\mu_{ij})_{n \times n}$ is the matrix

$$\left(\sum_{k=1}^{n} \lambda_{ik} \mu_{kj} \right)_{n \times n}.$$

These algebras form the most familiar reasonably representative family of algebras in which calculation is easy. It is surprising how often confusion or an ill-founded conjecture can be cleared up by checking the 2×2 matrix algebra. We urge the beginning reader, who is probably already acquainted with the basic facts, to become facile with such calculations. Of course, M_n has dimension n^2, has an identity element (which we denote by I) and is noncommutative unless n equals 1.

1.6.1 Matrix Units The usual choice of a linear space basis for M_n is the set $\{E^{k\ell} : 1 \leq k, \ell \leq n\}$ of *matrix units* where $E^{k\ell}$ has 1 in the $k\ell$-position and zero elsewhere (*i.e.*, the ij-entry is given by

$$(E^{k\ell})_{ij} = \delta_{ik}\delta_{j\ell},$$

where δ_{ik} is the Kronecker delta). The matrix units satisfy the product law:

$$E^{ij} \cdot E^{k\ell} = \delta_{jk} E^{i\ell} \qquad \forall\ i,j,k,\ell.$$

This relation provides an abstract description of a matrix algebra: An algebra is isomorphic to M_n if and only if it contains a linear space basis $\{E^{k\ell} : 1 \le k, \ell \le n\}$ satisfying the above product law.

1.6.2 The Trace We may define a linear map $\mathrm{Tr} \colon M_n \to \mathbb{C}$ by describing its value on the above basis by $\mathrm{Tr}(E^{ij}) = \delta_{ij}$ for all i,j. This map is, of course, just the usual *trace*, defined explicitly by

$$\mathrm{Tr}((a_{ij})_{n \times n}) = \sum_{i=1}^{n} a_{ii}.$$

From the product law for matrix units we see $\mathrm{Tr}(E^{ij} E^{k\ell}) = \delta_{jk} \delta_{i\ell} = \mathrm{Tr}(E^{k\ell} E^{ij})$. Hence we have the *trace identity*

$$\mathrm{Tr}(AB) = \mathrm{Tr}(BA) \qquad \forall\ A,\ B \in M_n.$$

By using the product law for matrix units, it is easy to show that any linear function satisfying the trace identity is a multiple of the trace as defined here. For some purposes, the *normalized trace* which equals $\frac{1}{n}\mathrm{Tr}$ on M_n is useful.

1.6.3 Matrix Representations of Linear Operators We briefly review the connection which exists between linear operators on a finite-dimensional linear space with dimension n and $n \times n$ matrices. Let $\mathcal{V}$ be a linear space of dimension n. Let V and W be two ordered bases for $\mathcal{V}$ with basis vectors $v^1, v^2, \ldots, v^n$ and $w^1, w^2, \ldots, w^n$, respectively. (An *ordered basis* is simply a basis listed in a fixed linear order.) For each $T \in \mathcal{L}(\mathcal{V})$, define n^2 complex numbers $\{\lambda_{ij} : 1 \le i,j \le n\}$ by

$$T(v^j) = \sum_{i=1}^{n} \lambda_{ij} v^i.$$

Denote the matrix $(\lambda_{ij})_{n \times n}$ by $M_V(T)$. For any other ordered basis such as W, $M_W(T)$ is defined similarly. The main fact about this construction is summarized in the following theorem.

Theorem. *Let V be an ordered basis for the n-dimensional vector space $\mathcal{V}$. Then M_V (as defined above) is an algebra isomorphism of $\mathcal{L}(\mathcal{V})$ onto M_n.*

1.6.4 Change of Bases The matrix corresponding to a given operator depends on the choice of ordered basis. For example, let $\mathcal{V}$ be the two-dimensional linear space consisting of all differentiable functions $f \colon \mathbb{R} \to \mathbb{C}$

such that $f'' = f$. Consider the two ordered bases V and W given by the functions $(\exp(\cdot), \exp(-(\cdot)))$ and $(\cosh(\cdot), \sinh(\cdot))$, respectively. Differentiation defines a linear operator D in $\mathcal{L}(V)$ and we have

$$M_V(D) = \begin{pmatrix} 1 & 0 \\ 0 & -1 \end{pmatrix}, \quad M_W(D) = \begin{pmatrix} 0 & 1 \\ 1 & 0 \end{pmatrix}.$$

We wish to recall the effect of changing bases. We continue the notation established above. Given the two ordered bases V and W, we can define an element C_{VW} of $\mathcal{L}(V)$ by

$$C_{VW} \sum_{i=1}^{n} \lambda_i v^i = \sum_{i=1}^{n} \lambda_i w^i \quad \forall \ \sum_{i=1}^{n} \lambda_i v^i \in V.$$

Note that C_{VW} is invertible with inverse C_{WV}. Denoting $M_W(T)$ by $(\mu_{ij})_{n \times n}$, we obtain

$$T C_{VW} v^j = T w^j = \sum_{i=1}^{n} \mu_{ij} w^i = C_{VW} \sum_{i=1}^{n} \mu_{ij} v^i.$$

Multiplying both sides by C_{VW}^{-1} and recalling the definition of $M_V(T)$, we get:

$$M_W(T) = M_V(C_{VW}^{-1} T C_{VW}) = M_V(C_{VW})^{-1} M_V(T) M_V(C_{VW}).$$

This shows that $M_W(T)$ and $M_V(T)$ are similar matrices. By reversing the above argument, we also see that any matrix similar to $M_V(T)$ will be $M_W(T)$ relative to a suitable ordered basis W.

The trace identity shows that similar matrices have the same trace. Thus the foregoing discussion shows that, for any finite-dimensional linear space V and any linear operator $T \in \mathcal{L}(V)$, we may define the trace of T by

$$\mathrm{Tr}(T) = \mathrm{Tr}(M_V(T))$$

where the choice of the ordered basis V is irrelevant.

1.6.5 Determinants   It is possible to define the trace and the determinant of a linear operator on a finite-dimensional space directly without introducing bases (*cf. e.g.*, Nickerson, Spencer and Steenrod [1959]), but this is sufficiently difficult that it is seldom done. We will not do it here. Hence we will assume that the reader is familiar with a definition of the determinant of an $n \times n$ matrix and the following three basic facts. We use $\mathrm{Det}(M)$ to denote the determinant of an $n \times n$ matrix M.

(1) $\mathrm{Det}(MN) = \mathrm{Det}(M)\mathrm{Det}(N) \quad \forall \ M, N \in M_n$.
(2) $\mathrm{Det}(M) = 0 \quad$ if and only if M is not invertible.

(3) $\mathrm{Det}(I) = 1$.

Properties (1) and (3) and the similarity of $M_V(T)$ and $M_W(T)$ show that $\mathrm{Det}(M_V(T))$ depends only on T and not on the choice of ordered basis V. Hence, from now on, we often consider Det to be a multiplicative map from $\mathcal{L}(\mathcal{V})$ into $\mathbb{C}$ (defined by $\mathrm{Det}(T) = \mathrm{Det}(M_V(T))$ where the choice of V is irrelevant).

Consideration of the extended left regular representation shows that every finite-dimensional algebra is isomorphic to a subalgebra of some matrix algebra. We will show later that every finite-dimensional prime algebra is isomorphic to M_n for some value of n. See Chapter 4 for terminology and Theorem 8.1.1 for a proof. Each M_n is a simple unital algebra. This means that there are no ideals except $\{0\}$ and M_n. Proposition 1.6.10 establishes a more general result.

Every finite-dimensional semiprime algebra is uniquely an internal direct sum of finitely many algebras, each of which is isomorphic to M_n for some n, and every such direct sum is strongly semisimple. Finally every finite-dimensional algebra is uniquely decomposable as the linear space internal direct sum of a (finite-dimensional) nilpotent ideal and a (finite dimensional) semiprime subalgebra.

1.6.6 Historical Remarks These last results were proved by Élie Cartan [1898] but are called the Wedderburn or Wedderburn–Artin structure theorems because of important generalizations due to Wedderburn [1907] and Emil Artin [1926]. They are proved at the beginning of Chapter 8.

Matrix algebras have an interesting history. They were invented and named by Arthur Cayley [1855], [1858]. However, quite independently, and from very different motives, they were re-invented by both Benjamin Peirce [1870] and his son Charles Saunders Peirce [1873]. James Joseph Sylvester stated [1884] that at Johns Hopkins University in 1881, he gave the first course on matrix theory ever given anywhere. For additional historical comments, particularly on the interesting separate viewpoints of Benjamin Peirce and C. S. Peirce, see Thomas Hawkins [1972] and Karen Hunger Parshall [1985].

1.6.7 Norms on M_n The algebra of $n \times n$ matrices can be made into a normed algebra in many ways. Since any norm on a finite-dimensional space is complete, M_n together with any of its algebra norms is a Banach algebra. Any two norms on a finite-dimensional space are equivalent, so that only when one is examining geometric rather than topological properties does it matter which algebra norm is used. The simplest norm for most calculations is the *Hilbert–Schmidt norm* defined by

$$\|(a_{ij})_{n \times n}\|_{HS} = \left(\sum_{i=1}^{n} \sum_{j=1}^{n} |a_{ij}|^2 \right)^{1/2}$$

for all $(a_{ij})_{n \times n} \in M_n$. The submultiplicative property is verified by an application of the Cauchy–Schwarz inequality.

Let $\mathbb{C}^n$ be the n-fold Cartesian product of $\mathbb{C}$ made into a linear space by the usual pointwise linear operations. When the elements of $\mathbb{C}^n$ are written as column vectors, matrix multiplication defines an isomorphism between M_n and the algebra $\mathcal{L}(\mathbb{C}^n)$ of linear operators on $\mathbb{C}^n$. If $\mathbb{C}^n$ is provided with a norm so that it is a Banach space, then (because of finite-dimensionality) the Banach algebra $\mathcal{B}(\mathbb{C}^n)$ of bounded linear operators is just $\mathcal{L}(\mathbb{C}^n)$ provided with the corresponding operator norm. When these operator norms are carried back to M_n through the isomorphism mentioned above, we get a large family of algebra norms on M_n. We will give the operator norms on M_2 corresponding to the three most common norms on $\mathbb{C}^2$. The operator norms for

$$a = \begin{pmatrix} \alpha & \beta \\ \gamma & \delta \end{pmatrix} \in M_2$$

corresponding to the ℓ^1-norm ($\|(\delta, \varepsilon)\| = |\delta| + |\varepsilon|$), the ℓ^2-norm ($\|(\delta, \varepsilon)\| = (|\delta|^2 + |\varepsilon|^2)^{1/2}$), and the ℓ^∞-norm ($\|(\delta, \varepsilon)\| = \max\{|\delta|, |\varepsilon|\}$) on $\mathbb{C}^2$ are, respectively,

$$\|a\|_{\ell^1} = \max\{|\alpha| + |\gamma|, |\beta| + |\delta|\}$$

$$\|a\|_{\ell^2} = 2^{-1/2}\{|\alpha|^2 + |\beta|^2 + |\gamma|^2 + |\delta|^2$$
$$+ [(|\alpha|^2 + |\beta|^2 + |\gamma|^2 + |\delta|^2)^2 - 4|\alpha\delta - \beta\gamma|^2]^{1/2}\}^{1/2}$$

$$\|a\|_{\ell^\infty} = \max\{|\alpha| + |\beta|, |\gamma| + |\delta|\}.$$

The first and last of the above norms are easily checked. The second norm can be most easily calculated from the properties of self-adjoint elements in C*-algebras discussed in Volume II.

1.6.8 Tensor Products of Matrices We wish to recall an old-fashioned concrete construction of the tensor product of two matrices. Let $a = (a_{ij})_{n \times n}$ and $b = (b_{ij})_{m \times m}$ be two matrices. Their tensor product $a \otimes b$ is the $nm \times nm$ matrix $(c_{ij})_{nm \times nm}$ defined as follows:

$$c_{ij} = a_{pq}b_{k\ell}, \quad \text{where} \quad \begin{cases} (k-1)n < i \le kn; \ p = i - (k-1)n \\ (\ell-1)n < j \le \ell n; \ q = j - (\ell-1)n. \end{cases}$$

Example:

$$\begin{pmatrix} a & b \\ c & d \end{pmatrix} \otimes \begin{pmatrix} e & f \\ g & h \end{pmatrix} = \begin{pmatrix} ae & be & af & bf \\ ce & de & cf & df \\ ag & bg & ah & bh \\ cg & dg & ch & dh \end{pmatrix}.$$

Another way to describe this product is to note

$$E^{ij} \otimes E^{k\ell} = E^{(kn-n+i)(\ell n-n+j)} \qquad \forall\, 1 \le i, j \le n;\ 1 \le k, \ell \le m$$

where E^{ij} is the ij-matrix unit in M_n and $E^{k\ell}$ is the $k\ell$-matrix unit in M_m. It is easy to check that this operation satisfies the following identities:

$$\begin{aligned} (\lambda a + \mu c) \otimes b &= \lambda(a \otimes b) + \mu(c \otimes b) \\ a \otimes (\lambda b + \mu d) &= \lambda(a \otimes b) + \mu(a \otimes d) \\ (a \otimes b)(c \otimes d) &= ac \otimes bd \end{aligned}$$

for all $a,\, c \in M_n$, $b,\, d \in M_m$ and $\lambda,\, \mu \in \mathbb{C}$. The formula for tensor products of matrix units shows that the set of all finite sums of tensor products $a \otimes b$ for $a \in M_n$ and $b \in M_m$ (which we would normally denote by $M_n \otimes M_m$) is all of M_{nm}. Hence $M_m \otimes M_n$ is isomorphic to $M_n \otimes M_m$.

The maps

$$\begin{aligned} a &\mapsto a \otimes I & a \in M_n \\ b &\mapsto I \otimes b & b \in M_m \end{aligned}$$

are unital algebra isomorphisms of M_n and M_m, respectively, into $M_{nm} \simeq M_n \otimes M_m$. Note that the images under these isomorphisms of M_n and M_m in M_{nm} are commuting unital subalgebras. Each is exactly the commutant of the other (Definition 1.1.4). Theorem 1.10.6 gives a general framework for this construction and show that the algebra $M_{nm} \simeq M_n \otimes M_m$ together with these embeddings is the solution to a universal mapping problem.

The Algebra $M_n(\mathcal{A})$ of $n \times n$ Matrices over an Algebra $\mathcal{A}$

1.6.9 Definition Let $\mathcal{A}$ be an algebra. For each i and j satisfying $1 \le i, j \le n$ let a_{ij} be an element of $\mathcal{A}$. Then $(a_{ij})_{n \times n}$ represents the matrix

$$\begin{pmatrix} a_{11} & a_{12} & \cdots & a_{1n} \\ a_{21} & a_{22} & \cdots & a_{2n} \\ \vdots & \vdots & \ddots & \vdots \\ a_{n1} & a_{n2} & \cdots & a_{nn} \end{pmatrix}.$$

We use $M_n(\mathcal{A})$ to denote the set of all $n \times n$ matrices with entries from $\mathcal{A}$. The sum of two matrices $(a_{ij})_{n \times n}$ and $(b_{ij})_{n \times n}$ in $M_n(\mathcal{A})$ is $(a_{ij} + b_{ij})_{n \times n}$, and the product of these matrices is

$$\left(\sum_{k=1}^{n} a_{ik} b_{kj} \right)_{n \times n}.$$

It is easy to check that $M_n(\mathcal{A})$ is an algebra. In Section 1.10, $M_n(\mathcal{A})$ will be identified as the algebra tensor product of M_n and $\mathcal{A}$.

The map which sends any element a of $\mathcal{A}$ to the diagonal matrix which has a at each position on the main diagonal (i.e., at each position with $i = j$) and 0's elsewhere is an isomorphism onto the subalgebra of diagonal matrices. Another way to embed $\mathcal{A}$ isomorphically into $M_n(\mathcal{A})$ is to send each $a \in \mathcal{A}$ onto the matrix with this element in its upper left corner (i.e. its 1,1-position) and 0's elsewhere. Clearly there are many other similar isomorphic embeddings.

If $\mathcal{A}$ is a normed algebra or a Banach algebra, then $M_n(\mathcal{A})$ is a normed algebra or a Banach algebra under various natural norms. For instance, the matrix $(a_{ij})_{n \times n}$ acts on the column vector $(b_1, b_2, \ldots, b_n)$ with entries from $\mathcal{A}^1$ by matrix multiplication

$$\begin{pmatrix} a_{11} & a_{12} & \cdots & a_{1n} \\ a_{21} & a_{22} & \cdots & a_{2n} \\ \vdots & \vdots & \ddots & \vdots \\ a_{n1} & a_{n2} & \cdots & a_{nn} \end{pmatrix} \begin{pmatrix} b_1 \\ b_2 \\ \vdots \\ b_n \end{pmatrix} = \begin{pmatrix} \sum_{k=1}^{n} a_{1k} b_k \\ \sum_{k=1}^{n} a_{2k} b_k \\ \vdots \\ \sum_{k=1}^{n} a_{nk} b_k \end{pmatrix}.$$

Clearly this action defines a bounded linear operator when the space $(\mathcal{A}^1)^n$ carries any one of several natural norms such as

$$\|(b_1, b_2, \ldots, b_n)\|_\infty = \sup\{\|b_1\|, \|b_2\|, \ldots, \|b_n\|\}.$$

The matrix $(a_{ij})_{n \times n}$ can now be given the operator norm for this bounded linear operator. If $\mathcal{A}$ is normed, $M_n(\mathcal{A})$ may also be given the Hilbert–Schmidt norm: $\|(a_{ij})_{n \times n}\|_{HS} = (\sum_{ij} \|a_{ij}\|^2)^{1/2}$ which is submultiplicative by the Cauchy–Schwarz inequality.

From now on, we will assume that $\mathcal{A}$ is unital. Then $M_n(\mathcal{A})$ contains various useful matrices in which all the entries are either 0 or 1. Let I be the identity matrix, and let $E^{k\ell}$ be the matrix which has 1 in the $k\ell$-position and zero elsewhere (i.e., the ij-entry is given by $(E^{k\ell})_{ij} = \delta_{ik}\delta_{j\ell}$, where δ_{ik} is the Kronecker delta, cf. §1.6.1).

1.6.10 Proposition *Let $\mathcal{A}$ be a unital algebra. Then $\mathcal{I} \mapsto M_n(\mathcal{I})$ is a bijection of the set of ideals of $\mathcal{A}$ onto the set of ideals of $M_n(\mathcal{A})$.*

Proof It is obvious that $\mathcal{I} \mapsto M_n(\mathcal{I})$ is an injection into the set of ideals of $M_n(\mathcal{I})$. Suppose $\tilde{\mathcal{I}}$ is an ideal in $M_n(\mathcal{I})$ and $A = (a_{ij})_{n \times n}$ is a matrix in $\tilde{\mathcal{I}}$. For any indices i, j, k and ℓ, $\tilde{\mathcal{I}}$ contains the matrix $E^{ki} A E^{j\ell}$ with a_{ij} at the $k\ell$-position and zeroes elsewhere. This shows that $\tilde{\mathcal{I}}$ has the form $M_n(\mathcal{I})$ for some ideal $\mathcal{I}$ of $\mathcal{A}$. $\qquad\square$

Let $\mathcal{A}$ be a unital algebra and let $\mathcal{B} = M_n(\mathcal{A})$. Let $e \in \mathcal{B}$ be the matrix E^{11} and let $w \in \mathcal{B}$ be the matrix with ones on the diagonal immediately below the main diagonal and in the upper right corner (the $1n$-position) and zeroes elsewhere. It is easy to check that e and w satisfy the conditions of the next theorem.

1.6.11 Theorem *Let $\mathcal{B}$ be a unital Banach algebra. Then $\mathcal{B}$ is isomorphic to $M_n(\mathcal{A})$ for some unital Banach algebra $\mathcal{A}$ and some $n \in \mathbb{N}$ if and only if there are elements e and w in $\mathcal{B}$ satisfying*

$$w^n = 1 = \sum_{j=1}^{n} w^{j-1} e w^{1-j} \quad and \quad e w^{j-1} e = \delta_{j1} e \qquad \forall\, j = 1, 2, \ldots, n. \quad (1)$$

In this case, $\mathcal{A}$ is isomorphic to the subalgebra $e\mathcal{B}e$ and $\mathcal{B}$ is isomorphic to $M_n(e\mathcal{B}e)$ under the isomorphism φ defined by

$$\varphi(b)_{jk} = e w^{1-j} b w^{k-1} e \qquad \forall\, b \in \mathcal{B};\ j, k = 1, 2, \ldots, n. \quad (2)$$

Furthermore, the map $\theta \colon e\mathcal{B}e \to \mathcal{B}$ defined by

$$\theta(ebe) = \sum_{j=1}^{n} w^{j-1} ebe w^{1-j} \qquad \forall\, b \in \mathcal{B} \quad (3)$$

is an isomorphism onto its range $\theta(e\mathcal{B}e) \subseteq \mathcal{B}$. Each element of $\theta(e\mathcal{B}e)$ commutes with e and w. Finally, the center of $\mathcal{B}$ is the image $\theta((e\mathcal{B}e)_Z)$ of the center of $e\mathcal{B}e$ under θ.

Proof The discussion preceding the statement of the theorem shows that if $\mathcal{B}$ is isomorphic to $M_n(\mathcal{A})$ then $\mathcal{B}$ contains elements satisfying (1). In this case $e\mathcal{B}e$ is obviously isomorphic to $\mathcal{A}$.

Suppose, conversely, that elements e and w satisfying (1) can be found in $\mathcal{B}$. For any $a, b \in \mathcal{B}$ and $j, k = 1, 2, \ldots, n$ we get

$$
\begin{aligned}
(\varphi(a)\varphi(b))_{jk} &= \sum_{i=1}^{n} \varphi(a)_{ji} \varphi(b)_{ik} \\
&= \sum_{i=1}^{n} e w^{1-j} a (w^{i-1} e^2 w^{1-i}) b w^{1-k} e = \varphi(ab)_{jk}
\end{aligned}
$$

which shows that φ is a homomorphism. To see that it is surjective, note

$$\varphi(w^{1-i} ebe w^{\ell-1})_{jk} = \delta_{ij} \delta_{\ell k} ebe \qquad \forall\, b \in \mathcal{B}.$$

If $b \in \ker(\varphi)$, then

$$b = \sum_{j=1}^{n} w^{j-1} e w^{1-j} b \sum_{k=1}^{n} w^{k-1} e w^{1-k} = \sum_{j=1}^{n} \sum_{k=1}^{n} w^{j-1} \varphi(b)_{jk} w^{1-k} = 0.$$

For any $a, b \in \mathcal{B}$ and $j, k = 1, 2, \ldots, n$ the calculation

$$\varphi(\theta(ebe))_{jk} = ew^{1-j}\left(\sum_{i=1}^{n} w^{i-1}ebew^{1-i}\right)w^{k-1} = \delta_{jk}ebe$$

shows that $\varphi(\theta(ebe))$ is the matrix $ebeI$ from which all the properties of θ follow.   □

Let $\mathcal{A}$ be a unital commutative algebra and let $M_n(\mathcal{A})$ be the algebra of all $n \times n$ matrices with entries from $\mathcal{A}$. The usual proof of Cramer's rule shows that a matrix in $M_n(\mathcal{A})$ is invertible in $M_n(\mathcal{A})$ if and only if its determinant is invertible in $\mathcal{A}$. Determinants do not work well in matrices over noncommutative algebras. Perhaps the easiest way to understand the problem is to consider the following matrix products

$$\begin{pmatrix} a & b \\ c & d \end{pmatrix}\begin{pmatrix} d & -b \\ -c & a \end{pmatrix} = \begin{pmatrix} ad - bc & -ab + ba \\ cd - dc & -cb + da \end{pmatrix}$$

$$\begin{pmatrix} d & -b \\ -c & a \end{pmatrix}\begin{pmatrix} a & b \\ c & d \end{pmatrix} = \begin{pmatrix} da - bc & db - bd \\ -ca + ac & -cb + ad \end{pmatrix}.$$

If a, b, c and d belong to a commutative algebra, the expressions on the right are just the determinant times the identity matrix, but if these elements do not commute, the right sides may be quite complicated. There are also four different ways to calculate the determinant. For specific examples, consider the algebra $M_2(M_2)$ of 2×2 matrices over M_2. If we choose

$$a = \begin{pmatrix} 2 & 0 \\ 0 & 1 \end{pmatrix}, \ b = \begin{pmatrix} 0 & 2 \\ 0 & 0 \end{pmatrix}, \ c = \begin{pmatrix} 0 & 0 \\ 2 & 0 \end{pmatrix} \ \text{and} \ d = \begin{pmatrix} 1 & 0 \\ 0 & 2 \end{pmatrix},$$

then the determinant calculated in any of the four possible orders is invertible but the matrix

$$A = \begin{pmatrix} a & b \\ c & d \end{pmatrix} \tag{4}$$

is not. In the other direction, if

$$a = \begin{pmatrix} 0 & 0 \\ 0 & 1 \end{pmatrix}, \ b = \begin{pmatrix} 0 & 1 \\ 0 & 0 \end{pmatrix}, \ c = \begin{pmatrix} 0 & 0 \\ 1 & 0 \end{pmatrix} \ \text{and} \ d = \begin{pmatrix} 1 & 0 \\ 0 & 0 \end{pmatrix},$$

then the element A defined in (4) satisfies $A^2 = I$ so that it is invertible even though its determinant is not invertible when calculated in any of the four orders. Criteria for invertibility and one-sided invertibility in terms of the determinant have been developed (cf. S. Š. Kesler, and Naum Ya. Krupnik [1967] or Krupnik [1987] Theorems 1.2 and 1.3).

The following criterion is occasionally useful for matrices in $M_2(\mathcal{A})$. If A has the form (4) and a is invertible, then A is invertible if and only if $d - ca^{-1}b$ is invertible since A satisfies

$$A = \begin{pmatrix} 1 & 0 \\ ca^{-1} & 1 \end{pmatrix} \begin{pmatrix} a & 0 \\ 0 & d - ca^{-1}b \end{pmatrix} \begin{pmatrix} 1 & a^{-1}b \\ 0 & 1 \end{pmatrix}.$$

Cf. the remarks following the proof of Theorem 2.2.14 and Danrun Huang [1992].

1.7   Operator Algebras

Algebras of operators on linear spaces are among the most important algebras to which the theory of this work applies. The extended left regular representation shows that every Banach algebra is isometrically isomorphic to a (necessarily closed) subalgebra of $\mathcal{B}(\mathcal{X})$ for some Banach space $\mathcal{X}$. The last section deals with the case of finite-dimensional vector spaces, so in this section we will usually assume that the vector spaces lack a finite basis and are therefore infinite-dimensional. Interesting results on infinite-dimensional linear spaces require some topology on the space and, in keeping with the general emphasis of this book, we will require a norm. In fact, the most common situation will be that in which the operators act on a Banach space.

At this point we do not have enough terminology for a detailed discussion of algebras of operators. Therefore, we will concentrate on introducing a few important ideals that occur in the algebras $\mathcal{B}(\mathcal{X})$. The theory of these ideals becomes somewhat messy unless the Banach space is fairly well behaved. From another viewpoint, we are merely acknowledging that a perfectly general Banach space may be wildly poorly behaved. Fortunately, most Banach spaces which appear in applications (other than as counterexamples) are relatively tame.

Hilbert spaces are the best behaved of all infinite-dimensional Banach spaces. However, the study of operator theory on Hilbert space is dominated by the extra structure provided by the Hilbert space adjoint which makes many interesting algebras of operators into *-algebras. Since the second volume of this work is devoted to the study of *-algebras, we prefer to postpone most of the study of operators on Hilbert space until that volume. (A short summary is included in 1.7.17 below.) Therefore we will state results for operators on Banach spaces which satisfy an array of desirable properties. There are many excellent books for further reading of which we mention a few. Dunford and Schwartz [1958] remains a basic reference but for newer developments Joseph Diestel and John Jerry Uhl, Jr. [1977], [1983] and Joram Lindenstrauss and Lior Tzafriri [1977], [1979] are particularly useful. See also Mahlon M. Day [1973], Diestel [1984], Graham

J. O. Jameson [1987], Vitali D. Milman and Gideon Schechtman [1986], Gilles Pisier [1986], Ivan Singer [1970], [1981], Nicole Tomczak-Jaegermann [1989], Dick van Dulst [1978] and Bernard Beauzamy [1982].

1.7.1 Reflexive Banach Spaces The most important distinction in Banach space theory is between reflexive and nonreflexive spaces. If $\mathcal{X}$ is a normed linear space, then its double dual space $\mathcal{X}^{**}$ is just the collection of continuous linear functionals on its dual space $\mathcal{X}^*$. Thus evaluation provides a natural linear map $\kappa : \mathcal{X} \to \mathcal{X}^{**}$ defined by

$$\kappa(x)(\omega) = \omega(x) \qquad \forall\, x \in \mathcal{X};\ \omega \in \mathcal{X}^*.$$

It is easy to see that this is an isometric linear isomorphism of $\mathcal{X}$ into $\mathcal{X}^{**}$. Thus it is often useful to think of $\mathcal{X}$ as a subspace of $\mathcal{X}^{**}$.

Definition A Banach space is called *reflexive* if the map κ, defined above, is surjective.

It is important to note that the definition requires that κ be the map which establishes a linear isometry between $\mathcal{X}$ and $\mathcal{X}^{**}$. Robert C. James [1951] provided an example of a (nonreflexive) Banach space $\mathcal{J}$ (called the *James space*) which is isometrically isomorphic to $\mathcal{J}^{**}$ but in which $\kappa(\mathcal{J})$ is a codimension one subspace of $\mathcal{J}^{**}$. (Alfred Andrew and William L. Green [1980] show that $\mathcal{J}$ is a Banach algebra under pointwise multiplication and that the Arens product on $\mathcal{J}^{**}$ is the unitization of $\mathcal{J}$. They determine the automorphism group of this algebra in detail. See also Shashi Kiran and Ajit Iqbal Singh [1988].)

Theorem *The following are equivalent for a Banach space $\mathcal{X}$:*
  (a) *$\mathcal{X}$ is reflexive.*
  (b) *$\mathcal{X}^*$ is reflexive.*
  (c) *The weak topology (i.e., $\mathcal{X}^{**}$-topology) on $\mathcal{X}^*$ equals the weak* topology (i.e., $\mathcal{X}$-topology).*
  (d) *The unit ball of $\mathcal{X}$ is compact in its weak topology.*
  (e) *The unit ball of $\mathcal{X}^*$ is compact in its weak topology.*

Proof (a)$\Rightarrow$(c): This is obvious.

(c)$\Rightarrow$(e): Alaoglu's theorem (Dunford and Schwartz [1958], V.4.2) shows this.

(e)$\Rightarrow$(b): Goldstine's theorem (Dunford and Schwartz [1958], V.4.5) asserts that the image under κ of the unit ball $\mathcal{Y}_1$ of any Banach space $\mathcal{Y}$ is dense in the unit ball $\mathcal{Y}_1^{**}$ of its double dual space in the weak* topology of that space. When this topology is transported back to $\mathcal{Y}_1$ by κ, it is just the weak topology. Hence if $\mathcal{X}_1^*$ is weakly compact, its image under κ must be all of $\mathcal{X}_1^{***}$. Thus κ is surjective.

(b)$\Rightarrow$(a): Let $\kappa_0 \colon \mathcal{X} \to \mathcal{X}^{**}$ and $\kappa_1 \colon \mathcal{X}^* \to \mathcal{X}^{***}$ be the natural maps. Condition (b) shows that any Ω in $\mathcal{X}^{***}$ has the form $\Omega = \kappa_1(\omega)$ for some

$\omega \in \mathcal{X}^*$. Therefore $\Omega(\kappa_0(x)) = \kappa_1(\omega)(\kappa_0(x)) = \kappa_0(x)(\omega) = \omega(x)$ for all $x \in \mathcal{X}$ shows that Ω is zero if it vanishes on $\kappa_0(\mathcal{X})$. The Hahn–Banach theorem then shows that $\kappa_0(\mathcal{X})$ is norm dense in $\mathcal{X}^{**}$. Since κ_0 is an isometry, its range is complete and thus norm closed. Therefore κ_0 is surjective.

Now the equivalence of (d) and (e) follows from the equivalence of (a) and (b). □

Relatively few Banach spaces are reflexive. The best known examples are L^p and ℓ^p for $1 < p < \infty$. Hilbert spaces are reflexive. In fact if $\mathcal{H}$ is a Hilbert space, then it is isometrically real-linearly isomorphic to $\mathcal{H}^*$ but the isomorphism is conjugate linear rather than complex linear.

In a nonreflexive space, the tower of dual space $\mathcal{X}$, $\mathcal{X}^*$, $\mathcal{X}^{**},\ldots$ usually grows very rapidly. (The space $\mathcal{J}$, mentioned above, is a weird exception.) In fact, nonreflexive Banach spaces $\mathcal{X}$ with $\mathcal{X}^*$ separable are comparatively rare and concrete representations for $\mathcal{X}^{***}$ are even rarer among nonreflexive spaces.

The theorem implies that every closed subspace and every quotient space of a reflexive Banach space is reflexive. We mention two additional interesting properties which characterize reflexive Banach spaces. The weak compactness of the unit ball shows that any continuous linear functional on a reflexive Banach space actually attains its norm at some point in the unit ball. (*I.e.* for each $\omega \in \mathcal{X}^*$, there is some $x \in \mathcal{X}_1$ satisfying $\omega(x) = \|\omega\|$.) This property characterizes reflexive Banach spaces (James [1972]). If a reflexive Banach space is given an equivalent norm, it is again reflexive by condition (c) of the last theorem. Thus, in particular, it is a dual Banach space. Again this property is characteristic. That is, any nonreflexive Banach space $\mathcal{X}$ has an equivalent norm in which it is not isometrically linearly isomorphic to a dual Banach space. (See van Dulst and Singer [1976] and van Dulst [1978], 4.8 for a slightly stronger result.)

The following result is due to Nicholas J. Young [1976].

Theorem *If $\mathcal{X}$ is a reflexive Banach space, then every continuous linear functional on $\mathcal{B}_K(\mathcal{X})$ is weakly almost periodic. Conversely, if $\mathcal{X}$ is a nonreflexive Banach space, then no non-zero continuous linear functional on $\mathcal{B}_F(\mathcal{X})$ (with its operator norm) is weakly almost periodic.*

Recall (from Theorem 1.4.11) that a normed algebra $\mathcal{A}$ is Arens regular if and only if every continuous linear functional on $\mathcal{A}$ is weakly almost periodic. Also, subalgebras and quotient algebras of Arens regular algebras are Arens regular. For a normed algebra $\mathcal{A}$ (such as $\mathcal{B}_F(\mathcal{X})$) which is not complete, Arens regularity and the weakly almost periodic character of linear functionals do not change when the algebra is replaced by its completion.

There are two easy and interesting corollaries.

Corollary *Let $\mathcal{X}$ be a Banach space. Then $\mathcal{B}_K(\mathcal{X})$ is Arens regular if and only if $\mathcal{X}$ is reflexive.*

Corollary *Let $\langle\, \mathcal{X} \,/\, \mathcal{Y} \,\rangle$ be a $\langle$ nonreflexive / reflexive $\rangle$ Banach space. Then there is no non-zero continuous homomorphism of $\mathcal{B}_F(\mathcal{X})$ into $\mathcal{B}(\mathcal{Y})$.*

Young shows that reflexive Banach spaces $\mathcal{X}$ exist for which $\mathcal{B}(\mathcal{X})$ is not Arens regular. In fact, he exhibits a reflexive Banach space $\mathcal{X}$ for which $\mathcal{B}_K(\mathcal{X})$ is Arens regular and $\mathcal{B}_K(\mathcal{X})^{**}$ with Arens multiplication is homeomorphically isomorphic to $\mathcal{B}(\mathcal{X})$ (*cf.* §1.7.14 below), but $\mathcal{B}(\mathcal{X})$ is not even Arens regular. Hence an algebra $\mathcal{A}$ may be Arens regular while its double dual $\mathcal{A}^{**}$ (with Arens multiplication) is not Arens regular. Pym [1983] gives an easier example. Ülger [1988b] proves a stronger form of Arens regularity for $\mathcal{B}_K(\mathcal{X})$ when $\mathcal{X}$ is reflexive.

1.7.2 Topological and Geometric Categories of Banach Spaces In Section 1.2 we mentioned the existence of two categories of Banach algebras. We have the same phenomenon in the present case also. We consider the *topological category*, in which the morphisms are continuous linear maps, to be more fundamental than the *geometric category*, in which the morphisms are contractive linear maps. In the first of these categories, objects are isomorphic if there is a homeomorphic linear isomorphism between them, and thus the objects are really topological algebras in which the topology is derived from a complete norm. In the geometric category, two objects are isomorphic if there exists an isometric linear isomorphism between them, and thus the precise value of the norm itself is part of the isomorphic structure of an object. Traditionally, two Banach spaces are called *isomorphic* if they are isomorphic in the topological category and *isometric* if they are isomorphic in the second category. We observe this distinction. We will stress properties of Banach spaces which are isomorphism invariants. We have already noted that reflexivity of a Banach space is an isomorphism invariant.

We wish to introduce four additional properties related to reflexivity. In two cases it requires care to define them in an isomorphism invariant way. We begin by considering a more restrictive condition.

1.7.3 Super-reflexive Banach Spaces Let us briefly discuss Hilbert space as a special case of a Banach space. We noted earlier that Hilbert space is usually defined in terms of an inner product and this gives rise to the Hilbert space adjoint of a bounded linear operator and thus to the *-algebra structure of algebras of operators. A Hilbert space inner product (*cf.* 1.7.17) can easily be introduced into any Banach space for which the norm satisfies the parallelogram law:

$$\|x + y\|^2 + \|x - y\|^2 = 2\|x\|^2 + 2\|y\|^2$$

in such a way that the inner product defines the given norm. (A proof is included in Volume II.) This condition may be considered as asserting the perfect roundness of the unit ball, and it turns out that reflexivity is related to the absence of infinite-dimensional flatness on the surface of the unit ball. James A. Clarkson [1936] introduced a condition which he called uniform convexity. D. P. Milman [1938] showed that it implied reflexivity although it is not enjoyed by all reflexive Banach spaces. (See Kôsaku Yosida [1965] V.2.2 for a proof.) It is a sort of weak form of the parallelogram law and, like it, is certainly not an isomorphism invariant. Uniform convexifiability, super-reflexivity and related properties were studied extensively by James, but the final characterization is due to Per Enflo [1972] (*cf.* van Dulst [1978]).

Definition A Banach space $\mathcal{X}$ is called *uniformly convex* if the following implication holds for any two sequences $\{x_n\}_{n\in\mathbb{N}}$ and $\{y_n\}_{n\in\mathbb{N}}$ of elements in the unit ball:

$$\left(\left\| \frac{x_n + y_n}{2} \right\| \to 1 \right) \Rightarrow \left(\|x_n - y_n\| \to 0 \right).$$

A Banach space $\mathcal{X}$ is called *uniformly smooth* if the norm is uniformly differentiable in the sense that the limit:

$$\lim_{t \to 0} \frac{\|x + ty\| - \|x\|}{t}$$

exists uniformly for $x, y \in \mathcal{X}_1, x \neq 0$. A Banach space $\mathcal{Y}$ is said to be *finitely representable* in a Banach space $\mathcal{X}$ if, for every finite-dimensional subspace $\mathcal{F}$ of $\mathcal{Y}$ and every $\varepsilon > 0$, there is a linear injection θ of $\mathcal{F}$ into $\mathcal{X}$ satisfying

$$(1 - \varepsilon)\|y\| \leq \|\theta(y)\| \leq (1 + \varepsilon)\|y\| \qquad \forall\, y \in \mathcal{F}.$$

Finally, a Banach space $\mathcal{X}$ is said to be *super-reflexive* if any Banach space $\mathcal{Y}$ which is finitely representable in it is reflexive.

Finite representability is not a demanding condition. For instance, for every Banach space, $\mathcal{X}^{**}$ is finitely representable in $\mathcal{X}$. More surprisingly, there is a reflexive Banach space $\mathcal{U}$ in which every Banach space is finitely representable. The space $\mathcal{U}$ is just the ℓ^2-direct sum of the spaces $\mathbb{C}^n$ for all $n \in \mathbb{N}$ where the individual pieces carry the sup norm: $\|(\lambda_1, \lambda_2, \ldots, \lambda_n)\|_\infty = \max\{|\lambda_1|, |\lambda_2|, \ldots, |\lambda_n|\}$. Obviously $\mathcal{U}$ is not super-reflexive. Moreover, ℓ^2 is finitely representable in any infinite-dimensional Banach space.

Theorem *A Banach space is uniformly convex if and only if its dual space is uniformly smooth. A uniformly convex or uniformly smooth Banach space is reflexive.*

The following conditions are equivalent for a Banach space $\mathcal{X}$:

(a) $\mathcal{X}$ *is super-reflexive.*

(b) $\mathcal{X}^*$ *is super-reflexive.*

(c) $\mathcal{X}$ *is uniformly convex in an equivalent norm (i.e., it is isomorphic to a uniformly convex space).*

(d) $\mathcal{X}$ *is uniformly smooth in an equivalent norm (i.e., it is isomorphic to a uniformly smooth space).*

(e) $\mathcal{X}$ *has an equivalent norm in which it is both uniformly convex and uniformly smooth.*

We prove only that a uniformly convex space is reflexive. By considering the converse of the definition, it is easy to see that a space is uniformly convex if and only if for every $\varepsilon > 0$, the number

$$\delta(\varepsilon) = \inf\{1 - \|\frac{x+y}{2}\| : x, y \in \mathcal{X}_1; \ \|x - y\| \geq \varepsilon\}$$

is positive. (A deeper and surprising result states that, if $\delta(\varepsilon)$ is positive for any single value of ε with $0 < \varepsilon < 2$, then the space is uniformly convex (van Dulst [1978] p. 188).) We have already noted Goldstine's result that the unit ball of a Banach space is dense in the unit ball of its double dual with respect to the weak* topology on the latter space. Thus any $x^{**} \in \mathcal{X}^{**}$ of norm 1 is the weak* limit of a net $\{x_\alpha\}_{\alpha \in A} \subseteq \mathcal{X}_1$. The net $\{(x_\alpha + x_\beta)/2\}_{(\alpha,\beta) \in A \times A}$ also converges to x^{**}. Hence the weak* lower semicontinuity of the norm implies that this net satisfies the condition in the uniform convexity criterion and thus $\lim_{(\alpha,\beta)} \|x_\alpha - x_\beta\| = 0$. Therefore the original net is a Cauchy net in the norm. Hence it must converge to an element of $\mathcal{X}$ which κ sends onto x^{**}. See van Dulst [1978] for a full proof.

We will mention other conditions below which are equivalent to super-reflexivity. However we will omit the large number of equivalent, more geometrical properties based on the growth of trees in the space. Those who wish to learn more should consult the very readable book by van Dulst [1978].

1.7.4 Weak Sequential Completeness The weak topology on a Banach space is seldom a metric topology (except in the comparatively rare case when its dual space is separable). Hence there is no reason to expect that, for example, a weak limit point of a sequence will be the weak limit of a subsequence. Nevertheless, since the very beginning of the subject, numerous results have been found showing that sequences are unusually useful in the weak topology. Probably the best known result is the Eberlein–Šmulian theorem which asserts that weak compactness of a subset $\mathcal{K}$ of a Banach space is equivalent to *weak countable compactness* (*i.e.*, every sequence has a weak limit point in $\mathcal{K}$) and to *weak sequential completeness* (*i.e.* every sequence has a subsequence converging weakly to an element in $\mathcal{K}$). For an easy proof see Robert Whitley [1967].

We now introduce an important sequential property which is enjoyed only by certain Banach spaces.

Definition A sequence $\{x_n\}_{n \in \mathsf{N}}$ in a Banach space $\mathcal{X}$ is said to be a weak Cauchy sequence if for each $\omega \in \mathcal{X}^*$ the sequence of numbers $\{\omega(x_n)\}_{n \in \mathsf{N}}$ is a Cauchy sequence. The Banach space $\mathcal{X}$ is said to be *weakly sequentially complete* if every weak Cauchy sequence has a weak limit in $\mathcal{X}$.

Proposition *A reflexive space is weakly sequentially complete.*

Proof A corollary of the uniform boundedness principle shows that a weak Cauchy sequence is bounded. Since the weak and weak* topologies agree on a reflexive space, the sequence must have a weak limit by the Alaoglu and Eberlein–Šmulian theorems. (This result can be established without the last theorem.)   □

In addition to reflexive spaces, most Banach spaces of measures and L^1- and ℓ^1-spaces are weakly sequentially complete. See Dunford and Schwartz [1958], pp. 374–379.

1.7.5 Dual Spaces Obviously a reflexive Banach space is always the dual of some other space, namely its own dual. We have already noted that reflexive spaces are the only Banach spaces which, when given any equivalent norm, are still isometric to a dual space. However, a space $\mathcal{X}$ which is a dual space in some norm has interesting properties. This is particularly true if it satisfies some other property such as separability or possession of a basis.

For any dual space, there is a canonical projection of $\mathcal{X}^{***}$ onto $\mathcal{X}^*$. However, this property does not characterize dual spaces–Banach spaces with this property are known which are not dual spaces in any equivalent norm (*e.g.* $L^1([0,1])$). We now give the interesting construction of this projection for a dual space.

Let $\mathcal{X}$ be a Banach space and let $\kappa_0 \colon \mathcal{X} \to \mathcal{X}^{**}$ and $\kappa_1 \colon \mathcal{X}^* \to \mathcal{X}^{***}$ be the canonical injections. Then for the dual $\kappa_0^* \colon \mathcal{X}^{***} \to \mathcal{X}^*$ of κ_0, the map

$$\kappa_0^* \circ \kappa_1 \colon \mathcal{X}^* \to \mathcal{X}^*$$

is the identity by an easy calculation. Hence

$$\kappa_1 \circ \kappa_0^* \colon \mathcal{X}^{***} \to \mathcal{X}^{***}$$

is a norm 1 projection onto the subspace $\kappa_1(\mathcal{X}^*)$. We will denote this projection by π. It is easy to see that the kernel of π is the closed complementary subspace $(I - \pi)(\mathcal{X}^{***}) = \kappa_0(\mathcal{X})^{\perp}$.

1.7.6 The Radon–Nikodym and Kreĭn–Milman Properties The Radon–Nikodym property has great importance in many areas but it arose in the study of vector measures. The excellent book by Diestel and Uhl [1977] presents many equivalent conditions, summarized on page 217 (*cf.* [1983] also). Because space does not permit a discussion of vector measures here, we will use one of the easy geometric conditions in our definition. Our chief interest lies in another consequence of the Radon–Nikodym property which will be introduced after certain ideals of operators have been defined.

Definition   Let $\mathcal{X}$ be a Banach space.

(a) Let $\mathcal{K}$ be a convex subset of $\mathcal{X}$. A point $x \in \mathcal{K}$ is said to be an extreme point of $\mathcal{K}$ if it is not the midpoint of a positive length line segment with end points in $\mathcal{K}$. Similarly, $x \in \mathcal{K}$ is a strongly exposed point of $\mathcal{K}$ if there is a linear functional $\omega \in \mathcal{X}^*$ with the following two properties: (1) $\omega(x)$ is strictly greater than $\omega(y)$ for any other $y \in \mathcal{K}$ and (2) if any sequence $\{x_n\}_{n \in \mathbb{N}}$ in $\mathcal{K}$ satisfies $\lim \omega(x_n) = \omega(x)$, then the sequence converges to x.

(b) $\mathcal{X}$ is said to satisfy the $\langle$ *Kreĭn–Milman / Radon–Nikodym* $\rangle$ property if every nonempty norm closed bounded convex subset of $\mathcal{X}$ is the norm closed convex hull of its $\langle$ extreme / strongly exposed $\rangle$ points.

It is not known whether the two properties introduced in (b) are actually equivalent. The classical Kreĭn–Milman theorem asserts that, in any locally convex topological linear space, every nonempty compact convex subset is the closed convex hull of its extreme points. Hence, in a dual Banach space, a nonempty norm closed bounded convex subset is the *weak** closed convex hull of its extreme points. If the Banach space were reflexive, so that the weak* and weak topologies agreed, this would make the set the weak closure and hence the norm closure of its convex hull. Obviously a strongly exposed point is an extreme point. These two remarks establish part of the next theorem. See Diestel and Uhl [1977] for a full proof.

Theorem   *A reflexive space or a separable dual space has the Radon–Nikodym property. Any space with the Radon–Nikodym property has the Kreĭn–Milman property and the two properties agree on dual spaces.*

Finally, a Banach space is super-reflexive if and only if every space finitely representable in it has either the Radon–Nikodym or the Kreĭn–Milman property.

If every Banach space finitely representable in a given space $\mathcal{X}$ has a certain property, then we say that $\mathcal{X}$ has the super-property. Thus the above theorem shows that super-reflexive is equivalent to super-Radon–Nikodym and to super-Kreĭn–Milman. It turns out that there are other properties for which the super version is equivalent to super-reflexive. We introduce some briefly.

First let us consider the girth of the unit ball in a Banach space. To understand this concept, consider first two-dimensional real space where our geometrical insight is good. We will introduce three possible unit balls and then measure the distance around them in the norm which they define on the space. A circular disc defines the ordinary Euclidean norm and its girth is 2π. A hexagon or parallelogram centered around the origin give girths of 6 and 8, respectively, and it turns out that these are the extreme values. In a higher dimensional space, we wish to measure the minimal distance of a closed curve on the surface of the unit sphere which contains the antipode $-x$ of any point x in the curve. It is easier to think of finding the shortest path from any point to its antipode and doubling the length of this path. The rather astounding fact is that, for some infinite-dimensional Banach spaces, the distance on the surface of the unit sphere from a point to its antipodal point can be as short as 2 (*i.e.*, the same as the straight line distance through the origin.) A Banach space in which this can happen is aptly called *flat* and a space where it does not is called *nonflat*.

We will show that $\mathcal{X} = C([0,1])$ is flat. Consider the path $x:[0,2] \to \mathcal{X}$ defined by $x(s) = f_s$ where

$$f_s(t) = \begin{cases} 1 + 2t - s & \text{if} \quad 0 \le t \le s/2, \\ 1 - 2t + s & \text{if} \quad s/2 < t \le 1. \end{cases} \tag{1}$$

It is easy to check that this path satisfies $\|f_s\| = 1$ and $\|f_s - f_r\| = |s - r|$ for all s and r in $[0,2]$ and $f_2 = -f_0$.

Theorem *If a Banach space $\mathcal{X}$ is flat, then so is its dual $\mathcal{X}^*$. In this case, $\mathcal{X}$ is not reflexive and its dual $\mathcal{X}^*$ is not separable.*

We can embed $C([0,1])$ isometrically into ℓ^∞ by simply enumerating the rational numbers in $[0,1]$ and mapping a function to the sequence of its values. Hence $\ell^\infty = c_0^{**}$ is flat, but c_0 is not since its dual ℓ^1 is separable. It turns out that ℓ^1 is not flat either.

We consider two more properties. A Banach space is said to have the *Banach–Saks property* if each of its sequences $\{x_n\}_{n \in \mathbb{N}}$ has a subsequence $\{x_{n_k}\}_{k \in \mathbb{N}}$ for which the sequence $\{n^{-1} \sum_{k=1}^n x_{n_k}\}_{n \in \mathbb{N}}$ of Cesáro means converges. Uniformly convex spaces have the Banach–Saks property and spaces with the property are reflexive, but the converses of both these implications fail. In particular, the Banach space $\mathcal{U}$, introduced in the discussion of super-reflexive spaces, has the Banach–Saks property. Finally, a Banach space $\mathcal{X}$ is said to be *ergodic* if for every linear isometry $T \in \mathcal{B}(\mathcal{X})$ the sequence $\{(n+1)^{-1} \sum_{k=0}^n T^k\}_{n \in \mathbb{N}}$ of Cesáro means converges to zero in the strong operator topology. Reflexive spaces are known to be ergodic but, again, the converse fails. Both of these concepts were introduced in order to state the following theorem.

Theorem *The following six properties of a Banach space are equivalent: super-reflexive, super-Radon–Nikodym, super-Kreĭn–Milman, super-nonflat, super-Banach–Saks and super-ergodic.*

All these results are proved in Diestel and Uhl [1977] and van Dulst [1978].

Before we discuss the final two properties of Banach spaces which we want to survey, the existence of bases and the approximation property, we will introduce several ideals of operators.

1.7.7 The Ideals of Approximable and Compact Operators

The ideal $\mathcal{B}_F(\mathcal{X})$ of finite-rank operators on a normed linear space $\mathcal{X}$ was introduced in Definition 1.1.17. Proposition 1.1.18 shows that $\mathcal{B}_F(\mathcal{X})$ is an ideal of $\mathcal{B}(\mathcal{X})$ which is included in every non-zero ideal of $\mathcal{B}(\mathcal{X})$. We recall the formulas. Let $\mathcal{W}$, $\mathcal{X}$, $\mathcal{Y}$ and $\mathcal{Z}$ be normed linear spaces. For elements $y \in \mathcal{Y}$ and $\omega \in \mathcal{X}^*$, define an operator $y \otimes \omega \in \mathcal{B}(\mathcal{X}, \mathcal{Y})$ by

$$(y \otimes \omega)(z) = \omega(z)y \qquad \forall\, z \in \mathcal{X}. \tag{2}$$

We denote the linear span of all such operators by $\mathcal{B}_F(\mathcal{X}, \mathcal{Y})$ and call operators in this space *finite-rank operators*. We simplify $\mathcal{B}_F(\mathcal{X}, \mathcal{X})$ to $\mathcal{B}_F(\mathcal{X})$. For arbitrary $S \in \mathcal{B}(\mathcal{W}, \mathcal{X})$, $T \in \mathcal{B}(\mathcal{Y}, \mathcal{Z})$, $y \in \mathcal{Y}$, $\omega \in \mathcal{X}^*$, $z \in \mathcal{Z}$ and $\tau \in \mathcal{Y}^*$ these operators satisfy

$$\begin{aligned}
(z \otimes \tau)(y \otimes \omega) &= \tau(y)z \otimes \omega \\
T(y \otimes \omega) &= T(y) \otimes \omega \\
(y \otimes \omega)S &= y \otimes S^*(\omega) \\
(x \otimes \omega)^* &= \omega \otimes \kappa(x).
\end{aligned} \tag{3}$$

We introduced tensor product notation in the last section in a preliminary way. The concept will be studied more systematically in Section 1.10. When this is done, it will be seen that $\mathcal{B}_F(\mathcal{X})$ is isomorphic as a linear space to the algebraic tensor product $\mathcal{X} \otimes \mathcal{X}^*$, and that the notation given for the operators $x \otimes \omega$ is consistent with the general notation. Furthermore, a matrix representation for the operators $x \otimes \omega$ can be constructed explicitly as a generalization of the explicit construction of the tensor product $M_n \otimes M_m$ given in the last section. To make things simpler, we will assume that $\mathcal{X}$ is finite-dimensional and that $\{x_1, x_2, \ldots, x_n\}$ is a basis for $\mathcal{X}$. Let $\{\omega_1, \omega_2, \ldots, \omega_n\}$ be the dual basis for $\mathcal{X}^*$ used in the proof of Proposition 1.1.18. Then the matrix for $x \otimes \omega$ is

$$(\lambda_i \mu_j)_{n \times n},$$

where $x = \sum_{i=1}^n \lambda_i x_i$ and $\omega = \sum_{j=1}^n \mu_j \omega_j$. When x is considered as an $n \times 1$ matrix (a column vector) and ω is considered as a $1 \times n$ matrix (a row vector) so that the evaluation $\omega(x)$ is just matrix multiplication, then the above formula for the matrix of $x \otimes \omega$ and the explicit formula for the matrix $a \otimes b$ (for $a \in M_n$ and $b \in M_m$) given in the last section agree.

Remark and Convention   We are about to introduce a number of ideals of $\mathcal{B}(\mathcal{X})$. However, all these ideals may be extended to be subspaces of the normed linear space $\mathcal{B}(\mathcal{X}, \mathcal{Y})$ of bounded linear operators from $\mathcal{X}$ to $\mathcal{Y}$. In this setting they still form a sort of generalized ideal in the sense indicated by equations (3) above. From now on, we will freely use our terminology for operators in $\mathcal{B}(\mathcal{X}, \mathcal{Y})$ when this is convenient even when the definition is given only for operators in $\mathcal{B}(\mathcal{X})$. In every case definitions and proofs are obvious generalizations.

In a Banach algebra, the closed ideals are usually more important than the nonclosed ideals. It is easy to see that $\mathcal{B}_F(\mathcal{X})$ is never closed unless $\mathcal{X}$ is finite-dimensional, in which case it equals $\mathcal{B}(\mathcal{X})$.

Definition   The closure of $\mathcal{B}_F(\mathcal{X})$ is called the ideal of *approximable operators* and denoted by $\mathcal{B}_A(\mathcal{X})$.

Obviously this ideal is included in each non-zero closed ideal of $\mathcal{B}(\mathcal{X})$. For some common Banach spaces $\mathcal{X}$, it is the only proper closed ideal of $\mathcal{B}(\mathcal{X})$.

Definition   An operator T in $\mathcal{B}(\mathcal{X})$ is said to be *compact* if the norm closure of $T(\mathcal{X}_1)$ is compact in the norm topology of $\mathcal{X}$. The set of compact operators in $\mathcal{B}(\mathcal{X})$ is denoted by $\mathcal{B}_K(\mathcal{X})$.

Proposition $\mathcal{B}_K(\mathcal{X})$ *is a non-zero closed ideal in* $\mathcal{B}(\mathcal{X})$ *which includes the ideal of approximable operators. In an infinite-dimensional Banach space it is always proper (i.e., not equal to* $\mathcal{B}(\mathcal{X})$).

Proof   To see that $\mathcal{B}_K(\mathcal{X})$ is closed, we use the fact that a subset of $\mathcal{X}$ has compact closure if and only if it is totally bounded. The sum of two compact operators is compact because the sum of two compact sets is compact. If K is a compact operator and T an arbitrary bounded operator, then $TK(\mathcal{X}_1)$ is the continuous image of a compact set and hence compact. Similarly $KT(\mathcal{X}_1)$ is included in $\|T\|K(\mathcal{X}_1)$ and hence is compact. Thus the compact operators form an ideal.

Any finite-rank operator is compact since the unit ball of a finite dimensional normed space is always compact. Hence $\mathcal{B}_K(\mathcal{X})$ is non-zero and includes the ideal of approximable operators.

If $\mathcal{X}$ is infinite-dimensional, we may choose a strictly increasing sequence of closed subspaces $\mathcal{X}_{(1)} \subseteq \mathcal{X}_{(2)} \subseteq \mathcal{X}_{(3)} \subseteq \cdots \subseteq \mathcal{X}$. Hence by the Riesz lemma, stated and proved below, we may choose a sequence of unit vectors $y_n \in \mathcal{X}_{(n)}$ satisfying $\|y_m - y_n\| \geq 1 - \varepsilon$ for $m \neq n$. This shows that $\mathcal{X}_1$ is not compact. Hence the identity operator I is not compact.   □

The following result was proved by Frigyes Riesz [1918].

Riesz Lemma *If $\mathcal{Y}$ is a proper closed subspace of a normed linear space $\mathcal{X}$, then for every $\varepsilon > 0$ there is a vector $x \in \mathcal{X} \setminus \mathcal{Y}$ satisfying $\|x\| = 1$ and $\|x + \mathcal{Y}\| = \inf\{\|x + y\| : y \in \mathcal{Y}\} \geq 1 - \varepsilon$.*

Proof We start with an arbitrary $z \in \mathcal{X} \setminus \mathcal{Y}$ and note that $\|z + \mathcal{Y}\|$ is positive since $\mathcal{Y}$ is closed. Hence there is a (non-zero) vector $y \in \mathcal{Y}$ satisfying $\|z + y\| \leq \|z + \mathcal{Y}\|(1 - \varepsilon)^{-1}$. An easy calculation shows that the vector $x = \|z + y\|^{-1}(z + y)$ has the desired property. □

The fact that $\mathcal{B}(\mathcal{X})$ has a nontrivial (*i.e.*, a non-zero, proper) closed ideal when $\mathcal{X}$ is infinite-dimensional contrasts with $\mathcal{B}(\mathcal{X}) \simeq M_n$ for a finite-dimensional normed linear space $\mathcal{X}$ of dimension n. In the last section we noted that M_n has no ideals except $\{0\}$ and M_n.

For all classical separable Banach spaces, the ideals of approximable and compact operators are equal. It was a long standing open question as to whether this was true for all separable spaces. Enflo [1973] settled the question by giving a counterexample.

Schauder's theorem (Dunford and Schwartz [1958] VI.5.2) states that an operator is compact if and only if its adjoint is compact. We prove the next proposition both for historical reasons and because we will use it later in this section.

Proposition *For any Banach space $\mathcal{X}$, a compact operator sends weakly convergent sequences into norm convergent sequences.*

Proof Suppose T is a compact operator and $\{x_n\}_{n \in \mathbb{N}} \subseteq \mathcal{X}$ is a sequence converging to x in the weak topology. For any $\omega \in \mathcal{X}^*$, the numbers $\omega T x_n - \omega T x = \omega(T x_n - T x) = T^*(\omega)(x_n - x)$ form a sequence converging to zero. Hence $\{T x_n\}_{n \in \mathbb{N}}$ converges weakly to $T x$. Suppose $\{T x_n\}_{n \in \mathbb{N}}$ does not converge to $T x$ in norm. Then there is a constant $\delta > 0$ and a subsequence which we again denote by $\{x_n\}_{n \in \mathbb{N}}$ such that $\{T x_n\}_{n \in \mathbb{N}}$ remains outside the δ-ball around $T x$. Since $\{x_n\}_{n \in \mathbb{N}}$ converges weakly, the uniform boundedness principle shows that it is a bounded sequence. Since T is compact, this shows that there is a subsequence of $\{T x_n\}_{n \in \mathbb{N}}$ which converges in norm to some element y outside the δ-ball around $T x$. However this would imply that the subsequence converges weakly to y which is impossible since it converges weakly to $T x$. This contradiction proves that $\{T x_n\}_{n \in \mathbb{N}}$ converges to $T x$ in norm. □

Corollary *For any Banach space $\mathcal{X}$, a norm continuous operator is also weakly continuous (i.e. any operator in $\mathcal{B}(\mathcal{X})$ is continuous as a function from $\mathcal{X}$ with its weak topology to $\mathcal{X}$ with its weak topology).*

Proof Use nets instead of sequences in the first step of the last proof. □

If $\mathcal{X}$ is reflexive, the property of compact operators proved in the above proposition characterizes them. This is an immediate consequence of the Eberlein–Šmulian theorem and the results on reflexive spaces given above. David Hilbert [1906] originally defined compact operators just on Hilbert space in terms of this characterization. The definition of compact operators that we have used is due to Frigyes Riesz [1918].

Calkin [1941], in one of the earliest papers dealing with the ideal theory of a Banach algebra, showed that when $\mathcal{X} = \mathcal{H}$ is a separable Hilbert space then every ideal of $\mathcal{B}(\mathcal{H})$ lies between $\mathcal{B}_F(\mathcal{H})$ and $\mathcal{B}_K(\mathcal{H})$. The result is known to hold for some additional Banach spaces (I. A. Feldman, Israel Tzutikovič Gohberg and A. S. Markus [1960]) and to fail for others (Dunford and Schwartz [1958] Chapter VI, and Seymour Goldberg [1966]). Calkin also classified all the ideals of $\mathcal{B}(\mathcal{H})$. We discuss several other ideals related to $\mathcal{B}_K(\mathcal{X})$ next. See Section 2.8 below for the spectral properties of compact operators.

1.7.8 Completely Continuous, Weakly Compact and Strictly Singular Operators An operator in $\mathcal{B}(\mathcal{X})$ is called *completely continuous* if it maps weakly convergent sequences into norm-convergent sequences. It is easy to see that the set $\mathcal{B}_{CC}(\mathcal{X})$ of completely continuous operators is a closed ideal of $\mathcal{B}(\mathcal{X})$. We have shown that this ideal includes $\mathcal{B}_K(\mathcal{X})$ and that it equals $\mathcal{B}_K(\mathcal{X})$ if $\mathcal{X}$ is reflexive. The term "completely continuous" with this meaning was used by Hilbert [1906] for operators on Hilbert space, where it has the same meaning as "compact", before the term "compact", was invented.

An operator T in $\mathcal{B}(\mathcal{X})$ is called *weakly compact* if the weak closure of $T\mathcal{X}_1$ is compact in the weak topology of $\mathcal{X}$. Because bounded linear operators are weakly continuous, it is easy to see that the set $\mathcal{B}_{WK}(\mathcal{X})$ of weakly compact operators is an ideal of $\mathcal{B}(\mathcal{X})$ containing the ideal of compact operators. Since a Banach space $\mathcal{Z}$ is reflexive if and only if its unit ball is weakly compact, it is also easy to see that $\mathcal{Z}$ is reflexive if and only if for all Banach spaces $\langle\, \mathcal{X} \,/\, \mathcal{Y}\, \rangle$, all operators $\langle\, R \in \mathcal{B}(\mathcal{X},\mathcal{Z}) \,/\, Q \in \mathcal{B}(\mathcal{Z},\mathcal{Y})\, \rangle$ are weakly compact. The fact that $\mathcal{B}_{WK}(\mathcal{X})$ is a closed ideal follows from the next theorem formally stated below. Gantmacher's theorem asserts that an operator is weakly compact if and only if its adjoint is. For a proof of this and other assertions given here, see Dunford and Schwartz [1958] VI-4. The following result, proved there, is often useful.

Theorem *An operator $T \in \mathcal{B}(\mathcal{X},\mathcal{Y})$ is weakly compact if and only if it satisfies:*

$$T^{**}(\mathcal{X}^{**}) \subseteq \kappa(\mathcal{Y}).$$

This theorem leads to an interesting exact sequence for any nonreflexive

Banach space $\mathcal{X}$

$$0 \to \mathcal{B}_{WK}(\mathcal{X}) \to \mathcal{B}(\mathcal{X}) \to \mathcal{B}(\mathcal{X}^{**}/\kappa(\mathcal{X}))$$

where any $T \in \mathcal{B}(\mathcal{X})$ is sent to $\tilde{T}$ defined by $\tilde{T}(F + \kappa(\mathcal{X})) = T^{**}(F) + \kappa(\mathcal{X})$ for $F \in \mathcal{X}^{**}$. This may be regarded as a faithful representation (*cf.* Definition 4.1.1 below) of the *weak generalized Calkin algebra* $\mathcal{B}(\mathcal{X})/\mathcal{B}_{WK}(\mathcal{X})$. If $\mathcal{X} = \mathcal{J}$ is the James space, this provides a homomorphism of $\mathcal{B}(\mathcal{J})$ onto $\mathbb{C}$ with kernel $\mathcal{B}_{WK}(\mathcal{J})$.

H. Garth Dales, Richard J. Loy and George A. Willis [1992] construct an interesting Banach space related to the above considerations. For each $p \geq 2$, they first construct an analogue $\mathcal{J}_p$ of the James space based on ℓ^p, so $\mathcal{J}_2$ is the ordinary James space. For $q > p$, they show $\mathcal{B}(\mathcal{J}_q, \mathcal{J}_p) = \mathcal{B}_K(\mathcal{J}_q, \mathcal{J}_p)$, and for $q \leq p$ every operator in $T \in \mathcal{B}(\mathcal{J}_q, \mathcal{J}_p)$ can be written uniquely in the form $\lambda + W$ for $\lambda \in \mathbb{C}$ and $W \in \mathcal{B}_{WK}(\mathcal{J}_q, \mathcal{J}_p)$. Let $\mathcal{W}$ be the ℓ^1 direct sum of the spaces $\mathcal{J}_{p_k}$ where $p_1 = 2$ and $\{p_k\}_{k \in \mathbb{N}}$ is a strictly increasing sequence. The operators on $\mathcal{W}$ have a matrix representation with matrix entries restricted as noted above: compact above the main diagonal and the sum of a scalar and a weakly compact operator on and below that diagonal. This allows the Banach algebra $\mathcal{B}(\mathcal{W})$ to be analyzed in detail. For instance, the operators for which the matrix entries on the main diagonal are weakly compact (*i.e.*, zero scalar part) form a closed ideal $\mathcal{I}$ which satisfies $\mathcal{B}(\mathcal{W}) = \mathcal{I} + \mathcal{D}$ where $\mathcal{D} \simeq \ell^\infty$ is the closed subalgebra of diagonal operators with purely scalar diagonal entries from ℓ^∞. Also $\mathcal{B}(\mathcal{W}) = \mathcal{K} + \mathcal{T}$ where $\mathcal{K}$ is the closed ideal of operators with the scalar part of each matrix entry zero, and $\mathcal{T}$ is the closed subalgebra of operators with purely scalar matrix entries. We refer the reader to the original paper for substantial further detail.

The first part of the next theorem is an important result of William J. Davis, Tadeusz Figiel, William B. Johnson and Aleksander Pełczynski [1974]. The last paragraph of the theorem comes from Kaijser [1976].

Theorem *An operator $P \in \mathcal{B}(\mathcal{X}, \mathcal{Y})$ is weakly compact if and only if there is a reflexive Banach space $\mathcal{Z}$ and linear operators $R \in \mathcal{B}(\mathcal{X}, \mathcal{Z})$ and $Q \in \mathcal{B}(\mathcal{Z}, \mathcal{Y})$ satisfying $P = QR$. Moreover, Q may be chosen injective and contractive, and S may be chosen to have the same norm and kernel as P and dense range in $\mathcal{Z}$.*

Suppose $T: \mathcal{A} \to \mathcal{B}(\mathcal{X})$ and $S: \mathcal{A} \to \mathcal{B}(\mathcal{Y})$ are continuous representations of a normed algebra $\mathcal{A}$ satisfying $PT_a(x) = S_a(Px)$ for all $a \in \mathcal{A}$ and $x \in \mathcal{X}$. If P is weakly compact and $\mathcal{Z}$, Q and R are as described above, then there is a continuous representation $U: \mathcal{A} \to \mathcal{B}(\mathcal{Z})$ satisfying $RT_a(x) = U_a(Rx)$ and $QU_a(z) = S_a(Qz)$ for all $a \in \mathcal{A}$, $x \in \mathcal{X}$ and $z \in \mathcal{Z}$.

Partial Proof For each $n \in \mathbb{N}$ define a new norm on $\mathcal{Y}$ by

$$\|y\|_n = \inf\{\|P\| \, 2^{-n/2} \|x\| + \|y - Px\| : x \in \mathcal{X}\} \qquad \forall \, y \in \mathcal{Y}. \tag{4}$$

Taking $x = 0$ in (4) gives

$$\|y\|_n \leq \|y\| \qquad \forall\, y \in \mathcal{Y};\ n \in \mathbb{N} \tag{5}$$

and $\|y\| \leq \|Px\| + \|y - Px\| \leq \|P\|\,\|x\| + 2^{n/2}\|y - Px\|$ implies

$$\|y\| \leq 2^{n/2}\|y\|_n \qquad \forall\, y \in \mathcal{Y};\ n \in \mathbb{N}. \tag{6}$$

Hence all these norms are equivalent. Note also

$$\|Px\|_n \leq 2^{-n/2}\|P\|\,\|x\| \qquad \forall\, x \in \mathcal{X}. \tag{7}$$

Define $\mathcal{Z}$ by $\mathcal{Z} = \{y \in \mathcal{Y} : \sum_{n=1}^{\infty} \|y\|_n^2 < \infty\}$, and define a norm on $\mathcal{Z}$ by

$$\||z\|| = \left(\sum_{n=1}^{\infty} \|z\|_n^2\right)^{1/2} \qquad \forall\, z \in \mathcal{Z}. \tag{8}$$

From (7) we see

$$\||Px\|| \leq \left(\sum_{n=1}^{\infty}(2^{-n/2}\|P\|\,\|x\|)_n^2\right)^{1/2} = \|P\|\,\|x\| \qquad \forall\, x \in \mathcal{X} \tag{9}$$

so $P\mathcal{X} \subseteq \mathcal{Z}$. Hence we may define $R \in \mathcal{B}(\mathcal{X}, \mathcal{Z})$ and $Q \in \mathcal{B}(\mathcal{Z}, \mathcal{Y})$ by

$$Rx = Px \qquad \forall\, x \in \mathcal{X}; \qquad\qquad Qz = z \quad \forall\, z \in \mathcal{Z}. \tag{10}$$

Inequality (9) shows $\|R\| \leq \|P\|$ and

$$\|Qz\| = \left(\sum_{n=1}^{\infty}(2^{-n/2}\|z\|)^2\right)^{1/2} \leq \left(\sum_{n=1}^{\infty}\|z\|_n^2\right)^{1/2} = \||z\|| \qquad \forall\, z \in \mathcal{Z} \tag{11}$$

shows $\|Q\| \leq 1$. Hence $P = QR$ implies $\|R\| = \|P\|$ and $\|Q\| = 1$. If $R\mathcal{X}$ is not dense in $\mathcal{Z}$, replace $\mathcal{Z}$ by the closure of $R\mathcal{X}$. This proves all the results except that $\mathcal{Z}$ is reflexive and the statements about representations. For the former fact we refer the reader to either Diestel and Uhl [1977], VIII.4.8 or Albrecht Pietsch [1980], 2.4.

We now turn to the proof of the last paragraph. The final equation in the theorem and our definition of Q show that we must define U by $U_a(z) = S_a(z)$ for all $a \in \mathcal{A}$ and $z \in \mathcal{Z}$. Hence we need to check that $S_a(z)$ is in $\mathcal{Z}$ for all $a \in \mathcal{A}$ and $z \in \mathcal{Z}$. Note

$$
\begin{aligned}
\|S_a y\|_n &\leq \inf\{\|P\|\,2^{-n/2}\,\|T_a x\| + \|S_a y - PT_a x\| : x \in \mathcal{X}\} \\
&\leq \inf\{\|T_a\|\,\|P\|\,2^{-n/2}\,\|T_a x\| + \|S_a\|\,\|y - Px\| : x \in \mathcal{X}\} \\
&\leq \max\{\|T\|,\ \|S\|\}\,\|a\|\,\|y\|_n \qquad \forall\, a \in \mathcal{A};\ y \in \mathcal{Y}
\end{aligned}
$$

where we used the intertwining equation $PT_a = S_a P$. This shows that S_a sends elements satisfying condition (7) into other elements satisfying that equation. If it is necessary to replace this original choice of Z by the closure of $P\mathcal{X}$, the intertwining equation shows again that $S_a z$ will still belong to Z if z does. Hence we are free to define $U: \mathcal{A} \to \mathcal{B}(Z)$ by $U_a(z) = S_a(z)$ for all $a \in \mathcal{A}$ and $z \in Z$. We also conclude that the norm of the representation U is at most $\max\{\|T\|, \|S\|\}$. □

A Banach space $\mathcal{X}$ is said to have the *Dunford–Pettis property* if every weakly compact operator from $\mathcal{X}$ into any Banach space $\mathcal{Y}$ is completely continuous. Obviously no infinite-dimensional reflexive space has this property, but most classical nonreflexive spaces do (Dunford and Schwartz [1958] p. 511). In fact, if $\mathcal{X}^*$ has the Dunford–Pettis property so does $\mathcal{X}$, but the converse fails.

The composition of a weakly compact operator followed by a completely continuous operator is compact by the Eberlein–Šmulian theorem, since the image of a bounded sequence will have a weakly convergent subsequence. Hence for any space with the Dunford–Pettis property, the square of the ideal $\mathcal{B}_{WK}(\mathcal{X})$ of weakly compact operators is the ideal of compact operators. (This is the first example we have given of a natural ideal for which the square is different and also a natural ideal. The example is rather unusual in that the square is closed in the norm in which the original ideal is complete.) If Ω is a compact Hausdorff topological space and $\mathcal{X} = C(\Omega)$ is the Banach space of continuous functions on Ω under the supremum norm, then the ideals of weakly compact and of completely continuous operators actually coincide and they are proper and properly include the ideal of compact operators unless $\mathcal{X}$ is finite-dimensional (or equivalently, Ω is finite). Weakly compact operators were introduced by Kōsaku Yosida [1938] and Shizuo Kakutani [1938].

An operator in $\mathcal{B}(\mathcal{X})$ is called *strictly singular* if it does not have a bounded inverse on any infinite-dimensional subspace. The set $\mathcal{B}_{SS}(\mathcal{X})$ of strictly singular operators in $\mathcal{B}(\mathcal{X})$ is a closed ideal which includes $\mathcal{B}_{CC}(\mathcal{X}) \cap \mathcal{B}_{WK}(\mathcal{X})$. This class of operators was introduced by Tosio Kato [1958].

We conclude this discussion by noting the inclusions which hold between these classes of operators from one Banach space to another. If $\mathcal{X}$ and $\mathcal{Y}$ are arbitrary, they satisfy

$$\mathcal{B}_K(\mathcal{X}, \mathcal{Y}) \subseteq \mathcal{B}_{CC}(\mathcal{X}, \mathcal{Y}) \cap \mathcal{B}_{WK}(\mathcal{X}, \mathcal{Y}) \subseteq \mathcal{B}_{SS}(\mathcal{X}, \mathcal{Y}).$$

When $\mathcal{Y}$ is reflexive, these inclusions can be improved:

$$\mathcal{B}_K(\mathcal{X}, \mathcal{Y}) \subseteq \mathcal{B}_{CC}(\mathcal{X}, \mathcal{Y}) \subseteq \mathcal{B}_{SS}(\mathcal{X}, \mathcal{Y}) \subseteq \mathcal{B}_{WK}(\mathcal{X}, \mathcal{Y}) = \mathcal{B}(\mathcal{X}, \mathcal{Y}).$$

Finally, if $\mathcal{X}$ is reflexive instead, we get even more:

$$\mathcal{B}_K(\mathcal{X}, \mathcal{Y}) = \mathcal{B}_{CC}(\mathcal{X}, \mathcal{Y}) \subseteq \mathcal{B}_{SS}(\mathcal{X}, \mathcal{Y}) \subseteq \mathcal{B}_{WK}(\mathcal{X}, \mathcal{Y}) = \mathcal{B}(\mathcal{X}, \mathcal{Y}).$$

For further results on these classes of ideals, see Goldberg [1966] and Gustafson [1968].

1.7.9 Bases It turns out that there is a fairly natural idea of a basis in a Banach space, but not all separable spaces have bases. A space with a basis is easily seen to be separable.

Definition A sequence $\{x_n\}_{n \in \mathbb{N}}$ in a Banach space $\mathcal{X}$ is called a *basis* or *Schauder basis* for $\mathcal{X}$ if for every x in $\mathcal{X}$ there is a unique sequence $\{\lambda_n\}_{n \in \mathbb{N}}$ of scalars so that $\sum_{n=1}^{\infty} \lambda_n x_n$ converges to x in the norm. For a basis $\{x_k\}_{k \in \mathbb{N}}$ and a positive integer n, define the nth *basis projection* $P_n \colon \mathcal{X} \to \mathrm{sp}\{x_1, x_2, \ldots, x_n\}$ by $P_n(\sum_{k=1}^{\infty} \lambda_k x_k) = \sum_{k=1}^{n} \lambda_k x_k$.

For a basis $\{x_n\}_{n \in \mathbb{N}}$, the sequence of linear functionals $\{x_n^*\}_{n \in \mathbb{N}}$ defined by

$$x_n^*(\sum_{j=1}^{\infty} \lambda_j x_j) = \lambda_n \quad \text{or} \quad x_n^*(x_m) = \delta_{mn}$$

is called the *biorthogonal system* defined by the basis.

A basis $\{x_n\}_{n \in \mathbb{N}}$ is said to be *shrinking* if for each $\omega \in \mathcal{X}^*$ the sequence $\{(I - P_n)^*(\omega)\}_{n \in \mathbb{N}}$ converges to zero in norm, and *boundedly complete* if for every sequence of scalars $\{\lambda_n\}_{n \in \mathbb{N}}$ for which the supremum of the set $\{\|\sum_{j=1}^{n} \lambda_j x_j\| : n \in \mathbb{N}\}$, is finite, the series $\sum_{n=1}^{\infty} \lambda_n x_n$ converges.

For very simple examples, consider the obvious bases for c_0 and ℓ^1 (defined by the Kronecker delta $\{\delta_n\}_{n \in \mathbb{N}}$) which are, respectively, shrinking and boundedly complete. Since all classical separable Banach spaces have bases, Banach asked whether this was true of all separable Banach spaces. The question was not settled until Enflo [1973] constructed a counterexample. Banach also noted that the uniform boundedness principle shows that the sequence of basis projections is norm bounded. Note that the basis projections have the expansion $P_n = \sum_{j=1}^{n} x_j \otimes x_j^*$ in terms of the biorthogonal system.

The next theorem gives some interesting properties of the concepts just introduced. Proofs can be found in Lindenstrauss and Tzafriri [1977], 1.b.

Theorem *If a Banach space $\mathcal{X}$ has a shrinking basis, then its biorthogonal system is a boundedly complete basis for $\mathcal{X}^*$.*

A space with a boundedly complete basis is isomorphic to a dual space. If a dual space $\mathcal{X}^*$ has a basis, then $\mathcal{X}$ has a shrinking basis so $\mathcal{X}^*$ has a boundedly complete basis.

A Banach space $\mathcal{X}$ with a basis is reflexive if and only if it satisfies one and hence all of the following equivalent conditions.

(a) *The given basis is both shrinking and boundedly complete.*
(b) *Every basis for $\mathcal{X}$ is shrinking.*
(c) *Every basis for $\mathcal{X}$ is boundedly complete.*

1.7.10 The Approximation Property In a space with a basis, the sequence $\{P_n\}_{n\in\mathbb{N}}$ of finite-rank projections converges to the identity operator in some sense, which we will now make explicit. Alexander Grothendieck [1955], [1956] discovered the right way to present the property involved.

Definition A Banach space is said to have the *approximation property* if its identity operator I can be uniformly approximated on every compact subset $\mathcal{K}$ by finite-rank operators (*i.e.*, for every $\varepsilon > 0$ there is a finite-rank operator T (depending on $\mathcal{K}$ and ε) satisfying $\|Tx - x\| \leq \varepsilon$ for all $x \in \mathcal{K}$.) It is said to have the *bounded approximation property* if there is some finite constant B so that the finite-rank operator T can be chosen to satisfy $\|T\| \leq B$ independent of $\mathcal{K}$ and ε. If $B = 1$, then the Banach space is said to have the *metric approximation property*.

A Banach space is said to have the *compact approximation property* if its identity operator can be uniformly approximated on every compact subset by compact operators. If the approximating compact operators can be chosen from a bounded set, the space is said to have the *bounded compact approximation property*.

As we just note, the bounded sequence of basis projections $\{P_n\}_{n\in\mathbb{N}}$ approximates the identity operator in this fashion for a space with a basis. Thus all Banach spaces which have a basis also enjoy the bounded approximation property. Enflo's counterexamples [1973], which we have already mentioned twice, were actually spaces which failed to have the approximation property. His examples were simplified and supplemented by Alexander M. Davie [1973b] and Andrzej Szankowski [1978] so that now subspaces of c_0 and ℓ^p for $1 \leq p < 2$ and $2 < p < \infty$ are known which lack the approximation property. In fact they do not even have the compact approximation property. Thus reflexive spaces may fail to have the compact approximation property. Spaces are known (even with separable duals) which have the approximation property but not the bounded approximation property and with the compact approximation property but not the bounded compact approximation property. Willis [1992d] has recently given an example of a separable reflexive Banach space with the bounded compact approximation property which does not have the approximation property. It is not known whether all separable spaces with the bounded approximation property have a basis (*cf.* Lindenstrauss and Tzafriri [1977] and [1979]). The only naturally occurring Banach space known to lack the approximation property is $\mathcal{B}(\mathcal{H})$ for a Hilbert space $\mathcal{H}$, Szankowski [1981]. In §5.1.12 we show the connection between these properties of $\mathcal{X}$ and the existence of approximate identities in various ideals of operators in $\mathcal{B}(\mathcal{X})$.

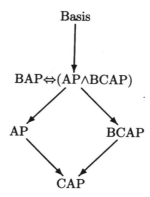

The diagram above summarizes the easy implications between these concepts (omitting the metric versions). We are about to note that the approximation property implies that the finite-rank operators are norm-dense in the compact operators. This establishes the equivalence BAP⇔(AP∧BCAP). All the other implications are obvious. The preceding remarks show that no implication arrow can be added to this diagram unless it turns out that BAP implies the existence of a basis (for separable spaces).

Here is an elementary result which applies only to the bounded case.

Theorem   Let $\mathcal{X}$ be a Banach space and let B be a finite constant. Then $\mathcal{X}$ has the ⟨ bounded / bounded compact ⟩ approximation property with bound B if and only if for every $\varepsilon > 0$ and every finite subset $x_1, x_2, \ldots, x_n$ of $\mathcal{X}$ there is an operator $T \in \langle\, \mathcal{B}_F(\mathcal{X})\ /\ \mathcal{B}_K(\mathcal{X})\,\rangle$ satisfying $\|T\| \le B$ and $\|Tx_j - x_j\| < \varepsilon$ for $j = 1, 2, \ldots, n$.

Proof   Let $\mathcal{K}$ be an arbitrary compact subset of $\mathcal{X}$. Cover $\mathcal{K}$ with open $\varepsilon/(B+1)$ balls centered at $x_1, x_2, \ldots, x_n$. If T is the operator guaranteed in the condition and $x \in \mathcal{K}$, then they satisfy $\|Tx - x\| \le \|T(x - x_j)\| + \|Tx_j - x_j\| + \|x_j - x\| \le 2\varepsilon$ for some j. The converse is immediate.   □

Before stating the next theorem, which is due to Grothendieck [1955], we mention two more ideas important in the proof. Compact subsets of a Banach space can be characterized as those included in the convex hull of some sequence converging to zero. Our definition of the approximation property mentions uniform approximation on compact sets. This corresponds to a locally convex topology on $\mathcal{B}(\mathcal{X}, \mathcal{Y})$, the *topology of uniform convergence on compact sets*, which is given by the collection of semi-norms:

$$\{\sigma_{\mathcal{K}} : \mathcal{K} \text{ is a compact subset of } \mathcal{X}\}, \text{ where } \sigma_{\mathcal{K}}(T) = \sup\{\|Tx\| : x \in \mathcal{K}\}.$$

From now on, we will call this the *Grothendieck topology* on $\mathcal{B}(\mathcal{X}, \mathcal{Y})$. The proof of the next theorem depends on showing that the linear functionals on $\mathcal{B}(\mathcal{X}, \mathcal{Y})$ which are continuous in the Grothendieck topology are exactly those given by:

$$T \mapsto \sum_{n=1}^{\infty} \omega_n(Tx_n) \text{ for } \{x_n\}_{n\in\mathbb{N}} \subseteq \mathcal{X}; \ \{\omega_n\}_{n\in\mathbb{N}} \subseteq \mathcal{Y}^*; \ \sum_{n=1}^{\infty} \|\omega_n\|\|x_n\| < \infty.$$

Theorem   *The following assertions are equivalent for any Banach space $\mathcal{X}$:*

(a) *$\mathcal{X}$ has the approximation property.*

(b) *For every Banach space $\mathcal{Y}$, $\mathcal{B}_F(\mathcal{X}, \mathcal{Y})$ is dense in $\mathcal{B}(\mathcal{X}, \mathcal{Y})$ in the Grothendieck topology.*

(c) *For every Banach space $\mathcal{Y}$, $\mathcal{B}_F(\mathcal{Y}, \mathcal{X})$ is dense in $\mathcal{B}(\mathcal{Y}, \mathcal{X})$ in the Grothendieck topology.*

(d) *Every choice of $\{x_n\}_{n\in\mathbb{N}} \subseteq \mathcal{X}$ and $\{\omega_n\}_{n\in\mathbb{N}} \subseteq \mathcal{X}^*$ such that*

$$\sum_{n=1}^{\infty} \|\omega_n\|\|x_n\| < \infty \text{ and } \sum_{n=1}^{\infty} \omega_n(x)x_n = 0 \text{ for all } x \in \mathcal{X}$$

satisfies $\sum_{n=1}^{\infty} \omega_n(x_n) = 0$.

(e) *For every Banach space $\mathcal{Y}$, $\mathcal{B}_F(\mathcal{Y}, \mathcal{X})$ is dense in $\mathcal{B}_K(\mathcal{Y}, \mathcal{X})$ in the norm topology.*

Moreover, $\mathcal{X}$ satisfies the $\langle$ metric / bounded $\rangle$ approximation property if and only if there is some finite constant B, so that in (b) or (c) finite-rank operators in the $\langle$ unit ball / B-ball $\rangle$ approximate operators in the unit ball and in (d) the last condition is replaced by $|\sum_{n=1}^{\infty} \omega_n(Tx_n)| \leq 1$ for all $T \in \mathcal{B}_F(\mathcal{X})_1$ and the conclusion is replaced by $\langle$ $|\sum_{n=1}^{\infty} \omega_n(x_n)| \leq 1$ / $|\sum_{n=1}^{\infty} \omega_n(x_n)| \leq B$ $\rangle$.

If $\mathcal{B}_K(\mathcal{Y}, \mathcal{X})$ and $\mathcal{B}_F(\mathcal{Y}, \mathcal{X})$ in condition (e) is replaced by $\mathcal{B}_K(\mathcal{X}, \mathcal{Y})$ and $\mathcal{B}_F(\mathcal{X}, \mathcal{Y})$, we get a characterization of when $\mathcal{X}^*$ has the approximation property. Thus we see that the ideal of compact operators equals the ideal of approximable operators in $\mathcal{B}(\mathcal{X})$ if either $\mathcal{X}$ or $\mathcal{X}^*$ has the approximation property. Attractive proofs of all these results can be found in Lindenstrauss and Tzafriri [1977], 1.e. With the ideas given in the last paragraph, they are all easy. In particular, note that (d) is just the statement that any linear functional, continuous in the Grothendieck topology, which vanishes on all rank one operators must vanish on the identity. Hence it is obviously equivalent to (a). Despite the next theorem, there is a space with a basis (and hence the approximation property) and a separable dual for which the dual space fails to have the approximation property.

Theorem   *A Banach space has the approximation property if its dual space does. Hence a reflexive space has the approximation property if and only if its dual space does. A reflexive space or a separable dual space with the approximation property has the metric approximation property.*

1.7.11 Nuclear Operators The closed ideals of operators considered above are all (possibly) larger than the ideal of compact operators. The ideal considered here lies between the compact operators and the finite-rank operators. Both the ideal of nuclear operators and the ideal of integral operators (considered next) are complete in their own norms which dominate the operator norm. We shall continue to make definitions for operators in $\mathcal{B}(\mathcal{X})$ (thus getting ordinary ideals, which is our main interest) but to use the fact that these are ideals in the extended sense when considered for operators from one space to another. The book by Pietsch [1980] provides an interesting view of these and other operator ideals.

Definition An operator $T \in \mathcal{B}(\mathcal{X})$ is said to be *nuclear* if there exist sequences $\{x_n\}_{n\in\mathbb{N}} \subseteq \mathcal{X}$ and $\{\omega_n\}_{n\in\mathbb{N}} \subseteq \mathcal{X}^*$ satisfying

$$\sum_{n=1}^{\infty} \|\omega_n\|\|x_n\| < \infty \text{ and } T(x) = \sum_{n=1}^{\infty} \omega_n(x)x_n \qquad \forall\, x \in \mathcal{X}. \qquad (12)$$

The set of all nuclear operators is denoted by $\mathcal{B}_N(\mathcal{X})$ and is provided with the *nuclear norm*

$$\|T\|_N = \inf\{\sum_{n=1}^{\infty} \|\omega_n\|\|x_n\| : T \text{ is represented by } \sum_{n=1}^{\infty} x_n \otimes \omega_n\}.$$

Of course, the representation mentioned in the definition of the norm is that described in equation (12). Obviously the adjoint of a nuclear operator is nuclear, but the converse fails (Pietsch [1980] 8.2.6) unless the dual space has the approximation property. The norm inequality in the next theorem is characteristic of ideals defined by norms.

Theorem *The collection $\mathcal{B}_N(\mathcal{X})$ of nuclear operators is an ideal in $\mathcal{B}(\mathcal{X})$. It is a Banach algebra in its own norm which is greater than or equal to the operator norm and satisfies the inequality:* $\|RTS\|_N \leq \|R\|\|T\|_N\|S\|$ *for $R, S \in \mathcal{B}(\mathcal{X})$ and $T \in \mathcal{B}_N(\mathcal{X})$. Moreover,* $\|x \otimes \omega\|_N = \|x\|\|\omega\|$ *for all $x \in \mathcal{X}$ and $\omega \in \mathcal{X}^*$ and $\|\cdot\|$ is the largest norm on $\mathcal{B}_F(\mathcal{X})$ satisfying this condition.*

All the above results are easy to prove. Nevertheless, the nuclear operators exhibit some mysterious behavior in perfectly arbitrary Banach spaces which was first identified by Grothendieck [1955], [1956]. This arises from considering their relationship to tensor products. Since we will put off a general discussion of tensor products until Section 1.10, we could avoid these complications here. However, we wish to introduce the duality theory involved in the two classes of ideals introduced in this and the following section, and that deeply involves their relationship with tensor products.

We begin by considering the ideal $\mathcal{B}_F(\mathcal{X})$ of finite-rank operators as the algebraic tensor product $\mathcal{X} \otimes \mathcal{X}^*$. We have already considered the operator norm on $\mathcal{B}_F(\mathcal{X})$ and its completion in this norm $\mathcal{B}_A(\mathcal{X})$. In the language of tensor products, this norm is called the *injective tensor norm* and the completion $\mathcal{B}_A(\mathcal{X})$ corresponds to the *injective tensor product*, denoted by $\mathcal{X} \check{\otimes} \mathcal{X}^*$, which carries this norm.

There is another natural norm for $T \in \mathcal{B}_F(\mathcal{X})$, the *projective tensor norm*:

$$\|T\|_p = \inf\{\sum_{k=1}^{n} \|x_k\|\|\omega_k\| : T = \sum_{k=1}^{n} x_k \otimes \omega_k\} \qquad (13)$$

where the infimum extends over all $n \in \mathbb{N}$; $x_k \in \mathcal{X}$; $\omega_k \in \mathcal{X}^*$. If we complete $\mathcal{X} \otimes \mathcal{X}^* \simeq \mathcal{B}_F(\mathcal{X})$ in this norm instead of the operator norm, we get the *projective tensor product* denoted by $\mathcal{X} \hat{\otimes} \mathcal{X}^*$. It is not hard to see that this norm is greater than or equal to the operator norm so that there is a natural contractive linear map $\eta : \mathcal{X} \hat{\otimes} \mathcal{X}^* \to \mathcal{B}_A(\mathcal{X})$. Also, every element of the projective tensor product can be represented in the form $\sum_{n=1}^{\infty} x_n \otimes \omega_n$ where $\{x_n\}_{n \in \mathbb{N}} \subseteq \mathcal{X}$ and $\{\omega_n\}_{n \in \mathbb{N}} \subseteq \mathcal{X}^*$ satisfy $\sum_{n=1}^{\infty} \|\omega_n\|\|x_n\| < \infty$. The infimum of the last sum over all such representations is the projective tensor norm. Thus it is obvious that the norm decreasing map η sends $\mathcal{X} \hat{\otimes} \mathcal{X}^*$ onto $\mathcal{B}_N(\mathcal{X})$. One would certainly hope that this map was an isometry, but in fact it may have a kernel. The remarks we have already made show that $\mathcal{B}_N(\mathcal{X})$ is isometrically isomorphic to $\mathcal{X} \hat{\otimes} \mathcal{X}^* / \ker(\eta)$, but for a finite-rank operator the two inequalities

$$\|\sum_{k=1}^{n} x_k \otimes \omega_k\| \leq \|\sum_{k=1}^{n} x_k \otimes \omega_k\|_N \leq \|\sum_{k=1}^{n} x_k \otimes \omega_k\|_p$$

are both strict in general. (The first norm is the operator norm which equals the injective tensor norm, the second is the nuclear norm which is calculated as the infimum over countable representations of the operator and the third is the projective norm which involves only finite representations of the finite-rank operator.) The approximation property for $\mathcal{X}$ is precisely the condition needed to ensure that η is injective and hence that the last two norms agree for every (nuclear) operator. This is the content of condition (d) of the main theorem on the approximation property which shows that a countable sum which vanishes as an operator is already zero in the projective tensor product.

In fact condition (d) has another interpretation. It asserts that the map

$$\sum_{n=1}^{\infty} x_n \otimes \omega_n \mapsto \sum_{n=1}^{\infty} \omega_n(x_n),$$

which is certainly well defined and therefore contractive in the projective tensor product, is also well defined (and therefore contractive) on the ideal of nuclear operators. Since this map is defined by the classical formula for a trace, we will call it the *trace*. Hence, when $\mathcal{X}$ has the approximation property and $T \in \mathcal{B}_N(\mathcal{X})$, we define

$$\mathrm{Tr}(T) = \sum_{n=1}^{\infty} \omega_n(x_n) \quad \text{where} \quad T = \sum_{n=1}^{\infty} x_n \otimes \omega_n,$$

$\{x_n\}_{n \in \mathbb{N}} \subseteq \mathcal{X}$, $\{\omega_n\}_{n \in \mathbb{N}} \subseteq \mathcal{X}^*$ and $\sum_{n=1}^{\infty} \|x_n\| \|\omega_n\| < \infty$. The definition makes the following trace identity and inequality clear for any $T \in \mathcal{B}_N(\mathcal{X})$ and $S \in \mathcal{B}(\mathcal{X})$,

$$\mathrm{Tr}(TS) = \mathrm{Tr}(ST)$$

$$|\mathrm{Tr}(TS)| \leq \|T\|_N \|S\|.$$

Hence nuclear operators induce continuous linear functionals on $\mathcal{B}(\mathcal{X})$. More is true as we will show in the next theorem.

When we examine the projective tensor product more closely, we will see that, for any Banach space, $\mathcal{B}(\mathcal{X}^*)$ is naturally isometrically isomorphic to the dual of $\mathcal{X} \hat{\otimes} \mathcal{X}^*$ under the map which sends an operator T onto the linear functional $\hat{T}$ defined by

$$\hat{T}(\tau) = \sum_{n=1}^{\infty} T(\omega_n)(x_n) \quad \text{where} \quad \tau = \sum_{n=1}^{\infty} x_n \otimes \omega_n.$$

When $\mathcal{X}$ has the approximation property and we apply this construction to the ideal of nuclear operators instead of the projective tensor product, we get the second statement of the following result.

Theorem *If $\mathcal{X}$ has the approximation property, then $\mathcal{B}_A(\mathcal{X})$ equals $\mathcal{B}_K(\mathcal{X})$. Also, the map $T \mapsto \hat{T}$ establishes an isometric linear isomorphism between $\mathcal{B}(\mathcal{X}^*)$ and the Banach space dual of $\mathcal{B}_N(\mathcal{X})$.*

Proof The first statement follows from (e) of the main theorem on the approximation property, and we have already outlined the proof of the second. □

If the Banach space $\mathcal{X}$ is a Hilbert space $\mathcal{H}$, then it has the approximation property since every separable subspace has an orthonormal basis which is a Schauder basis. Furthermore, the nuclear operators can be identified with the trace class operators, and $\mathcal{H}^*$ can be identified with $\mathcal{H}$ (by a conjugate linear isometry). Hence the second result states that the dual of the ideal of trace class operators is the algebra of all bounded linear operators. This special case is better known and has proved more useful (so far) than the more general result. It will be discussed in Volume II.

1.7.12 Integral Operators Next we identify the dual space of $\mathcal{B}_A(\mathcal{X})$. This will turn out to give a new ideal of operators on $\mathcal{X}$ in a natural way.

Any operator T in $\mathcal{B}(\mathcal{X})$ defines a linear functional on $\mathcal{B}_F(\mathcal{X}) = \mathcal{X} \otimes \mathcal{X}^*$ by the action $S \mapsto \mathrm{Tr}(ST)$ for any $S \in \mathcal{B}_F(\mathcal{X})$. There is no reason to suppose that this linear functional is continuous with respect to the operator norm on $\mathcal{B}_F(\mathcal{X})$, and for many bounded operators T it is not. When the linear functional is continuous in the operator norm on $\mathcal{B}_F(\mathcal{X})$, we say that T is an *integral operator* and write $\check{T}$ for the continuous extension to an element of $\mathcal{B}_A(\mathcal{X})^*$. We denote the set of integral operators on $\mathcal{X}$ by $\mathcal{B}_I(\mathcal{X})$ and give this space the norm $\|T\|_I = \|\check{T}\|$, the norm of the linear functional. This definition is due to Grothendieck. Unfortunately the term "integral operator" is also used with other meanings. (Diestel and Uhl [1977], VIII.2.10 shows that Pietsch integral operators coincide with integral operators in Grothendieck's terminology when $\mathcal{X}^{**}$ is the ℓ^1-direct sum of $\kappa(\mathcal{X})$ and a complementary closed subspace. Hence, this is true when $\mathcal{X}$ is a dual space.)

It is obvious that each nuclear operator is integral and that $\|T\|_I \leq \|T\|_N$ for these operators. (The embedding of ℓ^1 into ℓ^∞ is an example of an integral operator which is not nuclear because it is not even compact.) Note that if T is an integral operator which is not nuclear, the formula $\check{T}(S) = \mathrm{Tr}(ST)$ would not make sense for all operators $S \in \mathcal{B}_A(\mathcal{X})$. The linear functionals on $\mathcal{B}_A(\mathcal{X})$ which arise from operators in $\mathcal{B}_I(\mathcal{X})$ are all continuous in a stronger topology than the operator norm topology, but if we consider operators in $\mathcal{B}_I(\mathcal{X}^*)$ we get the following theorem.

Theorem   *The set $\mathcal{B}_I(\mathcal{X})$ is an ideal in $\mathcal{B}(\mathcal{X})$. It is a Banach algebra in its own norm which is greater than or equal to the operator norm and satisfies $\|STR\|_I \leq \|S\|\|T\|_I\|R\|$ for all $S, R \in \mathcal{B}(\mathcal{X})$ and $T \in \mathcal{B}_I(\mathcal{X})$. Moreover $\|x \otimes \omega\|_I = \|x\|\|\omega\|$ for all $x \in \mathcal{X}$ and $\omega \in \mathcal{X}^*$. Finally, the map $T \mapsto \check{T} \circ \delta$ (where δ is the map which takes an operator to its adjoint) establishes an isometric linear isomorphism between $\mathcal{B}_I(\mathcal{X}^*)$ and $\mathcal{B}_A(\mathcal{X})^*$.*

Proof All these statements are obvious except the claim that every continuous linear functional on $\mathcal{B}_A(\mathcal{X})$ comes from an integral operator in $\mathcal{B}_I(\mathcal{X}^*)$. If τ belongs to $\mathcal{B}_A(\mathcal{X})^*$, we can use it to define an operator T in $\mathcal{B}(\mathcal{X}^*)$ as follows. It is easy to see that for $\omega \in X^*$ the map $x \mapsto \tau(x \otimes \omega)$ is an element of $\mathcal{X}^*$ which we will denote by $T(\omega)$. Then the map T which takes ω into $T(\omega)$ is a well defined element of $\mathcal{B}(\mathcal{X}^*)$. It is easy to see that τ is just $\check{T} \circ \delta$. Diestel and Uhl [1977] VIII.2.12 show that T is an integral operator and this completes the proof.   □

We know that the finite-rank operators are included in any ideal (such as the ideal of integral operators), and by construction they are dense in the ideal of nuclear operators, but they are not always dense in $\mathcal{B}_I(\mathcal{X})$. The relationship between the nuclear and integral operators is obviously close, but in general the two ideals are distinct and the two norms differ on nuclear operators. Theorem VIII.3.8 of the reference cited in the last proof gives the easy result that $\mathcal{X}$ has the metric approximation property if and only if the nuclear and integral norms are equal for all operators in $\mathcal{B}_N(\mathcal{X})$. We combine these results to get the representation we need of the dual and bidual of the compact operators.

Theorem *If $\mathcal{X}^*$ has the metric approximation property and finite-rank operators are dense in the integral operators on $\mathcal{X}^*$, then $\mathcal{B}_N(\mathcal{X}^*)$ is isometrically linearly isomorphic to the Banach space dual of $\mathcal{B}_K(\mathcal{X})$ under the action*

$$\check{T}(S) = \mathrm{Tr}(TS^*) \qquad \forall\ T \in \mathcal{B}_N(\mathcal{X}^*);\ S \in \mathcal{B}_K(\mathcal{X}).$$

Hence $\mathcal{B}_K(\mathcal{X})^{**}$ is isometrically linearly isomorphic to $\mathcal{B}(\mathcal{X}^{**})$ under a map Θ which satisfies $\Theta(\kappa(K)) = K^{**}$ for all $K \in \mathcal{B}_K(\mathcal{X})$.

Corollary *If $\mathcal{X}$ is reflexive and has the approximation property, then $\mathcal{B}_K(\mathcal{X})^*$ is isometrically linearly isomorphic to $\mathcal{B}_N(\mathcal{X}^*)$ as in the theorem and $\mathcal{B}_K(\mathcal{X})^{**}$ is isometrically linearly isomorphic to $\mathcal{B}(\mathcal{X})$ under a map Θ which satisfies $\Theta(\kappa(K)) = K^{**}$ for all $K \in \mathcal{B}_K(\mathcal{X})$.*

We close by quoting two related results of Grothendieck [1955] on integral operators from Diestel and Uhl [1977]. Theorem VIII.2.11 shows the nontrivial fact that an operator is integral if and only if its adjoint is. Theorem VIII.4.6 shows that if $\mathcal{X}^*$ has the approximation property, then it has the Radon–Nikodym property if and only if $\mathcal{B}_I(\mathcal{X}, \mathcal{Y})$ and $\mathcal{B}_N(\mathcal{X}, \mathcal{Y})$ and their norms are equal for all Banach spaces $\mathcal{Y}$.

1.7.13 The Arens Products on $\mathcal{B}_A(\mathcal{X})^{**}$ We now wish to calculate the first Arens representation (*cf.* Section 1.4) for the double dual of the algebra $\mathcal{B}_K(\mathcal{X})$ of compact operators. This was first done in the case of Hilbert space. We will do it here for a wide class of Banach spaces, but not for arbitrary ones, since the exact nature of the Arens products on the double dual of $\mathcal{B}_K(\mathcal{X})$ or $\mathcal{B}_A(\mathcal{X})$ is not known for perfectly general Banach spaces. The results are due to Palmer [1985].

Let us assume that the Banach space $\mathcal{X}^*$ has the metric approximation property and that the finite-rank operators are norm dense in the integral operators on it, so that $\mathcal{X}$ satisfies the last theorem. We will calculate the first Arens representation for $\mathcal{A} = \mathcal{B}_K(\mathcal{X})$ and the representation T which is just the given action of the operators in $\mathcal{B}_K(\mathcal{X})$ on $\mathcal{X}$. This Arens representation is an algebra homomorphism of $\mathcal{B}_K(\mathcal{X})^{**}$ with its first Arens product into $\mathcal{B}(\mathcal{X}^{**})$. It is satisfying that this homomorphism is the same map as the isometric linear bijection Θ just developed from purely Banach space considerations. Before stating a comprehensive theorem, we give the meaning of the various steps in the construction of the first Arens representation and Arens product using the notation introduced above.

An arbitrary linear functional on $\langle\, \mathcal{A} \,/\, \mathcal{A}^* \,\rangle$ has the form $\langle\, \check{S} \,/\, \hat{T} \,\rangle$ for $\langle\, S \in \mathcal{B}_N(\mathcal{X}^*) \,/\, T \in \mathcal{B}(\mathcal{X}^{**}) \,\rangle$. Using the above notation, for $\tau \in \mathcal{X}^*$, $K \in \mathcal{B}_K(\mathcal{X})$ and $F \in \mathcal{X}^{**}$, Definition 1.4.14 gives

$$\tau_K = K^*(\tau) \qquad {}_F\tau = (\tau \otimes F)^\vee \qquad T^1_{\hat{T}}(F) = T(F).$$

Hence the first Arens representation $T^1_{\hat{T}}$ of $\hat{T} \in \mathcal{A}^{**}$ is just the natural action of T on $\mathcal{X}^{**}$. Since this is essentially an independent calculation, it allows us to give a converse to the last theorem of the previous section.

We conclude by stating the main result from the cited paper, which shows that if $\mathcal{B}_A(\mathcal{X})^{**}$ is even a little like $\mathcal{B}(\mathcal{X}^{**})$, it is isometrically algebraically isomorphic to it when the former carries its first Arens product. See the original paper for additional details, including a formula for the second Arens product.

Theorem *The following are equivalent for a Banach space $\mathcal{X}$:*

(a) *The dual $\mathcal{X}^*$ of $\mathcal{X}$ has the metric approximation property and $\mathcal{B}_F(\mathcal{X}^*)$ is dense in $\mathcal{B}_I(\mathcal{X}^*)$.*

(b) *There is a left identity element of norm one for $\mathcal{B}_A(\mathcal{X})^{**}$ with its first Arens product.*

(c) *The first Arens representation of $\mathcal{B}_A(\mathcal{X})^{**}$ is an isometry onto $\mathcal{B}(\mathcal{X}^{**})$.*

(d) *All the following are natural isometric linear isomorphisms which are also algebra isomorphisms when both spaces have natural algebra structure:*

$$\mathcal{B}_A(\mathcal{X}) = \mathcal{B}_K(\mathcal{X}) \simeq \mathcal{X} \check{\otimes} \mathcal{X}^*$$

$$\mathcal{B}_A(\mathcal{X})^* = \mathcal{B}_K(\mathcal{X})^* \simeq \mathcal{B}_I(\mathcal{X}^*) = \mathcal{B}_N(\mathcal{X}^*) \simeq \mathcal{X}^* \hat{\otimes} \mathcal{X}^{**}$$

$$\mathcal{B}_A(\mathcal{X})^{**} \simeq \mathcal{B}(\mathcal{X}^{**}).$$

Many common Banach spaces satisfy these conditions. See Lindenstrauss and Tzafriri [1977] and Diestel and Uhl [1977] for examples. In contrast, $\mathcal{B}_A(\mathcal{X})$ is Arens regular only for reflexive Banach spaces. For reflexive Banach spaces and Hilbert spaces, we have more.

Corollary *The following are equivalent for a reflexive Banach space $\mathcal{X}$:*

(a) *$\mathcal{X}$ has the approximation property.*

(b) *The two Arens products on $\mathcal{B}_K(\mathcal{X})^{**}$ agree and the natural injection κ of $\mathcal{B}_K(\mathcal{X})$ into $\mathcal{B}_K(\mathcal{X})^{**}$ can be extended to an isometric algebra isomorphism of $\mathcal{B}(\mathcal{X})$ onto $\mathcal{B}_K(\mathcal{X})^{**}$ which is compatible with the inclusion of $\mathcal{B}_K(\mathcal{X})$ in $\mathcal{B}(\mathcal{X})$. Moreover $\mathcal{B}_K(\mathcal{X})^*$ is isometrically isomorphic to $\mathcal{B}_N(\mathcal{X}^*)$.*

Corollary *If $\mathcal{H}$ is a Hilbert space, then (b) of the last corollary holds (with $\mathcal{X} = \mathcal{H}$). Moreover the ideal $\mathcal{B}_N(\mathcal{H})$ of nuclear operators is precisely the ideal $\mathcal{B}_{Tr}(\mathcal{H})$ of trace class operators and is included in $\mathcal{B}_K(\mathcal{H})$. The isometric linear isomorphisms*

$$\mathcal{B}_K(\mathcal{H})^* \simeq \mathcal{B}_{Tr}(\mathcal{H}) \quad \text{and} \quad \mathcal{B}_{Tr}(\mathcal{H})^* \simeq \mathcal{B}(\mathcal{H})$$

are both implemented by the pairing

$$(T, S) = \text{Tr}(TS) \qquad \forall\, T \in \mathcal{B}(\mathcal{H});\ S \in \mathcal{B}_{Tr}(\mathcal{H}).$$

1.7.14 Double Centralizers of Ideals of Operators In the last example, we calculated the Arens product on the double dual of $\mathcal{B}_K(\mathcal{X})$ for special Banach spaces and identified it with $\mathcal{B}(\mathcal{X})$ in the reflexive case. In this example, we will show that $\mathcal{B}(\mathcal{X})$ can also be identified with the double centralizer algebra of $\mathcal{B}_K(\mathcal{X})$ for any Banach space $\mathcal{X}$. Actually our results hold for a general class of ideals in $\mathcal{B}(\mathcal{X})$

Let $\mathcal{X}$ be a Banach space, and let $\mathcal{A}$ be a non-zero ideal in $\mathcal{B}(\mathcal{X})$ which is complete in its own norm $|||\cdot|||$ (which need not be the restriction of the operator norm on $\mathcal{B}(\mathcal{X})$). Proposition 1.1.18 shows that $\mathcal{A}$ must include $\mathcal{B}_F(\mathcal{X})$. The ideals $\mathcal{B}(\mathcal{X})$, $\mathcal{B}_K(\mathcal{X})$, $\mathcal{B}_{CC}(\mathcal{X})$, $\mathcal{B}_{WK}(\mathcal{X})$, $\mathcal{B}_{SS}(\mathcal{X})$, $\mathcal{B}_N(\mathcal{X})$ and $\mathcal{B}_I(\mathcal{X})$ (among others) satisfy all the requirements which we will subsequently make for $\mathcal{A}$. Proposition 1.2.6 gives a homomorphism θ of $\mathcal{B}(\mathcal{X})$ into $\mathcal{D}(\mathcal{A})$ defined by $\theta(T) = (L_T|\mathcal{A}, R_T|\mathcal{A})$ for all $T \in \mathcal{B}(\mathcal{X})$. The inclusion $\mathcal{B}_F(\mathcal{X}) \subseteq \mathcal{A}$ shows that θ is injective. Theorem 1.2.4 shows that the operators $L_T|\mathcal{A}$ and $R_T|\mathcal{A}$ are continuous. We will now investigate when θ is surjective. This will occur if and only if every left centralizer has the form $L_T|\mathcal{A}$.

From now on, we will assume that the norm $|||\cdot|||$ on $\mathcal{A}$ is a reasonable norm when restricted to $\mathcal{B}_F(\mathcal{X}) \subseteq \mathcal{A}$ (*i.e.*, it lies between the injective and projective tensor norms *cf.* Section 1.10) and satisfies

$$\max\{|||TS|||, |||ST|||\} \le |||S||| \cdot \|T\| \qquad \forall\, S \in \mathcal{A};\ T \in \mathcal{B}(\mathcal{X}).$$

To simplify notation, we will replace $L_T|\mathcal{A}$ and $R_T|\mathcal{A}$ by $T^{\#}$ and $T^{\flat}$. Thus $(\#)$ is a contractive injective homomorphism of $\mathcal{B}(\mathcal{X})$ into the subalgebra $\mathcal{L} \subseteq \mathcal{B}(\mathcal{A})$ of continuous left centralizers on $\mathcal{A}$, and $(\flat)$ is a contractive injective anti-homomorphism of $\mathcal{B}(\mathcal{X})$ into the subalgebra $\mathcal{R} \subseteq \mathcal{B}(\mathcal{A})$ of continuous right centralizers on $\mathcal{A}$.

We claim that $(\#)$ is an isometric isomorphism of $\mathcal{B}(\mathcal{X})$ onto $\mathcal{L}$. To prove this, let $L \in \mathcal{L}$, $x,\, y \in \mathcal{X}$ and $\omega,\, \tau \in \mathcal{X}^*$ satisfying $\omega(y) = 1$ be arbitrary, and define $z \in \mathcal{X}$ by $z = L(x \otimes \omega)y$. Then we get $L(x \otimes \tau) = \omega(y)L(x \otimes \tau) = L((x \otimes \omega)(y \otimes \tau)) = L(x \otimes \omega)y \otimes \tau = z \otimes \tau$. Clearly z depends linearly and continuously on x and is independent of τ. Hence the map $x \mapsto z$ belongs to $\mathcal{B}(\mathcal{X})$; if we denote it by T then the restriction of L to $\mathcal{B}_F(\mathcal{X})$ satisfies $L|\mathcal{B}_F(\mathcal{X}) = L_T|\mathcal{B}_F(\mathcal{X})$. Since $|||\cdot|||$ is a reasonable norm, we get $\|T\| \le \|L\|$. Hence for any $S \in \mathcal{A}$, $x \in \mathcal{X}$ and $\tau \in \mathcal{X}^*$ we have

$$L(S)x \otimes \tau = L(Sx \otimes \tau) = L((Sx) \otimes \tau) = (TSx) \otimes \tau = T^{\#}(S)x \otimes \tau.$$

This implies $L = T^{\#}$. Therefore $\#$ is an isometric isomorphism as claimed. (Note the curious fact that the restriction of L to $\mathcal{B}_F(\mathcal{X})$ determines L

even though $\mathcal{B}_F(\mathcal{X})$ need not be dense in $\mathcal{A}$ (*e.g.*, $\mathcal{A} = \mathcal{B}(\mathcal{X})$).) For any $(L, R) \in \mathcal{D}(\mathcal{A})$, $L = T^{\#}$ implies $R = T^{\flat}$ so that $T \mapsto (T^{\#}, T^{\flat})$ is an isometric isomorphism of $\mathcal{B}(\mathcal{X})$ onto $\mathcal{D}(\mathcal{A})$.

When we apply similar calculations to the homomorphism $(^{\flat})$, the answers are surprisingly different. Before doing so, we note that the Banach space adjoint $(^{*})$ is an anti-homomorphism of $\mathcal{B}(\mathcal{X})$ into $\mathcal{B}(\mathcal{X}^{*})$ which is surjective if and only if $\mathcal{X}$ is reflexive (Dunford and Schwartz [1958], VI.9.13).

Let $R \in \mathcal{R}$ be arbitrary. Choose $\omega \in \mathcal{X}^{*}$ and $y \in \mathcal{X}$ satisfying $\omega(y) = 1$. Then, for any $\tau \in \mathcal{X}^{*}$, define $\sigma \in \mathcal{X}^{*}$ by $\sigma = (R(y \otimes \tau))^{*}(\omega)$. For any $x \in \mathcal{X}$, the calculation $R(x \otimes \tau) = \omega(y)R(x \otimes \tau) = R((x \otimes \omega)(y \otimes \tau)) = x \otimes \omega R(y \otimes \tau) = x \otimes \sigma$ shows that σ is independent of the choice of ω and y. Clearly it depends linearly and continuously on τ. Hence the map $\tau \mapsto \sigma$ belongs to $\mathcal{B}(\mathcal{X}^{*})$. If we call this map $\tilde{R}$, then we have $R(x \otimes \tau) = x \otimes \tilde{R}(\tau)$ for all $x \in \mathcal{X}$ and $\tau \in \mathcal{X}^{*}$ from which we conclude $\|\tilde{R}\| \leq \|R\|$ since $\||| \cdot \|||$ is a reasonable norm. Hence for any $S \in \mathcal{A}$, $x \in \mathcal{X}$ and $\tau \in \mathcal{X}^{*}$ we have

$$x \otimes (R(S))^{*}\tau = x \otimes \tau R(S) = R(x \otimes \tau S) = R(x \otimes S^{*}\tau) = x \otimes \tilde{R}S^{*}\tau.$$

Therefore $R(S)^{*}$ equals $\tilde{R}S^{*}$ for any $S \in \mathcal{A}$. If $\mathcal{X}$ is reflexive, $\tilde{R}$ may be written as T^{*} for some $T \in \mathcal{B}(\mathcal{X})$ so we have $R = T^{\flat}$. Thus the map $T \mapsto T^{\flat}$ is an isometric anti-isomorphism of $\mathcal{B}(\mathcal{X})$ onto $\mathcal{R}$ when $\mathcal{X}$ is reflexive.

We conclude by briefly considering the case in which $\mathcal{X}$ is not reflexive and $\mathcal{A}$ is either $\mathcal{B}_K(\mathcal{X})$ or $\mathcal{B}_{WK}(\mathcal{X})$. Then well known results which can be found in Dunford and Schwartz [1958] VI.4.2, VI.4.8, and VI.5.2 show that, for any $T \in \mathcal{B}(\mathcal{X}^{**})$ and any $S \in \mathcal{A}$, the restriction $S^{**}T|_{\mathcal{X}}$ belongs to $\mathcal{A} \subseteq \mathcal{B}(\mathcal{X})$. This, together with the results obtained above, shows that the map $T \mapsto \tilde{R}_{T^{*}}$ defined by

$$\tilde{R}_{T^{*}}(S) = S^{**}T^{*}|_{\mathcal{X}} \qquad \forall\, T \in \mathcal{B}(\mathcal{X}^{*});\ S \in \mathcal{A}$$

is an isometric isomorphism of $\mathcal{B}(\mathcal{X}^{*})$ onto $\mathcal{R}$. The above example is due to Johnson [1964a], but the methods used here (and the generality obtained) appear to be novel.

1.7.15 Norms and Isomorphisms of $\mathcal{B}(\mathcal{X})$ In one of the earliest papers on Banach algebra theory, Meier Eidelheit [1940a] proved several remarkable theorems about the algebra $\mathcal{B}(\mathcal{X})$ of all bounded linear maps of a Banach space $\mathcal{X}$ into itself. He showed that an algebraic isomorphism between $\mathcal{B}(\mathcal{X})$ and $\mathcal{B}(\mathcal{Y})$ must come from a homeomorphic linear isomorphism between $\mathcal{X}$ and $\mathcal{Y}$. He also showed that every automorphism of $\mathcal{B}(\mathcal{X})$ is an inner automorphism and that any Banach algebra norm on $\mathcal{B}(\mathcal{X})$ must be equivalent to the operator norm. (We will enormously generalize this last result in Chapter 6.) Here we will prove slightly stronger statements using his simple but powerful ideas. We also give a related result based on an argument of Bertram Yood [1958].

Theorem *Let $\langle\, \mathcal{X} \,/\, \mathcal{Y}\,\rangle$ be a Banach space and let $\langle\, \mathcal{A} \,/\, \mathcal{B}\,\rangle$ be a norm closed subalgebra of $\langle\, \mathcal{B}(\mathcal{X})\widetilde{/}\,\mathcal{B}(\mathcal{Y})\,\rangle$ including $\langle\, \mathcal{B}_F(\mathcal{X})\,/\,\mathcal{B}_F(\mathcal{Y})\,\rangle$. If $\Phi\colon \mathcal{A} \to \mathcal{B}$ is an algebra isomorphism, then there is a homeomorphic linear isomorphism $P\colon \mathcal{X} \to \mathcal{Y}$ satisfying*

$$\Phi(S) = PSP^{-1} \qquad \forall\, S \in \mathcal{A}.$$

Proof Choose a non-zero element $z \in \mathcal{X}$ and a linear functional $\tau \in \mathcal{X}^*$ satisfying $\tau(z) = 1$. Then $z \otimes \tau$ is an idempotent operator of rank one. Hence $\Phi(z \otimes \tau)$ is also idempotent. Several easy arguments show that $\Phi(z \otimes \tau)$ also has rank one. For instance, note that $(z \otimes \tau)\mathcal{A}(z \otimes \tau)$ is the one-dimensional space $\mathbb{C}z \otimes \tau$. Hence $\Phi(z \otimes \tau)\mathcal{B}\Phi(z \otimes \tau)$ must also be one-dimensional. Since the image includes $\Phi(z \otimes \tau)\mathcal{B}_F(\mathcal{Y})\Phi(z \otimes \tau)$, this proves that $\Phi(z \otimes \tau)$ is a rank 1 idempotent. Hence we may write $\Phi(z \otimes \tau) = w \otimes \sigma$ for some $w \in \mathcal{Y}$ and $\sigma \in \mathcal{Y}^*$ satisfying $\sigma(w) = 1$.

We can now define the operator P. For any $x \in \mathcal{X}$, $\Phi(x \otimes \tau) = \Phi(x \otimes \tau)\Phi(z \otimes \tau) = \Phi(x \otimes \tau)(w \otimes \sigma) = y \otimes \sigma$ for some $y \in \mathcal{Y}$ which depends linearly on x. Define P to be the operator which sends x to y. We can write P as

$$P(x) = y \otimes \sigma(w) = \Phi(x \otimes \tau)(w) \qquad \forall\, x \in \mathcal{X}$$

and its inverse as

$$P^{-1}(y) = x \otimes \tau(z) = \Phi^{-1}(y \otimes \sigma)(z) \qquad \forall\, y \in \mathcal{Y}.$$

For any $S \in \mathcal{A}$ and $x \in \mathcal{X}$, we have $\Phi(S)P(x) \otimes \sigma = \Phi(S(x \otimes \tau)) = P(Sx) \otimes \sigma$, proving the formula.

Thus it only remains to show that Φ and its inverse are continuous. This follows from the next lemma, since any isomorphism obviously preserves the criterion for boundedness. ☐

Eidelheit's Lemma *Let $\mathcal{A}$ be a closed subalgebra of $\mathcal{B}(\mathcal{X})$ which includes $\mathcal{B}_F(\mathcal{X})$. Then a subset S is bounded if and only if for every $T \in \mathcal{B}_F(\mathcal{X})$ there exists an $\varepsilon > 0$ such that $I - \lambda TS$ is invertible for all $\lambda \in \varepsilon\mathbb{C}_1$ and all $S \in \mathcal{S}$.*

Proof The identity operator is the only operator which could serve as an identity element in $\mathcal{A}$. If it is not in $\mathcal{A}$, put it in. Suppose $\mathcal{S}$ is bounded by M. Choose $\varepsilon = (M\|T\| + 1)^{-1}$. Then $|\lambda| \le \varepsilon$ implies $\|\lambda TS\| < 1$. Hence the series $\sum_{n=0}^{\infty}(\lambda TS)^n$ converges in norm, so the calculation

$$(I - \lambda TS)\sum_{n=0}^{\infty}(\lambda TS)^n = \sum_{n=0}^{\infty}(\lambda TS)^n - \sum_{n=0}^{\infty}(\lambda TS)^{n+1} = I$$

is justified. It shows that $I - \lambda TS$ has a right inverse. Similarly, it has a left inverse and hence is invertible.

Conversely, suppose that $\mathcal{S}$ is not bounded. By a corollary of the uniform boundedness principle and the Hahn–Banach theorem (Dunford and Schwartz [1958], II.3.21), there exist $x \in \mathcal{X}$ and $\omega \in \mathcal{X}^*$ such that $\{\omega(Sx) : S \in \mathcal{S}\}$ is not bounded. Let T be the operator $x \otimes \omega$. Then for any $\varepsilon > 0$, there is an $S \in \mathcal{S}$ satisfying $|\omega(Sx)| > \varepsilon^{-1}$. Hence we see that the condition fails by taking $\lambda = (\omega(Sx))^{-1}$ and noting $(I - \lambda TS)x = 0$. □

Corollary 1 *If $\mathcal{X}$ and $\mathcal{Y}$ are Banach spaces, any algebraic isomorphism from $\mathcal{B}(\mathcal{X})$ to $\mathcal{B}(\mathcal{Y})$ (or even between closed subalgebras which include $\mathcal{B}_F(\mathcal{X})$ and $\mathcal{B}_F(\mathcal{Y})$, respectively) is continuous.*

Corollary 2 *Every algebra automorphism of $\mathcal{B}(\mathcal{X})$ is an inner automorphism.*

Corollary 3 *Let $\mathcal{A}$ be a norm closed subalgebra of $\mathcal{B}(\mathcal{X})$ which includes $\mathcal{B}_F(\mathcal{X})$. Then any two complete algebra norms on $\mathcal{A}$ are equivalent.*

Proof This follows from the lemma. Suppose a sequence $\{T_n\}_{n \in \mathbb{N}}$ converges to zero in another complete algebra norm. Then we can find a sequence $\{t_n\}_{n \in \mathbb{N}}$ of positive numbers converging to ∞ so that $\{t_n T_n\}_{n \in \mathbb{N}}$ still converges to zero and hence is bounded. The first half of the proof of the lemma depends only on the fact that we have a Banach algebra, so this sequence satisfies Eidelheit's boundedness criterion. Hence the sequence $\{t_n T_n\}_{n \in \mathbb{N}}$ is bounded in the operator norm and thus $\{T_n\}_{n \in \mathbb{N}}$ converges to zero in the operator norm. By the inverse boundedness theorem, the identity map between the algebra in these two complete norms must be a homeomorphism since we have just shown that it is continuous in one direction. □

The following related result is based on the proof of Theorem 2.3 in Yood [1958]. Theorem 2.6 in the same paper will be derived from this result as Theorem 2.5.17 in Chapter 2.

Theorem *Let $\mathcal{X}$ be a Banach space and let $\mathcal{A}$ be a non-zero ideal of $\mathcal{B}(\mathcal{X})$ or any subalgebra of $\mathcal{B}(\mathcal{X})$ which includes $\mathcal{B}_F(\mathcal{X})$. Any (not necessarily complete) algebra norm on $\mathcal{A}$ dominates the operator norm.*

Proof If $\mathcal{A}$ is an ideal, it includes $\mathcal{B}_F(\mathcal{X})$ by Proposition 1.1.18. The proof is by contradiction. Let the algebra norm on $\mathcal{A}$ be $||| \cdot |||$. Suppose it does not dominate the operator norm. Then we may find a sequence $\{T_n\}_{n \in \mathbb{N}}$ satisfying $|||T_n||| = 1$ for all $n \in \mathbb{N}$ and $\|T_n\| \to \infty$. Now let $x \in \mathcal{X}$ and $\omega \in \mathcal{X}^*$ be arbitrary. Then $|\omega(T_n(x))| \, |||x \otimes \omega||| = |||x \otimes \omega T_n x \otimes \omega||| \leq |||x \otimes \omega|||^2$. Hence, in particular, $\{|\omega(T_n(x))| : n \in \mathbb{N}\}$ is bounded for each $x \in \mathcal{X}$ and $\omega \in \mathcal{X}^*$. The uniform boundedness principle thus implies that $\{\|T_n\| : n \in \mathbb{N}\}$ is bounded, contradicting our assumption. □

1.7.16 Generalized Calkin Algebras and Fredholm Operators In the earliest American paper on Banach algebras with an abstract "modern" viewpoint, Calkin [1941] introduced an algebra which has come to be known as the *Calkin algebra*. It is the quotient algebra $\mathcal{B}(\mathcal{H})/\mathcal{B}_K(\mathcal{H})$, where $\mathcal{H}$ is a separable Hilbert space. This algebra is simple (*i.e.*, it has no nontrivial ideals) and unital. In Volume II we will further discuss this algebra and Calkin's representation of it.

Here, we generalize to the case in which $\mathcal{X}$ is a Banach space by defining $\mathcal{Q}(\mathcal{X}) = \mathcal{B}(\mathcal{X})/\mathcal{B}_K(\mathcal{X})$ to be its *generalized Calkin algebra*. We have seen Banach spaces in which $\mathcal{B}_{WK}(\mathcal{X})$ lies strictly between $\mathcal{B}_K(\mathcal{X})$ and $\mathcal{B}(\mathcal{X})$ so that the generalized Calkin algebra is not always simple. Furthermore, in some of these spaces (*e.g.*, those with the Dunford–Pettis property, including most classical nonreflexive spaces) the square of the ideal of weakly compact operators is the ideal of compact operators. In that case, the generalized Calkin algebra has a closed non-zero ideal with square zero. This is a type of bad behavior studied in more detail in Chapter 4 below. See also Selwyn R. Caradus, William E. Pfaffenberger and Bertram Yood [1974], John J. Buoni and Albert J. Klein [1979] and Michael J. Meyer [1992c].

An operator in $\mathcal{B}(\mathcal{X})$ is called *Fredholm* if its image in $\mathcal{Q}(\mathcal{X})$ is invertible, *Riesz* if its image in $\mathcal{Q}(\mathcal{X})$ is topologically nilpotent, and *inessential* if its image in $\mathcal{Q}(\mathcal{X})$ belongs to the Jacobson radical $\mathcal{Q}(\mathcal{X})_J$ of $\mathcal{Q}(\mathcal{X})$. The terms "topologically nilpotent" and "Jacobson-radical" are defined in the next chapter. The literature on Fredholm operators is enormous (*cf.* Caradus, Pfaffenberger and Yood [1974], Kato [1966], Martin Schechter [1971] and Section 8.5 below). Riesz operators are discussed briefly in Section 2.8 below, where the name is explained, and in great detail in Bruce A. Barnes, Gerard J. Murphy, M. R. F. Smyth, and Trevor T. West [1982]. For the other notions introduced above, see Anthony F. Ruston [1954], David C. Kleinecke [1963], West [1966], and Pfaffenberger [1970].

1.7.17 Hilbert Spaces and C*-algebras Although intensive study of these topics is postponed until Volume II, a brief overview will be useful here. All details can be found in Volume II.

Let $\mathcal{H}$ be a linear space. An *inner product* on $\mathcal{H}$ is a map

$$(\cdot,\cdot)\colon \mathcal{H} \times \mathcal{H} \to \mathbb{C}$$

which is linear in the first factor, conjugate linear in the second and satisfies

$$(x,x) > 0 \qquad \forall\, x \in \mathcal{H} \setminus \{0\}.$$

An inner product defines a norm on $\mathcal{H}$:

$$\|x\| = (x,x)^{1/2} \qquad \forall\, x \in \mathcal{H}.$$

A *Hilbert space* is a linear space with a fixed inner product such that the corresponding norm is complete. The most important property of Hilbert spaces is that for any continuous linear functional ω on a Hilbert space $\mathcal{H}$ there is a unique element $y \in \mathcal{H}$ satisfying

$$\omega(x) = (x, y) \qquad \forall\, x \in \mathcal{H}.$$

Furthermore, y also satisfies $\|y\| = \|\omega\|$.

Hilbert spaces are reflexive and even super-reflexive. They have *orthonormal bases* which are always both shrinking and boundedly complete bases in the terminology of §1.7.9 above. Hence Hilbert spaces have the bounded approximation property with bound 1. Moreover if $\mathcal{H}$ is a Hilbert space, the ideal of compact operators is the only nontrivial closed ideal in $\mathcal{B}(\mathcal{H})$ and includes all proper ideals. There is an interesting continuum of ideals C^p for all $0 < p \leq \infty$ in $\mathcal{B}(\mathcal{H})$ deeply connected with the ℓ^p spaces. Each ideal is a Banach algebra in its own norm $\|\cdot\|_p$. and is the completion of the ideal of finite-rank operators in this norm. Here C^∞ is just $\mathcal{B}_K(\mathcal{H})$ and C^p is continuously embedded in C^q if $p \leq q$. Furthermore C^1 is the ideal of nuclear operators which is also the ideal of *trace class* operators (with a different, but equivalent, definition), and C^2 is the ideal $\mathcal{B}_{HS}(\mathcal{H})$ of *Hilbert–Schmidt* operators. This last set is itself a Hilbert space in its own inner product and norm. It can be defined either abstractly or in terms of very concrete integral operators.

The inner product on $\mathcal{H}$ gives rise to the *Hilbert space adjoint* in $\mathcal{B}(\mathcal{H})$: for any $T \in \mathcal{B}(\mathcal{H})$, there is a unique operator T^* in $\mathcal{B}(\mathcal{H})$ satisfying

$$(T^*x, y) = (x, Ty) \qquad \forall\, x, y \in \mathcal{H}.$$

The map $T \mapsto T^*$ from $\mathcal{B}(\mathcal{H})$ to $\mathcal{B}(\mathcal{H})$ satisfies the next definition.

Definition   An *involution* on an algebra $\mathcal{A}$ is a map $(\,^*\,)\colon \mathcal{A} \to \mathcal{A}$ satisfying

$$
\begin{aligned}
(a + b)^* &= a^* + b^* & \text{(additive)} \\
(\lambda a)^* &= \lambda^* a^* & \text{(conjugate homogeneous)} \\
(ab)^* &= b^* a^* & \text{(anti-multiplicative)} \\
(a^*)^* &= a & \text{(involutive)}
\end{aligned}
\tag{14}
$$

for all $a, b \in \mathcal{A}$ and $\lambda \in \mathbb{C}$. An algebra with a fixed involution is called a **-algebra*, and a Banach algebra with an involution is called a *Banach *-algebra*. The following condition is called the *C*-condition*:

$$\|a^*a\| = \|a\|^2 \qquad \forall\, a \in \mathcal{A} \quad \text{(C*-condition)}. \tag{15}$$

A Banach algebra in which the norm and involution are related by the C*-condition is called a *C*-algebra*.

The simple definition of a C*-algebra does not suggest the profound effect the C*-condition has on the structure of an algebra. In Chapter 3 we will show that commutative unital C*-algebras can be characterized as those algebras isometrically isomorphic to the algebra $C(\Omega)$ of all continuous complex-valued functions on some compact Hausdorff space Ω (*cf.* §§1.5.1, 3.2.12 and 3.4.17). The remarks preceding the above definition show that $\mathcal{B}(\mathcal{H})$ is a C*-algebra when $\mathcal{H}$ is a Hilbert space. Hence any subalgebra of $\mathcal{B}(\mathcal{H})$ which is closed under the involution and also norm-closed is a C*-algebra. The fundamental Gelfand–Naĭmark theorem (Corollary 9.5.5 in Volume II) asserts the converse: any C*-algebra is isometrically isomorphic to a norm-closed and involution-closed subalgebra of $\mathcal{B}(\mathcal{H})$ for some Hilbert space $\mathcal{H}$. All the ideals of $\mathcal{B}(\mathcal{H})$ mentioned above are closed under the involution, but only the ideal of compact operators is closed in the operator norm.

A number of formulas are known to calculate the norm of a C*-algebra from its *-algebra structure. Hence, in particular, every homomorphism φ between C*-algebras which respects the *-algebra structure (*i.e.*, satisfies $\varphi(a^*) = (\varphi(a))^*$ for all a) is continuous (even contractive), and every isomorphism which respects the involution is an isometry.

These results show that, in the category of *-algebras, the norm of a C*-algebra is a consequence of the other structure. Similarly in the geometric category of Banach algebras (where morphisms are contractive algebra homomorphisms and isomorphisms are isometric), the involution in a C*-algebra is a consequence of the shape of the unit ball. See Section 2.6 and especially Proposition 2.6.8 for this viewpoint. Even in the topological category of Banach algebras (where morphisms are continuous algebra homomorphisms and hence the norm is only determined up to equivalence), the involution can be derived from consideration of bounded subgroups. All this will be discussed further in Volume II.

Propositions 2.3.17 and 2.3.18 show that every C*-algebra is semisimple (*cf.* Definition 2.3.2 and Section 4.3). Even the Banach space structure of a C*-algebra is strongly restricted by the definition. Any Banach algebra with the underlying Banach space of a C*-algebra is Arens regular. Also if a C*-algebra is linearly isometric to the dual of some Banach space, that Banach space is uniquely defined in a strong sense.

In Chapter 6 we need to know that the double centralizer of any C*-algebra is again a C*-algebra and that the regular homomorphism is an isometric isomorphism which preserves the involution. We will give the easy proof here. Let $\mathcal{A}$ be a C*-algebra. Recall that the norm of a bounded double centralizer is given by

$$\|(L, R)\| = \max(\|L\|, \|R\|) \qquad \forall \ (L, R) \in \mathcal{D}(\mathcal{A}).$$

The C*-condition shows directly that the hypotheses of Theorem 1.2.4 are satisfied (so every double centralizer is bounded) and that the regular homomorphism is an isometric isomorphism

$$\|a\|^2 = \|aa^*\| = \|L_a(a^*)\| \le \|L_a\|\|a\| \le \|a\|^2 \qquad \forall \, a \in \mathcal{A}$$

and similarly for the right regular representation R. The involution on $\mathcal{D}(\mathcal{A})$ is given by

$$(L, R)^* = (R^*, L^*) \qquad \forall \, (L, R) \in \mathcal{D}(\mathcal{A})$$

where R^* and L^* are defined by

$$T^*(a) = (T(a^*))^* \qquad \forall \, T \in \mathcal{L}(\mathcal{A}); \, a \in \mathcal{A}.$$

It is easy to check that this does define an involution. Theorem 1.2.4 shows that $\mathcal{D}(\mathcal{A})$ is a Banach algebra under its norm. In the present case each $(L, R) \in \mathcal{D}(\mathcal{A})$ satisfies

$$
\begin{aligned}
\|L\| &= \sup\{\|L(a)\| : a \in \mathcal{A}_1\} = \sup\{\|L(a)b\| : a, \, b \in \mathcal{A}_1\} \\
&= \sup\{\|aR(b)\| : a, \, b \in \mathcal{A}_1\} = \sup\{\|R(b)\| : b \in \mathcal{A}_1\} \\
&= \|R\|, \\
\|(L, R)^*\| &= \|(R^*, L^*)\| = \|R^*\| = \sup\{\|R^*(a)\| : a \in \mathcal{A}_1\} \\
&= \sup\{\|(R(a^*))^*\| : a \in \mathcal{A}_1\} = \|R\| \\
&= \|(L, R)\| \quad \text{and} \\
\|(L, R)^*(L, R)\| &= \|R^*L\| = \sup\{\|R^*L(a)b\| : a, \, b \in \mathcal{A}_1\} \\
&= \sup\{\|L(a)L^*(b)\| : a, \, b \in \mathcal{A}_1\} \\
&\ge \sup\{\|L(a)L(a)^*\| : a \in \mathcal{A}_1\} = \sup\{\|L(a)\|^2 : a \in \mathcal{A}_1\} \\
&= \|L\|^2 = \|(L, R)\|^2 \ge \|(L, R)^*(L, R)\|.
\end{aligned}
$$

Hence $\mathcal{D}(\mathcal{A})$ is a C*-algebra.

1.8 Group Algebras on T

This section provides elementary concrete realizations of many phenomena that we will discuss more abstractly and in much more generality later. See particularly Sections 1.9 and 3.6. Here we will look in detail at the various spaces of functions on the circle group which are algebras under convolution multiplication or are in some other way related to harmonic analysis in this setting.

1.8.1 Basic Facts and the Haar Integral The symbol $\mathbb{T}$ will always represent the set of complex numbers of absolute value one, considered as a group under multiplication. ($\mathbb{T}$ stands for torus since this group is sometimes called the one-dimensional torus.) This group will always carry the relative topology it inherits as a subset of the complex plane. Locally this is also the topology of the line as we will explain in greater detail below. The group operations (*i.e.* the multiplication map $(\lambda, \mu) \mapsto \lambda\mu$ from $\mathbb{T} \times \mathbb{T}$ to $\mathbb{T}$ and the inversion map $\lambda \mapsto \lambda^{-1}$ from $\mathbb{T}$ to $\mathbb{T}$) are obviously continuous so that $\mathbb{T}$ is an abelian, compact, connected topological group.

The map $h: \mathbb{R} \to \mathbb{T}$ defined by

$$h(t) = e^{it}$$

is a group homomorphism of the additive group of $\mathbb{R}$ onto $\mathbb{T}$ which is also a local homeomorphism. Thus h identifies $\mathbb{T}$ with the quotient of $\mathbb{R}$ modulo the subgroup $2\pi\mathbb{Z}$. From another viewpoint this makes it clear that $\mathbb{T}$ is an abelian, real, compact, connected, Lie group which has $h: \mathbb{R} \to \mathbb{T}$ as its simply connected covering group with fiber $\pi_1(\mathbb{T}) \approx \mathbb{Z}$.

When $f: \mathbb{T} \to \mathbb{C}$ is any function, it will be convenient to denote $f \circ h$ by adding a tilde: $\tilde{f}$. This identifies complex-valued functions on $\mathbb{T}$ with 2π-periodic functions on $\mathbb{R}$. Another useful viewpoint is to consider the restriction of $\tilde{f}$ to the closed interval $[-\pi, \pi]$. This identifies complex-valued functions on $\mathbb{T}$ with functions on $[-\pi, \pi]$ which have the same value at $-\pi$ and π.

We use the map h to define integration and differentiation of complex-valued functions on $\mathbb{T}$. For any suitable function $f: \mathbb{T} \to \mathbb{C}$, we define

$$\int_{\mathbb{T}} f(\lambda)d\lambda = \frac{1}{2\pi} \int_{-\pi}^{\pi} f \circ h(t)dt = \frac{1}{2\pi} \int_{-\pi}^{\pi} \tilde{f}(t)dt = \frac{1}{2\pi} \int_{-\pi}^{\pi} f(e^{it})dt.$$

(*N. B.* The integral defined here is not a contour integral although on other occasions we will use the notation $\int_{\Gamma} f(\lambda)d\lambda$ to denote a contour integral on a contour Γ in $\mathbb{C}$.) If $\tilde{f}$ is Riemann integrable, this integral may be interpreted as a Riemann integral, but usually we consider it as a Lebesgue integral. It has the following remarkable properties

$$\int_{\mathbb{T}} f(\mu\lambda)d\lambda = \int_{\mathbb{T}} f(\lambda)d\lambda \qquad \forall\, \mu \in \mathbb{T}, \tag{1}$$

$$\int_{\mathbb{T}} f(\lambda^{-1})d\lambda = \int_{\mathbb{T}} f(\lambda)d\lambda. \tag{2}$$

Translation invariance, as given in equation (1), is the characteristic property of a Haar integral to be defined in Section 1.9 below. Equation (2) is a property of unimodular Haar integrals.

Whenever $f: \mathbb{T} \to \mathbb{C}$ is a function such that $\tilde{f}$ has at least an almost everywhere defined derivative, we will define f' to be the function satisfying $(f')^{\tilde{}} = \tilde{f}'$. We will transport other properties of functions on $[-\pi, \pi]$ to functions on $\mathbb{T}$ in a similar fashion. Frequently we prefer to work with $\tilde{f}$ instead of f because of the more elementary appearance of the computations.

1.8.2 Some Banach Function Spaces We wish to define a number of Banach algebras of complex-valued functions on $\mathbb{T}$. Let us begin by discussing the Banach spaces involved. Their linear operations are always pointwise. Readers not already familiar with these Banach spaces can work out their properties as exercises after consulting Dunford and Schwartz [1958], Edwin Hewitt and Kenneth A. Ross [1963], [1970], or Walter Rudin [1962]. We will make some clarifying remarks after listing the Banach spaces of complex-valued functions on $\mathbb{T}$ and their norms.

Banach Space	Norm
$L^1 = \{$absolutely integrable functions$\}$	$\|f\|_1 = \int_{\mathbb{T}} \|f(\lambda)\| d\lambda$
$L^2 = \{$square integrable functions$\}$	$\|f\|_2 = \left(\int_{\mathbb{T}} \|f(\lambda)\|^2 d\lambda\right)^{1/2}$
$L^\infty = \{$essentially bounded functions$\}$	$\|f\|_\infty = \text{ess sup}\{\|f(\lambda)\| : \lambda \in \mathbb{T}\}$
$C = \{$continuous functions$\}$	$\|f\|_\infty = \sup\{\|f(\lambda)\| : \lambda \in \mathbb{T}\}$
$CBV = \{$continuous functions of bounded variation$\}$	$\|f\|_{CBV} = \|f\|_\infty + \|f\|_V$ where $\|f\|_V = \sup\{\sum_{j=1}^{n} \|\tilde{f}(t_j) - \tilde{f}(t_{j-1})\|\}$
$AC = \{$absolutely continuous functions$\}$	$\|f\|_{AC} = \|f\|_\infty + \|f'\|_1$
$C^1 = \{$continuously differentiable functions$\}$	$\|f\|_{C^1} = \|f\|_\infty + \|f'\|_\infty$

The supremum in the definition of $\|f\|_V$ is extended over all $n \in \mathbb{N}$ and all sets $\{t_0, t_1, \ldots, t_n\}$ satisfying $t_0 = -\pi < t_1 < \ldots < t_n = \pi$. If $\{f_n\}_{n \in \mathbb{N}}$ is a Cauchy sequence in CBV, then its limit f in C actually belongs to CBV and is its limit in CBV, so that CBV is complete. (To see this let $k = \sup\{\|f_n\|_V : n \in \mathbb{N}\}$ and, for any $\varepsilon > 0$, choose p so that $\|f_p - f_m\|_V \leq \varepsilon$ for all $m \geq p$. Then show that each set $\{t_0, t_1, \ldots, t_n\}$ satisfies both

$$\sum_{j=1}^{n} \|\tilde{f}(t_j) - \tilde{f}(t_{j-1})\| \leq k \quad \text{and} \quad \sum_{j=1}^{n} \|(\tilde{f}_p - \tilde{f})(t_j) - (\tilde{f}_p - \tilde{f})(t_{j-1})\| \leq \varepsilon.)$$

A function $f: \mathbb{T} \to \mathbb{C}$ is defined to be *absolutely continuous* if $\tilde{f}$ is absolutely continuous on $[-\pi, \pi]$. Hence, f belongs to AC if $\tilde{f}$ is differentiable almost everywhere and its derivative belongs to L^1 and satisfies

$$\tilde{f}(t) = \tilde{f}(-\pi) + \int_{-\pi}^{t} \tilde{f}'(r) dr \qquad \forall\, t \in \mathbb{R}.$$

This observation will be used several times. In particular, it gives the completeness of AC. (If $\{f_n\}_{n\in\mathbb{N}}$ is a Cauchy sequence in AC, then there is a $g \in L^1$ such that $\{f'_n\}_{n\in\mathbb{N}}$ converges to g in L^1. Hence if we define $f(-\pi)$ by $\lim f_n(-\pi)$ and define $f \in AC$ by $f(t) = f(-\pi) + \int_{-\pi}^{t} g(r)dr$, then $\{f_n\}_{n\in\mathbb{N}}$ converges to f in AC.) Alternatively, it is not particularly hard to see (*e.g.*, Hewitt and Karl R. Stromberg [1965], 18.1) that the norms $\|\cdot\|_{CBV}$ and $\|\cdot\|_{AC}$ are actually equal on AC and that AC is a closed subspace of CBV, so that it is complete. The proof of the completeness of C^1 is similar to the (first) proof of the completeness of AC.

The inclusions

$$C^1 \subseteq AC \subseteq CBV \subseteq C \subseteq L^\infty \subseteq L^2 \subseteq L^1 \tag{3}$$

are obvious from the definitions. (Of course the inclusion $C \subseteq L^\infty$ needs some interpretation, since the elements of L^∞ are equivalence classes of almost everywhere equal functions. If one of these equivalence classes contains a continuous function f, it contains only one and the two interpretations of $\|f\|_\infty$ agree. This gives meaning to the inclusion $C \subseteq L^\infty$ and shows that C is closed in L^∞.) The fact that $\mathbb{T}$ has measure one, and other elementary considerations, give the inequalities

$$\|f\|_1 \le \|f\|_2 \le \|f\|_\infty \le \|f\|_{CBV} = \|f\|_{AC} \le \|f\|_{C^1}. \tag{4}$$

Therefore each inclusion in (3) expresses a continuous embedding of the smaller Banach space into the larger. The embeddings $C \to L^\infty$ and $AC \to CBV$ are actually isometries.

1.8.3 Trigonometric Polynomials We will next discuss the linear space of trigonometric polynomials. It is dense in all of the above Banach spaces except L^∞ and CBV. For each $n \in \mathbb{Z}$, define $h_n \colon \mathbb{T} \to \mathbb{C}$ by $h_n(\lambda) = \lambda^n$ so that $\tilde{h}_n(t)$ is just e^{int}. The set $\mathcal{T}$ of *trigonometric polynomials* is the set of finite linear combinations of functions in $\mathbf{H} = \{h_n \ : \ n \in \mathbb{Z}\}$. If $p = \sum_{k=-n}^{n} \alpha_k h_k$ is a trigonometric polynomial, then we call the smallest possible n in this expression the *degree* of p.

Clearly $\mathcal{T}$ is an algebra under pointwise multiplication, which is closed under complex conjugation, separates the points of the compact space $\mathbb{T}$ and contains the constant functions. Since these are the hypotheses of the complex form of the Stone–Weierstrass theorem (*cf.* 6.1), we conclude that $\mathcal{T}$ is dense in C (*i.e.*, dense in the uniform norm $\|\cdot\|_\infty$). From the regularity of the measure on $\mathbb{T}$, one can conclude that C is dense in both L^1 and in L^2 in their respective norms. This fact and the inequalities $\|\cdot\|_1 \le \|\cdot\|_\infty$ and $\|\cdot\|_2 \le \|\cdot\|_\infty$ show that $\mathcal{T}$ is dense in L^1 and L^2 (*e.g.*, for any $\varepsilon > 0$ and $f \in L^2$, find $g \in C$ and $h \in T$ satisfying $\|f - g\|_2 < \frac{\varepsilon}{2}$ and $\|g - h\|_\infty < \frac{\varepsilon}{2}$, and thus conclude

$$\|f - h\|_2 \le \|f - g\|_2 + \|g - h\|_2 < \frac{\varepsilon}{2} + \|g - h\|_\infty < \varepsilon.)$$

Clearly $\mathcal{T}$ is not dense in L^∞ since it is included in C which is closed in L^∞. Similarly $\mathcal{T}$ is not dense in CBV since it is included in the closed subspace AC. The density of the trigonometric polynomials in AC can be established as follows. For any $f \in AC$ and any $\varepsilon > 0$, choose $g \in \mathcal{T}$ satisfying $\|g - f'\|_1 < \frac{\varepsilon}{2}$. Next define $h \in \mathcal{T}$ by $\tilde{h}(t) = \tilde{f}(-\pi) + \frac{1}{2\pi}\int_{-\pi}^t g(r)dr$. Then we have

$$\|f - h\|_{AC} = \sup |\frac{1}{2\pi}\int_{-\pi}^t (\tilde{f}'(r) - g(r))dr| + \|g - f'\|_1$$
$$< \frac{1}{2\pi}\int_{-\pi}^\pi |\tilde{f}'(r) - g(r)|dr + \frac{\varepsilon}{2} < \varepsilon.$$

An entirely similar argument proves that the trigonometric polynomials are dense in C^1.

1.8.4 Banach Algebra Structure   So far we have discussed these seven spaces as Banach spaces. It is easy to check that the five Banach spaces L^∞, C, CBV, AC and C^1 are all Banach algebras under pointwise multiplication. This is not the Banach algebra structure in which we are primarily interested here. We wish to consider all seven Banach spaces as Banach algebras under convolution multiplication, which we are about to define. To distinguish these two Banach algebra structures, we will use the symbols L^1, L^2, $L^\infty(*)$, $C(*)$, $CBV(*)$, $AC(*)$ and $C^1(*)$ for the Banach algebras with convolution multiplication and use L^∞, C, CBV, AC and C^1 to denote either the Banach spaces or the Banach algebras with pointwise multiplication.

1.8.5 Convolution Products For any two functions f, $g \in L^1$ we denote their *convolution product* by $f * g$ and define it by

$$f * g(\mu) = \int_{\mathbb{T}} f(\lambda)g(\lambda^{-1}\mu)d\lambda \qquad \forall \mu \in \mathbb{T}$$

$$\widetilde{f * g}(s) = \frac{1}{2\pi}\int_{-\pi}^\pi \tilde{f}(t)\tilde{g}(s - t)dt \qquad \forall s \in \mathbb{R}. \tag{5}$$

(§1.9.1 below gives an elementary explanation of the origin of these formulas.) Application of equations (1) and (2) allows us to rewrite this as

$$f * g(\mu) = \int_{\mathbb{T}} f(\mu\lambda)g(\lambda^{-1})d\lambda = \int_{\mathbb{T}} f(\mu\lambda^{-1})g(\lambda)d\lambda$$
$$= \int_{\mathbb{T}} f(\lambda^{-1})g(\lambda\mu)d\lambda \qquad \forall \mu \in \mathbb{T} \tag{6}$$

which implies

$$f * g = g * f. \tag{7}$$

An application of Fubini's theorem and equation (1) shows

$$\|f * g\|_1 = \left(\frac{1}{2\pi}\right)^2 \int_{-\pi}^{\pi} |\int_{-\pi}^{\pi} \tilde{f}(t)\tilde{g}(s-t)dt|ds \tag{8}$$

$$\leq \left(\frac{1}{2\pi}\right)^2 \int_{-\pi}^{\pi} \int_{-\pi}^{\pi} |\tilde{f}(t)\tilde{g}(s-t)|ds dt = \|f\|_1\|g\|_1.$$

So the norm on L^1 is submultiplicative. (Hereafter we omit the proof that various functions are measurable and other measure theoretic subtleties. Many of these results are checked explicitly in Hewitt and Ross [1963] or Hewitt and Stromberg [1965].) We could now check the other properties necessary to show that L^1 is an algebra by arguments similar to those already given. (Only associativity is nontrivial.) Actually we prefer to postpone this until we have introduced the Fourier transform at which point these results become obvious. Anticipating this discussion, we will regard L^1 as a commutative Banach algebra and prove that the other convolution algebras are subalgebras or ideals in L^1. We will also derive the submultiplicative property for each norm.

1.8.6 L^2 is an Ideal in L^1 For $f \in L^1$ and $g \in L^2$, the Cauchy–Schwarz inequality gives

$$\left(\frac{1}{2\pi}\int_{-\pi}^{\pi} |\tilde{f}(t)| \, |\tilde{g}(s-t)|dt\right)^2 = \left(\frac{1}{2\pi}\int_{-\pi}^{\pi} |\tilde{f}(t)|^{1/2}(|\tilde{f}(t)| \, |\tilde{g}(s-t)|^2)^{1/2}dt\right)^2$$

$$\leq \frac{1}{2\pi}\int_{-\pi}^{\pi} |\tilde{f}(t)| \, dt \frac{1}{2\pi}\int_{-\pi}^{\pi} |\tilde{f}(t)| \, |\tilde{g}(s-t)|^2 dt = \|f\|_1 \frac{1}{2\pi}\int_{-\pi}^{\pi} |\tilde{f}(t)| \, |\tilde{g}(s-t)|^2 dt.$$

This result together with Fubini's theorem gives

$$\|f * g\|_2{}^2 = \frac{1}{2\pi}\int_{-\pi}^{\pi} |(f * g)\tilde{\,}(s)|^2 ds \leq \left(\frac{1}{2\pi}\right)^3 \int_{-\pi}^{\pi} \left(\int_{-\pi}^{\pi} |\tilde{f}(t)\tilde{g}(s-t)|dt\right)^2 ds$$

$$\leq \|f\|_1\left(\frac{1}{2\pi}\right)^2 \int_{-\pi}^{\pi}\int_{-\pi}^{\pi} |\tilde{f}(t)| \, |\tilde{g}(s-t)|^2 ds \, dt = \|f\|_1{}^2\|g\|_2{}^2.$$

Hence we have established

$$\|f * g\|_2 \leq \|f\|_1\|g\|_2 \qquad \forall \, f \in L^1; \, g \in L^2, \tag{9}$$

so that L^2 is an ideal in L^1. Inequalities (4) and (9) show that the norm on L^2 is an algebra norm.

1.8.7 $L^\infty(*)$ **is an ideal in** L^1 It is even easier to show that $L^\infty(*)$ is also an ideal of L^2. For any $f \in L^1$, $g \in L^\infty$ and $s \in [-\pi, \pi]$, we have

$$|\widetilde{f * g}(s)| \le \frac{1}{2\pi} \int_{-\pi}^{\pi} |\tilde{f}(t)\tilde{g}(s-t)|dt \le \frac{1}{2\pi} \int_{-\pi}^{\pi} |\tilde{f}(t)| \, \|g\|_\infty dt = \|f\|_1 \|g\|_\infty.$$

This establishes

$$\|f * g\|_\infty \le \|f\|_1 \|g\|_\infty \qquad \forall \, f \in L^1; \, g \in L^\infty. \tag{10}$$

so that $L^\infty(*)$ is an ideal in L^1 and in L^2. This inequality and (4) show

$$\|f * g\|_\infty \le \|f\|_\infty \|g\|_\infty \qquad \forall \, f, g \in L^\infty.$$

So $L^\infty(*)$ is a Banach algebra.

1.8.8   **The Convolution Product of Functions from L^1 and L^∞ is Continuous**   To establish this and several related results we need a lemma. For any $f \colon \mathbb{T} \to \mathbb{C}$ and any $s \in \mathbb{R}$ we will define $f_s \colon \mathbb{T} \to \mathbb{C}$ by

$$f_s(\lambda) = f(\lambda e^{is}) \qquad \forall \, \lambda \in \mathbb{T} \quad \text{or} \quad \tilde{f}_s(t) = \tilde{f}(t+s) \qquad \forall \, t \in \mathbb{R}. \tag{11}$$

In all of the Banach spaces under consideration, note that the norm satisfies $\|f\| = \|f_s\|$ for all f in the space and all $s \in \mathbb{R}$.

Lemma   *Let $\mathcal{X}$ be one of the Banach spaces $L^1, L^2, C, AC,$ or C^1. Then for each $f \in \mathcal{X}$, the map $t \mapsto f_t$ from $\mathbb{R}$ to $\mathcal{X}$ is continuous.*

Proof   For any $s, r \in \mathbb{R}$, any $f \in \mathcal{X}$ and any $\varepsilon > 0$, we choose a $p \in \mathcal{T}$ satisfying $\|f - p\| < \frac{\varepsilon}{3}$, where $\| \cdot \|$ is the norm in $\mathcal{X}$. Then we have

$$\begin{aligned} \|f_s - f_r\| &\le \|f_s - p_s\| + \|p_s - p_r\| + \|p_r - f_r\| \\ &= \|(f-p)_s\| + \|p_s - p_r\| + \|(f-p)_r\| < \frac{2\varepsilon}{3} + \|p_s - p_r\|. \end{aligned}$$

Hence it is enough to show that any given $p \in \mathcal{T}$ and $r \in \mathbb{R}$ satisfy $\lim_{s \to r} \|p_s - p_r\| = 0$ in the norm of each of the Banach spaces under discussion. After choosing an n such that p can be written as $p = \sum_{k=-n}^{n} \alpha_k h_k$ for suitable coefficients α_k, one can easily establish these continuity properties. $\square$

This lemma easily establishes the inclusion

$$L^1 * L^\infty \subseteq C. \tag{12}$$

This implies that $L^\infty(*)$ and $C(*)$ are both ideals in L^1 and hence in L^2. For $f \in L^1$, $g \in L^\infty$ and r, $s \in \mathbb{R}$ we have

$$|\widetilde{(f*g)}(s) - \widetilde{(f*g)}(r)| = \frac{1}{2\pi}|\int_{-\pi}^{\pi}\left(\tilde{f}(s+t)\tilde{g}(-t) - \tilde{f}(r+t)\tilde{g}(-t)\right)dt|$$

$$\leq \frac{1}{2\pi}\int_{-\pi}^{\pi}|(\tilde{f}_s - \tilde{f}_r)(t)|\,|\tilde{g}(-t)|dt$$

$$\leq \|f_s - f_r\|_1\|g\|_\infty.$$

Since the first factor on the right can be made arbitrarily small, $f*g$ is continuous. Note that equation (12) also implies that the ideal $(L^\infty(*))^2$ in $L^\infty(*)$ is included in C. This ideal clearly contains T (each h_n is idempotent under convolution multiplication so $T^2 = T$) and hence we have

$$((L^\infty(*))^2)^- = C \neq L^\infty(*).$$

1.8.9 $CBV(*)$ **is an Ideal in** $L^\infty(*)$ This follows from the inequality

$$\|f*g\|_{CBV} \leq \|f\|_\infty\|g\|_{CBV} \qquad \forall\, f \in L^\infty;\ g \in CBV \qquad (13)$$

which we now prove. Given f and g as above, we have $\|f*g\|_\infty \leq \|f\|_\infty\|g\|_1 \leq \|f\|_\infty\|g\|_\infty$ by previous results. For any partition $t_0 = -\pi < t_1 < \cdots < t_n = \pi$ we have:

$$\sum_{j=1}^{n}|\widetilde{(f*g)}(t_j) - \widetilde{(f*g)}(t_{j-1})|$$

$$\leq \sum_{j=1}^{n}\frac{1}{2\pi}\int_{-\pi}^{\pi}|\tilde{f}(t)|\,|\tilde{g}(t_j - t) - \tilde{g}(t_{j-1} - t)|dt$$

$$\leq \|f\|_\infty\frac{1}{2\pi}\int_{-\pi}^{\pi}\sum_{j=1}^{n}|\tilde{g}_{-t}(t_j) - \tilde{g}_{-t}(t_{j-1})|dt$$

$$\leq \|f\|_\infty\frac{1}{2\pi}\int_{-\pi}^{\pi}\|g\|_V\,dt = \|f\|_\infty\|g\|_V.$$

Hence (13) follows from

$$\|f*g\|_{CBV} = \|f*g\|_\infty + \|f*g\|_V \leq \|f\|_\infty(\|g\|_\infty + \|g\|_V) = \|f\|_\infty\|g\|_{CBV}.$$

Combining again with (4) shows that the norm on CBV is submultiplicative.

1.8.10 $AC(*)$ **is an Ideal in** L^1 To complete the discussion of the relation of these convolution algebras we will show

$$(f * g)' = f * g' \qquad \forall \, f \in L^1; \; g \in AC \tag{14}$$

which implies $f * g \in AC$ and

$$\|f * g\|_{AC} \leq \|f\|_1 \|g\|_{AC} \qquad \forall \, f \in L^1; \; g \in AC. \tag{15}$$

Thus $AC(*)$ is an ideal in L^1 and hence in the other convolution algebras which contain it. Furthermore (14) and (10) show that $C^1(*)$ is an ideal in L^1. As in the earlier cases, we conclude that the norms on $AC(*)$ and $C^1(*)$ are submultiplicative.

To prove (14), assume $f \in L^1$ and $g \in AC$. Then the equation $\tilde{g}(s) = \tilde{g}(-\pi) + \int_{-\pi}^s \tilde{g}'(r)dr$ for all $s \in \mathbb{R}$ implies

$$\widetilde{f * g}(s) - \widetilde{f * g}(-\pi) = \frac{1}{2\pi} \int_{-\pi}^{\pi} \tilde{f}(t)(\tilde{g}(s - t) - \tilde{g}(-\pi - t))dt$$

$$= \frac{1}{2\pi} \int_{-\pi}^{\pi} \tilde{f}(t) \int_{-\pi-t}^{s-t} \tilde{g}'(r)dr \, dt = \frac{1}{2\pi} \int_{-\pi}^{\pi} \int_{-\pi}^{s} \tilde{f}(t)\tilde{g}'(u - t)du \, dt$$

$$= \frac{1}{2\pi} \int_{-\pi}^{s} \int_{-\pi}^{\pi} \tilde{f}(t)\tilde{g}'(u - t)dt \, du = \int_{-\pi}^{s} (f * (g'))\widetilde{}(u)du.$$

Differentiating this equation gives (14). The equation also shows directly that $f * g$ is an indefinite integral and therefore belongs to AC.

1.8.11 L^2 **as a Hilbert space** In order to introduce the Fourier transform in a natural way, we will discuss the Hilbert space L^2 in more detail. Its inner product is

$$(f, g) = \int_{\mathbb{T}} f(\lambda)g(\lambda)^* d\lambda = \frac{1}{2\pi} \int_{-\pi}^{\pi} \tilde{f}(t)\tilde{g}(t)^* dt.$$

The functions h_n, introduced above, satisfy

$$(h_n, h_m) = \frac{1}{2\pi} \int_0^{2\pi} e^{i(n-m)t} dt = \begin{cases} 0, & n \neq m \\ 1, & n = m. \end{cases}$$

Hence $\mathbf{H} = \{h_n : n \in \mathbb{Z}\}$ is an *orthonormal set* in L^2 since this is just the defining property of such a set. Recall that an orthonormal set $\mathbf{H}$ in a

Hilbert space $\mathcal{H}$ is called an *orthonormal basis* if it satisfies one (and hence all) of the following equivalent conditions:

(a) **H** is a maximal orthonormal set.

(b) If $x \in \mathcal{H}$ satisfies $(x, h) = 0$ for all $h \in \mathbf{H}$, then $x = 0$.

(c) $\mathcal{H}$ is the closed linear span of **H**.

(d) For all $x \in \mathcal{H}$, the unordered series

$$x = \sum_{h \in \mathbf{H}} (x, h)h$$

converges to x in the norm of $\mathcal{H}$.

(e) For all $x, y \in \mathcal{H}$, the unordered series

$$(x, y) = \sum_{h \in \mathbf{H}} (x, h)(h, y)$$

converges to (x, y).

(f) For all $x \in \mathcal{H}$, the unordered series

$$\|x\|^2 = \sum_{h \in \mathbf{H}} |(x, h)|^2$$

converges to $\|x\|^2$.

The series in (d) is called the *Fourier expansion* of x, and each of the last two equations is sometimes called *Parseval's identity*. (The easy proof of the equivalence of these properties can be found in any book discussing Hilbert spaces.) Since the set $\mathcal{T} = \text{span}(\mathbf{H})$ of trigonometric polynomials is dense in L^2, the orthonormal set $\mathbf{H} = \{h_n : n \in \mathbf{Z}\}$ is an orthonormal basis.

1.8.12 Fourier–Plancherel Transforms   For each $f \in L^2$, let us define a function $\hat{f} : \mathbf{Z} \to \mathbb{C}$ by

$$\hat{f}(n) = (f, h_n) = \int_{\mathbf{T}} f(\lambda)h_n(\lambda)^* d\lambda = \frac{1}{2\pi} \int_{-\pi}^{\pi} \tilde{f}(t)e^{-int} dt. \qquad (16)$$

Condition (d) for an orthonormal basis shows

$$f = \sum_{n \in \mathbf{Z}} \hat{f}(n)h_n \qquad \forall\, f \in L^2 \qquad (17)$$

where convergence is in the norm of L^2. Condition (f) shows

$$\|f\|_2{}^2 = \sum_{n \in \mathbf{Z}} |\hat{f}(n)|^2 \qquad \forall\, f \in L^2. \qquad (18)$$

Hence the map $f \mapsto \hat{f}$ is an isometric linear map of L^2 into the Hilbert space $\ell^2 = \ell^2(\mathbb{Z}) = \{$square summable, doubly-infinite sequences$\}$. Since $\sum_{n \in \mathbb{Z}} \alpha_n h_n$ converges to an element f in L^2 for any $\alpha \in \ell^2$, we see that the map $f \mapsto \hat{f}$ is surjective and hence is a unitary isomorphism of L^2 onto ℓ^2. We will call this map the *Fourier–Plancherel transform*.

1.8.13 Ideal Structure of the Banach Algebra L^2 Next we wish to examine the algebraic structure of L^2 and of the Fourier–Plancherel transform. A trivial calculation gives

$$h_n * h_m = \begin{cases} 0, & n \neq m \\ h_n, & n = m \end{cases} \qquad \forall \, n, m \in \mathbb{Z}. \tag{19}$$

Hence the orthonormal basis **H** consists of orthogonal idempotents so that

$$L^2 = \bigoplus_{n \in \mathbb{Z}} \mathbb{C} h_n \tag{20}$$

is both an orthogonal direct sum decomposition of the Hilbert space L^2 and a Banach algebra internal direct sum decomposition of the algebra. The formula

$$f * h_n = \hat{f}(n) h_n \qquad \forall \, f \in L^2; \, n \in \mathbb{Z} \tag{21}$$

provides an easy proof that $\mathcal{T}$ is an ideal in L^2 (and in all our other convolution algebras when the function $f \mapsto \hat{f}$ has been extended to L^1) and that the closed ideals of L^2 are exactly the subspaces of the form

$$S^\perp = \{f \in L^2 : (f, h_n) = 0 \qquad \forall \, n \in S\}$$

where S is an arbitrary subset of $\mathbb{Z}$. (The ideal $\mathcal{T}$ is the socle of all of these algebras in terminology that will be introduced in Chapter 8.)

Finally we come to the most interesting result. For any f and g in L^2 and any $n \in \mathbb{Z}$, Fubini's theorem and translation invariance, equation (1), give

$$\begin{aligned}
\widehat{f * g}(n) &= \left(\frac{1}{2\pi}\right)^2 \int_{-\pi}^{\pi} \int_{-\pi}^{\pi} \tilde{f}(t)\tilde{g}(s-t)dt \, e^{-ins}ds \\
&= \left(\frac{1}{2\pi}\right)^2 \int_{-\pi}^{\pi} \tilde{f}(t)e^{-int} \int_{-\pi}^{\pi} \tilde{g}(s-t)e^{-in(s-t)}ds \, dt \\
&= \frac{1}{2\pi} \int_{-\pi}^{\pi} \tilde{f}(t)e^{-int}\hat{g}(n)dt = \hat{f}(n)\hat{g}(n). \tag{22}
\end{aligned}$$

The Cauchy–Schwarz inequality shows that the pointwise product of two functions in ℓ^2 is in

$$\ell^1 = \ell^1(\mathbb{Z}) = \{\text{doubly-infinite summable sequences}\} \subseteq \ell^2.$$

Hence we may, and will, consider ℓ^2 to be an algebra under pointwise multiplication. Thus, we have shown that the Fourier–Plancherel transform is not only a unitary isomorphism but also an algebra isomorphism of L^2 (with the convolution product) onto ℓ^2 (with the pointwise product).

1.8.14 Fourier Transform on L^1 Some important parts of the above discussion can be generalized. Except for the inner product, each of the expressions for $\hat{f}(n)$ given in (16) makes sense for any $f \in L^1$. Hence we use them to define a linear map $f \mapsto \hat{f}$ from L^1 into the linear space of complex sequences. This map is called the *Fourier transform*. It is immediate that all $n \in \mathbb{Z}$ satisfy $|\hat{f}(n)| \leq \|f\|_1$, so that $\hat{f}$ is always a bounded sequence. In fact a stronger result is easy for us to prove now, since we know that $\mathcal{T}$ is dense in L^1.

Riemann–Lebesgue Lemma *The Fourier transform $\hat{f}$ of any function $f \in L^1$ belongs to*

$$c_0 = \{\text{doubly-infinite sequence converging to zero at } \infty \text{ and } -\infty\}.$$

Proof For any $f \in L^1$ and any $\varepsilon > 0$, we can find a trigonometric polynomial p satisfying $\|f - p\|_1 < \varepsilon$. Then for n greater than the degree of p, we have $|\hat{f}(n)| = |(\hat{f} - \hat{p})(n)| \leq \|f - p\|_1 < \varepsilon$. Hence $\hat{f}$ belongs to the space c_0. □

The space c_0 is always regarded as a Banach algebra under pointwise multiplication. The calculation (22) remains valid for $f, g \in L^1$ and shows that the Fourier transform is a contractive homomorphism of L^1 into c_0.

Our next task is to show that the Fourier transform is injective and hence an isomorphism onto its range. The analogous result for the Fourier–Plancherel transform was easy since we could reconstruct any $f \in L^2$ from its Fourier–Plancherel transform via the series $f = \sum_{n \in \mathbb{Z}} \hat{f}(n) h_n$ which converges in L^2. The formal series $\sum_{n \in \mathbb{Z}} \hat{f}(n) h_n$ is called the *Fourier series* of f, whether or not it converges in any sense.

1.8.15 The Fejér Kernel Unfortunately the Fourier series for $f \in L^1$ seldom converges in the norm of L^1. However the series is summable by various standard summability methods. We will investigate only one— Cesáro summability. For a (doubly-infinite) series $\sum_{n \in \mathbb{Z}} \alpha_n$, we will use S_n to denote the nth symmetric partial sum $S_n = \sum_{k=-n}^{n} \alpha_k$, and we will use σ_n for the nth term in the sequence of *Cesáro means*: $\sigma_n = \frac{1}{n+1} \sum_{k=0}^{n} S_k$. It is easy to see that, if the sequence $\{S_n\}_{n \in \mathbb{N}}$ converges (*i.e.*, the series converges), then the sequence $\{\sigma_n\}_{n\in\mathbb{N}}$ converges, but that the converse fails in many interesting cases. Since the limit of the latter always agrees with the limit of the former when that limit exists, it is a value that can be assigned in place of a sum for some nonconvergent series.

Suppose f belongs to L^1. Formula (21) shows that the Cesáro mean

$$\sigma_n(f) = \frac{1}{n+1} \sum_{k=0}^{n} \sum_{j=-k}^{k} \hat{f}(j)h_j$$

of the series $\sum_{n\in\mathbb{Z}} \hat{f}(n)h_n$ is given by

$$\sigma_n(f) = K_n * f, \tag{23}$$

where K_n is the trigonometric polynomial

$$K_n = \sum_{k=-n}^{n} \left(1 - \frac{|k|}{n+1}\right) h_k \tag{24}$$

which is called the *Fejér kernel*. Clearly K_n satisfies

$$\frac{1}{2\pi} \int_{-\pi}^{\pi} \tilde{K}_n(t)dt = 1. \tag{25}$$

It also satisfies

$$K_n \geq 0 \qquad \forall\, n \in \mathbb{N} \tag{26}$$

and

$$\lim_{n\to\infty} \frac{1}{2\pi} \int_{\delta}^{2\pi-\delta} \tilde{K}_n(t)dt = 0 \qquad \forall\, \delta > 0. \tag{27}$$

Both of these properties are immediate consequences of the identity

$$\tilde{K}_n(t) = \frac{1}{n+1} \left(\frac{\sin(\frac{n+1}{2}t)}{\sin(\frac{t}{2})} \right)^2$$

which in turn follows from the calculation

$$4(n+1)\left(\sin\left(\frac{t}{2}\right)\right)^2 \tilde{K}_n(t) = (-e^{it} + 2 - e^{-it}) \sum_{k=-n}^{n} (n+1-|k|)e^{ikt}$$

$$= -e^{i(n+1)t} + 2 - e^{-i(n+1)t} = 4\left(\sin\left(\frac{n+1}{2}t\right)\right)^2.$$

We will now show that if f belongs to one of the Banach spaces L^1, C, AC or C^1, the sequence $\{\sigma_n(f)\}_{n\in\mathbb{N}}$ converges to f in the norm of that space. Of course this Cesáro series also converges in L^2 since the Fourier series itself actually converges there.) The only properties of $\sigma_n(f)$ we will use are those given by (23), (25), (26) and (27). Denote L^1, C, AC or C^1 by $\mathcal{X}$ and the norm of $\mathcal{X}$ by $\|\cdot\|$. We will use only the following properties of these Banach spaces: for any $f \in \mathcal{X}$, the function $s \mapsto f_s$ is 2π-periodic and continuous from $\mathbb{R}$ to $\mathcal{X}$ and satisfies $\|f_s\| = \|f\|$ for all $s \in \mathbb{R}$.

In the proof of the next theorem we will use Riemann integrals of certain continuous $\mathcal{X}$-valued functions. The definition of these integrals, the proof that they exist, and the proof of the other basic properties which we use can be copied from any careful calculus textbook discussion using Riemann sums.

Fejér's Theorem   *Let $(\mathcal{X}, \|\cdot\|)$ be a Banach space such that for each $f \in \mathcal{X}$ and $s \in \mathbb{R}$ an element $f_s \in \mathcal{X}$ is defined satisfying:*

$$\|f_s\| = \|f\|, \qquad s \mapsto f_s \text{ is continuous from } \mathbb{R} \text{ to } (\mathcal{X}, \|\cdot\|)$$

and f_s is 2π-periodic in s. Let $\{K_n\}$ be a sequence of continuous functions on $\mathbb{T}$ satisfying (25), (26), and (27). Define $\sigma_n(f)$ by (23). Then the sequence $\{\sigma_n(f)\}$ converges to f in the norm of $\mathcal{X}$.

Corollary   *If a function f belongs to L^1, C, AC or C^1, then the Cesáro mean $\sigma_n(f)$ converges to f in the norm of the given space.*

Since the corollary is immediate, we prove the theorem.

Proof   For any $f \in \mathcal{X}$, (23) and (25) give

$$f - \sigma_n(f) = \frac{1}{2\pi} \int_{-\pi}^{\pi} \tilde{K}_n(t)(\tilde{f} - \tilde{f}_t) dt$$

$$= \frac{1}{2\pi} \int_{-\delta}^{\delta} \tilde{K}_n(t)(\tilde{f} - \tilde{f}_t) dt + \frac{1}{2\pi} \int_{\delta}^{2\pi-\delta} \tilde{K}_n(t)(\tilde{f} - \tilde{f}_t) dt.$$

For the first integral, (26) and (25) give

$$\|\frac{1}{2\pi} \int_{-\delta}^{\delta} \tilde{K}_n(t)(\tilde{f} - \tilde{f}_t) dt\| \leq \frac{1}{2\pi} \int_{-\pi}^{\pi} \tilde{K}_n(t) dt \max_{|t| \leq \delta} \{\|f - f_t\|\} = \max_{|t| \leq \delta} \{\|f - f_t\|\}$$

which approaches zero with δ, by our assumption on the continuity of $t \mapsto f_t$. For the second integral, after the choice of δ, the assumption $\|f_t\| = \|f\|$ for all $t \in \mathbb{R}$ gives

$$\|\frac{1}{2\pi} \int_{\delta}^{2\pi-\delta} \tilde{K}_n(t)(\tilde{f} - \tilde{f}_t) dt\| \leq 2\|f\| \frac{1}{2\pi} \int_{\delta}^{2\pi-\delta} \tilde{K}_n(t) dt$$

which also approaches zero with n, this time by (27).   □

This result shows that the Fourier transform is injective on L^1 since $\hat{f} = 0$ implies $\sigma_n(f) = 0$ for all n. We summarize our results on transforms:

Corollary *The Fourier–Plancherel transform is a unitary map and an algebra isomorphism of L^2 onto ℓ^2. The Fourier transform is a contractive injective homomorphism of L^1 into c_0.*

1.8.16 The Inverse Fourier Transform on ℓ^1 We will now reverse our point of view briefly. Consider the Banach space $\ell^1 = \ell^1(\mathbb{Z})$ of doubly-infinite absolutely summable complex sequences (*i.e.*, functions α from $\mathbb{Z}$ to $\mathbb{C}$ satisfying $\sum_{n\in\mathbb{Z}} |\alpha_n| < \infty$) with norm $\alpha \mapsto \|\alpha\|_1 = \sum_{n\in\mathbb{Z}} |\alpha_n|$. We can make ℓ^1 into a Banach algebra by defining a convolution product

$$(\alpha * \beta)(m) = \sum_{n\in\mathbb{Z}} \alpha_n \beta_{m-n} \quad \forall\, \alpha, \beta \in \ell^1.$$

To see that $\alpha * \beta$ belongs to ℓ^1 again and satisfies $\|\alpha * \beta\|_1 \leq \|\alpha\|_1 \|\beta\|_1$, we use the sequential form of Fubini's theorem:

$$\|\alpha * \beta\|_1 \leq \sum_{m\in\mathbb{Z}} \sum_{n\in\mathbb{Z}} |\alpha_n \beta_{m-n}| = \sum_{n\in\mathbb{Z}} |\alpha_n| \sum_{m\in\mathbb{Z}} |\beta_{m-n}| = \|\alpha\|_1 \|\beta\|_1.$$

The rest of the axioms for a Banach algebra are now most easily checked by reference to (29) below. We will follow the common practice of denoting this Banach algebra by ℓ^1 even though ℓ^1 could be considered a Banach algebra under pointwise multiplication.

For any sequence $\alpha \in \ell^1$, we define $\check{\alpha}: \mathbb{T} \to \mathbb{C}$ by

$$\check{\alpha}(\lambda) = \sum_{n\in\mathbb{Z}} \alpha_n \lambda^n \quad \forall\, \lambda \in \mathbb{T}, \qquad \text{or} \qquad \check{\alpha} = \sum_{n\in\mathbb{Z}} \alpha_n h_n \tag{28}$$

where the first series converges absolutely and uniformly, and therefore the second series converges in the norm of C to an element of C. The map $\alpha \mapsto \check{\alpha}$ is called the *inverse Fourier transform*. Since uniformly convergent series are integrable term-by-term, we find

$$(\check{\alpha})\hat{} = \alpha \quad \forall\, \alpha \in \ell^1.$$

1.8.17 The Fourier Algebra In fact, it is obvious that a function f in C has an absolutely convergent Fourier series if and only if $\hat{f}$ belongs to ℓ^1, in which case it has a uniformly convergent Fourier series. Let us denote the space of such functions by $A = A(\mathbb{T})$ and make A into a Banach space by transporting the norm of ℓ^1 *via* (˘) (or (ˆ)), *i.e.*,

$$\|f\|_A = \sum_{n\in\mathbb{Z}} |\hat{f}(n)| \quad \forall\, f \in A.$$

In fact, A is a Banach algebra under pointwise multiplication. Its closure under pointwise multiplication is a consequence of the identity

$$\widehat{fg} = \hat{f} * \hat{g} \qquad \forall\, f, g \in A \tag{29}$$

which, of course, can be rewritten as $\check{\alpha}\check{\beta} = (\alpha * \beta)\check{}$ for any $\alpha,\ \beta \in \ell^1$. Since the proof of this is essentially the same as (22) above, it is omitted. This algebra A is called the *Fourier algebra* of $\mathbb{T}$. The restriction of the Fourier transform to A is again called the Fourier transform. We have proved:

Proposition   *The Fourier transform is an isometric algebra isomorphism of A onto ℓ^1.*

(Of course the isometry part of this proposition is a bit of a fraud since the norm of A was defined to make it true.)

1.8.18 Other Algebras on $\mathbb{T}$ under the Pointwise Product   We have already remarked that C, CBV, AC and C^1 are, like A, Banach algebras under pointwise operations. We will close this section by showing that C^1 is included in A and some related results. This will depend on the following simple identity:

$$(f')\check{}(n) = in\hat{f}(n) \qquad \forall\, f \in AC;\ n \in \mathbb{Z} \tag{30}$$

which is established by integration by parts:

$$
\begin{aligned}
(f')\check{}(n) &= \frac{1}{2\pi} \int\limits_{-\pi}^{\pi} \tilde{f}'(t)e^{-int}\,dt \\
&= -\frac{1}{2\pi} \int\limits_{-\pi}^{\pi} \tilde{f}(t)(-in)e^{-int}\,dt + \tilde{f}(t)e^{-int}\Big|_{-\pi}^{\pi} = in\hat{f}(n).
\end{aligned}
$$

If f belongs to C^1, then f' belongs to $C \subseteq L^2$. Hence the Cauchy–Schwarz inequality gives

$$
\begin{aligned}
\sum_{n\in\mathbb{Z}\backslash\{0\}} |\hat{f}(n)| &= \sum_{n\in\mathbb{Z}\backslash\{0\}} |\tfrac{1}{n}(f')\check{}(n)| \\
&\leq \left(\sum_{n\in\mathbb{Z}\backslash\{0\}} (\tfrac{1}{n})^2 \sum_{n\in\mathbb{Z}} ((f')\hat{}(n))^2 \right)^{1/2} = \frac{\pi}{\sqrt{6}}\|f'\|_2 < \infty.
\end{aligned}
$$

Hence f belongs to A as claimed.

We have already shown that the convolution product of a function in L^1 and one in L^∞ gives a continuous function. Thus convolution is a smoothing operation. This is important in the more classical applications. Here is another amusing indication of the same thing. Consider $f \in AC$ and $g \in A$. Results (14) and (15) show $f * g \in AC$ and $(f * g)' = f' * g$. Hence $((f * g)')^\wedge = (f')^\wedge \hat{g}$ is the pointwise product of a sequence in c_0 and a sequence in ℓ^1 and hence belongs to ℓ^1. Therefore $(f * g)'$ belongs to A and hence to C. Thus we have proved the inclusion $AC * A \subseteq C^1$. Moreover, the convolution product of two functions without everywhere defined derivatives can have a continuous derivative.

1.8.19   Historical Remarks   The study of Fourier series was one of the main sources of functional analysis. Daniel Bernoulli (1700-1782) and Leonard Euler (1707-1783) first used what we would now call Fourier series. They did not develop their ideas, and not much progress was made on this approach until Jean Baptiste Joseph Fourier (1768-1830) used trigonometric series in a problem on heat. His methods were severely criticized by some of his contemporaries, but this motivated him to further develop his ideas for a decade until he wrote his famous book *Theorie analytique de la chaleur* [1822]. In this book he claimed that "any function" could be expanded in a convergent Fourier series. At that time, mathematicians were often not very precise about convergence and were extremely vague about the meaning of the word "function". Fourier's claim spurred a closer examination of the concept of a function. Peter Lejeune Dirichlet (1805-1859) stated essentially our modern view of functions and in [1829] he introduced his kernel and used it to prove the first rigorous convergence theorem for Fourier series. Since his theorem and method of proof are still important we will review them briefly.

A function $f: \mathbb{R} \to \mathbb{C}$ is said to be *piecewise smooth* on an interval $[a, b]$ if there is some partition $a = t_0 < t_1 < \cdots < t_n = b$ of $[a, b]$ so that f is continuously differentiable on each subinterval $]t_{j-1}, t_j[$. (This requires that the one-sided derivatives at the end points of each subinterval exist and equal the limits of the derivative in the interior.) Denote the limits from the right and left, respectively, for f at x by $f(x^+)$ and $f(x^-)$.

Dirichlet's Theorem   *Let $f: \mathbb{R} \to \mathbb{C}$ be piecewise smooth on $[-\pi, \pi]$ and 2π-periodic. Then the Fourier series converges pointwise to $f(x)$ at every point of continuity and it converges to*

$$\frac{f(x^+) + f(x^-)}{2}$$

at any point of discontinuity. If f is also continuous, then the series converges to f absolutely and uniformly on $\mathbb{R}$.

Dirichlet's proof was based on the following function known as *Dirich-*

let's kernel

$$D_n(t) = \sum_{k=-n}^{n} e^{ikt} = 1 + 2\sum_{k=1}^{n} \cos(kt) = \frac{\sin((n+\frac{1}{2})t)}{\sin(\frac{1}{2}t)}. \qquad (31)$$

The first expression in (31) shows that the convolution product with any function in L^1 is the nth partial sum of the Fourier series:

$$f * D_n(t) = \sum_{k=-n}^{n} \hat{f}(k)e^{ikt}.$$

The middle expression in (31) shows the same thing for Fourier series written in the standard real form. The last expression in (31) allows an elementary evaluation of the limit of $\{f * D_n(t)\}_{n \in \mathbb{N}}$ as n increases.

Around the beginning of this century, functional analysis began to develop around problems dealing with integral equations, many of which had originally arisen in the context of Fourier series (*e.g.*, Eric Ivar Fredholm [1900], David Hilbert [1904], [1906], Erhard Schmidt [1907], [1908]). Just before the Second World War increasingly algebraic methods, which arose naturally in the study of Fourier series and its lineal descendent, harmonic analysis, were a main stimulus to the definition of Banach algebras (*e.g.*, Norbert Wiener [1932], Arne Beurling [1938]).

1.9   Group Algebras

We will introduce some of the most important group algebras here and give a few of their basic properties. Since we are postponing the discussion of involutions to Volume II, and involutions play a prominent role in the theory of group algebras, we will not go into the theory of group algebras very deeply here. However, additional information on group algebras will be developed later in the present volume as new topics are explored. See particularly Section 3.6 (Commutative Group Algebras) and Example 5.1.9 (approximate identities in group algebras). We will denote the identity element in a group by e unless some other notation is established. For more details we recommend Hewitt and Ross [1963] and Hans Reiter [1968]. Jean-Paul Pier [1990] is a fascinating historical survey.

1.9.1   The Algebraic Group Algebra   Fundamentally, most group algebras are devices for introducing linear operations into a group. We will call the simplest possible group algebra of a group G the *algebraic group algebra*. This algebra, which we will denote by $\ell_0^1(G)$, consists of all complex-valued functions on G each of which is non-zero at only finitely many elements of G. Each of these functions f may be thought of as a formal sum

$$f \sim \sum_{u \in G} f(u)u$$

of finitely many complex multiples of elements of G. If two elements f and g of the algebraic group algebra are interpreted in this way, then the familiar rules of computation suggest the following definitions of addition, scalar multiplication and multiplication:

$$\sum_{u \in G} f(u)u + \sum_{u \in G} g(u)u = \sum_{u \in G}(f(u) + g(u))u$$

$$\lambda \sum_{u \in G} f(u)u = \sum_{u \in G} \lambda f(u)u$$

$$\sum_{u \in G} f(u)u \sum_{v \in G} g(v)v = \sum_{u \in G} \sum_{v \in G} f(u)g(v)uv.$$

Thus we define the linear operations pointwise. If we collect terms in the expression for the product, we get

$$\sum_{w \in G} \left(\sum_{u \in G} f(u)g(u^{-1}w) \right) w.$$

Hence we are led to the definition of the *convolution product*

$$f * g(u) = \sum_{v \in G} f(v)g(v^{-1}u) \qquad \forall\, u \in G. \tag{1}$$

It is obvious that these operations make the algebraic group algebra into an algebra. (Note that the above discussion would not need to be changed if another field (or even ring) were substituted for the complex field. Algebraic group algebras with respect to other rings and fields play an important role in the theory of finite groups. See Charles W. Curtis and Irving Reiner [1981, 1987]. However, we will consider only complex scalars.)

The algebraic group algebra has a number of disadvantages. It ignores the topology of G when G has a nontrivial topology, as we will often suppose. However even for finite groups (which are necessarily discrete in this discussion, since we will insist that topological groups have a Hausdorff topology) the algebraic group algebra is not a complete set of invariants. As a matter of fact the dihedral group of order eight (with two generators u and v satisfying $u^4 = v^2 = e$ and $uv = vu^3$) and the quarternion group (with two generators u and v satisfying $u^4 = e$, $u^2 = v^2$ and $uv = vu^3$) have isomorphic group algebras. (An explicit isomorphism is obtained by using the bases $(e, u, u^2, u^3, v + u^2v, uv + u^3v, w, uw)$ where in the first case w is $v - u^2v$ and in the second case w is $i(uv - u^3v)$.)

As we will show below, the normed algebra $(\ell_0^1(G), \|\cdot\|_1)$ obtained by endowing the algebraic group algebra with its natural ℓ^1-norm

$$\|f\|_1 = \sum_{u \in G} |f(u)| \qquad \forall\, f \in \ell_0^1(G) \tag{2}$$

is a complete set of invariants for the group G. (This norm is submultiplicative as shown in (4) below, thus verifying that $(\ell_0^1(G), \|\cdot\|_1)$ is a normed algebra.)

When thinking about topological groups we will call groups without topology, *discrete groups*. For discrete groups, this normed algebra is a fairly satisfactory group algebra. It is complete, and hence a Banach algebra, if and only if G is finite. In order to be able to make use of the theory of Banach algebras when G is not finite, it is useful to replace $(\ell_0^1(G), \|\cdot\|_1)$ by its completion. This completion is the most natural and useful group algebra for discrete groups. We will now introduce it formally.

1.9.2 The ℓ^1-Group Algebra
We begin with a formal definition.

Definition   Let G be a group. Let $\ell^1(G)$ be the set of functions $f: G \to \mathbb{C}$ satisfying

$$\|f\|_1 = \sum_{u \in G} |f(u)| < \infty. \tag{3}$$

Let $\ell^1(G)$ have the structure of a normed algebra with pointwise linear operations, the norm $\|\cdot\|_1$ defined above and the convolution product given in (1) above. Then $\ell^1(G)$ is called the ℓ^1-*group algebra of G*.

With the above definition, $\ell^1(G)$ is a Banach algebra. It is easy to check that $\|\cdot\|_1$ is a complete norm. To show that the product is well-defined, we check that the infinite series (1) converges absolutely. This calculation also shows that the norm is submultiplicative. We have

$$
\begin{aligned}
\|f * g\|_1 &= \sum_{u \in G} \Big| \sum_{v \in G} f(v)g(v^{-1}u) \Big| \\
&\leq \sum_{u \in G} \sum_{v \in G} |f(v)g(v^{-1}u)| = \sum_{v \in G} \sum_{u \in G} |f(v)|\,|g(v^{-1}u)| \\
&= \sum_{v \in G} |f(v)| \sum_{w \in G} |g(w)| = \|f\|_1 \|g\|_1 \qquad \forall\, f, g \in \ell^1(G)
\end{aligned}
\tag{4}
$$

where we have used the fact that $G = \{v^{-1}u : u \in G\}$ holds for any fixed $v \in G$. Now it is straightforward to check that $\ell^1(G)$ is an algebra (and hence a Banach algebra) under this multiplication.

For each $u \in G$, define δ_u to be the function on G which takes the value one at u and zero elsewhere. Then it is easy to see that δ_e is a multiplicative identity element for $\ell^1(G)$ and that each δ_u is invertible and satisfies

$$(\delta_u)^{-1} = \delta_{u^{-1}} \quad \delta_u \delta_v = \delta_{uv} \quad \forall\, u,\, v \in G. \tag{5}$$

Hence the map $u \mapsto \delta_u$ is a group isomorphism of G into the multiplicative group of invertible elements of $\ell^1(G)$. Each element of $\ell^1(G)$ may be expressed uniquely as an absolutely norm-convergent series of complex multiples of these elements:

$$f = \sum_{u \in G} f(u)\delta_u \quad \forall\, f \in \ell^1(G). \tag{6}$$

This is a rigorous interpretation of the formal sums introduced above in the discussion of algebraic group algebras.

Since $u \mapsto \delta_u$ is an isomorphism of G into the multiplicative group of $\ell^1(G)$, we see that $\ell^1(G)$ is commutative only when G is abelian. Conversely, the absolute convergence of the expansion (6) for $f \in \ell^1(G)$ gives an easy proof that $\ell^1(G)$ is commutative when G is abelian. Thus $\ell^1(G)$ is commutative if and only if G is abelian.

If f belongs to the center of $\ell^1(G)$, then it satisfies $\delta_u * f * \delta_{u^{-1}} = f$ for all $u \in G$. In fact, the expansion above shows that this is a necessary and sufficient condition for f to belong to the center of G. Hence $f \in \ell^1(G)$ belongs to the center of $\ell^1(G)$ if and only if it is constant on conjugacy classes. (A *conjugacy class* in a group G is a set of the form $C_v = \{uvu^{-1} : u \in G\}$ for some $v \in G$.) An ℓ^1 function can have a constant non-zero value only on finite sets. Hence $\ell^1(G)$ has a nontrivial center (*i.e.*, different from $\mathbb{C}\delta_e$ which is always included in the center) if and only if it has some finite conjugacy classes besides the trivial class $\{e\}$. Groups which have only infinite conjugacy classes (except for $\{e\}$) are called *Infinite Conjugacy Class* (abbreviated ICC) *groups*. They provide important examples of von Neumann algebra factors as we shall show subsequently.

1.9.3 Recovering G from $\ell^1(G)$　We now show how a discrete group G may be recovered from the Banach algebra $\ell^1(G)$. Our previous remarks show that the norm must play an essential role in the process. Note that the extreme points of the unit ball $\ell^1(G)_1$ of $\ell^1(G)$ have the form $\zeta\delta_u$ for some $\zeta \in \mathbb{T}$ and some $u \in G$. This follows immediately from the expansion (6). Thus the set E of extreme points forms a group. Let $h \colon G \to \mathbb{T}$ be any homomorphism. Then the set

$$\widetilde{H}_h = \{h(u)\delta_u : u \in G\}$$

is a subgroup of E which has the property

$$\|f + g\| = \|f\| + \|g\| \quad \forall\, f,\, g \in \widetilde{H}_h. \tag{7}$$

Furthermore $\ell^1(G)$ is the closed linear span of $\tilde{H}_h$. Conversely, suppose $\tilde{H}$ is a subgroup of E satisfying (7). Then $\tilde{H}$ contains at most one element of the form $\zeta\delta_u$ for each fixed element $u \in G$. Hence $\tilde{H}$ equals $\{h(u)\delta_u : u \in H\}$, where H is some subset of G and $h: H \to \mathbb{T}$ is some function. Since $\tilde{H}$ is a subgroup, H must be a subgroup of G and h must be a homomorphism of H into $\mathbb{T}$. If in addition $\ell^1(G)$ is the closed linear span of $\tilde{H}$, then H must be G. Hence $\tilde{H}$ is isomorphic to G. Therefore we can recover G, up to isomorphism, as any subgroup $\tilde{H}$ of E which satisfies (7) and which has $\ell^1(G)$ as its closed linear span. In fact, in Volume II we will use the assignment $G \mapsto \ell^1(G)$ to construct an isomorphism between the category of groups and group homomorphisms and a certain restricted category of Banach algebras. The following result is immediate from what we have proved.

Theorem   *Two groups G and H are isomorphic as groups if and only if $\ell^1(G)$ and $\ell^1(H)$ are isometrically isomorphic as Banach algebras.*

1.9.4 Introduction to Representations   Representations of algebras will be described systematically in Chapter 4. We introduce some of the ideas informally now in conjunction with representations of groups. A *representation* of a group G on a linear space $\mathcal{V}$ is a group homomorphism $V: G \to \mathcal{L}(\mathcal{V})_G$ of G into the group of invertible linear operators on $\mathcal{V}$. If V is a representation of G on $\mathcal{V}$, then it is clear that V can be "extended" to a *representation* (*i.e.*, an algebra homomorphism) T^V of the algebraic group algebra $\ell_0^1(G)$ of G on $\mathcal{V}$. We simply write

$$T_f^V(u) = \sum_{u \in G} f(u)V_u \qquad \forall\, f \in \ell_0^1(G)$$

where there is no problem of convergence since only finitely many terms of the sum are non-zero. It is clear that T^V is an algebra homomorphism and hence a representation of $\ell_0^1(G)$ on $\mathcal{V}$.

A *normed representation* of a group G on a Banach space $\mathcal{X}$ is a group homomorphism

$$V: G \to B(\mathcal{X})_G$$

of G into the group of invertible bounded linear operators on $\mathcal{X}$. A normed representation V is said to be *bounded* if there is a common bound B for all the operators V_u (*i.e.*, $\|V_u\| \le B$ for all $u \in G$). Given a bounded representation of a group G on a Banach space $\mathcal{X}$, we can "extend" V to a *continuous representation* T^V of $\ell^1(G)$ (*i.e.*, a norm-continuous algebra homomorphism $T^V: \ell^1(G) \to \mathcal{B}(\mathcal{X})$) simply by writing

$$T_f^V = \sum_{u \in G} f(u)V_u \qquad \forall\, f \in \ell^1(G). \tag{8}$$

Clearly the sum converges absolutely and T^V is a continuous homomorphism of $\ell^1(G)$ satisfying $\|T^V\| \leq B$ if B is a bound for V. Note that the continuity of T^V has nothing to do with any continuity property of V.

If G is a group and T is a unital representation of $\ell_0^1(G)$ or $\ell^1(G)$, then we may define a representation V^T of G by

$$V_u^T = T_{\delta_u} \qquad \forall\, u \in G. \tag{9}$$

If T is a continuous representation of $\ell^1(G)$ on a Banach space, then V^T is bounded. Clearly $T \mapsto V^T$ and $V \mapsto T^V$ are inverse constructions. Hence we can identify the class of all representations of a group G with the class of representations of $\ell_0^1(G)$, and we can identify the class of bounded representations of G on Banach spaces with the class of all continuous representations of $\ell^1(G)$ on Banach spaces. These identifications preserve all the important properties of the representations. For instance, if V is a representation of G on $\mathcal{V}$, then a linear subspace $\mathcal{W}$ of $\mathcal{V}$ is invariant with respect to V (*i.e.*, $V_u x \in \mathcal{W}$ for all $u \in G$ and all $x \in \mathcal{W}$) if and only if it is invariant with respect to T^V (*i.e.*, $T_f^V x \in \mathcal{W}$ for all $f \in \ell_0^1(G)$ and all $x \in \mathcal{W}$). Similarly if V is a bounded representation of G on a Banach space $\mathcal{X}$, then a closed linear subspace $\mathcal{W}$ is invariant with respect to V if and only if it is invariant with respect to T^V.

Here is a natural and important example of these ideas. (Others will be discussed in Volume II.) Let G be a group and, for any $u \in G$, let $V_u \in \mathcal{B}(\ell^1(G))$ be defined by

$$V_u(f) = \delta_u * f = {}_u f \qquad \forall\, f \in \ell^1(G).$$

It is clear that this is a bounded representation of G with bound 1. It is also clear that T^V is the left regular representation of $\ell^1(G)$ on itself.

1.9.5 Introduction to Topological Groups We now extend the foregoing considerations to topological groups. First we will define topological groups and give a few basic facts about them. The important work by Edwin Hewitt and Kenneth A. Ross [1963, 1970] is the best reference for additional facts. Unfortunately, our notation does not always agree with theirs.

A *topological group* is a group which is also a Hausdorff topological space in which the multiplication map from $G \times G$ to G and the inversion map from G to G, defined by $(u, v) \mapsto uv$ and $u \mapsto u^{-1}$, respectively, are continuous. Many authors do not require that a topological group be Hausdorff but we shall. Since inversion is its own inverse, inversion is actually a homeomorphism. These conditions ensure that, for any fixed $v \in G$, the maps $u \mapsto uv$ and $u \mapsto vu$ are homeomorphisms with inverse maps $u \mapsto uv^{-1}$ and $u \mapsto v^{-1}u$, respectively. Each of these maps will carry a neighborhood system at the identity e into a neighborhood system at v. The inverse maps have the inverse effect. Thus the topology of a topological group is completely determined by the neighborhoods of e and can be described by giving a neighborhood base at e. An abstract description of such a neighborhood base is given in Hewitt and Ross [1963], Theorem 4.5. Many properties of topological groups can be proved in terms of these bases. A subset U in a group G is called *symmetric* if it satisfies $U = U^{-1}$, where U^{-1} is just $\{u^{-1} : u \in U\}$. It is useful to know that any neighborhood V of e contains a symmetric neighborhood e.g., $V \cap V^{-1}$. Hence a base of neighborhoods at e may always be chosen to consist of symmetric neighborhoods.

Space forbids extensively developing the fundamentals of the theory of topological groups. We shall assume the most basic properties tacitly since they are almost self-evident. Complete proofs of all these facts, and some less basic ones, will be found in Hewitt and Ross [1963], Sections 4 through 8. For less obvious facts, we will give references. In particular, we assume the following results which are more or less obvious: the closure of a subgroup is a subgroup which is abelian or normal if the original subgroup has the corresponding property; the coset space $G/N = \{uN : u \in G\}$ of a closed normal subgroup N of a topological group G can be given the structure of a topological group in such a way that the natural map $G \to G/N$ is a continuous open homomorphism; the kernel of a continuous homomorphism is a closed normal subgroup; the product $US = \{uv : u \in U, v \in S\}$ of an open subset U with any subset S is open; $\bigcup_{n=1}^{\infty} U^n$ is an open subgroup if U is a symmetric neighborhood of e.

Here are some slightly less evident basic facts with references to Hewitt and Ross [1963]. If K is a compact subset and C is a closed subset, then KC is closed (Theorem 4.4); if V is a neighborhood of e, then there is a neighborhood U of e with $\overline{U} \subseteq V$ (Corollary 4.7) and if K is a compact subset, then there is a neighborhood U of e satisfying $uUu^{-1} \subseteq V$ for all $u \in K$ (Theorem 4.9); a subgroup is open if and only if it has nonvoid interior, and open subgroups are closed (Theorem 5.5). If $\varphi: G \to H$ is a continuous homomorphism with the closed subgroup N as kernel, then there is a continuous isomorphism of G/N onto H. This need not be a homeomorphism. For example, consider the identity map of $\mathbb{R}_d$ (the additive group of real numbers with the discrete topology) onto $\mathbb{R}$ (the same group with its usual topology).

1.9.6 Locally Compact Groups A *locally compact group* is a topological group which is locally compact as a topological space. Such groups have very special properties which we will briefly discuss later. However, to begin with, we only need the definition. Note that any group considered as a discrete topological group is a locally compact group. The additive group $\mathbb{R}$ and the multiplicative group $\mathbb{T} = \{\zeta \in \mathbb{C} : |\zeta| = 1\}$ are nondiscrete locally compact groups under their usual topologies. Also, any closed multiplicative group G of $n \times n$ real or complex matrices is a locally compact group under the relative topology it inherits as a subset of $\mathbb{R}^{n^2}$ or $\mathbb{C}^{n^2}$.

For a nondiscrete locally compact group G, we can of course construct $\ell^1(G)$, but it ignores the topology. (In fact in this case we will consider $\ell^1(G)$ to be the group algebra of G_d — the discrete group which is isomorphic to G as a group.) In order to find a natural group algebra which involves the topology of G, we first note that $\ell^1(G)$ is isometrically isomorphic to the dual space of the space $c_0(G)$ (the space of all complex-valued functions on G vanishing at infinity, $i.e.$, those functions $f; G \to \mathbb{C}$ such that for every $\varepsilon > 0$ there is a finite set off which the values of f are less than ε in absolute value). The duality is implemented by

$$g \mapsto \left(f \mapsto \sum_u f(u)g(u) \right) \qquad \forall\, g \in \ell^1(G);\ f \in c_0(G). \tag{10}$$

The following general considerations allow us to define a convolution product in a much wider context. (For historical comments on the development given here, see Hewitt and Ross [1963], §19 Notes].)

1.9.7 Convolution on Dual Spaces Let G be a group and let $f: G \to \mathbb{C}$ be a function. For any element $u \in G$, we may define two new functions $_uf: G \to \mathbb{C}$ and $f_u: G \to \mathbb{C}$ by

$$_uf(v) = f(u^{-1}v) \quad \text{and} \quad f_u(v) = f(vu^{-1}) \qquad \forall\, v \in G. \tag{11}$$

We also define a function $\tilde{f}: G \to \mathbb{C}$ by

$$\tilde{f}(u) = f(u^{-1}) \qquad \forall\, u \in G. \tag{12}$$

It is easy to check that these definitions satisfy

$$\begin{aligned} _u(_vf) &= \ _{uv}f,\ (f_u)_v = f_{uv},\ \tilde{\tilde{f}} = f \\ (_uf)^\sim &= \ (\tilde{f})_{u^{-1}} \quad \text{and} \quad (f_u)^\sim = \ _{u^{-1}}(\tilde{f}) \end{aligned} \tag{13}$$

for all $u,\, v \in G$. The function $\langle\, _uf\, /\, f_u\, \rangle$ is called the $\langle$ *left / right* $\rangle$ *translate of f by u*, and the function $\tilde{f}$ is called the *reverse of f*. (Unfortunately there is no standard universally accepted notation for these important concepts. Although our main reference for locally compact groups is Hewitt and Ross [1963], we have reluctantly chosen notation which is at odds with the notation used there.) For the given function $f: G \to \mathbb{C}$ we also define a function $\overline{f}: G \to \mathbb{C}$ by

$$\overline{f}(u) = f(u)^* \qquad \forall\, u \in G \tag{14}$$

where $f(u)^*$ is simply the complex conjugate of $f(u)$. (The notation f^* would be more natural here, and we use this notation in some other contexts for the function defined by conjugating the values. However, it will be convenient to reserve f^* for another concept when we deal with most convolution algebras of functions on groups.)

Let G be a group and let $\mathcal{X}$ be a linear space of complex-valued functions on G. Let $\mathcal{A}$ be a linear space of linear functionals on $\mathcal{X}$. We will define a very general notion of convolution on $\mathcal{A}$ when $\mathcal{A}$ and $\mathcal{X}$ satisfy certain conditions. First we assume that $\mathcal{X}$ is closed under left translation:

$$_uf \in \mathcal{X} \qquad \forall\, f \in \mathcal{X}; u \in G. \tag{15}$$

For any $\omega \in \mathcal{A}$ and $f \in \mathcal{X}$, we define a function ${}^\omega f \colon G \to \mathbb{C}$ by

$$ {}^\omega f(u) = \omega({}_{u^{-1}}f) \qquad \forall\, u \in G. \tag{16}$$

Second we assume

$$ {}^\omega f \in \mathcal{X} \qquad \forall\, f \in \mathcal{X};\; \omega \in \mathcal{A}. \tag{17}$$

Note that ${}^\omega f(e) = \omega(f)$ shows that $\omega \mapsto (f \mapsto {}^\omega f)$ is a linear injection of $\mathcal{A}$ into $\mathcal{L}(\mathcal{X})$. Given any two linear functionals $\omega,\, \tau \in \mathcal{A}$, we define their convolution product to be the function $\omega * \tau \colon \mathcal{X} \to \mathbb{C}$ given by

$$ \omega * \tau(f) = \omega({}^\tau f) \qquad \forall\, f \in \mathcal{X}. \tag{18}$$

Clearly $\omega * \tau$ is a linear functional, and our third assumption is:

$$ \omega * \tau \in \mathcal{A} \qquad \forall\, \omega,\, \tau \in \mathcal{A}. \tag{19}$$

A routine calculation now gives

$$ {}^{\omega * \tau}f = {}^\omega({}^\tau f) \qquad \forall\, \omega,\, \tau \in \mathcal{A};\; f \in \mathcal{X}. \tag{20}$$

This implies that $\mathcal{A}$ is an algebra under convolution multiplication and that $\omega \mapsto (f \mapsto {}^\omega f)$ is an injective algebra homomorphism of $\mathcal{A}$ into $\mathcal{L}(\mathcal{X})$.

Retaining our previous assumptions, let us further assume that $\mathcal{X}$ is a normed linear space under the supremum norm:

$$ \|f\|_\infty = \sup\{|f(u)| : u \in G\} $$

and that $\mathcal{A}$ is included in the space $\mathcal{A}^*$ of continuous linear functionals on $\mathcal{X}$. It follows immediately that any $f \in \mathcal{X}$, $u \in G$ and $\omega \in \mathcal{A}$ satisfy:

$$ \|{}_uf\| = \|f\| \quad \text{and} \quad \|{}^\omega f\| \le \|\omega\|\,\|f\|. $$

From this we conclude

$$ \|\omega * \tau\| \le \|\omega\|\,\|\tau\| \qquad \forall\, \omega,\, \tau \in \mathcal{A} \tag{21}$$

so that the norm is submultiplicative and $\mathcal{A}$ is a normed algebra.

If we apply the above theory to $\mathcal{X} = c_0(G)$ and $\mathcal{A} = c_0(G)^*$ we obtain a Banach algebra structure on $c_0(G)^*$ which is isometrically isomorphic to the convolution algebra $\ell^1(G)$ defined in §1.9.2. The isomorphism is given by the isometric linear isomorphism of $\ell^1(G)$ onto $c_0(G)^*$ given in (10) above.

1.9.8 The Group Measure Algebra $M(G)$ If G is an arbitrary locally compact group, then the Banach space $C_0(G)$ of all continuous functions from G to $\mathbb{C}$ which vanish at infinity is an obvious candidate to serve in the role of $\mathcal{X}$. This choice of $\mathcal{X}$ with the choice $\mathcal{A} = \mathcal{X}^*$ satisfies all of the assumptions made above. This is easily checked when it is noted that for each function f in the dense set of continuous functions which vanish off a (variable) compact set the map $u \mapsto {}_u f$ of G into $C_0(G)$ is continuous. Hence $C_0(G)^*$ is a Banach algebra under convolution. For the rest of this example we will assume that G is locally compact and that $C_0(G)^*$ has the Banach algebra structure just defined.

It is customary to use the Riesz representation theorem (Rudin [1974] Theorem 6.19 or Hewitt and Ross [1963], Chapter 3) to identify $C_0(G)^*$ with the Banach space $M(G)$ of all complex regular Borel measures on G, and then to call this convolution algebra *the measure algebra of G*. Of course, a measure $\mu \in M(G)$ corresponds to the linear functional

$$f \mapsto \int_G f \, d\mu \qquad \forall \, f \in C_0(G). \tag{22}$$

When this identification is made, the convolution product takes on the attractive form

$$\int_G f \, d\mu * \nu = \int_G \int_G f(uv) d\mu(u) \, d\nu(v) \qquad \forall \, f \in C_0(G) \tag{23}$$

where μ and ν are any two measures in $M(G)$. The norm on $M(G)$, which agrees with the norm on $C_0(G)^*$ under the above identification, is just

$$\|\mu\| = |\mu|(G) \qquad \forall \, \mu \in M(G)$$

where $|\mu|$ is the total variation positive measure associated with the complex measure μ. Note that for a discrete group G, the Radon–Nikodym theorem identifies $M(G)$ with $\ell^1(G)$, in such a way as to make our present discussion a direct generalization of that given in §1.9.2.

Since the present discussion uses no measure theory, we will often continue to denote our convolution algebra by $C_0(G)^*$ and think of its elements as continuous linear functionals on $C_0(G)$. We believe this approach is conceptually simpler. However, in other sections of the book we will usually adopt common practice and write $M(G)$ for this algebra. We note that much of any proof of the Riesz representation theorem is algebraic in nature and makes sense in a wider context than just that of algebras of continuous functions. This will be explored further in Volume II of this work.

Next we wish to prove that $G \mapsto C_0(G)^*$ is a covariant functor. In order to do this, we must construct a continuous algebra homomorphism

$$\overline{\varphi}: C_0(G)^* \to C_0(H)^*$$

for each continuous group homomorphism $\varphi: G \to H$, where G and H are locally compact topological groups. It is natural to expect that $\overline{\varphi}$ will be the dual of the map $\check{\varphi}: C_0(H) \to C_0(G)$ defined by

$$\check{\varphi}(f) = f \circ \varphi \qquad \forall \, f \in C_0(H).$$

However, unfortunately, the map $\check{\varphi}$ just defined does not usually map $C_0(H)$ into $C_0(G)$. (Consider φ to be the exponential map $t \mapsto e^{it}$ of $\mathbb{R}$ onto $\mathbb{T} = \{\lambda \in \mathbb{C} : |\lambda| = 1\}$.) At first sight, this appears to be a fatal flaw. Fortunately it is not. The map $\check{\varphi}$ actually maps $C_0(H)$ into $C(G)$. As we noted in §1.5.1, $C(G)$ is the double centralizer algebra of $C_0(G)$. R. Creighton Buck [1958] shows that $C_0(G)^*$ can be canonically identified as a Banach space with the closed subspace of $C(G)^*$ which consists of linear functionals continuous in the strict topology of $C(G)$ with respect to $C_0(G)$ (see Definition 5.1.5). Under this identification, it is now possible to define $\overline{\varphi}$ by the formula

$$\overline{\varphi}(\omega)f = \overline{\omega}(f \circ \varphi) \qquad \forall \, \omega \in C_0(G)^*; \, f \in C_0(H)$$

where $\overline{\omega}$ is ω considered as a strictly continuous linear functional on $C(G)$. Fortunately, all of this agrees with the identification of $C_0(G)^*$ with $M(G)$: the action of a measure $\mu \in M(G)$ on a function $f \in C(G)$, given by formula (22), is just the strictly continuous extension of its action on $C_0(G)$. Then formula (23) shows that the map $\overline{\varphi}$, which we have just defined, is a homomorphism. All this will be established in a wider setting in Volume II. For now, we have proved:

Theorem *Let G and H be locally compact groups. If $\varphi: G \to H$ is a continuous group homomorphism, then $\overline{\varphi}: M(G) \to M(H)$ is a contractive algebra homomorphism where*

$$\int_H f \, d\overline{\varphi}(\mu) = \int_G f \circ \varphi \, d\mu \qquad \forall \, f \in C_0(H); \, \mu \in M(G).$$

Equivalently, $\overline{\varphi}(\mu)(E) = \mu(\varphi^{\leftarrow}(E))$ for each Borel subset $E \subseteq H$, when we think of $\mu \in M(G)$ as a measure.

If K is another locally compact group and $\psi: H \to K$ is another continuous group homomorphism, then they satisfy $\overline{\psi \circ \varphi} = \overline{\psi} \circ \overline{\varphi}$.

1.9.9 Recovering G from $M(G)$ We now show that the abstract Banach algebra $C_0(G)^* \simeq M(G)$ determines G up to homeomorphic isomorphism. Thus the functor $G \mapsto C_0(G)^*$ provides a complete set of invariants for locally compact groups.

For each $u \in G$, let δ_u be the corresponding evaluation linear functional in $C_0(G)^*$ defined by

$$\delta_u(f) = f(u) \qquad \forall\, f \in C_0(G). \tag{24}$$

Then it is easy to see that the map

$$u \mapsto \delta_u \qquad \forall\, u \in G$$

is an injective group homomorphism of G into the group of invertible elements of $C_0(G)^*$ under convolution multiplication. It is injective since $C_0(G)$ separates the points of a locally compact space G. The convolution algebra $\ell^1(G)$ is obviously isometrically isomorphic to the closed linear span of $\{\delta_u : u \in G\}$ in $C_0(G)^*$ under the map

$$f \mapsto \sum_{u \in G} f(u)\delta_u \qquad \forall\, f \in \ell^1(G).$$

(Note that this gives a more concrete interpretation of formula (6) above.) Since G can be embedded in $C_0(G)^*$, it is obvious that G is abelian (*i.e.*, commutative) if $C_0(G)^*$ is commutative. The reverse implication holds as we will soon see, or the reader may consult Hewitt and Ross [1963], Theorem 19.6, for a proof based on Fubini's Theorem.

Now suppose we are presented with $\mathcal{A} \simeq C_0(G)^* \simeq M(G)$ as an abstract Banach algebra and wish to recover the topological group G from it. The extreme points of the unit ball are just the elements of the form $\zeta\delta_u$ for $\zeta \in \mathbb{T}$ and $u \in G$. This is easy to see when considering $M(G)$. Hence the set of extreme points of $\mathcal{A}_1$ form a group which is isomorphic to $\mathbb{T} \times G$. When starting from an abstract Banach algebra, in order to get a specific isomorphism, and thus identify G as a subgroup, we must find an element γ in the Gelfand space of $\mathcal{A}$. In the case of $M(G)$, this can be thought of as the map $\gamma(\mu) = \int_G d\mu$. In terms of γ, we define $G_\mathcal{A}$ to be the subgroup of $\mathcal{A}_G$ consisting of those extreme points which satisfy $\gamma(a) = 1$. We note that, as a Banach space, $\mathcal{A}$ is a dual space (namely of $C_0(G)$). Furthermore the product in $\mathcal{A}$ is separately continuous with respect to both variables in this topology. When $G_\mathcal{A}$ is provided with the relativized weak* topology, it is a locally compact group which is homeomorphically isomorphic to G. Hence we have established a procedure for recovering G as a topological group from the abstract Banach algebra $\mathcal{A}$.

Obviously the above procedure can be used to characterize those Banach algebras which are isomorphic to $M(G)$ for some locally compact group

G. This was done by Roger Rigelhof in his thesis [1969b], but it seems likely that further improvements may be possible. Proofs or references can be found there for remarks in the previous paragraph which need proof. In [1965] and [1966] Marc A. Rieffel characterized the group algebras of commutative and finite groups, respectively. For commutative groups the characterization makes heavy use of the order structure on the group algebra. Frederick P. Greenleaf [1965] characterized the group algebras of compact groups. This characterization also relies on the special nature of the group in an essential way and involves the existence of minimal ideals (see Chapter 8).

The Banach algebra $C_0(G)^*$ is so large and complicated that it is poorly understood. Hence another smaller Banach algebra, which is less natural than $C_0(G)^*$ in a number of respects, is usually taken as the group algebra of a nondiscrete locally compact group G. This is the L^1-group algebra. In order to define it we need to introduce Haar measure.

1.9.10 The Haar Functional Let G be a locally compact group and let $C_{00}(G)$ be the linear space (under pointwise operations) of all continuous functions f from G to $\mathbb{C}$ which have compact support, where the support of f is defined by:

$$\mathrm{supp}(f) = \{u \in G : f(u) \neq 0\}^-.$$

The cone of non-negative real-valued functions in $C_{00}(G)$ will be denoted by $C_{00}(G)_+$. If h is a non-zero function in $C_{00}(G)_+$, it is natural to think of its translates $_uh$ for $u \in G$ as all having the same size as h. Hence the size of a linear combination

$$\sum_{k=1}^{n} t_k \,_{u_k} h \qquad t_k \in \mathbb{R}_+; u_k \in G$$

with non-negative coefficients should be $\sum_{k=1}^{n} t_k$ times the size of h. Using this idea, the order properties of $C_{00}(G)_+$ and an elaborate limiting procedure which allows the comparison function h to be taken with smaller and smaller support, it is possible to find a linear functional on $C_{00}(G)$ which precisely measures the notion of size crudely described above. We refer the reader to Hewitt and Ross [1963], §15 Notes, for the long and interesting history of these ideas. (Attractive proofs of the theorem are also given in: Glen E. Bredon [1963], Erik M. Alfsen [1963] and Leopoldo Nachbin [1965].) Since the decisive step in generalization was taken by Alfred Haar [1933], the resulting linear functional is called the *left Haar functional* or *left Haar integral*. We state its properties formally.

Theorem *Let G be a locally compact group. There exist maps $\Lambda: C_{00}(G) \to \mathbb{C}$ and $\Delta: G \to \mathbb{R}_+^{\bullet}$ satisfying:*

(a) Λ *is linear.*

(b) $\Lambda(f) > 0$ *for all non-zero* $f \in C_{00}(G)_+$.

(c) $\Lambda({}_uf) = \Lambda(f)$ *for all* $f \in C_{00}(G)$; $u \in G$.

(d) *If* $\Lambda': C_{00}(G) \to \mathbb{C}$ *satisfies* (a), (c) *and*

       (b') $\Lambda(f) \geq 0$ *for all* $f \in C_{00}(G)_+$,

then there is some $c \in \mathbb{R}_+$ *satisfying* $\Lambda' = c\Lambda$.

(e) Δ *is a continuous group homomorphism into the multiplicative group of positive real numbers.*

(f) $\Lambda(f) = \tilde{\Delta}(u)\Lambda(f_u) = \Lambda(\tilde{f}\tilde{\Delta})$ *for all* $f \in C_{00}(G)$; $u \in G$.

Note that (d) asserts that the Haar functional is unique up to multiplication by a positive real number. The homomorphism $\Delta: G \to \mathbb{R}_+^{\bullet}$ is called the *modular function* of G. Its existence follows directly from the uniqueness result in (d). Its continuity depends on the uniform continuity of translation in the supremum norm

$$\|f\|_{\infty} = \sup\{|f(u)| : u \in G\}$$

on $C_{00}(G)$.

If G is a discrete group, then (d) shows that

$$\Lambda(f) = \sum_{u \in G} f(u) \qquad \forall\, f \in C_{00}(G) = \ell_0^1(G)$$

is the Haar functional, so that Δ is the constant function with value 1. The modular function is also the constant function with value 1 if G is abelian (use (f), (c) and the identity $f_u = {}_uf$) or compact (since $\Delta(G)$ is a compact multiplicative subgroup of $\mathbb{R}_+^{\bullet}$). Locally compact groups for which the modular function is identically equal to 1 are said to be *unimodular*.

The uniqueness result (d) shows that if one can guess a functional Λ satisfying (a), (b), and (c), then this is the Haar integral. Thus we easily find

$$\Lambda(f) = \int_{-\infty}^{\infty} f(t)\, dt \qquad \forall\, f \in C_{00}(\mathbb{R})$$

for the additive group $\mathbb{R}$ and

$$\Lambda(f) = \frac{1}{2\pi} \int_0^{2\pi} f(e^{it})\, dt \qquad \forall\, f \in C_{00}(\mathbb{T}) = C(\mathbb{T})$$

for the multiplicative group $\mathbb{T}$. Of course any positive multiple of these linear functionals could also serve as the Haar functional, but in the case of compact groups such as $\mathbb{T}$, it is customary to normalize the Haar integral so that it satisfies $\Lambda(1) = 1$. Hewitt and Ross [1963] have many more examples in §§15, 16.

Property (b) of the theorem shows that Λ is a positive linear functional on $C_{00}(G)$. A standard construction in measure theory, which is carried out in sufficient generality in Hewitt and Ross [1963] §11, associates a unique positive regular Borel measure on G with each positive linear functional on $C_{00}(G)$. We will denote the completion of the measure on G associated with Λ by λ and call it the *(left) Haar measure of G*. We will often continue to emphasize the functional Λ rather than the measure λ, but we record here the basic properties of left Haar measure. Paul R. Halmos [1950] gives a converse (essentially due to Weil [1940]) to this theorem.

Theorem *Let G be a locally compact group. There is a complete, extended real-valued, positive measure λ defined on a σ-algebra $\mathcal{B}_\lambda$ of subsets of G and satisfying:*

(a) $\mathcal{B}_\lambda$ *includes the σ-algebra $\mathcal{B}$ of Borel subsets of G.*

(b) $\lambda(U) > 0$ *for all nonempty open subsets U of G.*

(c) $\lambda(K) < \infty$ *for all compact subsets K of G.*

(d) $\lambda(uS) = \lambda(S)$ *for all $S \in \mathcal{B}_\lambda; u \in G$ where $uS = \{uv : v \in S\}$.*

(e) $\lambda(U) = \sup\{\lambda(K) : K \subseteq U \text{ and } K \text{ compact}\}$ *for all open subsets U of G.*

(f) $\lambda(S) = \inf\{\lambda(U) : S \subseteq U \text{ and } U \text{ open}\}$ *for all $S \in \mathcal{B}$.*

(g) *If λ' is any other measure satisfying* (a), (b), (c), (d), (e), (f), *then there is a positive constant c satisfying*

$$\lambda'(S) = c\lambda(S) \qquad \forall \, S \in \mathcal{B}_\lambda.$$

(h) $\Lambda(f) = \int f \, d\lambda$ *for all $f \in C_{00}(G)$ is a Haar functional.*

1.9.11 The Group Algebra $L^1(G)$ For each locally compact group which we consider, we make a choice once and for all of a particular Haar functional Λ and related Haar measure λ from among all possible positive multiples. We denote by $L^1(G)$ the usual Banach space $L^1(\lambda)$ of (equivalence classes of) functions absolutely integrable with respect to λ with norm

$$\|f\|_1 = \int_G |f| \, d\lambda. \tag{25}$$

It becomes a Banach algebra when convolution multiplication is defined by

$$g * h(u) = \int g(v)h(v^{-1}u) \, dv \qquad \forall \, g, h \in L^1(G); \ u \in G. \tag{26}$$

To prove that the integrand is $\mathcal{B}_\lambda$-measurable and that $g * h$ belongs to $L^1(G)$ requires a careful argument and Fubini's theorem.

We may interpret $L^1(G)$ as a subalgebra, indeed an ideal, in $M(G)$ by considering the Radon–Nikodym theorem. Left Haar measure on G is finite, and hence an element of $M(G)$, if and only if G is compact, but we may always consider the subspace $M_a(G)$ of $M(G)$ consisting of those measures absolutely continuous with respect to Haar measure. The Radon–Nikodym theorem provides a bijective isometric linear isomorphism between this subspace and $L^1(G)$. Calculations then show that this is an algebra isomorphism and the subspace is an ideal. From now on we will regard $L^1(G)$ as an ideal in $M(G)$ under this embedding, whenever that is convenient. See Hewitt and Ross [1963] for proofs.

Table 1, on the next page, records all the standard facts about $L^1(G)$ and $M(G)$ with the conventions and notation we use in this work.

We have already noted that the map $u \mapsto {}_u f$ is continuous from G into $C_0(G)$ for any locally compact group G and any $f \in C_0(G)$. It is frequently useful to know that $u \mapsto {}_u f$ is continuous from G into $L^1(G)$ for any $f \in L^1(G)$. The proof is similar to the one outlined in §1.9.7 for the continuous case but relies on the regularity of Haar measure. A converse was noted by Walter Rudin [1959] and rediscovered by Irving Glicksberg [1973]. This allows $M_a(G) \simeq L^1(G)$ to be defined independent of Haar measure.

Proposition *Let G be a locally compact group. A measure μ in $M(G)$ belongs to $M_a(G)$ if and only if $v \mapsto \mu(v^{-1}E)$ is continuous from G into $\mathbb{C}$) for any Borel set E.*

Proof Suppose μ satisfies the criterion and E is a Borel set with Haar measure zero. Let g be the characterisitic function of a Borel set A with finite measure so that g, and hence $g * \mu$, belong to $L^1(G)$. Then

$$\int_A \mu(v^{-1}E)dv = g * \mu(E) = \int_E g * \mu d\lambda = 0$$

shows that $\mu(v^{-1}E)$ is zero almost everywhere with respect to Haar measure. Continuity assures $\mu(E) = 0$. □

Elsewhere (Palmer, [1973]), we have fully developed another simple construction of the L^1-group algebra which involves no measure theory. Each $g \in C_{00}(G)$ determines an element Λ_g in $C_0(G)^*$ by the formula

$$\Lambda_g(f) = \Lambda(gf) \qquad \forall f \in C_0(G).$$

To see that Λ_g is continuous, we note

$$|\Lambda_g(f)| \leq \Lambda(|gf|) \leq \Lambda(\|f\|_\infty |g|) \leq \Lambda(|g|)\|f\|_\infty$$

for any $f \in C_0(G)$. By considering $\Lambda_g(\bar{g})$, it is easy to see that this is an injection of $C_{00}(G)$ into $C_0(G)^*$. A little more work (involving $g/(|g| + \varepsilon)$) shows that $g \mapsto \Lambda_g$ is an isometry with respect to the norm

$$\|g\|_1 = \Lambda(|g|) \qquad \forall g \in C_{00}(G).$$

Table 1: Formulas for a Locally Compact Group G

For any function $f: G \to \mathbb{C}$ and any $u,\, v \in G$ we define:

$$\tilde{f}(u) = f(u^{-1}); \qquad \overline{f}(u) = f(u)^*;$$

$$_uf(v) = f(u^{-1}v); \qquad f_u(v) = f(vu^{-1}).$$

Then this notation satisfies:

$$_u(\tilde{f}) = \widetilde{f_{u^{-1}}}; \qquad (\tilde{f})_u = \widetilde{_{u^{-1}}f};$$

$$_u(_vf) = {}_{uv}f; \quad _u(f_v) = (_uf)_v; \quad (f_u)_v = f_{uv}.$$

Denote left Haar measure by $d\lambda$, du or dv and the modular function by Δ. For $u \in G$ and $\mu,\, \nu \in M(G)$, define δ_u, $\tilde{\mu}$, $\overline{\mu}$ and $\mu * \nu$ in $M(G)$ by

$$\delta_u(E) = \begin{cases} 1 & u \in E \\ 0 & u \notin E; \end{cases} \quad \tilde{\mu}(E) = \mu(E^{-1}) \quad \overline{\mu}(E) = \mu(E)^* \quad \forall \text{ Borel sets } E;$$

$$\int h\, d\mu * \nu = \int h(uv)d\mu(u)d\nu(v) \qquad \forall\ h \in C_0(G).$$

Then any $f,\, g \in L^1(G)$, $u,\, v \in G$, $\mu,\, \nu \in M(G)$ and $h \in C_0(G)$ satisfy:

$$\int {}_uf d\lambda = \int f d\lambda; \quad \int f_u d\lambda = \Delta(u) \int f d\lambda; \quad \tilde{\Delta} = \Delta^{-1}$$

$$\int \tilde{f}\tilde{\Delta} d\lambda = \int f d\lambda; \quad \int \tilde{f} d\lambda = \int f \tilde{\Delta} d\lambda \text{ (when defined)}.$$

So right Haar measure is given by $\tilde{\Delta}\, d\lambda$.

$$\delta_u * f = {}_uf; \quad f * \delta_u = \tilde{\Delta}(u)f_u; \quad \tilde{\delta}_u = \delta_{u^{-1}}; \quad \delta_{uv} = \delta_u\delta_v;$$

$$\mu * f(u) = \int f(v^{-1}u)d\mu(v); \qquad \mu * f = \int {}_uf d\mu(u);$$

$$f * \mu(u) = \int \Delta(v^{-1})f(uv^{-1})d\mu(v); \qquad f * \mu = \int \tilde{\Delta}f_u d\mu(u);$$

$$
\begin{aligned}
f * g(u) &= \int f(v)g(v^{-1}u)dv & f * g &= \int f(u)\, _ug du \\
&= \int f(uv)g(v^{-1})dv & &= \int f_{u^{-1}}\, \tilde{g}(u)du \\
&= \int \Delta(v^{-1})f(uv^{-1})g(v)dv & &= \int \Delta(u)f_u\, g(u)du \\
&= \int \Delta(v^{-1})f(v^{-1})g(vu)du; & &= \int \tilde{\Delta}(u)\tilde{f}(u)\, _{u^{-1}}g du.
\end{aligned}
$$

For $h \in C_0(G)$, $k \in L^\infty(G)$, $\mu,\, \nu \in M(G)$ and $f,\, g \in L^1(G)$ define

$$\omega_h(\mu) = \int \tilde{h} d\mu \quad \text{and} \quad \omega_k(f) = \int \tilde{k} f d\lambda.$$

Then this notation satisfies:

$$\omega_h(\mu * \nu) = \omega_{\nu * h}(\mu) = \omega_{h * \Delta\mu}(\nu); \qquad \omega_k(f * g) = \omega_{g * k}(f) = \omega_{k * \Delta f}(g).$$

Then $L^1(G)$ is just the completion of $C_{00}(G)$ with respect to this norm. It may also be identified with the closure of $\Lambda_{C_{00}(G)}$ in $C_0(G)^*$.

To see the way in which this embedding is related to the multiplication on $C_0(G)^*$ we need a technical result on Λ. Let K be a compact subset of G and let $h \in C_{00}(G)_+$ be a function satisfying $h(u) = 1$ for all $u \in K$. Any $f \in C_{00}(G)$ with support in K satisfies

$$|\Lambda(f)| \leq \Lambda(|f|) \leq \Lambda(\|f\|_\infty h) \leq \Lambda(h)\|f\|_\infty,$$

where $\Lambda(h)$ depends only on K and not on the particular choice of f. Let us call a linear functional Λ' on $C_{00}(G)$ *tame* if for each compact K there is a constant B_K satisfying

$$|\Lambda'(f)| \leq B_K\|f\|_\infty \qquad \forall\, f \in C_{00}(G) \quad \text{with} \quad \text{supp}(f) \subseteq K.$$

In addition to Λ, the restriction to $C_{00}(G)$ of each $\omega \in C_0(G)^*$ is obviously tame. Tame linear functionals satisfy a weak form of Fubini's theorem which is strong enough for our present purposes. The simple proof is based on using the Stone–Weierstrass theorem to show that any function f in $C_{00}(G \times G)$ can be uniformly approximated by expressions of the form

$$\sum_{j=1}^n f_j(u)g_j(v) \quad \text{for} \quad n \in \mathbb{N} \quad \text{and} \quad f_j, g_j \in C_{00}(U)$$

where U is an open set with compact closure related to the support of f. Armed with this result, we can now define and calculate all the properties of $L^1(G)$ as an ideal in $C_0(G)^*$.

A large number of results are known for G or $L^1(G)$, both when G is abelian and when it is compact. We have already noted that G is unimodular in both cases. In §3.2.6 we note the easy fact that G is a [MAP] group when it is either compact or abelian. $L^1(G)$ is strongly semisimple (Section 4.5) whenever G belongs to [MAP]. The Proposition in §5.1.9 shows that G has small invariant neighborhoods (or equivalently, $L^1(G)$ has a central approximate identity) in both these cases. In §7.4.10 we will note that G belongs to [FIA]$^-$ and hence $L^1(G)$ is completely regular under either hypothesis. Finally, in Volume II we will show that G is amenable if it is either abelian or compact. The relations between these and many other similar classes of locally compact group are studied in detail in Palmer [1978]. Some of these results are also given in Volume II.

1.9.12 Functorial Properties of $L^1(G)$ The functorial properties of $L^1(G)$ are rather meager. Let G and H be locally compact groups and let $\varphi\colon G \to H$ be a continuous homomorphism. In §1.9.8 above, we constructed a contractive algebra homomorphism $\overline{\varphi}\colon M(G) \to M(H)$ and this, of course, carries $L^1(G)$ into $M(H)$. An easy computation gives the meaning of $\overline{\varphi}(f)$

for $f \in L^1(G)$ when $\varphi: G \to H$ is the continuous homomorphism $\langle \, \mathbb{Z} \to \mathbb{R}$ / $\mathbb{R} \to \mathbb{R} \times \mathbb{R}$ / $\mathbb{R}_d \to \mathbb{R} \, \rangle$ defined by $\langle \, n \mapsto n$ / $t \mapsto (t,0)$ / $t \mapsto t \, \rangle$. The image of $L^1(G)$ has trivial intersection with $L^1(H)$ in each of these three cases. Good results seem to depend on having φ open as well as continuous. We shall discuss this situation under two cases: (a) G is an open subgroup of H; (b) H is the quotient G/N of G with respect to some closed normal subgroup. In the first case, it is easy to see that $C_{00}(G)$ is embedded in $C_{00}(H)$ by extending functions in $C_{00}(G)$ to be zero on all of H outside G. Furthermore, the restriction of the Haar functional on $C_{00}(H)$ to $C_{00}(G)$ is obviously a Haar functional on $C_{00}(G)$. Hence $L^1(G)$ is naturally isometrically isomorphically embedded in $L^1(H)$.

The second case is a little more complicated. If N is any closed subgroup of G, N has its own Haar functional Λ^N. Hence for each $f \in C_{00}(G)$, we can define a function

$$\varphi_N^*(f)(uN) = \Lambda_v^N f(uv) \qquad \forall \, uN \in G/N \tag{27}$$

where the subscript v on Λ^N indicates that Λ is acting on $f(uv)$ considered as a function of $v \in N$ only. The left invariance of Λ^N shows that $\varphi_N^*(f)$ is well defined on the topological space $G/N = \{uN : u \in G\}$. Let $\varphi_N: G \to G/N$ be the natural (continuous and open) quotient map. Then the support of $\varphi_N^*(f)$ is included in the compact set $\varphi_N(\text{supp}(f))$ and an easy argument shows that $u \mapsto \Lambda_v^N f(uv)$ is a continuous map on G so that $\varphi_N^*(f)$ is continuous on G/N, i.e., $\varphi_N^*(f)$ belongs to $C_{00}(G/N)$.

Suppose now that N is a closed normal subgroup so that G/N is a locally compact group with Haar functional $\Lambda^{G/N}$. The functional

$$f \mapsto \Lambda^{G/N} \varphi_N^*(f) = \Lambda_{uN}^{G/N} \Lambda_v^N f(uv) \qquad \forall \, f \in C_{00}(G)$$

satisfies the hypotheses of (d) in the first theorem in §1.9.10 so that it is a multiple of Haar functional on $C_{00}(G)$. By a suitable normalization, we have Weil's formula [1940], 2nd ed., pp. 42-45:

$$\Lambda^G(f) = \Lambda_{uN}^{G/N} \Lambda_v^N f(uv) \qquad \forall \, f \in C_{00}(G) \quad \text{or} \tag{28}$$

$$\int_G f(u)du = \int_{G/N} \int_N f(uv)dv \, d(uN) \qquad \forall \, f \in C_{00}(G).$$

We can now state the main result for case (b).

Theorem *Let G be a locally compact group and let N be a closed normal subgroup. Let the Haar functionals of G, N and G/N be normalized so that Weil's formula holds. Let $\varphi_N: G \to G/N$ be the natural map and let $\varphi_N^*: C_{00}(G) \to C_{00}(G/N)$ be as defined in (27). Then φ_N^* induces a surjective algebra homomorphism $\overline{\varphi_N^*}$ of $L^1(G)$ onto $L^1(G/N)$ which establishes an isometric isomorphism between $L^1(G)/\ker(\overline{\varphi_N^*})$ and $L^1(G/N)$.*

To prove this, we state a lemma from which the theorem follows by the most general abstract considerations.

Lemma *Under the hypotheses of the theorem, φ_N^* maps $C_{00}(G)$ onto $C_{00}(G/N)$ and satisfies*

$$\|\varphi_N^*(f)\|_1 = \min\{\|g\|_1 : \varphi_N^*(g) = \varphi_N^*(f)\}.$$

Proof For any $k \in C_{00}(G/N)$, Hewitt and Ross [1963] 5.24 (b) assert that we can find a compact set K in G so that $\varphi_N(K)$ is the support of k. Choose a function $h \in C_{00}(G)_+$ which is identically equal to 1 on K. Define $f: G \to \mathbb{C}$ by

$$f(u) = \begin{cases} k \circ \varphi_N(u) \dfrac{h(u)}{\varphi_N^*(h) \circ \varphi_N(u)}, & \text{if } u \in \varphi_N^{-1}(\mathrm{supp}(k)) \\ 0, & \text{if } u \notin \varphi_N^{-1}(\mathrm{supp}(k)). \end{cases}$$

Then f is obviously continuous on $\varphi_N^{\leftarrow}(K)$ and is zero on the open complement of this set. Its support is included in $\mathrm{supp}(h)$. Hence f belongs to $C_{00}(G)$ and satisfies $\varphi_N^*(f) = k$. Since h is positive, Weil's formula gives $\|k\|_1 = \|f\|_1$. Weil's formula also shows that φ_N^* is a contraction with respect to the L^1-norms, giving the asserted equality.   □

If G and H are any locally compact groups and $\varphi: G \to H$ is a continuous open homomorphism, then φ can be written as the composition of two homomorphisms $G \to G/\ker(\varphi) \to H$ satisfying the two cases just discussed. Comparing these two cases to the construction of §1.9.8 and then combining them gives the following functorial result.

Theorem *Let G and H be locally compact groups and let $\varphi: G \to H$ be a continuous open group homomorphism. Then there is a contractive algebra homomorphism $\overline{\varphi}: L^1(G) \to L^1(H)$ which is the restriction to $L^1(G)$ of the natural map $\overline{\varphi}: M(G) \to M(H)$ defined by*

$$\int_H f \, d\overline{\varphi}(\mu) = \int_G f \circ \varphi \, d\mu \qquad \forall \, f \in C_0(H); \; \mu \in M(G).$$

If K is another locally compact group and $\varphi: H \to K$ is another continuous open group homomorphism, then $\overline{\psi \circ \varphi}: L^1(G) \to L^1(K)$ satisfies $\overline{\psi \circ \varphi} = \overline{\psi} \circ \overline{\varphi}$.

For a related result see Antoine Derighetti [1978].

1.9.13 Wendel's Theorem Next we give the result of James G. Wendel [1952] which represents $M(G)$ as the double centralizer algebra $\mathcal{D}(L^1(G))$ of $L^1(G)$ whenever G is a locally compact group. For each $\mu \in M(G) \simeq C_0(G)^*$, define $\tilde{L}_\mu$ and $\tilde{R}_\mu$ in $\mathcal{B}(L^1(G))$ by

$$\tilde{L}_\mu f = \mu * f \quad \tilde{R}_\mu f = f * \mu \qquad \forall \, f \in L^1(G)$$

so that $\tilde{L}$ and $\tilde{R}$ are merely the restrictions of the left and right regular representations of $M(G)$ to the ideal $L^1(G)$. Clearly we have $\|\tilde{L}_\mu\| \leq \|\mu\|$ and $\|\tilde{R}_\mu\| \leq \|\mu\|$ for all $\mu \in M(G)$ and we will soon see that equality actually holds in these inequalities.

Theorem 1.2.4 shows that for any double centralizer (L, R) of $L^1(G)$, L and R are bounded operators if the annihilator ideals $L^1(G)_{LA}$ and $L^1(G)_{RA}$ are both zero. We will show more than this: any $f, g \in L^1(G)$ satisfy $\|f\| = \sup\{\|f * g\| : g \in L^1(G)_1\}$ and $\|g\| = \sup\{\|f * g\| : f \in L^1(G)_1\}$. Hence when $\mathcal{D}(L^1(G))$ is given its natural norm

$$\|(L, R)\| = \max\{\|L\|, \|R\|\} \qquad \forall\, (L, R) \in \mathcal{D}(L^1(G)),$$

the embedding of $L^1(G)$ into $\mathcal{D}(L^1(G))$ is isometric. Obviously, the map $\theta: M(G) \to \mathcal{D}(L^1(G))$ defined by $\theta(\mu) = (\tilde{L}_\mu, \tilde{R}_\mu)$ is a contractive homomorphism, but even more is true.

Theorem   *If G is a locally compact group, the map $\theta: M(G) \to \mathcal{D}(L^1(G))$ is an isometric isomorphism of $M(G)$ onto $\mathcal{D}(L^1(G))$.*

Proof It is enough to show that each $(L, R) \in \mathcal{D}(L^1(G))$ can be expressed as $(\tilde{L}_\mu, \tilde{R}_\mu)$ for some $\mu \in M(G)$ satisfying $\|\mu\| \leq \|L\| \leq \|(L, R)\|$.

Let $\mathcal{V}$ be the family of all compact neighborhoods of e in G. Order $\mathcal{V}$ by reverse inclusion (*i.e.*, for $V, W \in \mathcal{V}$ $V \leq W$ means $W \subseteq V$). This makes $\mathcal{V}$ into a directed set. For each $V \in \mathcal{V}$, let e_V be $\lambda(V)^{-1}$ times the characteristic function of V. Example 5.1.9 below shows that the net $\{e_V\}_{V \in \mathcal{V}}$ satisfies:

$$\lim_{V \in \mathcal{V}} e_V * f = f; \quad \lim_{V \in \mathcal{V}} f * e_V = f \qquad \forall\, f \in L^1(G), \tag{29}$$

proving both $\|f\| = \sup\{\|f * g\| : g \in L^1(G)_1\}$ and also $\|g\| = \sup\{\|f * g\| : f \in L^1(G)_1\}$ for all $f, g \in L^1(G)$.

Let $(L, R) \in \mathcal{D}(L^1(G))$ and $f \in L^1(G)$ be given. Here is another consequence of (29) which uses the continuity of L also:

$$Lf = L \lim_{V \in \mathcal{V}} e_V * f = \lim_{V \in \mathcal{V}} L(e_V * f) = \lim_{V \in \mathcal{V}} h_V * f, \text{ where } h_V = L(e_V).$$

Clearly $\{h_V\}_{V \in \mathcal{V}}$ is a net in the closed ball $\|L\| M(G)_1$. Also $\|L\| M(G)_1 = \|L\| C_0(G)_1^*$ is compact in the weak*- or $C_0(G)$-topology by Alaoglu's theorem. Therefore some subnet $\{h_V\}_{V \in \mathcal{W}}$ converges to some $\mu \in \|L\| M(G)_1$ in the $C_0(G)$-topology.

Let f and g be arbitrary elements of $C_{00}(G)$ so that $f * \tilde{g}$ also belongs to $C_{00}(G)$. Given any $\varepsilon > 0$, there is some $V_0 \in \mathcal{W}$ such that $V \geq V_0$ implies

$$\left| \int f * \tilde{g}(u)\, d\mu(u) - \int f * \tilde{g}(u)\, h_V(u)\, du \right| < \varepsilon.$$

By Fubini's theorem we have

$$\left| \int f(v) \int \tilde{g}(v^{-1}u)d\mu(u)dv - \int f(v) \int \tilde{g}(v^{-1}u)h_V(u)du\,dv \right| < \varepsilon.$$

Taking the limit of $V \geq V_0$ in $\mathcal{W}$ gives

$$\left| \int f(v)[\tilde{L}_\mu(g)(v) - L(g)(v)]dv \right|$$

$$= \left| \int f(v)[\int g(u^{-1}v)d\mu(u) - \lim \int g(u^{-1}v)h_V(u)du]dv \right| \leq \varepsilon.$$

Since $f \in C_{00}(G)$ and $\varepsilon > 0$ were arbitrary, we see that $\tilde{L}_\mu(g) = L(g)$ holds for each $g \in C_{00}(G)$. Since both $\tilde{L}_\mu$ and L are continuous and $C_{00}(G)$ is dense in $L^1(G)$, we conclude $L = \tilde{L}_\mu$ where $\|\mu\| = \|L\|$. Furthermore, for any $f, g \in L^1(G)$ we have $R(f) * g = f * L(g) = f * \mu * g = \tilde{R}_\mu(f) * g$. Hence we have shown that (L, R) equals $(\tilde{L}_\mu, \tilde{R}_\mu)$ for some $\mu \in \|L\|M(G)_1$ as we wished to do. $\qquad\square$

Note that the fact that the left centralizer L came from a pair $(L, R) \in \mathcal{D}(L^1(G))$ was only invoked to justify the use of Theorem 1.2.4 to prove that L was continuous. Under the present circumstances, Proposition 5.2.6 below can be used in place of Theorem 1.2.4 to prove the continuity of L. In fact, we have shown that any left centralizer is a bounded linear map of the form $\tilde{L}_\mu$ for some $\mu \in M(G)$. Wendel [1952] used the above result to prove:

Corollary *A map* $: L^1(G) \to L^1(G)$ *is an isometric left centralizer if and only if it has the form*

$$L(f) = \zeta\,_uf \qquad \forall\, f \in L^1(G)$$

for some $u \in G$ *and* $\zeta \in \mathbb{T}$.

From this he concluded:

Theorem *Let* G *and* H *be locally compact groups and let* $T: L^1(G) \to L^1(H)$ *be a contractive algebra isomorphism. Then there is a homeomorphic group isomorphism* $\varphi: H \to G$ *and a continuous group homomorphism* $\chi: H \to \mathbb{T}$ *satisfying*

$$Tf(u) = c\chi(u)f(\varphi(u)) \qquad \forall\, u \in H$$

where c *is the constant value of the ratio* $\lambda_H\varphi(E)/\lambda_G(E)$ *for each measurable set* E *with finite non-zero measure. In particular* G *and* H *are isomorphic as topological groups and* T *is actually an isometry.*

The proof is based on the obvious fact that TLT^{-1} is a left centralizer of $L^1(H)$ when L is a left centralizer of $L^1(G)$. If L is isometric, then it can be shown that TLT^{-1} is also. Thus, with a little work, the previous result gives a map $\varphi \colon G \to H$ and a map $\chi \colon H \to \mathbb{T}$ defined by

$$T\tilde{L}_{\varphi(u)}T^{-1} = \chi(u)\tilde{L}'_u \qquad \forall\, u \in H$$

where $\tilde{L}_{\varphi(u)}$ and $\tilde{L}'_u$ are defined by $\tilde{L}_{\varphi(u)}f = {}_{\varphi(u)}f$ and $\tilde{L}'_u g = {}_u g$ for all $u \in H$, $f \in L^1(G)$ and $g \in L^1(H)$. For details, see the original reference. (Contractivity can be relaxed; see Nigel J. Kalton and Geoffrey V. Wood [1976] and Wood [1983], [1991]. Also see Roger P. Rigelhof [1969a], Adegoke Olubummo [1979].)

The major consequence of this result is that the Banach algebra $L^1(G)$ (with its norm), considered as an abstract Banach algebra, is a complete set of invariants for G. In [1966] Greenleaf described a procedure for recovering the topological group from the abstract Banach algebra $\mathcal{A} = L^1(G)$. First construct the double centralizer algebra $\mathcal{D}(\mathcal{A})$. (We have just seen that $\mathcal{D}(\mathcal{A})$ is isometrically isomorphic to $M(G)$.) Let $\tilde{G}(\mathcal{A})$ be the subgroup of the multiplicative group of $\mathcal{D}(\mathcal{A})$ consisting of those pairs (L, R) in which both L and R are surjective isometries and let γ_0 be some multiplicative linear functional on $\mathcal{D}(\mathcal{A})$. Under the identification of $\mathcal{D}(\mathcal{A})$ with $M(G)$, we have seen that $\tilde{G}(L^1(G))$ is the subgroup $\{\zeta \delta_u : \zeta \in \mathbb{T}, u \in G\}$ and that there is always a multiplicative linear functional. Let $G(\mathcal{A})$ be the subgroup

$$G(\mathcal{A}) = \{u \in \tilde{G}(\mathcal{A}) : \gamma_0(u) = \|u\|\}$$

of $\mathcal{D}(\mathcal{A})$ and endow $G(\mathcal{A})$ with the relativized strict topology (Definition 5.1.5) of $\mathcal{D}(\mathcal{A})$ (induced by its ideal $\mathcal{A}$). Greenleaf [1966] shows that $G(L^1(G))$ is homeomorphically isomorphic to G under the map of G into $M(G)$ given by

$$u \mapsto \gamma_0(\delta_u)^* \delta_u.$$

This construction is then used to characterize those Banach algebras isometrically isomorphic to $L^1(G)$ for some compact group G.

Jyunji Inoue [1971] introduces an interesting group algebra intermediate between $L^1(G)$ and $M(G)$. It is the closure in $M(G)$ of the sum of $L^1(G_\tau)$ where τ varies over all locally compact group topologies stronger than the given topology and $L^1(G_\tau)$ is just the corresponding group algebra. This algebra plays an important role in Joseph L. Taylor's [1973a] structure theory of $M(G)$ for G a locally compact abelian group. See 3.6.17 for related results.

1.9.14 Representations of Topological Groups A *representation V* of a group G on a linear space $\mathcal{X}$ is a group homomorphism of G into the group $\mathcal{L}(\mathcal{X})_G$ of invertibles in $\mathcal{L}(\mathcal{X})$. If $\mathcal{X}$ is normed and the representation is into $\mathcal{B}(\mathcal{X})_G$, it is called a *normed* representation. If G is a topological group, a normed representation V of G on a Banach space $\mathcal{X}$ is said to be $\langle$ *norm / weakly / strongly* $\rangle$ *continuous* if the map $V : G \to \mathcal{B}(\mathcal{X})$ is continuous with respect to the given topology of G and the $\langle$ norm / weak operator / strong operator $\rangle$ topology on $\mathcal{B}(\mathcal{X})$. Note that boundedness and continuity of representations are completely independent, with neither implying the other. Obviously norm continuity implies strong continuity which implies weak continuity. It turns out that norm continuity is too strong a concept to be of much interest. (Except of course when $\mathcal{X}$ is finite dimensional so that the various topologies on $\mathcal{B}(\mathcal{X})$ agree.) As we will show, weak continuity frequently implies strong continuity.

Note that the reconstruction of G from $L^1(G)$ carried out above depended on the representation $V : G \to \mathcal{B}(L^1(G))$ given by

$$V_u(f) = \delta_u * f = {}_uf \qquad \forall \, u \in G; \; f \in L^1(G).$$

This representation is strongly continuous. (To prove this, note that $C_{00}(G)$ is dense in $L^1(G)$ and that $u \mapsto f_u$ is continuous with respect to the supremum norm on $C_{00}(G)$.)

The following results will frequently be useful. Let G be a locally compact group and let $V : G \to \mathcal{B}(\mathcal{X})$ be a representation of G on a Banach space $\mathcal{X}$ such that $u \mapsto V_u z$ is bounded and weakly measurable for any fixed $z \in \mathcal{X}$. (For formal definitions of these almost self explanatory terms, see the appendix to this section where the Bochner integral and its basic properties are also defined and discussed.) For any $f \in L^1(G)$, we may use the Bochner integral to define

$$T_f^V z = \int f(u) V_u z \, du \qquad \forall \, z \in \mathcal{X}$$

as an element of $\mathcal{X}$. (The integrand is almost separably valued since $\mathrm{supp}(f)$ is σ-compact.) Clearly the map T_f^V (*i.e.*, $z \mapsto T_f^V z$) is linear. We now show that it belongs to $\mathcal{B}(\mathcal{X})$ and that the resulting map $T^V : L^1(G) \to \mathcal{B}(\mathcal{X})$ is a weakly continuous representation of $L^1(G)$ on $\mathcal{X}$. The uniform boundedness principle shows that $\{V_u : u \in G\}$ is a bounded set of operators in $\mathcal{B}(\mathcal{X})$. Let B be a bound for V and let $z \in \mathcal{X}$ and $\omega \in \mathcal{X}^*$ be arbitrary. We get

$$\begin{aligned}
|\omega(T_f^V z)| &= \left| \int f(u) \omega(V_u(z)) \, du \right| \le \int |f(u)| \; |\omega(V_u(z))| \, du \\
&\le \|f\|_1 \|\omega\| B \|z\|.
\end{aligned}$$

Weak continuity follows by replacing f by $f - g$ in the above expression. Fubini's Theorem gives

$$\omega(T^V_{f*g}(z)) = \int\int f(v)g(v^{-1}u)dv\,\omega(V_u(z))du$$

$$= \int\int f(v)g(v^{-1}u)\omega(V_u(z))du\,dv$$

$$= \int\int f(v)g(u)\omega(V_{vu}(z))du\,dv$$

$$= \int\int f(v)\omega(V_vV_u(z))dv\,g(u)du$$

$$= \int g(u)\omega(T^V_f V_u(z))du$$

$$= \int g(u)(T^V_f)^*(\omega)(V_u(z))du$$

$$= (T^V_f)^*(\omega)(T^V_g(z)) = \omega(T^V_f T^V_g(z)).$$

For later use, we need to note also that if V is weakly continuous at $e \in G$ (*i.e.*, if V is a continuous function at $e \in G$ into $\mathcal{B}(\mathcal{X})$ with its weak operator topology), then each $z \in \mathcal{X}$ satisfies $z \in \{T^V_f z : f \in L^1(G)\}^-$. If this were to fail, then the Hahn–Banach theorem would give an element $\omega \in \mathcal{X}^*$ satisfying $\omega(z) \neq 0$ but $\omega(T^V_f z) = 0$ for all $f \in L^1(G)$. However, this implies

$$\int f(u)\omega(V_u(z))du = \omega(T^V_f z) = 0 \qquad \forall\, f \in L^1(G)$$

which in turn implies that $\omega(V_u(z))$ is zero almost everywhere and hence is zero at e since it is continuous there. This gives the contradiction $0 \neq \omega(z) = \omega(V_e(z)) = 0$, which establishes $z \in \{T^V_f z : f \in L^1(G)\}^-$. A weak consequence is that $\{T^V_f z : f \in L^1(G), z \in \mathcal{X}\}$ is dense in $\mathcal{X}$. (Actually Theorem 5.2.2 and Example 5.1.9 below show that

$$\{T^V_f z : f \in L^1(G); z \in \mathcal{X}\}$$

equals $\mathcal{X}$ and that each $z \in \mathcal{X}$ may be written as $T^V_f y$ for some $f \in L^1(G)$ and $y \in \{T^V_f z : f \in L^1(G)\}^-$. In fact in the latter expression y may be chosen as close to z as desired and f may be chosen with $\|f\|_1 \leq 1$.)

The definition of the Bochner integral makes it clear that a closed V-invariant subspace of $\mathcal{X}$ is T^V-invariant. In order to prove that a closed T^V-invariant subspace is V-invariant, it seems necessary to make some continuity assumption on V. In the terminology of Definition 4.2.1, these two results will show that V is topologically irreducible if and only if T^V is.

From now on, in addition to the other assumptions made above, we assume that V is continuous with respect to the weak operator topology on $\mathcal{B}(\mathcal{X})$. Then we may use the Bochner integral to define a continuous representation $\overline{T}^V$ of $M(G)$ on $\mathcal{X}$ by

$$\overline{T}^V_\mu z = \int V_u z \, d\mu(u) \quad \mu \in M(G); \qquad \forall \, z \in \mathcal{X}.$$

The fact that this defines a continuous representation of $M(G)$ is proved by calculations similar to those above using results from the appendix as before. Note the identities $T^V_f = \overline{T}^V_{f\lambda}$ (where λ is Haar measure and $f\lambda(B) = \int_B f d\lambda$ for each Borel set B) and $V_u = \overline{T}^V_{\delta_u}$ for all $f \in L^1(G)$ and $u \in G$. The latter identity shows immediately that any $\overline{T}^V$-invariant subspace is V-invariant. Suppose $\mathcal{Y}$ is a closed T^V invariant subspace. We showed above that any $z \in \mathcal{X}$ can be written as $\lim_{n\to\infty} T^V_{f_n} z$ for a suitable sequence $\{f_n\}_{n\in\mathbb{N}}$ in $L^1(G)$. Hence for any $\mu \in M(G)$, we have

$$\overline{T}^V_\mu z = \overline{T}^V_\mu \lim T^V_{f_n} z = \lim T^V_{\mu*f_n} z.$$

Hence if z belongs to any T^V-invariant subspace, then $\overline{T}^V_\mu z$ does also. We have shown:

Theorem *If V is a weakly continuous, bounded representation of a locally compact group G on a Banach space $\mathcal{X}$ such that $u \mapsto V_u z$ is almost separably valued for each $z \in \mathcal{X}$ then V is strongly continuous. Moreover, if $T^V \colon L^1(G) \to \mathcal{B}(\mathcal{X})$ and $\overline{T}^V \colon M(G) \to \mathcal{B}(\mathcal{X})$ are defined by the Bochner integrals*

$$T^V_f z = \int f(u) V_u z \, dx \qquad \forall \, f \in L^1(G); \ z \in \mathcal{X}$$

$$\overline{T}^V_\mu z = \int V_u z \, d\mu(u) \qquad \forall \, \mu \in M(G); \ z \in \mathcal{X}$$

then the following are equivalent for a closed subspace $\mathcal{Y}$ of $\mathcal{X}$:
  (a) *$\mathcal{Y}$ is V-invariant;*
  (b) *$\mathcal{Y}$ is T^V-invariant;*
  (c) *$\mathcal{Y}$ is $\overline{T}^V$-invariant.*

It only remains to show that V is strongly continuous. It is enough to show continuity at e since V is a group homomorphism. Suppose $z \in \mathcal{X}$ and $\varepsilon > 0$ are given. Choose $f \in L^1(G)$ satisfying $\|T^V_f z - z\| < \varepsilon/(2 + 2B)$ (where, as before, B is a bound for $\{\|V_u\| : u \in G\}$). By the continuity of the map $u \mapsto {}_u f$ (Hewitt and Ross [1963], Theorem 20.4), there is a neighborhood N of e such that $u \in N$ implies $\|f - {}_u f\| < \varepsilon/2B$. Then, for any $u \in N$, we have

$$\|V_u z - z\| \leq \|V_u z - V_u T^V_f z\| + \|V_u T^V_f z - T^V_f z\| + \|T^V_f z - z\|$$
$$\leq B\|z - T^V_f z\| + \|T_{{}_u f - f} z\| + \|T^V_f z - z\| < \varepsilon.$$

Hence V is strongly continuous. Note that this conclusion is automatic if V is a unitary representation on Hilbert space since the weak operator and strong operator topologies agree on the set of unitary elements as we will show in Volume II.

Corollary *In the last theorem, the hypothesis that $u \mapsto V_u z$ is almost separably valued is unnecessary if V is assumed strongly continuous or if $\mathcal{X}$ is separable or reflexive.*

Proof We only used the hypothesis to ensure that the Bochner integral was defined. Hence this hypothesis may be dropped if the Banach space $\mathcal{X}$ is reflexive, for then the simpler integral defined in the appendix may be employed. Furthermore, if V is strongly continuous, then the map $u \mapsto V_u z$ will be strongly measurable for each $z \in \mathcal{X}$ with respect to measures in $M(G)$. Thus in this case also the above argument is valid. Finally if $\mathcal{X}$ is separable, the hypothesis mentioned is automatically satisfied.   □

1.9.15 Fourier Algebras, Beurling Algebras and Semigroup Algebras We wish to briefly introduce some variations on group algebras. Fourier algebras and Fourier–Stieltjes algebras will be studied in Volume II. Here we merely mention that they are commutative algebras of complex-valued functions with dual spaces which are convolution algebras. The functions can be interpreted as belonging to certain representation spaces. Pierre Eymard [1964] gave a systematic theory in the noncommutative case. The whole theory can be extended to Kac algebras introduced by L. I. Vaĭnerman and Grigoriĭ Isaakovič Kac [1974], Michel Enock and Jean-Marie Schwartz [1979], John Ernest [1967] and Masamichi Takesaki [1972]. See Jean De Cannière [1979], De Cannière, Enock and Schwartz [1979], De Cannière and Ronny J. E. Rousseau [1984]. See also Martin E. Walter [1986], [1989].

Definition A *weight function* ω on a locally compact group G is a locally bounded, measurable, positive-valued function $\omega : G \to \mathbb{R}_+^\bullet$ satisfying $\omega(e) = 1$ and $\omega(uv) \leq \omega(u)\omega(v)$ for all $u, v \in G$. If ω is a weight function, then the *Beurling algebra on G* defined by ω is the Banach space

$$L^1(G, \omega) = \{ f : G \to \mathbb{C} : \|f\|_\omega = \int_G |f(u)|\omega(u)du < \infty \},$$

provided with convolution multiplication.

It is easy to check that $L^1(G, \omega)$ is the subalgebra of $L^1(G)$ consisting of functions f satisfying $f\omega \in L^1(G)$. It is a Banach algebra in its own norm $\|\cdot\|_\omega$. If we replace ω by $\tilde{\omega}(u) = \inf_{\mathcal{K}} \{ \sup\{ \omega(uv) : v \in N \} : N \in \mathcal{K} \}$ (where $\mathcal{K}$ is the family of compact neighborhoods of e in G), then $\tilde{\omega}$ is an upper semi-continuous weight function with $L^1(G, \tilde{\omega})$ equal to $L^1(G, \omega)$ as a subset of $L^1(G)$ and having an equivalent norm. Hence we may further restrict weight functions to be upper semi-continuous without essential loss of generality.

Beurling algebras share many properties with L^1-group algebras, but their weight functions can be chosen to introduce subtle differences. For instance, Beurling algebras are approximately unital (see Chapter 5) and unital if the group is discrete. If N is a normal subgroup, the quotient algebra induced by the homomorphism $G \to G/N$ is a Beurling algebra on G/N. Arne Beurling essentially introduced these algebras into classical harmonic analysis in [1938]. Yngve Domar [1956] obtained many results for general locally compact abelian groups. Beurling algebras have also been studied on noncommutative locally compact groups. See also Bruce A. Barnes [1971b], Ian G. Craw and Nicholas J. Young [1974], John Liukkonen and Richard Mosak [1977], B. A. Rogozin and M. S. Sgibnev [1980], Niels Grønbaek [1983], Wilfried Hauenschild, Eberhard Kaniuth and Ajay Kumar [1983], William G. Bade, Philip C. Curtis, Jr. and H. Garth Dales [1987]. For other group algebras and related notions, see Satoru Igari and Yuichi Kanjin [1979], John G. Romo [1980a], [1980b], Klaus Hartmann [1981], S. A. Antonyan [1983] and David Gurarie [1984].

If S is a semigroup, we can introduce convolution multiplication on $\ell^1(S)$ by

$$f * g(u) = \sum_{\{v,w \in S \ : \ u=vw\}} f(v)g(w) \qquad \forall \ f,g \in \ell^1(S).$$

This makes $\ell^1(S)$ into a Banach algebra which we will call the *semigroup algebra of S*. This algebra can have quite different properties depending on the structure of S. If S is a unital semigroup (*i.e.*, if it has an identity element e for its multiplication), then δ_e is an identity element for $\ell^1(S)$. Example 4.8.5 below deals with the semigroup algebra of a free semigroup. If $\omega \colon S \to \mathbb{R}_+^*$ is a weight function satisfying the inequality of the last definition, then the Banach algebra $\ell^1(S, \omega)$ can be defined. The very special semigroups which have an analogue of Haar measure have been studied, and for them analogues of $L^1(G)$ and $L^1(G, \omega)$ can be defined. For further information on semigroup algebras see Young [1973b], John F. Berglund, Hugo D. Junghenn and Paul Milnes [1978], [1989], Ghahramani [1980], [1982], [1983], [1989], Lau [1983], Ghahramani and McClure [1990], Bade [1983] Grønbaek [1988], [1989a], Roderick G. McLean and Hans Kummer [1988] John Walter Baker and Ali Rejali [1989] and John Duncan and Alan L. T. Paterson [1990].

1.9.16 Appendix: Integration of Banach Space Valued Functions

In this appendix we will define and state the major results on: (1) a relatively elementary Lebesgue integral for reflexive Banach space valued functions, (2) the Bochner integral, and (3) other closely related matters. Proofs will not be given. They can be found in Hille and Phillips [1957], 3.7–3.9, Yosida [1965], Chapter V, 4 and 5 or Diestel and Uhl [1977], Chapter II. The first and third references contain much more material while the second gives the major results more concisely. Dunford and Schwartz [1958], Chapter 3, and other parts of Diestel and Uhl's book give closely related theories.

Throughout this appendix, let $(S, \mathcal{B}, \mu)$ be a measure space, where μ is either a non-negative extended real-valued, or a complex-valued, countably additive set function with the σ-field $\mathcal{B}$ of subsets of S as domain. Let $\mathcal{X}$ be a Banach space and let $f: S \to \mathcal{X}$ be a function. Then f is said to be *weakly $\mathcal{B}$-measurable* if the numerical valued function $\omega \circ f$ is measurable for each $\omega \in \mathcal{X}^*$.

Let f be a weakly $\mathcal{B}$-measurable function, and assume moreover that $\omega \circ f$ belongs to $L^1(\mu)$ for each $\omega \in \mathcal{X}^*$. We can always define an $\mathcal{X}^{**}$-valued integral of f with respect to μ by the following procedure. Consider the map $W: \mathcal{X}^* \to L^1(\mu)$ defined by $W(\omega) = \omega \circ f$. This is clearly linear and closed. Hence it is bounded by the closed graph theorem. Now consider the map

$$\omega \mapsto \int \omega(f(s)) \, d\mu(s) = \int W(\omega)(s) d\mu(s) \qquad \forall \, \omega \in \mathcal{X}^*.$$

This is also a linear map which must be bounded since W is. Therefore it is an element of $\mathcal{X}^{**}$ which could be denoted by $\int f(s) d\mu(s)$. This integral (which was defined independently by Gelfand [1938] and Dunford [1938]) has most of the usual properties of an integral. Its great defect is that the integral of f belongs to a larger space than the range of f. To remedy this defect, we will only use this integral for reflexive spaces $\mathcal{X}$ where we naturally interpret the integral of f as an element of $\mathcal{X}$ again. The easily obtained elementary properties of this integral are given, for instance, in Hille and Phillips [1957], 3.7.

A function $f: S \to \mathcal{X}$ (where as before $(S, \mathcal{B}, \mu)$ is a measure space and $\mathcal{X}$ is a Banach space) is said to be *μ-simple* if it has only a finite number of values and assumes each non-zero value on a $\mathcal{B}$-measurable set of finite μ-measure. A function $f: S \to \mathcal{X}$ is said to be *strongly μ-measurable* if in the norm of $\mathcal{X}$ it is the pointwise μ-almost everywhere limit of μ-simple functions. A function $f: S \to \mathcal{X}$ is said to be *μ-almost separably valued* if there is a μ-null set E in $\mathcal{B}$ such that $f(S \setminus E)$ is a separable set in the relativized norm topology of $\mathcal{X}$.

Clearly strong μ-measurability would be difficult to check directly. Hence the following theorem of Billy J. Pettis [1938] is important. It is proved in Yosida [1965], V.4.

Theorem *A function $f\colon S \to \mathcal{X}$ is strongly μ-measurable if and only if it is weakly $\mathcal{B}$-measurable and μ-almost separably valued.*

Next we turn to the Bochner integral which is defined in non-reflexive spaces, but is less general than our previous integral when that integral is defined. When both integrals are defined they agree.

A function $f\colon S \to \mathcal{X}$ is said to be *Bochner μ-integrable* if there exists a sequence $\{f_n\}_{n \in \mathbb{N}}$ of μ-simple functions from S to $\mathcal{X}$ which converges in the norm of $\mathcal{X}$ pointwise μ-almost everywhere to f in such a way that it satisfies

$$\lim_{n \to \infty} \int \|f(s) - f_n(s)\| \, d|\mu|(s) = 0.$$

The *Bochner integral* of a μ-simple function $f = \sum_{k=1}^{n} \chi_{E_k} x_k$ (where $E_1, E_2, \ldots, E_n$ is a collection of disjoint sets in $\mathcal{B}$ with finite μ-measure, χ_{E_k} is the characteristic function of E_k and $x_1, x_2, \ldots, x_n$ belong to $\mathcal{X}$) is

$$\int f \, d\mu = \sum_{k=1}^{n} \mu(E_k) x_k.$$

The *Bochner integral* with respect to μ of a Bochner μ-integrable function $f\colon S \to \mathcal{X}$ is

$$\int f \, d\mu = \lim_{n \to \infty} \int f_n \, d\mu$$

where $\{f_n\}_{n \in \mathbb{N}}$ is any sequence of μ-simple functions satisfying the defining property of Bochner integrability. Of course one must check that this integral is well defined since it apparently depends on the choice of a sequence. This is checked in both references given above. The usefulness of this integral depends largely on the following criterion for Bochner integrability first obtained by Salomon Bochner [1933].

Theorem *A function $f\colon S \to \mathcal{X}$ is Bochner integrable if and only if it is strongly $\mathcal{B}$-measurable and satisfies $\int \|f(s)\| d|\mu|(s) < \infty$.*

By combining the last two theorems we can obtain criteria for Bochner integrability which are easily checked in many situations.

The Bochner integral is easily seen to be a linear function of the integrand which satisfies

$$\left\| \int f(s) \, d\mu(s) \right\| \le \int \|f(s)\| d|\mu|(s).$$

Furthermore if $\mathcal{Y}$ is another Banach space, $T\colon \mathcal{X} \to \mathcal{Y}$ is a bounded linear function and $f\colon S \to \mathcal{X}$ is Bochner integrable, then $T \circ f$ is Bochner-integrable and satisfies

$$T\left(\int f(s) \, d\mu(s) \right) = \int T(f(s)) \, d\mu(s).$$

These results and a number of others are proved in our references.

Finally we make a convention. When G is a locally compact group, we will use all of the above terms referring to the measure space $(G, \mathcal{B}, \lambda)$, where $\mathcal{B}$ is the σ-field of Borel sets and λ is left Haar measure without specifying $\mathcal{B}$ and λ. Thus we speak of weakly measurable functions or Bochner integrable functions rather than weakly $\mathcal{B}$-measurable and Bochner λ-integrable functions.

1.10   Tensor Products

A number of the previous examples have involved tensor products of one sort or another. Perhaps the most interesting example is the set $\mathcal{B}_F(\mathcal{X}, \mathcal{Y})$ of bounded finite-rank operators from a normed linear space $\mathcal{X}$ to another normed linear space $\mathcal{Y}$ as presented in Definition 1.1.17. This can be identified with the algebraic tensor product $\mathcal{X} \otimes \mathcal{Y}^*$ defined below. In §1.6.8 we also noted that the matrix algebra M_{nm} of $nm \times nm$-complex matrices could be identified with $M_n \otimes M_m$ even as an algebra.

Here we shall give a more systematic, but still brief, development of this important topic. Other useful discussions can be found in Robert Schatten [1950], [1960], Grothendieck [1954], [1955], Joseph Diestel and John Jerry Uhl, Jr. [1977] Johann Cigler, Viktor Losert and Peter Michor [1979] and Alexandr Yakovlevič Helemskiĭ [1989b], [1993]. The latter three references concentrate on tensor products of Banach modules, a subject which we postpone until we treat cohomology theory. Schatten [1943a], [1943b], [1946] and Schatten and John von Neumann [1946] and [1948] began the study of tensor products of Banach spaces but Grothendieck's work, cited above, profoundly advanced the subject. A number of useful subsequent papers have presented in detail ideas and results which were implicit but far from explicit in his work.

In this section we give basic facts about tensor products, with some emphasis on tensor products of Banach algebras. We usually state our results for n-fold tensor products. Later in this work we will present additional results on tensor products of Banach algebras which involve concepts not yet introduced. See §§3.2.18, 4.2.25, 4.2.26, 5.1.4 and 7.1.19.

Tensor products are a device for reducing multilinear phenomena to linear ones. We are most interested in those tensor products of Banach spaces which are again Banach spaces. The two most important products of this kind are the projective tensor product and the injective tensor product. All other complete tensor products are intermediate between these two. We consider the purely algebraic case first. This is an area in which an unusual number of errors have been published. This probably stems from the fact that many proofs are utterly trivial, but in the Banach space theory some results which seem "obvious" are actually false.

Algebraic Tensor Products

1.10.1 Definition Let $\mathcal{X}^1, \mathcal{X}^2, \ldots, \mathcal{X}^n$ and $\mathcal{Y}$ be linear spaces. A map f of $\mathcal{X}^1 \times \mathcal{X}^2 \times \cdots \times \mathcal{X}^n$ into $\mathcal{Y}$ is called *multilinear* if it satisfies

$$f(x_1, \ldots, x_{k-1}, \lambda x_k' + \mu x_k'', x_{k+1}, \ldots, x_n)$$
$$= \lambda f(x_1, \ldots, x_{k-1}, x_k', x_{k+1}, \ldots, x_n) + \mu f(x_1, \ldots, x_{k-1}, x_k'', x_{k+1}, \ldots, x_n)$$

for all $k \in \{1, 2, \ldots, n\}$, $\lambda, \mu \in \mathbb{C}$, $x_k', x_k'' \in \mathcal{X}^k$ and $x_j \in \mathcal{X}^j$ for $j \neq k$. Denote the set of all multilinear maps of $\mathcal{X}^1 \times \mathcal{X}^2 \times \cdots \times \mathcal{X}^n$ into $\mathcal{Y}$ by

$$\mathcal{L} = \mathcal{L}(\mathcal{X}^1, \mathcal{X}^2, \ldots, \mathcal{X}^n; \mathcal{Y}).$$

An *algebraic tensor product* of $\mathcal{X}^1, \mathcal{X}^2, \ldots, \mathcal{X}^n$ is a pair $(\mathcal{Z}, p)$ consisting of a linear space $\mathcal{Z}$ and a map $p \in \mathcal{L}(\mathcal{X}^1, \mathcal{X}^2, \ldots, \mathcal{X}^n; \mathcal{Z})$ called the *tensor map* such that, if $\mathcal{Y}$ is any linear space and f is any map in $\mathcal{L}(\mathcal{X}^1, \mathcal{X}^2, \ldots, \mathcal{X}^n; \mathcal{Y})$, then there is a unique linear map $h \colon \mathcal{Z} \to \mathcal{Y}$ satisfying $f = h \circ p$. As usual, we will denote $p(x_1, x_2, \ldots, x_n)$ by $x_1 \otimes x_2 \otimes \cdots \otimes x_n$ and call such elements *elementary tensors*.

The universal mapping property of this definition and the uniqueness property in the next theorem can be illustrated by commutative diagrams.

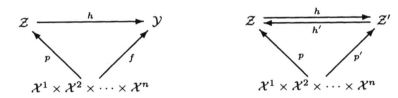

1.10.2 Theorem *If $\mathcal{X}^1, \mathcal{X}^2, \ldots, \mathcal{X}^n$ are linear spaces, then an algebraic tensor product $(\mathcal{Z}, p)$ exists. Each $t \in \mathcal{Z}$ can be written in the form*

$$t = \sum_{j=1}^{m} x_{1,j} \otimes x_{2,j} \otimes \cdots \otimes x_{n,j} \quad where \quad x_{k,j} \in \mathcal{X}^k \tag{1}$$

for some $m \in \mathbb{N}$. Both $\mathcal{Z}$ and $(\mathcal{Z}, p)$ are unique in the sense that if $(\mathcal{Z}', p')$ is another pair satisfying the definition, then $\mathcal{Z}$ and $\mathcal{Z}'$ are linearly isomorphic under an inverse pair of unique linear maps

$$h \colon \mathcal{Z} \to \mathcal{Z}' \text{ and } h' \colon \mathcal{Z}' \to \mathcal{Z} \text{ satisfying } p' = h \circ p \text{ and } p = h' \circ p'. \tag{2}$$

Proof We construct a linear space $\mathcal{Z}$ and a map p of $\mathcal{X}^1 \times \mathcal{X}^2 \times \cdots \times \mathcal{X}^n$ into $\mathcal{Z}$ such that the range of p generates $\mathcal{Z}$ (giving uniqueness) and such that any element of $\mathcal{L}(\mathcal{X}^1, \mathcal{X}^2, \ldots, \mathcal{X}^n; \mathcal{Y})$ factors through p uniquely. Let

$\mathcal{Z}'$ be the huge linear space of formal linear combinations of the elements of the Cartesian product $\mathcal{X}^1 \times \mathcal{X}^2 \times \cdots \times \mathcal{X}^n$. Let $\mathcal{Z}''$ be the linear subspace generated by the set of elements of the form

$$\{ \ (x_1, x_2, \ldots, x_{n-1}, x_k + x'_k, x_{k+1}, \ldots, x_n) - (x_1, x_2, \ldots, x_n)$$
$$-(x_1, x_2, \ldots, x_{k-1}, x'_k, x_{k+1}, \ldots, x_n), \tag{3}$$
$$(x_1, x_2, \ldots, x_{k-1}, \lambda x_k, x_{k+1}, \ldots, x_n) - \lambda(x_1, x_2, \ldots, x_n) :$$
$$\forall \ k \in \{1, 2, \ldots, n\}; \ x_k, x'_k \in \mathcal{X}^k; \ \lambda \in \mathbb{C} \ \}.$$

Let $\mathcal{Z}$ be the quotient $\mathcal{Z}'/\mathcal{Z}''$ and let $p: \mathcal{X}^1 \times \mathcal{X}^2 \times \cdots \times \mathcal{X}^n \to \mathcal{Z}$ be defined by $p(x_1, x_2, \ldots, x_n) = (x_1, x_2, \ldots, x_n) + \mathcal{Z}''$. Denote this element by $x_1 \otimes x_2 \otimes \cdots \otimes x_n$. Hence each element in $\mathcal{Z}$ can be written in the form (1) for suitable $m \in \mathbb{N}$ and $x_{k,j} \in \mathcal{X}^k$. It is easy to see that this construction does provide an algebraic tensor product: given $f \in \mathcal{L}(\mathcal{X}^1, \mathcal{X}^2, \ldots, \mathcal{X}^n; \mathcal{Y})$ as in the definition, we define h by $h(x_1 \otimes x_2 \otimes \cdots \otimes x_n) = f(x_1, x_2, \ldots, x_n)$ which is well defined by multilinearity.

In order to see that an algebraic tensor product is unique up to the natural notion of isomorphism, let $(\mathcal{Z}, p)$ and $(\mathcal{Z}', p')$ both be algebraic tensor products of $\mathcal{X}^1, \mathcal{X}^2, \ldots, \mathcal{X}^n$. Then by Definition 1.10.1 there are unique linear maps satisfying condition (2). The uniqueness of the identity maps $i: \mathcal{Z} \to \mathcal{Z}$ and $i': \mathcal{Z}' \to \mathcal{Z}'$ show $i = h' \circ h$ and $i' = h \circ h'$. Thus $\mathcal{Z}$ and $\mathcal{Z}'$ are isomorphic by a pair of inverse isomorphisms h and h' which intertwine with the tensor maps in the sense that they satisfy (2). This is the proper definition of an isomorphism between the pairs $(\mathcal{Z}, p)$ and $(\mathcal{Z}', p')$. $\qquad\qquad\qquad\qquad\qquad\qquad\qquad\qquad\qquad\qquad\qquad\qquad$ □

From now on we will speak of "the algebraic tensor product" rather than "an algebraic tensor product" and denote $\mathcal{Z}$ by $\mathcal{X}^1 \otimes \mathcal{X}^2 \otimes \cdots \otimes \mathcal{X}^n$. Note that the definition of the algebraic tensor product establishes a linear isomorphism

$$\eta: \mathcal{L}(\mathcal{X}^1, \mathcal{X}^2, \ldots, \mathcal{X}^n; \mathcal{Y}) \to \mathcal{L}(\mathcal{X}^1 \otimes \mathcal{X}^2 \otimes \cdots \otimes \mathcal{X}^n, \mathcal{Y}) \tag{4}$$

where $\eta(f) = h$, in the notation above. Clearly, η is the inverse of the linear isomorphism $g \mapsto g \circ p$ of $\mathcal{L}(\mathcal{X}^1 \otimes \mathcal{X}^2 \otimes \cdots \otimes \mathcal{X}^n, \mathcal{Y})$ onto $\mathcal{L}(\mathcal{X}^1, \mathcal{X}^2, \ldots, \mathcal{X}^n; \mathcal{Y})$.

If $m < n$, it is easy to construct a linear isomorphism

$$\mathcal{X}^1 \otimes \mathcal{X}^2 \otimes \cdots \otimes \mathcal{X}^n \tag{5}$$
$$\simeq (\mathcal{X}^1 \otimes \mathcal{X}^2 \otimes \cdots \otimes \mathcal{X}^m) \otimes (\mathcal{X}^{m+1} \otimes \mathcal{X}^{m+2} \otimes \cdots \otimes \mathcal{X}^n).$$

Hence any algebraic tensor product can be built up as a succession of algebraic tensor products between pairs of linear spaces:

$$\mathcal{X}^1 \otimes \mathcal{X}^2 \otimes \cdots \otimes \mathcal{X}^n \simeq ((\cdots((\mathcal{X}^1 \otimes \mathcal{X}^2) \otimes \mathcal{X}^3) \otimes \cdots) \otimes \mathcal{X}^{n-1}) \otimes \mathcal{X}^n. \tag{6}$$

Because of this construction, it is usually enough to deal with the tensor product of two spaces.

The definition of *bilinear maps* (*i.e.*, multilinear maps defined on the product of two spaces) gives obvious canonical linear isomorphisms

$$\mathcal{L}(\mathcal{X}^1, \mathcal{X}^2; \mathcal{Y}) \simeq \mathcal{L}(\mathcal{X}^1, \mathcal{L}(\mathcal{X}^2, \mathcal{Y})) \simeq \mathcal{L}(\mathcal{X}^2, \mathcal{L}(\mathcal{X}^1, \mathcal{Y})). \tag{7}$$

If $m < n$, this generalizes to the dual expression of (5)

$$\mathcal{L}(\mathcal{X}^1, \mathcal{X}^2, \ldots, \mathcal{X}^n; \mathcal{Y}) \simeq \mathcal{L}(\mathcal{X}^1, \mathcal{X}^2, \ldots, \mathcal{X}^m; \mathcal{L}(\mathcal{X}^{m+1}, \ldots, \mathcal{X}^n; \mathcal{Y}))$$

but the special case (7) is more important.

We need the following simple results.

1.10.3 Lemma *Every non-zero element t in $\mathcal{X}^1 \otimes \mathcal{X}^2$ can be written as*

$$t = \sum_{j=1}^m x_{1,j} \otimes x_{2,j} \tag{8}$$

where $\{x_{k,1}, x_{k,2}, \ldots, x_{k,m}\}$ is a linearly independent set for both $k = 1$ and $k = 2$.

Proof We may assume the expression (1) (with $n = 2$) has been chosen with m minimal. In order to show that this expression satisfies the conditions given for (8), by symmetry, it is enough to show that $x_{1,m}$ is not equal to a linear combination: $x_{1,m} = \sum_{j=1}^{m-1} \lambda_j x_{1,j}$. If this were true, we would have

$$t = \sum_{j=1}^{m-1} x_{1,j} \otimes (x_{2,j} + \lambda_j x_{2,m})$$

which contradicts the minimality of m.　　　　　□

As we have noted in previous examples, there is a linear map of $\mathcal{X} \otimes \mathcal{Y}^\dagger$ into $\mathcal{L}(\mathcal{Y}, \mathcal{X})$ given by

$$x \otimes \omega(z) = \omega(z)x \qquad \forall\, x \in \mathcal{X};\ \omega \in \mathcal{Y}^\dagger;\ z \in \mathcal{Y}. \tag{9}$$

Its range is the set $\mathcal{L}_F(\mathcal{Y}, \mathcal{X})$ of all (not necessarily bounded) finite-rank linear maps. The lemma implies that this map is injective as we will show in the proof of Proposition 1.10.4. There is also a linear map of $\mathcal{X} \otimes \mathcal{Y}^\dagger$ into $\mathcal{L}(\mathcal{X}, \mathcal{Y})^\dagger$ given by

$$x \otimes \omega(T) = \omega(Tx) \qquad \forall\, x \in \mathcal{X};\ \omega \in \mathcal{Y}^\dagger;\ T \in \mathcal{L}(\mathcal{X}, \mathcal{Y}). \tag{10}$$

We wish to highlight a special case of (9).

1.10.4 Proposition *Let $\mathcal{X}$ be a linear space and let $\mathcal{Y}$ be a subspace of $\mathcal{X}^\dagger$. Then $\mathcal{X} \otimes \mathcal{Y}$ is an algebra under the multiplication*

$$(x \otimes y)(z \otimes w) = y(z) x \otimes w \qquad \forall\ x, z, \in \mathcal{X};\ y, w \in \mathcal{Y}.$$

Equation (9) above establishes an algebra isomorphism of $\mathcal{X} \otimes \mathcal{Y}$ onto a subalgebra of the ideal $\mathcal{L}_F(\mathcal{X})$ of not necessarily bounded finite-rank operators on $\mathcal{X}$. If $\mathcal{Y} = \mathcal{X}^\dagger$, the isomorphism is onto $\mathcal{L}_F(\mathcal{X})$.

Proof Injectivity of the map defined in (9) is the only property which may not be clear. We may suppose that any non-zero $t = \sum_{j=1}^{m} x_j \otimes y_j \in \mathcal{X} \otimes \mathcal{Y}$ satisfies the lemma. Since y_1 is not in the linear span of $y_2, y_3, \ldots, y_n$, there is some $z \in \mathcal{X}$ satisfying $y_1(z) \neq 0$ and $y_k(z) = 0$ for $k = 2, 3, \ldots, n$. Thus $t(z) = x_1 \neq 0$.

The proof of Proposition 1.1.18 above shows that $\mathcal{L}_F(\mathcal{X})$ is the range of the isomorphism when $\mathcal{Y} = \mathcal{X}^\dagger$.   □

There is a linear map of $\theta: \mathcal{X}^1 \otimes \mathcal{X}^2 \otimes \cdots \otimes \mathcal{X}^n \to \mathcal{L}(\mathcal{X}^{1\dagger}, \ldots, \mathcal{X}^{n\dagger}; \mathbb{C})$ given on elementary tensors by

$$\theta(x_1 \otimes x_2 \otimes \cdots \otimes x_n)(\omega_1, \omega_2, \ldots, \omega_n) = \prod_{k=1}^{n} \omega_k(x_k) \qquad \forall\ x_k \in \mathcal{X}^k;\ \omega_k \in \mathcal{X}^{k\dagger}.$$

$$(11)$$

The existence of each of these three maps follows immediately from the definition of the algebraic tensor product. We will presently show that they are all injective. Hence they could be used to construct the algebraic tensor product.

We can also write (11) in a dual fashion. If $\omega_k \in \mathcal{X}^{k\dagger}$ for $k = 1, 2, \ldots, n$, we can define $\omega_1 \otimes \omega_2 \otimes \cdots \otimes \omega_n \in \mathcal{L}(\mathcal{X}^1, \mathcal{X}^2, \ldots, \mathcal{X}^n; \mathbb{C})$ by

$$\omega_1 \otimes \omega_2 \otimes \cdots \otimes \omega_n(x_1, x_2, \ldots, x_n) = \prod_{k=1}^{n} \omega_k(x_k) \qquad \forall\ x_k \in \mathcal{X}^k;\ \omega_k \in \mathcal{X}^{k\dagger}.$$

$$(12)$$

This notation is consistent with a variant of equation (4) above.

1.10.5 Proposition *Let $\mathcal{X}^1, \mathcal{X}^2, \ldots, \mathcal{X}^n$ be linear spaces. For each $k = 1, 2, \ldots, n$, let $\mathcal{Y}^k$ be a linear subspace of $\mathcal{X}^{k\dagger}$ which agrees with $\mathcal{Z}^\dagger$ when restricted to any finite-dimensional subspace $\mathcal{Z}$ of $\mathcal{X}^k$. The map*

$$\theta: \mathcal{X}^1 \otimes \mathcal{X}^2 \otimes \cdots \otimes \mathcal{X}^n \to \mathcal{L}(\mathcal{Y}^1, \mathcal{Y}^2, \ldots, \mathcal{Y}^n; \mathbb{C})$$

defined on elementary tensors by (11) is a canonical well defined injective linear map. Hence, for each non-zero tensor $t \in \mathcal{X}^1 \otimes \mathcal{X}^2 \otimes \cdots \otimes \mathcal{X}^n$, there are $\omega_k \in \mathcal{Y}^k$ for $k = 1, 2, \ldots, n$ satisfying $\theta(t)(\omega_1, \omega_2, \ldots, \omega_n) \neq 0$.

Furthermore, if B_k is a Hamel basis for $\mathcal{X}^k$ for $k = 1, 2, \ldots, n$ then $\{x_1 \otimes x_2 \otimes \cdots \otimes x_n : x_k \in B_k; \ k = 1, 2, \ldots, n\}$ is a Hamel basis for $\mathcal{X}^1 \otimes \mathcal{X}^2 \otimes \cdots \otimes \mathcal{X}^n$.

Proof It is obvious from multilinearity that θ is well defined and its injectivity will follow from the next argument.

Since any algebraic tensor product may be built up from successive algebraic tensor products of pairs, we may use the lemma for a proof by induction on n. The case $n = 1$ is obvious (taking the algebraic tensor product of one linear space to be the space itself). Suppose the result has been established for $n - 1$ and write $\mathcal{X}^1 \otimes \mathcal{X}^2 \otimes \cdots \otimes \mathcal{X}^n$ as $\mathcal{Y} \otimes \mathcal{X}^n$ where $\mathcal{Y} = \mathcal{X}^1 \otimes \mathcal{X}^2 \otimes \cdots \otimes \mathcal{X}^{n-1}$. By the lemma, we may write $t = \sum_{j=1}^m t_j \otimes x_{n,j}$ where $\{t_1, t_2, \ldots, t_m\} \subseteq \mathcal{Y}$ and $\{x_{n,1}, x_{n,2}, \ldots, x_{n,m}\} \subseteq \mathcal{X}^n$ are linearly independent. The induction hypothesis gives us $\omega_k \in \mathcal{Y}^k$ for $k = 1, 2, \ldots, n - 1$ with $\theta(t_1)(\omega_1, \omega_2, \ldots, \omega_{n-1})$ non-zero. Choose ω_n to be non-zero at $x_{n,1}$ and zero at $x_{n,j}$ for $j = 2, 3, \ldots, m$.

It is obvious from our construction that the proposed Hamel basis for the algebraic tensor product is a spanning set. The injectivity of θ shows that it is linearly independent and hence a Hamel basis. $\qquad\square$

Because of the injectivity of the map θ we could have defined the algebraic tensor product $\mathcal{X}^1 \otimes \mathcal{X}^2 \otimes \cdots \otimes \mathcal{X}^n$ to be the linear span in $\mathcal{L}(\mathcal{X}^{1\dagger}, \mathcal{X}^{2\dagger}, \ldots, \mathcal{X}^{n\dagger}; \mathbb{C})$ of all the maps defined by (11) for each elementary tensor. We will see below that the analogous construction for Banach spaces leads to a different Banach space tensor product from that given by the universal property of Definition 1.10.1 applied in the category of Banach spaces.

If $T_k \colon \mathcal{X}^k \to \mathcal{Y}^k \quad k = 1, 2, \ldots, n$ is a set of linear maps then the map

$$T_1 \otimes T_2 \otimes \cdots \otimes T_n \colon \mathcal{X}^1 \otimes \mathcal{X}^2 \otimes \cdots \otimes \mathcal{X}^n \to \mathcal{Y}^1 \otimes \mathcal{Y}^2 \otimes \cdots \otimes \mathcal{Y}^n \qquad (13)$$

defined on elementary tensors by

$$T_1 \otimes T_2 \otimes \cdots \otimes T_n(x_1 \otimes x_2 \otimes \cdots \otimes x_n) = T_1(x_1) \otimes T_2(x_2) \otimes \cdots \otimes T_n(x_n) \quad (14)$$

is a well defined linear map. Hence the construction of the algebraic tensor product is a functorial concept. In fact this functor partially preserves exactness (*cf.* Nicolas Bourbaki [1974], II, 3.4).

Algebra Tensor Products

If $\mathcal{A}^{(1)}, \mathcal{A}^{(2)}, \ldots, \mathcal{A}^{(n)}$ are algebras, then we can make their algebraic tensor product $\mathcal{A}^{(1)} \otimes \mathcal{A}^{(2)} \otimes \cdots \otimes \mathcal{A}^{(n)}$ into an algebra. The formal statement will be as long as its proof.

1.10.6　Theorem *Let $\mathcal{A}^{(1)}, \mathcal{A}^{(2)}, \ldots, \mathcal{A}^{(n)}$ be algebras. Their algebraic tensor product $\mathcal{A}^{(1)} \otimes \mathcal{A}^{(2)} \otimes \cdots \otimes \mathcal{A}^{(n)}$ becomes an algebra $\mathcal{A}$ when the* *product of elementary tensors is defined by*

$$(a_1 \otimes a_2 \otimes \cdots \otimes a_n)(b_1 \otimes b_2 \otimes \cdots \otimes b_n) = a_1 b_1 \otimes a_2 b_2 \otimes \cdots \otimes a_n b_n \quad (15)$$

for all a_k, $b_k \in \mathcal{A}^{(k)}$. If each $\mathcal{A}^{(k)}$ is unital, then the map $\varphi_k \colon \mathcal{A}^{(k)} \to \mathcal{A}$ defined by $\varphi_k(a) = 1 \otimes 1 \otimes \cdots \otimes 1 \otimes a \otimes 1 \otimes \ldots \otimes 1$ (with $a \in \mathcal{A}^{(k)}$ in the kth position) is an injective unital algebra homomorphism for each $k = 1, 2, \ldots, n$. These homomorphisms satisfy

$$\varphi_k(a)\varphi_\ell(b) = \varphi_\ell(b)\varphi_k(a) \qquad \forall\; k, \ell = 1, 2, \ldots, n;\; a \in \mathcal{A}^{(k)};\; b \in \mathcal{A}^{(\ell)}. \quad (16)$$

Finally, the $(n+1)$-tuple $(\mathcal{A}, \varphi_1, \varphi_2, \ldots, \varphi_n)$ is uniquely determined up to the natural notion of isomorphism by the following universal mapping property:

> *For any $(n+1)$-tuple $(\mathcal{B}, \psi_1, \psi_2, \ldots, \psi_n)$, consisting of a unital algebra $\mathcal{B}$ and n unital algebra homomorphisms $\psi_k \colon \mathcal{A}^{(k)} \to \mathcal{B}$ satisfying the analogue of (16), there is a unique unital algebra homomorphism $\theta \colon \mathcal{A} \to \mathcal{B}$ which satisfies $\theta \circ \varphi_k = \psi_k$ for each $k = 1, 2, \ldots, n$.*

For $n = 2$, the proposition is illustrated by the following commutative diagrams. The second one shows the isomorphism result.

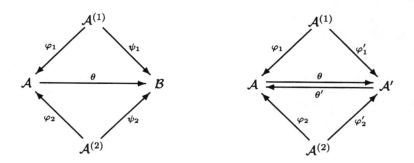

Proof The easiest way to see that the multiplication (15) is well defined is to note that in the explicit construction (in the proof of Theorem 1.10.2) of the algebraic tensor product as a quotient $\mathcal{Z}'/\mathcal{Z}''$, $\mathcal{Z}''$ is an ideal when $\mathcal{Z}'$ is given the obvious multiplication which induces this product. Hence the algebraic tensor product is an algebra under this definition. (The universal mapping property of algebraic tensor products can also be used to give a simple proof that this product is well defined, cf. Kjeld B. Laursen [1969a], p. 470.) Since $1 \otimes 1 \otimes \cdots \otimes 1$ is an identity element, $\mathcal{A}$ is unital. The maps

φ_k are obviously injective unital homomorphisms satisfying (16) so it only remains to establish that the map $h: \mathcal{A} \to \mathcal{B}$ of Definition 1.10.1 is a unital algebra homomorphism θ in the present setting. By induction on n (using (6)), we may restrict attention to the case $n = 2$ where we find

$$
\begin{aligned}
\theta((a \otimes b)(c \otimes d)) &= h(ac \otimes bd) = \psi_1(ac)\psi_2(bd) = \\
&= \psi_1(a)\psi_1(c)\psi_2(b)\psi_2(d) = \psi_1(a)\psi_2(b)\psi_1(c)\psi_2(d) \\
&= \theta(a \otimes b)\theta(c \otimes d) \qquad \forall\, a, c \in \mathcal{A}^{(1)};\ b, d \in \mathcal{A}^{(2)}. \quad \square
\end{aligned}
$$

When the algebraic tensor product is considered as an algebra in this way, we will call it the *algebra tensor product*. (Distinguish this from algebraic tensor product and tensor algebra which have different meanings.) The universal mapping property above indicates that within the category of unital commutative algebras and unital algebra homomorphisms, the algebra tensor product defines a *coproduct*. Note also that $M_n \otimes M_m$, introduced in §1.6.8, is an algebra tensor product and the notation associated with it is consistent with the present notations. If $\mathcal{A}$ is an arbitrary algebra, then the algebra $M_n(\mathcal{A})$ of all $n \times n$ matrices with elements of $\mathcal{A}$ as entries can be identified with $M_n \otimes \mathcal{A}$ as defined above. If $\mathcal{A}$ is unital, the algebra homomorphisms φ_1 and φ_2 are obvious.

It is sometimes useful to replace the universal mapping property displayed above by a purely internal algebraic property:

$\mathcal{A}$ is the linear span of

$$
\{\varphi_1(a_1)\varphi_2(a_2)\cdots\varphi_n(a_n) : a_k \in \mathcal{A}^{(k)};\ k = 1, 2, \ldots, n\}
$$

and for any finite subsets $\{a_{k,1}, a_{k,2}, \ldots, a_{k,m}\} \in \mathcal{A}^{(k)}$ each of which is linearly independent except for $k = \ell$ then

$$
\sum_{j=1}^{m} \varphi_1(a_{1,j})\varphi_2(a_{2,j})\cdots\varphi_n(a_{n,j}) = 0
$$

$$
\text{implies} \quad a_{\ell,1} = a_{\ell,2} = \ldots, a_{\ell,m} = 0.
$$

This condition is satisfied by the tensor product since for each i with $1 \le i \le m$ and each $k \ne \ell$ we can find $\omega_k \in \mathcal{A}^{(k)\dagger}$ satisfying $\omega_k(a_{k,j}) = 0$ for $j \ne i$ and $\omega_k(a_{k,i}) = 1$. If $\omega_\ell \in \mathcal{A}^{(\ell)\dagger}$ is arbitrary, this gives

$$
0 = \omega_1 \otimes \omega_2 \otimes \cdots \otimes \omega_n \left(\sum_{j=1}^{m} a_{1,j} \otimes a_{2,j} \otimes \cdots \otimes a_{n,j} \right) = \omega_\ell(a_{\ell,i})
$$

which implies $a_{\ell,i} = 0$ since $\omega_\ell \in \mathcal{A}^{(\ell)\dagger}$ was arbitrary. Also, i was arbitrary.

The present condition implies the universal mapping property since it ensures that the obvious definition of θ on products can be extended to sums of products.

The Projective Tensor Product of Banach Spaces

For finite-dimensional linear spaces and algebras, the above theory is quite complete and satisfying. However for infinite-dimensional spaces we naturally wish to introduce topological restrictions. To begin with these restrictions cause no problem, but if we wish to have the tensor product of Banach spaces be a Banach space, more complications arise.

We can form the algebraic tensor product of Banach spaces, and sometimes this construction is useful. For instance, $\mathcal{B}_F(\mathcal{X}, \mathcal{Y})$ was defined that way. These algebraic tensor products can be provided with natural norms in several ways, but they will essentially never be complete (unless one factor is finite-dimensional). We usually wish to have the tensor product of Banach spaces be a Banach space. There are three main ways to accomplish this as we will show in more detail here. First we may apply Definition 1.10.1 to the category of Banach spaces, producing the projective tensor product. We may apply the construction noted after Proposition 1.10.5 to get the injective tensor product. Finally we may consider various norms (satisfying mild and natural restrictions) on the algebraic tensor product and then complete it. It turns out that the projective tensor product and injective product are special cases of the third approach using the largest and smallest suitable norms, respectively.

1.10.7 Definition Let $\mathcal{X}^1, \mathcal{X}^2, \ldots, \mathcal{X}^n$, and $\mathcal{Y}$ be normed linear spaces. A multilinear function $f: \mathcal{X}^1 \times \mathcal{X}^2 \times \cdots \times \mathcal{X}^n \to \mathcal{Y}$ is said to be *bounded* if there is a number B satisfying

$$\|f(x_1, x_2, \ldots, x_n)\| \le B \|x_1\| \, \|x_2\| \cdots \|x_n\| \qquad \forall \ x_k \in \mathcal{X}^k.$$

The set of bounded multilinear maps of $\mathcal{X}^1 \times \mathcal{X}^2 \times \cdots \times \mathcal{X}^n$ into $\mathcal{Y}$ is denoted by $\mathcal{B}(\mathcal{X}^1, \mathcal{X}^2, \ldots, \mathcal{X}^n; \mathcal{Y})$. For each $f \in \mathcal{B}(\mathcal{X}^1, \mathcal{X}^2, \ldots, \mathcal{X}^n; \mathcal{Y})$ let $\|f\|$ be the infimum of the numbers B satisfying the above inequality.

The *projective tensor product* of $\mathcal{X}^1, \mathcal{X}^2, \ldots, \mathcal{X}^n$ is a pair $(\mathcal{Z}, p)$ where $\mathcal{Z} = \mathcal{X}^1 \hat{\otimes} \mathcal{X}^2 \hat{\otimes} \cdots \hat{\otimes} \mathcal{X}^n$ is a Banach space and the tensor map p belongs to $\mathcal{B}(\mathcal{X}^1, \mathcal{X}^2, \ldots, \mathcal{X}^n; \mathcal{Z})_1$ and satisfies the following universal property:

> If $\mathcal{Y}$ is any Banach space and f is any map in $\mathcal{B}(\mathcal{X}^1, \mathcal{X}^2, \ldots, \mathcal{X}^n; \mathcal{Y})$, then there is a unique continuous linear map $h: \mathcal{Z} \to \mathcal{Y}$ satisfying $f = h \circ p$ and $\|h\| \le \|f\|$.

It is easy to see that $f \mapsto \|f\|$ is a norm on $\mathcal{B}(\mathcal{X}^1, \mathcal{X}^2, \ldots, \mathcal{X}^n; \mathcal{Y})$ which is complete if $\mathcal{Y}$ is complete. Note that the projective tensor product of (possibly incomplete) normed linear spaces is a Banach space. It is naturally equivalent to the projective tensor product of their completions. We need an analogue of Theorem 1.10.2. The proof of uniqueness is no different, but to establish existence we will show how the projective tensor product can be constructed as the completion of the algebraic tensor product relative to a norm which we will call the *projective tensor norm*. The proof that this is a norm rather than just a semi-norm is particularly interesting.

1.10.8 Theorem *If $\mathcal{X}^1, \mathcal{X}^2, \ldots, \mathcal{X}^n$ are normed linear spaces, then their projective tensor product exists and is unique in the natural sense. The projective tensor product $\mathcal{X}^1 \hat{\otimes} \mathcal{X}^2 \hat{\otimes} \cdots \hat{\otimes} \mathcal{X}^n$ may be realized as the completion of the algebraic tensor product $\tilde{\mathcal{Z}} = \mathcal{X}^1 \otimes \mathcal{X}^2 \otimes \cdots \otimes \mathcal{X}^n$ with respect to the following expression which is a norm on $\tilde{\mathcal{Z}}$*

$$\|t\|_p = \inf \left\{ \sum_{j=1}^m \prod_{k=1}^n \|x_{k,j}\| : t = \sum_{j=1}^m x_{1,j} \otimes x_{2,j} \otimes \cdots \otimes x_{n,j} \right\} \quad \forall\, t \in \tilde{\mathcal{Z}}$$

(17)

where the infimum is extended over all of the representations for t with $m \in \mathbb{N}$ and $x_{k,j} \in \mathcal{X}^k$.

For any $t \in \mathcal{X}^1 \hat{\otimes} \mathcal{X}^2 \hat{\otimes} \cdots \hat{\otimes} \mathcal{X}^n$ and any $\varepsilon > 0$, we can find sequences $\{x_{k,m}\}_{m \in \mathbb{N}} \subseteq \mathcal{X}^k$ converging to zero for each $k = 1, 2, \ldots, n$ and satisfying $t = \sum_{j=1}^\infty x_{1,j} \otimes x_{2,j} \otimes \cdots \otimes x_{n,j}$ and

$$\|t\|_p \leq \sum_{j=1}^\infty \prod_{k=1}^n \|x_{k,j}\| \leq \|t\|_p + \varepsilon.$$

(18)

Proof As noted above, the uniqueness of the projective tensor product is defined and proved in a way analogous to that given in Theorem 1.10.2.

The expression (17) is easily seen to be a semi-norm on $\tilde{\mathcal{Z}}$. Let

$$\theta \colon \tilde{\mathcal{Z}} = \mathcal{X}^1 \otimes \mathcal{X}^2 \otimes \cdots \otimes \mathcal{X}^n \to B(\mathcal{X}^{1*}, \mathcal{X}^{2*}, \ldots, \mathcal{X}^{n*}; \mathbb{C})$$

be the canonical linear map given by the analogue of equation (11). Proposition 1.10.5 applies to the present situation directly so θ is injective. For any $t \in \tilde{\mathcal{Z}}$, the inequality $\|\theta(t)\| \leq \|t\|_p$ is obvious and shows that $\|\cdot\|_p$ is a norm not merely a semi-norm. Hence we may define $\mathcal{Z}$ to be the completion of $\tilde{\mathcal{Z}}$ in this norm, leaving the tensor map p unchanged. Clearly $p \in B(\mathcal{X}^1, \mathcal{X}^2, \ldots, \mathcal{X}^n; \mathcal{Z})$ is bounded with norm 1 when $\mathcal{Z}$ is defined this way. Furthermore, if $\mathcal{Y}$ and f satisfy the conditions in the definition, then the definition of the algebraic tensor product gives a map h of $\mathcal{X}^1 \otimes \mathcal{X}^2 \otimes \cdots \otimes \mathcal{X}^n$ into $\mathcal{Y}$ satisfying $f = h \circ p$. Any elementary tensor satisfies $\|h(x_1 \otimes x_2 \otimes \cdots \otimes x_n)\| = \|f(x_1, x_2, \ldots, x_n)\| \leq \|f\| \, \|x_1\| \, \|x_2\| \cdots \|x_n\|$. Hence h satisfies $\|h\| \leq \|f\|$ relative to the projective tensor norm. Thus h can be extended by continuity to be a linear map with the same norm defined on all of the projective tensor product and having the desired properties. Hence $(\mathcal{Z}, p)$ satisfies the universal mapping property.

In order to prove (18), choose a sequence $\{t_m\}_{m \in \mathbb{N}}$ in the algebraic tensor product satisfying $\|t - t_m\|_p < \varepsilon/2^{m+1}$ for each m. Then $\|t_1\|_p < \|t\|_p + \varepsilon/4$ and $\|t_{m+1} - t_m\|_p < 3\varepsilon/2^{m+2}$. Setting $t_0 = 0$, we can write $t = \sum_{m=0}^{\infty} (t_{m+1} - t_m)$. Each term in this series can be written as a finite sum of elementary tensors with the sum of the products of their norms smaller than the estimates just derived. The infinite series obtained by combining these finite sums satisfies (18). However, since the size of $\prod_{k=1}^{n} \|x_{k,j}\|$ tells us nothing about the size of the individual factors, the individual sequences $\{x_{k,m}\}_{m \in \mathbb{N}}$ need not converge to zero. Note that for any $q \in \mathbb{N}$, $x_{1,j} \otimes x_{2,j} \otimes \cdots \otimes x_{n,j} = \sum_{i=1}^{q^n} q^{-1} x_{1,j} \otimes q^{-1} x_{2,j} \otimes \cdots \otimes q^{-1} x_{n,j}$ (all terms equal). This trick allows us to write each term in the original infinite series as a sum of terms for which $\{x_{k,m}\}_{m \in \mathbb{N}}$ does converge to zero for each k. $\square$

The similarity of Definitions 1.10.1 and 1.10.7 yield a number of parallel results. Equations (5) and (6) above remain obvious when each $\otimes$ is replaced by a $\hat{\otimes}$ and hence n-fold projective tensor products can be built up from tensor products of pairs. Similarly, equations (4), (7) and (9)–(12) can each be extended to projective tensor products, but we will treat them more formally. We begin with (4) and (7) (for the case $\mathcal{Y} = \mathbb{C}$).

1.10.9 Proposition *For any normed linear spaces $\mathcal{X}^1, \mathcal{X}^2, \ldots, \mathcal{X}^n$, the Banach space dual of $\mathcal{X}^1 \hat{\otimes} \mathcal{X}^2 \hat{\otimes} \cdots \hat{\otimes} \mathcal{X}^n$ is naturally isometrically isomorphic to $\mathcal{B}(\mathcal{X}^1, \mathcal{X}^2, \ldots, \mathcal{X}^n; \mathbb{C})$.*

In particular, for any normed linear spaces $\mathcal{X}$ and $\mathcal{Y}$, $(\mathcal{X} \hat{\otimes} \mathcal{Y})^*$ is naturally isometrically linearly isomorphic to $\mathcal{B}(\mathcal{X}, \mathcal{Y}^*)$.

Proof The definition gives a contractive map $(f \mapsto h)$ of the second space into the first which has the contractive map $h \mapsto h \circ p$ as inverse.

For $\langle \omega \in (\mathcal{X} \hat{\otimes} \mathcal{Y})^* / T \in \mathcal{B}(\mathcal{X}, \mathcal{Y}^*) \rangle$ define $\langle T_\omega \in \mathcal{B}(\mathcal{X}, \mathcal{Y}^*) / \omega_T \in (\mathcal{X} \hat{\otimes} \mathcal{Y})^* \rangle$ by

$$\langle T_\omega(x)(y) = \omega(x \otimes y) / \omega_T(x \otimes y) = T(x)(y) \rangle \qquad \forall x \in \mathcal{X}; \ y \in \mathcal{Y}.$$

These yield inverse contractive linear isomorphisms. $\square$

The above construction of a natural predual Banach space for any Banach space of bounded linear operators into a dual space is fundamental. It is the precursor of Shoichiro Sakai's characterization [1956] of von Neumann algebras as C*-algebras which have preduals. See Volume II.

We can now explain the use of the name "projective".

1.10.10 Proposition *Let $\mathcal{X}^k$ and $\mathcal{Y}^k$ be Banach spaces and let $T_k \in \mathcal{B}(\mathcal{X}^k, \mathcal{Y}^k)$ be a bounded linear map for $k = 1, 2, \ldots, n$. Then there is a well defined continuous linear map*

$$T_1 \otimes T_2 \otimes \cdots \otimes T_n \colon \mathcal{X}^1 \hat{\otimes} \mathcal{X}^2 \hat{\otimes} \cdots \hat{\otimes} \mathcal{X}^n \to \mathcal{Y}^1 \hat{\otimes} \mathcal{Y}^2 \hat{\otimes} \cdots \hat{\otimes} \mathcal{Y}^n \qquad (19)$$

defined on elementary tensors by

$$T_1 \otimes T_2 \otimes \cdots \otimes T_n (x_1 \otimes x_2 \otimes \cdots \otimes x_n) = T_1(x_1) \otimes T_2(x_2) \otimes \cdots \otimes T_n(x_n) \quad (20)$$

with norm $\|T_1 \otimes T_2 \otimes \cdots \otimes T_n\|_p = \prod_{k=1}^{n} \|T_k\|$.

For $k = 1, 2, \ldots, n$, let T_k be the natural projection of $\mathcal{X}^k$ onto $\mathcal{X}^k / \mathcal{Z}^k$ where $\mathcal{Z}^k$ is a closed linear subspace of $\mathcal{X}^k$. Then $T_1 \otimes T_2 \otimes \cdots \otimes T_n$ is a projection of $\mathcal{X}^1 \hat{\otimes} \mathcal{X}^2 \hat{\otimes} \cdots \hat{\otimes} \mathcal{X}^n$ onto the quotient Banach space $\mathcal{X}^1 \hat{\otimes} \mathcal{X}^2 \hat{\otimes} \cdots \hat{\otimes} \mathcal{X}^n / \ker(T_1 \otimes T_2 \otimes \cdots \otimes T_n)$.

Proof In (13) we have already remarked that $T_1 \otimes T_2 \otimes \cdots \otimes T_n$ is a well defined linear map between the algebraic tensor products. Hence Theorem 1.10.8 and the following calculation, where $t = \sum_{j=1}^{\infty} x_{1,j} \otimes x_{2,j} \otimes \cdots \otimes x_{n,j}$, show that it is well defined with the asserted norm as a continuous linear map (with the asserted norm) between the projective tensor products:

$$\|T_1 \otimes T_2 \otimes \cdots \otimes T_n(t)\|_p \leq \sum_{j=1}^{m} \prod_{k=1}^{n} \|T_k x_{k,j}\| \leq \prod_{k=1}^{n} \|T_k\| \sum_{j=1}^{m} \prod_{k=1}^{n} \|x_{k,j}\|.$$

We simplify notation in the proof of the second claim by setting $n = 2$. It is enough to consider the case in which only the first subspace is nonzero, because the product of quotient maps is a quotient map. We write $\mathcal{X}^1 = \mathcal{X}$ and $\mathcal{Z}^1 = \mathcal{Z}$ and $\mathcal{X}^2 = \mathcal{Y}$. Let $\varepsilon > 0$ be arbitrary. Apply (18) and the definition of the quotient norm to $t \in (\mathcal{X}/\mathcal{Z}) \otimes \mathcal{Y}$ to find $\{x_m\}_{m \in \mathbb{N}} \subseteq \mathcal{X}$ and $\{y_m\}_{m \in \mathbb{N}} \subseteq \mathcal{Y}$ satisfying $t = \sum_{j=1}^{\infty}(x_j + \mathcal{Z}) \otimes y_j$ and $\|t\|_p \leq \sum_{j=1}^{\infty} \|x_j\|\|y_j\| \leq \|t\|_p + \varepsilon$. This shows that $T_1 \otimes T_2$ is a projection onto a quotient space with its quotient norm. $\quad \Box$

The following is an important case of the projective tensor product for our purposes. For a suitable product measure theory see Hewitt and Ross [1963], pp. 150-157.

1.10.11 Proposition *For $k = 1, 2, \ldots, n$ let (M^k, μ_k) be a pair consisting of a locally compact topological space M^k provided with a positive regular Borel measure μ_k. Then*

$$f_1 \otimes f_2 \otimes \cdots \otimes f_n(m_1, m_2, \ldots, m_n)$$
$$= f_1(m_1) f_2(m_2) \cdots f_n(m_n) \quad \forall f_k \in L^1(\mu_k); \; m_k \in M^k$$

defines an isometric linear isomorphism of $L^1(\mu_1) \hat{\otimes} L^1(\mu_2) \hat{\otimes} \cdots \hat{\otimes} L^1(\mu_n)$ onto $L^1(\mu_1 \times \mu_2 \times \cdots \times \mu_n)$.

Let $\mathcal{X}$ be a Banach space. Then

$$f \otimes x(m) = f(m)x \quad \forall f \in L^1(\mu_1); \; x \in \mathcal{X}$$

defines an isometric linear isomorphism of $L^1(\mu_1) \hat{\otimes} \mathcal{X}$ onto the space $L^1(\mu_1, \mathcal{X})$ of Bochner μ_1-integrable $\mathcal{X}$-valued functions on M^1.

Proof Consider the second case first and replace (M^1, μ_1) by (M, μ). The map $(f, x) \mapsto fx$ (where $fx(m) = f(m)x$ for all $m \in M$) is a bilinear map of $L^1(\mu) \times \mathcal{X}$ into $L^1(\mu, \mathcal{X})$, so there is a contractive linear map τ from the projective tensor product to $L^1(\mu, \mathcal{X})$. It is not hard to show that the space of continuous functions of compact support from M to $\mathcal{X}$ is dense in $L^1(\mu, \mathcal{X})$ and hence the image of τ is dense. The proof will be complete when we prove that the map τ is an isometry. We are reduced to showing that each $t = \sum_{j=1}^m f_j \otimes x_j \in L^1(\mu) \otimes \mathcal{X}$ satisfies $\|\tau(t)\| = \|t\|_p$. Theorem 1.10.9 shows that we may identify $(L^1(\mu) \otimes \mathcal{X})^*$ with $\mathcal{B}(L^1(\mu), \mathcal{X}^*)$. Since the f_j are supported on a σ-finite portion of (M, μ) and the x_j are included in a finite-dimensional subspace of $\mathcal{X}$, we may apply the Dunford–Pettis theorem (Dunford and Schwartz [1958], V1.8.6) to show that any element $\omega \in (L^1(\mu) \otimes \mathcal{X})^*$ has the form $\omega(f \otimes x) = \int f(m)x^*(m)(x)d\mu(m)$ for some function $x^*: M \to \mathcal{X}^*$ satisfying ess $\text{supp}\{\|x^*\|\} = \|\omega\|$. Since this can be rewritten as

$$\omega(t) = \int x^*(m)(\tau(t)(m))d\mu(m),$$

we find

$$\|t\|_p \le \sup\{|\omega(t)| : \omega \in (L^1(\mu) \otimes \mathcal{X})_1^*\} \le \|\tau(t)\|$$

as we wished to show.

Now the first result with $n = 2$ follows by setting $\mathcal{X} = L^1(\mu_2)$. As usual, the general case follows by taking successive tensor products as in (6). $\square$

We will briefly introduce the trace on certain projective tensor products. Consider the projective tensor product $\mathcal{X} \hat{\otimes} \mathcal{X}^*$. Any element can be expressed as a series $t = \sum_{n=1}^\infty x_n \otimes \omega_n$ satisfying $\sum_{n=1}^\infty \|x_n\| \|\omega_n\| < \infty$. It is easy to see that the expression

$$Tr(t) = \sum_{n=1}^\infty \omega_n(x_n) \tag{21}$$

(which we will call the *trace*) is well defined and satisfies

$$|Tr(t)| \le \|t\|_p. \tag{22}$$

When $\mathcal{X}$ is finite-dimensional, the matrix construction in §1.7.7 shows that this trace agrees with the familiar trace on square matrices. In §1.7.11 above and §1.10.25 below, we will note that it also agrees with other notions of traces of certain operators on infinite-dimensional Banach spaces.

Projective Tensor Products of Banach Algebras

We will now extend the results of Theorem 1.10.6 to Banach algebras. These ideas were first explored by Bernard R. Gelbaum [1959] and Jun Tomiyama [1960]. The projective tensor product is the only one of the Banach space tensor products which always preserves Banach algebra structure, but we will discuss some other special cases below. Other results on tensor products of general Banach algebras are contained in T. Keith Carne [1981]; Gelbaum [1959], [1961], [1962],[1965], [1970]; Jesús Gil de Lamadrid [1963], [1965a], [1965b], [1967]; Kjeld B. Laursen [1969b], [1970]; Arnold Lebow [1968]; Richard J. Loy [1970b]; Anna Maria Mantero and Andrew Tonge [1979] and [1980]; R. D. Mehta and Mahavirendra Vasavada [1985]; Jan Okniński [1982]; U. B. Tewari, M. Dutta, and Shobha Madan [1982]; Tomiyama [1960], [1972]; Ali Ülger [1988a]. This list omits all of the many papers relating to tensor products of C*-algebras and W*-algebras.

1.10.12 Proposition *Let $A^{(1)}, A^{(2)}, \ldots, A^{(n)}$ be Banach algebras. Then the projective tensor product $A^{(1)} \hat{\otimes} A^{(2)} \hat{\otimes} \cdots \hat{\otimes} A^{(n)}$ is a Banach algebra under the natural product and the projective tensor norm.*

Proof The obvious calculation shows that the projective tensor norm is submultiplicative with respect to the multiplication on the algebraic tensor product. Hence the multiplication introduced in Theorem 1.10.6 can be extended to the projective tensor product by continuity. □

When the projective tensor product of Banach algebras is regarded as a Banach algebra under this multiplication, we will call it the *projective Banach algebra tensor product*. We will note below that the other Banach tensor products we introduce later do not necessarily satisfy the above proposition. Combining Theorems 1.10.6 and 1.10.8 and previous arguments, we get the following result.

1.10.13 Corollary *Let $A^{(1)}, A^{(2)}, \ldots, A^{(n)}$ be unital Banach algebras with identity elements of norm 1 and let A be their projective Banach algebra tensor product. For $k = 1, 2, \ldots, n$, let $\varphi_k \colon A^{(k)} \to A$ be the natural contractive injective unital algebra homomorphisms defined in Theorem 1.10.6. Then the ordered $(n + 1)$-tuple $(A, \varphi_1, \varphi_2, \ldots, \varphi_n)$ is determined up to the natural notion of isometric isomorphism as the unique ordered $(n+1)$-tuple consisting of a unital Banach algebra and n contractive unital algebra homomorphisms $\varphi_k \colon A^{(k)} \to A$ with mutually commuting ranges, such that if $(B, \psi_1, \psi_2, \ldots, \psi_n)$ is any other such $(n + 1)$-tuple, then there is a unique contractive unital algebra homomorphism $\theta \colon A \to B$ satisfying $\psi_k = \theta \circ \varphi_k$ for $k = 1, 2, \ldots, n$. In particular the algebra projective tensor product defines a coproduct in the category of unital commutative Banach algebras with identity elements of norm 1 and contractive unital algebra homomorphisms.*

The following important result is a corollary of Proposition 1.10.11 and Corollary 1.10.13.

1.10.14 Corollary *Let $G^1, G^2, \ldots, G^n$ be locally compact groups. Then $L^1(G^1)\hat{\otimes}L^1(G^2)\hat{\otimes}\cdots\hat{\otimes}L^1(G^n)$ is isometrically algebra isomorphic to $L^1(G^1 \times G^2 \times \cdots \times G^n)$ under the map described in* Proposition 1.10.11.

Proof It only remains to check that the products agree. We may reduce to the case of two locally compact groups G and H. Then $f, g \in L^1(G)$ and $h, k \in L^1(H)$ satisfy

$$(f \otimes h)*(g \otimes k)(u, v) = \int\int f \otimes h(w, z)g \otimes k(w^{-1}u, z^{-1}v)dw\, dz$$

$$= \int f(w)g(w^{-1}u)dw \int h(z)k(z^{-1}v)dz = f*g \otimes h*k(u, v).$$

Hence the product in the algebra tensor product agrees with convolution in $L^1(G \times H)$. $\square$

Somewhat similar results are given by Jesús Gil de Lamadrid [1965b] for the projective tensor product of the Banach space $M(\Omega)$ of finite regular Borel measures on a locally compact space Ω with another Banach space $\mathcal{X}$. Then $M(\Omega)\hat{\otimes}\mathcal{X}$ is given as the space of certain $\mathcal{X}$-valued measures on Ω at least when Ω satisfies a fairly general condition. In particular, $M(G^1)\hat{\otimes}M(G^2)\hat{\otimes}\cdots\hat{\otimes}M(G^n)$ is isometrically algebra isomorphic to $M(G^1 \times G^2 \times \cdots \times G^n)$. Tewari, Dutta and Madan [1982] prove interesting partial converses to these results. They show that if $\mathcal{A}\hat{\otimes}\mathcal{B}$ is isometrically isomorphic to $L^1(G)$ or $M(G)$ for some locally compact abelian group G, then both $\mathcal{A}$ and $\mathcal{B}$ have the same form. The proof depends on Marc Rieffel's characterization [1965] of these group algebras for locally compact abelian groups.

We remark on another consequence of our definition of projective tensor products. Note that a Banach algebra consists of a Banach space $\mathcal{A}$ together with a multiplication map $M: \mathcal{A} \times \mathcal{A} \to \mathcal{A}$ defined by $M(a, b) = ab$. Clearly M belongs to $\mathcal{B}(\mathcal{A}, \mathcal{A}; \mathcal{A})$ and has norm at most one. Hence we may regard it as a contraction in $\mathcal{B}(\mathcal{A}\hat{\otimes}\mathcal{A}, \mathcal{A})$. The associativity of multiplication is then expressed by the equation $M\circ(M\otimes I) = M\circ(I\otimes M)$, where $I: \mathcal{A} \to \mathcal{A}$ is the identity map. Clearly any contractive linear map $M: \mathcal{A}\hat{\otimes}\mathcal{A} \to \mathcal{A}$ satisfying this condition determines a Banach algebra structure on $\mathcal{A}$. Hence all of Banach algebra theory can be stated in these terms. We will seldom be tempted to do this. However when confronted with a Banach algebra $\mathcal{A}$ and a Banach algebra structure on $\mathcal{A}^*$, the above procedure may be used to give $\mathcal{A}^*$ the structure of a Hopf algebra. See Theorem 4.2.26 for an interesting related result concerning *injective algebras* in which multiplication is well defined and continuous from the injective tensor product.

The Varopoulos Algebra

The projective tensor product of $C(\Omega_1)$ and $C(\Omega_2)$ may be embedded in $C(\Omega_1 \times \Omega_2)$ but it is not the whole algebra unless Ω_1 or Ω_2 is finite. We will prove this only in the following interesting case first considered by Nicholas Th. Varopoulos [1965], [1967]. The elegant presentation is due in part to the review of [1965] by Carl S. Herz. See also Jean-Pierre Kahane [1970]. These introductory comments will be generalized in Section 3.6.

Let G be a compact abelian group with normalized Haar measure (*i.e.*, $\lambda(G) = 1$) and let $\hat{G}$ be the set of all continuous group homomorphisms from G into $\mathbb{T}$. We will regard $\hat{G}$ as a discrete group under pointwise multiplication. The *Fourier algebra* $A(G)$ on G is the subalgebra of $C(G)$ consisting of all those functions which can be expressed in the form

$$f(u) = \sum_{h \in \hat{G}} g(h)h(u) \qquad \forall\, u \in G \quad \text{for some } g \in \ell^1(\hat{G}). \qquad (23)$$

In Section 3.6 we will give the easy proof that if $f \in A(G)$ satisfies (23), then g is uniquely determined and satisfies $g(h) = \hat{f}(h) = \int_G f(u)h(u)^* du$ for each $h \in \hat{G}$. Thus we may give $A(G)$ the norm $\|f\|_A = \|\hat{f}\|_1$. This makes $A(G)$ into a Banach algebra since any two functions $f, g \in A(G)$ satisfy $\widehat{fg} = \hat{f} * \hat{g}$ where $\hat{f} * \hat{g}(h) = \sum_{k \in \hat{G}} \hat{f}(k)g(k^{-1}h)$ and hence $\|fg\|_A \leq \|f\|_A \|g\|_A$.

Let $V(G \times G)$ be the projective tensor product $C(G) \hat{\otimes} C(G)$ considered as a subalgebra of $C(G \times G)$ by setting $(f \otimes g)(u, v) = f(u)g(v)$ for any $u, v \in G$ as usual. Any $F \in A(G \times G)$ belongs to $V(G \times G)$, since it can be written in the form $F(u, v) = \sum_{h, k \in \hat{G}} \hat{F}(h \otimes k)h(u)k(v)$ where we have simply noted that each element of $\widehat{G \times G}$ has the form $(h \otimes k)(u, v) = h(u)k(v)$ for some $h, k \in \hat{G}$. Hence we have continuous inclusions

$$A(G \times G) \to V(G \times G) \to C(G \times G).$$

1.10.15 Varopoulos' Theorem *Let G be a compact abelian group with normalized Haar measure. Define*

$$M: C(G) \to C(G \times G) \quad \text{and} \quad P: C(G \times G) \to C(G) \quad by$$

$$Mf(u, v) \;=\; f(uv) \qquad \forall\, f \in C(G);\ u, v \in G \quad and$$

$$PF(u) \;=\; \int_G F(v, v^{-1}u)dv \qquad \forall\, F \in C(G \times G);\ u \in G.$$

Then $P \circ M$ is the identity map on $C(G)$ and the restriction of $\langle\, M\,/\,P\,\rangle$ defines an isometric linear isomorphism of $\langle$ the Fourier algebra $A(G)\,/$ the

*Varopoulos algebra $V(G \times G) \rangle$ onto $\langle V(G \times G) / A(G) \rangle$. If $A(G)$ is en-
dowed with convolution multiplication and $V(G \times G)$ with the algebra ten-
sor product multiplication arising from convolution multiplication on $C(G)$,
then these maps are algebra isomorphisms.*

Proof The following elementary calculation shows $P \circ M = I$:

$$P \circ M(f)(u) = \int_G Mf(v, v^{-1}u)dv = \int_G f(u)dv = f(u) \quad \forall f \in C(G); \ u \in G.$$

For any $h \in \hat{G} \subseteq C(G)$, we get $Mh(u,v) = h(uv) = h(u)h(v)$. Hence M
sends any finite linear combination $f = \sum \hat{f}(h)h$ in $A(G)$ into $V(G \times G)$
and satisfies $\|Mf\|_p \leq \|f\|_A$. Therefore, M is a contractive linear map of
$A(G)$ into $V(G \times G)$.

For $f, g \in C(G)$, $f \otimes g$ belongs to $V(G \times G)$ and satisfies $P(f \otimes g)(u) = \int_G f(v)g(v^{-1}u)dv = f * g(u)$ for all $u \in G$. This gives

$$\|P(f \otimes g)\|_A = \|f * g\|_A = \|\hat{f}\hat{g}\|_1 \leq \|\hat{f}\|_2\|\hat{g}\|_2 = \|f\|_2\|g\|_2 \leq \|f\|_\infty\|g\|_\infty.$$

Hence the definition of the projective tensor norm shows that P is contractive from $V(G \times G)$ to $A(G)$. Since both M and P are contractive and
$P \circ M = I$, they are both isometries and they send $A(G)$ onto $V(G \times G)$
and conversely.

Now suppose $C(G)$ and $A(G)$ are endowed with convolution multiplication rather than their usual pointwise multiplication. They are both
Banach algebras under convolution. For $C(G)$, this follows from

$$|f * g(u)| \leq \int_G |f(v)g(v^{-1}u)|dv \leq \|f\|_\infty\|g\|_\infty \quad \forall f, g \in C(G); \ u \in G$$

since $\lambda(G) = 1$. For $A(G)$ we note

$$\|f * g\|_A = \|\widehat{f * g}\|_1 \leq \|\hat{f}\|_1\|\hat{g}\|_\infty \leq \|\hat{f}\|_1\|\hat{g}\|_1 \quad \forall f, g \in A(G).$$

Now give $V(G \times G) = C(G)\hat{\otimes}C(G)$ the multiplication induced by the convolution product in $C(G)$. We need only check that P is a homomorphism,
and it is enough to consider elementary tensors:

$$\begin{aligned}
P((f \otimes g)(h \otimes k)) &= P((f * h) \otimes (g * k)) \\
&= (f * h) * (g * k) = (f * g) * (h * k) \\
&= P(f \otimes g) * P(h \otimes k) \quad \forall f, g, h, k \in C(G). \quad \square
\end{aligned}$$

Injective Tensor Products of Banach Spaces

In the present setting, because of the Hahn–Banach theorem, Propositions 1.10.5 and 1.10.9 can be restated.

1.10.16 Corollary *Let $\mathcal{X}^1, \mathcal{X}^2, \ldots, \mathcal{X}^n$ be normed linear spaces. The map*

$$\theta: \mathcal{X}^1 \otimes \mathcal{X}^2 \otimes \cdots \otimes \mathcal{X}^n \to \mathcal{B}(\mathcal{X}^{1*}, \mathcal{X}^{2*}, \ldots, \mathcal{X}^{n*}; \mathbb{C}) = (\mathcal{X}^{1*} \hat{\otimes} \mathcal{X}^{2*} \hat{\otimes} \cdots \hat{\otimes} \mathcal{X}^{n*})^*$$

defined on elementary tensors by equation (11) is well defined and injective.

We have embedded the algebraic tensor product into the Banach space $\mathcal{B}(\mathcal{X}^{1*}, \mathcal{X}^{2*}, \ldots, \mathcal{X}^{n*}; \mathbb{C})$. It is easy to check that the corresponding norm on the algebraic tensor product is the one given in the next definition.

1.10.17 Definition *Let $\mathcal{X}^1, \mathcal{X}^2, \ldots, \mathcal{X}^n$ be normed linear spaces. The injective tensor norm on their algebraic tensor product is given by*

$$\|t\|_w = \sup\{|\sum_{j=1}^{m} \prod_{k=1}^{n} \omega_k(x_{k,j})| : \omega_k \in (\mathcal{X}^k)_1^*\} \qquad (24)$$

where $t = \sum_{j=1}^{m} x_{1,j} \otimes x_{2,j} \otimes \cdots \otimes x_{n,j}$. (The subscript w comes from the alternate name *weak tensor norm*.) The completion of the algebraic tensor product $\mathcal{X}^1 \otimes \mathcal{X}^2 \otimes \cdots \otimes \mathcal{X}^n$ in this norm is called the *injective tensor product* and denoted by $\mathcal{X}^1 \check{\otimes} \mathcal{X}^2 \check{\otimes} \cdots \check{\otimes} \mathcal{X}^n$.

The remarks above show that the injective tensor product may be considered to be the closure of the range of θ in $\mathcal{B}(\mathcal{X}^{1*}, \mathcal{X}^{2*}, \ldots, \mathcal{X}^{n*}; \mathbb{C})$. If $\mathcal{X}$ and $\mathcal{Y}$ are normed linear spaces, we have already noted that we may identify the algebraic tensor product $\mathcal{Y} \otimes \mathcal{X}^*$ with the space $\mathcal{B}_F(\mathcal{X}, \mathcal{Y})$ of bounded finite-rank linear operators from $\mathcal{X}$ to $\mathcal{Y}$. The identification is given by

$$\left(\sum_{j=1}^{m} y_j \otimes \omega_j \right)(z) = \sum_{j=1}^{m} \omega_j(z) y_j \qquad \forall \ m \in \mathbb{N}; \ y_j \in \mathcal{Y}; \ \omega_j \in \mathcal{X}^*; \ z \in \mathcal{X}.$$

Let us check the operator norm on $T = \sum_{j=1}^{m} y_j \otimes \omega_j$:

$$\begin{aligned} \|T\|_{op} &= \sup\{|\omega(T(z))| : z \in \mathcal{X}_1; \omega \in \mathcal{Y}_1^*\} \\ &= \sup\{|\sum_{j=1}^{m} \omega(y_j)\omega_j(z)| : z \in \mathcal{X}_1; \omega \in \mathcal{Y}_1^*\} = \|T\|_w. \end{aligned}$$

We have used the fact that the supremum of $|\sum_{j=1}^{m} \omega(y_j)\omega_j(z)|$ over $z \in \mathcal{X}_1$ is the same as the supremum of $|F(\sum_{j=1}^{m} \omega(y_j)\omega_j)|$ over $F \in \mathcal{X}_1^{**}$ since they both equal $\|\sum_{j=1}^{m} \omega(y_j)\omega_j\|$. We have proved the following.

1.10.18 Proposition *Let $\mathcal{X}$ and $\mathcal{Y}$ be normed linear spaces with at least $\mathcal{Y}$ complete. Then the space $\mathcal{B}_A(\mathcal{X}, \mathcal{Y}) = \mathcal{B}_F(\mathcal{X}, \mathcal{Y})^-$ of approximable operators from $\mathcal{X}$ to $\mathcal{Y}$ is isometrically linearly isomorphic to $\mathcal{Y} \check{\otimes} \mathcal{X}^*$ under the map indicated above.*

The reader can find many similar results. For instance there is a natural isometric linear embedding of $\mathcal{X} \check{\otimes} \mathcal{Y}$ onto a closed linear subspace of $\mathcal{B}(\mathcal{X}^*, \mathcal{Y})$. Next we note the functorial property of this tensor product and explain its name.

1.10.19 Proposition *Let $\mathcal{X}^k$ and $\mathcal{Y}^k$ be Banach spaces and let $T_k \in \mathcal{B}(\mathcal{X}^k, \mathcal{Y}^k)$ be a bounded linear map for $k = 1, 2, \ldots, n$. Then there is a well defined continuous linear map*

$$T_1 \otimes T_2 \otimes \cdots \otimes T_n \colon \mathcal{X}^1 \check{\otimes} \mathcal{X}^2 \check{\otimes} \cdots \check{\otimes} \mathcal{X}^n \to \mathcal{Y}^1 \check{\otimes} \mathcal{Y}^2 \check{\otimes} \cdots \check{\otimes} \mathcal{Y}^n$$

defined on elementary tensors by

$$T_1 \otimes T_2 \otimes \cdots \otimes T_n(x_1 \otimes x_2 \otimes \cdots \otimes x_n) = T_1(x_1) \otimes T_2(x_2) \otimes \cdots \otimes T_n(x_n)$$

with norm $\|T_1 \otimes T_2 \otimes \cdots \otimes T_n\|_w = \prod_{k=1}^{n} \|T_k\|$.

For each $k = 1, 2, \ldots, n$, let T_k be the natural injection of a closed linear subspace $\mathcal{X}^k$ into $\mathcal{Y}^k$. Then $T_1 \otimes T_2 \otimes \cdots \otimes T_n$ is an isometric injection of $\mathcal{X}^1 \check{\otimes} \mathcal{X}^2 \check{\otimes} \cdots \check{\otimes} \mathcal{X}^n$ into $\mathcal{Y}^1 \check{\otimes} \mathcal{Y}^2 \check{\otimes} \cdots \check{\otimes} \mathcal{Y}^n$.

Proof We have already noted (twice) that $T_1 \otimes T_2 \otimes \cdots \otimes T_n$ is a well defined linear map between the algebraic tensor products. Hence Definition 1.10.17 and the following calculation, where $t = \sum_{j=1}^{m} x_{1,j} \otimes x_{2,j} \otimes \cdots \otimes x_{n,j}$, show that it is well defined as a continuous linear map (with the given norm) between the projective tensor products:

$$\|T_1 \otimes T_2 \otimes \cdots \otimes T_n(t)\|_w = \sup\{|\sum_{j=1}^{m} \prod_{k=1}^{n} \omega_k(T_k(x_{k,j}))| : \omega_k \in (\mathcal{Y}^k)_1^*\}$$

$$= \sup\{|\sum_{j=1}^{m} \prod_{k=1}^{n} T_k^*(\omega_k)(x_{k,j})| : \omega_k \in (\mathcal{Y}^k)_1^*\}$$

$$\leq \prod_{k=1}^{n} \|T_k\| \sup\{|\sum_{j=1}^{m} \prod_{k=1}^{n} \omega_k(x_{k,j})| : \omega_k \in (\mathcal{X}^k)_1^*\} = \prod_{k=1}^{n} \|T_k\| \, \|t\|_w.$$

In order to see the injectivity result, consider the obvious map $\tilde{\theta}$ of the algebraic tensor product $\mathcal{Z} = \mathcal{X}^1 \otimes \mathcal{X}^2 \otimes \cdots \otimes \mathcal{X}^n$ into $\mathcal{B}(\mathcal{Y}^{1*}, \mathcal{Y}^{2*}, \ldots, \mathcal{Y}^{n*}; \mathbb{C})$. Because of the Hahn–Banach theorem, the norm on the image of $\mathcal{Z}$ is no different from the norm on $\theta(\mathcal{Z})$ in $\mathcal{B}(\mathcal{X}^{1*}, \mathcal{X}^{2*}, \ldots, \mathcal{X}^{n*}; \mathbb{C})$. But the former norm is that of $\mathcal{Z}$ considered as a subspace of $\mathcal{Y}^1 \check{\otimes} \mathcal{Y}^2 \check{\otimes} \cdots \check{\otimes} \mathcal{Y}^n$ and the latter is its norm as a subspace of $\mathcal{X}^1 \check{\otimes} \mathcal{X}^2 \check{\otimes} \cdots \check{\otimes} \mathcal{X}^n$. Since these are the respective injective tensor norms, $T_1 \otimes T_2 \otimes \cdots \otimes T_n$ is injective and isometric.   □

There are Banach spaces for which the projective tensor product does not satisfy the second paragraph of Proposition 1.10.19 and others for which the injective tensor product does not satisfy the last paragraph of Proposition 1.10.10. It is seldom true that the injective and projective tensor products of two Banach spaces agree, but Gilles Pisier [1981], [1983] contradicted a conjecture of Grothendieck by giving an example of a Banach space $\mathcal{X}$ satisfying $\mathcal{X} \hat{\otimes} \mathcal{X} = \mathcal{X} \check{\otimes} \mathcal{X}$.

We will briefly discuss the dual space of the injective tensor product, but we refer the reader to Chapter 8 of Diestel and Uhl [1977] for a more complete discussion. We restrict attention to two Banach spaces $\mathcal{X}$ and $\mathcal{Y}$. In the weak* topology, $\Omega = \mathcal{X}_1^* \times \mathcal{Y}_1^*$ is compact (by Alaoglu's theorem), and it is easy to see that an element of the algebraic tensor product $\mathcal{X} \otimes \mathcal{Y}$ defines a continuous function on Ω:

$$\sum_{j=1}^{m} x_j \otimes y_j(\omega, \tau) = \sum_{j=1}^{m} \omega(x_j)\tau(y_j) \qquad \forall \, (\omega, \tau) \in \Omega.$$

Definition 1.10.17 shows that this injection into $C(\Omega)$ is an isometry.

1.10.20 Theorem *Let $\mathcal{X}$ and $\mathcal{Y}$ be Banach spaces, and let $\Omega = \mathcal{X}_1^* \times \mathcal{Y}_1^*$ be the compact set defined by the weak* topology. Then $M(\Omega)$ is isometrically isomorphic to the dual space $(\mathcal{X} \check{\otimes} \mathcal{Y})^*$ of the injective tensor product under the map $\mu \mapsto \tilde{\mu}$ defined by*

$$\tilde{\mu}\left(\sum_{j=1}^{m} x_j \otimes y_j\right) = \sum_{j=1}^{m} \int_{\Omega} \omega(x_j)\tau(y_j) d\mu(\omega, \tau) \qquad \forall \, \mu \in M(\Omega); x_j \in \mathcal{X}; y_j \in \mathcal{Y}.$$

Proof It is immediate that $\tilde{\mu}$ is a continuous linear functional with norm bounded above by $\|\mu\|$.

Suppose, on the other hand, that a continuous linear functional on $\mathcal{X} \check{\otimes} \mathcal{Y}$ is given. The Hahn–Banach theorem shows that it can be extended from the isometric image of $\mathcal{X} \check{\otimes} \mathcal{Y}$ in $C(\Omega)$ to all of $C(\Omega)$ without increasing its norm. Then the Riesz representation theorem shows that it has the form $\tilde{\mu}$ for some $\mu \in M(\Omega)$ with $\|\mu\|$ no bigger than the norm of the original linear functional. $\qquad\qquad \square$

Injective Tensor Products of Algebras

We will give two related examples of injective tensor products of algebras. Let $\Omega_1, \Omega_2, \ldots, \Omega_n$ be sets and, for each k, let $\mathcal{A}(\Omega_k)$ be a linear space of complex valued functions on Ω_k. For each k, let $f_k \in \mathcal{A}(\Omega_k)$ be given and define

$$f_1 \otimes f_2 \otimes \cdots \otimes f_n(\omega_1, \omega_2, \ldots, \omega_n) = f_1(\omega_1)f_2(\omega_2)\cdots f_n(\omega_n). \qquad (25)$$

Then $f_1 \otimes f_2 \otimes \cdots \otimes f_n$ is a function on $\Omega_1 \times \Omega_2 \times \cdots \times \Omega_n$. Since the map $(f_1, f_2, \ldots, f_n) \mapsto f_1 \otimes f_2 \otimes \cdots \otimes f_n$ is clearly multilinear, this defines a linear map of $\mathcal{A}(\Omega_1) \otimes \mathcal{A}(\Omega_2) \otimes \cdots \otimes \mathcal{A}(\Omega_n)$ into the set $\mathcal{F} = \mathcal{F}(\Omega_1 \times \Omega_2 \times \cdots \times \Omega_n)$ of all complex functions on $\Omega_1 \times \Omega_2 \times \cdots \times \Omega_n$. If each $\mathcal{A}(\Omega_k)$ is an algebra under pointwise operations, then this map is a homomorphism when $\mathcal{F}$ is also considered as an algebra under pointwise operations. We state the most interesting case formally. Let Ω be a locally compact Hausdorff space. Recall the Banach algebra $C_0(\Omega)$ defined in §1.5.1.

1.10.21 Proposition *If $\Omega_1, \Omega_2, \ldots, \Omega_n$ are locally compact Hausdorff spaces, then the map indicated above establishes an isometric algebra isomorphism of $C_0(\Omega_1) \check\otimes C_0(\Omega_2) \check\otimes \cdots \check\otimes C_0(\Omega_n)$ onto $C_0(\Omega_1 \times \Omega_2 \times \cdots \times \Omega_n)$.*

Proof It is enough to consider the case $n = 2$ and we do so in order to simplify notation. The map indicated is obviously an algebra homomorphism of $C_0(\Omega_1) \otimes C_0(\Omega_2)$ onto a subalgebra of $C_0(\Omega_1 \times \Omega_2)$ which separates points, does not vanish identically at any point and is closed under complex conjugation. Furthermore, any element $f = \sum_{j=1}^m f_{1,j} \otimes f_{2,j}$ satisfies

$$
\begin{aligned}
\|f\|_w &= \sup\{|\sum_{j=1}^m \tau_1(f_{1,j})\tau_2(f_{2,j})| : \tau_k \in C_0(\Omega_k)_1^*\} \\
&= \sup\{\|\sum_{j=1}^m \tau_1(f_{1,j})f_{2,j}\|_\infty : \tau_1 \in C_0(\Omega_1)_1^*\} \\
&= \sup\{|\sum_{j=1}^m \tau_1(f_{1,j})f_{2,j}(\omega_2)| : \tau_1 \in C(\Omega_1)_1^*; \omega_2 \in \Omega_2\} \\
&= \sup\{\|\sum_{j=1}^m f_{2,j}(\omega_2)f_{1,j}\|_\infty : \omega_2 \in \Omega_2\} \\
&= \sup\{|\sum_{j=1}^m f_{2,j}(\omega_1)f_{2,j}(\omega_2)| : \omega_k \in \Omega_k\} = \|f\|_\infty.
\end{aligned}
$$

Hence the map is an isometry. Thus it can be extended to an isometry of $C_0(\Omega_1) \check\otimes C_0(\Omega_2)$ onto a closed subspace of $C_0(\Omega_1 \times \Omega_2)$. Since multiplication and complex conjugation are continuous, the image of this map is a subalgebra which is closed under complex conjugation and in the norm. It also separates points and does not vanish identically at any point. The Stone–Weierstrass theorem asserts that such a subalgebra is all of $C_0(\Omega_1 \times \Omega_2)$. $\square$

If Ω is a locally compact Hausdorff space and $\mathcal{A}$ is a Banach algebra, then the set $C_0(\Omega, \mathcal{A})$ of continuous functions from Ω to $\mathcal{A}$ which vanish at infinity is a Banach algebra under pointwise operations and the supremum norm. The elements of $C_0(\Omega) \otimes \mathcal{A}$ can be interpreted as elements of $C_0(\Omega, \mathcal{A})$ through the definition

$$(f \otimes a)(\omega) = f(\omega)a \qquad \forall \ f \in C_0(\Omega); \ a \in \mathcal{A}; \ \omega \in \Omega.$$

1.10.22 Proposition *Let Ω be a locally compact space and let $\mathcal{A}$ be a Banach algebra. Then the map described above induces an isometric algebra isomorphism of $C_0(\Omega) \check{\otimes} \mathcal{A}$ onto $C_0(\Omega, \mathcal{A})$.*

Proof The proof is similar to that of the last proposition except for showing that the image of the tensor product is dense. Clearly the set $C_{00}(\Omega, \mathcal{A})$ of continuous functions from Ω to $\mathcal{A}$ which have compact support is dense. Hence we may assume that Ω is compact. Let $f \in C(\Omega, \mathcal{A})$ and $\varepsilon > 0$ be given. Choose $f(\omega_1), f(\omega_2), \ldots, f(\omega_m)$ in $f(\Omega)$ so that for each $f(\omega)$ with $\omega \in \Omega$ there is some j satisfying $\|f(\omega) - f(\omega_j)\| < \varepsilon$. For each j, define $V_j \subseteq \Omega$ by $V_j = \{\omega \in \Omega : \|f(\omega) - f(\omega_j)\| < \varepsilon\}$ so that $V_1, V_2, \ldots, V_m$ is an open cover for Ω. There is a partition of unity $e_1, e_2, \ldots, e_m$ subordinate to this cover. By this we mean that each e_j is a continuous function of Ω into $[0,1]$ satisfying $e_j(\omega) = 0$ for $\omega \notin V_j$ and $\sum_{j=1}^{m} e_j(\omega) = 1$ for all $\omega \in \Omega$. (For the existence of such a partition of unity, see Bourbaki [1966], IX.4.3.) Thus we have

$$\left\| f(\omega) - \sum_{j=1}^{m} e_j \otimes f(\omega_j)(\omega) \right\| = \left\| \sum_{j=1}^{m} (f(\omega) - f(\omega_j)) e_j(\omega) \right\|$$

$$\leq \sum_{\{j : \omega \in V_j\}} e_j(\omega) \|f(\omega) - f(\omega_j)\| < \varepsilon \qquad \forall \ \omega \in \Omega.$$

This proves that the image of the tensor product is dense as desired. □

Other Complete Tensor Products

There are many similarities between projective and injective tensor products. Their most remarkable property is that each is defined independent of the particular Banach spaces occurring in the tensor product. Both Schatten [1950] and Grothendieck [1954], [1955] investigated this phenomenon and the latter found 14 equivalence classes of such norms. Their study is based on constructing norms on the tensor products of finite-dimensional subspaces, but we will give no further details.

Both the projective and injective tensor products are the completion of the algebraic tensor product with respect to a suitable norm. Both norms satisfy the following definition. The term "reasonable" and the next theorem are due to Grothendieck, while the term "uniform" was used already be Schatten. We will not use his term "cross norm".

We are giving the next definition in the geometric category of Banach spaces, as is most common and often suitable. However the reader should be aware that modifications suitable for the topological category of Banach algebras have been studied. Gil de Lamadrid [1967] called a norm on the algebraic tensor product *admissible* if it dominates the injective tensor norm and is dominated by the projective tensor norm (*cf.* Theorem 1.10.24(d) below). This terminology has been adopted by others, but we are not aware that his name *"modular"* for a uniform norm up to topological equivalence has seen recent use. In the cited paper he offers a more general setting in which modular norms can be defined.

1.10.23 Definition Let $\mathcal{X}^1, \mathcal{X}^2, \ldots, \mathcal{X}^n$ be normed linear spaces. Let $\|\cdot\|_\alpha$ be a norm on the algebraic tensor product $\mathcal{Z} = \mathcal{X}^1 \otimes \mathcal{X}^2 \otimes \cdots \otimes \mathcal{X}^n$. The norm induced on $\tilde{\mathcal{Z}} = \mathcal{X}^{1*} \otimes \mathcal{X}^{2*} \otimes \cdots \otimes \mathcal{X}^{n*}$ by its embedding into the dual space of $(\mathcal{Z}, \ \|\cdot\|_\alpha)$ defined on elementary tensors by

$$\omega_1 \otimes \omega_2 \otimes \cdots \otimes \omega_n(x_1, x_2, \ldots, x_n) = \prod_{k=1}^{n} \omega_k(x_k) \qquad \forall \ x_k \in \mathcal{X}^k; \ \omega_k \in (\mathcal{X}^k)^*$$

is called the *dual norm* and denoted by $\|\cdot\|_{\alpha^*}$.

A norm on $\mathcal{Z}$ is called a *reasonable* norm if it satisfies

$$\|x_1 \otimes x_2 \otimes \cdots \otimes x_n\|_\alpha \ \leq \ \|x_1\| \, \|x_2\| \cdots \|x_n\| \ \ \forall \ x_k \in \mathcal{X}^k \quad \text{and} \quad (26)$$
$$\|\omega_1 \otimes \omega_2 \otimes \cdots \otimes \omega_n\|_{\alpha^*} \ \leq \ \|\omega_1\| \, \|\omega_2\| \cdots \|\omega_n\| \ \ \forall \ \omega_k \in (\mathcal{X}^k)^* \qquad (27)$$

The completion of $\mathcal{Z}$ with respect to a reasonable norm $\|\cdot\|_\alpha$ on $\mathcal{Z}$ is denoted by $\mathcal{X}^1 \overset{\alpha}{\otimes} \mathcal{X}^2 \overset{\alpha}{\otimes} \cdots \overset{\alpha}{\otimes} \mathcal{X}^n$.

A reasonable norm on $\mathcal{Z}$ is called *uniform* if all choices of $T_k \in \mathcal{B}(\mathcal{X}^k)$ satisfy

$$\|T_1 \otimes T_2 \otimes \cdots \otimes T_n\| = \|T_1\| \, \|T_2\| \cdots \|T_n\|$$

where the norm on the left is the operator norm in $\mathcal{B}(\mathcal{Z})$. A reasonable norm $\|\cdot\|_\alpha$ on $\mathcal{Z}$ is called *nuclear* if the natural projection of $\mathcal{X}^1 \overset{\alpha}{\otimes} \mathcal{X}^2 \overset{\alpha}{\otimes} \cdots \overset{\alpha}{\otimes} \mathcal{X}^n$ onto $\mathcal{X}^1 \check{\otimes} \mathcal{X}^2 \check{\otimes} \cdots \check{\otimes} \mathcal{X}^n$ is injective.

Explicitly, the dual norm is $\|u\|_{\alpha^*} = \sup\{|u(t)| \ : \ t \in \mathcal{Z}; \|t\|_\alpha \leq 1\}$ where $u = \sum_{j=1}^{m} \omega_{1,j} \otimes \omega_{2,j} \otimes \cdots \otimes \omega_{n,j} \in \tilde{\mathcal{Z}}$, $t = \sum_{i=1}^{p} x_{1,i} \otimes x_{2,i} \otimes \cdots \otimes x_{n,i}$ and hence $u(t) = \sum_{j=1}^{m} \sum_{i=1}^{p} \prod_{k=1}^{n} \omega_{k,j}(x_{k,i})$.

1.10.24 Theorem *Let $\mathcal{X}^1, \mathcal{X}^2, \ldots, \mathcal{X}^n$ be normed linear spaces and denote $\mathcal{X}^1 \otimes \mathcal{X}^2 \otimes \cdots \otimes \mathcal{X}^n$ and $\mathcal{X}^{1*} \otimes \mathcal{X}^{2*} \otimes \cdots \otimes \mathcal{X}^{n*}$ by $\mathcal{Z}$ and $\tilde{\mathcal{Z}}$, respectively.*

(a) For reasonable norms, the inequalities (26) and (27) are actually equalities.

(b) The dual of a reasonable norm is a reasonable norm.

(c) Both the projective tensor norm and the injective tensor norm on $\mathcal{Z}$ are uniform reasonable norms. When either is applied to $\tilde{\mathcal{Z}}$ it is the dual of the other on $\mathcal{Z}$.

(d) A norm $\|\cdot\|_\alpha$ on $\mathcal{Z}$ is reasonable if and only if it satisfies

$$\|t\|_w \leq \|t\|_\alpha \leq \|t\|_p \qquad \forall \ t \in \mathcal{Z}. \qquad (28)$$

Thus the injective and projective tensor norms are, respectively, the smallest and largest reasonable norms.

Proof (a): Given any $x_k \in \mathcal{X}^k$, choose $\omega_k \in (\mathcal{X}^k)_1^*$ satisfying $\omega_k(x_k) = \|x_k\|$ for each k. Then (27) shows $\|\omega_1 \otimes \omega_2 \otimes \cdots \otimes \omega_n\|_{\alpha^*} \leq 1$ which implies $\|x_1 \otimes x_2 \otimes \cdots \otimes x_n\|_\alpha \geq \prod_{k=1}^n \|x_k\|$, as we wished to show.

The proof for (27) is the dual of the one just given except that this time we must consider sequences $\{x_{kn}\}_{n \in \mathbb{N}} \subseteq \mathcal{X}^k$ satisfying $\|x_{kn}\| = 1$ and $\lim \omega_k(x_{kn}) = \|\omega_k\|$.

(b): Let $\|\cdot\|_\alpha$ be a reasonable norm. Since (26) for $\|\cdot\|_{\alpha^*}$ is just (27) for $\|\cdot\|_\alpha$, we need only establish (27) for $\|\cdot\|_{\alpha^*}$. By induction, it is enough to consider the product of two normed linear spaces $\mathcal{X}$ and $\mathcal{Y}$. Let $\langle\, F \in \mathcal{X}^{**} \,/\, G \in \mathcal{Y}^{**} \,\rangle$ be arbitrary. Goldstine's theorem (Dunford and Schwartz [1958], V.4.5) allows us to choose a sequence $\langle\, \{x_\gamma\}_{\gamma \in \Gamma} \,/\, \{y_\gamma\}_{\gamma \in \Gamma} \,\rangle$ satisfying $\langle\, x_\gamma \to F \,/\, y_\gamma \to G \,\rangle$ and $\langle\, \|x_\gamma\| \leq \|F\| \,/\, \|y_\gamma\| \leq \|G\| \,\rangle$. An arbitrary tensor $u = \sum_{j=1}^m \omega_j \otimes \tau_j$ in $\mathcal{X}^* \otimes \mathcal{Y}^*$ satisfies

$$|(F \otimes G)(u)| = \left| \sum_{j=1}^m F(\omega_j) G(\tau_j) \right| = \lim \left| \sum_{j=1}^m \omega_j(x_\beta) \tau_j(y_\beta) \right|$$
$$\leq \limsup \|x_\beta\| \|y_\beta\| \|u\|_{\alpha^*} \leq \|F\| \|G\| \|u\|_{\alpha^*},$$

establishing $\|F \otimes G\|_{\alpha^{**}} \leq \|F\| \|G\|$.

(c), (d): The first paragraph of Proposition $\langle$ 1.10.10 / 1.10.19 $\rangle$ shows that the $\langle$ projective / injective $\rangle$ tensor norm is uniform.

First we will prove inequality (28) for any reasonable norm $\|\cdot\|_\alpha$. Any $\omega_k \in (\mathcal{X}^k)_1^*$ and any $t \in \mathcal{Z}$ satisfy

$$|\omega_1 \otimes \omega_2 \otimes \cdots \otimes \omega_n(t)| \leq \|\omega_1 \otimes \omega_2 \otimes \cdots \otimes \omega_n\|_{\alpha^*} \|t\|_\alpha \leq \|t\|_\alpha.$$

Since the weak tensor norm of t is defined to be the supremum over all the expressions on the left hand side, we conclude $\|t\|_w \leq \|t\|_\alpha$.

Any $t = \sum_{j=1}^m x_{1,j} \otimes x_{2,j} \otimes \cdots \otimes x_{n,j} \in \mathcal{Z}$ satisfies

$$\|t\|_\alpha \leq \sum_{j=1}^m \|x_{1,j} \otimes x_{2,j} \otimes \cdots \otimes x_{n,j}\|_\alpha = \sum_{j=1}^m \prod_{k=1}^n \|x_{k,j}\|$$

Since $\|t\|_p$ is defined to be the infimum over all the expressions on the right hand side, we get $\|t\|_\alpha \leq \|t\|_p$, completing the proof of (28).

Next we show that $\|\cdot\|_w$ and $\|\cdot\|_p$ are reasonable. Equation (26) is clear for both norms. The definition of the weak tensor norm shows

$$|\omega_1 \otimes \omega_2 \otimes \cdots \otimes \omega_n(t)| \leq \|\omega_1\| \|\omega_2\| \cdots \|\omega_n\| \|t\|_w \qquad \forall\, \omega_k \in (\mathcal{X}^k)^*;\ t \in \mathcal{Z}$$

which implies (27) for $\|\cdot\|_w$. These results imply

$$\|u\|_w \leq \|u\|_{p^*} \leq \|u\|_{w^*} \leq \|u\|_p \qquad \forall\, u \in \tilde{\mathcal{Z}}$$

which gives (27) for $||\cdot||_p$ when u is an elementary tensor.

Let us calculate the dual of the projective tensor norm for any $u \in \tilde{\mathcal{Z}}$:

$$
\begin{aligned}
||u||_{p^*} &= \sup\{|u(t)| : t \in \mathcal{Z};\ ||t||_p \leq 1\} \\
&= \sup\left\{\left|\sum_{j=1}^m u(x_{1,j} \otimes x_{2,j} \otimes \cdots \otimes x_{n,j})\right| : \right. \\
&\qquad\qquad\qquad\qquad \left. x_{k,j} \in \mathcal{X}^k;\ \sum_{j=1}^m \prod_{k=1}^n ||x_{k,j}|| \leq 1\right\} \\
&\leq \sup\left\{\sum_{j=1}^m |u(x_{1,j} \otimes x_{2,j} \otimes \cdots \otimes x_{n,j})| : \right. \\
&\qquad\qquad\qquad\qquad \left. x_{k,j} \in \mathcal{X}^k;\ \sum_{j=1}^m \prod_{k=1}^n ||x_{k,j}|| \leq 1\right\} \\
&= \sup\left\{|u(x_1 \otimes x_2 \otimes \cdots \otimes x_n)| : x_k \in \mathcal{X}^k;\ \prod_{k=1}^n ||x_k|| \leq 1\right\} \\
&= ||u||_w \leq ||u||_{p^*}.
\end{aligned}
$$

Finally we calculate the dual of the injective tensor norm for any $u = \sum_{j=1}^m \omega_{1,j} \otimes \omega_{2,j} \otimes \cdots \otimes \omega_{n,j} \in \tilde{\mathcal{Z}}$:

$$
\begin{aligned}
||u||_{w^*} &= \sup\{|u(t)| : t \in \mathcal{Z};\ ||t||_w \leq 1\} \\
&\leq \sup\left\{\sum_{j=1}^m |\omega_{1,j} \otimes \omega_{2,j} \otimes \cdots \otimes \omega_{n,j}(t)| : t \in \mathcal{Z};\ ||t||_w \leq 1\right\} \\
&\leq \sum_{j=1}^m \sup\{|\omega_{1,j} \otimes \omega_{2,j} \otimes \cdots \otimes \omega_{n,j}(t)| : t \in \mathcal{Z};\ ||t||_w \leq 1\} \\
&= \sum_{j=1}^m \prod_{k=1}^n ||\omega_{k,j}||.
\end{aligned}
$$

The projective tensor norm of u is just the infimum over all the expressions occurring on the last line, giving the desired inequality and equality.

Finally suppose $||\cdot||_\alpha$ is a norm on $\mathcal{Z}$ satisfying (28). When t is an elementary tensor, (28) becomes (26) with an equality sign. By the definition of the dual norm, (28) implies

$$
||u||_w = ||u||_{p^*} \leq ||u||_{\alpha^*} \leq ||u||_{w^*} = ||u||_p \qquad \forall\, u \in \tilde{\mathcal{Z}}.
$$

When u is an elementary tensor, this gives (27) with an equality sign. □

Because of inequality (28), it is obvious that there are contractive linear maps between the completions of $\mathcal{X}^1 \otimes \mathcal{X}^2 \otimes \cdots \otimes \mathcal{X}^n$ in the corresponding norms:

$$\mathcal{X}^1 \hat{\otimes} \mathcal{X}^2 \hat{\otimes} \cdots \hat{\otimes} \mathcal{X}^n \to \mathcal{X}^1 \overset{\alpha}{\otimes} \mathcal{X}^2 \overset{\alpha}{\otimes} \cdots \overset{\alpha}{\otimes} \mathcal{X}^n \to \mathcal{X}^1 \check{\otimes} \mathcal{X}^2 \check{\otimes} \cdots \check{\otimes} \mathcal{X}^n.$$

Caution: Even though these maps are extensions by continuity of the identity map on the algebraic tensor product, they may have nontrivial kernels. In particular, a formula which makes sense on the space $\mathcal{B}_A(\mathcal{X}, \mathcal{Y}) \simeq \mathcal{Y} \check{\otimes} \mathcal{X}^*$ of approximable operators is not well defined on the projective tensor product.

1.10.25 General Tensor Products of Banach Algebras It is natural to ask when a reasonable norm $|| \cdot ||_\alpha$ on the algebraic tensor product (considered as an algebra) $\mathcal{A} = \mathcal{A}^{(1)} \otimes \mathcal{A}^{(2)} \otimes \cdots \otimes \mathcal{A}^{(n)}$ of Banach algebras $\mathcal{A}^{(1)}, \mathcal{A}^{(2)}, \ldots, \mathcal{A}^{(n)}$ defines a Banach algebra structure on the completed tensor product $\overline{\mathcal{A}}^\alpha = \mathcal{A}^{(1)} \overset{\alpha}{\otimes} \mathcal{A}^{(2)} \overset{\alpha}{\otimes} \cdots \overset{\alpha}{\otimes} \mathcal{A}^{(n)}$. The obvious answer, that this happens exactly when the norm $|| \cdot ||_\alpha$ is submultiplicative, is useful, since submultiplicativity may be as easy to check directly as some of the properties we mention below. This question was first considered by Gelbaum [1962] and Gil de Lamadrid [1963], [1965a]. Subsequently, T. Keith Carne [1978] provided another type of answer, although he seems to have been unaware of the previous work. We will summarize some of this work without giving full details.

If $|| \cdot ||_\alpha$ is a uniform norm, then $\mathcal{B} = \mathcal{B}(\mathcal{A}^{(1)}) \otimes \mathcal{B}(\mathcal{A}^{(2)}) \otimes \cdots \otimes \mathcal{B}(\mathcal{A}^{(n)})$ clearly acts on $(\mathcal{A}, || \cdot ||_\alpha)$ in a bounded way:

$$
\begin{aligned}
||u(t)||_\alpha &\leq \sum_{j=1}^m ||T_{1,j} \otimes T_{2,j} \otimes \cdots \otimes T_{n,j}(t)||_\alpha \\
&\leq \sum_{j=1}^m \prod_{k=1}^n ||T_{k,j}|| \, ||t||_\alpha \quad \forall \, u \in \mathcal{B}; \, t \in \mathcal{A}.
\end{aligned}
$$

where $u = \sum_{j=1}^m T_{1,j} \otimes T_{2,j} \otimes \cdots \otimes T_{n,j}$ Hence we may define the operator norm $|| \cdot ||_{op\alpha}$ on $\mathcal{B}$ relative to $(\mathcal{A}, || \cdot ||_\alpha)$ and this is bounded above by the projective tensor norm. The dual action of $u^* = \sum_{j=1}^m T_{1,j}^* \otimes T_{2,j}^* \otimes \cdots \otimes T_{n,j}^*$ on $\mathcal{A}^*$ also satisfies $||u^*||_{op\alpha} \leq ||u^*||_p$, showing that $|| \cdot ||_{op\alpha}$ is reasonable. This operator norm is obviously submultiplicative. Gil de Lamadrid's main result essentially follows from the above remarks.

Proposition Let $\mathcal{A} = \mathcal{A}^{(1)} \otimes \mathcal{A}^{(2)} \otimes \cdots \otimes \mathcal{A}^{(n)}$ be Banach algebras such that the left regular representation L of each is an isometry. Let $|| \cdot ||_\alpha$ be a uniform reasonable norm on $\mathcal{A} = \mathcal{A}^{(1)} \otimes \mathcal{A}^{(2)} \otimes \cdots \otimes \mathcal{A}^{(n)}$. For each $t \in \mathcal{A}$, define $||t||_\beta$ to be $||L_t||_{op\alpha}$ where $|| \cdot ||_{op\alpha}$ is defined as indicated above on $\mathcal{B} = \mathcal{B}(\mathcal{A}^{(1)}) \otimes \mathcal{B}(\mathcal{A}^{(2)}) \otimes \cdots \otimes \mathcal{B}(\mathcal{A}^{(n)})$. Then $|| \cdot ||_\beta$ is an algebra norm on $\mathcal{A}$, and $|| \cdot ||_\alpha$ is submultiplicative if and only if $|| \cdot ||_\alpha = || \cdot ||_\beta$.

If $t = \sum_{j=1}^{m} a_{1,j} \otimes a_{2,j} \otimes \cdots \otimes a_{n,j} \in \mathcal{A}$, note that L_t is simply $\sum_{j=1}^{m} L_{a_{1,j}} \otimes L_{a_{2,j}} \otimes \cdots \otimes L_{a_{n,j}} \in \mathcal{B}$. Now it is easy to see that if $||\cdot||_\alpha$ is submultiplicative, then the two norms are equal. The converse is immediate. Note, however that the very ease of this proof shows that proving $||\cdot||_\alpha = ||\cdot||_\beta$ is likely to involve checking submultiplicativity directly. As Gil de Lamadrid noted, the main force of this theorem is to prove the existence of a reasonable number of algebra norms on tensor product algebras such as $\mathcal{A}$.

Carne [1978] considers the natural tensor norms in Grothendieck's sense mentioned above. He gives several necessary and sufficient conditions for such a norm to be submultiplicative. These conditions depend on a sort of mixed associativity of the metric tensor products. Using these criteria, he is able to show that representatives of three of the fourteen equivalence classes of natural tensor norms which Grothendieck defined and studied are submultiplicative.

It is obvious that the tensor product algebra of two unital algebras $\mathcal{A}^{(1)}$ and $\mathcal{A}^{(2)}$ is unital, even when the algebraic tensor product $\mathcal{A}^{(1)} \otimes \mathcal{A}^{(1)}$ is completed under some norm. Richard J. Loy [1970b] proves a very general converse for Banach algebras: if $\mathcal{A} = \mathcal{A}^{(1)} \overset{\alpha}{\otimes} \mathcal{A}^{(1)}$ is the completion of $\mathcal{A}^{(1)} \otimes \mathcal{A}^{(1)}$ under some reasonable (or even just admissible) algebra norm, the $\mathcal{A}$ is unital if and only if $\mathcal{A}^{(1)}$ and $\mathcal{A}^{(2)}$ are. Gelbaum [1962] contains a similar but more restricted result. For an extension of these results to approximate identities see Proposition 5.1.4.

A number of the papers cited before Theorem 1.10.12 above deal with the ideal theory of algebra tensor products. In particular, Gelbaum [1959], [1961], [1962], [1970], Tomiyama [1962], Lebow [1968] and Laursen [1970] fall in this category with the two 1970 papers correcting errors in the first pair. See also Carne [1981]. Some will be discussed in subsequent chapters.

1.10.26 Reasonable Hilbert Space Norms and C*-Norms §1.7.17 introduced Hilbert spaces and C*-norms on *-algebras. They will be studied much more intensively in Volume II. However, they present an important natural situation in which reasonable norms arise which differ from the projective and injective tensor norms.

If $\mathcal{H}^1, \mathcal{H}^2, \ldots, \mathcal{H}^n$ are Hilbert spaces, it is natural that we should wish to complete their algebraic tensor product to be a Hilbert space again. Since the norm on a Hilbert space is most often defined in terms of the inner product, we give $\mathcal{Z} = \mathcal{H}^1 \otimes \mathcal{H}^2 \otimes \cdots \otimes \mathcal{H}^n$ an inner product and then use that to define a norm. For elementary tensors we define

$$(x_1 \otimes x_2 \otimes \cdots \otimes x_n, y_1 \otimes y_2 \otimes \cdots \otimes y_n) = \prod_{k=1}^{n} (x_k, y_k) \quad \forall\, x_k, y_k \in \mathcal{H}^k.$$

Clearly this expression can be extended by bilinearity to $\mathcal{H}^1 \otimes \mathcal{H}^2 \otimes \cdots \otimes \mathcal{H}^n$ and satisfies the properties of an inner product. The only question which might arise is whether the inner product is definite, *i.e.*, whether $t \in \mathcal{Z}$ and $(t, u) = 0$ for all $u \in \mathcal{Z}$ implies $t = 0$. However, the semi-norm given by this inner product is certainly a reasonable norm, so it is larger than the injective tensor norm and really is a norm rather than just a semi-norm. Explicitly the norm is

$$\|t\| = \left(\sum_{j=1}^{m} \sum_{i=1}^{m} \prod_{k=1}^{n} (x_{k,j}, x_{k,i}) \right)^{1/2} \qquad \forall \, t = \sum_{j=1}^{m} x_{1,j} \otimes x_{2,j} \otimes \cdots \otimes x_{n,j} \in \mathcal{Z}.$$

Now let $\mathcal{A}^{(1)}, \mathcal{A}^{(2)}, \ldots, \mathcal{A}^{(n)}$ be C*-algebras and denote their algebraic tensor product (considered as an algebra) by $\mathcal{A}$. We may introduce an involution in $\mathcal{A}$ by setting a^* equal to $\sum_{j=1}^{m} a_{1,j}^* \otimes a_{2,j}^* \otimes \cdots \otimes a_{n,j}^*$ when $a = \sum_{j=1}^{m} a_{1,j} \otimes a_{2,j} \otimes \cdots \otimes a_{n,j}$. It is easy to see that this is well defined and satisfies the four properties of an involution given in §1.7.17. Thus we have made $\mathcal{A}$ into a *-algebra. In Volume II we will consider a number of classes of *-algebras in which there is a largest C*-norm, called the Gelfand–Naĭmark norm, for each algebra. It turns out that $\mathcal{A}$ has this property and that the Gelfand–Naĭmark norm $\|\cdot\|_\gamma$ is a reasonable norm on it. We denote the completion of $\mathcal{A}$ with respect to $\|\cdot\|_\gamma$ by $\overline{\mathcal{A}}^\gamma = \mathcal{A}^{(1)} \overset{\gamma}{\otimes} \mathcal{A}^{(2)} \overset{\gamma}{\otimes} \cdots \overset{\gamma}{\otimes} \mathcal{A}^{(n)}$.

The unitization of any C*-algebra has an algebra norm satisfying the C*-condition. Thus it is again a C*-algebra. Example 5.1.11 below shows that C*-algebras are approximately unital. This can be used to show that any C*-norm on $\mathcal{A}$ can be extended to the tensor product of the unitizations of the $\mathcal{A}^{(k)}$. Hence we may suppose all these algebras were already unital. In the category of unital C*-algebras (in which morphisms are *-homomorphisms which are necessarily contractive), $\overline{\mathcal{A}}^\gamma$ satisfies exactly the same universal mapping property as the projective tensor product of unital Banach algebras does in the category of unital Banach algebras.

When we introduced C*-algebras, we noted that they were characterized by the property that each one has at least one isometric isomorphism into $\mathcal{B}(\mathcal{H})$ (for a suitable Hilbert space $\mathcal{H}$) which sends the involution in the C*-algebra onto the Hilbert space adjoint map in $\mathcal{B}(\mathcal{H})$. Suppose that for each $k = 1, 2, \ldots, n$, $\varphi_k \colon \mathcal{A}^{(k)} \to \mathcal{B}(\mathcal{H}^k)$ is such a map. Then for each elementary tensor $a_1 \otimes a_2 \otimes \cdots \otimes a_n \in \mathcal{A}$, equation (15) shows how to interpret $\varphi_1(a_1) \otimes \varphi_2(a_2) \otimes \cdots \otimes \varphi_n(a_n)$ as an element of $\mathcal{B}(\mathcal{H})$ where $\mathcal{H}$ is the Hilbert space tensor product of the $\mathcal{H}^k$. This extends to an injective homomorphism $\varphi \colon \mathcal{A} \to \mathcal{B}(\mathcal{H})$. It turns out, though it is far from obvious, that the operator norm on $\varphi(a)$ is independent of the choice of the φ_k. This norm is obviously a C*-norm and turns out to be a reasonable norm on $\mathcal{A}$. In fact it is easy to see that it is the smallest C*-norm on $\mathcal{A}$. Since both the largest and the smallest C*-norm on $\mathcal{A}$ are reasonable, all C*-norms on $\mathcal{A}$ are reasonable.

The largest and smallest C*-norms do not always agree on the tensor product of C*-algebras. A C*-algebra $\mathcal{B}$ is called *nuclear* if they do agree on any tensor product of $\mathcal{B}$ with any other C*-algebra. It is also known that the smallest C*-norm on the tensor product of two C*-algebras is strictly larger than the injective tensor norm unless at least one of the factors is commutative. Since $C_0(\Omega)$ is the typical example of a commutative C*-algebra, this shows that Propositions 1.10.21 and 1.10.22 cannot be extended. See Kadison and Ringrose [1986] for further details.

2

The Spectrum

Introduction

The spectrum is undoubtedly the most important concept in the theory of algebras. In fact, it is primarily the spectrum which differentiates the theory of algebras from the general theory of rings. For an element a in an algebra $\mathcal{A}$, the spectrum is a collection of complex numbers, denoted by $Sp(a)$, which is a sort of shadow of the element. (This is the origin of the name.) In finite-dimensional algebras, it is always nonempty and finite (cf. Section 2.7). However, in perfectly general algebras it may be empty or unbounded and thus too intractable to be useful.

In this introduction to the chapter we will only define the *spectrum* of an element a in a unital algebra $\mathcal{A}$. In this case

$$Sp(a) = \{\lambda \in \mathbb{C} : \lambda 1 - a \text{ has no inverse in } \mathcal{A}\}.$$

Thus the spectrum of a matrix in the algebra M_n of all $n \times n$-matrices is the set of eigenvalues, and the spectrum of a function f in the algebra of all continuous functions on the unit interval [0,1] is the range of the function: $\{f(t) : t \in [0,1]\}$. In general, the spectrum offers a way to study the phenomenon of invertibility. Note that it is an isomorphism and automorphism invariant; under an isomorphism between algebras the spectrum of an element and its image are the same.

After this introduction, we will begin the detailed portion of the chapter by extending this definition to nonunital algebras. This will involve the concepts of *quasi-multiplication* and the *quasi-inverse*. In perfectly general algebras the spectrum has several useful properties including the *spectral mapping theorem* (Theorem 2.1.10). We also explore its possible pathologies with examples and a few simple results. Zero plays a special role in the spectrum and must be treated separately in the statement of many theorems. The problem is that the spectrum does not behave well with respect to homomorphisms or subalgebras unless they are unital. It behaves much better in the category of unital algebras where all algebras and homomorphisms are unital.

The *spectral radius* $\rho(a)$ of an element $a \in \mathcal{A}$ is defined by

$$\rho(a) = \sup\{|\lambda| : \lambda \in Sp(a)\}.$$

We prove (Theorem 2.2.2) that any element in an algebra $\mathcal{A}$ with a non-trivial algebra semi-norm σ satisfies the inequality:

$$\lim_{n\to\infty} \sigma(a^n)^{1/n} \leq \rho(a) \qquad \forall\, a \in \mathcal{A}.$$

This shows, in particular, that the spectrum of such an element is not empty (since the spectral radius would be $-\infty$ in that case). The *Gelfand–Mazur theorem* (Corollary 2.2.3), which asserts that the complex field is the only normed division algebra, is an immediate consequence of this result.

Next we introduce the concept of *spectral semi-norms*. Many properties characterize these semi-norms. In particular they are the algebra semi-norms σ for which there is some constant C (in fact we may always choose $C = 1$) satisfying

$$\rho(a) \leq C\sigma(a) \qquad \forall\, a \in \mathcal{A}.$$

Alternatively, they are the algebra semi-norms satisfying *Gelfand's spectral radius formula*:

$$\rho(a) = \lim_{n\to\infty} \sigma(a^n)^{1/n} \qquad \forall\, a \in \mathcal{A}.$$

This is the limit denoted by σ^∞ in Theorem 1.1.10. Israel Moiseevič Gelfand [1941a] proved this formula for complete algebra norms, which are therefore the best known examples of spectral semi-norms. Furthermore, spectral semi-norms are exactly those algebra semi-norms relative to which the multiplicative group of invertible elements is open. Additional characterizations are contained in Theorems 2.2.5, 2.2.14, and 2.5.7, Propositions 2.2.7 and 2.5.15 and Corollary 2.4.8. Many other properties of spectral semi-norms are developed. The following list (with specific references to results in this work) shows that they arise frequently in analysis. They will play a prominent role in the rest of the work.

Examples of Spectral Semi-norms

(a) Any complete algebra norm (Gelfand [1941a]; Corollary 2.2.8).

(b) Any algebra norm (not necessarily complete) on a completely regular semisimple commutative Banach algebra (Rickart [1953]; Theorem 3.2.10).

(c) Any algebra norm (not necessarily complete) on a C*-algebra (Rickart [1953], cf. Cleveland [1963]; Theorem 6.1.16).

(d) Any algebra norm (not necessarily complete) on any two-sided ideal of $\mathcal{B}(\mathcal{X})$ or on any closed subalgebra of $\mathcal{B}(\mathcal{X})$ which includes all finite-rank operators, where $\mathcal{B}(\mathcal{X})$ is the algebra of all bounded linear operators on a Banach space $\mathcal{X}$ (Yood, [1958]; Theorem 2.5.17).

(e) Any algebra norm (not necessarily complete) on a modular annihilator algebra (Yood [1958] *cf.* Bonsall [1954a]; Theorem 7.4.12).

(f) The operator norm on a full Hilbert algebra (Rieffel [1969b]; Volume II).

(g) The Gelfand–Naĭmark semi-norm on a hermitian Banach *-algebra (Palmer [1972], Raĭkov [1946] and Pták [1970], [1972]; Theorem 10.4.11).

(h) Any algebra semi-norm on a Jacobson-radical algebra (Corollary 2.3.4).

The Jacobson Radical and Fundamental Theorem

Newburgh's theorem (Theorem 2.2.15) shows that the spectrum and spectral radius are upper semi-continuous functions with respect to any spectral semi-norm on an algebra.

In order to formulate one of the most fundamental results on spectral semi-norms, we introduce in a preliminary fashion the *Jacobson radical*, which will be fully discussed in Chapter 4. It has a special role in connection with the spectrum and spectral semi-norms. The Jacobson radical can be described as the largest ideal on which the spectral radius is identically 0 or, alternatively, as the largest ideal such that the spectrum of any element in the algebra agrees with the spectrum of the image of this element modulo the ideal. However in order to prove that any algebra has such a largest ideal, and for other reasons, we introduce *algebraically irreducible representations* and define the Jacobson radical of an algebra as the intersection of the kernels of all such representations.

After a brief introduction of these concepts, we are able to prove the *fundamental theorem of spectral semi-norms* (Theorem 2.3.6) which states that the limits of any sequence with respect to different spectral semi-norms differ by an element in the Jacobson radical. The proof uses a recent lemma due to Thomas J. Ransford [1989]. This is a rather spectacular result since spectral semi-norms are not at all unique. It has a number of corollaries, among which Barry E. Johnson's [1967a] *uniqueness of norm theorem* (Corollary 2.3.10) is the best known. This theorem asserts that any two complete norms on a semisimple algebra must be equivalent.

Commutativity and Modular Ideals

Theorem 1.1.10 and the Gelfand spectral radius formula show that, on a commutative algebra with a spectral semi-norm, the spectral radius is both subadditive and submultiplicative. It turns out that it is enough to assume that the algebra is commutative modulo the Jacobson radical. Commutative algebras exist where the spectral radius is always finite-valued but is neither subadditive nor submultiplicative (Example 2.1.7). We will show (Theorem 2.4.11) that if the spectral radius has either property, then it has both, and that then the algebra must be commutative modulo its Jacobson radical. (This is a full converse of the first statement above, since the spectral radius is a spectral semi-norm under these hypotheses.)

We briefly introduce *modular ideals*. They will be further discussed in Chapter 4. Here they are used to characterize spectral semi-norms as those algebra semi-norms in which all maximal modular ideals are closed (Corollary 2.4.8) and to describe the kernels of multiplicative linear functionals on an algebra, *i.e.*, algebra homomorphisms into the complex number field (Theorem 2.4.12).

Spectral Algebras and Spectral Subalgebras

A *spectral algebra* is an algebra on which some spectral semi-norm can be defined. We emphasize that it is not an algebra with a particular choice of spectral semi-norm, but merely an algebra on which it is possible to define at least one spectral semi-norm. Most spectral algebras have many spectral semi-norms, and no particular one of them is preferred above the others. Proposition 2.4.2 gives a purely algebraic description of spectral algebras. These algebras play a fundamental role throughout this work. Note that any Banach algebra is a spectral algebra, and Theorem 2.5.18 characterizes spectral algebras in terms of the existence of homomorphisms into Banach algebras.

The *Gleason–Kahane–Żelazko* theorem (Theorem 2.4.13) shows that a linear functional $\varphi \colon \mathcal{A} \to \mathbb{C}$ on a spectral algebra is a homomorphism if and only if it satisfies

$$\varphi(a) \in Sp(a) \qquad \forall\, a \in \mathcal{A}.$$

In the next chapter we will show that a commutative algebra satisfies the Gelfand theory if and only if it is a spectral algebra. Chapter 4 contains the proof that any irreducible representation of a spectral algebra is strictly dense, generalizing results of Nathan Jacobson [1945b] and Charles E. Rickart [1950].

The chapter concludes with a thorough discussion of *spectral subalgebras*. These are the subalgebras such that an element of the subalgebra has the same non-zero spectrum whether calculated in the subalgebra or in the larger algebra. N. Bourbaki [1967] calls unital subalgebras full (*pleine*) if they are spectral subalgebras, but many interesting cases of spectral subalgebras are not unital subalgebras, as the following list of spectral subalgebras shows. (If not otherwise indicated, the statements have easy proofs which are gathered in Proposition 2.5.3.)

Examples of Spectral Subalgebras

(a) Any one- or two-sided ideal.

(b) The commutant of any subset. Hence the center of an algebra and any maximal commutative subalgebra.

(c) The "corner" subalgebra $e\mathcal{A}e$, for any idempotent $e \in \mathcal{A}$.

(d) Any finite-dimensional subalgebra or any subalgebra in which every element satisfies some polynomial equation.

(e) Any modular annihilator subalgebra of a normed algebra (Corollary 2.5.9).

(f) Any closed subalgebra of a Banach algebra in which each element of the subalgebra has nowhere dense spectrum relative to the subalgebra (Corollary 2.5.11).

(g) Any closed *-subalgebra of a hermitian Banach *-algebra (Theorem 10.4.19).

(h) The range of the left or right regular or extended regular representation.

(i) The image $\kappa(\mathcal{A})$ of $\mathcal{A}$ in $\mathcal{A}^{**}$ with either Arens product.

(j) Any intersection of spectral subalgebras or any spectral subalgebra of a spectral subalgebra.

As one would expect, a spectral subalgebra of a spectral algebra is again a spectral algebra in its own right. (However, a spectral subalgebra of a nonspectral algebra need not be a spectral algebra. Despite this, we believe the name "spectral subalgebra", is appropriate, since spectral subalgebras are the appropriate kind of subobject in the category of spectral algebras. The morphisms in this category are just algebra homomorphisms.) This gives a long list of spectral algebras. For instance, any (not necessarily closed) ideal in a Banach algebra is a spectral algebra.

The theory of *topological divisors of zero* extends to spectral semi-norms, and hence its consequences are available for spectral algebras, even though they do not carry a particular spectral semi-norm. Many results on spectral subalgebras arise in this way. Two propositions (Propositions 2.5.12 and 2.5.13) give conditions under which an algebra without topological quasi-divisors of zero is isomorphic to the complex field.

The last result (Theorem 2.5.18) characterizes spectral algebras as those algebras $\mathcal{A}$ for which $\mathcal{A}/\mathcal{A}_J$ can be embedded in a Banach algebra. This characterization remains valid even if the subalgebra is required to be dense and the Banach algebra is required to be semisimple. It gives insight into the reason for the similarity between the spectral theory of spectral algebras and Banach algebras.

Historical Notes

We conclude this introduction with a few historical remarks on the origin of the concept of spectrum. David Hilbert can be credited with the introduction of the words "spectrum", and "spectral theory", and with the beginning of the subject [1904, 1906; cf. 1912]. It arose out of his work on integral equations. These were first used by Niels Henrik Abel, but their systematic study was begun by Eric Ivar Fredholm [1900]. Fredholm's work inspired Hilbert to spend most of the first decade of the twentieth century working on integral equations, and this work was the origin of many of the fundamental ideas of functional analysis. In particular, Hilbert exploited a numerical parameter in the integral equations he wished to solve to understand the behavior in terms of the various values of this parameter. Although his formulation does not agree with the modern placement of the parameter, his spectrum is closely related to our modern concept. Hilbert even gave a fairly sophisticated spectral theorem which expresses the integral operator as a generalized integral over its spectrum. From our modern viewpoint, Hilbert, and later his student Erhard Schmidt, dealt mainly with the case of compact operators, where the spectrum resembles the spectrum of a matrix much more closely. Indeed, Fredholm, Hilbert and Schmidt used limits of matrix approximations as one of their main tools. However, Hilbert did begin with a more general spectral theorem involving a treatment of continuous spectrum.

It is a great irony of history that the word "spectrum", was first used in twentieth century science to describe the discrete lines in the light emitted by heated gases when that light was bent through a prism. Hilbert was well aware of this use of the word. These lines represent the energy of jumps between quantized states of the atoms involved. After considerable development of quantum mechanics, in which Hilbert played a role, it finally became clear that the physical spectrum of an atom could best be described mathematically in terms of the spectrum of an operator on Hilbert space. Hilbert is on record as saying that he had no hint of this connection when he introduced the concept of the spectrum in analysis. The connection received an early definitive treatment in John von Neumann's influential book [1927]. For more historical remarks see Lynn A. Steen [1973] and Constance Reid [1970].

We will not discuss continued fractions in Banach algebras. The reader will find a bibliography in Hans Denk and Max Riederle [1982].

2.1   Definition of the Spectrum

Quasi-multiplication and Quasi-inverses

We now begin the detailed portion of this chapter by extending the concept of spectrum to nonunital algebras. Recall that for any element a in a unital algebra $\mathcal{A}$ we define its *spectrum* and *spectral radius* by:

$$
\begin{aligned}
Sp(a) &= \{\lambda \in \mathbb{C} : \lambda 1 - a \text{ has no inverse in } \mathcal{A}\}, \qquad \text{and} \\
\rho(a) &= \sup\{|\lambda| : \lambda \in Sp(a)\},
\end{aligned}
\tag{1}
$$

respectively. If the spectrum is empty or unbounded, then the spectral radius will have the value $-\infty$ or ∞, respectively. When no element has empty or unbounded spectrum, the spectral radius is a non-negative real-valued function on the algebra—a beginning for analysis and geometry.

We can extend this definition to a not necessarily unital algebra $\mathcal{A}$ by saying that the spectrum of an element in $\mathcal{A}$ is the spectrum (as just defined) of its image in the unitization $\mathcal{A}^1$ of $\mathcal{A}$. According to this definition, 0 will belong to the spectrum of an element a unless $\mathcal{A}$ is already unital and a is invertible in $\mathcal{A}$. This is the accepted definition and should always be kept in mind. However, for both technical and theoretical reasons it is desirable to give a logically equivalent definition of spectrum entirely in terms of $\mathcal{A}$ rather than its unitization $\mathcal{A}^1$. In order to do this we must introduce a concept to take the place of invertibility in nonunital algebras. Of course such algebras have no invertible elements, but by recalling that we really wish to describe the conditions under which an element in $\mathcal{A}$ has an inverse in $\mathcal{A}^1$, we are led to the correct concept. It was first used by Sam Perlis [1942] (cf., Jacobson [1945a] and Irving Kaplansky [1947a]). The particular form of the quasi-product which is now standard was introduced in Einar Hille's influential book [1948].

2.1.1 Definition   The *quasi-product* of any two elements a and b in an algebra is denoted by $a \circ b$ and defined by

$$a \circ b = a + b - ab. \tag{2}$$

We shall also call this binary operation *quasi-multiplication*. (Note that 0 acts as an identity element under quasi-multiplication.) If $a \circ b = 0$ holds, then a is said to be a *left quasi-inverse* for b, and b is said to be a *right quasi-inverse* for a. An element which is both a left and a right quasi-inverse for a is called a *quasi-inverse* for a, and is denoted by a^q. The terms *right quasi-invertible*, *left quasi-invertible*, and *quasi-invertible* are used in the usual way. An element which does not have a (two-sided) quasi-inverse is said to be *quasi-singular*.

The set of quasi-invertible elements in an algebra $\mathcal{A}$ is called the *quasi-group of $\mathcal{A}$* and is denoted by $\mathcal{A}_{qG}$. Similarly, in a unital algebra the set of invertible elements is called the *group of $\mathcal{A}$* and is denoted by $\mathcal{A}_G$.

The first result in the next proposition explains the quasi-product. In particular, it implies that a quasi-inverse is unique if it exists, that an element which is both left and right quasi-invertible is quasi-invertible and that the quasi-inverse of a is in the double commutant of a: $a^q \in \{a\}''$.

If $\mathcal{A}$ is unital, it shows that the simple map

$$a \mapsto 1 - a \qquad \forall\, a \in \mathcal{A} \tag{3}$$

(which is its own inverse) is a semigroup isomorphism of the $\langle$ multiplicative semigroup of $\mathcal{A}$ / quasi-multiplicative semigroup of $\mathcal{A}$ $\rangle$ onto the $\langle$ quasi-multiplicative semigroup of $\mathcal{A}$ / multiplicative semigroup of $\mathcal{A}$ $\rangle$. Hence it is also a group isomorphism of the $\langle$ group $\mathcal{A}_G$ / quasi-group $\mathcal{A}_{qG}$ $\rangle$ onto the $\langle$ quasi-group $\mathcal{A}_{qG}$ / group $\mathcal{A}_G$ $\rangle$. If $\mathcal{A}$ is nonunital, the situation is more complicated: the above map establishes group isomorphisms between the normal subgroup $\mathcal{A}_{G(1)}^1 = \{\lambda + a \in \mathcal{A}_G^1 : \lambda = 1\}$ of the group $\mathcal{A}_G^1$ of $\mathcal{A}^1$ and the quasi-group $\mathcal{A}_{qG}$ of $\mathcal{A}$. The group $\mathcal{A}_G^1$ is homeomorphically isomorphic to the direct product of the multiplicative group $\mathbb{C}^\bullet$ of non-zero complex numbers and $\mathcal{A}_{qG}$ under the isomorphism

$$\lambda - a \mapsto (\lambda,\ \lambda^{-1}a).$$

Obviously, the inverse of this group isomorphism sends the second factor onto the normal subgroup of $\mathcal{A}_G^1$ just described.

Warning: Although quasi-multiplication is associative (and commutative if and only if $\mathcal{A}$ is commutative), it does *not* behave well with respect to the linear structure. Instead of the distributive law we get:

$$(a+b)\circ c = a\circ c + b\circ c - c \quad \text{and} \quad c\circ(a+b) = c\circ a + c\circ b - c \qquad (4)$$

and scalar multiplication gives:

$$\lambda a\circ b = \lambda(a\circ b) + (1-\lambda)b \quad \text{and} \quad a\circ\lambda b = \lambda(a\circ b) + (1-\lambda)a \qquad (5)$$

as short calculations will show.

2.1.2 Proposition *Let $\mathcal{A}$ be an algebra.*

(a) *The following are equivalent for any a, b, $c \in \mathcal{A}$:*

$$a\circ b = c \quad \text{and} \quad (1-a)(1-b) = 1 - c. \qquad (6)$$

(b) *Quasi-multiplication is associative and the quasi-group is a group under quasi-multiplication, with 0 as its identity element.*

(c) *If $\mathcal{A}$ is embedded as a one- or two-sided ideal in some larger unital algebra $\mathcal{B}$ (with identity element 1), then the following are equivalent for any non-zero $\lambda \in \mathbb{C}$ and $a \in \mathcal{A}$:*

$$\lambda^{-1}a \in \mathcal{A}_{qG} \quad \text{and} \quad (\lambda 1 - a) \in \mathcal{B}_G.$$

(d) *If $\mathcal{A}$ is not unital, then the quasi-invertible elements in $\mathcal{A}^1$ are given by:*

$$(\mathcal{A}^1)_{qG} = \{1 - \lambda + \lambda a : \lambda \in \mathbb{C}\setminus\{0\}; a \in \mathcal{A}_{qG}\}$$

and the quasi-inverse for such an element of $\mathcal{A}^1$ is given by:

$$(1 - \lambda + \lambda a)^q = 1 - \lambda^{-1} + \lambda^{-1}a^q.$$

Proof The first two claims are immediate. If $\lambda^{-1}a$ has b as a quasi-inverse, then $(\lambda^{-1}1 - \lambda^{-1}b)$ is an inverse for $(\lambda 1 - a)$ in $\mathcal{B}$. Conversely, if c is an inverse for $(\lambda 1 - a)$ in $\mathcal{B}$, then $-ac = -ca$ is a quasi-inverse for $\lambda^{-1}a$ in $\mathcal{A}$. The last claim is also routine.   □

The next proposition shows that the quasi-product and quasi-inverse have utility beyond just defining the spectrum in not necessarily unital algebras. No result similar to the following holds for the group of invertible elements in a unital algebra or for inverses, since a homomorphism between algebras does not necessarily respect inverses or invertibility. Only unital homomorphisms do so. However, perfectly arbitrary algebra homomorphisms respect quasi-inverses and quasi-invertibility.

2.1.3 Proposition *Let $\mathcal{A}$ and $\mathcal{B}$ be algebras and let $\varphi\colon \mathcal{A} \to \mathcal{B}$ be a homomorphism. Then they satisfy*

$$\varphi(\mathcal{A}_{qG}) \subseteq \mathcal{B}_{qG} \quad and \quad \varphi(a^q) = \varphi(a)^q \qquad \forall\, a \in \mathcal{A}_{qG}.$$

Thus if $\mathcal{B}$ is a subalgebra of $\mathcal{A}$, $\mathcal{B}_{qG}$ is included in $\mathcal{A}_{qG}$.
If $\mathcal{A}$, $\mathcal{B}$ and φ are unital, they also satisfy

$$\varphi(\mathcal{A}_G) \subseteq \mathcal{B}_G \quad and \quad \varphi(a^{-1}) = \varphi(a)^{-1} \qquad \forall\, a \in \mathcal{A}_G.$$

If $\mathcal{B}$ is a unital subalgebra of the unital algebra $\mathcal{A}$, $\mathcal{B}_G$ is included in $\mathcal{A}_G$.

Proof An easy calculation gives the first statement which immediately implies the second. The unital case is also straightforward. □

It is clear that quasi-multiplication is at least separately continuous in a semi-topological algebra. In a semi-normed algebra, the quasi-group is a topological group by the following result of Kaplansky [1947a].

2.1.4 Proposition *In a semi-normed algebra $\mathcal{A}$, quasi-multiplication is jointly continuous and the map which sends an element to its quasi-inverse is a homeomorphism.*

Proof The first statement is obvious. In order to establish the second, let a and b be two quasi-invertible elements and denote their difference by h. Then we get

$$b \circ a^q = a + h + a^q - (a + h)a^q = h(1 - a^q)$$

which implies

$$\begin{aligned}(1 - b^q)h(1 - a^q) &= h(1 - a^q) - b^q h(1 - a^q) \\ &= b^q \circ h(1 - a^q) - b^q = b^q \circ b \circ a^q - b^q = a^q - b^q.\end{aligned}$$

Denote the difference $a^q - b^q$ by k, so we have

$$k = k(-h(1 - a^q)) + (1 + a^q)h(1 - a^q).$$

Hence if σ is an algebra semi-norm and $\sigma(h)$ is sufficiently small, we get

$$\sigma(k) \leq [1 + \sigma(a^q)]^2[1 - \sigma(h)(1 + \sigma(a^q))]^{-1}\sigma(h).$$

This proves that the quasi-inverse is a continuous and hence homeomorphic map on the quasi-group. □

Definition and Basic Properties of the Spectrum

With these preliminaries out of the way, we are in a position to define the spectrum in a not necessarily unital algebra in terms of the quasi-inverse without reference to the unitization. Proposition 2.1.2(c) applied to the embedding of an algebra into its unitization shows how to do this. The present definition agrees completely with the informal definition offered at the very beginning of this section.

2.1.5 Definition Let a be an element in an algebra $\mathcal{A}$. Consider the set of complex numbers

$$\{\lambda \in \mathbb{C} : \lambda \neq 0 \text{ and } \lambda^{-1}a \notin \mathcal{A}_{qG}\}.$$

The *spectrum* $Sp(a)$ of a equals this set if $\mathcal{A}$ is unital and a is invertible. Otherwise $Sp(a)$ is the union of this set and $\{0\}$. The *spectral radius* $\rho(a)$ is defined by

$$\rho(a) = \sup\{|\lambda| : \lambda \in Sp(a)\}. \tag{7}$$

If the algebra $\mathcal{A}$ intended might not be clear, we write $Sp_{\mathcal{A}}(a)$ and $\rho_{\mathcal{A}}(a)$. The complement in $\mathbb{C}$ of the spectrum of a is called the *resolvent set* of a.

In perfectly general algebras the spectrum of an element may be empty or unbounded, as we will now see by example. In these two cases the spectral radius is $-\infty$ or ∞, respectively. However, in the types of algebras which are of most interest in this work the spectrum will be nonempty, closed and bounded and hence compact.

2.1.6 Example Let $\mathbb{C}(z)$ be the function field in one variable, consisting of all rational functions in z. Then any non-constant function has empty spectrum. Theorem 2.2.2 below shows that no element in a normed algebra can have empty spectrum. (Hence no submultiplicative norm can be defined on $\mathbb{C}(z)$.)

Many normed (even commutative) algebras contain elements with unbounded spectrum. For instance, consider the unital, commutative, normed algebra $\mathbb{C}[z]$ of all polynomials with the norm $\|p\| = \sup\{|p(\lambda)| : \lambda \in \mathbb{C}, |\lambda| \leq 1\}$. In this algebra, every element has closed spectrum since

$$Sp(p) = \begin{cases} \mathbb{C} & \text{if } p \text{ is not constant} \\ \{\lambda\} & \text{if } p \equiv \lambda \in \mathbb{C}. \end{cases}$$

There also exist unital, normed (even commutative) algebras in which some elements have unbounded and nonclosed spectrum. Consider $\mathcal{R} = \{$rational functions with poles outside the open unit disc, $\{\lambda \in \mathbb{C} : |\lambda| < 1\}$. We may give this the norm $\|f\| = \sup\{|f(\lambda)| : |\lambda| \leq \frac{1}{2}\}$. The spectrum of each rational function $f \in \mathcal{R}$ is just $\{f(\lambda) : |\lambda| < 1\}$.

Proposition 2.1.11 imparts some order to the situation illustrated by these examples.

If every element in an algebra has nonempty, bounded spectrum, then the spectral radius is a non-negative real-valued function on the algebra. We will show later that in a commutative spectral algebra the spectral radius is also subadditive and submultiplicative and hence is an algebra norm. Commutativity is not enough by itself to give this result, as the next example shows. Theorem 2.4.11 will show that if the spectral radius is subadditive or submultiplicative, then it enjoys both properties and is thus an algebra semi-norm.

2.1.7 Example There are unital, commutative algebras in which the spectrum of every element is nonempty and bounded (hence compact), so that the spectral radius is a non-negative real-valued function, but on which the spectral radius is neither subadditive nor submultiplicative. In $\mathbb{C} \oplus \mathbb{C}(z)$ the spectrum of an arbitrary element is given by:

$$Sp(\lambda, f) = \begin{cases} \{\lambda\} & \text{if } f \text{ not constant} \\ \{\lambda, \mu\} & \text{if } f \equiv \mu. \end{cases}$$

So we have

$$\rho(0,1) = 1 > 0 + 0 = \rho(0,z) + \rho(0,1-z)$$
$$\rho(0,1) = 1 > 0 \cdot 0 = \rho(0,z)\rho(0,z^{-1}).$$

This algebra even has a nontrivial algebra semi-norm: $\sigma(\lambda, f) = |\lambda|$.

These examples begin to show how badly behaved the spectrum is in perfectly general algebras. Nevertheless, there are some results which hold in any algebra. In Corollary 2.1.15 we will improve on inclusion (10) in the case of Banach algebras.

2.1.8 Proposition (a) *Any algebra $\mathcal{A}$ satisfies*

$$Sp(ab) \cup \{0\} = Sp(ba) \cup \{0\} \qquad \forall\, a,b \in \mathcal{A}. \tag{8}$$

(b) *Any algebras $\mathcal{A}$ and $\mathcal{B}$ and any homomorphism $\varphi \colon \mathcal{A} \to \mathcal{B}$ satisfies*

$$Sp_\mathcal{B}(\varphi(a)) \subseteq \bigcap_{b \in \ker(\varphi)} Sp_\mathcal{A}(a+b) \cup \{0\} \qquad \forall\, a \in \mathcal{A}. \tag{9}$$

If $\mathcal{A}, \mathcal{B}$ and φ are all unital or if $\mathcal{A}$ is not unital, this can be simplified to

$$Sp_\mathcal{B}(\varphi(a)) \subseteq Sp_\mathcal{A}(a) \qquad \forall\, a \in \mathcal{A}. \tag{10}$$

(c) *If $\mathcal{B}$ is a subalgebra of $\mathcal{A}$, then*

$$Sp_\mathcal{A}(b) \subseteq Sp_\mathcal{B}(b) \cup \{0\} \qquad \forall\, b \in \mathcal{B}. \tag{11}$$

If $\mathcal{B}$ is a unital subalgebra of $\mathcal{A}$ or if $\mathcal{B}$ is not a unital algebra, then we have

$$Sp_{\mathcal{A}}(b) \subseteq Sp_{\mathcal{B}}(b) \qquad \forall\, b \in \mathcal{B}. \tag{12}$$

Proof (a): If c is a quasi-inverse for $\lambda^{-1}ab$, then $\lambda^{-1}b(c-1)a$ is a quasi-inverse for $\lambda^{-1}ba$.

(b): This is an immediate consequence of Proposition 2.1.3.

(c): Apply (b) to the embedding map of $\mathcal{B}$ into $\mathcal{A}$. $\square$

2.1.9 Example In the above proposition it is really essential that 0 be treated differently from non-zero numbers in the spectrum. To see this for statement (a), consider the shift S and backward shift S^* on the space $\mathcal{X} = \{f \colon \mathbb{N} \to \mathbb{C}\}$ of all bounded complex sequences. These operators are defined by

$$S(f)(n) = \begin{cases} f(n-1) & \text{if}\quad n > 1 \\ 0 & \text{if}\quad n = 1 \end{cases} \quad ; \quad S^*(f)(n) = f(n+1) \qquad \forall\, n \in \mathbb{N}.$$

Thus S^*S is the identity operator, so its spectrum is $\{1\}$, but SS^* annihilates any sequence which has its only non-zero value in the first position and thus has spectrum $\{0, 1\}$. This example is intrinsically infinite-dimensional and it is easy to see that in a finite-dimensional algebra the symbols "$\cup\{0\}$", may be removed from the statement of (a) in the above proposition.

To see the special role of 0 in parts (b) and (c) of the proposition, consider $\mathbb{C}$ embedded as the upper left entry in the 2×2 matrix ring M_2. Then $1 \in \mathbb{C}$ has spectrum $\{1\}$ in $\mathbb{C}$ but spectrum $\{0, 1\}$ in M_2.

Elements with only one-sided inverses have special properties: Bernard H. Aupetit and L. Terrell Gardner [1981], James W. Rowell [1984], Scott H. Hochwald and Bernard Morrel [1987].

The Spectral Mapping Theorem

The next important result is called the *spectral mapping theorem*. This theorem includes the following kinds of results: $Sp(a^2) = Sp(a)^2$; $Sp(a^{-1}) = \{\lambda^{-1} : \lambda \in Sp(a)\}$ for invertible a; and $Sp(a^q) = \{\lambda/(\lambda - 1) : \lambda \in Sp(a)\}$ for quasi-invertible $a \in \mathcal{A}$. A *rational function* is, of course, simply a ratio of polynomials (which may be chosen relatively prime).

Let p be the polynomial $p(\lambda) = \sum_{k=0}^{n} \alpha_k \lambda^k$. (We denote the coordinate function on $\mathbb{C}$ by z:

$$z(\lambda) = \lambda \qquad \forall\, \lambda \in \mathbb{C},$$

so this polynomial may be written as $p = \sum_{k=0}^{n} \alpha_k z^k$.) Then for an element a in any algebra $\mathcal{A}$ we write $p(a) = \sum_{k=0}^{n} \alpha_k a^k$. This agrees with the notation of elementary algebra. (Here a^0 denotes the identity in $\mathcal{A}^1$, so that in a nonunital algebra, if the polynomial does not vanish at 0, is in $\mathcal{A}^1$

rather than $\mathcal{A}$.) Clearly, the function $p \mapsto p(a)$ is an algebra homomorphism of the algebra of polynomials into $\mathcal{A}^1$. If q is another polynomial, we also have $(q \circ p)(a) = q(p(a))$. This action of polynomials on elements of the algebra is called the *polynomial functional calculus*.

If f is a rational function with no poles in $Sp(a)$, then we may write $f = p/q$ where p and q are relatively prime polynomials and q has no zeroes in $Sp(a)$. As we show in more detail in the proof, $q(a)$ is then invertible, so we may define $f(a)$ to be $p(a)q(a)^{-1}$. This interpretation is so basic that we give no formal definition. Since this *rational functional calculus* is well defined in any algebra, the following theorem appears to be the natural form of the universal spectral mapping theorem. Surprisingly, the theorem is usually restricted to the polynomial functional calculus.

2.1.10 Theorem *If a is any element in any algebra $\mathcal{A}$ and f is any rational function with no poles in $Sp(a)$, then $f(a)$ is defined in $\mathcal{A}^1$. The function $f \mapsto f(a)$ is an algebra homomorphism. Unless $Sp(a)$ is empty and f is a constant, $f(a)$ satisfies:*

$$Sp(f(a)) = f(Sp(a))$$

where the last symbol is defined by $\{f(\lambda) : \lambda \in Sp(a)\}$. If g is a rational function and either side of the equation

$$(g \circ f)(a) = g(f(a))$$

is defined by the above remarks, then both sides are defined and they are equal.

Furthermore, for each $a \in \mathcal{A}$ the map $f \mapsto \varphi(f) = f(a)$ is the only algebra homomorphism φ from the algebra of rational functions with no poles in $Sp(a)$ into $\mathcal{A}^1$ which satisfies $\varphi(1) = 1$ and $\varphi(z) = a$.

Proof If $Sp(a)$ is nonempty and f is a constant λ_0, then they satisfy $f(a) = \lambda_0$ and $Sp(f(a)) = \{\lambda_0\} = f(Sp(a))$.

Henceforth we assume that f is not a constant and work in $\mathcal{A}^1$. Let p and q be relatively prime polynomials satisfying $f = p/q$. In order to determine the invertibility of $\lambda_0 - f(a)$ in $\mathcal{A}^1$, we factor q and $\lambda_0 q - p$:

$$q(\lambda) = \alpha(\lambda_1 - \lambda)(\lambda_2 - \lambda) \cdots (\lambda_n - \lambda)$$

$$\lambda_0 q(\lambda) - p(\lambda) = \beta(\mu_1 - \lambda)(\mu_2 - \lambda) \cdots (\mu_m - \lambda)$$

where α and β are not 0 (the latter since $f = p/q$ is not a constant). Since q has no zeroes in $Sp(a)$,

$$q(a) = \alpha(\lambda_1 - a)(\lambda_2 - a) \cdots (\lambda_n - a)$$

is invertible, so $f(a) = p(a)q(a)^{-1}$ is defined. No μ_j appears among the λ_k's since p and q are relatively prime. So

$$\lambda_0 - f(a) = \beta(\mu_1 - a)(\mu_2 - a)\cdots(\mu_m - a)[\alpha(\lambda_1 - a)(\lambda_2 - a)\cdots(\lambda_n - a)]^{-1}$$

is invertible unless some μ_j is in the spectrum of a; but the μ_j's are just the roots of $\lambda_0 q(\mu) - p(\mu) = 0$. Another way to say the same thing is that λ_0 is in the spectrum of $f(a)$ if and only if an element of the spectrum of a is a zero of $\lambda_0 q(\mu) - p(\mu)$. But this is what we wished to show. The other remarks are now obvious.   □

We will see other functional calculi with corresponding versions of the spectral mapping theorem later in this work. There are a number of interesting situations where a class of functions f acts on a class of elements a to give elements $f(a)$. If $f \mapsto f(a)$ is an algebra homomorphism, composition of functions acts appropriately and a version of the spectral mapping theorem $(Sp(f(a)) = \{f(\lambda) : \lambda \in Sp(a)\})$ holds, then the action is called a functional calculus. There is an intimate association between the spectral mapping theorem and the action of composite functions in all cases. In the next chapter we will define the most important functional calculus for arbitrary elements in arbitrary Banach algebras: the *holomorphic functional calculus* Theorem 3.3.7. (In that case we must consider an algebra of germs of functions rather than functions themselves in order to preserve the desired algebraic properties.) All our functional calculi are consistent with each other and unique under appropriate conditions.

The present spectral mapping theorem produces some order among the examples of poorly behaved spectra given in Example 2.1.6.

2.1.11 Proposition *If every element in an algebra has bounded spectrum, then every element has closed (hence, compact) spectrum.*

Proof Suppose the algebra $\mathcal{A}$ satisfies this hypothesis but $a \in \mathcal{A}$ has non-closed spectrum. Choose λ in the closure of $Sp(a)$ but not in $Sp(a)$. Then $Sp((\lambda - a)^{-1}) = \{(\lambda - \mu)^{-1} : \mu \in Sp(a)\}$ is unbounded.   □

The Exponential Function and Exponential Spectrum

We now introduce a variant of the spectrum which was first considered by Robin Harte [1976]. We begin by deriving the basic properties of the exponential function in a unital Banach algebra. We define this function in terms of its power series. (The next chapter will contain a systematic exposition of a much more general notion of functions of elements in a Banach algebra.)

In a topological group, the connected component which contains the identity is called the *principal component*. It is always a closed normal subgroup (*cf.* Hewitt and Ross [1963], 7.1). We will show that the set of finite products of exponentials is an open and closed normal subgroup of $\mathcal{A}_G$ which equals its connected component. Recall that a group is called *torsion-free* if it has no elements of finite order.

In the following theorem the second statement under (d) is a result of Mitio Nagumo [1936], and the last statement under (f) is due to Edgar R. Lorch [1943]. In a recent article, Alessandro Di Bucchianico [1991], like Nagumo, emphasizes that the principal component is arc connected and that the easiest way to show that an element is in this component is to find an arc connecting it to the identity. Theorem 2.1.12 below, follows Lorch in showing that the kernel of the exponential map (considered as a group homomorphism from a commutative Banach algebra) is just the additive group generated by $\{2\pi i e : e$ is an idempotent in $\mathcal{A}\}$. In contrast to (f), Vern I. Paulsen [1982] shows that every finitely generated group arises as $\mathcal{A}_G/\mathcal{A}_{Ge}$ for some (possibly noncommutative) Banach algebra $\mathcal{A}$.

2.1.12 Theorem *Let $\mathcal{A}$ be a unital Banach algebra. Define the exponential function* $\exp\colon \mathcal{A} \to \mathcal{A}$ *by* $\exp(a) = e^a$, *where*

$$e^a = \sum_{n=0}^{\infty} \frac{a^n}{n!} \qquad \forall\, a \in \mathcal{A}. \tag{13}$$

(a) *Any commuting elements* $a,\, b \in \mathcal{A}$ *satisfy:*

$$e^a\, e^b = e^{a+b}.$$

(b) *For any* $a \in \mathcal{A}$, e^a *is invertible and satisfies*

$$(e^a)^{-1} = e^{-a}.$$

(c) *For any* $a \in \mathcal{A}$, *the map* $t \mapsto e^{ta}$ *is a continuous group homomorphism of the additive group of real numbers into* $\mathcal{A}_G$.

(d) *The set* $\exp(\mathcal{A}) = \{e^a : a \in \mathcal{A}\}$ *includes the open unit ball around the identity element* 1. *Furthermore*, $c \in \mathcal{A}_G$ *belongs to* $\exp(\mathcal{A})$ *if and only if* c *belongs to some connected abelian subgroup of* $\mathcal{A}_G$.

(e) *The principal component of* $\mathcal{A}_G$ *is the subgroup*

$$\mathcal{A}_{Ge} = \{e^{a_1} e^{a_2} \cdots e^{a_n} : n \in \mathbb{N}; a_1, a_2, \ldots a_n \in \mathcal{A}\}$$

of $\mathcal{A}_G$ *generated by* $\exp(\mathcal{A})$. *It is open in* $\mathcal{A}$. *Thus* $\mathcal{A}_{Ge}$ *is an open and closed connected normal subgroup of* $\mathcal{A}_G$.

(f) *Let* $\mathcal{A}$ *be commutative. Then* $\exp(\mathcal{A})$ *is the principal component of* $\mathcal{A}_G$, *and the quotient group* $\mathcal{A}_G/\exp(\mathcal{A})$ *is torsion-free. Hence* $\mathcal{A}_G$ *has infinitely many components unless it is connected.*

Proof (a): Since the expansion (13) converges unconditionally, the series for $\exp(a)$ and $\exp(b)$ can be rearranged (or equivalently, multiplied by ordinary Cauchy multiplication) when a and b commute:

$$e^a e^b = \sum_{n=0}^{\infty} \frac{a^n}{n!} \sum_{m=0}^{\infty} \frac{b^n}{m!}$$

$$= \sum_{k=0}^{\infty} \frac{1}{k!} \sum_{n=0}^{k} \binom{k}{n} a^n b^{k-n} = e^{a+b}.$$

(b): This follows from (a).

(c): Statement (a) shows that $t \mapsto e^{ta}$ is a group homomorphism. For any $s, t \in \mathbb{R}$, the estimate

$$\|e^{sa} - e^{ta}\| \leq \|e^{sa}\| \, \|1 - e^{(t-s)a}\|$$

$$\leq \exp \|sa\| \sum_{n=1}^{\infty} \frac{\|(t-s)a\|^n}{n!}$$

$$\leq \exp \|sa\| \, |1 - \exp(|t-s| \, \|a\|)|$$

proves the continuity.

(d): If c satisfies $\|c - 1\| < 1$, then the series

$$a = \sum_{m=1}^{\infty} \frac{(1-c)^m}{m}$$

converges unconditionally in norm. Thus the double series

$$e^a = \sum_{n=0}^{\infty} \frac{1}{n!} \Big(\sum_{m=1}^{\infty} \frac{(1-c)^m}{m} \Big)^n$$

can be rearranged. Since the series for a is just the Taylor series expansion of the logarithm, rearrangement leads to the equation $e^a = c$. Result (c) shows that any exponential belongs to a connected abelian subgroup. We defer proof of the converse until we have established the first results of (f).

(e): Let $\mathcal{E}$ be the displayed set, which is obviously the subgroup generated by $\exp(\mathcal{A})$, and let $\mathcal{A}_{Ge}$ be the principal component of $\mathcal{A}_G$. Then $\mathcal{E}$ is connected since $t \mapsto e^{ta_1} e^{ta_2} \cdots e^{ta_n}$ for $t \in [0, 1]$ is an arc connecting an arbitrary element $e^{a_1} e^{a_2} \cdots e^{a_n}$ to 1. Hence $\mathcal{E}$ is a subset of $\mathcal{A}_{Ge}$.

Next we show that $\mathcal{E}$ is open in $\mathcal{A}$ and hence closed in $\mathcal{A}_G$. If $b \in \mathcal{E}$ and $a \in \mathcal{A}$ satisfy $\|b - a\| < \|b^{-1}\|^{-1}$, then they also satisfy $\|1 - b^{-1}a\| \leq \|b^{-1}\| \, \|b - a\| < 1$. Hence (d) shows $b^{-1}a = \exp(c)$ for some $c \in \mathcal{A}$. Thus $a = b \exp(c)$ belongs to $\mathcal{E}$, so $\mathcal{E}$ is open in $\mathcal{A}$ and hence in $\mathcal{A}_G$. However an open subgroup is always closed since its complement is a union of (necessarily open) cosets.

By connectedness this implies $\mathcal{A}_{Ge} = \mathcal{E}$. It only remains to show that $\mathcal{E}$ is a normal subgroup. For any $a \in \mathcal{A}$ and $b \in \mathcal{A}_G$, we have

$$b\, e^a\, b^{-1} = \sum_{n=0}^{\infty} (n!)^{-1} b a^n b^{-1} = \sum_{n=0}^{\infty} (n!)^{-1} (bab^{-1})^n = \exp(bab^{-1}).$$

Hence $\exp(\mathcal{A})$ is invariant under all inner automorphisms. Thus $\mathcal{A}_{Ge}$ is also invariant so it is a normal subgroup.

(f): When $\mathcal{A}$ is commutative, (a) and (b) show that $\exp(\mathcal{A})$ is a group. Hence (e) implies that $\exp(\mathcal{A})$ is the principal component of $\mathcal{A}_G$.

We now complete the proof of (d). Suppose $c \in \mathcal{A}$ belongs to a connected abelian subgroup H of $\mathcal{A}_G$. Let $\mathcal{B}$ be the (necessarily commutative) closed subalgebra of $\mathcal{A}$ generated by H. Then c is in the principal component of $\mathcal{B}_G$ and hence is an exponential by what we just proved.

Next we show that $\mathcal{A}_G / \exp(\mathcal{A})$ is torsion-free. Suppose $a \in \mathcal{A}_G$ and $n \in \mathbb{N}$ satisfy $a^n \in \exp(\mathcal{A})$. If b satisfies $a^n = \exp(b)$, then $c = a \exp(-n^{-1}b)$ satisfies $c^n = 1$. Hence the spectral mapping theorem (Theorem 2.1.10) shows that $Sp(c)$ is included in the set of nth roots of 1. The set $\{\lambda \in \mathbb{C} : 1 - \lambda + \lambda c \in \mathcal{A}_G\}$ is connected since its complement is finite. However this set contains 0 (where $1 - \lambda + \lambda c$ is 1) and 1 (where $1 - \lambda + \lambda c$ is c) so that c belongs to the principal component of $\mathcal{A}_G$. Thus c belongs to $\exp(\mathcal{A})$ by what we have just shown. Hence $a = \exp(n^{-1}b)c$ belongs to $\exp(\mathcal{A})$. Thus $\mathcal{A}_G / \exp(\mathcal{A})$ is torsion free. Since $\exp(\mathcal{A})$ is open, $\mathcal{A}_G$ has infinitely many components if it is not connected. $\square$

The following concept introduced by Harte [1976] is chiefly useful because of the improvement of inclusion (9) of Proposition 2.1.8 which we give in Theorem 2.1.14. We forego giving a definition in the nonunital case and note that the name is only appropriate in the case of Banach algebras. See Murphy and West [1992] for further results.

2.1.13 Definition Let $\mathcal{A}$ be a unital algebra. The *exponential spectrum* $eSp_{\mathcal{A}}(a) = eSp(a)$ of an element $a \in \mathcal{A}$ is the set of complex numbers

$$eSp(a) = \{\lambda \in \mathbb{C} : \lambda - a \notin \mathcal{A}_{Ge}\}.$$

The next theorem gives the fundamental properties of the exponential spectrum in a Banach algebra before we have derived the corresponding results for the spectrum itself. For the proof of non-emptiness, we depend on Theorem 2.2.2 (or Corollary 2.2.8) in the next section. Recall that we use ∂ to denote the boundary of a set in a topological space. The *polynomially convex hull* of a compact subset K of $\mathbb{C}$ is just the complement of the unbounded component of the complement of K. Thus it is obtained by "filling in all the holes" of K. We denote it by $pc(K)$.

2.1.14 Theorem (a) *Let A be a unital Banach algebra. For each $a \in A$, $eSp(a)$ is a nonempty compact subset of $\mathbb{C}$ satisfying*

$$\partial eSp(a) \subseteq Sp(a) \subseteq eSp(a) \subseteq pc(Sp(a)). \tag{14}$$

(b) *Let B be a closed unital subalgebra of a unital Banach algebra A. Then*

$$\partial eSp_B(b) \subseteq Sp_A(b) \; ; \quad eSp_B(b) \subseteq pc(Sp_A(b)) \qquad \forall \, b \in B.$$

(c) *Let A and B be unital Banach algebras and let $\varphi \colon A \to B$ be a continuous unital surjective homomorphism. Then they satisfy*

$$\varphi(A_{Ge}) = B_{Ge};$$

$$eSp_B(\varphi(a)) = \bigcap_{b \in \ker(\varphi)} eSp(a + b) \qquad \forall \, a \in A.$$

Proof (a): From the definition it is obvious that $eSp(a)$ includes $Sp(a)$, and we will show the latter is nonempty in Corollary 2.2.8. Since A_{Ge} is open, $eSp(a)$ is closed. It is bounded since $|\lambda| > \|a\|$ implies that $\lambda - a = \lambda(1 - \lambda^{-1}a)$ belongs to $\exp(A)$ by Theorem 2.1.12(d).

We have just noted the middle inclusion of (14). The last inclusion will follow from the first, since it is obvious that if the boundary of a compact set K in $\mathbb{C}$ is in L, then K is in $pc(L)$.

Let λ belong to the boundary of $eSp(a)$. Then we can find a sequence $\{\lambda_n\}_{n \in \mathbb{N}}$ of elements in the complement of $eSp(a)$ converging to λ. We conclude $|\lambda_n - \lambda| \, \|(\lambda_n - a)^{-1}\| \geq 1$, since otherwise

$$(\lambda - a)(\lambda_n - a)^{-1} = 1 - (\lambda_n - \lambda)(\lambda_n - a)^{-1}$$

would be in A_{Ge}, implying $\lambda - a \in A_{Ge}$, contrary to fact. This gives

$$\|(\lambda - a)(\lambda_n - a)^{-1}\| \leq 1 + |\lambda_n - \lambda| \, \|(\lambda_n - a)^{-1}\| \leq 2|\lambda_n - \lambda| \, \|(\lambda_n - a)^{-1}\|.$$

If $\lambda - a$ had a bounded inverse, this would imply the contradiction $1 \leq 2|\lambda_n - \lambda| \, \|(\lambda - a)^{-1}\|$. Hence λ is in $Sp(a)$.

(b): In the last paragraph, if $\lambda - a$ had a bounded inverse in any larger algebra, the argument would not change. Apply this to $b \in B$.

(c): Continuity implies $\varphi(e^a) = e^{\varphi(a)}$ and hence $\varphi(A_{Ge}) \subseteq B_{Ge}$. Conversely, any $b \in B_{Ge}$ satisfies $b = e^{c_1} e^{c_2} \cdots e^{c_n} = \varphi(e^{a_1} e^{a_2} \cdots e^{a_n})$ where $\varphi(a_j) = c_j$ for $j = 1, 2, \ldots, n$, giving the opposite inclusion.

In the second displayed equation in (c), it is obvious that the left-hand side is included in the right-hand side. Conversely if $\lambda \notin eSp_B(\varphi(a))$, then $\lambda - \varphi(a)$ equals $\varphi(c)$ for some $c \in A_{Ge}$. Hence $b = \lambda - a - c$ is in $\ker(\varphi)$ and $\lambda \notin eSp_A(a + b)$. $\qquad \square$

2.1.15 Corollary *Let $\mathcal{A}$ and $\mathcal{B}$ be unital Banach algebras. If $\varphi\colon \mathcal{A} \to \mathcal{B}$ is a continuous unital homomorphism with closed range, the spectrum satisfies:*

$$Sp_{\mathcal{B}}(\varphi(a)) \subseteq \bigcap_{b \in \ker(\varphi)} Sp_{\mathcal{A}}(a+b) \subseteq \mathrm{pc}(Sp_{\mathcal{B}}(\varphi(a))) \qquad \forall\, a \in \mathcal{A}. \qquad (15)$$

Proof The first inclusion was already obtained in Proposition 2.1.8(b). Let $\mathcal{C}$ be the closed, hence complete, range of φ. The theorem shows

$$\bigcap_{b \in \ker(\varphi)} Sp_{\mathcal{A}}(a+b) \subseteq \bigcap_{b \in \ker(\varphi)} eSp_{\mathcal{A}}(a+b) = eSp_{\mathcal{C}}(\varphi(a)) \subseteq \mathrm{pc}(Sp_{\mathcal{B}}(\varphi(a)))$$

which concludes the proof.   □

Ransford [1984a] derives most of spectral theory from very simple axioms without any multiplicative structure. He starts with a normed linear space $\mathcal{A}$, a fixed non-zero element $1 \in \mathcal{A}$ and an open subset Ω in $\mathcal{A}$ containing 1, excluding 0 and closed under multiplication by non-zero complex numbers. If $\mathcal{A}$ is a unital spectral normed algebra, we may take Ω to be either $\mathcal{A}_G$ or $\mathcal{A}_{Ge}$. Ransford's "spectrum" is defined to be $Sp_\Omega(a) = \{\lambda \in \mathbb{C} : a - \lambda \notin \Omega\}$. Other geometrical conditions can be imposed on Ω. See the reference above, Cătălin Badea [1991] and Ignacio Zalduendo [1989].

Relatively Regular Elements and Algebras

We conclude this section with a brief introduction to some generalized inverses. John von Neumann [1936b] first introduced regular algebras in connection with his studies of orthomodular lattices and continuous geometries. Kaplansky [1948d] studied weakly regular algebras. We have inserted "relatively" into both names in order to avoid confusion with other uses of the ubiquitous term "regular". In particular, we use "relatively regular element" in place of "regular element", since the latter term is used as a synonym for "invertible". We use weakly regular for another concept below.

2.1.16 Definition An element a in an algebra $\mathcal{A}$ is said to be $\langle$ *relatively regular / weakly relatively regular* $\rangle$ if there is a non-zero element $b \in \mathcal{A}$ satisfying $\langle$ $aba = a$ / $bab = b$ $\rangle$. In this case, b is called a $\langle$ *relative inverse / weak relative inverse* $\rangle$ for a. An algebra or ideal is called $\langle$ *relatively regular / weakly relatively regular* $\rangle$ if all its non-zero elements are $\langle$ *relatively regular / weakly relatively regular* $\rangle$.

Relative inverses are also called generalized inverses (particularly in operator theory) and pseudo-inverses. The next proposition sums up their most elementary properties.

2.1.17 Proposition *Let a and b be elements in an algebra $\mathcal{A}$. Define p and q by $p = ab$ and $q = ba$.*

(a) If b is a $\langle$ relative inverse / weak relative inverse $\rangle$ for a, then p and q are idempotents satisfying $\langle\ pa = a = aq\ /\ pa = paq = aq\ \rangle$.

(b) Any relatively regular element is weakly relatively regular.

(c) Any relatively regular element a with relative inverse b satisfies $a\mathcal{A} = p\mathcal{A}$ and $\mathcal{A}a = \mathcal{A}q$.

(d) If the algebra $\mathcal{A}$ satisfies $\langle\ \mathcal{A}_{LA} = 0\ /\ \mathcal{A}_{RA} = 0\ \rangle$, then any $a \in \mathcal{A}$ is relatively regular if and only if $\langle\ a\mathcal{A}\ /\ \mathcal{A}a\ \rangle$ is generated as a principal $\langle$ right / left $\rangle$ ideal by some idempotent.

(e) Any one-sided ideal which contains some weakly relatively regular element contains a non-zero idempotent.

(f) An algebra is weakly relatively regular if and only if every non-zero $\langle$ right / left $\rangle$ ideal contains a non-zero idempotent.

Proof (a): Multiplying $\langle\ aba = a\ /\ bab = b\ \rangle$ first on one-side and then on the other by $\langle\ b\ /\ a\ \rangle$ gives these results.

(b): Consider $(bab)a(bab) = bab$.

(c): Immediate from (a).

(d): Consider the case in which $a\mathcal{A}$ has the form $p\mathcal{A}$ for some idempotent p. Since $p = p^2$ is in $p\mathcal{A}$, we can write $p = ab$ for some $b \in \mathcal{A}$. For any $c \in \mathcal{A}$, $ac \in a\mathcal{A}$ can be written as $ac = pac = abac$, so $a - aba = 0$ under our hypothesis. The other case is similar.

(e): Suppose a belongs to a $\langle$ right ideal $\mathcal{R}$ / left ideal $\mathcal{L}$ $\rangle$ and satisfies $bab = b$. Then $\langle\ p = ab\ /\ q = ba\ \rangle$ belongs to $\langle\ \mathcal{R}\ /\ \mathcal{L}\ \rangle$.

(f): Result (e) shows that any non-zero one-sided ideal contains a non-zero idempotent. Suppose the $\langle$ right ideal $a\mathcal{A}$ / left ideal $\mathcal{A}a$ $\rangle$ contains the non-zero idempotent $\langle\ p = ab\ /\ q = ba\ \rangle$. Then a is weakly relatively regular by the equation given in the proof of (b) above. $\square$

Note that the conditions on the algebra $\mathcal{A}$ in (d) are satisfied if $\mathcal{A}$ is unital, semiprime (Definition 4.4.1) or approximately unital (Definition 5.1.1). Von Neumann's interest in (d) stemmed from the elementary fact that a one-sided ideal $\mathcal{I}$ in a unital algebra $\mathcal{A}$ is complemented (in the sense that there exists another one-sided ideal $\mathcal{J}$ of the same type satisfying $\mathcal{I} \cap \mathcal{J} = \{0\}$ and $\mathcal{I} + \mathcal{J} = \mathcal{A}$) if and only if $\mathcal{I}$ is the principal one-sided ideal generated by an idempotent. (In particular if $\mathcal{I} = p\mathcal{A}$, then $\mathcal{J} = (1 - p)\mathcal{A}$, and the idempotent p is uniquely determined by $\mathcal{I}$ and $\mathcal{J}$.)

We conclude this brief discussion by quoting Kaplansky's [1948d] main result without proof. For a proof and an accessible survey of related results see Selwyn R. Caradus [1978]. See also Harro G. Heuser [1982].

2.1.18 Theorem *A relatively regular Banach algebra is finite-dimensional.*

Other notions of generalized inverses are included in S. T. M. Ackermans and A. M. H. Gerards [1982], Harte [1987], Ackermans [1982], Pietro Aiena [1983], Vladimir Rakočević [1988] and Danrun Huang [1992].

Weak Invertibility and the Strong Spectrum

The following ideas have recently been introduced by Michael J. Meyer [1992d]. The definition is motivated by the observation that an element a in a unital algebra $\mathcal{A}$ is invertible exactly when $\mathcal{A}$ is both the smallest left ideal and the smallest right ideal containing a.

2.1.19 Definition   Let $\mathcal{A}$ be a unital algebra. An element $a \in \mathcal{A}$ is called *weakly invertible* if $\mathcal{A}$ is the smallest ideal including a. We denote the set of weakly invertible elements in $\mathcal{A}$ by $\mathcal{A}_{wG}$. The *strong spectrum* of $a \in \mathcal{A}$ is the subset $Sp_{\mathcal{A}}^{S}(a) = Sp^{S}(a)$ of $\mathbb{C}$ consisting of all λ such that $\lambda - a$ is not weakly invertible. The *strong spectral radius* is defined by $\rho_{\mathcal{A}}^{S}(a) = \rho^{S}(a) = \sup\{|\lambda| : \lambda \in Sp^{S}(a)\}$. If $\mathcal{A}$ is not necessarily unital, for $a \in \mathcal{A}$ we define $Sp_{\mathcal{A}}^{S}(a)$ and $\rho_{\mathcal{A}}^{S}(a)$ to be the strong spectrum $Sp_{\mathcal{A}^{1}}^{S}(a)$ and strong spectral radius $\rho_{\mathcal{A}^{1}}^{S}(a)$ of a in the unitization.

Clearly $a \in \mathcal{A}$ is weakly invertible if and only if there exist finite subsets $\{b_1, b_2, \ldots, b_n\}$ and $\{c_1, c_2, \ldots, c_n\}$ of $\mathcal{A}$ satisfying $1 = \sum_{j=1}^{n} b_j a c_j$. A weak analogue of quasi-invertibility will be introduced in Section 4.5.

2.1.20 Proposition   *Let $\mathcal{A}$ be a unital algebra and let $\Xi_{\mathcal{A}}$ be the set of all maximal proper ideals of $\mathcal{A}$.*

(a) *The set $\mathcal{A}_{wG}$ of weakly invertible elements includes all one- or two-sided invertible elements so all $a \in \mathcal{A}$ satisfy $Sp^{S}(a) \subseteq Sp(a)$. It is the complement of $\bigcup_{\mathcal{M} \in \Xi_{\mathcal{A}}} \mathcal{M}$. Hence each element $a \in \mathcal{A}$ satisfies*

$$Sp^{S}(a) = \{\lambda \in \mathbb{C} : \lambda - a \in \mathcal{M} \text{ for some } \mathcal{M} \in \Xi_{\mathcal{A}}\} \subseteq Sp(a).$$

(b) *The center $\mathcal{A}_{Z}$ of $\mathcal{A}$ satisfies $\mathcal{A}_{wG} \cap \mathcal{A}_{Z} = \mathcal{A}_{G} \cap \mathcal{A}_{Z}$ so any $a \in \mathcal{A}_{Z}$ satisfies $Sp^{S}(a) = Sp(a)$ and $\rho^{S}(a) = \rho(a)$.*

(c) *For any $a, b \in \mathcal{A}$ if ab is weakly invertible, so are both a and b. If $\langle a \, / \, b \rangle$ has a $\langle$ left $/$ right $\rangle$ inverse and $\langle b \, / \, a \rangle$ is weakly invertible, then ab is weakly invertible.*

(d) *If $\mathcal{B}$ is a unital algebra and $\varphi: \mathcal{A} \to \mathcal{B}$ is a unital homomorphism, then $\varphi(\mathcal{A}_{wG}) \subseteq \mathcal{B}_{wG}$ implies $Sp^{S}(\varphi(a)) \subseteq Sp^{S}(a)$ for all $a \in \mathcal{A}$.*

(e) *If $f: \mathbb{C} \to \mathbb{C}$ is a rational function with poles off $Sp(a)$, then $f(Sp^{S}(a))$ is a subset of $Sp^{S}(f(a))$. They are not always equal.*

Proof (a): If $a \in \mathcal{A}$ has a left inverse, then $\mathcal{A}a = \mathcal{A}$ so $\mathcal{A}a\mathcal{A} = \mathcal{A}$, a fortiori. The other cases are similar. An element $a \in \mathcal{A}$ is weakly invertible if and only if it belongs to no proper ideal. Theorem 2.4.6(d) contains the simple Zorn's lemma argument that each proper ideal in a unital algebra is included in a maximal proper ideal. The last sentence is an immediate consequence.

(b): If $a \in \mathcal{A}_{Z}$ satisfies $1 = \sum_{j=1}^{n} b_j a c_j$, then $\sum_{j=1}^{n} b_j c_j$ is an inverse.

(c): The equations $1 = \sum_{j=1}^{n} b_j(ab)c_j = \sum_{j=1}^{n} b_j a(bc_j) 1 = \sum_{j=1}^{n}(b_j a)bc_j$ prove the first statement. Considering the first case in the second sentence with c as a left inverse for a, we get $1 = \sum_{j=1}^{n} b_j bc_j = \sum_{j=1}^{n}(b_j c)(ab)c_j$.

(d): Any $a \in \mathcal{A}_{wG}$ satisfies $1 = \sum_{j=1}^{n} b_j ac_j$ and hence $\varphi(a)$ satisfies $1 = \sum_{j=1}^{n} \varphi(b_j)\varphi(a)\varphi(c_j)$.

(e): Given a, f and $\lambda \in Sp^S(a)$, (a) guarantees a maximal proper ideal $\mathcal{M}$ containing $\lambda - a$. Let $\varphi \colon \mathcal{A} \to \mathcal{A}/\mathcal{M}$ be the natural map. Then $f(\lambda) = f(\varphi(a)) = \varphi(f(a))$ is clear and establishes the positive result. Meyer [1992d] contains a simple counterexample to equality. □

If $\varphi_{\mathcal{M}} \colon \mathcal{A} \to \mathcal{A}/\mathcal{M}$ is the natural map, the equation in (a) can be written $Sp^S(a) = \{\varphi_{\mathcal{M}}(a) \in \mathbb{C} : \mathcal{M} \in \Xi_{\mathcal{A}}\}$. In Section 3.3 we will establish a meaning for $f(a)$ for larger classes of functions f when a belongs to certain special types of algebras. The proof of (e) above remains valid in this wider setting.

If $\{0\}$ is the only proper ideal of $\mathcal{A}$, then $\mathcal{A}_{wG}$ is just $\mathcal{A}\backslash\{0\}$. Similarly, if $\mathcal{A}$ (like $\mathcal{B}(\mathcal{H})$) has a unique largest ideal $\mathcal{I}$ then $\mathcal{A}_{wG}$ is $\mathcal{A}\backslash\mathcal{I}$. In both cases many elements will have empty strong spectrum. On the other extreme, if $\mathcal{A}$ is commutative, (b) shows $\mathcal{A}_{wG} = \mathcal{A}_G$. For the algebra $\mathcal{A} = C(\Omega, \mathcal{B})$ of continuous functions from a compact set Ω into an algebra $\mathcal{B}$, $\mathcal{A}_{wG}$ is the collection of functions f satisfying $f(\omega) \in \mathcal{B}_{wG}$ for all $\omega \in \Omega$. For additional results see 2.2.17 and 4.5.7.

2.2   Spectral Semi-norms

In this section we introduce the most important type of algebra semi-norms, namely spectral semi-norms. They are characterized by a large number of interesting equivalent properties which show their intimate connection with the algebraic structure. In the following section we will show that in certain well-behaved algebras all examples of spectral semi-norms are related.

Properties of the Spectral Radius

We need the following results on the spectral radius repeatedly.

2.2.1 Proposition *Let $\mathcal{A}$ be an algebra.*
(a) *Any $a \in \mathcal{A}$ and any $\lambda \in \mathbb{C}$ satisfy*

$$\rho(\lambda a) = |\lambda|\rho(a); \qquad \rho(a^n) = \rho(a)^n; \qquad \rho(1) = 1 \qquad (1)$$

where we are assuming that $\mathcal{A}$ is unital in the third case.
(b) *Any $a, b \in \mathcal{A}$ satisfy:*

$$\rho(ab) = \rho(ba) \qquad (2)$$

unless one product has spectrum $\{0\}$ and the other has empty spectrum.

(c) *Let $\mathcal{A}$ and $\mathcal{B}$ be algebras and let $\varphi \colon \mathcal{A} \to \mathcal{B}$ be a homomorphism. Then any element $a \in \mathcal{A}$ with nonempty spectrum satisfies*

$$\rho_{\mathcal{B}}(\varphi(a)) \leq \inf\{\rho_{\mathcal{A}}(a+b) : b \in \ker(\varphi)\} \leq \rho_{\mathcal{A}}(a). \tag{3}$$

(d) *If $\mathcal{B}$ is a subalgebra of $\mathcal{A}$, then any $b \in \mathcal{B}$ with nonempty spectrum in $\mathcal{B}$ satisfies*

$$\rho_{\mathcal{A}}(b) \leq \rho_{\mathcal{B}}(b).$$

(e) *If $\mathcal{A}$ is not unital, then any $\lambda + a \in \mathcal{A}^1$ satisfies*

$$\rho(\lambda + a) \leq |\lambda| + \rho(a) \leq 3\rho(\lambda + a).$$

Proof (a): These results all follow from the spectral mapping theorem or even more elementary considerations.

(b): Proposition 2.1.8(a) shows this.

(c) and (d): The corresponding results in Proposition 2.1.8 give these.

(e): From the spectral mapping theorem (or the definition), we see $Sp(\lambda + a) = \lambda + Sp(a)$ in $\mathcal{A}^1$. (Hence if either spectrum is unbounded, they both are and the inequality holds in a trivial way.) Since $\mathcal{A}$ is not unital, 0 belongs to $Sp(a)$. (Hence $Sp(a)$ is not empty, and $Sp(\lambda + a)$ is not empty either since the unitization of a nonunital algebra cannot be a division algebra.) Thus we see $|\lambda| \leq \rho(\lambda + a) \leq |\lambda| + \rho(a)$. For any $\varepsilon > 0$, we can find a $\mu \in Sp(a)$ with absolute value larger than $\rho(a) - \varepsilon$. Hence we get $\rho(a) - \varepsilon < |\mu| \leq |\lambda + \mu| + |\lambda| \leq \rho(\lambda + a) + |\lambda|$. The result follows by combining the inequalities and letting ε approach 0. $\qquad\square$

The reader might compare this result with Theorem 1.1.10. In spectral algebras, introduced below, the spectral radius has all the properties described in that theorem.

It appears reasonable that any two elements in an arbitrary algebra satisfy equation (2), but we have neither been able to rule out the possible restriction mentioned in the statement, nor have we been able to construct an algebra in which this actually occurs. Hence we do not know whether this equation is true in complete generality. All elements in algebras with a nontrivial semi-norm satisfy equation (2), as the Gelfand–Mazur theorem (Corollary 2.2.3) will show.

Let us restate inequality (3) above in a slightly different form involving an ideal $\mathcal{I}$ of $\mathcal{A}$:

$$\rho_{\mathcal{A}/\mathcal{I}}(a + \mathcal{I}) \leq \inf\{\rho_{\mathcal{A}}(a+b) : b \in \mathcal{I}\} \qquad \forall\, a \in \mathcal{A}. \tag{4}$$

In this form one may study the class of Banach algebras for which this inequality becomes an equality for all closed ideals $\mathcal{I}$ and all elements $a \in \mathcal{A}$. This has been examined by M. R. F. Smyth and Trevor T. West [1975],

Gert K. Pedersen [1976] and Gerald J. Murphy and West [1979]. Algebras
with this property are called *SR-algebras*. The last paper shows that C*-
algebras (*cf.* Volume II), commutative algebras $\mathcal{A}$ for which the algebra
$\hat{\mathcal{A}}$ of Gelfand transforms is dense in $C_0(\Gamma_\mathcal{A})$ (*cf.* Section 3.1) and Riesz
algebras (*cf.* Chapter 7) are SR-algebras, but that the disc algebra (§1.5.2)
is not. It seems likely that this property is fairly common.

The Spectral Radius and Algebra Norms

The following extremely important result was proved for the algebra of
absolutely convergent Fourier series by A. Beurling [1938] and for normed
algebras by Gelfand [1941a]. The elementary proof which we give is due
to Rickart [1958]. Theorem 2.2.5 below shows a case with equality rather
than the inequality obtained here. Recall that an algebra semi-norm σ on
an algebra $\mathcal{A}$ is nontrivial unless $\mathcal{A}$ is unital and σ is identically 0. Recall
also that if σ is an algebra semi-norm, then σ^∞ is defined by $\sigma^\infty(a) = \lim_{n \to \infty} \sigma(a^n)^{1/n}$.

2.2.2 Theorem *Let σ be a nontrivial algebra semi-norm on an algebra $\mathcal{A}$.
Then any element a in $\mathcal{A}$ has nonempty spectrum and the spectral radius
satisfies*

$$\sigma^\infty(a) \le \rho(a) \qquad \forall\, a \in \mathcal{A}.$$

Proof First we show that if $\sigma^\infty(a)$ is 0, then 0 belongs to $Sp(a)$, so there
is nothing to prove. Suppose 0 does not belong to $Sp(a)$. Then $\mathcal{A}$ is unital
and a is invertible. Thus Theorem 1.1.10(f) and (g) show

$$1 = \sigma^\infty(1) \le \sigma^\infty(a)\sigma^\infty(a^{-1})$$

so $\sigma^\infty(a)$ is non-zero.

Hence we may assume that $\sigma^\infty(a)$ is positive. The absolute homogeneity
of σ^∞ and ρ (Theorem 1.1.10(c) and Proposition 2.2.1(a)) allows us to
assume $\sigma^\infty(a) = 1$. Suppose the theorem is false, so that λa is quasi-
invertible for all λ in the closed unit disc $\mathbb{C}_1$. Proposition 2.1.4 shows that
$f(\lambda) = (\lambda a)^q$ is a uniformly continuous function on the compact set $\mathbb{C}_1$.
(In fact, Proposition 2.2.9 below shows that this function is analytic when
interpreted in the normed algebra $\mathcal{A}/\mathcal{A}_\sigma$. A simple, but non-elementary,
proof based on this fact and a little analytic function theory is given there.
We now continue with the elementary proof due to Rickart.) Let n be a
positive integer and let $\zeta_1, \zeta_2, \ldots, \zeta_n$ be the nth roots of unity. For any
$\lambda \in \mathbb{C}_1$, define λ_j by $\zeta_j \lambda$ for $j = 1, 2, \ldots, n$. Then

$$1 - (\lambda z)^n = (1 - \lambda_1 z)(1 - \lambda_2 z) \cdots (1 - \lambda_n z)$$

implies (*cf.* Proposition 2.1.2(a))

$$(\lambda a)^n = (\lambda_1 a) \circ (\lambda_2 a) \circ \cdots \circ (\lambda_n a).$$

Thus $(\lambda a)^n$ is quasi-invertible. Define $b_j \in \mathcal{A}$ by $b_j = -[\lambda_j a + (\lambda_j a)^2 + \cdots + (\lambda_j a)^{n-1}]$, so that for all j it satisfies $(\lambda a)^n = (\lambda_j a) \circ b_j$. By noting $\sum_{j=1}^{n} \zeta_j = 0$ and rearranging the double sum we also note $b_1 + b_2 + \cdots + b_n = 0$. Since $(\lambda a)^n$ and $\lambda_j a$ are quasi-invertible, we can write $f(\lambda_j) = b_j \circ ((\lambda a)^n)^q$. Thus equation (4) before Proposition 2.1.2 gives

$$
\begin{aligned}
n^{-1} \sum_{j=1}^{n} f(\lambda_j) &= n^{-1}\Big((\sum_{j=1}^{n} b_j) \circ ((\lambda a)^n)^q + (n-1)((\lambda a)^n)^q \Big) \\
&= ((\lambda a)^n)^q.
\end{aligned}
$$

By uniform continuity, for any positive ε we can find a number t close to 1 in the open interval $]0, 1[$ for which $\sigma(f(t_j) - f(\zeta_j)) < \varepsilon$ holds for all $n \in \mathbb{N}$ and all $j = 1, 2, \ldots, n$. Substituting t and 1 for λ in the last displayed equation gives

$$
\sigma(((ta)^n)^q - (a^n)^q) < \varepsilon
$$

independent of n. However $\sigma((ta)^n) = t^n \sigma(a^n)$ approaches 0 as n increases, since $\sigma^{\infty}(a) = 1$. Thus Proposition 2.1.4 shows that the sequence $((ta)^n)^q$ also converges to 0. This shows that for large n we will have $\sigma((a^n)^q) < 2\varepsilon$. Since $\varepsilon > 0$ was arbitrary, the sequence $\sigma((a^n)^q)$ and hence the sequence $\sigma(a^n)$ both converge to 0 with n. However this contradicts $\sigma^{\infty}(a) = 1$, which implies $\sigma(a^n) \geq 1$ for all n. Thus the theorem is established by contradiction. $\qquad \Box$

The next amazingly simple and powerful result is basic to much of the Gelfand theory, which will be presented in the next chapter. This result was announced without proof by Stanisław Mazur [1938] and a proof was supplied by Gelfand [1941a]. (Angus E. Taylor had used essentially the same proof for another theorem in [1938].) It is now generally called the Gelfand–Mazur theorem. This result generalizes the theorem of George Frobenius [1878, p. 63] which asserts that the complex numbers are the only finite-dimensional (complex) division algebra. The unpublished original proof of Mazur can be found in the book of Wiesław Żelazko [1973b]. For variant statements and proofs see Georgi Evgenyevič Šilov [1940a], Richard Arens [1947], S. Kametani [1952], L. Tornheim [1952], Marshall Harvey Stone [1953], Rickart [1958], V. K. Srinivasan [1979], Seth Warner [1979], Nolio Okada [1983], A. Cedilnik [1983], Nicola Rodinò [1983] and Propositions 2.5.12 and 2.5.13 below.

Recall that a *division algebra* is a unital algebra (not necessarily commutative) in which every non-zero element has an inverse.

2.2.3 Gelfand–Mazur Theorem *Any division algebra on which a nontrivial algebra semi-norm can be defined is isomorphic to the complex numbers. (In fact, it is enough to assume that every non-zero element of the*

algebra has a ⟨ left / right ⟩inverse.) The isomorphism is implemented by
$a \mapsto \lambda$ *where* λ *is the unique complex number in* $Sp(a)$.

Proof Suppose every non-zero element of a unital algebra $\mathcal{A}$ has a left
inverse. Let b be the left inverse of a. By assumption, b has a left inverse
which must be a. Thus b and a are invertible and $\mathcal{A}$ is a division algebra.
The theorem shows that any element a in an algebra with a nontrivial
algebra semi-norm has nonempty spectrum. If $\lambda \in Sp(a)$, then $\lambda 1 - a$ is
0 since it is not invertible. Thus the map $a = \lambda 1 \mapsto \lambda$ is a well-defined
isomorphism.   □

Spectral semi-norms

We now come to one of the most basic concepts of this work: *spectral
semi-norms*. The next theorem will show numerous conditions which are
equivalent for an algebra semi-norm. Logically it does not matter which of
these is chosen as the definition of a spectral semi-norm. We will select one
which is easy to remember and to check. We postpone historical remarks
until after the theorem.

2.2.4 Definition A *spectral semi-norm* on an algebra $\mathcal{A}$ is an algebra
semi-norm σ which is larger than or equal to the spectral radius, *i.e.*,

$$\rho(a) \leq \sigma(a) \qquad \forall\, a \in \mathcal{A}.$$

If a spectral semi-norm is actually a norm, then we call it a *spectral norm*.
An algebra together with a particular spectral semi-norm or spectral norm
is called a *spectral semi-normed algebra* or a *spectral normed algebra*.

2.2.5 Theorem *The following are equivalent for an algebra semi-norm σ
on an algebra $\mathcal{A}$.*
  (a) *σ is a spectral semi-norm.*
  (b) *For $a \in \mathcal{A}$, $\sigma(a) < 1$ implies that a is quasi-invertible in $\mathcal{A}$.*
  (c) *The set of quasi-invertible elements of $\mathcal{A}$ has nonempty interior with
respect to σ.*
  (d) *The set of quasi-invertible elements of $\mathcal{A}$ is open with respect to σ.*
  (e) *There is a finite constant C satisfying*

$$\rho(a) \leq C\sigma(a) \qquad \forall\, a \in \mathcal{A}.$$

(f) *σ satisfies*

$$\rho(a) = \lim_{n \to \infty} \sigma(a^n)^{1/n} \qquad \forall\, a \in \mathcal{A}.$$

*If $\mathcal{A}$ is unital, the quasi-group may be replaced by the group in conditions
(c) and (d).*

Proof (a) $\Rightarrow$ (b): Immediate since $\rho(a) < 1$ implies $1 \notin Sp(a)$.

(b) $\Rightarrow$ (c): Zero is in the interior of $\mathcal{A}_{qG}$.

(c) $\Rightarrow$ (d): Let $c \in \mathcal{A}_{qG}$ be arbitrary, and let $d \in \mathcal{A}_{qG}$ be an interior point of $\mathcal{A}_{qG}$. It is enough to show that c is an interior point. Define a map $L: \mathcal{A} \to \mathcal{A}$ by $L(a) = d \circ c^q \circ a$. Clearly L is continuous and maps both c onto d and $\mathcal{A}_{qG}$ onto $\mathcal{A}_{qG}$. Hence $\mathcal{A}_{qG} = L^{-1}(\mathcal{A}_{qG})$ is a neighborhood of c.

(d) $\Rightarrow$ (e): Since 0 is quasi-invertible, there is some $\varepsilon > 0$ such that $\sigma(a) < \varepsilon$ implies $a \in \mathcal{A}_{qG}$. Hence $\sigma(a) < \varepsilon$ implies $\lambda^{-1}a \in \mathcal{A}_{qG}$ for all $\lambda \in \mathbb{C}$ with $|\lambda| \geq 1$, which in turn implies $\rho(a) < 1$. Hence we may choose C to be ε^{-1}.

(e) $\Rightarrow$ (f): Proposition 2.2.1(a) established $\rho(a^n) = \rho(a)^n$ for all $a \in \mathcal{A}$ and $n \in \mathbb{N}$. Thus (e) implies

$$\rho(a) = \rho(a^n)^{1/n} \leq C^{1/n}\sigma(a^n)^{1/n} \qquad \forall \, a \in \mathcal{A}.$$

Combining this with Theorem 2.2.2 gives the desired result.

(f) $\Rightarrow$ (a): Immediate since $\lim \sigma(a^n)^{1/n} \leq \sigma(a)$.

The final remark is obvious. $\square$

Since 1 is never quasi-invertible, every spectral semi-norm is a nontrivial algebra semi-norm. Algebra norms with property (d) were first studied by Kaplansky [1947a], who called algebras provided with such a norm, Q-algebras. (He actually considered rings which were not necessarily algebras.) Ernest A. Michael [1952] and Bertram Yood [1958] first noted the equivalence of several of these properties.

Other characterizations of spectral semi-norms (or norms) will be given below in Theorems 2.2.14 and 2.5.7, Propositions 2.2.7 and 2.5.15 and Corollary 2.4.8. Many examples of spectral semi-norms are known and will be discussed in this work. Most were listed in the introduction to this chapter.

In a spectral semi-normed algebra $(\mathcal{A}, \sigma)$, an element a has spectrum $\{0\}$ if and only it satisfies $\lim_{n \to \infty} \sigma(a^n)^{1/n} = 0$. We have already called such elements *topologically nilpotent*. We now introduce the symbol $\mathcal{A}_{tN}$ for the class of such elements in any spectral algebra $\mathcal{A}$.

2.2.6 A Non-spectral Algebra Norm Let $\mathcal{A}$ be the disc algebra: the algebra under pointwise operations of functions continuous on the closed unit disc in $\mathbb{C}$ and holomorphic on its interior D. Let S be any subset of D with a limit point in D. Then $\| \cdot \|_S$ defined by $\|f\|_S = \sup\{|f(\lambda)| : \lambda \in S\}$ is a norm on $\mathcal{A}$. It is not spectral unless the closure of S includes the boundary of the disc. Note that in this example the algebra is a commutative, semisimple (Definition 2.3.2) Banach algebra and thus non-spectral norms can occur on quite well-behaved algebras. Nevertheless, Proposition 2.5.16 and examples (b), (c), (d), (e) and (h) in the introduction show a number of cases in which every norm on an algebra is spectral.

Geometric Series and Spectral Semi-norms

When dealing with algebra norms rather than semi-norms, Fuster and Marquina [1984] noted the following equivalence, which can be stated informally as: An algebra norm is spectral if and only if all the geometric series which should converge have limits. We remark that this result shows a connection between spectral semi-normed algebras and the more restrictive notion of functional algebras which we will introduce in the next chapter after the holomorphic functional calculus is discussed.

2.2.7 Proposition *The following conditions are equivalent for an algebra norm σ on an algebra $\mathcal{A}$.*

(a) *σ is a spectral norm.*

(b) *If the sequence $\sigma(a^n)$ converges to 0, then the geometric series $\sum_{n=1}^{\infty} a^n$ has a sum in $(\mathcal{A}, \sigma)$.*

(c) *If $\sum_{n=1}^{\infty} \sigma(a^n)$ is finite, then the geometric series $\sum_{n=1}^{\infty} a^n$ has a sum in $(\mathcal{A}, \sigma)$.*

(d) *If $\sigma(a) < 1$ holds, the geometric series $\sum_{n=1}^{\infty} a^n$ has a sum in $(\mathcal{A}, \sigma)$.*

Proof (a) $\Rightarrow$ (b): For a spectral norm σ, the equation $\lim \sigma(a^n) = 0$ implies that $a \circ (-\sum_{n=1}^{N} a^n) = (-\sum_{n=1}^{N} a^n) \circ a = a^{N+1}$ is quasi-invertible in $\mathcal{A}$ for large enough N. Hence a is quasi-invertible in $\mathcal{A}$. We will show that $-a^q$ is the sum of the series $\sum_{n=1}^{\infty} a^n$. This follows from the estimate:

$$
\begin{aligned}
\sigma\left(a^q + \sum_{n=1}^{N} a^n\right) &= \sigma\left((1 - a^q)(1 - a)(a^q + \sum_{n=1}^{N} a^n)\right) \\
&\leq (1 + \sigma(a^q))\sigma(a^q - aa^q + a - a^{N+1}) \\
&= (1 + \sigma(a^q))\sigma(a^{N+1}).
\end{aligned}
$$

(b) $\Rightarrow$ (c) $\Rightarrow$ (d): Immediate.

(d) $\Rightarrow$ (a): If the series converges, then

$$
a \circ \left(-\sum_{n=1}^{\infty} a^n\right) = \left(-\sum_{n=1}^{\infty} a^n\right) \circ a = a - \sum_{n=1}^{\infty} a^n + \sum_{n=1}^{\infty} a^{n+1} = 0,
$$

so a is quasi-invertible, verifying (b) of the previous theorem. $\qquad\square$

Note that the proof of the implication (a) $\Rightarrow$ (b) remains valid when σ is merely a spectral semi-norm rather than a norm. However, the proof of the opposite implication establishes only

$$
\sigma\left(a \circ \left(-\sum_{n=1}^{\infty} a^n\right)\right) = 0.
$$

Since the set of elements on which any spectral semi-norm vanishes is included in the Jacobson radical (Corollary 2.3.4), this implies that $a \circ$

$(-\sum_{n=1}^{\infty} a^n)$ is quasi-invertible when σ is a spectral semi-norm. Hence the second condition of Proposition 2.2.7 could be included in Theorem 2.2.5 by restating it as: For $a \in \mathcal{A}$, $\sigma(a) = 0$ implies $a \in \mathcal{A}_{qG}$ (or alternatively, $a \in \mathcal{A}_J$) and $\sigma(a) < 1$ implies that there is a sum in $\mathcal{A}$ for $\sum_{n=1}^{\infty} a^n$. This was omitted because of its inelegance.

We use the last result to establish the oldest and still the most important example of a spectral semi-norm. The result is due to Gelfand [1941a]. If an algebra or linear space is complete with respect to a norm we will say that the norm is *complete*.

2.2.8 Corollary *Any complete algebra norm is a spectral norm. Thus the norm of a Banach algebra is a spectral norm. In particular, the spectrum of any element in a Banach algebra is nonempty closed and bounded, hence compact and the spectral radius satisfies Theorem 2.2.5(f).*

Any norm on a finite-dimensional algebra is complete and hence is spectral if it is an algebra norm. For a strong converse see Peter D. Johnson, Jr. [1978] and Robert Grone and Johnson [1982].

Proposition 2.2.7 shows that geometric series play a special role with respect to spectral semi-norms. For future reference, it is useful to list the various geometric series which are commonly used. We do this in the next proposition.

2.2.9 Proposition *Let $(\mathcal{A}, \sigma)$ be a spectral semi-normed algebra.*
(a) For any $a \in \mathcal{A}_{qG}$ and $b \in \mathcal{A}$, either

$$\sigma(b - a) < [1 + \sigma(a^q)]^{-1} \quad or \quad \sigma^{\infty}(a^q \circ b) < 1$$

implies $b \in \mathcal{A}_{qG}$ and

$$b^q = a^q - \sum_{n=1}^{\infty} (a^q \circ b)^n (1 - a^q).$$

Also, $\sigma(a^q \circ b) < 1$ implies

$$\sigma(b^q - a^q) \le [1 - \sigma(a^q \circ b)]^{-1} [1 + \sigma(a^q)]^2 \sigma(b - a).$$

(b) If $\mathcal{A}$ is unital, for any $a \in \mathcal{A}_G$ and $b \in \mathcal{A}$, either

$$\sigma(b - a) < \sigma(a^{-1})^{-1} \quad or \quad \sigma^{\infty}(1 - a^{-1}b) < 1$$

implies $b \in \mathcal{A}_G$ and

$$b^{-1} = \sum_{n=0}^{\infty} (1 - a^{-1}b)^n a^{-1}.$$

Also, $\sigma(1 - a^{-1}b) < 1$ implies

$$\sigma(b^{-1} - a^{-1}) \le [1 - \sigma(1 - a^{-1}b)]^{-1} \sigma(a^{-1})^2 \sigma(b - a).$$

(c) *For any* $a \in \mathcal{A}$, $\mu \in \mathbb{C} \setminus (Sp(a) \cup \{0\})$ *and* $\lambda \in \mathbb{C} \setminus \{0\}$, *the inequality* $|\lambda - \mu| < |\lambda|[\sigma((\mu^{-1}a)^q)]^{-1}$ *implies* $\lambda \notin Sp(a)$,

$$(\lambda^{-1}a)^q = \mu(\lambda - \mu)^{-1} \sum_{n=1}^{\infty} (\lambda^{-1}(\lambda - \mu)(\mu^{-1}a)^q)^n \qquad and$$

$$\sigma((\lambda^{-1}a)^q - (\mu^{-1}a)^q) \leq |\lambda - \mu| \; |\lambda|^{-1}s[1 + |\mu| \; |\lambda|^{-1}s(1 - |\lambda - \mu| \; |\lambda|^{-1}s)^{-1}]$$

where s *is* $\sigma((\mu^{-1}a)^q)$.

(d) *If* $\mathcal{A}$ *is unital, then for any* $a \in \mathcal{A}$, $\mu \in \mathbb{C} \setminus Sp(a)$, *and* $\lambda \in \mathbb{C}$, *the inequality* $|\lambda - \mu| < \sigma((\mu - a)^{-1})^{-1}$ *implies* $\lambda \notin Sp(a)$,

$$(\lambda - a)^{-1} = \sum_{n=0}^{\infty} (\mu - \lambda)^n (\mu - a)^{-(n+1)}, \qquad and$$

$$\sigma((\lambda - a)^{-1} - (\mu - a)^{-1}) \leq (1 - |\mu - \lambda|\sigma((\mu - a)^{-1}))^{-1}\sigma((\mu - a)^{-1})^2|\mu - \lambda|.$$

(e) *For any* $a \in \mathcal{A}$ *and any* $\lambda \in \mathbb{C}$, $|\lambda| > \sigma^{\infty}(a)$ *implies* $\lambda \notin Sp(a)$ *and*

$$(\lambda^{-1}a)^q = - \sum_{n=1}^{\infty} (\lambda^{-1}a)^n.$$

Also, $|\lambda| > \sigma(a)$ *implies*

$$\sigma((\lambda^{-1}a)^q) \leq \sigma(a)(|\lambda| - \sigma(a))^{-1}.$$

(f) *If* $\mathcal{A}$ *is unital, then for any* $a \in \mathcal{A}$ *and* $\lambda \in \mathbb{C}$, $|\lambda| > \sigma^{\infty}(a)$ *implies* $\lambda \notin Sp(a)$, *and*

$$(\lambda - a)^{-1} = \sum_{n=0}^{\infty} \lambda^{-(n+1)}a^n.$$

Also, $|\lambda| > \sigma(a)$ *implies*

$$\sigma((\lambda - a)^{-1}) \leq |\lambda|^{-1}(| - |\lambda|^{-1}\sigma(a))^{-1}.$$

Proof The idea of the proof is already given in the proof of (a) $\Rightarrow$ (b) in Proposition 2.2.7. Note that (c) and (d) are special cases of (a) and (b), respectively. For (a), prove the simple identities $(1 - a)a^q = -a$, $(1 - a)(a^q \circ b) = b - a$ and $a^q \circ b = (1 - a^q)(b - a)$. The identity $(\mu^{-1}a)^q \circ \lambda^{-1}a = (\lambda - \mu)\lambda^{-1}(\mu^{-1}a)^q$ is useful in deriving (c) from (a).   □

Analyticity of the Resolvent Function

The last proposition shows that even though a spectral normed algebra need not be complete, it does have a rich supply of convergent geometric series. We wish to use this to prove one of the basic properties of the

resolvent function $\lambda \mapsto (\lambda - a)^{-1} \in \mathcal{A}^1$ defined on the *resolvent set* $\mathbb{C} \backslash Sp(a)$
for any element a in a spectral normed algebra.

2.2.10 Definition   Let $(\mathcal{X}, \| \cdot \|)$ be a normed linear space and let U
be an open (not necessarily connected) subset of $\mathbb{C}$. A function $f : U \to \mathcal{X}$
is said to be $\langle$ *weakly analytic / analytic* $\rangle$ on U if $\langle$ for each $\omega \in \mathcal{X}^*$ the
function $\omega \circ f : U \to \mathbb{C}$ is analytic on U / for each $\lambda_0 \in U$ there is a sequence
$\{x_n\}_{n \in \mathbb{N}^0} \subseteq \mathcal{X}$ and a positive ε satisfying

$$f(\lambda) = \sum_{n=0}^{\infty} (\lambda - \lambda_0)^n x_n \qquad \forall \, \lambda \text{ such that } |\lambda - \lambda_0| < \varepsilon \, \rangle.$$

If the complement of U is bounded, a function $f : U \to \mathcal{X}$ is said to be *ana-
lytic at infinity* if there is a positive constant R and a sequence $\{x_n\}_{n \in \mathbb{N}} \subseteq \mathcal{X}$
satisfying

$$f(\lambda) = \sum_{n=1}^{\infty} \lambda^{-n} x_n \qquad \forall \, \lambda \text{ such that } |\lambda| > R.$$

We insert the obvious result called for by this definition, even though it
is not needed for Corollary 2.2.12 which depends on Proposition 2.2.9.

2.2.11 Theorem   *Let $\mathcal{X}$ be a Banach space and let U be an open (not
necessarily connected) subset of $\mathbb{C}$. A function $f : U \to \mathcal{X}$ is analytic on U
if and only if it is weakly analytic on U.*

Proof Analyticity clearly implies weak analyticity, so let us assume that f
is weakly analytic on U and $\lambda_0 \in U$. Choose $\varepsilon > 0$ so that the closed disc
of radius ε around λ_0 is included in U. Let Γ be the positively oriented
boundary of this disc and let D be the open disc of radius $\varepsilon / 2$ centered at
λ_0. Consider the formula

$$\frac{1}{\lambda_1 - \lambda_2} \left(\frac{f(\lambda_1) - f(\lambda_0)}{\lambda_1 - \lambda_0} - \frac{f(\lambda_2) - f(\lambda_0)}{\lambda_2 - \lambda_0} \right) \tag{5}$$
$$= \frac{1}{2\pi i} \int_\Gamma \frac{f(\lambda) \, d\lambda}{(\lambda - \lambda_0)(\lambda - \lambda_1)(\lambda - \lambda_2)}).$$

If we knew that f was continuous so that the right side existed as a Riemann
integral, then this equation would follow from weak analyticity and the
Hahn–Banach theorem for any $\lambda_1, \lambda_2 \in D$. Since the norm of the right side
is bounded above by $\varepsilon \sup\{\|f(\lambda)\| : \lambda \in \Gamma\}/(2/\varepsilon)^3$, the left side is bounded
proving the existence of

$$f'(\lambda_0) = \lim_{\lambda \to \lambda_0} \frac{f(\lambda) - f(\lambda_0)}{\lambda - \lambda_0}$$

in the Banach space $\mathcal{X}$.

In order to avoid assuming the continuity of f or the existence and boundedness of the right side of (5), we replace f by $\omega \circ f$ for an arbitrary $\omega \in \mathcal{X}^*$ throughout (5). Both sides exist and are equal, and the right side is bounded by the same expression as before with $\sup\{\|f(\lambda)\| : \lambda \in \Gamma\}$ replaced by $\sup\{\|\omega \circ f(\lambda)\| : \lambda \in \Gamma\}$. By the uniform boundedness principle, the original right side is bounded in absolute value by some constant. Hence f is not only continuous but actually differentiable.

The definition of weak analyticity makes it clear that f' is also weakly analytic. Hence the argument can be repeated. For higher derivatives, the usual Cauchy formula and the Hahn–Banach theorem imply

$$f^{(n)}(\lambda_0) = \frac{n!}{2\pi i} \int_\Gamma \frac{f(\lambda)}{(\lambda - \lambda_0)^{n+1}} d\lambda.$$

The obvious estimate then shows that $f(\lambda) = \sum_{n=0}^\infty (\lambda - \lambda_0)^n f^{(n)}(\lambda_0)/n!$ converges for $|\lambda - \lambda_0| < r$. □

2.2.12 Corollary *Let $\mathcal{A}$ be a spectral normed algebra. For any $a \in \mathcal{A}$, the resolvent function*

$$\lambda \mapsto (\lambda - a)^{-1} \in \mathcal{A}^1 \qquad \forall \lambda \in \mathbb{C} \setminus Sp(a)$$

is analytic on the resolvent set $\mathbb{C} \setminus Sp(a)$ and at infinity.

Proof Results (d) and (f) in Proposition 2.2.9 give this. □

This result will be exploited in the next chapter, where we will show how to use contour integrals in the resolvent set of elements in a Banach algebra to define a functional calculus involving analytic functions. For more details on vector valued analytic functions, see Hille–Phillips [1957] pp. 92–116. We do not need anything beyond Corollary 2.2.12 for the following result which Bonsall and Duncan [1973b] call the *abstract Runge theorem*. In the next chapter we will use it to prove the classical approximation theorem of Runge.

2.2.13 Theorem *Let $\mathcal{A}$ be a unital Banach algebra. Let a be an element of $\mathcal{A}$, and let $\mathcal{B}$ be a closed unital subalgebra of $\mathcal{A}$ which contains a. Then the set $Sp_\mathcal{B}(a) \setminus Sp_\mathcal{A}(a)$ is a (possibly empty) union of bounded components of the $\mathcal{A}$-resolvent set of a.*

If $\tilde{S}$ contains at least one point from each bounded component of the $\mathcal{A}$-resolvent set of a and $\mathcal{B}$ is the closed unital subalgebra generated by a and $\{(\lambda - a)^{-1} : \lambda \in \tilde{S}\}$, then the resolvent sets of $\mathcal{A}$ with respect to $\mathcal{A}$ and $\mathcal{B}$ are equal.

Proof Let S be the resolvent set of a relative to $\mathcal{B}$, which can also be described as the set $\{\lambda \in \mathbb{C} \setminus Sp_\mathcal{A}(a) : (\lambda - a)^{-1} \in \mathcal{B}\}$. Proposition 2.2.9(d) shows that S is open. Let ω be an arbitrary element of $\mathcal{A}^*$ which vanishes on $\{(\lambda - a)^{-1} : \lambda \in S\}$. If $\lambda \in \mathbb{C} \setminus Sp_{\mathcal{A}}(a)$ can be approximated by elements of S, the expansion of Proposition 2.2.9(d) shows that ω must also vanish on $(\lambda - a)^{-1}$. Since $\omega \in \mathcal{A}^*$ was arbitrary, the Hahn–Banach theorem shows that S is also a relatively closed subset of the $\mathcal{A}$-resolvent set $\mathbb{C} \setminus Sp_{\mathcal{A}}(a)$ and hence must be a union of components. Proposition 2.2.9(f) shows that S includes every complex number with absolute value greater than the spectral radius of a, so it contains the unbounded component of the resolvent set of a in $\mathcal{A}$. Hence the two resolvent sets can only differ by some collection of bounded components of $\mathbb{C} \setminus Sp_{\mathcal{A}}(a)$. The second statement is now immediate.   □

The above result shows that when the spectrum of an element is changed by calculating it in smaller closed unital subalgebras, it increases only by adding a whole bounded component of the resolvent set all at once. In the other direction, if the spectrum of an element is changed by calculating it in a larger algebra in which the original algebra is embedded unitally and isometrically, the theorem shows that the boundary of the spectrum can only decrease by losing the whole boundary of a bounded component all at once. Thus for $1, a \in \mathcal{B} \subseteq \mathcal{A}$ we have

$$\partial Sp_{\mathcal{B}}(a) \subseteq \partial Sp_{\mathcal{A}}(a) \subseteq Sp_{\mathcal{A}}(a) \subseteq Sp_{\mathcal{B}}(a).$$

We will extend this observation in Corollary 2.5.8, which is based on an entirely different argument.

Stability of Spectral Semi-norms

Next we mention the stability properties of spectral semi-norms. We have collected these results together for convenience of reference even though they involve two definitions which will not be given formally until later in this chapter. Spectral subalgebras and the Jacobson radical are defined in the introduction and more formally in Definitions 2.5.1 and 2.3.3, respectively. The introduction and Proposition 2.5.3 give long lists of spectral subalgebras. Properties (d) and (e) are due to Kaplansky [1948b] and (h) is due to Richard G. Swan [1977].

As we have noted, there are many ways to extend an algebra norm from $\mathcal{A}$ to $\mathcal{A}^1$ if $\mathcal{A}$ is not unital, but all the reasonable ones satisfy condition (g) in the following theorem. The situation is similar for matrix algebras: the condition in (h) is satisfied by all reasonable extensions of an algebra norm.

2.2.14 Theorem *The following are equivalent for an algebra semi-norm σ on an algebra $\mathcal{A}$.*

(a) *σ is spectral.*

(b) *The restriction of σ to each spectral subalgebra is spectral.*

(c) *The restriction of σ to each maximal commutative subalgebra is spectral.*

(d) *The quotient semi-norm induced by σ on $\mathcal{A}/\mathcal{I}$ is spectral for each ideal $\mathcal{I}$ of $\mathcal{A}$.*

(e) *There is some ideal $\mathcal{I}$ for which both σ and its quotient semi-norm are spectral on $\mathcal{I}$ and $\mathcal{A}/\mathcal{I}$, respectively. (When this condition holds for some ideal it holds for all ideals, of course.)*

(f) *The quotient norm induced by σ on $\mathcal{A}/\mathcal{A}_J$ is spectral, where $\mathcal{A}_J$ is the Jacobson radical.*

(g) *Any extension σ^1 of σ to be an algebra semi-norm on $\mathcal{A}^1$ in which $\mathcal{A}$ is closed in $\mathcal{A}^1$ is a spectral semi-norm.*

(h) *Any extension of the semi-norm to be an algebra semi-norm on the $n \times n$-matrix algebra over $\mathcal{A}$ is spectral if the topology of this extension is that defined by the linear space semi-norm $\max\{\sigma(a_{ij}) : 1 \leq i, j \leq n\}$.*

(i) *σ is equivalent to a spectral semi-norm.*

Proof (a) $\Rightarrow$ (b): Since the non-zero spectrum of an element in a spectral subalgebra is the same whether calculated in the subalgebra or in the full algebra, this is immediate from the definition.

(b) $\Rightarrow$ (c): Maximal commutative subalgebras are spectral subalgebras by Proposition 2.5.3(b).

(c) $\Rightarrow$ (a): Each element in the algebra belongs to some maximal commutative subalgebra, so this also follows from the definition of a spectral semi-norm.

(a) $\Rightarrow$ (d): The inequality $\rho_{\mathcal{A}/\mathcal{I}}(a + \mathcal{I}) \leq \rho_{\mathcal{A}}(a)$ for all $a \in \mathcal{A}$ from Proposition 2.2.1(c) and the definition give this.

(d) $\Rightarrow$ (f) $\Rightarrow$ (e): Immediate since $\mathcal{A}_J$ is always spectral.

(e) $\Rightarrow$ (a): It is enough to show $\sigma(a) < 1/3$ implies that a is quasi-invertible in $\mathcal{A}$. This inequality implies $\sigma(a + \mathcal{I}) < 1/3$ and hence that the quasi-inverse $b + \mathcal{I}$ of $a + \mathcal{I}$ satisfies $\sigma(b + \mathcal{I}) < (1/3)/(1 - 1/3) = 1/2$. Hence the coset representative b can be chosen in $\mathcal{A}$ to satisfy $\sigma(b) < 1/2$, so $\sigma(a \circ b) \leq \sigma(a) + \sigma(b) + \sigma(ab) < 1$. Thus $a \circ b$ is quasi-invertible in $\mathcal{I}$ (where σ is spectral) so $a \circ b$ and a are quasi-invertible in $\mathcal{A}$.

(a) $\Rightarrow$ (g): The condition shows that the quotient semi-norm of σ^1 on $\mathbb{C} = \mathcal{A}^1/\mathcal{A}$ is spectral. If the quotient norm of a non-zero λ is less than $|\lambda|$, then there is a $b \in \mathcal{A}$ satisfying $\sigma^1(1 - b) < 1$, which implies that 1 is the limit of $1 - (1 - b)^n \in \mathcal{A}$. Hence (a) $\Rightarrow$ (g) follows from condition (e).

(g) $\Rightarrow$ (a): Condition (g) implies that the extension $\sigma^1(\lambda + a) = |\lambda| + \sigma(a)$ satisfies the condition. Hence the original semi-norm is spectral on $\mathcal{A}$ by (b), since $\mathcal{A}$ is an ideal in $\mathcal{A}^1$ and hence a spectral subalgebra by Proposition 2.5.3(a).

(a) $\Rightarrow$ (h): If σ is spectral, we will show that the set of quasi-invertible elements is a neighborhood of 0. Choose a neighborhood $\mathcal{U}$ of 0 in $M_n(\mathcal{A})$ so small that $\sigma(a_{ij}) < 2^{-n}$ holds for all matrices (a_{ij}) in $\mathcal{U}$ and all i and j. Let

$a = (a_{ij})$ be a matrix in $\mathcal{U}$. Then, in particular, by Theorem 2.2.5(b) all its elements are quasi-invertible and the elements satisfy $\sigma(a_{i1}(1-a_{11})^{-1}a_{1j}) < 2^{-n}$. Multiply the matrix $I - a$ on the right by all the elementary matrices which differ from the identity matrix by the insertion of $(1 - a_{11})^{-1}a_{1j}$ in the $1j$ position for $j > 1$. Multiply on the left by all the elementary matrices which differ from the identity matrix by the insertion of $a_{i1}(1 - a_{11})^{-1}$ in the $i1$ position for $i > 1$. The result will be that there are only zeroes in the first row and column except in the $1\,1$ position and that the remaining non-zero elements are perturbed by addition of elements of the form $a_{i1}(1 - a_{11})^{-1}a_{1j}$ and thus satisfy $\sigma(b_{ij}) < 2^{-n+1}$. Hence the process can be continued inductively until $I - a$ has been shown to be similar to a matrix of the form $I - b$ where b is diagonal with each diagonal entry quasi-invertible.

(h) $\Rightarrow$ (a): We can extend σ to the matrix algebra as the Hilbert–Schmidt semi-norm $\sigma'((a_{ij})_{1 \leq i,j \leq n}) = [\sum_{i,j=1}^{n} \sigma(a_{ij})^2]^{1/2}$. It is submultiplicative by the Cauchy–Schwarz inequality. Hence this implication follows from (b), because $\mathcal{A}$ is embedded in the matrix algebra in the upper left-hand entry as a corner subalgebra which is a spectral subalgebra by Proposition 2.5.3(c) below.

(i) $\Leftrightarrow$ (a): This follows immediately from Theorem 2.2.5(f). $\qquad\square$

It is worth noting that conclusion (i) can be combined with Theorem 1.1.10(i) to show that for any element a in a spectral normed algebra $(\mathcal{A}, \sigma)$ and any $\varepsilon > 0$ we can always find an equivalent spectral norm σ' satisfying

$$\rho(a) > \sigma'(a) - \varepsilon.$$

Here is an alternative proof for (h) which may have independent interest. Let $\mathcal{A}$ be a unital topological algebra in which the set of invertible elements is open and inversion is continuous. Let (a_{ij}) be a 2×2 matrix over $\mathcal{A}$ which is close to the identity matrix in the sense that the differences of all corresponding entries is in the same small neighborhood of 0. Then its inverse (b_{ij}) is given by

$$b_{ij} = \delta(a_{ii} - a_{ik}a_{kk}^{-1}a_{ki})^{-1}a_{ij}a_{jj}^{-1}$$

where k always has the other value from i in $\{0, 1\}$ and δ is positive if $i = j$ and negative otherwise. All the inverses written exist if the matrix is close enough to the identity. Thus $M_2(\mathcal{A})$ also has an open set of invertible elements and inversion is continuous. The method works for arbitrary size matrices, but we stop after giving the result for 3×3 matrices. Use the same notation except this time we assume that $\{i, m, n\}$ is the set $\{1, 2, 3\}$ in some order and that m differs from both i and j if they are distinct. Then

$$b_{ij} = \delta[a_{ii} + (a_{in}a_{nn}^{-1}a_{nm} - a_{im})(a_{mm} - a_{mn}a_{nn}^{-1}a_{nm})^{-1}a_{mi}$$
$$+ (a_{im}a_{mm}^{-1}a_{mj} - a_{ij})(a_{nn} - a_{nm}a_{mm}^{-1}a_{mn})^{-1}a_{ni}]^{-1}$$
$$(a_{im}a_{mm}^{-1}a_{mj} - a_{ij})(a_{nn} - a_{nm}a_{mm}^{-1}a_{mn})^{-1}.$$

Continuity of the Spectrum

The next result, which is due to John D. Newburgh [1951], shows that for a spectral semi-normed algebra $\mathcal{A}$, $Sp(a) \subseteq \mathbb{C}$ is an upper semi-continuous function of $a \in \mathcal{A}$ when we use the Hausdorff metric on the subsets of $\mathbb{C}$. Example 2.2.16 shows that there exists a sequence of nilpotent (hence topologically nilpotent) elements in a Banach algebra which converges to an element with spectrum larger than $\{0\}$. However, Corollary 3.1.6 shows that $Sp(a) \subseteq \mathbb{C}$ is a continuous function of $a \in \mathcal{A}$ when $\mathcal{A}$ is also commutative. For further results see Newburgh [1951], Aupetit [1975] and [1979a], Gerard J. Murphy [1981] and Laura Burlando [1986], [1990] and [1991]. The last paper gives a criterion for the spectrum to be continuous at an element a in a Banach algebra. The condition is much weaker than Newburgh's sufficient condition that $Sp(a)$ be totally disconnected. (The notation $\varepsilon\mathbb{C}_1$ used in the theorem denotes the closed disk of radius ε in accordance with our convention on closed balls in normed linear spaces.)

2.2.15 Theorem *Let $(\mathcal{A}, \sigma)$ be a spectral semi-normed algebra. For any $a \in \mathcal{A}$ and any $\varepsilon > 0$, there is a $\delta > 0$ so that for all $b \in \mathcal{A}$, $\sigma(a - b) < \delta$ implies*

$$Sp(b) \subseteq Sp(a) + \varepsilon\mathbb{C}_1 \qquad and \qquad \rho(b) \leq \rho(a) + \varepsilon.$$

Hence the spectrum and spectral radius are upper semi-continuous functions.

Proof Suppose the inclusion fails for a given $a \in \mathcal{A}$ and $\varepsilon > 0$. We can find a sequence $\{a_n\}_{n\in\mathbb{N}} \subseteq \mathcal{A}$ converging to a and a sequence $\{\lambda_n\}_{n\in\mathbb{N}} \subseteq \mathbb{C}$ satisfying $\lambda_n \in Sp(a_n) \setminus (Sp(a) + \varepsilon\mathbb{C}_1)$. Since $\{a_n : n \in \mathbb{N}\}$ is bounded, $\{\lambda_n : n \in \mathbb{N}\}$ is bounded. Hence (by taking a subsequence if necessary) we may assume $\{\lambda_n\}_{n\in\mathbb{N}}$ converges to some $\lambda \notin Sp(a)$. If λ is 0, then $\mathcal{A}$ is unital and a is invertible. Since $\lambda_n - a_n$ converges to $-a$, $\lambda_n - a_n$ is eventually invertible, contradicting the choice of λ_n. If λ is not 0, then $\lambda^{-1}a$ is quasi-invertible. Eventually λ_n is non-zero and since $\lambda_n^{-1}a_n$ converges to $\lambda^{-1}a$, $\lambda_n^{-1}a_n$ is eventually quasi-invertible, contradicting the choice of λ_n again. Hence the inclusion is proved by contradiction. The inequality follows immediately.   □

2.2.16 Example This example is due to Kaplansky but was first published by Rickart [1960]. It shows that the results of Newburgh's theorem are restricted to semicontinuity and cannot be extended to give continuity.

We produce a sequence $\{T_n\}_{n\in\mathbb{N}}$ of nilpotent operators which converges to an operator having spectrum different from $\{0\}$.

Let $\mathcal{H}$ be $\ell^2(\mathbb{N})$ with $\{z_n\}_{n\in\mathbb{N}}$ denoting the natural orthonormal basis. The operators which we consider are all weighted shift operators defined by

$$S(z_m) = s_m z_{m+1} \quad m \in \mathbb{N}$$

for some bounded sequence $\{s_m\}_{m\in\mathbb{N}}$ of weights. The norm of such an operator is easily seen to be

$$\|S\| = \sup\{|s_m| : m \in \mathbb{N}\}.$$

For each m, $n \in \mathbb{N}$, define t_m and t_{mn} to be e^{-k} and $(1 - \delta_{nk})e^{-k}$, respectively, where k is defined by $m = 2^k(2j + 1)$. Define T and T_n by

$$T(z_m) = t_m z_{m+1} \quad \text{and} \quad T_n(z_m) = t_{mn} z_{m+1} \quad m \in \mathbb{N}.$$

Since $T_n - T$ satisfies $(T_n - T)(z_m) = \delta_{nk}e^{-k}z_{m+1}$ for all n, $m \in \mathbb{N}$ and for k as defined previously, we see that the sequence $\{T_n\}_{n\in\mathbb{N}}$ does converge to T. It is easy to see that each T_n satisfies $T_n^{2^{n+1}} = 0$, since 2^{n+1} shifts of the index will send any basis vector z_m past an index of the form by $2^n(2j+1)$ where t_{mn} is 0. Hence each T_n is nilpotent and satisfies $Sp(T_n) = \{0\}$ and $\rho(T_n) = 0$.

Finally, we show that T satisfies $\rho(T) \geq e^{-1}$ and hence has non-zero spectrum. The norm of T^n is given by

$$\|T^n\| = \sup\{\prod_{m=k}^{k+n-1} t_m : k \in \mathbb{N}\}.$$

For $n = 2^j - 1$, the definition of the weights gives

$$\prod_{m=1}^{n} t_m = \prod_{k=1}^{j-1} \exp(-k2^{j-k-1})$$

and hence

$$\|T^n\|^{1/n} > \exp(-\sum_{k=1}^{j-1} \frac{k}{2^{k+1}}).$$

Since the right-hand side converges to e^{-1} with increasing j (or n), Corollary 2.2.8 gives the desired result.

2.2.17 Weakly Spectral Norms   The strong spectrum $Sp^S(a)$ and strong spectral radius $\rho^S(a)$ of an element a in an algebra $\mathcal{A}$ were introduced in Definition 2.1.19. Meyer [1992d] defined an algebra norm $\|\cdot\|$ on $\mathcal{A}$ to be *weakly spectral* if the set $\mathcal{A}_{wG}$ of weakly invertible elements is open relative to it. We note some of the parallels with spectral norms.

Proposition   *Let $\| \cdot \|$ be an algebra norm on a unital algebra $\mathcal{A}$.*
    (a) *If $\| \cdot \|$ is spectral, it is weakly spectral.*
The following conditions are equivalent:
    (b) *The norm $\| \cdot \|$ is weakly spectral.*
    (c) *All $a \in \mathcal{A}$ satisfy $\rho_S(a) \le \|a\|$ so $Sp^S(a)$ is compact.*
    (d) *The identity element 1 is in the interior of $\mathcal{A}_{wG}$.*
    (e) *Every maximal ideal of $\mathcal{A}$ is closed.*
    (f) *For any $a \in \mathcal{A}$, $\|1 - a\| < 1$ implies $a \in \mathcal{A}_{wG}$.*

We omit the proofs which are similar to the spectral case. In a nonunital algebra $\mathcal{A}$, the definition and proposition can be modified using the set $\mathcal{A}_{wqG}$ of weakly quasi-invertible elements in place of $\mathcal{A}_{wG}$. (This set is defined in Section 4.5.) In the nonunital case, (e) refers to all maximal modular ideals and 0 replaces 1 in (d) and (f).

2.3   The Jacobson Radical and the Fundamental Theorem

We begin by showing that spectral semi-norms are not unique even on the best behaved algebras. Our next major goal is to prove that if a sequence converges to two possibly different limits in two different spectral semi-norms, these limits differ by an element of the Jacobson radical. Hence we introduce a preliminary treatment of the Jacobson radical. A few remarks about representations are needed in order to define this concept. Representations and the Jacobson radical will be discussed much more fully in Chapter 4, where all of the current arguments will be elaborated and the reader will find historical remarks.

2.3.1 Spectral Semi-norms are not Unique   This is true even in unital, semisimple commutative Banach algebras. Consider $\ell^1(\mathbb{Z})$ with convolution multiplication. Both ρ and $\| \cdot \|_1$ are spectral norms. Another easy noncommutative example is the finite-rank operators on a Banach space. In Section 1.7 we described several norms on this set, in addition to the operator norm. Theorem 2.2.14(b) shows that all of these norms are spectral since the finite-rank operators form an ideal in the various completions.

Despite these examples, Theorem 2.3.6 below shows that if a sequence converges in two different spectral norms, then the two limits differ by an element of the Jacobson radical (cf. Definition 2.3.2 and Theorem 2.3.3). Hence in semisimple algebras such limits are unique.

Introduction to the Jacobson Radical

2.3.2 Definition   A *representation* of an algebra $\mathcal{A}$ on a linear space $\mathcal{X}$ is an algebra homomorphism T from $\mathcal{A}$ into the algebra $\mathcal{L}(\mathcal{X})$ of all linear maps (also called *operators*) of $\mathcal{X}$ into itself. For $a \in \mathcal{A}$, we write T_a for the operator associated with a. A subspace $\mathcal{Y}$ of $\mathcal{X}$ is said to be *invariant* (or *T-invariant* in cases where the representation might not be clear) if every representing operator T_a maps $\mathcal{Y}$ into itself. A representation is called *irreducible* if it has some non-zero representing operators, and $\mathcal{X}$ is its only non-zero invariant subspace. The *Jacobson radical* $\mathcal{A}_J$ of an algebra $\mathcal{A}$ is the intersection of the kernels of all the irreducible representations of $\mathcal{A}$. An algebra is called *semisimple* if $\mathcal{A}_J$ is $\{0\}$ and *Jacobson-radical* if $\mathcal{A}_J$ is the whole algebra (*i.e.*, if there are no irreducible representations at all).

In order to prove the next theorem we need to know that if T is an irreducible representation of $\mathcal{A}$ on $\mathcal{X}$ and z is a non-zero vector in $\mathcal{X}$, then any vector x in $\mathcal{X}$ can be written in the form $T_a z$ for a suitable $a \in \mathcal{A}$. To see this, note that $T_{\mathcal{A}} z = \{T_a z : a \in \mathcal{A}\}$ is clearly an invariant subspace. If it were the zero subspace, then the set of all w for which $T_{\mathcal{A}} w$ equals 0 would be a non-zero invariant subspace and hence equal to $\mathcal{X}$. Since this would contradict the existence of a non-zero representing operator, $T_{\mathcal{A}} z$ must be all of $\mathcal{X}$ for each non-zero z, which gives the desired result.

2.3.3 Theorem *In any algebra $\mathcal{A}$ the following sets are equal.*

(a) *The Jacobson radical $\mathcal{A}_J$ of $\mathcal{A}$.*

(b) $\{b \in \mathcal{A} : \rho(ab) = 0$ *for all* $a \in \mathcal{A}\}$.

(c) *The largest ideal $\mathcal{I}$ satisfying any (hence all) of the following equivalent properties:*

$$Sp_{\mathcal{A}/\mathcal{I}}(a + \mathcal{I}) \subseteq Sp_{\mathcal{A}}(a) \subseteq Sp_{\mathcal{A}/\mathcal{I}}(a + \mathcal{I}) \cup \{0\} \qquad \forall\, a \in \mathcal{A};$$

$$\mathcal{A}_{qG} = \{a \in \mathcal{A} : a + \mathcal{I} \in (\mathcal{A}/\mathcal{I})_{qG}\};$$

$$Sp(a + b) = Sp(a) \qquad \forall\, a \in \mathcal{A};\ b \in \mathcal{I};$$

$$Sp(b) = \{0\} \qquad \forall\, b \in \mathcal{I};$$

$$\rho(a + b) = \rho(a) \qquad \forall\, a \in \mathcal{A};\ b \in \mathcal{I};$$

$$\rho(b) = 0 \qquad \forall\, b \in \mathcal{I}.$$

Proof Let us call the conditions in (c), (c_1), (c_2), (c_3), (c_4), (c_5) and (c_6), respectively. Note $(c_4) \Leftrightarrow (c_6)$. All ideals satisfy the first inclusion in (c_1). The definition of the non-zero elements in the spectrum shows $(c_1) \Leftrightarrow (c_2)$. The following implications are also easy: $(c_3) \Rightarrow (c_4)$, $(c_3) \Rightarrow (c_5)$, $(c_4) \Rightarrow (c_6)$ and $(c_5) \Rightarrow (c_6)$. Furthermore (c_1) almost implies (c_3). Since we may choose an arbitrary representative of the coset $a + \mathcal{I}$ in (c_1), it is obvious that the spectra of a and $a + b$ differ by at most zero in (c_3). We will temporarily denote the set in (b) by $\mathcal{B}$.

Suppose $\mathcal{J}$ is an ideal (not necessarily the largest) with the property described in (c_1) and let $\mathcal{I}$ be the largest ideal satisfying (c_6). Then, in particular, each element of $b \in \mathcal{J}$ satisfies $Sp_{\mathcal{A}}(b) = \{0\}$. Hence $\mathcal{J}$ is included in $\mathcal{I}$ which is included in $\mathcal{B}$ (since $\mathcal{I}$ is an ideal).

We will show that any element b in $\mathcal{B}$ is in the kernel of every irreducible representation, which implies $\mathcal{J} \subseteq \mathcal{I} \subseteq \mathcal{B} \subseteq \mathcal{A}_J$. Suppose $b \in \mathcal{B}$ is not in the kernel of some irreducible representation $T : \mathcal{A} \to \mathcal{L}(\mathcal{X})$. Then we can choose $z \in \mathcal{X}$ so that $T_b z \neq 0$. By irreducibility we may choose $a \in \mathcal{A}$ satisfying $T_a T_b z = z$. Since $\rho(ab)$ is 0, we may find a quasi-inverse c for ab. But this gives the contradiction:

$$z = z + T_c z - T_c z = T_{ab} z + T_c z - T_c T_{ab} z = T_0 z = 0.$$

We will complete the proof by showing that the Jacobson radical has the property of $\mathcal{J}$ described in (c_2) and satisfies (c_3) even at zero. In order to do the former it is enough to show that any element which is quasi-invertible modulo the Jacobson radical is quasi-invertible. We will begin by showing that such an element e has a left quasi-inverse. Suppose not. Then e does not belong to the left ideal $\{ae - a : a \in \mathcal{A}\}$ denoted by $\mathcal{A}(e-1)$. Hence we may use Zorn's lemma to find a left ideal $\mathcal{M}$ which is maximal (under the inclusion order) and includes $\mathcal{A}(e-1)$ but does not contain e. Consider the natural action of $\mathcal{A}$ on $\mathcal{A}/\mathcal{M}$ defined by

$$T_a(b + \mathcal{M}) = ab + \mathcal{M} \qquad \forall\, b \in \mathcal{M}.$$

The maximality of $\mathcal{M}$ shows that this is an irreducible representation. (Any left ideal properly including $\mathcal{M}$ would have to contain $\{e\}$ and thus be the whole algebra $\mathcal{A}$.) Hence the Jacobson radical is included in the kernel of T. Furthermore any element b of this kernel belongs to $\mathcal{M}$ since $T_b(e+\mathcal{M}) = \mathcal{M}$ implies $b = be - (be - b) \in \mathcal{M}$. Since we assumed that e was quasi-invertible modulo the Jacobson radical, there is an element $c \in \mathcal{A}$ with $e + c - ce$ in the Jacobson radical and hence in $\mathcal{M}$. But this implies that e is in $\mathcal{M}$ contrary to the construction of $\mathcal{M}$.

We have now shown that each element $e \in \mathcal{A}$ which becomes quasi-invertible in $\mathcal{A}/\mathcal{A}_J$ has a left quasi-inverse c in $\mathcal{A}$. Then $c + \mathcal{A}_J$ must be the quasi-inverse of $e + \mathcal{A}_J$. Thus $c + \mathcal{A}_J$ is quasi-invertible so c has a left quasi-inverse which must be e. Thus e is quasi-invertible, as we wished to show.

Finally, we show that $\mathcal{A}_J$ satisfies (c_3) even for 0 in the spectra. If a is invertible and b belongs to the Jacobson radical, then $-a^{-1}b$ has a quasi-inverse c so $a^{-1} - ca^{-1}$ is an inverse for $a + b$. Since (c_1) implies (c_3) for non-zero values, the Jacobson radical satisfies (c_3).   □

2.3.4 Corollary *The set on which any spectral semi-norm vanishes is included in the Jacobson radical. Hence, any spectral semi-norm on a semi-simple algebra is a norm, and the Jacobson radical is closed in any spectral semi-norm. Any algebra norm on a Jacobson-radical algebra is spectral.*

Proof If a spectral semi-norm σ vanishes on $b \in \mathcal{A}$, then it vanishes on ab for all $a \in \mathcal{A}$, so $\rho(ab)$ is 0. Hence the set on which σ vanishes is included in the Jacobson radical by condition (b) of the theorem. The second statement is an immediate consequence, and the third follows by considering the quotient semi-norm on $\mathcal{A}/\mathcal{A}_J$, which is spectral by Theorem 2.2.14(d). Since the quotient semi-norm is a norm, $\mathcal{A}_J$ is closed. The last sentence follows from the definition of spectral semi-norms, since the spectral radius is identically 0 on the Jacobson radical.   □

We will eventually see several other proofs that the Jacobson radical is closed in spectral semi-norms.

The Fundamental Theorem of Spectral Semi-norms

Theorem 2.3.6 below is derived from Proposition 1.3 of Angel Rodriguez-Palacio [1985]. It was first stated explicitly in the present form as Proposition 3.6 in Palmer [1992] (submitted in 1985). The present proof (Palmer [1989]) is a simple transformation of Ransford's beautiful proof [1989] (based on Aupetit [1982a]) of Johnson's uniqueness of norm theorem. The next lemma, which we call *Ransford's three circles theorem*, is a slight extension of his proof [1989] based on Jacques Hadamard. A spectral algebra (Definition 2.4.1) is simply an algebra on which some spectral semi-norm can be defined. We prefer to give this lemma here where it is needed, with its natural hypothesis, rather than waiting until after the formal definition.

2.3.5 Lemma *Let p be a polynomial of a complex variable with coefficients from a spectral algebra $\mathcal{A}$. Then any $R > 0$ satisfies:*

$$\sup\{\rho(p(\zeta))^2 : |\zeta| = 1\} \leq \sup\{\rho(p(\lambda)) : |\lambda| = R\} \sup\{\rho(p(\lambda)) : |\lambda| = R^{-1}\}.$$

Proof Choose a spectral semi-norm σ on $\mathcal{A}$, and let $\mathcal{A}_\sigma$ be the ideal on which σ vanishes. Then the formula $\|a + \mathcal{A}_\sigma\| = \sigma(a)$ defines a spectral norm on the quotient algebra $\mathcal{A}/\mathcal{A}_\sigma$ (Theorem 2.2.14(d)). Let $\mathcal{B}$ be the completion of $\mathcal{A}/\mathcal{A}_\sigma$ in this norm. By Theorem 2.2.5(f) and Corollary 2.2.8, the spectral radii on $\mathcal{A}$ and $\mathcal{B}$ satisfy:

$$\rho_{\mathcal{A}}(a) = \lim \sigma(a^n)^{1/n} = \lim \|a^n + \mathcal{A}_\sigma\|^{1/n} = \rho_{\mathcal{B}}(a + \mathcal{A}_\sigma) \qquad \forall \, a \in \mathcal{A}.$$

Let n be an arbitrary positive integer and ζ be an arbitrary complex number of modulus 1. Choose a norm-one continuous linear functional ω on $\mathcal{B}$ satisfying $\omega(p(\zeta)^n + \mathcal{A}_\sigma) = \|p(\zeta)^n + \mathcal{A}_\sigma\| = \sigma(p(\zeta)^n)$. Denote $\omega(p(\lambda)^n + \mathcal{A}_\sigma)$ by $q(\lambda) = \sum_{k=0}^{nd} q_k \lambda^k$ where d is the degree of p and each q_k is a complex number. Then the Cauchy–Schwarz–Bunyakowski inequality gives:

$$\sigma(p(\zeta)^n)^2 \ \leq \ \left(\sum |q_k|\right)^2 \leq (nd+1)\left(\sum |q_k|^2\right)$$

$$\leq \ (nd+1)\left(\sum |q_k|^2 R^{2k}\right)^{1/2}\left(\sum |q_k|^2 R^{-2k}\right)^{1/2}$$

$$\leq \ (nd+1)\left(\frac{1}{2\pi}\int_0^{2\pi} |q(Re^{it})|^2 dt \ \frac{1}{2\pi}\int_0^{2\pi} |q(R^{-1}e^{it})|^2 dt\right)^{1/2}$$

$$\leq \ (nd+1)\sup\{\sigma(p(\lambda)^n) : |\lambda| = R\}\sup\{\sigma(p(\lambda)^n) : |\lambda| = R^{-1}\}.$$

We get the desired result by taking nth roots and letting n approach infinity through powers of two. This makes the convergence monotone decreasing, so that a slight variant of Dini's theorem (e. g., Stromberg [1981]) shows that the supremum and limit may be interchanged. $\qquad\square$

2.3.6 Fundamental Theorem *Let σ_1 and σ_2 be two spectral semi-norms on an algebra $\mathcal{A}$. If a sequence has limits with respect to both semi-norms, then these limits differ by an element of the Jacobson radical.*

Proof Let $\{a_n\}$ be such a sequence. Without loss of generality we may assume the σ_1-limit is 0, and we will denote the other by a. Let b in $\mathcal{A}$ be arbitrary. Denote the supremum of $\{\sigma_2(ba_n)\}$ by B. We shall apply Ransford's three circles theorem with $\zeta = 1$ to the first degree polynomials:

$$p_n(\lambda) = \lambda ba_n + ba - ba_n.$$

This gives

$$\rho(ba)^2 \ \leq \ \sup\{\rho(p_n(\lambda)) : |\lambda| = R\}\sup\{\rho(p_n(\lambda)) : |\lambda| = R^{-1}\}$$

$$\leq \ \sup\{\sigma_1(p_n(\lambda)) : |\lambda| = R\}\sup\{\sigma_2(p_n(\lambda)) : |\lambda| = R^{-1}\}$$

$$\leq \ [(R+1)\sigma_1(ba_n) + \sigma_1(ba)][R^{-1}\sigma_2(ba_n) + \sigma_2(ba - ba_n)].$$

Letting n approach infinity, we get the upper bound $\sigma_1(ba)R^{-1}B$. Now let R approach infinity, in order to conclude $\rho(ba) = 0$. Thus by Theorem 2.3.3, a is in the Jacobson radical, as we wished to show. $\qquad\square$

In order to express the next corollary elegantly, we introduce a concept due to Bohdan J. Tomiuk and Bertram Yood [1989] and Michael J. Meyer [1992a].

2.3.7 Definition Two (not necessarily complete) norms $\|\cdot\|$ and $\|\|\cdot\|\|$ on a normed linear space $\mathcal{A}$ are said to be *consistent* if the identity map between them has a closed graph. More explicitly, this means that if a sequence $\{a_n\}_{n\in\mathbb{N}} \subseteq \mathcal{A}$ converges to a in $\|\cdot\|$ and to b in $\|\|\cdot\|\|$, then $a = b$.

Note that if two norms dominate a common norm, or if the closed unit ball in one is closed in the other, then they are consistent. In particular if one norm dominates another, they are consistent. For two complete norms, the closed graph theorem shows that consistency is the same as equivalence. Corollary 2.3.4 shows that the next obvious corollary captures the full force of the fundamental theorem on semisimple algebras.

2.3.8 Corollary *Any two spectral norms on a semisimple algebra are consistent.*

2.3.9 Corollary *Every surjective homomorphism from a Banach algebra onto a semisimple Banach algebra is continuous. In fact, every surjective homomorphism from a spectral semi-normed algebra onto a semisimple spectral normed algebra is closed.*

Proof We explain the second result first. If $\varphi \colon A \to B$ satisfies these hypotheses with σ as the spectral semi-norm on A, apply the theorem to B with σ_1 as the given norm on B and σ_2 as the spectral quotient semi-norm transported from $A/\ker(\varphi)$, i.e., $\sigma_2(b) = \inf\{\sigma(a) : a \in \varphi^{-1}(b)\}$ for all $b \in B$. Thus if a sequence converges in σ and its image under φ converges in σ_1, then its image under φ surely converges in both σ_2 and σ_1, so the limits are the same by the theorem. For the first result, apply the closed graph theorem.   □

2.3.10 Corollary (Johnson's uniqueness of norm theorem) *Any two complete norms on a semisimple Banach algebra are equivalent. Thus a semisimple Banach algebra has a unique complete norm topology. Furthermore, if an algebra A is complete with respect to two submultiplicative norms, then they induce equivalent quotient norms on A/A_J.*

Proof Apply the last corollary with the identity map on the algebra bearing its two different norms.   □

The final corollary above is one of the most important results in the whole theory of Banach algebras. Johnson's original proof [1967a] was based in part on work of Rickart [1950]. The original difficult proof, which uses the principle of accumulation of singularities, will be given to establish Theorem 4.2.15. The present proof is derived from ideas of Aupetit [1982a], Rodriguez-Palacio [1985] and Ransford [1989]. A more complete discussion, both historical and mathematical, is postponed until Chapter 6.

Subharmonic Functions

Edoardo Vesentini [1968] [1970] showed that the spectral radius and its logarithm were both subharmonic functions (Definition 2.3.11). He used these results to establish some maximum modulus results involving the spectrum (*cf.* Hille-Phillips [1957] Theorem 3.13.1, Vasile I. Istrǎţescu [1969] and Bernd Schmidt [1970]). In a long series of papers beginning with [1974], Bernard H. Aupetit enormously extended the range and power of these methods. Aupetit summarized the first five years of his research in

[1979a] and gave a simplified account in [1991]. In many respects the culmination of this technique was his influential proof [1982a] [1983] of Johnson's uniqueness of norm theorem (Corollary 2.3.8). A previous version of the present chapter used subharmonic functions extensively, deriving the fundamental theorem by Aupetit's proof. Ransford's ingenious proof [1989] has allowed us to return to more standard parts of complex variable theory. We wish to briefly summarize this powerful theory, but the reader should consult Aupetit's books and papers for more detail.

2.3.11 Definition Let U be an open subset of $\mathbb{C}$. We call a function $f: U \to \mathbb{R} \cup \{-\infty\}$ *subharmonic* if it is upper semi-continuous and for any closed disc $D(\lambda_0, r)$ included in U it satisfies

$$f(\lambda_0) \leq \frac{1}{2\pi} \int_0^{2\pi} f(\lambda_0 + re^{i\theta})\, d\theta. \tag{1}$$

The appendix to Aupetit [1991] is a good short list of basic results on subharmonic functions; Hayman and Kennedy [1976] contains full proofs. Obviously if both f and $-f$ are subharmonic, then f is harmonic (*Note* , f is continuous and equality holds in (1)). It is classical that if $f: U \to \mathbb{C}$ is analytic, then $\lambda \mapsto \log |f(\lambda)|$ is subharmonic. In order to extend this result to Banach algebra valued functions, one uses a technical property of subharmonic functions to show that $\lambda \mapsto \log \|f(\lambda)\|$ is subharmonic for any vector valued analytic function f (Definition 2.2.10). Then Gelfand's spectral radius formula (Corollary 2.2.8) and obvious properties of subharmonic functions give Vesentini's result:

2.3.12 Theorem *Let $\mathcal{A}$ be a Banach algebra and let $U \subseteq \mathbb{C}$ be open. If a function $f: U \to \mathcal{A}$ is analytic, then $\lambda \mapsto \rho(f(\lambda))$ and $\lambda \mapsto \log \rho(f(\lambda))$ are both subharmonic.*

The following generalizations of the maximum modulus principle and Liouville's Theorem are almost immediate consequences of basic properties of subharmonic functions. Recall that $\mathrm{pc}(S)$ is the polynomially convex hull of S.

2.3.13 Theorem *Let $\mathcal{A}$ be a Banach algebra, let $U \subseteq \mathbb{C}$ be open and connected and let $f: U \to \mathcal{A}$ be analytic. If $\lambda_0 \in U$ satisfies $Sp(f(\lambda)) \subseteq Sp(f(\lambda_0))$ for all $\lambda \in U$, then $\partial Sp(f(\lambda_0)) \subseteq \partial Sp(f(\lambda))$ and $\mathrm{pc} Sp(f(\lambda)) = \mathrm{pc} Sp(f(\lambda_0))$ hold for all $\lambda \in U$. If, in addition, $Sp(f(\lambda_0))$ has empty interior or $\mathbb{C} \setminus Sp(f(\lambda))$ has no bounded component, then $Sp(f(\lambda))$ is constant for $\lambda \in U$.*

2.3.14 Theorem *If $\mathcal{A}$ is a Banach algebra, $f: \mathbb{C} \to \mathcal{A}$ is analytic and $\bigcup_{\lambda \in \mathbb{C}} Sp(f(\lambda))$ is bounded, then $\mathrm{pc} Sp(f(\lambda))$ is constant.*

In [1991], Aupetit uses these and similar results to derive numerous results describing or characterizing the center, the Jacobson radical, commutativity, elements with finite spectrum, etc. See also Aupetit [1982b], Vesentini [1984a] [1984b], Gong Ning Chen [1987] and O. H. Cheikh and Mohamed Oudadess [1988].

Capacity in Banach Algebras

Paul R. Halmos [1971] introduced this idea which has close links to subharmonic functions. He was motivated by the desire to find a class of elements related to algebraic elements (*i.e.*, those satisfying a one-variable polynomial) the way topologically nilpotent elements are related to nilpotent elements. This motivates the definition.

2.3.15 Definition   Let $\mathcal{A}$ be a Banach algebra. For any element $a \in \mathcal{A}$, define $\mathrm{cap}_n(a) = \inf\{\|p(a)\| : p$ is a monic polynomial of degree $\leq n\}$ and $\mathrm{cap}(a) = \lim_{n \to \infty} \mathrm{cap}_n(a)^{1/n}$.

It turns out that the capacity of an element depends only on the spectrum of the element and can be calculated by the following attractive (but not easy) alternative formula: $\mathrm{cap}(a) = \lim_{n \to \infty} \delta_n(Sp(a))$ where δ_n is the nth spectral diameter defined on a compact set K by

$$\delta_n(K) = \max\left\{ \left(\prod_{0 \leq j < k \leq n} |\lambda_j - \lambda_k| \right)^{2/(n(n+1))} : \lambda_0, \lambda_1, \ldots, \lambda_n \in K \right\}.$$

Elements with capacity zero do behave like generalizations of algebraic elements. It is obvious that elements with finite spectrum have capacity zero, and it is true (but not obvious) that elements with countable spectrum also have capacity zero. See also Aupetit and Wermer [1978]. For algebraic elements of Banach algebras see Nathan Jacobson [1945d], Zemánek [1978] Bernard R. Gelbaum [1980], Sołtysiak [1978], [1982], [1984], Young [1980], Corach [1982], Aupetit and Zemánek [1983b], Raúl Rodríguez Macías [1983] and Barnes [1985].

The Approximate $B^\#$-condition

There is a very simple concept which forces semisimplicity and has other consequences for spectral norms. The approximate $B^\#$-condition was introduced by Smiley [1955] generalizing a definition of Frank F. Bonsall [1954a]. It captures a portion of the power of the C*-condition which was introduced in §1.7.17.

2.3.16 Definition   An algebra norm $\|\cdot\|$ on an algebra $\mathcal{A}$ is said to satisfy the *approximate* $B^\#$-*condition* if, for each $a \in \mathcal{A}$ and each $\varepsilon > 0$, there is a (non unique) non-zero element $a^\# \in \mathcal{A}$ satisfying

$$\|a^\# a\|^\infty \geq \|a^\#\| \, \|a\|(1 - \varepsilon).$$

This metric condition is often satisfied by algebras of operators. Some examples are covered by the following proposition.

2.3.17 Proposition *Let $\mathcal{X}$ be any normed linear space and let $\mathcal{A}$ be a non-zero ideal of $\mathcal{B}(\mathcal{X})$ or any subalgebra of $\mathcal{B}(\mathcal{X})$ including $\mathcal{B}_F(\mathcal{X})$. The operator norm on $\mathcal{A}$ satisfies the approximate $B^\#$-condition.*

If $\mathcal{H}$ is a Hilbert space, the operator norm on any subalgebra of $\mathcal{B}(\mathcal{H})$ which is closed under the Hilbert space adjoint satisfies the approximate $B^\#$-condition.

Proof Proposition 1.1.18 shows that any non-zero ideal includes the finite-rank operators. If $T \in \mathcal{B}$ and $\varepsilon > 0$ are given, we can find a non-zero $x \in \mathcal{X}$ satisfying $\|Tx\| \geq \|T\| \, \|x\|(1-\varepsilon)$ and a non-zero $\omega \in \mathcal{X}^*$ satisfying $\omega(Tx) = \|\omega\| \, \|Tx\|$. Define $T^\#$ to be $x \otimes \omega$ so that we have $\|T^\#\| = \|x\| \, \|\omega\|$; and $T^\# T(x) = \omega(Tx)x = \|\omega\| \, \|Tx\|x$. The last equation shows that $\|\omega\| \, \|Tx\|$ is a characteristic value of $T^\# T$ so that we have $\|T^\# T\|^\infty = \rho_{\mathcal{B}(\mathcal{X})}(T^\# T) \geq \|\omega\| \, \|Tx\| \geq \|\omega\| \, \|T\| \, \|x\|(1 - \varepsilon) = \|T^\#\| \, \|T\|(1 - \varepsilon)$.

The closure of a subalgebra of $\mathcal{B}(\mathcal{H})$ satisfying this condition is a C*-algebra.   □

The next proposition, which is essentially due to Bonsall [1954a], shows the power of this apparently weak condition.

2.3.18 Proposition *A normed algebra $(\mathcal{A}, \|\cdot\|)$ which satisfies the approximate $B^\#$-condition is semisimple. Furthermore, if σ is a spectral norm on $\mathcal{A}$ which satisfies: $\sigma(a) \leq \|a\|$ for all $a \in \mathcal{A}$, then in fact $\sigma(a)$ equals $\|a\|$ for all $a \in \mathcal{A}$. Hence a spectral norm which satisfies the approximate $B^\#$-condition is minimal among all spectral norms on the algebra.*

Proof For any $a \in \mathcal{A}_J$, Theorems 2.2.2 and 2.3.3 show $0 = \|a^\# a\|^\infty \geq \|a\|2^{-1}$ for a suitable element $a^\#$ of norm one. Hence a is zero, proving that $\mathcal{A}$ is semisimple. To prove the last assertion, let $\varepsilon > 0$ and $a \in \mathcal{A}$ be arbitrary. Choose $a^\#$ satisfying the approximate $B^\#$-condition and note that Theorem 2.2.2 gives $(1-\varepsilon)\|a^\#\| \, \|a\| \leq \|a^\# a\|^\infty \leq \rho(a^\# a) \leq \sigma(a^\# a) \leq \sigma(a^\#)\sigma(a) \leq \|a^\#\| \, \|a\|$. Since $\varepsilon > 0$ was arbitrary, the desired conclusion follows.   □

For other results on approximate $B^\#$-norms, see Bonsall and Duncan [1967], [1973b], Smiley [1955] as well as Proposition 8.7.4 below. Bonsall [1954a] and Antonio Fernández López and Ángel Rodríguez-Palacios [1985] together show that a Banach space $\mathcal{X}$ is reflexive if and only if $\mathcal{B}_A(\mathcal{X})$ satisfies the $B^\#$-condition.

2.4   Spectral Algebras

In this section we introduce a type of abstract algebra (*i.e.*, one without any definite norm or topology) which provides a natural setting for many results originally proved for Banach algebras. These algebras, which we call spectral algebras, will be used extensively in the rest of this work. The section also contains the proof of a remarkable theorem which asserts that the spectral radius is subadditive if and only if it is submultiplicative, and that both conditions hold exactly when the algebra is a spectral algebra which is commutative modulo its Jacobson radical. In the next chapter, this result will be further strengthened by the addition of Gelfand theory which applies to precisely the class of spectral algebras which are commutative modulo their Jacobson radicals.

After introducing modular ideals, the section concludes with a few interesting results using this new machinery.

Spectral Algebras

The concept introduced here is a key one in this work. It is a purely algebraic concept which nevertheless implies many basic results in the theory of Banach algebras.

2.4.1 Definition An algebra is called *spectral* if some spectral semi-norm can be defined on it.

We wish to emphasize that a spectral algebra is not an algebra together with some particular spectral semi-norm, but is rather an abstract algebra on which some spectral semi-norm can be defined. Most spectral algebras have a vast diversity of spectral semi-norms, and no one spectral semi-norm is preferred over another.

Obviously, Banach algebras are spectral algebras (since the complete norm is a spectral norm), but we will see many other examples of spectral algebras. However, the spectrum in spectral algebras behaves the way it does in Banach algebras. In fact, the main difference between the theory of Banach algebras and that of spectral algebras is simply the absence of the holomorphic functional calculus in the latter case. (This functional calculus for Banach algebras will be described in Chapter 3.) Before exploring these ideas, we express the definition in a more geometric way.

2.4.2 Proposition *An algebra is spectral if and only if there is some balanced, convex, absorbing semigroup included in the set of quasi-invertible elements.*

Proof The open unit ball of a spectral semi-norm would be such a set, and conversely given such a set, its Minkowski functional would be a spectral semi-norm.   □

Theorem 1.1.10 above shows that for any algebra semi-norm σ $\lim \sigma(a^n)^{1/n}$ has many desirable properties. Theorem 2.2.5(f) shows that the spectral radius in any spectral algebra has all these properties. (This is a nontrivial observation since we have already shown by example that subadditivity and submultiplicativity are not properties of the spectral radius (even when it is finite-valued) in arbitrary commutative algebras.) We record this and another similar observation about spectral algebras in the next theorem.

2.4.3 Theorem *Let A be a spectral algebra.*

(a) *The spectrum of every element of A is closed and bounded, hence compact.*

(b) *The spectral radius in A satisfies all the properties listed in Theorem 1.1.10.*

Proof Both results follow from Theorem 2.2.5 applied to any spectral semi-norm on the algebra.   □

We will need some simple technical results.

2.4.4 Proposition *An algebra is spectral if and only if its ⟨ unitization / quotient modulo its Jacobson radical ⟩ is spectral. The quotient of any spectral algebra by any ideal is a spectral algebra.*

Proof If A is nonunital and σ is a spectral semi-norm on A, the semi-norm $\sigma^1 \colon A^1 \to \mathbb{R}_+$ defined by $\sigma^1(\lambda + a) = |\lambda| + \sigma(a)$ on A^1 is also spectral since $\rho(\lambda + a) \leq |\lambda| + \rho(a)$ is obvious. If σ is a spectral semi-norm on A^1, then its restriction to A is spectral since the non-zero spectrum of an element in A is the same whether computed in A or A^1 (Proposition 2.1.8).

The other statements follow directly from Theorem 2.2.13.   □

Some Questions

Theorem 2.2.13(c) asserts that an algebra semi-norm is spectral if its restriction to each maximal commutative subalgebra is spectral. We do not know whether an algebra is spectral if each of its maximal commutative subalgebras is spectral. There is some evidence (and several plausible but fallacious "proofs" based on the isomorphism between $(\mathcal{I}_1 + \mathcal{I}_2)/\mathcal{I}_1$ and $\mathcal{I}_2/(\mathcal{I}_1 \cap \mathcal{I}_2)$ and Theorem 2.2.13(d)) that in any algebra, or at least in any commutative algebra, the sum of two ideals is a spectral algebra if each of the ideals is a spectral algebra in its own right. A variant of this possible result relates to a semi-normed algebra (A, σ) and asks whether σ is spectral on the sum of two ideals when it is spectral on each ideal separately. This last result would imply that each semi-normed algebra has a largest ideal on which the algebra semi-norm is spectral. The result for commutative algebras implies the existence of a largest spectral ideal in any commutative algebra. One may also inquire whether an algebra is spectral if there is an ideal $\mathcal{I}$ in it which is spectral and for which $\mathcal{A}/\mathcal{I}$ is also spectral. We do not know whether any of these results is true.

Modular Ideals

Modular ideals play an important role in the project of extending results from unital to nonunital algebras. Maximal modular left ideals are also closely related to the Jacobson radical and irreducible representations. Modular ideals were first used (with the name "regular ideal") by Irving E. Segal [1947a]. The term "modular ideal" was introduced by Jacobson [1956]. Even in nonunital algebras we use the common notations $\mathcal{A}(1-e)$ and $(1-e)\mathcal{A}$ to denote, respectively, the sets $\{a - ae : a \in \mathcal{A}\}$ and $\{a - ea : a \in \mathcal{A}\}$.

2.4.5 Definition Let $\mathcal{A}$ be an algebra and let $\mathcal{L}$, $\mathcal{R}$, and $\mathcal{I}$ be, respectively, a left ideal, a right ideal and an ideal of $\mathcal{A}$. An element $e \in \mathcal{A}$ is called a $\langle$ right / left / two-sided $\rangle$ *relative identity* for $\langle \mathcal{L} / \mathcal{R} / \mathcal{I} \rangle$ if it satisfies $\langle \mathcal{A}(1-e) \subseteq \mathcal{L} / (1-e)\mathcal{A} \subseteq \mathcal{R} / \mathcal{A}(1-e) \cup (1-e)\mathcal{A} \subseteq \mathcal{I} \rangle$. A one- or two-sided ideal is called *modular* if it has a relative identity of the type described above. A one- or two-sided ideal of $\mathcal{A}$ is called *proper* if it is a proper subset of $\mathcal{A}$ and is called *maximal* if it is maximal under inclusion among all proper ideals of its type.

Obviously if $\mathcal{A}$ is unital, every one- or two-sided ideal is modular, with the identity element as a relative identity. As a partial converse, a proper (two-sided) ideal $\mathcal{I}$ is modular if and only if $\mathcal{A}/\mathcal{I}$ is unital. If $\mathcal{I}$ is an ideal of $\mathcal{A}$ and e and f are a right and a left relative identity for $\mathcal{I}$, then $\mathcal{A}/\mathcal{I}$ has both a right and a left identity so that these must agree and be an identity. Hence $\mathcal{I}$ is a modular ideal with both e and f as two-sided relative identities. It is also easy to see directly that $e \circ f$ is a two-sided relative identity for $\mathcal{I}$. For any algebra, the whole algebra is always a modular ideal of itself, but we will be interested in proper modular ideals.

It is not obvious *a priori* that a maximal modular ideal is a maximal ideal, but the easy proof is included with the following results due to Segal [1947a].

2.4.6 Theorem *Let $\mathcal{A}$ be an algebra.*

(a) *A one- or two-sided ideal which includes a modular ideal of the same type is also modular.*

(b) *No proper one- or two-sided modular ideal contains its relative identity. If e is the $\langle$ right / left $\rangle$ relative identity for a modular $\langle$ left / right $\rangle$ ideal $\langle \mathcal{L} / \mathcal{R} \rangle$, then $\{e - b : b \in \langle \mathcal{L} / \mathcal{R} \rangle\}$ is disjoint from $\mathcal{A}_{qG}$.*

(c) *A maximal modular one- or two-sided ideal is a maximal ideal of the same type. In fact, if $a \in \mathcal{A}$ does not belong to a maximal modular $\langle$ left / right $\rangle$ ideal $\langle \mathcal{L} / \mathcal{R} \rangle$, then $\langle \mathcal{A}a + \mathcal{L} / a\mathcal{A} + \mathcal{R} \rangle$ equals $\mathcal{A}$.*

(d) *Any proper modular one- or two-sided ideal is included in a maximal modular ideal of the same type.*

(e) *An element $e \in \mathcal{A}$ is a $\langle$ right / left $\rangle$ relative identity for some maximal modular $\langle$ left / right $\rangle$ ideal if and only if e has no $\langle$ left / right $\rangle$ quasi-inverse.*

Proof We consider only left ideals, since the other cases are similar.

(a): The relative identity for the included ideal obviously serves the same role for the larger ideal.

(b): If the right relative identity e of the modular left ideal $\mathcal{L}$ belongs to $\mathcal{L}$, then any $a \in \mathcal{A}$ satisfies $a = ae - (ae - a) \in \mathcal{L}$. For $b \in \mathcal{L}$, if $e - b$ had a quasi-inverse c, then the following calculation would show $e \in \mathcal{L}$, contrary to what we have just proved:

$$e = c(e - b) - c + b = (ce - c) - cb + b \in \mathcal{L}.$$

(c): The first statement follows immediately from (a). In order to prove the second, let e be a right relative identity for the maximal modular left ideal $\mathcal{L}$. Consider the set $\{b \in \mathcal{A} : \mathcal{A}b \subseteq \mathcal{L}\}$. It is easy to see that this is a left ideal which includes $\mathcal{L}$ but does not contain e. Hence by the maximality of $\mathcal{L}$, it equals $\mathcal{L}$. Thus $\mathcal{A}a + \mathcal{L}$ properly contains $\mathcal{L}$. By maximality again it equals $\mathcal{A}$.

(d): The following standard application of Zorn's lemma proves this. Consider the family of all left ideals which include a given modular left ideal $\mathcal{L}$ but do not contain its right relative identity e. When ordered by inclusion, the union of any linearly ordered subfamily is an upper bound for the subfamily. Hence the partially ordered set has a maximal element, which is a maximal modular left ideal including $\mathcal{L}$.

(e): First we show the necessity of the condition. If e is a right relative identity for $\mathcal{L}$ and satisfies $b \circ e = 0$, then $e = be - b \in \mathcal{L}$ holds, contradicting (b) above. Conversely if e has no left quasi-inverse, then e does not belong to the modular left ideal $\mathcal{A}(1 - e)$. Thus by (d), e is the right relative identity for a maximal modular left ideal including $\mathcal{A}(1 - e)$. $\square$

2.4.7 Theorem *Let σ be a spectral semi-norm on an algebra $\mathcal{A}$. The closure with respect to σ of every proper modular one- or two-sided ideal of $\mathcal{A}$ is proper. Hence, every maximal modular one-or two-sided ideal of $\mathcal{A}$ is closed with respect to σ.*

Proof Since all cases are similar, we consider only left ideals. Let e be a right relative identity for the proper modular left ideal $\mathcal{L}$. We will show that the σ-distance from e to any element b of $\mathcal{L}$ is at least 1. If not, $\sigma(e - b) < 1$ implies that $e - b$ has a quasi-inverse c. But this contradicts Theorem 2.4.6(b). Hence e is not in the closure of $\mathcal{L}$ with respect to σ, and thus this closure is proper. The second sentence is an immediate consequence of the first. $\square$

2.4.8 Corollary *An algebra semi-norm is spectral if and only if all maximal modular one-sided ideals are closed with respect to it.*

Proof If σ is a spectral semi-norm on $\mathcal{A}$, the theorem shows that all maximal modular one-sided ideals are σ-closed. Conversely, suppose that σ is an algebra semi-norm on $\mathcal{A}$ which is not a spectral semi-norm. Theorem 2.2.5(b) shows that we can find a quasi-singular element $e \in \mathcal{A}$ satisfying $\sigma(e) < 1$. By symmetry we may suppose that e has no left quasi-inverse so that, by Theorem 2.4.6, $\mathcal{A}(1-e)$ is a modular left ideal included in some maximal modular left ideal $\mathcal{M}$ which does not contain e. However, $e - e^n = (\sum_{k=1}^{n-1} e^k)(1-e) \in \mathcal{M}$ together with $\sigma(e) < 1$ show that e belongs to the σ-closure of $\mathcal{M}$. $\square$

2.4.9 Arens' Algebra L^ω The last two results show that for spectral algebras (and *a fortiori* for Banach algebras) all maximal modular ideals are closed. In [1946] Arens introduced a unital complete metric algebra which has no closed ideals, in contrast to the above results. The algebra which is denoted by L^ω is the intersection of all the L^p spaces $1 \le p < \infty$ on the unit interval $I = [0, 1]$. The multiplication is pointwise, so that L^ω is commutative and semisimple as well as unital. Thus it has many maximal (necessarily modular) ideals. For each $p \in \mathbb{N}$, let

$$\|f\|_p = \left(\int_0^1 |f(x)|^p dx \right)^{1/p} \qquad \forall\, f \in L^\omega$$

be the usual L^p-norm. Define a complete metric

$$d(f, g) = \sum_{p=1}^\infty 2^{-p} \min\{1, \|f - g\|_p\} \qquad \forall\, f, g \in L^\omega$$

on L^ω in the usual way so that convergence in this metric is simultaneous convergence in all L^p norms $1 \le p < \infty$. Hence L^ω has a complete invariant metric (*i.e.*, $d(f, g) = d(0, f - g)$ for all f, g) and is thus a Fréchet space. Kaplansky [1948c] showed that L^ω has no closed maximal ideals. This contrasts with the situation in unital semisimple spectral algebras, in which every maximal ideal is closed and there are enough such ideals to have intersection $\{0\}$. In [1964] Yood proved the stronger statement that L^ω contains no closed ideal which is maximal among closed ideals.

In the terminology introduced in Chapter 7, this is a topological annihilator algebra which is not a modular annihilator algebra and which has socle zero. This shows that topological annihilator algebras need not resemble normed annihilator algebras, which are automatically spectral. See also Mohamed Oudadess [1985].

Commutativity and the Spectral Radius as an Algebra Semi-norm

The next theorem shows that if the spectral radius is either subadditive or submultiplicative, then it is both and thus is a spectral semi-norm. This happens if and only if the algebra is commutative modulo the Jacobson radical. In this generality, the result is due to Palmer [1992] based on arguments of Jaroslav Zemánek [1977a], [1977b] and Vlastimil Pták and Zemánek [1977a] [1977b]. For the proof we need the following interesting result of Claude Le Page [1967]. The elegant proof is derived from apparently independent proofs of similar results due to Isadore M. Singer and John Wermer [1955], Ivan Vidav [1955], F. V. Širokov [1956] and Marvin Rosenblum [1958]. See also Victor A. Belfi, and Robert S. Doran [1980], Paula Vrbová [1981], Oudadess [1983b], [1987], Gerd Niestegge [1984], M. I. Karakhanyan, M. I. [1984], Yiannis Tsertos [1986], Zalduendo [1989], Bertram Yood [1990] and Badea [1991].

2.4.10 Lemma *A unital normed algebra $\mathcal{A}$ is commutative if and only if there is a constant C satisfying*

$$\|ab\| \leq C\|ba\| \qquad \forall\, a, b \in \mathcal{A}.$$

Proof If $\mathcal{A}$ is commutative, we may choose $C = 1$. To prove the converse, first note that we may replace $\mathcal{A}$ by its completion and thus assume that $\mathcal{A}$ is complete. Define $e^{\lambda a}$ by $e^{\lambda a} = \sum_{n=0}^{\infty} (\lambda a)^n / n!$, so that by the usual elementary calculation $e^{\lambda a}$ has inverse $e^{-\lambda a}$. (See the end of Section 2.1 for a discussion of the exponential function.) Let $a, b \in \mathcal{A}$ be arbitrary but fixed for the rest of this argument. For any fixed $\omega \in \mathcal{A}^*$ (where $\mathcal{A}^*$ is the Banach space dual of $\mathcal{A}$), define $f: \mathbb{C} \to \mathbb{C}$ by

$$f(\lambda) = \omega(e^{\lambda a} b e^{-\lambda a}).$$

Clearly, for a suitable choice of $c_n \in \mathcal{A}$, we can write $f(\lambda) = \sum_{n=0}^{\infty} \omega(c_n)\lambda^n$ as an entire function satisfying

$$|f(\lambda)| \leq \|\omega\| \, \|e^{\lambda a}(be^{-\lambda a})\| \leq C\|\omega\| \, \|be^{-\lambda a}e^{\lambda a}\| = C\|\omega\| \, \|b\|.$$

Hence, by Liouville's theorem, f is a constant function. Thus the derivative f' of f is identically 0. However

$$f'(\lambda) = \omega(ae^{\lambda a}be^{-\lambda a} + e^{\lambda a}b(-a)e^{-\lambda a})$$

implies $f'(0) = \omega(ab - ba)$. Since $\omega \in \mathcal{A}^*$ was arbitrary, we conclude $ab = ba$ by the Hahn–Banach theorem.   □

2.4.11 Theorem *The following are equivalent for any algebra $\mathcal{A}$ in which the spectral radius is finite-valued:*

(a) *There is a constant C satisfying*

$$\rho(a + b) \leq C(\rho(a) + \rho(b)) \qquad \forall\, a,\, b \in \mathcal{A}.$$

(b) *There is a constant C satisfying*

$$\rho(ab) \leq C\rho(a)\rho(b) \qquad \forall\, a,\, b \in \mathcal{A}.$$

(c) $\mathcal{A}$ *is a spectral algebra and is commutative modulo its Jacobson radical.*

(d) *The spectral radius is an algebra semi-norm and hence the unique smallest spectral semi-norm.*

Proof Note that the constants C necessarily satisfy $C \geq 1$ unless $\mathcal{A}$ is a radical algebra. Since that case needs no further consideration, we assume this inequality.

(a) $\Rightarrow$ (b): For $a,\, b \in \mathcal{A}$, let $\lambda_0 \in \mathbb{C}$ satisfy $|\lambda_0| > 9C^2\rho(a)\rho(b)$. Then we may write $\lambda_0 = \mu\nu$ with $|\mu| > 3C\rho(a)$ and $|\nu| > 3C\rho(b)$. Hence we get

$$\begin{aligned}
\rho((\mu - a)^{-1}a) &= \sup\{|\lambda/(\mu - \lambda)| : \lambda \in Sp(a)\} \\
&\leq \sup\{|\lambda/(\mu - \lambda)| : |\lambda| \leq \rho(a)\} < (2C)^{-1}
\end{aligned}$$

and similarly $\rho((\nu - b)^{-1}b) < (2C)^{-1}$. Thus

$$\lambda_0 - ab = (\mu - a)(1 + (\mu - a)^{-1}a + (\nu - b)^{-1}b)(\nu - b)$$

is invertible by the spectral mapping theorem, and the following inequality

$$\rho((\mu - a)^{-1}a + (\nu - b)^{-1}b) \leq C[\rho((\mu - a)^{-1}a) + \rho((\nu - b)^{-1}b)] < 1$$

implies that the middle factor is invertible. We conclude $\rho(ab) \leq 9C^2\rho(a)\rho(b)$.

(b) $\Rightarrow$ (a): For $a,\, b \in \mathcal{A}$, let $\lambda_0 \in \mathbb{C}$ satisfy $|\lambda_0| > \rho(a) + C\rho(b)$. We must consider the unital and nonunital cases separately. If $\mathcal{A}$ is unital,

$$\lambda_0 - (a + b) = (\lambda_0 - a)(1 - (\lambda_0 - a)^{-1}b)$$

is invertible, since

$$\rho((\lambda_0 - a)^{-1}b) \leq C\rho((\lambda_0 - a)^{-1})\rho(b) < C(C\rho(b))^{-1}\rho(b) = 1$$

shows $1 \notin Sp((\lambda_0 - a)^{-1}b)$. If $\mathcal{A}$ is nonunital, $(\lambda_0 - a)^{-1}$ has the decomposition $\lambda_0^{-1} + c$ with $c \in \mathcal{A}^1$. Hence

$$\begin{aligned}
\lambda_0 - (a + b) &= (\lambda_0 - a)(1 - (\lambda_0 - a)^{-1}b) \\
&= (\lambda_0 - a)(1 - \lambda_0^{-1}b - cb) \\
&= (\lambda_0 - a)(1 - cb(1 - \lambda_0^{-1}b)^{-1})(1 - \lambda_0^{-1}b)
\end{aligned}$$

is invertible in $\mathcal{A}^1$, since

$$\rho(cb(1 - \lambda_0^{-1}b)^{-1}) \le C\rho(c)\rho(b(1 - \lambda_0^{-1}b)^{-1}) < 1/C \le 1$$

follows from

$$
\begin{aligned}
\rho(c) &= \rho((\lambda_0 - a)^{-1} - \lambda_0^{-1}) = \rho(a\lambda_0^{-1}(\lambda_0 - a)^{-1}) \\
&\le |\lambda_0^{-1}|\rho(a)/(|\lambda_0| - \rho(a)) < |\lambda_0^{-1}|\rho(a)(C\rho(b))^{-1}
\end{aligned}
$$

and

$$
\begin{aligned}
\rho(b(1 - \lambda_0^{-1}b)^{-1}) &= |\lambda_0|\rho(b(\lambda_0 - b)^{-1}) \\
&\le |\lambda_0^{-1}|\rho(b)/(|\lambda_0| - \rho(b)) < |\lambda_0|\rho(b)\rho(a)^{-1}.
\end{aligned}
$$

We conclude:

$$\rho(a + b) \le \rho(a) + C\rho(b) \le C(\rho(a) + \rho(b)).$$

(b) $\Rightarrow$ (c): By the last argument we may assume $\rho(a+b) \le \rho(a)+C\rho(b)$ for all $a, b \in \mathcal{A}$. Define

$$p(a) = C^2 \sup\{\rho(a + b) - \rho(b) : b \in \mathcal{A}\} \qquad \forall\, a \in \mathcal{A}.$$

Then p satisfies

$$
\begin{aligned}
p(a + c) &= C^2 \sup\{(\rho(a + c + b) - \rho(c + b)) + (\rho(c + b) - \rho(b)) : b \in \mathcal{A}\} \\
&\le p(a) + p(c) \qquad \forall\, a, c \in \mathcal{A}; \\
C^2\rho(a) &\le p(a) \le C^3\rho(a) \qquad \forall\, a \in \mathcal{A}; \\
p(ab) &\le C^3\rho(ab) \le C^4\rho(a)\rho(b) \le p(a)p(b) \qquad \forall\, a, b \in \mathcal{A}
\end{aligned}
$$

and thus p is a spectral semi-norm which also satisfies

$$p(a)^2 \le C^6\rho(a)^2 = C^6\rho(a^2) \le C^4 p(a^2) \qquad \forall\, a \in \mathcal{A}.$$

Hence the ideal on which p vanishes is the largest ideal on which the spectral radius vanishes, so that by Corollary 2.3.4 it is the Jacobson radical. Thus p induces a norm $\|\cdot\|$ on $\mathcal{A}/\mathcal{A}_J$ defined by $\|a + \mathcal{A}_J\| = p(a)$ for all $a \in \mathcal{A}$. However, this implies

$$
\begin{aligned}
\|(a + \mathcal{A}_J)(b + \mathcal{A}_J)\| &= \|ab + \mathcal{A}_J\| = p(ab) \le C^3\rho(ab) \\
&= C^3\rho(ba) \le Cp(ba) = C\|ba + \mathcal{A}_J\| \\
&= C\|(b + \mathcal{A}_J)(a + \mathcal{A}_J)\|,
\end{aligned}
$$

so that $\mathcal{A}/\mathcal{A}_J$ is commutative by the last lemma.

(c) $\Rightarrow$ (d): Theorem 2.2.5(f) and Theorem 1.1.10(g) applied to the spectral norm induced on $\mathcal{A}/\mathcal{A}_J$ show that ρ is an algebra semi-norm there.

However, Theorem 2.3.3 shows that the spectral radius of a coset $a + \mathcal{A}_J$ in $\mathcal{A}/\mathcal{A}_J$ equals that of a in $\mathcal{A}$, so that the spectral radius is also an algebra semi-norm there. Hence, it is clearly the unique smallest spectral semi-norm. (We may, therefore, take C as 1, so the definition of p shows that it is just ρ.)

(d) $\Rightarrow$ (a) and (b): Since ρ is an algebra semi-norm, we may choose $C = 1$. $\square$

In the next chapter, Theorem 3.1.5 will add additional information to this theorem. It will show that the conditions of Theorem 2.4.5 are also equivalent to the existence of a complete Gelfand theory. In Chapter 4, Theorem 4.2.18 will show that these conditions are also equivalent to uniform continuity of the spectral radius.

Yu. V. Turovskiĭ [1985] shows that certain closed subalgebras are commutative modulo their Jacobson radical. In any associative algebra $\mathcal{A}$ we may introduce the *Lie product* $[a, b] = ab - ba$ for all $a, b \in \mathcal{A}$. A linear subspace closed under this product is called a *Lie subalgebra* and a Lie subalgebra $\mathcal{B}$ is said to be *Lie nilpotent* if there is a positive integer n such that any list $b_0, b_1, \ldots, b_n$ in $\mathcal{B}$ satisfies $[b_n, [b_{n-1}, \cdots, [b_1, b_0] \cdots]] = 0$. The cited paper shows that the norm closure of the associative subalgebra generated by any Lie nilpotent Lie subalgebra is commutative modulo its Jacobson radical.

Numerous properties and subsets of Banach algebras have been characterized in terms of spectral radius criteria. We list the following papers which deal with commutativity, the center, the Jacobson radical, finite dimensionality, etc. Since the spectral radius is unchanged by dividing by the Jacobson radical, these properties and subsets are actually determined modulo the Jacobson radical. Aupetit [1979a], [1991]; Aupetit and Zemánek [1978], [1981], [1983a], [1983b] and Zemánek [1977a], [1977b], [1980] and [1982].

Homomorphisms onto the Complex Field

The following result, which describes the kernels of homomorphisms onto the complex numbers, plays a crucial role in the next chapter which establishes the Gelfand theory for commutative spectral algebras.

2.4.12 Theorem *The following are equivalent for a modular ideal $\mathcal{I}$ in a spectral algebra $\mathcal{A}$.*

(a) $\mathcal{A}/\mathcal{I}$ *is isomorphic to the complex field.*

(b) $\mathcal{I}$ *is maximal as a left ideal.*

(c) $\mathcal{I}$ *is maximal as a right ideal.*

Hence a modular ideal in a commutative algebra is the kernel of a homomorphism into the complex field if and only if it is maximal.

Proof Clearly (a) implies the other two properties since in this case $\mathcal{I}$

has codimension 1. Suppose (b) holds. Choose a spectral semi-norm σ on $\mathcal{A}$. Theorem 2.4.7 shows that $\mathcal{I}$ is closed, so $\mathcal{A}/\mathcal{I}$ is a (spectral) normed algebra. We wish to show that every non-zero element of $\mathcal{A}/\mathcal{I}$ has a left inverse so that $\mathcal{A}/\mathcal{I}$ is isomorphic to the complex field by the Gelfand–Mazur theorem (Corollary 2.2.3). Let e be a relative identity for $\mathcal{I}$, so that $e+\mathcal{I}$ is an identity element for $\mathcal{A}/\mathcal{I}$. Let a be an arbitrary element not in $\mathcal{I}$, so that $a+\mathcal{I}$ is an arbitrary non-zero element of $\mathcal{A}/\mathcal{I}$. The left ideal $\mathcal{A}a+\mathcal{I}$ contains $a = ea - (ea - a)$ which does not belong to $\mathcal{I}$, so $\mathcal{A}a + \mathcal{I} = \mathcal{A}$. Thus $a+\mathcal{I}$ has a left inverse, as we wished to show. The case of right ideals is similar.   □

This last proposition identifies all the kernels of homomorphisms of a spectral algebra onto the complex number field. The next result characterizes the homomorphisms themselves. For commutative Banach algebras this theorem was proved by Andrew M. Gleason [1967] and by Jean-Pierre Kahane and Żelazko [1968], independently. Żelazko [1968] extended the original result to noncommutative algebras, and Moshe Roitman and Yitzak Sternfield [1981] gave essentially the present statement and proof. Relevant counterexamples are contained in the last cited article and in Żelazko [1973a], [1973b]. For other variations see Alberto Arosio and Artur V. Ferreira [1978a] [1978b], Aupetit [1979b], Sergiusz Kowalski and Zbigniew Słodkowski [1980], Chang Pao Chen [1983], S. H. Kulkarni [1984], Regina Wille [1985], Arens [1987], Badea [1991] and Krzysztof Jarosz [1991]. A linear map satisfying condition (b) is called a Jordan homomorphism, and such maps are studied systematically in Section 6.3 below.

2.4.13 Theorem *Let $\mathcal{A}$ be a unital algebra and let $\varphi \colon \mathcal{A} \to \mathbb{C}$ be linear and unital (i.e., $\varphi(1) = 1$). The first four conditions below are equivalent and imply the last two equivalent conditions. If $\mathcal{A}$ is a spectral algebra, then all six conditions are equivalent.*

  (a) $\varphi(a) = 0$ *implies* $\varphi(a^2) = 0$ *for* $a \in \mathcal{A}$.
  (b) $\varphi(a^2) = \varphi(a)^2$ *for* $a \in \mathcal{A}$.
  (c) $\varphi(a) = 0$ *implies* $\varphi(ba) = 0$ *for* $a, b \in \mathcal{A}$ *(i.e.,* $\ker(\varphi)$ *is a left ideal).*
  (d) φ *is an algebra homomorphism (i.e.,* $\varphi(ab) = \varphi(a)\varphi(b)$ *for* $a, b \in \mathcal{A}$*).*
  (e) $\ker(\varphi)$ *contains no invertible element.*
  (f) *For each* $a \in \mathcal{A}$, $\varphi(a) \in Sp(a)$.

Proof (a) $\Rightarrow$ (b): Since $\varphi(a - \varphi(a))$ is 0, we conclude

$$0 = \varphi((a - \varphi(a))^2) \Rightarrow \varphi(a^2) = \varphi(a)^2.$$

(b) $\Rightarrow$ (c): First note:

$$\varphi(bc + cb) = \varphi((b + c)^2) - \varphi(b^2) - \varphi(c^2) = 2\varphi(b)\varphi(c).$$

Hence we see $\varphi(ab + ba) = 0$. By (b) we conclude $\varphi((ab + ba)^2) = 0$. Now the identity $(bc - cb)^2 = 2((bcb)c + c(bcb)) - (bc + cb)^2$, together with our previous results gives

$$
\begin{aligned}
(\varphi(ba - ab))^2 &= \varphi((ba - ab)^2) \\
&= 2\varphi((bab)a + a(bab)) = 4\varphi(bab)\varphi(a) = 0.
\end{aligned}
$$

By adding the resulting equations we get:

$$2\varphi(ba) = \varphi(ab + ba) + \varphi(ba - ab) = 0.$$

(c) $\Rightarrow$ (d): Since $\varphi(a - \varphi(a))$ is 0, (c) gives

$$0 = \varphi(b(a - \varphi(a))) = \varphi(ba) - \varphi(b)\varphi(a) \qquad \forall\, a, b \in \mathcal{A}.$$

(d) $\Rightarrow$ (a): Trivial.

(d) $\Rightarrow$ (e): If a is invertible and $\varphi(a)$ is 0, we get the contradiction $0 = \varphi(a^{-1}a) = \varphi(1) = 1$.

(e) $\Rightarrow$ (f): This follows from $\varphi(\varphi(a) - a) = 0$.

(f) $\Rightarrow$ (e): If a is invertible, 0 is not in its spectrum.

(f) $\Rightarrow$ (a) (For a spectral algebra $\mathcal{A}$.): For any fixed $n \in \mathbb{N}$, consider the polynomial $p(\lambda) = \varphi((\lambda - a)^n)$ and its roots λ_j for $1 \le j \le n$. The equation $\varphi((\lambda_j - a)^n) = 0$ implies that $(\lambda_j - a)^n$ is not invertible. Hence $\lambda_j - a$ is also not invertible. This is equivalent to $\lambda_j \in Sp(a)$ for $1 \le j \le n$. By expanding

$$p(\lambda) = \Pi_{j=1}^n (\lambda - \lambda_j) = \lambda^n - n\varphi(a)\lambda^{n-1} + \frac{n(n-1)}{2}\varphi(a^2)\lambda^{n-2} + \cdots$$

and comparing coefficients, we get:

$$\sum_{j=1}^n \lambda_j = n\varphi(a) = 0 \text{ and } \sum_{j<k} \lambda_j \lambda_k = \frac{n(n-1)}{2}\varphi(a^2).$$

However,

$$0 = (\sum_{j=1}^n \lambda_j)^2 = \sum_{j=1}^n \lambda_j^2 + 2\sum_{j<k} \lambda_j \lambda_k$$

gives

$$n(n-1)|\varphi(a^2)| = 2|\sum_{j<k} \lambda_j \lambda_k| = |\sum_{j=1}^n \lambda_j^2| \le \sum_{j=1}^n |\lambda_j^2| \le n\rho(a)^2.$$

Since $n \in \mathbb{N}$ was arbitrary and $\rho(a)$ is finite in a spectral algebra, this gives $\varphi(a^2) = 0$. $\qquad\square$

In the above theorem, weaken condition (f) to $\varphi(a) \in \mathrm{co}Sp(a)$. It is easy to see that this is equivalent to $|\varphi(a)| \leq \rho(a)$ for all $a \in \mathcal{A}$ (where we are still assuming linearity and $\varphi(1) = 1$). Bonsall and Duncan [1971] call such linear functionals *spectral states*. Obviously the set Ω of spectral states forms a weak* compact convex subset of $\mathcal{A}^*$. Each spectral state obviously vanishes on $\mathcal{A}_J \subseteq \mathcal{A}_{tN}$ and on all commutators by the argument in the proof of Lemma 2.4.10. Every algebra homomorphism $\varphi: \mathcal{A} \to \mathbb{C}$ belongs to Ω and Niestegge [1983] shows that if $\mathcal{A}$ is a unital commutative Banach algebra, then the extreme points of Ω are algebra homomorphisms. From this he proves a stronger form of the following theorem of Nagasawa [1959]: *If $\mathcal{A}$ and $\mathcal{B}$ are unital commutative Banach algebras with $\mathcal{B}$ semisimple and $\varphi: \mathcal{A} \to \mathcal{B}$ is a unital linear map satisfying $\rho_{\mathcal{B}}(\varphi(a)) = \rho_{\mathcal{A}}(a)$ for all $a \in \mathcal{A}$, then φ is an algebra isomorphism.* See also George Maltese [1979], Dénes Petz [1983], Przemysław Kajetanowicz [1984] and Krzysztof Jarosz [1985b].

Norms on Algebras of Continuous Functions

The following results are due to Kaplansky [1949a]. Theorem 3.5.17 below is a considerable generalization of Theorem 2.4.15, but its proof is far more difficult, relying on the Šilov idempotent theorem. The algebras $C_0(\Omega)$ were defined in §1.5.1.

2.4.14 Lemma *Let $\mathcal{A}$ be a commutative normed algebra and let $\bar{\mathcal{A}}$ be its completion. If $a \in \mathcal{A}$ is quasi-invertible modulo every maximal modular closed ideal of $\mathcal{A}$, then it is quasi-invertible in $\bar{\mathcal{A}}$.*

Proof Theorem 2.4.6(e) shows that if a is not quasi-invertible in $\bar{\mathcal{A}}$, then a will be the relative identity for a maximal modular ideal $\mathcal{M}$ of $\bar{\mathcal{A}}$. Theorem 2.4.7 shows that $\mathcal{M}$ is closed so that $\mathcal{M} \cap \mathcal{A}$ is a closed maximal modular ideal of $\mathcal{A}$. Clearly a is not quasi-invertible modulo this ideal. $\qquad\square$

2.4.15 Theorem *Let $(\mathcal{A}, \|\cdot\|)$ be a commutative Banach algebra isometrically isomorphic to $C_0(\Omega)$ for some locally compact space Ω. (It is equivalent to describe $(\mathcal{A}, \|\cdot\|)$ as any commutative Banach algebra satisfying Theorem 3.2.12.) Then any (not necessarily complete) algebra norm $\|\|\cdot\|\|$ on $\mathcal{A}$ satisfies*

$$\|a\| \leq \|\|a\|\| \qquad \forall \, a \in \mathcal{A}.$$

Proof For the proof, we may assume that $\mathcal{A}$ actually is $C_0(\Omega)$. Let Ω_C be the set of points $\omega \in \Omega$ such that the ideal $\mathcal{Z}(\omega) = \{f \in \mathcal{A} : f(\omega) = 0\}$ is closed in the norm $\|\|\cdot\|\|$. We will show that Ω_C is dense in Ω and this will prove the theorem since $\|\|f\|\| \geq \sup\{|f(\omega)| : \omega \in \Omega_C\}$. (Otherwise, we get $\lim_{n\to\infty}(f - f(\omega))\sum_{k=0}^{n}(f/f(\omega))^n = -f$ for some $f \notin \mathcal{Z}(\omega)$.)

In order to give a proof by contradiction, suppose that Ω_C is not dense. Let U be a nonempty open subset of Ω disjoint from Ω_C. The discussion in §1.5.1 shows that there is a function $f: \Omega \to [0, 1]$ in $\mathcal{A}$ which assumes the value 1 on some nonempty open subset V of U and vanishes off U. By the last lemma, f is quasi-invertible in the completion of $\mathcal{A}$ under $||| \cdot |||$. However, this is impossible since there is a non-zero function $g : \Omega \to [0, 1]$ in $\mathcal{A}$ with $g(\omega_0) = 1$ for some $\omega_0 \in V$ and vanishing outside V satisfying $g \circ f = f$ and hence $g = g \circ (f \circ f^q) = (g \circ f) \circ f^q = f \circ f^q = 0.$ □

2.5   Spectral Subalgebras and Topological Divisors of Zero

In the introduction we listed a number of sufficient conditions for a subalgebra to be a spectral subalgebra. The easy conditions will be collected in Proposition 2.5.3. The more difficult ones all depend on the theory of topological quasi-divisors of zero, introduced by Kaplansky [1949a] following Šilov [1940b]. Topological quasi-divisors of zero cannot be defined in a spectral algebra, since the algebra has no particular spectral semi-norm. However the most useful consequences of this concept are independent of the particular choice of norm, so that they are available in the theory of spectral algebras. We begin with a formal definition of spectral subalgebras, which were informally defined in the introduction to this chapter.

Fundamental Properties of Spectral Subalgebras

2.5.1 Definition   A subalgebra $\mathcal{B}$ of an algebra $\mathcal{A}$ is said to be a *spectral subalgebra* if it satisfies

$$\mathcal{B}_{qG} = \mathcal{B} \cap \mathcal{A}_{qG}. \tag{1}$$

In the next proposition we characterize spectral subalgebras in terms of conditions on the spectrum of elements in the subalgebra. Zero plays an exceptional role here as it does whenever nonunital subalgebras are under consideration. Thus we include two additional conditions on the spectra of elements of the subalgebra, each of which is more stringent than the conditions shown to be equivalent to the subalgebra being spectral.

2.5.2 Proposition   *Let $\mathcal{B}$ be a subalgebra of an algebra $\mathcal{A}$.*

(a) *Each of the following two displayed conditions is equivalent to $\mathcal{B}$ being a spectral subalgebra of $\mathcal{A}$:*

$$Sp_{\mathcal{B}}(b) \cup \{0\} = Sp_{\mathcal{A}}(b) \cup \{0\} \qquad \forall\, b \in \mathcal{B}; \tag{2}$$

$$Sp_{\mathcal{B}}(b) \subseteq Sp_{\mathcal{A}}(b) \cup \{0\} \qquad \forall\, b \in \mathcal{B}. \tag{3}$$

Thus any element b in a spectral subalgebra $\mathcal{B}$ of $\mathcal{A}$ satisfies

$$\rho_{\mathcal{B}}(b) = \rho_{\mathcal{A}}(b) \tag{4}$$

(unless b has spectrum $\{0\}$ in one algebra and empty spectrum in the other).

In particular, a spectral subalgebra of a spectral algebra is a spectral algebra.

(b) *The condition*

$$Sp_{\mathcal{B}}(b) = Sp_{\mathcal{A}}(b) \quad \forall\, b \in \mathcal{B} \tag{5}$$

holds if and only if $\mathcal{B}$ is a spectral subalgebra of $\mathcal{A}$ which also satisfies one of the following two mutually exclusive conditions:

(b_1) $\mathcal{B}$ *is a unital subalgebra of $\mathcal{A}$, or*

(b_2) $\mathcal{B}$ *is a nonunital algebra and none of its elements is invertible in $\mathcal{A}$.*

(c) *If $\mathcal{B}$ is a spectral subalgebra of $\mathcal{A}$, then the inclusion*

$$Sp_{\mathcal{B}}(b) \subseteq Sp_{\mathcal{A}}(b) \quad \forall\, b \in \mathcal{B} \tag{6}$$

follows from any of the following circumstances:

(c_0) *Every element of $\mathcal{B}$ invertible in $\mathcal{A}$ is also invertible in $\mathcal{B}$.*

(c_1) $\mathcal{A}$ *is not unital.*

(c_2) $\mathcal{B}$ *is a unital algebra.*

(c_3) $\mathcal{A}$ *and $\mathcal{B}$ satisfy condition* (a), (b), (c), (d), (e) *or* (f) *of* Proposition 2.5.3.

Proof (a): The equivalence of equations (1) and (2) follows from the definition of the spectrum. Conditions (2) and (3) are equivalent by Proposition 2.1.8(c). The equality for the spectral radii is an immediate consequence of (2), and it implies the last sentence. (We do not know any example in which an element $b \in \mathcal{B}$ has spectrum $\{0\}$ in $\mathcal{B}$ and empty spectrum in $\mathcal{A}$ although the other possibility is easy to achieve, e. g., in the algebra of Example 2.1.7.)

(b): Assume equation (5). This implies equation (2) and thus that $\mathcal{B}$ is a spectral subalgebra. Furthermore, if $\mathcal{B}$ is a unital algebra with identity element e, then e must be the identity of $\mathcal{A}$ (rather than a nontrivial idempotent) since otherwise it would have spectrum $\{1\}$ in $\mathcal{B}$ and $\{1,0\}$ in $\mathcal{A}$. Therefore either $\mathcal{B}$ is a unital subalgebra or it is a nonunital algebra. In the latter case, none of its elements could be invertible in $\mathcal{A}$, since they all have 0 in their spectra relative to $\mathcal{B}$ and we are assuming (5).

Conversely if $\mathcal{B}$ is a spectral subalgebra satisfying either (b_1) or (b_2), then (2) implies (5).

(c): Condition (c_0) is just a restatement of the condition necessary to derive inclusion (6) from inclusion (3) and thus is both necessary and sufficient. Condition (c_1) implies that 0 is in the spectrum in $\mathcal{A}$ of any element and thus that (3) implies (6).

Suppose next that $\mathcal{B}$ is a unital algebra (although not necessarily a unital subalgebra) with identity element e. Then $b \in \mathcal{B}$ is invertible in $\mathcal{B}$ if and only if $e - b$ is quasi-invertible in $\mathcal{B}$ and thus if and only if it is quasi-invertible in $\mathcal{A}$. However if b has an inverse b^{-1} in $\mathcal{A}$, then $e - eb^{-1}e$ is a quasi-inverse for $e - b$. Thus inclusion (6) holds.

In the proof of Proposition 2.5.3 we will actually establish (6) rather than just (1), (2) or (3) and thereby establish (c_3). $\square$

The following example, suggested by Larry B. Schweitzer [1993], shows that not every spectral subalgebra satisfies conditions (4) and (6) above. Let $\mathcal{F}$ be the algebra of formal power series defined in §2.9.7 below. Every $f \in \mathcal{F}$ with non-zero constant term f_0 has an inverse in $\mathcal{F}$ with constant term $1/f_0$. Let $\mathcal{A}$ be the algebra of all formal power series which start at some (possibly negative) index. Thus every non-zero element $f \in \mathcal{A}$ can be written uniquely as $f = \sum_{n=m}^{\infty} f_n z^n = \alpha z^m \sum_{n=0}^{\infty} \beta_n z^n$ with $m \in \mathbb{Z}$ and $\beta_0 = 1$ (so $\alpha = f_m$, $\beta_n = f_{n+m}/f_m$). Hence $\mathcal{A}$ is a field and can be identified with the field of quotients of $\mathcal{F}$. Let $\mathcal{B}$ be the subalgebra of $\mathcal{A}$ consisting of all power series beginning at a positive index. Note that $\mathcal{B}^1$ can be naturally identified with $\mathcal{F} \subseteq \mathcal{A}$. It is easy to see that every non-zero element $g \in \mathcal{B}$ satisfies $Sp_{\mathcal{B}}(g) = \{0\}$ and $Sp_{\mathcal{A}}(g) = \emptyset$. Hence g satisfies neither (4) nor (6).

2.5.3 Proposition *Let $\mathcal{A}$ be an algebra. The subsets indicated in* (a), (b), (c), (d) *and* (g) *below are spectral subalgebras of $\mathcal{A}$ satisfying inclusion* (6) *above. The sets indicated in* (e) *and* (f) *are spectral subalgebras satisfying equation* (5) *(which implies inclusion* (6)*) above relative to the indicated larger algebras.*

(a) *Any one- or two-sided ideal of $\mathcal{A}$, any hereditary subalgebra or indeed any subalgebra $\mathcal{B}$ for which there is some $n \in \mathbb{N}$ satisfying $\mathcal{B}^n \mathcal{A} \mathcal{B}^n \subseteq \mathcal{B}$;*

(b) *The commutant of any subset; hence also the center of an algebra and any maximal commutative subalgebra;*

(c) *For any idempotent $e \in \mathcal{A}$ the "corner" subalgebra $e\mathcal{A}e$;*

(d) *Any finite-dimensional subalgebra or any subalgebra in which each element satisfies a polynomial identity;*

(e) *The range of the left or right regular or extended regular representation in $\mathcal{L}(\mathcal{A})$;*

(f) *The image $\kappa(\mathcal{A})$ of $\mathcal{A}$ in $\mathcal{A}^{**}$ with either Arens product;*

(g) *Any intersection of spectral subalgebras or any spectral subalgebra of a spectral subalgebra of $\mathcal{A}$.*

Proof (a), (c): We prove the last portion of (a) first since that establishes all of (a) and (c). If $b \in \mathcal{B}$ has a quasi-inverse $a \in \mathcal{A}$, then they satisfy $b^n a b^{n-1} = b^n(ab - b)b^{n-1} \in \mathcal{B}$. At the nth step in a reduction process we get $b^n a = b^n(ab - b) \in \mathcal{B}$. But we may continue $b^{n-1}a = b^{n-1}(ba - b) \in \mathcal{B}$ until we conclude $a = (ba - b) \in \mathcal{B}$. If $b \in \mathcal{B}$ is invertible in $\mathcal{A}$, then $b^n b^{-2n} b^n = 1$ shows that $\mathcal{B}$ is a unital subalgebra so (6) holds.

(b): The inverse or quasi-inverse of any element in the commutant of a set is obviously in the commutant also.

(d): Suppose $\mathcal{B}$ is finite-dimensional. Then each of its elements satisfies a polynomial equation of degree less than or equal to its dimension. Suppose $b \in \mathcal{B}$ satisfies some polynomial equation. Since $\mathcal{A}$ or $\mathcal{B}$ or both may be nonunital, we will choose $p(\lambda) = \sum_{k=1}^{n} \alpha_k \lambda^k$ to be the monic polynomial, having constant term 0 and lowest degree, which satisfies $p(b) = 0$. If b has a quasi-inverse $c \in \mathcal{A}$, then the equation $(1-c)b^m = -(c+\sum_{k=1}^{m-1} b^k)$ for all $m \in \mathbb{N}$ allows us to rewrite the equation $(1 - c)p(b) = 0$ as $(\sum_{k=1}^{n} \alpha_k)c = -q(b)$, where q is a monic polynomial having constant term 0 and degree at most $m - 1$. Hence both sides of the equation are non-zero, so c is in $\mathcal{B}$. Thus $\mathcal{B}$ is a spectral subalgebra.

If $b \in \mathcal{B}$ has an inverse $b^{-1} \in \mathcal{A}$, then multiplication of $p(b)$ by a suitable non-negative power of b^{-1} allows us to write $1 \in \mathcal{A}$ as a linear combination of positive powers of b, so 1 is in $\mathcal{B}$. Hence $\mathcal{B}$ is a unital subalgebra and its elements must satisfy (6).

(e): We will consider the case of the left regular representation, since the right regular representation is similar and the extended regular representations are easier. Suppose that $I - T \in \mathcal{L}(\mathcal{A})$ is a quasi-inverse for L_a so that T is an inverse for $I - L_a$. Then $b = T(-a)$ satisfies $b - ab = (I - L_a)b = (I - L_a)T(-a) = -a$. Hence L_b is a right quasi-inverse for L_a, but in fact it is a quasi-inverse, since L_a is quasi-invertible. Thus $L_{\mathcal{A}}$ is a spectral subalgebra of $\mathcal{L}(\mathcal{A})$.

Suppose next that for some $a \in \mathcal{A}$ the operator L_a has an inverse S in $\mathcal{L}(\mathcal{A})$. Note that L_a must be injective since it has an inverse. Now define $e = S(a)$ and $c = S(e)$. Any $b \in \mathcal{A}$ satisfies

$$L_a(eb - b) = L_aS(a)b - L_a(b) = ab - ab = 0,$$

hence $eb - b$ is 0 and finally

$$L_a(L_c(b) - S(b)) = L_aS(e)b - L_aS(b) = eb - b = 0.$$

We conclude $I = L_e$ and $S = L_c$. Thus equation (5) (with $\mathcal{L}(\mathcal{A})$ and $L_{\mathcal{A}}$ as the algebra and subalgebra) holds in this case, since $L_{\mathcal{A}}$ is a unital subalgebra if it is a unital algebra.

(f): This proof is similar to the last one. Suppose $\kappa(a)$ has $f \in \mathcal{A}^{**}$ as a quasi-inverse. (This does not depend on the choice of an Arens product.) The equation $f - \kappa(a)f = -\kappa(a)$ shows that any $g \in \mathcal{A}^{**}$ satisfies $(I - L_a)^{**}(g - fg) = g$. Thus $(I - L_a)^{**}$ maps $\mathcal{A}^{**}$ onto itself, and hence $(I - L_a)$ must map $\mathcal{A}$ onto itself. We conclude that there is some $b \in \mathcal{A}$ satisfying $b - ab = -a$. But this just shows that b is a left quasi-inverse for a. Hence $\kappa(b)$ equals f since f was a two-sided quasi-inverse for $\kappa(a)$. Therefore a is quasi-invertible in $\mathcal{A}$.

Finally, we deal with the case in which $\kappa(a)$ is invertible in $\mathcal{A}^{**}$. Since multiplication on the left by $\kappa(a)$ in either Arens product is just $(L_a)^{**}$, this implies that L_a is a bijection on $\mathcal{A}$. Let e be the element which L_a

maps onto a. Then we have $L_a(eb - b) = ab - ab = 0$ for all $b \in \mathcal{A}$. Hence $L_e = I$ so $(L_e)^{**} = I$ and $\kappa(e)$ equals the identity in $\mathcal{A}^{**}$. Thus, if any element of $\kappa(\mathcal{A})$ is invertible, then this algebra is a unital subalgebra of $\mathcal{A}^{**}$, which implies equation (5), as we wished to show.

(g): This is obvious. $\square$

We caution the reader that in (e) above we have shown that an element in the range of the left regular representation has precisely the same spectrum whether calculated in this range or in the bigger algebra $\mathcal{L}(\mathcal{A})$. We do not claim, and it is not true, that an element $a \in \mathcal{A}$ has the same spectrum in $\mathcal{A}$ as L_a has in $\mathcal{L}(\mathcal{A})$. The problem is that the left regular representation has a nontrivial kernel unless the left annihilator $\mathcal{A}_{LA}$ of $\mathcal{A}$ is $\{0\}$. The two-dimensional algebra with basis a, b satisfying $a^2 = ab = 0$, $ba = a$ and $b^2 = b$ shows that the two spectra just mentioned may differ without some hypothesis on the algebra $\mathcal{A}$. We have established the following facts. Any algebra $\mathcal{A}$ satisfies

$$Sp(a) = Sp(L_a^1) \quad \forall \, a \in \mathcal{A} \tag{7}$$

and

$$Sp(L_a) \subseteq Sp(a) \subseteq Sp(L_a) \cup \{0\} \quad \forall \, a \in \mathcal{A}. \tag{8}$$

If the left annihilator $\mathcal{A}_{LA}$ of $\mathcal{A}$ is $\{0\}$, then we also have

$$Sp(L_a) = Sp(a) \quad \forall \, a \in \mathcal{A}. \tag{9}$$

2.5.4 A Non-spectral Subalgebra Proposition 2.5.3 shows that many kinds of subalgebras are spectral subalgebras. The introduction to this chapter listed additional examples. (See also Barnes [1983].) We now give an example of a non-spectral, closed, unital subalgebra which is a generic in some sense.

Let $\mathcal{A}$ be a unital algebra and let $\mathcal{B}$ be a subalgebra containing the identity element 1 of $\mathcal{A}$. If $\mathcal{B}$ is not spectral, there is some element $b \in \mathcal{B}$ with an inverse b^{-1} in $\mathcal{A}$ but not in $\mathcal{B}$. Thus, the subalgebra generated by b and 1 is also a non-spectral subalgebra of $\mathcal{A}$ and this shows that every unital non-spectral subalgebra includes one of this type. If $\mathcal{A}$ is normed and $\mathcal{B}$ is closed, then an element b of the type just described can be characterized by the fact that b^{-1} exists in $\mathcal{A}$ but is not the limit of any sequence of polynomials in b (*i.e.*, expressions of the form $\sum_{k=1}^n \alpha_k b^k$).

Here is an explicit example. Let $\mathcal{H}$ be a separable Hilbert space and choose an orthonormal basis $\{\varphi_n : n \in \mathbb{Z}\}$ parameterized by the set of all integers. Consider the operator $U \in \mathcal{B}(\mathcal{H})$ defined by

$$U\left(\sum_{j=-\infty}^{\infty} \lambda_j \varphi_j\right) = \sum_{j=-\infty}^{\infty} \lambda_j \varphi_{j+1}$$

which is called the *bilateral shift.* Clearly U has an inverse defined by

$$U^{-1}\left(\sum_{j=-\infty}^{\infty} \lambda_j \varphi_j\right) = \sum_{j=-\infty}^{\infty} \lambda_j \varphi_{j-1}.$$

We claim that U^{-1} cannot be approximated by polynomials in U. Perhaps the easiest way to see this is to consider the subalgebra $\mathcal{B}$ of $\mathcal{B}(\mathcal{H})$ defined by

$$\mathcal{B} = \{T \in \mathcal{B}(\mathcal{H}) : T\mathcal{H}_n \subseteq \mathcal{H}_n \quad n \in \mathbb{Z}\}$$

where

$$\mathcal{H}_n = \text{closed span } \{\varphi_j : j \geq n\}.$$

Obviously, $\mathcal{B}$ is a closed subalgebra which contains U but not U^{-1}.

Topological Divisors of Zero

The next concepts come in versions related to both invertibility and quasi-invertibility. We will give the two definitions in parallel. As usual, we will find the second version more useful, but the first version may be easier to grasp on first reading. We will routinely give proofs for the second version only, leaving it to the reader to reformulate them for the first version.

2.5.5 Definition An element a in any algebra $\mathcal{A}$ with an algebra seminorm σ is called a *left topological $\langle$ divisor / quasi-divisor $\rangle$ of zero* if there exists a sequence $\{b_n\}_{n\in\mathbb{N}} \subseteq \mathcal{A}$ satisfying both:

$$\sigma(b_n) = 1 \quad n \in \mathbb{N} \text{ and } \langle \sigma(ab_n) \to 0 \ / \ \sigma(b_n - ab_n) \to 0 \ \rangle.$$

A *right topological $\langle$ divisor / quasi-divisor $\rangle$of zero* are defined similarly. An element which is either a right or a left topological $\langle$ divisor / quasi-divisor $\rangle$ of zero is called a *topological $\langle$ divisor / quasi-divisor $\rangle$of zero* and an element which belongs to both sets is called a *joint topological $\langle$ divisor / quasi-divisor $\rangle$of zero* if the same sequence $\{b_n\}_{n\in\mathbb{N}}$ can be used on both sides. These sets are denoted by $\langle \mathcal{A}_{LZ} / \mathcal{A}_{LqZ} \rangle$, $\langle \mathcal{A}_{RZ} / \mathcal{A}_{RqZ} \rangle$, $\langle \mathcal{A}_Z / \mathcal{A}_{qZ} \rangle$ and $\langle \mathcal{A}_{JZ} / \mathcal{A}_{JqZ} \rangle$, respectively.

It may help the reader to know that an element a in a semi-normed algebra $(\mathcal{A}, \sigma)$ is called a *left quasi-divisor of zero* if there is some $b \in \mathcal{A}$ satisfying $\sigma(b) = 1$ and $b - ab = (1 - a)b = 0$. Hence if the sequence in the above definition of a left topological quasi-divisor of zero had a limit a, then a would be a left quasi-divisor of zero. The utility of these concepts arises from two facts. First, even a (one-sided) topological quasi-divisor of zero cannot have a quasi-inverse in any larger semi-normed algebra, so that it is what has been called a *permanently quasi-singular element.* Second, in spectral semi-normed algebras the boundary of the set of quasi-invertible elements consists of joint topological quasi-divisors of zero, so that such elements are fairly common. These two results will be recorded in the first portion of the next proposition and the following theorem, respectively. The rest of the proposition contains other easy results needed later. Arens [1958], [1960] shows (among other results) that, for any element which is not a topological divisor of zero in a commutative Banach algebra, there is a larger algebra in which the element is invertible. Bonsall and Duncan [1973b] give an example due to Peter G. Dixon showing that an element may be both a right and a left divisor of zero without being a joint divisor of zero. See also Example 3.4.17 below and Zofia Balsam [1980].

2.5.6 Proposition *Let* $(\mathcal{A}, \sigma)$ *be a semi-normed algebra.*

(a) *If* $a \in \mathcal{A}$ *is a topological* ⟨ *divisor / quasi-divisor* ⟩ *of zero, then it is not* ⟨ *invertible / quasi-invertible* ⟩ *in any larger algebra to which the algebra semi-norm can be extended.*

(b) *The following set inclusions hold:*

$$\mathcal{A} \circ \mathcal{A}_{LqZ} \subseteq \mathcal{A}_{LqZ} \qquad \mathcal{A}_{RqZ} \circ \mathcal{A} \subseteq \mathcal{A}_{RqZ}.$$

(c) *If* $a, c \in \mathcal{A}$ *satisfy* $a \circ c \in \langle \mathcal{A}_{LqZ} / \mathcal{A}_{RqZ} \rangle$*, then either* a *or* c *belongs to* ⟨ $\mathcal{A}_{LqZ}$ / $\mathcal{A}_{RqZ}$ ⟩.

Proof (a): If the sequence $\{b_n\}_{n \in \mathbb{N}} \subseteq \mathcal{A}$ satisfies the definition for $a \in \mathcal{A}_{LqZ}$ and a has a quasi-inverse in $\mathcal{B}$, then we get the contradiction:

$$1 = \sigma(b_n) = \sigma(a^q \circ a \circ b_n) \leq \sigma(b_n - ab_n) + \sigma(a^q)\sigma(b_n - ab_n) \to 0.$$

The other cases are similar or easier.

(b): We consider only the first inclusion. Again let $\{b_n\}_{n \in \mathbb{N}} \subseteq \mathcal{A}$ satisfy the definition for $a \in \mathcal{A}_{LqZ}$ and let c belong to $\mathcal{A}$. Then we get

$$\sigma((1 - (c \circ a))b_n) = \sigma((1 - c)(1 - a))b_n) \leq (1 + \sigma(c))\sigma((1 - a)b_n) \to 0.$$

(c): If the sequence $\{b_n\}_{n \in \mathbb{N}} \subseteq \mathcal{A}$ satisfies the definition for $a \circ c \in \mathcal{A}_{LqZ}$, then we get

$$\sigma((1 - (a \circ c))b_n) = \sigma((1 - a)(1 - c)b_n) \to 0.$$

If $\sigma((1 - c)b_n) \to 0$, then c is a left topological quasi-divisor of zero. If not, there is some subsequence of $\{b_n\}$ for which this expression is bounded away from zero. We will denote the subsequence by $\{b_n\}$ again. Define a new sequence $\{b'_n\}$ by

$$b'_n = \sigma((1 - c)b_n)^{-1}(1 - c)b_n.$$

This new sequence shows that a is a left topological quasi-divisor of zero, as we wished. □

Richard Arens [1958] showed that in any unital commutative Banach algebra an element was permanently singular if and only if it was a topological divisor of zero. In fact for any finite set of nontopological divisors of zero he constructed a unital super-algebra in which they all become invertible. Recently Antonio Fernández López, Miguel Florencio, Pedro J. Paúl and Vladimir Müller [1990] extended this result to commutative topological algebras. (In this book, multiplication in a topological algebra is only required to be singly continuous. The result is false for the class of topological algebras with jointly continuous multiplication, since it is harder to produce super-algebras in this situation.) See also Wiesław Żelazko [1984], [1986], [1987].

The next theorem has many important corollaries. In it the (less useful) sufficiency of the condition was first noted by Vania D. Mascioni in [1987], but the necessity of the condition really goes back to Šilov [1950].

2.5.7 Theorem *Let $\mathcal{A}$ be an algebra and let σ be an algebra semi-norm on $\mathcal{A}$. Then σ is a spectral semi-norm if and only if every element in the boundary of the quasi-group $\mathcal{A}_{qG}$ is a joint topological quasi-divisor of zero. Similarly, in a unital algebra an algebra semi-norm is spectral if and only if every element in the boundary of the group $\mathcal{A}_G$ is a joint topological divisor of zero.*

Proof To establish the necessity, suppose σ is a spectral semi-norm and $\{a_n\}_{n\in\mathbb{N}} \subseteq \mathcal{A}_{qG}$ converges to $a \in \mathcal{A}$. Then $a_n^q \circ a = (1 - a_n^q)(a - a_n)$ implies

$$\sigma(a_n^q \circ a) \leq (1 + \sigma(a_n^q))\sigma(a - a_n).$$

Thus if $\{a_n^q\}_{n\in\mathbb{N}}$ has a bounded subsequence, $a_n^q \circ a$ will belong to $\mathcal{A}_{qG}$ for some n, since 0 is in the σ-interior of $\mathcal{A}_{qG}$. In this case a has a left quasi-inverse. Similarly it has a right quasi- inverse and thus belongs to $\mathcal{A}_{qG}$. Therefore if a belongs to the boundary of the open set $\mathcal{A}_{qG}$ and $\{a_n\}_{n\in\mathbb{N}} \subseteq \mathcal{A}_{qG}$ converges to a, then $\sigma(a_n^q)$ approaches infinity. Thus $b_n = \sigma(a_n^q)^{-1}a_n^q$ satisfies the requirements of Definition 2.5.5 on both sides, since, for instance, we get

$$\begin{aligned}
\sigma(a\sigma(a_n^q)^{-1}a_n^q - \sigma(a_n^q)^{-1}a_n^q) &= \sigma(a_n^q)^{-1}\sigma((a - a_n)a_n^q + a_na_n^q - a_n^q) \\
&\leq \sigma(a_n^q)^{-1}[\sigma(a - a_n)\sigma(a_n^q) + \sigma(a_n)] \to 0.
\end{aligned}$$

Thus if a is in the boundary of $\mathcal{A}_{qG}$, then a is a joint topological quasi-divisor of zero.

If a is in the boundary of the set of invertible elements, then $\mathcal{A}$ is unital and $1 - a$ is in the boundary of the set of quasi-invertible elements. Hence by what we have already proved, $1 - a$ is a joint topological quasi-divisor of zero. Therefore, a is a joint topological divisor of zero.

For the sufficiency, simply note that if σ is not a spectral semi-norm, then $\mathcal{A}_{qG}$ is not σ-open, so some element a of the boundary of $\mathcal{A}_{qG}$ belongs to $\mathcal{A}_{qG}$. Since we are assuming that a is a joint topological divisor of zero, this contradicts Proposition 2.5.6 without even extending the algebra. □

The Boundary of the Spectrum and the Permanent Spectrum

We remark that the proof of Theorem 2.1.14(a) above shows that if λ is on the boundary of the exponential spectrum of an element a in a unital Banach algebra $\mathcal{A}$, then $\lambda - a$ is a joint topological divisor of zero. Let us specialize to a unital Banach algebra $\mathcal{A}$ and call the *permanent spectrum* of an element $a \in \mathcal{A}$ the set

$$per\,Sp_{\mathcal{A}}(a) = \bigcap\{Sp_{\mathcal{B}}(a) : \mathcal{B} \text{ is a Banach algebra including } \mathcal{A}\} \qquad (10)$$

where inclusion means *isometric inclusion*, *i.e.*, the norm of $\mathcal{A}$ is just the restriction of the norm of $\mathcal{B}$. Thus we have just shown $\partial e Sp(a) \subseteq per\,Sp(a)$ and

$$\partial Sp(a) \subseteq per\,Sp(a) \quad \forall\, a \in \mathcal{A}. \qquad (11)$$

The question of when inclusion (11) is proper has been explored by Šilov [1947b], Murphy and West [1980a] and others. The latter paper simplifies an example from the former which shows that proper inclusion can occur. However, it seems to be much more common for the two sets to agree. Results and references are contained in the latter paper.

In the next result, note that we do not need to require that the semi-norm be spectral on the larger algebra nor that the subalgebra be a spectral subalgebra. Any complete subalgebra of a normed algebra and in particular any closed subalgebra of a Banach algebra (*cf.* Corollary 2.5.10) satisfies the corollary. (Recall that pc$S$ is the polynomially convex hull of S.)

2.5.8 Corollary *Let σ be an algebra semi-norm on an algebra $\mathcal{A}$. If $\mathcal{B}$ is a subalgebra on which σ is a spectral semi-norm, then it satisfies:*

$$\partial Sp_{\mathcal{B}}(b) \subseteq Sp_{\mathcal{A}}(b) \subseteq Sp_{\mathcal{B}}(b) \cup \{0\} \qquad \forall\, b \in \mathcal{B}$$

$$\partial Sp_{\mathcal{B}}(b) \subseteq \partial Sp_{\mathcal{A}}(b) \qquad \forall\, b \in \mathcal{B}$$

and hence

$$Sp_{\mathcal{A}}(b) \subseteq Sp_{\mathcal{B}}(b) \subseteq pc\,Sp_{\mathcal{A}}(b) \qquad \forall\, b \in \mathcal{B}.$$

Proof Proposition 2.1.8(c) asserts the second inclusion in the first line for any subalgebra. The first inclusion in the same line follows from the theorem and Proposition 2.5.6(a). We are using the fact that if $\langle$ 0 is / λ is a non-zero element $\rangle$ in the boundary of the spectrum of b in $\mathcal{B}$, then $\langle b / \lambda^{-1}b \rangle$ is in the boundary of the set of $\langle$ invertible / quasi-invertible $\rangle$ elements. For the inclusion in the second line, note that if λ is any number in the boundary of the spectrum of $Sp_{\mathcal{B}}(a)$, it is the limit of a sequence $\{\lambda_n\}_{n\in\mathbb{N}}$ not in $Sp_{\mathcal{B}}(a)$ and hence surely not in $Sp_{\mathcal{A}}(a)$. Hence the first inclusion shows λ is in the boundary of $Sp_{\mathcal{A}}(a)$. The final display is an immediate consequence.   □

The corollary above can be used to show that certain types of subalgebras are always spectral subalgebras. A number of similar results could be formulated, and we offer this one as an example. Modular annihilator algebras are discussed in Chapter 7.

2.5.9 Corollary *Any subalgebra of a normed algebra which is a modular annihilator algebra is a spectral subalgebra.*

Proof The restriction of the norm to the subalgebra is spectral by Theorem 7.4.12. Furthermore, by Theorem 7.5.5 the spectrum of each element of the subalgebra is nowhere dense and hence equal to its own boundary. The result now follows from the previous corollary.   □

The next result is particularly important for its application to Banach algebras. Note that while the subalgebra $\mathcal{B}$ is a spectral algebra, it need not be a spectral subalgebra of $\mathcal{A}$. We call it a *feebly spectral subalgebra*.

2.5.10 Corollary *Let $\mathcal{A}$ be a spectral algebra. Then the following are equivalent for any subalgebra $\mathcal{B}$.*

(a) $\rho_{\mathcal{A}}(b) = \rho_{\mathcal{B}}(b) \qquad \forall \, b \in \mathcal{B}$.

(b) $\mathcal{B}$ *is a spectral algebra satisfying*

$$\partial Sp_{\mathcal{B}}(b) \subseteq \partial Sp_{\mathcal{A}}(b) \subseteq Sp_{\mathcal{A}}(b) \subseteq Sp_{\mathcal{B}}(b) \cup \{0\} \qquad \forall \, b \in \mathcal{B}.$$

(c) *The restriction to $\mathcal{B}$ of some spectral semi-norm on $\mathcal{A}$ is a spectral semi-norm on $\mathcal{B}$.*

(d) *The restriction to $\mathcal{B}$ of every spectral semi-norm on $\mathcal{A}$ is a spectral semi-norm on $\mathcal{B}$.*

A closed subalgebra of a Banach algebra satisfies these conditions.

Proof (a) $\Rightarrow$ (d): Immediate from the definition of a spectral semi-norm.

(d) $\Rightarrow$ (c): Obvious, since $\mathcal{A}$ is a spectral algebra.

(c) $\Rightarrow$ (b): This follows from Corollary 2.5.8 and the fact that the spectrum is closed in a spectral algebra.

(b) $\Rightarrow$ (a): Obvious.

The final remark is obvious by considering conditions (a) or (c).   □

Recall the discussion following Theorem 2.2.12. It dealt only with the case of closed unital subalgebras of Banach algebras. However, by applying Proposition 2.5.15 below to the completion of $\mathcal{A}/\mathcal{A}_J$ in some spectral norm and to the closure of the image of $\mathcal{B}$ in this completion we can extend the results given there to the setting of Corollary 2.5.10. Hence, the only difference between the spectrum of an element $b \in \mathcal{B}$ when calculated relative to $\mathcal{A}$ and $\mathcal{B}$ is that certain whole bounded components of the resolvent set in the former case may belong to the spectrum in the latter case, or zero may appear if b is invertible in $\mathcal{A}$ but $\mathcal{B}$ does not contain the inverse of b. Adding a bounded component of the resolvent set to the spectrum will remove its boundary from the boundary of the spectrum.

The next result is similar to Corollary 2.5.9.

2.5.11 Corollary *A closed subalgebra of a Banach algebra is a spectral subalgebra if every element in the subalgebra has nowhere dense spectrum in the subalgebra.*

Proof Apply the final remark of the last corollary. If $Sp_\mathcal{B}(b)$ (which is closed) is nowhere dense, then it is equal to its own boundary; consequently condition (b) of the last corollary shows that $\mathcal{B}$ is a spectral subalgebra when every element of $\mathcal{B}$ has this property.   □

Algebras without Divisors of Zero

The following two results show two circumstances under which an algebra without topological divisors or quasi-divisors of zero must be the complex field. In this respect they are related to the Gelfand–Mazur theorem (Corollary 2.2.3). The first is due to Żelazko [1967] improving on Kaplansky [1949a].

2.5.12 Proposition *A non-zero normed algebra $\mathcal{A}$ is isomorphic to $\mathbb{C}$ if and only if it contains no non-zero joint topological divisors of zero.*

Proof Clearly an algebra isomorphic to $\mathbb{C}$ contains no non-zero topological divisors of zero.

Suppose every commuting subalgebra of $\mathcal{A}$ is isomorphic to $\mathbb{C}$. Then there is certainly no commuting pair $a, b \in \mathcal{A} \setminus \{0\}$ satisfying $ab = 0$. However, if these elements do not commute, then they satisfy $ba \neq 0$ and $(ba)^2 = 0$, which is also impossible. Any two idempotents $e, f \in \mathcal{A}$ satisfy $e(ef - fe)e = f(ef - fe)f = 0$, so they commute. By hypothesis any maximal commutative subalgebra $\mathcal{C}$ of $\mathcal{A}$ has the form $\mathbb{C}e$ for some idempotent e. Similarly, any element a of $\mathcal{A}$ has the form λf for some idempotent f, but since e and f commute, a belongs to $\mathcal{C}$ and hence actually has the form λe. Hence e is an identity element for $\mathcal{A} = \mathbb{C}e$ which is isomorphic to $\mathbb{C}$. Therefore, in the rest of this proof we may assume that $\mathcal{A}$ is commutative, so that any topological divisor of zero is a joint topological divisor of zero.

Case (1): Suppose that $\mathcal{A}$ is unital without non-zero topological divisors of zero. If $a \in \mathcal{A}$ is a topological divisor of zero in the completion $\mathcal{B}$ of $\mathcal{A}$, then it is easy to see (by approximating the sequence $\{b_n\}_{n \in \mathbb{N}} \subseteq \mathcal{B}$ by a sequence in $\mathcal{A}$) that a is a topological divisor of zero in $\mathcal{A}$. Theorem 2.2.2 shows that the boundary of the spectrum in $\mathcal{B}$ of each element $a \in \mathcal{A}$ contains some $\lambda \in \mathbb{C}$. Theorem 2.5.7 shows that $\lambda - a$ is a topological divisor of zero. Hence by assumption $\lambda - a$ is 0. Since $a \in \mathcal{A}$ was arbitrary, we have $\mathcal{A} = \mathbb{C}$.

Case (2): Next suppose $\mathcal{A}$ is nonunital. Give $\mathcal{A}^1$ the norm $\|\lambda + a\| = |\lambda| + \|a\|$ for all $\lambda + a \in \mathcal{A}^1$. Then $\mathcal{A}^1$ is not isomorphic to $\mathbb{C}$ since $\mathcal{A}$ is not zero. Thus $\mathcal{A}^1$ has a topological divisor of zero $\lambda + a$. Let $\{\lambda_n + b_n\}_{n \in \mathbb{N}}$ satisfy $\|\lambda_n + b_n\| = 1$ for all $n \in \mathbb{N}$ and $(\lambda + a)(\lambda_n + b_n) \to 0$. Since $\{\lambda_n\}_{n \in \mathbb{N}}$ is a bounded sequence we may choose a convergent subsequence and hence assume $\lambda_n \to \lambda_0$. If both λ and λ_0 are 0, then $\mathcal{A}$ has a topological divisor of zero and there is nothing more to prove.

Case (2a): Next suppose $\lambda = 0$ but $\lambda_0 \neq 0$. If some subsequence of $\{b_n\}_{n \in \mathbb{N}}$ converges to some $b \in \mathcal{A}$, then b satisfies $a(\lambda_0 + b) = 0$. Since $-\lambda_0^{-1}b$ is not an identity element in $\mathcal{A}$, there is some $c \in \mathcal{A}$ satisfying $(\lambda_0 + b)c \neq 0$. Hence $a(\lambda_0 + b)c = 0$ shows that a is a divisor of zero and *a fortiori* a topological divisor of zero. If no subsequence of $\{b_n\}_{n \in \mathbb{N}}$ converges, we can (by selecting a subsequence) assume that $\|b_{n+1} - b_n\|$ is bounded away from zero. Hence the sequence $\{\|b_{n+1} - b_n\|^{-1}(b_{n+1} - b_n)\}_{n \in \mathbb{N}}$ shows that a is topological divisor of zero.

Case (2b): Finally suppose $\lambda \neq 0$ but $\lambda_0 = 0$. As before, the fact that $-\lambda^{-1}a$ is not an identity element in $\mathcal{A}$ gives some $c \in \mathcal{A}$ satisfying $c(\lambda + a) \neq 0$. Hence $c(\lambda + a)b_n \to 0$ shows that c is a topological divisor of zero.   □

We now restate the result above for topological quasi-divisors of zero.

2.5.13 Proposition *A spectral semi-normed algebra $\mathcal{A}$ which contains no topological quasi-divisors of zero (except possibly 1) is either a Jacobson-radical algebra or isomorphic to the complex field, $\mathbb{C}$. The latter case occurs if and only if some element of $\mathcal{A}$ has non-zero spectrum or if and only if $\mathcal{A}$ is unital.*

Proof Suppose some element $a \in \mathcal{A}$ has spectrum not equal to $\{0\}$. Then $\partial Sp(a)$ must contain a non-zero complex number λ, since the spectrum is bounded and nonempty (by Theorem 2.4.3). Theorem 2.5.7 shows that $\lambda^{-1}a$ is a topological quasi-divisor of zero and hence equals 1. Now let $b \in \mathcal{A}$ be arbitrary and choose $t > \sigma(b)$, so that $Sp(t - b)$ contains a non-zero element μ in its boundary. The argument just given for $\lambda^{-1}a$ implies $\mu^{-1}(t - b) = 1$. Hence the map $b = (t - \mu)1 \mapsto t - \mu$ is an isomorphism of $\mathcal{A}$ onto $\mathbb{C}$.

If every element of $\mathcal{A}$ has spectrum $\{0\}$, then $\mathcal{A}$ is a Jacobson-radical algebra by Theorem 2.3.3 (b) and thus has no identity element.   □

The following result is due to Rickart [1950]. Unlike most of the other results in this chapter, there does not seem to be an interesting formulation in terms of spectral algebras.

2.5.14 Theorem *Let $\mathcal{A}$ be a Banach algebra. Let a be an element of $\mathcal{A}$ having zero in the boundary of its spectrum. Then either there is a non-zero proper idempotent e (i.e., $0 \neq e \neq 1$) such that a is invertible in $e\mathcal{A}e$ or a is a joint topological divisor of zero.*

Proof As an element of $\mathcal{A}^1$, a belongs to the boundary of $(\mathcal{A}^1)_G$. Hence by Theorem 2.5.7, a is a joint topological divisor of zero in $\mathcal{A}^1$ (with the norm $\|\lambda + b\| = |\lambda| + \|b\|$ if $\mathcal{A}^1 \neq \mathcal{A}$). Thus if $\mathcal{A} = \mathcal{A}^1$, the proof is complete. (We have only used the fact that the norm is a spectral norm.)

Now suppose $\mathcal{A}$ is nonunital and a is not a joint topological divisor of zero. By symmetry we may assume that a is not a left topological divisor of zero. We will prove that there is a left identity e for $\mathcal{A}$ such that a is invertible in $e\mathcal{A}e$.

By what we have just proved, there is a sequence $\{\lambda_n + c_n\}_{n \in \mathbb{N}} \subseteq \mathcal{A}^1$ with $|\lambda_n| + \|c_n\| = 1$ such that $a(\lambda_n + c_n)$ converges to 0. Since $\{\lambda_n\}_{n \in \mathbb{N}}$ is a bounded sequence, it has a convergent subsequence. By a change of notation we assume $\{\lambda_n\}_{n \in \mathbb{N}}$ converges. It does not converge to 0, since in that case $b_n = (1 - |\lambda_n|)^{-1} c_n$ (omitting any n with $|\lambda_n| = 1$) would put a in $\mathcal{A}_{LZ}$. Thus λ_n converges to some $\lambda \neq 0$. Then $a(-\lambda^{-1} c_n)$ converges to a. The sequence $\{-\lambda^{-1} c_n\}_{n \in \mathbb{N}}$ must be a Cauchy sequence, since otherwise $b_n = \|\lambda^{-1}(c_{p_n} - c_{q_n})\|^{-1} \lambda^{-1}(c_{p_n} - c_{q_n})$ for suitable subsequences $\{c_{p_n}\}$ and $\{c_{q_n}\}$ puts a into $\mathcal{A}_{LZ}$. Thus $-\lambda^{-1} c_n$ converges to a non-zero element $e \in \mathcal{A}$ which satisfies $ae = a$. Then any $b \in \mathcal{A}$ satisfies $0 = a(eb - b)$. If there were any b with $eb - b$ non-zero, a would be a left divisor of zero. Hence e is a left identity for $\mathcal{A}$ and *a fortiori* an idempotent. In particular $ea = a = ae$ implies that a belongs to $e\mathcal{A}e$. In this spectral subalgebra (Proposition 2.5.3(c)), 0 is still in the boundary of the spectrum unless a is invertible in $e\mathcal{A}e$. Since $e\mathcal{A}e$ is unital, the first step of this proof shows that a would be a joint topological divisor of zero in $e\mathcal{A}e$ (and hence *a fortiori* in $\mathcal{A}$, contrary to assumption) if it were not invertible in $e\mathcal{A}e$. Since a is non-zero, e is a proper idempotent. $\square$

The Spectral Radius in Subalgebras

The following criterion is sometimes useful and in particular it will be used in the next two results.

2.5.15 Proposition *Let $\| \cdot \|$ be an algebra norm on an algebra $\mathcal{A}$ and let $\overline{\mathcal{A}}$ be the completion of $\mathcal{A}$ with respect to $\| \cdot \|$. Then the following are equivalent:*

    (a) $\| \cdot \|$ *is a spectral norm on $\mathcal{A}$.*
    (b) $\rho_{\overline{\mathcal{A}}}(a) = \rho_{\mathcal{A}}(a)$ $a \in \mathcal{A}$.
    (c) $\mathcal{A}$ *is a spectral subalgebra of $\overline{\mathcal{A}}$.*

Proof (a) $\Rightarrow$ (c): Suppose $a \in \mathcal{A}$ has a quasi-inverse $b \in \overline{\mathcal{A}}$. Choose a sequence $\{b_n\}_{n \in \mathbb{N}}$ in $\mathcal{A}$ converging to b. This sequence satisfies $a \circ b_n \to$

$a \circ b = 0$ and $b_n \circ a \to b \circ a = 0$. Hence $a \circ b_n$ and $b_n \circ a$ are eventually quasi-invertible in $\mathcal{A}$. This implies that a is quasi-invertible in $\mathcal{A}$.

(c) $\Rightarrow$ (b): $\Rightarrow$ (a): Immediate.   □

In the list of spectral semi-norms given in the introduction, (b), (c), (d), (e) and (h) assert that all algebra norms on certain algebras are spectral. Yood [1958] called algebras with this property permanent Q-algebras and noted some of the following equivalent conditions. We call a spectral algebra *permanently spectral* if all algebra norms on it are spectral. See Michael J. Meyer [1991], [1992b] and Theorem 6.1.5(g) below.

2.5.16 Proposition *The following are equivalent for an algebra $\mathcal{A}$:*

(a) *$\mathcal{A}$ is permanently spectral.*

(b) *If $\mathcal{B}$ is a Banach algebra and $\varphi\colon \mathcal{A} \to \mathcal{B}$ is an injective homomorphism, then*

$$\rho_{\mathcal{B}}(\varphi(a)) = \rho_{\mathcal{A}}(a) \qquad \forall\, a \in \mathcal{A}.$$

(c) *If $\mathcal{B}$ is a Banach algebra and $\varphi\colon \mathcal{A} \to \mathcal{B}$ is an injective homomorphism with dense range, then $\varphi(\mathcal{A})$ is a spectral subalgebra of $\mathcal{B}$.*

Proof (a) $\Rightarrow$ (c): Proposition 2.5.15 applies since we can think of $\mathcal{B}$ as the completion of $\varphi(\mathcal{A})$ in the norm induced from $\mathcal{B}$.

(c) $\Rightarrow$ (b): Let $\|\cdot\|$ be the complete norm on $\mathcal{B}$ and denote the closure of $\varphi(\mathcal{A})$ in $\mathcal{B}$ by $\mathcal{C}$. Then any $a \in \mathcal{A}$ satisfies

$$\rho_{\mathcal{A}}(a) = \rho_{\mathcal{C}}(\varphi(a)) = \lim \|\varphi(a)^n\|^{1/n} = \rho_{\mathcal{B}}(\varphi(a)).$$

(b) $\Rightarrow$ (a): Given an algebra norm $\|\cdot\|$ on $\mathcal{A}$, let $\mathcal{B}$ be the completion of $\mathcal{A}$ in this norm, and let $\varphi\colon \mathcal{A} \to \mathcal{B}$ be the natural embedding. Each $a \in \mathcal{A}$ satisfies $\rho_{\mathcal{A}}(a) = \rho_{\mathcal{B}}(a) \le \|a\|$, so $\|\cdot\|$ is a spectral norm.   □

The following result, due to Bertram Yood [1958], provides the simplest examples of infinite-dimensional algebras satisfying the above proposition.

2.5.17 Theorem *Let $\mathcal{X}$ be a Banach space and let $\mathcal{A}$ be a subalgebra of $\mathcal{B}(\mathcal{X})$ which includes $\mathcal{B}_F(\mathcal{X})$. If $\mathcal{A}$ is either closed or a spectral subalgebra of $\mathcal{B}(\mathcal{X})$, then any algebra norm on $\mathcal{A}$ is a spectral norm.*

Proof The last result in §1.7.15 shows that any algebra norm on $\mathcal{A}$ dominates the operator norm. Either hypothesis implies that the spectral radius on $\mathcal{A}$ is less than or equal to the operator norm. Hence Theorem 2.2.5(e) shows that any algebra norm is a spectral norm on $\mathcal{A}$.   □

We have noted that the spectrum in a spectral algebra enjoys the properties of the spectrum in a Banach algebra. The next theorem and example show that an algebra is spectral if and only if its semisimple quotient can be embedded as a spectral subalgebra of a Banach algebra, but that the spectral algebra itself may not have an embedding into a Banach algebra.

2.5.18 Theorem *The following are equivalent for an algebra $\mathcal{A}$:*

(a) $\mathcal{A}$ *is spectral.*

(b) *There is some Banach algebra $\mathcal{B}$ and some homomorphism $\varphi: \mathcal{A} \to \mathcal{B}$ with the Jacobson radical as kernel satisfying*

$$\rho_{\mathcal{B}}(\varphi(a)) = \rho_{\mathcal{A}}(a) \qquad \forall \, a \in \mathcal{A}.$$

(c) $\mathcal{A}$ *modulo its Jacobson radical can be embedded as a dense spectral subalgebra of a semisimple Banach algebra.*

Proof (a) $\Rightarrow$ (c): Let σ' be an arbitrary spectral semi-norm on $\mathcal{A}$. Corollary 2.3.4 and Theorem 2.2.13(d) show that σ' induces a spectral norm $||| \cdot |||$ on $\mathcal{A}/\mathcal{A}_J$. Let $(\overline{\mathcal{A}}, ||| \cdot |||)$ be the completion of $\mathcal{A}/\mathcal{A}_J$ in this norm. Note that $\mathcal{A}/\mathcal{A}_J$ is a spectral subalgebra of $\overline{\mathcal{A}}$ by Proposition 2.5.15. The Jacobson radical $\overline{\mathcal{A}}_J$ of $\overline{\mathcal{A}}$ has intersection $\{0\}$ with $\mathcal{A}/\mathcal{A}_J \subseteq \overline{\mathcal{A}}$, since the intersection is an ideal of the semisimple algebra $\mathcal{A}/\mathcal{A}_J$ and in it each element has spectral radius zero. Note that the quotient spectral norm $\| \cdot \|$ on $\overline{\mathcal{A}}/\overline{\mathcal{A}}_J$ when restricted to the isomorphic image of $\mathcal{A}/\mathcal{A}_J$ induces a semi-norm σ on $\mathcal{A}/\mathcal{A}_J$ which is spectral, since we have:

$$\begin{aligned}
\rho_{\mathcal{A}/\mathcal{A}_J}(a + \mathcal{A}_J) &= \rho_{\overline{\mathcal{A}}}(a + \mathcal{A}_J) = \rho_{\overline{\mathcal{A}}/\overline{\mathcal{A}}_J}((a + \mathcal{A}_J) + \overline{\mathcal{A}}_J) \\
&\leq \|(a + \mathcal{A}_J) + \overline{\mathcal{A}}_J)\| = \sigma(a + \mathcal{A}_J)
\end{aligned}$$

where we have used Theorem 2.3.3 and Proposition 2.5.15. Therefore $\overline{\mathcal{A}}/\overline{\mathcal{A}}_J$ can be identified with the completion of $(\mathcal{A}/\mathcal{A}_J, \| \cdot \|)$ and $\mathcal{A}/\mathcal{A}_J$ is a dense subalgebra. Hence Proposition 2.5.15 shows that $\mathcal{A}/\mathcal{A}_J$ is a spectral subalgebra of $\overline{\mathcal{A}}/\overline{\mathcal{A}}_J$.

(c) $\Rightarrow$ (b): Immediate.

(b) $\Rightarrow$ (a): Corollary 2.5.10 shows that $\mathcal{A}/\mathcal{A}_J$ is spectral and then Proposition 2.4.4 shows that $\mathcal{A}$ itself is spectral. $\square$

2.5.19 Not Every Spectral Algebra is a Subalgebra of a Banach Algebra   Since any Jacobson-radical algebra is spectral by Corollary 2.3.4, the example of H. Garth Dales [1981b] provides the desired example. See also J. Jacobus Grobler and Heinrich Raubenheimer [1991] and Lenore Groenewald and Raubenheimer [1988].

2.5.20 A Simple Unital Banach Algebra with a Non-spectral Norm   We give a Banach algebra $\mathcal{A}$ (due to Schweitzer [1993]) which is simple (*i.e.*, it has no non-zero proper ideals), unital and a dense non-spectral subalgebra of a C*-algebra $\mathcal{B}$ called the *irrational rotation algebra*. Thus the C*-norm on $\mathcal{A}$ is a non-spectral norm.

For any function $F : \mathbb{Z} \times \mathbb{T} \to \mathbb{C}$ and any $n \in \mathbb{Z}$, define $F_n : \mathbb{T} \to \mathbb{C}$ by $F_n(\zeta) = F(n, \zeta)$ for all $\zeta \in \mathbb{T}$. Consider the Banach space $\mathcal{A}$ of all functions

$F : \mathbb{Z} \times \mathbb{T} \to \mathbb{C}$ such that $F_n \in C(\mathbb{T})$ for each $n \in \mathbb{Z}$ and

$$||F|| = \sum_{n \in \mathbb{Z}} e^{|n|} ||F_n||_\infty < \infty \qquad \forall \, F \in \mathcal{A}.$$

This is a Banach algebra under the following twisted convolution product where α is an irrational multiple of π

$$F * G(n, \zeta) = \sum_{m \in \mathbb{Z}} F(m, \zeta) G(n - m, e^{i\alpha m} \zeta) \qquad \forall \, F, G \in \mathcal{A}; \; n \in \mathbb{Z}; \; \zeta \in \mathbb{T}.$$

A Banach algebra defined in this way is called the *crossed product* of $\mathbb{Z}$ with $C(\mathbb{T})$ induced by the automorphism $\theta(f)(\zeta) = f(e^{i\alpha}\zeta)$ of $C(\mathbb{T})$. Then $E(n, \zeta) = \delta_{n0}$ (Kronecker delta) is an identity element for $\mathcal{A}$. Consider $C(\mathbb{T})$ as the subalgebra of all functions $F \in \mathcal{A}$ with $F_n = 0$ for $n \neq 0$. Furthermore, $\mathcal{A}$ is a Banach *-algebra under the involution

$$F^*(n, \zeta) = F(-n, e^{i\alpha n}\zeta)^* \qquad F \in \mathcal{A}; \; n \in \mathbb{Z}; \; \zeta \in \mathbb{T}.$$

Theorem 10.2.8 shows that any Banach *-algebra has a largest C*-semi-norm. It is a norm in this case. Let $\mathcal{B}$ be the completion of $\mathcal{A}$ in this norm.

Let $P : \mathcal{A} \to C(\mathbb{T}) \subseteq \mathcal{A}$ be the contractive linear projection defined by $P(F) = F_0$. If $\mathcal{I}$ were a proper ideal in $\mathcal{A}$, its intersection with $C(\mathbb{T})$ would have to be $\{0\}$. Schweitzer shows that this implies $P(\mathcal{I}) = \{0\}$ and then $\mathcal{I} = \{0\}$, so $\mathcal{A}$ is simple.

Consider the following homomorphism T of $\mathcal{A}$ into $\mathcal{B}(C(\mathbb{T}))$:

$$T_F(g)(\zeta) = \sum_{n \in \mathbb{Z}} F(n, \zeta) e^n g(e^{in}\zeta) \qquad \forall \, F \in \mathcal{A}; \; g \in C(\mathbb{T}); \; \zeta \in \mathbb{T}.$$

Theorem 4.2.10 below and the theory of *-representations of crossed product C*-algebras show that $\mathcal{A}$ is not a spectral subalgebra of $\mathcal{B}$ since this irreducible representation is not the restriction of a *-representation of $\mathcal{B}$.

2.6 Numerical Range in Banach Algebras

In this section we will call a normed algebra *norm-unital* if it contains an identity element which has norm one. We will deal almost exclusively with norm-unital Banach algebras although some of the results also hold for incomplete norm-unital normed algebras. The numerical range of an element in such an algebra is a nonempty, compact, convex subset of the complex plane which contains the spectrum of the element. The numerical range is chiefly useful because it is better behaved than the spectrum under addition and limits. The elements of the algebra which have numerical range included in the real line also have important special properties.

The numerical range of an element depends on the *precise* value of the norm on the two-dimensional space spanned by the element and the identity element. Therefore, it is not preserved under homeomorphic isomorphism of the algebra, but only under isometric isomorphism. On the other hand, the numerical range is preserved under unital isometric linear maps which are not homomorphisms. The geometric character of the numerical range has advantages and disadvantages. The properties of Banach algebras in which we are usually most interested are either purely algebraic (preserved under isomorphism) or topologically algebraic (preserved under homeomorphic isomorphism). The numerical range cannot reflect these two types of properties precisely. On the other hand, one might sometimes be able to choose a norm from among several equivalent ones, in order to apply numerical range considerations to best effect.

The excellent expository treatment by Frank F. Bonsall and John Duncan [1971] and [1973a] allows us to limit this section to those topics actually used in subsequent chapters. Thus we omit entirely a discussion of the numerical range of operators on normed linear spaces and of elements in normed algebras which are not norm-unital, despite the historical importance of these topics. See also Vasile I. Istrăţescu [1981] and its review.

A numerical range was first defined by Otto Toeplitz in [1918] for operators on Hilbert space. We define this concept and derive some useful results about it in Section 9.1 of Volume II. The definition we treat in this section was first given explicitly in full generality by Bonsall [1969a]. However, many of the results and methods trace back to a paper by Gunter Lumer [1961], which was in turn based on work by Ivan Vidav [1956] and H. Frederic Bohnenblust and Samuel Karlin [1955]. Other early developments related to numerical range were given by Friedrich L. Bauer [1962] and Palmer [1965], [1968a], [1968b] and [1970]. (The letter W in our notation is from the German name *Wertbereich* for the numerical range.)

Definitions and Basic Properties

2.6.1 Definition   Let $\mathcal{A}$ be a norm-unital normed algebra. The set $\{\omega \in \mathcal{A}^* : \omega(1) = \|\omega\| = 1\}$ is denoted by $\mathcal{A}_W^*$. For any element $a \in \mathcal{A}$ the *numerical range* $W(a)$ of a is the following set:

$$W(a) = \{\omega(a) : \omega \in \mathcal{A}_W^*\}. \tag{1}$$

The number

$$\|a\|_W = \sup\{|\lambda| : \lambda \in W(a)\} \tag{2}$$

is called the *numerical radius* of a.

The Hahn–Banach theorem shows that $\mathcal{A}_W^*$ is never empty. Clearly it is convex and compact in the $\mathcal{A}$-topology (*i.e.*, the weak* topology) on $\mathcal{A}^*$. If $\overline{\mathcal{A}}$ is the completion of $\mathcal{A}$, then $\overline{\mathcal{A}}_W^*$ and $\mathcal{A}_W^*$ can be naturally identified.

Hence for any $a \in \mathcal{A}$ the numerical range is the same whether calculated in $\mathcal{A}$ or in $\overline{\mathcal{A}}$. Thus Banach algebra techniques can be used to obtain results on the numerical range of elements in a norm-unital normed algebra.

In the next proposition we formally record some more observations which are basic to all that follows. The last result is essentially due to James P. Williams [1967].

2.6.2 Proposition *Let $\mathcal{A}$ be a norm-unital normed algebra.*

(a) *The numerical range of any element of $\mathcal{A}$ is a nonempty, compact, convex subset of $\mathbb{C}$.*

(b) $W(a + b) \subseteq W(a) + W(b) \qquad \forall\, a, b \in \mathcal{A}.$

(c) $W(1) = \{1\}.$

(d) $W(\lambda a) = \lambda W(a) \qquad \forall\, \lambda \in \mathbb{C};\ a \in \mathcal{A}.$

(e) *If $\{a_n\}_{n \in \mathbb{N}} \subseteq \mathcal{A}$ converges to $a \in \mathcal{A}$, then*

$$W(a) \subseteq \{\lim \lambda_n : \lambda_n \in W(a_n)\} \subseteq \cap_{n=1}^{\infty} \overline{\cup_{k=n}^{\infty} W(a_n)}.$$

(f) *If $\mathcal{A}$ is complete (i.e., a Banach algebra), then*

$$Sp(a) \subseteq W(a) \subseteq \|a\| \mathbb{C}_1.$$

Proof The map $\omega \mapsto \omega(a)$ is continuous and affine from the nonempty, compact, convex set $\mathcal{A}_W^*$ with the $\mathcal{A}$-topology. Since $W(a)$ is the range of this map, $W(a)$ is a nonempty, compact, convex subset of $\mathbb{C}$. Conditions (b), (c), (d), and (e) are obvious from the definition of the numerical range. In condition (e) we can even take convergence to mean weak convergence.

(f) By Corollary 2.2.8, the norm of $\mathcal{A}$ is spectral. Suppose λ belongs to $Sp(a)$. Then $\lambda - a$ lacks either a left or a right inverse, so that either $\mathcal{A}(\lambda - a)$ or $(\lambda - a)\mathcal{A}$ (or both) is a proper one-sided ideal of $\mathcal{A}$. Let S denote one of these which is a proper one-sided ideal. Each element of S is singular. Hence Proposition 2.2.9(b) (with $a = 1$) shows $\|1 - b\| \geq 1$ for all $b \in S$. Thus the Hahn–Banach theorem shows that we may choose an $\omega \in \mathcal{A}^*$ such that $\omega(1) = 1 = \|\omega\|$ (i.e., $\omega \in \mathcal{A}_W^*$) but $\omega(b) = 0$ for all $b \in S$. Therefore $\omega(\lambda - a) = \lambda - \omega(a) = 0$ implies λ belongs to $W(a)$. The last inclusion follows from $|\omega(a)| \leq \|\omega\|\,\|a\| = \|a\|$ for all $\omega \in \mathcal{A}_W^*$. □

The next lemma is essentially due to Bohnenblust and Karlin [1955]. Note that the first three expressions in equation (3) are equal even if $\mathcal{A}$ is just a normed algebra. This follows from our previous remark on the numerical range in the completion of $\mathcal{A}$. The last two expressions in equation (3) are not defined in general unless $\mathcal{A}$ is complete. The exponential function in a Banach algebra was introduced in Theorem 2.1.12

2.6.3 Lemma *Any element a of a norm-unital Banach algebra $\mathcal{A}$ satisfies the following equations.*

$$\sup\{\operatorname{Im}(\lambda) : \lambda \in W(a)\} = \inf\{t^{-1}(\|1 - ita\| - 1) : t > 0\}$$
$$= \lim_{t \to 0^+} t^{-1}(\|1 - ita\| - 1) \tag{3}$$
$$= \lim_{t \to 0^+} t^{-1} \log \|e^{-ita}\|$$
$$= \sup\{t^{-1} \log \|e^{-ita}\| : t > 0\};$$

$$\|a\|_W = \sup\{|\lambda|^{-1} \log \|e^{\lambda a}\| : \lambda \in \mathbb{C}, \lambda \neq 0\}. \tag{4}$$

Proof If a is 0 there is nothing to prove so we assume a is not 0. Let s be $\sup\{\operatorname{Im}(\lambda) : \lambda \in W(a)\}$. The last proposition shows that this is finite and satisfies $|s| \leq \|a\|$. For any $t > 0$ and any $\omega \in \mathcal{A}_W^*$, we have $-i\omega(a) = t^{-1}(\omega(1 - ita) - 1)$. Hence we see $\operatorname{Im}(\omega(a)) = \operatorname{Re}(-i\omega(a)) \leq t^{-1}(\|1 - ita\| - 1)$ so

$$s \leq \inf\{t^{-1}(\|1 - ita\| - 1) : t > 0\}. \tag{5}$$

For any $t \in \mathbb{R}$ with $0 < t < \|a\|^{-1}$, the Hahn–Banach theorem gives an $\omega \in \mathcal{A}^*$ satisfying $\omega(1-ita) = \|\omega\| \|1-ita\| = 1$. The map $b \mapsto \omega(b(1-ita))$ from $\mathcal{A}$ into $\mathbb{C}$ clearly belongs to $\mathcal{A}_W^*$. Using this map we get

$$1 + t^2\|a\|^2 \geq \|1 + t^2 a^2\| = \|(1 + ita)(1 - ita)\|$$
$$\geq \|1 - ita\|\operatorname{Re}(\omega((1 + ita)(1 - ita)))$$
$$\geq \|1 - ita\|(1 - ts) \geq (1 - t\|a\|)^2 > 0.$$

Rearranging the inequality $1 + t^2\|a\|^2 \geq \|1 - ita\|(1 - ts)$ we get

$$\frac{\|1 - ita\| - 1}{t} \leq \frac{s + t\|a\|^2}{1 - ts}$$

for all $0 < t < \|a\|^{-1}$. Hence, using (5), we conclude

$$s \leq \inf\{t^{-1}(\|1 - ita\| - 1) : t > 0\} \leq \limsup_{t \to 0^+} t^{-1}(\|1 - ita\| - 1) \leq s.$$

This confirms the first two equalities of (3).

It is easy to see that there is a constant B satisfying

$$|\|e^{-ita}\| - \|1 - ita\|| \leq Bt^2 \quad \text{for all } |t| < \|a\|^{-1}.$$

Since the second limit is already known to exist, this gives

$$\lim_{t \to 0^+} t^{-1}(\|e^{-ita}\| - 1) = \lim_{t \to 0^+} t^{-1}(\|1 - ita\| - 1).$$

Similarly, we can easily find a constant C satisfying $|\|e^{-ita}\| - 1| \leq Ct$ for all $|t| < \|a\|^{-1}$. Expanding the logarithm function around 1 gives

$$\lim_{t \to 0^+} t^{-1} \log \|e^{-ita}\| = \lim_{t \to 0^+} t^{-1}(\|e^{-ita}\| - 1).$$

These two equations confirm the third equality of (3).

To obtain the last equality of (3), we note that the function $f(t) = \log \|e^{-ita}\|$ is subadditive for $t > 0$. Hence for $t > 0$ and $n \in \mathbb{N}$, we have $f(t) \leq nf(t/n)$. This implies $t^{-1}f(t) \leq (t/n)^{-1}f(t/n)$. Thus, given any value of $t^{-1}f(t)$, there are arbitrarily small values of $s > 0$ such that $s^{-1}f(s) \geq t^{-1}f(t)$. Since the limit is known to be finite and not greater than the supremum, we conclude $\lim_{t\to 0+} t^{-1}f(t) = \sup\{t^{-1}f(t) : t > 0\}$.

Equation (4) can be derived as follows:

$$\|a\|_W = \sup\{|\lambda| : \lambda \in W(a)\} = \sup\{\mathrm{Im}(\lambda) : \lambda \in W(\zeta a); \zeta \in \mathbb{T}\}$$
$$= \sup\{t^{-1}\log\|e^{-i\zeta ta}\| : t > 0, \zeta \in \mathbb{T}\}. \quad \Box$$

This lemma can be used to justify the remark, in the introduction to this section, that $W(a)$ depends only on the norm on the two-dimensional subspace generated by 1 and a. For any $\lambda \in \mathbb{C}$ and $\zeta \in \mathbb{T}$, Proposition 2.6.2 shows $W(\lambda + \zeta a) = \lambda + \zeta W(a)$. Thus, we can freely translate and rotate $W(a)$ in the complex plane. Therefore, the second or third expression for $\sup\{\mathrm{Im}(\lambda) : \lambda \in W(a)\}$ gives a complete description of the support lines of $W(a)$, and thus a complete description of the closed convex set $W(a)$.

The next theorem is also essentially due to Bohnenblust and Karlin [1955].

2.6.4 Theorem *Let $\mathcal{A}$ be a norm-unital Banach algebra. The numerical radius is a (linear space) norm on $\mathcal{A}$ which is equivalent to the original norm of $\mathcal{A}$ and satisfies*

$$\rho(a) \leq \|a\|_W \leq \|a\| \leq e\|a\|_W. \tag{6}$$

Proof Proposition 2.6.2(f) shows that the numerical radius is a semi-norm satisfying $\rho(a) \leq \|a\|_W \leq \|a\|$.

In order to obtain the final inequality, we let ζ_n be a primitive n^{th} root of unity (e.g., $\zeta_n = \exp(2\pi i/n)$). Then $\sum_{k=1}^{n} \zeta_n^{kj} = 0$ for any integer j which is not divisible by n. Let r be a positive number. Then we have

$$a = \frac{1}{n} \sum_{k=1}^{n} \frac{r}{\zeta_n^k} \sum_{j=0}^{n} \frac{[(\frac{\zeta_n^k}{r})a]^j}{j!} \tag{7}$$

However, by choosing n sufficiently large we can make the right side as close as we please to $n^{-1} \sum_{k=1}^{n} (\zeta_n^k/r)^{-1} e^{(\zeta_n^k/r)a}$ (e. g., to make the difference less than $\varepsilon > 0$, choose n so that $\sum_{j=n+1}^{\infty} (j!)^{-1}(\|a\|/r)^n < \varepsilon/r$). Using equations (7) and (4), we conclude $\|a\| \leq re^{r^{-1}\|a\|_W}$ for all $r > 0$. Hence $\|a\|_W = 0$ implies $\|a\| = 0$. If $\|a\|_W$ is non-zero, we choose $r = \|a\|_W$ to obtain the desired inequality. $\quad \Box$

We are now in a position to give an easy proof to the final statement of Proposition 1.1.13 which we left unproved before. The reader is referred to the original statement for notation. We use the numerical range and numerical radius for $(\mathcal{A}^1,\ \|\cdot\|_R)$. Since 0 is in the closure of the numerical range of any $a \in A$, we get the elementary estimate $\|\lambda+a\|_R \geq \|\lambda+a\|_W \geq \max\{|\lambda|,\ \|a\|_W/2\}$ which gives

$$\frac{\|\lambda + a\|_1}{\|\lambda + a\|_R} \leq \frac{|\lambda| + e\|a\|_W}{\max\{|\lambda|,\ \|a\|_W/2\}} \leq 1 + 2e \qquad \forall\ \lambda + a \in \mathcal{A}^1.$$

This completes the proof of Proposition 1.1.13.

Symmetric Elements

We turn now to the theory of those elements in a norm-unital Banach algebra which have real numerical range. Here the terminology is not quite settled. Vidav [1956] called these elements self adjoint. Lumer [1961] and Earl Berkson [1963] called them hermitian. Palmer called them symmetric (and with another definition, self-conjugate) [1965; 1968b]. The term "hermitian" was adopted in the exposition by Bonsall and Duncan [1971], and therefore this term should probably be considered standard. However, in this work we use the word "hermitian", in a different sense. Therefore, we have retained the term "symmetric" in the next definition.

2.6.5 Definition An element h in a norm-unital Banach algebra is called *symmetric* if its numerical range is included in the real axis. The set of symmetric elements in a norm-unital Banach algebra $\mathcal{A}$ is denoted by $\mathcal{A}_S$.

Note that $\mathcal{A}_S$ is a real linear space which is closed in the weak topology and hence in the norm topology of $\mathcal{A}$.

In order to keep the proof of Theorem 2.6.7 elementary we include a lemma with an elementary proof. The result could also be derived from a (non-elementary) application of the functional calculus (Theorem 3.3.7). This lemma is based on one first given by Bonsall and Michael J. Crabb [1970].

In the following lemma we need to know that the power series for $\arcsin(a)$ converges unconditionally when a belongs to a norm-unital Banach algebra $\mathcal{A}$ and $\sup\{\|a^{2n+1}\| : n \in \mathbb{N}\}$ is finite. The easiest way to show this here is by comparison to the binomial series for $(1 + x)^{1/2}$, which we show to converge absolutely for $|x| \leq 1$ in the discussion preceding Theorem 3.4.5. We have

$$\arcsin(x) = \sum_{n=0}^{\infty} \frac{(2n)!x^{2n+1}}{2^{2n}(n!)^2(2n+1)}.$$

Denoting the nth coefficient by r_n, we see $r_n = \frac{2n+2}{2n+1}\left|\binom{1/2}{n+1}\right|$. Thus the power series $\sum_{n=0}^{\infty} r_n x^{2n+1}$ converges absolutely and uniformly for $|x| \leq 1$ so that its sum is continuous, which implies $\sum_{n=0}^{\infty} r_n = \arcsin(1) = \frac{\pi}{2}$.

Hence if $a \in \mathcal{A}$ satisfies $\|a^{2n+1}\| \leq B$ for all $n \in \mathbb{N}$, then the series satisfies

$$\sum_{n=0}^{\infty} \|r_n a^{2n+1}\| \leq \sum_{n=0}^{\infty} \frac{2n+2}{2n+1}\left|\binom{1/2}{n+1}\right|B \leq 2B\sum_{n=1}^{\infty}\left|\binom{1/2}{n}\right| = 2B,$$

where the last equality depends on equation (4) before Theorem 3.4.5.

We use an elementary portion of Gelfand theory in the following proof, even though that theory is not formally introduced until the next chapter.

2.6.6 Lemma *Let $\mathcal{A}$ be a complex unital Banach algebra. Let h be an element of $\mathcal{A}$ with spectrum contained in the open interval $]-\frac{\pi}{2}, \frac{\pi}{2}[$ of the real axis. Then we get $h = \arcsin(\sin(h))$, where the arcsine and sine functions are defined by their (convergent) power series expansions.*

Proof We may assume that $\mathcal{A}$ is commutative by replacing it with a maximal commutative subalgebra if necessary (cf. Proposition 2.5.3(a)). Let Γ be the Gelfand space of $\mathcal{A}$. Let s and t be real numbers with $|s|\rho(h) < \frac{\pi}{2}$. Thus we have

$$\begin{aligned}
\rho(\sin((s+it)h)) &= \sup_{\gamma\in\Gamma}\{|\gamma(\sin((s+it)h))|\} \\
&= \sup_{\gamma\in\Gamma}\{|\sin((s+it)\gamma(h))|\} \\
&= \sup_{\gamma\in\Gamma}\{|\sin(s\gamma(h))\cosh(t\gamma(h)) \\
&\qquad\qquad + i\cos(s\gamma(h))\sinh(t\gamma(h))|\} \\
&\leq |\sin(s\rho(h))|\cosh(t\rho(h)) + |\sinh(t\rho(h))|.
\end{aligned}$$

Therefore there is an open set U containing $[-1,1]$ consisting of $\lambda = s + it$ in the complex plane for which $\rho(\sin(\lambda h)) < 1$. As we have seen, the power series $\sum_{n=0}^{\infty} r_n h^{2n+1}$ for $\arcsin(h)$ converges unconditionally for any $h \in \mathcal{A}$ with $\{\|h^{2n+1}\| : n \in \mathbb{N}\}$ bounded, and hence for any $h \in \mathcal{A}$ with $\rho(h) < 1$. Thus $f(\lambda) = \arcsin(\sin(\lambda h))$ is defined by a convergent double series for $\lambda \in U$, so $f(\lambda)$ is the composition of two analytic functions and hence is analytic for $\lambda \in U$. However, if $|\lambda|$ is sufficiently small, $\sum_{n=0}^{\infty} \frac{\|\lambda h\|^{2n+1}}{(2n+1)!}$ is less than 1 so the double series for $f(\lambda)$ converges unconditionally and can be rearranged to give $f(\lambda) = \lambda h$. Thus by the uniqueness of power series expansions, $f(\lambda) = \lambda h$ for all $\lambda \in U$. Since $1 \in U$, the lemma is proved. $\square$

The equivalence of (a), (b) and (c) in the next theorem is essentially due to Bohnenblust and Karlin [1955] but was first explicitly given recognition by Lumer [1961]. Results (f), (g) and (h) were discovered by Vidav [1956].

Condition (e) has a more complicated history. The first equality was proved by Vidav [1956]. At the same time he derived the full equation in a very special case. Allan M. Sinclair [1971] and Andrew Browder [1971] each obtained the full result independently. The present proof is based on a proof by Bonsall and Crabb [1970]. Condition (d) is merely a restatement of some of the other results.

2.6.7　Theorem *Let A be a norm-unital Banach algebra. The following three conditions are equivalent for an element $h \in A$:*

(a) $h \in A_S$.
(b) $\|e^{ith}\| = 1 \quad \forall\, t \in \mathbb{R}$.
(c) $\lim\limits_{t \to 0} |t|^{-1}(\|1 - ith\| - 1) = 0$.

Furthermore A_S and any h, $k \in A_S$ satisfy:

(d) $co(Sp(h)) = W(h) \subseteq \mathbb{R}$.
(e) $\rho(h) = \|h\|_W = \|h\|$.
(f) $A_S \cap iA_S = \{0\}$ *and* $A_S + iA_S$ *is closed.*
(g) $i(hk - kh) \in A_S$.
(h) $\|h - ik\| \le 2\|h + ik\|$.

Proof The equivalence of (a), (b) and (c) is an immediate consequence of Lemma 2.6.3. For the rest of this proof let h and k be elements of A_S. Theorem 2.6.4 shows that (e) follows from the inequality $\|h\| \le \rho(h)$ which we will now establish. Let $t \in \mathbb{R}_+$ satisfy $\rho(t^{-1}h) < \frac{\pi}{2}$, so Proposition 2.6.2(f) implies $Sp(t^{-1}h) \subseteq (-\pi/2, \pi/2)$. Since $t^{-1}h$ is symmetric, we also get $\|\sin(t^{-1}h)\| = 2^{-1}\|e^{it^{-1}h} - e^{-it^{-1}h}\| \le 1$. Lemma 2.6.6 then shows $t^{-1}\|h\| = \|\arcsin(\sin(t^{-1}h))\| \le \sum_{n=0}^{\infty} r_n \|\sin(t^{-1}h)\| \le \frac{\pi}{2}$, where $\{r_n\}$ is the sequence of coefficients which was defined before Lemma 2.6.6. We have shown that $\rho(h) < t\frac{\pi}{2}$ implies $\|h\| \le t\frac{\pi}{2}$. Hence we conclude $\|h\| \le \rho(h)$, so that (e) is proved.

(d): Since $W(h)$ is a nonempty compact convex subset of $\mathbb{R}$, it has the form $[s, t]$ for suitable s, $t \in \mathbb{R}$. Thus (e) shows $\rho(h - s) = \|h - s\|_W = t - s = \|h - t\|_W = \rho(h - t)$. This shows that both s and t belong to the spectrum. Since $Sp(h) \subseteq W(h)$ was already proved in Proposition 2.6.2, we conclude $co(Sp(h)) = W(h) \subseteq \mathbb{R}$.

(f): The first result follows from (d) and (e). We must show that $A_S + iA_S$ is closed. Suppose $\{h_n + ik_n\}_{n \in \mathbb{N}}$ converges to $a \in A$ with h_n, $k_n \in A_S$. For any n, $m \in \mathbb{N}$, Theorem 2.6.4 and (e) show

$$
\begin{aligned}
\|h_n - h_m\| &\le \|h_n - h_m\|_W = \sup\{|\omega(h_n - h_m)| : \omega \in A_W^*\} \\
&\le \sup\{|\omega((h_n + ik_n) - (h_m + ik_m))| : \omega \in A_W^*\} \\
&= \|(h_n + ik_n) - (h_m + ik_m)\|_W \\
&\le \|(h_n + ik_n) - (h_m + ik_m)\|.
\end{aligned}
$$

Hence $\{h_n\}_{n \in \mathbb{N}}$ is a Cauchy sequence in the real Banach space A_S. Let its limit be $h \in A_S$. Similarly $\{k_n\}_{n \in \mathbb{N}}$ has a limit $k \in A_S$. Thus $a = h + ik$ belongs to $A_S + iA_S$, which is therefore closed.

(g): Condition (b) implies $\|e^{isk}e^{ish}e^{-isk}e^{-ish}\| = 1$ for all $s \in \mathbb{R}$. By expanding the exponentials, we see that there is a constant B satisfying

$$|\|1 + s^2(hk - kh)\| - 1| \leq Bs^3 \qquad \forall\, s;\ 0 \leq s \leq 1.$$

Letting t be s^2 we see

$$\lim_{t \to 0^+} t^{-1}(\|1 - it(i(hk - kh))\| - 1) = 0.$$

Lemma 2.6.3 shows that no number in the numerical range of $i(hk - kh)$ has positive imaginary part. By interchanging h and k we also exclude numbers with negative imaginary part. Hence $i(hk - kh)$ belongs to $\mathcal{A}_S$.

(h): We have

$$\begin{aligned}
\|h\| = \|h\|_W &= \sup\{|\omega(h)| : \omega \in \mathcal{A}_W^*\} \\
&\leq \sup\{(\omega(h)^2 + \omega(k)^2)^{1/2} : \omega \in \mathcal{A}_W^*\} \\
&= \sup\{|\omega(h + ik)| : \omega \in \mathcal{A}_W^*\} \\
&= \|h + ik\|_W \leq \|h + ik\|.
\end{aligned}$$

Similarly we get $\|k\| \leq \|h + ik\|$, which implies $\|h - ik\| \leq \|h\| + \|k\| \leq 2\|h + ik\|$. $\square$

Symmetric Elements in Algebras

The following two propositions will be used in Theorem 9.5.8 of Volume II to characterize C*-algebras (cf. §1.7.17). The equivalence of (a) and (b) is proved by an argument in Palmer [1968a], but was first explicitly stated by Ellen Torrance [1970]. The involution in part (c) was first studied by Vidav [1956]. When $\mathcal{A}_V$ is an algebra, it is a C*-algebra under the involution $*: \mathcal{A}_V \to \mathcal{A}_V$.

2.6.8 Theorem *Let A be a norm-unital Banach algebra. Denote the set $\mathcal{A}_S + i\mathcal{A}_S$ by $\mathcal{A}_V$. Then the following are equivalent:*

(a) *$\mathcal{A}_V$ is an algebra.*

(b) *$h \in \mathcal{A}_S$ implies $h^2 \in \mathcal{A}_V$.*

(c) *$\mathcal{A}_V$ is a closed subalgebra in which each element has a unique expression of the form $h + ik$ for $h, k \in \mathcal{A}_S$. Furthermore the map $*: \mathcal{A}_V \to \mathcal{A}_V$ defined by $(h + ik)^* = h - ik$ for all $h, k \in \mathcal{A}_S$ is a real-linear homeomorphism with norm at most 2 (and an isometry with respect to the numerical radius) which satisfies: $(a + b)^* = a^* + b^*$, $(\lambda a)^* = \lambda^* a^*$, $(ab)^* = b^* a^*$ and $(a^*)^* = a$ for all $a, b \in \mathcal{A}_V$ and $\lambda \in \mathbb{C}$.*

Proof (c) $\Rightarrow$ (a) $\Rightarrow$ (b): Obvious.

(b) $\Rightarrow$ (c): Theorem 2.6.7(f) shows that $\mathcal{A}_V$ is a closed linear subspace. Let h^2 equal $k + ij$ where h, k and j belong to $\mathcal{A}_S$. The identity $h(k + ij) =$

$(k+ij)h$ implies $hk - kh = i(jh - hj)$. Theorem 2.6.7(g) shows that $hk - kh$ belongs to both $\mathcal{A}_S$ and $i\mathcal{A}_S$. Theorem 2.6.7(f) shows $hk - kh = 0$. Thus h commutes with k and hence with j. Proposition 3.1.6 and Theorems 2.1.10 and 2.6.7(d) give $iSp(j) \subseteq Sp(h^2) - Sp(k) \subseteq Sp(h)^2 - Sp(k) \subseteq \mathbb{R}$ and $Sp(j) \subseteq \mathbb{R}$, so j is 0 by Theorem 2.6.7(f). Therefore h^2 belongs to $\mathcal{A}_S$ for all $h \in \mathcal{A}_S$.

Let h and k be arbitrary elements of $\mathcal{A}_S$; then $h + k$ and therefore $hk + kh = (h + k)^2 - h^2 - k^2$ belong to $\mathcal{A}_S$. This, in addition to Theorem 2.6.7(g), gives $2hk = hk + kh + i[i(kh - hk)] \in \mathcal{A}_S + i\mathcal{A}_S = \mathcal{A}_V$. Since $\mathcal{A}_V$ is the linear span of $\mathcal{A}_S$, it is an algebra.

Theorem 2.6.4 shows $\mathcal{A}_S \cap i\mathcal{A}_S = \{0\}$ so that the expression $h + ik$ with $h, k \in \mathcal{A}_S$ is unique. Thus, as a real-linear space $\mathcal{A}_V$ has a direct sum decomposition $\mathcal{A}_V = \mathcal{A}_S \oplus i\mathcal{A}_S$. Hence the map $*: \mathcal{A}_V \to \mathcal{A}_V$ is well-defined. It is obviously real-linear and it is a homeomorphism (with norm at most 2) by Theorem 2.6.7(h) and an isometry in the numerical radius by the definition. Clearly it satisfies the first, second, and third equations asserted in this theorem. To see that it satisfies the fourth equation, it is enough to consider $h, k \in \mathcal{A}_S$ because of the conjugate linearity of $*$. However, the results we have already obtained show $2(hk)^* = hk + kh - i[i(hk - kh)]^* = hk + kh + i[i(hk - kh)] = 2kh = 2k^*h^*$ so this fourth equation is also satisfied.　　　□

The next proposition is an adaptation to the present setting of arguments used by Kaplansky (cf. Joseph A. Schatz [1953]). We will draw the natural conclusion from this proposition in Theorem 9.5.8 of Volume II, after introducing several additional concepts.

2.6.9 Proposition *Let $\mathcal{A}$ be a norm-unital Banach algebra satisfying the equivalent conditions of* Theorem 2.6.8. *Define $\mathcal{A}_V$ and $*: \mathcal{A}_V \to \mathcal{A}_V$ as in that theorem and let $\mathcal{A}_Q$ denote $\{h \in \mathcal{A} : W(h) \subseteq \mathbb{R}_+\}$.*

(a) *For each $h \in \mathcal{A}_S$ there are unique elements $p, q \in \mathcal{A}_Q \cap \{h\}''$ satisfying $h = p - q$ and $pq = qp = 0$. Hence we have*

$$\mathcal{A}_S = \mathcal{A}_Q - \mathcal{A}_Q \quad \text{and} \quad \mathcal{A}_Q \cap (-\mathcal{A}_Q) = \{0\}.$$

(b) *For each $a \in \mathcal{A}_V, a^*a$ belongs to $\mathcal{A}_Q$ and satisfies $\|a^*a\| \geq \|a\|_W^2$.*

Proof (a): It is enough to consider $h \in \mathcal{A}_S$ satisfying $\|h\| \leq 1$. Let h be such an element and let $\mathcal{C}$ denote $\{h\}''$. Then $Sp_\mathcal{A}(h^2) = Sp_\mathcal{C}(h^2) \subseteq [0, 1]$ and Theorem 2.6.7(e) show $W(h^2) \subseteq [0, 1]$ and $\|h^2 - 1\| \leq 1$, since h^2 is symmetric. Hence $k = 1 + \sum_{n=1}^{\infty} \binom{1/2}{n}(h^2 - 1)^n$ belongs to $\mathcal{A}_S \cap \mathcal{C}$ and satisfies $W(k) \subseteq [0, 1]$ and $k^2 = h^2$, since the series converges unconditionally (cf. the discussion before Theorem 3.4.5). Define p and q by $p = 2^{-1}(k + h)$ and $q = 2^{-1}(k - h)$. (Again we use a simple argument based on the Gelfand theory introduced in Section 3.1.) Each $\gamma \in \Gamma_C$ satisfies

$$
\begin{aligned}
\gamma(p) &= 2^{-1}[\sum_{n=0}^{\infty} \binom{1/2}{n}(\gamma(h^2-1)^n) + \gamma(h)] \\
&= 2^{-1}((\gamma(h^2))^{1/2} + \gamma(h)) = 2^{-1}(|\gamma(h)| + \gamma(h)).
\end{aligned}
$$

Theorems 3.1.5 and 2.6.7(d) imply $W(p) = co(Sp_{\mathcal{A}}(p)) = co(Sp_C(p)) \subseteq \mathbb{R}_+$. Similarly $W(q) \subseteq \mathbb{R}_+$ implies that p and q belong to $\mathcal{A}_Q$. We conclude $4pq = (k+h)(k-h) = k^2 - h^2 = 0 = (k-h)(k+h) = 4qp$. Theorem 2.6.7(d) shows the uniqueness of p and q.

(b): The identities $W(h) = co(Sp(h))$ and $\|h\| = \|h\|_W = \rho(h)$ for a symmetric element h in $\mathcal{A}$ from Theorem 2.6.7, which are used repeatedly in the rest of this proof, will not be cited again explicitly.

Suppose $b \in \mathcal{A}_V$ satisfies $b = h + ik$ with h, $k \in \mathcal{A}_S$ and $-bb^* \in \mathcal{A}_Q$. Theorem 2.1.8(a) shows

$$
W(-b^*b) \subseteq co(Sp(-bb^*) \cup \{0\}) = co(Sp(-b^*b) \cup \{0\}) \subseteq \mathbb{R}_+.
$$

The identity $b^*b + bb^* = 2(h^2 + k^2)$, Theorem 2.1.10 and Proposition 2.6.2 show

$$
\begin{aligned}
W(b^*b) &\subseteq 2W(h^2) + 2W(k^2) + W(-bb^*) \\
&= 2co(Sp(h))^2 + 2co(Sp(k))^2 + W(-bb^*) \subseteq \mathbb{R}_+.
\end{aligned}
$$

Hence $W(b^*b) \subseteq (-\mathbb{R}_+) \cap \mathbb{R}_+$, so b^*b is 0. Similarly bb^* and, in turn, $h^2 + k^2$ are 0. For any $\omega \in \mathcal{A}_W^*$, $\omega(h^2) \in co(Sp(h))^2$ and $\omega(k^2) \in co(Sp(k))^2$ imply $0 \le \omega(h^2) \le \omega(h^2 + k^2) = 0$. Therefore $\rho(h) = \rho(h^2)^{1/2} = 0$ implies that h is 0. Similarly k, and therefore b, are 0.

For any $b \in \mathcal{A}_V$, conclusion (a) shows $b^*b = p - q$ with p, $q \in \mathcal{A}_Q$ and $pq = qp = 0$. Hence $-(bq)^*bq = -q(p-q)q = q^3 \in \mathcal{A}_Q$. We have just shown that this implies $bq = 0$. Hence we get $q^2 = -b^*bq = 0$. As before, this implies $q = 0$. Thus $b^*b = p$ belongs to $\mathcal{A}_Q$.

For $h \in \mathcal{A}_S \cap \mathcal{A}_V$, $t \in \mathbb{R}$ and $\omega \in \mathcal{A}_W^*$, the result just proved implies $t^2 + 2t\omega(h) + \omega(h^2) = \omega((t+h)^2) \ge 0$. Thus the discriminant of this quadratic is nonpositive which implies $\omega(h)^2 \le \omega(h^2)$. We will use this result with an arbitrary $a \in \mathcal{A}_V$ which can be written as $h + ik$ for $h, k \in \mathcal{A}_S \cap \mathcal{A}_V$. Thus we find

$$
\begin{aligned}
\|a^*a\| &= \rho(a^*a) = (\rho(a^*a) + \rho(aa^*))/2 = (\|a^*a\| + \|aa^*\|)/2 \\
&\ge \|a^*a + aa^*\|/2 = \|h^2 + k^2\| \ge \sup\{\omega(h^2 + k^2) : \omega \in \mathcal{A}_W^*\} \\
&= \sup\{\omega(h)^2 + \omega(k)^2 : \omega \in \mathcal{A}_W^*\} = \|a\|_W^2. \qquad \square
\end{aligned}
$$

See also James P. Williams [1967], Crabb and Duncan [1979], Husain and Anitha Srinivasan [1979], Subhash J. Bhatt [1983], Bonsall and Duncan

[1980], Roger R. Smith [1979], [1981], J. Martínez-Moreno, A. Mojtar-Kaidi and Ángel Rodríguez-Palacios [1981], Martínez-Moreno and Rodríguez-Palacios [1985], Thanassis Chryssakis [1986], Crabb and McGregor [1988] and A. K. Gaur and Taqdir Husain [1989].

Power Bounded Elements

We wish to use Crabb's Lemma 2.6.6 to give some results on power bounded elements. Theorem 2.6.12(a) is contained in Gelfand [1941c] but the very simple proof using Crabb's lemma is due to Allan and Ransford [1989]. Part (b) was first proved by Katznelson and Tzafriri [1986], but its proof is based on the above paper and Allan [1989]. For extensions of these results and other proofs, see the above papers Allan, O'Farrell and Ransford [1987], Esterle, Strouse and Zouakia [1990] and Vũ Quốc Phóng [1992].

2.6.10 Definition Let $\mathcal{A}$ be a normed algebra. An element $\langle\, a \in \mathcal{A}\,/\,a \in \mathcal{A}_G\,\rangle$ is said to be $\langle$ *power bounded / doubly power bounded* $\rangle$ if there is some finite constant B satisfying $||a||^n \leq B$ for all $\langle\, n \in \mathbb{N}\,/\,n \in \mathbb{Z}\,\rangle$.

Gelfand's spectral radius formula shows that the spectrum of a $\langle$ power bounded / doubly power bounded $\rangle$ element a in a Banach algebra is included in $\langle\, \mathbb{D}\,/\,\mathbb{T}\,\rangle$. In either case, Proposition 1.1.9 shows that there is an equivalent norm with $B = 1$.

2.6.11 Lemma *Let $\mathcal{A}$ be a unital commutative Banach algebra and let $a \in \mathcal{A}$ be power bounded and satisfy $\rho(a) = 1$. Then there is a Banach algebra $(\mathcal{B}, |||\cdot|||)$ and a unital algebra homomorphism $\varphi\colon \mathcal{A} \to \mathcal{B}$ satisfying:*

(a) $|||\varphi(b)||| = \limsup_{n\to\infty} ||a^n b||$ *for all $b \in \mathcal{A}$.*

(b) $\varphi(a)$ *is invertible in $\mathcal{B}$ and satisfies $|||\varphi(a)||| = |||\varphi(a)^{-1}||| = 1$.*

(c) $Sp_{\mathcal{B}}(\varphi(a)) \subseteq Sp_{\mathcal{A}}(a) \cap \mathbb{T}$.

Proof Proposition 1.1.9 shows that we may assume that $||a^n|| = 1$ for all $n \in \mathbb{N}^0$ without loss of generality. Define an algebra semi-norm σ on $\mathcal{A}$ by $\sigma(b) = \limsup_{n\to\infty} ||a^n b||$ for all $b \in \mathcal{A}$. Note $\sigma(1) = 1$ and $\sigma(ab) = \sigma(b)$ for all $b \in \mathcal{A}$. Let $\tilde{\mathcal{A}}$ be the normed algebra $\mathcal{A}/\mathcal{A}_\sigma$ (where, as usual, $\mathcal{A}_\sigma = \{b \in \mathcal{A} : \sigma(b) = 0\}$) with norm $|||b + \mathcal{A}_\sigma||| = \sigma(b)$ for all $b \in \mathcal{A}$. Denote $b + \mathcal{A}_\sigma \in \tilde{\mathcal{A}}$ by $\tilde{b}$ for all $b \in \mathcal{A}$. Let $\tilde{\mathcal{B}}$ be the algebra of pairs $\{(\tilde{b}, n) : b \in \mathcal{A}, n \in \mathbb{N}^0\}$ with $(\tilde{b}, n)$ interpreted as $\tilde{b}\tilde{a}^{-n}$ so that $\lambda(\tilde{b}, n) = (\lambda\tilde{b}, n)$, $(\tilde{b}, n) + (\tilde{c}, m) = (\tilde{a}^m\tilde{b} + \tilde{a}^n\tilde{c}, n + m)$, $(\tilde{b}, n)(\tilde{c}, m) = (\tilde{b}\tilde{c}, n + m)$ for all $b, c \in \mathcal{A}$, $\lambda \in \mathbb{C}$ and $n, m \in \mathbb{N}^0$. It is easy to see that this is a well defined algebra and that $|||(\tilde{b}, n)|||$, defined by $|||\tilde{b}|||$ for all $b \in \mathcal{A}$ and $n \in \mathbb{N}$, is an algebra norm on it. Let $\mathcal{B}$ be the completion of $\tilde{\mathcal{B}}$ in this norm, and for each $b \in \mathcal{A}$ define $\varphi(b)$ to be $(\tilde{b}, 0) \in \mathcal{B}$. Conditions (a) and (b) and $Sp_{\mathcal{B}}(\varphi(a)) \subseteq Sp_{\mathcal{A}}(a) \cap \mathbb{T}$ are obvious. For any λ with $|\lambda| < 1$, the convergent series $\sum_{n=0}^{\infty}(\lambda^n, n + 1)$ in $\mathcal{B}$ is an inverse for $\lambda - a$. $\qquad\square$

2.6.12 Theorem *Let $\mathcal{A}$ be a unital Banach algebra.*

(a) *If $a \in \mathcal{A}_G$ is doubly power bounded, then $Sp(a) = \{1\}$ is equivalent to $a = 1$.*

(b) *If $a \in \mathcal{A}$ is power bounded, then $Sp(a) \cap \mathbb{T} \subseteq \{1\}$ is equivalent to $\lim_{n \to \infty} a^n(1 - a) = 0$.*

Proof Proposition 1.1.9 allows us to assume $||a^n|| \leq 1$ for $\langle\, n \in \mathbb{Z} \,/\, n \in \mathbb{N}^0 \,\rangle$ in case $\langle\, (a) \,/\, (b) \,\rangle$. We may also assume that $\mathcal{A}$ is commutative by replacing the original algebra by one of its maximal commutative (necessarily unital) subalgebras including a.

(a): Obviously, $a = 1$ implies $Sp(a) = \{1\}$. Conversely, $Sp(a) = \{1\}$ implies $Sp(1-a) = \{0\}$ and $1-a \in \mathcal{A}_J$. Define h to be $i \sum_{k=1}^{\infty} (1-a)^k/k \in \mathcal{A}_J$ so $a = e^{ih}$. For any integer m, $\sin(mh) \in \mathcal{A}_J$ implies $Sp(\sin(mh)) = \{0\}$. Lemma 2.6.6 shows $h = \arcsin(\sin(h))$ and the bound on $||a^n||$ gives

$$||(\sin(mh))^k|| = \left|\left|\left(\frac{a^m - a^{-m}}{2i}\right)^k\right|\right| \leq 1.$$

Before Lemma 2.6.6 we noted that the coefficients r_k in the Taylor expansion of the arcsine satisfy $\sum_{k=1}^{\infty} |r_k| = \pi/2$. Hence we get

$$||mh|| = ||\arcsin(\sin(mh))|| \leq \sum_{k=1}^{\infty} |r_k|\,||(\sin(mh))^k|| \leq \frac{\pi}{2}.$$

Since $m \in \mathbb{N}$ is arbitrary, this implies $h = 0$ and $a = 1$.

(b): For any $\lambda \in Sp(a) \cap \mathbb{T}$, $\lambda^n(1 - \lambda) \in Sp(a^n(1-a))$ implies $|1 - \lambda| \leq ||a^n(1-a)||$. Hence $\lim_{n \to \infty} a^n(1-a) = 0$ certainly implies $Sp(a) \cap \mathbb{T} \subseteq \{1\}$. Suppose, conversely, that this inclusion holds. If the intersection is empty, $\lim_{n \to \infty} a^n = 0$ so we are reduced to considering the case $Sp(a) \cap \mathbb{T} = \{1\}$. Apply the last lemma. Then $\varphi(a)$ satisfies part (a), so it equals $1 \in \mathcal{B}$. Equivalently, $\varphi(1 - a) = 0$ which implies $\lim_{n \to \infty} a^n(1 - a) = 0$, as we wished to show.   □

2.7   The Spectrum in Finite-Dimensional Algebras

In this section we will study the spectral theory of finite-dimensional algebras. Actually the whole section will be devoted to the spectral theory of linear operators on finite-dimensional linear spaces. This accomplishes the goal stated in the first sentence since any finite-dimensional algebra $\mathcal{A}$ may be isomorphically embedded as a subalgebra of an algebra $\mathcal{L}(\mathcal{A})$ of all linear operators on a finite-dimensional linear space $\mathcal{V}$ in such a way that the spectrum of an element $a \in \mathcal{A}$ is the same as the spectrum of its image in $\mathcal{L}(\mathcal{A})$.

Let $\mathcal{A}$ be an arbitrary algebra. Then the extended left regular representation L^1 is always an isomorphism and the left regular representation L is an isomorphism if and only if the annihilator ideal $\mathcal{A}_{LA}$ is $\{0\}$. After Proposition 2.5.3, we noted that any element a in any algebra $\mathcal{A}$ satisfies

$$Sp_{\mathcal{A}}(a) = Sp_{\mathcal{L}(\mathcal{A}^1)}(L_a^1) \tag{1}$$

and if $\mathcal{A}_{LA}$ is $\{0\}$, then it also satisfies

$$Sp_{\mathcal{A}}(a) = Sp_{\mathcal{L}(\mathcal{A})}(L_a). \tag{2}$$

Thus all spectral questions about an algebra $\mathcal{A}$ may be translated into similar questions about the algebra $\mathcal{L}(\mathcal{V})$ where $\mathcal{V}$ is a vector space. When $\mathcal{A}$ is finite-dimensional, $\mathcal{V}$ may be chosen finite-dimensional, since it can be chosen as either $\mathcal{A}$ or $\mathcal{A}^1$.

We will give an extended discussion of spectral questions about $\mathcal{L}(\mathcal{V})$ where $\mathcal{V}$ is a finite-dimensional linear space. In particular we will show that all operators in $\mathcal{L}(\mathcal{V})$ are spectral in the sense of Nelson Dunford [1954] and that the proof of this fact is the major step in the construction of the Jordan canonical form.

2.7.1 Convention For the rest of this section $\mathcal{V}$ will be a finite-dimensional linear space with dimension n. The word "operator" will always denote an element of $\mathcal{L}(\mathcal{V})$.

In Section 1.7 we reviewed the well-known isomorphism between $\mathcal{L}(\mathcal{V})$ and the algebra M_n of all $n \times n$ matrices, and we will refer to this isomorphism when convenient.

2.7.2 The Characteristic Polynomial In Section 1.7 we noted that the determinant defines a multiplicative map

$$\mathrm{Det}\colon \mathcal{L}(\mathcal{V}) \to \mathbb{C}$$

which takes the value 0 at an operator $T \in \mathcal{L}(\mathcal{V})$ if and only if T is not invertible. Hence we conclude

$$Sp(T) = \{\lambda \in \mathbb{C} : \mathrm{Det}(\lambda - T) = 0\}.$$

Any of the usual procedures for calculating determinants in terms of a matrix representation shows that for a fixed operator T, $\mathrm{Det}(\lambda - T)$ is a monic polynomial of degree n. We define the *characteristic polynomial* of T, denoted by ch_T, by

$$ch_T(\lambda) = \mathrm{Det}(\lambda - T). \tag{3}$$

Since the complex field is algebraically closed, the characteristic polynomial always has at least one and at most n roots. Hence $Sp(T)$ is a nonempty set of n or fewer complex numbers.

2.7.3 Rank and Nullity   Any operator $T \in \mathcal{L}(\mathcal{V})$ satisfies

$$\mathrm{rank}(T) + \mathrm{null}(T) = n \tag{4}$$

where $\mathrm{rank}(T)$ is the dimension of the subspace $T(\mathcal{V})$ and $\mathrm{null}(T)$ is the dimension of $\ker(T)$. (This is easily proved by extending a basis for $\ker(T)$ to a basis for $\mathcal{V}$ and then considering the image of this basis under T.) In particular an operator is non-invertible if and only if it has a nontrivial kernel. Thus a complex number λ belongs to the spectrum of T if and only if there is at least one non-zero vector x satisfying

$$(\lambda - T)x = 0.$$

This motivates the following terminology.

2.7.4 Characteristic Values and Characteristic Vectors   A *characteristic vector* for a linear operator $T \in \mathcal{L}(\mathcal{V})$ is a non-zero vector $x \in \mathcal{V}$ for which there is a complex number λ satisfying

$$Tx = \lambda x.$$

A complex number λ satisfying this relationship with a characteristic vector x of T is called a *characteristic value* of T, and λ and x are said to correspond to each other.

     The terms "eigenvector", and "proper vector" are synonyms for "characteristic vector" and the terms "eigenvalue", and "proper value" are synonyms for "characteristic value".

     In terms of this definition the spectrum of any operator T can be described as the set of characteristic values of T and these are the roots of the characteristic polynomial of T.

     Since the complex field is algebraically closed, the characteristic polynomial factors into linear factors

$$ch_T(\lambda) = \prod_{k=1}^{m} (\lambda - \lambda_k)^{n_k} \tag{5}$$

where $\lambda_1, \lambda_2, \ldots, \lambda_m$ is a list of distinct characteristic values of T. The exponent n_k corresponding to the characteristic value λ_k is called the *algebraic multiplicity* of λ_k. The *geometric multiplicity* of a characteristic value λ_j is defined to be the dimension of $\ker(\lambda_j - T)$. Later in our discussion it will be obvious that the geometric multiplicity is less than or equal to the algebraic multiplicity, and they are equal only for the diagonalizable operators we are about to discuss.

2.7.5 Diagonalizable Operators   The simplest class of matrices to deal with (outside of complex multiples of the identity) is the class of *diagonal* *matrices* (*i.e.*, those matrices which have all their non-zero entries on the main diagonal). The product of two diagonal matrices is their pointwise product and hence any two diagonal matrices commute. An operator S on a finite-dimensional space $\mathcal{V}$ is called *diagonalizable* if it is represented by a diagonal matrix relative to some ordered basis.

Note than an operator S has a diagonal matrix relative to an ordered basis if and only if each basis vector is a characteristic vector of S. Hence S is diagonalizable if and only if $\mathcal{V}$ has a basis of characteristic vectors for S. When S is represented by a diagonal matrix, the characteristic values of S, repeated according to their algebraic multiplicity, are just the diagonal entries of the matrix.

If S is diagonalizable, the choice of an ordered basis of characteristic vectors is highly nonunique. The order of the basis vectors is clearly indeterminate. Of course, any basis vector can be replaced by any of its non-zero multiples. If any characteristic value λ has algebraic multiplicity greater than 1, then $\ker(\lambda - S)$ may be more than 1-dimensional, and the choice of basis vectors within it is completely arbitrary. Most of this indeterminacy is removed by substituting a list of suitable canonical projection operators onto the linear subspaces of characteristic vectors of S for the choice of an ordered basis of characteristic vectors. In order to study this approach, we need to introduce the concept of the minimal polynomial of a linear operator T on a finite-dimensional linear space $\mathcal{V}$.

2.7.6 Minimal Polynomials We could begin this discussion by proving the Cayley-Hamilton theorem which asserts

$$ch_T(T) = 0 \tag{6}$$

for all $T \in \mathcal{L}(\mathcal{V})$. (What is wrong with the simple argument : "$ch_t(\lambda) = \text{Det}(\lambda - T)$ so $ch_T(T) = \text{Det}(t - t) = 0$"?) However it is logically simpler to note that $\mathcal{L}(\mathcal{V})$ has dimension n^2 (since it is isomorphic to the algebra M_n). Hence the powers $I, T, T^2, \ldots, T^{n^2}$ of T are linearly dependent so that there is some monic polynomial p satisfying $p(T) = 0$. The well-ordering of the integers allows us to choose a monic polynomial mi_T of smallest degree satisfying

$$mi_T(T) = 0. \tag{7}$$

This is called the *minimal polynomial* of T. It is denoted simply by mi when T is understood. If p is any polynomial satisfying $p(T) = 0$, then mi_T divides p. (To see this, note that polynomial division gives polynomials q and r satisfying $p = (mi)q + r$, with the degree of r strictly less than the degree of mi. Hence by the minimality of the degree of mi we have $r = 0$.) This shows that mi_T is well-defined.

2.7.7 Proposition *If* $\{\lambda_1, \lambda_2, \ldots, \lambda_m\}$ *is the set of distinct characteristic values of an operator* T, *then the minimal polynomial of* T *has the* *factorization*

$$mi_T(\lambda) = \prod_{j=1}^{m}(\lambda - \lambda_j)^{a_j} \tag{8}$$

where each a_j is positive.

Proof In order to prove this, we will assume the truth of (8) with each a_j positive and then prove that $\{\lambda_1, \lambda_2, \ldots, \lambda_m\}$ is the set of characteristic values. First we have $0 = mi_T(T) = (T - \lambda_j)q_j(T)$ for each j and a suitable polynomial q_j. Since $q_j(T)$ is not 0 (by the minimality of the degree of mi_T), we see that there is a non-zero vector $x = q_j(T)y$ satisfying $(\lambda_j - T)x = 0$. Hence λ_j is a characteristic value of T. Conversely, if λ_0 is a characteristic value of T with corresponding characteristic vector x, and if p is any polynomial, then $p(T)x = p(\lambda_0)x$. Since this equals 0 if and only if $(\lambda - \lambda_0)$ divides $p(\lambda)$, we see that $\lambda - \lambda_0$ divides mi_T, for each characteristic value λ_0 of T. □

2.7.8 Ascent of an Operator In (8) the exponent a_j corresponding to the characteristic value λ_j is called the *ascent* of λ_j. (The term comes from the fact that the sequence of subspaces $\ker(\lambda_j - T) \subseteq \ker((\lambda_j - T)^2) \subseteq \cdots \subseteq \ker((\lambda_j - T)^k)$ is strictly ascending until $k = a_j$ and is stationary thereafter. This can be proved by arguments similar to those just given.) If T is diagonalizable, it is obvious that each characteristic value has ascent one. The converse is easily established after some remarks on projection operators.

2.7.9 Projection Operators Recall that an operator $E \in \mathcal{L}(\mathcal{V})$ is called a *projection operator* if it is idempotent (*i.e.*, if it satisfies $E^2 = E$). In this case $I - E$ is also a projection operator and E is said to project $\mathcal{V}$ *onto the linear subspace*

$$E\mathcal{V} = \{x \in \mathcal{V} : Ex = x\} = \ker(I - E)$$

along the linear subspace

$$\ker(E) = \{x \in \mathcal{V} : Ex = 0\} = (I - E)\mathcal{V}.$$

Clearly these linear subspaces satisfy

$$E\mathcal{V} \cap \ker(E) = \{0\}; \quad E\mathcal{V} + \ker(E) = \mathcal{V}$$

so that E determines, and is determined by, the direct sum decomposition $\mathcal{V} = E\mathcal{V} \oplus \ker(E)$. Two projection operators E and F are said to be *orthogonal* if they satisfy $EF = FE = 0$. This is equivalent to the condition $E\mathcal{V} \cap F\mathcal{V} = \{0\}$.

2.7.10 Invariant and Decomposing Subspaces If E is idempotent, then the subspace $E\mathcal{V}$ is *invariant* under an operator S if and only if E satisfies

$$ESE = SE.$$

Hence the direct sum decomposition $E\mathcal{V} \oplus \ker(E)$ *decomposes* S (*i.e.*, both $E\mathcal{V}$ and $\ker(E) = (I - E)\mathcal{V}$ are invariant under S) if and only if E and S commute. This is shown by the calculation:

$$ES = ESE + ES(I - E) = SE + E(I - E)S(I - E) = SE.$$

We are now ready to determine the detailed structure of the diagonalizable operators. This is done here to prepare the reader for the more complicated structure theorem we will later present for arbitrary operators. Any of the implications in the next theorem is easy to establish directly, but we will use a fairly efficient circle of implications.

2.7.11 Theorem *The following conditions for $S \in \mathcal{L}(\mathcal{V})$ are equivalent.*

(a) *S is diagonalizable.*
(b) *$\mathcal{V}$ has a basis consisting of characteristic vectors of S.*
(c) *The minimal polynomial of S has no multiple roots.*
(d) *There exist distinct complex numbers $\lambda_1, \lambda_2, \ldots, \lambda_m$ and mutually orthogonal non-zero projection operators $E_1, E_2, \ldots, E_m$ satisfying*

$$I = \sum_{j=1}^{m} E_j \quad and \quad SE_j = \lambda_j E_j. \tag{9}$$

(e) *Condition (d) holds and $\lambda_1, \lambda_2, \ldots, \lambda_m$ are the distinct characteristic values of S. The minimal polynomial of S is given by:*

$$mi_S(\lambda) = \prod_{j=1}^{m} (\lambda - \lambda_j) \tag{10}$$

and the projections $E_1, E_2, \ldots, E_m$ are given by the following polynomials in S

$$E_j = q_j(S), \quad where \quad q_j(\lambda) = \prod_{\substack{k=1 \\ k \neq j}}^{m} ((\lambda - \lambda_k)/(\lambda_j - \lambda_k)) \tag{11}$$

and satisfy

$$S = \sum_{j=1}^{m} \lambda_j E_j \quad and \quad E_j \mathcal{V} = \ker(\lambda_j - S) \tag{12}$$

in addition to (9).

Proof (a)⇔(b): Each vector in the basis relative to which the operator has a diagonal matrix must be a characteristic vector, and conversely the matrix relative to a basis of characteristic vectors would be diagonal.

(b)⇒(c): As usual, we denote the distinct characteristic values of S by $\lambda_1, \lambda_2, \ldots, \lambda_m$. For each characteristic value λ_j of S, $\lambda_j - S$ will annihilate each basis vector which is a characteristic vector with λ_j as its characteristic value. The other characteristic vectors will simply be multiplied by some non-zero complex number. Hence, applying the polynomial $\prod_{j=1}^{m}(\lambda - \lambda_j)$ to S, gives 0. Since the minimal polynomial must therefore divide this one (it is easy to see that they are actually equal but we do not need this here), the minimal polynomial has no repeated roots.

(c)⇒(e): Let the minimal polynomial be given by (10) which determines the set $\lambda_1, \lambda_2, \ldots, \lambda_m$. For each j consider the polynomial q_j and the operator E_j in equation (11). Since the degree of q_j is $m - 1$, E_j is not 0. However by factoring the minimal polynomial we see that $(\lambda_j - S)E_j$ is 0. Hence each λ_j is a characteristic value of S. If λ is a characteristic value of S, with characteristic vector x, then $mi_S(S)x = mi_S(\lambda)x$, so λ is one of the roots of the minimal polynomial and $\{\lambda_1, \lambda_2, \ldots, \lambda_m\}$ coincides with the spectrum of S. Similarly, the minimal polynomial shows that $E_j E_k$ is 0 if j and k are distinct.

The polynomial $1 - \sum_{j=1}^{m} q_j$ is 0 since it has degree $m - 1$ but has the m roots $\lambda_1, \lambda_2, \ldots, \lambda_m$. Hence we conclude $I = \sum_{j=1}^{m} E_j$ and $SE_j = \lambda_j E_j$. If x is a characteristic vector of S, with characteristic value λ_k, then $q_j(S)x$ is 0 if $k \neq j$ and is x if $k = j$. Hence $E_j = q_j(S)$ is precisely the projection onto the subspace of $\mathcal{V}$ consisting of characteristic vectors with characteristic value λ_j. Thus all the statements of (e) are established.

(e)⇒(d): Obvious.

(d)⇒(b): A basis for $\mathcal{V}$ consists of the union of the bases for the subspaces $E_j \mathcal{V}$.   □

The unique representation $S = \sum_{j=1}^{m} \lambda_j E_j$ of condition (e) above is called the *spectral resolution* of S. In order to partially extend this representation to non-diagonalizable operators, we need one more result on commuting diagonalizable operators. This result is true for an arbitrary family of commuting operators, but we prove it only for a pair.

2.7.12 Proposition *If S and S' are commuting diagonalizable operators in $\mathcal{L}(\mathcal{V})$, then there is an ordered basis for $\mathcal{V}$ relative to which both S and S' are represented by diagonal matrices.*

Proof A proof by induction on the dimension of $\mathcal{V}$ starting from the definition is not hard, but we prefer to use the spectral resolution. Let $S = \sum_{j=1}^{m} \lambda_j E_j$ and $S' = \sum_{k=1}^{p} \lambda_k' E_k'$ be the spectral resolutions. Since each E_j is a polynomial in S, it commutes with S' and hence with each E_k'. Therefore $E_j E_k' = E_k' E_j$ is a projection operator for each j and k.

Choose a basis for each linear space $E_j E'_k \mathcal{V}$. (If $E_j E'_k$ is 0, then the basis for $E_j E'_k \mathcal{V} = \{0\}$ is the empty set.) It is easy to see that the union of all these bases is a basis for $\mathcal{V}$ with the desired property.   □

We now generalize, as far as possible, the spectral decomposition to non-diagonalizable operators. The following elementary result shows that any linear operator on a finite-dimensional space is a spectral operator in the terminology of Dunford [1954], which we will introduce later.

2.7.13 Theorem   *Let $\mathcal{V}$ be a finite-dimensional space and let T be a linear operator on $\mathcal{V}$. Then there exist positive integers m, $a_1, a_2, \ldots, a_m$, distinct complex numbers $\lambda_1, \lambda_2, \ldots, \lambda_m$, operators S and N and non-zero operators $E_1, E_2, \ldots, E_m$ satisfying:*

(a) $T = S + N$.
(b) $SN = NS$.
(c) S is diagonalizable.
(d) N is nilpotent.
(e) $E_1, E_2, \ldots, E_m$ are projection operators.
(f) $I = \sum_{j=1}^m E_j$.
(g) $S = \sum_{j=1}^m \lambda_j E_j$.
(h) $E_j E_k = 0$ whenever $j \neq k$.
(i) *The minimal polynomial for T is $\prod_{j=1}^m (\lambda - \lambda_j)^{a_j}$, $\lambda_1, \lambda_2, \ldots, \lambda_m$ is the list of distinct characteristic values of T and $a_1, a_2, \ldots, a_m$ is the list of corresponding ascents.*
(j) *For each $j = 1, 2, \ldots, m$ the minimal polynomial for the restriction of T to $E_j \mathcal{V}$ is $(\lambda - \lambda_j)^{a_j}$.*
(k) *Each $j = 1, 2, \ldots, m$ satisfies $E_j \mathcal{V} = \ker((\lambda_j - T)^{a_j})$ and the restriction of $(\lambda_j - T)$ to $(I - E_j)\mathcal{V} = (\lambda_j - T)^{a_j} \mathcal{V}$ is invertible.*
(l) $S, N, E_1, E_2, \ldots, E_m$ *are all polynomials in T.*

Furthermore, S and N are uniquely determined by conditions (a), (b), (c), and (d). If S and N are defined by (g) and (a), then (b), (d), (f), and (h) determine $\lambda_1, \lambda_2, \ldots, \lambda_m$ (provided the numbers in this list are required to be distinct) and $E_1, E_2, \ldots, E_m$ uniquely up to a permutation of $\{1, 2, \ldots, m\}$. If condition (j) is also assumed in the last sentence, then $a_1, a_2, \ldots, a_m$ are also determined up to the permutation of indices.

Proof Let

$$mi_T(\lambda) = \prod_{j=1}^m (\lambda - \lambda_j)^{a_j}$$

be the minimal polynomial for T, so that $\lambda_1, \lambda_2, \ldots, \lambda_m$ is the list of distinct characteristic values with associated ascents $a_1, a_2, \ldots, a_m$ according to Proposition 2.7.7. Then the polynomials

$$p_j(\lambda) = \prod_{\substack{k=1 \\ k \neq j}}^{m} (\lambda - \lambda_k)^{a_k} \quad j = 1, 2, \ldots, m$$

have no common divisor, so the Euclidean algorithm guarantees the existence of polynomials h_j satisfying

$$1 = \sum_{j=1}^{m} h_j p_j. \tag{13}$$

Define q_j by $q_j = h_j p_j$ and define $E_j \in \mathcal{L}(\mathcal{V})$ by

$$E_j = q_j(T) \quad j = 1, 2, \ldots, m.$$

Then (13) shows

$$\sum_{j=1}^{m} E_j = I. \tag{14}$$

Since the minimal polynomial mi_T divides $q_j q_k$ whenever $j \neq k$, we have

$$E_j E_k = 0 \quad \text{when} \quad j \neq k.$$

Hence we get

$$E_j^2 = E_j \sum_{k=1}^{m} E_k = E_j I = E_j$$

for all j, so that each E_j is a projection operator. Thus we have established (e), (f), (h), (i), and part of (l). Define S and N by (g) and (a); then (a), (g), (c), the rest of (l) and therefore (b) are also satisfied.

For each j the definition $E_j = q_j(T)$ makes it clear that if q_0 is a polynomial, then $q_0(T)E_j$ will be 0 if and only if $(\lambda - \lambda_j)^{a_j}$ divides q_0. Hence, in particular,

$$(\lambda_j - T)^p E_j = 0$$

is equivalent to $p \geq a_j$, so that (j) is satisfied. Since $(\lambda - \lambda_j)^{a_j}$ divides each q_k for $k \neq j$, we conclude $\ker((\lambda_j - T)^{a_j})$ is included in $\ker(p_k(T))$ for each $k \neq j$. Hence (14) shows the first equality of (k). This in turn shows that $\lambda_j - T$ is injective when restricted to $(I - E_j)\mathcal{V}$. Since E_j is a polynomial in T we have

$$(\lambda_j - T)(I - E_j)\mathcal{V} = (I - E_j)(\lambda_j - T)\mathcal{V} \subseteq (I - E_j)\mathcal{V}.$$

Hence $(\lambda_j - T)$ restricted to $(I - E_j)\mathcal{V}$ is invertible, since it is an injective linear map of a finite-dimensional space into, and therefore onto, itself.

Again using that $(\lambda - \lambda_j)^{a_j}$ divides each q_k for $k \neq j$, we get

$$(\lambda_j - T)^{a_j}\mathcal{V} \supseteq (\sum_{\substack{k=1 \\ k \neq j}}^{m} E_k)\mathcal{V} = (I - E_j)\mathcal{V}.$$

The opposite inclusion follows from

$$(I - E_j)(\lambda_j - T)^{a_j}\mathcal{V} = (\lambda_j - T)^{a_j}(I - E_j)\mathcal{V} = (\lambda_j - T)^{a_j}\mathcal{V}.$$

This completes the proof of (k).

Equation (g) shows that we may write

$$N = T - S = T - \sum_{j=1}^{m}\lambda_j E_j = \sum_{j=1}^{m}(T - \lambda_j)E_j.$$

Hence for any positive integer p, we can establish the equation

$$N^p = \sum_{j=1}^{m}(T - \lambda_j)^p E_j$$

by induction, using (h) and the commutativity of T and E_j (which follows from (a) and (b) or from (l)). If p satisfies $p \geq \max\{a_j : j = 1, 2, \ldots, m\}$, then N^p equals 0 since N^p is expressed as a polynomial in T which is divisible by the minimal polynomial of T. This proves (d) and completes the proof of the twelve properties (a), (b), ...,(l).

Suppose S' and N' are also operators satisfying (a), (b), (c), and (d). Then (a) implies $S - S' = N' - N$. Properties (a) and (b) for S' and N' imply $TS' = S'T$ and $TN' = N'T$. Since S and N are polynomials in T, we conclude $S'S = SS'$ and $N'N = NN'$. Hence if p equals the sum of the indices of nilpotence for N and N', we get

$$(N' - N)^p = \sum_{k=0}^{p}\binom{n}{k}(N')^k N^{p-k} = 0.$$

Thus $S - S' = N' - N$ is nilpotent. On the other hand, by choosing an ordered basis relative to which both S and S' are represented by diagonal matrices, we see that $S - S'$ can be represented by a diagonal matrix. Clearly no power of a diagonal operator is 0 unless each of the diagonal entries (and hence the operator) is 0. Hence $S - S' = N' - N = 0$, so that S and N are uniquely determined by (a), (b), (c), and (d).

Suppose another family of distinct $\lambda_1', \lambda_2', \ldots, \lambda_r' \in \mathbb{C}$ and non-zero operators $E_1', E_2', \ldots, E_r'$ also satisfy (b), (d), (f), and (h). Define S and N by the analogues of (g) and (a) relative to this data. We have already given the argument which shows that (f) and (h) imply (e). Hence S satisfies (c). Thus the argument of the last paragraph shows that S and N have their original meaning. Hence we have $\sum_{j=1}^{m} \lambda_j E_j = S = \sum_{k=1}^{r} \lambda_k' E_k'$. By Theorem 2.7.11 we see that $\lambda_1, \lambda_2, \ldots, \lambda_m$ and $\lambda_1', \lambda_2', \ldots, \lambda_r'$ are both the list of distinct characteristic values of S, and that $E_1, E_2, \ldots, E_m$ and $E_1', E_2', \ldots, E_r'$ are both the list of corresponding projections onto the subspaces of characteristic vectors. Hence $\lambda_1, \lambda_2, \ldots, \lambda_m$ and $E_1, E_2, \ldots, E_m$ are unique up to simultaneous permutations of both lists. Clearly (j) determines $a_1, a_2, \ldots, a_m$.   □

We remark on one elementary but interesting consequence of this theorem using terminology introduced in Section 2.3 above. Suppose $\mathcal{A}$ is a commutative subalgebra of $\mathcal{L}(\mathcal{V})$. Then the decomposition $T = S + N$ is easily seen to define a direct sum decomposition of $\mathcal{A}$. The second direct summand is an ideal consisting of nilpotent operators and hence, by Theorem 2.3.3, is included in the Jacobson radical. The first direct summand is a subalgebra which contains no non-zero nilpotent operator (since every operator in it is a linear combination of projections) and is thus semisimple. Hence this direct sum decomposition is just the direct sum decomposition of $\mathcal{A}$ into a semisimple subalgebra isomorphic to $\mathcal{A}/\mathcal{A}_J$ and the ideal $\mathcal{A}_J$ guaranteed by the Feldman decomposition theorem(Theorem 7.1.3 below).

In order to improve on Theorem 2.7.13 we need to ascertain the structure of the nilpotent operator N. The next proposition deals with this question. Actually this proposition will not be applied directly to N, but rather to the restriction of N to each of the invariant subspaces $E_k \mathcal{V}$ in the terminology of Theorem 2.7.13.

2.7.14 Proposition   Let $\mathcal{V}$ be a finite-dimensional linear space with dimension n. Let $N \in \mathcal{L}(\mathcal{V})$ be nilpotent. Let r and d be the index of nilpotence of N and the nullity of N, respectively. Then there are positive integers $n_1, n_2, \ldots, n_d$ satisfying $r = n_1 \geq \cdots \geq n_d \geq 1$ and $\sum_{k=1}^{d} n_k = n$, and vectors $x_1, x_2, \ldots, x_d$ satisfying $N^{n_k} x_k = 0$ for $k = 1, 2, \ldots, d$, and such that

$$V = \{x_1, Nx_1, \ldots, N^{n_1-1}x_1, x_2, Nx_2,$$
$$\ldots, N^{n_2-1}x_2, \ldots, x_d, Nx_d, \ldots, N^{n_d-1}x_d\}$$

is a basis for $\mathcal{V}$.

Proof The proof is by induction on n. If $n = 1$ there is nothing to prove. Suppose $n > 1$ and that the result holds for all $n' < n$. Since N^{r-1} is non-zero we can find $x_1 \in \mathcal{V}$ satisfying $N^{n_1-1}x_1 \neq 0$, where we have set n_1 equal to r. The space $\mathcal{W}$ spanned by $W = \{x_1, Nx_1, \ldots, N^{n_1-1}x_1\}$ is clearly invariant and W is a basis for it. (Linear independence follows from the fact that λ^r is the minimal polynomial for N.) The restriction $N_\mathcal{W}$ of

N to $\mathcal{W}$ and the operator $N^{\mathcal{W}}$ induced by N on $\mathcal{V}/\mathcal{W}$ ($N^{\mathcal{W}}$ is defined by $N^{\mathcal{W}}(x+\mathcal{W}) = Nx+\mathcal{W}$) both satisfy $(N_{\mathcal{W}})^r = 0$ and $(N^{\mathcal{W}})^r = 0$. Let $\tilde{r}$ and $\tilde{d} - 1$ be the index of nilpotence and nullity of $N^{\mathcal{W}}$. Clearly we have $\tilde{r} \leq r$. By our induction hypothesis we may find positive integers $n_2, n_3, \ldots, n_{\tilde{d}}$ and vectors $\tilde{x}_2, \ldots, \tilde{x}_{\tilde{d}}$ satisfying $r \geq \tilde{r} = n_2 \geq n_3 \geq \cdots \geq n_{\tilde{d}}, \sum_{k=2}^{\tilde{d}} n_k = n - r, N^{n_k}\tilde{x}_k + \mathcal{W} = \mathcal{W}$ for $k = 2, \ldots, \tilde{d}$ and such that

$$\tilde{V} = \{\tilde{x}_2 + \mathcal{W}, N\tilde{x}_2 + \mathcal{W}, \ldots, N^{n_2-1}\tilde{x}_2 + \mathcal{W}, \tilde{x}_3 + \mathcal{W}, N\tilde{x}_3 + \mathcal{W},$$
$$\ldots, N^{n_3-1}\tilde{x}_3 + \mathcal{W}, \ldots, \tilde{x}_{\tilde{d}} + \mathcal{W}, N\tilde{x}_{\tilde{d}} + \mathcal{W}, \ldots, N^{n_{\tilde{d}}-1}\tilde{x}_{\tilde{d}} + \mathcal{W}\}$$

is a basis for $\mathcal{V}/\mathcal{W}$.

For any fixed k satisfying $2 \leq k \leq \tilde{d}$, let $\lambda_0, \lambda_1, \ldots, \lambda_{n_1-1}$ satisfy $N^{n_k}\tilde{x}_k = \sum_{j=0}^{n_1-1} \lambda_j N^j x_1$. Then $N^{n_1}\tilde{x}_k = \sum_{j=0}^{n_1-1} \lambda_j N^{n_1-n_k+j}x_1$ vanishes. Hence $N^{n_1}x_1 = 0$ and the linear independence of $x_1, Nx_1, \ldots, N^{n_1-1}x_1$ imply $\lambda_j = 0$ for $j < n_k$. If we define x_k by $x_k = \tilde{x}_k - \sum_{j=n_k}^{n_1-1} \lambda_j N^{j-n_k}x_1$, then it therefore satisfies $N^{n_k}x_k = 0$ and $N^j x_k + \mathcal{W} = N^j \tilde{x}_k + \mathcal{W}$ for $j = 0, 1, \ldots, n_k - 1$ and for $k = 2, 3, \ldots, \tilde{d}$. By the linear independence of $\tilde{V}$ and of W, there is no nontrivial relation $\sum_{k=1}^{\tilde{d}} \sum_{j=0}^{n_k} \lambda_{jk} N^j x_k = 0$ of linear dependence. Thus $\tilde{d}$ is d and the set V is a basis for $\mathcal{V}$ as claimed. $\square$

Use the notation of the last proposition. For each $k = 1, 2, \ldots, d$ the restriction of N to the subspace spanned by $x_k, Nx_k, \ldots, N^{n_k-1}x_k$ has matrix

$$\begin{pmatrix} 0 & & & & & & \\ 1 & 0 & & & & \mathbf{0} & \\ & 1 & 0 & & & & \\ & & 1 & \ddots & & & \\ & & & \ddots & \ddots & & \\ \mathbf{0} & & & & \ddots & 0 & \\ & & & & & 1 & 0 \end{pmatrix}$$

relative to this ordered basis. Clearly the vectors $x_1, \ldots, x_d$ of Proposition 2.7.14 are not uniquely defined in any sense at all. Nevertheless, it is easy to see that the form of the matrix of N relative to any ordered basis satisfying the conditions on V is completely unique.

We now adopt the notation of Theorem 2.7.13. Let λ_j be a characteristic value of T and let E_j be the corresponding projection. Apply Theorem 2.7.13 to the restriction of N (in the notation of Theorem 2.7.13) to $E_j\mathcal{V}$. Then the matrix of the restriction of N to $E_j\mathcal{V}$ is the direct sum of blocks of the form

$$\begin{pmatrix} \lambda_j & & & & & & \\ 1 & \lambda_j & & & & 0 & \\ & 1 & \lambda_j & & & & \\ & & 1 & \ddots & & & \\ & & & \ddots & \ddots & & \\ & 0 & & & \ddots & \lambda_j & \\ & & & & & 1 & \lambda_j \end{pmatrix}.$$

Such a matrix is called a *Jordan block*. A matrix is said to be the direct sum of Jordan blocks when it has the form

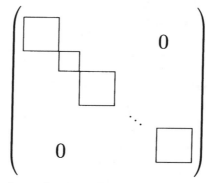

where each square along the main diagonal represents a Jordan block.

Combining Theorem 2.7.13 and Proposition 2.7.14 gives the following *Jordan canonical form* for an operator. Note that the choice of ordered basis is generally far from unique, but the form of the matrix is completely determined by the description given.

2.7.15 Theorem *Let T be a linear operator on a finite-dimensional linear space $\mathcal{V}$ with dimension n. Let $\lambda_1, \lambda_2, \ldots, \lambda_m$ be a list of the distinct characteristic values of T. Then we can choose an ordered basis for T such that the matrix J_T of T relative to this basis is the direct sum of Jordan blocks. Furthermore we may assume that for $1 \leq k < j \leq m$ the Jordan blocks with diagonal entries λ_k all occur before any of the blocks with diagonal entries λ_j, and that the blocks with diagonal entries λ_k occur in order of decreasing size. These conditions completely determine J_T once the ordering of $\lambda_1, \lambda_2, \ldots, \lambda_m$ is chosen.*

Note that we can find all of the spectral properties of T from inspection of its Jordan matrix $J = J_T$. The set of characteristic values of T is the set of diagonal entries of J, and the algebraic multiplicity of any characteristic value λ_j of T is the number of times λ_j occurs on the diagonal of J. The geometric multiplicity of λ_j is just the number of Jordan blocks with diagonal entries λ_j (since the last basis vector corresponding to any such block is a characteristic vector of T corresponding to λ_j). The ascent of any characteristic value λ_j is the length of the longest Jordan block with diagonal entries λ_j. Finally, for any characteristic value λ_j of T and any positive integer p, the dimension of $\ker((\lambda_j - T)^p)$ is $k_0 p + \sum_{k=k_0+1}^{d} n_k$, where $n_1, n_2, \ldots, n_d$ is the list of block lengths in descending order of blocks with diagonal entries equal to λ_j, and n_{k_0} is the last block length which is not less than p.

We have reviewed the details of spectral theory for linear operators on finite-dimensional spaces as a guide to what follows. However, the theory can be applied directly to certain situations. Let T be an operator on an infinite-dimensional Banach space $\mathcal{X}$. Suppose that $\lambda_1, \lambda_2, \ldots, \lambda_m$ is a list of distinct characteristic values of T which are isolated points in $Sp(T)$. Suppose that each λ_j has finite ascent (*i.e.*, there exists an integer p satisfying $\ker((\lambda_j - T)^p) = \ker((\lambda_j - T)^{p+1})$, so that the ascent a_j of λ_j can be defined as the smallest such integer) and that the dimension of $\ker((\lambda_j - T)^k)$ is finite for each j and k. These conditions are satisfied if $\lambda_1, \lambda_2, \ldots, \lambda_m$ is a list of distinct non-zero characteristic values of a compact operator T. Let Γ be a contour in the resolvent set of T surrounding $\{\lambda_1, \lambda_2, \ldots, \lambda_m\}$ and no other points in $Sp(T)$. For each j let Γ_j be a contour in the resolvent set surrounding λ_j and no other point of $Sp(T)$. Then (anticipating the detailed discussion in the next chapter) for $j = 1, 2, \ldots, m$

$$E_j = \frac{1}{2\pi i} \int_{\Gamma_j} (\lambda - T)^{-1} d\lambda \quad \text{and} \quad E = \frac{1}{2\pi i} \int_{\Gamma} (\lambda - T)^{-1} d\lambda \qquad (15)$$

are commuting projection operators onto finite-dimensional subspaces satisfying

$$E = \sum_{j=1}^{m} E_j; \quad E_j E_k = 0 \quad \text{for} \quad j \neq k$$

$$E_j \mathcal{X} = \ker((\lambda_j - T)^{a_j}).$$

Furthermore the restriction of $\lambda - T$ to $(I - E)\mathcal{X} = (\prod_{j=1}^{m} (\lambda_j - T)^{a_j})\mathcal{X}$ is invertible for all $\lambda \in \mathbb{C}$ which are surrounded by Γ. All these remarks are easily verified when integrals such as those occurring in (15) are discussed in the next chapter. Now Theorems 2.7.13 and 2.7.15 can be applied to the restriction of T to $E\mathcal{X}$ and give a complete description of this operator, hence also of the spectral behavior of T on the portion of the complex plane inside Γ.

It is, of course, not true that the spectral theory of finite-dimensional operators is a reliable guide to typical spectral problems for operators on infinite-dimensional Banach or Hilbert spaces. As a matter of fact, it is reasonable to say that the class of compact operators on a Banach space is the most general class of operators which have spectral behavior which is an infinite-dimensional analogue of the spectral behavior of matrices. Even among compact operators there occur operators T (called *transcendental topologically nilpotent operators*) which are not nilpotent but satisfy $Sp(T) = \{0\}$.

2.8   Spectral Theory of Operators.

In this section we will introduce some generalities on the spectral theory of bounded linear operators and then discuss the case of compact operators in more detail. Let $\mathcal{X}$ be a Banach space. The elements of $\mathcal{B}(\mathcal{X})$ (Definition 1.1.16) will simply be called operators in this example. There are many sources of additional information, *e.g.*, Dunford and Schwartz [1958], [1963], Henry R. Dowson [1978] and Robin Harte [1988]

We begin by giving the classical terminology for the classification of the points of $Sp(T)$ for $T \in \mathcal{B}(\mathcal{X})$.

2.8.1 Definition   The ⟨ *point spectrum / continuous spectrum / residual spectrum / approximate point spectrum* ⟩ of $T \in \mathcal{B}(\mathcal{X})$ is the set:

$$
\begin{aligned}
\langle\, pSp(T) &= \{\lambda \in \mathbb{C} : \lambda - T \text{ is not injective}\} \,/ \\
cSp(T) &= \{\lambda \in \mathbb{C} : \lambda - T \text{ is injective and } (\lambda - T)\mathcal{X} \\
&\quad \text{is dense in } \mathcal{X} \text{ but does not equal } \mathcal{X}\} \,/ \\
rSp(T) &= \{\lambda \in \mathbb{C} : \lambda - T \text{ is injective but } (\lambda - T)\mathcal{X} \\
&\quad \text{is not dense in } \mathcal{X}\} \,/ \\
apSp(T) &= \{\lambda \in \mathbb{C} : \text{there is a sequence } \{x_n\}_{n\in\mathbb{N}} \subseteq \mathcal{X} \\
&\quad \text{satisfying } \|x_n\| = 1 \text{ for all } n \text{ and} \\
&\quad \lim(\lambda - T)(x_n) = 0\} \,\rangle.
\end{aligned}
$$

In the terminology introduced in the last section, the point spectrum is just the set of characteristic values. Thus the spectrum of an operator on a finite-dimensional space equals its point spectrum. Obviously each of the sets described in this definition is a subset of the spectrum and the first three are disjoint. A point λ is in the approximate point spectrum of T if and only if $\lambda - T$ is a left topological divisor of zero. (For sufficiency, consider the one-dimensional operators $x_n \otimes x^*$ which are of norm 1 if $x^* \in \mathcal{X}^*$ is. For necessity, choose a bounded sequence $\{y_n\}_{n\in\mathbb{N}}$ such that each $x_n = T_n(y_n)$ has norm 1.) In addition to these subsets, we also wish to consider the boundary of the spectrum, denoted by $\partial Sp(\cdot)$.

2.8.2 Proposition *The spectrum of any operator $T \in \mathcal{B}(\mathcal{X})$ satisfies:*

$$Sp(T) = pSp(T) \cup cSp(T) \cup rSp(T) \quad \text{(disjoint union).} \qquad (1)$$

The approximate point spectrum is a closed subset satisfying:

$$pSp(T) \cup cSp(T) \cup \partial Sp(T) \subseteq apSp(T). \qquad (2)$$

Proof The open mapping theorem (or its corollary the inverse boundedness theorem) shows that a point not in $pSp(T) \cup cSp(T) \cup rSp(T)$ belongs to the resolvent set of T, proving (1).

The point spectrum is included in the approximate point spectrum because we may choose $x_n = x$ for all n where $Tx = 0$ and $\|x\| = 1$. Since the approximate point spectrum comes from topological divisors of zero, Theorem 2.5.7 shows that it includes the boundary of the spectrum. We may easily prove:

(a) the approximate point spectrum is the set: $\{\lambda \in \mathbb{C} :$ there is no $M \in \mathbb{R}$ satisfying $\|x\| \leq M\|(\lambda - T)x\|$ for all $x \in \mathcal{X}\}$.

(b) $(\lambda - T)\mathcal{X}$ is closed if and only if there is an $M \in \mathbb{R}$ such that for each $y \in (\lambda - T)(\mathcal{X})$ there is some $x \in \mathcal{X}$ satisfying $y = (\lambda - T)x$ and $\|x\| \leq M\|y\|$.

Hence the approximate point spectrum includes the continuous spectrum. A standard diagonalization shows that $apSp(T)$ is closed for any operator T. $\qquad\square$

Note that conditions (a) and (b) stated in the proof have independent interest. Defining distance by $\text{dist}(\lambda, Sp(T)) = \inf\{|\lambda - \mu| : \mu \in Sp(T)\}$, Proposition 2.2.9 shows

$$\|(\lambda - T)^{-1}\| \geq \text{dist}(\lambda, Sp(T))^{-1} \quad \forall\, \lambda \notin Sp(T).$$

Dual maps were introduced following Definition 1.1.15. The *dual* $T^* \in \mathcal{B}(\mathcal{X}^*)$ of an operator $T \in \mathcal{B}(\mathcal{X})$ is defined by

$$T^*(x^*)(x) = x^*(Tx) \quad \forall\, x^* \in \mathcal{X}^*;\ x \in X. \qquad (3)$$

It is easy to see that T^* does belong to $\mathcal{B}(X^*)$ and satisfies:

$$\|T^*\| = \|T\| \quad \forall\, T \in \mathcal{B}(\mathcal{X}) \qquad (4)$$

so that the map $T \mapsto T^*$ is an isometric anti-isomorphism of $\mathcal{B}(\mathcal{X})$ into $\mathcal{B}(\mathcal{X}^*)$. The natural isometric injection κ of $\mathcal{X}$ into $\mathcal{X}^{**}$ given by

$$\kappa(x)(x^*) = x^*(x) \quad \forall\, x \in \mathcal{X};\ x^* \in \mathcal{X}^*$$

was also introduced following Definition 1.1.15. Usually we will regard $\mathcal{X}$ as a subset of $\mathcal{X}^{**}$ under this embedding. When this is done, an operator

T may be regarded as a subset of (the graph of) T^{**}. This is a valuable viewpoint as we will now show.

2.8.3 Proposition *The dual of any operator satisfies*

$$Sp(T) = Sp(T^*) \quad T \in \mathcal{B}(\mathcal{X}). \tag{5}$$

Proof For any $\lambda \in \mathbb{C}$ and $T \in \mathcal{B}(\mathcal{X})$, if $(\lambda - T)^{-1}$ exists in $\mathcal{B}(\mathcal{X})$ then it is trivial to check that $((\lambda - T)^{-1})^*$ is the inverse of $(\lambda - T)^* = \lambda - T^*$. Suppose, conversely, that $(\lambda - T^*)^{-1}$ exists in $\mathcal{B}(\mathcal{X}^*)$, then by what we have just shown, $(\lambda - T^{**})^{-1}$ exists in $\mathcal{B}(\mathcal{X}^{**})$. Hence $\lambda - T$ is injective since it is a restriction of $\lambda - T^{**}$. Also $\lambda - T$ is a linear homeomorphism of $\mathcal{X}$ onto $(\lambda - T)\mathcal{X}$ which is therefore a closed subspace. If $(\lambda - T)\mathcal{X}$ were not all of $\mathcal{X}$, then the Hahn–Banach theorem would provide a non-zero $x^* \in \mathcal{X}^*$ satisfying $0 = x^*(\lambda - T)\mathcal{X} = (\lambda - T^*)(x^*)(\mathcal{X})$. This contradicts the injectivity of $(\lambda - T)^*$ and thus proves that $\lambda - T$ is invertible. □

For any subsets $\mathcal{Y}$ of $\mathcal{X}$ and $\mathcal{Z}$ of $\mathcal{X}^*$, we denote the *annihilator sets* of $\mathcal{Y}$ in $\mathcal{X}^*$ and of $\mathcal{Z}$ in $\mathcal{X}$ by:

$$\mathcal{Y}^\perp = \{x^* \in \mathcal{X}^* : x^*(x) = 0 \quad \forall \, x \in \mathcal{Y}\}$$

and

$$^\perp\mathcal{Z} = \{x \in \mathcal{X} : x^*(x) = 0 \quad \forall \, x^* \in \mathcal{Z}\}.$$

These are clearly closed linear subspaces with the former even closed in the $\mathcal{X}$- or weak*-topology. For any $T \in \mathcal{B}(\mathcal{X})$ it is easy to show:

$$\begin{aligned} \ker(T^*) &= (T\mathcal{X})^\perp; \quad T\mathcal{X}^- = {}^\perp(\ker(T^*)) \\ \ker(T) &= {}^\perp(T^*\mathcal{X}^*) : \quad T^*\mathcal{X}^{*-} \subseteq (\ker(T))^\perp. \end{aligned} \tag{6}$$

These results imply:

$$\begin{aligned} pSp(T) &\subseteq pSp(T)^* \cup rSp(T^*) \tag{7} \\ rSp(T) &\subseteq pSp(T^*) \subseteq pSp(T) \cup rSp(T). \end{aligned}$$

It is a little harder to show that if $T\mathcal{X}$ is closed then $T^*\mathcal{X}^*$ is closed and satisfies $T^*\mathcal{X}^* = \ker(T)^\perp$, and it is still harder to show that if $T^*\mathcal{X}^*$ is closed then $T\mathcal{X}$ is closed and satisfies $T\mathcal{X} = {}^\perp \ker(T^*)$. (*Cf.* Dunford and Schwartz [1958], pp. 487, 8.) See also Seymour Goldberg [1966], II.3.14 and II.4.11 where more detailed results are obtained for not necessarily continuous linear operators defined on dense linear subspaces of a normed linear space $\mathcal{X}$ with values in a normed linear space $\mathcal{Y}$ and the duals of these operators defined on a linear subspace of $\mathcal{Y}^*$.

Any $T \in \mathcal{B}(\mathcal{X})$ and $n \in \mathbb{N}$ obviously satisfy

$$\ker(T^n) \subseteq \ker(T^{n+1}).$$

Moreover, $\ker(T^{n_0}) = \ker(T^{n_0+1})$ implies $\ker(T^n) = \ker(T^{n_0})$ for all $n \geq n_0$. This suggests defining the *ascent* $\alpha(T)$ of $T \in \mathcal{S}(\mathcal{X})$ to be the smallest $n \in \mathbb{N}$ satisfying

$$\ker(T^n) = \ker(T^{n+1}),$$

if such an n exists, and ∞ otherwise. Similarly it is obvious that we get $T^n\mathcal{X} \supseteq T^{n+1}\mathcal{X}$ and that $T^{n_0}\mathcal{X} = T^{n_0+1}\mathcal{X}$ implies $T^n\mathcal{X} = T^{n_0}\mathcal{X}$ for all $n \geq n_0$. This in turn suggests defining the *descent* $\delta(T)$ of T to be the smallest $n \in \mathbb{N}$ satisfying $T^n\mathcal{X} = T^{n+1}\mathcal{X}$, if such an n exists, and ∞ otherwise. A nontrivial standard result, which we will prove below in Lemma 7.5.2, shows $\alpha(T) = \delta(T)$ whenever both $\alpha(T)$ and $\delta(T)$ are finite. In this case, moreover, there is a direct sum decomposition

$$\mathcal{X} = T^n\mathcal{X} \oplus \ker(T^n) \quad n = \alpha(T) = \delta(T) < \infty, \tag{8}$$

and T restricted to $T^n\mathcal{X}$ is a bijection onto $T^n\mathcal{X}$.

Proposition 3.4.1 in the next chapter shows how to construct spectral idempotents with the following properties. Suppose $T \in \mathcal{B}(\mathcal{X})$ has a disconnected spectrum with $\sigma \subseteq Sp(T)$ both open and closed in $Sp(T)$. Then there is a projection operator (*i.e.*, an idempotent in $\mathcal{B}(\mathcal{X})$) $E = E(\sigma)$ satisfying:

(a) $E \in \{T\}''$.

(b) $Sp(T|E\mathcal{X}) \subseteq \sigma$ and $Sp(T|(I - E)\mathcal{X}) \subseteq Sp(T) \setminus \sigma$.

This may be a good place to remark that any projection $E \in \mathcal{B}(\mathcal{X})$ satisfies:

(1) $E(\mathcal{X}) = \{x \in \mathcal{X} : x = Ex\} = \ker(I - E)$;

(2) $I - E$ is a projection, so $(I - E)(\mathcal{X}) = \{x \in \mathcal{X} : x = (I - E)x\} = \ker(E)$;

(3) $\mathcal{X} = E(\mathcal{X}) \oplus (I - E)\mathcal{X}$, where $E\mathcal{X}$ and $(I - E)\mathcal{X}$ are closed. Conversely if a Banach space $\mathcal{X}$ is the direct sum of two closed subspaces $\mathcal{Y}$ and $\mathcal{Z}$, then there is a projection $E \in \mathcal{B}(\mathcal{X})$ satisfying $E\mathcal{X} = \mathcal{Y}$ and $(I - E)\mathcal{X} = \mathcal{Z}$.

The Spectral Theory of Compact Operators.

Compact operators were defined and a few of their basic properties were proved in §1.7.7. We start our discussion by quoting a basic result due to Juliusz Schauder [1930].

2.8.4 Proposition *Let $\mathcal{X}$ be a Banach space. Then $T \in \mathcal{B}(\mathcal{X})$ is compact if and only if its Banach space dual $T^* \in \mathcal{B}(\mathcal{X}^*)$ is compact.*

For a proof, see Dunford and Schwartz [1958], VI.5.2. If $\mathcal{X}$ is a Hilbert space $\mathcal{H}$, the usual conjugate linear identification of $\mathcal{H}^*$ with $\mathcal{H}$ shows that this is equivalent to saying that $\mathcal{B}_K(\mathcal{H})$ is invariant under the Hilbert space adjoint. This result is also an immediate consequence of the identity $\mathcal{B}_K(\mathcal{H}) = \mathcal{B}_F(\mathcal{H})^-$. Still another proof comes from noting that $T \in$

$\mathcal{B}(\mathcal{H})$ is compact if and only if it transforms sequences converging weakly to zero into sequences converging to zero in norm. Hence if K is compact and $\{x_n\}_{n\in\mathbb{N}}$ converges weakly to zero, then we have $\|K^*x_n\|^2 = (KK^*x_n, x_n) \leq \|K(K^*x_n)\|\,\|x_n\| \to 0$ since $\{K^*x_n\}_{n\in\mathbb{N}}$ also converges weakly and $\{\|x_n\|\}_{n\in\mathbb{N}}$ is bounded by the uniform boundedness principle. Thus K^* is compact.

2.8.5 Proposition *Let K be a compact operator on a Banach space $\mathcal{X}$. Let λ be a non-zero complex number and let n be a positive integer. Then $\ker((\lambda - K)^n)$ is finite-dimensional and $(\lambda - K)^n\mathcal{X}$ is closed.*

Proof Note that $(\lambda - K)^n$ may be written as $\lambda^n - KP$, where P is a polynomial in K. Hence KP is compact and it suffices to prove both statements for the case $n = 1$.

In order to show that $\ker(\lambda - K)$ is finite-dimensional, it is enough to show that every bounded sequence $\{x_n\}_{n\in\mathbb{N}}$ in $\ker(\lambda - K)$ contains a convergent subsequence. Since we may write $x_n = \lambda^{-1}(K + (\lambda - K))x_n = K(\lambda^{-1}x_n)$, this follows immediately from the compactness of K.

Suppose $(\lambda - K)\mathcal{X}$ is not closed. Then we may find a sequence $\{x_n\}_{n\in\mathbb{N}} \subseteq \mathcal{X}$ such that $\{(\lambda-K)x_n\}_{n\in\mathbb{N}}$ converges to $y \notin (\lambda-K)\mathcal{X}$. Replacing $\{x_n\}_{n\in\mathbb{N}}$ by a subsequence, we may assume that no x_n lies in $\ker(\lambda-K)$. Define d_n by $d_n = \operatorname{dist}(x_n, \ker(\lambda-K))$ and choose $y_n \in \ker(\lambda-K)$ to satisfy $\|y_n - x_n\| < 2d_n$.

Suppose $\|y_n - x_n\|$ is a bounded sequence. Then we may replace it by a subsequence so that $\{K(y_n - x_n)\}_{n\in\mathbb{N}}$ converges. In this case

$$
\begin{aligned}
y_n - x_n &= \lambda^{-1}[(\lambda - K) + K](y_n - x_n) \\
&= -\lambda^{-1}(\lambda - K)x_n + \lambda^{-1}K(y_n - x_n)
\end{aligned}
$$

converges to some z which satisfies $y = \lim(\lambda - K)(y_n - x_n) = (\lambda - K)z$. Since this contradicts our choice of y, we may assume that $\|y_n - x_n\|$ is an unbounded sequence. Replacing it by a subsequence, we may assume $\|y_n - x_n\| \to \infty$. Define z_n by

$$
z_n = \|y_n - x_n\|^{-1}(y_n - x_n)
$$

so that we have $\|z_n\| = 1$,

$$
(\lambda - K)z_n = \|y_n - x_n\|^{-1}(\lambda - K)(x_n) \to 0y = 0
$$

and

$$
z_n = \lambda^{-1}((\lambda - K)z_n + Kz_n).
$$

Since K is compact, these facts show that we may replace $\{z_n\}_{n\in\mathbb{N}}$ by a subsequence which has a limit z. Clearly $(\lambda - K)z = \lim(\lambda - K)z_n = 0$ implies that z belongs to $\ker(\lambda - K)$. Define w_n by $w_n = y_n + \|y_n - $

$x_n \| z \in \ker(\lambda - K)$. Thus we find $\|x_n - w_n\| \geq \text{dist}(x_n, \ker(\lambda - K)) = d_n$. However, $x_n - w_n = x_n - y_n - \|y_n - x_n\| z = \|y_n - x_n\|(z_n - z)$ shows that $\|x_n - w_n\| < 2d_n \|z_n - z\|$. However this contradicts $z = \lim z_n$. Thus we conclude that $(\lambda - K)\mathcal{X}$ is closed. $\square$

2.8.6 Proposition *Let K be a compact operator on the Banach space $\mathcal{X}$. If $\{\lambda_n\}_{n \in \mathbb{N}}$ is a sequence of distinct characteristic values of K, then the sequence $\{\lambda_n\}_{n \in \mathbb{N}}$ converges to zero.*

Proof Assume the sequence does not converge to zero. Then we may replace it by a subsequence which satisfies $|\lambda_n| \geq \varepsilon$ for all n where ε is some fixed positive number.

For each n, choose $x_n \in \mathcal{X}$ satisfying $\|x_n\| = 1$ and $Kx_n = \lambda_n x_n$. We claim that $\{x_n : n \in \mathbb{N}\}$ is a linearly independent set. If not, let n be the smallest integer for which $x_1, x_2, \ldots, x_n$ is not linearly independent so that we can write

$$x_n = \alpha_1 x_1 + \alpha_2 x_2 + \cdots + \alpha_{n-1} x_{n-1}$$

for $\alpha_j \in \mathbb{C}$. Applying $K - \lambda_n$ gives

$$0 = \alpha_1(\lambda_1 - \lambda_n)x_1 + \alpha_2(\lambda_2 - \lambda_n)x_2 + \cdots + \alpha_{n-1}(\lambda_{n-1} - \lambda_n)x_{n-1}.$$

The linear independence of $x_1, x_2, \ldots, x_{n-1}$ and the non-zero character of $\lambda_j - \lambda_n$ for $j = 1, 2, \ldots, n-1$ imply $0 = \alpha_1 = \alpha_2 = \cdots = \alpha_{n-1}$ which is inconsistent with $\|x_n\| = 1$. Hence $\{x_n : n \in \mathbb{N}\}$ is linearly independent.

Define $\mathcal{Y}_n$ to be the closed subspace span $\{x_1, x_2, \ldots, x_n\}$ and apply the Riesz Lemma (see §1.7.7) to $\mathcal{Y}_{n-1} \not\subseteq \mathcal{Y}_n$ to obtain $y_n \in \mathcal{Y}_n$ satisfying $\|y_n\| = 1$ and $\text{dist}(y_n, \mathcal{Y}_{n-1}) > 1/2$. For $n > m$ we have

$$\|K(\lambda_n^{-1} y_n) \quad - \quad K(\lambda_m^{-1} y_m)\|$$
$$= \quad \|y_n - (\lambda_n^{-1}(\lambda_n - K)y_n + \lambda_m^{-1} K y_m)\| > 1/2$$

since $(\lambda_n - K)y_n$ and Ky_m both belong to $\mathcal{Y}n - 1$. However this contradicts the fact that $\{K(\lambda_n^{-1} y_n)\}_{n \in \mathbb{N}}$ must have a convergent subsequence since K is compact and $\{\lambda_n^{-1} y_n\}_{n \in \mathbb{N}}$ is a bounded sequence. Thus our assumption that $\{\lambda_n\}_{n \in \mathbb{N}}$ did not converge to zero is untenable. $\square$

2.8.7 Theorem *Let K be a compact operator on a Banach space $\mathcal{X}$ which is not finite-dimensional. Then $Sp(K)$ contains zero and $Sp(K) \setminus \{0\}$ is either finite or a countable set which converges to zero whenever ordered as a sequence. Any $\lambda_0 \in Sp(K) \setminus \{0\}$ satisfies:*

(a) *λ_0 is an isolated point in $Sp(K)$.*

(b) *There is a projection $E_{\lambda_0} \in \{K\}''$ such that the subspace $E_{\lambda_0}\mathcal{X}$ is finite-dimensional, $Sp(K|E_{\lambda_0}\mathcal{X}) = \{\lambda_0\}$ and $\lambda_0 \notin Sp(K|(I - E_{\lambda_0})\mathcal{X})$.*

(c) *λ_0 is a characteristic value of both K and $K^* \in \mathcal{B}(\mathcal{X}^*)$ satisfying:*

$$0 < \dim(\ker(\lambda_0 - K)^n) = \dim(\ker(\lambda_0 - K^*)^n) < \infty \qquad \forall\, n \in \mathbb{N};$$

$$0 < \alpha(\lambda_0 - K) = \alpha(\lambda_0 - K^*) = \delta(\lambda_0 - K) = \delta(\lambda_0 = K)^*) < \infty;$$

$$E_{\lambda_0}(\mathcal{X}) = \ker((\lambda_0 - K)^a); \quad E_{\lambda_0}^*(\mathcal{X}^*) = \ker((\lambda_0 - K^*)^a);$$

$$(I - E_{\lambda_0})\mathcal{X} = (\lambda_0 - K)^a \mathcal{X}; \quad (I - E_{\lambda_0}^*)\mathcal{X}^* = (\lambda_0 - K)^* \mathcal{X},$$

where $a = \alpha(\lambda_0 - K)$ is the ascent of $(\lambda_0 - K)$.

Proof Suppose $Sp(K)$ has $\lambda_0 \neq 0$ as a limit point. Choose a sequence $\{\lambda_n\}_{n \in \mathbb{N}} \subseteq Sp(K) \setminus \{0\}$ of distinct numbers converging to λ_0. The last proposition shows that $\lambda_n - K$ is injective except for a finite set of indices. If $(\lambda_n - K)\mathcal{X} = \mathcal{X}$, the inverse boundedness theorem shows that any injective $\lambda_n - K$ is invertible, contradicting $\lambda_n \in Sp(K)$. Since $(\lambda_n - K)\mathcal{X}$ is always closed, this shows that $(\lambda_n - K)\mathcal{X}$ cannot be dense. However if $(\lambda_n - K)\mathcal{X}$ is not dense, then the Hahn–Banach theorem gives an $x^* \in \mathcal{X}^*$ satisfying $x^*((\lambda_n - K)\mathcal{X}) = (\lambda_n - K^*)(x^*)\mathcal{X} = 0$ and hence belonging to $\ker(\lambda_n - K^*)$. However, since K^* is compact by Schauder's theorem, the last proposition shows that $\lambda_n - K^*$ is injective for all but finitely many indices. This contradiction shows that $Sp(K)$ has no limit points except possibly zero. Hence $Sp(K) \setminus \{0\}$ contains only isolated points and is either finite or a sequence converging to zero.

Let λ_0 be an element of $Sp(K) \setminus \{0\}$ and let Γ be a circle centered at λ_0 and of such small radius that the intersection of $Sp(K)$ and the closed disc bounded by Γ is just λ_0. Consider Γ as a positively oriented contour (in the usual sense of complex analysis) and define an operator E_{λ_0} by

$$E_{\lambda_0} = \frac{1}{2\pi i} \int_\Gamma (\lambda - K)^{-1} d\lambda$$

as described in Definition 3.3.6. The theorem just mentioned shows that E_{λ_0} is a projection operator (*i.e.*, an idempotent) satisfying $E_{\lambda_0} \in \{K\}''$ and $E_{\lambda_0} = K T_{\lambda_0}$ where T_{λ_0} is defined by

$$T_{\lambda_0} = \frac{1}{2\pi i} \int_\Gamma \lambda^{-1}(\lambda - K)^{-1} d\lambda.$$

Hence E_{λ_0} is a compact operator so that $E_{\lambda_0}(\mathcal{X})$ (on which E_{λ_0} acts as the identity element) is finite-dimensional. Furthermore, the spectrum of the restriction $K_{\lambda_0} = K|_{E_{\lambda_0}\mathcal{X}}$ of K to $E_{\lambda_0}\mathcal{X}$ is exactly λ_0. The spectral theory of Theorem 2.7.13 may now be applied to K_{λ_0}. It shows $E_{\lambda_0}\mathcal{X} = \ker((\lambda_0 - K_{\lambda_0})^a)$ where a is the index of nilpotence of $\lambda_0 - K_{\lambda_0}$ which is the ascent of $\lambda_0 - K$. Note that Theorem 3.3.7 shows that the restriction of

$\lambda_0 - K$ to $(I - E_{\lambda_0})\mathcal{X}$ is invertible which implies $(I - E_{\lambda_0})\mathcal{X} \subseteq (\lambda_0 - K)^a \mathcal{X}$. This inclusion is an equation since any $x \in \mathcal{X}$ satisfies $E_{\lambda_0}(\lambda_0 - K)^a x = (\lambda_0 - K)^a E_{\lambda_0} x = 0$. If n is strictly less than a, then $(\lambda_0 - K)^n \mathcal{X}$ has nontrivial intersection with $E_{\lambda_0}\mathcal{X}$ so that a is also the descent of $\lambda_0 - K$.

Since the adjoint operation $*: \mathcal{B}(\mathcal{X}) \to \mathcal{B}(\mathcal{X}^*)$ is a linear isometry, we find

$$E_{\lambda_0}^* = \frac{1}{2\pi i} \int_{\Gamma} (\lambda - K^*)^{-1} d\lambda.$$

Since K^* is also a compact operator, the results already established show that the common value a of ascent $(\lambda_0 - K^*)$ and descent $(\lambda_0 - K^*)$ satisfies

$$E_{\lambda_0}^* = \ker((\lambda_0 - K^*)^a); \quad (I - E_{\lambda_0}^*)\mathcal{X}^* = (\lambda_0 - K^*)^a \mathcal{X}^*.$$

If x^* belongs $(E_{\lambda_0}\mathcal{X})^*$ and satisfies $x^*(\lambda_0 - K)E_{\lambda_0}\mathcal{X} = \{0\}$, then $x^* \circ E_{\lambda_0}$ belongs to $\ker(\lambda_0 - K^*)$. If d_{λ_0} and n_{λ_0} represent the dimension of $E_{\lambda_0}\mathcal{X}$ and of $\ker(\lambda_0 - K_{\lambda_0})$, respectively, then Theorem 2.7.13 shows that $(\lambda_0 - K_{\lambda_0})E_{\lambda_0}\mathcal{X}$ has dimension $n_{\lambda_0} - d_{\lambda_0}$ so that the linear space of $x^* \in (E_{\lambda_0}\mathcal{X})^*$ vanishing on $(\lambda_0 - K_{\lambda_0})E_{\lambda_0}\mathcal{X}$ has a basis $x_1^*, x_2^*, \ldots, x_d^*$. Clearly $B = \{x_1^* \circ E_{\lambda_0}, x_2^* \circ E_{\lambda_0}, \ldots, x_d^* \circ E_{\lambda_0}\}$ is a linearly independent set in $\ker(\lambda_0 - K^*)$. However we claim that this set spans $\ker(\lambda_0 - K^*)$ so that we have

$$\dim \ker(\lambda_0 - K) = d_{\lambda_0} = \dim \ker(\lambda_0 - K^*).$$

To see this, suppose $x^* \in \ker(\lambda_0 - K^*)$. Since $\lambda_0 - K$ is invertible on $I - E_{\lambda_0}$, we have $x^* = x^* \circ E_{\lambda_0} + x^* \circ I - E_{\lambda_0} = x^* \circ E_{\lambda_0} + x^* \circ (\lambda_0 - K) \circ T_{\lambda_0} = x^* \circ E_{\lambda_0}$ for a suitable T_{λ_0}. Hence x^* can be written as $x^* \circ E_{\lambda_0}$ for some $x^* \in (E_{\lambda_0}\mathcal{X})^*$ which must satisfy $x^*(\lambda_0 - K)E_{\lambda_0} = 0$. This shows that B spans $\ker(\lambda_0 - K)$. The equation $(\lambda_0 - K)^n = \lambda_0^n - K$ again shows that

$$\dim \ker(\lambda_0 - K)^n) = \dim \ker((\lambda_0 - K^*)^n)$$

follows for all $n \in \mathbb{N}$ from the case $n = 1$ which we have just proved. This of course shows $\dim(E_{\lambda_0}\mathcal{X}) = \dim((\lambda_0 - K^*)^a) = \dim(E_{\lambda_0}^*\mathcal{X})$ and $\alpha(\lambda_0 - K) = \alpha(\lambda_0 - K^*)$. □

Hyperinvariant Subspaces

The spectral results proved above are more than enough to settle a basic question about compact operators. When considering bounded linear operators on Banach spaces, one naturally wonders whether such an operator necessarily has a nontrivial closed invariant subspace (*i.e.*, for $T \in \mathcal{B}(\mathcal{X})$, does there exist a closed subspace $\mathcal{Y}$ of the Banach space $\mathcal{X}$ satisfying $\mathcal{Y} \neq \{0\}, \mathcal{Y} \neq \mathcal{X}$ and $T\mathcal{Y} \subseteq \mathcal{Y}$?). Of course one assumes that $\mathcal{X}$ is not one-dimensional. Results from the last section give an easy affirmative answer in the finite-dimensional case. If T is an operator on a finite-dimensional space and T is not a constant multiple of the identity (in which case every subspace is invariant), then Theorem 2.7.13 shows that we need only consider the case in which T has a single eigenvalue and a non-zero nilpotent summand N. Then ker N is a nontrivial subspace (closed with respect to any norm) invariant under not only T but any operator which commutes with T.

Infinite-dimensional Banach spaces have been constructed with non-zero operators on them which have no closed invariant subspaces: Per Enflo [1976], [1987], Charles J. Read [1984], [1985], Bernard Beauzamy [1985]. Before these difficult constructions were discovered, Nachman Aronszajn and Kennan T. Smith [1954] showed that any compact operator on a Banach space, not of dimension one, also has a nontrivial closed invariant subspace. The Aronszajn–Smith result was extended somewhat but its proof remained very difficult for twenty years until a Russian mathematician, Victor I. Lomonosov [1974], introduced a new and radically simpler method of proof. It was quickly realized that the new method of proof would give much stronger results. The following easy, elementary result is from some unpublished notes of Carl M. Pearcy and Alan L. Shields which incorporate an idea of Hugh M. Hilden.

2.8.8 Definition   An operator K on a Banach space $\mathcal{X}$ is said to have a *hyperinvariant subspace* if there is a closed subspace $\mathcal{Y}$ of K satisfying $\mathcal{Y} \neq \{0\}, \mathcal{Y} \neq \mathcal{X}$, and $T\mathcal{Y} \subseteq \mathcal{Y}$ for all $T \in \{K\}'$.

2.8.9 Theorem   *Any non-zero compact operator on an infinite-dimensional Banach space has a hyperinvariant subspace.*

Proof   Let the compact operator K act on an infinite-dimensional Banach space $\mathcal{X}$. If K has a non-zero complex number λ in its spectrum, then λ is a characteristic value and the finite-dimensional subspace $\ker(\lambda - K)$ is hyperinvariant. Hence we may assume that K is topologically nilpotent. We may also assume $\|K\| = 1$. Choose $x_0 \in \mathcal{X}$ satisfying $\|Kx_0\| > 1$. Define $\mathcal{S} \subseteq \mathcal{X}$ and $\mathcal{T} \subseteq \mathcal{X}$ by $\mathcal{S} = \{x \in \mathcal{X} : \|x - x_0\| \leq 1\} = x_0 + \mathcal{X}_1$ and $\mathcal{T} = K\mathcal{S}^-$. Neither $\mathcal{S}$ nor $\mathcal{T}$ contain zero and $\mathcal{T}$ is compact.

The proof is by contradiction. Suppose that K has no hyperinvariant subspace. Then for each non-zero $x \in \mathcal{X}$, the non-zero $\{K\}'$-invariant closed subspace $\{K\}'x^-$ must equal $\mathcal{X}$. Hence $\{K\}'x$ is dense for each non-zero $x \in \mathcal{X}$. For each $T \in \{K\}'$, define an open set $\mathcal{N}_T$ by

$$\mathcal{N}_T = \{x \in \mathcal{X} : \|Tx - x_0\| < 1\}.$$

Since $\{K\}'x$ is dense for each non-zero x, we have

$$\bigcup_{T \in \{K\}'} \mathcal{N}_T = \mathcal{X} \setminus \{0\}.$$

Therefore there is a finite subset $\mathcal{F} \subseteq \{K\}'$ such that

$$\bigcup_{T \in \mathcal{F}} \mathcal{N}_T$$

includes the compact set $\mathcal{T}$. We will now construct a sequence $\{T_n\}_{n \in \mathbb{N}}$ inductively where each T_n is chosen from $\mathcal{F}$. Since Kx_0 belongs to $\mathcal{T}$, we can find $T_1 \in \mathcal{F}$ such that $T_1 K x_0$ belongs to $\mathcal{S}$. Hence $K T_1 K x_0$ belongs to $\mathcal{T}$ so that we can find $T_2 \in \mathcal{F}$ such that $T_2 K T_1 K x_0$ belongs to $\mathcal{S}$. Continuing in this way, for any n we find

$$x_n = T_n K T_{n-1} K \cdots T_2 K T_1 K x_0$$

in $\mathcal{S}$. Since each T_j belongs to $\mathcal{F} \subseteq \{K\}'$, we can rewrite this as

$$x_n = T_n T_{n-1} \cdots T_2 T_1 K^n x_0.$$

Define d and M by $d = \text{dist}(\mathcal{S}, 0)$ and $M = \max\{\|T\| : T \in \mathcal{F}\}$. Then we have

$$d \leq \|x_n\| \leq \left(\sum_{j=1}^{n} \|T_j\| \right) \|K^n\| \; \|x_0\| \leq M^n \|K^n\| \; \|x_0\|.$$

Dividing by $\|x_0\|$ and taking nth roots, we get

$$\left(\frac{d}{\|x_0\|} \right)^{1/n} \leq M \|K^n\|^{1/n}.$$

This is a contradiction since K is topologically nilpotent so that the right side approaches 0. This contradiction shows that K has a hyperinvariant subspace.   □

2.8.10 Essential spectrum   We wish to mention this concept briefly and give some references. Related results can be found in §1.7.16 and Chapter 8. Let $\mathcal{X}$ be a Banach space. The *essential spectrum* of an operator $T \in \mathcal{B}(\mathcal{X})$ is the spectrum of its image $T + \mathcal{B}_K(\mathcal{X})$ in the generalized Calkin algebra $\mathcal{Q}(\mathcal{X}) = \mathcal{B}(\mathcal{X})/\mathcal{B}_K(\mathcal{X})$. It is closely related to the sometimes slightly larger *Weyl spectrum* which is the intersection of $Sp(T + K)$ for all $K \in \mathcal{B}_K(\mathcal{X})$. These smaller spectra overlook spectral points where the singularity is not "too bad". A *Riesz operator* in $\mathcal{B}(\mathcal{X})$ is one with essential spectrum $\{0\}$. Thus they are the operators with topologically nilpotent image in the Calkin algebra. Trevor T. West has shown that on Hilbert space each Riesz operator can actually be written as the sum of a compact and a topologically nilpotent operator. These terms are sometimes used in a more generalized setting with another ideal playing the role of the compact operators or even another algebra playing the role of $\mathcal{B}(\mathcal{X})$. See the references in §1.7.16, Murphy and West [1980b] and Jaroslav Zemánek [1983].

Local Spectral Theory and Decomposable Operators

Local spectral theory associates subsets of the spectrum of an operator $T \in \mathcal{B}(\mathcal{X})$ on a Banach space $\mathcal{X}$ with subspaces of $\mathcal{X}$. In [1954] Nelson Dunford introduced the theory of spectral operators in order to extend the theory of normal operators to somewhat more general operators on Hilbert space and to Banach spaces. We have already shown in Theorem 2.7.13 that every finite dimensional operator is spectral, and it was originally hoped that many operators of analysis (both bounded and unbounded), which were known to have rich spectral theories, would turn out to be spectral operators. Despite the substantial success recorded in Dunford and Schwartz [1971], this important pioneering theory is too restrictive to encompass the most interesting examples. However, Dunford's concept of the single valued extension property has been basic to all more recent theories. The books by Ion Colojoară and Ciprian Foiaş [1968] and Florian Horia Vasilescu [1982] record other successful attempts at generalization. In [1963], Foiaş defined decomposable operators. To date, this concept has achieved the best balance between strong enough hypotheses to give a rich theory and weak enough hypotheses to contain the most important and interesting examples. We treat only the case of bounded operators.

The following definition is a simplification, due to Ernst Albrecht [1979], of the original definition of decomposable operators. Property (β) and an early version of (δ) are due to Errett Bishop [1959].

2.8.11 Definition Let $\mathcal{X}$ be a Banach space and let T be an operator in $\mathcal{B}(\mathcal{X})$. If U and V are open subsets of $\mathbb{C}$ with union $\mathbb{C}$, then two closed T-invariant subspaces $\mathcal{Y}$ and $\mathcal{Z}$ of $\mathcal{X}$ are said to *decompose* T relative to U and V if they satisfy

$$Sp(T|_{\mathcal{Y}}) \subseteq U \qquad Sp(T|_{\mathcal{Z}}) \subseteq V \qquad \text{and} \qquad \mathcal{Y} + \mathcal{Z} = \mathcal{X}.$$

The operator T is said to be *decomposable* if it has decomposition relative to any open cover $\{U, V\}$ of $\mathbb{C}$.

For any open subset U of $\mathbb{C}$, let $\mathcal{O}(U, \mathcal{X})$ be the linear space of all analytic functions $f: U \to \mathcal{X}$ (*cf.* Definition 2.2.10). Give this linear space the topology of pointwise convergence which is uniform on all compact subsets of U, which makes it a Fréchet space. Each $T \in \mathcal{B}(\mathcal{X})$ defines a continuous linear map $\tilde{T}_U : \mathcal{O}(U, \mathcal{X}) \to \mathcal{O}(U, \mathcal{X})$. If $\tilde{T}_U$ is injective and has closed range for each open set U, then T is said to have *property* (β).

For each closed set F in $\mathbb{C}$, let $\mathcal{X}_F$ be the set of vectors $x \in \mathcal{X}$ that are in the range of $T_{\mathbb{C}\backslash F}$ (in the sense that the constant function x on $\mathbb{C} \backslash F$ is in this range). The operator T is said to have the *decomposition property* (δ) provided that $\mathcal{X}_{\overline{U}} + \mathcal{X}_{\overline{V}}$ equals $\mathcal{X}$ for each open cover $\{U, V\}$ of $\mathbb{C}$.

The class of decomposable operators contains not only normal operators on a Hilbert space but also (on any Banach space) compact operators (indeed, all operators with totally disconnected spectra), spectral operators, generalized scalar operators and surjective isometries. Many interesting partial differential operators are unbounded decomposable operators

2.8.12 Theorem *Let $\mathcal{X}$ be a Banach space and let $T \in \mathcal{B}(\mathcal{X})$ be an operator.*

(a) *T has property $\langle\ (\beta)\ /\ (\delta)\ \rangle$ if and only if $T^* \in \mathcal{B}(\mathcal{X}^*)$ has property $\langle\ (\delta)\ /\ (\beta)\ \rangle$.*

(b) *T has property $\langle\ (\beta)\ /\ (\delta)\ \rangle$ if and only if T is similar to the $\langle$ restriction / quotient $\rangle$ of a decomposable operator $\overline{T}$ $\langle$ to / by $\rangle$ a $\overline{T}$-invariant subspace.*

(c) *T is decomposable if and only if it has both properties (β) and (δ).*

For proofs and further details see Albrecht and Jórge Eschmeier [1993], Albrecht, Eschmeier and Michael M. Neumann [1986], Kjeld B. Laursen and Neumann [1992], Neumann [1992] and Vivien G. Miller [1993].

2.9 Topological Algebras

This work discusses abstract algebras (*i.e.*, those without topology), normed algebras and Banach algebras in some detail. It has been a deliberate decision to exclude any general discussion of the many classes of topological algebras which have been studied, since otherwise the book would have grown beyond any bound. Many results in the book extend routinely from normed algebras to some wider class or classes of topological algebras, but we will seldom point this out. However, some important applications involve non-normed topological algebras, so we need to give an outline of certain parts of their theory. We shall concentrate on those portions which are closest to Banach algebra theory. The discussion here will be brief and without proofs. It was postponed to this chapter because at several points it involves the hypothesis that the group of quasi-invertible elements should be open.

2.9.1 Definition A *Fréchet algebra*, sometimes abbreviated *F-algebra*, is a topological algebra in which the topology arises from a complete metric.

Although we only require multiplication to be continuous in each variable separately in a topological algebra, Richard Arens [1947] proved that multiplication is actually jointly continuous in a Fréchet algebra. Furthermore Stefan Banach [1948], with improvements by Wiesław Żelazko [1960] and by Andrew M. Gleason in his *Mathematical Reviews* review of that paper, showed that the quasi-inversion map ($a \mapsto a^q$ for all $a \in \mathcal{A}_{qG}$) is continuous if and only if the set of quasi-invertible elements is a G_δ-set. We state these results as a formal theorem.

2.9.2 Theorem *Let $\mathcal{A}$ be a Fréchet algebra. Then multiplication is jointly continuous and the topology may be defined by a norm-like function $\|\cdot\|: \mathcal{A} \to \mathbb{R}_+$ satisfying:*

$$\|a\| = 0 \text{ if and only if } a = 0 \qquad \forall\, a \in \mathcal{A} \qquad (1)$$

$$\|a + b\| \leq \|a\| + \|b\| \qquad \forall\, a, b \in \mathcal{A} \qquad (2)$$

$$\|\lambda a\| = |\lambda|\,\|a\| \text{ if } \lambda \in \mathbb{C} \text{ and } |\lambda| = 1 \qquad \forall\, a \in \mathcal{A} \qquad (3)$$

$$\|\lambda_n a\| \to 0 \text{ if } \{\lambda_n\}_{n \in \mathbb{N}} \subseteq \mathbb{C} \text{ and } |\lambda_n| \to 0 \qquad \forall\, a \in \mathcal{A} \qquad (4)$$

$$\|\lambda a_n\| \to 0 \text{ if } \{a_n\}_{n \in \mathbb{N}} \subseteq \mathcal{A} \text{ and } \|a_n\| \to 0 \qquad \forall\, \lambda \in \mathbb{C} \qquad (5)$$

$$\mathcal{A} \text{ is complete in the metric: } \delta(a, b) = \|a - b\| \qquad \forall\, a, b \in \mathcal{A}. \qquad (6)$$

The quasi-inversion map (and therefore the inversion map in the unital case) is continuous if and only if the set $\mathcal{A}_{qG}$ of quasi-invertible elements is a G_δ set. Finally, a metric algebra may be completed to be a Fréchet algebra if and only if multiplication is jointly continuous.

Following the important and still useful early exposition by Ernest A. Michael [1952], many authors restrict the term Fréchet algebra to the case where the complete topology arises from a countable family of algebra seminorms. We will introduce this class briefly below, but first we discuss another property. A good source for some of this material is Żelazko [1965].

2.9.3 Definition In a topological linear space, a set $\mathcal{B}$ is said to be *bounded* if for any neighborhood $\mathcal{U}$ of 0 there is some non-zero scalar λ such that $\lambda \mathcal{B} \subseteq \mathcal{U}$. A topological linear space is said to be *locally bounded* if there exists some neighborhood of 0 which is bounded.

In a normed or metrizable topological linear space, balls around the origin are obviously bounded neighborhoods, so such spaces are locally bounded. In the complete case the converse is true, and more surprisingly, the metric can always be chosen to be subadditive and to have a strong substitute for homogeneity. If $\mathcal{X}$ is a locally bounded topological linear space and $\mathcal{U}$ is a bounded neighborhood of 0, then

$$\left\{\frac{1}{n}\mathcal{U} \ : \ n \in \mathbb{N}\right\} \qquad (7)$$

is a countable basis of neighborhoods of 0 so the linear space is metrizable. In fact, Stefan Rolewicz [1957] shows that the topology can be given by a p-homogeneous norm, which is a norm-like function $\|\cdot\|: \mathcal{X} \to \mathbb{R}_+$ satisfying:

$$\|a\| \quad = \quad 0 \text{ if and only if } a = 0 \qquad \forall\, a \in \mathcal{X}$$
$$\|a + b\| \quad \leq \quad \|a\| + \|b\| \qquad\qquad\qquad \forall\, a, b \in \mathcal{X}$$
$$\|\lambda a\| \quad = \quad |\lambda|^p \|a\| \qquad\qquad\qquad\quad \forall\, \lambda \in \mathbb{C}, a \in \mathcal{X}$$

for some particular $0 < p \leq 1$. Obviously a p-homogeneous norm is a norm (in the usual sense) if and only if p is 1. Locally bounded linear spaces resemble normed linear spaces rather closely, and for complete ones the resemblance to Banach spaces is even closer. We will now characterize the complete locally bounded algebras, beginning with a brief consideration of general Fréchet algebras.

2.9.4 Theorem *The following are equivalent for a Fréchet algebra $\mathcal{A}$:*
(a) *There is a metric δ defining the topology and satisfying*

$$\delta(ab, 0) \leq \delta(a, 0)\delta(b, 0) \qquad \forall\, a, b \in \mathcal{A}; \tag{8}$$

(b) *$\mathcal{A}$ is locally bounded;*
(c) *For some $0 < p \leq 1$ there is a submultiplicative p-homogeneous norm defining the topology.*

Algebras satisfying these hypotheses share most of the properties of Banach algebras except that, since they are not necessarily locally convex, the space of linear functionals on the algebra may be too small to be of use. The topology of a topological linear space which is both locally bounded and locally convex arises from a norm. We now explore the locally convex topological algebras.

Locally convex topological linear spaces were introduced briefly prior to Definition 1.1.8. Topological algebras in which the underlying topological linear space is locally convex do not have exactly the properties which we desire. Thus, we specialize the concept a bit further.

2.9.5 Definition A *locally multiplicatively convex algebra* is a topological algebra in which the topology is given by some collection of (ordinary) algebra semi-norms. If P is a set of algebra semi-norms on an algebra $\mathcal{A}$, then a net $\{a_\delta\}_{\delta \in \Delta}$ in $\mathcal{A}$ converges to $a \in \mathcal{A}$ with respect to this set if

$$\sigma(a_\delta - a) \to 0 \qquad \forall\, \sigma \in P. \tag{9}$$

Locally multiplicatively convex algebras are often called *lmc-algebras*. Obviously, multiplication is jointly continuous and quasi-inversion is continuous in these algebras. There are many cases in which the collection of algebra semi-norms can be taken to be countable. As noted previously for linear spaces, if $\{\sigma_n : n \in \mathbb{N}\}$ is a countable collection of algebra semi-norms, then the same topology is given by the metric defined by

$$\delta(a, b) = \sum_{n=1}^{\infty} \frac{2^{-n}\sigma_n(a - b)}{1 + \sigma_n(a - b)} \qquad \forall\, a, b \in \mathcal{A}. \tag{10}$$

The following important result was obtained by Michael in his seminal memoir [1952], where it was used to show that this class of algebras has some of the behavior usually associated with algebras having an open set of quasi-invertible elements.

2.9.6 Theorem *Every locally multiplicatively convex Fréchet algebra is the projective limit of a projective system of Banach algebras. An element in the algebra is quasi-invertible if and only if its representative from each of these Banach algebras is quasi-invertible.*

2.9.7 The Algebra $\mathcal{F} = \mathbb{C}[[z]]$ of Formal Power Series   An element of $\mathcal{F}$ is a formal power series $\sum_{n=0}^{\infty} \alpha_n z^n$ without any question of convergence, so $\{\alpha_n\}_{n \in \mathbb{N}^0} \subseteq \mathbb{C}$ is an arbitrary sequence. The algebraic operations are the usual ones for power series:

$$\sum_{n=0}^{\infty} \alpha_n z^n \sum_{n=0}^{\infty} \beta_n z^n = \sum_{n=0}^{\infty} \sum_{k=0}^{n} \alpha_k \beta_{n-k} z^n.$$

We will give $\mathcal{F}$ the *topology of coordinatewise convergence*. The following family of semi-norms shows that $\mathcal{F}$ is locally multiplicatively convex:

$$\| \sum_{n=0}^{\infty} \alpha_n z^n \|_m = \sum_{n=0}^{m} |\alpha_n| \qquad \forall \, m \in \mathbb{N}^0.$$

Theorem   *For each $m \in \mathbb{N}^0$, $\| \cdot \|_m$ is an algebra semi-norm on $\mathcal{F}$. With the topology induced by this family of semi-norms, $\mathcal{F}$ is a locally multiplicatively convex Fréchet algebra. Furthermore, this is the only Fréchet algebra topology on $\mathcal{F}$.*

The last sentence is due to Graham R. Allan [1972]. Of course this implies that $\mathcal{F}$ is not a Banach algebra under any norm. Nevertheless its group of invertible elements $\mathcal{F}_G = \{\sum_{n=0}^{\infty} \alpha_n z^n : \alpha_0 \neq 0\}$. is obviously open. The algebra $\mathcal{F}$ also has a decreasing sequence of closed ideals $\mathcal{F}_m$ with intersection $\{0\}$:

$$\mathcal{F}_m = \{\sum_{n=0}^{\infty} \alpha_n z^n : \alpha_n = 0 \text{ for all } n \leq m\} \qquad \forall \, m \in \mathbb{N}^0.$$

Note that the whole algebra is the disjoint union of $\mathcal{F}_G$ and $\mathcal{F}_0$. In the terminology of Section 6.4 below, it is known that all derivations of $\mathcal{F}$ into itself are continuous.

3

Commutative Algebras and Functional Calculus

Introduction

In this work we are relatively little concerned with commutative algebras. The involution in the *-algebras, which are the main subject of study here, is a sort of weak substitute for commutativity. However, some commutative theory is essential in dealing with the more complicated noncommutative theory. In particular, commutativity is intimately associated with the various functional calculi which we will introduce in this chapter and later.

Section 3.1 presents the remarkably simple theory, created by Israel Moiseevič Gelfand [1941a], which replaces any commutative Banach algebra by a homomorphic image (modulo its Jacobson radical) which is an algebra of complex-valued continuous functions vanishing at infinity on a locally compact topological space. This canonically constructed space is called the Gelfand space. (In many applications the homomorphism is an isomorphism.) We show that all the, now classical, results in this theory hold for an algebra if and only if it is an almost commutative spectral algebra where almost commutative means commutative modulo its Jacobson radical. A number of other equivalent conditions are given in Theorem 3.1.5. They can be briefly summarized by saying that if the spectrum or spectral radius of an algebra has any similarity to the spectrum or spectral radius of an algebra of functions, then the algebra has a full Gelfand theory. Theorem 3.1.5 and Proposition 3.1.7 give some results which imply that a Banach algebra is either commutative or almost commutative. We postpone others to Chapter 4, where we can use Sinclair's density theorem to give an elegant proof (Theorem 4.2.18) due to Jaroslav Zemánek [1982]. Theorem 3.1.11, due to Georgi Evgenyevič Šilov [1947a], shows that any homomorphism from a Banach algebra into a semisimple commutative Banach algebra is automatically continuous. The Gelfand theory for double centralizer algebras, double duals with Arens multiplication and singly generated unital Banach algebras are all explored.

Section 3.2 introduces the Šilov boundary, hulls, kernels and the hull–kernel topology on the Gelfand space. Completely regular commutative Banach algebras are shown to be those commutative algebras in which the hull–kernel topology agrees with the original Gelfand topology of the

Gelfand space. Theorem 3.2.12 gives the Gelfand–Naĭmark characterization of algebras of bounded continuous functions on a topological space. This is used to construct the Stone–Čech compactification of a topological space and the Bohr compactification of a locally compact group. It is shown that the latter construction is injective if and only if the group has enough finite-dimensional representations to separate its points.

Much of the remainder of the chapter is devoted to functional calculi. These will be discussed at greater length when introduced. We treat two cases in depth in Sections 3.3 and 3.5, respectively. Theorem 3.3.7 details the theory (due to Gelfand [1941a]) in which a meaning is assigned to $f(a)$ when a is any element in any Banach algebra and f is a function which is analytic on some open set including the spectrum of a. The uniqueness of this construction is explored, and it is extended to semisimple spectral algebras in which the algebra of functions which act may be smaller. A spectral normed algebra with a full functional calculus is called a functional algebra.

Section 3.4 contains examples and various applications of the theory developed in the previous section to the construction of idempotents, square roots, exponentials and logarithms. Actually, we show how more elementary methods can be applied in many of these situations.

Section 3.5 treats the case in which $(a_1, a_2, \ldots, a_n)$ is an ordered n-tuple of commuting elements in a unital Banach algebra $\mathcal{A}$ and f is a function of n variables which is analytic in a neighborhood of the "joint spectrum" of $(a_1, a_2, \ldots, a_n)$ defined in some suitable sense. The proper definition in the most general case is not clear. In the easiest case, $\mathcal{A}$ is commutative and the proper meaning of joint spectrum (Definition 3.5.4) is obvious. Then $f(a_1, a_2, \ldots, a_n)$ also has a standard definition and very useful properties. These results are essentially due to Richard Arens and Alberto P. Calderon [1955] and Lucien Waelbroeck [1954] based on Šilov's proof [1953] for finitely generated commutative algebras. The construction of $f(a_1, a_2, \ldots, a_n)$ is still technically complicated enough that we do not prove the more recent results and present our proof in such a way that the result of an exterior algebra construction can be taken as an axiom, and its proof omitted by the reader who wishes to do so. (We do, however, give a complete proof which should be accessible without any previous knowledge of differential geometry.) Theorem 3.5.9 contains the main results. An implicit function theorem (Theorem 3.5.12, essentially due to Arens and Calderon [1955]) and the Šilov idempotent theorem (Corollary 3.5.13 and 3.5.15) are among a number of applications. Theorem 3.5.17 is an important result of Charles E. Rickart [1953] which shows that any (possibly incomplete) algebra norm on a semisimple completely regular Banach algebra is spectral. The Arens–Royden theorem (Theorem 3.5.19, proved by Arens [1963] and Halsey L. Royden [1963], independently) completes our investigation of the exponential function begun in Theorem 2.1.12. The section ends with a brief introduction to Joseph L. Taylor's construction [1970b], [1970c], [1972a] and [1972b] for commuting n-tuples of operators.

Section 3.6 gives an overview of the duality theory for locally compact abelian groups.

3.1 Gelfand Theory

We now turn to Gelfand theory. In Section 1.5 we briefly introduced algebras of complex-valued function under pointwise operations. This is the easiest sort of commutative algebra to understand. Gelfand theory allows us to study any commutative spectral algebra $\mathcal{A}$ in terms of such an algebra which is a homomorphic image of the original algebra with kernel equal to the Jacobson radical $\mathcal{A}_J$ of $\mathcal{A}$.

Let $\mathcal{A}$ be a linear space of complex-valued functions on a set Ω. Assume that $\mathcal{A}$ is closed under pointwise multiplication. In Section 1.5 we agreed to consider a space of functions such as $\mathcal{A}$ to be an algebra under pointwise operations, unless we specifically defined some other multiplication. We also noted there that, if each function in $\mathcal{A}$ is bounded, then

$$\|f\|_\infty = \sup\{|f(\omega)| : \omega \in \Omega\}$$

is an algebra norm on $\mathcal{A}$. Convergence relative to this norm is uniform convergence, so it is called the *uniform norm*. Note that the range $f(\Omega)$ of f is included in $Sp_{\mathcal{A}}(f)$ (since $\lambda - f$ assumes the value zero at $\omega \in \Omega$ if $\lambda = f(\omega)$). Thus $\rho_{\mathcal{A}}(f) \geq \|f\|_\infty$ holds for all $f \in \mathcal{A}$ so that no spectral semi-norm on $\mathcal{A}$ can be smaller than $\|\cdot\|_\infty$. Thus, if $\|\cdot\|_\infty$ is a spectral norm, then it equals the spectral radius.

Two specific Banach algebras of bounded continuous functions with the uniform norm as their complete norm were introduced in Section 1.5 and will be used again here. If Ω is a locally compact space, $\langle\, C(\Omega) \,/\, C_0(\Omega) \,\rangle$ is the algebra of all continuous complex-valued functions on Ω which $\langle$ are bounded / vanish at infinity $\rangle$. Obviously $C_0(\Omega)$ is always an ideal of $C(\Omega)$, and the two are equal if and only if Ω is compact.

3.1.1 Definition Let $\mathcal{A}$ be an algebra. Define $\Gamma_{\mathcal{A}}$ to be the set:

$$\Gamma_{\mathcal{A}} = \{\text{all non-zero algebra homomorphisms of } \mathcal{A} \text{ into } \mathbb{C}\}.$$

If $\Gamma_{\mathcal{A}}$ is nonempty, for each $a \in \mathcal{A}$, define $\hat{a}\colon \Gamma_{\mathcal{A}} \to \mathbb{C}$ by evaluation:

$$\hat{a}(\gamma) = \gamma(a) \qquad \forall\ \gamma \in \Gamma_{\mathcal{A}}.$$

The map $(\hat{\ })$ is called the *Gelfand homomorphism* and, for any $a \in \mathcal{A}$, $\hat{a}$ is called the *Gelfand transform of a*. The *Gelfand topology* is the weakest topology on $\Gamma_{\mathcal{A}}$ in which the maps $\hat{a}$ are continuous for all $a \in \mathcal{A}$. If it is nonempty, the set $\Gamma_{\mathcal{A}}$ together with this topology is called the *Gelfand space of* $\mathcal{A}$. The subset $\mathcal{A}_{\Gamma}$ of $\mathcal{A}$ on which the Gelfand map vanishes (or equivalently on which all $\gamma \in \Gamma_{\mathcal{A}}$ vanish) is called the *Gelfand radical of* $\mathcal{A}$. The Gelfand radical of an algebra $\mathcal{A}$ for which $\Gamma_{\mathcal{A}}$ is empty is $\mathcal{A}$ itself. We call $\mathcal{A}$ ⟨ *Gelfand-radical / Gelfand-semisimple* ⟩ if $\mathcal{A}_{\Gamma}$ equals ⟨ $\mathcal{A}$ / $\{0\}$ ⟩. Finally we call $\mathcal{A}$ *almost commutative* if it is commutative modulo its Jacobson radical (*i.e.*, if $\mathcal{A}/\mathcal{A}_J$ is commutative).

Note that an algebra homomorphism of an algebra $\mathcal{A}$ into $\mathbb{C}$ is either identically zero or surjective. The term Gelfand homomorphism is justified in Theorem 3.1.2.

In the Gelfand topology on $\Gamma_{\mathcal{A}}$, a neighborhood base at an element γ_0 is defined by $\{\mathcal{U}_{\mathcal{F}} : \mathcal{F}$ is a finite subset of $\mathcal{A}\}$ where

$$\mathcal{U}_{\mathcal{F}} = \{\gamma \in \Gamma_{\mathcal{A}} : |\gamma(a) - \gamma_0(a)| < 1 \text{ for all } a \in \mathcal{F}\}.$$

(There is no need to consider small positive ε's since the elements in $\mathcal{F}$ can be chosen of arbitrarily large norm.) A net $\{\gamma_\alpha\}_{\alpha \in A} \subseteq \Gamma_{\mathcal{A}}$ will converge to $\gamma \in \Gamma_{\mathcal{A}}$ in the Gelfand topology if and only if $\gamma_\alpha(a) = \hat{a}(\gamma_\alpha)$ converges to $\hat{a}(\gamma) = \gamma(a)$ for all $a \in \mathcal{A}$. If $(\mathcal{A}, \|\cdot\|)$ is a spectral normed algebra, we will see below that $\Gamma_{\mathcal{A}}$ is a subset of the unit ball of the dual space $\mathcal{A}^*$ of $(\mathcal{A}, \|\cdot\|)$ and the Gelfand topology is just the relativized weak* topology

The Gelfand space of $\mathcal{A}$ is called the *structure space*, the *spectrum*, and the *character space* by various authors. When $\mathcal{A}$ is a spectral algebra which is almost commutative, Proposition 3.1.3 below shows that the map which assigns to each homomorphism in $\Gamma_{\mathcal{A}}$ its kernel is a bijection onto the set of all maximal modular ideals of $\mathcal{A}$. Thus, the structure space is also called the *maximal ideal space*. Many authors refer to the homomorphisms of $\mathcal{A}$ into $\mathbb{C}$ as *multiplicative linear functionals* on $\mathcal{A}$.

Section 3.6 shows that when $\mathcal{A}$ is the L^1 group algebra of a locally compact abelian group, then the Gelfand homomorphism corresponds to the Fourier transform. Hence, the Gelfand homomorphism is frequently called the Gelfand transform. We restrict this terminology to calling $\hat{a}$ the Gelfand transform of a. The Gelfand homomorphism is also often called the *Gelfand representation* since it represents the algebra (modulo its Gelfand radical) as a more concrete algebra of continuous functions.

We note in Theorem 3.1.5(b) that in an almost commutative spectral algebra the Gelfand radical equals the Jacobson radical. In Chapter 4, where we introduce primitive ideals, Theorem 4.1.9 will show that in an almost commutative algebra primitive ideals coincide with maximal modular ideals. In Section 4.5, where the Brown–McCoy radical is introduced, Theorem 4.5.9 will show that this radical also coincides with the Gelfand radical in almost commutative spectral algebras. Hence, from now on we will speak of an almost commutative spectral algebra as *semisimple* if it is

Gelfand-semisimple. It is interesting to note that Nathan Jacobson [1945a] cites the Gelfand radical as one of the (several) examples which show that his new concept (the Jacobson radical) is useful.

In order to prove the following result, and for later use, we need a symbol for the collection of all (not necessarily non-zero) homomorphisms of an algebra $\mathcal{A}$ into $\mathbb{C}$. We will denote this set by $\Gamma^0_{\mathcal{A}}$. Note that $\Gamma^0_{\mathcal{A}}$ is never empty because it always contains the trivial, identically zero, homomorphism which we will denote by 0. In fact, this *augmented Gelfand space* always satisfies:

$$\Gamma^0_{\mathcal{A}} = \Gamma_{\mathcal{A}} \cup \{0\}. \tag{1}$$

When it is convenient, we extend a Gelfand transform $\hat{a}$ to $\hat{a}^0$ defined on the augmented Gelfand space by:

$$\hat{a}^0(\gamma) = \begin{cases} \gamma(a) & \gamma \in \Gamma_{\mathcal{A}} \\ 0 & \gamma = 0. \end{cases} \tag{2}$$

The augmented Gelfand space has a Gelfand topology defined by requiring each $\hat{a}^0$ to be continuous. It is easy to see that the Gelfand topology on $\Gamma_{\mathcal{A}}$ is just the relativized Gelfand topology of $\Gamma^0_{\mathcal{A}}$. We have little need for these concepts in the unital case where $\hat{1}$ shows that 0 is an isolated point in the augmented Gelfand space.

3.1.2 Theorem *Let $\mathcal{A}$ be an algebra in which the set $\Gamma_{\mathcal{A}}$ is not empty.*

(a) *The Gelfand space is Hausdorff. If $\mathcal{A}$ is nonunital, restriction of each $\gamma \in \Gamma_{\mathcal{A}^1}$ to $\mathcal{A}$ defines a homeomorphism of $\Gamma_{\mathcal{A}^1}$ onto $\Gamma^0_{\mathcal{A}}$. The Gelfand homomorphism is an algebra homomorphism of $\mathcal{A}$ into the algebra of all continuous complex-valued functions on $\Gamma_{\mathcal{A}}$. If $\mathcal{A}$ is a unital algebra, then the Gelfand homomorphism is a unital homomorphism.*

(b) *The range of the Gelfand transform $\hat{a}$ of any element $a \in \mathcal{A}$ is included in the spectrum of a:*

$$\hat{a}(\Gamma_{\mathcal{A}}) \subseteq Sp(a) \qquad \forall\, a \in \mathcal{A}. \tag{3}$$

(c) *The Gelfand radical includes the Jacobson radical and contains any commutator:*

$$\mathcal{A}_J \subseteq \mathcal{A}_\Gamma \quad and \quad ab - ba \in \mathcal{A}_\Gamma \qquad \forall\, a, b \in \mathcal{A}.$$

Hence $\Gamma_{\mathcal{A}}$, $\Gamma_{\mathcal{A}/\mathcal{A}_J}$ and $\Gamma_{\mathcal{A}/\mathcal{A}_D}$ are naturally homeomorphic where $\mathcal{A}_D$ is the smallest ideal containing all commutators.

(d) *If σ is a spectral semi-norm on $\mathcal{A}$, then any $\gamma \in \Gamma_{\mathcal{A}}$ satisfies:*

$$|\gamma(a)| \leq \rho(a) \leq \sigma(a) \qquad \forall\, a \in \mathcal{A}. \tag{4}$$

Furthermore, the Gelfand radical is closed with respect to σ and includes the set $\mathcal{A}_\sigma$ on which σ vanishes.

(e) *If $\mathcal{A}$ is a spectral algebra, then $\Gamma^0_{\mathcal{A}}$ is compact and $\Gamma_{\mathcal{A}}$ is locally compact in the Gelfand topology. The Gelfand transform $\hat{a}$ of any element $a \in \mathcal{A}$ is a bounded continuous function vanishing at infinity:*

$$(\hat{\ }): \mathcal{A} \to C_0(\Gamma_{\mathcal{A}}).$$

If $\mathcal{A}$ or $\mathcal{A}/\mathcal{A}_J$ is also unital, then $\Gamma_{\mathcal{A}}$ is compact in the Gelfand topology so $C_0(\Gamma_{\mathcal{A}})$ equals $C(\Gamma_{\mathcal{A}})$.

Proof (a): For any $a, b \in \mathcal{A}$, any $\lambda \in \mathbb{C}$, and any $\gamma \in \Gamma_{\mathcal{A}}$, we have:

$$(\lambda a)^{\wedge}(\gamma) = \gamma(\lambda a) = \lambda \gamma(a) = \lambda \hat{a}(\gamma);$$
$$(a + b)^{\wedge}(\gamma) = \gamma(a + b) = \gamma(a) + \gamma(b) = (\hat{a} + \hat{b})(\gamma);$$
$$(ab)^{\wedge}(\gamma) = \gamma(ab) = \gamma(a)\gamma(b) = (\hat{a}\hat{b})(\gamma).$$

Hence the Gelfand homomorphism is indeed a homomorphism. The topology of the Gelfand space was chosen to make all the Gelfand transforms $\hat{a}$ continuous functions. If γ_1 and γ_2 are distinct elements in $\Gamma_{\mathcal{A}}$, then there is some $a \in \mathcal{A}$ with $\hat{a}(\gamma_1) \neq \hat{a}(\gamma_2)$. If we define ε to be the positive distance between these two numbers, then $\{\gamma \in \Gamma_{\mathcal{A}} : |\hat{a}(\gamma) - \hat{a}(\gamma_1)| < \varepsilon/2\}$ and $\{\gamma \in \Gamma_{\mathcal{A}} : |\hat{a}(\gamma) - \hat{a}(\gamma_2)| < \varepsilon/2\}$ are disjoint neighborhoods of γ_1 and γ_2, respectively. Hence $\Gamma_{\mathcal{A}}$ is Hausdorff.

Suppose $\mathcal{A}$ is nonunital. Then $\gamma_0 : \mathcal{A}^1 \to \mathbb{C}$ defined by $\gamma_0(\lambda + a) = \lambda$ belongs to $\Gamma_{\mathcal{A}^1}$ and its restriction to $\mathcal{A}$ is the zero homomorphism in $\Gamma^0_{\mathcal{A}}$. If the restriction to $\mathcal{A}$ of any $\gamma \in \Gamma_{\mathcal{A}^1}$ is not zero, then the restriction belongs to $\Gamma_{\mathcal{A}}$. This restriction map is a homeomorphism since it is obviously a continuous bijection of the compact Hausdorff space $\Gamma_{\mathcal{A}^1}$ onto the compact Hausdorff space $\Gamma^0_{\mathcal{A}}$. This proves the second sentence. (Note that the inverse homeomorphism of $\Gamma^0_{\mathcal{A}}$ onto $\Gamma_{\mathcal{A}^1}$ is given by $\gamma \mapsto \bar{\gamma}$ defined by $\bar{\gamma}(\lambda + a) = \lambda + \gamma(a)$ since each $\gamma \in \Gamma_{\mathcal{A}^1}$ is unital.)

If $\mathcal{A}$ is unital, then $\hat{1}$ is the function constantly equal to 1 because the idempotent number $\gamma(1) = \gamma(1)^2$ must be 1 rather than 0 since γ is not identically zero. Hence the Gelfand homomorphism is unital.

(b): If b is an inverse for $\gamma(a) - a$ in $\mathcal{A}^1$, then

$$1 = \gamma(1) = \gamma(b(\gamma(a) - a)) = \gamma(b)(\gamma(a) - \gamma(a)) = 0,$$

where γ has been extended (if necessary) to $\mathcal{A}^1$ by $\gamma(\lambda + a) = \lambda + \gamma(a)$.

(c): If an element belongs to the Jacobson radical then its spectrum is $\{0\}$, so by (b) the element is in the Gelfand radical. Any commutator $ab - ba$ for $a, b \in \mathcal{A}$ and any element of $\gamma \in \Gamma_{\mathcal{A}}$ satisfy

$$\gamma(ab - ba) = \gamma(a)\gamma(b) - \gamma(b)\gamma(a) = 0.$$

If φ is the natural map of $\mathcal{A}$ onto $\langle\ \mathcal{A}/\mathcal{A}_J\ /\ \mathcal{A}/\mathcal{A}_D\ \rangle$, then $\gamma \mapsto \gamma \circ \varphi$ is a natural homeomorphism of $\Gamma_{\mathcal{A}}$ onto $\langle\ \Gamma_{\mathcal{A}/\mathcal{A}_J}\ /\ \Gamma_{\mathcal{A}/\mathcal{A}_D}\ \rangle$.

(d): The inequality (4) is an immediate consequence of (b). It shows that the kernel of any $\gamma \in \Gamma_{\mathcal{A}}$ is closed with respect to σ. The Gelfand radical is just the intersection of these sets for all $\gamma \in \Gamma_{\mathcal{A}}$. The last statement follows from the inequality.

(e): We begin by showing that $\Gamma_{\mathcal{A}}^0$ is compact. Inequality (4) of (d) shows that for each $a \in \mathcal{A}$, $\rho(a)$ is a finite bound for both $|\hat{a}(\gamma)|$ and $|\hat{a}^0(\gamma)|$. Consider $\langle \, \Gamma_{\mathcal{A}} \, / \, \Gamma_{\mathcal{A}}^0 \, \rangle$ as a subset of the product space $K = \prod_{a \in \mathcal{A}} \rho(a)\mathbb{C}_1$. The Gelfand topology on $\langle \, \Gamma_{\mathcal{A}} \, / \, \Gamma_{\mathcal{A}}^0 \, \rangle$ is precisely the relativization of the product topology on K. In this topology, K is Hausdorff and compact by the Tychonoff theorem. Thus if $\langle \, \Gamma_{\mathcal{A}} \, / \, \Gamma_{\mathcal{A}}^0 \, \rangle$ is closed in K, it is compact. However, each of the equations which defines whether $\gamma \colon \mathcal{A} \to \mathbb{C}$ is an algebra homomorphism involves at most three elements of $\mathcal{A}$. Therefore, $\Gamma_{\mathcal{A}}^0$ is closed since membership in it is determined by these equations. (The set of elements satisfying any equation involving only a finite number of elements of $\mathcal{A}$ is closed by definition of the product topology. In the case of three elements, an $(\varepsilon/3)$-argument will show this.) Furthermore, each $\hat{a}$ belongs to $C_0(\Gamma_{\mathcal{A}})$, since each $\hat{a}^0$ vanishes at $0 \in \Gamma_{\mathcal{A}}^0$. It follows that $\Gamma_{\mathcal{A}}^0$ is just the one point compactification of $\Gamma_{\mathcal{A}}$, so the latter is locally compact. In the unital case the additional restriction $\gamma(1) = 1$ (where 1 is an identity element at least modulo the Jacobson radical) is enough to determine membership in $\Gamma_{\mathcal{A}}$, and thus the latter set is closed and therefore compact by our previous remark. □

Note that if $(\mathcal{A}, \, \| \cdot \|)$ is a spectral normed algebra, result (d) shows that $\Gamma_{\mathcal{A}}$ is a subset of the unit ball of $\mathcal{A}^*$. In this case, the Gelfand topology is just the relativized weak* topology. This situation will be the most common in the rest of this work. In particular this shows that any homomorphism $\gamma \colon \mathcal{A} \to \mathbb{C}$ on a Banach algebra $\mathcal{A}$ is continuous. Whether this extends to arbitrary Fréchet algebras is a question of some interest often called Michael's problem from his Memoir [1952] (*cf.* Jean Esterle [1984], Mohamed Oudadess [1983a], Taqdir Husain [1983], Angel Larotonda and Ignacio M. Zalduendo [1983]).

So far we have provided no conditions to force the Gelfand space to be nonempty. Since the Gelfand transform of any commutator is zero, we cannot expect the Gelfand space to be large enough to be useful unless $\mathcal{A}$ is "almost" commutative. In this case, that turns out to mean that $\mathcal{A}$ is commutative modulo its Jacobson radical, and we have already introduced *almost commutative* as a name for this condition. Also, $\Gamma_{\mathcal{A}}$ would be empty if any element of $\mathcal{A}$ had empty spectrum. In fact, we only get a useful theory when there are enough elements in the Gelfand space so that inclusion (3) of Proposition 3.1.2(b) can be reversed. (As usual, zero plays a special role in the spectrum here, so that zero must be omitted from the reversed inclusions.) If $\mathcal{A}$ is almost commutative, the reverse inclusion follows from the assumption that $\mathcal{A}$ is a spectral algebra, as we will now show. In fact, these two conditions are necessary and sufficient ones for a number of interesting results. Before proving this, we establish a simple but powerful consequence of results on maximal modular ideals from the last chapter, particularly Theorem 2.4.12.

3.1.3 Proposition *Let $\mathcal{A}$ be an almost commutative algebra. Then a subset $\mathcal{M}$ of $\mathcal{A}$ is the kernel of a non-zero homomorphism onto $\mathbb{C}$ if and only if it is a maximal modular ideal.*

Proof Clearly the kernel of a homomorphism onto $\mathbb{C}$ is a maximal modular ideal since it has codimension one and a unital image. Conversely, suppose $\mathcal{M}$ is a maximal modular ideal. First we will show that it includes the Jacobson radical. (This will follow from basic results in Chapter 4, but we need a proof now.) Theorem 2.4.6 shows that $\mathcal{M}$ is included in a maximal modular left ideal $\mathcal{L}$. This means that $\mathcal{M}$ is included in the kernel of the irreducible representation of $\mathcal{A}$ on $\mathcal{A}/\mathcal{L}$ defined by $T_a(b + \mathcal{L}) = ab + \mathcal{L}$ for all $a, b \in \mathcal{A}$. Hence by maximality, $\mathcal{M}$ is this kernel. Since it is the kernel of an irreducible representation, it includes the Jacobson radical by Definition 2.3.2. Now the image $\mathcal{M}/\mathcal{A}_J$ of $\mathcal{M}$ in the commutative algebra $\mathcal{A}/\mathcal{A}_J$ obviously satisfies Theorem 2.4.12 and thus is the kernel of a homomorphism of $\mathcal{A}/\mathcal{A}_J$ onto $\mathbb{C}$. Hence $\mathcal{M}$ is the kernel of a homomorphism of $\mathcal{A}$ onto $\mathbb{C}$.   □

3.1.4 Lemma *Let $\mathcal{A}$ be an almost commutative spectral algebra. Let a be an element in $\mathcal{A}$. Then a non-zero complex number λ belongs to the spectrum of a if and only if there is a non-zero homomorphism $\gamma \in \Gamma_\mathcal{A}$ satisfying:*

$$\gamma(a) = \lambda.$$

Furthermore, 0 belongs to the spectrum of a if there is a non-zero homomorphism $\gamma \in \Gamma$ satisfying $\gamma(a) = 0$, and this condition is also necessary if the algebra $\mathcal{A}$ is unital.

Proof Suppose $\gamma: \mathcal{A} \to \mathbb{C}$ is a homomorphism onto $\mathbb{C}$ satisfying $\gamma(a) = \lambda$. Theorem 3.1.2(b) shows $\lambda \in Sp(a)$.

Suppose, conversely, that a non-zero $\lambda \in \mathbb{C}$ belongs to $Sp(a)$. Theorem 2.3.3(c) shows that λ belongs to $Sp_{\mathcal{A}/\mathcal{A}_J}(a + \mathcal{A}_J)$. Since this quotient algebra is commutative, we may assume $\mathcal{A}$ is commutative for the rest of the proof. Then $\lambda^{-1}a$ is not quasi-invertible. Theorem 2.4.6 and the commutativity of $\mathcal{A}$ show that $\lambda^{-1}a$ is the relative identity for a maximal modular ideal $\mathcal{M}$. Thus Proposition 3.1.3 asserts that $\mathcal{M}$ is the kernel of a homomorphism γ of $\mathcal{A}$ onto $\mathbb{C}$. Let $b \in \mathcal{A}$ be any element satisfying $\gamma(b) = 1$. Then $0 = \gamma(b\lambda^{-1}a - b) = \gamma(\lambda^{-1}a) - 1$ holds and implies $\gamma(a) = \lambda$.

If $\mathcal{A}$ is unital and 0 belongs to $Sp_\mathcal{A}(a)$, then $\mathcal{A}a$ is included in a maximal modular ideal $\mathcal{I}$ with relative identity 1. Thus, by the same argument, there is a homomorphism γ of $\mathcal{A}$ onto $\mathbb{C}$ with $\gamma(a) = 0$.   □

The following theorem combines results from Theorem 2.4.11 and Theorem 3.1.2 to give our definitive statement on Gelfand theory. It gives necessary and sufficient conditions for this theory to apply and shows that an algebra which has even the slightest resemblance to an algebra of functions (modulo its Jacobson radical) in fact enjoys a complete Gelfand theory.

3.1.5 Theorem *The following are equivalent for an algebra $\mathcal{A}$.*

(a) *$\mathcal{A}$ is a spectral algebra which is almost commutative.*

(b) *$\mathcal{A}$ is a spectral algebra in which the Jacobson radical equals the Gelfand radical.*

(c) *The spectral radius on $\mathcal{A}$ is an algebra semi-norm.*

(d) *The spectral radius is finite-valued and there is a constant C satisfying:*
$$\rho(a+b) \leq C(\rho(a) + \rho(b)) \qquad \forall \, a, b \in \mathcal{A}.$$

(e) *The spectral radius is finite-valued and there is a constant C satisfying:*
$$\rho(ab) \leq C\rho(a)\rho(b) \qquad \forall \, a, b \in \mathcal{A}.$$

(f) *Every element of $\mathcal{A}$ has nonempty bounded spectrum and any two elements satisfy:*
$$Sp(a+b) \subseteq Sp(a) + Sp(b).$$

(g) *Every element of $\mathcal{A}$ has nonempty bounded spectrum and any two elements satisfy:*
$$Sp(ab) \subseteq Sp(a)Sp(b).$$

(h) *Either the Gelfand space is empty (which occurs if and only if $\mathcal{A}$ is a Jacobson-radical algebra), or the Gelfand space $\Gamma_{\mathcal{A}}$ is a nonempty locally compact Hausdorff space and the algebra $\hat{\mathcal{A}} = \{\hat{a} : a \in \mathcal{A}\}$ of Gelfand transforms is a point-separating, spectral subalgebra of $C_0(\Gamma_{\mathcal{A}})$ satisfying*
$$\hat{a}(\Gamma_{\mathcal{A}}) \subseteq Sp_{\mathcal{A}}(a) \subseteq \hat{a}(\Gamma_{\mathcal{A}}) \cup 0 \qquad \forall \, a \in \mathcal{A}. \tag{5}$$

Moreover, for each $a \in \mathcal{A}$, the supremum $\|\hat{a}\|_\infty = \rho(a)$ of $\{\, |\hat{a}(\gamma)| : \gamma \in \Gamma_{\mathcal{A}} \,\}$ is attained at some $\gamma \in \Gamma_{\mathcal{A}}$.

If any of conditions (a), (b), (c), (d), (e), (f), (g) or (h) hold and $\mathcal{A}$ is also unital, then $\Gamma_{\mathcal{A}}$ is compact and each element $a \in \mathcal{A}$ also satisfies:
$$\hat{a}(\Gamma_{\mathcal{A}}) = Sp(a). \tag{6}$$

Proof (a)$\Rightarrow$(h): If $\mathcal{A}$ is Jacobson-radical, $\Gamma_{\mathcal{A}}$ is empty since each of its elements would define an irreducible representation of $\mathcal{A}$ on $\mathbb{C}$. If $\mathcal{A}$ is not Jacobson-radical, it contains an element with a non-zero complex number in its spectrum. Hence by Lemma 3.1.4, the Gelfand space of $\mathcal{A}$ is not empty. The rest of the statements follow from Theorem 3.1.2(b) and (e) and Lemma 3.1.4 except the claims that $\hat{\mathcal{A}}$ is a spectral subalgebra and that the spectral radius is a maximum value.

For the latter we show that the supremum $\rho(a) = \sup\{\,|\hat{a}(\gamma)| : \gamma \in \Gamma_{\mathcal{A}}\,\}$ is actually attained. For each $n \in \mathbb{N}$, choose $\gamma_n \in \mathcal{A}$ satisfying $|\hat{a_n}| > \rho(a) - 1/n$. There is a subnet converging in the compact space $\Gamma_{\mathcal{A}}^0$ but it cannot converge to 0, so its limit γ_0 is in $\Gamma_{\mathcal{A}}$ and satisfies $|\hat{a}(\gamma_0)| = \rho(a)$.

It only remains to show the image $\hat{\mathcal{A}}$ is a spectral subalgebra of $C_0(\Gamma_{\mathcal{A}})$. If $\hat{a} \in \hat{\mathcal{A}}$ is quasi-invertible in $C_0(\Gamma_{\mathcal{A}})$, then 1 does not belong to $Sp_{C_0(\Gamma_{\mathcal{A}})}(\hat{a})$. Thus 1 does not belong to $\hat{a}(\Gamma_{\mathcal{A}})$. Hence, the second inclusion in (5) implies $1 \notin Sp_{\mathcal{A}}(a)$. Thus, a has a quasi-inverse b in $\mathcal{A}$, so $\hat{b}$ is a quasi-inverse in $\hat{\mathcal{A}}$ for $\hat{a}$.

(h)$\Rightarrow$(f) and (g): This is immediate except that it appears at first that we need to add a "$\cup\{0\}$" to the right side of these inclusions. If $\mathcal{A}$ is unital, we do not need to add "$\cup\{0\}$" since we can use the final equation of (h), and if $\mathcal{A}$ is not unital, then zero belongs to the spectrum of every element.

The implications (f)$\Rightarrow$(d) and (g)$\Rightarrow$(e) are obvious. Theorem 2.4.11 shows the equivalence of conditions (a), (c), (d) and (e).

(h)$\Rightarrow$(b): The inclusion $\mathcal{A}_J \subseteq \mathcal{A}_\Gamma$ holds for all algebras by Theorem 3.1.2(c), hence it is enough to establish the other inclusion. If $\mathcal{A}$ is Jacobson-radical, there is nothing to prove. Otherwise, the inclusions of (h) show that the Gelfand radical is an ideal in which every element has spectrum $\{0\}$. Thus the Gelfand radical is included in the Jacobson radical by Theorem 2.3.3.

(b)$\Rightarrow$(a): Theorem 3.1.2(c) shows this. $\square$

Some Consequences of the Gelfand Theory

This is a very powerful theorem with many obvious consequences. For instance, note that condition (b) shows that an almost commutative spectral algebra $\mathcal{A}$ is semisimple (and therefore commutative) if and only if it satisfies one (and hence all) of the following equivalent conditions: (1) the Gelfand homomorphism is injective; (2) the spectral radius is a norm; or, (3) the elements of $\Gamma_{\mathcal{A}}$ separate the points of $\mathcal{A}$. The inclusions in conditions (f) and (g) (which will be extended below to commuting elements in an arbitrary spectral algebra) are powerful aids in computation. The only related results for noncommuting elements involve the numerical radius introduced in Section 2.6 rather than the generally more useful spectrum. Corollary 3.5.14 will show that if $\mathcal{A}$ is a semisimple commutative Banach algebra and $\Gamma_{\mathcal{A}}$ is compact, then $\mathcal{A}$ is unital, thus providing a partial converse for the last statement in (h).

We now extend several of our results from commutative algebras to commuting elements in possibly non-commutative algebras. Theorem 2.2.14 showed that the spectrum and the spectral radius were upper semicontinuous functions. The following corollary shows that the spectrum and spectral radius are uniformly continuous when restricted to commuting sets.

3.1.6 Corollary *Let $\mathcal{A}$ be a spectral algebra. Then any commuting elements $a, b \in \mathcal{A}$ satisfy the following results:*

$$
\begin{aligned}
Sp(a + b) &\subseteq Sp(a) + Sp(b); \\
Sp(ab) &\subseteq Sp(a)Sp(b); \\
\rho(a + b) &\leq \rho(a) + \rho(b); \\
\rho(ab) &\leq \rho(a)\rho(b); \\
Sp(a) &\subseteq Sp(b) + \rho(a - b)\mathbb{C}_1 \subseteq Sp(b) + \sigma(a - b)\mathbb{C}_1; \\
|\rho(a) - \rho(b)| &\leq \rho(a - b) \leq \sigma(a - b)
\end{aligned}
$$

where σ is some spectral semi-norm in the last two results. Thus the spectrum (in the Hausdorff metric for subsets of $\mathbb{C}$) and spectral radius are uniformly continuous on commuting elements with respect to any spectral semi-norm.

Proof Propositions 2.5.2 and 2.5.3(c) show that we may replace $\mathcal{A}$ by any maximal commutative subalgebra that contains a and b. Theorem 3.1.5 establishes the first four results. The other results follow easily. $\qquad\square$

Characterizations of Commutativity

Theorem 3.1.5 gives a number of characterizations of almost commutative algebras among spectral algebras. It also shows, as we have just emphasized, that an almost commutative spectral normed algebra $(\mathcal{A}, \|\cdot\|)$ satisfies:

$$|\rho(a) - \rho(b)| \leq \|a - b\| \qquad \forall\, a, b \in \mathcal{A}.$$

This certainly implies that the spectral radius is a uniformly continuous function on such an algebra. Theorem 4.2.15 in the next chapter will give a converse.

We now give additional conditions which imply that a normed algebra is commutative. Note that condition (c) below (which was already obtained in Lemma 2.4.10) is necessary as well as sufficient. The other two conditions are considered again in Proposition 3.1.8. These results are due to Claude Le Page [1967], Rudolf A. Hirschfeld and Wiesław Żelazko [1968] and John W. Baker and John S. Pym [1971]. See also Gh. Mocanu [1967], [1971], O. H. Cheikh and M. Oudadess [1988], We give related results in Theorem 4.2.15.

3.1.7 Proposition *A normed algebra $(\mathcal{A}, \|\cdot\|)$ is commutative if there is a constant C satisfying any of the following three conditions.*

(a) $\|a\|^2 \leq C\|a^2\| \qquad \forall\, a \in \mathcal{A}.$

(b) $\|a\| \leq C\|a\|^\infty \qquad \forall\, a \in \mathcal{A}.$

(c) $\|ab\| \leq C\|ba\|$ $\forall\, a,\, b \in \mathcal{A}^1$

where the norm on $\mathcal{A}^1$ is defined by $\|\lambda + a\| = |\lambda| + \|a\|$ for all $\lambda \in \mathbb{C}$ and $a \in \mathcal{A}$.

Proof Inequality (a) implies

$$
\begin{aligned}
\|a\| &\leq\ C^{2^{-1}}\|a^2\|^{2^{-1}} \leq C^{2^{-1}}C^{2^{-2}}\|a^{2^2}\|^{2^{-2}} \leq\ \cdots \\
&\leq\ C^{2^{-1}}C^{2^{-2}}\cdots C^{2^{-n}}\|a^{2^n}\|^{2^{-n}}
\end{aligned}
$$

for any n. Taking the limit as n increases we get (b).

Suppose (b) holds. Let $\overline{\mathcal{A}}$ be the completion of $\mathcal{A}$. For any $a \in \overline{\mathcal{A}}$, let $\{a_n\}_{n\in\mathbb{N}}$ be a sequence in $\mathcal{A}$ converging to a. Theorem 1.1.10(h) implies

$$
\|a\| = \lim \|a_n\| \leq C \limsup \|a_n\|^\infty \leq C\|a\|^\infty.
$$

Then for any $\lambda + a \in \overline{\mathcal{A}}^1$, Proposition 2.2.1(e) implies

$$
\begin{aligned}
\|\lambda + a\| &=\ |\lambda| + \|a\| \leq |\lambda| + C\|a\|^\infty \\
&\leq\ C(|\lambda| + \rho(a)) \leq 3C\rho(\lambda + a) = 3C\|\lambda + a\|^\infty.
\end{aligned}
$$

Hence for any $a,\, b \in \overline{\mathcal{A}}^1$, Theorem 1.1.10(d) implies

$$
\|ab\| \leq 3C\|ab\|^\infty = 3C\|ba\|^\infty \leq 3C\|ba\|.
$$

Thus we have shown (a) $\Rightarrow$ (b) $\Rightarrow$ (c), (with a different choice of constant in (c)). Now Lemma 2.4.10 completes the proof. $\square$

3.1.8 Proposition *The following conditions are equivalent for a normed algebra* $(\mathcal{A}, \|\cdot\|)$.

(a) *The norm is complete and satisfies* $\|a\|^2 \leq C\|a^2\|$ *for all* $a \in \mathcal{A}$.

(b) *The norm is complete and satisfies* $\|a\| \leq C\|a\|^\infty$ *for all* $a \in \mathcal{A}$.

(c) *The spectral radius is a complete norm on* $\mathcal{A}$.

(d) *The Gelfand homomorphism is injective and* $\hat{\mathcal{A}}$ *is closed in* $C_0(\Gamma_{\mathcal{A}})$.

(e) *The algebra* $\mathcal{A}$ *is isomorphic to a closed subalgebra of* $C_0(\Omega)$ *for some locally compact space* Ω.

Proof (a)$\Rightarrow$(b): The proof of the last proposition establishes this.

(b)$\Rightarrow$(c): The last proposition shows that $\mathcal{A}$ is a commutative Banach algebra, so the spectral radius is an algebra norm. Since the norm in a Banach algebra is spectral, it dominates the spectral radius. Hence, (b) shows that they are equivalent norms.

(c)$\Rightarrow$(d): Theorem 3.1.5 shows that if the spectral radius is a norm, then the Gelfand homomorphism is injective. However, the Gelfand homomorphism is an isometry with respect to the supremum norm in $C_0(\Gamma_{\mathcal{A}})$. Hence, the image $\hat{\mathcal{A}}$ of this map is closed.

(d)$\Rightarrow$(e): Immediate.

(e)$\Rightarrow$(c): If the isomorphism is φ, then $\rho_{\mathcal{A}}(a) = \rho_{C_0(\Omega)}(\varphi(a)) = \|\varphi(a)\|_\infty$ holds for all $a \in \mathcal{A}$.

(c)$\Rightarrow$(a): The identity map from $\mathcal{A}$ with its own norm to $\mathcal{A}$ with the spectral radius norm is a contraction. Hence by the inverse boundedness theorem, the two complete norms are equivalent. Thus, for a suitable constant C we get $\|a\|^2 \le C\rho(a)^2 = C\rho(a^2) \le C\|a^2\|$. $\square$

Algebras satisfying the equivalent conditions of the above proposition are usually called *uniform algebras*. For an introduction to the theory of uniform algebras, including many examples, see the excellent book of Edgar Lee Stout [1971]. Theorem 4.2.26 gives a surprising characterization of uniform algebras. Irving Glicksberg [1964] and Sergei Vitalyevič Kislyakov [1989] show that most uniform algebras are uncomplemented subspaces of the Banach space of all continuous functions. Quotients of uniform algebras have been characterized by A. M. Davie [1973c]. They are sometimes called Q-algebras, but that term is also used for topological or normed algebras with an open set of invertible elements. For a lattice theoretic approach see Egon Scheffold [1988] and previous papers.

Dual Maps of Homomorphisms

Let $\mathcal{A}$ and $\mathcal{B}$ be algebras and let φ be a homomorphism of $\mathcal{A}$ into $\mathcal{B}$. Then the dual map $\varphi^\dagger \colon \mathcal{B}^\dagger \to \mathcal{A}^\dagger$ maps $\Gamma_{\mathcal{B}}^0$ into $\Gamma_{\mathcal{A}}^0$ since $\varphi^\dagger(\gamma) = \gamma \circ \varphi$ is a homomorphism into $\mathbb{C}$ whenever γ is a homomorphism into $\mathbb{C}$. Clearly, $\varphi^\dagger$ need not necessarily take a non-zero $\gamma \in \Gamma_{\mathcal{B}}$ into a non-zero homomorphism in $\Gamma_{\mathcal{A}}$, so we must work with the augmented Gelfand spaces in this instance. We will also use $\varphi^\dagger$ to represent the restriction to $\Gamma_{\mathcal{B}}^0$ of the map $\varphi^\dagger \colon \mathcal{B}^\dagger \to \mathcal{A}^\dagger$, and we will call this restricted map the *dual of* φ.

3.1.9 Proposition *Let $\mathcal{A}$ and $\mathcal{B}$ be algebras and let φ be a homomorphism of $\mathcal{A}$ into $\mathcal{B}$. Then the dual map $\varphi^\dagger \colon \Gamma_{\mathcal{B}}^0 \to \Gamma_{\mathcal{A}}^0$ is continuous relative to the Gelfand topologies.*

Proof This is immediate. $\square$

In Section 3.3 we will use the completeness of a Banach algebra to define a functional calculus. It will then be important to know the relationship between the Gelfand theory of a commutative spectral normed algebra and that of its completion. The next proposition gives a very satisfactory answer: they can be naturally identified.

3.1.10 Proposition *Let $(\mathcal{A}, \|\cdot\|)$ be a spectral normed algebra with completion $\overline{\mathcal{A}}$. If φ is the inclusion map of $\mathcal{A}$ into $\overline{\mathcal{A}}$, then $\varphi^\dagger \colon \Gamma_{\overline{\mathcal{A}}} \to \Gamma_{\mathcal{A}}$ is a homeomorphism of $\Gamma_{\overline{\mathcal{A}}}$ onto $\Gamma_{\mathcal{A}}$.*

If $\mathcal{A}$ is ⟨ commutative / almost commutative ⟩ then $\overline{\mathcal{A}}$ is also ⟨ commutative / almost commutative ⟩.

Proof Proposition 2.1.5 shows that $\mathcal{A}$ (or, equivalently, $\varphi(\mathcal{A})$) is a dense spectral subalgebra of $\overline{\mathcal{A}}$ and Proposition 3.1.9 shows that $\varphi^\dagger$ is a continuous map into $\Gamma_{\mathcal{A}}$. First we show that $\varphi^\dagger$ is surjective. For each $\gamma \in \Gamma_{\mathcal{A}}$, let $\overline{\gamma}: \overline{\mathcal{A}} \to \mathbb{C}$ be the extension by continuity. It is easy to see that $\overline{\gamma}$ belongs to $\Gamma_{\overline{\mathcal{A}}}$ and that it is mapped to γ by $\varphi^\dagger$. It only remains to show that $\varphi^\dagger$ is an open map or, equivalently, that the map $\gamma \mapsto \overline{\gamma}$ is continuous. Let $\mathcal{U} = \{\gamma \in \Gamma_{\overline{\mathcal{A}}} : |\gamma(a_j) - \overline{\gamma}_0(a_j)| < 1 \text{ for } j = 1, 2, \ldots, n\}$ be an arbitrary neighborhood of the extension $\overline{\gamma}_0$ of some $\gamma_0 \in \Gamma_{\mathcal{A}}$ where $a_1, a_2, \ldots, a_n$ are elements of $\overline{\mathcal{A}}$. For each j, we can choose $\tilde{a}_j$ in $\mathcal{A}$ so that $\|a_j - \tilde{a}_j\| < 1/3$. Then $\overline{\gamma}$ belongs to $\mathcal{U}$ whenever γ satisfies $|\gamma(\tilde{a}_j) - \gamma_0(\tilde{a}_j)| < 1/3$ holds for $j = 1, 2, \ldots, n$. This proves the continuity of $\gamma \mapsto \overline{\gamma}$.

If $\mathcal{A}$ is commutative, it is obvious that $\overline{\mathcal{A}}$ is also. If $\mathcal{A}$ is almost commutative, then the first part of the proof shows that the Gelfand transform of $ab - ba$ is zero for all $a, b \in \overline{\mathcal{A}}$. Hence, $\overline{\mathcal{A}}$ is also almost commutative. $\square$

The next result, which is due to Šilov [1947a], is related to Corollary 2.3.9. Both results show that certain homomorphisms into Banach algebras are automatically continuous (*i.e.* they are continuous without requiring continuity *a priori*). The former result requires surjectivity but does not restrict the target algebra, while the present one makes no restriction on the "size" of the range of the homomorphism, except that it should be in a semisimple commutative algebra. Chapter 6 will contain a number of related results. The last statement in the theorem was obtained earlier by Gelfand [1941a] and is a special case of the Johnson-Rickart uniqueness of norm theorem (Corollary 2.3.10 and Theorem 6.1.1).

3.1.11 Theorem *Let $(\mathcal{A}, \sigma)$ be a spectral semi-normed algebra and let $(\mathcal{B}, \tau)$ be a commutative, semisimple spectral semi-normed algebra. Then τ is a norm and (the graph of) any homomorphism*

$$\varphi: \mathcal{A} \to \mathcal{B}$$

is closed. If $\mathcal{A}$ and $\mathcal{B}$ are Banach algebras (with the latter still commutative and semisimple), then φ is continuous.

In particular, any two complete norms on a commutative semisimple algebra are equivalent.

Proof Corollary 2.3.4 shows that τ is a norm. (Theorem 3.1.2(d) also gives the same result since it implies $\|\hat{b}\|_\infty \leq \rho_{\mathcal{B}} \leq \tau$ for all $b \in \mathcal{B}$ and $\|\cdot\|_\infty$ is a norm on a semisimple commutative algebra.)

Let $\{a_n\}_{n \in \mathbb{N}} \subseteq \mathcal{A}$ converge to $a \in \mathcal{A}$, and let $\{\varphi(a_n)\}_{n \in \mathbb{N}}$ converge to $b \in \mathcal{B}$. For any $\gamma \in \Gamma_{\mathcal{B}}$, inequality (4) in Theorem 3.1.2(d) applied to $\gamma \circ \varphi$ implies that $\gamma \circ \varphi(a) = \lim \gamma \circ \varphi(a_n) = \gamma \lim \varphi(a_n)$. Hence $(\varphi(a))^\wedge = \hat{b}$ implies $\varphi(a) = b$ by semisimplicity. Thus φ has a closed graph and is continuous by the closed graph theorem if $\mathcal{A}$ and $\mathcal{B}$ are Banach algebras.

We can also give a simple proof which does not use the Gelfand theory. Theorem 2.4.3(b) shows that the spectral radius in $\mathcal{B}$ is an algebra norm. Let $\{a_n\}_{n\in\mathbb{N}}$ and b have the same meaning as in the first proof and let $c \in \mathcal{B}$ be arbitrary. Then these elements satisfy

$$
\begin{aligned}
\rho(cb) &\leq \rho(c)\rho(\varphi(a_n)) + \rho(c)\rho(\varphi(a_n) - b) \leq \rho(c)[\rho(a_n) + \rho(\varphi(a_n) - b)] \\
&\leq \rho(c)[\|a_n\|_{\mathcal{A}} + \|\varphi(a_n) - b\|_{\mathcal{B}}] \to 0
\end{aligned}
$$

where we used Proposition 2.2.1(c) to replace $\rho(\varphi(a_n))$ by $\rho(a_n)$ in the third inequality. Hence b is in the Jacobson radical of $\mathcal{B}$ by Theorem 2.3.3 and thus equals 0 as we wished to show.   □

Double Centralizer Algebras

Next we turn to showing the connection between Gelfand theory and both double centralizer algebras and Arens multiplication on the double dual of an algebra. Some of the following results were obtained by Ju Kwei Wang [1961] and László Máté [1965]. Pietro Aiena [1990] contains related specialized results.

3.1.12 Theorem *Let $\mathcal{A}$ be an algebra.*
   (a) *For any $\gamma \in \Gamma_{\mathcal{A}}$, there is an element $\overline{\gamma} \in \Gamma_{\mathcal{D}(\mathcal{A})}$ satisfying*

$$
\overline{\gamma}(L_a, R_a) = \gamma(a) \qquad \forall \, a \in \mathcal{A}.
$$

Hence, $\Gamma_{\mathcal{D}(\mathcal{A})}$ is nonempty if $\Gamma_{\mathcal{A}}$ is. If $\Gamma_{\mathcal{A}}$ is nonempty, restriction of the Gelfand transforms of elements of $\mathcal{D}(\mathcal{A})$ to $\{\overline{\gamma} : \gamma \in \Gamma_{\mathcal{A}}\}$ defines a homomorphism ψ of $\mathcal{D}(\mathcal{A})$ into an algebra of continuous functions on $\Gamma_{\mathcal{A}}$ with kernel $\{(L,R) \in \mathcal{D}(\mathcal{A}) : L(\mathcal{A}) + R(\mathcal{A}) \subseteq \mathcal{A}_\Gamma\}$.
   (b) *If $\mathcal{A}$ is a commutative, semisimple spectral algebra, then $\mathcal{D}(\mathcal{A})$ is commutative and semisimple. If $\|\cdot\|$ is a spectral norm on $\mathcal{D}(\mathcal{A})$, then each map in the following commutative diagram is contractive and injective.*

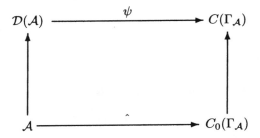

Proof (a): Choose $(L,R) \in \mathcal{D}(\mathcal{A})$ and $\gamma \in \Gamma_{\mathcal{A}}$ and $a, b, \in \mathcal{A}$ with $\gamma(a) \neq 0$. The double centralizer equation gives:

$$
\gamma(a)\gamma(L(b)) = \gamma(aL(b)) = \gamma(R(a)b) = \gamma(R(a))\gamma(b).
$$

By setting b equal to a, we conclude $\gamma(L(a)) = \gamma(R(a))$. If $\gamma(b) \neq 0$, then

$$\gamma(L(a))/\gamma(a) = \gamma(R(a))/\gamma(a) = \gamma(L(b))/\gamma(b) = \gamma(R(b))/\gamma(b).$$

Therefore, for each $\gamma \in \Gamma_{\mathcal{A}}$ and $(L, R) \in \mathcal{D}(\mathcal{A})$ we can define

$$\overline{\gamma}\Big((L, R)\Big) = \gamma(L(a))/\gamma(a)$$

for some $a \in \mathcal{A}$ with $\gamma(a) \neq 0$. It is easy to check $\overline{\gamma} \in \Gamma_{\mathcal{D}(\mathcal{A})}$. Since the same a may be used in the definition of $\overline{\gamma}$ for all γ in some neighborhood of each γ_0, the restriction of $(L, R)^{\wedge}$ will be continuous on $\{\overline{\gamma} : \gamma \in \Gamma_{\mathcal{A}}\}$. The kernel of ψ is obvious.

(b): The vertical maps in the diagram are just the regular homomorphisms. Since the hypotheses imply $\mathcal{A}_{\Gamma} = \{0\}$, ψ is injective, as are all the other maps. Hence $\mathcal{D}(\mathcal{A})$ is commutative and semisimple. It is also clear that all the maps are contractive.   □

When $\mathcal{A}$ is a commutative Banach algebra, it would be interesting to know the size of the subalgebra $\hat{\mathcal{A}}$ of Gelfand transforms of elements in $\mathcal{A}$ in the algebra $C_0(\Gamma_{\mathcal{A}})$. An interesting answer is given in Graham R. Allan [1968]: a function f on $\Gamma_{\mathcal{A}}$ belongs to $\hat{\mathcal{A}}$ if and only if it is the restriction of a locally bounded analytic function on a weak* neighborhood of $\Gamma_{\mathcal{A}}$ in $\mathcal{A}^*$. The word "analytic" in this definition has a variety of natural meanings.

Sin-ei Takahasi and Osamu Hatori [1990] have recently found answers to the related question of the size of the subalgebra of the algebra $C(\Gamma_{\mathcal{A}})$ induced by the double centralizer algebra $\mathcal{D}(\mathcal{A})$. In a number of interesting cases, but not in every case, this is exactly the algebra of functions which we describe in the next definition. The condition in the definition was first used by Salomon Bochner [1934], Isaac J. Schoenberg [1934] and William F. Eberlein [1955] to characterize those continuous functions on the dual group of a locally compact group induced by measures on the original group. This explains the name. We will show that this algebra is precisely the algebra of all continuous functions which arise by restricting elements of the double dual $\mathcal{A}^{**}$ to $\Gamma_{\mathcal{A}} \subseteq \mathcal{A}^*$.

The Image of the Double Dual Algebra

3.1.13 Definition Let $\mathcal{A}$ be a Banach algebra with nonempty Gelfand space. A function $f \in C(\Gamma_{\mathcal{A}})$ is said to satisfy the *Bochner–Schoenberg–Eberlein condition* if there is a finite constant C so that for any finite list $\alpha_1, \alpha_2, \ldots, \alpha_n$ of complex numbers and corresponding list $\gamma_1, \gamma_2, \ldots, \gamma_n$ of elements of $\Gamma_{\mathcal{A}}$, f satisfies:

$$|\sum_{j=1}^{n} \alpha_j f(\gamma_j)| \leq C \|\sum_{k=1}^{n} \alpha_j \gamma_j\|_{\mathcal{A}^*}. \tag{7}$$

For such a function f, the infimum of all possible constants C is called the *Bochner–Schoenberg–Eberlein norm* of f and is denoted by $\|f\|_{\mathrm{BSE}}$. The normed linear space of all Bochner–Schoenberg–Eberlein functions is denoted by $C_{\mathrm{BSE}}(\Gamma_{\mathcal{A}})$.

It is straightforward to show that $C_{\mathrm{BSE}}(\Gamma_{\mathcal{A}})$ is a subalgebra of $C(\Gamma_{\mathcal{A}})$ including $\hat{\mathcal{A}}$ and that $\|\cdot\|_{\mathrm{BSE}}$ is a complete algebra norm on $C_{\mathrm{BSE}}(\Gamma_{\mathcal{A}})$ which satisfies $\|\hat{a}\|_{\infty} \leq \|\hat{a}\|_{\mathrm{BSE}}$ for all $a \in \mathcal{A}$. Some of these results are checked in the cited reference. The following result stems from Paul Civin and Bertram Yood [1961] 3.6, 3.7 and 3.16 as well as Takahasi and Hatori [1990], Theorem 4. See Section 1.4 for background on Arens multiplication.

3.1.14 Theorem *Let $\mathcal{A}$ be a Banach algebra and let $\kappa_0 \colon \mathcal{A} \to \mathcal{A}^{**}$ and $\kappa_1 \colon \mathcal{A}^* \to \mathcal{A}^{***}$ be the canonical isometric linear injections.*

(a) When $\mathcal{A}^{**}$ is provided with either Arens product, κ_1 maps $\Gamma_{\mathcal{A}} \subseteq \mathcal{A}^*$ into $\Gamma_{\mathcal{A}^{**}} \subseteq \mathcal{A}^{***}$.

(b) If $\Gamma_{\mathcal{A}}$ is nonempty, the set $(\Gamma_{\mathcal{A}})^{\perp}$ is an ideal in $\mathcal{A}^{**}$ with either Arens product and the two resulting quotient algebras coincide and are commutative. Moreover, there is a contractive algebra homomorphism ψ which makes the following sequence exact:

$$\{0\} \longrightarrow (\Gamma_{\mathcal{A}})^{\perp} \xrightarrow{\ \varphi\ } \mathcal{A}^{**} \xrightarrow{\ \psi\ } B(\Gamma_{\mathcal{A}})$$

where φ is the inclusion map and $B(\Gamma_{\mathcal{A}})$ is the Banach algebra (under pointwise multiplication and the supremum norm) of all bounded functions on $(\Gamma_{\mathcal{A}})$.

(c) If $\mathcal{A}$ is almost commutative, then

$$\psi(\mathcal{A}^{**}) \cap C(\Gamma_{\mathcal{A}}) = C_{\mathrm{BSE}}(\Gamma_{\mathcal{A}})$$

where ψ is the homomorphism of (b) above.

(d) If $\mathcal{A}$ is commutative, then any two elements $f, g \in \mathcal{A}^{**}$ satisfy $fg = g \cdot f$. Hence, κ_0 maps $\mathcal{A}$ into the center of $\mathcal{A}^{**}$ with respect to either Arens product.

Proof (a): For any $\gamma \in \Gamma_{\mathcal{A}}$, $a \in \mathcal{A}$ and $f \in \mathcal{A}^{**}$, the three step definition of the Arens products gives $_a\gamma = \gamma(a)\gamma = \gamma_a$, $_f\gamma = f(\gamma)\gamma = \gamma_f$, and $fg(\gamma) = f(\gamma)g(\gamma) = f \cdot g(\gamma)$, proving this result.

(b): This follows easily from the last result in the proof of (a).

(c): It is easy to see that the left side is a subset of the right side. Suppose f is a function in $C_{\mathrm{BSE}}(\Gamma_{\mathcal{A}})$ with norm $\|f\|_{\mathrm{BSE}} = F$. Let B be the collection of all finite subsets of $\Gamma_{\mathcal{A}}$ ordered by inclusion so that it is a directed set. By Helly's theorem (Yôsida [1965], Theorem 6.5, p.109), for each $\beta \in B$ we may find an element a_β satisfying $f(\gamma) = \gamma(a_\beta)$ for all $\gamma \in \beta$ and $\|a_\beta\| \leq F + 1/n$ where n is the cardinal of β. Thus $\{\kappa(a_\beta)\}_{\beta \in B}$ is a bounded net in $\mathcal{A}^{**}$. By Alaoglu's theorem, some subnet converges in the weak* topology to an element $g \in \mathcal{A}^{**}$ which satisfies $g(\gamma) = f(\gamma)$ for all $\gamma \in \Gamma_{\mathcal{A}}$ and $\|g\| = \|f\|_{\text{BSE}}$.

(d): These immediate consequences of the definition of the Arens products are noted here for completeness.   □

A unital normed algebra $\mathcal{A}$ is said to be *generated* by one of its elements a if it is the smallest closed unital subalgebra of itself containing a. Note that this is equivalent to saying that polynomials in a are dense in $\mathcal{A}$. In this case, $\mathcal{A}$ is said to be *singly generated*. A compact subset K of $\mathbb{C}$ is said to be *polynomially convex* if it is equal to its polynomially convex hull pc$K$. This is equivalent to saying that there are no bounded components of $\mathbb{C} \setminus K$. The next theorem shows the relationship of these concepts. It will be generalized to finitely generated algebras in Section 3.5.

3.1.15 Theorem *Let $(\mathcal{A}, \| \cdot \|)$ be a unital spectral normed algebra which is generated by one of its elements a. Then $\hat{a}$ is a homeomorphism of $\Gamma_{\mathcal{A}}$ onto $Sp(a)$ and the latter set is polynomially convex. Furthermore, for any compact polynomially convex subset K of $\mathbb{C}$ there is a singly generated algebra $\mathcal{A}$ with generator a satisfying $Sp(a) = K$.*

Proof If two elements γ_1 and γ_2 of $\Gamma_{\mathcal{A}}$ satisfy $\gamma_1(a) = \gamma_2(a)$, then they agree on every polynomial in a and hence on $\mathcal{A}$. Thus, $\hat{a}$ is a one-to-one continuous map from the compact space $\Gamma_{\mathcal{A}}$ into the Hausdorff space $\mathbb{C}$. Any such map is a homeomorphism.

If there were a bounded component U of $\mathbb{C} \setminus Sp(a)$, we could choose an element λ_0 from it. For any polynomial p, the maximum modulus principle implies

$$\begin{aligned}
|p(\lambda_0)| &\leq \max\{|p(\lambda)| : \lambda \in \partial U\} \leq \max\{|p(\lambda)| : \lambda \in Sp(a)\} \\
&\leq \max\{|\lambda| : \lambda \in Sp(p(a))\} = \rho(p(a)) \leq \|p(a)\|.
\end{aligned}$$

Given any element b in the smallest (not necessarily closed) unital subalgebra $\mathcal{A}_0$ of $\mathcal{A}$ containing a, we can find a polynomial p satisfying $p(a) = b$. The above inequality shows that if two polynomials both satisfy this condition then they agree at λ_0. Hence, we can define a homomorphism $\varphi \colon \mathcal{A}_0 \to \mathbb{C}$ by $b \mapsto p(\lambda_0)$. Since φ is contractive by the same inequality, it can be extended to all of $\mathcal{A}$. However, this implies the contradiction $\lambda_0 \in Sp(a)$ since $\lambda_0 = \varphi(a)$. Thus $Sp(a)$ is polynomially convex.

Finally, if K is a compact polynomially convex subset of $\mathbb{C}$, we can consider the unital closed subalgebra $\mathcal{A}$ of $C(K)$ generated by z. Then $Sp_{\mathcal{A}}(z) = Sp_{C(K)}(z) = K$ follows from Corollary 2.5.8 and proves that the pair $(\mathcal{A}, z)$ has the desired property.   □

3.2 Šilov Boundary, Hulls and Kernels

In this section we will generally assume that the algebras considered are commutative and spectral. It is of some interest that the definitions of the last section were made in such a way that this assumption is not really necessary. Similarly, we sometimes assume that the Gelfand spaces of the algebras involved are nonempty. This is only to avoid triviality, and if carefully construed, the statements remain true without this restriction.

The Šilov Boundary

We introduce the notion of boundaries and prove the existence of a minimum closed boundary for any commutative spectral algebra. The terminology is suggested by algebras such as the disc algebra $A(\mathbb{D})$ introduced in 1.5.2 and further explained in §3.2.13. For this algebra, the Gelfand transform $\hat{z}$ of the coordinate function z on $\mathbb{C}$ (*i.e.*, the function satisfying $z(\lambda) = \lambda$ for all $\lambda \in \mathbb{C}$) sets up a homeomorphism between Γ_A and the closed unit disc $\mathbb{C}_1$. Under this homeomorphism, the Šilov boundary, which we are about to define, is mapped onto the boundary $\mathbb{T}$ of the disc. We will use the maximum modulus principle below to show that the inverse image of $\mathbb{T}$ under $\hat{z}$ satisfies the defining properties of the Šilov boundary. As the name suggests, the concept is due to Šilov [1940a].

3.2.1 Definition   Let A be a commutative spectral algebra with Γ_A nonempty. A *boundary* for A is a subset B of Γ_A such that for each element $a \in A$ there is a point $\gamma \in B$ satisfying

$$|\hat{a}(\gamma)| = \|\hat{a}\|_\infty = \rho(a).$$

The intersection of all closed boundaries for A is called the *Šilov boundary* for A and is denoted by ∂_A.

Note that a closed subset B of Γ_A is a boundary if and only if it satisfies $\|\hat{a}|_B\|_\infty = \|\hat{a}\|_\infty$. The argument is the same as that given in the proof of Theorem 3.1.5 (a)$\Rightarrow$(h) which showed that $|\hat{a}|$ attains its supremum on Γ_A.

It is not obvious that the intersection ∂_A of all closed boundaries is non-empty and certainly not obvious that it is a boundary. However, the next theorem, which is due to Šilov [1940a], proves these results and thus justifies the terminology "Šilov boundary". The proof we give is due to Lars Hormander [1966] when A is a unital Banach algebra. The first inclusion in (1) gives another interpretation of the term boundary. In relation to the second inclusion, it is of some interest that $Sp(a)$ includes any component of the complement of $\hat{a}(\partial_A) \cup \{0\}$ with which it has nonempty intersection. Otherwise the component would include points in the boundary of $Sp(a)$.

3.2.2 Theorem   *Let A be a commutative spectral algebra. The Šilov boundary ∂_A for A is a nonempty boundary for A and hence is the smallest* *closed boundary for* $\mathcal{A}$. *Each* $a \in \mathcal{A}$ *satisfies*

$$\partial Sp(a) \subseteq \hat{a}(\partial_{\mathcal{A}}) \cup \{0\} \subseteq Sp(a) \cup \{0\} \tag{1}$$

where " $\cup\{0\}$ *" can be omitted if* $\mathcal{A}$ *is unital. If* $\mathcal{A}$ *is nonunital, then the map which restricts each element of* $\Gamma_{\mathcal{A}^1}$ *to* $\mathcal{A}$ *maps* $\partial_{\mathcal{A}^1}$ *bijectively onto either* $\partial_{\mathcal{A}}$ *or* $\partial_{\mathcal{A}} \cup \{0\}$.

Proof First we assume that $\mathcal{A}$ is unital. We begin by obtaining a particularly convenient base for the topology of $\Gamma_{\mathcal{A}}$. Since the topology is the weakest one making each $\hat{a}$ in $\hat{\mathcal{A}}$ continuous, a base is given by the sets of the form $\{\gamma \in \Gamma_{\mathcal{A}} : |\hat{b}_j(\gamma) - \hat{b}_j(\varphi)| < \varepsilon_j;\ j = 1, 2, \ldots, n\}$ where $\varphi \in \Gamma_{\mathcal{A}}, n \in \mathbb{N}, b_j \in \mathcal{A}$ and $\varepsilon_j > 0$ vary. For each $j = 1, 2, \ldots, n$, define an element $a_j \in \mathcal{A}$ by $a_j = \varepsilon_j^{-1}(b_j - \varphi(b_j))$. Then the basic open set above is described by

$$\{\gamma \in \Gamma_{\mathcal{A}} : |\hat{a}_j(\gamma)| < 1 \qquad \forall\, j = 1, 2, \ldots, n\}.$$

Call the latter set $V_{a_1, a_2, \ldots, a_n}$. A base for the topology of $\Gamma_{\mathcal{A}}$ is given by sets of this form. We will call them special basic open sets.

Next we show that if $V = V_{a_1, a_2, \ldots, a_n}$ is one of these special basic sets, then either $B \setminus V$ is a boundary for each closed boundary B, or V has a non-empty intersection with each boundary of $\mathcal{A}$. Suppose B is a closed boundary but $B \setminus V$ is not a boundary. Then, there is an element $b \in \mathcal{A}$ such that $\|\hat{b}|_{(B \setminus V)}\|_{\infty} < 1$ while $\|\hat{b}\|_{\infty} = 1$. For a large enough positive integer m, we can achieve

$$|(b^m a_j)\hat{\ }(\gamma)| = |\hat{b}(\gamma)^m \hat{a}_j(\gamma)| < 1 \qquad \forall\, j = 1, 2, \ldots, n;\ \gamma \in B \setminus V.$$

Since $|(b^m a_j)^{\wedge}(\gamma)| < 1$ also holds on V, compactness of B gives

$$\|(b^m a_j)^{\wedge}|_B\|_{\infty} < 1.$$

Since B is a boundary, this implies $\|(b^m a_j)^{\wedge}\|_{\infty} < 1$. Thus, $|\hat{b}(\gamma)|$ must achieve its maximum 1 only when each $|\hat{a}_j(\gamma)|$ is strictly less than one. That is, $|\hat{b}|$ achieves its maximum only in V. Hence, V must have nontrivial intersection with each boundary of $\mathcal{A}$.

Now we prove that $\partial_{\mathcal{A}}$ is a boundary. If not, there is an element $a \in \mathcal{A}$ satisfying $\|\hat{a}|_{\partial_{\mathcal{A}}}\|_{\infty} < 1$ and $\|\hat{a}\|_{\infty} = 1$. Let C be the set $\{\gamma \in \Gamma_{\mathcal{A}} : |\hat{a}(\gamma)| = 1\}$. Then, any $\gamma \in C$ does not belong to $\partial_{\mathcal{A}} = \cap\{B : B$ is a closed boundary for $\mathcal{A}\}$. Hence there is a special basic open neighborhood V_{γ} of γ which has empty intersection with some closed boundary. By the result proved in the last paragraph, $B \setminus V_{\gamma}$ is a boundary for each closed boundary B. However, $\{V_{\gamma} : \gamma \in C\}$ is an open cover for the compact set C. Let $\{V_1, V_2, \ldots, V_n\}$ be a finite subcover. Since $\Gamma_{\mathcal{A}}$ is a closed boundary, $\Gamma_{\mathcal{A}} \setminus V_1$ is a closed boundary by the choice of V_1 as one of the V_γs. Repeating this argument $n - 1$ more times, we conclude that $\Gamma_A \setminus \cup_{j=1}^n V_j \subseteq \Gamma_A \setminus C$ is a closed boundary. However, the element a contradicts this assertion. Hence ∂_A is a boundary. Clearly, it is the smallest closed boundary.

Now suppose that there is an $a \in A$ and a $\lambda_0 \in \partial Sp(a)$ which does not belong to $\hat{a}(\partial_A)$. Since $Sp(a)$ is closed, Lemma 3.1.4 shows $\lambda_0 = \gamma_0(a)$ for some $\gamma_0 \in \Gamma_A$. Since ∂_A is compact, the distance from λ_0 to $\hat{a}(\partial_A)$ is positive. Call it ε. Choose $\lambda \in \mathbb{C} \setminus Sp(a)$ satisfying $|\lambda - \lambda_0| < \varepsilon/2$. Then the distance from λ to $\hat{a}(\partial_A)$ is at least $\varepsilon/2$. Since $\lambda - a$ is invertible, $((\lambda - a)^{-1})^\wedge = (\lambda - \hat{a})^{-1}$ belongs to $\hat{A}$. Since ∂_A is a boundary, we must have $\|(\lambda - \hat{a})^{-1}\|_\infty = \|(\lambda - \hat{a})^{-1}|_{\partial_A}\|_\infty \leq 2/\varepsilon$. However, the inequality $|(\lambda - \gamma_0(a))^{-1}| = |\lambda - \lambda_0|^{-1} > 2/\varepsilon$ contradicts the previous inequality, thus providing the desired inclusion.

Now we turn to the case in which A is not unital. We will establish that the Šilov boundary is a boundary and the last statement of the theorem first. Suppose B is a closed boundary for A. For each $\gamma \in \Gamma_A$, let $\gamma^1 \in \Gamma_{A^1}$ be defined by $\gamma^1(\lambda + a) = \lambda + \gamma(a)$ so that γ^1 is the unique element of Γ_{A^1}, the restriction of which to A equals γ. Define $B^1 \subseteq \Gamma_{A^1}^0$ by $B^1 = \{\gamma^1 : \gamma \in B\} \cup \{\gamma_0\}$ where γ_0 is defined by $\gamma_0(\lambda + a) = \lambda$. We will prove by contradiction that B^1 includes ∂_{A^1}. Suppose not. Then, there is an element $b \in A^1$ and a number t satisfying

$$\|\hat{b}|_{B^1}\|_\infty < t < \|\hat{b}\|_\infty.$$

Let $\lambda + a$ be such a high power of $t^{-1}b$ that it satisfies

$$\|(\lambda + a)^\wedge|_{B^1}\|_\infty < 1 < 3 < \|(\lambda + a)^\wedge\|_\infty.$$

Then $\lambda = \gamma_0(\lambda + a)$ satisfies $|\lambda| < 1$. Since B is a boundary for A, $a \in A \subseteq A^1$ satisfies $\|\hat{a}\|_\infty = \|\hat{a}|_{B^1}\|_\infty < 2$. This implies $\|(\lambda+a)^\wedge\|_\infty \leq |\lambda|+\|\hat{a}\|_\infty < 3$, which is a contradiction.

The inclusion we have just established implies that

$$\partial_A \cup \{0\} = \cap\{B : B \text{ is a closed boundary for } A\} \cup \{0\}$$

includes the set of restrictions to A of elements of ∂_{A^1}. Hence ∂_A is a boundary for A since we have already shown that ∂_{A^1} is a boundary for A^1. Since the set of non-zero restrictions to A of elements in ∂_{A^1} is a closed boundary for A, this set includes ∂_A. This completes the proof of the last statement in the theorem. Inclusion (1) for nonunital algebras is now an immediate consequence. □

Theorems 3.2.10 and 3.2.12 will show two important classes of commutative Banach algebras in which the Šilov boundary is the whole Gelfand space. We now give the simplest criterion for this to happen.

3.2.3 Proposition *Let $\mathcal{A}$ be a commutative spectral algebra. If the algebra $\hat{\mathcal{A}}$ of Gelfand transforms is dense in $C_0(\Gamma_\mathcal{A})$, then the Šilov boundary $\partial_\mathcal{A}$ is the whole Gelfand space $\Gamma_\mathcal{A}$. In particular, this happens if for each $a \in \mathcal{A}$ there is some $b \in \mathcal{A}$ satisfying $\hat{b}(\gamma) = \hat{a}(\gamma)^*$ for all $\gamma \in \Gamma_\mathcal{A}$.*

Proof Suppose $\gamma \in \Gamma_\mathcal{A}$ were not in $\partial_\mathcal{A}$. Then we could find $f \in C_0(\Gamma_\mathcal{A})$ satisfying $f(\gamma) = 1$ and $f|_{\partial_\mathcal{A}} \equiv 0$. Any element $a \in \mathcal{A}$ satisfying $\|\hat{a} - f\|_\infty < 1/2$ would contradict the fact that $\partial_\mathcal{A}$ is a boundary.

If $\hat{\mathcal{A}}$ is closed under complex conjugation, it satisfies all the hypotheses of the Stone–Weierstrass theorem stated in 1.5.1.   □

Šilov [1940b] introduced his boundary in order to determine which homomorphisms in the Gelfand space of a closed subalgebra could be extended to be elements of the Gelfand space of a larger commutative Banach algebra. Recall the results on dual maps in Proposition 3.1.9. In the following theorem, note that if $\mathcal{A}$ is any closed subalgebra of a commutative Banach algebra $\mathcal{B}$, then the inclusion map satisfies all the given hypotheses. See also Meyer [1991].

3.2.4 Theorem *Let $\mathcal{A}$ and $\mathcal{B}$ be commutative spectral algebras and let $\varphi \colon \mathcal{A} \to \mathcal{B}$ be a homomorphism satisfying*

$$\rho_\mathcal{B}(\varphi(a)) = \rho_\mathcal{A}(a) \qquad \forall\, a \in \mathcal{A}.$$

Then $\varphi^\dagger(\partial_\mathcal{B})$ includes $\partial_\mathcal{A}$. Furthermore, the Šilov boundary $\partial_\mathcal{A}$ is the largest subset of $\Gamma_\mathcal{A}$ with this property for all choices of $\mathcal{B}$ and φ.

If $\mathcal{A}$ is also semisimple, $\partial_\mathcal{A}$ is exactly the set of homomorphisms in $\Gamma_\mathcal{A}$ which can be extended to any commutative spectral algebra $\mathcal{B}$ which includes $\mathcal{A}$ as a subalgebra satisfying $\rho_\mathcal{B}(a) = \rho_\mathcal{A}(a)$ for all $a \in \mathcal{A}$.

Proof Note that $\partial_\mathcal{B} \cup \{0\}$ is a compact subset of $\Gamma_\mathcal{B}^0$. Hence its image in $\Gamma_\mathcal{A}^0$ is closed. Thus $\varphi^\dagger(\partial_\mathcal{B})$ is closed in $\Gamma_\mathcal{B}$. It is also a boundary since any $a \in \mathcal{A}$ satisfies

$$\|\hat{a}|_{\varphi^\dagger(\partial_\mathcal{B})}\|_\infty = \|(\varphi(a))^\wedge|_{\partial_\mathcal{B}}\|_\infty = \rho_\mathcal{B}(\varphi(a)) = \rho_\mathcal{A}(a).$$

Thus, this set includes the Šilov boundary of $\mathcal{A}$.

Consider the Gelfand homomorphism of $\mathcal{A}$ into $\mathcal{B} = C_0(\Gamma_\mathcal{A})$. This satisfies the hypotheses of the theorem, and $\Gamma_\mathcal{B} = \partial_\mathcal{B}$ can obviously be identified with $\Gamma_\mathcal{A}$ under the dual of the Gelfand homomorphism. Hence, this shows that $\partial_\mathcal{A}$ is the largest subset of $\Gamma_\mathcal{A}$ covered by $\varphi^\dagger(\partial_\mathcal{B})$ for all choices of $\mathcal{B}$ and φ.

For the final result, consider φ to be just the inclusion map. Hence, in the first paragraph of the proof we have shown that each homomorphism γ in the Šilov boundary of $\mathcal{A}$ can be extended to be an element of the Gelfand space (indeed the Šilov boundary) of $\mathcal{B}$. The second paragraph shows that the Gelfand homomorphism (which may now be regarded as an embedding map) is one in which no element of $\Gamma_{\mathcal{A}}$ outside of the Šilov boundary can be extended.   □

The Hull-Kernel Topology

We introduce hulls, kernels and the hull–kernel topology. This last concept is due to Gelfand and Šilov [1941] in the case of commutative Banach algebras. Marshall Harvey Stone [1937] used a closely related concept for Boolean rings earlier. In Chapter 7 we will study essentially the same concept in the case of noncommutative Banach algebras. We note here that the construction we are about to give depends only on the fact that the kernel of an element of $\Gamma_{\mathcal{A}}$ is a prime ideal. (This term is defined in Chapter 4.) The hull–kernel topology on the set of prime ideals is called the *Zariski topology* in Grothendieck's theory of schemes in algebraic geometry. The hull–kernel topology for primitive ideals in a general ring (also defined in Chapter 4) is usually called the *Jacobson topology* (*cf.* Jacobson [1945b]).

3.2.5 Definition   Let $\mathcal{A}$ be a commutative spectral algebra. The *hull* $h(\mathcal{S})$ of a subset $\mathcal{S}$ of $\mathcal{A}$ is the set:

$$h(\mathcal{S}) = \{\gamma \in \Gamma_{\mathcal{A}} : \gamma(a) = 0 \quad \forall\, a \in \mathcal{S}\}.$$

The *kernel* $k(B)$ of a subset B of $\Gamma_{\mathcal{A}}$ is the set:

$$k(B) = \{a \in \mathcal{A} : \gamma(a) = 0 \quad \forall\, \gamma \in B\}.$$

A subset B of $\Gamma_{\mathcal{A}}$ which has the form $h(\mathcal{S})$ for some $\mathcal{S} \subseteq \mathcal{A}$ is called a *hull*.

It is sometimes useful to think of $h(\mathcal{S})$ and $k(B)$ described in the following fashion:

$$h(\mathcal{S}) \;=\; \bigcap_{a \in \mathcal{S}} \{\gamma \in \Gamma_{\mathcal{A}} : \hat{a}(\gamma) = 0\} \tag{2}$$

$$k(B) \;=\; \bigcap_{\gamma \in B} \ker(\gamma). \tag{3}$$

These descriptions make clear the first assertion of the next proposition, which records some elementary consequences of the definition.

3.2.6 Proposition   *Let $\mathcal{A}$ be a commutative spectral algebra with $\Gamma_{\mathcal{A}}$ nonempty. Consider $\mathcal{S}_1 \subseteq \mathcal{S}_2 \subseteq \mathcal{A}$ and $B_1 \subseteq B_2 \subseteq \Gamma_{\mathcal{A}}$. Then $h(\mathcal{S}_1)$ is a closed set in $\Gamma_{\mathcal{A}}$ and $k(B_1)$ is a closed ideal of $\mathcal{A}$ when $\mathcal{A}$ is furnished with any spectral semi-norm. Moreover*

$$h(\mathcal{S}_2) \;\subseteq\; h(\mathcal{S}_1) \qquad\qquad k(B_2) \subseteq k(B_1) \tag{4}$$

$$\mathcal{S}_1 \;\subseteq\; hk(\mathcal{S}_1) \qquad\qquad B_1 \subseteq hk(B_1) \tag{5}$$

$$hkh(\mathcal{S}_1) \;=\; h(\mathcal{S}_1) \qquad\qquad khk(B_1) = k(B_1). \tag{6}$$

Hence, a subset B of Γ_A is a hull if and only if it satisfies $B = hk(B)$.

If S_2 is the ideal generated by S_1, then $h(S_2) = h(S_1)$. Similarly, if B_2 is the closure of B_1 in the Gelfand topology, then $k(B_2) = k(B_1)$.

Proof The first assertions follow immediately from the intersection results which precede the proposition. The four inclusions are direct consequences of the definitions. Together they give the two displayed equations. The first of these equations implies the statement about hulls. The last two assertions are immediate consequences of the original definition. $\square$

3.2.7 Theorem *Let A, B and C be commutative spectral algebras. Let*

$$0 \longrightarrow A \overset{\varphi}{\longrightarrow} B \overset{\psi}{\longrightarrow} C \longrightarrow 0$$

be a short exact sequence. Let

$$\Gamma_C^0 \overset{\psi^\dagger}{\longrightarrow} \Gamma_B^0 \overset{\varphi^\dagger}{\longrightarrow} \Gamma_A^0$$

be the associated dual maps. Then $\psi^\dagger$ maps Γ_C homeomorphically onto $h(\varphi(A))$ and $\varphi^\dagger$ maps $\Gamma_B \setminus h(\varphi(A))$ homeomorphically onto Γ_A.

Proof For $\gamma \in \Gamma_C$, $\psi^\dagger(\gamma) = \gamma \circ \psi$ vanishes on $\varphi(A)$, so that $\psi^\dagger$ maps Γ_C into $h(\varphi(A))$. Since ψ is onto, $\psi^\dagger$ is injective. Suppose γ belongs to $h(\varphi(A)) \subseteq \Gamma_B$. For any $c \in C$, define $\tilde{\gamma}(c)$ to be $\gamma(b)$ where $b \in B$ and $\psi(b) = c$. This is well defined since γ belongs to $h(\ker(\psi))$. Now it is immediate that $\tilde{\gamma}$ belongs to Γ_C and satisfies $\psi^\dagger(\tilde{\gamma}) = \gamma$. Hence $\psi^\dagger$ is a surjection onto $h(\varphi(A))$. For a net $\{\gamma_\alpha\}_{\alpha \in A}$ and an element γ in $\Gamma_C, \psi^\dagger(\gamma_\alpha)$ converges to $\psi^\dagger(\gamma)$ if and only if γ_α converges to γ, since $\psi^\dagger(\gamma_\alpha)(b) = \gamma_\alpha \circ \psi(b) = \gamma_\alpha(\psi(b))$ and $\psi^\dagger(\gamma)(b) = \gamma \circ \psi(b) = \gamma(\psi(b))$ for all $b \in B$. Thus, $\psi^\dagger$ is a homeomorphism.

We simplify notation by abbreviating $\Gamma_B \setminus h(\varphi(A))$ to S. It is clear that $\varphi^\dagger$ maps $h(\varphi(A))$ onto $\{0\}$ and S into Γ_A. Suppose that $\varphi^\dagger(\gamma_1) = \varphi^\dagger(\gamma_2)$ holds for some $\gamma_1, \gamma_2, \in \Gamma_B$. If γ_1 does not equal γ_2, then there is some $b \in B$ satisfying $\gamma_1(b) \neq \gamma_2(b)$. If $\gamma_1(b)$ or $\gamma_2(b)$ is zero, we may assume without loss of generality that it is $\gamma_1(b)$. Then, $\gamma_1(d) = 0$ and $\gamma_2(d) = 1$ hold when d is defined to be $(\gamma_2(b))^{-1}b$. If neither $\gamma_1(b)$ nor $\gamma_2(b)$ is zero, then by choosing d to be $\lambda b + \mu b^2$ we can achieve $\gamma_1(d) = 0$ and $\gamma_2(d) = 1$ for a suitable choice of $\lambda, \mu \in \mathbb{C}$. Choose d as indicated. Then for any $a \in A$, we have $0 = \gamma_1(d\varphi(a)) = \gamma_2(d\varphi(a)) = \gamma_2(a)$ since $d\varphi(a)$ belongs to $\varphi(A)$ and $\gamma_1 \circ \varphi$ equals $\gamma_2 \circ \varphi$. This would imply that γ_2 (and hence γ_1) belongs to $h(\varphi(A))$. Thus $\varphi^\dagger$ is injective when restricted to S.

Next we show that $\varphi^\dagger$ maps S onto Γ_A. Suppose γ belongs to Γ_A. Let $e \in A$ satisfy $\gamma(e) = 1$. Define $\overline{\gamma} \colon B \to \mathbb{C}$ by $\overline{\gamma}(b) = \gamma(\varphi(e)b)$ for all $b \in B$. Then $\overline{\gamma}$ is clearly linear and $\overline{\gamma}(bd) = \gamma(\varphi(e)bd)\gamma(\varphi(e)) = \gamma(\varphi(e)\varphi(e)bd) = \gamma(\varphi(e)b)\gamma(\varphi(e)d) = \overline{\gamma}(b)\overline{\gamma}(d)$ holds for all $b, d \in B$. Thus $\overline{\gamma}$ belongs to Γ_B^0.

For all $a \in \mathcal{A}$, we get $\overline{\gamma}(a) = \gamma(\varphi(ea)) = \gamma(e)\gamma(a) = \gamma(a)$. Hence $\overline{\gamma}$ belongs to $\mathcal{S}$ and $\varphi^{\dagger}(\overline{\gamma})$ is γ. Therefore $\varphi^{\dagger}$ maps $\mathcal{S}$ onto $\Gamma_{\mathcal{A}}$.

Finally, we must show that $\varphi^{\dagger}$ is a homeomorphism when restricted to $\mathcal{S}$. It is obviously continuous. For $\{\gamma_{\alpha}\}_{\alpha \in A} \subseteq \mathcal{S}$ and $\gamma \in \mathcal{S}$, we will show that γ_{α} converges to γ if $\varphi^{\dagger}(\gamma_{\alpha})$ converges to $\varphi^{\dagger}(\gamma)$. Choose $e \in \mathcal{A}$ satisfying $\varphi^{\dagger}(\gamma)(e) = 1$. For any $b \in \mathcal{B}$ and any $\varepsilon > 0$, we can find $\alpha_0 \in A$ so that $\alpha \geq \alpha_0$ implies $|\gamma_{\alpha}(\varphi(e)b) - \gamma(\varphi(e)b)| < \varepsilon/2(1 + \rho(b))$, and $|\gamma_{\alpha}(\varphi(e)) - \gamma(\varphi(e))| < \varepsilon/2$. Then for $\alpha \geq \alpha_0$, we have

$$
\begin{aligned}
|\gamma_{\alpha}(b) - \gamma(b)| &= |\gamma_{\alpha}(b)\gamma(\varphi(e)) - \gamma(b)\gamma(\varphi(e))| \\
&\leq |\gamma_{\alpha}(b)| \, |\gamma(\varphi(e)) - \gamma_{\alpha}(\varphi(e))| + |\gamma_{\alpha}(b\varphi(e)) - \gamma(b\varphi(e))| \\
&< \varepsilon.
\end{aligned}
$$

This shows that $\varphi^{\dagger}$ is a homeomorphism.   □

The next theorem is frequently proved by an appeal to Kuratowski's closure axioms, but we will give a direct proof. Recall that a hull is any subset H of $\Gamma_{\mathcal{A}}$ which can be written as $h(\mathcal{I})$ for some subset $\mathcal{I}$ of $\mathcal{A}$. Consequently, a hull H satisfies $H = hk(H)$.

3.2.8 Theorem *Let $\mathcal{A}$ be an algebra with $\Gamma_{\mathcal{A}}$ nonempty. The family*

$$\{\Gamma_{\mathcal{A}} \setminus H : H \text{ is a hull in } \Gamma_{\mathcal{A}}\}$$

is a topology for $\Gamma_{\mathcal{A}}$. This topology is included in the Gelfand topology for $\Gamma_{\mathcal{A}}$.

Proof Let $\mathcal{T}$ be the family of sets which is asserted to be a topology. Clearly $h(\mathcal{A}) = \emptyset$ and $h(\{0\}) = \Gamma_{\mathcal{A}}$ hold, so that $\Gamma_{\mathcal{A}}$ and $\emptyset$ belong to $\mathcal{T}$. The definition shows that any family $\{\mathcal{I}_{\alpha} : \alpha \in A\}$ of subsets of $\mathcal{A}$ satisfies

$$\bigcap_{\alpha \in A} h(\mathcal{I}_{\alpha}) = \{\gamma \in \Gamma_{\mathcal{A}} : \gamma(a) = 0 \quad \forall \, a \in \bigcup_{\alpha \in A} \mathcal{I}_{\alpha}\} = h(\bigcup_{\alpha \in A} \mathcal{I}_{\alpha}).$$

This implies that $\mathcal{T}$ is closed under arbitrary unions. Finally, we must show that $\mathcal{T}$ is closed under the intersection of two of its elements. Let H_1 and H_2 be two hulls. Then

$$H_1 \cup H_2 \subseteq hk(H_1 \cup H_2) = h(k(H_1) \cap k(H_2))$$

is obvious. Suppose γ does not belong to $H_1 \cup H_2$. Then for some a_j in $k(H_j)$, we have $\gamma(a_j) \neq 0$ for $j = 1, 2$. Since the product $a_1 a_2$ belongs to $k(H_1) \cap k(H_2)$ and $\gamma(a_1 a_2)$ is non-zero, we see $\gamma \notin h(k(H_1) \cap k(H_2))$. Thus, $H_1 \cup H_2 = hk(H_1 \cup H_2)$ is a hull.

If $H = hk(H)$ is any hull, then $H = \{\gamma \in \Gamma_{\mathcal{A}} : \hat{a}(\gamma) = 0, \ a \in k(H)\}$ is closed in the Gelfand topology. Hence the topology $\mathcal{T}$ is weaker than the Gelfand topology.   □

The topology described in Theorem 3.2.8 is called the *hull–kernel topology*. We will now show some conditions which are equivalent to the equality of the hull–kernel topology and the Gelfand topology. The definition is due to Šilov [1947a]. who used the term *regular*. Charles E. Rickart [1960] introduced the term *completely regular* which is now standard.

3.2.9 Definition   Let Ω be a topological space. A subalgebra $\mathcal{A}$ of $C(\Omega)$ is said to be *completely regular* on Ω if, for any closed subset K of Ω and any $\omega_0 \in \Omega \setminus K$, there is a function $f \in \mathcal{A}$ satisfying

$$f(\omega_0) = 1 \quad \text{and} \quad f|_K \equiv 0.$$

A commutative spectral algebra $\mathcal{A}$ is said to be *completely regular* if $\hat{\mathcal{A}} \subseteq C(\Gamma_{\mathcal{A}})$ is completely regular on $\Gamma_{\mathcal{A}}$.

Note that a Hausdorff topological space Ω will be completely regular (in the topological sense) if and only if $C(\Omega)$ is completely regular on Ω. An easy compactness argument shows that any compact Hausdorff topological space is normal, and Urysohn's lemma shows that a normal space is completely regular. It is clear from the definition that any subspace of a completely regular space is completely regular. Hence, we conclude that any subspace of a compact Hausdorff space is completely regular. (In particular, locally compact Hausdorff spaces are completely regular.) §3.2.15 will show that, conversely, any completely regular Hausdorff space is a subspace of a compact Hausdorff space.

The first portion of the following result is due to Šilov [1947a] for commutative unital Banach algebras. Condition (c) will be used again in a more general setting in Chapter 7. Corollary 3.5.14 will show that a hull in the Gelfand space of a commutative Banach algebra is compact in the Gelfand (and hence in the hull–kernel) topology if and only if its kernel is modular. This makes condition (c) a bit clearer.

3.2.10 Theorem   *The following two conditions are equivalent for any commutative spectral algebra $\mathcal{A}$:*
   (a) *$\mathcal{A}$ is completely regular.*
   (b) *The Gelfand topology and the hull–kernel topology coincide.*
   The following condition implies (a) *and* (b) *and is equivalent to them if $\mathcal{A}$ is either unital or a Banach algebra.*
   (c) *The hull–kernel topology is Hausdorff and each point in $\Gamma_{\mathcal{A}}$ has a hull–kernel neighborhood with a modular kernel.*
   Any of these conditions imply that the Šilov boundary of $\mathcal{A}$ equals $\Gamma_{\mathcal{A}}$.

Proof (a) $\Rightarrow$ (b): Since each hull is closed in the Gelfand topology, we need only show that each subset $B \subseteq \Gamma_{\mathcal{A}}$ which is closed in the Gelfand topology is a hull. Let $\gamma \in \Gamma_{\mathcal{A}}$ not belong to B. Then, complete regularity implies that there is an $a \in \mathcal{A}$ satisfying $\hat{a}(\gamma) = 1$ and $\hat{a}|_B = 0$. Hence, γ does not belong to $hk(B)$. Since $\gamma \notin B$ was arbitrary, this implies $hk(B) \subseteq B$. The opposite inclusion always holds as shown in Proposition 3.1.4. Hence each Gelfand closed set B is a hull, so the two topologies agree.

(b) $\Rightarrow$ (a): Suppose B is a Gelfand closed set, and hence a hull, in $\Gamma_{\mathcal{A}}$ and $\gamma \in \Gamma_{\mathcal{A}}$ does not belong to B. Since B equals $hk(B)$, there is some element $b \in k(B)$ satisfying $\hat{b}(\gamma) \neq 0$. Then $a = \hat{b}(\gamma)^{-1}b$ satisfies $\hat{a}(\gamma) = 1$ and $\hat{a}(B) \subseteq \{0\}$.

(b) $\Leftrightarrow$ (c) (Under extra hypothesis.): Theorem 3.1.2(a) asserts that the Gelfand topology is always Hausdorff, so (b) implies that the hull–kernel topology is Hausdorff.

If $\mathcal{A}$ is unital, any ideal of $\mathcal{A}$ is modular so that the second condition of (c) is vacuous. Again, if $\mathcal{A}$ is unital and (c) holds, then the identity map from $\Gamma_{\mathcal{A}}$ with its Gelfand topology onto $\Gamma_{\mathcal{A}}$ with its hull–kernel topology will be a homeomorphism since it is a continuous bijection of a compact space onto a Hausdorff space. This completes the proof of the equivalence of (b) and (c) when $\mathcal{A}$ is unital.

Now suppose (c) holds. (We do not need to assume completeness.) It is enough to show that each closed set $B \subseteq \Gamma_{\mathcal{A}}$ is a hull, and for this it is enough to show that any $\gamma_0 \in \Gamma_{\mathcal{A}}$ which does not belong to B does not belong to $hk(B)$. Let V be a hull kernel neighborhood of γ_0 with $k(V)$ modular. Let $\varphi: \mathcal{A} \to \mathcal{A}/k(V)$ be the natural map. Theorem 3.2.7 shows that $\varphi^{\dagger}$ is a homeomorphism of $\Gamma_{\mathcal{A}/k(V)}$ onto $hk(V)$ (in the Gelfand topologies). Since $\mathcal{A}/k(V)$ is unital, Theorem 3.1.2(e) shows that $\Gamma_{\mathcal{A}/k(V)}$ is compact. Hence $B_0 = B \cap hk(V)$ is compact in the Gelfand topology. Since the hull–kernel topology is Hausdorff, for each $\gamma \in B_0$ we can find a hull–kernel, and hence Gelfand, neighborhood U_{γ} of γ satisfying $\gamma_0 \notin hk(U_{\gamma})$. Then $\{U_{\gamma} : \gamma \in B_0\}$ is an open cover for B_0. Let $\{U_1, U_2, \ldots, U_n\}$ be a finite subcover. For each $j = 1, 2, \ldots, n$, $\gamma_0 \notin hk(U_j)$ implies the existence of an element $b_j \in k(U_j)$ satisfying $\gamma_0(b_j) \neq 0$. Similarly, $\gamma_0 \notin hk(\Gamma_{\mathcal{A}} \setminus V)$ implies the existence of an element $b_0 \in k(\Gamma_{\mathcal{A}} \setminus V)$ satisfying $\gamma_0(b_0) \neq 0$. Then $b = b_0 b_1 \cdots b_n$ satisfies $\gamma_0(b) \neq 0$ and belongs to

$$k(\Gamma_{\mathcal{A}} \setminus V) \cap \bigcap_{j=1}^{n} k(U_j) = k(\Gamma_{\mathcal{A}} \setminus V \cup B_0) \subseteq k(B).$$

Hence γ_0 does not belong to $hk(B)$. This concludes the proof of $(c) \Rightarrow (b)$

Now suppose $\mathcal{A}$ is a Banach algebra and (b) holds. We will construct a hull–kernel neighborhood V for an arbitrary $\gamma \in \Gamma_{\mathcal{A}}$ such that $k(V)$ is modular. First choose an element $b \in \mathcal{A}$ such that $\gamma_0(b)$ is non-zero, and denote $|\gamma_0(b)|/3$ by ε. Then we claim that $k(V)$ is modular where V is the neighborhood of γ_0, $V = \{\gamma : |\gamma(b)| \geq 2\varepsilon\}$. Considering V as a subset of $\prod_{a \in \mathcal{A}} \rho(a)\mathbb{C}_1$, we see easily that V is compact. Hence, Theorem 3.2.7 shows that $\Gamma_{\mathcal{A}/k(V)}$ is compact. (Thus, we could refer to Corollary 3.5.14 below which implies that the semisimple algebra $\mathcal{A}/k(V)$ is unital so that $k(V)$

is modular. Instead of using this deep result, we will give an elementary proof, due to Rickart [1960], that $k(V)$ is modular.) Denote $\mathcal{A}/k(V)$ by $\mathcal{A}'$ and $b + k(V) \in \mathcal{A}'$ by a. We take over the notation from the proof of Theorem 2.2.2 replacing $\mathcal{A}$ by $\mathcal{A}'$ and λ by ε^{-1}. Then the long displayed equation in the proof becomes

$$((\varepsilon^{-1}a)^n)^q = n^{-1} \sum_{j=1}^{n} f(\varepsilon^{-1}\zeta_j)$$

where $n \in \mathbb{N}$ is arbitrary, $\{\zeta_1 = 1, \zeta_2, \ldots, \zeta_n\}$ is a full set of nth-roots of unity, and f is an $\mathcal{A}'$-valued function which is continuous on the circle of radius ε^{-1}. The right hand side of this expression is a Riemann sum for the Riemann integral

$$\int_0^1 f(\varepsilon^{-1}e^{2\pi it})\,dt,$$

which exists since f is continuous. Hence, there is an element $e' \in \mathcal{A}'$ satisfying

$$e' = \lim_{n\to\infty} ((\varepsilon^{-1}a)^n)^q.$$

Theorem 3.2.7 shows that any element of $\Gamma_{\mathcal{A}'}$ corresponds to an element of $h(k(V)) = V$. Hence, for any $\gamma \in \Gamma_{\mathcal{A}'}$ we get

$$\gamma(e') = \lim_{n\to\infty}((\varepsilon^{-1}\gamma(a))^n)^q = \lim_{n\to\infty} \frac{(\varepsilon^{-1}\gamma(a))^n}{(\varepsilon^{-1}\gamma(a))^n - 1} = 1.$$

Therefore $k(V)$ is modular, since e' is an identity element for the semisimple algebra $\mathcal{A}' = \mathcal{A}/k(V)$. Hence (b) and (c) are equivalent if $\mathcal{A}$ is a Banach algebra.

Finally, we prove the last assertion in the theorem by contradiction. If the closed set $\partial_{\mathcal{A}}$ is not all of $\Gamma_{\mathcal{A}}$, complete regularity guarantees an element $a \in \mathcal{A}$ with nonvanishing Gelfand transform satisfying $\hat{a}(\partial_{\mathcal{A}}) = \{0\}$. This contradicts the fact that $\partial_{\mathcal{A}}$ is a boundary. $\square$

The next result is due to Ernst Albrecht [1982a] and this simple proof was discovered by Jyunji Inoue and Sin-ei Takahasi [1992].

3.2.11 Corollary *Every commutative unital Banach algebra $\mathcal{A}$ has a largest closed completely regular subalgebra $\mathcal{A}_{\mathrm{reg}}$.*

Proof Let $\mathcal{A}_{\mathrm{reg}}$ (with Gelfand space Γ) be the closed subalgebra of $\mathcal{A}$ generated by all closed completely regular subalgebras. Let b belong to one of these subalgebras $\mathcal{B}$, which we may assume to be unital. It is clear that the map which restricts $\gamma \in \Gamma$ to $\mathcal{B}$ is continuous in the hull–kernel topology from Γ to $\Gamma_{\mathcal{B}}$. The theorem shows that $\hat{b}$ is hull–kernel continuous on Γ_B, so it is also hull–kernel continuous when considered on Γ. Hence any element of $\mathcal{A}_{\mathrm{reg}}$ has a hull–kernel continuous Gelfand transform since it is the norm limit of finite algebraic expressions of elements like b. Thus $\mathcal{A}_{\mathrm{reg}}$ verifies condition (b) of the theorem.   □

Next we turn to a condition which is much stronger than complete regularity. Algebras satisfying the conditions of the next theorem are called *commutative C*-algebras*. They will be dealt with at greater length in Volume II. The next result was a lemma in the fundamental paper of Gelfand and Mark A. Naǐmark [1943], but is often referred to as the commutative Gelfand–Naǐmark theorem. The complete regularity of the locally compact Hausdorff space Γ_C implies immediately that an algebra satisfying the conditions of this theorem is completely regular. The following result has many consequences some of which will not be considered until Volume II.

3.2.12 Theorem *Let $(\mathcal{A}, \|\cdot\|)$ be a commutative Banach algebra. Then, the Gelfand homomorphism is an isometric map of $(\mathcal{A}, \|\cdot\|)$ onto $(C_0(\Gamma_{\mathcal{A}}),$ $\|\cdot\|_{\infty})$ if and only if a map $*\colon \mathcal{A} \to \mathcal{A}$ (called an involution) can be defined satisfying*

$$(a+b)^* = a^* + b^* \tag{7}$$

$$(\lambda a)^* = \lambda^* a^* \tag{8}$$

$$(ab)^* = a^* b^* \tag{9}$$

$$(a^*)^* = a \tag{10}$$

$$\|a^* a\| = \|a\|^2 \tag{11}$$

for all $a, b \in \mathcal{A}$ and $\lambda \in \mathbb{C}$. When such a map $*\colon \mathcal{A} \to \mathcal{A}$ is defined, it satisfies

$$\widehat{a^*} = (\hat{a})^* \qquad \forall\, a \in \mathcal{A}. \tag{12}$$

Proof If the Gelfand homomorphism ($\widehat{\ }$) is an isometric isomorphism of $(\mathcal{A}, \|\cdot\|)$ onto $(C_0(\Gamma_{\mathcal{A}}), \|\cdot\|_{\infty})$, then we can define $*\colon \mathcal{A} \to \mathcal{A}$ by

$$\widehat{a^*}(\gamma) = (\hat{a}(\gamma))^* \qquad \forall\, \gamma \in \Gamma_{\mathcal{A}}; \quad a \in \mathcal{A}.$$

It is easy to check that ($*$) has the stated properties.

Suppose, conversely, that a map $*\colon \mathcal{A} \to \mathcal{A}$ exists satisfying (7) through (11). For any $h \in \mathcal{A}$ such that $h^* = h$, $\|h\| = \|h^2\|^{1/2} = \|h^{2^2}\|^{2^{-2}} = \cdots = \|h^{2^n}\|^{2^{-n}}$ can be derived by repeated application of (11). Thus Corollary 2.2.8 implies

$$\|h\| = \|h\|^{\infty} = \rho(h) \qquad \forall\, h = h^* \in \mathcal{A}. \tag{13}$$

Next we will use this equality to show that any $h \in \mathcal{A}$ satisfiying $h = h^*$ also satisfies $Sp(h) \subseteq \mathbb{R}$. Suppose not. Then for some s, t in $\mathbb{R}$, $k = sh^2 + th$

satisfies $i \in Sp(k)$ and $k = k^*$. We will derive a contradiction from this. Let $r \in \mathbb{R}$ satisfy $1 + 2r > \rho(k)^2$. For any positive integer n, define $c_n \in \mathcal{A}$ by $c_n = (k + ri)^n k$. Theorem 2.1.10 shows $(1 + r)^n i^{n+1} \in Sp(c_n)$. Thus, Theorem 3.1.5(c) shows $(1 + r)^{2n} \leq \rho(c_n)^2 \leq \|c_n\|^2 = \|c_n^* c_n\| = \rho(c_n^* c_n) = \rho((k^2 + r^2)^n k^2) \leq \sum_{j=0}^{n} \binom{n}{j} r^{2(n-j)} \rho(k)^{2j+2} = (\rho(k^2) + r^2)^n \rho(k^2)$. Taking the nth root and then the limit as n increases, we obtain $\rho(k)^2 + r^2 < 1 + 2r + r^2 \leq \rho(k)^2 + r^2$. This contradiction proves that $h = h^*$ implies $Sp(h) \subseteq \mathbb{R}$ for any $h \in \mathcal{A}$. Thus by Theorem 3.1.5, $h = h^*$ implies that $\hat{h}$ is real-valued.

Now we show that (12) holds. For $a \in \mathcal{A}$, define $h, k \in \mathcal{A}$ by $h = 2^{-1}(a + a^*)$ and $k = (2i)^{-1}(a - a^*)$. These elements satisfy $h = h^*, k = k^*, a = h + ik$ and $a^* = h - ik$. Thus for any $\gamma \in \Gamma_\mathcal{A}$, we get $\widehat{a^*}(\gamma) = \hat{h}(\gamma) - i\hat{k}(\gamma) = (\hat{h}(\gamma) + i\hat{k}(\gamma))^* = \hat{a}(\gamma)^*$, establishing (12).

We can now show easily that the Gelfand transform maps $\mathcal{A}$ isometrically and isomorphically onto $C_0(\Gamma_\mathcal{A})$. For any $a \in \mathcal{A}$, we have

$$
\begin{aligned}
\|a\| &= \|a^* a\|^{1/2} = \rho(a^* a)^{1/2} = \sup\{|\widehat{a^* a}(\gamma)|^{1/2} : \gamma \in \Gamma_\mathcal{A}\} \\
&= \sup\{(\widehat{a^*}(\gamma)\hat{a}(\gamma))^{1/2} : \gamma \in \Gamma_\mathcal{A}\} = \sup\{|\hat{a}(\gamma)| : \gamma \in \Gamma_\mathcal{A}\}.
\end{aligned}
$$

Thus (^) is isometric, and this together with Theorem 3.1.5 shows that it is an isomorphism into $C_0(\Gamma_\mathcal{A})$. Since (^) is an isometry, $\hat{\mathcal{A}}$ is complete and hence is closed in $C_0(\Gamma_\mathcal{A})$. However, $\hat{\mathcal{A}}$ is dense by Proposition 3.2.3 and hence actually equals $C_0(\Gamma_\mathcal{A})$, so that the Gelfand transform maps $\mathcal{A}$ onto $C_0(\Gamma_\mathcal{A})$.   □

We now turn to a few examples to illustrate the theory introduced so far in this chapter.

3.2.13 The Disc Algebra This algebra $A(\mathbb{D})$ was introduced in §1.5.2, which the reader should consult. It is the algebra of all functions $f: \mathbb{D} \to \mathbb{C}$ which are continuous on the closed unit disc $\mathbb{D} = \{\lambda \in \mathbb{C} : |\lambda| \leq 1\}$ and analytic on its interior $\mathbb{D}^\circ = \{\lambda \in \mathbb{C} : |\lambda| < 1\}$. The algebraic operations are pointwise and the norm is simply the supremum norm, which is complete since the uniform limit of analytic functions is analytic. Hence $A(\mathbb{D})$ is a unital commutative Banach algebra.

The first step in studying any commutative Banach algebra is the determination of its Gelfand space. It is often easy to guess the homomorphisms in the Gelfand space but harder to prove that one has guessed all of them. The present example follows this pattern. For each $\lambda \in \mathbb{D}$ the map

$$\gamma_\lambda(f) = f(\lambda) \qquad \forall\ f \in A(\mathbb{D})$$

belongs to $\Gamma_{A(\mathbb{D})}$. There are several ways to see that the set $\{\gamma_\lambda : \lambda \in \mathbb{D}\}$ is actually the whole Gelfand space. One instructive way is to note two isomorphic constructions for $A(\mathbb{D})$.

Let $P(\mathbb{D})$ be the algebra under pointwise operations of all functions $f: \mathbb{D} \to \mathbb{C}$ which are uniform limits of polynomials. The norm on $P(\mathbb{D})$ is, of course, again the supremum norm which gives the topology of uniform convergence. Let $P(\mathbb{T})$ be the algebra under pointwise operations of all continuous functions $f: \mathbb{T} \to \mathbb{C}$ which have vanishing negative Fourier coefficients (cf. §1.8.14). The norm is just the supremum norm again. Here we regard $\mathbb{T} = \{\lambda \in \mathbb{C} : |\lambda| = 1\}$ as the boundary of $\mathbb{D}$ and we recall (from §1.8.14) that the nth Fourier coefficient of $f \in C(\mathbb{T})$ is given by

$$\hat{f}(n) = \frac{1}{2\pi} \int_{-\pi}^{\pi} f(e^{it})e^{-int}dt.$$

We claim that $A(\mathbb{D})$ and $P(\mathbb{D})$ coincide and that restriction to $\mathbb{T} = \partial\mathbb{D}$ maps them isometrically isomorphically onto $P(\mathbb{T})$.

Fejér's Theorem (§1.8.15) shows that the sequence $\{\sigma_n f\}_{n\in\mathsf{N}}$ of Cesáro means of the partial sums of the Fourier series of a function $f \in C(\mathbb{T})$ converges uniformly (i.e., in the norm of $C(\mathbb{T})$) to f. Hence if f belongs to $P(\mathbb{T})$, then f is the uniform limit on $\mathbb{T}$ of the sequence $\{\sigma_n f\}_{n\in\mathsf{N}}$ of polynomial functions of $\lambda = e^{it} \in \mathbb{T}$. Now the maximum modulus principle shows that the sequence $\{\sigma_n f\}_{n\in\mathsf{N}}$ of polynomials converges uniformly on all of $\mathbb{D}$. (Explicitly: $|\sigma_n f(\lambda) - \sigma_m f(\lambda)| \le \|\sigma_n f - \sigma_m f\|_\infty \to 0$ holds for all $\lambda \in \mathbb{D}$, where the supremum norm is defined on $\mathbb{T}$.) This shows that each $f \in P(\mathbb{T})$ is the restriction to $\mathbb{T}$ of some function $\tilde{f}$ in $P(\mathbb{D})$ satisfying $\|\tilde{f}\|_\infty = \|f\|_\infty$. In fact, the construction of the Fejér kernel given in §1.8.15 can be modified to give $\tilde{f}$ explicitly:

$$\tilde{f}(re^{i\theta}) = \frac{1}{2\pi} \int_{-\pi}^{\pi} f(e^{it})P_r(\theta - t)dt$$

where P_r is the classical Poisson kernel

$$P_r(s) = \frac{1 - r^2}{1 - 2r\cos(s) + r^2}.$$

The completeness of $A(\mathbb{D})$ shows immediately that $P(\mathbb{D})$ is included in $A(\mathbb{D})$. Hence, in order to show that $A(\mathbb{D})$ equals $P(\mathbb{D})$ and that restriction to $\mathbb{T}$ maps $A(\mathbb{D})$ isometrically isomorphically onto $P(\mathbb{T})$, it only remains to show that restriction maps $A(\mathbb{D})$ into $P(\mathbb{T})$. This is obvious since we have

$$
\begin{aligned}
\hat{f}(-n) &= \lim_{r\to 1-} \frac{1}{2\pi} \int_{-\pi}^{\pi} f(re^{it})e^{int}dt \\
&= \lim_{r\to 1-} \frac{1}{2\pi i} \int_{|\lambda|=r} f(\lambda)\lambda^{n-1}d\lambda = 0
\end{aligned}
$$

for each $n \in \mathbb{N}$ and $f \in A(\mathbb{D})$. The last equality holds because the integrand is an analytic function.

(Note that for any $f \in A(\mathbb{D})$ the Taylor series expansion for f around zero converges pointwise to f on $\mathbb{D}^\circ$. However, it need not converge uniformly on $\mathbb{D}^\circ$ nor even pointwise at each point of $\mathbb{D}$. This is why the proof of the inclusion $A(\mathbb{D}) \subseteq P(\mathbb{D})$ depended on Cesáro summability.)

However, the coordinate function z is a single generator for $P(\mathbb{D})$, so Theorem 3.1.15 shows that $\{\gamma_\lambda : \lambda \in \mathbb{D}\}$ is the Gelfand space of $P(\mathbb{D})$ and hence of $A(\mathbb{D})$. The same result shows that $\lambda \mapsto \gamma_\lambda$ and $\gamma \mapsto \gamma(z)$ are inverse homeomorphisms of $\mathbb{D}$ onto $\Gamma_{A(\mathbb{D})}$ and of $\Gamma_{A(\mathbb{D})}$ onto $\mathbb{D}$, respectively.

The Šilov boundary of the disc algebra is also easy to identify. The maximum modulus principle shows that each $f \in A(\mathbb{D})$ must assume its maximum absolute value somewhere on $\mathbb{T} = \partial\mathbb{D}$. Conversely for each $\lambda \in \mathbb{T}$, the function $1 + \lambda^* z$ assumes its maximum absolute value at λ and only at λ. Hence, under the identification of $\Gamma_{A(\mathbb{D})}$ with $\mathbb{D}$, the Šilov boundary is identified with the topological boundary $\mathbb{T}$. This is, of course, a main reason for the name "boundary".

The last remark and the final result in Theorem 3.2.10 both show that the disc algebra is *not* completely regular. Hence, the hull–kernel topology on $\Gamma_{A(\mathbb{D})}$ is different from the Gelfand topology. In fact, it is clear that no proper subset of $\mathbb{D}$ can correspond to a hull if it contains a limit point in $\mathbb{D}^\circ$ since no non-zero function in $A(\mathbb{D})$ can vanish on such a subset.

Yood [1963b] noted the following examples of non-spectral algebra norms on $A(\mathbb{D})$. For any $f \in A(\mathbb{D})$, define

$$|||f||| = \sup\{|f(\lambda)| : \lambda \in K\}$$

where K is some closed subset of $\mathbb{D}$ which has nonempty interior and does not include $\mathbb{T}$. This is a norm since an analytic function which vanishes on K must be zero throughout $\mathbb{D}$. If λ is in $\mathbb{T} \setminus K$, then the function $g = (1 + \lambda^* z)/2$ satisfies the inequality $|||g||| < 1 = \rho(g)$ which shows that $||| \cdot |||$ is not a spectral norm. Furthermore, the maximal modular ideal $\mathcal{M} = \ker(\gamma_\lambda) = \{f \in A(\mathbb{D}) : f(\lambda) = 0\}$ is not closed and is even dense. To see this, note that for any $f \in A(\mathbb{D})$, $f - g^n f$ belongs to $\mathcal{M}$ but in the norm $||| \cdot |||$, $\lim_{n \to \infty} f - g^n f = f$ holds since $|||g^n||| \to 0$.

A larger algebra related to the disc algebra is

$$H^\infty = \{f : \mathbb{D}^\circ \to \mathbb{C} : f \text{ is holomorphic and bounded}\}.$$

Multiplication is pointwise and $|| \cdot ||_\infty$ is the norm, so $A(\mathbb{D})$ is a closed subalgebra of H^∞. It is not hard to see that for each $\zeta \in \mathbb{T}$ the radial limit $\tilde{f}(\zeta) = \lim_{r \to 1^-} f(r\zeta)$ exists. Then $\tilde{f}$ belongs to $L^\infty(\mathbb{T})$ and satisfies $||\tilde{f}||_\infty = ||f||_\infty$ for all $f \in H^\infty$ by the maximum modulus principle. The map $f \mapsto \tilde{f}$ identifies H^∞ with the subalgebra of $L^\infty(\mathbb{T})$ (with convolution multiplication) consisting of those functions $\tilde{f}$ satisfying

$$\frac{1}{2\pi} \int_{-\pi}^{\pi} \tilde{f}(e^{it})e^{nt}dt = 0 \qquad \forall\, n \in \mathbb{N}.$$

Note that these are exactly the functions in $L^\infty(\mathbb{T})$ for which all the negative Fourier coefficients vanish. Any $f \in H^\infty$ can be reconstructed from the corresponding $\tilde{f}$ by the Poisson formula given above. Since $C(\mathbb{T})$ is closed in $L^\infty(\mathbb{T})$, the polynomials are not dense in H^∞. In fact, H^∞ is not even separable. Thus the Gelfand space of H^∞ is much larger than $\mathbb{D}^\circ$. The portion of the Gelfand space corresponding to each point on the boundary of the disc is enormous.

3.2.14 Example Many variations of the last example have been studied. For instance, one could take any compact set Ω in the complex plane, other than the closed unit disc $\mathbb{D}$, and consider the algebras $A(\Omega)$ and $P(\Omega)$ defined in a way analogous to $A(\mathbb{D})$ and $P(\mathbb{D})$. (Note that the proof given above shows that the algebra $P(\mathbb{T})$, introduced above, agrees with this notation.) Similarly, one may consider the commutative Banach algebra $R(\Omega)$ (under pointwise operations and the supremum norm) of all uniform limits on Ω of rational functions with their poles off Ω. Any compact $\Omega \subset \mathbb{C}$, for which the interior of every component is connected (in order to assure the middle inclusion), clearly satisfies $P(\Omega) \subseteq R(\Omega) \subseteq A(\Omega) \subseteq C(\Omega)$. We have just shown $P(\mathbb{D}) = R(\mathbb{D}) = A(\mathbb{D}) \neq C(\mathbb{T})$ and the case $\Omega = \mathbb{T}$ satisfies $P(\mathbb{T}) \neq R(\mathbb{T}) = A(\mathbb{T}) = C(\mathbb{T})$ where the equality of $R(\mathbb{T})$ and $C(\mathbb{T})$ follows again from Fejér's theorem. (*N. B.* The algebra $A(\mathbb{T})$ equals $C(\mathbb{T})$ for the trivial reason that $\mathbb{T}$ has no interior. The notation here conflicts with the meaning assigned to $A(\mathbb{T})$ in Section 1.8 where it represents the functions with absolutely convergent Fourier series. Each notation is standard in its own specialized area.)

Since $R(\Omega)$ is generated as a Banach algebra by the set of functions $\{z\} \cup \{(z-\lambda)^{-1} : \lambda \notin \Omega\}$, each γ in the Gelfand space of $R(\Omega)$ is determined by the number $\lambda_\gamma = \gamma(z)$. Since $z - \lambda_\gamma$ is never invertible in $R(\Omega)$, each λ_γ belongs to Ω. Hence, as in the argument given before for $P(\mathbb{D})$, we see that $\gamma \mapsto \lambda_\gamma$ is a homeomorphism of $\Gamma_{R(\Omega)}$ onto Ω. A similar argument shows that the Šilov boundary of $R(\Omega)$ is always mapped onto the topological boundary of Ω under this homeomorphism.

For any compact subset Ω of $\mathbb{C}$, let $\mathrm{pc}(\Omega)$ represent the polynomially convex hull of Ω defined before Theorem 2.1.14. Thus $\mathrm{pc}(\Omega)$ is the union of Ω with all the bounded components of the complement of Ω. (More graphically, $\mathrm{pc}(\Omega)$ is just Ω with all its holes filled in.) The maximum modulus principle shows that restriction to Ω maps $P(\mathrm{pc}(\Omega))$ isometrically isomorphically onto $P(\Omega)$. Runge's approximation theorem (Theorem 3.3.3) shows that $(z-\lambda)^{-1}$ for $\lambda \notin \mathrm{pc}(\Omega)$ belongs to $P(\mathrm{pc}(\Omega))$, so that

$P(\mathrm{pc}(\Omega))$ equals $R(\mathrm{pc}(\Omega))$. Mergelyan's theorem (mentioned after Theorem 3.3.3) shows that $P(\mathrm{pc}(\Omega))$ actually equals $A(\mathrm{pc}(\Omega))$. Hence, we have $P(\Omega) \simeq P(\mathrm{pc}(\Omega)) = R(\mathrm{pc}(\Omega)) = A(\mathrm{pc}(\Omega))$. Note that we have proved the special case where Ω and $\mathrm{pc}\Omega$ are $\mathbb{T}$ and $\mathbb{D}$, respectively. Note also that Runge's theorem shows that any function in $C(\Omega)$ which is the restriction of a function analytic in some neighborhood of Ω actually belongs to $R(\Omega)$.

We have seen examples (*i.e.*, $\mathbb{T}$ and $\mathbb{D}$, respectively) in which the inclusions $P(\Omega) \subseteq R(\Omega)$ and $A(\Omega) \subseteq C(\Omega)$ are proper. Certain choices of Ω called Swiss cheeses show that $R(\Omega) \subseteq A(\Omega)$ may be proper (*cf.* Theodore W. Gamelin [1969], p.24). Since $A(\Omega)$ satisfies $R(\Omega) \subseteq A(\Omega) \subseteq C(\Omega)$ where the inclusions are isometries, and since the Gelfand spaces of both $R(\Omega)$ and $C(\Omega)$ can be identified with Ω under the map $\gamma \mapsto \lambda_\gamma = \gamma(z)$, it is not surprising that this map is also a homeomorphism of the Gelfand space of $A(\Omega)$ onto Ω for all compact Ω in $\mathbb{C}$. Furthermore, the Šilov boundary of $A(\Omega)$ is always identified by this map with the topological boundary of Ω (*cf.* Gamelin [1969], p. 31).

Before leaving these examples we also mention that if Ω is a compact subset of $\mathbb{C}$ with connected complement and connected interior, then restriction to the boundary $\partial\Omega$ of Ω maps $P(\Omega)$ isometrically isomorphically onto a maximal closed subalgebra of $C(\Omega)$. (Errett Bishop [1960]).

The classes of examples discussed above can be extended in various ways to algebras of analytic functions of several complex variables. The theory, of course, becomes far more complicated in this setting. For further results, we refer the reader to the extensive contemporary literature, starting with the following books: Andrew Browder [1969], Gamelin [1969], Stout [1971], and John Wermer [1976].

An interesting but unrelated example was given by Eva Kallin [1963]. It is a unital commutative Banach algebra $\mathcal{A}$ generated by four elements. There is a continuous function f on $\Gamma_{\mathcal{A}}$ such that for each point γ in $\Gamma_{\mathcal{A}}$, there is a neighborhood U of γ and an $a \in \mathcal{A}$ such that f equals $\hat{a}$ on U, but $f \notin \hat{\mathcal{A}}$. This contradicted a published theorem of Šilov (*cf.* Arens [1961] for background). The algebra also contains elements a and b such that $\hat{b}$ vanishes in the neighborhood of each zero of $\hat{a}$, but b does not belong to the ideal generated by a.

3.2.15 The Stone–Čech Compactification of a Topological Space

Our construction is based on the abstract characterization of the algebras $C_0(\Omega)$ given in Theorem 3.2.12, which shows that a commutative Banach algebra is isometrically isomorphic to $C_0(\Omega)$ for some locally compact Hausdorff space Ω if and only if it is a commutative C*-algebra. Note that Ω is compact if and only if $C_0(\Omega)$ is a unital algebra, in which case $C_0(\Omega)$ equals the algebra $C(\Omega)$ of all bounded continuous functions on Ω. Hence, Theorem 3.2.12 also shows that a unital commutative Banach algebra $\mathcal{A}$ is isometrically isomorphic to $C(\Omega)$ for some compact Hausdorff space Ω if and only if it is a unital commutative C*-algebra.

Let Ω be any topological space and let $C(\Omega)$ be the unital commutative Banach algebra of all bounded continuous functions $f\colon \Omega \to \mathbb{C}$ under pointwise operations and the supremum norm. It is obvious that

$$f^*(\omega) = f(\omega)^* \qquad \forall\, f \in C(\Omega);\ \omega \in \Omega$$

defines an involution which makes $C(\Omega)$ into a unital commutative C*-algebra. Hence, the Gelfand homomorphism (ˆ) is an isometric isomorphism of $C(\Omega)$ onto $C(\Gamma_{C(\Omega)})$. We will denote $\Gamma_{C(\Omega)}$ by $\beta\Omega$. For each $\omega \in \Omega$ the map $\gamma_\omega\colon C(\Omega) \to \mathbb{C}$ defined by

$$\gamma_\omega(f) = f(\omega) \qquad \forall\, f \in C(\Omega)$$

clearly belongs to $\beta\Omega$. Furthermore the map γ (*i.e.*, the map $\omega \mapsto \gamma_\omega$) is clearly a continuous map of Ω into $\beta\Omega$. In fact, the range γ_Ω of this map is dense in $\beta\Omega$. To see this, note that a function $f \in C(\Gamma_{C(\Omega)})$ which vanishes on γ_Ω is zero since it has the form $\hat{g}$ for some $g \in C(\Omega)$ which must be identically zero.

Assume next that Ω is a completely regular Hausdorff topological space (*i.e.*, a Tychonoff space). Then we claim that the map γ is a homeomorphism of Ω onto its image $\gamma_\Omega \subseteq \beta\Omega$ in its relative topology. Since the functions $f \in C(\Omega)$ clearly separate the points of a completely regular Hausdorff space, γ is injective. The definition of complete regularity shows that for each $\omega_0 \in \Omega$, a neighborhood base at ω_0 is given by

$$\{\{\omega \in \Omega\ :\ |f(\omega)| > 0\} : f \in C(\Omega);\ f(\omega_0) = 1\}.$$

Since γ transforms this into a neighborhood base for γ_{ω_0} in γ_Ω, γ is a homeomorphism as claimed. Hence, we have shown that any completely regular Hausdorff space Ω can be homeomorphically embedded as a dense subset of a compact Hausdorff space $\beta\Omega$ so that each bounded continuous function on Ω extends to a continuous function on $\beta\Omega$ with the same bound. In the next paragraph we will show that the embedding into $\beta\Omega$ satisfying these conditions is unique up to an obvious notion of equivalence which we will define there. The essentially unique space $\beta\Omega$ (or more precisely the embedding of Ω into $\beta\Omega$) is called the *Stone–Čech compactification* of Ω.

If Ω is any topological space, then a *compactification* of Ω is a pair (ψ, Ψ) consisting of a compact topological space Ψ and a continuous map ψ of Ω onto a dense set in Ψ. Note that we are not requiring that ψ be injective or a homeomorphism onto its image with the relativized topology of Ψ as some authors do. Two compactifications (ψ, Ψ) and (ψ', Ψ') are called *equivalent* if there is a homeomorphism $\theta\colon \Psi \to \Psi'$ satisfying $\theta \circ \psi = \psi'$.

To see that the Stone–Čech compactification is unique up to this notion of equivalence, suppose $\psi\colon \Omega \to \Psi$ satisfies the conditions for the Stone–Čech compactification. For each $f \in C(\Omega)$, let $\tilde{f} \in C(\Psi)$ be its extension to Ψ. (That is, $\tilde{f}$ satisfies $\tilde{f} \circ \psi = f$.) Then $\theta: \Psi \to \Gamma_{C(\Omega)}$, defined by $\theta(\tau)(f) = \tilde{f}(\tau)$ for all $\tau \in \Psi$, defines an equivalence between $\psi: \Omega \to \Psi$ and $\gamma: \Omega \to \Gamma_{C(\Omega)} = \beta\Omega$. These results, together with the remarks preceding Theorem 3.2.10, show that a topological space Ω has a compactification (ψ, Ψ) in which ψ is a homeomorphism if and only if Ω is completely regular.

We remind the reader of the algebraic analogue of the Stone–Čech compactification given in §1.5.1. There we showed that, for any locally compact Hausdorff space, there was a natural identification of $C(\Omega)$ with the double centralizer algebra $\mathcal{D}(C_0(\Omega))$ of $C_0(\Omega)$.

If (ψ, Ψ) is a compactification of Ω, then $\mathcal{A} = \{f \circ \psi : f \in C(\Psi)\}$ is a closed, self-adjoint (i.e., $\mathcal{A}$ is closed under complex conjugation of values), unital subalgebra of $C(\Omega)$. Conversely, any such subalgebra $\mathcal{A} \subseteq C(\Omega)$ is a unital commutative C*-algebra under complex conjugation as involution, so that the Gelfand transform is an isometric isomorphism of $\mathcal{A}$ onto $C(\Gamma_{\mathcal{A}})$. Clearly, the map $\gamma: \Omega \to \Gamma_{\mathcal{A}}$ defined by

$$\gamma_\omega(f) = f(\omega) \qquad \forall\, f \in \mathcal{A}; \ \omega \in \Omega$$

is a continuous map. The same argument as before shows that γ_Ω is dense in $\Gamma_{\mathcal{A}}$. Hence, we have established a bijection between the set of equivalence classes of compactification of Ω and the closed self-adjoint unital subalgebras $\mathcal{A}$ of $C(\Omega)$. Note that the algebra $\mathcal{A}$ corresponding to (ψ, Ψ) is the set of functions $f \in C(\Omega)$ which can be extended to Ψ in the sense that there is a function $g \in C(\Psi)$ satisfying $f = g \circ \psi$. Thus, $C(\Omega)$ itself is the algebra associated with the Stone–Čech compactification. Similarly, if Ω is locally compact and (ψ, Ψ) is the one-point compactification of Ω, then the associated algebra is $C_0(\Omega)^1 \subseteq C(\Omega)$.

We will give another even more interesting example of a compactification in the next example.

3.2.16 The Bohr Compactification This is a compactification of a locally compact group which takes into account both the topological and the group structure. Let G be a locally compact group. Recall from Section 1.10 that if $f: G \to \mathbb{C}$ is any function, then its left and right translates are defined by

$$_uf(v) = f(u^{-1}v) \quad \text{and} \quad f_u(v) = f(vu^{-1}) \qquad \forall\, v \in G,$$

respectively. A function $f \in C(G)$ is said to be $\langle$ left / right / two-sided $\rangle$ *almost periodic* if the set

$$\langle \{_uf : u \in G\} \,/\, \{f_u : u \in G\} \,/\, \{_uf_v : u, v \in G\} \,\rangle$$

has compact closure in $C(G)$. If f and g are left almost periodic, then $\{_u(f+g) : u \in G\}^-$ is included in the (continuous) image under summation of the compact set $\{_uf : u \in G\}^- \times \{_uf : u \in G\}^-$ and hence is compact. This shows that the set of left almost periodic functions on G is closed under addition. Similar arguments show that it is a subalgebra of $C(G)$ which is even closed under complex conjugation. It is also easy to see that if $\{f_n\}_{n \in \mathbb{N}}$ is a sequence of left almost periodic functions which converge to f in $C(G)$, then $\{_uf : u \in G\}$ is totally bounded so that f is almost periodic. Thus, the set of left almost periodic functions on G is a closed self-adjoint unital subalgebra of $C(G)$. It is obvious that similar arguments could be applied to the sets of right or two-sided almost periodic functions. However, it turns out that there is an elementary proof showing that left (or right) almost periodic functions are necessarily two-sided almost periodic. (For a proof see Lynn H. Loomis [1953], p.167, where the present approach to the Bohr compactification was first given.) Hence, we may speak of the unital commutative C*-*algebra of all almost periodic functions* and denote this set by $AP(G)$.

Our previous remarks show that $AP(G) \subseteq C(G)$ determines a compactification $\gamma: G \to bG$ of G. Thus γ is a continuous map of G into bG. Even though we are assuming that G is locally compact, and therefore completely regular, γ need not be injective nor a homeomorphism.

It is not clear *a priori* that bG has any more structure than that of a compact Hausdorff space. In fact, it turns out that the preimage under $\gamma: G \to bG$ of the point $\gamma(e)$ is a closed normal subgroup of G which we naturally denote by $\ker(\gamma)$ and the preimage of each point in the dense set γ_G is a coset of $\ker(\gamma)$. Hence, γ_G carries the structure of a group. The group operations on γ_G are uniformly continuous in the natural uniformity of bG and hence can be extended to all of bG. This makes bG into a compact topological group. (Proof of these facts may be found in the same reference.) Hence, for any locally compact group G, we have constructed a homomorphism $\gamma: G \to bG$ onto a dense subgroup of a compact group bG such that each almost periodic function f on G can be extended to be a continuous function $\overline{f}$ on bG (*i.e.*, f satisfies $\overline{f} \circ \gamma = f$). The extension process is an isometric isomorphism of $AP(G)$ onto $C(bG)$. The same argument which showed the uniqueness of the Stone–Čech compactification up to equivalence shows that $\gamma: G \to bG$ is determined up to equivalence by this extension property. The present compactification is called the *Bohr-compactification* since Harald Bohr [1932] first studied it in the special case in which the group G is the group $\mathbb{R}$ of real numbers.

The Haar measure on the compact group bG defines a linear functional on $C(bG)$ which can be pulled back to $AP(G)$ through the isometric isomorphic extension map $f \mapsto \overline{f}$ of $AP(G)$ onto $C(bG)$. This gives a linear functional $M: AP(G) \to \mathbb{C}$ satisfying

$$
\begin{aligned}
M(1) &= 1 \\
|M(f)| &\leq \|f\|_\infty \qquad \forall\, f \in AP(G)
\end{aligned}
$$

$$M(f^*) \;=\; M(f)^* \quad \forall \; f \in AP(G)$$
$$M(h) \;\geq\; 0 \qquad\quad \forall \; h \in AP(G); \; h \geq 0$$
$$M(_uf_v)) \;=\; M(f) \quad \forall \; f \in AP(G); \; u, \, v \in G$$

which is called the *von Neumann mean*. John von Neumann [1934] proved the existence of this mean by an elementary argument showing that if the set of left translates of a function $f \colon G \to \mathbb{C}$ (not necessarily continuous) has compact closure in the uniform norm, then its closed convex hull includes exactly one constant function, namely $M(f)$.

For Bohr's classical case $G = \mathbb{R}$, the von Neumann mean is given by

$$M(f) = \lim_{T \to \infty} (2T)^{-1} \int\limits_{-T}^{T} f(u)du \quad \forall \; f \in AP(\mathbb{R}).$$

In this case, there are also two important characterizations of $AP(\mathbb{R})$ in addition to the definition which we have given (which is due to Bochner [1927]). A function $f \in C(\mathbb{R})$ is almost periodic if and only if for each $\varepsilon > 0$ there is a $T > 0$ such that in each interval of length T there is an s satisfying $\|f - f_s\|_\infty < \varepsilon$. (This explains the name "almost periodic function".) Bohr showed that the functions satisfying the above criterion were precisely the uniform limits of functions of the form $t \mapsto \sum_{j=1}^{n} \alpha_j e^{is_j t}$ for $\alpha_j \in \mathbb{C}, s_j \in \mathbb{R}$. Since these functions obviously separate the points of $\mathbb{R}$, the Bohr compactification $\gamma \colon \mathbb{R} \to b\mathbb{R}$ is an injection.

3.2.17 Some Classes of Groups The continuous group homomorphism $\gamma \colon G \to bG$ of the Bohr compactification of an arbitrary locally compact group G is not necessarily an injection. Our description of its construction shows that it is an injection if and only if the functions in $AP(G)$ separate the points of G. A locally compact group for which this holds is said to be *maximally almost periodic* and the collection of all such groups is denoted by [MAP].

Let $U \colon G \to \mathcal{B}(\mathcal{H})$ be a continuous group homomorphism of G onto a bounded subgroup of the group of invertible elements in $\mathcal{B}(\mathcal{H})$ where $\mathcal{H}$ is a finite-dimensional normed linear space. Such a homomorphism is called a *finite-dimensional, continuous, bounded representation* of G. (An important special case arises when $\mathcal{H}$ is an inner product space and each representing operator U_u for $u \in G$ is unitary. Then U is called a *finite-dimensional, continuous, unitary representation*.) By choosing a basis for $\mathcal{H}$, we can write each representing operator $U_u(u \in G)$ as a matrix $(U_{u,i,j})_{n \times n}$ where n is the dimension of $\mathcal{H}$. The multiplicativity of U shows that the translates of each matrix entry function $u \mapsto U_{u,i,j}$ are linear combinations of the other such functions with a universal bound for the coefficients. Hence, the set of all such translates is a bounded subset of the finite-dimensional space spanned by $\{u \mapsto U_{u,i,j} : 1 \leq i, j \leq n\}$. Thus, each function in this space is almost periodic. This gives a sufficient condition for a locally compact group to belong to [MAP], and in fact this condition is also necessary.

Theorem. *The following are equivalent for a locally compact group G:*

(a) $G \in$ [MAP], *i.e., the algebra $AP(G)$ separates the points of G.*

(b) *In the Bohr compactification $\gamma: G \to bG$ the homomorphism γ is injective.*

(c) *There is some injective continuous group homomorphism of G into a compact group.*

(d) *There are enough continuous finite-dimensional bounded representations to separate the points of G.*

(e) *There are enough continuous finite-dimensional unitary representations to separate the points of G.*

Proof From previous remarks, the implications (e)$\Rightarrow$(d) $\Rightarrow$(a)$\Leftrightarrow$(b)$\Rightarrow$(c) are obvious. If K is any compact group, then the set $\mathcal{U}$ of continuous finite-dimensional unitary representations of K separates the points of K as we will show later by several different arguments. Hence if (c) holds and $\gamma \to K$ is a continuous homomorphism, then the set $\{U \circ \gamma : U \in \mathcal{U}\}$ shows that (e) holds.   □

In the above theorem, the implication (d)$\Rightarrow$(c) is also obvious directly. (If (d) holds and $\mathcal{U}$ is a point separating set of continuous finite-dimensional bounded representations of G, then $K = \prod_{U \in \mathcal{U}} U_G$ is a compact group, and the map which sends an element of G to its representing operator in each representation of $\mathcal{U}$ is an injective continuous group homomorphism of G into K.) Also, the implication (d)$\Rightarrow$(e) follows directly from the simple argument outlined in the next paragraph.

Let G be a subgroup of the multiplicative group of $\mathcal{L}(\mathcal{H})$, where $\mathcal{H}$ is some finite-dimensional linear space. Choose a basis $v_1, v_2, \ldots, v_n$ for $\mathcal{H}$ and define an inner product on $\mathcal{H}$ by

$$\left(\sum_{j=1}^{n} \lambda_j v_j, \sum_{j=1}^{n} \mu_j v_j \right)_0 = \sum_{j=1}^{n} \lambda_j \mu_j^*.$$

Assume that G is bounded with respect to the corresponding norm. Since any two norms on $\mathcal{H}$ are equivalent, and hence give rise to equivalent norms on $\mathcal{L}(\mathcal{H}) = \mathcal{B}(\mathcal{H})$, the particular choice of the basis is unimportant. Then G is a compact topological group in the relative topology of $\mathcal{B}(\mathcal{H})$. For any two vectors $x, y \in \mathcal{H}$, define a second inner product by

$$(x, y) = \int_{U \in G} (Ux, Uy)_0 dU,$$

where $\int \cdots dU$ represents the Haar integral on G. Then, it is clear that each $U \in G$ is a unitary operator with respect to this new inner product.

Thus for any bounded group G in $\mathcal{L}(\mathcal{H})$ (where $\mathcal{H}$ is finite-dimensional and G is bounded with respect to any norm on $\mathcal{H}$), there is an inner product on $\mathcal{H}$ relative to which G is a unitary group. Note that the last portion of this argument shows that any compact group in $\mathcal{B}(\mathcal{H})$ where $(\mathcal{H}, (\cdot, \cdot)_0)$ is an infinite-dimensional Hilbert space is a unitary group relative to an equivalent inner product. If G was originally the image of another locally compact group under a homomorphism continuous with respect to any norm on $\mathcal{H}$, then the homomorphism is still continuous with respect to the new inner product.

3.2.18 Gelfand Theory and Šilov Boundaries for Projective Tensor Products of Commutative Banach Algebras
The following results are essentially due to Jun Tomiyama [1960]. See also Bernard R. Gelbaum [1962] and Arnold Lebow [1968]. More general results are obtained by considerably more complicated arguments in Gelbaum [1959], [1961], [1970] and Kjeld B. Laursen [1970]. Some of these will be given in Example 7.1.19.

Theorem. *Let $\mathcal{A}$ and $\mathcal{B}$ be commutative Banach algebras. Then $\mathcal{A}\hat{\otimes}\mathcal{B}$ is a commutative Banach algebra and its Gelfand space is naturally homeomorphic to $\Gamma_\mathcal{A} \times \Gamma_\mathcal{B}$.*

Proof Proposition 1.10.12 shows that $\mathcal{A}\hat{\otimes}\mathcal{B}$ is a commutative Banach algebra. For $\gamma \in \Gamma_\mathcal{A}$ and $\delta \in \Gamma_\mathcal{B}$, the continuous bilinear functional

$$(a, b) \mapsto \gamma(a)\delta(b)$$

obviously induces a well defined function $\gamma \otimes \delta \colon \mathcal{A}\hat{\otimes}\mathcal{B} \to \mathbb{C}$, which is given by $\gamma \otimes \delta(\sum_{k=1}^{\infty} a_k \otimes b_k) = \sum_{k=1}^{\infty} \gamma(a_k)\delta(b_k)$ when $\sum_{k=1}^{\infty} \|a_k\| \, \|b_k\| < \infty$ holds. This provides a continuous map of $\Gamma_\mathcal{A} \times \Gamma_\mathcal{B}$ into $\Gamma_{\mathcal{A}\hat{\otimes}\mathcal{B}}$. (This map is just the tensor product of the functions γ and δ composed with the multiplication map $M \colon \mathbb{C} \otimes \mathbb{C} \to \mathbb{C}$.)

Next we wish to define a map from $\Gamma_{\mathcal{A}\hat{\otimes}\mathcal{B}}$ into $\Gamma_\mathcal{A} \times \Gamma_\mathcal{B}$ which will be the inverse of the above map. For each $\theta \in \Gamma_{\mathcal{A}\hat{\otimes}\mathcal{B}}$, choose an element $c \otimes d$ with $\theta(c \otimes d) \neq 0$. Replacing this element by a suitable multiple allows us to assume $\theta(c \otimes d) = 1$. Now define $\langle\, \gamma_\theta \colon \mathcal{A} \to \mathbb{C} \, / \, \delta_\theta \colon \mathcal{B} \to \mathbb{C} \,\rangle$ by $\langle\, \gamma_\theta(a) = (ac \otimes d) \, / \, \delta_\theta(b) = (c \otimes bd) \,\rangle$ for any $\langle\, a \in \mathcal{A} \, / \, b \in \mathcal{B} \,\rangle$. These maps satisfy

$$\gamma_\theta \otimes \delta_\theta(a \otimes b) = \theta(ac \otimes d)\theta(c \otimes bd) = \theta(ac^2 \otimes bd^2) = \theta(a \otimes b),$$

$$\gamma_\theta(aa') = \theta(aa'c \otimes d)\theta(c \otimes d) = \theta(ac \otimes d)\theta(a'c \otimes d) = \gamma_\theta(a)\gamma_\theta(a'),$$

and similarly $\delta_\theta(bb') = \delta_\theta(b)\delta_\theta(b')$ for all a, $a' \in \mathcal{A}$ and b, $b' \in \mathcal{B}$. Both maps are non-zero since they satisfy $1 = \theta(c \otimes d) = \gamma_\theta \otimes \delta_\theta(c \otimes d) = \gamma_\theta(c)\delta_\theta(d)$. Finally, $\gamma_{\gamma \otimes \delta} = \gamma$ and $\delta_{\gamma \otimes \delta} = \delta$ are obvious for all $\gamma \in \Gamma_\mathcal{A}$

and $\delta \in \Gamma_{\mathcal{B}}$. Hence, we have inverse bijections of $\Gamma_{\mathcal{A}} \times \Gamma_{\mathcal{B}}$ onto $\Gamma_{\mathcal{A} \hat{\otimes} \mathcal{B}}$. In fact, it is obvious that $\theta \mapsto (\gamma_{\theta}, \delta_{\theta})$ is also continuous so these bijections are homeomorphisms. □

Proposition *Let $\mathcal{A}$ and $\mathcal{B}$ be commutative Banach algebras. Then $\mathcal{A} \hat{\otimes} \mathcal{B}$ is semisimple if and only if:*

(a) *Both $\mathcal{A}$ and $\mathcal{B}$ are semisimple, and*

(b) *The natural map $\theta: \mathcal{A} \hat{\otimes} \mathcal{B} \to \mathcal{A} \check{\otimes} \mathcal{B}$ is injective. (This will hold if either $\mathcal{A}$ or $\mathcal{B}$ has the approximation property.)*

Proof If $\mathcal{A} \hat{\otimes} \mathcal{B}$ is semisimple, then its spectral radius is a norm. By the last proposition, this means that

$$t \mapsto \sup\{|\gamma \otimes \delta(t)| : \gamma \in \Gamma_{\mathcal{A}}; \ \delta \in \Gamma_{\mathcal{B}}\} \qquad \forall\, t \in \mathcal{A} \hat{\otimes} \mathcal{B}$$

is a norm on $\mathcal{A} \hat{\otimes} \mathcal{B}$. This clearly implies that $\rho_{\mathcal{A}}(a) = \sup\{|\gamma(a)| : \gamma \in \Gamma_{\mathcal{A}}\}$, $\rho_{\mathcal{B}}(b) = \sup\{|\delta(b)| : \delta \in \Gamma_{\mathcal{B}}\}$ and $\|t\|_{\omega} = \sup\{|\omega \otimes \omega'(t)| : \omega \in \mathcal{A}_1^* : \omega' \in \mathcal{B}_1^*\}$ are all norms. Hence, $\mathcal{A}$ and $\mathcal{B}$ are semisimple and θ is injective.

Suppose, conversely, that conditions (a) and (b) hold. Then any $t = \sum_{k=1}^{\infty} a_k \otimes b_k$ which satisfies $\sum_{k=1}^{\infty} \|a_k\|\, \|b_k\| < \infty$ and is in the Gelfand radical of $(\mathcal{A} \hat{\otimes} \mathcal{B})$ also satisfies $\gamma \otimes \delta(t) = \sum_{k=1}^{\infty} \gamma(a_k)\delta(b_k) = 0$. Since $\mathcal{B}$ is semisimple, this implies that $\sum_{k=1}^{\infty} \gamma(a_k)b_k$ is zero for each $\gamma \in \Gamma_{\mathcal{A}}$. Hence for each $\gamma \in \Gamma_{\mathcal{A}}$ and $\omega' \in \mathcal{B}^*$, $\gamma(\sum_{k=1}^{\infty} \omega'(b_k)a_k) = \gamma \otimes \omega'(t)$ is zero. Since $\mathcal{A}$ is semisimple, this implies that $\omega \otimes \omega'(t) = \omega(\sum_{k=1}^{\infty} \omega'(b_k)a_k)$ is zero. Condition (b) now shows that t is zero as we wished. □

Note that this last proposition contradicts Corollary 1 of Lebow [1968]. Gelbaum [1962], p. 533 notes that the above function $\theta \mapsto (\gamma_{\theta}, \delta_{\theta})$ sends the Šilov boundary of $\mathcal{A} \hat{\otimes} \mathcal{B}$ onto the products of the Šilov boundaries of $\mathcal{A}$ and $\mathcal{B}$. Tomiyama [1960], Theorem 5 and Gelbaum [1962], p. 530 show that $\mathcal{A}$ and $\mathcal{B}$ are completely regular if and only if the same is true of $\mathcal{A} \hat{\otimes} \mathcal{B}$. Several other similar results are proved in one or the other of these papers (*e.g.*, concerning anti-symmetry, analyticity, Wiener's property.) Finally, in Tomiyama's paper it is noted that if $\mathcal{A}$ and $\mathcal{B}$ are semisimple commutative Banach algebras, there is always a cross norm on $\mathcal{A} \otimes \mathcal{B}$ such that the completion of $\mathcal{A} \otimes \mathcal{B}$ relative to this cross norm is semisimple. (Just pull the norm of $\mathcal{A} \hat{\otimes} \mathcal{B} / (\mathcal{A} \hat{\otimes} \mathcal{B})_J$ (where $(\mathcal{A} \hat{\otimes} \mathcal{B})_J$ is the Gelfand radical of $\mathcal{A} \hat{\otimes} \mathcal{B}$) back to $\mathcal{A} \otimes \mathcal{B}$, where our previous results show that it is a norm.)

3.3 Functional Calculus

Introduction to Various Functional Calculi

A functional calculus is a way of assigning a meaning to "function $f(a)$ of an element a" for some algebra of functions $f: \mathbb{C} \to \mathbb{C}$ and some set of elements a in some class of algebras. We will use several functional calculi in this work. As the set of functions increases in size, the classes of algebras and the sets of elements within these algebras to which the functions can be applied will naturally decrease in size. The assignment $f \mapsto f(a)$ will always be an algebra homomorphism. The constant function 1

$$1(\lambda) = 1 \qquad \forall\, \lambda \in \mathbb{C}$$

will be mapped onto the identity element in $\mathcal{A}^1$, and the identity function

$$z(\lambda) = \lambda \qquad \forall\, \lambda \in \mathbb{C}$$

will be mapped onto a. (We will use z with this meaning throughout the remainder of this section.) These two requirements will determine the homomorphism, perhaps under mild additional continuity assumptions. Furthermore, $f(g(a))$ will be defined exactly when $f \circ g(a)$ is defined (where $f \circ g$ is the composition of f and g) and we will have $f(g(a)) = f \circ g(a)$. For each given functional calculus we consider, the question of whether $f(a)$ is defined for a given element a (of the proper type) will depend only on the spectrum of a. Furthermore, some form of the spectral mapping theorem which asserts

$$Sp(f(a)) = \{f(\lambda) \colon \lambda \in Sp(a)\}$$

will hold. Finally, the various functional calculi will agree when more than one is defined for a given f and a.

We will also consider two functional calculi in which functions are applied to ordered n-tuples of elements rather than to single elements. These will be taken up separately, since they are technically much more difficult.

The two most elementary functional calculi, $i.e.$, those involving polynomials and rational functions, were already considered in conjunction with the spectral mapping theorem (Theorem 2.1.10). We briefly review the facts established there. They are defined for any element in any algebra. The functions involved in the first are polynomials. Let a be an element in the algebra $\mathcal{A}$. Let p be a polynomial with complex coefficients, say $p = \sum_{j=0}^{n} \alpha_j z^j$ ($i.e.$, $p(\lambda) = \sum_{j=0}^{n} \alpha_j \lambda^j$ for all $\lambda \in \mathbb{C}$). Then define $p(a)$ by $p(a) = \sum_{j=0}^{n} \alpha_j a^j \in \mathcal{A}^1$ where $a^0 = 1$. It is clear that the map $p \mapsto p(a)$ is a homomorphism which also respects composition. Obviously it is the only algebra homomorphism of the algebra of polynomials into $\mathcal{A}^1$ which takes 1 onto 1 and z onto a.

We can easily extend the polynomial functional calculus to a functional calculus in which any rational function can be applied to an element $a \in \mathcal{A}$ if it has no poles in $Sp(a)$. (Of course, this is not a proper extension if $Sp(a)$ is the whole plane. No element in a spectral algebra has this problem.) Let $a \in \mathcal{A}$ be given and let p and q be polynomials such that q has no roots in $Sp(a)$. The spectral mapping theorem shows that zero does not belong to the spectrum of $q(a)$ in $\mathcal{A}^1$. Hence, $q(a)$ is invertible in $\mathcal{A}^1$ and its inverse commutes with $p(a)$. Thus, for any $a \in \mathcal{A}$ and any rational function $r = p/q$ where p and q are polynomials and q has no roots in $Sp(a)$, we can define $r(a)$ by

$$r(a) = p(a)q(a)^{-1} \in \mathcal{A}^1.$$

It is immediately clear that the map $r \mapsto r(a)$ does not depend on the particular choice of p and q (satisfying $r = p/q$ and q has no roots in $Sp(a)$) and is a homomorphism. The spectral mapping theorem

$$Sp(r(a)) = r(Sp(a))$$

was proved as Theorem 2.1.10. If s is another rational function, $s(r(a))$ is defined if and only if the rational function $s \circ r$ has its poles off $Sp(a)$. It is easy to check that $s \circ r(a) = s(r(a))$ by factoring s and checking the equation for each of the factors $\lambda - z$ and $(\lambda' - z)^{-1}$ which occur. It is also easy to check that $r \mapsto r(a)$ is the only homomorphism of this set of rational functions into $\mathcal{A}^1$ which takes 1 onto 1 and z onto a. Thus we have another well-behaved functional calculus.

Let $\mathcal{A}$ be a Banach algebra, and let a be an element in $\mathcal{A}$. Let $f = \sum_{n=0}^{\infty} \alpha_n z^n$ be a power series with complex coefficients having radius of convergence strictly greater than $\rho(a)$. Then, $\sum_{n=0}^{\infty} \alpha_n a^n$ converges in norm unconditionally for any extension of the norm of $\mathcal{A}$ to $\mathcal{A}^1$. The set of functions with power series with radius of convergence strictly greater than $\rho(a)$ is an algebra and the map $f \mapsto f(a)$ is a homomorphism, since the series converges unconditionally and thus can be multiplied. It is also obvious that this functional calculus extends the polynomial calculus. (We used a special case of this idea to define the exponential function in Theorem 2.1.12.) Rather than discuss uniqueness, composition of functions or the spectral mapping theorem in the present setting, we go on to a more inclusive functional calculus for analytic functions.

The Analytic Functional Calculus

We turn to the construction of the most important functional calculus for general Banach algebras: that based on analytic functions. This functional calculus was first used by Frigyes Riesz in [1911] for spectral projections of eigenvalues of compact operators but was developed more fully by Angus E. Taylor [1938b], Gelfand [1941a], Edgar R. Lorch [1942] and especially Nelson Dunford [1943], who gave the general spectral mapping theorem. Norbert Wiener's [1923] use of Banach space valued analytic functions and Marshall Harvey Stone's book [1932] provided a stimulus to some of this work. Similar ideas were also used in a more concrete setting by Arne Beurling [1938]. The particular construction we will use derives from the first edition (1971) of Wermer [1976] but the idea is probably due to Nicolaas Govert de Bruijn [1958]. It was further refined by Bonsall and Duncan [1973b], Walter Rudin [1973] and Robert B. Burckel [1994].

This functional calculus applies to any element a in any Banach algebra $\mathcal{A}$, and the functions for which $f(a)$ is defined are those which are analytic in some neighborhood of the spectrum of a. Actually, the set of all functions analytic in some (small, variable) neighborhood of $Sp(a)$ is not an algebra, so that we must turn to the algebra of germs of functions analytic in some neighborhood of $Sp(a)$. (The problem arises because we must define the domain of a sum to be the intersection of the domains of the individual functions, so that $(f + g) - g$ will not usually be the same function as f. For the power series with positive radius of convergence considered above, it is not necessary to intersect the domain of a sum since we may always choose the largest open disc on which the series converges.) We postpone consideration of this complication until our basic construction is complete. The construction depends on certain contour integrals of functions with values in a Banach algebra. We introduce our terminology and notation before explaining the details of the construction.

3.3.1 Definition Let $\mathcal{A}$ be a Banach algebra and let a be an element of $\mathcal{A}$. Let $\tilde{\mathcal{F}}_a$ be the collection of all functions f each of which is defined and analytic on some (possibly disconnected) open set D_f which includes $Sp(a)$. We will say a contour Γ in the complex plane is *suitable for the pair* $(a,\ f)$ if Γ consists of a finite set $\{\Gamma_1, \Gamma_2, \ldots, \Gamma_p\}$ of disjoint, rectifiable, piecewise smooth, simple closed curves in $D_f \setminus Sp(a)$ such that the winding number of Γ satisfies

$$\frac{1}{2\pi i} \int_\Gamma \frac{1}{(\lambda - \lambda_0)}\, d\lambda = \begin{cases} 1 & \text{if } \lambda_0 \in Sp(a) \\ 0 & \text{if } \lambda_0 \in \mathbb{C} \setminus D_f. \end{cases}$$

If Γ is suitable for $(a,\ f)$, we define an element $I_a^\Gamma(f)$ of $\mathcal{A}^1$ by

$$I_a^\Gamma(f) = \frac{1}{2\pi i} \int_\Gamma f(\lambda)(\lambda - a)^{-1}\, d\lambda. \tag{1}$$

First we note a way to construct a suitable contour Γ for a pair $(a,\ f)$. Let $\varepsilon > 0$ be $\mathrm{dist}(Sp(a), \mathbb{C} \setminus D_f)$. Choose $R > \rho(a)$. Cover the compact set $R\mathbb{C}_1 \setminus D_f$ by a finite number of open discs $D_1, D_2, \ldots, D_q$ of radius $\varepsilon/2$ centered at points in the set. If the boundaries of some of these discs are tangent, their centers may be moved slightly to remove this problem. Let V be the neighborhood of $Sp(a)$ given by

$$V = R\mathbb{C}_1 \setminus \bigcup_{k=1}^q D_k.$$

Let Γ be the boundary of V oriented so that any arc of $\partial R\mathbb{C}_1$ in Γ is counterclockwise and any arc of ∂D_k is clockwise for $k = 1, 2, \ldots, q$. It is easy to check that this contour is suitable.

The contour integral $I_a^\Gamma(f) = (2\pi i)^{-1} \int_\Gamma f(\lambda)(\lambda - a)^{-1} \, d\lambda$ is defined as an ordinary Riemann–Stieltjes integral in the unitization $\mathcal{A}^1$ converging in the norm topology. Explicitly, for each of the finitely many Jordan closed curves Γ_ℓ, the union of which is Γ, the integral is the limit of Riemann–Stieltjes sums of the form

$$\frac{1}{2\pi i} \sum_{k=1}^{n} f(\lambda_k)(\lambda_k - a)^{-1}(\gamma_k - \gamma_{k-1})$$

where the λ_k's and γ_k's belong to Γ_k, $\gamma_0 = \gamma_n$, and one positive circuit around Γ_ℓ starting at γ_0 visits the λ_k's and γ_k's in the order

$$\gamma_0, \lambda_1, \gamma_1, \lambda_2, \gamma_2, \ldots, \gamma_{n-1}, \lambda_n, \gamma_n.$$

The limit is taken as $\max\{d(\gamma_{k-1}, \gamma_k) : k = 1, 2, \ldots, n\}$ approaches zero, where $d(\gamma_{k-1}, \gamma_k)$ is the distance along Γ_ℓ in the positive direction from γ_{k-1} to γ_k. Because the integrand is continuous, the integral exists by exactly the same proof given in any careful discussion of the Riemann integral for numerical-valued functions. Moreover, the definition shows that any continuous linear functional $\omega \in (\mathcal{A}^1)^*$ can be moved through the integral sign to give:

$$\omega(I_a^\Gamma(f)) = \frac{1}{2\pi i} \int_\Gamma f(\lambda)\omega((\lambda - a)^{-1}) \, d\lambda. \tag{2}$$

Using the analyticity of the resolvent function $\lambda \mapsto (\lambda - a)^{-1}$ (Corollary 2.2.12) on the resolvent set $\mathbb{C} \setminus Sp(a)$ of a, we could easily use classical complex analysis, equation (2) and the Hahn–Banach theorem to show that $I_a^\Gamma(f)$ is actually independent of the choice of suitable contour Γ, but we proceed differently.

3.3.2 Lemma *Let a be an element of a Banach algebra $\mathcal{A}$, f a function in $\tilde{\mathcal{F}}_a$ and Γ a suitable contour for (a, f).*

(a) *If f is identically equal to 1, then $I_a^\Gamma(1)$ equals $1 \in \mathcal{A}^1$.*

(b) *If $f = r$ is a rational function with its poles off $Sp(a)$, then*

$$I_a^\Gamma(r) = r(a).$$

(c) *If $\{f_n\}_{n \in \mathbb{N}}$ is a sequence of analytic functions on D_f which converges pointwise to f and uniformly on compact subsets of D_f, then*

$$I_a^\Gamma(f) = \lim_{n \to \infty} I_a^\Gamma(f_n). \tag{3}$$

(d) *If $\gamma \in \Gamma_{\mathcal{A}^1}$, then $\gamma(I_a^\Gamma(f)) = f(\gamma(a))$. Hence, if $\Gamma_{\mathcal{A}^1}$ is nonempty, the Gelfand transform satisfies*

$$\widehat{I_a^\Gamma(f)} = f \circ \hat{a}.$$

Proof (a): Choose R strictly larger than $\max\{|\lambda| : \lambda \in \Gamma\}$ and let Γ_0 be the circle of radius R centered at 0 and oriented counterclockwise. Let $\tilde{\Gamma}$ be the contour consisting of Γ_0 together with Γ with its orientation reversed. Corollary 2.2.12 shows that the resolvent $(\lambda - a)^{-1}$ is analytic at every point where the winding number of $\tilde{\Gamma}$ is non-zero. Hence, for any $\omega \in (\mathcal{A}^1)^*$ equation (2) shows that $\omega(I_a^\Gamma(f)) = (2\pi i)^{-1} \int_{\Gamma_0} f(\lambda)\omega((\lambda - a)^{-1}) \, d\lambda$. Proposition 2.2.9(f) shows that the resolvent can be expanded in a uniformly and absolutely convergent Laurent series on Γ_0, so we have

$$\omega(I_a^\Gamma(1)) = \frac{1}{2\pi i} \int_{\Gamma_0} \sum_{n=0}^{\infty} \omega(a^n)\lambda^{-n-1}d\lambda = \frac{1}{2\pi i} \int_{\Gamma_0} \omega(1)\lambda^{-1}d\lambda = \omega(1).$$

Since $\omega \in (\mathcal{A}^1)^*$ was arbitrary, (a) is established by the Hahn Banach theorem.

(b): Let p and q be polynomials satisfying $r = p/q$ and with q having no roots in $Sp(a)$. By expanding the left side, we see that there are polynomials $g_0, g_1, \ldots, g_n$ satisfying

$$p(\lambda)q(\zeta) - p(\zeta)q(\lambda) = (\lambda - \zeta)\sum_{k=0}^{n} g_k(\lambda)\zeta^k.$$

Hence, we may write

$$(r(\lambda) - r(a))(\lambda - a)^{-1} = \sum_{k=0}^{n} h_k(\lambda)a_k$$

where $h_k = g_k/q$ and $a_k = a^k/q(a)$. Since each complex-valued function h_k is analytic inside Γ, its integral is zero. Hence, conclusion (a) gives

$$I_a^\Gamma(r) = r(a)\frac{1}{2\pi i} \int_\Gamma (\lambda - a)^{-1} \, d\lambda = r(a).$$

(c): The point set of the contour Γ is a compact subset of D_f. Hence if $\{f_n\}_{n \in \mathbb{N}}$ is any sequence of functions satisfying the hypotheses, then it converges uniformly to f on Γ. The same proof that works for numerical-valued functions shows that the limit of a uniformly convergent sequence of functions with values in $\mathcal{A}^1$ can be interchanged with a Riemann integral. This gives the desired result.

(d): We have already noted that a continuous linear functional like γ can be moved through the integral sign, so we see

$$\gamma(I_a^\Gamma(f)) = \frac{1}{2\pi i} \int_\Gamma f(\lambda)(\lambda - \gamma(a))^{-1}) \, d\lambda = f(\gamma(a)).$$

The remark about Gelfand transforms is an immediate consequence. □

Runge Approximation Theorem

Lemma 3.3.2 contains enough information about the map $I_a^\Gamma \colon \tilde{\mathcal{F}}_a \to \mathcal{A}^1$ so that we may use it to prove the classical Runge approximation theorem [1885]. That theorem will be central to the next step of our construction, which will make the rest of the properties of our functional calculus, including its uniqueness, quite easy.

3.3.3 Runge's Approximation Theorem *Let K be a compact subset of $\mathbb{C}$. Then any function analytic in a neighborhood of K can be uniformly approximated on K by rational functions with poles off K. In fact, if Λ is any subset of the complement of K which includes at least one point from each bounded component of that complement, then only rational functions with poles in Λ are needed.*

Hence, if the complement of K is connected, then any function analytic in the neighborhood of K can be uniformly approximated on K by polynomials.

Proof Let $R(K)$ be the closed subalgebra of $C(K)$ generated by the rational functions with poles in Λ. The abstract Runge theorem (Theorem 2.2.13) asserts that every element of $R(K)$ has the same spectrum in $R(K)$ as it does in $C(K)$. The identity function z belongs to $R(K)$ and its spectrum is exactly K since $(\lambda - z)^{-1}$ belongs to $C(K)$ (and hence to $R(K)$) exactly when $\lambda \notin K$. If f is analytic in a neighborhood of K, then f belong to $\tilde{\mathcal{F}}_z$. Choose a contour Γ suitable for (a, f). Then, $I_z^\Gamma(f)$ belongs to $R(K)$ and satisfies $\widehat{I_z^\Gamma(f)} = f \circ \hat{z}$ by (d). However, the Gelfand homomorphism is injective in this case (since for any $\lambda \in K$ we can define a homomorphism $\gamma_\lambda \in \Gamma_{R(K)}$ by $\gamma_\lambda(f) = f(\lambda)$ for all $f \in R(K)$). Hence $f = I_z^\Gamma(f)$ belongs to $R(K)$. This proves the first statement of the theorem.

If the complement of K is connected, it has no bounded components so that Λ may be chosen empty. This proves the final statement. □

The final statement of Runge's theorem has two extremely important extensions. One is K. Oka's theorem [1936] which we state as Theorem 3.5.3 below. The other is S. N. Mergelyan's theorem [1951] which we will not prove. Mergelyan's theorem asserts that a function f is the uniform limit of polynomials on a compact subset K of $\mathbb{C}$ which has connected complement if f is continuous on K and analytic on the interior of K. Many proofs of this theorem have been given: *cf.* Rudin [1974], Stout, [1971].

Runge's theorem greatly increases the importance of equation (3) in Lemma 3.3.2(c).

3.3.4 Corollary *Let a be an element of a Banach algebra $\mathcal{A}$ and let f be a function in $\tilde{\mathcal{F}}_a$. Then, it is possible to find a suitable contour Γ for* *(a, f) and a sequence $\{r_n\}_{n\in\mathbb{N}}$ of rational functions satisfying*

$$I_a^\Gamma(f) = \lim_{n\to\infty} r_n(a). \tag{4}$$

Moreover, any sequence of rational functions, defined and converging to f uniformly on each compact subset of some neighborhood of $Sp(a)$, satisfies this equation.

Proof Equation (4) follows from result (a) and equation (3) of (c) in the lemma if the sequence satisfies the last sentence in the corollary. We use Runge's theorem to establish the existence of such approximating sequences. First choose Λ with exactly one point in each bounded component of $\mathbb{C} \setminus Sp(a)$. Then choose an open neighborhood V of $Sp(a)$ with compact closure $\bar{V}$ included in $D_f \setminus \Lambda$. Apply Runge's theorem to f and $\bar{V}$. $\square$

We will use equation (4) to define the functional calculus. The equality of the two expressions shows that each is independent of the various choices involved. First, we introduce the algebra of functions (or rather, of equivalence classes of functions) on which the functional calculus is based.

The Topological Algebra of Germs of Analytic Functions

Let $\mathcal{A}$ be a Banach algebra and let a be an element of $\mathcal{A}$ which will remain fixed throughout this discussion. We will define the algebra $\mathcal{F}_a$ of germs of functions analytic in a neighborhood of $Sp(a)$. Let $\mathcal{N}_a$ be the family of all open subsets of $\mathbb{C}$ which include $Sp(a)$. Note that the elements of $\mathcal{N}_a$ need not be connected. For each $U \in \mathcal{N}_a$, let $\mathcal{F}_U$ be the algebra (under pointwise operations) of all analytic functions $f : U \to \mathbb{C}$. We give $\mathcal{F}_U$ the topology in which a net $\{f_\alpha\}_{\alpha\in A}$ converges to f if $f_\alpha(\lambda)$ converges to $f(\lambda)$ for each $\lambda \in U$ and the convergence is uniform on each compact subset of U. We have already used $\tilde{\mathcal{F}}_a$ to represent the union of all the algebras $\mathcal{F}_U$ for $U \in \mathcal{N}_a$ and D_f to represent the domain of $f \in \tilde{\mathcal{F}}_a$. Our functional calculus is defined for each function in $\tilde{\mathcal{F}}_a$ but, since $\tilde{\mathcal{F}}_a$ is not an algebra, we wish to replace it with an algebra. In order to do this, we introduce an equivalence relation:

$$f \sim g \quad \Leftrightarrow \quad f|_U = g|_U \text{ for some } U \in \mathcal{N}_a. \tag{5}$$

We will denote the set of equivalence classes by $\mathcal{F}_a$. For each $U \in \mathcal{N}_a$, define $\rho_U : \mathcal{F}_U \to \mathcal{F}_a$ as the natural map. For $f, g \in \tilde{\mathcal{F}}_a$, fg and $f + g$ are the usual pointwise product and sum defined on $D_f \cap D_g$. It is easy to see that these operations are compatible with the equivalence relation. Hence, they give $\mathcal{F}_a$ the structure of an algebra. We will not distinguish in notation between a function $f \in \tilde{\mathcal{F}}_a$ and its equivalence class in $\mathcal{F}_a$.

We wish to give $\mathcal{F}_a$ the strongest locally convex linear topology such that each map $\rho_U : \mathcal{F}_U \to \mathcal{F}_a$ will be continuous. To do this we define a neighborhood system at the origin consisting of all convex, balanced, absorbing subsets N of $\mathcal{F}_a$ such that $\rho_U^{-1}(N)$ is open in $\mathcal{F}_U$ for each $U \in \mathcal{N}_a$. (In other words, we have defined $\mathcal{F}_a$ to be the inductive limit (Definition 1.3.4) of $\{\mathcal{F}_U : U \in \mathcal{N}_a\}$ under the maps $\rho_{UV} : \mathcal{F}_U \to \mathcal{F}_V$ defined by restriction to V for each pair U, $V \in \mathcal{N}_a$ with $V \subseteq U$.)

3.3.5 Definition The topological algebra $\mathcal{F}_a$ which we have just constructed is called the *algebra of germs of functions analytic in a neighborhood of $Sp(a)$*.

The Homomorphism I_a

For each $U \in \mathcal{N}_a$, we will now define a continuous homomorphism $I_a^U : \mathcal{F}_U \to \mathcal{A}^1$. It will be easy to see that if $f, g \in \tilde{\mathcal{F}}_a$ are equivalent, then $I_a^{D_f}(f) = I_a^{D_g}(g)$. Hence, the family $\{I_a^U : U \in \mathcal{N}_a\}$ defines a homomorphism I_a from $\mathcal{F}_a$ into $\mathcal{A}^1$. Moreover, it is easy to see that $I_a : \mathcal{F}_a \to \mathcal{A}^1$ is continuous.

Let $U \in \mathcal{N}_a$ be given. Then, a contour Γ which is suitable for (a, f) for one $f \in \mathcal{F}_U$ is suitable for all of them, so we define $I_a^U : \mathcal{F}_U \to \mathcal{A}^1$ to be I_a^Γ. Corollary 3.3.4 shows that this is independent of the choice of Γ among suitable contours. Clearly, I_a^U is linear and Lemma 3.3.2(c) shows that it is continuous. Equation (4) of Corollary 3.3.4 shows that it is multiplicative, since the functional calculus for rational functions is multiplicative. We have already noted that $I_a^U(f)$ is invariant under the equivalence relation used to construct $\mathcal{F}_a$, so for $f \in \mathcal{F}_a$ we may define $I_a(f)$ to be $I_a^{D_f}(f)$ for some representative f.

3.3.6 Definition Let a be an element in a Banach algebra $\mathcal{A}$ and let f be a function in $\mathcal{F}_a$. In the *analytic functional calculus*, $f(a)$ has the meaning $I_a(f) \in \mathcal{A}^1$ as defined above.

We are now ready to state a comprehensive theorem. The statement of the theorem is slightly complicated by the fact that we do not require an identity element. In fact, the reader may wish to note how both the statement and the proof are simplified if $\mathcal{A}$ is unital.

3.3.7 Theorem *Let $\mathcal{A}$ be a Banach algebra. For each $a \in \mathcal{A}$, the map $I_a : f \mapsto f(a)$ constructed above is the unique continuous homomorphism $I_a : \mathcal{F}_a \to \mathcal{A}^1$ satisfying:*

(a) $I_a(1) = 1$ $I_a(z) = a$.

Furthermore, for any $a \in \mathcal{A}$ and $f \in \mathcal{F}_a$, these maps also satisfy:

(b) $Sp(f(a)) = f(Sp(a))$.

(c) *For any $g \in \mathcal{F}_{f(a)}$, $g \circ f$ belongs to $\mathcal{F}_a$ and satisfies*

$$g \circ f(a) = g(f(a)).$$

(d) *If either 0 is not in $Sp(a)$ or $f(0)$ is zero, then $f(a)$ belongs to the smallest closed spectral subalgebra of $\mathcal{A}$ containing a. Otherwise $f(a)$*

belongs to the smallest closed spectral subalgebra of $\mathcal{A}^1$ containing a and 1. If a representative of f can be chosen in $\tilde{\mathcal{F}}_a$ so that D_f includes a compact neighborhood of $Sp(a)$ with connected complement, then the word "spectral" can be removed from each of the last two statements.

(e) *For any Banach algebra $\mathcal{B}$ and continuous homomorphism $\varphi\colon \mathcal{A} \to \mathcal{B}$, both $\mathcal{F}_a \subseteq \mathcal{F}_{\varphi(a)}$ and*

$$\varphi(f(a)) = f(\varphi(a))$$

hold when either (1) $\mathcal{A}$, $\mathcal{B}$ and φ are all unital; or (2) $Sp(a)$ does not include zero; or (3) $f(0)$ is zero.

(f) *If $\mathcal{C}$ is a maximal commutative subalgebra of $\mathcal{A}$ and the element a belongs to $\mathcal{C}$, then*

$$\widehat{f(a)} = f \circ \hat{a} \tag{6}$$

holds, where $(\hat{\ })$ denotes the Gelfand homomorphism on $\mathcal{C}$ unless $\mathcal{A}$ is not unital and also $f(0)$ is not zero in which case $(\hat{\ })$ denotes the Gelfand homomorphism on $\mathcal{C}^1$ (embedded in $\mathcal{A}^1$).

(g) *If f is a rational function, $f(a)$ has the same meaning in the analytic functional calculus as it does in the rational functional calculus.*

(h) *If f has a power series expansion $f = \sum_{n=0}^{\infty} \alpha_n z^n$ with radius of convergence strictly greater than $\rho(a)$, then f belongs to $\mathcal{F}_a$ and satisfies*

$$f(a) = \sum_{n=0}^{\infty} \alpha_n a^n.$$

(i) *For each $f \in \tilde{\mathcal{F}}_a$, there is a neighborhood $\mathcal{D}$ of a such that f belongs to $\mathcal{F}_b$ for each $b \in \mathcal{D}$ and $f(b)$ is a continuous function of $b \in \mathcal{D}$.*

Proof Throughout this proof, $a \in \mathcal{A}$ will be arbitrary but fixed. We have already noted that I_a is a well defined continuous homomorphism which agrees with the rational functional calculus if the function is rational. Hence, in particular, conditions (a) and (g) hold.

Let $I\colon \mathcal{F}_a \to \mathcal{A}^1$ be any other continuous homomorphism satisfying (1) and (2). Theorem 2.1.10 states that I must agree with the rational functional calculus for rational functions. Corollary 3.3.4 now shows $I = I_a$, establishing uniqueness.

(b): Consider $\mu \in \mathbb{C} \setminus f(Sp(a))$. Then, $g = (\mu - f)^{-1}$ is analytic on the open neighborhood $D_f \setminus \{\lambda : f(\lambda) = \mu\}$. Hence, g belongs to $\mathcal{F}_a$ and $g(a)$ is an inverse for $\mu - f(a)$, so $Sp(f(a)) \subseteq f(Sp(a))$.

Conversely, consider $\mu \in Sp(a)$. For any λ, there is an analytic function g on D_f satisfying $(\mu - \lambda)g(\lambda) = f(\mu) - f(\lambda)$. Therefore $(\mu - a)g(a) = f(\mu) - f(a)$ holds. However, this shows that if $f(\mu) - f(a)$ had an inverse then $\mu - a$ would have one, contrary to assumption. This establishes the reverse inclusion.

(c): By (b), we see that $g \circ f$ belongs to $\mathcal{F}_a$ if g belongs to $\mathcal{F}_{f(a)}$. The function $I: \mathcal{F}_{f(a)} \to \mathcal{A}^1$ defined by $I(g) = g \circ f(a)$ is a continuous homomorphism satisfying $I(1) = 1$ and $I(z) = f(a)$. Hence by the uniqueness of this functional calculus (applied to $f(a)$ instead of a), we see $f \circ g(a) = f(g(a))$.

(d) and (e): Either side of equation (4) of Corollary 3.3.4 shows that $f(a)$ belongs to the smallest closed spectral subalgebra of $\mathcal{A}^1$ which contains a and 1. The last sentence follows from the last sentence of Runge's theorem. We combine the proof of the first sentence with the similar case for (e).

Let $\mathcal{B}$ be another Banach algebra and let $\varphi: \mathcal{A} \to \mathcal{B}$ be a continuous homomorphism. Assume first that $\mathcal{A}$, $\mathcal{B}$ and φ are all unital. Then, $Sp_{\mathcal{A}}(a)$ includes $Sp_{\mathcal{B}}(\varphi(a))$ by Proposition 2.1.8. Hence, $\mathcal{F}_a$ is included in $\mathcal{F}_{\varphi(a)}$ (in a natural sense). Again the desired equation is an immediate consequence of either side of equation (4) of Corollary 3.3.4. Hence, (e) holds under hypothesis (1).

Next we will establish (e) under the alternative hypotheses that either $f(0) = 0$ or $0 \notin Sp(a)$ holds. At the same time we will establish the first sentence of (d).

For $a \in \mathcal{A}$ and any $f \in \mathcal{F}_a$, let Γ be a suitable contour for (a, f) that has winding number 0 around zero if zero is not in $Sp(a)$. For any $\lambda \in \Gamma$ we can write

$$(\lambda - a)^{-1} = \lambda^{-1} - \lambda^{-1}(\lambda^{-1}a)^q. \tag{7}$$

Thus we obtain

$$f(a) = \frac{1}{2\pi i} \int_{\Gamma} f(\lambda)\lambda^{-1}d\lambda - \frac{1}{2\pi i} \int_{\Gamma} f(\lambda)\lambda^{-1}(\lambda^{-1}a)^q d\lambda. \tag{8}$$

The first integral is zero by the Cauchy integral formula. The second belongs to any closed spectral subalgebra containing a since all of its Riemann sums do. This proves the first sentence of (d). Furthermore, if $\varphi: \mathcal{A} \to \mathcal{B}$ is any continuous homomorphism, we have

$$
\begin{aligned}
\varphi(f(a)) &= -\frac{1}{2\pi i} \int_{\Gamma} f(\lambda)\lambda^{-1}(\lambda^{-1}\varphi(a))^q d\lambda \\
&= \frac{1}{2\pi i} \int_{\Gamma} f(\lambda)(\lambda - \varphi(a))^{-1}d\lambda = I_{\varphi(a)}(f)
\end{aligned}
$$

where we have used the analogue of (7) stated for $\varphi(a)$ in place of a. This proves condition (e) under hypotheses (2) or (3).

(f): Suppose first that $\mathcal{A}$ is unital. By (e) with hypothesis (1), the injection of a maximal commutative subalgebra $\mathcal{C}$ into $\mathcal{A}$ will preserve the functional calculus for $f(a)$ for $a \in \mathcal{C}$ and $f \in \mathcal{F}_a$. (Of course, $\mathcal{F}_a$ will be defined relative to $\mathcal{C}$. However, Proposition 2.1.8 shows $Sp_{\mathcal{A}}(a) = Sp_{\mathcal{C}}(a)$, so $\mathcal{F}_a$ has the same meaning whether defined relative to $\mathcal{A}$ or $\mathcal{C}$.) Thus, we may work exclusively in $\mathcal{C}$. Now (e) with hypothesis (1) applied to the Gelfand homomorphism of $\mathcal{C}$ give the desired conclusion.

Next suppose $\mathcal{A}$ is not unital, so that zero belongs to $Sp(a)$. If $f(0) = 0$ holds, condition (d) implies $f(a) \in \mathcal{C}$. Hence, we can again apply (e) (with either hypothesis depending on whether or not $\mathcal{C}$ is unital) to the Gelfand homomorphism of $\mathcal{C}$ to obtain the desired conclusion. Finally if $\mathcal{A}$ is not unital and $f(0)$ is not zero, then we can apply (e) with hypothesis (1) to the Gelfand homomorphism of $\mathcal{C}^1$ to prove the last portion of (f).

(h): (Recall that (g) was established earlier.) The partial sums of the series $\sum_{n=0}^{\infty} \alpha_n z^n$ converge to $f = \sum_{n=0}^{\infty} \alpha_n z^n$ in $\mathcal{F}_a$ under the assumptions of (h). Hence, the conclusion of (h) follows from applying (g) and the continuity of I_a to the sequence of partial sums.

(i): Choose a contour Γ which is suitable for (a, f). Theorem 2.2.13 shows that there is a neighborhood $\mathcal{D}$ of a such that each $b \in \mathcal{D}$ has its spectrum in the open set $\{\lambda \in \mathbb{C}: \lambda \notin \Gamma$ and the winding number of Γ around λ is $+1\}$. Hence, each $b \in \mathcal{D}$ satisfies $f \in \mathcal{F}_b$. Use the chosen contour Γ to define $I_b(f) = \frac{1}{2\pi i} \int_{\Gamma} f(\lambda)(\lambda - b)^{-1} d\lambda$ for each $b \in \mathcal{D}$. Proposition 2.1.4 shows that $(\lambda - b)^{-1}$ is a jointly uniformly continuous function of $\lambda \in \Gamma$ and $b \in \mathcal{D}$. Hence, the map $b \mapsto I_b(f)$ is continuous by the usual proof from calculus applied to the integral above. □

We mention explicitly three consequences of condition (d) of Theorem 3.3.7. If $a \in \mathcal{A}$ belongs to some proper closed ideal $\mathcal{I}$ of $\mathcal{A}$ (so 0 belongs to $Sp(a)$) and if $f(0) = 0$ holds, then $f(a)$ belongs to $\mathcal{I}$. Also,

$$f(a) \in \{a\}'' \qquad \forall \, a \in \mathcal{A}; \; f \in \mathcal{F}_a \tag{9}$$

holds (where $\{a\}''$ is the double commutant of $a \in \mathcal{A}$). Finally, if $\mathcal{A}$ is a closed subalgebra of $\mathcal{B}(\mathcal{X})$ for some Banach space $\mathcal{X}$ and $\mathcal{D}$ is a closed subspace of $\mathcal{X}$ satisfying $a\mathcal{X} \subseteq \mathcal{X}$ for some $a \in \mathcal{A}$, then $f(a)\mathcal{X} \subseteq \mathcal{X}$ holds for all $f \in \mathcal{F}_a$ for which a representative in $\mathcal{F}_a$ can be chosen so that D_f includes a compact neighborhood of $Sp(a)$ with connected complement.

For a long time it had been hoped that for each $a \in \mathcal{A}$ the homomorphism I_a was unique even without the assumption of continuity. This was disproved by an interesting construction of Graham R. Allan [1972]. See 3.4.8 through 3.4.16 below for a more detailed discussion of these results.

See also Florian Horia Vasilescu [1982]. The paper [1989] by John B. Conway, Domingo A. Herrero and Bernard B. Morrel discusses some interesting extensions. Mohammed Akkar [1978] defines a functional calculus in certain locally convex algebras.

We have examined this functional calculus in considerable detail. This is motivated in part by the applications which we wish to make of it, but it is also motivated by our desire to compare this functional calculus with two others which we shall introduce in succeeding chapters. These are the functional calculi of continuous functions of normal elements in C*-algebras and Borel functions of normal elements in W*-algebras.

Functional Calculus for Spectral Algebras

We would like to extend the functional calculus to spectral algebras but we are only able to do so for Jacobson-semisimple spectral algebras. There are two related problems. Spectral algebras have no topology and we have just noted that there do exist discontinuous functional calculi in some Banach algebras which differ from the standard functional calculus. Thus we need to know that the functional calculus obtained by completing in different spectral semi-norms is always the same. The next proposition removes this difficulty in the semisimple case. In a non-semisimple spectral algebra, we would also face the problem that only spectral semi-norms might be available, so that the completion would actually be a quotient algebra and thus finding preimages in the algebra itself of the function of the image of one of its element would cause difficulty.

3.3.8 Proposition *Let A be a Jacobson-semisimple spectral algebra. Let a be an element of A and f a function in $\mathcal{F}_a$. Let σ_1 and σ_2 be any two spectral norms on A. Let b and c be the element $f(a)$ defined by the functional calculus in the completions of A with respect to σ_1 and σ_2, respectively. If both b and c are in A, then they are equal.*

Proof This is a slight variation on the uniqueness portion of the proof of Theorem 3.3.7. Recall from Proposition 2.5.15 that the spectrum of a does not change in any completion with respect to a spectral semi-norm. Choose a contour Γ suitable for defining $f(a)$ and let K be the compact set $\Gamma \cup \{\lambda \in \mathbb{C} : \Gamma$ has winding number $+1$ around $\lambda\}$. By Runge's Theorem (Theorem 3.5.3), we can find a sequence $\{f_n\}$ of rational functions with poles off $K \supseteq Sp(a)$ which approximates f uniformly on K. Each $f_n(a)$ is unambiguously defined in A and the sequence $\{f_n(a)\}$ converges to b in σ_1 and to c in σ_2, so the Fundamental theorem (Theorem 2.3.6) shows $b = c$ since A is semi-simple. $\qquad\qquad\qquad\qquad\qquad\qquad\qquad\qquad\qquad\qquad\Box$

This proposition shows that the following definition is unambiguous and does not depend on the choice of spectral norm.

3.3.9 Definition Let A be a Jacobson-semisimple spectral algebra. Choose a spectral norm on A and let $\overline{A}$ be the completion. For any a in A and any $f \in \mathcal{F}_a$ if $f(a) \in \overline{A}^1$ actually belongs to A^1, then we say that $f(a)$ is defined in A.

As already noted in the proof of Proposition 3.3.8, the spectrum of a does not change in the completion with respect to any spectral norm. Hence, if $f(a)$ is defined relative to such a completion, and that element is in $\mathcal{A}$, then it agrees with $f(a)$ defined relative to the completion with respect to any other spectral norm. Thus, the restricted functional calculus in a spectral algebra $\mathcal{A}$ is a homomorphic image of whatever class of functions is involved and satisfies the rules on composition of functions of any normal functional calculus. It also satisfies the spectral mapping theorem. Of course, we can use the above procedure to define a partial functional calculus in any spectral normed algebra without any restriction of semisimplicity.

3.3.10 Example The proposition does not say anything about the existence of $f(a)$. If $\mathcal{A}$ is the disc algebra and $\mathcal{B}$ is the subalgebra of rational functions with poles off the closed disc, it is clear that $\mathcal{B}$ is a spectral subalgebra of $\mathcal{A}$ and thus a spectral algebra. For the identity function z in $\mathcal{B} \subseteq \mathcal{A}$, the functional calculus image $f(z)$ belongs to $\mathcal{B}$ if and only if f is rational with poles off the spectrum of z which is just the closed disc. Thus, the algebra of functions involved in the functional calculus is just the algebra of these rational functions and is therefore as small as it possibly could be. It would be interesting to know what other subalgebras of $\mathcal{F}_a$ can arise in different spectral algebras. We will now define and study the class of spectral normed algebras in which the functional calculus is defined for every $f \in \mathcal{F}_a$.

Functional Algebras and Subalgebras

3.3.11 Definition A *functional algebra* is a spectral normed algebra $\mathcal{A}$ such that for each $a \in \mathcal{A}$ and each function $f \in \mathcal{F}_a$ the element $f(a)$ defined by the functional calculus relative to the completion $\overline{\mathcal{A}}$ of $\mathcal{A}$ is already in $\mathcal{A}^1$. A subalgebra of a Banach algebra or functional algebra is called a *functional subalgebra* if it contains $f(a)$ for each of its elements a and each $f \in \mathcal{F}_a$. A normed inductive limit of Banach algebras is called a *pseudo-Banach algebra*.

The first class of algebras introduced above may in time prove to be of fundamental importance, but we are not sure that the current definition is exactly correct. The class of local algebras, which seems to have been first formally developed by Bruce Blackadar in [1986], is our model. Similar ideas have been used chiefly in the K-theory of C*-algebras. The term "local Banach algebra" was intended to suggest that each element individually has a property (*viz.* a functional calculus) associated with completeness even though the algebra as a whole need not be complete. Since "local algebra" has an entirely different meaning stemming from algebraic geometry, we will not use it. Note that a functional subalgebra is necessarily a spectral subalgebra and that a functional subalgebra of a functional algebra is again a functional algebra.

The term "pseudo-Banach algebra" was introduced by Allan, Dales and

J. Peter McClure [1971]. (The cited paper actually considered only the commutative case.) Their formal definition was in terms of bounded sets, but they showed that their pseudo-Banach algebras are exactly those algebras which occur as normed inductive limits of Banach algebras. The normed inductive limit construction was described in Definition 1.3.4 and Example 1.3.5. For unital commutative Banach algebras, the reference already cited characterizes these algebras among several well known classes of topological algebras (*e.g.*, locally multiplicatively convex, Fréchet algebras, etc.) and shows that their Gelfand theory is well behaved. We now note the connection between these two classes below. This is a fundamental result from Blackadar [1986].

3.3.12 Theorem *A normed inductive limit of functional algebras is a functional algebra. Hence, in particular, a pseudo-Banach algebra is a functional algebra.*

Proof We will use the notation from Definition 1.3.4. First we note that the norm on $\mathcal{A}$ is spectral. If $a \in \mathcal{A}$ satisfies $\|a\| < 1$, then there is some α and $a_\alpha \in \mathcal{A}_\alpha$ with $a = \varphi_\alpha(a_\alpha)$ and $\|a_\alpha\| < 1$. (Much more is true, of course, but this is all we need.) So a_α is quasi-invertible since a Banach algebra is spectral by Corollary 2.2.8. Since a is a homomorphic image, it is also quasi-invertible so the norm is spectral by the same corollary. Let $a \in \mathcal{A}$ and $f \in \mathcal{F}_a$ be given. Suppose α is such that $a = \varphi_\alpha(b)$ for some $b \in \mathcal{B} = \mathcal{A}_\alpha$. Let D_f be the domain of f. Then we have $Sp_\mathcal{A}(a) \subseteq Sp_\mathcal{B}(b)$, so $C = Sp_\mathcal{B}(b) \setminus D_f$ may be empty but is certainly compact. If it is not empty, then for each $\lambda \in C \setminus \{0\}$ we can find a $\beta \geq \alpha$ and a quasi-inverse for $\lambda^{-1}\varphi_{\beta\alpha}(b)$ (since $\lambda^{-1}a$ has a quasi-inverse in $\mathcal{A}$). In fact, there is a neighborhood of λ in which such quasi-inverses exist. If zero is in C, then $\mathcal{A}$ is unital and there is a neighborhood of zero and a β so that $\varphi_{\beta\alpha}(b)$ is invertible in $\mathcal{A}_\beta$. Cover C with a finite number of such neighborhoods and then choose γ larger than all the β which correspond to the neighborhoods in this collection. It is clear that $f(\varphi_{\gamma\alpha}(b))$ is defined in $\mathcal{A}_\gamma$. Define $f(a)$ to be $\varphi_\gamma(f(\varphi_{\gamma\alpha}(b)))$.   □

We wish to examine the inductive limit of matrix algebras of increasing size over a functional algebra. It is apparently unknown whether a matrix algebra over a functional algebra is again a functional algebra, but the following recent theorem of Larry B. Schweitzer [1992] covers a case which includes many applications. The applications are of the following type: Let $\mathcal{A}$ be the L^1-algebra or C^*-algebra of a Lie group G. It may be easier to calculate the K-theory of a dense functional subalgebra consisting of functions which are, for instance, C^∞ on G. The results in this example show that the machinery all works smoothly.

3.3.13   Theorem   *Let $\mathcal{A}$ be a spectral normed algebra which is also a Fréchet algebra in a topology such that the embedding of $\mathcal{A}$ with its Fréchet algebra topology into $\mathcal{A}$ with its norm is continuous. Let $\overline{\mathcal{A}}$ be the completion of $\mathcal{A}$ in its norm. Then $\mathcal{A}$ is a functional subalgebra of $\overline{\mathcal{A}}$ and hence a functional algebra. Furthermore, any matrix algebra $M_n(\mathcal{A})$ over $\mathcal{A}$ is also a functional subalgebra of $M_n(\overline{\mathcal{A}})$ (when this matrix algebra is given any algebra norm which defines the topology pointwise in terms of its entries) and hence a functional algebra.*

Proof   In Theorem 2.9.2 we have noted that the inverse is continuous in a Fréchet algebra. Thus, the integrand $f(\lambda)(\lambda - a)^{-1}$ in the integral defining $f(a)$ is continuous. Hence, its Riemann sums converge in the Fréchet topology of $\mathcal{A}$ just as in the discussion of the functional calculus in a Banach algebra. Because $\mathcal{A}$ (with its Fréchet topology) is continuously embedded in $\overline{\mathcal{A}}$ (with its complete norm), the image of $f(a)$ in $\mathcal{A}$ is just $f(a)$ in $\overline{\mathcal{A}}$. Therefore, $\mathcal{A}$ is a functional subalgebra of $\overline{\mathcal{A}}$.

Theorem 2.2.14(h) shows that the extension of the norm of $\mathcal{A}$ to $M_n(\mathcal{A})$ is still spectral. Hence the image of $M_n(\mathcal{A})$ in $M_n(\overline{\mathcal{A}})$ is a spectral subalgebra by Proposition 2.5.15. Also, this image is still a continuously embedded Fréchet algebra. Thus, it is a functional subalgebra by the same argument as before.   □

Blackadar's [1986] Definition 3.1.1 actually requires that $M_n(\mathcal{A})$ be functional (in our terminology) for all $n \in \mathbb{N}$ before he calls it a local algebra. This was necessary for the applications to K-theory, but the theorem above is sufficient for most applications. In terms of Blackadar's definition, Lothar M. Schmitt [1991] shows that $\mathcal{A}/\mathcal{I}$ is local whenever $\mathcal{A}$ is local and $\mathcal{I}$ is a closed ideal. We do not know whether matrix algebras and quotients of functional algebras are always functional without any further restrictions, but this whole theory is in its infancy. See Schweitzer [1991], [1992] and [1993] for examples and discussion.

3.4   Examples and Applications of Functional Calculus

Our careful study of the various elementary functional calculi and the analytic functional calculus is justified partly by their important applications and partly by our desire to compare them with two additional functional calculi which will be defined for certain elements in certain *-algebras. We introduce one of these at the end of this section. We will also explore the situations under which the analytic functional calculus is, or is not, unique when the continuity assumption is removed. However, most of this section will be devoted to some of the important applications of the analytic functional calculus. In the interest of giving elementary proofs, we will sometimes avoid using this functional calculus in situations to which it clearly applies. We will do this whenever elementary methods yield the same results with comparatively little extra effort. In addition to polynomials and rational functions, the functions which occur most often in practice are the exponential, the logarithm, and the square root. Also, for elements with disconnected spectrum the characteristic function of a portion of the spectrum is frequently useful. Except for this last case, these applications can usually be treated as easily by more elementary methods.

Remark. We often state theorems about the functional calculus just for Banach algebras, even though they hold for functional algebras also. The reader can easily check whether some small change in wording is needed.

Existence of Idempotents

In Section 3.5 we will prove the Šilov idempotent theorem (Corollary 3.5.15) which shows the existence of idempotents in a commutative Banach algebra if its Gelfand space is disconnected. The proof of that result depends on the powerful theory developed in Section 3.5. However, we can now prove an easier analogue of this result which is a trivial consequence of the analytic functional calculus we have already constructed. The next result is due to Gelfand [1941a].

3.4.1 Proposition *Let $\mathcal{A}$ be a unital, Banach algebra, and let a be an element of $\mathcal{A}$. If $Sp(a)$ is the union of a family $\{K_j : j = 1, 2, \ldots, n\}$ ($n > 1$) of disjoint, nonempty, closed sets, then there is a set $\{e_j : j = 1, 2, \ldots, n\}$ of orthogonal non-zero idempotents satisfying $\sum_{j=1}^{n} e_j = 1$ and $Sp(ae_j) = K_j \cup \{0\}$ for $j = 1, 2, \ldots, n$. Moreover, each of these idempotents is contained in the smallest unital closed spectral subalgebra $\mathcal{B}$ of $\mathcal{A}$ which contains a, so $e_j \in \{a\}''$. Considering Gelfand transforms relative to the commutative Banach algebra $\mathcal{B}$, we see that $\widehat{e_j}$ is the characteristic function of $\hat{a}^{\leftarrow}(K_j)$.*

Proof For each $j = 1, 2, \ldots, n$, let f_j be a function in $\mathcal{F}_a$ which is equal to 1 in a neighborhood of K_j and is equal to 0 in a neighborhood of $Sp(a) \setminus K_j$. Define e_j by $e_j = f_j(a)$ for $j = 1, 2, \ldots, n$. Theorem 3.3.7 gives all of the stated results.   □

The above proof is an example of using a disconnected neighborhood of the spectrum and a function which is analytic in the chosen neighborhood but is not globally analytic. It is interesting to write down the idempotents produced in the last proposition explicitly. Let Γ be a suitable contour in $\mathbb{C} \setminus Sp(a)$ which has winding number $+1$ around each point of K_j and winding number 0 around each other point in $Sp(a)$. Then we have:

$$e_j = \frac{1}{2\pi i} \int_\Gamma (\lambda - a)^{-1} \, d\lambda.$$

The Kernel of the Exponential Function

Theorem 2.1.12 gave the basic results on the exponential function $\exp\colon \mathcal{A} \to \mathcal{A}$ in a unital Banach algebra $\mathcal{A}$ where

$$\exp(a) = e^a = \sum_{n=0}^{\infty} \frac{a^n}{n!} \qquad \forall\, a \in \mathcal{A}.$$

If $\mathcal{A}$ is commutative, that theorem shows that $\exp$ is a group homomorphism from the additive group $\mathcal{A}$ to the multiplicative group $\mathcal{A}_G$. Following Edgar R. Lorch [1942] and [1943], we now find the kernel of this homomorphism.

3.4.2 Theorem *Let $\mathcal{A}$ be a commutative unital Banach algebra. The kernel of the exponential map $\exp\colon \mathcal{A} \to \mathcal{A}_G$ is the additive group generated by the set $\{2\pi i e\colon e$ is an idempotent in $\mathcal{A}\}$.*

Proof Suppose e is an idempotent in $\mathcal{A}$. Then

$$\exp(2\pi i e) = 1 + \sum_{n=1}^{\infty} (n!)^{-1}(2\pi i)^n e = 1 + (\exp(2\pi i) - 1)e = 1.$$

Hence, the kernel of the group homomorphism $\exp\colon \mathcal{A} \to \mathcal{A}_G$ contains the additive group generated by $\{2\pi i e : e$ is an idempotent in $\mathcal{A}\}$. Conversely suppose $a \in \mathcal{A}$ satisfies $\exp(a) = 1$. Then Theorem 3.3.7(b) shows that $Sp(a)$ is a finite subset of the set $\{2\pi i n\colon n \in \mathbb{Z}\}$. Proposition 3.3.1 gives a finite set $\{e_j\colon j = 1, 2, \ldots, m\}$ of orthogonal idempotents satisfying $\sum_{j=1}^{m} e_j = 1$ and $Sp(ae_j) = \{2\pi i n_j\}$ for some integer n_j. Define b by $b = \sum_{j=1}^{m} 2\pi i n_j e_j$. Then $\hat{a} - \hat{b}$ is zero so there is an element r in the Gelfand radical of $\mathcal{A}$ satisfying $a + r = b$. Furthermore, the absolute convergence of the series involved allow us to write $0 = \exp(b) - \exp(a) = \sum_{n=0}^{\infty}(n!)^{-1}((a + r)^n - a^n) = r \sum_{n=1}^{\infty}(n!)^{-1} \sum_{k=1}^{n} \binom{n}{k} r^{k-1} a^{n-k}$. The Gelfand transform of the infinite series is just $\sum_{n=1}^{\infty}(n!)^{-1} n \hat{a}^{n-1} = \exp(\hat{a})$. Hence, the infinite series is invertible so r is zero. Thus $a = b$ has the desired form. $\square$

Logarithms

We have just determined the kernel of the exponential function on a commutative Banach algebra $\mathcal{A}$. Theorem 2.1.12 shows that, in the commutative case, the range of the exponential function is the connected component of the identity in $\mathcal{A}_G$. However, it is harder to determine whether a given element is an exponential in the noncommutative case. We will consider not necessarily commutative Banach algebras. Clearly any element in the range of the exponential function is invertible. We are about to show that if 0 is in the unbounded component of $\mathbb{C} \setminus Sp(a)$ then a is the exponential of an element $\log(a)$. However, the next example shows that this sufficient condition is not necessary even in the commutative case. A more subtle condition will be given in the next section.

3.4.3 Example　Consider the commutative unital Banach algebra $C(\mathbb{T})$ of continuous functions on the 1-torus $\mathbb{T}$. Let $g\colon \mathbb{T} \to \mathbb{C}$ be defined by

$$
g(e^{i\theta}) = \begin{cases} 2i\theta & 0 \le \theta < \pi \\ i(4\pi - 2\theta) & \pi \le \theta < 2\pi. \end{cases}
$$

Clearly, g is continuous so it belongs to our algebra. Just as clearly, $\exp(g)$ has the whole of $\mathbb{T}$ as its spectrum.

3.4.4 Theorem　*Let $\mathcal{A}$ be a unital Banach algebra and let $a \in \mathcal{A}$ be an element which has 0 in the unbounded component of $\mathbb{C} \setminus Sp(a)$. Then in the smallest unital closed spectral subalgebra of $\mathcal{A}$ containing a, there is an element $\log(a)$ which satisfies*

$$
\exp(\log(a)) = a.
$$

Hence for any $n \in \mathbb{N}$, the element $a^{1/n} = \exp(\frac{1}{n}\log(a))$ satisfies

$$
(a^{1/n})^n = a.
$$

Any element $a \in \mathcal{A}$ which satisfies $\|1 - a\| < 1$ satisfies the previous condition and we can write

$$
\log(a) = -\sum_{n=1}^{\infty} \frac{(1 - a)^n}{n}.
$$

Proof　Since the unbounded component which contains 0 is arc connected, we can construct a continuous curve connecting 0 to some point on the negative real axis beyond the spectral radius of a. Using this curve, we can choose a branch of the logarithm function which is analytic in a neighborhood of $Sp(a)$ and use it to define $\log(a)$. Theorem 3.3.7 then gives the results in the first paragraph. The alternative hypothesis ensures that the series converges, so Theorem 3.3.7(h) gives the final result.　□

Note that if $Sp(a)$ is disconnected, then $\log(a)$ and $a^{1/n}$ depend on the choice of the contour in the proof of the theorem. Obviously, any invertible matrix $A \in M_n$ for any $n \in \mathbb{N}$ satisfies the theorem and thus has a logarithm and arbitrary roots.

Square Roots

Let a be an element in a Banach algebra and let b be a square root for a in the sense that $b^2 = a$ holds. Then a belongs to any closed commutative subalgebra $\mathcal{C}$ containing b, and $\hat{b}$ is a square root for $\hat{a}$ in $\hat{\mathcal{C}}$. The implicit function theorem proved in the next section (Theorem 3.5.12) will give a strong converse. If $a \in \mathcal{A}$ is invertible in a closed commutative subalgebra $\mathcal{C}$ which is a unital algebra, and if $\hat{a}$ has a square root h in $C(\Gamma_{\mathcal{C}})$, then a has a square root in $\mathcal{C}$ and hence in $\mathcal{A}$. Furthermore, this square root is unique in $\mathcal{C}$ subject to the condition that $\hat{c} = h$. This result is, of course, obtained by considering the function $F(w, z) = w^2 - z$. The main disadvantage of this result is that it only deals with elements which are invertible, at least in some closed commutative subalgebra.

Assume now that $\mathcal{A}$ is a unital Banach algebra and a is an invertible element in $\mathcal{A}$ such that the whole negative real line lies in $\mathbb{C} \setminus Sp(a)$. Then, the usual principal branch of the square root function belongs to $\mathcal{F}_a$ so that the analytic functional calculus gives a square root $a^{1/2}$ in $\{a\}'' \subseteq \mathcal{A}$ with spectrum in the open right half plane. We will now show that this is the only square root with spectrum in the open right half plane. If b is any other square root, $a^{1/2}$ and b belong to a common closed commutative subalgebra $\mathcal{C}$. If b has spectrum in the right half plane, then $a^{1/2}$ and b must have the same Gelfand transform in $\hat{\mathcal{C}}$ since they satisfy $\widehat{(a^{1/2})}^2 = (\hat{b})^2$. Hence, there is some r in the Gelfand radical of $\mathcal{C}$ satisfying $b = a^{1/2} + r$. Then, $2a^{1/2} + r$ is invertible in $\mathcal{C}$ since it has a nonvanishing Gelfand transform. Thus, $0 = b^2 - (a^{1/2})^2 = (2a^{1/2} + r)r$ implies $r = 0$.

The uniqueness of square roots with spectrum in the right half plane was first obtained explicitly by Einar Hille [1958]. It was rediscovered and given an attractive proof by L. Terrell Gardner [1966]. However, an immediate corollary of this result, which plays a decisive role in the theory of Banach *-algebras was not known to most workers in that field until discovered by James W. M. Ford in his dissertation [1966], [1967].

We now derive an elementary version of the above result which does apply to certain noninvertible elements. Actually, we will deal with elements b satisfying $b \circ b = a$. Such an element is called *a quasi-square root of a*. This is more convenient, particularly in nonunital algebras. Obviously such elements satisfy $(1 - b)^2 = 1 - a$ in $\mathcal{A}^1$.

We will give a simple derivation of the binomial series expansion for $(1 + \lambda)^{1/2}$. First, assume the existence of a power series expansion with undetermined coefficients:

$$(1 + \lambda)^{1/2} = y(\lambda) = \sum_{n=0}^{\infty} a_n \lambda^n. \tag{1}$$

Clearly, a_0 must be 1. The first equality implies

$$2(1 + \lambda)y'(\lambda) = y(\lambda). \tag{2}$$

Substituting the series expansion gives

$$\sum_{n=0}^{\infty} [2(n + 1)a_{n+1} + 2na_n]\lambda^n = \sum_{n=0}^{\infty} a_n \lambda^n.$$

By equating the coefficients, we get the recurrence relation $a_{n+1} = (1 - 2n)(2n + 2)^{-1}a_n$, from which

$$a_n = \binom{1/2}{n} = \frac{(-1)^{n+1}(2n - 2)!}{2^{2n-1}n!(n - 1)!} \qquad \forall\, n > 0$$

follows. The ratio test shows that the power series of (2) converges for $|\lambda| < 1$. Thus, for $|\lambda| < 1$ we conclude

$$(1 + \lambda)^{1/2} = 1 + \sum_{n=1}^{\infty} \binom{1/2}{n}\lambda^n. \tag{3}$$

The identity

$$\sum_{n=1}^{N}\left|\binom{1/2}{n}\right| = 1 - \frac{(2N)!}{2^{2N}(N!)^2} < 1 \tag{4}$$

is easily proved by induction. It shows that

$$\sum_{n=1}^{\infty}\left|\binom{1/2}{n}\right| \tag{5}$$

converges so that the power series of (3) converges absolutely and uniformly for $|\lambda| \leq 1$. Since it represents a continuous function there, the power series satisfies (3) for all $|\lambda| \leq 1$, and in particular (5) converges to 1.

3.4.5 Theorem   *Let $\mathcal{A}$ be a Banach or a functional algebra. Let $a \in \mathcal{A}$ satisfy either $\rho(a) < 1$ or $\|a\|^n \leq 1$ for all sufficiently large n. Then the series*

$$-\sum_{n=1}^{\infty} \binom{1/2}{n}(-a)^n \tag{6}$$

converges unconditionally to an element $b \in a\mathcal{A} \cap \mathcal{A}a \cap \{a\}''$ satisfying

$$b \circ b = a \quad and \quad \rho(b) \leq 1. \tag{7}$$

If a satisfies $\rho(a) < 1$, then b is the only element in $\mathcal{A}$ satisfying (7) and, in fact, satisfies $1 - \rho(b) \leq \sqrt{1 - \rho(a)} < 1$.

Proof If a satisfies $\rho(a) < 1$, Theorem 2.2.5(f) shows $\|a^n\| \leq 1$ for all sufficiently large n. Thus, under either hypothesis on $a \in \mathcal{A}$, the remarks preceding this theorem show that $\sum_{n=1}^{\infty} \|\binom{1/2}{n}a^{n-1}\|$ converges so that (6) converges unconditionally to an element $b \in a\mathcal{A} \cap \mathcal{A}a \cap \{a\}''$ which satisfies $b \circ b = a$. Corollary 3.1.6 shows that each partial sum satisfies

$$\rho\left(-\sum_{n=1}^{N}\binom{1/2}{n}(-a)^n\right) \leq \sum_{n=1}^{N}\left|\binom{1/2}{n}\right|\rho(a)^n$$

$$\leq -\sum_{n=1}^{\infty}\binom{1/2}{n}(-\rho(a))^n = 1 - \sqrt{1 - \rho(a)}.$$

The last part of Corollary 3.1.6 then shows $\rho(b) \leq 1 - \sqrt{1 - \rho(a)}$ since each partial sum commutes with b.

The inequality $\rho(a) < 1$ then implies $\rho(b) < 1$. When this holds, suppose $c \in \mathcal{A}$ satisfies $c \circ c = a$ and $\rho(c) \leq 1$. Then $ca = 2c^2 - c^3 = ac$ holds so c commutes with a, and hence with b also. Thus $\rho((b+c)/2) \leq (1+\rho(b))/2 < 1$ implies that $(b+c)/2$ has a quasi-inverse which we denote by d. Hence, we get $b - c = 0 \circ (b-c) = d \circ ((b+c)/2) \circ (b-c) = d \circ [(b+c+2b-2c-(b^2-c^2))/2] = d \circ [(b + c + b \circ b - c \circ c)/2] = d \circ ((b + c)/2) = 0$.   □

Continuity of the Spectrum

The following two results are due to John D. Newburgh [1951]. Lemma 3.4.6 will be used both in Proposition 3.4.7 and in Proposition 6.1.10. Proposition 3.4.7 gives the same result as the last one in Corollary 3.1.6 with completely different hypotheses. It is also related to Newburgh's theorem (Theorem 2.2.15).

3.4.6 Lemma *Let a be an element in a Banach algebra $\mathcal{A}$. Let V be a bounded open subset of $\mathbb{C}$ satisfying $\overline{V} \cap Sp(a) = V \cap Sp(a) \neq \emptyset$. Then there is a $\delta > 0$ such that $\|b - a\| < \delta$ implies $Sp(b) \cap V \neq \emptyset$.*

Proof Let f be the characteristic function of V. By assumption, f is analytic in a neighborhood of $Sp(a)$. Theorem 3.3.7 shows that $f(a)$ is a non-zero idempotent. Hence, Theorem 3.3.7(i) shows that there is a neighborhood a, which we may take to be an open disc of radius δ, such that $f(b)$ is defined for each b in the neighborhood and satisfies $\|f(b) - f(a)\| < \|f(a)\|$. This implies $f(b) \neq 0$ and thus, by Theorem 3.3.7(b), $Sp(b) \cap V \neq \emptyset$ for each b satisfying $\|b - a\| < \delta$.   □

3.4.7 Proposition *Let $\mathcal{A}$ be a Banach algebra. Let $a \in \mathcal{A}$ have totally disconnected spectrum. For any $\varepsilon > 0$ there is a $\delta > 0$ such that $\|b - a\| < \delta$ implies*

$$Sp(a) \subseteq Sp(b) + \varepsilon\mathbb{C}_1, \qquad Sp(b) \subseteq Sp(a) + \varepsilon\mathbb{C}_1 \qquad and$$

$$|\rho(a) - \rho(b)| \leq \varepsilon.$$

Proof Since $Sp(a)$ is compact and totally disconnected, there are a finite number of disjoint closed discs D_j $j = 1, 2, \ldots, n$ each containing some point of $Sp(a)$, of radius less than $\varepsilon/2$ and such that $Sp(a)$ is included in the interior of $D = \bigcup_{j=1}^{n} D_j$. Theorem 2.2.15 shows that we can find a $\delta_0 > 0$ so that $\|b - a\| < \delta_0$ implies $Sp(b) \subseteq D^\circ$. For each $j = 1, 2, \ldots, n$ the lemma gives a $\delta_j > 0$ so that $Sp(b) \cap D_j \neq \emptyset$. Define δ to be the minimum of $\{\delta_j : j = 0, 1, \ldots, n\}$. Then any b satisfying $\|b - a\| < \delta$ has all the desired properties since each point of $Sp(a)$ is in some D_j° of diameter less than ε, which also contains a point of $Sp(b)$.   □

Discontinuous Functional Calculi

We will now explore the existence of discontinuous functional calculi. Theorem 3.3.7 states that for any element a in a Banach algebra $\mathcal{A}$ there is a unique continuous homomorphism $I_a\colon \mathcal{F}_a \to \mathcal{A}^1$ satisfying

$$I_a(1) = 1 \qquad I_a(z) = a. \tag{8}$$

We wish to investigate conditions on the Banach algebra $\mathcal{A}$ which will ensure that for $a \in \mathcal{A}$, any (not necessarily continuous) homomorphism $J_a\colon \mathcal{F}_b \to \mathcal{A}^1$ satisfying (8) equals I_a. When this is true for a particular $a \in \mathcal{A}$, we say that a *has a unique functional calculus* and when it is true for all $a \in \mathcal{A}$, we say that $\mathcal{A}$ *has a unique functional calculus*. Since I_a is unique among continuous homomorphisms satisfying (8), a will have a unique functional calculus if and only if every homomorphism $J_a\colon \mathcal{F}_a \to \mathcal{A}^1$ satisfying (8) is continuous. (Thus, this is a type of automatic continuity result similar to those discussed in Chapter 6.) The main results here are due to Allan [1972], Dales [1973] and Thomas [1978].

First, we will show that not all algebras have a unique functional calculus. Let $\mathcal{F} = \mathbb{C}[[z]]$ be the algebra of formal power series in z introduced in Example 2.9.7. Let $\mathcal{A}$ be a Banach algebra and let a be an element in $\mathcal{A}$ satisfying $Sp(a) = \{0\}$. Then the algebra $\mathcal{F}_a$ of germs of functions analytic in the neighborhood of $Sp(a) = \{0\}$ can be identified with the subalgebra $\mathcal{F}_{(0)}$ of $\mathcal{F}$ consisting of those series with a non-zero radius of convergence. Hence if we can embed $\mathcal{F}$ into $\mathcal{A}^1$ by two distinct unital homomorphisms φ and ψ which agree at z, but disagree at some element of $\mathcal{F}_{(0)}$, then we can be sure that the functional calculus is not unique. Thus the search for non-unique functional calculi can begin by studying unital homomorphisms of $\mathcal{F}$ into a unital Banach algebra $\mathcal{A}$. We quote, without proofs, the main theorems from Allan [1972]. Recall that the Jacobson radical and Gelfand radical coincide for commutative Banach algebras by Theorem 3.1.5.

3.4.8 Theorem *Let $\mathcal{A}$ be a unital Banach algebra and let $\varphi\colon \mathcal{F} \to \mathcal{A}$ be an injective unital homomorphism. Then $\varphi(z)$ belongs to the Jacobson radical of $\mathcal{A}$ and the descending sequence $\{[\mathcal{A}(\varphi(z))^n]^-\}_{n\in\mathbb{N}}$ of closed left ideals eventually becomes constant at a non-zero left ideal.*

3.4.9 Theorem *Let $\mathcal{A}$ be a unital, commutative Banach algebra. Let there be an element b in the Gelfand radical of $\mathcal{A}$ such that the descending sequence $\{(\mathcal{A}b^n)^-\}_{n\in\mathbb{N}}$ of closed ideals eventually becomes constant at a non-zero ideal denoted by $\overline{\mathcal{I}}$. Then $\mathcal{I} = \cap_{n=1}^{\infty} \mathcal{A}b^n$ is dense in $\overline{\mathcal{I}}$. Furthermore for any $c \in \mathcal{I} \setminus \{0\}$ and any $f \in \mathcal{F}$ which is transcendental over the polynomials, there are injective unital homomorphisms $\varphi\colon \mathcal{F} \to \mathcal{A}$ and $\psi\colon \mathcal{F} \to \mathcal{A}$ satisfying $\varphi(z) = b = \psi(z)$ and $\varphi(f) \neq \varphi(f) + c = \psi(f)$.*

Hence, any algebra (such as those given in Examples 3.4.15 and 3.4.16) satisfying the conditions of Theorem 3.4.9 will fail to have a unique functional calculus and therefore will have a discontinuous functional calculus in addition to its unique continuous functional calculus. We now give the results due to Marc P. Thomas [1978]. They require some preparation. See also Bade, Curtis and Laursen [1980].

3.4.10 Definition   Let $\mathcal{A}$ be a Banach algebra and let a be an element in $\mathcal{A}$. A subspace $\mathcal{D}'$ is said to be a-*divisible* if it satisfies

$$(\lambda - a)\mathcal{D}' = \mathcal{D}' \qquad \forall \, \lambda \in \mathbb{C}.$$

If $\mathcal{X}$ is a subspace of $\mathcal{A}$, denote the largest a-divisible subspace of $\mathcal{X}$ by $\mathcal{D}(a, \mathcal{X})$.

The definition means that $\mathcal{D}(a, \mathcal{X})$ is an a-divisible subspace of $\mathcal{X}$ which includes all other such subspaces. Hence it is unique if it exists. Since $\{0\}$ is a-divisible and the sum of any family of a-divisible subspaces is a-divisible, it is obvious that $\mathcal{D}(a, \mathcal{X})$ always exists. We incorporate this observation in the next proposition. We are primarily interested in the case in which $\mathcal{A}$ is commutative and $a\mathcal{X} \subseteq \mathcal{X}$ already holds.

3.4.11 Proposition   *Let $\mathcal{A}$ be a Banach algebra. Let a be an element in $\mathcal{A}$ and let $\mathcal{X}$ be a subspace of $\mathcal{A}$. Then $\mathcal{D}(a, \mathcal{X})$ exists and is unique. It is the largest subspace $\mathcal{D}'$ of $\mathcal{X}$ satisfying $(\lambda - a)\mathcal{D}' = \mathcal{D}'$ for all $\lambda \in Sp(a)$.*

If $\mathcal{A}$ is commutative, then $\mathcal{D}(a, \mathcal{X})$ is included in the Jacobson radical of $\mathcal{A}$. Also for any a, $\mathcal{D}(a, \mathcal{X})$ is an ideal of $\mathcal{A}$ if $\mathcal{X}$ is.

Proof We have already shown existence and uniqueness. Suppose $\mathcal{D}'$ is the largest subspace of $\mathcal{X}$ satisfying $(\lambda - a)\mathcal{D}' = \mathcal{D}'$ for all $\lambda \in Sp(a)$. This implies $a\mathcal{D}' \subseteq \mathcal{D}'$ since the spectrum is nonempty. It is enough to show $(\mu - a)\mathcal{D}' = \mathcal{D}'$ for all $\mu \in \mathbb{C} \setminus Sp(a)$. Any $\lambda \in Sp(a)$ and $\mu \in \mathbb{C} \setminus Sp(a)$ satisfy $(\lambda - a)(\mu - a)^{-1}\mathcal{D}' = (\mu - a)^{-1}(\lambda - a)\mathcal{D}' = (\mu - a)^{-1}\mathcal{D}'$. Hence the maximum nature of $\mathcal{D}'$ gives $(\mu - a)^{-1}\mathcal{D}' \subseteq \mathcal{D}'$. However, this implies $\mathcal{D}' = (\mu - a)(\mu - a)^{-1}\mathcal{D}' \subseteq (\mu - a)\mathcal{D}' \subseteq \mathcal{D}'$ as we wished to show.

If $\mathcal{A}$ is commutative, let $\gamma \in \Gamma_{\mathcal{A}}$ be arbitrary. For any $d \in \mathcal{D}(a, \mathcal{X})$, there is an $e \in \mathcal{D}(a, \mathcal{X})$ satisfying $d = (\gamma(a) - a)e$. Applying γ shows $\gamma(d) = 0$. Since $\gamma \in \Gamma_{\mathcal{A}}$ was arbitrary, Theorem 3.1.5 shows $\mathcal{D}(a, \mathcal{X}) \subseteq \mathcal{A}_J$. If $\mathcal{X}$ is an ideal of $\mathcal{A}$ and $\mathcal{D}' \subseteq \mathcal{X}$ satisfies $(\lambda - a)\mathcal{D}' = \mathcal{D}'$ for all $\lambda \in \mathbb{C}$, then for any $b \in \mathcal{A}$, $b\mathcal{D}'$ satisfies the same conditions. Hence $\mathcal{D}(a, \mathcal{X})$ is an ideal. $\square$

This proposition shows that the ideal $\mathcal{I}$ of Theorem 3.4.9 is just $\mathcal{D}(b, \mathcal{A})$.

3.4.12 Proposition   *Let $\mathcal{A}$ be a Banach algebra. For $a \in \mathcal{A}$ let $I_a: \mathcal{F}_a \to \mathcal{A}^1$ be the usual functional calculus homomorphism and let $J_a: \mathcal{F}_a \to \mathcal{A}^1$ be any other homomorphism also satisfying (8). Define $\delta_J: \mathcal{F}_a \to \mathcal{A}$ by $\delta_J(f) = J_a(f) - I_a(f)$ for all $f \in \mathcal{F}_a$. Then δ_J is linear and satisfies*

(a) $\delta_J(fg) = I_a(f)\delta_J(g) + \delta_J(f)I_a(g) + \delta_J(f)\delta_J(g)$ $\forall f, g \in \mathcal{F}$.

(b) $\delta_J(p) = 0$ $\forall$ polynomials p.

(c) $\delta_J(\mathcal{F}_a) \subseteq \mathcal{D}(a, \mathcal{A})$.

(d) $\delta_J(e) = 0$ whenever e is idempotent.

(e) If $\mathcal{A}$ is commutative, $\delta_J(\mathcal{F}_a) \subseteq \mathcal{D}(a, \mathcal{A}_J) \subseteq \mathcal{A}_J$.

Proof Properties (a) and (b) follow from easy computations.

(c) and (e): For $f \in \mathcal{F}_a$ and $\lambda \in Sp(a)$, we can write $f = f(\lambda) + (\lambda - z)g$ for some $g \in \mathcal{F}_a$. By applying δ_J, we get $\delta_J(f) = (\lambda - a)\delta_J(f)$. This shows $(\lambda - a)\delta_J(\mathcal{F}_a) = \delta_J(\mathcal{F}_a)$ for all $\lambda \in Sp(a)$. The last proposition now gives $\delta_J(\mathcal{F}_a) \subseteq \mathcal{D}(a, \mathcal{A})$. If $\mathcal{A}$ is commutative, the final result of the last proposition gives (e).

(d): The elements $I_a(e)$ and $J_a(e)$ are idempotents. Since $I_a(e)$ belongs to the double commutant of $\{a\}$ and $aJ_a(e) = J_a(ze) = J_a(e)a$ holds, there is a maximal commutative subalgebra $\mathcal{C}$ which includes both $J_a(\mathcal{F}_a)$ and $I_a(\mathcal{F}_a)$. This implies $\delta_J(e) = \delta_J(e)^3 = \cdots = \delta_J(e)^{3^n}$ for all $n \in \mathbb{N}$. Gelfand's spectral radius formula now gives $||\delta_J(e)|| = \rho(\delta_J(e))$. We may replace $\mathcal{A}$ by $\mathcal{C}$ and then apply (e) to conclude $\delta_J(e) = 0$. .   □

Note that $e \in \mathcal{F}_a$ is idempotent if and only if $e|_{Sp(a)}$ is the characteristic function of a clopen subset S of $Sp(a)$. For S clopen in $Sp(a)$ we write e_S^a for such an element of $\mathcal{F}_a$.

3.4.13 Theorem *Let $\mathcal{A}$ be a Banach algebra. Any $a \in \mathcal{A}$ for which $\mathcal{D}(a, \mathcal{A})$ is $\{0\}$ has a unique functional calculus. This holds for all $a \in \mathcal{A}$ if $\mathcal{A}$ is commutative and its Jacobson radical is finite-dimensional.*

If $\mathcal{A}$ is commutative and $Sp(a)$ has at most countably many connected components, then the following are equivalent:

(a) *a has a unique functional calculus.*

(b) *$\mathcal{D}(a, \mathcal{A}) = \{0\}$.*

(c) *Each connected component S of $Sp(a)$ satisfies*

$$\mathcal{D}(a, \bigcap_{\substack{T \text{ clopen} \\ S \subseteq T}} e_T^a \mathcal{A}) = \{0\}.$$

Partial Proof We have already established the first paragraph except the last sentence. Suppose $\mathcal{A}$ is commutative and $\mathcal{A}_J$ is finite-dimensional. Then there is a non-zero polynomial p such that the restriction of the left regular representation L_a to $\mathcal{A}_J$ satisfies $L_{p(a)} = p(L_a) = 0$. Factoring $p(a) = \alpha(\lambda_1 - a)(\lambda_2 - a) \cdots (\lambda_n - a)$ shows $\{0\} = p(a)\mathcal{D}(a, \mathcal{A}) = \mathcal{D}(a, \mathcal{A})$.

For a proof of the second paragraph of the theorem see the original paper Thomas [1978] or Dales [1973] and [1994]. It is naturally hard to prove the existence of a discontinuous functional when some component of $Sp(a)$ fails to satisfy (c). The unusual (and not very restrictive) hypothesis on the topology of the spectrum is used to show that one component S of $Sp(a)$ is clopen. This is crucial in the proof that (c) implies (b).   □

We also quote Dales' main result [1973]. The reader is directed to that paper for further details and for reference to previous work.

3.4.14 Theorem *Let $\mathcal{A}$ and $\mathcal{B}$ be unital commutative Banach algebras with $\mathcal{B}$ semisimple. Then, $\mathcal{A}$ has a unique functional calculus if it has an ideal $\mathcal{I}$ containing its Gelfand radical and satisfying:*
   (a) *There is a continuous, unital homomorphism θ of $\mathcal{B}$ into $\mathcal{B}(\mathcal{I})$.*
   (b) *Each a, $c \in \mathcal{I}$ and $b \in \mathcal{B}$ satisfy:*

$$\theta(b)(ac) = a\theta(b)(c).$$

   (c) *There is a finite set $\{c_1, c_2, \ldots, c_n\} \subseteq \mathcal{I}$ such that each $c \in \mathcal{I}$ has a unique expression*

$$c = \sum_{j=1}^{n} \theta(b_j)(c_j) \quad \text{with} \quad b_j \in \mathcal{B}.$$

If we take $\mathcal{B}$ to be the complex field in the above theorem, we see again that a unital commutative Banach algebra with a finite-dimensional radical has a unique functional calculus. (This covers Feldman's example discussed in Example 7.1.4.) Arens–Hoffman extensions of semisimple commutative Banach algebras, which are discussed in Example 3.4.17, are also covered by the theorem. A simple argument (given on p. 642 of Dales [1973]) shows that any algebra satisfying the hypotheses of the theorem must have a nilpotent radical. This property of Arens–Hoffman extensions of semisimple commutative Banach algebras was first obtained by John A. Lindberg, Jr. [1964]. Unfortunately it seems difficult to derive any criterion for general Banach algebras to have a unique functional calculus from this criterion for commutative Banach algebras.

3.4.15 Example We will exhibit an algebra satisfying Theorem 3.4.9 which also has other interesting properties. Let $\mathcal{A}$ be the Banach space $L^1([0, 1])$ with Lebesgue measure on $[0, 1]$. This can be made into a Banach algebra by defining convolution multiplication:

$$f * g(t) = \int_0^t f(s)g(t - s)ds \qquad \forall\, f,\, g \in \mathcal{A}; t \in [0, 1].$$

An obvious change of variables shows that $\mathcal{A}$ is commutative. It is easy to see that the convolution powers of the constant function 1 are $1^n = x^{n-1}/(n - 1)!$ and have norm $(n!)^{-1}$. Hence, the constant function 1 is topologically nilpotent and generates the dense subalgebra $\mathcal{P}$ of polynomials in $\mathcal{A}$. (The Stone–Weierstrass theorem (cf. 1.5.1) shows that the algebra $\mathcal{P}$ of polynomials is dense in $C([0,1])$ in its norm. A standard argument, based on the regularity of Lebesgue measure, then shows that $\mathcal{P}$ is dense in $\mathcal{A}$.) Therefore, the continuity of the spectral radius in commutative spectral normed algebras (Corollary 3.1.6) shows that $\mathcal{A}$ is equal to its own Gelfand radical. Another way to see that $\mathcal{A}$ is Gelfand-radical is to note that any function which vanishes almost everywhere in some neighborhood of zero is nilpotent, and that the ideal of such functions is dense in $\mathcal{A}$.

Consider the sequence of functions $\{e_n\}_{n \in \mathbb{N}}$ where $e_n(s)$ equals n for $s \leq 1/n$ and equals zero for $s > 1/n$. These functions all have norm one. Any $f \in C([0,1])$ and $t \geq 1/n$ satisfies

$$
\begin{aligned}
|f(t) - e_n * f(t)| &= |f(t) - \int_0^t e_n(s)f(t-s)ds| \\
&= n \int_0^{1/n} (f(t) - f(t-s))ds \\
&\leq n \int_0^{1/n} |f(t) - f(t-s)|ds \\
&\leq \max\{|f(t) - f(r)| : t - (1/n) \leq r \leq t\}.
\end{aligned}
$$

Hence, the uniform continuity of functions in $C([0,1])$ shows that

$$
f = \lim e_n * f \tag{9}
$$

holds pointwise on $]0,1]$. The Lebesgue dominated convergence theorem shows that equation (9) holds in the norm of $\mathcal{A}$ for any $f \in C([0,1])$. Since (as remarked above) $C([0,1])$ is dense in $\mathcal{A}$, (9) holds for all $f \in \mathcal{A}$. In the terminology introduced in Section 5.1, $\{e_n\}_{n \in \mathbb{N}}$ is an approximate identity of norm one in $\mathcal{A}$. Let $\{f_n\}_{n \in \mathbb{N}}$ be a countable dense sequence in $\mathcal{A}$ (e.g., an ordering of the polynomials with complex rational coefficients). Corollary 5.2.4 (the Cohen factorization theorem) shows that there is some $b \in \mathcal{A}$ and some sequence $\{g_n\}_{n \in \mathbb{N}} \subseteq \mathcal{A}$ satisfying $\|b\| \leq 1$ and $f_n = g_n * b$. Hence $\mathcal{A}b^-$ equals $\mathcal{A}$. Therefore, $(\mathcal{A}b^n)^-$ equals $\mathcal{A}$ for any $n \in \mathbb{N}$. Hence the unital algebra $\mathcal{A}^1$ and the element b satisfy Theorem 3.4.9, so $\mathcal{A}$ fails to have a unique functional calculus.

Note that $\mathcal{A}b = \mathcal{A}$ would imply the existence of an element e to satisfying $eb = b$ and hence $ea = ecb = ceb = cb = a$ for any $a = cb$ in $\mathcal{A}$. Since we have shown that $\mathcal{A}$ is Gelfand radical, it cannot be unital, so $\mathcal{A}b \neq \mathcal{A}$. Hence, the ideal $\mathcal{I} = \cap_{n=1}^\infty \mathcal{A}b^n$ is a proper (dense) ideal.

Before continuing the discussion of uniqueness of the functional calculus, we will mention some other interesting properties of the algebra

$\mathcal{A} = L^1([0,1])$. Since the constant function 1 generates $\mathcal{A}$ as a Banach algebra, the closed ideals of $\mathcal{A}$ are just the subspaces invariant under the Volterra operator

$$(Tf)(t) = \int_0^t f(s)ds \qquad \forall\, f \in \mathcal{A};\ t \in [0,1]. \tag{10}$$

William F. Donoghue, Jr. [1957] showed that these are exactly the subspaces

$$\mathcal{I}_t = \{f \in \mathcal{A}\colon f(s) = 0 \ \text{ for almost all } s \in [0,t]\}$$

for $t \in [0,1]$. Obviously, $f \in \mathcal{I}_t$ and $g \in \mathcal{I}_r$ imply $f*g \in \mathcal{I}_{t+r}$. A convolution theorem of Titchmarsh [1926] shows that, conversely, $f \in \mathcal{I}_t$ and $f*g \in \mathcal{I}_{t+r}$ imply $g \in \mathcal{I}_r$ (for $t, r, t+r \in [0,1]$). Each ideal $\mathcal{I}_t$ $t > 0$ is clearly nilpotent (see Definition 4.4.1 for this and other terms not yet introduced), so $\mathcal{A}$ is not semi-prime and, indeed, the dense ideal $\cup_{a>0}\mathcal{I}_a$ is included in its Baer radical by Theorem 4.4.6. Herbert Kamowitz and Stephen Scheinberg [1969] have obtained explicit formulas for all the continuous derivations of $\mathcal{A}$ and all the automorphisms of $\mathcal{A}$ which are in the connected component of the group of automorphisms. As we will show in Example 6.1.19, the algebra $\mathcal{A}$ satisfies the hypotheses of Theorems 6.1.18 and 6.4.20 so that all its automorphisms and derivations are continuous. Examples 4.8.3, 4.8.4 and 5.1.10 present somewhat similar convolution algebras.

We note here that the disc algebra $A(D)$ can obviously be identified with a subalgebra of $\mathcal{F}$. Since $L^1([0,1])$ satisfies Theorem 3.4.9 and at least one of φ and ψ is discontinuous, there is a discontinuous injective homorphism of $A(D)$ into $L^1([0,1])$. Thus, in particular, there is an incomplete algebra norm on $A(D)$ which dominates its usual norm $\|\cdot\|_\infty$. The results which we are about to discuss show that such a norm can even be chosen so that the Gelfand radical of the completion of $A(D)$ is nilpotent. See Allan [1980] for a variation.

3.4.16 Example We discuss more briefly another example of a discontinuous functional calculus. Details will be found in Dales [1973]. Let $\mathcal{U}$ be the algebra of analytic functions with power series which converge uniformly on the closed unit disc $\mathbb{D}$. Then $\mathcal{U}$ is a Banach algebra under the norm:

$$\|f\| = \sum_{n=0}^\infty |\alpha_n| \ \text{ when } f(z) = \sum_{n=0}^\infty \alpha_n z^n.$$

Let $\mathcal{R}$ be the Banach space $C([0,1])$ and let T be the Volterra operator defined by (10). Since $Sp(T)$ is $\{0\}$, we may define $f(T)$ for any $f \in \mathcal{U}$ by

$$f(T) = \sum_{n=0}^\infty \alpha_n T^n \ \text{ for } \ f(z) = \sum_{n=0}^\infty \alpha_n z^n \in \mathcal{U}$$

and we will just have the usual functional calculus. Now let $\mathcal{A}$ be the Banach space $\mathcal{U} \oplus \mathcal{R}$ with the norm $\|f \oplus x\| = \|f\| + \|x\|$ and multiplication

$$(f \oplus x)(g \oplus y) = (fg, f(T)y + g(T)x) \qquad \forall\, f,\, g \in \mathcal{U};\ x,\, y \in \mathcal{R}.$$

Clearly, $\mathcal{A}$ is a commutative Banach algebra with Gelfand radical $\{0\} \oplus \mathcal{R}$. Furthermore, the isometric isomorphism $\theta: \mathcal{U} \to \mathcal{A}$ defined by $\theta(f) = f \oplus 0$ shows that $\mathcal{A}$ has a strong Wedderburn decomposition (*cf.* Example 7.1.4). Nevertheless a discontinuous functional calculus for $\mathcal{A}$ is constructed in the reference already cited. The construction is closely related to a construction given in Theorem 6.4.23, which leads to a Banach algebra with two inequivalent Banach algebra norms. These give rise to two different functional calculi—each continuous with respect to one of the two norms. Even the exponential function has a different effect in these two functional calculi.

Miscellaneous Examples

3.4.17 Extensions and Arens–Hoffman Extensions   In this example $\mathcal{B}$ is said to be an *extension* of $\mathcal{A}$ if $\mathcal{A}$ is a subalgebra of $\mathcal{B}$. Unfortunately there is considerable variation in terminology. Some authors insist that the embedding map be continuous, homeomorphic or even isometric, while others make no continuity restrictions at all. Obviously the theories derived depend strongly on these assumptions.

If $a \in \mathcal{A}$ is sufficiently well behaved, there may be a unital algebra $\mathcal{B}$ including $\mathcal{A}$ in which there is a joint solution to the equations $ax = 1$ and $xa = 1$. Similar ideas have been investigated in great detail usually when the algebras are assumed to be commutative: Arens [1958], [1960], Ian G. Craw [1978], Vladimir Müller [1979], [1982], [1983], [1988], Gerard J. Murphy and Trevor T. West [1980a], Craw and Susan Ross [1983], Souren Arshakovič Grigoryan [1984], Zame [1984], and Michael J. Meyer [1991].

More generally, let $\mathcal{A}$ be a unital commutative algebra and let $\mathcal{A}[x]$ be the algebra of all polynomials in an indeterminant x with coefficients in $\mathcal{A}$. Let $p \in \mathcal{A}[x]$ be a monic polynomial of degree at least 2. Let $\mathcal{B} = \mathcal{A}[x]/p\mathcal{A}[x]$ be the quotient algebra of $\mathcal{A}$ modulo the principal ideal $p\mathcal{A}[x]$, and let $\theta: \mathcal{A}[x] \to \mathcal{B}$ be the natural map. Clearly, $\mathcal{B}$ is a unital commutative algebra and the map $a \mapsto \theta(a)$ embeds $\mathcal{A}$ as a subalgebra of $\mathcal{B}$. Furthermore, $\theta(x) \in \mathcal{B}$ is a solution of the equation $p(x) = 0$. When $\mathcal{A}$ is a Banach algebra, Arens and Kenneth M. Hoffman [1956] show how to provide $\mathcal{B}$ with a complete algebra norm making $\mathcal{B}$ into a Banach algebra. The key idea is the following simple result.

Proposition.   *Let $a_0, a_1, \ldots, a_{n-1}$ be elements in a commutative unital Banach algebra. Then there is a positive number s satisfying*

$$\|a_0\| + \|a_1\| s + \cdots + \|a_{n-1}\| s^{n-1} \leq s^n.$$

For any such number s, the expression

$$\|a_0 + a_1 x + \cdots + a_{n-1} x^{n-1} + p\mathcal{A}[x]\|_s = \sum_{j=0}^{n-1} \|a_j\| s^j$$

defines a complete algebra norm on $\mathcal{B}$.

David T. Brown [1966] and Lindberg [1971] show that if $\mathcal{A}$ is semisimple and s satisfies the condition above, then any complete algebra norm on $\mathcal{B}$ is equivalent to $\|\cdot\|_s$. The norms $\|\cdot\|_s$ defined above for different values of s (which satisfy the condition) are all equivalent. It has also been shown that all derivations of an Arens–Hoffman extension of $\mathcal{A}$ are continuous if $\mathcal{A}$ has the same property (Lindberg [1971], Richard J. Loy [1970a], Julian M. Cusack [1976]). In particular, this holds if $\mathcal{A}$ is semisimple.

3.4.18 Banach Algebras of Power Series   The earliest references to the subject matter discussed here appear to be Gelfand, Raĭkov and Šilov [1946], Lorch [1944] and Šilov [1947a], [1947b]. Important more recent discussions are Loy [1974b], Sandy Grabiner [1974] and Bade, Dales and Laursen [1984]. (See also Dales [1994] and Bade [1983].) The terminology in this field is unfortunately not standardized. We use $\mathcal{F} = \mathbb{C}[[z]]$ to represent the *algebra of all formal power series* with complex coefficients (*cf.* 2.9.7). If $\sum_{n=0}^{\infty} \alpha_n \lambda^n$ happens to converge for some $f = \sum_{n=0}^{\infty} \alpha_n z^n \in \mathcal{F}$ and some $\lambda \in \mathbb{C}$, we denote it by $\check{f}(\lambda)$.

Definition.   A *Banach algebra of power series* is a Banach space $\mathcal{A}$ which is a subalgebra of $\mathcal{F}$ containing z such that the coefficient functional

$$C_n\left(\sum_{m=1}^{\infty} \alpha_m z^m\right) = \alpha_n \qquad \forall \; \sum_{m=1}^{\infty} \alpha_m z^m \in \mathcal{A}$$

is continuous for each $n \in \mathbb{N}^0$.

If $\mathcal{A}$ is a Banach algebra of power series, the closed graph theorem shows that multiplication is separately continuous for each $a \in \mathcal{A}$ so Proposition 1.1.9 asserts that there is an equivalent norm on $\mathcal{A}$ in which it is a Banach algebra. Henceforth, we assume that $\mathcal{A}$ is a Banach algebra whenever it is convenient to do so, although the norm given on certain examples may not be submultiplicative until an equivalent renorming is performed. We always consider $\mathcal{A}^1$ to be embedded in $\mathcal{F}$ in the obvious way. Some authors require that polynomials be dense in a Banach algebra of power series and call algebras that lack this property generalized Banach algebras of power series. Banach algebras of power series arise naturally in at least five contexts and also provide a simple format for the construction of counterexamples. Many such algebras can be profitably considered in more than one context. They are often Jacobson-radical.

Let U be an open connected region of the complex plane containing zero. Let $\mathcal{A}$ be a linear space of functions analytic on U and assume that $\mathcal{A}$

is closed under pointwise multiplication and is complete in some submultiplicative norm. Two obvious examples are the disc algebra $A(\mathbb{D})$ and the algebra $H^\infty(\mathbb{D})$. (These two algebras are fundamentally different in this theory, because the polynomials are dense in the first case and not in the second.) Since the functions in $\mathcal{A}$ are completely determined by their power series expansion at zero, any such algebra is a Banach algebra of power series. Evaluation at any point of U defines an element of the Gelfand space $\Gamma_\mathcal{A}$, and since such evaluations obviously separate the points of $\mathcal{A}$ these algebras are semisimple. This example is not typical of the type of Banach algebras of power series usually studied.

Now suppose the region U is star shaped with respect to zero. Again let $\mathcal{A}$ be a linear space of functions analytic on U, but this time define multiplication by convolution

$$f * g(z) = \int\limits_0^z f(w)g(z-w)dw$$

and assume that $\mathcal{A}$ is a Banach algebra relative to this multiplication. Obviously, this does not become a Banach algebra of power series when the Taylor series at zero is associated with each function. Instead, associate with each function

$$f(z) = \sum_{n=0}^\infty \alpha_n z^n$$

the formal series

$$\hat{f}(z) = \sum_{n=1}^\infty f^{(n-1)}(0)z^n = \sum_{n=1}^\infty \alpha_{n-1}(n-1)!z^n.$$

It is not hard to show

$$(f * g)\check{} = \hat{f} \cdot \hat{g} \quad f, g \in \mathcal{A},$$

so that $\mathcal{A}$ is a Banach algebra of power series provided it contains the constants (since $\hat{1}(z) = z$) and meets the topological conditions. Grabiner [1974], Section 13, shows that many well-known spaces of analytic functions (e.g., H^p spaces) do become radical Banach algebras of power series. The ideal structure and other properties of these spaces are determined explicitly. See also Example 4.8.3.

As a third example of a naturally defined class of Banach algebras of power series, we consider a topologically nilpotent but not nilpotent operator T on a Banach space $\mathcal{X}$. Let $\mathcal{A}$ be the algebra of all power series $\sum_{n=1}^\infty \alpha_n z^n$ such that the series $\sum_{n=1}^\infty \alpha_n T^n$ converges in the weak operator topology of $\mathcal{B}(\mathcal{X})$, and provide $\mathcal{A}$ with the operator norm of $\mathcal{B}(\mathcal{X})$. Under suitable conditions on T, $\mathcal{A}$ is a Banach algebra of power series. See Grabiner [1974], Section 14, for further details and references.

Our fourth natural example of Banach algebras of power series arises from considering shift operators. Let $\mathcal{X}$ be a Banach space. A *biorthogonal system* in $\mathcal{X}$ is a pair of sequences $\{z_n\}_{n\in\mathbb{N}} \subseteq \mathcal{X}$ and $\{\omega_n\}_{n\in\mathbb{N}} \in \mathcal{X}^*$ satisfying $\omega_j(z_k) = \delta_{jk}$. The biorthogonal system $(\{z_n\}_{n\in\mathbb{N}}, \{\omega_n\}_{n\in\mathbb{N}})$ is called a *generalized basis* if the sequence of functionals separates the points of $\mathcal{X}$ and is called a *Markushevich basis* if, in addition, the sequence of vectors has a dense range. Relative to such a basis, an operator $T \in \mathcal{B}(\mathcal{X})$ is called a *weighted shift* determined by the sequence $\{\lambda_n\}_{n\in\mathbb{N}}$ of non-zero complex numbers if each $x \in \mathcal{X}$ satisfies

$$\omega_{n+1}(Tx) = (\lambda_{n+1}/\lambda_n)\omega_n(x) \quad n \geq 0$$

$$\omega_0(Tx) = 0.$$

The power series representation determined by T is the map which associates with each $x \in \mathcal{X}$ the formal power series $\sum_{n=0}^{\infty} \omega_n(x)\lambda_n^{-1}z^n$. The space $\mathcal{A}^1$ of all such power series is a Banach space under the inherited norm. Under this representation of $\mathcal{X}$, T corresponds to the operator of multiplication by z, $\lambda_n z_n$ corresponds to T^n, and $\lambda_n^{-1}\omega_n$ corresponds to the nth coefficient functional C_n on $\mathcal{A}^1$. The collection $\mathcal{A}$ of power series in $\mathcal{A}^1$ without constant terms is frequently a Banach algebra of power series.

We wish to examine our fifth class of examples, the weighted semigroup algebras on the semigroup $\mathbb{N}^0$ of non-negative integers, more closely. These algebras were introduced in §1.9.15. Recall that a weight function on $\mathbb{N}^0$ is a positive valued sequence $\{\omega_n\}_{n\in\mathbb{N}^0}$ satisfying $\omega_0 = 1$ and $\omega_m\omega_n \leq \omega_{m+n}$ for all $m, n \in \mathbb{N}^0$. We will denote the semigroup algebra $\ell^1(\mathbb{N}^0, \omega)$ by $\mathcal{A}$. Thus $\mathcal{A}$ is the algebra of sequences $f: \mathbb{N}^0 \to \mathbb{C}$ with finite norm $\|f\|_\omega = \sum_{n=0}^{\infty} |f(n)|\omega_n$ under convolution multiplication $f * g(n) = \sum_{m=0}^{n} f(m)g(n-m)$. Clearly $\mathcal{A}$ is a singly generated unital commutative Banach algebra which is a subalgebra of $\mathcal{F}$ and in which sequences with finite support (*i.e.*, polynomials in δ_1) are dense. The next simple result, published in Gelfand, Raĭkov and Šilov [1946], is fundamental and follows from Theorem 3.1.15.

Theorem   Let $\rho = \rho(\ell^1(\mathbb{N}^0, \omega))$ be $\inf\{\omega_n^{1/n}\} = \lim \omega_n^{1/n}$.

(a) If ρ is positive, then $\mathcal{A}$ is a unital semisimple commutative Banach algebra and $\lambda \mapsto \gamma_\lambda$ is a homeomorphism of $\{\lambda \in \mathbb{C} : |\lambda| \leq \rho\}$ onto $\Gamma_\mathcal{A}$ where $\gamma_\lambda(f) = \check{f}(\lambda)$.

(b) If ρ is zero, then $\mathcal{A}$ is isomorphic to the unitization of its radical $\mathcal{A}_J = \{f \in \mathcal{A} : f(0) = 0\}$.

This obviously suggests calling a weight function ω *radical* if ρ is zero. Recall from Example 2.9.7 that $\mathcal{F}$ has a decreasing sequence of closed ideals $\mathcal{F}_m = \{f \in \mathcal{F} : f(n) = 0 \text{ for all } n \leq m\}$. When we consider $\mathcal{A}$ as embedded in $\mathcal{F}$, this gives a sequence of closed ideals called *standard ideals*. We will call $\mathcal{A}$ or its weight sequence ω *unicelullar* if all the closed ideals of $\mathcal{A}$

are standard. During the 1940s, Šilov asked whether every radical weight function is unicellular. Thomas [1983], [1984a], [1984b] finally proved that nonunicellular ones exist. This raises questions about the structure of $\mathcal{A}/\mathcal{I}$ when $\mathcal{I}$ is not standard.

We introduce several technical properties of weight functions and briefly record their logical relations. It is convenient to use η_n for $-\log \omega_n$ so that $\rho = 0$ if and only if $\lim \eta_n/n = \infty$. A weight sequence $\{\omega_n\}_{n\in\mathbb{N}^0}$ is called $\langle$ *convex / star-shaped* $\rangle$ if the sequence $\langle \eta_{n+1} - \eta_n / \{\eta_n/n\}_{n\in\mathbb{N}^0} \rangle$ is increasing. It is called *basic* if $\sup\{\omega_{m+n+k}/\omega_{m+k}\omega_{n+k} : m, n \in \mathbb{N}^0\}$ is finite for each $k \in \mathbb{N}^0$, and *ordinary* if, for each $a \in \mathcal{A}$, there is some $k \in \mathbb{N}^0$ satisfying $z^k \in a\mathcal{A}$.

Theorem   *There are radical weight functions on $\mathbb{N}^0$ which are not unicellular. All the arrows in the following diagram indicate known implications between properties of radical weight functions and none are logical equivalences.*

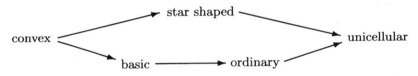

For proofs and related results, see Dales [1994]. The counterexamples needed to show that none of the implications can be reversed are from Bade, Dales and Laursen [1984] and and Thomas [1984b]. See also Grabiner [1975], [1974] Section 14, [1984c]; Grabiner and Thomas [1985], Yngve Domar [1983], [1987] Fereidoun Ghahramani and McClure [1990], [1992], Ghahramani, McClure and Grabiner [1990].

3.4.19 Example   Our next example is a unital, completely regular, semi-simple, commutative Banach algebra with a non-maximal proper closed prime ideal. It is a refinement due to Cusack [1976] of an example of Lorch [1944]. See also Dales and Alexander M. Davie [1973].

Let $K = \{K_n\}_{n\in\mathbb{N}}$ be a sequence of strictly positive real numbers satisfying $K_0 = 1$ and

$$K_n \geq \binom{n}{k} K_k K_{n-k} \tag{11}$$

for all $n \in \mathbb{N}^0$ and all k satisfying $0 \leq k \leq n$. Let $\mathcal{A}(K)$ be the Banach algebra of all $f \in C^\infty([0,1])$ (the infinitely differentiable functions under pointwise multiplication) which are finite with respect to the norm

$$\|f\|_K = \sum_{n=0}^{\infty} K_n^{-1} \|f^{(n)}\|_\infty.$$

Loy [1974a] shows that these algebras are semisimple if $\lim(n!/K_n)^{1/n}$ is positive. Dales and Davie [1973] show that, if K satisfies

$$\sum_{n=2}^{\infty}(\sum_{j=1}^{n-1}\binom{n}{j}K_jK_{n-j}K_n^{-1}) < \infty, \tag{12}$$

then $[0,1]$ is naturally identified with the Gelfand space of $\mathcal{A}(K)$.

For all r, s satisfying $0 < r < s < 1$ define $f_{rs} \in C^{\infty}[0,1]$ by

$$f_{rs}(t) = \begin{cases} 0 & t \leq r \text{ and } t \geq s \\ \exp((r-t)^{-1} + (t-s)^{-1}) & r < t < s \end{cases}$$

Then for any closed set $C \in [0,1]$ and any $t \notin C$, we can find an f_{rs} which vanishes on C but is non-zero at t. Hence, if we can choose K satisfying (12) so that $\mathcal{A}(K)$ contains each function f_{rs}, then $\mathcal{A}(K)$ will be completely regular. An elementary but tedious estimate shows that $k_n = \sup\{\|f_{rs}^{(n)}\|_{\infty} : 0 < r < s < 1\}$ is finite for each n. Hence, it remains to show that for any sequence $\{k_n\}_{n\in\mathbb{N}}$ of positive numbers, we can find a sequence $K = \{K_n\}_{n\in\mathbb{N}}$ satisfying both (12) and also $\sum_{n=0}^{\infty}K_n^{-1}k_n < \infty$. We may assume $k_n \geq 2^n$ for all $n \in \mathbb{N}$. Define K inductively by $K_0 = K_1 = 1$ and for $n \geq 2$

$$K_n = 2^n(\sum_{j=1}^{n-1}\binom{n}{j}K_jK_{n-j}) + k_n^2.$$

Clearly, we have $K_n \geq \binom{n}{j}K_jK_{n-j}$ $0 \leq j \leq n$,

$$\sum_{n=2}^{\infty}\sum_{j=1}^{n-1}\binom{n}{j}K_jK_{n-j}K_n^{-1} \leq \sum_{n=2}^{\infty}2^{-1} < \infty,$$

and

$$\sum_{n=2}^{\infty}K_n^{-1}k_n \leq \sum_{n=2}^{\infty}k_n^{-2}k_n \leq \sum_{n=2}^{\infty}2^{-n} < \infty.$$

With this choice of the sequence K, $\mathcal{A}(K)$ is a unital, completely regular, semisimple commutative Banach algebra.

For any choice of the sequence K satisfying (11),

$$\mathcal{P} = \{f \in \mathcal{A}(K) : f^{(n)}(0) = 0 \quad n \in \mathbb{N}^0\}$$

is a non-maximal closed prime ideal. It is non-maximal since it has infinite codimension. It is prime since it is the kernel of the homomorphism

$$f \mapsto \sum_{n=0}^{\infty}(n!)^{-1}f^{(n)}(0)z^n \quad f \in \mathcal{A}(K)$$

of $\mathcal{A}(K)$ into the integral domain $\mathcal{F}$.

3.4.20 Point Derivations Let $\mathcal{A}$ be a unital algebra and let γ be a point in its Gelfand space which is arbitrary but will remain fixed throughout this discussion.

Definition A *point derivation* of $\mathcal{A}$ at γ is a linear functional $\delta \colon \mathcal{A} \to \mathbb{C}$ satisfying

$$\delta(ab) = \delta(a)\gamma(b) + \gamma(a)\delta(b) \qquad \forall \, a, b \in \mathcal{A}.$$

If $\mathcal{A}$ is an algebra of differentiable functions on $\mathbb{R}$ and if γ is evaluation at a point $x \in \mathbb{R}$, then evaluation of the derivative of a function at x would be a point derivation at γ. This is obviously the origin of the terminology.

Let $\mathcal{M}$ be the codimension one ideal which is the kernel of γ and let $\mathcal{M}^2$ denote the linear span of all products of elements in $\mathcal{M}$, as usual. It is easy to see that a linear functional on $\mathcal{A}$ is a point derivation at γ if and only if it vanishes on $\mathcal{M}^2 + \mathbb{C}1$. Consequently, the linear space of point derivations at γ is in natural bijective correspondence with the algebraic dual of the quotient space $\mathcal{M}/\mathcal{M}^2$. Hence, there are no point derivations at γ if $\mathcal{M} = \mathcal{M}^2$.

If $\mathcal{A}$ is a normed algebra and the closure of $\mathcal{M}^2$ equals $\mathcal{M}$, then there are no continuous point derivations at γ. Perhaps of greater interest is the case in which $\mathcal{M}/\mathcal{M}^2$ is infinite-dimensional in a normed algebra. For then there are surely discontinuous point derivations at γ. This has the following consequence, which is the easiest way to construct discontinuous algebra homomorphisms.

Proposition *Let $\mathcal{A}$ be a unital normed algebra with a discontinuous point derivation δ at $\gamma \in \Gamma_{\mathcal{A}}$. Let $\mathcal{B}$ be a unital normed algebra (such as the matrix algebra M_2) with some non-zero nilpotent element. Then there is a discontinuous unital homomorphism of $\mathcal{A}$ into $\mathcal{B}$.*

Proof If $\mathcal{B}$ has a non-zero nilpotent element, it has a non-zero element b satisfying $b^2 = 0$. Define the homomorphism φ to be

$$\varphi(a) = \gamma(a)1 + \delta(a)b \qquad \forall \, a \in \mathcal{A}. \qquad \square$$

It is obvious that for all elements of the Gelfand space of the algebra $C(\Omega)$ for a compact Hausdorff space Ω, $\mathcal{M}^2 = \mathcal{M}$ in the notation introduced above, so there are no non-zero point derivations. We will mention the difficult construction of discontinuous homomorphisms on $C(\Omega)$ in Section 6.1. In Volume II we will show that there are no non-zero point derivations on noncommutative C*-algebras either. If $\mathcal{A}$ is the disc algebra (§3.2.13), there are no non-zero point derivations at points of the Gelfand space corresponding to the boundary points of the disc. The only point derivations at points of Γ_A corresponding to interior points of the disc are those arising from differentiation and they are continuous point derivations. Hence, all the point derivations are continuous even though we will show that there are discontinuous homomorphisms from the disc algebra. In Volume II we will show that there are no non-zero point derivations on $L^1(G)$ for an amenable group or on other amenable Banach algebras.

3.4.21 A Functional Calculus for Continuous Functions

We will use Theorem 3.2.12 to construct a functional calculus in which continuous functions act on normal elements in a C*-algebra. This topic will be pursued in much greater depth in Volume II, but the comparison with the functional calculus of this section is instructive. C*-algebras were defined and introduced in §1.7.17. An element $\langle\, h\, /\, c\, \rangle$ in a C*-algebra is said to be $\langle$ hermitian / normal $\rangle$ if it satisfies $\langle\, h^* = h\, /\, c^*c = cc^*\, \rangle$. The set of $\langle$ hermitian / normal $\rangle$ elements in a C*-algebra A is denoted by $\langle\, A_H\, /\, A_N\, \rangle$.

Volume II will include many examples of C*-algebras and a much more complete discussion of their properties. The commutative examples are characterized in Theoreom 3.2.12. We need two easy results now. First, note that the inequality $\|a^*a\| \geq \|a\|^2$ for all $a \in A$ is equivalent to the C*-condition and implies that the involution is an isometry. To see this, note that $\|a\|^2 \leq \|a^*a\| \leq \|a^*\|\,\|a\|$ implies $\|a\| \leq \|a^*\|$, $\|a^*\| \leq \|(a^*)^*\| = \|a\|$ and $\|a^*a\| \leq \|a^*\|\,\|a\| = \|a\|^2 \leq \|a^*a\|$. Second, we note that if A is a C*-algebra, then the unitization A^1 is also a C*-algebra when we define

$$(\lambda + a)^* = \lambda^* + a^* \text{ and } \|\lambda + a\| = \sup\{\|(\lambda + a)b\| : b \in A_1\}.$$

The C*-inequality follows from

$$\begin{aligned}
\|(\lambda + a)^*(\lambda + a)\| &= \sup\{\|(\lambda + a)^*(\lambda + a)b\| : b \in A_1\} \\
&\geq \sup\{\|b^*(\lambda + a)^*(\lambda + a)b\| : b \in A_1\} \\
&= \sup\{\|(\lambda + a)b\|^2 : b \in A_1\} = \|\lambda + a\|^2.
\end{aligned}$$

The functional calculus is established in the next theorem.

Theorem. Let A be a C*-algebra. For each $c \in A_N$, let $A(c)$ be the smallest norm-closed unital subalgebra of A^1 which includes c and c^*. Each element d of $A(c)$ is normal and satisfies $Sp_{A^1}(d) = Sp_{A(c)}(d)$. Furthermore, there is an isometric isomorphism, denoted by $f \mapsto f(c)$, of $C(Sp_A(c))$ onto $A(c)$ which satisfies:

  (a) $z(c) = c$
  (b) $1(c) = 1 \in A^1$
  (c) $\widehat{f(c)} = f \circ \hat{c}$
  (d) $Sp_{A^1}(f(c)) = f(Sp_A(c))$
  (e) $f^*(c) = f(c)^*$

(f) $g(f(c)) = g \circ f(c)$
for all $f \in C(Sp(c))$ and $g \in C(Sp(f(c)))$.

Proof For simplicity, we will write Γ, Sp and $C(c)$ for $\Gamma_{A(c)}, Sp_{A(c)}$ and $C(Sp_{A(c)}(c))$, respectively. Since $A(c)$ clearly satisfies the hypotheses of Theorem 3.2.12, the Gelfand homomorphism is an isometric isomorphism of $A(c)$ onto $C(\Gamma)$. Theorem 3.1.5 shows that $\hat{c}\colon \Gamma \to Sp(c)$ is a continuous surjection. Next, we will show that $\hat{c}$ is an injection. Suppose $\hat{c}(\gamma_1) = \hat{c}(\gamma_2)$ holds for $\gamma_1, \gamma_2 \in \Gamma$. The last remark of Theorem 3.2.12 asserts $\widehat{c^*}(\gamma_1) = \widehat{c^*}(\gamma_2)$, which implies $p(c,c^*)\hat{\;}(\gamma_1) = p(c,c^*)\hat{\;}(\gamma_2)$ for all polynomials p in two variables. Since elements of the form $p(c,c^*)$ are clearly dense in $A(c)$, and ($\hat{\;}$) is continuous, this implies $\gamma_1(a) = \hat{a}(\gamma_1) = \hat{a}(\gamma_2) = \gamma_2(a)$ for all $a \in A(c)$. Hence γ_1 equals γ_2. Thus $\hat{c}\colon \Gamma \to Sp(c)$ is a continuous bijection and hence a homeomorphism. Therefore, we may identify Γ with $Sp(c)$, and interpret ($\hat{\;}$) as an isometric isomorphism of $A(c)$ onto $C(Sp(c))$. The map $f \mapsto f(c)$ is just the inverse map. All the stated results will now follow from Theorem 3.2.12 when we establish $Sp_{A^1}(d) = Sp(d)$ for all $d \in A(c)$.

Proposition 2.1.8(c) shows $Sp_{A^1}(d) \subseteq Sp_{A(c)}(d) = Sp(d)$ for any $d \in A(c)$. Suppose $\lambda \in Sp(d)$. Then $\lambda - \hat{d}$ vanishes at some $\gamma_0 \in \Gamma$. Given any $\varepsilon > 0$, there is a neighborhood N of γ_0 satisfying $|(\lambda - \hat{d})(\gamma)| < \varepsilon$ for all $\gamma \in N$. Since ($\hat{\;}$) is surjective, Urysohn's lemma gives some $a \in A(c) \subseteq A^1$ satisfying $\|a\| = \|\hat{a}\|_\infty = 1$ and $\mathrm{supp}(\hat{a}) \subseteq N$. We conclude $\|(\lambda - d)a\| = \|(\lambda - \hat{d})\hat{a}\|_\infty < \varepsilon$. Since $\varepsilon > 0$ was arbitrary, $\lambda - d$ is not invertible in A^1. We have established $Sp_{A^1}(d) = Sp(d)$ for all $d \in A(c)$. $\square$

We indicate a few applications of this powerful functional calculus. Others will be given in Section 5.1. Any normal element c in a C*-algebra satisfies $\|c\| = \|\hat{c}\|_\infty = \rho(c)$ (where ($\hat{\;}$) is defined relative to $A(c)$). Any hermitian element h in a C*-algebra has its spectrum included in $\mathbb{R}$. Conversely, if $h \in A_N$ has real spectrum it is hermitian.

If $p \in A_N$ has spectrum included in $\mathbb{R}_+$ then the square root function operates on p so there is a normal element q in p satisfying $p = q^2$. The spectrum of q is also non-negative and hence q is hermitian. For $h \in A_H$ define h_+ and h_- by $h_+ = f(h)$ and $h_- = g(h)$, where f and g are defined on $\mathbb{R}$ by

$$f(t) = \begin{cases} t & t \geq 0 \\ 0 & t < 0 \end{cases}$$

$$g(t) = \begin{cases} 0 & t \geq 0 \\ -t & t < 0. \end{cases}$$

Then h_+ and h_- are hermitian elements with spectrum in $\mathbb{R}_+$ satisfying:

$$h = h_+ - h_- \qquad h_+ h_- = h_- h_+ = 0.$$

For an example of a Banach algebra with a remarkably large functional calculus see Gilles Pisier [1979].

Algebra Norms on C*-algebras

We now use Theorem 3.2.12 again, this time to get a genuinely noncommutative result. The last argument in the following proof is due to Bonsall [1954a]. Related results will be discussed in Theorem 6.1.5.

3.4.22 Theorem *Let $(\mathcal{A}, \|\cdot\|)$ be a C*-algebra. Any (possibly incomplete) algebra norm $\|\|\cdot\|\|$ on $\mathcal{A}$ satisfies*

$$\|a\|^2 \leq \||a^*a\|| \leq \||a\|| \, \||a^*\|| \qquad \forall\, a \in \mathcal{A} \tag{13}$$

and is spectral. The C*-norm $\|\cdot\|$ is minimal in the sense that if $\|\|\cdot\|\|$ satisfies $\||a\|| \leq \|a\|$ for all $a \in \mathcal{A}$ then the two norms are actually equal.

Proof For any $a \in \mathcal{A}$ let $\mathcal{C}$ be the closed subalgebra generated by a^*a. Since $\mathcal{C}$ satisfies Theorem 3.2.12, Theorem 2.4.15 gives the inequality. Let $\mathcal{B}$ be the completion of $(\mathcal{A}, \|\|\cdot\|\|)$ and recall that Proposition 3.2.2 implies $\rho_{\mathcal{B}}(a) \leq \rho_{\mathcal{A}}(a)$ for all $a \in \mathcal{A}$. Replace a by a^n in inequality (13), take the nth root, and then let n approach ∞. Since $a \mapsto a^*$ is a real algebra anti-isomorphism which sends each complex numer to its conjugate, this gives $\rho_{\mathcal{A}}(a)^2 \leq \rho_{\mathcal{B}}(a)\rho_{\mathcal{B}}(a^*) \leq \rho_{\mathcal{B}}(a)\rho_{\mathcal{A}}(a^*) = \rho_{\mathcal{B}}(a)\rho_{\mathcal{A}}(a)$. We conclude $\rho_{\mathcal{A}}(a) = \rho_{\mathcal{B}}(a) \leq \||a\||$ for all $a \in \mathcal{A}$ so that $\|\|\cdot\|\|$ is a spectral norm.

Now suppose that $\|\|\cdot\|\|$ satisfies $\||a\|| \leq \|a\|$ for all $a \in \mathcal{A}$. If there were any $a \in \mathcal{A}$ satisfying $\||a\|| < \|a\|$ then we would get the contradiction $\|a\|^2 \leq \||a\|| \, \||a^*\|| < \|a\|\|a^*\| = \|a\|^2$. □

3.5   Multivariable Functional Calculus

Introduction

This section describes functional calculi for finite ordered sets of commuting elements in unital Banach algebras. Let $a = (a_1, a_2, \ldots, a_n)$ be an ordered n-tuple of commuting elements in a unital Banach algebra $\mathcal{A}$. We wish to define a notion of "joint spectrum" $Sp(a)$ of a which will be a subset of $\mathbb{C}^n$. If f is a function of n complex variables which is analytic on a neighborhood of $Sp(a)$, we hope to define an element $f(a)$ of $\mathcal{A}$ with the usual properties of a functional calculus. Thus we want $f(a)$ to satisfy: (1) polynomials act appropriately, (2) for each $a \in \mathcal{A}$, $f \mapsto f(a)$ is a unital homomorphism from the algebra of functions analytic in a fixed neighborhood U of $Sp(a)$ to $\mathcal{A}$, (3) a spectral mapping theorem holds, (4) composition of functions satisfies $f \circ g(a) = f(g(a))$ when either side is defined, (5) $f \mapsto f(a)$ is continuous and unique among continuous functions satisfying (1) and (2). If $Sp(a)$ is chosen quite large, the definition of $f(a)$ becomes easy, but the spectral mapping theorem and uniqueness will probably fail, while the class of functions which act will be too restricted. We do not know a completely satisfactory definition in the most general case, so we will give only a partial treatment. See also Wiesław Żelazko [1979], where the class of all joint spectra satisfying the spectral mapping theorem at least for polynomials is studied, Robin Harte [1972], [1973], [1977], [1981], [1988], Arne Kokk [1989] and Yu. V. Turovskiĭ [1984a].

Historically the first case to be studied, and the easiest case, is that of a commutative unital Banach algebra $\mathcal{A}$. Here the results are essentially due to Richard Arens and Alberto P. Calderon [1955] and Lucien Waelbroeck [1954], based on Šilov's proof [1953] for finitely generated commutative algebras. Only Waelbroek obtains a homomorphism. The most natural statements about the uniqueness of the resulting functional calculus were proved much later by William R. Zame [1979] and Stephen H. Schanuel and Zame [1982]. Bonsall and Duncan [1973b], Chapter 20 gives a beautifully simple treatment of a functional calculus which is not shown to be a homomorphism or unique. We will construct a system of homomorphisms with the weaker uniqueness property similar to Bourbaki [1967], Theodore W. Gamelin [1969] and Edgar Lee Stout [1971]. We separate the most technical portion of the construction and give an elementary exposition suitable to readers unacquainted with the exterior differential calculus. See also Graham R. Allan [1968], [1970], [1971] and Waelbroeck [1982a], [1982b], [1983a] and [1983b].

If the Banach algebra $\mathcal{A}$ is noncommutative, the situation is much more complicated. The difficulty arises first because the proper definition of the joint spectrum is no longer obvious. In [1970a] Joseph L. Taylor gave a complicated but very natural definition of the joint spectrum for the special case in which $\mathcal{A}$ is the noncommutative Banach algebra $\mathcal{B}(\mathcal{X})$ of all bounded operators on a Banach space $\mathcal{X}$. This is now called the Taylor spectrum. In [1970b], [1970c], [1972a] and [1972b] he constructed the functional calculus associated with his notion of spectrum. Both the definition and construction depend on Koszul complexes and are too technical to describe here in detail. Furthermore, basic properties of functional calculi remain unclear and difficult in this case. Alexandr Yakovlevič Helemskiĭ [1981] and [1989b], Chapter VI give a somewhat simplified version based on Taylor's [1972b]. See also Vasilescu [1982]. Mihai Putinar [1980], [1982], [1983a], [1983b], [1984a], [1984b], [1984c] has apparently solved these problems and given a coherent theory but at the expense of the further complication of extending the definition from functions analytic in a neighborhood of the spectrum to Stein algebras and less trivial applications of sheaf theory. (See also Jörg Eschmeier and Putinar [1984].) The most detailed exposition of this theory is still available only in the Rumanian language, but we understand (personal communication) that Putinar is currently preparing a detailed book-length exposition in English.

Given these two special cases (ordered n-tuples in commutative algebras and commuting ordered n-tuples in $\mathcal{B}(\mathcal{X})$), there are some obvious ways to try to define the joint spectrum and functional calculus for commuting ordered n-tuples in arbitrary unital Banach algebras $\mathcal{A}$. If $\mathcal{C}$ is any commutative unital subalgebra of $\mathcal{A}$ including $\{a_1, a_2, \ldots, a_n\}$, then the commutative theory can be applied to $a = (a_1, a_2, \ldots, a_n)$ considered relative to $\mathcal{C}$. Obviously the theory depends on the choice of $\mathcal{C}$. In order to have as many functions as possible in the functional calculus, one must choose $\mathcal{C}$ as large as possible. The only natural choice is the double commutant $\{a_1, a_2, \ldots, a_n\}''$ of $\{a_1, a_2, \ldots, a_n\}$. This procedure is satisfactory in the very special case in which $a_1, a_2, \ldots, a_n$ belong to the center $\mathcal{A}_Z$ of $\mathcal{A}$ and $\mathcal{A}_Z$ is chosen as $\mathcal{C}$. Unfortunately, in essentially all other cases examples are known in which no choice will give a satisfactory spectrum or one as small as Taylor's spectrum (see Taylor [1970a], Zbigniew Słodkowski and Żelazko [1979]). This procedure is also incompatible with a satisfactory spectral mapping theorem. Another approach would be to apply Taylor's theory to the left regular representing operators $\{L_{a_1}, L_{a_2}, \ldots, L_{a_n}\}$. This also presents many problems in most cases, although not for C*-algebras (Raul E. Curto [1982]).

Before introducing our notation and proceeding with the detailed exposition, we wish to mention, but not pursue, the idea of a functional calculus for several noncommuting elements. One could begin such a discussion by considering some kind of generalization of polynomials in several noncommuting indeterminants. This approach has been pursued by Taylor [1972a], [1972b] and [1973b], Harte [1972] and [1973] and Alan James Pryde [1988].

Another approach is to abandon the homomorphism property of a functional calculus and use ordinary analytic functions of several variables, symmetry considerations and a suitable idea of joint spectrum. Albrecht has developed that approach. His valuable survey [1982b] cites the relevant literature. For some related ideas see N. C. Luong and Árpád Száz [1979] and Denis Luminet [1986] for a special case.

We now return to the main subject of this chapter: a functional calculus for several elements in a commutative algebra. We need a little notation. An element of $\mathbb{C}^n$ will be denoted by a single lower case Greek letter such as λ and its components will then be denoted by subscripts, $e.g.$,

$$\lambda = (\lambda_1, \lambda_2, \ldots, \lambda_n). \tag{1}$$

For each $j = 1, 2, \ldots, n$ the function $z_j \colon \mathbb{C}^n \to \mathbb{C}$ will be defined by

$$z_j(\lambda) = z_j(\lambda_1, \lambda_2, \ldots, \lambda_n) = \lambda_j \qquad \forall \, \lambda \in \mathbb{C}^n. \tag{2}$$

Then $z = (z_1, z_2, \ldots, z_n)$ is the identity function or the coordinate function on $\mathbb{C}^n$:

$$z(\lambda) = \lambda \qquad \forall \, \lambda \in \mathbb{C}^n. \tag{3}$$

Polynomials and Polynomial Convexity

Let $\mathcal{P}(\mathbb{C}^n)$ be the algebra, under pointwise operations, of polynomial functions with domain $\mathbb{C}^n$ and range in $\mathbb{C}$. Of course, $\mathcal{P}(\mathbb{C}^n)$ is generated as a unital algebra by its subset $\{z_1, z_2, \ldots, z_n\}$. Let $p \in \mathcal{P}(\mathbb{C}^n)$ be expressed by

$$p = \sum_{0 \leq j_1 + j_2 + \cdots + j_n \leq N} \alpha_{j_1, j_2, \ldots, j_n} z_1^{j_1} z_2^{j_2} \cdots z_n^{j_n}.$$

If $a = (a_1, a_2, \ldots, a_n)$ is any ordered n-tuple of commuting elements in any algebra $\mathcal{A}$, we define $p(a)$ by

$$p(a) = \sum_{0 \leq j_1 + j_2 + \cdots + j_n \leq N} \alpha_{j_1, j_2, \ldots, j_n} a_1^{j_1} a_2^{j_2} \cdots a_n^{j_n}. \tag{4}$$

For any $n \in \mathbb{N}$, this clearly defines a homomorphism of $\mathcal{P}(\mathbb{C}^n)$ into $\mathcal{A}$. Furthermore if $q = (q_1, q_2, \ldots, q_m)$ is an ordered m-tuple of polynomials from $\mathcal{P}(\mathbb{C}^n)$ and p belongs to $\mathcal{P}(\mathbb{C}^m)$, then we have $p(q(a)) = p \circ q(a)$. In order to state a spectral mapping theorem, we need to define the joint spectrum of an ordered n-tuple of commuting elements in an algebra. The same problem, along with many others, arises in trying to extend this functional calculus to some class of analytic functions. We will now turn to the preliminaries needed before the construction of a functional calculus for analytic functions of ordered n-tuples of commuting elements.

3.5.1 Definition The *polynomially convex hull* of a compact subset K of $\mathbb{C}^n$ is the set $\mathrm{pc}(K) \supseteq K$ of all $\lambda \in \mathbb{C}^n$ satisfying

$$|p(\lambda)| \leq \| \, p|_K \, \|_\infty \qquad \forall \, p \in \mathcal{P}(\mathbb{C}^n).$$

A set $K \subseteq \mathbb{C}^n$ is said to be *polynomially convex* if it satisfies $K = \mathrm{pc}(K)$. A *compact polynomial polyhedron* in $\mathbb{C}^n$ is a set of the form

$$\{\lambda \in \mathbb{C}^n : |\lambda_j| \leq M, |p_k(\lambda)| \leq M \text{ for } j = 1, 2, \ldots, n$$

$$\text{and } k = 1, 2, \ldots, m\}$$

where M is a constant and each p_k belongs to $\mathcal{P}(\mathbb{C}^n)$.

For any compact $K \subseteq \mathbb{C}^n$, we can write

$$\mathrm{pc}(K) = \bigcap_{p \in \mathcal{P}(\mathbb{C}^n)} \{\lambda \in \mathbb{C}^n : |p(\lambda)| \leq \| \, p|_K \, \|_\infty\}.$$

Hence, $\mathrm{pc}(K)$ is again compact. It is also immediate that a compact polynomial polyhedron is polynomially convex. We now show that the converse is almost true in a certain sense.

3.5.2 Lemma *If K is a compact polynomially convex set in $\mathbb{C}^n$ and $U \subseteq \mathbb{C}^n$ is a neighborhood of K, then there is some compact polynomial polyhedron included in U and including K in its interior.*

Proof Choose M sufficiently large so that the compact set K is included in the interior of the closed unit ball B of radius M centered at the origin. For each $\zeta \in B \setminus U$, the polynomial convexity of $K \subseteq U$ allows us to choose a polynomial p^ζ which will satisfy $|p^\zeta(\zeta)| > M > \|p^\zeta|_K\|_\infty$. For each $\zeta \in B \setminus U$, let U^ζ be the open set $\{\lambda \in \mathbb{C}^n : |p^\zeta(\lambda)| > M\}$. Then $\{U^\zeta : \zeta \in B \setminus U\}$ is an open cover for the compact set $B \setminus U$. Let $U^{\zeta_1}, U^{\zeta_2}, \ldots, U^{\zeta_m}$ be a finite subcover. Then $\{\lambda \in \mathbb{C}^n : |\lambda_j| \leq M;\ |p^{\zeta_k}(\lambda)| \leq M; j = 1, 2, \ldots, n;\ k = 1, 2, \ldots, m\}$ satisfies the conclusion of this lemma.   □

In addition to polynomial polyhedra, certain other types of compact subsets of $\mathbb{C}^n$ are known to be polynomially convex (Eva Kallin [1965]). Any compact convex set is polynomially convex. Hence, the polynomially convex hull of a compact set is included in the convex hull. Surprisingly, the union of any two disjoint compact convex sets is polynomially convex, but the union of three disjoint closed *polydiscs* (*i.e.*, sets of the form $\{\lambda \in \mathbb{C}^n : |\lambda_j - \zeta_j| \leq r_j, j = 1, 2, \ldots, n\}$ where $\zeta \in \mathbb{C}^n$ and $r = (r_1, r_2, \ldots, r_n)$ is an ordered n-tuple of positive numbers called the *polyradius* of the polydisc) is not always polynomially convex. The union of up to three disjoint balls $\{\lambda \in \mathbb{C}^n : (\sum_{j=1}^n |\lambda_j - \zeta_j|^2)^{1/2} < r\}$ is polynomially convex but the situation for four balls is unknown. For $n > 1$, polynomial convexity is not very well understood. However, for $n = 1$ it is not hard to show that a compact set $K \subseteq \mathbb{C}$ is polynomially convex if and only if $\mathbb{C} \setminus K$ is connected. At this point we will state an important approximation theorem due to Oka [1936].

3.5.3 Theorem *Let $K \subseteq \mathbb{C}^n$ be a compact polynomially convex set. Any function analytic in a neighborhood of K can be uniformly approximated on K by polynomials.*

The proof of this theorem will be deferred. It will be proved in the course of establishing the existence of the multivariable functional calculus just as Runge's approximation theorem, Theorem 3.3.3, was proved in the course of establishing the one-variable functional calculus. The special case of Theorem 3.5.3 for $n = 1$ is just the last sentence of Theorem 3.3.3.

The Joint Spectrum

We now introduce a notation for the linear space of ordered n-tuples of elements of an algebra and define the joint spectrum.

3.5.4 Definition For any algebra $\mathcal{A}$, let $\mathcal{A}^{(n)}$ be the set of ordered n-tuples of elements in $\mathcal{A}$. We will consider $\mathcal{A}^{(n)}$ as a vector space under pointwise operations, but never as an algebra (unless $n = 1$ in which case we identify $\mathcal{A}^{(1)}$ and $\mathcal{A}$). For any unital algebra $\mathcal{A}$ and for any $a =$

$(a_1, a_2, \ldots, a_n) \in \mathcal{A}^{(n)}$, the *joint spectrum* of a, denoted by $Sp_{\mathcal{A}}(a) = Sp(a)$, is the set of $\{\lambda \in \mathbb{C}^n\}$ such that either the left or the right ideal generated by $\{\lambda_1 - a_1, \lambda_2 - a_2, \ldots, \lambda_n - a_n\}$ is proper. For $a \in \mathcal{A}^{(n)}$ let $\mathcal{N}_a$ be the collection of all open subsets of $\mathbb{C}^n$ which include $Sp(a)$.

When $n = 1$, this definition of $Sp(a)$ agrees with our previous definition. The next lemma shows that the joint spectrum is nonempty, compact and satisfies the spectral mapping theorem for the polynomial functional calculus if $\mathcal{A}$ is a unital, commutative, spectral algebra.

3.5.5 Lemma *Let $\mathcal{A}$ be a unital, commutative, spectral algebra. Any element a of $\mathcal{A}^{(n)}$ satisfies:*

$$
\begin{aligned}
Sp(a) &= \{\lambda \in \mathbb{C}^n \colon \{\lambda_1 - a_1, \lambda_2 - a_2, \ldots, \lambda_n - a_n\} \\
&\qquad \text{is included in some proper ideal } \mathcal{I} \text{ of } \mathcal{A}\} \\
&= \{\lambda \in \mathbb{C}^n \colon \text{ no element } b \text{ in } \mathcal{A}^{(n)} \text{ satisfies } 1 = \sum_{j=1}^{n}(\lambda_j - a_j)b_j\} \\
&= \{\gamma(a) : \gamma \in \Gamma_{\mathcal{A}}\}.
\end{aligned}
$$

Proof The first two sets obviously equal $Sp(a)$. If λ belongs to the second set, then Theorem 2.4.6 gives a maximal modular ideal $\mathcal{A}$ which includes $\{\lambda_1 - a_1, \lambda_2 - a_2, \ldots, \lambda_n - a_n\}$. Proposition 3.1.3 and Lemma 3.1.4 then show that there is a homomorphism $\gamma \in \Gamma_{\mathcal{A}}$ satisfying $\gamma(a) = \lambda$. Thus, the second set is included in $Sp(a)$. The opposite inclusion is established by choosing $\mathcal{I} = \ker(\gamma)$ where $\gamma \in \Gamma_{\mathcal{A}}$ satisfies $\lambda - \gamma(a)$. □

The first statement in the following result is a generalization of a special case of Corollary 2.5.8. The rest of this result is a partial generalization of Theorem 3.1.15 on singly generated unital spectral normed algebras. Let $(\mathcal{A}, \|\cdot\|)$ be a spectral normed algebra. Let $a_1, a_2, \ldots, a_n$ belong to $\mathcal{A}$. If $\mathcal{A}$ is the smallest unital closed subalgebra of itself which includes $\{a_1, a_2, \ldots, a_n\}$, then this set is said to *generate $\mathcal{A}$*. If such a finite set of generators can be found, then $\mathcal{A}$ is said to be *finitely generated*.

3.5.6 Proposition *Let $(\mathcal{A}, \|\cdot\|)$ be a unital commutative spectral normed algebra. Let $a = (a_1, a_2, \ldots, a_n)$ belong to $\mathcal{A}^{(n)}$, and let $\mathcal{C}$ be the unital closed subalgebra of $\mathcal{A}$ generated by $\{a_1, a_2, \ldots, a_n\}$. Then a satisfies*

$$Sp_{\mathcal{A}}(a) \subseteq Sp_{\mathcal{C}}(a) \subseteq \mathrm{pc}(Sp_{\mathcal{A}}(a)). \tag{5}$$

If $\mathcal{A}$ itself is generated by $\{a_1, a_2, \ldots, a_n\}$, then the map $\hat{a}: \Gamma_{\mathcal{A}} \to \mathbb{C}^n$ defined by

$$\hat{a}(\gamma) = (\gamma(a_1), \gamma(a_2), \ldots, \gamma(a_n))$$

is a homeomorphism onto its range $Sp_{\mathcal{A}}(a)$ which is polynomially convex.

Proof The following trivial observation proves the first inclusion

$$Sp_{\mathcal{A}}(a) = \{\gamma(a): \gamma \in \Gamma_{\mathcal{A}}\} = \{(\gamma|_C)(a): \gamma \in \Gamma_{\mathcal{A}}\}$$
$$\subseteq \{\gamma(a): \gamma \in \Gamma_C\} = Sp_C(a).$$

The second inclusion is established by the following computation where $\lambda \in Sp_C(a)$ and $p \in \mathcal{P}(\mathbb{C}^n)$ are arbitrary:

$$|p(\lambda)| \leq \max\{|p(\gamma(a))|: \gamma \in \Gamma_C\} = \max\{|\gamma(p(a))|: \gamma \in \Gamma_C\} = \rho_C(p(a))$$
$$= \rho_{\mathcal{A}}(p(a)) = \max\{|\gamma(p(a))|: \gamma \in \Gamma_{\mathcal{A}}\} = \max\{|p(\gamma(a))|: \gamma \in \Gamma_{\mathcal{A}}\}$$
$$= \|p|_{Sp_{\mathcal{A}}(a)}\|_\infty.$$

Suppose $\mathcal{A}$ is generated by $\{a_1, a_2, \ldots, a_n\}$. Any element $\gamma \in \Gamma_{\mathcal{A}}$ is determined by its values on these elements so that $\hat{a}$ is one-to-one. Thus, $\hat{a}$ is a homeomorphism since it is a continuous map from a compact space to a Hausdorff space.

Any polynomial $p \in \mathcal{P}(\mathbb{C}^N)$ and $\lambda_0 \in \mathrm{pc}Sp(a)$ satisfy

$$|p(\lambda_0)| \leq \max\{|p(\lambda)| : \lambda \in Sp(a)\} = \max\{|p(\gamma)| : \gamma \in \Gamma_{\mathcal{A}}\} \leq \|p(a)\|.$$

Hence, just as in the proof of Theorem 3.1.15, we can define a homomorphism $\varphi: \mathcal{A}_0 \to \mathbb{C}$ by $\varphi(p(a)) = p(\lambda_0)$ where $\mathcal{A}_0$ is the dense (not necessarily closed) unital subalgebra of $\mathcal{A}$ generated by $\{a_1, a_2, \ldots, a_n\}$. Since φ is contractive, it can be extended to all of $\mathcal{A}$ into $\mathbb{C}$. Hence, $\lambda_0 \in Sp(a)$ so $Sp(a)$ is polynomially convex. □

The Algebra $P(K)$

The algebra $P(K)$ of all uniform limits of polynomials on a compact subset K of $\mathbb{C}^n$, with pointwise algebraic operations and the supremum norm, is an example of a finitely generated unital commutative Banach algebra. This algebra is generated as a unital Banach algebra by the components of the coordinate function $z = (z_1, z_2, \ldots, z_n)$. Thus, the last proposition shows that the map $\hat{z}$ is a homeomorphism of $\Gamma_{P(K)}$ onto the polynomially convex set $Sp(z)$. Each point $\lambda \in K$ clearly belongs to $Sp(z)$. Hence $Sp(z)$ includes $\mathrm{pc}(K)$. However, any λ not contained in $\mathrm{pc}(K)$ satisfies $|p(\lambda)| > \|p|_K\|_\infty$ for some polynomial p. The right side of this inequality is simply the norm of p considered as an element of $P(K)$. If λ belongs to $Sp(z)$, this contradicts Theorem 3.1.2(d). Thus, $Sp(z)$ equals $\mathrm{pc}(K)$ which is homeomorphic to $\Gamma_{P(K)}$ under the map $\lambda \mapsto \gamma_\lambda$ where $\gamma_\lambda(f) = f(\lambda)$.

Now suppose K is already polynomially convex. For any polynomial p, considered as an element of $P(K)$, we have

$$\hat{p} = p(\hat{z}) = p \circ \hat{z}$$

since ($\hat{}$) is a homomorphism. Since $\|f\| = \|f^n\|^{n^{-1}}$ holds for any $f \in P(K)$ and any $n \in \mathbb{N}$, the Gelfand map ($\hat{}$) is an isometry and hence injective. Thus, the density of the polynomials in $P(K)$ proves

$$\hat{f} = f \circ \hat{z} \text{ for all } f \in P(K).$$

We summarize these results and a few additional comments in a proposition.

3.5.7 Proposition *Let K be a compact subset of $\mathbb{C}^n$ and let z be the coordinate function on $\mathbb{C}^n$ considered as an element in $P(K)^{(n)}$. Then $\hat{z}$ is a homeomorphism of $\Gamma_{P(K)}$ onto $\mathrm{pc}(K)$. If K is polynomially convex, $\hat{z}$ satisfies*

$$\hat{f} = f \circ \hat{z} \qquad f = \hat{f} \circ (\hat{z})^{\leftarrow} \qquad \forall \ f \in P(K).$$

Furthermore, a unital Banach algebra $\mathcal{A}$ is isometrically isomorphic to $P(K)$ for some polynomially convex set K if and only if $\mathcal{A}$ is finitely generated and satisfies $\|a\|^2 = \|a^2\|$ for all $a \in \mathcal{A}$.

Proof We have just proved the first two statements of this proposition. Clearly any Banach algebra which is isometrically isomorphic to $P(K)$ satisfies the stated conditions.

Suppose $\mathcal{A}$ is a finitely generated unital Banach algebra and satisfies $\|a\|^2 = \|a^2\|$ for all $a \in \mathcal{A}$. Then $\mathcal{A}$ is commutative by Proposition 3.1.8 and satisfies $\|\hat{a}\|_\infty = \rho(a) = \|a\|^\infty = \|a\|$ for all $a \in \mathcal{A}$. Hence, the Gelfand map is an isometric isomorphism of $\mathcal{A}$ onto $\hat{\mathcal{A}} \subseteq \mathbb{C}(\Gamma_{\mathcal{A}})$.

Let $\mathcal{A}$ be generated by $\{a_1, a_2, \ldots, a_n\}$ and let $a \in \mathcal{A}^{(n)}$ be the element $a = (a_1, a_2, \ldots, a_n)$. Then Proposition 3.5.6 shows that $\hat{a}$ is a homeomorphism of $\Gamma_{\mathcal{A}}$ onto the polynomially convex set $Sp(a)$. Hence, the map

$$c \to \hat{c} \circ (\hat{a}^{\leftarrow}) \qquad c \in \mathcal{A}$$

is an isometric isomorphism of $\mathcal{A}$ onto $P(Sp(a))$. $\qquad\qquad\square$

The Arens–Calderon Theorem

We need a little more technical notation. For any unital commutative Banach algebra $\mathcal{A}$, let $\mathcal{A}^{(\infty)}$ be the disjoint union of $\mathcal{A}^{(n)}$ for $n \in \mathbb{N}$. We regard each $\mathcal{A}^{(n)}$ as a vector space under the obvious pointwise operations. For $a = (a_1, a_2, \ldots, a_n) \in \mathcal{A}^{(n)}$ and $b = (b_1, b_2, \ldots, b_m) \in \mathcal{A}^{(m)}$, we define (a, b) to be the element $(a_1, a_2, \ldots, a_n, b_1, b_2, \ldots, b_m)$ in $\mathcal{A}^{(n+m)}$. This binary operation on $\mathcal{A}^{(\infty)}$ is associative. For any $n < m$, let $\pi_{m,n} \colon \mathcal{A}^{(m)} \to \mathcal{A}^{(n)}$ be defined by

$$\pi_{m,n}(a) = \pi_{m,n}(a_1, a_2, \ldots, a_m) = (a_1, a_2, \ldots, a_n)$$

and let $\pi'_{m,n} \colon \mathcal{A}^{(m)} \to \mathcal{A}^{(m-n)}$ be defined by

$$\pi'_{m,n}(a) = \pi'_{m,n}(a_1, a_2, \ldots, a_m) = (a_{n+1}, a_{n+2}, \ldots, a_m).$$

We regard $\mathbb{C}^n$ as embedded in $\mathcal{A}^{(n)}$ so that for any $\lambda \in \mathbb{C}^m$, we have

$$\pi_{m,n}(\lambda) = \pi_{m,n}(\lambda_1, \lambda_2, \dots, \lambda_m) = (\lambda_1, \lambda_2, \dots, \lambda_n), \quad \text{and}$$

$$\pi'_{m,n}(\lambda) = \pi'_{m,n}(\lambda_1, \lambda_2, \dots, \lambda_m) = (\lambda_{n+1}, \lambda_{n+2}, \dots, \lambda_m).$$

The following proposition is due to Arens and Calderon [1955]. It will play a key role in our proof that the multivariable functional calculus has the desired properties.

3.5.8 Arens–Calderon Theorem *Let $\mathcal{A}$ be a unital commutative Banach algebra. Any $m, n \in \mathbb{N}$ (satisfying $m > n$), $a \in \mathcal{A}^{(n)}$ and $b \in \mathcal{A}^{(m-n)}$ also satisfy*

$$\pi_{m,n}Sp(a,b) = Sp(a), \qquad \text{and} \tag{6}$$

$$Sp(a) \subseteq \pi_{m,n}\mathrm{pc}(Sp(a,b)) \subseteq \mathrm{pc}(Sp(a)). \tag{7}$$

Furthermore if $n \in \mathbb{N}, a \in \mathcal{A}^{(n)}$ and a neighborhood U of $Sp(a)$ in $\mathbb{C}^n$ are given, it is always possible to find $m \in \mathbb{N}$ and $b \in \mathcal{A}^{(m-n)}$ satisfying

$$\pi_{m,n}\mathrm{pc}(Sp(a,b)) \subseteq U. \tag{8}$$

Proof The first claim follows from the observation

$$Sp(a) = \{\gamma(a)\colon \gamma \in \Gamma_{\mathcal{A}}\} = \pi_{m,n}\{\gamma(a,b)\colon \gamma \in \Gamma_{\mathcal{A}}\} = \pi_{m,n}Sp(a,b)$$

where n, m, a and b satisfy the hypotheses. From this result we can conclude $Sp(a) = \pi_{m,n}(Sp(a,b)) \subseteq \pi_{m,n}\mathrm{pc}(Sp(a,b))$. Now for any compact set $K \subseteq \mathbb{C}^m$, we have

$$\begin{aligned}
\pi_{m,n}\mathrm{pc}(K) &= \pi_{m,n}\{\lambda \in \mathbb{C}^m\colon |p(\lambda)| \le \|p|_K\|_\infty \text{ for all } p \in \mathcal{P}(\mathbb{C}^m)\} \\
&\subseteq \pi_{m,n}\{\lambda \in \mathbb{C}^m\colon |p(\lambda)| \le \|p|_K\|_\infty \\
&\qquad\qquad \text{for all } p = q \circ \pi_{m,n} \text{ for } q \in \mathcal{P}(\mathbb{C}^n)\} \\
&= \mathrm{pc}(\pi_{m,n}(K)).
\end{aligned}$$

Consideration of the case $K = Sp(a,b)$ gives

$$\pi_{m,n}\mathrm{pc}(Sp(a,b)) \subseteq \mathrm{pc}(\pi_{m,n}Sp(a,b)) = \mathrm{pc}(Sp(a)).$$

We now prove the last assertion of the proposition. Let $n \in \mathbb{N}, a \in \mathcal{A}^{(n)}$ and $U \in \mathcal{N}_a$ be arbitrary. By (7) we need only choose b to exclude each $\zeta \in \mathrm{pc}(Sp(a)) \setminus U$ from $\pi_{m,n}\mathrm{pc}(Sp(a,b))$. Suppose such a ζ is given. Since ζ does not belong to U, it does not belong to $Sp(a)$. Hence, Lemma 3.5.5 shows that there is an element $b^\zeta \in \mathcal{A}^{(n)}$ satisfying

$$\sum_{j=1}^{n} (\zeta_j - a_j) b_j^\zeta = 1.$$

Hence, ζ does not belong to the spectrum of a in the unital subalgebra C generated by $\{a_1, a_2, \ldots, a_n, b_1^\zeta, b_2^\zeta, \ldots, b_n^\zeta\}$. This, together with the first inclusion of (5) and (6), implies $\zeta \notin \pi_{2n,n} Sp_C(a, b^\zeta)$. However, the second paragraph of Proposition 3.5.6 shows $Sp_C(a, b^\zeta) = \mathrm{pc}(Sp_A(a, b^\zeta))$ so ζ is not contained in $\pi_{2n,n} \mathrm{pc}(Sp_A(a, b^\zeta))$. This set is closed since $\pi_{2n,n}$ is continuous and $\mathrm{pc}(Sp_A(a, b^\zeta))$ is compact. Thus the complement U^ζ of this set is a neighborhood of ζ. The family $\{U^\zeta : \zeta \in \mathrm{pc}(Sp(a)) \setminus U\}$ is an open cover for the compact set $\mathrm{pc}(Sp(a)) \setminus U$. Choose a finite subcover $\{U^{\zeta_j} : j = 1, 2, \ldots, p\}$. We will now show that $m = n(p+1)$ and $b = (b^{\zeta_1}, b^{\zeta_2}, \ldots, b^{\zeta_p}) \in \mathcal{A}^{(m-n)}$ satisfy (8).

Any $\zeta \in \mathrm{pc}(Sp(a)) \setminus U$ belongs to some U^{ζ_k}. Thus, ζ does not belong to $\pi_{2n,n} \mathrm{pc}(Sp(a, b^{\zeta_k}))$. Let C_k be the unital closed subalgebra of $\mathcal{A}$ generated by $\{a_1, a_2, \ldots, a_n, b_1^{\zeta_k}, b_2^{\zeta_k}, \ldots, b_n^{\zeta_k}\}$ and let C be the unital closed subalgebra of $\mathcal{A}$ generated by $\{a_1, a_2, \ldots, a_n, b_1, b_2, \ldots, b_{m-n}\}$. Then, ζ does not belong to $\pi_{2n,n} Sp_{C_k}(a, b^{\zeta_k}) = Sp_{C_k}(a)$ which includes $Sp_C(a) = \pi_{m,n} Sp_C(a, b) = \pi_{m,n} \mathrm{pc}(Sp(a, b))$. Thus, no $\zeta \in \mathrm{pc}(Sp(a, b)) \setminus U$ belongs to $\pi_{m,n} \mathrm{pc}(Sp(a, b))$. $\square$

Multivariable Functional Calculus

We generalize the notation introduced in Section 3.3. For each $a \in \mathcal{A}^{(\infty)}$, let $\widetilde{\mathcal{F}}_a$ be the set $\{f : f$ is analytic on its domain D_f which belongs to $\mathcal{N}_a\}$. Note that if $a \in \mathcal{A}^{(\infty)}$ belongs to $\mathcal{A}^{(n)}$, then $Sp(a)$ is a compact subset of $\mathbb{C}^n$. Define a relation $\sim$ on $\widetilde{\mathcal{F}}_a$ by:

$$f \sim g \Leftrightarrow \text{ there exists a } U \in \mathcal{N}_a \text{ satisfying } f|_U = g|_U.$$

This is an equivalence relation and the set of equivalence classes is denoted by $\mathcal{F}_a$. As before, we do not distinguish in notation between $f \in \mathcal{F}_a$ and a representative of f in $\widetilde{\mathcal{F}}_a$. It is again easy to check that $\mathcal{F}_a$ is an algebra under the operations inherited from the following operations in $\widetilde{\mathcal{F}}_a$:

$$
\begin{aligned}
f + g &= f|_{D_f \cap D_g} + g|_{D_f \cap D_g} \\
fg &= (f|_{D_f \cap D_g})(g|_{D_f \cap D_g})
\end{aligned}
$$

where the right hand side has the customary (pointwise) meaning.

Finally, we make $\mathcal{F}_a$ into a topological algebra by defining $f_\alpha \to f$ to mean that there is a neighborhood U of $Sp(a)$ such that for some α_0 and some representatives f and f_α in $\widetilde{\mathcal{F}}_a$

$$U \subseteq D_f \cap \bigcap_{\alpha > \alpha_0} D_{f_\alpha}$$

holds and f_α converges to f pointwise on U and uniformly on any compact subset of U.

The following theorem is essentially due to Waelbroeck [1954]. A theory with some of the same features was developed by Šilov [1953] and Arens and Calderon [1955]. However, their construction was not shown to define a homomorphism except in the case of semisimple Banach algebras. The proof that we give is based upon Arens [1961], Nicolas Bourbaki [1967], and Gamelin [1969].

Note that we have not generalized the first two properties of (d) in Theorem 3.3.7. Moreover, the uniqueness proved here is definitely a property of the family $\{I_a : a \in \mathcal{A}^\infty\}$, whereas in Theorem 3.3.7 each separate map I_a was proved unique. Zame [1979] has shown that the individual maps are unique in this multivariable functional calculus, but the proof is not included here since it requires substantial preparation. If we restrict attention to the functional calculus in a semisimple commutative Banach algebra $\mathcal{A}$, any homomorphism from $\mathcal{F}_a$ into $\mathcal{A}$ composed with an element of $\Gamma_\mathcal{A}$ is necessarily continuous, so the closed graph theorem shows that the original homomorphism was continuous. Zame [1979] introduces a transfinite construction to extend this uniqueness result somewhat. Angel Larotonda and Ignacio Zalduendo [1984] give an easier construction. See also Schanuel and Zame [1982] for a functorial interpretation.

3.5.9 Theorem *For any unital commutative Banach algebra $\mathcal{A}$, there is a unique family $\{I_a : a \in \mathcal{A}^{(\infty)}\}$ of continuous linear maps*

$$I_a : \mathcal{F}_a \to \mathcal{A}$$

satisfying

(a) $I_a(p) = p(a) \qquad \forall\, n \in \mathbb{N};\ a \in \mathcal{A}^{(n)};\ p \in \mathcal{P}(\mathbb{C}^n).$

(b) $I_{\pi_{m,n}(a)}(f) = I_a(f \circ \pi_{m,n}) \qquad \forall\, \begin{cases} m,\, n \in \mathbb{N} & \text{with } m > n; \\ a \in \mathcal{A}^{(m)};\ f \in \mathcal{F}_{\pi_{m,n}(a)}. \end{cases}$

Furthermore, for each $n \in \mathbb{N}$, $a \in \mathcal{A}^{(n)}$ and $f \in \mathcal{F}_a$, this family also satisfies:

(c) I_a *is a continuous homomorphism.*

(d) *If $\mathcal{C}$ is another unital commutative Banach algebra, then any continuous unital homomorphism $\varphi : \mathcal{A} \to \mathcal{C}$ satisfies*

$$\varphi(I_a(f)) = I_{\varphi(a)}(f).$$

(e) $\widehat{I_a(f)} = f \circ \hat{a}.$

(f) $Sp(I_a(f)) = f(Sp(a)).$

(g) *For all $h \in \mathcal{F}_a^{(m)}$ (cf. Definition 3.5.4) and $g \in \mathcal{F}_{I_a(h)}$, $g \circ h$ belongs to $\mathcal{F}_a$ and satisfies $I_a(g \circ h) = I_{I_a(h)}(g).$*

(h) *If a representative of f can be chosen in $\widetilde{\mathcal{F}}_a$ so that D_f includes a polynomially convex compact neighborhood K of $Sp(a)$, then $I_a(f)$ belongs to the unital closed subalgebra generated by $\{a_1, a_2, \ldots, a_n\}$. If, in addition,*

K contains the origin 0 *in* $\mathbb{C}^n$ *and* $f(0) = 0$, *then* $I_a(f)$ *belongs to the unital closed subalgebra generated by* $\{a_1, a_2, \ldots, a_n\}$.

(i) *Each* $m \in \mathbb{N}$, $b \in \mathcal{A}^{(m)}$ *and* $g \in \mathcal{F}_b$ *satisfy*

$$I_{(a,b)}(h) = I_a(f)I_b(g)$$

where $h(\lambda_1, \lambda_2, \ldots, \lambda_{n+m}) = f(\lambda_1, \lambda_2, \ldots, \lambda_n)g(\lambda_{n+1}, \lambda_{n+2}, \ldots, \lambda_{n+m})$.

(j) *Each* $b \in \mathcal{A}$ *and* $g \in \mathcal{F}_b$ *satisfy*

$$I_b(g) = g(b).$$

Proof For this proof we fix $n \in \mathbb{N}$, a Banach algebra $\mathcal{A}$ satisfying the hypotheses, and $a \in \mathcal{A}^{(n)}$. For any neighborhood U of $Sp(a)$, we make the proof depend on the existence of a continuous (actually C^∞) $\mathcal{A}$-valued function v with compact support in U, which is independent of U and of the (many) choices made in its construction in the following weak sense: If $\tilde{U}$ is another neighborhood of $Sp(a)$ with $\tilde{U} \subseteq U$ and if $\tilde{v}$ is constructed from $\tilde{U}$ by the same procedure (which includes various choices) by which v was constructed from U, then

$$\int_{\mathbb{C}^n} fv\,dV = \int_{\mathbb{C}^n} f\tilde{v}\,dV \tag{9}$$

holds for all functions f analytic on U where the product fv is the pointwise scalar product and dV represents $2n$-dimensional Lebesgue measure. We then define $I_a: \mathcal{F}_a \to \mathcal{A}$ by

$$I_a(f) = \int_{\mathbb{C}^n} fv\,dV \tag{10}$$

where we have chosen a representative $f \in \tilde{\mathcal{F}}_a$ of $f \in \mathcal{F}_a$ and constructed v from D_f. This is well-defined (as a function of $f \in \mathcal{F}_a$) by (9). It obviously makes I_a a continuous linear map. The construction of v shows that I_a satisfies (e), (i), and (j). The construction of v and the verification of (e), (i), and (j) depend on Stokes' theorem for exterior differential forms on $\mathbb{C}^n$ and lie entirely outside the realm of Banach algebra theory. We postpone these steps. Readers who wish to avoid them may simply assume the existence of a function v with the above properties. We shall now prove Theorem 3.5.9 starting from the following axiom to be verified later.

Axiom. *For each* $a \in \mathcal{A}^{(\infty)}$, *there is at least one continuous linear map* $I_a: \mathcal{F}_a \to \mathcal{A}$ *which satisfies properties* (e), (i) *and* (j).

Note first that, by definition of the joint spectrum, (f) is an immediate consequence of (e). Furthermore, (j) implies $I_{a_j}(1) = 1$ for $j = 1, 2, \ldots, n$.

This result together with repeated application of (i) implies $I_a(1) = 1$. From this, we can conclude (b) as the special case of (i) when the function g is 1. Condition (j) also implies $I_{a_j}(z) = a_j$ for $j = 1, 2, \ldots, n$ where z is the coordinate function on $\mathbb{C}$. This result, our previous results, and (i) imply

$$I_a(z_j) = a_j$$

for $j = 1, 2, \ldots, n$ where $z = (z_1, z_2, \ldots, z_n)$ is the coordinate function on $\mathbb{C}^n$. Now, this result together with (i) implies

$$I_a(p) = p(a) \tag{11}$$

where p is any monomial. The linearity of I_a now gives (11) for any polynomial $p \in \mathcal{P}(\mathbb{C}^n)$. We have proved (a).

We have now proved or assumed that $\{I_a : a \in \mathcal{A}^{(\infty)}\}$ is a family of continuous linear maps satisfying (a), (b), (e), (f), (i), and (j). It only remains to show the uniqueness of the family $\{I_a : a \in \mathcal{A}^{(\infty)}\}$, and properties (c), (d), (g), and (h).

We will turn to the proof of uniqueness as soon as we have proved the Oka approximation theorem which was stated as Theorem 3.5.3 above.

Proof of Theorem 3.5.3. Let $K \subseteq \mathbb{C}^n$ be compact and polynomially convex. We wish to show that any complex-valued function f which is analytic in a neighborhood of K belongs to the Banach algebra $P(K)$ of uniform limits of polynomials on K. The components $z_1, z_2, \ldots, z_n$ of the coordinate function generate $P(K)$ as a unital Banach algebra. Hence, Proposition 3.5.7 shows that $Sp(z)$ equals K and that $\hat{z}$ is a homeomorphism of $\Gamma_{\mathcal{A}}$ onto K. Thus, property (e) shows $\widehat{I_z(f)} = f \circ \hat{z}$. Proposition 3.5.7 now shows that $f = \widehat{I_z(f)} \circ (\hat{z}^{\leftarrow})$ belongs to $P(K)$. This completes the proof of Theorem 3.5.3. □

Proof of Theorem 3.5.9, continued. We now prove the uniqueness assertion. Suppose $\{\tilde{I}_a : a \in \mathcal{A}^{(\infty)}\}$ is another family of continuous linear maps $\tilde{I}_a : \mathcal{F}_a \to \mathcal{A}$, satisfying (a) and (b). Let $f \in \mathcal{F}_a$ be arbitrary but fixed. Choose a representative of f in $\tilde{\mathcal{F}}_a$. By Theorem 3.5.8, we can choose $b \in \mathcal{A}^{(m-n)}$ satisfying

$$Sp(a) \subseteq \pi_{m,n}\mathrm{pc}(Sp(a,b)) \subseteq D_f.$$

By Lemma 3.5.2 we can find a polynomial convex set K in $\mathbb{C}^m$ satisfying $\mathrm{pc}(Sp(a,b)) \subseteq K^0$ and $\pi_{m,n}(K) \subseteq D_f$. By Theorem 3.5.3, we can find a sequence $\{p_k\}_{k \in \mathbb{N}}$ of polynomials converging to $f \circ \pi_{m,n}$ uniformly on K. Continuity and (b) imply

$$\tilde{I}_a(f) = \tilde{I}_{\pi_{m,n}(a,b)}(f) = \tilde{I}_a(f \circ \pi_{m,n}) = \lim \tilde{I}_a(p_k) = \lim I_a(p_k)$$
$$= I_a(f \circ \pi_{m,n}) = I_a(f).$$

Hence, $\{I_a : a \in \mathcal{A}^{(\infty)}\}$ is unique, subject to the restriction that each $I_a : \mathcal{F}_a \to \mathcal{A}$ is a continuous linear map satisfying (a) and (b).

(d): The family of maps $\{\varphi \circ I_a : a \in \mathcal{A}^{(\infty)}\}$ satisfies the properties of $\{I_{\varphi(a)} : a \in \mathcal{A}^{(\infty)}\}$ except that each

$$\varphi \circ I_a : \mathcal{F}_a \to \mathcal{A}$$

is defined on the algebra $\mathcal{F}_a$ which is possibly smaller than the algebra $\mathcal{F}_{\varphi(a)}$ on which $I_{\varphi(a)}$ is defined. Since $\mathcal{F}_a$ includes the polynomials, the above uniqueness argument proves (d).

(c): Of course, only the multiplicative property of each I_a remains to be proved. Let $f, g \in \mathcal{F}_a$ be given. Choose representatives of f and g in $\tilde{\mathcal{F}}_a$. By Theorem 3.5.8, choose $b \in \mathcal{A}^{(m-n)}$ satisfying $\pi_{m,n}\mathrm{pc}(Sp(a,b)) \subseteq D_f \cap D_g$. Use Lemma 3.5.2 to choose a polynomially convex set K in $\mathbb{C}^m$ satisfying $\mathrm{pc}(Sp(a,b)) \subseteq K^\circ$ and $\pi_{m,n}(K) \subseteq D_f \cap D_g$. Now, Theorem 3.5.3 shows that we can choose sequences $\{p_k\}_{k \in \mathbb{N}}$ and $\{q_k\}_{k \in \mathbb{N}}$ of polynomials so that $\{p_k\}_{k \in \mathbb{N}}$ converges to $f \circ \pi_{m,n}$ and $\{q_k\}_{k \in \mathbb{N}}$ converges to $g \circ \pi_{m,n}$ uniformly on K. Then we obtain

$$
\begin{aligned}
I_a(fg) &= I_{\pi_{m,n}(a,b)}(fg) = I_{(a,b)}(fg \circ \pi_{m,n}) \\
&= I_{(a,b)}((f \circ \pi_{m,n})(g \circ \pi_{m,n})) = \lim_{k \to \infty} I_{(a,b)}(p_k q_k) \\
&= \lim_{k \to \infty} I_{(a,b)}(p_k) \lim_{k \to \infty} I_{(a,b)}(q_k) = I_{(a,b)}(f \circ \pi_{m,n}) I_{(a,b)}(g \circ \pi_{m,n}) \\
&= I_{\pi_{m,n}(a,b)}(f) I_{\pi_{m,n}(a,b)}(g) = I_a(f) I_a(g).
\end{aligned}
$$

This proves (c).

(h): The first sentence is an immediate consequence of Oka's approximation theorem and the continuity of I_a. The second sentence follows since its hypotheses guarantee that the approximating polynomials can be chosen with constant terms equal to zero.

(g): Let $h \in \mathcal{F}_a^{(m)}$ and $g \in \mathcal{F}_{I_a(h)}$ be given. For the present argument we will use π and π' to denote the maps $\pi_{n+m,n}$ and $\pi'_{n+m,n}$. Let $k : \mathbb{C}^{n+m} \to \mathbb{C}$ be defined by

$$k = g \circ h \circ \pi - g \circ \pi',$$

i.e., $k(\lambda_1, \lambda_2, \ldots, \lambda_{n+m}) = g(h(\lambda_1, \lambda_2, \ldots, \lambda_n)) - g(\lambda_{n+1}, \lambda_{n+2}, \ldots, \lambda_{n+m})$. Then k belongs to $\mathcal{F}_{(a,I_a(h))}$ and satisfies

$$k \circ (z, h \circ \pi) = g \circ h \circ \pi - g \circ h \circ \pi = 0 \qquad (12)$$

where z is the coordinate function on $\mathbb{C}^n$. For $j = 0, 1, 2, \ldots, m$, let $k_j : \mathbb{C}^{n+m} \to \mathbb{C}$ be defined by

$$
\begin{aligned}
k_j &= k \circ (z_1, z_2, \ldots, z_{n+j}, h_{j+1} \circ \pi, h_{j+2} \circ \pi, \ldots, h_m \circ \pi) \\
&= g \circ h \circ \pi - g \circ (z_{n+1}, z_{n+2}, \ldots, z_{n+j}, h_{j+1} \circ \pi, h_{j+2} \circ \pi, \ldots, h_m \circ \pi).
\end{aligned}
$$

Note $k_0 = k \circ (z, h \circ \pi) = 0$ and $k_m = k$. Thus, we can write

$$k = \sum_{j=1}^{m}(k_j - k_{j-1}) = \sum_{j=1}^{m}(z_{n+j} - h_j \circ \pi)\left(\frac{k_j - k_{j-1}}{z_{n+j} - h_j \circ \pi}\right),$$

where each fraction is actually analytic in a neighborhood of $Sp(a, I_a(h))$ since it can be written as

$$(z_{n+j} - h_j \circ \pi)^{-1}[-g \circ (z_{n+1}, z_{n+2}, \ldots, z_{n+j}, h_{j+1} \circ \pi, h_{j+2} \circ \pi, \ldots, h_m \circ \pi)$$

$$+g \circ (z_{n+1}, z_{n+2}, \ldots, z_{n+j-1}, h_j \circ \pi, h_{j+1} \circ \pi, \ldots, h_m \circ \pi)]$$

in which the two terms in the brackets differ only in the $n+j$ position with z_{n+j} in one term and $h_j \circ \pi$ in the other. We already know that $I_{(a,I_a(h))}$ is a homomorphism and that it satisfies

$$\begin{aligned} I_{(a,I_a(h))}(z_{n+j} - h_j \circ \pi) &= I_{(a,I_a(h))}(z_{n+j}) - I_{(a,I_a(h))}(h_j \circ \pi) \\ &= I_a(h_j) - I_a(h_j) = 0 \end{aligned}$$

where we have used (a) on the first term and (b) on the second. Hence, we conclude $0 = I_{(a,I_a(h))}(k) = I_{(a,I_a(h))}(g \circ h \circ \pi) - I_{(a,I_a(h))}(g \circ \pi') = I_a(g \circ h) - I_{I_a(h)}(g)$, where on the first term we have used (b) and on the second term an analogue of (b), which can be proved from (i) just as (b) was. This concludes the proof of (g), and hence of the whole theorem, except for the construction of the function v satisfying (10) and the verification of the properties (e), (i), and (j).   □

Construction of v: Exterior Differential Calculus

We turn now to the construction of the function v associated with a neighborhood U of $Sp(a)$ in $\mathbb{C}^n$ which will allow us to establish the Axiom on which the proof of Theorem 3.5.9 has depended up to now. The next 13 pages are essentially an elementary appendix and can be skipped with little loss of continuity. We begin with a brief review of the differential and integral calculus of C^∞ exterior differential forms on an open subset V of $\mathbb{C}^n$. As a source for further information on this subject at about the level needed here, we mention the elementary book by Nickerson, Spencer, and Steenrod [1959].

Let $C^\infty(V)$ be the set of all complex-valued functions on the open subset V of $\mathbb{C}^n$ which have continuous partial derivatives of all orders with respect to the $2n$ real coordinates in V. The identity function on V will be denoted by $z = (z_1, z_2, \ldots, z_n)$, where $z(\lambda) = \lambda$ for all $\lambda = (\lambda_1, \lambda_2, \ldots, \lambda_n) \in V \subseteq \mathbb{C}^n$. (Thus for $j = 1, 2, \ldots, n$, $z_j: \mathbb{C}^n \to \mathbb{C}$ satisfies $z_j(\lambda) = \lambda_j$ for all $\lambda \in V$.) The partial derivatives with respect to the real and imaginary parts of z_j will be denoted by $\frac{\partial}{\partial x_j}$ and $\frac{\partial}{\partial y_j}$, respectively.

A *derivation* of $C^\infty(V)$ is a linear map $D\colon C^\infty(V) \to C^\infty(V)$ which satisfies

$$D(fg) = D(f)g + fD(g) \qquad \forall\, f, g \in C^\infty(V).$$

The set of all derivations on $C^\infty(V)$ will be denoted by $\Delta(V)$. Pointwise operations on $\Delta(V)$ make it into a linear space. For $f \in C^\infty(V)$ and $D \in \Delta(V)$, we define fD by

$$(fD)(g) = f(D(g)) \qquad \forall\, g \in C^\infty(V).$$

This makes $\Delta(V)$ into a $C^\infty(V)$-module. It can be shown that the set $\{\frac{\partial}{\partial x_j}, \frac{\partial}{\partial y_j} : j = 1, 2, \ldots, n\}$ is a free module basis for $\Delta(V)$ over $C^\infty(V)$. However, we will usually use another module basis. For $j = 1, 2, \ldots, n$, define elements $\frac{\partial}{\partial z_j}$ and $\frac{\partial}{\partial \bar{z}_j}$ of $\Delta(V)$ by $\frac{\partial}{\partial z_j} = \frac{1}{2}(\frac{\partial}{\partial x_j} + \frac{1}{i}\frac{\partial}{\partial y_j})$ and $\frac{\partial}{\partial \bar{z}_j} = \frac{1}{2}(\frac{\partial}{\partial x_j} - \frac{1}{i}\frac{\partial}{\partial y_j})$. If f is analytic on V, an easy calculation shows $\frac{\partial f}{\partial z_j}$ is the ordinary complex derivative with respect to the jth-coordinate. Furthermore, the equations $\frac{\partial f}{\partial \bar{z}_j} = 0$ for $j = 1, 2, \ldots, n$ are equivalent to the Cauchy–Riemann equations. Hence, these equations hold if and only if f is analytic on V by Hartog's theorem (which asserts that a function on $\mathbb{C}^n$ is analytic if and only if it is analytic in each variable separately, *cf.* Hormander, [1966]). From previous remarks $\{\frac{\partial}{\partial z_j}, \frac{\partial}{\partial \bar{z}_j} : j = 1, 2, \ldots, n\}$ is clearly a free module basis for $\Delta(V)$ over $C^\infty(V)$.

The space $\mathcal{D}^1(V)$ of *differential 1-forms* on V is by definition the module dual of $\Delta(V)$. That is, an element σ of $\mathcal{D}^1(V)$ is a linear function $\sigma\colon \Delta(V) \to C^\infty(V)$ satisfying,

$$\sigma(fD) = f\sigma(D) \qquad \forall\, f \in C^\infty(V);\ D \in \Delta(V).$$

For any $f \in C^\infty(V)$, the map

$$D \mapsto D(f) \qquad \forall\, D \in \Delta(V)$$

obviously belongs to $\mathcal{D}^1(V)$. It is denoted by df. Hence the set $\{dz_j, d\bar{z}_j : j = 1, 2, \ldots, n\}$ is a free module basis for $\mathcal{D}^1(V)$ which is dual to the module basis $\{\frac{\partial}{\partial z_j}, \frac{\partial}{\partial \bar{z}_j} : j = 1, 2, \ldots, n\}$ of $\Delta(V)$ (*i.e.*, these bases satisfy $dz_j(\frac{\partial}{\partial \bar{z}_k}) = d\bar{z}_j(\frac{\partial}{\partial z_k}) = 0, dz_j(\frac{\partial}{\partial z_k}) = d\bar{z}_j(\frac{\partial}{\partial \bar{z}_k}) = \delta_{jk}$ for all $j, k \in \{1, 2, \ldots, n\}$). Sometimes it will also be convenient to use the module basis $\{dx_j, dy_j : j = 1, 2, \ldots, n\}$ for $\mathcal{D}^1(V)$. These basis vectors are related by the usual equations: $dz_j = dx_j + idy_j, d\bar{z}_j = dx_j - idy_j$. Hence, this second module basis is dual to the basis $\{\frac{\partial}{\partial x_j}, \frac{\partial}{\partial y_j} : j = 1, 2, \ldots, n\}$ for $\Delta(V)$. In terms of these bases the differential df of any $f \in C^\infty(V)$ can be written as

$$df = \sum_{j=1}^{n}(\frac{\partial}{\partial z_j}dz_j + \frac{\partial}{\partial \bar{z}_j}d\bar{z}_j) = \sum_{j=1}^{n}(\frac{\partial f}{\partial x_j}dx_j + \frac{\partial f}{\partial y_j}dy_j).$$

The next step in our review is quite abstract. For $q > 1$, an *elementary q-form* is an expression of the form $f\sigma_1 \wedge \sigma_2 \wedge \cdots \wedge \sigma_q$ where f is a C^∞

function on V and each σ_j is a 1-form on V. For $q > 1$, a q-*form* on V is a finite sum of elementary q-forms. Two q-forms are identical if one can be transferred into the other by a finite sequence of identities of the following two types:

$$f\sigma_1 \wedge \sigma_2 \wedge \cdots \wedge \sigma_{j-1} \wedge (g\sigma_j + h\sigma_j') \wedge \sigma_{j+1} \wedge \cdots \wedge \sigma_q$$

$$= fg\sigma_1 \wedge \sigma_2 \wedge \cdots \wedge \sigma_{j-1} \wedge \sigma_j \wedge \sigma_{j+1} \wedge \cdots \wedge \sigma_q$$

$$+ fh\sigma_1 \wedge \sigma_2 \wedge \cdots \wedge \sigma_{j-1} \wedge \sigma_j' \wedge \sigma_{j+1} \wedge \cdots \wedge \sigma_q;$$

$$f\sigma_1 \wedge \sigma_2 \wedge \cdots \wedge \sigma_j \wedge \sigma_{j+1} \wedge \cdots \wedge \sigma_q = -f\sigma_1 \wedge \sigma_2 \wedge \cdots \wedge \sigma_{j+1} \wedge \sigma_j \wedge \cdots \wedge \sigma_q$$

where f, g and h belong to $C^\infty(V)$ and σ_j' and each σ_k belong to $\mathcal{D}^1(V)$. Thus, we require $\sigma_1 \wedge \sigma_2 \wedge \cdots \wedge \sigma_q$ to be linear over $C^\infty(V)$ in each factor separately and to be anti-commutative in each pair of adjacent positions. In particular, the q-form $\sigma_1 \wedge \sigma_2 \wedge \cdots \wedge \sigma_q$ is zero if σ_j equals σ_k for any distinct integers j and k. Let $\mathcal{D}^q(V)$ be the $C^\infty(V)$ module of q-forms on V. Since $\mathcal{D}^1(V)$ had a free module basis with $2n$ elements, and since an elementary q-form $\sigma_1 \wedge \sigma_2 \wedge \cdots \wedge \sigma_q$ is zero unless $\{\sigma_1, \sigma_2, \ldots, \sigma_q\}$ is independent over $C^\infty(V)$, an elementary computation shows that $\mathcal{D}^q(V)$ has a free module basis with $\binom{2n}{q}$ elements for $1 \leq q \leq n$. For q greater than n, $\mathcal{D}^q(V)$ is zero.

For the elementary q-form $\sigma = f\sigma_1 \wedge \sigma_2 \wedge \cdots \wedge \sigma_q$ and the elementary p-form $\rho = g\rho_1 \wedge \rho_2 \wedge \cdots \wedge \rho_p$, we define the exterior product, $\sigma \wedge \rho$, of σ and ρ to be the elementary $(q+p)$-form $fg\sigma_1 \wedge \sigma_2 \wedge \cdots \wedge \sigma_q \wedge \rho_1 \wedge \rho_2 \wedge \cdots \wedge \rho_p$. Extend this product to all q-forms and p-forms by requiring the exterior product to satisfy the distributive laws. Thus, the exterior product is a binary operation on the set $\cup_{q=1}^\infty \mathcal{D}^q(V)$ which satisfies the associative and distributive laws. Furthermore,

$$\sigma \wedge \rho = (-1)^{pq} \rho \wedge \sigma \qquad \forall \, \sigma \in \mathcal{D}^q(V); \rho \in \mathcal{D}^p(V)$$

also follows. Clearly, the notation $\wedge$ for the exterior product is compatible with our previous notation for q-forms.

It is convenient to denote the regular $C^\infty(V)$-module $C^\infty(V)$ by $\mathcal{D}^0(V)$ and to call the functions which belong to it 0-forms. The *exterior derivative* d is a collection of linear maps

$$d: \mathcal{D}^q(V) \to \mathcal{D}^{q+1}(V) \qquad q = 0, 1, 2, \ldots$$

defined as follows. For each $f \in \mathcal{D}^0(V) = C^\infty(V)$, df has the meaning already assigned as an element of $\mathcal{D}^1(V)$. For the elementary q-form

$$\sigma = f d\sigma_1 \wedge \cdots \wedge d\sigma_q,$$

$d\sigma$ has the meaning

$$d\sigma = df \wedge d\sigma_1 \wedge \cdots \wedge d\sigma_q.$$

The definition for other q-forms is obtained by the postulated linearity. It is easy to check that for any q-form σ and any p-form ρ we get

$$d(\sigma \wedge \rho) = (d\sigma) \wedge \rho + (-1)^q \sigma \wedge d\rho.$$

Another important property of exterior differentiation is expressed by $d^2 = 0$, (*i.e.*, $d(d\sigma) = 0$ for all p-forms σ and all $p \in \mathbb{N}^0$).

Stokes' Theorem

These are all the results we need on the exterior differential calculus of forms. Note that all the results are of a computational nature. The integral calculus of forms gives a little more substance to the theory. The main result is Stokes' theorem which asserts that any "$(q+1)$-dimensional", compact subset K of $\mathbb{C}^n$ with suitably smooth "q-dimensional" boundary ∂K and any q-form σ satisfy

$$\int_K d\sigma = \int_{\partial K} \sigma. \tag{13}$$

To make sense of this formula, we must define what a "q-dimensional", compact subset of $\mathbb{C}^n$ is and define the integral of a q-form over such a set. In general, it is not necessary to require that a "q-dimensional", compact subset actually be a q-dimensional submanifold of $\mathbb{C}^n$, but merely that it be an image under a (possibly highly singular) C^∞-mapping of some piecewise smoothly bounded compact subset of $\mathbb{R}^q$ which is the closure of its interior. This can be systematically expressed in terms of C^∞-singular simplices and chains. (This leads to deRham's theorem, which identifies the cohomology groups based on the exterior derivative as the duals (under integration) of the singular homology groups.) However, we will only need to apply Stoke's theorem to $(2n-1)$-forms on the boundary of a $2n$-ball B centered at the origin, so we can use simpler definitions. An elementary $2n$-form on V can be written as $\sigma = f dx_1 \wedge dy_1 \wedge dx_2 \wedge dy_2 \wedge \cdots \wedge dx_n \wedge dy_n$ for a suitable C^∞-function f. (Note that the order is important in order to get the sign correct.) Then $\int_B \sigma$ is defined to be the ordinary $2n$-fold integral $\int_B f \, dx_1 dy_1 dx_2 dy_2 \cdots dx_n dy_n$. An easy calculation shows

$$d\bar{z}_1 \wedge dz_1 \wedge d\bar{z}_2 \wedge dz_2 \wedge \cdots \wedge d\bar{z}_n \wedge dz_n$$
$$= (2i)^n dx_1 \wedge dy_1 \wedge dx_2 \wedge dy_2 \wedge \cdots \wedge dx_n \wedge dy_n.$$

A $(2n-1)$-form can be written as $\sum_{j=1}^{2n} f_j \sigma_j$ where each σ_j is of the form $dx_1 \wedge dy_1 \wedge dx_2 \wedge dy_2 \wedge \cdots \wedge dx_n \wedge dy_n$ with the jth factor removed. In order to define the integral of this form over the boundary ∂B of B, we need a suitable C^∞ parameterization of σB, say $F: R \to \sigma B$ where R is a compact rectangular parallelepiped in $\mathbb{R}^{2n-1}$. Then we define

$$\int_{\partial B} f_j \sigma_j = \int_R f_j \circ F \frac{\partial(x_1, y_1, \ldots, x_n, y_n)}{\partial(t_1, t_2, \ldots, t_{2n-1})} dt_1 dt_2 \cdots dt_{2n-1},$$

where in the numerator of the Jacobian the jth component is omitted. The orientation of the parameterization is important in general, and we could describe this by giving an explicit parameterization (in terms of trigonometric functions). However, since all the boundary integrals we need will be zero, we omit this.

We have described a standard form of the calculus of forms. However, in fact, the functions and forms we use will mostly have values in $\mathcal{A}$ or $\mathcal{A}^{(n)}$ rather than in $\mathbb{C}$. The partial derivatives of an $\mathcal{A}$- or $\mathcal{A}^{(n)}$-valued function are defined in the usual way. We use $C^\infty(\mathbb{C}^n, \mathcal{A})$ and $C^\infty(\mathbb{C}^n, \mathcal{A}^{(n)})$ to denote the $C^\infty(\mathbb{C}^n, \mathbb{C})$-module of C^∞ functions on $\mathbb{C}^n$ with values in $\mathcal{A}$ and in $\mathcal{A}^{(n)}$, respectively. There is no difficulty in extending the rest of the theory. The reader who wishes to check some particular point will find that evaluation of elements of $\mathcal{A}^*$ or $(\mathcal{A}^{(n)})^*$ on both sides of an equation usually reduces the problem to the classical complex-valued setting.

Construction of v in the One-dimensional Case

Before constructing the function v in the n-dimensional case, we will show how the one-variable functional calculus could have been defined in terms of differential forms. This result will be used to check property (j) in the theorem. Recall that for any $a \in \mathcal{A}$ and $f \in \mathcal{F}_a$ we defined $I_a(f)$ by

$$I_a(f) = \frac{1}{2\pi i} \int_\Gamma f(\lambda)(\lambda - a)^{-1} d\lambda$$

where Γ was an admissible contour. We would now drop the dummy variable λ and write this as

$$I_a(f) = \frac{1}{2\pi i} \int_\Gamma (z - a)^{-1} f \, dz$$

where z is the coordinate function on $\mathbb{C}$. Let U be the open neighborhood of $Sp(a)$ defined by $U = \{\lambda \in \mathbb{C} : \Gamma$ has winding number $+1$ with respect to $\lambda\}$, and let K be the compact set $K = \overline{U} = U \cup \Gamma$. Choose a function w in $C^\infty(\mathbb{C})$ satisfying $w = 1$ in a neighborhood of $Sp(a)$ and having compact support in U. Define a function $u \in C^\infty(\mathbb{C}, \mathcal{A})$ by

$$u(\lambda) = \begin{cases} (1 - w(\lambda))(\lambda - a)^{-1} & \lambda \notin Sp(a) \\ 0 & \lambda \in Sp(a). \end{cases}$$

Stokes' theorem and the other properties of forms give

$$
\begin{aligned}
I_a(f) &= \frac{1}{2\pi i} \int_\Gamma (z-a)^{-1} f dz = \frac{1}{2\pi i} \int_\Gamma f u\, dz \\
&= \frac{1}{2\pi i} \int\int_K d(fu) \wedge dz = \frac{1}{2\pi i} \int\int_K \frac{\partial(fu)}{\partial \bar z} d\bar z \wedge dz \\
&= \frac{1}{2\pi i} \int\int_K f \frac{\partial u}{\partial \bar z} d\bar z \wedge dz = \int\int_C f\Big(\frac{1}{\pi}\frac{\partial u}{\partial \bar z}\Big) dV.
\end{aligned}
$$

(Here we have used the fact that the integral $\int_\Gamma \cdot dz$ corresponds to the boundary integral in Stokes' theorem.) This shows that we could have used the last expression to define $I_a(f)$.

An Auxiliary Function u

We now wish to generalize this construction to elements $a \in \mathcal{A}^{(n)}$ and functions of n-variables which are analytic in a neighborhood of $Sp(a)$. To do this, we note that we could have started from the penultimate line of the last equation to get

$$
I_a(f) = \frac{1}{2\pi i} \int\int_K f\, du \wedge dz = \frac{1}{2\pi i} \int_B f\, du \wedge dz
$$

where B is a closed ball centered at the origin and satisfying $K \subseteq B$.

To handle the general situation, we start out with a technical lemma about the construction of an auxiliary function $u \in C^\infty(\mathbb{C}^n, \mathcal{A}^{(n)})$ and a form based upon this function.

3.5.10 Lemma *Let $n \in \mathbb{N}, a \in \mathcal{A}^{(n)}$ and $U \in \mathcal{N}_a$ be given. Then there are:*

(a) *a C^∞-function $w\colon \mathbb{C}^n \to \mathbb{R}$ with compact support in U such that $w = 1$ holds in a neighborhood of $Sp(a)$, and*

(b) *a C^∞-function $u\colon \mathbb{C}^n \to \mathcal{A}^{(n)}$ satisfying*

$$
\sum_{j=1}^n u_j(z_j - a_j) = 1 - w. \tag{14}
$$

When w and u satisfy these conditions, the $2n$-form $\mu = du_1 \wedge dz_1 \wedge du_2 \wedge dz_2 \wedge \cdots \wedge du_n \wedge dz_n$ has compact support in U. Moreover, if $\tilde U \in \mathcal{N}_a$ with $\tilde U \subseteq U$ is given and $\tilde w, \tilde u, \tilde \mu$ are any elements satisfying the same conditions relative to each other and to $\tilde U$ as w, u and μ satisfy relative to each other and to U, then there is an $(n-1)$-form τ supported on U such that

$$
\mu - \tilde\mu = d\tau \wedge dz_1 \wedge dz_2 \wedge \cdots \wedge dz_n.
$$

(For the *cognescenti* we remark that we are claiming that μ is uniquely determined as an element in the Dolbeault cohomology group $H^{n,n}_{\bar{\partial}}$ of (the germ of neighborhoods of) $Sp(a)$ with coefficients in $\mathcal{A}^{(n)}$.)

Proof The existence of a function with the properties of w is well-known. Choose such a function.

We turn to the construction of the functions u_j for $j = 1, 2, \ldots, n$. For each $\zeta \in \mathbb{C}^n$ which does not belong to $Sp(a)$, choose $b^\zeta \in \mathcal{A}^{(n)}$ satisfying

$$\sum_{j=1}^{n} b_j^\zeta(\zeta_j - a_j) = 1.$$

Choose an open neighborhood U^ζ of ζ on which the continuous affine function

$$\sum_{j=1}^{n} b_j^\zeta(z_j - a_j)$$

has invertible values. For each $k = 1, 2, \ldots, n$ let u_k^ζ be the function defined by

$$u_k^\zeta = b_k^\zeta (\sum_{j=1}^{n} b_j^\zeta(z_j - a_j))^{-1}$$

on U^ζ. Then each u_k^ζ is analytic on U^ζ.

The family $\{U^\zeta : \zeta \in \mathbb{C}^n \setminus Sp(a)\}$ is an open cover for the paracompact space $\mathbb{C}^n \setminus Sp(a)$. Therefore, we can choose a locally finite C^∞ partition of unity subordinate to this cover. That is, we choose a family $\{h_\alpha : \alpha \in A\}$ of functions so that each function h_α belongs to $C^\infty(\mathbb{C}^n \setminus Sp(a))$, has range in $[0, 1]$ and has support in U^{ζ_α} for some $\zeta_\alpha \in \mathbb{C}^n \setminus Sp(a)$. Furthermore, near any point $\lambda \in \mathbb{C}^n \setminus Sp(a)$ there are only finitely many non-zero members of the family $\{h_\alpha : \alpha \in A\}$ and these satisfy $\sum_{\alpha \in A} h_\alpha(\lambda) = 1$. Now choose u_k for $k = 1, 2, \ldots, n$ to be the functions defined by

$$u_k(\lambda) = \begin{cases} \displaystyle\sum_{\alpha \in A} (1 - w)(\lambda) h_\alpha(\lambda) u_k^{\zeta_\alpha}(\lambda) & \text{if } \lambda \in \mathbb{C}^n \setminus Sp(a) \\ 0 & \text{if } \lambda \in Sp(a) \end{cases}$$

The sum is a finite sum for any particular $\lambda \in \mathbb{C}^n \setminus Sp(a)$. Clearly, each u_k belongs to $C^\infty(\mathbb{C}^n, \mathcal{A})$. Hence $u = (u_1, u_2, \ldots, u_n)$ satisfies the condition specified for $u \in C^\infty(\mathbb{C}^n, \mathcal{A}^{(n)})$.

We check that the form μ behaves as desired. Since $\sum_{j=1}^{n} u_j(z_j - a_j) = 1$ holds on the complement V of $\text{supp}(w)$, exterior differentiation gives

$$\sum_{j=1}^{n}(z_j - a_j)du_j + \sum_{j=1}^{n} u_j dz_j = 0 \tag{15}$$

on V. Taking the exterior product with $du_1 \wedge du_2 \wedge \cdots \wedge du_{k-1} \wedge du_{k+1} \wedge \cdots \wedge du_n \wedge dz_1 \wedge dz_2 \wedge \cdots \wedge dz_n$, we get

$$(z_k - a_k)du_1 \wedge du_2 \wedge \cdots \wedge du_n \wedge dz_1 \wedge dz_2 \wedge \cdots \wedge dz_n = 0$$

on V. Multiplying by u_k, summing and applying (14), we find that μ is zero on V. Hence, μ has compact support in U.

Now let $\tilde{U} \subseteq U$ be another neighborhood of $Sp(a)$. Let $\tilde{w}, \tilde{u}$ and $\tilde{\mu}$ satisfy the same conditions relative to each other and to $\tilde{U}$ that w, u and μ satisfy relative to each other and to U. In particular, we have

$$\sum_{j=1}^{n} \tilde{u}_j(z_j - a_j) = 1 - \tilde{w}. \tag{16}$$

We want to show that $\tilde{\mu} - \mu$ has a special form. To simplify our discussion, we will write σ for the n-form $dz_1 \wedge dz_2 \wedge \cdots \wedge dz_n$. Then we want to find an $(n-1)$-form τ supported on U such that

$$\tilde{\mu} - \mu = d\tau \wedge \sigma.$$

(The point is that Stokes' theorem will then show that the integrals with respect to μ and $\tilde{\mu}$ of an analytic function over a ball which includes U will have the same values.)

For each $k = 1, 2, \ldots, n$, we have

$$\begin{aligned}
\tilde{u}_k - u_k &= \left(\sum_{j=1}^{n} u_j(z_j - a_j) + w\right)\tilde{u}_k - \left(\sum_{j=1}^{n} \tilde{u}_j(z_j - a_j) + \tilde{w}\right)u_k \\
&= \sum_{j=1}^{n}(z_j - a_j)(u_j\tilde{u}_k - \tilde{u}_j u_k) + w\tilde{u}_k - \tilde{w}u_k.
\end{aligned}$$

Defining $p_{j,k}$ and q_k for $j, k = 1, 2, \ldots, n$ by $p_{j,k} = u_j\tilde{u}_k - \tilde{u}_j u_k$ and $q_k = w\tilde{u}_k - \tilde{w}u_k$, we get

$$\tilde{u}_k - u_k = \sum_{j=1}^{n}(z_j - a_j)p_{j,k} + q_k \quad k = 1, 2 \ldots, n. \tag{17}$$

Similarly we get

$$\begin{aligned}
\tilde{w} - w &= -\sum_{k=1}^{n}(z_k - a_k)(\tilde{u}_k - u_k) \\
&= -\sum_{k=1}^{n}(z_k - a_k)\left(\sum_{j=1}^{n}(z_j - a_j)p_{j,k} + q_k\right) = -\sum_{k=1}^{n}(z_k - a_k)q_k
\end{aligned}$$

since $p_{j,k} = -p_{k,j}$ for all $j, k = 1, 2, \ldots, n$.

Now define $\tilde{\tilde{u}} \in C^\infty(\mathbb{C}^n, \mathcal{A}^{(n)})$ by

$$\tilde{\tilde{u}}_k = u_k + \sum_{j=1}^n (z_j - a_j) p_{j,k} \quad k = 1, 2, \ldots, n.$$

This function satisfies

$$\sum_{j=1}^n (\tilde{\tilde{u}}_j - a_j) = 1 - w$$

since $p_{j,k} = -p_{k,j}$ for all $k, j = 1, 2, \ldots, n$. Let $\tilde{\tilde{\mu}}$ be the $2n$-form

$$\tilde{\tilde{\mu}} = d\tilde{\tilde{u}}_1 \wedge dz_1 \wedge d\tilde{\tilde{u}}_2 \wedge dz_2 \wedge \cdots \wedge d\tilde{\tilde{u}}_n \wedge dz_n.$$

Obviously, it is enough to show that both $\tilde{\mu} - \tilde{\tilde{\mu}}$ and $\tilde{\tilde{\mu}} - \mu$ can be written in the form $d\tau \wedge \sigma$ for suitable choices of an $(n-1)$-form τ supported on U. We shall write each of the n-forms $\tilde{\mu} - \tilde{\tilde{\mu}}$ and $\tilde{\tilde{\mu}} - \mu$ as a sum of n-forms which we will show explicitly can be written as $d\tau \wedge \sigma$ with τ supported in U.

For each $k = 0, 1, 2, \ldots, n$, let $u^k \in C^\infty(\mathbb{C}^n, \mathcal{A}^{(n)})$ be the function

$$u_j^k = \begin{cases} \tilde{\tilde{u}}_j + q_j & \text{if } j \leq k \\ \tilde{\tilde{u}} & \text{if } j > k \end{cases}$$

so that $u^0 = \tilde{\tilde{u}}$ and $u^n = \tilde{u}$ hold where in the last case we have used (17). For each $k = 0, 1, 2, \ldots, n$, define μ^k by

$$\mu^k = du_1^k \wedge dz_1 \wedge du_2^k \wedge dz_2 \wedge \cdots \wedge du_n^k \wedge dz_n$$

so that $\mu^0 = \tilde{\tilde{\mu}}$ and $\mu^n = \tilde{\mu}$ both hold. Then we get $\mu^k - \mu^{k-1} = d\tilde{\tilde{u}}_1 \wedge dz_1 \wedge d\tilde{\tilde{u}}_2 \wedge dz_2 \wedge \cdots \wedge dz_{k-1} \wedge dq_k \wedge dz_k \wedge \cdots d\tilde{\tilde{u}}_n \wedge dz_n = (d\tau) \wedge \sigma$ where τ is $\pm q_k d\tilde{\tilde{u}}_1 \wedge d\tilde{\tilde{u}}_2 \wedge \cdots \wedge d\tilde{\tilde{u}}_{k-1} \wedge d\tilde{\tilde{u}}_{k+1} \wedge \cdots \wedge d\tilde{\tilde{u}}_n$. Thus, $\tilde{\mu} - \tilde{\tilde{\mu}} = \sum_{k=1}^n (\mu^k - \mu^{k-1})$ has the desired form since $q_k = w\tilde{\tilde{u}}_k - \tilde{w}u_k$ has support in $\text{supp}(w) \cup \text{supp}((\tilde{w})) \subseteq U$.

For distinct integers i and k from $\{1, 2, \ldots, n\}$, let $u^{i,k} \in C^\infty(\mathbb{C}^n, \mathcal{A}^{(n)})$ be defined by

$$u_j^{i,k} = \begin{cases} u_j & \text{if } j \neq i, j \neq k \\ u_k + (z_i - a_i) p_{i,k} & \text{if } j = k \\ u_i + (z_k - a_k) p_{k,i} & \text{if } j = i. \end{cases}$$

Then $p_{i,k} = -p_{k,i}$ implies

$$\sum_{j=1}^n u_j^{i,k} (z_j - a_j) = 1 - w. \tag{18}$$

Define a $2n$-form $\mu^{i,k}$ to be $\mu^{i,k} = du_1^{i,k} \wedge dz_1 \wedge du_2^{i,k} \wedge dz_2 \wedge \cdots \wedge du_n^{i,k} \wedge dz_n$ and define a $(2n-2)$-form ρ to be the form $du_1 \wedge du_2 \wedge \cdots \wedge du_n$ with du_i and du_k omitted. We get

$$
\begin{aligned}
du_k^{i,k} &= du_k + p_{i,k}dz_i + (z_i - a_i)dp_{i,k} \\
du_i^{i,k} &= du_i - p_{i,k}dz_k - (z_k - a_k)dp_{i,k}
\end{aligned}
$$

which implies

$$
\begin{aligned}
&du_k^{i,k} \wedge du_i^{i,k} \wedge dz_k \wedge dz_i - du_k \wedge du_i \wedge dz_k \wedge dz_i \\
&= (z_i - a_i)dp_{i,k} \wedge du_i \wedge dz_k \wedge dz_i - (z_k - a_k)du_k \wedge dp_{i,k} \wedge dz_k \wedge dz_i.
\end{aligned}
$$

For a suitable choice of sign, this in turn implies

$$
\mu^{i,k} - \mu = \pm((z_i - a_i)dp_{i,k} \wedge du_i - (z_k - a_k)du_k \wedge dp_{i,k}) \wedge \rho \wedge \sigma.
$$

Note also

$$
-dw = \sum_{j=1}^{n}(z_j - a_j)du_j + \sum_{j=1}^{n} u_j dz_j
$$

which implies

$$
-dp_{i,k} \wedge dw \wedge \rho \wedge \sigma = dp_{i,k} \wedge ((z_i - a_i)du_i + (z_k - a_k)du_k) \wedge \rho \wedge \sigma
$$

$$
= ((z_i - a_i)dp_{i,k} \wedge du_i - (z_k - a_k)du_k \wedge dp_{i,k}) \wedge \rho \wedge \sigma.
$$

Hence, we can write $\mu^{i,k} - \mu$ as $\pm d\tau \wedge \sigma$ where $\tau = p_{i,k}dw \wedge \rho$ is supported on U since w is supported there.

Now equation (18) shows that $u^{i,k}$ and $\mu^{i,k}$ could be substituted for u and μ at the beginning of this argument. Then the argument could be repeated for another choice of two distinct indices i', k'. After $n(n-1)/2$ steps we would arrive at $\tilde{u}$ and $\tilde{\mu}$. Thus $\tilde{\mu} - \mu$ has the desired form. This completes the proof of this lemma. $\qquad\square$

Note that when each du_j is expanded $\frac{n!}{(2\pi i)^n}\mu$ has the form vdV where v is some C^∞-function (a Jacobian determinant), and where dV is $2n$-dimensional Lebesgue measure. This is the function v mentioned at the beginning of the proof of Theorem 3.5.9. We assume $vdV = \frac{n!}{(2\pi i)^n}\mu$ from now on.

3.5.11 Lemma *Let* U, w, u, μ *and* v *be as defined in* Lemma 3.5.10 *and above. For each* $f \in \mathcal{F}_a$, *define* $I_a(f)$ *to be*

$$
I_a(f) = \frac{n!}{(2\pi i)^n} \int_B f\, d\mu = \int_{\mathbb{C}^n} fv\, dV \tag{19}
$$

where f is some representative of f in $\mathcal{F}_a$, μ and v are constructed relative to $U = D_f$, and B is a closed ball centered at the origin which includes D_f. Then $I_a: \mathcal{F}_a \to \mathcal{A}$ is a well-defined continuous linear map satisfying (e), (i), and (j) of Theorem 3.5.9.

Remark. Despite the fact that f may not be defined on all of B or $\mathbb{C}^n$, the integrals are unambiguous since μ and v are zero off the domain of f.

Proof We check first that $I_a(f)$ is well-defined (*i.e.*, it depends only on $a \in \mathcal{A}$ and $f \in \mathcal{F}_a$). Suppose f and $\tilde{f}$ are two representatives in $\tilde{\mathcal{F}}_a$ of $f \in \mathcal{F}_a$. Without loss of generality, we may assume $D_{\tilde{f}} \subseteq D_f$. Let μ and $\tilde{\mu}$ be any two $2n$-forms constructed as in Lemma 3.5.10 from D_f and $D_{\tilde{f}}$, respectively. Then Lemma 3.5.10 asserts that there is an $(n-1)$-form τ with support in D_f such that $\mu = \tilde{\mu} + d\tau \wedge \sigma$ where $\sigma = dz_1 \wedge dz_2 \wedge \cdots \wedge dz_n$. For an analytic function $f, d(f\tau \wedge \sigma) = df \wedge \tau \wedge \sigma + f d\tau \wedge \sigma = f d\tau \wedge \sigma$ holds. Hence Stokes' theorem shows

$$\int_B f d\mu - \int_B \tilde{f} d\tilde{\mu} = \int_B f d\tau \wedge \sigma = \int_B d(f\tau \wedge \sigma) = \int_{\partial B} f\tau \wedge \sigma = 0$$

since $\tau = 0$ in a neighborhood of ∂B. Thus $I_a: \mathcal{F}_a \to \mathcal{A}$ is well-defined by (19). Clearly I_a is a continuous linear map. The remarks given before Lemma 3.5.10 show that I_a satisfies (j) (when n is 1).

(i): Let $n, m \in \mathbb{N}$, $a \in \mathcal{A}^{(n)}$, $b \in \mathcal{A}^{(m)}$, $f \in \mathcal{F}_a$ and $g \in \mathcal{F}_b$ be given, and let h be the function specified in condition (i). To simplify notation, let π and π' represent the maps $\pi_{n+m,n}$ and $\pi'_{n+m,n}$. Then the function h of condition (i) is simply $(f \circ \pi)(g \circ \pi')$. Let $w \in C^\infty(\mathbb{C}^n), u \in C^\infty(\mathbb{C}^n, \mathcal{A}^{(n)})$ and μ be constructed as in Lemma 3.5.10 so that they are suitable for defining $I_a(f)$. Let $\tilde{w} \in C^\infty(\mathbb{C}^m), \tilde{u} \in C^\infty(\mathbb{C}^m, \mathcal{A}^{(m)})$ and $\tilde{\mu}$ be constructed as in Lemma 3.5.10 so that they are suitable for defining $I_b(g)$. To simplify notation, from now on we will write $f, w, u, \tilde{g}, \tilde{w}$ and $\tilde{u}$ to represent the functions $f \circ \pi, w \circ \pi, u \circ \pi, g \circ \pi', \tilde{w} \circ \pi'$ and $\tilde{u} \circ \pi'$, respectively. Similarly, we use μ to represent the $2n$-form on $\mathbb{C}^{n+m}$, which has the same formal expression as μ on $\mathbb{C}^n$, and we use $\tilde{\mu}$ to represent the $2m$-form

$$\tilde{\mu} = d\tilde{u}_1 \wedge dz_{n+1} \wedge d\tilde{u}_2 \wedge dz_{n+2} \wedge \cdots \wedge d\tilde{u}_m \wedge dz_{n+m}$$

on $\mathbb{C}^{n+m}$ (where we are using the convention on $\tilde{u} \in C^\infty(\mathbb{C}^{n+m}, \mathcal{A}^{(m)})$ just established). With these notational conventions, we have

$$I_a(f) I_b(g) = \frac{n! m!}{(2\pi i)^{n+m}} \int f\tilde{g}\mu \wedge \tilde{\mu}. \tag{20}$$

The function $w\tilde{w}$ is supported on

$$(D_f, D_g) = \{(\lambda, \tilde{\lambda}): \lambda \in D_f, \tilde{\lambda} \subseteq D_g\} \in \mathbb{C}^{n+m}.$$

Also the equations

$$\sum_{j=1}^{n} u_j(z_j - a_j)\, 1 - w \quad \text{and} \quad \sum_{j=1}^{m} \tilde{u}_j(z_{n+j} - b_j) = 1 - \tilde{w}$$

imply

$$\sum_{j=1}^{m} u_j(z_j - a_j) + \sum_{j=1}^{m} w\tilde{u}_j(z_{j+n} - b_j) = 1 - w + w(1 - \tilde{w}) = 1 - w\tilde{w}.$$

Hence, we can use $w\tilde{w} \in C^{\infty}(\mathbb{C}^{n+m})$ and $(u, w\tilde{u}) \in C^{\infty}(\mathbb{C}^{n+m}, \mathcal{A}^{(n+m)})$ to define $I_{(a,b)}(h) = I_{(a,b)}(f\tilde{g})$. Thus, we have

$$I_{(a,b)}(h) = \frac{(n+m)!}{(2\pi i)^{n+m}} \int f\tilde{g}\mu \wedge d(w\tilde{u}_1) \wedge dz_{n+1} \wedge d(w\tilde{u}_2) \wedge dz_{n+2} \wedge \cdots \wedge d(w\tilde{u}_n) \wedge dz_{n+m}.$$

However, for each $j = 1, 2, \ldots, m$,

$$d(w\tilde{u}_j) = \tilde{u}_j dw + w d\tilde{u}_j = \tilde{u}_j\left(-\sum_{k=1}^{n}(z_k - a_k)du_k - \sum_{k=1}^{n} u_k dz_k\right) + w d\tilde{u}_j$$

implies

$$\mu \wedge d(w\tilde{u}_j) = w(\mu \wedge d\tilde{u}_j).$$

Thus, we conclude

$$I_{(a,b)}(h) = \frac{(n+m)!}{(2\pi i)^{n+m}} \int f\tilde{g}w^m \mu \wedge \tilde{\mu}.$$

Stokes' theorem and (20) will give the desired equality if we can find an $(n-1)$-form τ supported on (D_f, D_g) and satisfying

$$[(n+m)!w^m - n!m!]\mu = d\tau \wedge \sigma \tag{21}$$

where σ is the form $\sigma = dz_1 \wedge dz_2 \wedge \cdots \wedge dz_n$.

We prove (21) by writing the left side as a sum of terms, each of which is shown to have the correct form. For $j = 1, 2, \ldots, n$, let μ_j be the $(2n-1)$-form which is equal to μ except that the factor du_j is removed. For $k = 0, 1, 2, \ldots, n$, let τ_k be the $(n-1)$-form satisfying

$$\tau_k \wedge \sigma = w^k \sum_{j=1}^{n} u_j\mu_j.$$

Since $du_j \wedge \mu_j = \mu$ for $j = 1, 2, \ldots, n$, we see

$$d\tau_k \wedge \sigma = kw^{k-1} \sum_{j=1}^{n} u_j dw \wedge \mu_j + nw^k \mu.$$

Using the now familiar expansion for $-dw$ and noting again that $du_j \wedge \mu_j = \mu$, we get

$$
\begin{aligned}
d\tau_k \wedge \sigma &= -kw^{k-1} \sum_{j=1}^{n} u_j(z_j - a_j)\mu + nw^k \mu \\
&= [-kw^{k-1}(1-w) + nw^k]\mu = [(n+k)w^k - kw^{k-1}]\mu.
\end{aligned}
$$

Thus, defining τ by

$$\tau = \sum_{k=1}^{m} \frac{(n+k-1)!m!}{k!} \tau_k,$$

we get

$$
\begin{aligned}
d\tau \wedge \sigma &= \sum_{k=1}^{m} \frac{(n+k-1)!m!}{k!} \left((n+k)w^k - kw^{k-1} \right) \mu \\
&= \sum_{k=1}^{m} \left(\frac{(n+k)!m!}{k!} w^k - \frac{(n+k-1)!m!}{(k-1)!} w^{k-1} \right) \mu \\
&= [(n+m)!w^m - n!m!]\mu.
\end{aligned}
$$

This concludes the proof of (i).

It only remains to show that our definition of I_a satisfies condition (e). This amounts to showing

$$\gamma(I_a(f)) = f(\gamma(a))$$

for each $\gamma \in \Gamma_A$ and $f \in \mathcal{F}_a$. Let w, u and μ be defined as in Lemma 3.5.10 so that they are suitable for defining $I_a(f)$. Let $\tilde{u} \in C^\infty(\mathbb{C}^n, \mathbb{C}^{(n)})$ be defined by $\tilde{u}_j = \gamma \circ u_j$, and let $\tilde{\mu}$ be defined from $\tilde{u}$ as μ was defined from u. Since γ is a continuous homomorphism, we get

$$\gamma(I_a(f)) = \frac{n!}{(2\pi i)^n} \int f d\tilde{\mu}.$$

This reduces the problem to a calculation relative to the algebra $\mathbb{C}$. We can apply the available results on the functional calculus to the case $\mathcal{A} = \mathbb{C}$. Let $\gamma(a)$ be the element $(\gamma(a_1), \gamma(a_2), \ldots, \gamma(a_n))$ in $\mathbb{C}^{(n)}$. Then f belongs to $\mathcal{F}_{\gamma(a)}$ and $w, \tilde{u}, \tilde{\mu}$ are suitable for defining $I_{\gamma(a)}(f)$. By our uniqueness result, we could also obtain $I_{\gamma(a)}(f)$ from functions $\tilde{w} \in C^\infty(\mathbb{C}^n), \tilde{u} \in$

$C^\infty(\mathbb{C}^n, \mathbb{C}^{(n)})$ and an associated form $\tilde{\tilde{\mu}}$ supported in a closed polydisc centered at $\gamma(a)$ on which f has a uniformly convergent power series expansion. Since we know that $I_{\gamma(a)}$ satisfies (j) and (i), the argument given previously establishes that $I_{\gamma(a)}(p) = p(\gamma(a))$ for any polynomial p. Hence the continuity of $I_{\gamma(a)}$ shows that $I_{\gamma(a)}(f)$ is the limit of the partial sums of the power series for f evaluated at $\gamma(a)$. Hence we obtain

$$\gamma(I_a(f)) = \frac{n!}{(2\pi i)^n} \int f d\tilde{\mu} = \frac{n!}{(2\pi i)^n} \int f d\tilde{\tilde{\mu}} = f(\gamma(a)).$$

This concludes the proof of this lemma, and hence, the complete proof of Theorem 3.5.9.   □

Having proved this theorem, we are justified in simplifying our notation by replacing $I_a(f)$ by

$$f(a) = I_a(f) \qquad \forall\, n \in \mathbb{N};\; a \in \mathcal{A}^{(n)};\; f \in \mathcal{F}_a.$$

We leave it to the reader to translate the various equations in the statement of Theorem 3.5.9 into this notation.

Implicit Function Theorem

The following implicit function theorem is essentially due to Arens and Calderon [1955]. However, they obtained the full result only for semisimple commutative Banach algebras or for functions f with well-behaved power series expansions. The particular version given here is found in Gamelin [1969]. Its proof is based on the results of Theorem 3.5.9. This theorem can in turn be developed into a functional calculus of multiple valued functions (*i.e.*, functions defined on a Riemann surface with more than one sheet). For this viewpoint, which we will not develop here, see Michel Bonnard [1969], Allan [1970], Edgar Lee Stout [1971], and V. Ja. Lin [1973].

3.5.12 Theorem *Let $\mathcal{A}$ be a commutative unital Banach algebra. Let $n \in \mathbb{N}$ and $a \in \mathcal{A}^{(n)}$ be given. Let F be a complex-valued function which is analytic on an open subset U of $\mathbb{C}^{n+1}$ satisfying $Sp(a) \subseteq \pi'_{n+1,1}U$. Suppose that $\frac{\partial F}{\partial z_1}$ never vanishes on U. If there is an $h \in C(\Gamma_\mathcal{A})$ satisfying*

$$F(h, \hat{a}) \equiv 0$$

(i.e. $F(h(\gamma), \hat{a}_1(\gamma), \hat{a}_2(\gamma), \ldots, \hat{a}_n(\gamma)) = 0$ for all $\gamma \in \Gamma_\mathcal{A}$), then there is a unique element $c \in \mathcal{A}$ satisfying

$$\hat{c} = h \quad and \quad F(c, a) = 0.$$

Proof Let z and w be the coordinate functions on $\mathbb{C}^n$ and on $\mathbb{C}$, respectively. The ordinary implicit function theorem for analytic functions (*cf. e.g.*,

Hormander [1966]) shows that for any $\psi \in \Gamma_A$ there will be constants $\varepsilon^\psi > 0$ and $\delta^\psi > 0$ so that the equation $F(\lambda, \mu) = 0$ has only one solution with λ in the disc $D^\psi = \{\lambda : |\lambda - h(\psi)| < \varepsilon^\psi\}$ for each μ in the polydisc $P^\psi = \{\mu \in \mathbb{C}^n : |\mu_j - \hat{a}_j(\psi)| < \delta^\psi$ for $j = 1, 2, \ldots, n\}$. Let $V^\psi \subseteq \Gamma_A$ be the open neighborhood of ψ defined by $V^\psi = \{\gamma \in \Gamma_A : |h(\gamma) - h(\psi)| < \varepsilon^\psi/2$ and $|\hat{a}_j(\gamma) - \hat{a}_j(\psi)| < \delta^\psi/2$ for $j = 1, 2, \ldots, n\}$. Choose a subcover $\{V^{\psi_1}, V^{\psi_2}, \ldots, V^{\psi_p}\}$. Define $\varepsilon > 0$ by $\varepsilon = \min\{\varepsilon^{\psi_j} : j = 1, 2, \ldots, p\}$, and define $V \subseteq \Gamma_A \times \Gamma_A$ by $V = \{(\gamma_1, \gamma_2) \in \Gamma_A \times \Gamma_A : |h(\gamma_1) - h(\gamma_2)| < \varepsilon/2\}$. Suppose $(\gamma_1, \gamma_2) \in V$ satisfies $\hat{a}(\gamma_1) = \hat{a}(\gamma_2)$. For some j, γ_1 belongs to V^{ψ_j} so that $h(\gamma_1)$ and $h(\gamma_2)$ both belong to D^{ψ_j} and $\hat{a}(\gamma_1) = \hat{a}(\gamma_2)$ belongs to P^{ψ_j}. Also, $F(h(\gamma_k), \hat{a}(\gamma_k))$ is zero for $k = 1, 2$. Thus, $h(\gamma_1)$ equals $h(\gamma_2)$ by the uniqueness of solutions of $F(\lambda, \mu) = 0$ with $\lambda \in D^{\psi_j}$ for $\mu \in P^{\psi_j}$. Therefore, V is an open neighborhood of the diagonal in $\Gamma_A \times \Gamma_A$ such that $(\gamma_1, \gamma_2) \in V$ and $\hat{a}(\gamma_1) = \hat{a}(\gamma_2)$ imply $h(\gamma_1) = h(\gamma_2)$. For any $(\gamma_1, \gamma_2) \in \Gamma_A \times \Gamma_A \setminus V$, choose $b \in \mathcal{A}$ satisfying $\hat{b}(\gamma_1) \neq \hat{b}(\gamma_2)$ and then choose a neighborhood V^b of (γ_1, γ_2) in $\Gamma_A \times \Gamma_A$ such that $(\psi_1, \psi_2) \in V^b$ implies $\hat{b}(\psi_1) \neq \hat{b}(\psi_2)$. The family of all such V^b forms an open cover for the compact set $\Gamma_A \times \Gamma_A \setminus V$. Let $\{V^{b_1}, V^{b_2}, \ldots, V^{b_{m-n}}\}$ be a finite subcover and define $b \in \mathcal{A}^{(m-n)}$ by $b = (b_1, b_2, \ldots, b_{m-n})$. Then $\gamma(a, b) = \psi(a, b)$ implies $h(\gamma) = h(\psi)$ for all $\gamma, \psi \in \Gamma_A$. Hence, we obtain a well-defined function H on $Sp(a, b)$ from the definition

$$H(\gamma(a, b)) = h(\gamma) \qquad \forall \, \gamma \in \Gamma_A. \tag{22}$$

We wish to show that this function H is continuous. Suppose $\gamma_k(a, b)$ converges to $\gamma_0(a, b)$ for some sequence $\{\gamma_k\}_{k \in \mathbb{N}} \subseteq \Gamma_A$. It is enough to show that $h(\gamma_k)$ converges to $h(\gamma_0)$. If this is false, we can choose a neighborhood U of $h(\gamma_0)$ and a subsequence of $\{h(\gamma_k)\}_{k \in \mathbb{N}}$ which remains completely outside U. Since Γ_A is compact, we may choose a subnet of $\{\gamma_k\}_{k \in \mathbb{N}}$ which converges to some $\gamma_0' \in \Gamma_A$. Then we have $\gamma_0'(a, b) = \gamma_0(a, b)$ so that $h(\gamma_0') = h(\gamma_0)$. This contradiction shows that $\{h(\gamma_k)\}_{k \in \mathbb{N}}$ converges to $h(\gamma_0)$ as we wished.

Now we claim that H can be extended to a function which is analytic in a neighborhood of $Sp(a, b)$. To simplify notation, let $\tilde{F}$ be $F \circ \pi_{m+1, n+1}$ defined on $(U, \mathbb{C}^{m-n})$. Since H satisfies

$$\tilde{F}(H, z_1, z_2, \ldots, z_m) \equiv 0 \tag{23}$$

on $Sp(a, b)$, and the partial derivative of $\tilde{F}$ with respect to its first variable never vanishes on $(U, \mathbb{C}^{m-n})$, the implicit function theorem for analytic functions shows that for each $\lambda \in Sp(a, b)$ there is an open polydisc $P^\lambda \subseteq \mathbb{C}^m$ centered at λ and an open disc $D^\lambda \subseteq \mathbb{C}$ centered at $H(\lambda)$, such that for each $\mu \in P^\lambda$ there is a unique solution $H^\lambda(\mu) \in D^\lambda$ of the equation $\tilde{F}(w, \mu) = 0$. Furthermore, the function $H^\lambda : P^\lambda \to D^\lambda$ defined in this way is analytic and is completely determined on any connected open subset of

P^λ by its value at a single point and the requirement that it be a continuous solution of $F(H^\lambda, z_1, z_2, \ldots, z_n) = 0$. Since H is continuous, we may assume that P^λ is chosen small enough so that H maps P^λ into D^λ. Assume that P^λ, D^λ, and H^λ satisfying these conditions have been chosen for each $\lambda \in Sp(a, b)$. If λ and μ belong to $Sp(a, b)$, the uniqueness of H^λ and H^μ show that either $\lambda \in P^\mu$ or $\mu \in P^\lambda$ implies $H^\lambda = H^\mu$ on $P^\lambda \cap P^\mu$. For any $\lambda \in Sp(a, b)$, let V^λ be the polydisc centered at λ with polyradius equal to half the polyradius of P^λ. If $V^\lambda \cap V^\mu$ is nonempty for any λ, $\mu \in Sp(a, b)$, then λ or μ belongs to $P^\lambda \cap P^\mu$ so that H^λ and H^μ agree on $V^\lambda \cap V^\mu$. Hence, we can extend H to be an analytic function on the neighborhood $V = \cup_{\lambda \in Sp(a,b)} V^\lambda$ of $Sp(a, b)$.

Let c be the element $H(a, b) \in \mathcal{A}$ defined by the multivariable functional calculus. Theorem 3.5.9(e) and equation (22) give $\hat{c} = H(a, b)\hat{} = h$. Theorem 3.5.9(b) and (g) and equation (23) give

$$F(c, a) = F \circ \pi_{m+1, n+1}(c, a, b) = \tilde{F}(H(a, b), a, b) = 0.$$

This proves the existence of the element c.

Suppose d were another element satisfying the same conditons as c. Then $\hat{d} = h = \hat{c}$ would imply the existence of an element r in the Gelfand radical of $\mathcal{A}$ satisfying

$$d = c + r.$$

We wish to show that r is zero. Let G be the function defined on $\mathbb{C}^{n+2}$ by

$$G(u, w, z) = \frac{1}{u^2}[F(w + u, z) - F(w, z) - u\frac{\partial F}{\partial w}(w, z)]$$

where z is the coordinate function on $\mathbb{C}^n$. This function is analytic on a neighborhood of $Sp(r, c, a)$. Also by assumption we have $F(c + r, a) = F(c, a) = 0$. Hence, when $u[uG(u, w, z) + \frac{\partial F}{\partial w}(w, z)] = F(w + u, z) - F(w, z)$ is applied to $(r, c, a) \in \mathcal{A}^{(n+2)}$, we obtain

$$r[rG(r, c, a) + \frac{\partial F}{\partial w}(c, a)] = 0.$$

However, $\frac{\partial F}{\partial w}$ does not vanish on $Sp(c, a)$ so the Gelfand transform of the expression in the bracket does not vanish on $\Gamma_{\mathcal{A}}$. Thus the expression in the bracket is invertible, so r must be zero.   □

If we interpret our notation suitably, we may even apply Theorem 3.5.12 in the case $n = 0$. Let $\mathcal{A}$ be as in Theorem 3.5.12. Suppose that f is a complex-valued analytic function on an open subset V of $\mathbb{C}$ and that f' does not vanish at any point of V. If there is an $h \in C(\Gamma_{\mathcal{A}})$ satisfying $f \circ h \equiv 0$ on $\Gamma_{\mathcal{A}}$, then we claim that there is a unique element $c \in \mathcal{A}$ satisfying

$$\hat{c} = h \quad \text{and} \quad f(c) = 0.$$

To prove this, we merely apply Theorem 3.5.12 to the case $n = 1, a = 0$, $U = V \times \mathbb{C}$ and $F = f \circ \pi_{2,1}$. It is this case $n = 0$ which we will use in our first application.

The next result and Corollary 3.5.15 are both known as the Šilov idempotent theorem. They were first proved by Šilov [1953]. We will use the implicit function theorem to give an extremely simple proof. Usually they are derived from the multivariable functional calculus directly. No proofs are known which avoid this heavy machinery, although Šilov's original proof depended only on his imperfect (although still highly nontrivial) multivariable functional calculus.

3.5.13 Corollary *Let $\mathcal{A}$ be a commutative Banach algebra. If K is a compact open subset of $\Gamma_{\mathcal{A}}$, then there is a unique idempotent $e \in \mathcal{A}$ such that $\hat{e}$ is the characteristic function of K.*

Proof If $\mathcal{A}$ is unital, apply the theorem to $F(w) = w^2 - w = 0$ and its solution h, the characteristic function of K in $C(\Gamma_{\mathcal{A}})$. If $\mathcal{A}$ is nonunital, replace $\mathcal{A}$ by $\mathcal{A}^1$. Theorem 3.1.2(a) shows that restriction to $\mathcal{A}$ defines a homeomorphism of $\Gamma_{\mathcal{A}^1}$ onto $\Gamma^0_{\mathcal{A}}$, which is the one point compactification of $\Gamma_{\mathcal{A}}$. Hence, the element e can be defined as before. It belongs to $\mathcal{A}$ since $\hat{e}$ clearly vanishes at infinity. □

We give two important corollaries of the Šilov idempotent theorem here. (We will also use the idempotent theorem in the proofs of Lemma 3.5.16 and Theorem 3.5.19.) The first gives a partial converse of the easy assertion that $\Gamma_{\mathcal{A}}$ is compact if $\mathcal{A}$ is a commutative unital Banach algebra. This shows that a semisimple commutative Banach algebra $\mathcal{A}$ is unital if and only if $\Gamma_{\mathcal{A}}$ is compact. Easy examples show that the hypothesis of semisimplicity cannot be dropped. This result is due to Šilov [1953].

3.5.14 Corollary *Let $\mathcal{A}$ be a commutative Banach algebra.*

(a) *The Gelfand space of $\mathcal{A}$ is nonempty and compact if and only if $\mathcal{A}$ modulo its Gelfand radical is unital. In this case $\mathcal{A}$ is algebra direct sum of of the closed ideals $e\mathcal{A}$ and $\mathcal{A}_J$.*

(b) *A hull H in $\Gamma_{\mathcal{A}}$ is compact if and only if its kernel $k(H)$ is a modular ideal.*

Proof (a): Denote the Gelfand radical of $\mathcal{A}$ by $\mathcal{A}_J$. Theorem 3.1.2(e) asserts that $\Gamma_{\mathcal{A}}$ is compact if $\mathcal{A}/\mathcal{A}_J$ is unital. In this case $\mathcal{A}_J$ is modular, so Theorem 1.2.24, Proposition 3.1.3 and Lemma 3.1.4 show that $\Gamma_{\mathcal{A}}$ is nonempty. Hence, $\Gamma_{\mathcal{A}}$ is nonempty and compact. Conversely, if $\Gamma_{\mathcal{A}}$ is nonempty and compact, Corollary 3.5.13 shows the existence of an idempotent e such that e is the identity function on $\Gamma_{\mathcal{A}}$. Hence, $e + \mathcal{A}_J$ is an identity element in $\mathcal{A}/\mathcal{A}_J$ and $e\mathcal{A}$ is an ideal satisfying $e\mathcal{A} \cap \mathcal{A}_J = \{0\}$.

(b): Theorem 3.2.7 now gives this. □

The next result is also due to Šilov [1953].

3.5.15 Corollary *The following are equivalent for a unital commutative Banach algebra $\mathcal{A}$.*

(a) $\Gamma_{\mathcal{A}}$ *is disconnected.*

(b) *There is an idempotent in $\mathcal{A}$ other than 0 and 1.*

(c) $\mathcal{A}$ *is the direct sum of two closed non-zero ideals.*

(d) $\mathcal{A}$ *is the direct sum of two non-zero ideals.*

Proof (a) $\Rightarrow$ (b): If $\Gamma_{\mathcal{A}}$ is disconnected, let it be the disjoint union of two nonempty open and closed sets K_1 and K_2. These sets are compact so Corollary 3.5.13 gives an idempotent e which is neither 0 nor 1 since its Gelfand transform is the characteristic function of K_1.

(b) $\Rightarrow$ (c): Let e be an idempotent not equal to zero or one. Then $e\mathcal{A} = \{a \in \mathcal{A} : ea = a\}$ and $(1 - e)\mathcal{A} = \{a \in \mathcal{A} : ea = 0\}$ are non-zero closed ideals satisfying $\mathcal{A} = e\mathcal{A} \oplus (1 - e)\mathcal{A}$.

(c) $\Rightarrow$ (d): Obvious.

(d) $\Rightarrow$ (b): If $\mathcal{I}_1$ and $\mathcal{I}_2$ are non-zero ideals satisfying $\mathcal{A} = \mathcal{I}_1 \oplus \mathcal{I}_2$, let $1 = e_1 \oplus e_2$ be the corresponding direct sum decomposition of 1. The identity $e_1 \oplus e_2 = 1 = 1^2 = e_1^2 \oplus e_2^2$ implies that e_1 and e_2 are idempotents. They are both non-zero since $\mathcal{I}_1$ and $\mathcal{I}_2$ are non-zero.

(b) $\Rightarrow$ (a): If e is a nontrivial idempotent, $\hat{e} \in \mathbb{C}(\Gamma_{\mathcal{A}})$ is a nontrivial continuous characteristic function, so $\Gamma_{\mathcal{A}}$ is disconnected. $\Box$.

As we have seen, the Šilov idempotent theorem lies very deep (at least in our present state of knowledge). However, in certain circumstances the existence of nontrivial idempotents can be proved in a much more elementary way. See Proposition 3.4.1.

We now wish to prove an important theorem of Charles E. Rickart [1953]. It depends on a lemma which uses the Šilov idempotent theorem. It also depends on the concepts of hulls and kernels and completely regular algebras introduced in Section 3.2. Irving Kaplansky [1949a] proved a result similar to Theorem 3.5.17 in which the algebra $\mathcal{A}$ was assumed to have the form $C(\Omega)$ for some locally compact Hausdorff space Ω.

3.5.16 Lemma *Let $\mathcal{A}$ be a commutative Banach algebra. Let $\mathcal{I}$ be an ideal of $\mathcal{A}$ and let H be a compact hull in $\Gamma_{\mathcal{A}}$ which is disjoint from $h(\mathcal{I})$. Then, there is an element $b \in \mathcal{I}$ which satisfies $\hat{b}(H) = \{1\}$.*

Proof Corollary 3.5.14 shows that $k(H)$ is a modular ideal. Hence, $\mathcal{I} + k(H)$ is also a modular ideal. It satisfies $h(\mathcal{I} + k(H)) = h(\mathcal{I}) \cap hk(H) = h(\mathcal{I}) \cap H = \emptyset$ and hence is not included in any maximal modular ideal. Theorem 2.4.6 thus shows $\mathcal{I} + k(H) = \mathcal{A}$. Corollary 3.5.14 then shows that there is some element $a \in \mathcal{A}$ satisfying $\gamma(a) = 1$ for all $\gamma \in H$. Let $a = b + c$ be a decomposition (guaranteed by $\mathcal{A} = \mathcal{I} + k(H)$) satisfying $b \in \mathcal{I}$ and $c \in k(H)$.

Since $\gamma(c) = 0$ holds for all $\gamma \in H$, the equation $\gamma(b) = \gamma(b+c) = \gamma(a) = 1$ shows that b has the desired property.   □

3.5.17 Theorem *Let $\mathcal{A}$ be a completely regular, semisimple commutative Banach algebra and let $\mathcal{B}$ be a commutative spectral algebra. Let $\varphi: \mathcal{A} \to \mathcal{B}$ be an injective homomorphism.*

(a) *The associated dual map $\varphi^\dagger: \Gamma_{\mathcal{B}}^0 \to \Gamma_{\mathcal{A}}^0$ maps $\Gamma_{\mathcal{B}}^0$ onto $\Gamma_{\mathcal{A}}^0$.*

(b) *Every element $a \in \mathcal{A}$ satisfies $Sp_{\mathcal{A}}(a) \cup \{0\} = Sp_{\mathcal{B}}(\varphi(a)) \cup \{0\}$. This result also holds if $\mathcal{B}$ is a normed algebra instead of a spectral algebra.*

(c) *Any (not necessarily complete) algebra norm $||| \cdot |||$ on $\mathcal{A}$ is spectral.*

Proof (a): Let B denote the image of $\Gamma_{\mathcal{B}}^0$ in $\Gamma_{\mathcal{A}}^0$ under $\varphi^\dagger$. Suppose B is not all of $\Gamma_{\mathcal{A}}^0$ and choose $\gamma_0 \in \Gamma_{\mathcal{A}}^0 \setminus B$. Then for each $\gamma \in \Gamma_{\mathcal{B}}^0$, there is an element $a_\gamma \in \mathcal{A}$ satisfying $\hat{a}_\gamma(\gamma_0) \neq \hat{a}_\gamma(\varphi^\dagger(\gamma))$. For each $\gamma \in \Gamma_{\mathcal{B}}^0$, define ε_γ and U_γ by $\varepsilon_\gamma = 2^{-1}|\hat{a}_\gamma(\gamma_0) - \hat{a}_\gamma(\varphi^\dagger(\gamma))|$ and $U_\gamma = \{\psi \in \Gamma_{\mathcal{B}}^0 : |\hat{a}_\gamma(\gamma_0) - \hat{a}_\gamma(\varphi^\dagger(\psi))| < \varepsilon_\gamma\}$ so that U_γ is an open set in $\Gamma_{\mathcal{B}}^0$ containing γ. Choose a finite subcover $U_{\gamma_1}, U_{\gamma_2}, \ldots, U_{\gamma_n}$ of the compact set $\Gamma_{\mathcal{B}}^0$ and denote each a_{γ_j} and ε_{γ_j} by a_j and ε_j, respectively. Let ε be $\min\{\varepsilon_1, \varepsilon_2, \ldots, \varepsilon_n\}$ and let V be the open neighborhood

$$V = \{\gamma \in \Gamma_{\mathcal{A}}^0 : |\hat{a}_j(\gamma_0) - \hat{a}_j(\gamma)| < \varepsilon \quad \forall \; j = 1, 2, \ldots, n\}$$

of γ_0 in $\Gamma_{\mathcal{A}}^0$ which is disjoint from B. Choose an open neighborhood W of γ_0 with compact closure in V. Lemma 3.5.16 shows that there are elements $b, c \in \mathcal{A}$ satisfying $\hat{b}(\gamma_0) = 1$, $\hat{b}(\gamma) = 0$ for all $\gamma \notin W$, $\hat{c}(\gamma) = 1$ for all $\gamma \in W$ and $\hat{c}(\gamma) = 0$ for all $\gamma \notin V$. Therefore, $\widehat{bc}$ equals $\hat{b}$, and since the Gelfand radical is zero this implies $bc = b$. Since $\widehat{\varphi(c)}$ vanishes on $\Gamma_{\mathcal{B}}^0$, $\varphi(c)$ has a quasi-inverse d in $\mathcal{B}$. However, this gives $\varphi(b) = \varphi(b) \circ \varphi(c) \circ d = \varphi(b \circ c) \circ d = \varphi(c) \circ d = 0$, which contradicts the condition $\hat{b}(\gamma_0) = 1$. Hence, the dual map is surjective.

(b): Theorem 3.1.5 shows that this follows from (a) when $\mathcal{B}$ is spectral. If $\mathcal{B}$ is normed, apply this result to the completion $\overline{\mathcal{B}}$ of $\mathcal{B}$ and note $Sp_{\mathcal{A}}(a) \cup \{0\} \subseteq Sp_{\mathcal{B}}(\varphi(a)) \cup \{0\} \subseteq Sp_{\overline{\mathcal{B}}}(\varphi(a)) \cup \{0\}$.

(c): If $||| \cdot |||$ is an algebra norm on $\mathcal{A}$, let $\varphi: \mathcal{A} \to \mathcal{B}$ be the injection of $\mathcal{A}$ into the completion $\mathcal{B}$ of $(\mathcal{A}, ||| \cdot |||)$. This result now follows from (b) and Proposition 2.5.15.   □

The special case of this theorem, which is due to Kaplansky [1949a], has a very simple proof which uses the relatively elementary Theorem 3.2.12 in place of Lemma 3.5.16 which depends on the Šilov idempotent theorem.

3.5.18 Corollary *Let $\mathcal{A}$ be a completely regular commutative Banach algebra and let σ be an algebra semi-norm on $\mathcal{A}$. Then σ is spectral if and only if $\mathcal{A}_J = \mathcal{A}_\Gamma$ is σ-closed.*

Proof If σ is spectral, Theorem 3.1.5(b) shows $\mathcal{A}_J = \mathcal{A}_\Gamma$ and Corollary 2.3.4 shows that this set is σ-closed. Conversely, suppose that $\mathcal{A}_J$ is σ-closed. Let $\bar{\sigma}$ be the norm induced on $\mathcal{A}/\mathcal{A}_J$. Since $\mathcal{A}/\mathcal{A}_J$ satisfies the hypotheses of Theorem 3.5.17, $\bar{\sigma}$ must be spectral. Thus, any $a \in \mathcal{A}$ satisfies $\rho(a) = \rho(a + \mathcal{A}_J) \leq \bar{\sigma}(a + \mathcal{A}_J) \leq \sigma(a)$. Hence σ is spectral. □

The Exponential Function and the Group $\mathcal{A}_G/\exp(\mathcal{A})$

The exponential function was introduced in Theorem 2.1.12 where it was shown that in a commutative unital Banach algebra $\mathcal{A}$, $\exp(\mathcal{A})$ is just the connected component of the identity in the group $\mathcal{A}_G$ of invertible elements. Theorem 3.4.2 gave the kernel of the homomorphism $\exp: \mathcal{A} \to \mathcal{A}_G$ and Example 3.4.3 showed that Theorem 3.4.4 could not completely describe the range. Now, with the help of the multivariable functional calculus, we can answer the resulting question and show relations to a number of interesting topological concepts.

Consider first the case $\mathcal{A} = C(X)$ where X is some compact Hausdorff space. We recall the definition of the first cohomotopy group of a topological space X. (For further information on this subject see Sze-Tsen Hu [1959].) The set G of continuous functions from X to $\mathbb{T}$ is obviously a group under pointwise multiplication. Two functions in G are said to be *homotopic in G* if one can be continuously deformed in G into the other. More precisely, we say f_0 and f_1 in G are homotopic if there is a continuous map $F: X \times [0, 1] \to \mathbb{T}$ (called a *homotopy*) satisfying $F(x, 0) = f_0(x)$ and $F(x, 1) = f_1(x)$ for all $x \in X$. Obviously, this is an equivalence relation on G. It is also obvious that it respects multiplication in G. Hence, the set of homotopy classes of G is again a group. It is denoted by $\pi^1(X)$ and is called the *first cohomotopy group* of X.

For any $f \in C(X)_G$, the function $f/|f|$ is defined and belongs to G. (In fact, $f(x, t) = f(x)/(1 - t + t|f(x)|)$ is a homotopy of f into $f/|f|$ in the set $C(X)_G$.) Let θ be the group homomorphism of $C(X)_G$ onto $\pi^1(X)$ defined by setting $\theta(f)$ equal to the homotopy class of $f/|f|$ in G. We claim that $\exp(C(X))$ is the kernel of θ so that $(C(X))_G/\exp(C(X))$ is isomorphic to $\pi^1(X)$.

For any $g \in C(X), \exp(g)/|\exp(g)|$ is just $\exp(i\mathrm{Im}(g))$. Thus, the homotopy

$$G(x, t) = \exp(i(1 - t)\mathrm{Im}(g)) \qquad \forall\, x \in X;\ t \in [0, 1]$$

between $\exp(g)/|\exp(g)|$ and the constant function 1 shows that $\theta(\exp(g))$ is the identity element in $\pi^1(X)$. Hence, $\exp(C(X)) \subseteq \ker(\theta)$ holds.

If, conversely, $\theta(f)$ is 1, then there is a homotopy $F: X \times [0, 1] \to \mathbb{T}$ satisfying $F(x, 0) = f(x)/|f(x)|$ and $F(x, 1) \equiv 1$. Since $t \mapsto F(\cdot, t)$ is a continuous map of $[0,1]$ into $C(X)_G$, $f/|f|$ belongs to the principal component of $C(X)_G$. Similarly, the homotopy between f and $f/|f|$ noted above shows that f belongs to the principal component of $C(X)_G$. Hence, Theorem 2.1.12 shows that f belongs to $\exp(C(X))$. This establishes that θ induces a group isomorphism

$$C(X)_G / \exp(C(X)) \simeq \pi^1(X). \tag{24}$$

We now consider $\mathcal{A}_G / \exp(\mathcal{A})$ for an arbitrary, unital, commutative Banach algebra. The Gelfand space is an invariant of a commutative spectral algebra. If the algebra is determined up to isomorphism, then the Gelfand space is determined up to homeomorphism. However, many nonisomorphic commutative Banach algebras have the same Gelfand space up to homeomorphism (*e.g.*, the disc algebra (§3.2.13) and the algebra of all continuous functions on the disc). Nevertheless, certain algebraic properties of a commutative Banach algebra depend only on its Gelfand space. Corollary 3.5.14 provides an example. It shows that a commutative Banach algebra is unital modulo its radical if and only if its Gelfand space is nonempty and compact. We will now show two more cases of this phenomenon. In fact, the next theorem shows that the zero and first order Čech cohomology groups of the Gelfand space with integer coefficients arise as natural algebraic invariants of the algebra. We will return to this interpretation after stating and proving the theorem.

The first half of the next theorem is a reformulation of the Šilov idempotent theorem. The second half is known as the Arens–Royden theorem. It was discovered by Arens [1963] and, partly independently, by Halsey L. Royden [1963]. See also Rickart [1969]. The proof we give is due to Gamelin [1969]. It is based on Oka's [1936] solution for open polynomial polyhedra of a problem solved by P. Cousin [1895] for polydiscs. We will now state this Cousin problem. For a proof, see Gunning and Rossi [1965].

Let $\{V_\alpha : \alpha \in A\}$ be an open cover for an open polynomial polyhedron $V \subseteq \mathbb{C}^n$. *Cousin data* for $\{V_\alpha : \alpha \in A\}$ is defined to be a family of functions $\{g_{\alpha,\beta} : (\alpha, \beta) \in A \times A\}$ such that each $g_{\alpha,\beta}$ is analytic on $V_\alpha \cap V_\beta$ and the family satisfies

$$g_{\alpha,\beta} + g_{\beta\gamma} + g_{\gamma\alpha} = 0 \quad \text{on} \quad V_\alpha \cap V_\beta \cap V_\gamma$$

for all $\alpha, \beta, \gamma \in A$. Oka's theorem asserts the existence of a family of functions $\{g_\alpha : \alpha \in A\}$ such that each g_α is analytic on V_α and the family satisfies

$$g_\alpha - g_\beta = g_{\alpha,\beta}$$

for all $\alpha, \beta \in A$.

3.5.19 Theorem *Let $\mathcal{A}$ be a unital commutative Banach algebra. Let $H^0(\mathcal{A})$ be the additive subgroup of $\mathcal{A}$ generated by the idempotents in $\mathcal{A}$*

and let $H^1(\mathcal{A})$ be the multiplicative group $\mathcal{A}_G/\exp(\mathcal{A})$. The Gelfand homomorphism induces an isomorphism of $H^0(\mathcal{A})$ onto $H^0(C(\Gamma_\mathcal{A}))$ and an isomorphism of $H^1(\mathcal{A})$ onto $H^1(C(\Gamma_\mathcal{A}))$.

Proof The idempotent elements in $C(\Gamma_\mathcal{A})$ are exactly the characteristic functions of open and closed subsets. Hence the Šilov idempotent theorem (Corollary 3.5.13) shows that the Gelfand homomorphism restricted to $H^0(\mathcal{A})$ is an isomorphism onto $H^0(C(\Gamma_\mathcal{A}))$.

It is obvious that the Gelfand homomorphism maps $\mathcal{A}_G$ into $C(\Gamma_\mathcal{A})_G$ and $\exp(\mathcal{A})$ into $\exp(\hat{\mathcal{A}})$. Hence, (ˆ) does induce a homomorphism of $H^1(\mathcal{A})$ into $H^1(C(\Gamma_\mathcal{A}))$. The implicit function theorem (Theorem 3.5.12) applied to $F(w,z) = e^w - z$, shows that this homomorphism is injective. It only remains to show that it is surjective. In other words, we must show that for any $f \in C(\Gamma_\mathcal{A})_G$, there is a $c \in \mathcal{A}_G$ such that $f/\hat{c}$ belongs to $\exp(C(\Gamma_\mathcal{A}))$.

Let $f \in C(\Gamma_\mathcal{A})_G$ be given. By the Stone–Weierstrass theorem (1.5.1), the algebra $\{\sum_{j=1}^n \hat{b}_j \hat{b}_{n+j} : n \in \mathbb{N}, b_j \in \mathcal{A}\}$ is dense in $C(\Gamma_\mathcal{A})$. Hence, we can find an element $b = (b_1, b_2, \ldots, b_{2n}) \in \mathcal{A}^{(2n)}$ satisfying

$$\|1 - f^{-1} \sum_{j=1}^n \hat{b}_j \hat{b}_{n+j}^*\|_\infty < 1.$$

Theorem 3.4.4 shows that there is a function $g \in C(\Gamma_\mathcal{A})$ satisfying

$$f^{-1} \sum_{j=1}^n \hat{b}_j \hat{b}_{n+j}^* = \exp(g).$$

Thus, it is enough to find elements $c \in \mathcal{A}_G$ and $h \in C(\Gamma_\mathcal{A})$ satisfying

$$\hat{c}^{-1} \sum_{j=1}^n \hat{b}_j \hat{b}_{n+j}^* = \exp(h), \tag{25}$$

since we will then have $f/\hat{c} = \exp(h - g)$.

Choose a neighborhood U of $Sp(b) \subseteq \mathbb{C}^{2n}$ such that

$$\tilde{p} = \sum_{j=1}^n z_j z_{n+j}^*$$

does not vanish on U where $z = (z_1, z_2, \ldots, z_{2n})$ is the coordinate function on $\mathbb{C}^{2n}$. Proposition 3.5.8 asserts the existence of an element $d \in \mathcal{A}^{(m-2n)}$ such that $\pi_{m,2n}\mathrm{pc}(Sp(b,d))$ is included in U and includes $Sp(b)$. Lemma 3.5.2 shows that we can find an open polynomial polyhedron $V \subseteq \mathbb{C}^m$ satisfying $\mathrm{pc}(Sp(b,d)) \subseteq V \subseteq \pi_{m,2n}^{\leftarrow}(U)$ and hence also satisfying $Sp(b) \subseteq \pi_{m,2n}(V) \subseteq U$. Then, p defined by $p = \tilde{p} \circ \pi_{m,2n}$ does not vanish on the open polynomial polyhedron V. Let $\{V_\alpha : \alpha \in A\}$ be an open cover of V such that, on each V_α, p has a continuous logarithm p_α which varies in absolute value by less than π. For each ordered pair (α, β), define $p_{\alpha,\beta}$ by $p_{\alpha,\beta} = p_\alpha - p_\beta$. Since $\exp(p_{\alpha,\beta}) = \exp(p_\alpha)\exp(-p_\beta) = 1$ holds, and $|p_\alpha - p_\beta|$ varies by less than 2π, each $p_{\alpha,\beta}$ is a constant function which is an integer multiple of $2\pi i$. Furthermore, we have

$$p_{\alpha,\beta} + p_{\beta,\gamma} + p_{\gamma,\alpha} = 0$$

for each triple α, β, γ. Hence, the $\{p_{\alpha,\beta} : \alpha, \beta \in A\}$ are Cousin data for the open covering $\{V_\alpha : \alpha \in A\}$ of the polynomially convex set V. Thus, by the theorem of Oka stated above, there are analytic functions $\{q_\alpha : \alpha \in A\}$ such that each q_α is defined on V_α and satisfies $q_\alpha - q_\beta = p_{\alpha,\beta}$ on $V_\alpha \cap V_\beta$ for each pair (α, β). Hence, there is a well-defined analytic function q satisfying $q|V_\alpha = \exp(q_\alpha)$ for $\alpha \in A$. Note that q is never zero on V since it is locally an exponential function. Furthermore, since $p_\alpha - q_\alpha = p_\beta - q_\beta$ holds on $V_\alpha \cap V_\beta$ for any pair (α, β), there is a continuous function $\tilde{h}$ on V satisfying $\tilde{h}|V_\alpha = p_\alpha - q_\alpha$ for all α. Therefore, $q^{-1}p = \exp(\tilde{h})$ holds on V since it holds on each V_α. Define $c \in \mathcal{A}$ to be $q(\hat{b}, \hat{d})$. Then c is invertible (since q was) and satisfies $\hat{c}^{-1} \sum_{j=1}^{n} \hat{b}_j \hat{b}_{n+j}^* = \hat{c}^{-1}\hat{p}(\hat{b}) = (q(\hat{b}, \hat{d})^{-1}p(\hat{b}, \hat{d})) = \exp(\tilde{h}(\hat{b}, \hat{d}))$. This is what we wished to show. $\square$

We now interpret $H^0(\mathcal{A})$ and $H^1(\mathcal{A})$ as cohomology groups. For details on sheaf theory, see Roger Godement [1958]. For any commutative unital Banach algebra $\mathcal{A}$, we have an exact sequence

$$0 \longrightarrow H^0(\mathcal{A}) \xrightarrow{2\pi i} \mathcal{A} \xrightarrow{\exp} \mathcal{A}_G \longrightarrow H^1(\mathcal{A}) \longrightarrow 0.$$

For the special case $\mathcal{A} = C(X)$, this sequence has a sheaf theoretic interpretation. (In fact a similar interpretation is valid at least whenever $\mathcal{A}$ is semisimple and completely regular.) Let $\tilde{Z}$ be the constant sheaf on X of integers under addition. Let $\tilde{C}$ and $\tilde{C}_G$ be the sheaves on X of germs of continuous functions under addition and of germs of nonvanishing continuous functions under multiplication. Then multiplication by $2\pi i$ and the exponential map induce a short exact sequence of sheaves

$$0 \longrightarrow \tilde{Z} \xrightarrow{2\pi i} \tilde{C} \xrightarrow{\exp} \tilde{C}_G \longrightarrow 0.$$

This gives rise to an exact sequence of cohomology groups

$$0 \longrightarrow H^0(X, \tilde{Z}) \xrightarrow{2\pi i} H^0(X, \tilde{C}) \xrightarrow{\exp} H^0(X, \tilde{C}_G)$$

$$\xrightarrow{\partial} H^1(X, \tilde{Z}) \longrightarrow H^1(X, \tilde{C}) \longrightarrow \cdots.$$

However, the zero order cohomology group with coefficients in a sheaf is just the group of global sections of the sheaf and $H^1(X, \tilde{C}) = 0$ holds, since $\tilde{C}$ is a fine sheaf. Clearly, the groups of global sections of $\tilde{Z}, \tilde{C}$, and $\tilde{C}_G$ are, respectively, $H^0(C(X)), C(X)$ and $C(X)_G$. Therefore, the cohomology exact sequence becomes

$$0 \longrightarrow H^0(X, \tilde{Z}) \xrightarrow{2\pi i} C(X) \xrightarrow{\exp} C(X)_G$$

$$\xrightarrow{\partial} H^1(X, \tilde{Z}) \longrightarrow 0$$

where we could substitute $H^0(C(X))$ and $H^1(C(X))$ for $H^0(X, \tilde{Z})$ and $H^1(X, \tilde{Z})$ if we wished. The Arens–Royden theorem gives the identification of $H^0(\mathcal{A})$ with $H^0(C(X)) = H^0(X, \tilde{Z})$ and of $H^1(\mathcal{A})$ with $H^1(C(X)) = H^1(X, \tilde{Z})$. Since we have already identified the cohomotopy group $\pi^1(X)$ as $C(X)_G / \exp(C(X))$, we have established a natural isomorphism of $H^1(X, \tilde{Z})$ onto $\pi^1(X)$ for any compact space X. Recall also that Theorem 2.1.12 shows that $C(X)_G / \exp(C(X))$ is torsion free.

Edward K. Blum [1953] gives two interesting arguments, which are independent of the Šilov idempotent theorem, showing that $H^0(\mathcal{A})$ is the fundamental group $\pi_1(G)$ of the principal component G of $\mathcal{A}_G$.

Taylor ([1971], [1972a], [1975] and [1976]) defines natural variants of $H^0(C(X))$ and $H^1(C(X))$ for a pair $Y \subseteq X$ of compact spaces and shows that the resulting structure satisfies the axioms of cohomology theory. From the uniqueness of structures satisfying these axioms, he infers that these groups are identical to the Čech cohomology groups with integer coefficients. The axioms allow him to identify $H^0(M(G))$ (the result is essentially Paul J. Cohen's idempotent theorem [1960a], [1959b]) and $H^1(M(G))$ where $M(G)$ is the measure algebra on a locally compact abelian group (cf. the next section). (This general result gave concrete new information on which invertible measures on $\mathbb{R}$ are exponentials.) Taylor also obtains results in the more general case of convolution measure algebras. Along similar lines, he defined and studied the K-theory of a commutative Banach algebra [1973a] using the results of Arens [1966].

For some related results the reader may wish to consult Rickart [1969], Iain Raeburn and Taylor [1977], and Kallin [1963]. The last named article contains an example related to the first article of Arens. The literature of commutative Banach algebras is enormous. For further general information see the following books: Browder [1969], Gamelin [1969], Stout [1971], and Wermer [1976].

Taylor Spectrum and Functional Calculus in $\mathcal{B}(\mathcal{X})$

We give only the briefest introduction to this subject and refer the reader to Helemskiĭ [1981] and [1989b] Chapter VI, the original articles

Taylor [1970a], [1970b], [1970c], [1972a] and [1972b], Vasilescu [1982] and the forthcoming book by Putinar for further details.

If $\mathcal{X}$ is a Banach space, $S = (S_1, S_2, \ldots, S_n)$ and $\lambda = (\lambda_1, \lambda_2, \ldots, \lambda_n)$ are ordered n-tuples of commuting operators in $\mathcal{B}(\mathcal{X})$ and of complex numbers, respectively, then λ belongs to the Taylor spectrum if the Koszul complex $\mathcal{K}(\mathcal{X}, S - \lambda)$ is not exact. For lack of space, we shall not define the Koszul complex (see Helemskiĭ [1989b], V.1.1). Its exactness implies $\bigcap_{1 \leq k \leq n} \ker(S_k) = \{0\}$ and $\sum_{k=1}^{n} S_k(\mathcal{X}) = \mathcal{X}$. The other conditions involved in proving exactness are of a similar nature. (For instance if $n = 2$, the only other condition is that $S_2 x_1 = S_1 x_2$ should imply the existence of a $z \in \mathcal{X}$ satisfying $x_1 = S_1 z$ and $x_2 = S_2 z$.) Thus, although the Taylor spectrum is hard to define, it is not hard to compute.

In spite of the fact that we have not even explained the definition of Taylor's joint spectrum adequately, we believe it will be worthwhile to state the resulting theorem which relates it to the functional calculus. The proof may be put together from Helemskiĭ [1989b] and Taylor [1970b]. We begin with a simple definition of a functional calculus.

3.5.20 Definition Let $a = (a_1, a_2, \ldots, a_n)$ be an ordered n-tuple of commuting elements in a unital Banach algebra $\mathcal{A}$. Let a'' denote the double commutant of the set $\{a_1, a_2, \ldots, a_n\}$ in $\mathcal{A}$. Let U be an open subset of $\mathbb{C}^n$. We say that a has a functional calculus on U if there is a continuous unital homomorphism $I_a^U : \mathcal{F}_U \to a'' \subseteq \mathcal{A}$ satisfying

$$I_a^U(z_k) = a_k \qquad \forall \, k = 1, 2, \ldots, n.$$

3.5.21 Theorem Let $\mathcal{X}$ be a Banach space and let $S = (S_1, S_2, \ldots, S_n)$ be an ordered n-tuple of commuting operators in $\mathcal{B}(\mathcal{X})$.

(a) The Taylor spectrum $Sp^T(S)$ of S is a nonempty compact subset of $\mathbb{C}^n$ satisfying $|\lambda_k| \leq \rho(S_k)$ for all $\lambda = (\lambda_1, \lambda_2, \ldots, \lambda_n) \in Sp^T(S)$ and $k = 1, 2, \ldots, n$.

(b) If $\mathcal{A}$ is any closed unital subalgebra of $\mathcal{B}(\mathcal{X})$ including $\{S_1, S_2, \ldots, S_n\}$ in its center, then $Sp^T(S)$ is included in $Sp_{\mathcal{A}}(S)$. There do exist examples where this inclusion is proper for all choices of $\mathcal{A}$.

(c) The ordered n-tuple S has a functional calculus on every open subset U of $\mathbb{C}^n$ which includes $Sp^T(S)$.

(d) If U is a domain of holomorphy, then S has a functional calculus on U if and only if $Sp^T(S) \subseteq U$. In this case the functional calculus for S on U is unique.

(e) Let $f = (f_1, f_2, \ldots, f_m)$ be an ordered m-tuple of functions analytic on an open set U which includes $Sp^T(S)$. Then $Sp^T(f(S)) = f(Sp^T(S))$ where we have written $f(S)$ for $I_S^U(f)$. Hence if U and $V \supseteq Sp^T(f(S))$ are domains of holomorphy and g is analytic on V, then $g \circ f(S)$ is defined and equals $g(f(S))$.

3.6   Commutative Group Algebras

Section 1.8 described the group algebras associated with the compact abelian group $\mathbb{T}$ in some detail. This involved some results about group algebras of the discrete abelian group $\mathbb{Z}$. Section 1.9 gave basic information about the Banach algebras $L^1(G)$ and $M(G)$ for an arbitrary locally compact group G. In particular, 1.9.13 identified $M(G)$ with the double centralizer algebra of $L^1(G)$. For locally compact abelian groups, 1.9.9 showed that both algebras are commutative. The reader may wish to consult those sections for background information and notation. All notation is summarized in Table 1, p. 144. In the present section we will show that the Gelfand space of $L^1(G)$ can be identified with the locally compact group $\hat{G}$ of all continuous homomorphisms of G into $\mathbb{T}$, with pointwise product and the compact open topology. The group $\hat{G}$ is called the *character group* of G. We then prove that the natural map of G into $\hat{\hat{G}}$ (the character group of $\hat{G}$) is a homeomorphic group isomorphism. The proof, which depends on showing that $L^1(G)$ is completely regular, is essentially due to David Abramovič Raĭkov [1945] (written in 1940), Henri Cartan and Roger Godement [1947]. We also give a brief survey of some results of Joseph L. Taylor [1965], [1973a] on the Gelfand space of $M(G)$. The section concludes by mentioning some other group algebras.

We begin with a simple special case of the result we will prove in complete generality below. Let G be a discrete abelian group and define $\hat{G}$ to be the set of group homomorphisms from G to $\mathbb{T}$. We will regard $\hat{G}$ as a group under pointwise multiplication:

$$hk(u) = h(u)k(u) \qquad \forall h, k \in \hat{G}; \ u \in G$$

and as a topological space under the topology of pointwise convergence. (If we identify each homomorphism $h \in \hat{G}$ with its graph in $\prod_{u \in G} \mathbb{T} = \mathbb{T}^G$, this becomes the product topology.) If $\{h_\alpha\}_{\alpha \in A}$ is a net in $\hat{G}$ converging in this topology to an element $h \in \mathbb{T}^G$, then h satisfies $h(uv) = \lim h_\alpha(uv) = \lim h_\alpha(u)h_\alpha(v) = \lim h_\alpha(u) \lim h_\alpha(v) = h(u)h(v)$ for all $u, v \in G$. Thus $\hat{G}$ is closed in $\mathbb{T}^G$ and hence is compact. We claim that $\hat{G}$ is a topological group. Continuing with the same notation, we find $\lim h_\alpha^{-1}(u) = \lim h_\alpha(u)^* = h(u)^* = h^{-1}(u)$. If $\{k_\alpha\}_{\alpha \in A}$ is another net in $\hat{G}$ converging to an element $k \in \hat{G}$, then it satisfies $\lim h_\alpha(u)k_\alpha(u) = \lim h_\alpha(u) \lim k_\alpha(u) = h(u)k(u) = hk(u)$. This verifies our claim, so $\hat{G}$ is a compact abelian group.

Theorem   *Let G be a discrete abelian group. Let $\hat{G}$ be the compact abelian group described above and let $\Gamma_{\ell^1(G)}$ be the Gelfand space of the commutative unital group algebra $\ell^1(G)$. Then the isometric linear isomorphism*

$\omega \colon \ell^\infty(G) \to \ell^1(G)^*$ *defined by*

$$\omega_h(f) = \sum_{u \in G} f(u)h(u^{-1}) \qquad \forall h \in \ell^\infty(G); \ f \in \ell^1(G)$$

sends $\hat{G} \subseteq \ell^\infty(G)$ *(with its topology of pointwise convergence) homeomorphically onto* $\Gamma_{\ell^1(G)} \subseteq \ell^1(G)^*$ *(with its Gelfand topology).*

Proof Only the last sentence remains to be proven. Any $h \in \hat{G}$ and any $f, g \in \ell^1(G)$ satisfy $h(u^{-1}) = h(u)^*$ and hence

$$\begin{aligned}
\omega_h(f * g) &= \sum_{u \in G} f * g(u)h(u)^* = \sum_{u \in G} \sum_{v \in G} f(v)g(v^{-1}u)h(u)^* \\
&= \sum_{v \in G} \sum_{u \in G} f(v)g(v^{-1}u)h(v)^*h(v^{-1}u)^* \\
&= \sum_{v \in G} f(v)h(v)^* \sum_{u \in G} g(v^{-1}u)h(v^{-1}u)^* = \omega_h(f)\omega_h(g).
\end{aligned}$$

Hence ω sends $\hat{G}$ into $\Gamma_{\ell^1(G)}$. The map ω is well known to be a linear isometry of $\ell^\infty(G)$ onto $\ell^1(G)^*$. Thus if γ belongs to the Gelfand space there is a unique $h \in \ell^\infty(G)_1$ satisfying $\gamma = \omega_h$. This function then satisfies $\gamma(\delta_u) = h(u)^*$ for any $u \in G$. For $u, v \in G$ we get $h(u)h(v) = \gamma(\delta_u)^*\gamma(\delta_u)^* = \gamma(\delta_u * \delta v)^* = h(uv)$, so h belongs to $\hat{G}$. This shows that ω is a bijection of $\hat{G}$ onto $\Gamma_{\ell^1(G)}$. Since both spaces are compact and Hausdorff, ω is a homeomorphism if either it or its inverse is continuous. It is not hard to check that ω is continuous, but we take this opportunity to note that the inverse map is given by the obviously continuous map $\gamma \mapsto h_\gamma$ defined by $h_\gamma(u) = \gamma(\delta_u)^*$ for all $\gamma \in \Gamma_{\ell^1(G)}$ and $u \in G$. □

Now let G be an arbitrary locally compact group. The Banach space dual $L^1(G)^*$ of $L^1(G)$ can be identified with $L^\infty(G)$. (See Hewitt and Ross [1963], Section 12.18, for a proof in the generality needed here.) With each $h \in L^\infty(G)$, we will associate the linear functional $\omega_h \in L^1(G)^*$ defined by

$$\omega_h(f) = \int_G f(u)h(u^{-1})du = \int f\tilde{h}d\lambda \qquad \forall f \in L^1(G) \tag{1}$$

(where $\tilde{h}$ is defined by $\tilde{h}(u) = h(u^{-1})$ for all $u \in G$). The twist built into this definition gives the following formulas

$$\omega_h(f * g) = \omega_{g*h}(f) = \omega_{h*\triangle_f}(g) \qquad \forall h \in L^\infty(G); \ f, g \in L^1(G) \tag{2}$$

where the convolution products are defined and belong to $L^\infty(G)$ even when G is a not necessarily unimodular group. In the unimodular case (*a fortiori*, in the abelian case), these reduce to

$$\omega_h(f * g) = \omega_{g*h}(f) = \omega_{h*f}(g) \qquad \forall h \in L^\infty(G); \ f, g \in L^1(G). \tag{3}$$

Of course, the proofs depend on Fubini's theorem.

Next suppose that G is a locally compact abelian group and note that any γ in the Gelfand space Γ_G of $L^1(G)$ belongs to the unit ball of $L^1(G)^*$ and hence can be associated with a function $h_\gamma \in L^\infty(G)_1$ satisfying

$$\gamma(f) = \int f(u)h_\gamma(u^{-1})du \tag{4}$$

and

$$\int f * g(u)h_\gamma(u^{-1})du = \int f(u)h_\gamma(u^{-1})du \int g(u)h_\gamma(u^{-1})du \tag{5}$$

for all $f, g \in L^1(G)$. It is easy to see (just use the definition of convolution multiplication and Fubini's theorem) that the second formula will hold if h_γ is multiplicative

$$h_\gamma(uv) = h_\gamma(u)h_\gamma(v) \qquad \forall\, u, v \in G$$

and the integrals exist. Thus we are led to consider bounded group homomorphism of G into the multiplicative group of $\mathbb{C}$. In fact, by considering $h(u^n) = h(u)^n$ for $u \in G$ and $n \in \mathbb{Z}$ we see that such a bounded group homomorphism h into the multiplicative group of $\mathbb{C}$ must be a group homomorphism into $\mathbb{T} = \{\lambda \in \mathbb{C}: |\lambda| = 1\}$. We give a formal definition.

3.6.1 Definition Let G be a group. A homomorphism of G into $\mathbb{T}$ is called a *character* of G. If G is a topological group, then the collection of all continuous characters of G is denoted by $\hat{G}$. This set is endowed with pointwise multiplication and the compact–open topology. It is called the *character group* or *dual group* of G.

It is obvious that $\hat{G}$ is a group under pointwise multiplication. The compact–open topology is defined by the basic open neighborhoods

$$N(h_0, K, \varepsilon) = \{h \in \hat{G}: |h(u) - h_0(u)| < \varepsilon \text{ for all } u \in K\} \tag{6}$$

where $h_0 \in \hat{G}$, K a compact subset of G and $\varepsilon > 0$ are arbitrary. It is easy to see that $\hat{G}$ is a topological group under this topology. Note that any character on any (not necessarily topological) group satisfies

$$h(e) = 1 \tag{7}$$
$$|h(u)| = 1 \qquad \forall\, u \in G \tag{8}$$
$$\tilde{h}(u) = h(u^{-1}) = h(u)^* = \overline{h}(u) \quad \forall\, u \in G \tag{9}$$
$$h(uv) = h(u)h(v) \qquad \forall\, u, v \in G. \tag{10}$$

So far we have noted that the map $h \mapsto \omega_h$ defined by

$$\omega_h(f) = \int f\tilde{h}d\lambda = \int f\overline{h}d\lambda = \int f(u)h(u)^*du \qquad \forall\, f \in L^1(G) \tag{11}$$

maps $\hat{G}$ into Γ_G. This map is injective since distinct continuous functions give rise to distinct elements of $L^\infty(G)$ and hence to distinct elements in $L^1(G)^*$. More is true.

3.6.2 Proposition *Let G be a locally compact abelian group. Then the map $h \mapsto \omega_h$ defined by (11) is a bijection of the set $\hat{G}$ onto the Gelfand space Γ_G of $L^1(G)$.*

Proof It only remains to show that the map is surjective. The best way to prove this is to show that for any $\gamma \in \Gamma_G$ the function $h_\gamma \in L^\infty(G)$ satisfying (4) actually belongs to $\hat{G}$. For any $g \in L^1(G)$, consider the function $u \mapsto \gamma(_ug)$ from G to $\mathbb{C}$. We have already noted in Section 1.9 that $u \mapsto {}_ug$ is a continuous map of G into $L^1(G)$ so that $u \mapsto \gamma(_ug)$ is continuous and is bounded by $\|g\|_1$. We claim that this function satisfies

$$\gamma(g)h_\gamma(u^{-1}) = \gamma(_ug) \tag{12}$$

as an element of $L^\infty(G)$ (*i.e.*, almost everywhere). To establish this, it is enough to show that $\int \gamma(g)h_\gamma(u^{-1})f(u)du = \int \gamma(_ug)f(u)du$ holds for all $f \in L^1(G)$. However, Fubini's theorem and (4) give

$$
\begin{aligned}
\int \gamma(g)h_\gamma(u^{-1})f(u)du &= \gamma(g)\gamma(f) = \gamma(f * g) \\
&= \int h_\gamma(v^{-1})\int f(u)g(u^{-1}v)du\,dv \\
&= \int f(u)\int h_\gamma(v^{-1}){}_ug(v)dv\,du \\
&= \int f(u)\gamma(_ug)du
\end{aligned}
$$

as we wished to show. Choose $g \in L^1(G)$ to satisfy $\gamma(g) \neq 0$ and change h_γ on a set of measure zero so that it satisfies (11) everywhere and hence is continuous on all of G. This gives

$$
\begin{aligned}
\gamma(g)h_\gamma(v^{-1}u^{-1}) &= \gamma(_{uv}g) = \gamma(_u(_vg)) \\
&= \gamma(_vg)h_\gamma(u^{-1}) = \gamma(g)h_\gamma(v^{-1})h_\gamma(u^{-1})
\end{aligned}
$$

for all u, $v \in G$, so that h_γ is indeed a continuous character on G.   □

We have identified $\hat{G}$ and Γ_G as sets. However Γ_G is a locally compact topological space under the Gelfand topology and $\hat{G}$ is a topological abelian group under pointwise multiplication and the compact–open topology. It is not immediately obvious that the group operations on $\hat{G}$ are continuous in the Gelfand topology transferred from Γ_G, nor is it clear that the compact–open topology on $\hat{G}$ is locally compact. However, we will prove both of these at once by showing that the bijection $h \mapsto \omega_h$ of $\hat{G}$ onto Γ_G is actually a homeomorphism.

3.6.3 Theorem *Let G be a locally compact abelian group. Then the map $h \mapsto \omega_h$ defined by (11) is a homeomorphism of $\hat{G}$ with its compact–open topology onto Γ_G with its Gelfand topology. Hence $\hat{G}$ is a locally compact abelian group and a net $\{h_\alpha\}_{\alpha \in A} \subseteq \hat{G}$ converges to $h_0 \in \hat{G}$ if and only if $\lim \int h_\alpha(u^{-1})f(u)du = \int h_0(u^{-1})f(u)du$ holds for all $f \in L^1(G)$. Furthermore the map*

$$(h, u) \mapsto h(u) \qquad \forall \, h \in \hat{G}; \; u \in G$$

of $\hat{G} \times G \to \mathbb{C}$ is jointly continuous.

Proof First we show that if $\{h_\alpha\}_{\alpha \in A} \subseteq \hat{G}$ converges to $h_0 \in \hat{G}$ then the last statement of the theorem holds. For any given $f \in L^1(G)$ and $\varepsilon > 0$ choose a compact set $K \subseteq G$ satisfying $\int_{G \backslash K} |f| d\lambda < \varepsilon/4$. Then $h_\alpha \in N(h_0, K^{-1}, \varepsilon/(2\|f\|_1 + 1))$ implies

$$\left| \int h_\alpha(u^{-1})f(u)du - \int h_0(u^{-1})f(u)du \right|$$

$$\leq \quad (\int\limits_{K} + \int\limits_{G \backslash K})|h_\alpha(u^{-1}) - h_0(u^{-1})| \, |f(u)|du$$

$$\leq \quad (\varepsilon/(2\|f\|_1 + 1))\|f\|_1 + 2(\varepsilon/4) < \varepsilon$$

as desired. Hence $h \mapsto \omega_h$ is continuous from $\hat{G}$ to Γ_G.

To complete the proof, we must show that the image in Γ_G of any compact–open neighborhood $N(h_0, K, \varepsilon) \subseteq \hat{G}$ includes a Gelfand neighborhood. In a moment we will show that $(\omega_h, u) \mapsto h(u)$ is jointly continuous from $\Gamma_G \times G$ to $\mathbb{C}$ with the Gelfand topology on Γ_G. Assuming this, we see that for each $u \in K$ we can choose a Gelfand neighborhood U_u of ω_{h_0} in Γ_G and a neighborhood W_u of u in G so that $\omega_h \in U_u$ and $v \in W_u$ imply $|h(v) - h_0(u)| < \varepsilon/2$. Since K is compact we can find a finite set $u_1, u_2, \ldots, u_n$ satisfying $K \subseteq \bigcup_{j=1}^{n} W_{u_j}$. Let U be the Gelfand neighborhood of ω_{h_0} defined by $U = \bigcap_{j=1}^{n} U_{u_j}$. Then any $\omega_h \in U$ and $v \in K$ satisfy

$$|h_0(v) - h(v)| \leq |h_0(v) - h_0(u_j)| + |h_0(u_j) - h(v)| < \varepsilon$$

for the j satisfying $v \in U_{u_j}$. This shows $U \subseteq N(h_0, K, \varepsilon)$ as we wished.

It remains to show the joint continuity of the map $(\omega_h, u) \mapsto h(u)$ for $h \in \hat{G}$ and $u \in G$. Let $\gamma_0 \in \Gamma_G$ and $u_0 \in G$ be given and choose an $f \in L^1(G)$ satisfying $\gamma_0(f) \neq 0$ and a Gelfand neighborhood U of γ_0 such that $\omega_h \in U$ implies $\omega_h(f) \neq 0$. Then all $\omega_h \in U$ satisfy

$$h(u) \quad = \quad h(u)\int f(v)h(v^{-1})dv/\omega_h(f) = \int f(v)h(uv^{-1})dv/\omega_h(f)$$

$$= \int f(vu^{-1})h(u)dv/\omega_h(f) = \omega_h(f_u)/\omega_h(f).$$

Hence it is enough to show that $(\gamma, u) \mapsto \gamma(f_u)$ is continuous on $U \times G$. For any $\varepsilon > 0$, choose a neighborhood W of $u_0 \in G$ satisfying $\|f_u - f_{u_0}\|_1 < \varepsilon/2$ and the Gelfand neighborhood $U_0 \subseteq U$ of γ_0 defined by $|\gamma(f_{u_0}) - \gamma_0(f_{u_0})| < \varepsilon/2$ for all $\gamma \in U_0$. Then any $(\gamma, u) \in U_0 \times W$ satisfies $|\gamma(f_u) - \gamma_0(f_{u_0})| \le |\gamma(f_u) - \gamma(f_{u_0})| + |\gamma(f_{u_0}) - \gamma_0(f_{u_0})| < \|f_u - f_{u_0}\|_1 + \varepsilon/2 < \varepsilon$. Hence the map is jointly continuous.   □

Given any locally compact abelian group G, we have constructed another locally compact abelian group $\hat{G}$. The construction can be repeated to construct the character group $\hat{\hat{G}}$ of $\hat{G}$. There is a natural map of G into $\hat{\hat{G}}$ defined by

$$\theta(u)(h) = h(u) \qquad \forall \, u \in G; \; h \in \hat{G}. \tag{13}$$

The joint continuity of $(h, u) \mapsto h(u)$ shows that $\theta(u)$ is a continuous character on $\hat{G}$. Obviously θ is a homomorphism. We now wish to prove the Pontryagin duality theorem which asserts that θ is a homeomorphic isomorphism of G onto $\hat{\hat{G}}$. This theorem establishes a complete duality between G and $\hat{G}$ in that each is the character group of the other (up to homeomorphic isomorphism). It is, perhaps, the most fundamental result of commutative harmonic analysis.

Since the duality map $\theta \colon G \to \hat{\hat{G}}$ (defined by (13)) is obviously a homomorphism, the duality theorem will be proved when the following results have been established:

(a) θ is injective.

(b) θ is a homeomorphism onto its image.

(c) $\theta(G)$ is closed in $\hat{\hat{G}}$.

(d) $\theta(G)$ is dense in $\hat{\hat{G}}$.

Establishing (a) is equivalent to showing that $\hat{G}$ separates the points of G. Since we have identified $\hat{G}$ with Γ_G, this is obviously related to (and in fact follows easily from) showing that Γ_G separates the points of $L^1(G)$. The latter is precisely the question of whether $L^1(G)$ is semisimple. This would be the next natural question to ask in the study of the commutative Banach algebra $L^1(G)$ after finding a representation of its Gelfand space. We will give an easy proof of the semisimplicity of $L^1(G)$ in Volume II, but here we will prove that $\hat{G}$ separates the points of G directly.

Result (b) can be proved by showing that when G is considered as a family of maps of $\hat{G}$ into $\mathbb{C}$ *via* the map θ, then the topology of G agrees with the compact–open topology. Then result (c) will follow from the elementary fact that any locally compact subgroup of a locally compact group must be closed (Hewitt and Ross [1963], 5.11).

Result (d) is an immediate consequence of any proof that $L^1(G)$ is a completely regular commutative Banach algebra whenever G is a locally compact abelian group. (See Jean Esterle and José E. Galé [1982] for an extension.) When this result is applied to $L^1(\hat{G})$, it easily shows that the image of $\theta(G)$ is dense in $\hat{\hat{G}}$ identified with the Gelfand space of $L^1(\hat{G})$. For the second time we see that a major step in establishing the Pontryagin duality theorem is a natural step in studying the commutative Banach algebra $L^1(G)$ (*viz.* determining whether it is completely regular).

The proofs of (a), (b), and (d) can all be based on a Fourier inversion theorem. First we introduce the Fourier–Stieltjes transform. Let G be a locally compact abelian group. The Gelfand transform is a contractive homomorphism of $L^1(G)$ into $C_0(\Gamma_G)$. The identification of Γ_G with $\hat{G}$ given by (11) allows us to make the following definition.

3.6.4 Definition Let G be a locally compact abelian group. The *Fourier transform* is the contractive homomorphism $(\hat{\ }): L^1(G) \to C_0(\hat{G})$ defined by

$$\hat{f}(h) = \omega_h(f) = \int f\tilde{h}d\lambda = \int f\overline{h}d\lambda = \int f(u)h(u)^*du \quad \forall\, f \in L^1(G); h \in \hat{G}.$$
(14)

The range of the Fourier transform is denoted by $A(\hat{G})$ and is called the *Fourier algebra* on $\hat{G}$. Similarly, the *inverse Fourier transform* is the contractive homomorphism $(\check{\ }): L^1(\hat{G}) \to C_0(G)$ defined by

$$\check{g}(u) = \int g\theta(u)d\hat{\lambda} = \int g(h)h(u)dh \quad \forall\, g \in L^1(\hat{G}); u \in G$$
(15)

where $d\hat{\lambda}$ and dh represent Haar measure on $\hat{G}$. Its range is denoted by $A(G)$ and called the *Fourier algebra* on G.

This definition generalizes the classical concept of (coefficients of) Fourier series introduced in Section 1.8 where $G = \mathbb{T}$ and $\hat{G} \simeq \mathbb{Z}$. It also generalizes the Fourier integral transform where $G = \mathbb{R}$ and $\hat{G} \simeq \mathbb{R}$. The calculation $(\tilde{\bar{f}})\hat{\ }(h) = \int f(u^{-1})^*h(u)^*du = \left(\int f(v)h(v^{-1})dv\right)^* = \hat{f}(h)$, which holds for all $f \in L^1(G)$ and all $h \in \hat{G}$, shows that the Fourier algebra is closed under complex conjugation.

Since $L^1(G)$ is frequently regarded as an ideal in the measure algebra $M(G)$ under the map $f \mapsto f d\lambda$ (Section 1.9) this suggests the following more general definition.

3.6.5 Definition For a locally compact abelian group G, the contractive homomorphism $\langle\, (\hat{\ }): M(G) \to C(\hat{G})\, /\, (\check{\ }): M(\hat{G}) \to C(G)\, \rangle$ defined by

$$\langle\; \hat{\mu}(h) = \int \tilde{h}d\mu = \int \overline{h}d\mu \quad \forall\, h \in \hat{G}; \; \mu \in M(G)\; /$$
(16)

$$\check{\mu}(u) = \int \theta(u)d\mu = \int h(u)d\mu(h) \quad \forall\, \mu \in M(\hat{G}); \; u \in G\; \rangle$$
(17)

is called the $\langle$ *Fourier–Stieltjes / inverse Fourier–Stieltjes* $\rangle$ *transform* and its range $\langle$ $B(\hat{G})$ / $B(G)$ $\rangle$ is the *Fourier–Stieltjes algebra on* $\langle$ $\hat{G}$ / G $\rangle$.

Note that the inverse Fourier and inverse Fourier–Stieltjes transforms differ from the direct transforms by the omission of a complex conjugation. This prevents the perfect symmetry which the duality theorem shows exists between G and $\hat{G}$ from being extended to the direct and inverse transforms. This difference is necessary if the inversion theorem is to hold. It is quite easy to show that the inverse Fourier–Stieltjes transform is an injection. The calculation given above to show that the Fourier algebra $A(\hat{G})$ is closed under complex conjugation easily extends to show that $B(\hat{G})$, $A(G)$ and $B(G)$ are all closed under complex conjugation. Now we may state a standard inversion theorem.

3.6.6 Inversion Theorem *Let G be a locally compact abelian group. If a function belongs to $B(G) \cap L^1(G)$, then its Fourier transform belongs to $L^1(\hat{G})$. Moreover the Haar measure on G and $\hat{G}$ can be chosen to satisfy*

$$\tilde{\hat{f}} = f \qquad \forall\, f \in B(G) \cap L^1(G).$$

Our proof of this theorem will depend on the theory of positive definite functions and on Bochner's Theorem.

3.6.7 Definition Let G be a group. A function $f: G \to \mathbb{C}$ is said to be *positive definite* if for any finite sets $\{u_1, u_2, \ldots, u_n\} \subseteq G$ and $\{\lambda_1, \lambda_2, \ldots, \lambda_n\} \subseteq \mathbb{C}$ it satisfies

$$\sum_{k=1}^{n} \sum_{j=1}^{n} \lambda_k \lambda_j^* f(u_k u_j^{-1}) \geq 0.$$

3.6.8 Proposition *Let G be a group and let $f: G \to \mathbb{C}$ be a function.*

(a) *Then f is positive definite if it it is a character or if G is a locally compact group and f has the form $g * \tilde{\bar{g}}$ for some $g \in L^1(G)$.*

(b) *If f is positive definite then it is bounded and satisfies $\tilde{f} = \bar{f}$, $\|f\|_\infty \leq f(e)$ and*

$$|f(u) - f(v)|^2 \leq 2f(e)\mathrm{Re}(f(e) - f(uv^{-1})) \qquad \forall\, u, v \in G. \qquad (18)$$

Proof (a): The two cases have very similar proofs. Suppose a character h, $\{u_1, u_2, \ldots, u_n\} \subseteq G$ and $\{\lambda_1, \lambda_2, \ldots, \lambda_n\} \subseteq \mathbb{C}$ are given. Then we get

$$\sum_{k=1}^{n} \sum_{j=1}^{n} \lambda_k \lambda_j^* h(u_k u_j^{-1}) = \sum_{k=1}^{n} \sum_{j=1}^{n} \lambda_k \lambda_j^* h(u_k) h(u_j)^*$$

$$= \left(\sum_{k=1}^{n} \lambda_k h(u_k) \right) \left(\sum_{j=1}^{n} \lambda_j h(u_j) \right)^* \geq 0.$$

Replace h by $g * \tilde{g}$ in the above calculation to get

$$\sum_{k=1}^{n}\sum_{j=1}^{n}\lambda_k\lambda_j^* g * \tilde{g}(u_k u_j^{-1}) = \sum_{k=1}^{n}\sum_{j=1}^{n}\lambda_k\lambda_j^* \int_G g(u_k u_j^{-1}v)g(v)^* dv$$

$$= \int_g \left(\sum_{k=1}^{n}\lambda_k g(u_k^{-1}v)^*\right)\left(\sum_{j=1}^{n}\lambda_j g(u_j^{-1}v)^*\right)^* dv \geq 0.$$

(b): For the first two results choose $n = 2$, $u_1 = e$, $u_2 = u$, $\lambda_1 = 1$ and $\lambda_2 = \lambda$ in the definition, to get

$$f(e) + |\lambda|^2 f(e) + \lambda f(u) + \lambda^* f(u^{-1}) \geq 0.$$

The choice $\langle \lambda = 1 \, / \, \lambda = i \,\rangle$ shows that $\langle \, f(u)+f(u^{-1}) \, / \, if(u)-if(u^{-1}) \,\rangle$ is real, proving $\check{f} = \bar{f}$. If λ is chosen to satisfy $\lambda f(u) = -|f(u)|$, the displayed inequality gives $|f(u)| \leq f(e)$ for any $u \in G$.

For (18), we may assume $f(u) \neq f(v)$ and choose $n = 3$, $u_1 = e$, $u_2 = u$, $u_3 = v$, $\lambda_1 = 1$ and for any real t, $\lambda_2 = -\lambda_3 = t|f(u) - f(v)|/(f(u) - f(v))$. This gives $2\mathrm{Re}(f(e) - f(uv^{-1}))t^2 + 2|f(u) - f(v)|t + f(e) \geq 0$. This quadratic has no real roots, so its discriminant is negative, giving (18).   □

For locally compact abelian groups we can characterize the positive definite functions. Readers will notice that the proof involves ideas from Hilbert space theory. When we look at unitary group representations in Volume II, we will see a close connection with positive definite functions. This will explain the origin of the following proof and the appearance of Hilbert space ideas in it.

3.6.9 Bochner's Theorem *Let G be a locally compact abelian group. The following assertions about a function $f: G \to \mathbb{C}$ are equivalent:*
   (a) *f is continuous and positive definite.*
   (b) *f belongs to $B(G)$ and there is a non-negative measure $\mu \in M(\hat{G})$ satisfying $f = \hat{\mu}$.*

Proof (b) $\Rightarrow$ (a): This "easy" direction is proved in the same way as (a) in Proposition 3.6.8:

$$\sum_{k=1}^{n}\sum_{j=1}^{n}\lambda_k\lambda_j^* f(u_k u_j^{-1}) = \sum_{k=1}^{n}\sum_{j=1}^{n}\lambda_k\lambda_j^* \int_{\hat{G}} h(u_k)h(u_j)^* d\mu$$

$$= \int_{\hat{G}} \left(\sum_{k=1}^{n}\lambda_k h(u_k)\right)\left(\sum_{j=1}^{n}\lambda_j h(u_j)\right)^* d\mu \geq 0.$$

(a) $\Rightarrow$ (b): Let $p: G \to \mathbb{C}$ be continuous and positive definite with $p(e) = 1$. Let $f \in C_{00}(G)$ have compact support K. Inequality (18) shows that K can be partitioned into a finite number of Borel sets $B_1, B_2, \ldots, B_n$ so that the function $(u, v) \mapsto f(u)f(v)^*p(uv^{-1})$ varies is little as we please on each $B_k \times B_j$. Hence the integral

$$\int_G \int_G f(u)f(v)^*p(uv^{-1})du\,dv \tag{19}$$

can be approximated as closely as we please by finite sums of the form

$$\sum_{k=1}^{n} \sum_{j=1}^{n} f(u_k)f(u_j)^*p(u_k u_j^{-1})\lambda(B_k)\lambda(B_j) \geq 0$$

which are non-negative by positive definiteness. This shows that the integral (19) is non-negative. In fact (19) is non-negative for any f in $L^1(G)$, since $C_{00}(G)$ is dense in $L^1(G)$.

We use the positive definite function p to define a linear functional η_p and a (possibly indefinite) inner product $(\cdot, \cdot)_p$ on $L^1(G)$:

$$\eta_p(f) = \int_G f(u)p(u)du \qquad (f, g)_p = \eta_p(f * \bar{g}) \qquad \forall\, f, g \in L^1(G) \tag{20}$$

Obviously $(f, g)_p$ is linear in the first variable and conjugate linear in the second. The positivity of the inner product follows from that of the integral (19):

$$
\begin{aligned}
(f, g)_p &= \int_G \int_G f(u)\tilde{g}(u^{-1}v)^*p(v)du\,dv = \int_G \int_G f(u)\tilde{g}(v)^*p(uv)dv\,du \\
&= \int_G \int_G f(u)g(v)^*p(uv^{-1})dv\,du. \tag{21}
\end{aligned}
$$

This is enough to derive the Cauchy–Schwarz inequality:

$$|(f, g)_p|^2 \leq (f, f)_p (g, g)_p \qquad \forall\, f, g \in L^1(G). \tag{22}$$

Next we consider the case in which g is the characteristic function of some compact symmetric neighborhood V of e in G multiplied by $\lambda(V)^{-1}$. Equation (21) gives

$$
\begin{aligned}
(f, g)_p - \tau_p(f) &= \int_G f(u) \int_V \lambda(V)^{-1}p(uv^{-1})dv\,du - \int_G f(u)p(u)du \\
&= \int_G f(u)\lambda(V)^{-1} \int_V (p(uv^{-1}) - p(u))dv\,du \\
(g, g)_p - 1 &= \lambda(V)^{-2} \int_V \int_V (p(uv^{-1}) - p(u))dv\,du.
\end{aligned}
$$

The continuity of p together with inequality (18) show that we can make these expressions as small as we please merely by requiring V to be small. Thus (22) yields

$$|\tau_p(f)|^2 \leq (f, f)_p = \tau_p(f * \tilde{\bar{f}}) \qquad \forall\, f \in L^1(G). \tag{23}$$

Let k represent $f * \tilde{\bar{f}}$ and let k^n be the nth convolution power of k. Then iteration of the last step gives

$$|\tau_p(f)|^2 \leq |\tau_p(k^2)|^{2^{-1}} \leq \cdots \leq |\tau_p(k^{2^n})|^{2^{-n}}.$$

Recalling $||p||_\infty = 1$ and applying Gelfand's spectral radius formula as n increases and our identification of the Fourier transform with the Gelfand transform, we get

$$|\tau_p(f)|^2 \leq \rho_{L^1(G)}(k) = ||\hat{k}||_\infty = ||\widehat{f * \tilde{\bar{f}}}||_\infty = ||\hat{f}||_\infty^2.$$

Thus τ_p is linear functional on the Fourier algebra $A(\hat{G})$ on $\hat{G}$ with respect to the supremum norm with bound 1. We can extend τ_p to a linear functional on $C_0(\hat{G})$ with norm 1. Hence the Riesz representation theorem gives a measure $\mu \in M(\hat{G})$ of norm 1 satisfying

$$\int_G f(u)p(u)du \;=\; \tau_p(f) = \int_{\hat{G}} \hat{f}(h)d\mu(h) = \int_{\hat{G}} \int_G f(u)h(u)du\,d\mu(h)$$

$$=\; \int_G f(u)\breve{\mu}(u)du \qquad \forall\, f \in L^1(G)$$

Since $C_{00}(G) \subseteq L^1(G) \cap C_0(G)$ is dense in $C_0(G)$, p equals $\breve{\mu}$ at least almost everywhere. Since both p and $\breve{\mu}$ are continuous, they are equal. It only remains to show that μ is positive. This follows from

$$1 = p(e) = \int_{\hat{G}} d\mu(h) = \mu(G) \leq ||\mu|| = 1. \qquad \square$$

Bochner's theorem and the Jordan decomposition theorem for measures shows that any function f in the Fourier–Stieltjes algebra $B(G)$ is a linear combination $f = f_1 - f_2 + i(f_3 - f_4)$ of four positive definite functions.

We can now prove the inversion theorem.

Proof of inversion theorem First we consider two functions f and g in $B(G) \cap L^1(G)$ and a function $k \in L^1(G)$. Let μ_f and μ_g be measures in $M(\hat{G})$ satisfying $f = \breve{\mu}_f$ and $g = \breve{\mu}_g$. This notation gives

$$k * f(e) \;=\; \int_G k(u^{-1})f(u)du = \int_G k(u^{-1}) \int_{\hat{G}} h(u)d\mu_f(h)du$$

$$=\; \int_{\hat{G}} \int_G k(u^{-1})h(u)du\,d\mu_f(h) = \int_{\hat{G}} \hat{k}(h)d\mu_f(h)$$

from which we derive

$$\int_{\hat{G}} \hat{k}(h)\hat{g}(h)d\mu_f(h) \; = \; \int_{\hat{G}} \widehat{k*g}(h)d\mu_f(h) = k*g*f(e) = k*f*g(e)$$

$$= \; \int_{\hat{G}} \widehat{k*f}(h)d\mu_g(h) = \int_{\hat{G}} \hat{k}(h)\hat{f}(h)d\mu_g(h).$$

Since $B(\hat{G}) = \{\hat{k} : k \in L^1(G)\}$ is dense in $C_0(\hat{G})$, this proves

$$\hat{g}d\mu_f = \hat{f}d\mu_g \qquad \forall \; f, g \in B(G) \cap L^1(G). \tag{24}$$

We will define a linear functional Ω on $C_{00}(\hat{G})$ which will turn out to be a multiple of the Haar functional. In this paragraph and the next let k belong to $C_{00}(\hat{G})$ and let K be the support of k. We want a function $f \in B(G) \cap L^1(G)$ such that its Fourier transform does not vanish on K. For any $h_0 \in K$ we can find some function $f_0 \in C_{00}(G)$ such that $\hat{f}_0(h_0)$ is non-zero. Then $\widehat{f_0 * \tilde{\tilde{f}}_0}(h) = \hat{f}_0(h)\hat{f}_0(h)^*$ is a non-negative-valued function which is positive at $h = h_0$. Since K is compact, we can find $f_1, f_2, \ldots, f_n \in L^1(G)$ such that the Fourier transform of $f = f_1 * \tilde{\tilde{f}}_1 + f_2 * \tilde{\tilde{f}}_2 + \cdots + f_n * \tilde{\tilde{f}}_n$ does not vanish on K. Proposition 3.6.9 shows that $f \in C_{00}(G) \subseteq L^1(G)$ is positive definite and hence belongs to $B(G)$ also. Thus we can define

$$\Omega(k) = \int_{\hat{G}} \frac{k(h)}{\hat{f}(h)} d\mu_f(h). \tag{25}$$

Equation (1) shows that this expression is actually independent of f (so long as the integrand is defined and $f \in B(G) \cap L^1(G)$). Therefore Ω is a linear functional on $C_{00}(\hat{G})$ since $f \in B(G) \cap L^1(G)$ can be chosen to be positive on the supports of any two functions k_1 and k_2 in $C_{00}(\hat{G})$ and used to define $\Omega(\alpha k_1 + \beta k_2)$ for $\alpha, \beta \in \mathbb{C}$. Since f in (2) is positive definite, $\hat{f}$ and μ_f are both positive. Hence Ω is a positive linear functional on $C_{00}(\hat{G})$.

Next we will show that Ω is left invariant so that it is a multiple of Haar measure on $\hat{G}$. Let $h_0 \in \hat{G}$ be arbitrary but fixed. Let k and f be as in the last paragraph except that we require $\hat{f}$ to be non-zero on both $K = \mathrm{supp}(k)$ and $h_0 K$. Let g be the pointwise product $g = h_0^{-1}f$ so we get $\hat{f}(h) = \hat{g}(h_0^{-1}h)$ and $d\mu_f(h) = d\mu_f(h_0^{-1}h)$. This gives

$$\Omega(_{h_0}k) \; = \; \int_{\hat{G}} \frac{k(h_0^{-1}h)}{\hat{f}(h)} d\mu_f(h) = \int_{\hat{G}} \frac{k(h_0^{-1}h)}{\hat{g}(h_0^{-1}h)} d\mu_g(h_0^{-1}h)$$

$$= \; \int_{\hat{G}} \frac{k(h)}{\hat{g}(h)} d\mu_g(h) = \Omega(k),$$

so Ω is left invariant. To see that Ω is not zero, we fix a non-zero continuous positive definite function f in $L^1(G)$ for a moment, and choose a function

$k \in C_{00}(\hat{G})$ so that $\int k d\mu_f$ is non-zero. This gives

$$\Omega(k\hat{f}) = \int_{\hat{G}} k\hat{f}/\hat{g} d\mu_g = \int_{\hat{G}} k d\mu_f \neq 0. \tag{26}$$

Thus we may choose Haar measure dh on $\hat{G}$ to satisfy

$$\Omega(k) = \int_{\hat{G}} k(h) dh \qquad \forall \, k \in C_{00}(\hat{G}). \tag{27}$$

Combining (3) and (4) gives

$$\int_{\hat{G}} k d\mu_f = \Omega(k\hat{f}) = \int_{\hat{G}} k(h)\hat{f}(h) dh \qquad \forall \, k \in C_{00}(\hat{G}); f \in B(G) \cap L^1(G) \tag{28}$$

and thus the measures satisfy

$$d\mu_f(h) = \hat{f}(h) dh \qquad \forall \, f \in B(G) \cap L^1(G). \tag{29}$$

Since μ_f is a finite measure, $\hat{f}$ belongs to $L^1(\hat{G})$. Furthermore for any $f \in B(G) \cap L^1(G)$ and $u \in G$ equation (6) gives

$$f(u) = (\breve{\mu}_f)(u) = \int_{\hat{G}} h(u) d\mu_f(h) = \int_{\hat{G}} \hat{f}(h) h(u) du = \breve{\hat{f}}(u)$$

as we wished to show.   □

The power of Bochner's theorem is the ability to produce a function in $B(G)$ merely by choosing a continuous positive definite function. By combining the inversion theorem with Bochner's theorem, we see that a continuous positive definite function in $L^1(G)$ is the inverse Fourier transform of its non-negative-valued Fourier transform. We illustrate the utility of this starting with a lemma. Theorem 3.6.11 is a special case of the Gelfand–Raĭkov theorem proved in Volume II.

3.6.10 Lemma *Let G be a locally compact abelian group and suppose U is a neighborhood of e. Then there is a continuous positive definite function g in $L^1(G) \cap B(G)$ satisfying $g(e) = 1$ and $g(u) = 0$ for all $u \in G \setminus U$.*

Proof Choose a symmetric compact neighborhood V of e satisfying $V^2 \subseteq U$. Define g by $u \mapsto \lambda(V)^{-1}\chi_V * \chi_V(u)$, where χ_V is the characteristic function of V. This expression shows that g is positive definite. It is easy to show $g(u) = \lambda(V)^{-1}\lambda(V \cap uV)$ which gives the continuity of g.   □

3.6.11 Theorem *Let G be a locally compact abelian group. Then $\hat{G}$ separates the points of G, so $\theta: G \to \hat{\hat{G}}$ is injective. If G is compact, the span of $\hat{G}$ is dense in $C(G)$.*

Proof It is enough to show that for any $u \neq e$ in G there is some $h \in \hat{G}$ satisfying $h(u) \neq 1$. By the lemma and the Hausdorff property of G, choose $g \in B(G) \cap L^1(G)$ satisfying $g(e) = 1$ and $g(u) = 0$. The inversion theorem gives

$$0 = g(u) = (\hat{g})\check{\ }(u) = \int \hat{g}(h)h(u)dh$$

and

$$1 = g(e) = (\hat{g})\check{\ }(e) = \int \hat{g}(h)h(e)dh = \int \hat{g}(h)dh$$

from which the existence of such an $h \in G$ is obvious.

If G is compact, then the span $\mathrm{sp}(\hat{G})$ of $\hat{G}$ is a point separating subalgebra of $C(\hat{G})$ which is closed under complex conjugation. Hence the Stone–Weierstrass theorem (§1.5.1) shows that $\mathrm{sp}(\hat{G})$ is dense.   □

We state a partial result.

3.6.12 Proposition *If G is a locally compact abelian group, then θ is a homeomorphism of G onto its range in $\hat{\hat{G}}$.*

Proof Recall that $\hat{\hat{G}}$ has the compact open topology as a set of maps from $\hat{G}$ into T. Hence it is enough to show that the sets

$$U(K, \varepsilon) = \{u \in G : |h(u) - 1| < \varepsilon \ \text{for all} \ h \in K\},$$

with $K \subseteq \hat{G}$ compact and ε positive, form a neighborhood base at e.

First we show that each of these sets is a neighborhood of e. The joint continuity of $(h, u) \mapsto h(u)$ shows that for each $k \in \hat{G}$ there are neighborhoods $U_k \subseteq G$ and $V_k \subseteq \hat{G}$ of e and k, respectively, satisfying $|h(u) - 1| < \varepsilon$ for all $h \in V_k$ and $u \in U_k$. Choose an open cover $V_{k_1}, V_{k_2}, \ldots, V_{k_n}$ of K. Then $\cap_{j=1}^n U_{k_j}$ is obviously a neighborhood of e included in $U(K, \varepsilon)$.

Next we show that if U is any neighborhood of e in G then there is some K and ε satisfying $U(K, \varepsilon) \subseteq U$. Use the lemma to find $g \in B(G) \cap L^1(G)$ satisfying $g(e) = 1, \hat{g} \geq 0$ and vanishing off U. The inversion theorem gives

$$1 = g(e) = (\hat{g})\check{\ }(e) = \int \hat{g}(h)dh.$$

Choose a compact set K satisfying

$$\int\limits_{\hat{G}\backslash K} |\hat{g}(h)|dh < \frac{1}{3}.$$

Then the neighborhood $U(K, \frac{1}{3})$ is included in U since the inversion theorem gives $|1 - g(u)| < 1$ for $u \in U(K, \frac{1}{3})$.   □

3.6.13 Plancherel Theorem *Let G be a locally compact abelian group. The map $f \mapsto \hat{f}$ is an isometry of $(L^1(G) \cap L^2(G), \|\cdot\|_2)$ into $(L^2(\hat{G}), \|\cdot\|_2)$. This map extends to an isometry of $L^2(G)$ onto $L^2(\hat{G})$.*

Proof For any $f \in L^1(G) \cap L^2(G)$, it is easy to check that $g = \tilde{f}^- * f$ is continuous and positive definite. Therefore we may apply the inversion theorem to get

$$\|f\|_2^2 = \int \overline{f}(u)f(u)du = \tilde{f}^- * f(e)$$

$$= (\hat{g})\check{}(e) = \int \hat{g}(h)h(e)dh$$

$$= \int \hat{g}d\hat{\lambda} = \int (\tilde{f}^- * f)\check{}d\hat{\lambda} = \int |\hat{f}|^2 d\hat{\lambda} = \|\hat{f}\|_2^2.$$

Hence ($\check{}$) is an isometry of $L^1(G) \cap L^2(G)$ into $L^2(\hat{G})$. We must still show that the image is dense. It is enough to show that any $g \in L^2(\hat{G})$ satisfying

$$\int \hat{f}g d\hat{\lambda} = 0 \qquad \forall\, f \in L^1(G) \cap L^2(G) \tag{30}$$

is zero. Note that $f \in L^1(G) \cap L^2(G)$ and $u \in G$ imply $f_u \in L^1(G) \cap L^2(G)$ and $\widehat{f_u}(h) = \hat{f}(h)h(u^{-1})$ so that any $g \in L^2(\hat{G})$ satisfying (18) also satisfies

$$(\hat{f}g)\check{}(u^{-1}) = \int \hat{f}(h)g(h)h(u^{-1})dh = \int \widehat{f_u}(h)\hat{g}(h)dh = 0$$

for all $f \in L^1(G)$ and $u \in G$. The injectivity of the inverse Fourier transform now gives $\hat{f}g = 0$ (as a function in $L^1(\hat{G})$) for all $f \in L^1(G) \cap L^2(G)$. Let f be the characteristic functions of open set V with compact closure. Then f belongs to $L^1(G) \cap L^2(G)$ and for any $h \in \hat{G}$, $\hat{f}(h)$ is just $\int_V h(u)^* du$. Letting V vary, we conclude $g = 0$ in $L^2(\hat{G})$.

For any $f \in L^2(G)$, we may define $\hat{f} \in L^2(\hat{G})$ by

$$\hat{f} = \lim \hat{f}_n$$

where $\{f_n\}_{n \in \mathbb{N}} \in L^1(G) \cap L^2(G)$ is any sequence converging to f in the L^2-norm. This gives an isometry $f \mapsto \hat{f}$ of $L^2(G)$ onto $L^2(\hat{G})$. $\qquad \square$

The map $f \mapsto \hat{f}$ of $L^2(G)$ onto $L^2(\hat{G})$ just defined is called the *Fourier–Plancherel transform*. The polarization identity applied to Plancherel's theorem now gives *Parseval's identity*

$$(f, g) = (\hat{f}, \hat{g}) \qquad \forall\, f, g \in L^2(G) \tag{31}$$

where the inner products are those of $L^2(G)$ and $L^2(\hat{G})$, respectively. Also many identities such as

$$
\begin{aligned}
\widehat{f_u}(h) &= h(u^{-1})\hat{f}(h) \\
\widehat{kf}(h) &= \hat{f}(k^{-1}h) \\
\widehat{f^-} &= \hat{f}^- \qquad (\overline{f})^{\hat{}} = (\hat{f})^{\,\overline{}}
\end{aligned}
\qquad
\begin{aligned}
&\forall\, f \in L^2(G);\ h, k \in \hat{G}; \\
&\qquad \forall\, u \in G
\end{aligned}
\qquad (32)
$$

immediately extend from functions in $L^1(G)$ to functions in $L^2(G)$. Parseval's identity also implies

$$\widehat{fg} = \hat{f} * \hat{g} \qquad \forall\, f, g \in L^2(G) \qquad (33)$$

which implies

$$A(\hat{G}) = L^2(\hat{G}) * L^2(\hat{G}). \qquad (34)$$

3.6.14 Theorem *For any locally compact abelian group G, the commutative Banach algebra $L^1(G)$ is completely regular. Hence the $\langle$ Fourier / Fourier–Stiltjes $\rangle$ algebra $\langle$ $A(\hat{G})$ / $B(\hat{G})$ $\rangle$ is dense in $\langle$ $C_0(\hat{G})$ / $C(\hat{G})$ $\rangle$.*

Proof Let k be a point in $\hat{G}$ and let U be a neighborhood of k. We must construct a function $f \in L^1(G)$ satisfying $\hat{f}(k) = 1$ and $\hat{f}(h) = 0$ for all $h \notin U$. Let V be a compact symmetric neighborhood of $1 \in \hat{G}$ satisfying $V^2k \subseteq U$. Then the characteristic functions χ_V and χ_{Vk} belong to $L^2(\hat{G})$ so that there is a function $f \in L^1(G)$ satisfying

$$\hat{f} = \hat{\lambda}(V)^{-1}\chi_V * \chi_{Vk}.$$

As before, we find

$$\hat{f}(h) = \hat{\lambda}(V)^{-1} \int \chi_V(hg)\chi_{Vk}(g^{-1})dg = \hat{\lambda}(V)^{-1}\hat{\lambda}(Vh \cap Vk)$$

which gives the desired result.

We have already noted that both $A(\hat{G})$ and $B(\hat{G})$ are closed under complex conjugation, so in both cases complete regularity gives all the hypotheses of the Stone–Weierstrass theorem.   □

As noted previously, the results we have just derived from the inversion theorem, together with Hewitt and Ross [1963], 5.11, establish the following.

3.6.15 Pontryagin Duality Theorem *For any locally compact abelian group G, the natural map $\theta \colon G \to \hat{\hat{G}}$ is a homeomorphic isomorphism onto $\hat{\hat{G}}$.*

Proof The only part of the proof which might not be perfectly clear is the density of $\theta(G)$ in $\hat{G}$. If $\theta(G)$ were not dense, the complete regularity of $L^1(\hat{G})$ would give a non-zero function $g \in L^1(G)$ satisfying

$$\hat{g}(\theta(u)) = 0 \qquad \forall\, u \in G.$$

By the uniqueness of the inverse Fourier transform, $\check{g}$ is non-zero. However this gives the contradiction

$$\begin{aligned}
\check{g}(u) &= \int h(u^{-1})^* g(h) dh = \int \theta(u^{-1})(h)^* g(h) dh \\
&= \hat{g}(\theta(u^{-1})) = 0 \qquad \forall\, u \in G. \qquad \square
\end{aligned}$$

The duality theorem now allows all of the results proved previously only for $\hat{G}$ to be reinterpreted as statements about arbitrary locally compact abelian groups. For instance (22) becomes

$$A(G) = L^2(G) * L^2(G). \tag{35}$$

We remark that it is easy to show that if G is a compact group then $\hat{G}$ is discrete, and if G is discrete then $\hat{G}$ is compact. The Haar integral on a discrete group G is simply summation

$$\int f d\lambda = \sum_{u \in G} f(u).$$

Then the Haar measure $\hat{\lambda}$ on $\hat{G}$ satisfying the inversion theorem and/or Plancherel's theorem is the one normalized to satisfy

$$\hat{\lambda}(\hat{G}) = 1.$$

3.6.16 Ideal Theory Throughout this subsection, G will denote a locally compact abelian group. We are interested in the closed ideal structure of $L^1(G)$. We will use the identification of $L^\infty(G)$ as the dual Banach space of the group algebra $L^1(G)$. We consider the elements of $\hat{G}$ as continuous characters which are elements of $L^\infty(G)$ and also as multiplicative linear functionals on $L^1(G)$ which are elements of $\Gamma_{L^1(G)}$. This identification agrees with our previous notation since when $h \in \hat{G}$ is thought of as a function in $L^\infty(G)$ the corresponding element of $\Gamma_{L^1(G)}$ is just $\omega_h \in L^1(G)^*$. We also use the complete regularity of $L^1(G)$. Since $L^1(G)$ is Gelfand semisimple, $\{0\}$ is the intersection of maximal modular ideals, a condition which will be defined as strong semisimplicity in Section 4.5 below.

Let $\mathcal{I}$ be a subset of $L^1(G)$ and $\mathcal{W}$ a subset of $L^\infty(G)$. We are interested in their *annihilators in* $L^\infty(G)$ *and* $L^1(G)$, respectively:

$$
\begin{aligned}
\mathcal{I}^\perp &= \{h \in L^\infty(G) : \omega_h(f) = 0 \ \text{for all} \ f \in \mathcal{I}\} \\
&= \{h \in L^\infty(G) : \int_G f(v)h(v^{-1})dv = 0 \ \text{for all} \ f \in \mathcal{I}\} \\
\mathcal{W}_\perp &= \{f \in L^1(G) : \omega_h(f) = 0 \ \text{for all} \ h \in \mathcal{W}\} \\
&= \{f \in L^1(G) : \int_G f(v)h(v^{-1})dv = 0 \ \text{for all} \ h \in \mathcal{W}\}.
\end{aligned}
$$

Clearly $\mathcal{I}^\perp$ is a weak* closed linear subspace of $L^\infty(G)$ and $\mathcal{W}_\perp$ is a (norm) closed linear subspace of $L^1(G)$. If $\mathcal{W}$ and $\mathcal{I}$ already satisfy these additional conditions, then the Hahn–Banach theorem shows

$$
(\mathcal{I}^\perp)_\perp = \mathcal{I} \quad \text{and} \quad (\mathcal{W}_\perp)^\perp = \mathcal{W}.
$$

Now suppose $\mathcal{I}$ is a norm closed linear subspace which is invariant under translation by elements of G. Then $\mathcal{I}^\perp$ is also translation invariant, but more interestingly it can be expressed as

$$
\begin{aligned}
\mathcal{I}^\perp &= \{h \in L^\infty(G) : \int_G f(uv)h(v^{-1})dv = 0 \ \text{for all} \ f \in \mathcal{I}; \ u \in G\} \\
&= \{h \in L^\infty(G) : f * h = 0 \ \text{for all} \ f \in \mathcal{I}\}.
\end{aligned}
$$

From this, it is not hard to see that a closed linear subspace of $L^1(G)$ is an ideal if and only if it is translation invariant.

We can apply the theory of hulls and kernels to $L^1(G)$ and $\hat{G}$ (considered as the Gelfand space of $L^1(G)$). The question of spectral synthesis is: When is a closed ideal $\mathcal{I}$ of $L^1(G)$ completely determined by its hull $h(\mathcal{I})$? We say that a (necessarily closed) subset S of $\hat{G}$ is a *set of spectral synthesis* if its kernel $k(S)$ is the only ideal with hull equal to S.

To explain this terminology further, let S be a subset of $\hat{G}$ and let $\mathcal{W}(S)$ be the weak* closed linear span of S in $L^\infty(G)$. Since sp$S$ is invariant under translation by element of G, $\mathcal{W}(S)$ is also invariant. For any weak* closed, translation invariant, linear subspace $\mathcal{W}$ of $L^\infty(G)$, call $S(\mathcal{W}) = \mathcal{W} \cap \hat{G}$ *the spectrum of* $\mathcal{W}$. Clearly, $\mathcal{W}(S(\mathcal{W}))$ is the smallest weakly closed, translation invariant subset $\mathcal{W}'$ of $\mathcal{W}$ with spectrum S. Obviously, $S(\mathcal{W})$ is always closed and, conversely, it is true that every closed subset of $\hat{G}$ is $S(\mathcal{W})$ for some $\mathcal{W}$ satisfying our conditions. This follows from complete regularity and the fact that if $\mathcal{W} = \mathcal{I}^\perp$, then $S(\mathcal{W}) = h(\mathcal{I})$. Then S is a set of spectral synthesis if and only if $\mathcal{W}(S)$ is the only weakly closed, translation invariant subset of $L^\infty(G)$ with spectrum S. In this case, any function in any $\mathcal{W}'$ with $S(\mathcal{W}') = S$ can be approximated by linear combinations of elements of S (*i.e.* *synthesized* from S).

If G is a compact abelian group so $\hat{G}$ is discrete, then every subset of $\hat{G}$ is a set of synthesis. However if G is locally compact abelian but not compact, Paul Malliavin [1959] showed that there are always subsets of $\hat{G}$ which are not sets of synthesis. Norbert Wiener's Tauberian theorem [1932] is essentially the statement that every closed ideal is included in some maximal modular ideal which is equivalent, in the present setting, to the density in $L^1(G)$ of the set of elements f for which their Gelfand transform $\hat{f}$ has compact support. Suppose S is a closed subset of $\hat{G}$. Let $\mathcal{J}(S)$ be the closure in $L^1(G)$ of the set of elements with compactly supported Gelfand transforms vanishing on some neighborhood of S. It is easy to see that $\mathcal{J}(S)$ is the smallest closed ideal of $L^1(G)$ with hull S just as $k(S)$ is the largest such ideal. Hence S is a set of synthesis if and only if these two ideals coincide. Stated another way, S is a set of synthesis if and only if every $f \in L^1(G)$ for which $\hat{f}$ vanishes on S can be approximated in $L^1(G)$ by elements with Gelfand transforms vanishing in a neighborhood of S. See Rudin [1962], Chapter 8 for examples of sets of synthesis and nonsynthesis. Chapter 7 will extend a number of these questions and results to noncommutative groups and noncommutative Banach algebras and will give proofs in that setting.

3.6.17 The Measure Algebra $M(G)$ The algebra $M(G)$ is so large that it is still not well understood even for locally compact abelian groups. Cohen's idempotent theorem [1959b], [1960a] identifies the idempotents in $M(G)$. Let λ be Haar measure on a compact abelian subgroup and let $\chi_1, \chi_2, \ldots, \chi_n$ be a finite list of distinct continuous characters on H. Then $\mu = (\chi_1 + \chi_2 + \cdots + \chi_n)\lambda$ is an idempotent measure on G and every such measure can be constructed by a finite number of operations of the form $\mu \mapsto 1 - \mu$ and $(\mu_1, \mu_2) \mapsto \mu_1 * \mu_2$ starting from idempotents of this special form. (Note $\mu_1 + \mu_2 - \mu_1 * \mu_2 = 1 - (1 - \mu_1) * (1 - \mu_2)$.) another theorem by Cohen [1960b] describes all homomorphisms from $L^1(G)$ into $M(H)$ where G and H are arbitrary locally compact abelian groups. Every such homomorphism can be extended to a homomorphism of $M(G)$ into $M(H)$, but not uniquely. See Jyunji Inoue [1971] for an extension.

Taylor [1965], [1973a] gives by far the most detailed information on the structure of $M(G)$. First he identifies an abstract class of commutative *convolution measure algebras* based on the theory of complex ordered L-spaces. A convolution measure algebra is simply a commutative Banach algebra $\mathcal{M}$ which is an L-space and for which multiplication is an L-homomorphism from $\mathcal{M} \otimes \mathcal{M}$ to $\mathcal{M}$. The dual of such an algebra has the form $C(\Omega)$ for a suitable (large) compact space Ω. Let $\mathcal{A}$ be the closed linear span of the Gelfand space $\Gamma_{\mathcal{M}}$ of $\mathcal{M}$ in $\mathcal{M}^* \simeq C(\Omega)$. Then $\mathcal{A}$ has the form $C(S)$ for a compact commutative semigroup S and convolution multiplication in $M(S)$ agrees with the original multiplication in $\mathcal{M}$. Thus his class of abstract convolution measure algebras really deserves its name. The Gelfand space $\Gamma_\mathcal{M}$ is also a semigroup which has its own natural topology in which it is a topological semigroup. Certain idempotents in this semigroup are identities for locally compact subgroups of $\Gamma_\mathcal{M}$. They are called *critical points* and from them much of the structure of $\mathcal{M}$ can be calculated. Taylor characterizes those convolution measure algebras which are isomorphic to $L^1(G)$ for some locally compact abelian group G as those semisimple algebras which include no proper *L*-subalgebra. He is able to describe the invertible and exponential elements rather concretely, and these results were new even for the case $G = \mathbb{R}$. See the second reference above for an expository presentation.

3.6.18 The Beurling Algebra $\ell^1(\mathbb{Z}, \omega)$ Beurling algebras were defined in §1.9.15 but we wish to give a simple example.

Definition   A *weight function* ω on $\mathbb{Z}$ is a positive valued function $\omega\colon \mathbb{Z} \to \mathbb{R}_+^0$ satisfying

$$\omega_0 = 1 \quad \text{and} \quad \omega_{m+n} \le \omega_m \omega_n \quad \forall\, m, n \in \mathbb{Z}.$$

If ω is a weight function, then the Banach space

$$\ell^1(\mathbb{Z}, \omega) = \{f\colon \mathbb{Z} \to \mathbb{C} \;:\; \|f\|_\omega = \sum_{n=-\infty}^{\infty} |f(n)|\omega_n < \infty\},$$

provided with convolution multiplication, is called the *Beurling algebra on* $\mathbb{Z}$ defined by ω. Denote $\lim_{n \to \pm\infty} \omega_n^{1/n}$ by $\rho_\pm = \rho_\pm(\ell^1(\mathbb{Z}, \omega))$.

The inequality for the weight function ω shows that the limits above exist and that $\rho_\pm$ agrees with $\pm \inf\{\pm\omega_n^{1/n}\}$ (*cf.* Theorem 1.1.10).

Proposition   *For any weight function ω, $\ell^1(\mathbb{Z}, \omega)$ is a unital semisimple commutative Banach algebra.*

For any $\lambda \in \mathbb{C}$ satisfying $\rho_- \le \lambda \le \rho_+$, the series $\sum_{n=-\infty}^{\infty} f(n)\lambda^n$ converges. If we denote its sum by $\check{f}(\lambda)$, then $\check{f}$ is a homeomorphism of the annulus $\{\lambda : \rho_- \le \lambda \le \rho_+\}$ onto $\Gamma_{\ell^1(\mathbb{Z}, \omega)}$.

Proof It is easy to show convergence, that $\ell^1(\mathbb{Z}, \omega)$ is a unital commutative Banach algebra and that $\check{f}$ is continuous into $\Gamma_{\ell^1(\mathbb{Z}, \omega)}$. Consideration of $\gamma(\delta_\pm)$ for any $\gamma \in \Gamma_{\ell^1(\mathbb{Z}, \omega)}$ (which satisfy $\gamma(\delta_-)\gamma(\delta_+) = 1$), shows that the map is onto and hence a homeomorphism. Thus it is obvious that $\ell^1(\mathbb{Z}, \omega)$ is semisimple.   □

4

Ideals, Representations and Radicals

Introduction

In any mathematical theory it is important to compare the general abstract structures which arise with comparatively well understood concrete examples. Representation theory is a systematic attempt to do this. We have already seen one example of a representation theory in the Gelfand theory for commutative spectral algebras given in Chapter 3. Any commutative spectral algebra modulo its Gelfand radical is isomorphic to a spectral subalgebra of the algebra of all continuous functions vanishing at infinity on a locally compact Hausdorff space. Since algebras of continuous functions are comparatively concrete and can perhaps be somewhat better understood *a priori* than general commutative spectral algebras, this is an important (and characteristic) representation theory. Note that the class of *all* commutative spectral algebras is too large to be successfully represented. Only the semisimple commutative spectral algebras can be faithfully represented. Fortunately the pathology of the non-semisimple algebras can be neatly excised by dividing out the Gelfand radical.

In this work we will study several other representation theories in considerable detail. This chapter deals with the representation of general algebras and part of the second volume is devoted to the representation theory of *-algebras. In both these cases the phenomena noted above occur. Not all of the objects (algebras or *-algebras) can be faithfully represented, but the pathological part can be divided out.

The first two sections of this chapter deal with the representations themselves, postponing until the later sections a discussion of how the pathological portion splits off. Sections 4.3 to 4.7 introduce the theory of radicals which explores the latter question.

In setting up a representation theory for general algebras, it is first necessary to choose a class of algebras which are concrete and comparatively well understood. Of course the algebras chosen must be sufficiently general to represent many of the most interesting examples. One's first choice might be the algebras of finite-dimensional matrices studied in Sections 1.7 and 2.7. Historically these were used through the first half of this century. However, for the type of algebras in which we are principally interested, this choice is much too restrictive. Typical and well-behaved algebras would have no non-zero representations. Thus it turns out that the class of algebras most suitable for our representation theory consists of the algebras of all linear maps of some (not necessarily finite-dimensional) linear space into itself.

We will study homomorphisms of general algebras into algebras of linear maps on linear spaces. These homomorphisms will be called *representations* of the algebra in question and the linear spaces will be called *representation spaces*. When the spaces are not finite-dimensional, we will usually need to consider norms on them and confine our attention to bounded linear maps.

Representations can be broken down into simpler pieces by finding invariant subspaces of their representation spaces. The atoms under this process of decomposition are called *irreducible representations*. They play a fundamental role. Unfortunately it is not usually possible to decompose a general representation into the sum of irreducible representations.

Much of the first two sections will deal with those ideals which are the kernels of irreducible representations. These are called *primitive ideals*. They are examples of a class of ideals called *prime ideals* which will also be studied. We also introduce ultraprime ideals.

The first section deals with ideals and with the purely algebraic theory of representations while the second section explores representations more fully and the phenomena related to norms. The case of spectral algebras and spectral normed algebras will be considered most intensively. It turns out that in order to study irreducible representations it is useful to consider a wider class of representations, called *cyclic representations*. Each cyclic representation can be constructed (up to equivalence) from the algebra itself. We will see this same situation again in the second volume when we consider representations of *-algebras. Representations on reflexive spaces play a role intermediate between the general theory of this volume and the more special theory of the next volume. Some basic results about them are given at the end of Section 4.2.

Historical Remarks

Theorem 4.2.13 asserts that any irreducible representation of a spectral algebra is strictly dense. For Banach algebras this result was proved by Charles E. Rickart [1950]. The concept of strict density and the principal techniques and results on this subject are due to Nathan Jacobson [1945c]. This paper brought the theory of irreducible representations to its present central position.

The theory of representations is quite old. For a discussion of its early history see Thomas Hawkins [1972]. Representations of algebras were considered by Charles Saunders Peirce [1881] and a more systematic study was begun by Theodor Molien [1893]. However, the representation theory of finite groups, which was also considered by Molien [1897] but was mainly developed by G. Frobenius, soon eclipsed the representation theory for algebras. Much of this theory is contained in the second edition of W. Burnside's classic book [1911]. Representations of algebras are mentioned in Leonard Eugene Dickson's book [1927]. They are treated in a quite modern way by Emmy Noether [1929] and in Adrian A. Albert's book [1939]. Since early discussions of algebras dealt with finite dimensional concrete algebras, the representation theory was not very explicit. However, even before algebras were generally thought of abstractly, their representations were studied implicitly since it was obvious that the given presentation of a hypercomplex system (an algebra presented as a linear space with a particular basis and a multiplication table for the basis elements) was not necessarily the most advantageous presentation.

Since the time when Noether [1929] began to develop the general theory of modules, the language of module theory has tended to supersede the entirely equivalent language of representation theory. In this work we will primarily retain the older language for two reasons. First, our emphasis will always be on the algebras rather than the representations themselves. Second, the appropriate modules for Banach algebras are rather special, and the added complication of a norm on the algebra makes the language of modules less convenient. Nevertheless, the reader should be aware that the theories of modules and of ⟨ normed / Banach ⟩ modules are entirely equivalent to the theories presented here of representations on linear spaces and of continuous representations of ⟨ normed / Banach ⟩ algebras on ⟨ normed / Banach ⟩ spaces. We adopt language of modules occasionally (*cf.* §4.1.17).

Radicals

In the rest of this chapter we will introduce various concepts of a radical for an algebra. By far the most important is the Jacobson radical which we will call simply the radical and explore in detail in Section 4.3. (It was briefly introduced already in Section 2.3.) In Sections 4.4 and 4.5 we will consider the Baer radical (or prime radical) and the Brown–McCoy or strong radical. Most of the material dealing with these three radicals could be developed for general rings, but we prefer to work in the setting of algebras.

In Section 4.6 subdirect product representations are considered corresponding to each of these radicals. Spectral algebras are also characterized in terms of their natural representation as a subdirect product relative to the Jacobson radical. We then discuss in Section 4.7 a theory of radicals in an arbitrary semi-abelian category. This theory provides a common abstract generalization of the three radicals studied in the previous sections and of the reducing ideal and Leptin radical of a *-algebra which will be introduced in Volume II. At the end of the chapter we will discuss examples and the history of radicals. We prefer to postpone the history until the technical notions involved have been introduced.

4.1   Ideals and Representations

As usual we start by defining the basic concepts. Irreducible representations were already briefly introduced in Section 2.3.

4.1.1 Definition   Let $\mathcal{A}$ be an algebra and let $\mathcal{X}$ be a linear space. A $\langle$ *representation / anti-representation* $\rangle$ T of $\mathcal{A}$ on $\mathcal{X}$ is a $\langle$ homomorphism / anti-homomorphism $\rangle$ $a \mapsto T_a$ of $\mathcal{A}$ into $\mathcal{L}(\mathcal{X})$. A subspace $\mathcal{Y}$ of $\mathcal{X}$ is said to be *T-invariant* if $T_a y$ belongs to $\mathcal{Y}$ for all $y \in \mathcal{Y}$ and all $a \in \mathcal{A}$. The $\langle$ representation / anti-representation $\rangle$ T of $\mathcal{A}$ on $\mathcal{X}$ is said to be:

(a) *Faithful* if the $\langle$ homomorphism / anti-homomorphism $\rangle$ is injective.

(b) *Trivial* if $T_a = 0$ for all $a \in \mathcal{A}$.

(c) *Irreducible* if $\{0\}$ and $\mathcal{X}$ are the only T-invariant subspaces, and T is not trivial.

(d) *Cyclic* if there is a vector $z \in \mathcal{X}$ (called a *cyclic vector*) satisfying $\mathcal{X} = \{T_a z : a \in \mathcal{A}\}$.

(e) *Equivalent* to a $\langle$ representation / anti-representation $\rangle$ S of $\mathcal{A}$ on a linear space $\mathcal{Y}$ if there is a linear bijection (called an *equivalence*) $U : \mathcal{X} \to \mathcal{Y}$ satisfying

$$S_a U = U T_a \qquad \forall\, a \in \mathcal{A}. \tag{1}$$

We make a few comments on the formal side of this definition before beginning to discuss the concepts systematically. We will sometimes use the word "representation" to cover both representations and anti-representations and will seldom explicitly state results for anti-representations. If trivial representations were not specifically excluded in the definition of irreducible representations, the trivial representations on zero- and one-dimensional spaces would satisfy the definition of an irreducible representation for an essentially trivial reason. However, the trivial representation on a one-dimensional space has quite different properties from those of irreducible representations, although some authors (possibly through inadvertence) allow it as an irreducible representation. The trivial representation on a zero-dimensional space is excluded simply for convenience. Its inclusion or exclusion makes little difference to the theory. Note that a (necessarily trivial) representation on a zero-dimensional space is cyclic.

Although the definition of equivalence in (e) is stated asymmetrically, note that $U^{-1} : \mathcal{Y} \to \mathcal{X}$ establishes an equivalence between S and T. Thus it is immediate that equivalence is an equivalence relation. Note also that equivalence of representations clearly preserves all of the other properties defined here.

If $\mathcal{A}$ is an algebra and $S : \mathcal{A} \to \mathcal{L}(\mathcal{X}), T : \mathcal{A} \to \mathcal{L}(\mathcal{Y})$ are two representations of $\mathcal{A}$, we will write $\mathrm{Equiv}(S, T)$ for the set of all equivalences of S and T. Hence $\mathrm{Equiv}(S, T) \subseteq \mathcal{L}(\mathcal{X}, \mathcal{Y})$ is nonempty if and only if S and

T are equivalent. Its elements are linear bijections $V: \mathcal{X} \to \mathcal{Y}$ satisfying $T_a V = V S_a$ for all $a \in \mathcal{A}$.

Whenever we speak of a representation T of an algebra $\mathcal{A}$ without specifying the linear space on which the maps $\{T_a : a \in \mathcal{A}\}$ act, we will denote this linear space by $\mathcal{X}^T$. From now on we will speak of the dimension of $\mathcal{X}^T$ as the *dimension* of T.

The left regular representation (cf. Definition 1.2.1) L of an algebra on its underlying linear space is an example of a representation. A subspace $\mathcal{L}$ of an algebra $\mathcal{A}$ is L-invariant (*i.e.*, invariant under the left regular representation) if and only if it is a left ideal. Our next results will show that this example is particularly important.

Construction of New Representations from Invariant Subspaces

If T is a representation of $\mathcal{A}$ on $\mathcal{X}$ and $\mathcal{Y}$ is a T-invariant subspace, then

$$a \mapsto T_a|_{\mathcal{Y}} \tag{2}$$

is a representation of $\mathcal{A}$ on $\mathcal{Y}$. This representation is called the *restriction* of T to $\mathcal{Y}$ and is denoted by $T^{\mathcal{Y}}$. (In this case $T^{\mathcal{Y}}$ is also called a *subrepresentation* of T.) There is another notion of restriction which one must carefully distinguish from this one. If $\mathcal{B}$ is a subalgebra of $\mathcal{A}$ and T is a representation of $\mathcal{A}$ on $\mathcal{X}$, then the homomorphism $a \mapsto T_a$ can be restricted to $\mathcal{B}$. When we need to distinguish these two notions of restriction we will call the first (where each T_a is restricted to $\mathcal{Y}$) *the restriction of the representation* T, and we call the second (where the map T itself is restricted to $\mathcal{B}$) *the restriction of the homomorphism* T.

Given a representation T of $\mathcal{A}$ on $\mathcal{X}$ and a T-invariant subspace $\mathcal{Y}$ there is another natural, and even more important, way to construct a new representation. For each $a \in \mathcal{A}$, define $T_a^{\mathcal{X}/\mathcal{Y}}$ by

$$T_a^{\mathcal{X}/\mathcal{Y}}(x + \mathcal{Y}) = T_a x + \mathcal{Y} \qquad \forall\, x + \mathcal{Y} \in \mathcal{X}/\mathcal{Y}. \tag{3}$$

Since $\mathcal{Y}$ is T-invariant, $T_a^{\mathcal{X}/\mathcal{Y}}$ is a well defined element of $\mathcal{L}(\mathcal{X}/\mathcal{Y})$. Thus $a \mapsto T_a^{\mathcal{X}/\mathcal{Y}}$ is a representation of $\mathcal{A}$ on $\mathcal{X}/\mathcal{Y}$. It is called the *reduction of* T *by* $\mathcal{Y}$. We will use $L^{\mathcal{A}/\mathcal{L}}$ to denote the reduction of the left regular representation L by a left ideal $\mathcal{L}$.

Internal Models for Cyclic Representations

The last construction gives a standard model for all cyclic representations as shown in the following proposition. This result seems to have first appeared explicitly in Jacobson [1956]. For unital algebras, the result can be stated without mentioning modular ideals and in that form it was known much earlier.

4.1.2 Proposition *Let $\mathcal{A}$ be an algebra and let T be a representation of $\mathcal{A}$ on $\mathcal{X}$. Then T is cyclic if and only if T is equivalent to $L^{\mathcal{A}/\mathcal{L}}$ for some modular left ideal $\mathcal{L}$ of $\mathcal{A}$. If z is any cyclic vector for T, then $\mathcal{L}$ can be chosen as*

$$\mathcal{L} = \{a \in \mathcal{A} : T_a z = 0\}.$$

Proof Suppose z is any vector in $\mathcal{X}$. Then $\mathcal{L} = \{a \in \mathcal{A} : T_a z = 0\}$ is a left ideal. Now assume that z is a cyclic vector. Then there is some $e \in \mathcal{A}$ satisfying $T_e z = z$. Clearly e is a right relative identity for $\mathcal{L}$, so that $\mathcal{L}$ is modular. For $a \in \mathcal{A}$, let $U(a + \mathcal{L}) = T_a z$. This is well-defined by the choice of $\mathcal{L}$ and hence is a linear injection of $\mathcal{A}/\mathcal{L}$ into $\mathcal{X}$. It is surjective since z is a cyclic vector. Finally it is easy to check that $T_a U = U L_a^{\mathcal{A}/\mathcal{L}}$ holds for all $a \in \mathcal{A}$. Thus U is an equivalence. This completes the proof of the necessity of the condition.

Now suppose $\mathcal{L}$ is a modular left ideal with right relative identity e. Then $L_a^{\mathcal{A}/\mathcal{L}}(e + \mathcal{L}) = a + \mathcal{L}$ holds for all $a \in \mathcal{A}$. Hence $e + \mathcal{L}$ is a cyclic vector for $L^{\mathcal{A}/\mathcal{L}}$. Since equivalence preserves cyclicity, this completes the proof of the sufficiency of the condition.   □

Irreducible Representations

Having obtained a model for each cyclic representation we now turn to the problem of picking out the irreducible representations. They are special cases of cyclic representations.

The next theorem, which is essentially due to Jacobson [1945a], shows that any irreducible representation is equivalent to $L^{\mathcal{A}/\mathcal{M}}$ for a maximal left ideal $\mathcal{M}$. However, modular ideals had not yet been introduced by Irving E. Segal [1947a] and the satisfactory result just quoted made it difficult to see that theory should be based on only certain distinguished maximal left ideals. (If $\mathcal{M}$ is a maximal left ideal, then it is easy to see that $L^{\mathcal{A}/\mathcal{M}}$ is either irreducible or trivial. The latter case is equivalent to $\mathcal{A}^2 \subseteq \mathcal{M}$. Thus if $L^{\mathcal{A}/\mathcal{M}}$ is irreducible, there is a maximal modular left ideal $\mathcal{M}'$ with $L^{\mathcal{A}/\mathcal{M}'}$ equivalent to $L^{\mathcal{A}/\mathcal{M}}$.) Although Segal [1947a] and Einar Hille [1948] use modular ideals cleverly in close proximity to their discussion of the irreducibility of $L^{\mathcal{A}/\mathcal{M}}$ for $\mathcal{M}$ a maximal left ideal, they seem to have overlooked the fact that $\mathcal{M}$ may always be chosen modular. Hence again it seems that the next theorem first appeared in full generality in Jacobson's book [1956].

4.1.3 Theorem *Let $\mathcal{A}$ be an algebra and let T be a representation of $\mathcal{A}$ on the linear space $\mathcal{X}$. Then the following are equivalent.*

(a) *T is irreducible.*

(b) *Every non-zero vector in $\mathcal{X}$ is a cyclic vector, and $\mathcal{X}$ is not zero-dimensional.*

(c) *There is a maximal modular left ideal $\mathcal{L}$ so that T is equivalent to* $L^{\mathcal{A}/\mathcal{L}}$.

Proof $(a) \Rightarrow (b)$: The set $\{z \in \mathcal{X} : T_{\mathcal{A}}z = \{0\}\}$ is a T-invariant subspace and hence equals $\{0\}$ or $\mathcal{X}$. Since T is nontrivial this subspace is $\{0\}$. For any $z \in \mathcal{X}, T_{\mathcal{A}}z = \{T_a z : a \in \mathcal{A}\}$ is a T-invariant subspace and hence equals $\{0\}$ or $\mathcal{X}$. For each non-zero $z \in \mathcal{X}, T_{\mathcal{A}}z$ equals $\mathcal{X}$ so z is cyclic.

$(b) \Rightarrow (c)$: Proposition 4.1.2 shows that T is equivalent to $L^{\mathcal{A}/\mathcal{L}}$ for some modular left ideal $\mathcal{L}$. If $\mathcal{L}$ is proper but not maximal choose a proper left ideal $\mathcal{K}$ with $\mathcal{K}$ properly including $\mathcal{L}$. Then $\mathcal{K}/\mathcal{L}$ is a non-zero and proper $L^{\mathcal{A}/\mathcal{L}}$-invariant subspace. This is impossible since every non-zero element of $\mathcal{K}/\mathcal{L}$ is a cyclic vector for $L^{\mathcal{A}/\mathcal{L}}$ by the equivalence of $L^{\mathcal{A}/\mathcal{L}}$ and T.

$(c) \Rightarrow (a)$: If $\mathcal{L}$ is a maximal left ideal and $\mathcal{Y}$ is an $L^{\mathcal{A}/\mathcal{L}}$-invariant subspace of $\mathcal{A}/\mathcal{L}$, then $\{a \in \mathcal{A} : a + \mathcal{L} \in \mathcal{Y}\}$ must be a left ideal of $\mathcal{A}$ which includes $\mathcal{L}$. Hence $\mathcal{Y}$ is either $\{0\}$ or $\mathcal{A}/\mathcal{L}$. Thus if $\mathcal{L}$ is modular, both $L^{\mathcal{A}/\mathcal{L}}$ and its equivalent representation T are irreducible.   □

Decomposition of Representations

Suppose T is a representation of an algebra $\mathcal{A}$ on a linear space $\mathcal{X}$ which satisfies

$$\mathcal{X} = \mathcal{Y} \oplus \mathcal{Z} \qquad (4)$$

where $\mathcal{Y}$ and $\mathcal{Z}$ are both T-invariant. When this happens the pair $(\mathcal{Y}, \mathcal{Z})$ is said to *decompose* T. In this case the reduction of T by $\mathcal{Y}$ is equivalent to the restriction of T to $\mathcal{Z}$. Given a T-invariant subspace $\mathcal{Y}$, only in very special cases will there be a T-invariant subspace $\mathcal{Z}$ so that $(\mathcal{Y}, \mathcal{Z})$ decomposes T. Obviously irreducible representations are indecomposable. The next theorem will state an interesting and useful intermediate condition. The theorem is essentially due to Issai Schur [1905] and is one of the oldest results in this whole theory. It was stated in the present form in Albert's book [1939].

4.1.4 Theorem *Let T be an irreducible representation of an algebra $\mathcal{A}$ on a linear space $\mathcal{X}$. Then*

$$(T_{\mathcal{A}})' = \{S \in \mathcal{L}(\mathcal{X}) : ST_a = T_a S, \quad a \in \mathcal{A}\} \qquad (5)$$

is a division algebra. If $(T_{\mathcal{A}})'$ is a division algebra then T is indecomposable.

Proof Suppose S belongs to $(T_{\mathcal{A}})'$. Then $S\mathcal{X}$ is a T-invariant subspace. Since $S\mathcal{X}$ is $\{0\}$ if and only if S is zero, $S\mathcal{X}$ equals $\mathcal{X}$ for all non-zero S. Similarly $\{x : Sx = 0\}$ is T-invariant and must equal $\{0\}$ unless S is zero. Thus every non-zero $S \in (T_{\mathcal{A}})'$ is invertible in $\mathcal{L}(\mathcal{X})$, and its inverse must belong to $(T_{\mathcal{A}})'$, so $(T_{\mathcal{A}})'$ is a division algebra.

Suppose the representation is decomposable. Then the projections onto the two subspaces both belong to $(T_{\mathcal{A}})'$ since the subspaces are invariant.

However these projection operators are nontrivial idempotents, and thus $(T_A)'$ is not a division algebra.   □

The usual representation on $\mathbb{C}^n (n \geq 2)$ of the algebra of strictly lower triangular $n \times n$ matrices is an example of a representation which has a nontrivial invariant subspace but no nontrivial decomposition. In this case $(T_A)'$ is not a division algebra. If one represents the algebra of all $\mathbb{N} \times \mathbb{N}$ matrices which have only finitely many non-zero entries on the vector space of all complex sequences, it is not hard to see that the commutant is just $\mathbb{C}$ and hence is a division algebra, but the subspace $\mathcal{Y}$ of all sequences with only finitely many non-zero entries is an invariant subspace (which does not give rise to a decomposition). Thus we have seen that none of the implications in the theorem are logical equivalences. In Volume II we will see a theory in which the corresponding concepts are all equivalent.

If T is a representation of an algebra $\mathcal{A}$ on $\mathcal{X}$, then the decomposition of $\mathcal{X}$ by the T-invariant subspaces $\mathcal{Y}$ and $\mathcal{Z}$ corresponds to an internal direct sum decomposition of the corresponding $\mathcal{A}$-module. We will briefly indicate the construction of the corresponding external direct sum. Suppose S and T are representations of $\mathcal{A}$ on $\mathcal{Y}$ and $\mathcal{Z}$, respectively. Then we define a representation $S \oplus T$ of $\mathcal{A}$ on $\mathcal{Y} \oplus \mathcal{Z}$ as the linear extension to all of $\mathcal{Y} \oplus \mathcal{Z}$ of the following expression:

$$(S \oplus T)_a(y \oplus z) = S_a y \oplus T_a z \qquad \forall \, a \in \mathcal{A}; \; y \in \mathcal{Y}; \; z \in \mathcal{Z}.$$

Clearly this can be extended to define the direct sum of a finite number of representations. If $\{T^\alpha : \alpha \in A\}$ is an infinite family of representations of an algebra $\mathcal{A}$ on linear spaces $\{\mathcal{X}^\alpha : \alpha \in A\}$, then we define an algebraic direct sum $\bigoplus_{\alpha \in A} T^\alpha$ of $\{T^\alpha : \alpha \in A\}$ on the direct sum (cf. Definition 1.4.2) $\bigoplus_{\alpha \in A} \mathcal{X}^\alpha$, by

$$(\oplus_{\alpha \in A} T^\alpha)_a(\oplus_{\alpha \in A} x^\alpha) = \oplus_{\alpha \in A} T^\alpha_a x^\alpha$$

where $a \in \mathcal{A}$ is arbitrary and $\alpha \mapsto x^\alpha$ is any function in $\prod_{\alpha \in A} \mathcal{X}_\alpha$ which is equal to zero except for a finite set of $\alpha \in A$.

If $\{T^\alpha : \alpha \in A\}$ is an infinite family of representations of $\mathcal{A}$ on Banach spaces $\{\mathcal{X}^\alpha : \alpha \in A\}$, then it is sometimes desirable to define the topological direct sum of the representations on the topological direct sum of the representation spaces $\{\mathcal{X}^\alpha : \alpha \in A\}$. This is not always possible but when we need the construction we will discuss it.

Intertwining Operators

We will also state a variant of the first portion of the last theorem. Let $\mathcal{A}$ be an algebra and let S and T be representations of $\mathcal{A}$ on linear spaces $\mathcal{X}$ and $\mathcal{Y}$, respectively. We wish to consider the set of *intertwining operators* of S and T consisting of all linear maps $V : \mathcal{X} \to \mathcal{Y}$ which satisfy:

$$T_a V x = V S_a x \qquad x \in \mathcal{X}; \; a \in \mathcal{A}.$$

This is obviously a linear space under pointwise operations.

4.1.5 Proposition *All non-zero intertwining operators between irreducible representations are equivalences.*

Proof We use the notation established just before the statement of the proposition. It is enough to show that each non-zero element is a bijection. However (just as in the proof of the last theorem) if V is a non-zero intertwining operator, then $V\mathcal{X}$ is a non-zero T-invariant subspace of $\mathcal{Y}$ which must equal $\mathcal{Y}$, and $\ker(V)$ is a proper S-invariant subspace of $\mathcal{X}$ which must equal $\{0\}$. □

Ideals and Irreducible Representations

Next we investigate the connection between irreducible representations and ideals. These results were at least essentially known to Jacobson when he wrote [1945a].

4.1.6 Theorem *Let T be an irreducible representation of an algebra $\mathcal{A}$ on a linear space $\mathcal{X}$ and let $\mathcal{I}$ be an ideal in $\mathcal{A}$.*

(a) *The restriction of the homomorphism T to $\mathcal{I}$ is either trivial or an irreducible representation of $\mathcal{I}$. Furthermore every irreducible representation of $\mathcal{I}$ arises in this way.*

(b) *If $\mathcal{I} \subseteq \ker(T)$ holds, then*

$$\tilde{T}_{a+\mathcal{I}} = T_a \qquad \forall\, a \in \mathcal{A} \tag{6}$$

defines an irreducible representation $\tilde{T}$ of $\mathcal{A}/\mathcal{I}$. Furthermore every irreducible representation of $\mathcal{A}/\mathcal{I}$ arises in this way.

Proof (a): The set $\{x \in \mathcal{X} : T_\mathcal{I}x = \{0\}\}$ is a T-invariant subspace and hence must be $\{0\}$ or $\mathcal{X}$. If it is $\mathcal{X}$, the restriction of the homomorphism T to $\mathcal{I}$ is trivial. If it is $\{0\}$, then for each non-zero $x \in \mathcal{X}$ the T-invariant subspace $T_\mathcal{I}x$ must be $\mathcal{X}$. Hence the restriction of the homomorphism T to $\mathcal{I}$ is an irreducible representation by Theorem 4.1.3(b).

Now suppose T is an irreducible representation of $\mathcal{I}$. If $z \in \mathcal{X}$, $b, d \in \mathcal{I}$ and $a \in \mathcal{A}$ satisfy $T_b z = 0$, then $T_d T_{ab} z = T_{dab} z = T_{da} T_b z = 0$ holds. Thus $\{T_{ab}z : a \in \mathcal{A}\}$ is included in the T-invariant subspace $\{x \in \mathcal{X} : T_\mathcal{I}x = \{0\}\}$, which must equal $\{0\}$. Hence we can choose any $z \in \mathcal{X}^T \setminus \{0\}$ and use

$$\overline{T}_a T_b z = T_{ab} z \qquad \forall\, a \in \mathcal{A}, \ b \in \mathcal{I}$$

to define $\overline{T}_a$ uniquely on $\mathcal{X} = T_\mathcal{I}z$. Clearly the restriction of the homomorphism $\overline{T}$ to $\mathcal{I}$ is T and $\overline{T}$ is an irreducible representation of $\mathcal{A}$. This proves that every irreducible representation of $\mathcal{I}$ arises by restricting the homomorphism of some irreducible representation of $\mathcal{A}$.

(b) Clearly $\tilde{T}$ is well-defined and hence is an irreducible representation of $\mathcal{A}/\mathcal{I}$ when $\mathcal{I} \subseteq \ker(T)$. On the other hand, if T is an irreducible representation of $\mathcal{A}/\mathcal{I}$ and $\varphi \colon \mathcal{A} \to \mathcal{A}/\mathcal{I}$ is the natural map, then T is obtained in the indicated way from the irreducible representation $T \circ \varphi$ of $\mathcal{A}$. □

Primitive and Prime Ideals

We comment on the history of the next definition. Jacobson [1945a] introduced the term "primitive algebra" defining it to be an algebra in which $\{0\}$ satisfies condition (a) of our Theorem 4.1.8. In the same place he introduced the concept and notation of the quotient $\mathcal{I} : \mathcal{A}$ of an ideal. The term "primitive ideal" was introduced by Jacobson also but in [1945b]. Prime ideals were defined by Wolfgang Krull [1928], cf. Neal H. McCoy [1949]. We have remarked earlier on the origin of the word "ideal" from the concept of an ideal number in an algebraic number field. These were introduced in order to restore unique factorization, and thus the concept of a prime ideal was associated with the very origin of ideal theory.

The reader should note that primitive ideals are defined in an asymmetrical fashion. Since representations correspond to multiplication on the left, a primitive ideal is not necessarily primitive (although it is an ideal) in the reverse of the algebra. This is the reason for the asymmetry of the notion of quotient introduced in the next definition.

4.1.7 Definition An ideal is called *primitive* if it is the kernel of some irreducible representation. A proper ideal $\mathcal{I}$ is called *prime* if whenever $\mathcal{I}_1$ and $\mathcal{I}_2$ are ideals satisfying $\mathcal{I}_1 \mathcal{I}_2 \subseteq \mathcal{I}$, then either $\mathcal{I}_1 \subseteq \mathcal{I}$ or $\mathcal{I}_2 \subseteq \mathcal{I}$. The set of primitive ideals of an algebra $\mathcal{A}$ will be denoted by $\Pi_{\mathcal{A}}$ and the set of prime ideals by $P_{\mathcal{A}}$.

If $\mathcal{L}$ is a left ideal, then the set

$$\mathcal{L} : \mathcal{A} = \{a \in \mathcal{A} : a\mathcal{A} \subseteq \mathcal{L}\} \tag{7}$$

is called the *quotient of $\mathcal{L}$.*

In Chapter 7 we will introduce a topology on the sets $\Pi_{\mathcal{A}}$ and $P_{\mathcal{A}}$, and thereafter they will be considered to be topological spaces. If $\mathcal{I}$ is any ideal of $\mathcal{A}$ the symbols $h^{\Pi}(\mathcal{I})$ and $h^P(\mathcal{I})$ denote the sets

$$h^{\Pi}(\mathcal{I}) = \{\mathcal{P} \in \Pi_{\mathcal{A}} : \mathcal{I} \subseteq \mathcal{P}\}$$

$$h^P(\mathcal{I}) = \{\mathcal{P} \in P_{\mathcal{A}} : \mathcal{I} \subseteq \mathcal{P}\}$$

which are called the *hulls* of $\mathcal{I}$ in $\Pi_{\mathcal{A}}$ and in $P_{\mathcal{A}}$, respectively. These hulls, which will be the closed sets in the topology introduced in Chapter 7, will be studied in more depth there.

It is easy to see that the quotient $\mathcal{L} : \mathcal{A}$ of $\mathcal{L}$ is a (two-sided) ideal. If $\mathcal{L}$ is a modular left ideal with right relative identity e, then any $a \in \mathcal{L} : \mathcal{A}$, satisfies $a = ae - (ae - a) \in \mathcal{L}$. Hence $\mathcal{L} : \mathcal{A} \subseteq \mathcal{L}$ holds. We refine this observation in the following theorem, which is essentially due to Jacobson [1945a], although of course he did not mention modular ideals.

4.1.8 Theorem *Let $\mathcal{A}$ be an algebra.*

(a) *An ideal $\mathcal{P}$ of $\mathcal{A}$ is primitive if and only if it has the form $\mathcal{L} : \mathcal{A}$ for some maximal modular left ideal $\mathcal{L}$. In this case $\mathcal{P}$ is the largest ideal included in $\mathcal{L}$.*

(b) *A primitive ideal is prime. (In fact, if $\mathcal{A}$ is the completion of a normed algebra $\mathcal{B}$, then the intersection of a primitive ideal of $\mathcal{A}$ with $\mathcal{B}$ is a prime ideal of $\mathcal{B}$.)*

(c) *A primitive ideal $\mathcal{P}$ satisfies $\mathcal{P} = \cap\{\mathcal{L} : \mathcal{L}$ is a maximal modular left ideal satisfying $\mathcal{P} = \mathcal{L} : \mathcal{A}\}$.*

(d) *An element $a \in \mathcal{A}$ is quasi-invertible if and only if it is quasi-invertible modulo each primitive ideal.*

(e) *If $\mathcal{I}$ is an ideal of $\mathcal{A}$, then the map $\mathcal{P} \mapsto \mathcal{P} \cap \mathcal{I}$ is a bijection of $\Pi_{\mathcal{A}} \setminus h^{\Pi}(\mathcal{I})$ onto $\Pi_{\mathcal{I}}$ and the map $\mathcal{P} \mapsto \mathcal{P}/\mathcal{I}$ is a bijection of $h^{\Pi}(\mathcal{I})$ onto $\Pi_{\mathcal{A}/\mathcal{I}}$.*

Proof (a): Suppose $\mathcal{P}$ is the kernel of the (non-trivial) irreducible representation $T: \mathcal{A} \to \mathcal{L}(\mathcal{X})$. Choose $z \in \mathcal{X} \setminus \{0\}$ and define $\mathcal{L}$ by $\mathcal{L} = \{a \in \mathcal{A} : T_a z = 0\}$. Then $\mathcal{L}$ is a maximal modular left ideal and T is equivalent to $L^{\mathcal{A}/\mathcal{L}}$ as shown in Proposition 4.1.2 and Theorem 4.1.3. Clearly $\mathcal{P}$ is the kernel of $L^{\mathcal{A}/\mathcal{L}}$ and this kernel is $\mathcal{L} : \mathcal{A}$. We have noted $\mathcal{L} : \mathcal{A} \subseteq \mathcal{L}$. Any ideal $\mathcal{I} \subseteq \mathcal{L}$ clearly satisfies $\mathcal{I}\mathcal{A} \subseteq \mathcal{L}$ and hence is included in $\mathcal{L} : \mathcal{A}$.

(c): Since $\mathcal{L} : \mathcal{A} \subseteq \mathcal{L}$ holds for any modular left ideal $\mathcal{L}$, we know that $\mathcal{P}$ is included in $\cap\{\mathcal{L} : \mathcal{L}$ is a modular left ideal with $\mathcal{P} = \mathcal{L} : \mathcal{A}\}$. To prove the opposite inclusion suppose $\mathcal{P}$ is the kernel of the irreducible representation $T: \mathcal{A} \to \mathcal{L}(\mathcal{X})$ and a belongs to $\mathcal{A}\setminus\mathcal{P}$. Then $T_a z$ is non-zero for some $z \in \mathcal{X}$. Thus a does not belong to $\mathcal{L} = \{a \in \mathcal{A} : T_a z = 0\}$. However, we have just noted that $\mathcal{L}$ satisfies $\mathcal{P} = \mathcal{L} : \mathcal{A}$.

(b): We will show that if $\mathcal{P}$ is a primitive ideal of $\mathcal{A}$ then $\mathcal{P} \cap \mathcal{B}$ satisfies a condition formally stronger than our definition of a prime ideal of $\mathcal{B}$. Suppose $\mathcal{L}_1$ and $\mathcal{L}_2$ are left ideals of $\mathcal{B}$ satisfying $\mathcal{L}_1\mathcal{L}_2 \subseteq \mathcal{P} = \mathcal{L} : \mathcal{A}$, where $\mathcal{L}$ is a maximal modular left ideal of $\mathcal{A}$. Note that $\mathcal{L}$ is closed by Theorem 2.4.7 and hence $\mathcal{P} = \mathcal{L} : \mathcal{A}$ is also closed. Since $\mathcal{B}$ is dense in $\mathcal{A}$, the closures $\mathcal{L}_1^-$ and $\mathcal{L}_2^-$ of $\mathcal{L}_1$ and $\mathcal{L}_2$ are left ideals in $\mathcal{A}$. If $\mathcal{L}_2$ is not included in $\mathcal{L} : \mathcal{A}$, then $\mathcal{L}_2\mathcal{A} + \mathcal{L}$ properly contains the maximal left ideal $\mathcal{L}$. This implies $\mathcal{A} = \mathcal{L}_2^-\mathcal{A} + \mathcal{L}$. Thus $\mathcal{L}_1^-\mathcal{A} = \mathcal{L}_1^-\mathcal{L}_2^-\mathcal{A} + \mathcal{L}_1^-\mathcal{L} \subseteq \mathcal{P}\mathcal{A} + \mathcal{L} = \mathcal{L}$ implies $\mathcal{L}_1 \subseteq \mathcal{L} : \mathcal{A} = \mathcal{P}$.

(d): If a is quasi-inveritible in $\mathcal{A}$ it is certainly quasi-invertible modulo each primitive ideal. In order to establish the converse, we will assume that $a \in \mathcal{A}$ is quasi-invertible modulo each primitive ideal of $\mathcal{A}$. First we show that a has a left quasi-inverse. Suppose not. Then Theorem 2.4.6 shows that a is the right relative identity for some maximal modular left ideal, $\mathcal{L}$. By assumption a is quasi-invertible modulo the primitive ideal $\mathcal{L} : \mathcal{A}$. Let b be a quasi-inverse for a modulo $\mathcal{L} : \mathcal{A}$ so that $b - ba + a = b \circ a$ belongs to $\mathcal{L} : \mathcal{A} \subseteq \mathcal{L}$. Since $b - ba$ belongs to $\mathcal{L}$ this implies $a \in \mathcal{L}$ which contradicts our choice of $\mathcal{L}$. Therefore a has a left quasi-inverse.

If we denote the left quasi-inverse of a by b, then $b + \mathcal{P}$ must equal $(a + \mathcal{P})^q$ for each primitive ideal $\mathcal{P}$. Hence b has a left quasi-inverse modulo each primitive ideal. The argument just given shows that b has a left quasi-inverse. Hence b is quasi-invertible and a is its right quasi-inverse, so $a = b^q$ is quasi-invertible.

(e): Theorem 4.1.6(a) shows that the map $\mathcal{P} \mapsto \mathcal{P} \cap \mathcal{I}$ is a surjection of $\Pi_{\mathcal{A}} \setminus h^\Pi(\mathcal{I})$ onto $\Pi_{\mathcal{I}}$. In order to show that it is an injection, suppose $\mathcal{P}_1, \mathcal{P}_2 \in \Pi_{\mathcal{A}} \setminus h^\Pi(\mathcal{I}) \subseteq P_{\mathcal{A}} \setminus h^P(\mathcal{I})$ satisfy $\mathcal{P}_1 \cap \mathcal{I} = \mathcal{P}_2 \cap \mathcal{I}$. Then $\mathcal{P}_2 \mathcal{I} \subseteq \mathcal{P}_1 \cap \mathcal{I} \subseteq \mathcal{P}_2$ and $\mathcal{I} \not\subseteq \mathcal{P}_2$ imply $\mathcal{P}_1 \subseteq \mathcal{P}_2$. A similar argument gives $\mathcal{P}_2 \subseteq \mathcal{P}_1$ so that the map is a bijection. Theorem 4.1.6(b) shows that the map $\mathcal{P} \mapsto \mathcal{P}/\mathcal{I}$ is a surjection of $h^\Pi(\mathcal{I})$ onto $\Pi_{\mathcal{A}/\mathcal{I}}$, and this map is obviously an injection.   □

Primitive and Maximal Modular Ideals

The next result was noted by Jacobson [1945b] for unital algebras.

4.1.9 Theorem *A maximal modular ideal is primitive. Hence every proper modular ideal is included in some primitive ideal. In a commutative algebra every primitive ideal is maximal modular.*

Proof If $\mathcal{M}$ is a maximal modular ideal, then by Theorem 2.4.6 there is a maximal modular left ideal $\mathcal{L}$ which includes $\mathcal{M}$. Then $\mathcal{M}\mathcal{A} \subseteq \mathcal{M} \subseteq \mathcal{L}$ implies $\mathcal{M} \subseteq \mathcal{L} : \mathcal{A}$. Hence maximality implies $\mathcal{M} = \mathcal{L} : \mathcal{A}$. Thus $\mathcal{M}$ is primitive by Theorem 4.1.8. The second statement now follows by Theorem 2.4.6. If $\mathcal{L}$ is a maximal modular left ideal in a commutative algebra, then it is a maximal ideal so that $\mathcal{L} \subseteq \mathcal{L} : \mathcal{A}$ (which is clear for any two-sided ideal $\mathcal{L}$) implies $\mathcal{L} : \mathcal{A} = \mathcal{L}$. Since any primitive ideal has the form $\mathcal{L} : \mathcal{A}$ for some maximal modular left ideal, this proves the last statement.   □

Prime Ideals

We will have much more to do with primitive ideals than with prime ideals, but the next theorem, which is due to McCoy [1949], will find a use. Note that it shows that the condition established for primitive ideals in the proof of Theorem 4.1.8(b) is not properly stronger than the definition of a prime ideal.

4.1.10 Theorem *The following conditions on a proper two-sided ideal $\mathcal{P}$ in an algebra $\mathcal{A}$ are equivalent.*
  (a) *$\mathcal{P}$ is prime.*
  (b) *For all a, $b \in \mathcal{A}$, $a\mathcal{A}b \subseteq \mathcal{P}$ implies $a \in \mathcal{P}$ or $b \in \mathcal{P}$.*

(c) *For all $\langle$ left / right $\rangle$ ideals $\mathcal{L}_1$ and $\mathcal{L}_2$ of $\mathcal{A}$, $\mathcal{L}_1\mathcal{L}_2 \subseteq \mathcal{P}$ implies*
$\mathcal{L}_1 \subseteq \mathcal{P}$ or $\mathcal{L}_2 \subseteq \mathcal{P}$.

Furthermore if $\mathcal{I}$ is an ideal of $\mathcal{A}$, then the map $\mathcal{P} \mapsto \mathcal{P} \cap \mathcal{I}$ is a bijection
of $P_{\mathcal{A}} \setminus h^P(\mathcal{I})$ onto $P_{\mathcal{I}}$ and the map $\mathcal{P} \mapsto \mathcal{P}/\mathcal{I}$ is a bijection of $h^P(\mathcal{I})$ onto
$P_{\mathcal{A}/\mathcal{I}}$.

Proof $(a) \Rightarrow (b)$: The inclusion $a\mathcal{A}b \subseteq \mathcal{P}$ implies $\mathcal{A}a\mathcal{A}\mathcal{A}b\mathcal{A} \subseteq \mathcal{P}$. Since $\mathcal{P}$ is
prime we conclude that either $\mathcal{A}a\mathcal{A} \subseteq \mathcal{P}$ or $\mathcal{A}b\mathcal{A} \subseteq \mathcal{P}$. If the first inclusion
holds, then $(\mathcal{A}^1 a \mathcal{A}^1)^3 \subseteq \mathcal{A}a\mathcal{A} \subseteq \mathcal{P}$. However, since $\mathcal{P}$ is prime this implies
$a \in \mathcal{A}^1 a \mathcal{A}^1 \subseteq \mathcal{P}$. If the second inclusion holds, the analogous argument
gives $b \in \mathcal{A}^1 b \mathcal{A}^1 \subseteq \mathcal{P}$.

$(b) \Rightarrow (c)$: Suppose $\mathcal{L}_1\mathcal{L}_2 \subseteq \mathcal{P}$ holds but $\mathcal{L}_2$ is not included in $\mathcal{P}$. Choose
$b \in \mathcal{L}_2 \setminus \mathcal{P}$. Then any $a \in \mathcal{L}_1$ satisfies $a\mathcal{A}b \subseteq \mathcal{L}_1\mathcal{L}_2 \subseteq \mathcal{P}$. Thus (b) implies
$a \in \mathcal{P}$ since $b \notin \mathcal{P}$. Hence $\mathcal{L}_1$ is included in $\mathcal{P}$.

$(c) \Rightarrow (a)$: Obvious.

Finally we prove the last sentence. If $\mathcal{P}$ is a prime ideal of $\mathcal{A}$ which
does not include $\mathcal{I}$, then $\mathcal{P} \cap \mathcal{I}$ is a proper ideal of $\mathcal{I}$. Suppose $\mathcal{I}_1$ and $\mathcal{I}_2$
are ideals in $\mathcal{I}$ satisfying $\mathcal{I}_1\mathcal{I}_2 \subseteq \mathcal{P} \cap \mathcal{I}$. Since $\mathcal{P}$ is a prime ideal of $\mathcal{A}$ and
satisfies $(\mathcal{I}\mathcal{I}_1)(\mathcal{I}\mathcal{I}_2) \subseteq \mathcal{P}$, either $\mathcal{I}\mathcal{I}_1 \subseteq \mathcal{P}$ or $\mathcal{I}\mathcal{I}_2 \subseteq \mathcal{P}$ must hold by (c).
Similarly since either $\mathcal{I}(\mathcal{I}_1\mathcal{A}^1) = (\mathcal{I}\mathcal{I}_1)\mathcal{A}^1 \subseteq \mathcal{P}$ or $\mathcal{I}(\mathcal{I}_2\mathcal{A}^1) = (\mathcal{I}\mathcal{I}_2)\mathcal{A}^1 \subseteq \mathcal{P}$
holds and $\mathcal{I} \subseteq \mathcal{P}$ does not, either $\mathcal{I}_1 \subseteq \mathcal{P} \cap \mathcal{I}$ or $\mathcal{I}_2 \subseteq \mathcal{P} \cap \mathcal{I}$ holds. Thus
$\mathcal{P} \cap \mathcal{I}$ is a prime ideal of $\mathcal{I}$.

In order to prove that $\mathcal{P} \mapsto \mathcal{P} \cap \mathcal{I}$ is a surjection onto the set of prime
ideals of $\mathcal{I}$, suppose that $\mathcal{P}$ is a prime ideal of $\mathcal{I}$. Let $\overline{\mathcal{P}}$ be the ideal of $\mathcal{A}$
defined by $\overline{\mathcal{P}} = \{a \in \mathcal{A} : \mathcal{I}a\mathcal{I} \subseteq \mathcal{P}\}$. Since $\mathcal{P}$ is prime we conclude $\overline{\mathcal{P}} = \{a \in \mathcal{A} : \mathcal{I}a \subseteq \mathcal{P}\}$. The inclusion $\mathcal{P} \subseteq \overline{\mathcal{P}} \cap \mathcal{I}$ is obvious. To obtain the opposite
inclusion, note that any $b \in \overline{\mathcal{P}} \cap \mathcal{I}$ satisfies $\mathcal{I}(b\mathcal{I}^1) = (\mathcal{I}b)\mathcal{I}^1 \subseteq \mathcal{P}\mathcal{I}^1 = \mathcal{P}$.
Since $\mathcal{P}$ is a prime ideal of $\mathcal{I}$ this implies $b \in b\mathcal{I}^1 \subseteq \mathcal{P}$. In order to see that
$\overline{\mathcal{P}}$ is a prime ideal of $\mathcal{A}$, let $\mathcal{I}_1$ and $\mathcal{I}_2$ be ideals of $\mathcal{A}$ satisfying $\mathcal{I}_1\mathcal{I}_2 \subseteq \overline{\mathcal{P}}$.
Then $(\mathcal{I}\mathcal{I}_1)(\mathcal{I}\mathcal{I}_2) \subseteq \mathcal{I}\mathcal{I}_1\mathcal{I}_2 \subseteq \mathcal{P}$ implies either $\mathcal{I}\mathcal{I}_1 \subseteq \mathcal{P}$ or $\mathcal{I}\mathcal{I}_2 \subseteq \mathcal{P}$ since $\mathcal{P}$
is a prime ideal of $\mathcal{I}$. Hence either $\mathcal{I}_1 \subseteq \overline{\mathcal{P}}$ or $\mathcal{I}_2 \subseteq \overline{\mathcal{P}}$ holds, so that $\overline{\mathcal{P}}$ is a
prime ideal of $\mathcal{A}$ which clearly does not include $\mathcal{I}$. The proof of Theorem
4.1.8(e) shows that $\mathcal{P} \mapsto \mathcal{P} \cap \mathcal{I}$ is injective on $P_{\mathcal{A}} \setminus h^P(\mathcal{I})$, so it is a bijection.

Let $\varphi: \mathcal{A} \mapsto \mathcal{A}/\mathcal{I}$ be the natural map. The map $\mathcal{P} \to \varphi(\mathcal{P})$ is clearly
an injection of $h^P(\mathcal{I})$ into the set of prime ideals of $\mathcal{A}/\mathcal{I}$ since $\mathcal{A}/\mathcal{I}$ and
$(\mathcal{A}/\mathcal{P})/\varphi(\mathcal{P})$ are isomorphic. Similarly it is immediate that the map $\mathcal{P} \mapsto$
$\varphi^{-1}(\mathcal{P})$ sends the set of prime ideals of $\mathcal{A}/\mathcal{I}$ into $h(\mathcal{I})$. Hence these two
inverse maps are bijections. $\qquad\qquad\qquad\qquad\qquad\qquad\qquad\qquad\qquad\qquad\qquad\quad$ $\square$

Ultraprime Algebras and Ideals

The concept we discuss next was introduced recently by Martin Mathieu [1989] and it is not yet clear what role it will play in the theory. It can
be thought of as a metric adaptation of criterion (b) of Theorem 4.1.10.
The term "ultraprime" comes from the theory of ultraproducts (*cf.* Defi-

nition 1.3.8) which is a branch of model theory sometimes associated with nonstandard analysis. We shall not investigate these roots but will use Mathieu's characterization as a definition. This stems from work of Gunter Lumer and Marvin Rosenblum [1959].

4.1.11 Definition A normed algebra $\mathcal{A}$ is called an *ultraprime algebra* if there exists a positive constant k satisfying

$$\|L_b R_c\| \geq k\|b\|\,\|c\| \qquad \forall\, b, c \in \mathcal{A}.$$

A constant k for which the above condition holds is called *an ultraprime constant for $\mathcal{A}$.*

A closed ideal $\mathcal{I}$ of a normed algebra $\mathcal{A}$ is said to be an *ultraprime ideal of $\mathcal{A}$* if $\mathcal{A}/\mathcal{I}$ is an ultraprime algebra in its quotient norm.

To paraphrase this definition, a normed algebra $\mathcal{A}$ is ultraprime with a (positive) ultraprime constant k if for any $b, c \in \mathcal{A}$ there is some $a \in \mathcal{A}$ satisfying

$$\|bac\| \geq k\|b\|\,\|a\|\,\|c\|.$$

To further simplify we often restrict a, or b and c, or all three to have norm 1. Note that the supremum of all ultraprime constants need not be an ultraprime constant. A (necessarily closed) ideal $\mathcal{P}$ is an ultraprime ideal of a normed algebra $\mathcal{A}$ if there is some positive constant k such that for any $b, c \in \mathcal{A}$ there is some $a \in \mathcal{A}$ satisfying

$$\|bac + \mathcal{P}\| \geq k\|b + \mathcal{P}\|\,\|a + \mathcal{P}\|\,\|c + \mathcal{P}\|.$$

Obviously an ultraprime ideal is a closed prime ideal. Unfortunately, not all primitive ideals are ultraprime ideals (Pere Ara and Mathieu [1991]). An example is the ideal $\{0\}$ in the algebra $\mathcal{B}_{HS}(\mathcal{H})$ of Hilbert–Schmidt operators. (Let $b = c$ be a projection onto an n-dimensional subspace.)

If $\mathcal{X}$ is a normed linear space, any (necessarily primitive) subalgebra $\mathcal{A}$ of $\mathcal{B}(\mathcal{X})$ which includes $\mathcal{B}_F(\mathcal{X})$ is an ultraprime algebra and any positive number less than 1 may be used as an ultraprime constant. To see this, let $S, T \in \mathcal{A}$ of norm 1 and $1 > \varepsilon > 0$ be arbitrary and choose $x \in \mathcal{X}_1$ and $x^* \in \mathcal{X}_1^*$ to satisfy $\|S(x)\| > 1 - \varepsilon$ and $\|T^*(x^*)\| > 1 - \varepsilon$. Then $\|S\,x \otimes x^*\,T\| = \|S(x) \otimes T^*(x^*)\| = \|S(x)\|\,\|T^*(x^*)\| > (1 - \varepsilon)^2$.

George A. Willis [1989b] shows that for a discrete group G, $\ell^1(G)$ is ultraprime if and only if G is an ICC-group (*i.e.*, a group in which all conjugacy classes except $\{e\}$ are infinite). As we shall see, any prime commutative normed algebra (not isomorphic to $\mathbb{C}$) is an example of a prime but not ultraprime algebra.

4.1.12 Theorem *Let $\mathcal{A}$ be a normed algebra.*

(a) *$\mathcal{A}$ is an ultraprime algebra if and only if one (hence all) of its dense subalgebras is an ultraprime algebra. Hence the completion of an ultraprime normed algebra is ultraprime.*

(b) *If $\mathcal{A}$ is an ultraprime algebra, then all of its ideals are ultraprime algebras.*

(c) *$\mathcal{A}$ is an ultraprime algebra if and only if $\mathcal{A}^1$ is an ultraprime algebra.*

(d) *If $\mathcal{A}$ is an ultraprime algebra, then its center is trivial.*

Proof (a): Let $\mathcal{B}$ be a dense subalgebra. If $\mathcal{B}$ is an ultraprime algebra with an ultraprime constant k and $b, c \in \mathcal{A}$ are arbitrary with norm 1, choose b', c' and a in $\mathcal{B}$ satisfying $\|b'\| = 1$, $\|c'\| = 1$, $\|b - b'\| \leq k/3$, $\|c - c'\| \leq k/3$ and $\|b'ac'\| \geq k\|a\|$. Then $bac = b'ac' + (b - b')ac + b'a(c - c')$ implies $\|bac\| \geq \|b'ac'\| - \|(b - b')ac\| - \|b'a(c - c')\| \geq k\|a\| - (k/3)\|a\| - (k/3)\|a\| = (k/3)\|a\|$. Hence $\mathcal{A}$ is ultraprime.

Conversely, if $\mathcal{A}$ is an ultraprime algebra with an ultraprime constant k and $b, c \in \mathcal{B}$ are arbitrary, choose $a \in \mathcal{A}$ satisfying $\|bac\| \geq k\|b\|\,\|a\|\,\|c\|$. By approximating a by an element in $\mathcal{B}$, we see that $\mathcal{B}$ is an ultraprime algebra.

(b): Since $\mathcal{A}$ is an ideal of itself, it is enough to consider an arbitrary ideal $\mathcal{I}$ in an ultraprime algebra with an ultraprime constant k. Let $b, c \in \mathcal{I}$ be arbitrary with norm 1. Then we can find $a \in \mathcal{A}$ of norm 1 satisfying $\|bac\| \geq k$. Again we can choose $d \in \mathcal{A}$ of norm 1 satisfying $\|bacdc\| \geq k\|bac\|$. Hence $acd \in \mathcal{I}$ satisfies $\|b(acd)c\| \geq k\|bac\| \geq k^2$. Hence $\mathcal{I}$, an arbitrary ideal, is an ultraprime algebra.

(c): If $\mathcal{A}^1$ is an ultraprime algebra, then $\mathcal{A}$ is also by (b). Suppose $\mathcal{A}$ is a nonunital ultraprime algebra with ultraprime constant k. We give $\mathcal{A}^1$ the norm $\|\lambda + a\| = |\lambda| + \|a\|$ as usual. Let $\mu + b$ and $\nu + c$ in $\mathcal{A}^1$ be arbitrary and choose $a \in \mathcal{A}$ to satisfy $\|bac\| \geq k\|b\|\,\|a\|\,\|c\|$. Then we also have $\|ba\|\,\|c\| \geq k\|b\|\,\|a\|\,\|c\|$ so $\|ba\| \geq k\|b\|\,\|a\|$ and similarly $\|ac\| \geq k\|a\|\,\|c\|$. Hence the inequality

$$\|(\mu + b)a(\nu + c)\| \geq \|\mu ac\| + \|\nu ba\| + \|bac\|$$
$$\geq k\|\mu + b\|\,\|a\|\,\|\nu + c\|$$

shows that $\mathcal{A}^1$ is ultraprime.

(d): The center $\mathcal{A}_Z$ is called trivial if it is zero- or one-dimensional. If $\mathcal{A}_Z$ is not trivial, Proposition 2.5.12 shows that it has a topological divisor of zero. Suppose $c \in \mathcal{A}_Z$ and the sequence $\{b_n\}_{n \in \mathbb{N}} \subseteq \mathcal{A}_Z$ satisfy $\|c\| = 1$, $\|b_n\| = 1$ for all $n \in \mathbb{N}$ and $b_n c \to 0$. Then any sequence $\{a_n\}_{n \in \mathbb{N}} \subseteq \mathcal{A}$ of norm 1 elements would satisfy $b_n a_n c \to 0$ which contradicts the ultraprime character of $\mathcal{A}$. $\qquad\square$

Mathieu [1989] shows that a normed algebra is ultraprime if and only if some, hence every, ultrapower is a prime (or, equivalently, an ultraprime) algebra. He uses this to prove several of the above results and also that every matrix algebra $M_n(\mathcal{A})$ over an ultraprime algebra is ultraprime.

We now indicate the construction of the *normed symmetric algebra of quotients* of an ultraprime normed algebra. For details, see Mathieu [1991]. We could carry out the construction as a direct limit, but we will proceed in another way. Mathieu [1989] constructs a larger one-sided algebra of quotients, but we judge the symmetric version to be of more general interest. It generalizes the double centralizer algebra.

Let $\mathcal{A}$ be an ultraprime normed algebra with k as an ultraprime constant. Consider the set $\mathcal{F}$ of triples $(\mathcal{I}, \tilde{L}, \tilde{R})$ where $\mathcal{I}$ is a non-zero ideal and $\tilde{L}, \tilde{R}: \mathcal{I} \to \mathcal{A}$ are continuous linear maps satisfying the double centralizer equation:

$$a\tilde{L}(b) = \tilde{R}(a)b \qquad \forall\, a, b \in \mathcal{I}. \tag{8}$$

Thus these maps satisfy $\tilde{L}(ac) = \tilde{L}(a)c$ and $\tilde{R}(ba) = b\tilde{R}(a)$ for all $a \in \mathcal{I}$ and $b, c \in \mathcal{A}$ by the argument of Theorem 1.2.4, since even in a prime algebra the annihilator ideals are zero. (Note that the results of Proposition 1.2.3 are also available.) We introduce an equivalence relation on $\mathcal{F}$ by

$$(\mathcal{I}, \tilde{L}, \tilde{R}) \sim (\mathcal{I}', \tilde{L}', \tilde{R}') \Leftrightarrow \tilde{L}|_{\mathcal{I} \cap \mathcal{I}'} = \tilde{L}'|_{\mathcal{I} \cap \mathcal{I}'} \text{ and } \tilde{R}|_{\mathcal{I} \cap \mathcal{I}'} = \tilde{R}'|_{\mathcal{I} \cap \mathcal{I}'}.$$

Because $\mathcal{A}$ is a prime algebra, no product or intersection of non-zero ideals is the zero ideal. Also if there is any non-zero ideal $\mathcal{K} \subseteq \mathcal{I} \cap \mathcal{I}'$ on which $\tilde{L}$ agrees with $\tilde{L}'$ and $\tilde{R}$ with $\tilde{R}'$, then $(\mathcal{I}, \tilde{L}, \tilde{R})$ and $(\mathcal{I}', \tilde{L}', \tilde{R}')$ are equivalent. This establishes the transitivity of the equivalence relation. Denote the equivalence class of $(\mathcal{I}, \tilde{L}, \tilde{R})$ by $[\mathcal{I}, \tilde{L}, \tilde{R}]$. As a set, let $\mathcal{B}$ be the set of these equivalence classes of triples. The algebra operations on $\mathcal{B}$ are defined by

$$[\mathcal{I}, \tilde{L}, \tilde{R}] + [\mathcal{I}', \tilde{L}', \tilde{R}'] = [\mathcal{I} \cap \mathcal{I}', \tilde{L} + \tilde{L}', \tilde{R} + \tilde{R}']$$
$$\lambda[\mathcal{I}, \tilde{L}, \tilde{R}] = [\mathcal{I}, \lambda\tilde{L}, \lambda\tilde{R}]$$
$$[\mathcal{I}, \tilde{L}, \tilde{R}] \cdot [\mathcal{I}', \tilde{L}', \tilde{R}'] = [\mathcal{I}\mathcal{I}' \cap \mathcal{I}'\mathcal{I}, \tilde{L}\tilde{L}', \tilde{R}'\tilde{R}].$$

For any triple $(\mathcal{I}, \tilde{L}, \tilde{R})$ in $\mathcal{F}$ and $b, c \in \mathcal{I}$, we can find $a \in \mathcal{A}$ satisfying

$$\|b\|\, \|\tilde{L}\|\, \|a\|\, \|c\| \geq \|b\tilde{L}(ac)\| = \|\tilde{R}(b)ac\| \geq k\|\tilde{R}(b)\|\, \|a\|\, \|c\|$$

which implies $\|\tilde{L}\| \geq k\|\tilde{R}\|$. Similarly we find $\|\tilde{R}\| \geq k\|\tilde{L}\|$. Again, for any pair of equivalent triples $(\mathcal{I}, \tilde{L}, \tilde{R}) \sim (\mathcal{I}', \tilde{L}', \tilde{R}')$ and any $b \in \mathcal{I}$ and $c \in \mathcal{I}'$, we can find an $a \in \mathcal{A}$ satisfying

$$\|\tilde{L}'\|\, \|b\|\, \|a\|\, \|c\| \geq \|\tilde{L}'(bac)\| = \|\tilde{L}(bac)\| = \|\tilde{L}(b)ac\|$$
$$\geq k\|\tilde{L}(b)\|\, \|a\|\, \|c\|$$

which implies $\|\tilde{L}'\| \geq k\|\tilde{L}\|$. Obviously, $\|\tilde{R}'\| \geq k\|\tilde{R}\|$, $\|\tilde{L}\| \geq k\|\tilde{L}'\|$ and $\|\tilde{R}\| \geq k\|\tilde{R}'\|$ also hold. Hence if we define a norm on $\mathcal{B}$ by

$$\| [\mathcal{I}, \tilde{L}, \tilde{R}] \| = \inf\{\max\{\|\tilde{L}'\|, \|\tilde{R}'\|\} : (\mathcal{I}, \tilde{L}, \tilde{R}) \sim (\mathcal{I}', \tilde{L}', \tilde{R}')\},$$

we do get an algebra norm rather than just a semi-norm. Finally, we can define a homeomorphic isomorphism of $\mathcal{A}$ into $\mathcal{B}$ by $a \mapsto (\mathcal{A}, L_a, R_a)$. We will state the resulting theorem, but leave the rest of the details to the reader who may consult Mathieu [1991].

4.1.13 Theorem *Let $\mathcal{B}$ be the normed symmetric algebra of quotients as defined above. Then $\mathcal{B}$ is a unital ultraprime normed algebra containing a homeomorphic and isomorphic image of $\mathcal{A}$. It is the largest normed algebra extension of $\mathcal{A}$ satisfying:*

(a) *For each $b \in \mathcal{B}$ there is a non-zero ideal $\mathcal{I}$ of $\mathcal{A}$ satisfying $b\mathcal{I} \subseteq \mathcal{A}$ and $\mathcal{I}b \subseteq \mathcal{A}$.*

(b) *If $b \in \mathcal{B}$ and a non-zero ideal $\mathcal{I}$ of $\mathcal{A}$ satisfy $b\mathcal{I} = \{0\}$ or $\mathcal{I}b = \{0\}$ then $b = 0$.*

If $\mathcal{A}$ is an algebra, then an operator $T: \mathcal{A} \to \mathcal{A}$ is an *elementary operator* if it has the form $T = \sum_{j=1}^{n} L_{a_j} R_{b_j}$ for some $n \in \mathbb{N}$ and $a_j, b_j \in \mathcal{A}$. There is an obvious homomorphism θ of $\mathcal{A} \otimes \mathcal{A}^R$ (where $\mathcal{A}^R$ is the reverse algebra of $\mathcal{A}$) onto the algebra $\mathcal{E}\ell(\mathcal{A})$ of elementary operators on $\mathcal{A}$. Usually this homomorphism has a nontrivial kernel. Ultraprime normed algebras provide an exception. The elementary proof is given in Mathieu [1989]. (See Erling Størmer [1980] for related interesting ideas.)

4.1.14 Proposition *Let $\mathcal{A}$ be an ultraprime normed algebra. With the notation introduced above, θ is an isomorphism of $\mathcal{A} \otimes \mathcal{A}^R$ onto $\mathcal{E}\ell(\mathcal{A})$.*

Ideals and Arens Multiplication

The following results of Paul Civin and Bertram Yood [1961] describe the behavior of ideals in relation to the Arens products on the double dual algebra. Nilpotent ideals are introduced in Definition 4.4.1.

4.1.15 Proposition *Let $\mathcal{A}$ be a Banach algebra and let $\mathcal{I}$ be a $\langle$ left / right / two-sided $\rangle$ ideal of $\mathcal{A}$. Let $\overline{\mathcal{I}}$ be the weak* closure of $\kappa(\mathcal{I})$ in $\mathcal{A}^{**}$.*

(a) *$\overline{\mathcal{I}}$ is a $\langle$ left / right / two-sided $\rangle$ ideal of $\mathcal{A}^{**}$ with respect to both Arens products.*

(b) *If $\mathcal{I}$ is a proper modular ideal, $\overline{\mathcal{I}}$ is also a proper modular ideal with respect to both Arens products.*

(c) *If $\mathcal{I}$ is a primitive ideal, $\overline{\mathcal{I}}$ is included in a primitive ideal of $\mathcal{A}^{**}$ with respect to both Arens products.*

(d) *If $\mathcal{I}$ is nilpotent, $\overline{\mathcal{I}}$ is nilpotent with respect to both Arens products.*

Proof (a): We consider only the case in which $\mathcal{I}$ is a left ideal since the other cases are similar. Let $\langle\, f = \lim \kappa(a_\alpha) \,/\, g = \lim \kappa(b_\beta) \,\rangle$ be an arbitrary element of $\langle\, \mathcal{A}^{**} \,/\, \overline{\mathcal{I}} \,\rangle$ with $\langle\, a_\alpha \,/\, b_\beta \,\rangle$ elements of $\langle\, \mathcal{A} \,/\, \mathcal{I} \,\rangle$ and the limit in the weak* topology. We use the continuity properties (14) and (15) from Theorem 1.4.2. Consider the first Arens product and a fixed α. Then

$\kappa(a_\alpha)g = \lim_\beta \kappa(a_\alpha b_\beta)$ belongs to $\overline{\mathcal{I}}$. Hence $fg = \lim_\alpha \kappa(a_\alpha)g$ also belongs to $\overline{\mathcal{I}}$. For the second Arens product, $f \circ \kappa(b_\beta) = \lim_\alpha \kappa(a_\alpha b_\beta)$ belongs to $\overline{\mathcal{I}}$, so $f \circ g = \lim_\beta f \circ \kappa(b_\beta)$ also belongs to $\overline{\mathcal{I}}$.

(b): Again we consider only the case in which $\mathcal{I}$ is a proper modular left ideal with right relative identity e. We may suppose $\|e\| = 1$. Clearly $\kappa(e)$ is a right relative identity for $\overline{\mathcal{I}}$, so it only remains to show that it is not in $\overline{\mathcal{I}}$. The proof of Theorem 2.4.7 shows that the distance from e to (the norm closure of) $\mathcal{I}$ is at least 1. Thus the Hahn–Banach theorem gives an $\omega \in \mathcal{A}_1^*$ satisfying $\omega(e) = 1$ and $\omega(\mathcal{I}) = \{0\}$. Thus $\kappa(e)$ is not in $\overline{\mathcal{I}}$ as we wished to show.

(c): Theorem 4.1.8(a) shows that there is a maximal modular left ideal $\mathcal{L}$ of $\mathcal{A}$ satisfying $\mathcal{I} = \mathcal{L} : \mathcal{A}$. By what we have just proved, there is a maximal modular left ideal $\mathcal{M}$ of $\mathcal{A}^{**}$ which includes $\overline{\mathcal{L}}$. The argument given in the proof of (a) shows that the primitive ideal $\mathcal{M} : \mathcal{A}$ includes $\overline{\mathcal{I}}$.

(d): Suppose $\mathcal{I}$ satisfies $\mathcal{I}^n = \{0\}$. We will show that $\overline{\mathcal{I}}$ satisfies $\overline{\mathcal{I}}^n = \{0\}$. Let $f_1, f_2, \ldots, f_n$ be arbitrary elements in $\overline{\mathcal{I}}$. For each $j = 1, 2, \ldots, n$, suppose $f_j = \lim a_{j\alpha}$ holds in the weak* topology with each $a_{j\alpha}$ in $\mathcal{A}$. Consider the first Arens product first. For any choice of indices $\kappa(a_{1\alpha(1)})\kappa(a_{2\alpha(2)}) \cdots \kappa(a_{(n-1)\alpha(n-1)})f_n$ is zero since it is the limit of $\kappa(a_{1\alpha(1)})\kappa(a_{2\alpha(2)}) \cdots \kappa(a_{(n-1)\alpha(n-1)})\kappa(a_{n\alpha}) = 0$. Similarly, for any choice of indices, $\kappa(a_{1\alpha(1)})\kappa(a_{2\alpha(2)}) \cdots \kappa(a_{(n-2)\alpha(n-2)})f_{n-1}f_n$ is zero since it is the limit of $\kappa(a_{1\alpha(1)})\kappa(a_{2\alpha(2)}) \cdots \kappa(a_{(n-2)\alpha(n-2)})\kappa(a_{(n-1)\alpha})f_n = 0$. Obviously we can proceed in this way until we finally conclude $f_1 f_2 \cdots f_n = 0$. For the second Arens product we simply work from left to right instead of from right to left. □

Irreducible Representations and Double Centralizer Algebras

4.1.16 Proposition *Any irreducible representation of an algebra $\mathcal{A}$ has a unique extension to a representation of the double centralizer algebra $\mathcal{D}(\mathcal{A})$. This extension is also irreducible.*

Proof Let $T: \mathcal{A} \to \mathcal{L}(\mathcal{X})$ be an irreducible representation. For $x, y \in \mathcal{X}$ and $b, c \in \mathcal{A}$, suppose $T_b x = T_c y$. Then any $a \in \mathcal{A}$ and $(L, R) \in \mathcal{D}(\mathcal{A})$ satisfy

$$T_a(T_{L(b)}x - T_{L(c)}y) = T_{aL(b)}x - T_{aL(c)}y = T_{R(a)b}x - T_{R(a)c}y$$
$$= T_{R(a)}(T_b x - T_c y) = 0.$$

The irreducibility of T implies $T_{L(b)}x = T_{L(c)}y$. Therefore we may define $\overline{T}: \mathcal{D}(\mathcal{A}) \to \mathcal{L}(\mathcal{X})$ by $\overline{T}_{(L,R)}x = T_{L(c)}y$ for any $x \in \mathcal{X}$ and $(L, R) \in \mathcal{D}(\mathcal{A})$ where $x = T_c y$. (When convenient, we could also write $\overline{T}_{(L,R)}x = T_{L(e)}x$ where $x = T_e x$.) This obviously defines a representation extending T, since it is well defined by the calculation above. It is also clear that any extension of T would have to agree with $\overline{T}$. Finally, $\overline{T}$ is irreducible since it extends an irreducible representation. □

Modules, Normed Modules and Bimodules

We close this section by formally defining modules and describing the identification between the theory of representations and the theory of modules mentioned in the introduction.

4.1.17 Definition Let $\mathcal{A}$ be an algebra and let $\mathcal{X}$ be a linear space. Then $\mathcal{X}$ is called a *left $\mathcal{A}$-module* if there is a fixed representation $T: \mathcal{A} \to \mathcal{L}(\mathcal{X})$ and $T_a(x)$ is denoted by ax for all $a \in \mathcal{A}$ and $x \in \mathcal{X}$. If $\mathcal{A}$ and $\mathcal{X}$ are both normed and the representation $T: \mathcal{A} \to \mathcal{B}(\mathcal{X})$ is continuous, then the corresponding module is called a *normed left $\mathcal{A}$-module*. The norm of T (as a bounded linear map from $\mathcal{A}$ to $\mathcal{B}(\mathcal{X})$) is called the *the bound of the normed module* $\mathcal{X}$. In the above situation, if both $\mathcal{A}$ and $\mathcal{X}$ are complete, the left module is called a *left Banach module*.

If $\mathcal{X}$ is a linear space and $S: \mathcal{A} \to \mathcal{L}(\mathcal{X})$ is a fixed anti-representation of $\mathcal{A}$ on $\mathcal{X}$ with its action denoted by $xa = S_a(x)$ for all $a \in \mathcal{A}$ and $x \in \mathcal{X}$, then $\mathcal{X}$ is called a *right $\mathcal{A}$-module*. *Normed right $\mathcal{A}$-modules* and *right Banach $\mathcal{A}$-modules* are defined analogously to left ones.

If $\mathcal{X}$ is both a left and a right $\mathcal{A}$-module, and if the two module actions satisfy

$$a(xb) = (ax)b \qquad \forall \, a, b \in \mathcal{A}; \; x \in \mathcal{X}$$

then it is called an *$\mathcal{A}$-bimodule*. (This means that the related representation and anti-representation have commuting actions.) If both module actions are normed module actions, then the bimodule is said to be *normed*. The larger of the bounds for the left and right module actions is called the *bound* for the bimodule action. If, in addition, both $\mathcal{A}$ and $\mathcal{X}$ are complete, the bimodule is called a *Banach bimodule*.

Note that a left $\mathcal{A}$-module is a linear space $\mathcal{X}$ with an action of $\mathcal{A}$, denoted by juxtaposition, and satisfying

$$(a + b)x = ax + bx; \quad a(x + y) = ax + ay;$$

$$(\lambda a)x = \lambda(ax) = a(\lambda x); \quad \text{and} \quad (ab)x = a(bx)$$

for all $a, b \in \mathcal{A}$, $x, y \in \mathcal{X}$ and $\lambda \in \mathbb{C}$. For right modules, the very last of these conditions must be replaced by $x(ab) = (xa)b$ for all $a, b \in \mathcal{A}$ and $x \in \mathcal{X}$.

If $T: \mathcal{A} \to \mathcal{L}(\mathcal{X})$ is any representation, it defines a left $\mathcal{A}$-module structure on $\mathcal{X}$. Conversely, if $\mathcal{X}$ is a left $\mathcal{A}$-module, then the map $T: \mathcal{A} \to \mathcal{L}(\mathcal{X})$ defined by

$$T_a(x) = ax \qquad \forall \, a \in \mathcal{A}; x \in \mathcal{X}$$

is a representation of $\mathcal{A}$ on $\mathcal{X}$. Clearly these are inverse constructions. All the terminology for representations is extended to modules in the obvious way.

A left $\mathcal{A}$-module action of a normed algebra $\mathcal{A}$ on a normed linear space $\mathcal{X}$ defines a normed left $\mathcal{A}$-module if there is some finite constant M satisfying

$$\|ax\| \leq M\|a\|\,\|x\| \qquad \forall\, a \in \mathcal{A};\ x \in \mathcal{X}.$$

The infimum of all constants M having the above property is the bound of the normed module $\mathcal{X}$.

4.2 Representations and Norms

Definitions

We now introduce norms into these purely algebraic considerations. The reader may wish to compare the following definitions with those given in Definition 4.1.1. Although both irreducible representations and topologically irreducible representations have been studied since the beginning of Banach algebra theory, sometimes no distinction in terminology was made. As we will see in Chapter 9 of Volume II topological irreducibility plays a key role in the representation theory of *-algebras while irreducibility is normally more important in other contexts. The terminology we adopt seems to have been first used systematically in Rickart's important book [1960]. Some authors use "algebraically irreducible" or "strictly irreducible" for what we call simply "irreducible". Sometimes they then use "irreducible" for what we are calling "topologically irreducible". The same remarks hold for "cyclic" and "topologically cyclic". The other terminology of this definition also seems to have been standardized by its use in Rickart's book.

4.2.1 Definition Let T be a representation of an algebra $\mathcal{A}$ on a normed linear space $\mathcal{X}$. Then T is called:

(a) *Normed* if $T_a \in \mathcal{B}(\mathcal{X})$ for each $a \in \mathcal{A}$.

(b) *Topologically cyclic* if there is a vector $z \in \mathcal{X}$ (called a *topologically cyclic vector*) such that $T_{\mathcal{A}}z = \{T_a z : a \in \mathcal{A}\}$ is dense in $\mathcal{X}$.

(c) *Topologically irreducible* if $\{0\}$ and $\mathcal{X}$ are the only closed T-invariant subspaces, and T is not trivial.

(d) *Topologically equivalent* to a representation S of $\mathcal{A}$ on a normed linear space $\mathcal{Y}$ if there is a homeomorphic linear bijection U (called a *topological equivalence*) of $\mathcal{X}$ onto $\mathcal{Y}$ which satisfies

$$S_a U = U T_a \qquad \forall\, a \in \mathcal{A}. \tag{1}$$

If $(\mathcal{A}, \sigma)$ is a semi-normed algebra then a representation T of $\mathcal{A}$ is called:

(e) *Continuous* if it is normed and continuous as a map from $(\mathcal{A}, \sigma)$ to $\mathcal{B}(\mathcal{X})$.

(f) *Strongly continuous* if $a \mapsto T_a x$ is continuous as a map from $(\mathcal{A}, \sigma)$ to $\mathcal{X}$ for each fixed $x \in \mathcal{X}$.

We will informally record a number of simple consequences of this definition. Note that it is easier for a representation on a normed linear space to be topologically cyclic or topologically irreducible than to be cyclic or irreducible, but that it is harder for it to be topologically equivalent to another representation than to be equivalent.

If T is a normed representation, it is obvious that the closure of every T-invariant subspace is T-invariant.

Any restriction of a normed, continuous or strongly continuous representation has the same property. Reduction by a closed T-invariant subspace also preserves each of these properties.

Let $\mathcal{A}$ be a spectral normed algebra and let $\mathcal{L}$ be a modular left ideal of $\mathcal{A}$. Then $L^{\mathcal{A}/\mathcal{L}}$ will be irreducible if and only if $\mathcal{L}$ is maximal. However it is also easy to see that $L^{\mathcal{A}/\mathcal{L}}$ will be topologically irreducible if and only if $\mathcal{L}$ is maximal among closed ideals. Since the two concepts (irreducibility and topological irreducibility for $L^{\mathcal{A}/\mathcal{L}}$) agree for any closed left ideal such as $\mathcal{L}$, then $L^{\mathcal{A}/\mathcal{L}}$ is topologically irreducible if and only if it is irreducible. We will find the same result in some other contexts, notably for *-representations of C*-algebras, but the two concepts do not usually agree.

Strongly Continuous Representations

Note that a continuous representation of a semi-normed algebra on a normed linear space is always normed and strongly continuous. The converse is not always true. However, the uniform boundedness principle gives a simple result in that direction for Banach algebras.

4.2.2 Proposition *A normed strongly continuous representation of a Banach algebra on a normed linear space is continuous.*

Proof Let T be a strongly continuous normed representation of a Banach algebra $\mathcal{A}$ on a normed linear space $\mathcal{X}$. Apply the uniform boundedness principle to $\{S_x : x \in \mathcal{X}_1\}$ where (as usual) $\mathcal{X}_1$ is the unit ball of $\mathcal{X}$ and

$$S_x(a) = T_a(x) \qquad \forall\, a \in \mathcal{A};\ x \in \mathcal{X}.$$

Strong continuity shows that each linear map S_x is continuous from $\mathcal{A}$ to $\mathcal{X}$. The fact that T is normed shows:

$$\|S_x(a)\| = \|T_a x\| \leq \|T_a\|\,\|x\| \leq \|T_a\| \qquad \forall\, a \in \mathcal{A};\ x \in \mathcal{X}.$$

Hence, there is some M satisfying

$$\|S_x\| \leq M \qquad \forall\, x \in \mathcal{X}_1$$

which implies

$$\|T_a\| = \sup_{x \in \mathcal{X}_1} \|T_a(x)\| = \sup_{x \in \mathcal{X}_1} \|S_x(a)\| \leq M\|a\|$$

so that T is continuous. □

We now obtain the analogue of Proposition 4.1.2 which first occurred explicitly in Rickart [1960]. Closely related results were discussed in Segal [1947a] and Hille [1948]. Note that the next two results deal with cyclic rather than topologically cyclic representations.

4.2.3 Theorem *Let T be a strongly continuous, cyclic representation of a semi-normed algebra $(\mathcal{A}, \sigma)$ on a normed linear space $\mathcal{X}$ with cyclic vector z. Then $L^{\mathcal{A}/\mathcal{L}}$ is equivalent (under a continuous equivalence) to T where $\mathcal{L}$ is the closed modular left ideal $\mathcal{L} = \{a \in \mathcal{A} : T_a z = 0\}$. If $\mathcal{A}$ is a Banach algebra and $\mathcal{X}$ is a Banach space then $L^{\mathcal{A}/\mathcal{L}}$ and T are topologically equivalent.*

Proof The proof of Proposition 4.1.2 shows that the map $U: \mathcal{A}/\mathcal{L} \to \mathcal{X}$ defined by $U(a + \mathcal{L}) = T_a z$ establishes an equivalence between $L^{\mathcal{A}/\mathcal{L}}$ and T. If T is strongly continuous then $\mathcal{L}$ is certainly closed so the quotient semi-norm σ' is a norm. Let M satisfy $\|T_a z\| \le M\sigma(a)$ for all $a \in \mathcal{A}$. Then $\|U(a + \mathcal{L})\| \le M\sigma'(a + \mathcal{L})$. Thus U is continuous. Suppose in addition that $\mathcal{A}$ is a Banach algebra and $\mathcal{X}$ is a Banach space. Then $\mathcal{A}/\mathcal{L}$ is a Banach space under its quotient norm. Since $U: \mathcal{A}/\mathcal{L} \to \mathcal{X}$ is a continuous bijection it is a homeomorphism and hence a topological equivalence. □

Proposition 4.2.2 recorded that the uniform boundedness principle shows that a normed strongly continuous representation of a Banach algebra is continuous. In the following corollary, which does not seem to have been explicitly noted before, we do not require that the representation be normed.

4.2.4 Corollary *Any strongly continuous cyclic representation of a Banach algebra on a Banach space is continuous. Any equivalence between two strongly continuous cyclic representations of a Banach algebra on Banach spaces is a topological equivalence.*

Proof The first statement follows from the last statement of Theorem 4.2.3 since $L^{\mathcal{A}/\mathcal{L}}$ is obviously continuous. To prove the second statement, suppose $T: \mathcal{A} \to \mathcal{L}(\mathcal{X})$ and $S: \mathcal{A} \to \mathcal{L}(\mathcal{Y})$ are strongly continuous cyclic representations of the Banach algebra $\mathcal{A}$ on the Banach spaces $\mathcal{X}, \mathcal{Y}$. Let $U: \mathcal{X} \to \mathcal{Y}$ be an equivalence. Let $z \in \mathcal{X}$ be a cyclic vector for T. Then Uz is a cyclic vector for S and we may define $\mathcal{L}$ by $\mathcal{L} = \{a \in \mathcal{A} : T_a z = 0\} = \{a \in \mathcal{A} : S_a U z = 0\}$. Theorem 4.2.3 asserts that the maps $a + \mathcal{L} \mapsto T_a z$ and $a + \mathcal{L} \mapsto S_a U z$ are topological equivalences. Thus $T_a z \mapsto S_a U z$ is a topological equivalence. However, this map is just U since $S_a U z = U T_a z$ holds for all $a \in \mathcal{A}$. □

In Corollary 4.2.16 results similar to those of the above corollary will be obtained again under different hypotheses using the much deeper Theorem 4.2.15.

Topologically Irreducible Representations

The next result contains a partial analogue of Theorem 4.1.3 and Proposition 4.1.5. Conclusion (a) is an exceptional result because it draws an algebraic conclusion from a partly topological hypothesis.

4.2.5 Proposition *Let T be a normed representation of an algebra $\mathcal{A}$ on a normed linear space $\mathcal{X}$. The representation T is topologically irreducible if and only if every non-zero vector $z \in \mathcal{X}$ is a topologically cyclic vector. If these conditions hold, then:*

(a) *The kernel of T is a prime ideal.*

(b) *The restriction of the homomorphism T to an ideal $\mathcal{I}$ of $\mathcal{A}$ is either trivial or topologically irreducible.*

(c) *If $\mathcal{I}$ is an ideal included in $\ker(T)$, then*

$$\tilde{T}_{a+\mathcal{I}} = T_a \qquad \forall\, a \in \mathcal{A}$$

defines a topologically irreducible representation of $\mathcal{A}/\mathcal{I}$. Every topologically irreducible representation of $\mathcal{A}/\mathcal{I}$ arises in this way.

Proof The equivalence of the two conditions is proved as in Theorem 4.1.3 using the easy fact that the closure of a T-invariant subspace is T-invariant.

(a): To prove this statement, suppose $\mathcal{L}_1$ and $\mathcal{L}_2$ are left ideals satisfying $\mathcal{L}_1\mathcal{L}_2 \subseteq \ker(T)$ and $\mathcal{L}_2 \nsubseteq \ker(T)$. Then $\mathcal{Y} = \text{span}(T_{\mathcal{L}_2}\mathcal{X})$ is a nontrivial T-invariant subspace. Hence $\mathcal{Y}$ is dense in $\mathcal{X}$. Let a belong to $\mathcal{L}_1$. Then $\mathcal{L}_1\mathcal{L}_2 \subseteq \ker(T)$ implies $T_a\mathcal{Y} = \{0\}$, and this implies $T_a = 0$ since $\mathcal{Y}$ is dense. Therefore $\mathcal{L}_1$ is included in $\ker(T)$. This proves that $\ker(T)$ is prime.

(b): The set $\{x \in \mathcal{X} : T_{\mathcal{I}}x = \{0\}\}$ is a closed T-invariant subspace and hence must be $\{0\}$ or $\mathcal{X}$. If it is $\mathcal{X}$, the restriction of the homomorphism T to $\mathcal{I}$ is trivial. If it is $\{0\}$, then for each non-zero $x \in \mathcal{X}$ the closed T-invariant subspace $(T_{\mathcal{I}}x)^-$ must be $\mathcal{X}$. Hence in this case the restriction of the homomorphism T to $\mathcal{I}$ is topologically irreducible.

(c): Clearly the two sets of operators $\{T_a : a \in \mathcal{A}\}$ and $\{\tilde{T}_{a+\mathcal{I}} : a + \mathcal{I} \in \mathcal{A}/\mathcal{I}\}$ are equal when T and $\tilde{T}$ satisfy the equation in (c). Hence if either set is topologically irreducible, the other is also. $\qquad\qquad\square$

Note that the proofs of (b) and (c) above are essentially the same as the proofs of (a) and (b) of Theorem 4.1.6. However, we cannot extend the second sentence of Theorem 4.1.6(a) to the present situation. Here is the trouble. Suppose $\mathcal{I}$ is an ideal of $\mathcal{A}$ and $T\colon\mathcal{I} \to \mathcal{L}(\mathcal{X})$ is a normed topologically irreducible representation of $\mathcal{I}$ on $\mathcal{X}$. Then $\{x \in \mathcal{X} : T_b(x) = 0 \text{ for } b \in \mathcal{I}\}$ is a closed T-invariant subspace which must be $\{0\}$. Suppose $b_1, b_2, \ldots, b_n \in \mathcal{I}$ and $z_1, z_2, \ldots, z_n \in \mathcal{X}$ satisfy $\sum_{j=0}^{n} T_{b_j} z_j = 0$. Then any $a \in \mathcal{A}$ and $d \in \mathcal{I}$

satisfy

$$T_d \sum_{j=0}^n T_{ab_j} z_j = \sum_{j=0}^n T_{dab_j} z_j = T_{da} \sum_{j=0}^n T_{b_j} z_j = 0.$$

Hence $\sum_{j=0}^n T_{ab_j} z_j$ belongs to the subspace which we have just shown to be $\{0\}$. Therefore we may define $\overline{T} \in \mathcal{L}(\mathcal{Y})$, where $\mathcal{Y} = \mathrm{span}(T_{\mathcal{I}}\mathcal{X})$, by

$$\overline{T}_a(\sum_{j=0}^n T_{b_j} z_j) = \sum_{j=0}^n T_{ab_j} z_j \qquad \forall \, a \in \mathcal{A}; \ n \in \mathbb{N}; \ b_j \in \mathcal{I}; \ z_j \in \mathcal{X}.$$

Clearly $\overline{T}$ is a topologically irreducible representation of $\mathcal{A}$ on $\mathcal{Y}$ which extends the restriction of the representation T to $\mathcal{Y}$. In general there seems to be no reason for $\overline{T}_a$ to be a bounded operator on $\mathcal{Y}$ and hence no way to extend T to be a normed representation of $\mathcal{A}$ on $\mathcal{X}$. (If $\mathcal{I}$ is an approximately unital Banach algebra, T is continuous and $\mathcal{X}$ is a Banach space then Corollary 5.2.3 gives an easy proof that the extension is possible.) However, we do not actually know of any topologically irreducible normed representation T of an ideal $\mathcal{I}$ in an algebra $\mathcal{A}$ which cannot be extended to a normed representation of the whole algebra $\mathcal{A}$. Of course, the most interesting counterexample would have $\mathcal{A}$ a Banach algebra, $\mathcal{I}$ closed, the representation space a Banach space and T continuous.

Topologically irreducible representations are not well understood and they can behave very badly. For instance let $T \in \mathcal{B}(\mathcal{X})$ be one of the operators without invariant subspaces discussed before Definition 2.8.8. Then the identity representation of the commutative unital subalgebra $\mathcal{A}$ of $\mathcal{B}(\mathcal{X})$ generated by T is a continuous topologically irreducible representation. The inclusion $\mathcal{A} \subseteq \mathcal{A}'$ shows that this representation has a nontrivial commutant which is not a division algebra by the Gelfand–Mazur Theorem (Corollary 2.2.3). Compare Schur's lemma (Theorem 4.1.4).

A normed representation T of an algebra $\mathcal{A}$ on a normed linear space $\mathcal{X}$ is said to be *topologically completely irreducible* if for any $S \in \mathcal{L}(\mathcal{X})$, any $\varepsilon > 0$ and any elements $x_1, x_2, \dots, x_n \in \mathcal{X}$ there exists an $a \in \mathcal{A}$ satisfying

$$\|T_a x_j - S x_j\| < \varepsilon \qquad \forall \, j = 1, 2, \dots, n.$$

It is not known whether continuous topologically irreducible representations of a Banach algebra on a Banach space are topologically completely irreducible. For some partial results in this direction involving the existence of finite-rank operators, which are due to J. M. G. Fell and J. Dixmier, see Example 4.8.8. See also Wiesław Żelazko [1991].

Spectral Semi-norms and Norms on Representation Spaces

We now investigate some of these concepts for spectral semi-normed algebras and spectral algebras. The following result is implicit in Irving Kaplansky [1947a], Segal [1947a] and Hille [1948].

4.2.6 Proposition *A primitive ideal in a spectral semi-normed algebra is closed.*

Proof Suppose $\mathcal{P}$ is a primitive ideal in a spectral semi-normed algebra $(\mathcal{A}, \sigma)$. Theorem 4.1.8(a) shows that there is some maximal modular left ideal $\mathcal{L}$ satisfying $\mathcal{P} = \mathcal{L} : \mathcal{A}$. Theorem 2.4.7 shows that $\mathcal{L}$ is closed. The definition of $\mathcal{L} : \mathcal{A}$ now shows that $\mathcal{P}$ is closed.   □

We next generalize Theorem 4.2.3 and also restate the conclusion in a way that will be useful. Theorem 4.2.8 will then show that this generalized theorem applies to irreducible representations of spectral semi-normed algebras. The following two results are due to Rickart [1950] in the case of Banach algebras.

4.2.7 Theorem *Let T be a cyclic representation of a semi-normed algebra $(\mathcal{A}, \sigma)$ on a linear space $\mathcal{X}$. If z is a cyclic vector for T and if $\{a \in \mathcal{A} : T_a z = 0\}$ is closed then*

$$\|x\|_z = \inf\{\sigma(a) : a \in \mathcal{A}, T_a z = x\} \qquad \forall\, x \in \mathcal{X} \qquad (2)$$

defines a norm on $\mathcal{X}$ such that T is a continuous representation on $(\mathcal{X}, \|\cdot\|_z)$ satisfying $\|T_a\|_z \leq \sigma(a)$ for all $a \in \mathcal{A}$. If $\mathcal{A}$ is a Banach algebra, then $(\mathcal{X}, \|\cdot\|_z)$ is a Banach space. If in addition $\mathcal{X}$ is already a Banach space under some norm $\|\cdot\|$ and T is strongly continuous with respect to $\|\cdot\|$, then $\|\cdot\|$ and $\|\cdot\|_z$ are equivalent.

Proof Let $\mathcal{L} = \{a \in \mathcal{A} : T_a z = 0\}$. Then $\|\cdot\|_z$ is simply the quotient norm of $\mathcal{A}/\mathcal{L}$ transferred to $\mathcal{X}$ by the equivalence established in Proposition 4.1.2. The inequality $\|T_a\|_z \leq \sigma(a)$ is a restatement of the fact that $L^{\mathcal{A}/\mathcal{L}}$ is a contractive representation. It follows immediately that $\|\cdot\|_z$ is complete if $\mathcal{A}$ is a Banach algebra. The last statement is now an immediate consequence of the second sentence of Corollary 4.2.4.   □

4.2.8 Theorem *Let $(\mathcal{A}, \sigma)$ be a spectral semi-normed algebra, and let T be an irreducible representation of $\mathcal{A}$ on a linear space $\mathcal{X}$. Any non-zero vector $z \in \mathcal{X}$ satisfies the hypotheses (and therefore the conclusion) of Theorem 4.2.7.*

Proof Theorem 4.1.3 shows that z is a cyclic vector and its proof shows that $\mathcal{L} = \{a \in \mathcal{A} : T_a z = 0\}$ is a maximal modular left ideal. Hence $\mathcal{L}$ is closed by Theorem 2.4.7. Thus Theorem 4.2.7 applies.   □

Segal proved essentially the next result in [1947a].

4.2.9 Corollary *If $\mathcal{A}$ is a Banach algebra, then each primitive ideal of $\mathcal{A}$ is the kernel of some continuous irreducible representation of $\mathcal{A}$ on a Banach space.*

The following result of Larry Schweitzer [1992] will be used later.

4.2.10 Theorem *Let A be a spectral normed algebra. Let B be a normed algebra and let $\varphi: B \to A$ be a continuous injective homomorphism onto a dense subalgebra of A. (A might be the completion of B.) Then the following are equivalent:*

(a) $\varphi(B)$ is a spectral subalgebra.

(b) The closure $\overline{\varphi(\mathcal{L})}$ in A of every maximal modular left ideal $\mathcal{L}$ in B satisfies:

$$\{b \in B : \varphi(b) \in \overline{\varphi(\mathcal{L})}\} = \mathcal{L}.$$

(c) For any irreducible representation T of B there is a representation S of A and a linear injection $V: \mathcal{X}^T \to \mathcal{X}^S$ satisfying:

$$VT_b x = S_{\varphi(b)} V x \qquad \forall\, b \in B, x \in \mathcal{X}^T.$$

When these conditions hold:

(d) The closure in A of the image under φ of each proper modular ideal of B is a proper modular ideal of A.

(e) The representation S can be taken to be irreducible.

(f) The linear map V is continuous and $V(\mathcal{X}^T)$ is dense in $\mathcal{X}^S$ in the norms introduced in Theorem 4.2.7 for the representation spaces $\mathcal{X}^T$ and $\mathcal{X}^S$.

Before giving the proof of this theorem we wish to point out explicitly that we will use weaker hypotheses than given in the formal statement of the theorem. For all the results except (f) it is enough to assume that A is a topological algebra with an open set of quasi-invertible elements. For the whole theorem it is enough to assume that B is a topological algebra. In (f) the norm on $\mathcal{X}^T$ can be considered as arising from either the norm giving the topology of B if there is such a norm, or from the norm induced by φ from A. We preferred to use norms as usual in the formal statement, but we will use these weaker hypotheses to generalize some results in Volume II. In many applications B is actually a dense subalgebra of A.

Proof We will actually prove this result under the above weaker hypotheses which were originally used by Schweitzer: Let A be a topological algebra with A_{qG} open and let B be a topological algebra. We will use $\bar{\mathcal{L}}$ to denote the closure in A of $\varphi(\mathcal{L})$.

$(a) \Rightarrow (b)$: Let e be a right relative identity for $\mathcal{L}$ in B. By Theorem 2.4.6(b) we conclude that $\{e - b : b \in \mathcal{L}\}$ is disjoint from B_{qG}. Since $A_{qG} \cap \varphi(B) = \varphi(B_{qG})$ and A_{qG} is open, A_{qG} is disjoint from $\{\varphi(e) - b : b \in \bar{\mathcal{L}}\}$. Hence $\bar{\mathcal{L}}$ is a proper subset of A. Also it is easy to check that $\tilde{\mathcal{L}} = \{b \in B : \varphi(b) \in \bar{\mathcal{L}}\}$ is a left ideal in B. It is also a proper subset by the density of $\varphi(B)$ in A. (Up to this point the argument does not depend on the maximality of $\mathcal{L}$.) Thus $\tilde{\mathcal{L}}$ is a proper left ideal in $\mathcal{B}$ including $\mathcal{L}$ which must therefore equal $\mathcal{L}$ by maximality.

(b) $\Rightarrow$ (c): (Condition (b) is now interpreted to say that $\tilde{\mathcal{L}}$ equals $\mathcal{L}$.) Let T be an irreducible representation of $\mathcal{B}$ on $\mathcal{X}$ and choose a non-zero $z \in \mathcal{X}$. Define $\mathcal{L} = \{b \in \mathcal{B} : T_b z = 0\}$. Then $\mathcal{L}$ is a maximal modular left ideal in $\mathcal{B}$ and the map $b + \mathcal{L} \mapsto T_b z$ provides an equivalence between $L^{\mathcal{B}/\mathcal{L}}$ and T by the proof of Proposition 4.1.2. Now use the notation already established. Note that $\bar{\mathcal{L}}$ is a proper modular left ideal in $\mathcal{A}$ by the density of $\varphi(\mathcal{B})$ in $\mathcal{A}$. Hence we can take S to be the representation $L^{\mathcal{A}/\bar{\mathcal{L}}}$ on $\mathcal{A}/\bar{\mathcal{L}}$ and V to be the (well-defined) map

$$T_d z \mapsto \varphi(d) + \bar{\mathcal{L}}$$

for all $d \in \mathcal{B}$.

(c) $\Rightarrow$ (a): Suppose $\varphi(e) \in \mathcal{A}_{qG} \cap \varphi(\mathcal{B})$ but the quasi-inverse c of $\varphi(e)$ in $\mathcal{A}$ is not in $\varphi(\mathcal{B})$. Then e is not left quasi-invertible in $\mathcal{B}$ since its left quasi-inverse would have to have c as its image under φ. Thus by Theorem 2.4.6 there is a maximal modular left ideal $\mathcal{L}$ in $\mathcal{B}$ with e as its right relative identity. Since $L^{\mathcal{B}/\mathcal{L}}$ is an irreducible representation of $\mathcal{B}$ there must be some representation S of $\mathcal{A}$ and an intertwining map $V: \mathcal{B}/\mathcal{L} \to \mathcal{X}^S$ as described in (c). Let $z = V(e + \mathcal{L})$ which is non-zero by construction and note:

$$S_{\varphi(e)} z = V L_e^{\mathcal{B}/\mathcal{L}}(e + \mathcal{L}) = V(ee + \mathcal{L}) = V(e + \mathcal{L}) = z.$$

Now we have a contradiction:

$$z = S_{\varphi(e)} z = S_c S_{\varphi(e)} z - S_c z = S_c z - S_c z = 0.$$

To prove the last statement in the theorem we may use all three conditions and the construction in the paragraph above. Thus conclusions (d), (e) and (f) follow easily. □

Strict Density of Irreducible Representations

The next theorem, which was proved in the case of Banach algebras by Rickart [1950], is an important consequence of Schur's lemma (Theorem 4.1.4) and the Gelfand–Mazur theorem (Corollary 2.2.3).

4.2.11 Theorem *Let T be an irreducible representation of a spectral algebra $\mathcal{A}$ on a linear space $\mathcal{X}$. Then*

$$(T_{\mathcal{A}})' = \{S \in \mathcal{L}(\mathcal{X}) : S T_a = T_a S \quad a \in \mathcal{A}\}$$

is the set of complex multiples of the identity map on $\mathcal{X}$.

Proof Choose a spectral semi-norm σ on $\mathcal{A}$. Let $z \in \mathcal{X}$ be non-zero and let $\| \cdot \|_z$ be as defined in Theorem 4.2.7. Suppose S belongs to $(T_{\mathcal{A}})'$.

Then for $x = T_a z$ we get $Sx = ST_a z = T_a Sz$ which implies $\|Sx\|_z \le$ $\|T_a\|_z \|Sz\|_z \le \sigma(a)\|Sz\|_z$. Taking the infimum over a with $x = T_a z$, we see that $\|Sx\|_z \le \|x\|_z \|Sz\|_z$ so S is continuous. Hence by Theorem 4.1.4, $(T_\mathcal{A})'$ is a normed division algebra with I as identity. Therefore Corollary 2.2.3 shows that $(T_\mathcal{A})'$ equals $\mathbb{C}I$.   □

4.2.12 Definition   A representation T of an algebra $\mathcal{A}$ on a linear space $\mathcal{X}$ is called *strictly dense* if whenever $x_1, x_2, \ldots, x_n$ is a finite list of linearly independent vectors in $\mathcal{X}$ and $y_1, y_2, \ldots, y_n$ is a list of vectors in $\mathcal{X}$ there is an element $a \in \mathcal{A}$ with $T_a x_j = y_j$ for $j = 1, 2, \ldots, n$.

The definition can be rephrased by saying that, for all $n \in \mathbb{N}, T_\mathcal{A}$ is n-fold transitive on linearly independent sets of vectors in $\mathcal{X}$. The left regular representation of the quarternions (a *real* rather than complex algebra) is irreducible but not 2-fold transitive, and hence not strictly dense. However, for (complex) spectral algebras we have the following remarkable theorem. This important result, which was mentioned in the introduction to this section, is due to Rickart [1950]. A version for a closed irreducible algebra of operators on a Banach space was obtained by Yood [1949]. The refinement stated second in the theorem is due to Allan M. Sinclair [1976] for Banach algebras. The result was known earlier for C*-algebras. Recall that a functional algebra is simply a spectral normed algebra in which the analytic functional calculus is preserved.

4.2.13 Theorem   *Any irreducible representation of a spectral algebra is strictly dense. If T is an irreducible representation of a functional algebra $\mathcal{A}$ and $x_1, x_2, \ldots, x_n$ and $y_1, y_2, \ldots, y_n$ are both finite lists of linearly independent vectors in $\mathcal{X}$, then there is an element $a \in \mathcal{A}$ satisfying $T_{e^a} x_j = y_j$ for $j = 1, 2, \ldots, n$.*

Proof Let T be an irreducible representation of a spectral algebra $\mathcal{A}$ on a linear space $\mathcal{X}$. We use induction on $n \ge 2$ in the following statement: For any list $x_1, x_2, \ldots, x_n$ of n linearly independent vectors, we can find an $a \in \mathcal{A}$ satisfying $T_a x_j = 0$ for $j = 1, 2, \ldots, n-1$ and $T_a x_n \ne 0$. First we show that the statement is true for $n = 2$. Suppose not. Then there are linearly independent vectors $y, z \in \mathcal{X}$ such that for all $a \in \mathcal{A}, T_a y = 0$ implies $T_a z = 0$. For any $x \in \mathcal{X}$, choose $b(x) \in \mathcal{A}$ so that $T_{b(x)} y = x$ and define $S: \mathcal{X} \to \mathcal{X}$ by $Sx = T_{b(x)} z$. Then S is well-defined by the assumption on y and z. Thus S is linear.

We now show that S belongs to $(T_\mathcal{A})'$. For any $a \in \mathcal{A}$ and $x \in \mathcal{X}$, $b(T_a x) = ab(x)$ follows from $T_a T_{b(x)} y = T_a x$. Hence, $S T_a x = T_{ab(x)} z = T_a T_{b(x)} z = T_a Sx$ shows $S \in (T_\mathcal{A})'$. Theorem 4.2.11 now shows that S must be just multiplication by some scalar, $\lambda \in \mathbb{C}$. Then $T_b z = S T_b y = \lambda T_{b(x)} y$ holds for all $b \in \mathcal{A}$. However this implies $T_\mathcal{A}(z - \lambda y) = \{0\}$ and hence $z = \lambda y$. This contradicts the independence of y and z and shows that the statement holds for $n = 2$.

Now assume the statement for some $n \geq 2$ and let $x_1, x_2, \ldots, x_{n+1}$ be a list of $n + 1$ linearly independent vectors. Choose $a \in \mathcal{A}$ satisfying $T_a x_1 = T_a x_2 = \ldots = T_a x_{n-1} = 0$ and $T_a x_{n+1} \neq 0$. If $T_a x_n$ is zero, then a has the desired property. If $T_a x_n$ and $T_a x_{n+1}$ are independent, choose $c \in \mathcal{A}$ so that $T_c T_a x_n = 0$ and $T_c T_a x_{n+1} \neq 0$. Then ca has the desired property.

Thus we may suppose $T_a x_{n+1} = \lambda T_a x_n$ for some $\lambda \in \mathbb{C}$. Choose b in $\mathcal{A}$ satisfying $T_b x_1 = T_b x_2 = \cdots = T_b x_{n-1} = 0$ and $T_b(x_{n+1} - \lambda x_n) \neq 0$. If $T_b x_n$ is zero, then b has the desired property (since $T_b x_{n+1} \neq 0$). If $T_b x_n$ and $T_b x_{n+1}$ are independent, then a $c \in \mathcal{A}$ can be chosen as before so that cb has the desired property. Thus we may suppose $T_b x_{n+1} = \mu T_b x_n$ for some $\mu \in \mathbb{C}$. Clearly μ does not equal λ. Choose $d \in \mathcal{A}$ satisfying $T_d T_b x_n = T_a x_n$. Then $a - db$ has the desired property since $T_{a-db} x_{n+1} = \lambda T_a x_n - T_d T_b x_{n+1} = \lambda T_a x_n - T_d T_b \mu x_n = (\lambda - \mu) T_a x_n \neq 0$. This completes the induction.

Now suppose $x_1, x_2, \ldots, x_n$ are linearly independent and $y_1, y_2, \ldots, y_n$ are arbitrary. By what we have just proved we can find $a_j \in \mathcal{A}$ satisfying $T_{a_j} x_j \neq 0$ and $T_{a_j} x_i = 0$ when $i \neq j$ for $j = 1, 2, \ldots, n$. Use the irreducibility of T to find $b_j \in \mathcal{A}$ satisfying $T_{b_j} T_{a_j} x_j = y_j$ for $j = 1, 2, \ldots, n$. Then $a = b_1 a_1 + b_2 a_2 + \cdots + b_n a_n$ satisfies $T_a x_j = y_j$ for $j = 1, 2, \ldots, n$. Thus T is strictly dense.

Finally we turn to the second statement. Let $\mathcal{Y}$ be the subspace spanned by $x_1, x_2, \ldots, x_n$ and $y_1, y_2, \ldots, y_n$. Since $\mathcal{Y}$ is finite-dimensional and the lists are each linearly independent, there is an invertible linear operator which takes each x_j into the corresponding y_j. But an invertible linear operator on a finite-dimensional space can be written in the form e^A for some linear operator A on $\mathcal{Y}$. The first part of this theorem shows that there is some $a \in \mathcal{A}$ with $T_a x_j = A x_j$ for $j = 1, 2, \ldots, n$. Theorem 4.2.8 shows that there is some norm on $\mathcal{X}$ relative to which the representation is normed and continuous. Hence T_{e^a} has the desired effect. $\qquad \square$

The following technical result will be used later.

4.2.14 Corollary *Let T be an irreducible representation of a complex Banach algebra $\mathcal{A}$ on a linear space $\mathcal{X}$. If $\{x_n\}_{n \in \mathbb{N}}$ is a sequence of linearly independent vectors in $\mathcal{X}$, then for any $p \in \mathbb{N}$ there is an element $a \in \mathcal{A}$ such that $T_a x_m$ is zero for $1 \leq m \leq p$ and the set $\{T_a x_q : q > p\}$ is linearly independent.*

Proof By Theorem 4.2.13, we can successively choose $a_{p+1}, a_{p+2}, \ldots$ such that $\|a_n\| < 2^{-n}$,

$$T_{a_n} x_1 = T_{a_n} x_2 = \cdots = T_{a_n} x_{n-1} = 0,$$

and

$$T_{a_n} x_n \notin \text{span}\{T_{b_n} x_1, T_{b_n} x_2, \ldots, T_{b_n} x_n\}$$

all hold where b_n is defined by $b_n = a_{p+1} + a_{p+2} + \cdots + a_{n-1}$. Let a be $\sum_{n=p+1}^{\infty} a_n$. Give $\mathcal{X}$ the complete norm described in Theorem 4.2.7 so that T becomes continuous. Then $T_a x_m = \sum_{n=p+1}^{\infty} T_{a_n} x_m = 0$ holds for $m \leq p$ and $T_a x_q = \sum_{n=p+1}^{\infty} T_{a_n} x_q = \sum_{n=p+1}^{q} T_{a_n} x_q = T_{a_q} x_q + T_{b_q} x_q$ holds for $q > p$. Also $T_a x_k = \sum_{n=p+1}^{\infty} T_{a_n} x_k = \sum_{n=p+1}^{m-1} T_{a_n} x_k = T_{b_m} x_k$ holds for any $p < k < q$. Hence $T_a x_q = T_{a_q} x_q + T_{b_q} x_q$ is independent of $\{T_a x_k : 1 \leq k \leq q-1\} = \{T_{b_q} x_k : 1 \leq k \leq q-1\}$ for all $q > p$.   □

Continuity of Irreducible representations

We now derive the most important consequence from this circle of ideas. The following result is a slight restatement of Barry E. Johnson's main result in [1967a]. This was the final step in his original proof of the uniqueness of the Banach algebra topology of a semisimple Banach algebra (Corollary 2.3.10). We will discuss this and other important consequences and extensions in Section 6.1. Our proof is essentially Johnson's original one.

4.2.15 Theorem *Let $\mathcal{A}$ be a Banach algebra. Any irreducible normed representation of $\mathcal{A}$ on a normed linear space is continuous.*

Proof Let $\mathcal{P}$ be the kernel of the irreducible normed representation T of $\mathcal{A}$ on a normed linear space $\mathcal{X}$. Then $\mathcal{P}$ is primitive and hence closed by Proposition 4.2.6. Thus we may, and do, replace $\mathcal{A}$ by $\mathcal{A}/\mathcal{P}$ and assume that T is faithful. To avoid double subscripts we then consider $\mathcal{A}$ as embedded in $\mathcal{B}(\mathcal{X})$ as an irreducible algebra of operators with elements A, B, C, etc. However we shall continue to use $\|A\|$ to represent the norm of A as an element of $\mathcal{A}$. By Proposition 4.2.2, it is enough to show that T is strongly continuous (*i.e.*, for each $x \in \mathcal{X}, A \to Ax$ is continuous on $\mathcal{A}$).

If $\mathcal{X}$ is finite-dimensional, then $\mathcal{A} \subseteq \mathcal{B}(\mathcal{X})$ is also finite dimensional. Hence the linear map $A \to Ax$ would be continuous on $\mathcal{A}$ for each $x \in \mathcal{X}$. Thus we may assume that $\mathcal{X}$ is infinite dimensional. Furthermore, the commutator $\mathcal{A}'$ of $\mathcal{A}$ in $\mathcal{B}(\mathcal{X})$ is $\mathbb{C}I$ by Theorem 4.2.11. Thus we may choose a sequence $\{x_n\}_{n \in \mathbb{N}}$ of vectors in $\mathcal{X}$ which are linearly independent over $\mathcal{A}'$ and satisfy $\|x_n\| = 1$ for all $n \in \mathbb{N}$.

Suppose $x \in \mathcal{X}$ is an element such that the map $A \mapsto Ax$ is continuous on $\mathcal{A}$. Then for each fixed $B \in \mathcal{A}$, the map $A \mapsto AB \mapsto ABx$ would be continuous for all $A \in \mathcal{A}$. However, the irreducibility of T shows $\{Bx : B \in \mathcal{A}\} = \mathcal{X}$ for any non-zero $x \in \mathcal{X}$. Thus either the map $A \mapsto Ax$ is continuous on $\mathcal{A}$ for all $x \in \mathcal{X}$ or it is discontinuous for all non-zero $x \in \mathcal{X}$. By deriving a contradiction from the second possibility we will establish the theorem.

Assume $A \mapsto Ax$ is discontinuous for each non-zero $x \in \mathcal{X}$. We first show that for each $K > 0, \varepsilon > 0$, and $m \in \mathbb{N}$ there is an $A \in \mathcal{A}$ satisfying:

(a) $\|A\| < \varepsilon$;

(b) $Ax_1 = Ax_2 = \cdots = Ax_{m-1} = 0$;

(c) $\|Ax_m\| > K$.

Let K, ε and m be fixed for the rest of this paragraph. Let $\mathcal{M}_n$ be the set $\{A \in \mathcal{A} : Ax_n = 0\}$. Then each $\mathcal{M}_n$ is a maximal modular left ideal of $\mathcal{A}$. Define $\mathcal{R}$ and $\mathcal{L}$ by $\mathcal{R} = \mathcal{M}_1 \cap \mathcal{M}_2 \cap \cdots \cap \mathcal{M}_{m-1}$ and $\mathcal{L} = \mathcal{R} + \mathcal{M}_m$. Theorem 4.2.13 shows that there is an element $B \in \mathcal{A}$ satisfying $Bx_1 = Bx_2 = \cdots = Bx_{m-1} = 0, Bx_m = x_m \neq 0$. This element B belongs to $\mathcal{R} \subseteq \mathcal{L}$ but not to $\mathcal{M}_m$ so that the maximality of $\mathcal{M}_m$ implies $\mathcal{L} = \mathcal{A}$. Thus addition defines a continuous linear map of $\mathcal{R} \oplus \mathcal{M}_m$ onto $\mathcal{A}$. By the open mapping theorem there is a constant $\delta > 0$ such that for any $B \in \mathcal{A}$ satisfying $\|B\| < \delta\varepsilon$ there will be elements $A \in \mathcal{R}$ and $C \in \mathcal{M}_m$ satisfying $B = A + C$ and $\|A\| < \varepsilon, \|C\| < \varepsilon$. Since $A \to Ax_m$ is discontinuous, we can choose an element $B \in \mathcal{A}$ satisfying $\|B\| < \delta\varepsilon$ and $\|Bx_m\| > K$. Now choose A and C as just indicated. Then A satisfies (a), (b), and (c) (since $Cx_m = 0$) as required.

By induction we can choose a sequence $\{A_n\}_{n\in\mathbb{N}}$ of elements of $\mathcal{A}$ satisfying:

(a') $\|A_n\| < 2^{-n}$;

(b') $A_n x_1 = A_n x_2 = \cdots = A_n x_{n-1} = 0$;

(c') $\|A_n x_n\| \geq n + \|A_1 x_n + \cdots + A_{n-1} x_n\|$.

Define $B_k \in \mathcal{A}$ by $B_k = \sum_{n>k} A_n$. Since each A_n belongs to the closed ideal $\mathcal{M}_k$ for all $n > k$, B_k belongs to $\mathcal{M}_k$. Thus for each $k \in \mathbb{N}$, we get

$$
\begin{aligned}
\|B_0 x_k\| &= \|A_1 x_k + A_2 x_k + \cdots + A_k x_k + B_k x_k\| \\
&\geq \|A_k x_k\| - \|A_1 x_k + \cdots + A_{k-1} x_k\| \\
&\geq k = k\|x_k\|.
\end{aligned}
$$

However, B_0 is a bounded operator since T is a normed representation. This contradiction establishes the theorem.   □

4.2.16 Corollary *Let $\mathcal{A}$ be a Banach algebra.*

(a) *If $T: \mathcal{A} \to \mathcal{L}(\mathcal{X})$ is an irreducible representation on a linear space, then $\mathcal{X}$ has a unique Banach space topology relative to which T is normed.*

(b) *An algebraic equivalence between two normed irreducible representations of $\mathcal{A}$ on Banach spaces is a topological equivalence.*

Proof (a): Theorem 4.2.8 gives such a Banach space norm on $\mathcal{X}$. If $\mathcal{X}$ has any Banach space norm relative to which T is normed, Theorem 4.2.15 shows that T is continuous. Hence Theorem 4.2.8 shows that the given complete norm is equivalent to the one assigned by construction (using any non-zero $z \in \mathcal{X}$).

(b): Theorem 4.2.15 shows that the representations are continuous. Hence Corollary 4.2.4 applies.   □

Characterizations of Commutative Banach Algebras

We will now show the power of Sinclair's refinement in Theorem 4.2.13 by giving Jaroslav Zemánek's remarkably simple proof [1982] of spectral criteria for commutativity (or more precisely, almost commutativity) in Banach algebras. A number of results equivalent to those in Theorem 4.2.18 were obtained in Theorems 2.4.11 and 3.1.5 for spectral algebras. Recall that if a functional algebra $\mathcal{A}$ is nonunital, the exponential function is defined in $\mathcal{A}^1$.

4.2.17 Theorem *Let $\mathcal{A}$ be a functional algebra. For any $a \in \mathcal{A}$, consider*

$$\zeta(a) = \sup\{\rho(a - e^{-b}ae^b) : b \in \mathcal{A}\}.$$

The following three conditions are equivalent for any $c \in \mathcal{A}$
  (a) $\zeta(c)$ *is finite.*
  (b) $\zeta(c) = 0.$
  (c) $c + \mathcal{A}_J$ *is in the center of $\mathcal{A}/\mathcal{A}_J$.*
If in addition to these conditions the spectral radius of c is zero, then c is in the Jacobson radical of $\mathcal{A}$.

Proof (a)$\Rightarrow$(c): We may assume $\zeta(c) < 1$. Let T be an arbitrary irreducible representation of $\mathcal{A}$. If $\mathcal{A}$ is nonunital, T can be extended to $\mathcal{A}^1$ by $T_{\lambda+a} = \lambda I + T_a$ for all $\lambda + a \in \mathcal{A}^1$. We want to show $T_c = \lambda T_1 = \lambda I$. If not, there is some vector x in the representation space with $T_c x = y$ linearly independent from x. By Theorem 4.2.13 we can therefore find $b \in \mathcal{A}$ satisfying $T_{e^b} x = x$ and $T_{e^b} y = x + y$. We conclude

$$T_{c - e^{-b}ce^b}x = T_{e^{-b}}(T_{e^b c - ce^b}(x)) = T_{e^{-b}}(x) = x \neq 0.$$

However, this implies that 1 belongs to the spectrum of $c - e^{-b}ce^b$, contradicting our assumption.

  (c)$\Rightarrow$(b): If (c) holds, then $c - e^{-b}ce^b$ belongs to the Jacobson radical so its spectral radius is zero.

  (b)$\Rightarrow$(a): Obvious.

The final remark follows since the Jacobson radical of the set $\{a \in \mathcal{A} : a + \mathcal{A}_J$ is in the center of $\mathcal{A}/\mathcal{A}_J\}$ is the subset of elements with spectral radius zero. $\square$

4.2.18 Theorem *The following conditions are equivalent for a functional algebra $\mathcal{A}$.*
  (a) *The spectral radius is uniformly continuous on $\mathcal{A}$.*
  (b) *$\mathcal{A}$ is almost commutative.*

Proof (a)$\Rightarrow$(b): Choose an $\varepsilon > 0$ so that for any two elements $a, c \in \mathcal{A}$, $\|a - c\| < \varepsilon$ implies $|\rho(a) - \rho(c)| < 1$. Then any $a, b \in \mathcal{A}$ satisfy

$$|\rho(a - e^{-b}ae^b) - \rho(a)| = |\rho(a - e^{-b}ae^b) - \rho(e^{-b}ae^b)| \leq \|a\|\varepsilon^{-1}$$

which implies $\zeta(a) \leq \rho(a) + \varepsilon^{-1}\|a\| < \infty$. Hence (b) holds by the last theorem.

(b)$\Rightarrow$(a): By Theorem 3.1.5, the spectral radius is an algebra semi-norm which is certainly less than or equal to the norm of the functional algebra $\mathcal{A}$, since that norm is spectral. Hence we conclude $|\rho(a) - \rho(b)| \leq \rho(a-b) \leq \|a - b\|$ for all $a, b \in \mathcal{A}$. $\qquad\square$

Zemánek's [1982] paper includes many other simple consequences of Theorem 4.2.17. Let us denote the set of topologically nilpotent elements in a Banach algebra $\mathcal{A}$ by $\mathcal{A}_{tN}$. We know this set includes the Jacobson radical of $\mathcal{A}$. If $\mathcal{A}_{tN}$ is closed under either sums or products then it is closed under both and thus equals the Jacobson radical which is also closed in the norm. Similarly an element $b \in \mathcal{A}$ belongs to the Jacobson radical if and only if it satisfies $b + \mathcal{A}_{tN} \subseteq \mathcal{A}_{tN}$.

We give another important consequence of Theorem 4.2.11, which can also be obtained from Theorem 4.1.9 and Proposition 2.4.12.

4.2.19 Theorem *Every irreducible representation of a commutative spectral algebra is one-dimensional.*

Proof (First Proof) Let T be an irreducible representation of the commutative spectral algebra $\mathcal{C}$ on a linear space $\mathcal{X}$. Then $T_{\mathcal{C}}$ is commutative and hence satisfies $T_{\mathcal{C}} \subseteq T_{\mathcal{C}}'$. Since Theorem 4.2.11 asserts that the latter space is $\mathbb{C}I$, and since $T_{\mathcal{C}}$ is irreducible, we have $T_{\mathcal{C}} = \mathbb{C}I$ and $\mathcal{X}$ is one-dimensional. $\qquad\square$

Proof (Second Proof) Use the same notation. Theorem 4.1.9 shows that $\ker(T)$ is a maximal modular ideal in $\mathcal{C}$. Proposition 2.4.12 shows that $\mathcal{C}/\ker(T)$ is isomorphic to $\mathbb{C}$. Hence again T is one-dimensional. $\qquad\square$

It seems to be unknown when the above result holds for topologically irreducible representations in place of algebraically irreducible representations. In Volume II we describe a special case in which this is true.

Representation on Reflexive Banach Spaces

The extended left regular representation shows that any Banach algebra has a faithful (even isometric) representation on a Banach space, and Corollary 4.2.9 shows that any primitive Banach algebra has a faithful continuous irreducible representation on a Banach space. Representations on reflexive Banach spaces are rarer, but we can elucidate when they occur. The results in this subsection are due to Nicholas J. Young [1976] and Sten Kaijser [1981]. We begin with a construction which will yield all normed irreducible representations of a Banach algebra on a reflexive Banach space. It generalizes the Gelfand–Naĭmark–Segal construction which will be studied in Section 9.4 of Volume II.

4.2.20 Definition Let $\mathcal{A}$ be a Banach algebra and let ω be a linear functional on $\mathcal{A}$. Let $\langle\, \mathcal{A}_\omega\, /\, {}_\omega\mathcal{A}\, \rangle$ be the $\langle$ left / right $\rangle$ ideal

$$\langle\, \mathcal{A}_\omega = \{b \in \mathcal{A} : \omega(ab) = 0 \quad a \in \mathcal{A}\}\, /\, {}_\omega\mathcal{A} = \{b \in \mathcal{A} : \omega(ba) = 0 \quad a \in \mathcal{A}\}\, \rangle.$$

Define a norm on $\langle\, \mathcal{A}/\mathcal{A}_\omega\, /\, \mathcal{A}/{}_\omega\mathcal{A}\, \rangle$ by

$$\langle\, \|b + \mathcal{A}_\omega\|_\omega \;=\; \sup\{|\omega(ab)| : a \in \mathcal{A}_1\}\, /$$
$$\|b + {}_\omega\mathcal{A}\|^\omega \;=\; \sup\{|\omega(ba)| : a \in \mathcal{A}_1\}\, \rangle.$$

Denote the contractive $\langle$ representation / anti-representation $\rangle$ of $\mathcal{A}$ on $\langle\, \mathcal{A}/\mathcal{A}_\omega\, /\, \mathcal{A}/{}_\omega\mathcal{A})\, \rangle$ induced by the $\langle$ left / right $\rangle$ representation of $\mathcal{A}$ on $\mathcal{A}$ by $\langle\, \tilde{L}^\omega\, /\, \tilde{R}^\omega\, \rangle$. Denote the completion of $\langle\, (\mathcal{A}/\mathcal{A}_\omega, \|\cdot\|_\omega)\, /\, (\mathcal{A}/{}_\omega\mathcal{A}, \|\cdot\|^\omega)\, \rangle$ by $\langle\, \mathcal{X}^\omega\, /\, \mathcal{Y}^\omega\, \rangle$ and the extension by continuity of each $\langle\, \tilde{L}^\omega_a\, /\, \tilde{R}^\omega_a\, \rangle$ to $\langle\, \mathcal{X}^\omega\, /\, \mathcal{Y}^\omega\, \rangle$ by $\langle\, L^\omega_a\, /\, R^\omega_a\, \rangle$. Finally, in order to obtain a representation from R^ω, we let $R^{\omega*}\colon \mathcal{A} \to \mathcal{B}((\mathcal{Y}^\omega)^*)$ be defined by $R^{\omega*}_a = (R^\omega_a)^*$.

Call ω $\langle$ left / right $\rangle$ *autoperiodic* if $\langle\, \mathcal{X}^\omega\, /\, \mathcal{Y}^\omega\, \rangle$ is reflexive. A linear functional which is either left or right autoperiodic is called *autoperiodic*.

Young shows that autoperiodic linear functionals are weakly almost periodic (defined before Theorem 1.4.11). He characterizes when $\omega \in \mathcal{A}^*_{WAP}$ is $\langle$ left / right $\rangle$ autoperiodic in terms of a geometric condition on the weak closure of $\langle\, \{{}_a\omega : a \in \mathcal{A}_1\}\, /\, \{\omega_a : a \in \mathcal{A}_1\}\, \rangle$ and in terms of the equality of the mixed limits of certain double sequences. He also gives a number of examples and determines the autoperiodic linear functionals in a few cases. For instance, if $\mathcal{A} = C(\Omega)$ where Ω is a locally compact Hausdorff space, the left (= right) autoperiodic functionals on $\mathcal{A}$ are those of the form: $f \mapsto \sum_{j=1}^{n} \alpha_j f(\omega_j)$ for $n \in \mathbb{N}, \alpha_j \in \mathbb{C}$ and $\omega_j \in \Omega$. In other words, the set of left autoperiodic functionals in this case is just the linear span of the Gelfand space $\Gamma_{\mathcal{A}}$. However, the most important examples of autoperiodic functionals arise as follows.

Using the notation of Definition 4.2.20, suppose that $\mathcal{A}$ is unital. Let $z \in \mathcal{X}^\omega$ be the element $1 + \mathcal{A}_\omega$. Then z is clearly a topologically cyclic vector for L^ω. Moreover, $a + \mathcal{A}_\omega \mapsto \omega(a)$ defines a continuous linear functional on $\mathcal{A}/\mathcal{A}_\omega$, so there is an element $z^* \in \mathcal{X}^{\omega*}$ satisfying

$$\omega(a) = z^*(L^\omega_a(z)) \qquad \forall\, a \in \mathcal{A}.$$

(These ideas were already included in the important paper Bonsall and Duncan [1967].) We now show that in the present context there is a much stronger converse.

4.2.21 Theorem *Let $\mathcal{A}$ be a Banach algebra and let $T\colon \mathcal{A} \to \mathcal{B}(\mathcal{X})$ be an irreducible normed representation of $\mathcal{A}$ on a reflexive Banach space $\mathcal{X}$. Choose non-zero vectors $z \in \mathcal{X}$ and $z^* \in \mathcal{X}^*$ and define $\omega \in \mathcal{A}^*$ by $\omega(a) =$*

$z^*(T_a(z))$ *for all* $a \in \mathcal{A}$. *Then* ω *is right autoperiodic and satisfies* $Y^\omega = \mathcal{A}/_\omega\mathcal{A}$ *(i.e., no completion is necessary). Furthermore, the map* $V \colon \mathcal{A}/_\omega\mathcal{A} \to \mathcal{X}^*$ *defined by*

$$V(a +\, _\omega\mathcal{A}) = T_a^*(z^*) \qquad \forall\, a \in \mathcal{A}$$

is a well-defined homeomorphic linear isomorphism and its adjoint is a topological equivalence between T *and* $R^{\omega*}$.

Hence every irreducible normed representation of a Banach algebra $\mathcal{A}$ on a reflexive Banach space is topologically equivalent to $R^{\omega*}$ for some right autoperiodic functional $\omega \in \mathcal{A}^*$.

Proof Theorem 4.2.15 shows that T is continuous. Let $\mathcal{L}$ be the left ideal $\{b \in \mathcal{A} : T_b(z) = 0\}$. In Theorem 4.2.3, we have already noted that $\mathcal{A}/\mathcal{L}$ with its quotient norm is homeomorphically linearly isomorphic to $\mathcal{X}$. Hence for $x^* \in \mathcal{X}^*$ the expression $x^* \mapsto \sup\{|x^*(T_b(z))| : b \in \mathcal{A}_1\}$ is equivalent to the original norm. So

$$\|a +\, _\omega\mathcal{A}\|^\omega = \sup\{|\omega(ab)| : b \in \mathcal{A}_1\} = \sup\{|V(a +\, _\omega\mathcal{A})(T_b(z))| : b \in \mathcal{A}_1\}$$

shows that V is a well-defined homeomorphic linear isomophism. Thus its range is norm closed. Therefore, its range (which is clearly a linear subspace) is both weakly closed and weak = weak* dense. This implies that $\mathcal{A}/_\omega\mathcal{A} = \mathcal{Y}^\omega$ is a reflexive Banach space, so ω is right autoperiodic.

It only remains to show that V^* is an equivalence between T and $R^{\omega*}$. The following calculation, which holds for all $a, b, c \in \mathcal{A}$, proves this:

$$\begin{aligned}
V^*(T_a(T_b(z)))(c +\, _\omega\mathcal{A}) &= V(c +\, _\omega\mathcal{A})(T_{ab}(z)) = z^*(T_{cab}(z)) \\
&= V(ca +\, _\omega\mathcal{A})(T_b(z)) \\
&= V^*(T_b(z))(R_a^\omega(c +\, _\omega\mathcal{A})) \\
&= R_a^{\omega*}(V^*(T_b(z)))(c +\, _\omega\mathcal{A}). \qquad \square
\end{aligned}$$

Essentially the following result was obtained by Kaijser [1981] as well as Young [1976]. Although their proofs differ, both involve the theorem on weakly compact operators of Davis, Figiel, Johnson and Pelczynski (Diestel and Uhl [1977], VIII.4.8). Because of the technicalities involved, we give only a sketch of the more difficult direction in Young's proof.

4.2.22 Theorem *A normed algebra* $\mathcal{A}$ *has a* ⟨ *faithful continuous / isometric* ⟩ *representation on some reflexive Banach space if and only if the weakly almost periodic functionals of norm 1 on* $\mathcal{A}$ ⟨ *separate the points of* $\mathcal{A}$ / *determine the norm of* $\mathcal{A}$ ⟩.

Proof Suppose $T \colon \mathcal{A} \to \mathcal{B}(\mathcal{X})$ is a ⟨ faithful continuous / isometric ⟩ representation of $\mathcal{A}$ on a reflexive Banach space $\mathcal{X}$. Choose norm 1 vectors $z \in \mathcal{X}$ and $z^* \in \mathcal{X}^*$. As before, define $\omega \in \mathcal{A}^*$ by $\omega(a) = z^*(T_a(z))$ for all $a \in \mathcal{A}$. Then ω is weakly almost periodic by direct application of the definition since bounded subsets of reflexive spaces are weakly compact. As $z \in \mathcal{X}_1$ and $z^* \in \mathcal{X}_1^*$ vary, this collection of weakly almost periodic functionals of $\langle$ bounded norm separates the points / norm at most one determines the norm $\rangle$ of $\mathcal{A}$.

(Sketch.) In order to prove the converse, Young shows how weakly almost periodic functionals can be approximated by autoperiodic ones using the construction of Davis, Figiel, Johnson and Pelczynski already cited. We have shown how to obtain a representation on a reflexive space from an autoperiodic linear functional. The direct sum of all these representations on the ℓ^2-direct sum of the reflexive spaces then provides the $\langle$ faithful continuous / isometric $\rangle$ representation needed to complete the proof. $\quad \square$

We now turn to Kaijser's approach in terms of factoring a weakly compact map through a reflexive space.

4.2.23 Theorem *Let $\mathcal{A}$ be a unital Banach algebra and let $\omega \in \mathcal{A}^*$ be a weakly almost periodic linear functional. Then there is a representation T^ω of $\mathcal{A}$ on a reflexive Banach space $\mathcal{Z}^\omega$, and topologically cyclic vectors $z^\omega \in \mathcal{Z}^\omega$ and $z^{\omega*} \in \mathcal{Z}^{\omega*}$ satisfying*

$$\|z^\omega\| \, \|z^{\omega*}\| \leq \|\omega\| \quad \text{and} \quad \omega(a) = z^{\omega*}(T_a^\omega z^\omega) \qquad \forall \, a \in \mathcal{A}. \qquad (3)$$

Proof Theorem 1.4.11 shows that the map $P_\omega \colon \mathcal{A} \to \mathcal{A}^*$, defined by $P_\omega(a) = {}_a\omega$ for $a \in \mathcal{A}$, is weakly compact. Hence the last theorem in 1.7.8 shows that there is a reflexive Banach space $\mathcal{Z}^\omega$, a map $S \in \mathcal{B}(\mathcal{A}, \mathcal{Z}^\omega)$ with range dense in $\mathcal{Z}^\omega$ and an injective map $Q \in \mathcal{B}(\mathcal{Z}^\omega, \mathcal{A}^*)$ satisfying $P_\omega = QS$, $\|S\| = \|P_\omega\| = \|\omega\|$ and $\|Q\| \leq 1$. If, as usual, L and R represent the left and right regular representations of $\mathcal{A}$ on itself, then P^ω intertwines L and the representation $a \to (R_a)^*$ of $\mathcal{A}$ on $\mathcal{A}^*$ in the sense that $P_\omega(L_a(b)) = (R_a)^*(P_\omega(b))$ holds. Hence by the result already cited, there is a representation T^ω of $\mathcal{A}$ on $\mathcal{Z}^\omega$ satisfying $S(ab) = S(L_a(b)) = T_a^\omega(S(b))$, $Q(T_a^\omega(x)) = (R_a)^*(Q(x))$ and hence ${}_{ab}\omega = P_\omega(ab) = Q(T_a^\omega(S(b)))$ for any $a, b \in \mathcal{A}$. Thus we have

$$\omega(cab) = {}_{ab}\omega(c) = Q^*(\kappa(c))(T_a^\omega(S(b))) \qquad \forall \, a, b, c \in \mathcal{A} \qquad (4)$$

Taking $z^\omega = S(1)$ and $z^{\omega*} = Q^*(\kappa(1))$, gives the desired result. The injective map Q is the dual of $Q^* \circ \kappa$ which therefore has dense range in $\mathcal{Z}^{\omega*}$. This shows that both these vectors are topologically cyclic. $\quad \square$

4.2.24 Corollary *Let $\mathcal{A}$ be an Arens regular unital Banach algebra. Then there is a homeomorphic faithful representation of $\mathcal{A}$ on a reflexive Banach space $\mathcal{Z}$. Moreover the Arens representation of $\mathcal{A}^{**}$ on $\mathcal{Z}$ is also faithful and homeomorphic.*

Proof For each $\omega \in \mathcal{A}^*$ of norm 1, construct the representation T^ω on $\mathcal{Z}^\omega$ as in the theorem. Consider the ℓ^2-direct sum of these representations on the ℓ^2-direct sum $\mathcal{Z}$ of the family $\mathcal{Z}^\omega$. For each $a \in \mathcal{A}$, there is some ω of norm 1 satisfying $\|a\| = \omega(a) = z^{\omega*}(T_a^\omega(z^\omega)) \leq \|T_a^\omega\| \leq \|T_a\|$. This shows that the continuous representation T is homeomorphic and faithful. For the Arens representation T^1 of Definition 1.4.14 and $f \in \mathcal{A}^{**}$, we find $\|T_f^1\| \geq \sup\{(T^\omega)_f^1 : \omega \in \mathcal{A}^*; \|\omega\| = 1\} \geq \sup\{|z^{\omega*}(T^\omega)_f^1(z^\omega)| : \omega \in \mathcal{A}^*; \|\omega\| = 1\} \geq \sup\{|f(\omega)| : \omega \in \mathcal{A}^*; \|\omega\| = 1\} = \|f\|$ where z^ω and $z^{\omega*}$ are defined as in the theorem relative to each particular ω. Thus T^1 is also homeomorphic and faithful.   □.

Injective Banach Algebras

  This notion was introduced by Grothendieck and has been put in a more modern setting by Kaijser [1976], which is our main reference. For each ω in the dual Banach space of a Banach algebra $\mathcal{A}$, the map $P_\omega\colon \mathcal{A} \to \mathcal{A}^*$ defined by $P_\omega(a) = {}_a\omega$ was introduced before Theorem 1.4.11, and uniform algebras were defined following Proposition 3.1.8. Integral operators were discussed in §1.7.12 and the injective tensor product in Section 1.10. See also Anna Maria Mantero and Andrew Tonge [1979].

4.2.25 Definition A Banach algebra $\mathcal{A}$ is said to be ⟨ *injective / geometrically injective* ⟩ if it is unital and its multiplication defines a ⟨ continuous / contractive ⟩ map from $\mathcal{A} \otimes \mathcal{A}$ with its injective tensor norm into $\mathcal{A}$.

  Of course, multiplication always defines a continuous map with respect to the projective tensor norm on $\mathcal{A} \otimes \mathcal{A}$. It is easy to see that $\mathcal{A}$ is ⟨ injective / geometrically injective ⟩ if and only if for each $\omega \in \mathcal{A}^*$ the map P_ω is an integral operator and the map $\omega \mapsto P_\omega$ from $\mathcal{A}^*$ to $\mathcal{B}_I(\mathcal{A}, \mathcal{A}^*)$ is ⟨ continuous / contractive ⟩. Kaijser gives a new proof of Grothendieck's criterion that $\mathcal{A}$ is injective if and only if for every Banach algebra $\mathcal{B}$ the injective tensor product $\mathcal{A} \check{\otimes} \mathcal{B}$ is a Banach algebra under the product induced by the products in $\mathcal{A}$ and $\mathcal{B}$.

4.2.26 Theorem *Every injective Banach algebra has a faithful representation as a norm-closed subalgebra of $\mathcal{B}(\mathcal{H})$ for some Hilbert space $\mathcal{H}$ and hence is Arens regular.*

  Every geometrically injective Banach algebra is a uniform algebra and hence is commutative and semisimple.

Proof (Sketch of the first portion) Injective maps such as P_ω can be factored through a Hilbert space, just as weakly compact ones can be factored through a reflexive space. The analogue of the last theorem in §1.7.8 still holds. Hence the proofs of Theorem 4.2.23 and Corollary 4.2.24 do not need any essential change.   □

4.3   The Jacobson Radical

This section studies the most important of all radicals, the Jacobson radical. In the introduction to this chapter we explained that representation theories ordinarily cannot provide information on perfectly general mathematical objects but that the pathological portion can often be neatly split off. The Jacobson radical is the intractable portion of an algebra relative to irreducible representations. This radical was briefly introduced in Chapter 2 (Definition 2.3.2) since it plays a fundamental role with respect to spectral theory as well as representation theory. It was also pointed out in Chapter 3 that the Gelfand radical introduced there equals the Jacobson radical, at least for spectral algebras which are commutative modulo either of these radicals.

The next definition is due to Jacobson [1945a], although he used property (b) of our Theorem 4.3.6 as his definition.

4.3.1   Definition   The *radical*, or *Jacobson radical*, of an algebra $\mathcal{A}$ is the intersection of the kernels of the irreducible representations of $\mathcal{A}$. It is denoted by $\mathcal{A}_J$. An algebra is said to be *semisimple* if its radical is zero and is said to be *Jacobson-radical* if it equals its Jacobson radical.

In this definition we are following the common convention that the intersection of an empty family of subsets of a set is the set itself. Thus an algebra is Jacobson-radical if and only if it has no irreducible representations at all. We could rephrase the definition by saying that $\mathcal{A}_J$ is the intersection of all the primitive ideals of $\mathcal{A}$. Of course the Jacobson radical is an ideal. We shall see that semisimple algebras are comparatively well behaved. (The fundamental theorem of spectral semi-norms (Theorem 2.3.6) was an early example of this.) The next theorem shows how to construct a (largest) semisimple quotient algebra from any algebra.

4.3.2   Theorem   *Any algebra $\mathcal{A}$ and any ideal $\mathcal{I}$ of $\mathcal{A}$ satisfy:*
(a) $\mathcal{I}_J = \mathcal{I} \cap \mathcal{A}_J$.
(b) $\mathcal{I} \subseteq \mathcal{A}_J$ *implies* $(\mathcal{A}/\mathcal{I})_J = \mathcal{A}_J/\mathcal{I}$.
(c) $\mathcal{A}/\mathcal{I}$ *semisimple implies* $\mathcal{A}_J = \mathcal{I}_J$.
Thus in particular $\mathcal{A}_J$ is radical and $\mathcal{A}/\mathcal{A}_J$ is semisimple.

Proof (a): Theorem 4.1.6(a) asserts that the irreducible representations of an ideal of $\mathcal{A}$ are precisely the restrictions to the ideal (of the homomorphisms) of the irreducible representations of $\mathcal{A}$. This implies $\mathcal{I}_J = \mathcal{I} \cap \mathcal{A}_J$ and hence $(\mathcal{A}_J)_J = \mathcal{A}_J$. (b): Theorem 4.1.6(b) asserts that if $\mathcal{I}$ is an ideal of $\mathcal{A}$ which is included in $\mathcal{A}_J$, then the irreducible representations of $\mathcal{A}$ are precisely those which can be written as $T \circ \varphi$ for some irreducible representation, T, of $\mathcal{A}/\mathcal{I}$ where $\varphi \colon \mathcal{A} \to \mathcal{A}/\mathcal{I}$ is the natural map. This shows $(\mathcal{A}/\mathcal{I})_J = \mathcal{A}_J/\mathcal{I}$ and hence $(\mathcal{A}/\mathcal{A}_J)_J = \{0\}$. (c): If $\mathcal{A}/\mathcal{I}$ is semisimple, Theorem 4.1.8(e) shows $\mathcal{I} = \cap\{\mathcal{P} : \mathcal{P} \in h^{\Pi}(\mathcal{I})\}$ which implies $\mathcal{I}_J = \mathcal{I} \cap \mathcal{A}_J = \cap\{\mathcal{P} : \mathcal{P} \in \Pi_A\} = \mathcal{A}_J$. □

An immediate consequence of this theorem is the following simple but useful result.

4.3.3 Corollary *Any algebra $\mathcal{A}$ satisfies:* $\mathcal{A}_J = (\mathcal{A}^1)_J$.

Proof The theorem gives $\mathcal{A}_J = (\mathcal{A}^1)_J \cap \mathcal{A}$. If the algebra is unital there is nothing to prove, so we assume otherwise. Since $\lambda + a \mapsto \lambda$ defines an irreducible representation of $\mathcal{A}^1$ on $\mathbb{C}$ with kernel $\mathcal{A}$, we conclude $(\mathcal{A}^1)_J \subseteq \mathcal{A}$ which implies $\mathcal{A}_J = (\mathcal{A}^1)_J \cap \mathcal{A} = (\mathcal{A}^1)_J$. □

Quasi-regular Ideals

The following definition is due to Reinhold Baer [1943]. The idea stems from Sam Perlis [1942].

4.3.4 Definition A one-sided ideal or ideal is called *quasi-regular* if it contains a quasi-inverse for each of its elements.

Because a one-sided ideal or ideal is a spectral subalgebra, it is quasi-regular if each of its elements has a quasi-inverse in any algebra in which it is embedded. Note also that a one-sided ideal or ideal $\mathcal{L}$ of an algebra $\mathcal{A}$ is quasi-regular if and only if it satisfies

$$Sp_A(a) = Sp_{\mathcal{L}}(a) = \{0\} \qquad \forall\, a \in \mathcal{L}.$$

The equations obviously imply that $\mathcal{L}$ is quasi-regular. To see the opposite implication note that

$$\lambda^{-1}a \in \mathcal{L} \subseteq \mathcal{A}_{qG} \qquad \forall\, a \in \mathcal{L}; \ \lambda \in \mathbb{C} \setminus \{0\}$$

implies $Sp_A(a) \subseteq Sp_{\mathcal{L}}(a) \subseteq \{0\}$. If $\mathcal{A}$ is not unital these inclusions are obviously equalities, but if $\mathcal{A}$ is unital then 1 does not belong to $\mathcal{L} \subseteq \mathcal{A}_{qG}$ so that $Sp_A(a)$ equals $\{0\}$ in this case also.

The first result in the following lemma is due to Baer [1943]. The second is due to Jacobson [1945a].

4.3.5 Lemma *A left ideal is quasi-regular if every element is left quasi-invertible. An element which belongs to a quasi-regular left ideal belongs to every primitive ideal.*

Proof Suppose $\mathcal{L}$ is a left ideal and every element of $\mathcal{L}$ is left quasi-invertible (*i.e.*, has a left quasi-inverse). Let b be the left quasi-inverse for $a \in \mathcal{A}$. Then $b = ba - a$ belongs to $\mathcal{L}$. Thus b has a left quasi-inverse c as well as its right quasi-inverse a. This implies $c = c \circ (b \circ a) = (c \circ b) \circ a = a$. Thus a is quasi-invertible with quasi-inverse b. Hence $\mathcal{L}$ is quasi-regular.

To prove the second statement suppose that an element a does not belong to some primitive ideal. Then there is an irreducible representation T satisfying $T_a \neq 0$. Choose $z \in \mathcal{X}^T$ satisfying $T_a z \neq 0$ and then choose $b \in \mathcal{A}$ satisfying $T_b T_a z = T_{ba} z = z$. If c is a left quasi-inverse for ba then $c \circ (ba) = 0$ holds and so $T_0 z = T_{c \circ (ba)} z = T_c z + T_{ba} z - T_c T_{ba} z = z$ is a contradiction. Thus if an element a does not belong to some primitive ideal, then ba is not quasi-invertible for some $b \in \mathcal{A}$ and hence a does not belong to any quasi-regular left ideal.   □

Diverse Characterizations of the Jacobson Radical

In the next theorem (a) and part of (d) are due to Hille [1948], while (b) and (c) are due to Jacobson [1945a]. In [1947a] Segal defined an algebra to be weakly semisimple if the maximal modular left ideals have intersection $\{0\}$. He was apparently unaware of Jacobson's paper [1945a]. Part of this theorem was already obtained in Theorem 2.3.3.

4.3.6 Theorem *Let $\mathcal{A}$ be an algebra.*

(a) *The Jacobson radical is the intersection of all maximal modular left (or right) ideals of $\mathcal{A}$. Hence it is the same in the reverse algebra.*

(b) *The Jacobson radical is a quasi-regular ideal which includes all quasi-regular one- or two-sided ideals of $\mathcal{A}$.*

(c) *The Jacobson radical is the set*

$$\{a \in \mathcal{A} : \mathcal{A}^1 a \subseteq \mathcal{A}_{qG}\}.$$

(d) *The Jacobson radical is the largest ideal $\mathcal{I}$ satisfying any (hence all) of the following equivalent conditions*

$$Sp_{\mathcal{A}/\mathcal{I}}(a + \mathcal{I}) \subseteq Sp_{\mathcal{A}}(a) \subseteq Sp_{\mathcal{A}/\mathcal{I}}(a + \mathcal{I}) \cup \{0\} \qquad \forall\, a \in \mathcal{A};$$

$$\mathcal{A}_{qG} = \{a \in \mathcal{A} : a + \mathcal{I} \in (\mathcal{A}/\mathcal{I})_{qG}\};$$

$$Sp(a + b) = Sp(a) \qquad \forall\, a \in \mathcal{A}; b \in \mathcal{I};$$

$$Sp(b) = \{0\} \qquad \forall\, b \in \mathcal{I};$$

$$\rho(a + b) = \rho(a) \qquad \forall\, a \in \mathcal{A}; b \in \mathcal{I}.$$

$$\rho(b) = 0 \qquad \forall\, b \in \mathcal{I}.$$

(e) *Any spectral subalgebra $\mathcal{B}$ of $\mathcal{A}$ satisfies $\mathcal{B} \cap \mathcal{A}_J \subseteq \mathcal{B}_J$.*

Proof We first prove (c), (b) and the first sentence of (a); the latter two only for left ideals. Theorem 4.1.8 shows that the intersection $\mathcal{A}_J$ of all the primitive ideals of $\mathcal{A}$ equals the intersection of all maximal modular left ideals of $\mathcal{A}$ (even if one, and hence both, of these families is empty). Lemma 4.3.5 shows that every quasi-regular left ideal is included in the intersection $\mathcal{A}_J$ of all the primitive ideals. If $e \in \mathcal{A}$ is not left quasi-invertible, then there is a maximal modular left ideal $\mathcal{L}$ with $e \notin \mathcal{L}$ by Theorem 2.4.6. Thus e does not belong to $\mathcal{A}_J$ by what we have already shown. Now the first sentence of Lemma 4.3.5 shows that $\mathcal{A}_J$ is quasi-regular. Clearly any element belongs to a quasi-regular left ideal if and only if it belongs to the set defined in condition (c). Thus we have established (a), (b) and (c) except that we have not yet shown that $\mathcal{A}_J$ is the intersection of all maximal modular right ideals and contains all right quasi-regular ideals. However, $(\lambda + b)a \in \mathcal{A}_{qG}$ is equivalent to $a(\lambda + b) \in \mathcal{A}_{qG}$ by Proposition 2.1.8. We conclude $\mathcal{A}_J = \{a \in \mathcal{A} : a\mathcal{A}^1 \subseteq \mathcal{A}_{qG}\} = \{a \in \mathcal{A} : a$ belongs to some quasi-regular right ideal$\}$. This completes the proof of (b) and shows that the radical of the reverse of an algebra is the same as the radical of the algebra itself. This fact completes the proof of (a) also.

(d): This was proved as Theorem 2.3.3(c).

(e): This follows from (b) since $\mathcal{B} \cap \mathcal{A}$ is a quasi-regular ideal of $\mathcal{B}$. $\square$

As remarked after the definition of quasi-regular ideals, an ideal or one-sided ideal is quasi-regular if and only if each element has spectrum equal to $\{0\}$. Hence if the set of elements with spectrum $\{0\}$ is an ideal, then that ideal is the Jacobson radical. (Usually, however, there are many elements with spectrum $\{0\}$ not contained in the Jacobson radical. The various nilpotent elements in a full matrix algebra are examples.) As noted below, commutative spectral algebras always have the property that the set of elements with spectrum $\{0\}$ is an ideal. An algebra, $\mathcal{A}$, with this property has the further remarkable property that any subalgebra, $\mathcal{B}$, satisfies $\mathcal{B}_J \subseteq \mathcal{A}_J \cap \mathcal{B}$ and any spectral subalgebra, $\mathcal{B}$, satisfies $\mathcal{B}_J = \mathcal{A}_J \cap \mathcal{B}$.

Algebras which have no non-zero elements with spectrum $\{0\}$ form an extreme case. Commutative semisimple spectral algebras are an example. For finite dimensional algebras the Wedderburn structure theorems show that any algebra satisfying this condition must be commutative. However, we will give a family of infinite-dimensional noncommutative Banach algebras satisfying this condition in Example 4.8.5. Any subalgebra of such an algebra is semisimple.

4.3.7 Corollary *Every one-sided identity in a semisimple algebra is a two-sided identity.*

Proof Suppose $\mathcal{A}$ is an algebra and e is a right identity for $\mathcal{A}$. Then the set $\mathcal{I} = \{ey - y : y \in \mathcal{A}\}$ satisfies $\mathcal{A}\mathcal{I} = \{0\}$ and $\mathcal{I}\mathcal{A} \subseteq \mathcal{I}$ so that it is a quasi-regular ideal in which each element satisfies $b^q = -b$. Hence $\mathcal{I}$ is included in $\mathcal{A}_J$. If $\mathcal{A}$ is semisimple, then $\mathcal{I}$ is zero. $\square$

We will now record the beautiful way in which norms, spectral norms and complete norms relate to the Jacobson radical. The next four corollaries are due to Jacobson [1945a] in the complete case. A one- or two-sided ideal $\mathcal{I}$ in a semi-normed algebra is called *topologically nil* if every element $a \in \mathcal{A}$

satisfies $\sigma^\infty(a) = 0$. (This concept is also due to Jacobson [1945a] who called these ideals generalized nil ideals.) In a spectral semi-normed algebra this is equivalent to $Sp(a) = \{0\}$ for all $a \in \mathcal{I}$. Thus in this case an ideal is topologically nil if it is quasi-regular.

4.3.8 Corollary *Let $(\mathcal{A}, \sigma)$ be a semi-normed algebra. Then $\sigma^\infty(a) = 0$ holds for each $a \in \mathcal{A}_J$ so that the Jacobson radical is a topologically nil ideal.*

Proof This follows immediately from Theorems 4.3.6(b), 4.1.6(b) and 2.2.2. □

The second statement in the next corollary was obtained independently by Max Zorn and Einar Hille about the same time (cf. Hille [1948], pp. 475, 476) as Jacobson [1945a].

4.3.9 Corollary *The Jacobson radical of a spectral semi-normed algebra $(\mathcal{A}, \sigma)$ is a topologically nil ideal which contains every topologically nil one-sided ideal or ideal. It also equals the set*

$$\{a \in \mathcal{A} : \lim_{n \to \infty} \sigma((ab)^n) = 0 \text{ for all } b \in \mathcal{A}^1\}.$$

Proof The first statement follows immediately from Theorems 4.3.6(b) and 2.2.5 and the remark following Definition 4.3.4. To prove the second statement we note that $a \in \mathcal{A}_J$ and $b \in \mathcal{A}^1$ imply $ab \in \mathcal{A}_J$ which implies $0 = \sigma^\infty(ab) = \lim_{n\to\infty}(\sigma((ab)^n))^{1/n}$. Hence $\lim_{n\to\infty} \sigma((ab)^n) = 0$ holds for any $a \in \mathcal{A}_J$ and $b \in \mathcal{A}^1$. Conversely, if $\sigma^\infty(a) > 0$ holds for $a \in \mathcal{A}$, then $\lim_{n\to\infty} \sigma(ab)^n = 0$ fails for any $b = \lambda \in \mathbb{C}$ satisfying $|\lambda| > \sigma^\infty(a)^{-1}$. Thus $a \notin \mathcal{A}_J$ implies that $\lim_{n\to\infty} \sigma(ab)^n = 0$ fails for some $b \in \mathcal{A}^1$. □

The following simple results are of tremendous importance, so we repeat them despite their earlier appearances in Chapters 2 and 3.

4.3.10 Corollary *The Jacobson radical of a spectral semi-normed algebra is closed. Hence a spectral semi-norm on a semisimple algebra is a norm.*

Proof Theorems 4.3.6(a) and 2.4.7 prove the first statement. The second is an immediate consequence of the first. □

4.3.11 Corollary *The Jacobson radical of an almost commutative spectral algebra equals the Gelfand radical.*

Proof Recall that an algebra is almost commutative if its quotient modulo its Jacobson radical is commutative. Theorem 3.1.5 shows that the statement of this corollary is equivalent to almost commutativity for a spectral algebra. We record this fact again here for completeness. □

Jacobson Radical and Idempotent Elements

In the next proposition we collect some results on the Jacobson radical and idempotent elements. Result (c) is due to Kaplansky [1949a]. A perusal of the proof of Theorem 2.5.14 shows that (c) remains true for unital spectral semi-normed algebras. The proof of (d) is a slight modification of an argument due to Joseph H. Maclagan Wedderburn [1907]. He made no topological assumption on $\mathcal{A}$ but assumed that $\mathcal{A}_J$ was a nilpotent ideal. (Nilpotent ideals and nil ideals will be defined below.) Whenever $\mathcal{A}_J$ is nil, the power series used in the proof terminate, so that no limits are involved in summing them. Since the existence of these limits is the only issue in the proof, this result holds for functional algebras as well as Banach algebras. Result (e) was obtained by Chester Feldman [1951]. It will be used later (Theorem 7.1.3) to prove one of the main results of his paper.

4.3.12 Proposition *Let $\mathcal{A}$ be an algebra.*
(a) *$\mathcal{A}_J$ contains no non-zero idempotents.*
(b) *Any idempotent e in $\mathcal{A}$ satisfies*

$$(e\mathcal{A}e)_J = e\mathcal{A}e \cap \mathcal{A}_J = e\mathcal{A}_J e.$$

Hence, the Jacobson radical $M_n(\mathcal{A})_J$ of an $n \times n$ matrix algebra over a unital algebra $\mathcal{A}$ is just $M_n(\mathcal{A}_J)$.

Let $\mathcal{A}$ be a Banach algebra.
(c) *Every element of $\mathcal{A}_J$ is a two-sided topological divisor of zero.*
(d) *Any idempotent in $\mathcal{A}/\mathcal{A}_J$ can be written as $e + \mathcal{A}_J$, where e is an idempotent in $\mathcal{A}$.*
(e) *Any finite or countable set of orthogonal idempotents in $\mathcal{A}/\mathcal{A}_J$ may be written as $\{e + \mathcal{A}_J : e \in \mathcal{E}\}$, where $\mathcal{E}$ is a set of orthogonal idempotents in $\mathcal{A}$.*

Proof (a): If e is an idempotent element in $\mathcal{A}_J$, then it is quasi-invertible and hence $e + ee^q = e(e + e^q) = e(ee^q) = ee^q$ implies $e = 0$.

(b): Proposition 2.5.3(e) gives $e\mathcal{A}e \cap \mathcal{A}_J \subseteq (e\mathcal{A}e)_J$. Let $a \in \mathcal{A}^1$ and $b \in (e\mathcal{A}e)_J$ be arbitrary. Then $ab = aeb$ holds and $eaeb \in (e\mathcal{A}e)_J$ has a quasi-inverse c in $e\mathcal{A}e$. We have $c \circ ab = ab + c - cab = aeb + c - ceaeb = (1 - e)aeb + c \circ eaeb = (1 - e)aeb$. However $(1 - e)aeb$ has square zero so that it is quasi-invertible. Hence ab is left quasi-invertible. Since $a \in \mathcal{A}^1$ was arbitrary we conclude that b belongs to $\mathcal{A}_J$ by Theorem 4.3.6(c). This concludes the proof of the first equality since $b \in (e\mathcal{A}e)_J$ was arbitrary. The inclusion $e\mathcal{A}_J e \subseteq e\mathcal{A}e \cap \mathcal{A}_J$ is obvious and any $b \in e\mathcal{A}e \cap \mathcal{A}_J$ satisfies $b = ebe \in e\mathcal{A}_J e$. The remark about matrix algebras now follows from Theorem 1.6.10.

(c): Corollary 4.3.8 and Theorem 2.5.14 show that each element of $\mathcal{A}_J$ is either a two-sided topological divisor of zero or is invertible in $e\mathcal{A}e$ for a suitable proper idempotent e. However, the latter condition is impossible since it would imply that $\mathcal{A}_J$ contains the idempotent e.

(d): Let $a + \mathcal{A}_J$ be idempotent. We will actually show that there is a (necessarily unique) idempotent e in the smallest closed spectral subalgebra of $\mathcal{A}$ containing a (e.g., it is in the double commutant $\{a\}''$ of $\{a\}$ in $\mathcal{A}$) satisfying $e + \mathcal{A}_J = a + \mathcal{A}_J$. Define j by $j = a^2 - a \in \mathcal{A}_J$. Then $-4j(1 - (-4j)^q) = b$ is defined and belongs to $\mathcal{A}_J \cap \{a\}''$. Thus b satisfies $\|b\|^\infty = 0$ so that the series

$$\frac{1}{2} \sum_{k=1}^{\infty} \binom{1/2}{k} b^k$$

converges absolutely to an element c which belongs to both $\mathcal{A}_J$ and any closed spectral subalgebra of $\mathcal{A}$ containing a, and which satisfies $4c^2 + 4c = b$. Define e by $e = a + 2ac - c \in a + \mathcal{A}_J$. Then we get

$$
\begin{aligned}
e^2 - e &= a^2 + 4a^2c^2 + c^2 + 4a^2c - 2ac - 4ac^2 - a - 2ac + c \\
&= j + 4jc^2 + c^2 + 4jc + c = j + jb + 4^{-1}b \\
&= j - (4j + 1)j(1 - (-4j)^q) = 0.
\end{aligned}
$$

(e): Suppose that $\{a_j + \mathcal{A}_J : 1 \leq j \leq n\}$ is a set of orthogonal idempotents in $\mathcal{A}/\mathcal{A}_J$. By induction using (d) we may assume that a set $\{e_j : a \leq j \leq n - 1\}$ of orthogonal idempotents have been found satisfying $e_j + \mathcal{A}_J = a_j + \mathcal{A}_J$. Define an idempotent $e \in \mathcal{A}$ by $e = \sum_{j=1}^{n-1} e_j$. Then the orthogonality of $\{a_j + \mathcal{A}_J : 1 \leq j \leq n\}$ implies $(1 - e)a_n(1 - e) + \mathcal{A}_J = a_n + \mathcal{A}_J$. Let e_n be the idempotent constructed as in (d) from $(1 - e)a_n(1 - e)$. The construction gives an element in the closed spectral subalgebra $(1 - e)\mathcal{A}(1 - e)$ of $\mathcal{A}$. Hence $\{e_j : 1 \leq j \leq n\}$ is a set of orthogonal idempotents satisfying $e_j + \mathcal{A}_J = a_j + \mathcal{A}_J$ for $1 \leq j \leq n$. Clearly the argument covers the case of a countably infinite set of orthogonal idempotents. $\square$

The *perturbation set* $\mathcal{P}(\mathcal{S})$ of a set $\mathcal{S}$ in an algebra $\mathcal{A}$ is the set $\{a \in \mathcal{A} : a + \mathcal{S} \subseteq \mathcal{S}\}$. Clearly this set is always closed under addition. Suppose $\mathcal{A}$ is unital and $\mathcal{S} = \mathcal{A}_G$. Since $\mathcal{A}_G$ is closed under multiplication by non-zero numbers, $a \in \mathcal{P}(\mathcal{A}_G)$, $b \in \mathcal{A}_G$ and $\lambda \neq 0$ implies $\lambda a + b = \lambda(a + \lambda^{-1}b) \in \mathcal{A}_G$. Hence $\mathcal{P}(\mathcal{A}_G)$ is a linear subspace. If every element of $\mathcal{A}$ can be written as the sum of invertible elements, then $\mathcal{P}(\mathcal{A}_G)$ is an ideal. To see this, consider $a \in \mathcal{P}(\mathcal{A}_G)$, $b \in \mathcal{A}_G$ and $c \in \mathcal{A}$ with $c = \sum_{j=1}^{n} c_j$ where the c_j are all invertible. Then $ac_j + b = (a + bc_j^{-1})c_j$ is invertible for each j, so ac_j and hence $ac = \sum_{j=1}^{n} ac_j$ belong to the perturbation set. Since multiplication from the left works similarly, $\mathcal{P}(\mathcal{A}_G)$ is an ideal of $\mathcal{A}$ as claimed. We wish to work in not necessarily spectral algebras, so we must consider the *perturbation set of $\mathcal{A}_G^1$ in $\mathcal{A}$*: $\mathcal{P}_\mathcal{A}(\mathcal{A}_G^1) = \mathcal{A} \cap \mathcal{P}(\mathcal{A}_G^1)$. The next result is due to Arnold Lebow and Martin Schecter [1971] for unital Banach algebras. See also Jürgen Schulz [1985].

4.3.13 Proposition *In any spectral algebra* $\mathcal{A}$, *the Jacobson radical is the perturbation set in* $\mathcal{A}$ *of the group* $\mathcal{A}_G^1$ *in the unitization:*

$$\mathcal{A}_J = \{a \in \mathcal{A} : a + \mathcal{A}_G^1 \subseteq \mathcal{A}_G^1\}.$$

Proof Denote the perturbation set by $\mathcal{P} = \mathcal{P}_\mathcal{A}(\mathcal{A}_G^1)$. The third condition under Theorem 4.3.6(d) shows $\mathcal{A}_J \subseteq \mathcal{P}$ in any algebra. Since the spectrum is bounded in a spectral algebra, any element is the sum of the two invertible elements λ and $-(\lambda - a)$ for any λ with absolute value strictly greater than $\rho(a)$. Hence the perturbation set $\mathcal{P}$ is an ideal by the argument given above. In this case, Theorem 4.3.6(d) shows $\mathcal{P} = \mathcal{A}_J$. □

T. Keith Carne [1981] studies irreducible representations of general algebra tensor products and gives conditions for their semisimplicity. The nuclearity (Definition 1.10.23) of the tensor norm is a key factor. This improves previous results of Bernard R. Gelbaum [1962], Kjeld B. Laursen [1970] and Jun Tomiyama [1972].

4.4 The Baer Radical

We will now turn to the study of another radical—the Baer radical. It is of much less importance in our theory mainly because it is not necessarily closed in spectral normed algebras. Nevertheless the algebras for which this radical is zero have a number of useful properties.

Definitions

The terms "nilpotent" and "nil" introduced in the next definition were used by Wedderburn [1907], and Gottfried Köthe [1930], respectively. Köthe provided an example which showed that they did not coincide. The concept of a nil ideal was exploited by Elie Cartan [1898]. The term "semiprime" was introduced by Masayoshi Nagata [1951], but the concept was already considered by Baer [1943].

It should be noted that, unlike primitive ideals, all the concepts presented here and for the rest of this section are invariant under reversal of the order of multiplication. Thus, anything proved for left ideals will hold (perhaps with an obvious reflection in the statement) for right ideals.

4.4.1 Definition Let $\mathcal{A}$ be an algebra. Then $\mathcal{A}$ is called:

(a) *nil* if each element is nilpotent (*i.e.*, for each $a \in \mathcal{A}$ there is some $n \in \mathbb{N}$ satisfying $a^n = 0$);

(b) *nilpotent* if there is some $n \in \mathbb{N}$ satisfying $\mathcal{A}^n = \{0\}$ (*i.e.*, there is some $n \in \mathbb{N}$ such that for any $a_1, a_2, \ldots, a_n \in \mathcal{A}$ the product $a_1 a_2 \cdots a_n$ is zero);

(c) *semiprime* if it has no non-zero nilpotent ideal.

An ideal $\mathcal{I}$ of $\mathcal{A}$ is called nil or nilpotent if it is *nil* or *nilpotent* when it is considered as an algebra in its own right. However, an ideal $\mathcal{I}$ of $\mathcal{A}$ is called *semiprime* if $\mathcal{A}/\mathcal{I}$ is semiprime.

The two systems for applying the above terms to algebras and to ideals are slightly confusing, but we are following well established tradition. The terms "prime" and "primitive" are applied to algebras and ideals according to the same system as "semiprime". One must be careful to note that a semiprime ideal of an algebra is not usually semiprime when considered as an algebra in its own right.

Nil and Nilpotent Ideals

In Theorem 4.4.11 below we will give results of Sandy Grabiner [1969] and Peter G. Dixon [1973b] which assert that a nil Banach algebra is nilpotent and a semiprime Banach algebra has no non-zero nil one-sided ideals. Most of the following results can be found in Cartan [1898] and Wedderburn [1907].

4.4.2 Proposition *Let $\mathcal{A}$ be an algebra.*

(a) *If $\mathcal{A}$ is nil or nilpotent, then every subalgebra of $\mathcal{A}$ and every homomorphic image of $\mathcal{A}$ has the corresponding property.*

(b) *If $\mathcal{A}$ has an ideal $\mathcal{I}$ such that $\mathcal{I}$ and $\mathcal{A}/\mathcal{I}$ are both nil or both nilpotent, then $\mathcal{A}$ has the corresponding property.*

(c) *The sum $\mathcal{M}$ of all the nil ideals is a nil ideal and $\mathcal{A}/\mathcal{M}$ has no non-zero nil ideals.*

(d) *The sum of a finite set of nilpotent ideals is nilpotent. The sum of all the nilpotent ideals is a nil ideal which contains every nilpotent one-sided ideal.*

(e) *If some ideal $\mathcal{I}$ of $\mathcal{A}$ has a non-zero nilpotent ideal, then there is a non-zero nilpotent ideal of $\mathcal{A}$. Hence any ideal of a semiprime algebra is a semiprime algebra.*

Proof Both (a) and (b) are obvious. (c) and (d): First we note that the sum of two nil or nilpotent ideals has the corresponding property. Suppose $\mathcal{I}_1$ and $\mathcal{I}_2$ are both nil or both nilpotent ideals in $\mathcal{A}$. Then $(\mathcal{I}_1 + \mathcal{I}_2)/\mathcal{I}_1$ is isomorphic to $\mathcal{I}_2/(\mathcal{I}_1 \cap \mathcal{I}_2)$ which is nil or nilpotent by (a). Thus (b) shows that $\mathcal{I}_1 + \mathcal{I}_2$ is nil or nilpotent since both $(\mathcal{I}_1 + \mathcal{I}_2)/\mathcal{I}_1$ and $\mathcal{I}_1$ have the corresponding property. By induction both these properties are preserved under finite sums of ideals. If an element $a \in \mathcal{A}$ belongs to the sum of all nil ideals, then a belongs to the sum of finitely many nil ideals and hence is nilpotent by what we have just proved. Hence the sum of all nil ideals is nil. Since the sum of all nilpotent ideals is an ideal included in the sum of all nil ideals it is a nil ideal also.

Let $\mathcal{M}$ be the sum of all nil ideals. We will show that $\mathcal{A}/\mathcal{M}$ has no non-zero nil ideals. If $\mathcal{I}$ is the preimage in $\mathcal{A}$ of a nil ideal of $\mathcal{A}/\mathcal{M}$, then

$\mathcal{I}/\mathcal{M}$. and $\mathcal{M}$ are both nil so $\mathcal{I}$ is nil by (b). Therefore $\mathcal{I}$ is included in $\mathcal{M}$ by the definition of $\mathcal{M}$. Thus $\mathcal{A}/\mathcal{M}$ has no non-zero nil ideals.

Finally, we note that the last claim of (d) follows from the fact that any nilpotent one-sided ideal is included in some nilpotent two-sided ideal. For instance, if $\mathcal{L}$ is a left ideal satisfying $\mathcal{L}^n = \{0\}$, then $\mathcal{L}$ is included in the ideal $\mathcal{L}\mathcal{A}^1$ which satisfies $(\mathcal{L}\mathcal{A}^1)^n = \mathcal{L}(\mathcal{A}^1\mathcal{L})^{n-1}\mathcal{A}^1 \subseteq \mathcal{L}^n\mathcal{A}^1 = \{0\}$.

(e) Let $\mathcal{N}$ be a non-zero nilpotent ideal of $\mathcal{I}$. The ideal $\mathcal{M} = \mathcal{A}^1\mathcal{N}\mathcal{A}^1$ of $\mathcal{A}$ satisfies $\mathcal{M}^3 \subseteq \mathcal{I}\mathcal{N}\mathcal{I} \subseteq \mathcal{N}$. Thus $\mathcal{M}$ is nilpotent, since $\mathcal{N}$ is nilpotent. The last sentence follows immediately from the first. □

The sum of all the nil ideals of an algebra is called the *nil radical*. This was the first radical considered for algebras. It was introduced by Cartan [1898]. We denote the nil radical of an algebra $\mathcal{A}$ by $\mathcal{A}_{\mathrm{nil}}$. Proposition 4.4.2(c) shows

$$(\mathcal{A}/\mathcal{A}_{\mathrm{nil}})_{\mathrm{nil}} = \{0\} \quad \text{and} \quad (\mathcal{A}_{\mathrm{nil}})_{\mathrm{nil}} = \mathcal{A}_{\mathrm{nil}}.$$

These are the two most basic properties which any radical should possess. For finite-dimensional algebras nil ideals are nilpotent. Wedderburn [1907] introduced and concentrated on the nilpotent ideals. Their theory is unfortunately even harder than the theory of nil ideals to generalize beyond finite-dimensional algebras and algebras with the descending chain condition on left ideals (where again nil ideals are nilpotent).

Semiprime Ideals

The next results are mainly due to Nagata [1951]. The only nontrivial step is the proof that each semiprime ideal satisfies (e). This theorem is very similar to Theorem 4.1.10.

4.4.3 Theorem *Let $\mathcal{I}$ be an ideal in an algebra $\mathcal{A}$. The following are equivalent:*

(a) $\mathcal{I}$ *is a semiprime ideal of $\mathcal{A}$.*

(b) $a\mathcal{A}a \subseteq \mathcal{I}$ *implies $a \in \mathcal{I}$ for all $a \in \mathcal{A}$;*

(c) *If $\mathcal{M}$ is an ideal of $\mathcal{A}$, then $\mathcal{M}^2 \subseteq \mathcal{I}$ implies $\mathcal{M} \subseteq \mathcal{I}$.*

(d) *If $\mathcal{L}$ is a one-sided ideal of $\mathcal{A}$ satisfying $\mathcal{L}^n \subseteq \mathcal{I}$ for some positive integer n, then $\mathcal{L}$ satisfies $\mathcal{L} \subseteq \mathcal{I}$.*

(e) $\mathcal{I}$ *is the intersection of some set of prime ideals of $\mathcal{A}$.*

Proof $(a) \Rightarrow (b)$: If $a\mathcal{A}a \subseteq \mathcal{I}$ holds, then $(\mathcal{A}^1a\mathcal{A}^1/\mathcal{I})^3$ equals $\{\mathcal{I}\}$ in $\mathcal{A}/\mathcal{I}$. Thus $(\mathcal{A}^1a\mathcal{A}^1)/\mathcal{I} = \{\mathcal{I}\}$ holds in the semiprime algebra $\mathcal{A}/\mathcal{I}$. However, this implies $a \in \mathcal{I}$.

$(b) \Rightarrow (c)$: If $\mathcal{M}^2 \subseteq \mathcal{I}$ and $a \in \mathcal{M}$ hold, then $a\mathcal{A}a \subseteq \mathcal{M}^2 \subseteq \mathcal{I}$ implies $a \in \mathcal{I}$. Since $a \in \mathcal{M}$ was arbitrary, $\mathcal{M}$ is included in $\mathcal{I}$.

$(c) \Rightarrow (d)$: Let $\mathcal{L}$ be a left ideal satisfying $\mathcal{L}^n \subseteq \mathcal{I}$. Then $\mathcal{L}\mathcal{A}^1$ is an ideal satisfying $(\mathcal{L}\mathcal{A}^1)^n \subseteq \mathcal{L}^n\mathcal{A}^1 \subseteq \mathcal{I}\mathcal{A}^1 = \mathcal{I}$. Suppose that m is the smallest integer satisfying $(\mathcal{L}\mathcal{A}^1)^m \subseteq \mathcal{I}$. If $m > 1$ held, then $((\mathcal{L}\mathcal{A}^1)^{m-1})^2 \subseteq \mathcal{I}$ would hold. Then (c) would imply $(\mathcal{L}\mathcal{A}^1)^{m-1} \subseteq \mathcal{I}$. This contradiction shows $m = 1$ so $\mathcal{L} \subseteq \mathcal{L}\mathcal{A}^1 \subseteq \mathcal{I}$ holds.

$(d) \Rightarrow (a)$: We must show that $\mathcal{A}/\mathcal{I}$ has no non-zero nilpotent ideals. If $(\mathcal{N}/\mathcal{I})^n$ is the zero ideal (i.e., it equals $\{\mathcal{I}\}$ in $\mathcal{A}/\mathcal{I}$), then $\mathcal{N}^n$ is included in $\mathcal{I}$. This implies $\mathcal{N} \subseteq \mathcal{I}$ so that $\mathcal{N}/\mathcal{I}$ already equals $\{\mathcal{I}\}$ in $\mathcal{A}/\mathcal{I}$.

$(e) \Rightarrow (c)$: If $\mathcal{M}^2 \subseteq \cap\{\mathcal{P}_\alpha : \alpha \in A\}$ holds for any set of prime ideals $\{\mathcal{P}_\alpha : \alpha \in A\}$, then for each $\alpha \in A, \mathcal{M}^2 \subseteq \mathcal{P}_\alpha$ implies $\mathcal{M} \subseteq \mathcal{P}_\alpha$ by the definition of a prime ideal. Hence $\mathcal{M}$ is a subset of $\cap\{\mathcal{P}_\alpha : \alpha \in A\}$.

$(b) \Rightarrow (e)$: If (e) is false there is an element a_1 which belongs to the intersection of all the prime ideals of $\mathcal{A}$ which include $\mathcal{I}\}$, but a_1 does not belong to $\mathcal{I}$. We can inductively choose $a_n \in a_{n-1}\mathcal{A}a_{n-1} \setminus \mathcal{I}$ by (b). Define $\mathcal{S}$ as the set $\{a_n : n \in \mathbb{N}\}$. Consider the family of ideals which include $\mathcal{I}$ but do not intersect $\mathcal{S}$. It contains $\mathcal{I}$. When this family is ordered by inclusion, Zorn's lemma shows that it has a maximal element which we will call $\mathcal{P}$. We want to show that $\mathcal{P}$ is prime. This will prove the implication $(b) \Rightarrow (e)$ by contradiction since the choice of a_1 then shows that $a_1 \in \mathcal{S}$ belongs to $\mathcal{P}$. Suppose $a, b \in \mathcal{A}$ do not belong to $\mathcal{P}$. We will show $a\mathcal{A}b \not\subseteq \mathcal{P}$ and thus establish that $\mathcal{P}$ is prime by Theorem 4.1.10(b). By maximality of $\mathcal{P}$ there are elements $a_i, a_j \in \mathcal{S}$ satisfying $a_i \in \mathcal{P} + \mathcal{A}^1 a \mathcal{A}^1$ and $a_j \in \mathcal{P} + \mathcal{A}^1 b \mathcal{A}^1$. By symmetry we may suppose $i \leq j$. This implies $a_{j+1} \in a_j \mathcal{A}a_j \subseteq (\mathcal{P} + \mathcal{A}^1 a \mathcal{A}^1)\mathcal{A}(\mathcal{P} + \mathcal{A}^1 b \mathcal{A}^1)$. If $\mathcal{P}$ includes $a\mathcal{A}b$, then the last expression would be included in $\mathcal{P}$. However, $a_{j+1} \in \mathcal{S}$ does not belong to $\mathcal{P}$. Hence $a\mathcal{A}b$ is not included in $\mathcal{P}$ so $\mathcal{P}$ is prime. As remarked above this shows $(b) \Rightarrow (e)$.   □

Note that property (d) in the last theorem shows that a semiprime algebra has no non-zero nilpotent one-sided ideals. This is one of the conditions originally considered by Baer [1943] in his definition of radical ideals.

We now prove another easy proposition which includes results needed later. Similar statements hold for prime ideals instead of semiprime ideals.

4.4.4 Proposition *Let $\mathcal{A}$ be an algebra and let $\mathcal{I}$ be an ideal of $\mathcal{A}$.*

(a) *If $\mathcal{P}$ is a semiprime ideal of $\mathcal{I}$, then $\mathcal{P}$ is an ideal of $\mathcal{A}$. If in addition $\mathcal{I}$ is a semiprime ideal of $\mathcal{A}$, then $\mathcal{P}$ is a semiprime ideal of $\mathcal{A}$.*

(b) *If $\mathcal{P}$ is a semiprime ideal of $\mathcal{A}$, then $\mathcal{P}\cap\mathcal{I}$ is a semiprime ideal of $\mathcal{I}$.*

(c) *If $\mathcal{P}$ is a semiprime ideal of $\mathcal{A}$ which includes $\mathcal{I}$, then $\mathcal{P}/\mathcal{I}$ is a semiprime ideal of $\mathcal{A}/\mathcal{I}$.*

(d) *If $\mathcal{P}$ is a semiprime ideal of $\mathcal{A}/\mathcal{I}$, then its complete preimage in $\mathcal{A}$ is a semiprime ideal of $\mathcal{A}$.*

Proof (a): If $\mathcal{P}$ is any ideal of $\mathcal{I}$, then $\mathcal{P}\mathcal{A}$ and $\mathcal{A}\mathcal{P}$ are ideals of $\mathcal{I}$. (The following calculations show this for $\mathcal{P}\mathcal{A} : \mathcal{P}\mathcal{A} \subseteq \mathcal{I}; \mathcal{I}(\mathcal{P}\mathcal{A}) = (\mathcal{I}\mathcal{P})\mathcal{A} \subseteq \mathcal{P}\mathcal{A}; (\mathcal{P}\mathcal{A})\mathcal{I} = \mathcal{P}(\mathcal{A}\mathcal{I}) \subseteq \mathcal{P}\mathcal{A}$. Note also $(\mathcal{P}\mathcal{A})\mathcal{I} = \mathcal{P}(\mathcal{A}\mathcal{I}) \subseteq \mathcal{P}\mathcal{I} \subseteq \mathcal{P}$.) However we have $(\mathcal{P}\mathcal{A})^2 = \mathcal{P}(\mathcal{A}\mathcal{P}\mathcal{A}) \subseteq \mathcal{P}\mathcal{I} \subseteq \mathcal{P}$ and similarly $(\mathcal{A}\mathcal{P})^2 \subseteq \mathcal{P}$.

Hence if $\mathcal{P}$ is a semiprime ideal of $\mathcal{I}$, then $\mathcal{P}$ includes $\mathcal{PA}$ and $\mathcal{AP}$ so $\mathcal{P}$ is an ideal of $\mathcal{A}$.

Now assume that $\mathcal{I}$ is a semiprime ideal of $\mathcal{A}$ and let $\mathcal{P}$ be a semiprime ideal of $\mathcal{I}$. Then we have just shown that $\mathcal{P}$ is an ideal of $\mathcal{A}$. We will use Theorem 4.4.3(c) to show that $\mathcal{P}$ is a semiprime ideal of $\mathcal{A}$. Suppose $\mathcal{N}$ is an ideal of $\mathcal{A}$ satisfying $\mathcal{N}^2 \subseteq \mathcal{P}$. Then $\mathcal{N}^2 \subseteq \mathcal{P} \subseteq \mathcal{I}$ implies $\mathcal{N} \subseteq \mathcal{I}$. Hence $\mathcal{N}$ is an ideal of $\mathcal{I}$. Since $\mathcal{P}$ is a semiprime ideal of $\mathcal{I}$ which includes $\mathcal{N}^2$ it includes $\mathcal{N}$. Hence $\mathcal{P}$ is a semiprime ideal of $\mathcal{A}$.

(b): We will use Theorem 4.4.3(c) and (d). Suppose $\mathcal{N}$ is an ideal of $\mathcal{I}$ satisfying $\mathcal{N}^2 \subseteq \mathcal{P} \cap \mathcal{I}$. Then $\mathcal{A}^1 \mathcal{N}$ is a left ideal of $\mathcal{A}$ satisfying $(\mathcal{A}^1 \mathcal{N})^3 = \mathcal{A}^1 \mathcal{N} (\mathcal{A}^1 \mathcal{N}) \mathcal{A}^1 \mathcal{N} \subseteq \mathcal{A}^1 \mathcal{N} \mathcal{I} \mathcal{A}^1 \mathcal{N} = \mathcal{A}^1 \mathcal{N} (\mathcal{I} \mathcal{A}^1) \mathcal{N} \subseteq \mathcal{A}^1 \mathcal{N} \mathcal{I} \mathcal{N} \subseteq \mathcal{A}^1 \mathcal{N}^2 \subseteq \mathcal{A}^1 (\mathcal{P} \cap \mathcal{I}) \subseteq \mathcal{P}$. Since $\mathcal{P}$ is a semiprime ideal of $\mathcal{A}$ it includes $\mathcal{N} \subseteq \mathcal{A}^1 \mathcal{N}$. Since $\mathcal{N}$ is an ideal of $\mathcal{I}$ we conclude $\mathcal{N} \subseteq \mathcal{P} \cap \mathcal{I}$. This proves that $\mathcal{P} \cap \mathcal{I}$ is a semiprime ideal of $\mathcal{I}$.

(c): Suppose $\mathcal{N}$ is an ideal of $\mathcal{A}/\mathcal{I}$ which satisfies $\mathcal{N}^2 \subseteq \mathcal{P}/\mathcal{I}$. Let $\mathcal{M}$ be its complete preimage in $\mathcal{A}$. Then $\mathcal{M}^2$ is included in $\mathcal{P} + \mathcal{I} = \mathcal{P}$. Since $\mathcal{P}$ is a semiprime ideal of $\mathcal{A}$ we conclude $\mathcal{M} \subseteq \mathcal{P}$ and hence $\mathcal{N} = \mathcal{M}/\mathcal{I} \subseteq \mathcal{P}/\mathcal{I}$. Theorem 4.4.3(c) shows that $\mathcal{P}/\mathcal{I}$ is semiprime.

(d): Let $\mathcal{M}$ be the complete preimage in $\mathcal{A}$ of $\mathcal{P}$ and let $\mathcal{N}$ be an ideal of $\mathcal{A}$ satisfying $\mathcal{N}^2 \subseteq \mathcal{M}$. Then $(\mathcal{N}/\mathcal{I})^2 \subseteq \mathcal{M}/\mathcal{I} = \mathcal{P}$ implies $\mathcal{N}/\mathcal{I} \subseteq \mathcal{P}$ since $\mathcal{P}$ is a semiprime ideal of $\mathcal{A}/\mathcal{I}$. Hence $\mathcal{N}$ is included in $\mathcal{M}$. Therefore $\mathcal{M}$ is semiprime. □

Reinhold Baer [1943] called any ideal $\mathcal{I}$ of an algebra $\mathcal{A}$ a radical ideal if $\mathcal{I}$ is a nil ideal and $\mathcal{A}/\mathcal{I}$ contains no non-zero nilpotent one-sided ideals. He proved that the intersection of all radical ideals is a radical ideal (now called the Baer lower radical) and the sum of all radical ideals is a radical ideal (now called the Baer upper radical). He made the easy observation that the Baer lower radical is (in our terminology) simply the intersection of the semiprime ideals since this intersection is automatically nil.

The Baer Radical

The following definition is due to McCoy [1949]. Jacob Levitzki [1951] and Nagata [1951] independently showed that the ideal defined here, which McCoy called the prime radical, coincides with the Baer lower radical. We follow Nathan J. Divinsky [1965] in calling this simply the Baer radical. Theorem 4.4.6 below shows that this agrees with Baer's original definition.

4.4.5 Definition The *Baer radical*, or *prime radical*, of an algebra, $\mathcal{A}$, is the intersection of all the prime ideals of $\mathcal{A}$. It is denoted by $\mathcal{A}_B$. The algebra is called *Baer-radical* if it equals its own Baer radical.

To agree with systematic usage we should call an algebra Baer-semisimple if its Baer radical is zero. However, the next result shows that the term semiprime already covers this situation. (Nagata [1951] invented the term semiprime to signify that such an algebra could be written as a subdirect product of prime algebras. He suggested the term semiprimitive to replace Jacobson semisimple. His systematic notation has not been adopted by others.)

The (easy) fact that the Baer radical (as defined here) is semiprime was noted by McCoy [1949]. The fact that it is the smallest such ideal was proved by Nagata [1951]. The following theorem shows that $\mathcal{A}_B$ is Baer's original lower radical.

4.4.6 Theorem *The Baer radical of an algebra is a semiprime nil-ideal which is included in every semiprime ideal. Hence the Baer radical is $\{0\}$ if and only if the algebra is semiprime. The Baer radical of a commutative algebra is the set of nilpotent elements and hence is the nil radical.*

Proof The fact that the Baer radical is a semiprime ideal which is included in every semiprime ideal is an immediate consequence of Theorem 4.4.3(e). Since $\{0\}$ is a semiprime ideal if and only if the algebra is semiprime, the second sentence follows from this. Proposition 4.4.2(c) shows that the sum, $\mathcal{M}$, of all nil ideals (the nil radical) is a nil, semi-prime ideal. Hence Proposition 4.4.2(a) shows that the Baer radical is nil since it is included in $\mathcal{M}$.

If $\mathcal{A}$ is commutative the set $\mathcal{N}$ of nilpotent elements is clearly an ideal. It includes $\mathcal{A}_B$ by the first result of this theorem. If this inclusion were proper, $\mathcal{A}/\mathcal{A}_B$ would include the non-zero nilpotent ideal $\mathcal{A}^1 b/\mathcal{A}_B$ for some $b \in \mathcal{N} \setminus \mathcal{A}_B$. Since $\mathcal{A}/\mathcal{A}_B$ is semiprime, we conclude that $\mathcal{N}$ equals the Baer radical. □

Having characterized the Baer radical as an intersection of ideals and characterized the algebras with the smallest possible Baer radicals, we turn our attention to the Baer radical itself and characterize it as a sum of ideals.

4.4.7 Theorem *An algebra is Baer radical if and only if every non-zero homomorphic image of it has a non-zero nilpotent ideal. The Baer radical of any algebra is a Baer radical ideal which includes all Baer radical ideals and all one-sided nilpotent ideals.*

Proof Let $\mathcal{A}$ be an algebra. By the last theorem $\mathcal{A}_B$ is the smallest ideal of $\mathcal{A}$ such that $\mathcal{A}/\mathcal{A}_B$ is semiprime. Hence $\mathcal{A} = \mathcal{A}_B$ holds if and only if every non-zero homomorphic image of $\mathcal{A}$ contains a non-zero nilpotent ideal (*i.e.*, is not semiprime).

Theorem 4.4.6 shows that $\mathcal{A}_B$ is a semiprime ideal of $\mathcal{A}$ and that $(\mathcal{A}_B)_B$ is a semiprime ideal of $\mathcal{A}_B$. Hence Proposition 4.4.4(a) shows that $(\mathcal{A}_B)_B$ is a semiprime ideal of $\mathcal{A}$. Theorem 4.4.6 then shows $\mathcal{A}_B \subseteq (\mathcal{A}_B)_B$. Since the opposite inclusion is trivial, $\mathcal{A}_B$ is a Baer radical ideal.

Now suppose $\mathcal{I}$ is any Baer radical ideal of $\mathcal{A}$. If $\mathcal{I}$ is not included in $\mathcal{A}_B$, then $\mathcal{I}/(\mathcal{I} \cap \mathcal{A}_B)$ is a non-zero homomorphic image which must contain a non-zero nilpotent ideal by the first statement of this theorem. However $\mathcal{I}/(\mathcal{I} \cap \mathcal{A}_B)$ is isomorphic to $(\mathcal{I} + \mathcal{A}_B)/\mathcal{A}_B$ which is an ideal of the semiprime algebra $\mathcal{A}/\mathcal{A}_B$ and hence contains no non-zero nilpotent ideal by Proposition 4.4.2(e). This contradiction shows that $\mathcal{I}$ (which was an arbitrary Baer radical ideal) is included in $\mathcal{A}_B$. Finally Proposition 4.4.2(d) shows that $\mathcal{A}_B$ includes all one-sided nilpotent ideals since it includes all Baer radical ideals and hence all nilpotent ideals. $\quad\square$

Note that the first sentence of the last theorem shows how to characterize Baer radical ideals in terms of nilpotent ideals. One might be tempted to continue the process and define a new concept by the same construction starting with Baer radical ideals. However, an application of Theorem 4.4.3(e) shows easily that we simply get the concept Baer radical again by this construction. This remark is due to Adam Suliński, Robert F. V. Anderson, and Divinsky [1966] whose work suggested the treatment of Theorem 4.4.7 above.

Next we prove the analogue of Theorem 4.3.2. The two assertions in the last sentence have each been proved already (in Theorems 4.4.6 and 4.4.7 respectively) but are included for comparison. Property (a) was established by McCoy [1949].

4.4.8 Theorem *Any algebra $\mathcal{A}$ and any ideal $\mathcal{I}$ of $\mathcal{A}$ satisfy:*
(a) $\mathcal{I}_B = \mathcal{I} \cap \mathcal{A}_B$.
(b) $\mathcal{I} \subseteq \mathcal{A}_B$ implies $(\mathcal{A}/\mathcal{I})_B = \mathcal{A}_B/\mathcal{I}$.
(c) $\mathcal{A}/\mathcal{I}$ semiprime implies $\mathcal{A}_B = \mathcal{I}_B$.
Thus in particular $\mathcal{A}/\mathcal{A}_B$ is semiprime and $\mathcal{A}_B$ is Baer radical.

Proof The proof of Theorem 4.3.2 may be copied, merely substituting "prime" for "primitive" and "the last sentence of Theorem 4.1.10" for "Theorem 4.1.8(e)". We give alternative proofs for (a) and (b) also.

(a): Proposition 4.4.4(b) shows that $\mathcal{I} \cap \mathcal{A}_B$ is a semiprime ideal of $\mathcal{I}$. Hence Theorem 4.4.6 implies $\mathcal{I}_B \subseteq \mathcal{I} \cap \mathcal{A}_B$.

To prove the opposite inclusion, suppose a is an element of $\mathcal{I} \setminus \mathcal{I}_B$. Then there is a prime ideal $\mathcal{M}$ of $\mathcal{I}$ which does not contain a. Define $\mathcal{S}$ to be the set $\mathcal{I} \setminus \mathcal{M}$ which contains a. Consider the family of ideals of $\mathcal{A}$ which do not intersect $\mathcal{S}$. This family is not empty since $\{0\}$ belongs to it. Zorn's lemma applies to this family when it is ordered by inclusion and gives a maximal element $\mathcal{P}$. We repeat the argument given in the proof of Theorem 4.4.3 (b)$\Rightarrow$(e), which shows that $\mathcal{P}$ is a prime ideal. Suppose b_1 and b_2 are any elements of $\mathcal{A}$ satisfying $b_1, b_2 \notin \mathcal{P}$. Then the maximality of $\mathcal{P}$ gives elements c_1 and c_2 in $\mathcal{S}$ satisfying $c_j \in \mathcal{P} + \mathcal{A}^1 b_j \mathcal{A}^1$ for $j = 1, 2$. This implies $c_1 \mathcal{A} c_2 \subseteq (\mathcal{P} + \mathcal{A}^1 b_1 \mathcal{A}^1) \mathcal{A} (\mathcal{P} + \mathcal{A}^1 b_2 \mathcal{A}^1) \subseteq \mathcal{P}$. However this contradicts the fact that $c_1 \mathcal{I} c_2$ has nontrivial intersection with $\mathcal{S}$. Hence $\mathcal{P}$ is a prime ideal which does not include a. Since a was an arbitrary element of $\mathcal{I} \setminus \mathcal{I}_B$ and since $\mathcal{I} \cap \mathcal{A}_B$ equals $\cap \{\mathcal{I} \cap \mathcal{P} : \mathcal{P}$ is a prime ideal of $\mathcal{A}\}$, we conclude that $\mathcal{I} \cap \mathcal{A}_B$ is included in $\mathcal{I}_B$. This proves (a).

(b): Proposition 4.4.4(c) and (d) prove

$$
\begin{aligned}
(\mathcal{A}/\mathcal{I})_B &= \cap \{\mathcal{P} : \mathcal{P} \text{ is a semiprime ideal of } \mathcal{A}/\mathcal{I}\} \\
&= \cap \{\mathcal{P}/\mathcal{I} : \mathcal{P} \text{ is a semiprime ideal of } \mathcal{A}\} \\
&= \mathcal{A}_B/\mathcal{I}. \qquad \square
\end{aligned}
$$

The following two conditions which imply that an algebra is semiprime may have some interest. The second condition significantly strengthens Yood [1972], Corollary 2.7.

4.4.9 Proposition *An algebra is semiprime if either:*

(a) *It has a separating family of topologically irreducible normed representations; or*

(b) *It is a dense subalgebra of a semiprime topological algebra.*

In fact, the intersection of a closed $\langle$ prime / semiprime $\rangle$ ideal of a topological algebra with a dense subalgebra is a $\langle$ prime / semiprime $\rangle$ ideal of the subalgebra.

Proof (a): This follows from Proposition 4.2.5 and Theorem 4.4.3(a).

(b): This follows from Theorem 4.4.3(b).

The last remark follows from Theorem $\langle$ 4.1.10 (b) / 4.4.3(b) $\rangle$. $\square$

The intersection of the kernels of the continuous topologically irreducible representations of an arbitrary Banach algebra $\mathcal{A}$ would seem to be an interesting object. It is a closed semiprime ideal, but not much more seems to be known about it.

The following result is chiefly of interest to us through its use in Theorem 4.4.11. It shows that a nil algebra with a uniform bound on the degree of nilpotency of its elements is nilpotent. It is due to Nagata [1952]. Graham Higman [1956] gave a more general version. The simpler proof which we use here is due to Philip J. Higgins and appears in Jacobson [1964].

4.4.10 Proposition *Let $\mathcal{A}$ be an algebra. Let n be an integer such that $a^n = 0$ holds for all $a \in \mathcal{A}$. Then $\mathcal{A}$ is nilpotent and satisfies $\mathcal{A}^N = \{0\}$ where $N = 2^n - 1$.*

Proof The proof is by induction on n. The case $n = 1$ is trivial. Let $a^n = 0$ hold for all $a \in \mathcal{A}$. For $a, b \in \mathcal{A}$ and $\lambda \in \mathbb{C}$, the polynomial $p(\lambda) = (a + \lambda b)^n$ vanishes identically. Hence the coefficient $q_n(a, b) = \sum_{j=0}^{n-1} a^j b a^{n-j-1}$ of λ is zero for $a, b \in \mathcal{A}$. For all $a, b, c \in \mathcal{A}$, we see

$$
0 = \sum_{j=0}^{n-1} q_n(a, cb^j) b^{n-j-1} = \sum_{k=0}^{n-1} \sum_{j=0}^{n-1} a^k c b^j a^{n-k-1} b^{n-j-1}
$$

$$= \sum_{k=0}^{n-1} a^k c q_n(b, a^{n-k-1}) = a^{n-1} c q_n(b, 1) = n a^{n-1} c b^{n-1}.$$

We conclude $a^{n-1} c b^{n-1} = 0$ for all a, b, $c \in \mathcal{A}$.

Let $\mathcal{N}$ be the ideal which is the linear span of $\{ab^{n-1}c : a, c \in \mathcal{A}^1, b \in \mathcal{A}\}$. Then we have just shown $\mathcal{N}\mathcal{A}\mathcal{N} = \{0\}$. Also each $a + \mathcal{N} \in \mathcal{A}/\mathcal{N}$ satisfies $(a + \mathcal{N})^{n-1} = \mathcal{N}$. By induction we may assume $(\mathcal{A}/\mathcal{N})^M = \mathcal{N}$ where $M = 2^{n-1} - 1$. Thus $\mathcal{A}^M \subseteq \mathcal{N}$ holds and this implies $\mathcal{A}^N = \mathcal{A}^M \mathcal{A} \mathcal{A}^M \subseteq \mathcal{N}\mathcal{A}\mathcal{N} = \{0\}$. This proves the proposition. $\qquad\square$

Nil and Nilpotent Ideals in Banach Algebras

In the next theorem (b) is due to Grabiner [1969]. Conclusion (c) and the first sentence of (a) are due to Dixon [1973b] with proofs based on that of the previous reference. The second sentence of (a) is due to Grabiner [1976]. Conclusion (d) is an easy consequence of (b). From (c) we conclude that the Baer radical is the smallest interesting radical for Banach algebras. Note that a Banach algebra is Jacobson semisimple if and only if it has no non-zero topologically nil ideals and it is semiprime if and only if it has no non-zero nil (equivalently, nilpotent) ideals.

4.4.11 Theorem *Let $\mathcal{A}$ be a Banach algebra.*

(a) *Every nil left ideal of $\mathcal{A}$ is included in a sum of nilpotent ideals. Every finitely generated nil left ideal is included in a nilpotent ideal.*

(b) *If $\mathcal{A}$ is nil, then it is nilpotent.*

(c) *The Baer radical of $\mathcal{A}$ equals both the sum of all the nilpotent ideals of $\mathcal{A}$ and the sum of all the nil left ideals of $\mathcal{A}$. Hence $\mathcal{A}_B$ is the nil radical of $\mathcal{A}$ and $\mathcal{A}/\mathcal{A}_B$ has no non-zero nil left ideals.*

(d) *The Baer radical is nilpotent if and only if it is closed.*

Proof (a): Without loss of generality we may assume that $\mathcal{A}$ is unital. For each $b \in \mathcal{A}$ let $\mathcal{A}b$ and $\mathcal{A}b\mathcal{A}$ be, respectively, the principal left ideal and the principal ideal generated by b. If $\mathcal{A}b$ is nil we will show that $\mathcal{A}b\mathcal{A}$ is nilpotent. Hence any nil left ideal $\mathcal{L}$ is included in the sum of the collection $\{\mathcal{A}b\mathcal{A} : b \in \mathcal{L}\}$ of nilpotent ideals. If $\mathcal{L} = \mathcal{A}b_1 + \mathcal{A}b_2 + \cdots + \mathcal{A}b_n$ is a finitely generated left ideal, then it is included in the nilpotent ideal $\mathcal{A}b_1\mathcal{A} + \mathcal{A}b_2\mathcal{A} + \cdots + \mathcal{A}b_n\mathcal{A}$ which is nilpotent by Proposition 4.4.2(d).

Assume that $\mathcal{A}b$ is nil. For each $n \in \mathbb{N}$ define $\mathcal{S}_n$ to be the closed set $\{a \in \mathcal{A} : (ab)^n = 0\}$. Since $\mathcal{A}b$ is nil, $\mathcal{A}$ equals $\cup_{n=1}^\infty \mathcal{S}_n$. The Baire category theorem shows that there is some $\mathcal{S}_n$ with nonempty interior. Thus for some $a_0 \in \mathcal{A}$ and some $\varepsilon > 0$, $((a_0 + a)b)^n = 0$ holds for all $a \in \mathcal{A}_\varepsilon$. Let $a \in \mathcal{A}_1$ and $\omega \in \mathcal{A}^*$ be arbitrary. Since the polynomial $\omega(((a_0 + \lambda a)b)^n)$ is zero for all $\lambda \in \mathbb{C}$ with $|\lambda| \leq \varepsilon$, it is zero for all $\lambda \in \mathbb{C}$. Since ω was arbitrary $((a_0 + \lambda a)b)^n$ is zero. Hence $a_0 + \lambda a$ belongs to $\mathcal{S}_n$ for all $a \in \mathcal{A}_1$ and $\lambda \in \mathbb{C}$. In other words $\mathcal{S}_n = \mathcal{A}$ and each element ab of the principal left ideal $\mathcal{A}b$ satisfies $(ab)^n = 0$. Proposition 4.4.10 now shows that $\mathcal{A}b$ is nilpotent. Since $(\mathcal{A}b)^N = \{0\}$ implies $(\mathcal{A}b\mathcal{A})^N \subseteq (\mathcal{A}b)^N\mathcal{A} = \{0\}$, $\mathcal{A}b\mathcal{A}$ is also nilpotent.

(b): For each $n \in \mathbb{N}$ let $\mathcal{S}_n$ be the closed set $\{a \in \mathcal{A} : a^n = 0\}$. Since $\mathcal{A}$ is nil, it equals $\cup_{n=1}^n \mathcal{S}_n$. Hence by the Baire category theorem there is some $a_0 \in \mathcal{A}$ and $\varepsilon > 0$ satisfying $(a_0 + a)^n = 0$ for all $a \in \mathcal{A}_\varepsilon$. Arguing as before, we conclude $\mathcal{A} = \mathcal{S}_n$. Proposition 4.4.10 shows again that $\mathcal{A}$ is nilpotent.

(c): Theorem 4.4.7 shows that $\mathcal{A}_B$ is a nil ideal which includes the sum of all nilpotent ideals. Hence (a) shows that $\mathcal{A}_B$ equals the sum of all nilpotent ideals which equals the sum of all nil left ideals. Proposition 4.4.2(c) shows that $\mathcal{A}/\mathcal{A}_B$ has no non-zero nil ideals.

(d): If $\mathcal{A}_B$ is closed, Theorem 4.4.6 and (b) above show that it is nilpotent. If it is nilpotent then there is some N satisfying $\mathcal{A}_B \subseteq \{a : a^N = 0\}$. Hence the closure of $\mathcal{A}_B$ is included in the same set and is thus nilpotent. Hence $\mathcal{A}_B$ is closed by Theorem 4.4.7.   □

4.5 The Brown-McCoy or Strong Radical

We will now discuss one more radical. This radical was defined by B. Brown and McCoy [1947] and is sometimes called the Brown–McCoy radical. Segal [1941] defined a unital algebra to be semisimple if its maximal ideals had intersection $\{0\}$. Later [1947a], he changed the term to strongly semisimple. Brown and McCoy originally defined this radical in terms of G-regular ideals which we introduce with a different name in Definition 4.5.5. Michael J. Meyer [1992d] establishes some deep analogies with the Jacobson radical.

This radical is closed in spectral algebras and is quite well behaved in Banach algebras. However, it is of less importance in functional analysis than the Jacobson radical. The Jacobson radical of familiar Banach algebras is usually a pathological portion. This is not true of the strong radical. For instance, if $\mathcal{H}$ is a separable Hilbert space, then the strong radical of $\mathcal{B}(\mathcal{H})$ is the ideal of compact operators.

4.5.1 Definition Let $\mathcal{A}$ be an algebra. The intersection of all maximal modular ideals of $\mathcal{A}$ will be called the *Brown–McCoy radical*, or *strong radical*, of $\mathcal{A}$ and will be denoted by $\mathcal{A}_M$. An algebra is called *strongly semisimple* if $\mathcal{A}_M$ is zero and is called *Brown–McCoy-radical* if $\mathcal{A}_M$ equals $\mathcal{A}$. The set of maximal modular ideals of an algebra $\mathcal{A}$ will be denoted by $\Xi_\mathcal{A}$. If $\mathcal{I}$ is a subset of $\mathcal{A}$, then $h^\Xi(\mathcal{I})$ will denote the *hull* of $\mathcal{I}$ in $\Xi_\mathcal{A}$: $h^\Xi(\mathcal{I}) = \{\mathcal{M} \in \Xi_\mathcal{A} : \mathcal{I} \subseteq \mathcal{M}\}$.

We follow our usual convention on the intersection of an empty family of sets, so that $\mathcal{A}_M$ is $\mathcal{A}$ if and only if $\mathcal{A}$ has no maximal modular ideals.

Then an algebra $\mathcal{A}$ is a Brown–McCoy radical algebra if and only if it has no maximal modular ideals and hence, by Theorem 2.4.6, if and only if it has no modular ideals at all.

We need a simple result analogous to Theorem 4.1.8(e) and the last sentence of Theorem 4.1.10.

4.5.2 Theorem   *Let $\mathcal{A}$ be an algebra and let $\mathcal{I}$ be an ideal of $\mathcal{A}$. The map*

$$\mathcal{M} \mapsto \mathcal{M} \cap \mathcal{I}$$

is a bijection of $\Xi_{\mathcal{A}} \setminus h^{\Xi}(\mathcal{I})$ onto $\Xi_{\mathcal{I}}$ and the map $\mathcal{M} \mapsto \mathcal{M}/\mathcal{I}$ is a bijection of $h^{\Xi}(\mathcal{I})$ onto $\Xi_{\mathcal{A}/\mathcal{I}}$.

Proof Let $\mathcal{M}$ be a maximal modular ideal of $\mathcal{A}$ which does not include $\mathcal{I}$. Then $\mathcal{I} + \mathcal{M}$ equals $\mathcal{A}$ by maximality. Thus $\mathcal{I}/(\mathcal{M} \cap \mathcal{I})$ which is isomorphic to $(\mathcal{I} + \mathcal{M})/\mathcal{M} = \mathcal{A}/\mathcal{M}$ is a simple unital algebra. Hence $\mathcal{M} \cap \mathcal{I}$ is a maximal modular ideal of $\mathcal{I}$. The last sentence of Theorem 4.1.10 shows that $\mathcal{M} \mapsto \mathcal{M} \cap \mathcal{I}$ is injective.

Now suppose $\mathcal{M}$ is a maximal modular ideal of $\mathcal{I}$. Let $\varphi \colon \mathcal{I} \to \mathcal{I}/\mathcal{M}$ be the natural map and let $e \in \mathcal{I}$ be a relative identity for $\mathcal{M}$ so that $\varphi(e)$ is the identity element in $\mathcal{I}/\mathcal{M}$. Consider the map $\psi \colon \mathcal{A} \to \mathcal{I}/\mathcal{M}$ defined by $\psi(a) = \varphi(eae)$ for all $a \in \mathcal{A}$. It is easy to check that ψ is a homomorphism. For instance for a, $b \in \mathcal{A}$, we have $\psi(ab) = \varphi(eabe) = \varphi(ea)\varphi(be) = \varphi(ea)\varphi(e)^2\varphi(be) = \varphi(eae)\varphi(ebe) = \psi(a)\psi(b)$. Also ψ maps $\mathcal{A}$ onto $\mathcal{I}/\mathcal{M}$ since any $a \in \mathcal{I}$ satisfies $\psi(a) = \varphi(eae) = \varphi(e)\varphi(a)\varphi(e) = \varphi(a)$. Thus the kernel of ψ, $\{a \in \mathcal{A} : eae \in \mathcal{M}\}$, is a maximal modular ideal of $\mathcal{A}$, and its intersection with $\mathcal{I}$ is just $\mathcal{M}$ since φ is just $\psi|_{\mathcal{I}}$. Therefore the map $\mathcal{M} \mapsto \mathcal{M} \cap \mathcal{I}$ is a surjection of $\Xi_{\mathcal{A}} \setminus h^{\Xi}(\mathcal{I})$ onto $\Xi_{\mathcal{I}}$.

For $\mathcal{M} \in h^{\Xi}(\mathcal{I})$ it is obvious that $\mathcal{M}/\mathcal{I}$ is a maximal modular ideal of $\mathcal{A}/\mathcal{I}$ with relative identity $e + \mathcal{I}$ when e is a relative identity for $\mathcal{M}$. If $\varphi \colon \mathcal{A} \to \mathcal{A}/\mathcal{I}$ is the natural map, then $\mathcal{N} \mapsto \varphi^{-1}(\mathcal{N})$ for $\mathcal{N} \in \Xi_{\mathcal{A}/\mathcal{I}}$ is obviously an inverse for $\mathcal{M} \mapsto \mathcal{M}/\mathcal{I}$ (for $\mathcal{M} \in h^{\Xi}(\mathcal{I})$) so that both maps are bijections.   □

The following result is the analogue of Theorems 4.3.2 and 4.4.8. It was essentially proved by Brown and McCoy [1947].

4.5.3 Theorem   *Any algebra $\mathcal{A}$ and any ideal $\mathcal{I}$ of $\mathcal{A}$ satisfy:*
(a) $\mathcal{I}_M = \mathcal{I} \cap \mathcal{A}_M$.
(b) $\mathcal{I} \subseteq \mathcal{A}_M$ *implies* $(\mathcal{A}/\mathcal{I})_M = \mathcal{A}_M/\mathcal{I}$.
(c) $\mathcal{A}/\mathcal{I}$ *strongly semisimple implies* $\mathcal{A}_M = \mathcal{I}_M$. *Thus in particular* $\mathcal{A}/\mathcal{A}_M$ *is strongly semisimple and* $\mathcal{A}_M$ *is Brown–McCoy radical.*

Proof The proof of Theorem 4.3.2 may be repeated, merely replacing "primitive" by "maximal modular" and "Theorem 4.1.8(e)" by "Theorem 4.5.2".
□

The following simple but important result was noted by Segal [1947a] for complete algebra norms.

4.5.4 Theorem *The strong radical is closed in any spectral or weakly spectral norm.*

Proof This follows from Theorem 2.4.7 and Proposition 2.2.17.   □

We now give some characterizations of the strong radical similar to the characterizations of the Jacobson radical in Theorem 4.3.6 and Proposition 4.3.13. They are due to Brown and McCoy [1947] and Michael J. Meyer [1992d]. Definition 2.1.19 introduced the notion of weakly invertible elements, but we deferred defining "weakly quasi-invertible elements". By the general convention on the use of the prefix "quasi-", an element $a \in \mathcal{A}$ should be weakly quasi-invertible if $1 - a$ is weakly invertible in $\mathcal{A}^1$. Weakly quasi-invertible elements were called G-regular by Brown and McCoy [1947] who used the criterion of Theorem 4.5.7 as the definition of their radical.

4.5.5 Definition An element a in an algebra $\mathcal{A}$ is called *weakly quasi-invertible* if the ideal

$$\mathcal{I}(a) = \{(b - ba) + (c - ac) + \sum_{i=1}^{n}(d_i e_i - d_i a e_i) :$$
$$n \in \mathbb{N}, d, c, d_i, e_i \in \mathcal{A} \quad \text{for} \quad i = 1, 2, \ldots, n\}$$

is all of $\mathcal{A}$. The set of such elements is denoted by $\mathcal{A}_{wqG}$. An ideal is called *weakly quasi-regular* if each of its elements is weakly quasi-invertible.

4.5.6 Proposition The following are equivalent for an element a in an arbitrary algebra $\mathcal{A}$.
  (a) a is weakly quasi-invertible.
  (b) a belongs to $\mathcal{I}(a)$.
  (c) $1 - a$ is weakly invertible in $\mathcal{A}^1$.
  (d) a is not the relative identity for any maximal modular ideal of $\mathcal{A}$.

Proof Let $\mathcal{J}$ be the ideal of $\mathcal{A}^1$ generated by $1 - a$ so (c) is equivalent to $\mathcal{J} = \mathcal{A}^1$. It is easy to check that $\mathcal{J} = \mathcal{I}(a) + \mathbb{C}(1 - a)$. Note also that a is a relative identity for an ideal if and only if the ideal includes $\mathcal{I}(a)$.
  (a) $\Rightarrow$ (b): Obvious.
  (b) $\Rightarrow$ (c): Since $1 - a \in \mathcal{J}$, (b) implies $1 = 1 - a + a \in \mathcal{J}$.
  (c) $\Rightarrow$ (d): If a is the relative identity for a maximal modular ideal $\mathcal{M}$ of $\mathcal{A}$, then $\mathcal{I}(a) \subseteq \mathcal{M}$ implies $\mathcal{J} \subseteq \mathcal{M} + \mathbb{C}(1 - a) \neq \mathcal{A}^1$.
  (d) $\Rightarrow$ (a): Theorem 2.4.6(d) shows that a is a relative identity for some maximal modular ideal which includes $\mathcal{I}(a)$ whenever $\mathcal{I}(a)$ is a proper subset of $\mathcal{A}$.   □

4.5.7 Theorem *Let $\mathcal{A}$ be an algebra. Then the strong radical $\mathcal{A}_M$ is a weakly quasi-regular ideal of $\mathcal{A}$ which includes all weakly quasi-regular ideals of $\mathcal{A}$. Hence it is the largest ideal $\mathcal{I}$ satisfying any (hence all) of the following equivalent conditions*

$$Sp^S_{\mathcal{A}/\mathcal{I}}(a+\mathcal{I}) \subseteq Sp^S_{\mathcal{A}}(a) \subseteq Sp^S_{\mathcal{A}/\mathcal{I}}(a+\mathcal{I}) \cup \{0\} \qquad \forall \ a \in \mathcal{A};$$

$$a + \mathcal{I} \in (\mathcal{A}/\mathcal{I})_{wqG} \Leftrightarrow a \in \mathcal{A}_{wqG} \qquad \forall \ a \in \mathcal{A};$$

$$Sp^S(b+a) = Sp^S(b) \qquad \forall \ b \in \mathcal{A}; a \in \mathcal{I};$$

$$Sp^S(a) = \{0\} \qquad \forall \ a \in \mathcal{I};$$

$$\rho^S(b+a) = \rho^S(b) \qquad \forall \ b \in \mathcal{A}; a \in \mathcal{I};$$

$$\rho^S(a) = 0 \qquad \forall \ a \in \mathcal{I}.$$

Proof If a is an element of a weakly quasi-regular ideal $\mathcal{I}$, then every element in $\mathcal{A}^1 a \mathcal{A}^1 \subseteq \mathcal{I}$ is weakly quasi-regular. Thus an element $a \in \mathcal{A}$ belongs to some weakly quasi-regular ideal if and only if $\mathcal{A}^1 a \mathcal{A}^1$ is a weakly quasi-regular ideal. Therefore the theorem will be proved if we can show that $a \in \mathcal{A}$ belongs to $\mathcal{A}_M$ if and only if $\mathcal{A}^1 a \mathcal{A}^1$ is a weakly quasi-regular ideal.

Suppose $\mathcal{A}^1 a \mathcal{A}^1$ is not weakly quasi-regular. Then there is some $b \in \mathcal{A}^1 a \mathcal{A}^1$ which does not belong to $\mathcal{I}(b)$. Thus $\mathcal{I}(b)$ is a proper modular ideal and by applying Zorn's lemma to the set of ideals including $\mathcal{I}(b)$ but not containing b we see that b is the relative identity of some maximal modular ideal $\mathcal{M}$. Thus b does not belong to $\mathcal{M}$ and hence b does not belong to $\mathcal{A}_M$. However this implies $a \notin \mathcal{A}_M$ since $a \in \mathcal{A}_M$ would imply $b \in \mathcal{A}^1 a \mathcal{A}^1 \subseteq \mathcal{A}_M$.

Conversely, suppose $\mathcal{A}^1 a \mathcal{A}^1$ is a weakly quasi-regular ideal of $\mathcal{A}$ and $\mathcal{M}$ is a maximal modular ideal of $\mathcal{A}$. Then $(\mathcal{A}/\mathcal{M})(a+\mathcal{M})(\mathcal{A}/\mathcal{M})$ is a weakly quasi-regular ideal of $\mathcal{A}/\mathcal{M}$. However since $\mathcal{A}/\mathcal{M}$ is simple, either a belongs to $\mathcal{M}$ or $(\mathcal{A}/\mathcal{M})(a+\mathcal{M})(\mathcal{A}/\mathcal{M})$ equals $\mathcal{A}/\mathcal{M}$. The latter is impossible since $1 \in \mathcal{A}/\mathcal{M}$ cannot be weakly quasi-regular. Thus a belongs to $\mathcal{M}$. Since $\mathcal{M}$ was an arbitrary maximal modular ideal of $\mathcal{A}$, a belongs to $\mathcal{A}_M$.

A non-zero number $\lambda \in \mathbb{C}$ belongs to the strong spectrum if and only if $\lambda^{-1}a$ is weakly quasi-regular. If an element a belongs to a weakly quasi-regular ideal, then λa is weakly quasi-regular for all $\lambda \in \mathbb{C}$, so $Sp^S(a) = \{0\}$. This, together with the arguments in the proof of Theorem 4.3.6, establishes the second sentence. □

4.5.8 Proposition *In any spectral algebra $\mathcal{A}$, the strong radical is the perturbation set in $\mathcal{A}$ of the set $\mathcal{A}^1_{wG}$ of weakly regular elements in the unitization:*

$$\mathcal{A}_M = \{a \in \mathcal{A} : a + \mathcal{A}^1_{wG} \subseteq \mathcal{A}^1_{wG}\}.$$

Proof Denote the perturbation set by $\mathcal{P}$. Proposition 2.1.20(c) shows that $\mathcal{A}^1_{wG}$ is invariant under left or right multiplication by invertible elements. Hence the proof given for Proposition 4.3.13 shows that $\mathcal{P}$ is an ideal. The third condition listed in Theorem 4.5.7 shows $\mathcal{A}_M \subseteq \mathcal{P}$ in any algebra. Since $\mathcal{P}$ is an ideal which satisfies this condition, Theorem 4.5.7 shows $\mathcal{P} \subseteq \mathcal{A}_M$ establishing their equality.   □

Comparison of the Three Radicals

We now compare the three radicals we have considered. The following inclusions were noted as soon as the various radicals were defined. It was also noted right away that these three radicals agree with the nil radical in any ring satisfying the descending chain condition on left ideals. All semisimple algebras satisfying the descending chain condition are finite-dimensional and, as shown by Elie Cartan [1898], are finite sums of full matrix algebras over the complex field.

4.5.9 Theorem *Let $\mathcal{A}$ be an algebra. The following inclusions hold:*

$$\mathcal{A}_B \subseteq \mathcal{A}_J \subseteq \mathcal{A}_M.$$

If $\mathcal{A}$ is commutative $\mathcal{A}_J, \mathcal{A}_M$ and $\{a \in \mathcal{A} : Sp(a) = \{0\}\}$ are equal.

Proof Theorem 4.1.9 asserts that every maximal modular ideal is primitive and every primitive ideal in a commutative algebra is maximal modular. Theorem 4.1.8 asserts that every primitive ideal is prime. Thus the present theorem follows since the Baer radical is the intersection of all prime ideals, the Jacobson radical is the intersection of all primitive ideals, and the strong radical is the intersection of all maximal modular ideals.   □

Let $\mathcal{A}$ be a semiprime algebra so that the ideal $\{0\}$ is the intersection of some set $B \subseteq P_\mathcal{A}$ of prime ideals of $\mathcal{A}$. Theorem 4.4.3 and the equations $(\mathcal{A}_{LA})^2 = (\mathcal{A}_{RA})^2 = \{0\}$ imply $\mathcal{A}_{LA} = \mathcal{A}_{RA} = \{0\}$. Hence the regular homomorphism of $\mathcal{A}$ into its double centralizer algebra $\mathcal{D}(\mathcal{A})$ is an injection so we may regard $\mathcal{A}$ as an ideal in $\mathcal{D}(\mathcal{A})$. For any prime ideal $\mathcal{P}$ in $\mathcal{A}$ the unique prime ideal $\overline{\mathcal{P}}$ of $\mathcal{D}(\mathcal{A})$ satisfying $\overline{\mathcal{P}} \cap \mathcal{A} = \mathcal{P}$ which is guaranteed by Theorem 4.1.10 is given by

$$\overline{\mathcal{P}} = \{(L, R) \in \mathcal{D}(\mathcal{A}) : \mathcal{A}(L, R)\mathcal{A} = \mathcal{A}L(\mathcal{A}) = R(\mathcal{A})\mathcal{A} \subseteq \mathcal{P}\}.$$

Hence if $\tilde{P}$ is any subset of $P_\mathcal{A}$ we have

$$\begin{aligned}
\bigcap_{\mathcal{P} \in \tilde{P}} \overline{\mathcal{P}} &= \{(L, R) \in \mathcal{D}(\mathcal{A}) : \mathcal{A}L(\mathcal{A}) = R(\mathcal{A})\mathcal{A} \subseteq \bigcap_{\mathcal{P} \in \tilde{P}} \mathcal{P}\} \\
&= \{(L, R) \in \mathcal{D}(\mathcal{A}) : L(\mathcal{A}) + R(\mathcal{A}) \subseteq \bigcap_{\mathcal{P} \in \tilde{P}} \mathcal{P}\}
\end{aligned}$$

where the last equality follows from Theorem 4.4.3 since $L(\mathcal{A})$ and $R(\mathcal{A})$ are one-sided ideals of $\mathcal{A}$ satisfying $L(\mathcal{A})^2 \subseteq \mathcal{R}$ and $R(\mathcal{A})^2 \subseteq \mathcal{R}$ for the semiprime ideal $\mathcal{R} = \bigcap_{\mathcal{P} \in \tilde{P}} \mathcal{P}$. In particular, if $\tilde{P}$ includes B, then we conclude $L(\mathcal{A}) \subseteq \mathcal{A}_{RA} = \{0\}$ and $R(\mathcal{A}) \subseteq \mathcal{A}_{LA} = \{0\}$ for each (L, R) in $\mathcal{D}(\mathcal{A})$ so that $\bigcap_{\mathcal{P} \in \tilde{P}} \overline{\mathcal{P}}$ is zero. If $\tilde{P}$ is $\langle\ P_{\mathcal{A}}\ /\ \Pi_{\mathcal{A}}\ /\ \Xi_{\mathcal{A}}\ \rangle$, then for each $\mathcal{P} \in \tilde{P}$, $\overline{\mathcal{P}}$ belongs to $\langle\ P_{\mathcal{D}(\mathcal{A})}\ /\ \Pi_{\mathcal{D}(\mathcal{A})}\ /\ \Xi_{\mathcal{D}(\mathcal{A})}\ \rangle$ according to Theorem $\langle$ 4.1.10 / 4.1.8(e) / 4.5.2 $\rangle$. Hence we have proved:

4.5.10 Theorem *If $\mathcal{A}$ is semiprime, semisimple, or strongly semisimple, then $\mathcal{D}(\mathcal{A})$ has the corresponding property.*

If $\mathcal{A}_{LA} \cap \mathcal{A}_{RA}$ is zero, then $\mathcal{D}(\mathcal{A})$ satisfies:

$$\mathcal{D}(\mathcal{A})_B \quad \subseteq \quad \{(L, R) \in \mathcal{D}(\mathcal{A}) : L(\mathcal{A}) + R(\mathcal{A}) \subseteq \mathcal{A}_B\};$$

$$\mathcal{D}(\mathcal{A})_J \quad \subseteq \quad \{(L, R) \in \mathcal{D}(\mathcal{A}) : L(\mathcal{A}) + R(\mathcal{A}) \subseteq \mathcal{A}_J\};$$

$$\mathcal{D}(\mathcal{A})_M \quad \subseteq \quad \{(L, R) \in \mathcal{D}(\mathcal{A}) : L(\mathcal{A}) + R(\mathcal{A}) \subseteq \mathcal{A}_M\}.$$

Some of these considerations are found in Johnson [1964a], Section 4.

4.6   Subdirect Products

The reader has surely noticed the many similarities between the three radicals we have discussed. The next two sections will be devoted to a discussion of these similarities. In Section 4.7 we will discuss a general abstract theory which includes not only the radicals considered here, but also the reducing ideal and the Leptin radical, which will play a fundamental role in Volume II. There are many other examples of radicals which we will not discuss. Before we turn to this very general setting we consider the subdirect product decomposition related to the three radicals which we have already discussed. Subdirect products were introduced in Section 1.3. They have a long history which is summarized in some detail by McCoy [1947].

The Jacobson radical is the intersection of primitive ideals, the Baer radical is the intersection of prime ideals, and the strong radical is the intersection of maximal modular ideals. These three intersection results lead to corresponding subdirect product decompositions. Recall that an algebra is called prime or primitive if the ideal $\{0\}$ is prime or primitive respectively. Hence an algebra is prime if and only if it has no non-zero ideals which divide $\{0\}$. (It is not hard to see that a commutative algebra is prime if and only if it has no non-zero elements which are zero divisors. Prime ideals in commutative algebras (over the rationals) were essentially considered by Ernst Eduard Kummer.) An algebra is primitive if and only if it has a faithful irreducible representation. Theorem 4.2.13 shows that a spectral algebra is primitive if and only if it has a faithful strictly dense representation. An algebra $\mathcal{A}$ is called *simple* if it has no ideals except $\{0\}$ and $\mathcal{A}$. Obviously an algebra is both unital and simple if and only if $\{0\}$ is a maximal modular ideal. Simple unital algebras are the constituents of the subdirect product decomposition of strongly semisimple algebras.

Recall that Corollary 4.3.10 asserts that any spectral semi-norm on a semisimple algebra is a norm. Theorem 4.5.7 shows that the same is true for a strongly semisimple algebra. Hence without loss of generality we speak of spectral norms instead of spectral semi-norms in the next theorem.

4.6.1 Theorem *An algebra $\mathcal{A}$ is semiprime, semisimple, or strongly semisimple if and only if it is isomorphic to a subdirect product of prime algebras, primitive algebras, or simple unital algebras, respectively. If $\mathcal{A}$ is a spectral normed algebra which is either semisimple or strongly semisimple, then the subdirect product representation may be chosen normed with the isomorphism contractive.*

Proof Let $\{\mathcal{I}^\alpha : \alpha \in A\}$ be the set of prime, primitive, or maximal modular ideals of $\mathcal{A}$ depending on whether $\mathcal{A}$ is semiprime, semisimple, or strongly semisimple. Then $\cap\{\mathcal{I}^\alpha : \alpha \in A\}$ is $\{0\}$. So the map $a \mapsto \hat{a}$ where $\hat{a}(\alpha) = a + \mathcal{I}^\alpha$ is an isomorphism of $\mathcal{A}$ onto a subdirect product of $\{\mathcal{A}/\mathcal{I}^\alpha : \alpha \in A\}$. This proves the necessity of all the purely algebraic conditions. If $\mathcal{A}$ is a spectral normed algebra, then each $\mathcal{I}^\alpha$ in the decomposition by primitive or maximal modular ideals is closed. Thus $a \mapsto \hat{a}$ is a contractive isomorphism onto the normed subdirect product when each $\mathcal{A}/\mathcal{I}^\alpha$ has its quotient norm.

Suppose $\mathcal{A}$ is a subdirect product of $\{\mathcal{A}^\alpha : \alpha \in A\}$, where each $\mathcal{A}^\alpha$ is a prime algebra, a primitive algebra or a unital simple algebra. Then the ideals $\mathcal{I}^\alpha = \{a \in \mathcal{A} : a(\alpha) = 0\}$ satisfy $\mathcal{A}^\alpha = \mathcal{A}/\mathcal{I}^\alpha$ and $\cap_{\alpha \in A}\mathcal{I}^\alpha = \{0\}$, so that $\mathcal{A}$ is semiprime, semisimple, or strongly semisimple respectively. This proves the sufficiency. □

A Subdirect Product Representation for Double Centralizer Algebras

We remark on an alternative direct product decomposition in terms of simple unital direct factors which is sometimes available. Let $\mathcal{A}$ be an algebra and let $\mathcal{I}$ be an ideal of $\mathcal{A}$. Theorem 4.5.2 shows that the map $\mathcal{M} \mapsto \mathcal{M} \cap \mathcal{I}$ is a bijection of $\Xi_\mathcal{A} \setminus h^\Xi(\mathcal{I})$ onto $\Xi_\mathcal{I}$. If $\mathcal{M}$ is any ideal in $\Xi_\mathcal{I}$ let $\overline{\mathcal{M}} \in \Xi_\mathcal{A} \setminus h^\Xi(\mathcal{I})$ be the corresponding ideal satisfying $\overline{\mathcal{M}} \cap \mathcal{I} = \mathcal{M}$. Since $\overline{\mathcal{M}}$ is maximal and does not include $\mathcal{I}$ we have $\mathcal{A}/\overline{\mathcal{M}} = (\overline{\mathcal{M}} + \mathcal{I})/\overline{\mathcal{M}} \simeq \mathcal{I}/\overline{\mathcal{M}} \cap \mathcal{I} = \mathcal{I}/\mathcal{M}$. Hence the direct factors in $\prod_{\overline{\mathcal{M}} \in \Xi_\mathcal{A} \setminus h^\Xi(\mathcal{I})}(\mathcal{A}/\overline{\mathcal{M}})$ are isomorphic to the corresponding direct factors in $\prod_{\mathcal{M} \in \Xi_\mathcal{I}}(\mathcal{I}/\mathcal{M})$. Therefore there is a homomorphism of $\mathcal{A}$ into $\prod_{\mathcal{M} \in \Xi_\mathcal{I}}(\mathcal{A}/\overline{\mathcal{M}})$ with kernel $\cap_{\mathcal{M} \in \Xi_\mathcal{I}}\overline{\mathcal{M}} = \cap_{\overline{\mathcal{M}} \in \Xi_\mathcal{A} \setminus h^\Xi(\mathcal{I})}\overline{\mathcal{M}}$. This is a subdirect product decomposition of $\mathcal{A}$ over a smaller space of maximal modular ideals. In general the kernel of the representation will be larger.

Now replace $\mathcal{A}$ and $\mathcal{I}$ in the last paragraph by $\mathcal{D}(\mathcal{A})$ and $\mathcal{A}$, respectively, where $\mathcal{A}$ is a strongly semisimple algebra. Theorem 4.5.8 shows $\cap_{\mathcal{M} \in \Xi_{\mathcal{A}}} \overline{\mathcal{M}} = \cap_{\overline{\mathcal{M}} \in \Xi_{\mathcal{D}(\mathcal{A})} \setminus h^{\Xi}(\mathcal{A})} \overline{\mathcal{M}} = \{0\}$ so that the subdirect product decomposition is now an isomorphism. Hence we have established the first portion of the next result.

4.6.2 Corollary *If $\mathcal{A}$ is strongly semisimple, then there is a natural isomorphism of $\mathcal{D}(\mathcal{A})$ onto a subdirect product included in $\Pi_{\mathcal{M} \in \Xi_{\mathcal{A}}}(\mathcal{A}/\mathcal{M})$. This subdirect product is given explicitly by:*

$$(L, R) \mapsto (\mathcal{M} \mapsto L(e_{\mathcal{M}})) \qquad \forall \ (L, R) \in \mathcal{D}(\mathcal{A}); \mathcal{M} \in \Xi_{\mathcal{A}}$$

where for each $\mathcal{M} \in \Xi_{\mathcal{A}}$ $e_{\mathcal{M}}$ is a relative identity for $\mathcal{M}$.

Proof It only remains to justify the explicit formula. The calculation $a((L, R) - L(e_{\mathcal{M}}))b = a(L(b) - L(e_{\mathcal{M}})b) = R(a)(b - e_{\mathcal{M}}b) \in \mathcal{M}$ for all $a, b \in \mathcal{A}$ shows that $(L, R) = L(e) + ((L, R) - L(e_{\mathcal{M}}))$ is a decomposition of $(L, R) \in \mathcal{D}(\mathcal{A})$ corresponding to $\mathcal{D}(\mathcal{A}) = \mathcal{A} + \overline{\mathcal{M}}$. This shows that the subdirect product decomposition is given by the indicated formula. $\qquad \square$

A Characterization of Spectral Algebras

We wish to improve Theorem 4.6.1 to give a characterization of spectral algebras. For this we need to introduce a definition.

4.6.3 Definition Let $\mathcal{A}$ be a subdirect product of algebras $\{\mathcal{A}^{\alpha} : \alpha \in A\}$. Then $\mathcal{A}$ is called *ample* if an element $a \in \mathcal{A}$ is quasi-invertible whenever $a(\alpha)$ is quasi-invertible in $\mathcal{A}^{\alpha}$ for each $\alpha \in A$.

4.6.4 Theorem *The following are equivalent for an algebra $\mathcal{A}$.*

(a) *$\mathcal{A}$ is a spectral algebra.*

(b) *$\mathcal{A}/\mathcal{A}_J$ can be embedded as a dense subalgebra of a semisimple Banach algebra $\mathcal{B}$ such that the spectral radius of $\mathcal{A}/\mathcal{A}_J$ is the restriction of the spectral radius of $\mathcal{B}$.*

(c) *$\mathcal{A}/\mathcal{A}_J$ is isomorphic to an ample, normed, subdirect product of strictly dense spectral normed algebras of bounded operators on normed linear spaces.*

Proof $(a) \Rightarrow (b)$: Let σ' be an arbitrary spectral semi-norm on $\mathcal{A}$. Theorem 2.2.14(d) and Corollary 4.3.10 show that σ' induces a spectral norm $|||\cdot|||$ on $\mathcal{A}/\mathcal{A}_J$. Let $(\overline{\mathcal{A}}, |||\cdot|||)$ be the completion of $(\mathcal{A}/\mathcal{A}_J, |||\cdot|||)$. Let $\varphi: \mathcal{A} \to \mathcal{A}/\mathcal{A}_J$ and $\psi: \overline{\mathcal{A}} \to \overline{\mathcal{A}}/\overline{\mathcal{A}}_J$ be the natural maps. Any $a \in \mathcal{A}$ satisfies $\rho_{\varphi(\mathcal{A})}\varphi(a) = |||\varphi(a)|||^{\infty} = \rho_{\overline{\mathcal{A}}}(\varphi(a))$ so we simply write $\rho(\varphi(a))$ for the spectral radius of $\varphi(a)$. Define $\sigma: \mathcal{A} \to \mathbb{R}_+$ by $\sigma(a) = \inf\{|||\varphi(a) + b||| : b \in \overline{\mathcal{A}}_J\}$. Clearly σ is an algebra semi-norm on $\mathcal{A}$. (In fact it is the quotient norm of $\overline{\mathcal{A}}/\overline{\mathcal{A}}_J$ pulled back to $\mathcal{A}$ through ψ and φ.) Each $a \in \mathcal{A}$ satisfies $\sigma(a) \geq \rho(\psi(\varphi(a))) = \rho(\varphi(a)) = \rho(a)$ by Theorem 4.3.6(d). Hence σ is a spectral semi-norm.

A second application of Corollary 4.3.10 shows that σ induces a norm on $\mathcal{A}/\mathcal{A}_J$. The completion of $\mathcal{A}/\mathcal{A}_J$ in this norm is clearly $\overline{\mathcal{A}}/\overline{\mathcal{A}}_J$. This is the semisimple algebra $\mathcal{B}$ of (b). By Theorem 4.3.6(d) and previous remarks we get $\rho(\varphi(a)) = \rho_{\overline{\mathcal{A}}/\overline{\mathcal{A}}_J}(\psi(\varphi(a))) = \rho_{\mathcal{B}}(\psi(\varphi(a)))$, where $\psi|_{\varphi(\mathcal{A})}$ is now the embedding of $\varphi(\mathcal{A}) = \mathcal{A}/\mathcal{A}_J$ into $\mathcal{B}$.

(b) $\Rightarrow$ (a): Corollary 2.5.10 shows that $\mathcal{A}/\mathcal{A}_J$ is a spectral algebra. Theorem 4.3.6(d) shows $\rho_{\mathcal{A}}(a) = \rho_{\mathcal{A}/\mathcal{A}_J}(a + \mathcal{A}_J) \leq \|a + \mathcal{A}_J\| \leq \|a\|$ for any $a \in \mathcal{A}$. Hence $\mathcal{A}$ is a spectral algebra by Theorem 2.2.5.

(a) $\Rightarrow$ (c): Consider the subdirect product decomposition for $\mathcal{A}/\mathcal{A}_J$ described in Theorem 4.6.1. Choose a spectral semi-norm σ on $\mathcal{A}$. For each primitive ideal $\mathcal{P}, \sigma$ induces a spectral norm on $\mathcal{A}/\mathcal{P}$ by Proposition 4.2.6 and Theorem 2.2.5. Let $\mathcal{P}$ be the kernel of the irreducible representation T on $\mathcal{X}$. Then T is strictly dense by Theorem 4.2.13 and $\mathcal{X}$ can be normed so that T is a continuous representation by Theorem 4.2.8. Theorem 4.1.8(d) shows that the subdirect product decomposition is ample.

(c) $\Rightarrow$ (a): The norm $\|a\| = \sup\{\|a(\alpha)\| : \alpha \in A\}$ of $\mathcal{A}$ defined by any ample, normed, subdirect product decomposition is a spectral norm by Theorem 2.2.5 since $\|a\| < 1$ implies $a \in \mathcal{A}_{qG}$. $\square$

4.7   Categorical Theory of Radicals

We will now investigate some more fundamental common properties of all radicals. A number of general theories of radicals have been advanced: e.g., Shimshon Amitsur [1952], Vladimir A. Andrunakievič [1961], Divinsky [1965], Gray [1970], Aleksandr Gennadievič Kuroš [1953], Adam Suliński [1958], Barry James Gardner [1989]. The theory which most conveniently covers the three radicals introduced in this section and the radicals which will be introduced in Volume II is that of Mary W. Gray [1970]. In order to give this theory we must introduce a small part of the language of categories and express some of the results of this section in that language. Many aspects of the subject will be ignored. However, we will show that several of the categories which we study in these notes are *semiabelian*. This type of category was introduced by Gray in part to give a setting for a very general theory of radicals [1967], [1970]. For further information on the more traditional approach to radicals of algebras and rings, see Divinsky [1965].

We give only a superficial introduction to category theory in functional analysis here. For more details see Boris Samuilovič Mitjagin and Albert Solomonovič Švarc [1964], Sten Kaijser [1976], A. W. M. Graven [1979], Yuriĭ Vasilyevič Selivanov [1978], Michael Grosser [1979a], Johann Cigler, Viktor Losert and Peter Michor [1979], Michor [1978] and Alexandr Yakovlevič Helemskiĭ [1989b].

4.7.1 Definition A *category* **C** consists of:

(a) A class **C** of *objects*.

(b) For each ordered pair $(\mathcal{A}, \mathcal{B})$ of objects from **C** a set $Hom_{\mathbf{C}}(\mathcal{A}, \mathcal{B})$.

(c) For each ordered triple $(\mathcal{A}, \mathcal{B}, \mathcal{C})$ of objects from **C** a map (called *composition*)

$$(\circ)\colon Hom_{\mathbf{C}}(\mathcal{B}, \mathcal{C}) \times Hom_{\mathbf{C}}(\mathcal{A}, \mathcal{B}) \to Hom_{\mathbf{C}}(\mathcal{A}, \mathcal{C}).$$

These are subject to the following restrictions:

(d) Each ordered quadruple $(\mathcal{A}, \mathcal{B}, \mathcal{C}, \mathcal{D})$ of objects from **C** and each $\varphi \in Hom_{\mathbf{C}}(\mathcal{A}, \mathcal{B}), \psi \in Hom_{\mathbf{C}}(\mathcal{B}, \mathcal{C})$ and $\theta \in Hom_{\mathbf{C}}(\mathcal{C}, \mathcal{D})$ satisfy $\theta \circ (\psi \circ \varphi) = (\theta \circ \psi) \circ \varphi$.

(e) For each object $\mathcal{A}$ in **C** there is an element $I_{\mathcal{A}} \in Hom_{\mathbf{C}}(\mathcal{A}, \mathcal{A})$ such that each object $\mathcal{B}$ in **C**, each $\varphi \in Hom_{\mathbf{C}}(\mathcal{A}, \mathcal{B})$ and each $\psi \in Hom_{\mathbf{C}}(\mathcal{B}, \mathcal{A})$ satisfy

$$I_{\mathcal{A}} \circ \psi = \psi \quad \text{and} \quad \varphi \circ I_{\mathcal{A}} = \varphi.$$

An object 0 in a category **C** is called a *zero object* if for each object $\mathcal{A}$ in **C**, $Hom_{\mathbf{C}}(\mathcal{A}, 0)$ and $Hom_{\mathbf{C}}(0, \mathcal{A})$ are both singletons. If $(\mathcal{A}, \mathcal{B})$ is an ordered pair of objects in a category **C**, an element $\theta \in Hom_{\mathbf{C}}(\mathcal{A}, \mathcal{B})$ is called an *isomorphism* if there exists an element $\theta^{-1} \in Hom_{\mathbf{C}}(\mathcal{B}, \mathcal{A})$ satisfying $\theta^{-1} \circ \theta = I_{\mathcal{A}}$ and $\theta \circ \theta^{-1} = I_{\mathcal{B}}$. Objects $\mathcal{A}$ and $\mathcal{B}$ are said to be *isomorphic* if $Hom_{\mathbf{C}}(\mathcal{A}, \mathcal{B})$ contains an isomorphism.

The elements of $Hom_{\mathbf{C}}(\mathcal{A}, \mathcal{B})$ will be called *morphisms*.

From now on we write $Hom(\mathcal{A}, \mathcal{B})$ instead of $Hom_{\mathbf{C}}(\mathcal{A}, \mathcal{B})$ if there is no danger of confusion. It is convenient to insist that the sets $Hom(\mathcal{A}, \mathcal{B})$ in a given category be disjoint. We make this requirement (although our notation for zero-maps is at variance with it). Thus for any $\varphi \in Hom(\mathcal{A}, \mathcal{B})$ and $\psi \in Hom(\mathcal{C}, \mathcal{D})$, $\varphi \circ \psi$ is defined (uniquely) if and only if $\mathcal{D} = \mathcal{A}$.

In all the categories which we will consider, the class of objects will be so large that it is a proper class rather than a set, the objects themselves will be sets (with additional structure) and each set $Hom(\mathcal{A}, \mathcal{B})$ will consist of maps with domain $\mathcal{A}$ and range in $\mathcal{B}$. Composition of functions will agree with composition as defined in the category. These are the most commonly considered types of categories. All our categories will also contain zero objects.

It is easy to see that for each object $\mathcal{A}$ in any category there is only one element in $Hom(\mathcal{A}, \mathcal{A})$ with the properties of $I_{\mathcal{A}}$. Furthermore, if $\mathcal{A}$ and $\mathcal{B}$ are objects and $\theta \in Hom(\mathcal{A}, \mathcal{B})$ is an isomorphism, then θ^{-1} is uniquely determined and is also an isomorphism.

Similarly any two zero-objects in a category are isomorphic. Thus there will be no confusion if we denote every zero object by 0 and also use the symbol 0 to denote the unique element of $Hom(0, \mathcal{A})$ and $Hom(\mathcal{A}, 0)$ for all $\mathcal{A}$ in the category. Finally, if $\mathcal{A}$ and $\mathcal{B}$ are any two objects in a category and $f \in Hom(\mathcal{A}, \mathcal{B})$ can be written as $f = 0 \circ 0$ for $0 \in Hom(\mathcal{A}, 0)$ and $0 \in Hom(0, \mathcal{B})$, then we denote f by 0 also. Then $0 \in Hom(\mathcal{A}, \mathcal{B})$ is uniquely determined independent of the choice of zero objects used in its definition. The notation has the advantage of making $f \circ 0 = 0$ and $0 \circ f = 0$ hold whenever the composition is defined.

We will now introduce the examples with which we are mainly concerned in this chapter. In each case we define $Hom(\mathcal{A}, \mathcal{B})$ for an ordered pair $(\mathcal{A}, \mathcal{B})$ of objects, merely by a qualitative description of the kind of maps which belong to $Hom(\mathcal{A}, \mathcal{B})$. If these maps are identified (as usual) with their graphs in the Cartesian product $\mathcal{A} \times \mathcal{B}$, then the sets $Hom(\mathcal{A}, \mathcal{B})$ will not be disjoint. Disjointness may be achieved by an easy indexing scheme but we ignore this technicality.

4.7.2 Definition   The following table describes seven categories.

Symbol	Objects	Morphisms
A	algebras	homomorphisms
SA	spectral algebras	homomorphisms
SSA	spectral semi-normed algebras	continuous homomorphisms
NA	normed algebras	continuous homomorphisms
BA	Banach algebras	continuous homomorphisms
GNA	normed algebras	contractive homomorphisms
GBA	Banach algebras	contractive homomorphisms

These categories are called, respectively, the *category of algebras, the category of spectral algebras, the category of spectral semi-normed algebras, the category of normed algebras, the category of Banach algebras, the geometric category of normed algebras, and the geometric category of Banach algebras.*

Example 4.8.11 will show a substantial difference between the topological and geometric categories of Banach algebras. We now introduce categorically defined kernels, cokernels, images, and coimages. The verifications which are left to the reader are all easy.

A *monomorphism* k in a category is simply a morphism which can be cancelled from the left of an equation (*i.e.*, $k \circ f = k \circ g \Rightarrow f = g$). A *subobject* in a category is an equivalence class of monomorphisms, where k_1 and k_2 are *equivalent* if there exist morphisms f and g satisfying $k_1 = k_2 \circ f$ and $k_2 = k_1 \circ g$. In particular a subobject is not an object in the category, and although it corresponds to a monomorphic image of an object it need not itself correspond to any object. For instance, the embedding of $C([0,1])$ in $L^1([0,1])$ defines a subobject of $L^1([0,1])$ in each of the two categories of Banach algebras listed in the last definition. However, since the image of $C([0,1])$ is not closed in $L^1([0,1])$ it does not correspond to any object in these categories. This complicates the application of this definition to such examples as the category of Banach algebras.

Let k_1 and k_2 be monomorphisms representing two subobjects $\mathcal{R}_1$ and $\mathcal{R}_2$. Then $\mathcal{R}_2$ is said to *include* $\mathcal{R}_1$ if there is a monomorphism k satisfying $k_2 \circ k = k_1$. A subobject is said to be a *zero subobject* if one (and hence all) of its representing monomorphisms is a zero morphism.

A morphism f is said to *have a kernel* if there is some monomorphism k satisfying:

(a) $f \circ k = 0$;

(b) If g is any morphism satisfying $f \circ g = 0$, then there is a unique morphism h satisfying $k \circ h = g$.

The subobject which is the class of all monomorphisms equivalent to k (which is also the class of all monomorphisms satisfying (a) and (b)) is called the *kernel of f*.

In all of the categories listed in Definition 4.7.2, every morphism has a kernel and the kernel of $f \in Hom(\mathcal{A}, \mathcal{B})$ is simply (the subobject represented by the injection into $\mathcal{A}$ of) the set theoretic kernel $\ker(f)$ of f. Thus in the first three categories any ideal is a kernel and in the last four categories any closed ideal is a kernel.

An *epimorphism u* in a category is a morphism which can be cancelled from the right of an equation (*i.e.*, $f \circ u = g \circ u \Rightarrow f = g$). Note that in the last five categories listed in Definition 4.7.2 any morphism with dense range is an epimorphism. A *quotient object* is an equivalence class of epimorphisms where u_1 and u_2 are equivalent if there exist morphisms f and g satisfying $u_1 = f \circ u_2$ and $u_2 = g \circ u_1$.

A morphism f is said to *have a cokernel* if there is an epimorphism u satisfying:

(a) $u \circ f = 0$;

(b) If g is any morphism satisfying $g \circ f = 0$, then there is a unique morphism h satisfying $h \circ u = g$.

The quotient object which is the class of all epimorphisms equivalent to u (which is also the class of all epimorphisms satisfying (a) and (b)) is called the *cokernel of f*.

In the categories **A** and **SA**, the cokernel of $f \in Hom(\mathcal{A}, \mathcal{B})$ is represented by

$$\mathcal{B} \mapsto \mathcal{B}/ \text{ ideal generated by } f(\mathcal{A}).$$

In the categories **SSA**, **NA**, **BA**, **GNA** and **GBA**, the cokernel of $f \in Hom(\mathcal{A}, \mathcal{B})$ is represented by

$$\mathcal{B} \mapsto \mathcal{B}/ \text{ closed ideal generated by } f(\mathcal{A}).$$

A morphism f is said to *have an image* if there is some monomorphism i satisfying:

(a) There is a morphism g satisfying $i \circ g = f$;

(b) If j is a monomorphism such that there is a morphism h satisfying $j \circ h = f$, then there is a monomorphism k satisfying $j \circ k = i$.

The subobject which is the class of all monomorphisms equivalent to i (which is also the class of all monomorphisms satisfying (a) and (b)) is called the *image of f*.

In all of the categories of Definition 4.7.2, the image of $f \in Hom(\mathcal{A}, \mathcal{B})$ is represented by the map

$$\mathcal{A}/\ker(f) \mapsto \mathcal{B}$$

induced by f.

A morphism f is said to *have a coimage* if there is an epimorphism u satisfying:

(a) There is a morphism g satisfying $g \circ u = f$;

(b) If v is an epimorphism such that there is a morphism h satisfying $h \circ v = f$, then there is an epimorphism w satisfying $w \circ v = u$.

The quotient object which is the class of all epimorphisms equivalent to u (which is also the class of all epimorphisms satisfying (a) and (b)) is called the *coimage of f*.

The coimage of $f \in Hom(\mathcal{A}, \mathcal{B})$ in all of the categories of Definition 4.7.2 is represented by the natural morphism

$$\mathcal{A} \mapsto \mathcal{A}/\ker(f).$$

4.7.3 Definition   A category with a zero object is called *semiabelian* if:

(a) Every morphism may be factored into a representative of its coimage followed by a representative of its image, and

(b) Every morphism has a cokernel.

A category with a zero object is called *cosemiabelian* if it satisfies (a) and

(c) Every morphism has a kernel.

4.7.4 Proposition   *All of the categories of Definition 4.7.2 are both semiabelian and cosemiabelian.*

Proof   This follows from the remarks above.   □

A category **B** is called a *subcategory* of a category **C** if every object of **B** is an object of **C**, any ordered pair $(\mathcal{A}, \mathcal{B})$ of objects of **B** satisfies $Hom_{\mathbf{B}}(\mathcal{A}, \mathcal{B}) \subseteq Hom_{\mathbf{C}}(\mathcal{A}, \mathcal{B})$ and $Hom_{\mathbf{B}}(\mathcal{A}, \mathcal{A})$ contains the identity $I_{\mathcal{A}} \in Hom_{\mathbf{C}}(\mathcal{A}, \mathcal{A})$. A subcategory **B** of **C** is called *full* if it satisfies $Hom_{\mathbf{B}}(\mathcal{A}, \mathcal{B}) = Hom_{\mathbf{C}}(\mathcal{A}, \mathcal{B})$ for all ordered pairs $(\mathcal{A}, \mathcal{B})$ of objects of **B**. For instance, **SBA** is a subcategory of **N, A, SNA**, and **BA**, and it is a full subcategory of **SNA**.

In the following theorem we use the notation $\mathcal{A}_{\mathcal{R}}$ to represent a subobject of an object, $\mathcal{A}$. Recall that by definition $\mathcal{A}_{\mathcal{R}}$ is not an object but an equivalence class of monomorphisms. However, in all of the cases under discussion here, $\mathcal{A}_{\mathcal{R}}$ is easily and naturally associated with a definite object.

4.7.5 Definition Let $\mathbf{C}$ be a semiabelian category. A *radical subcategory* of $\mathbf{C}$ is a full subcategory $\mathbf{R}$ satisfying:

(a) If $\mathcal{A}$ is an object in $\mathbf{R}$ and $i \in Hom(\mathcal{I}, \mathcal{B})$ represents the image of $f \in Hom(\mathcal{A}, \mathcal{B})$, then $\mathcal{I}$ is an object in $\mathbf{R}$;

(b) If $\mathcal{A}$ is an object in $\mathbf{R}$ and $k \in Hom(\mathcal{R}, \mathcal{A})$ represents the kernel of $f \in Hom(\mathcal{A}, \mathcal{B})$, then $\mathcal{R}$ is an object in $\mathbf{R}$;

(c) For each $\mathcal{A} \in \mathbf{C}$, there is unique subobject $\mathcal{A}_{\mathbf{R}}$ of $\mathcal{A}$ which satisfies:

   (c$_1$) $\mathcal{A}_{\mathbf{R}}$ is a kernel;

   (c$_2$) $\mathcal{A}_{\mathbf{R}}$ is represented by a monomorphism with an object of $\mathbf{R}$ as domain;

   (c$_3$) $\mathcal{A}_{\mathbf{R}}$ includes any subobject of $\mathcal{A}$ which is a kernel and is also represented by a monomorphism with an object of $\mathbf{R}$ as domain;

(d) If $u \in Hom(\mathcal{A}, \mathcal{B})$ is a representative of the cokernel of a representative of $\mathcal{A}_{\mathbf{R}}$, then the subobject $\mathcal{B}_{\mathbf{R}}$ is the zero-subobject of $\mathcal{B}$.

What we call a radical subcategory here (following Gray [1967]) corresponds to a hereditary radical in the terminology of some writers. See, for instance, the excellent book [1965] by Divinsky.

4.7.6 Theorem *The full subcategory defined by the condition that all its objects satisfy* ($\mathcal{A} = \mathcal{A}_J$; $\mathcal{A} = \mathcal{A}_M$) *is a radical subcategory of each of the categories* **A**, **SA**, **SSA**, **BA**, *and* **SBA**. *The full subcategory defined by the condition* $\mathcal{A} = \mathcal{A}_B$ *is a radical subcategory of each of the categories* **A**, **SA**, *and* **SSA**.

Proof To prove (a) in the definition of a radical subcategory, note that $\mathcal{I}$ can be chosen as $\mathcal{A}/\ker(f)$ for all of the categories we are considering. Hence (a) follows from Theorems 4.3.2, 4.4.8, and 4.5.3. Similarly, to prove (b) one notes that $\mathcal{R}$ may be taken as $\ker(f)$ which is an ideal of $\mathcal{A}$. The same three theorems complete the argument. The subobject $\mathcal{A}_{\mathbf{R}}$ of (c) is represented by $\mathcal{A}_J \to \mathcal{A}, \mathcal{A}_M \to \mathcal{A}$ and $\mathcal{A}_B \to \mathcal{A}$ in the three cases under discussion. Thus (c$_1$) follows from Corollary 4.3.10 and Theorem 4.5.4 and the fact that $\mathcal{A}_J, \mathcal{A}_M$, and $\mathcal{A}_B$ are always ideals. Finally (c$_2$), (c$_3$), and (d) follow from Theorems 4.3.2, 4.4.8, and 4.5.3. $\quad\square$

In **NA** and **GNA**, $\mathcal{A}_J$, $\mathcal{A}_B$ and $\mathcal{A}_M$ are not necessarily closed so that $\mathcal{A}/\mathcal{A}_J, \mathcal{A}/\mathcal{A}_B$ and $\mathcal{A}/\mathcal{A}_M$ are not normed algebras.

4.8 History of Radicals and Examples

4.8.1 History of Radicals

Because of the central importance of radicals in the theory of algebras, we will discuss their history in detail. In order to simplify the discussion as much as possible, we will concentrate on radicals of algebras, mentioning only briefly the important advances which related solely to the extension of the concepts from algebras to more general rings, and we will omit any discussion of radicals other than those considered in this work. For information on them, see Divinsky [1965]. We will also transpose notation and terminology to agree with that used in this work whenever it is convenient. In particular, we make operators act on the left and speak of left modules although many authors have used the opposite convention.

Again we emphasize the root idea of radicals. When trying to establish a structure theory for a wide class of abstract mathematical objects, such as algebras, it is usually necessary to discard certain pathological cases. It frequently turns out, as it does in the case of algebras, that in each object the pathological part can be excised by judiciously choosing the kernel of a morphism which contains the pathology and studying only the image of that morphism which is healthy. The trick is to identify the pathology so that: (a) the above construction is possible; (b) a significant theory can be developed for the images; and (c) the excluded kernels are not too large, *i.e.*, they do not contain the really interesting examples.

It is a curious historical fact that the strategy just outlined was first applied to a class of non-associative algebras—the Lie algebras. These algebras arise naturally in the study of differentiable groups initiated by Sophus Lie. They were first seriously studied by Wilhelm Killing [1888] although the term "Lie algebra" was not introduced until 1933 in lectures by Hermann Weyl at the Institute for Advanced Study. Killing's work, although very powerful, was not rigorous and his proofs were not accepted by most contemporary mathematicians. In his thesis [1898], Elie Cartan gave rigorous proofs of Killing's ideas and extended them to a theory in which the radical of a Lie algebra plays a central and very explicit role. (For a Lie algebra, the radical is the sum of all solvable ideals which is itself a solvable ideal. The quotient modulo this ideal has no non-zero solvable ideals.) One of his most notable contributions was the discovery of what has come to be called "Cartan's criterion"—that a Lie algebra is semisimple if and only if its Killing form is nondegenerate. The thesis gave a complete classification of complex simple Lie algebras (slightly correcting Killing's results) and showed that any complex Lie algebra in which the radical is zero is the direct sum of (finitely many) simple Lie algebras. The term semisimple (*halbeinfach*) had been introduced by Killing to describe a Lie algebra which is the direct sum of simple Lie algebras.

In [1898] Cartan applied the theory of radicals, which had been so successful in his treatment of Lie algebras, to the treatment of finite dimensional real or complex associative algebras. The distinction between algebras which contained quarternionic components and those that did not was at the forefront of his discussion and makes it look somewhat odd to the modern reader. Furthermore, Cartan did not use the recently developed idea due to Theodor Molien [1893] of an algebra $\mathcal{A}/\mathcal{I}$ defined as the quotient of an algebra $\mathcal{A}$ modulo an ideal $\mathcal{I}$. However, except for these points the results will look familiar to any mathematics graduate student. He shows ([1898b], pp. 57-59) that every algebra decomposes into a direct sum of a nil ideal and a semisimple subalgebra where again the latter term signifies a direct sum of finitely many simple unital algebras. Furthermore, all of the simple summands are full matrix algebras over $\mathbb{R}$, $\mathbb{C}$ or $\mathbf{H}$ (the quarternions). (The last part is not expressed quite so explicitly, and the quarternions cause considerable difficulty.) This is the full set of structure theorems usually called the "Wedderburn structure theorems". Cartan showed the power of the concept of a radical for associative algebras and also set the theory of finite-dimensional associative algebras over the real and complex fields on a very firm footing indeed.

Joseph H. Maclagan Wedderburn's fundamental paper [1907] generalized the structure theorem just given to finite-dimensional algebras over arbitrary fields. (Wedderburn had just completed a definitive study of the finite fields.) The paper is important for many reasons, one of which is that it seems to have been much more widely read than Cartan's paper. It is a very modern paper which can be read without any culture shock by a present-day mathematician. The methods are much more general and abstract than those used previously, and he uses Molien's [1893] idea of quotient algebras very effectively. He shows quite directly that there is a largest nilpotent ideal (this result is of course special to finite-dimensional algebras) and that the quotient modulo this ideal has no non-zero nilpotent ideals. He uses these properties to define the term semisimple.

Wedderburn does not use the term radical, but it was apparently standard by the time of Leonard Eugene Dickson's book [1927]. Of course the term arose from the fact that the radical either consists of the set of roots of zero (if it is thought of as a nil ideal) or is the ideal root of the zero ideal (if it is thought of as a nilpotent ideal).

These results were so impressive that many mathematicians tried to extend their validity. After Emmy Noether had introduced the idea of chain conditions on ideals of a ring, Emil Artin [1926] proved that a ring satisfying both the ascending and descending chain conditions on left ideals satisfies the Wedderburn structure theorems. Later C. Hopkins [1939] showed that the descending chain condition implies the ascending chain condition so that only the former is needed. Artin's result, with this improvement, is frequently called the Artin–Wedderburn theorem. This success further encouraged mathematicians to try to extend these results to more general rings. For instance, see Köthe [1930]. It gradually became apparent that the definition of the radical for more general rings was the central problem. For rings satisfying the descending chain condition, the largest nil ideal (called the nil radical) is nilpotent and satisfies various other properties which make it the obvious choice to play the role of a radical. However, for more general rings, including infinite-dimensional algebras, the various desirable properties of the nil ideal belong to different ideals, each one of which can be considered as a radical in some situations.

Reinhold Baer [1943] proposed to call an ideal $\mathcal{I}$ of a ring $\mathcal{A}$ a radical ideal if $\mathcal{I}$ were nil and $\mathcal{A}/\mathcal{I}$ had no non-zero nilpotent left ideals. He showed that any ring contained a largest and a smallest radical ideal. (The latter is the Baer lower radical introduced in Definition 4.4.5.) He showed by example that not every intermediate ideal has to be a radical ideal. He also observed (although he attributes this to the referee) that the largest or smallest radical ideal of $\mathcal{A}/\mathcal{I}$ is zero when $\mathcal{I}$ is, respectively, the largest or smallest radical ideal of $\mathcal{A}$. From his definition it is obvious that $\mathcal{I}$ itself is its own largest or smallest radical ideal. Baer's largest radical ideal had been considered previously by Köthe [1930a] and Hans Fitting [1935].

Sam Perlis [1942] showed that the nil radical of a finite-dimensional algebra is the largest quasi-regular ideal of the algebra. Baer [1943] introduced the term "quasi-regular ideal" and showed that in any ring the sum of all quasi-regular ideals is a quasi-regular ideal, and that the quotient modulo this ideal has no non-zero quasi-regular ideals. However, Nathan Jacobson [1945a], in a paper the great importance of which was immediately recognized, showed that a significant theory could be based on the use of this ideal as a radical. Indeed, except for the role played by modular ideals, Jacobson's original paper, together with his paper [1945b], contains the complete theory very much as it is known today. Jacobson noted that his radical agrees with the nil-radical in rings satisfying the descending chain condition on left ideals, and agrees with the Gelfand radical for commutative Banach algebras.

Several of the results on the Jacobson radical were obtained independently and simultaneously by Max Zorn and Einar Hille, cf. Hille ([1948], pp. 475-6). Hille had access to preliminary versions of Irving E. Segal's paper [1947a] in which modular ideals were first defined (with the name regular ideals). He seems to have been the first mathematician to recognize the important connections between modular ideals and the theory of the Jacobson radical. Most of these connections are presented in his book (Hille [1948]). However, an exposition of the theory containing all elements of our present day theory does not seem to have been given until the book of Jacobson [1956].

We have already mentioned the important references relating Baer's definition of his lower radical to the prime radical defined by McCoy [1949]. These are Jacob Levitzki [1951] and Masayoshi Nagata [1951]. A transfinite construction of the Baer radical due to Baer [1943] was considerably simplified and generalized in Sulinsky, Anderson and Divinsky [1966]. In ring theory the Baer radical is associated with a successful structure theory due to Alfred W. Goldie [1960] for rings satisfying the ascending chain condition.

The strong radical was defined by B. Brown and Neal H. McCoy [1947] and [1948], while strong semisimplicity was independently introduced by Segal [1941], [1947a].

The discovery of these and many other radicals for rings suggested creation of a general theory of radicals. This was carried out in different ways in a number of papers which were mentioned preceding Definition 4.7.1. The book by Divinsky [1965] summarizes a number of these theories. However, we have found that the theory due to Mary W. Gray [1970] is most suitable to the application which we wish to make throughout this work. The radicals satisfying her definitions are called hereditary radicals by most authors. Some non-hereditary radicals for rings have been widely considered. The distinction relates to property (a) of Theorems 4.3.2, 4.4.8, and 4.5.3 which is not enjoyed by non-hereditary radicals. For further discussion see Divinsky [1965], especially page 125.

Examples

We collect several interesting examples which illustrate the theory of the chapter. Our first examples show that the Jacobson, Baer, and strong radicals are all distinct.

4.8.2 Example   We have already remarked that when $\mathcal{X}$ is an infinite-dimensional Banach space, the algebra $\mathcal{B}_A(\mathcal{X})$ of approximable operators is a semisimple (even primitive) Banach algebra which is Brown–McCoy-radical. This algebra has many other desirable properties illustrating the general fact that Brown–McCoy-radical algebras, unlike Jacobson-radical algebras, are often well behaved from the analysts' viewpoint. Similarly $\mathcal{B}(\mathcal{X})$ is semisimple and primitive, but $\mathcal{B}_A(\mathcal{X})$ is included in $\mathcal{B}(\mathcal{X})_M$.

Apart from a few examples, Jacobson-radical algebras received little attention until the 1980s. Even now, little seems to be known about noncommutative Jacobson-radical algebras. However, the 1981 conference held in Long Beach, California sparked renewed interest. Examples 3.4.15 and 3.4.18 deal with radical Banach algebras, but we will have relatively more to say about them, so we will simply list a number of additional references: Grabiner [1971], [1973], [1974], [1981], [1983], Jean Esterle [1981a], [1981b], Grønbaek [1982a], [1983], Bachelis [1983], Bade [1983], Laursen [1983b], Grabiner and Thomas [1985].

We mention two more examples of semiprime Banach algebras which are Jacobson-radical.

4.8.3 The Disc Algebra under Convolution   This example and some variants are mentioned in Georgi Evgenyevič Šilov ([1947a], p. 104) and Hille ([1948], p. 478). Let the underlying Banach space of $\mathcal{A}$ be the same as the disc algebra (§3.2.13), but let multiplication be defined by convolution

$$f * g(\lambda) = \int\limits_0^\lambda f(\lambda - \mu)g(\mu)d\mu = \int\limits_0^1 f(\lambda(1 - s))g(\lambda s)\lambda ds.$$

Then any $f \in \mathcal{A}$ satisfies

$$|f^2(\lambda)| = |f * f(\lambda)| \le \int\limits_0^1 \|f\|^2 |\lambda| ds = \|f\|^2 |\lambda|$$

and hence by induction satisfies

$$
\begin{aligned}
|f^n(\lambda)| &\le \int\limits_0^1 \|f\| \, \|f\|^{n-1} |\lambda|^{n-2}((n-2)!)^{-1} |\lambda| s^{n-2} ds \\
&= \frac{\|f\|^n |\lambda|^{n-1}}{(n-1)!}.
\end{aligned}
$$

This shows that each $f \in \mathcal{A}$ is topologically nilpotent which implies $\mathcal{A} = \mathcal{A}_J$. On the other hand the Titchmarsh convolution theorem ([1926], p. 286) shows that $\mathcal{A}$ has no non-zero divisors of zero so that it has no non-zero nil (or nilpotent) ideals and hence is semiprime. For further details and some generalizations see 3.4.18 above and Grabiner [1974].

4.8.4 Another Jacobson-radical Semiprime Algebra This example is due to Dixon [1973b]. It is separable and possesses a bounded approximate identity (see Chapter 5 for the definition) so that it satisfies Graham R. Allan's [1972] criterion for possession of a discontinuous functional calculus (Theorem 3.4.9 above). The algebra $\mathcal{A}$ is the set of (almost everywhere equal equivalence classes of) measurable functions on $[0, \infty[$ satisfying

$$\|f\| = \int\limits_0^\infty e^{-te^t} |f(t)| dt < \infty$$

with convolution multiplication

$$f * g(t) = \int\limits_0^t f(s)g(t - s)ds.$$

Arguments similar to those given above show that each element is topologically nilpotent and that there are no non-zero divisors of zero, so that $\mathcal{A}$ is Jacobson-radical and semiprime. The sequence of functions $\{x\chi_{[0,n^{-1}]}\}_{n \in \mathbb{N}}$

(where χ_S is the characteristic function of some subset S of the real line) form an approximate identity for $\mathcal{A}$.

As noted earlier, sets of nonsynthesis in commutative algebras give rise to radical quotients: Gregory F. Bachelis [1983].

4.8.5   The Algebra of the Free Semigroup on Two Generators

Let X be an arbitrary nonempty set. A word formed from X is a finite sequence $w = x_1 x_2 \cdots x_p$ of elements x_j from X where the x_j are not necessarily distinct. The *length* of a word w, denoted by $n(w)$, is the length of the sequence of symbols (*i.e.*, $n(x_1 x_2 \cdots x_p) = p$). We allow the empty word $\emptyset$ with length $n(\emptyset) = 0$. The unital free semigroup $FS(X)$ generated by X is the collection of all words from X made into a semigroup by the juxtaposition product:

$$(x_1 x_2 \cdots x_p)(y_1 y_2 \cdots y_q) = x_1 x_2 \cdots x_p y_1 y_2 \cdots y_q.$$

Clearly the empty word is an identity element in this semigroup and the length function is additive

$$n(wz) = n(w) + n(z) \qquad \forall \; w, z \in FS(X).$$

The set X is called the alphabet for $FS(X)$. It is clear that the cardinal of X determines $FS(X)$ up to isomorphism of semigroups. Thus we write $FS(n)$ for $FS(X)$ if X has cardinal $n \in \mathbb{N}$. For $n > 1$, $FS(n)$ is noncommutative. Clearly $FS(X)$ is countable if and only if X is finite or countably infinite. For $n > 1$, $FS(n)$ includes a subsemigroup isomorphic to any other countable unital free semigroup. (To see this, one chooses two distinct elements x, y from the alphabet of $FS(n)$ and then considers the subsemigroup generated by a suitable subset of $\{xy^k x : k \in \mathbb{N}\}$.) Hence, in a certain sense, all countable unital free semigroups are equally complicated. For simplicity and definiteness we will concentrate attention on $FS(2)$ although the reader can extend most of the results to other cases without difficulty.

For the rest of this example (except Proposition 2), we use $FS(2)$ to denote $FS(\{x, y\})$. It is sometimes convenient to have a particular sequential ordering for $FS(2)$. We define $\{w_p\}_{p \in \mathbb{N}}$ to be the sequence

$$\emptyset, x, y, x^2, xy, yx, y^2, x^3, x^2 y, xyx, xy^2, yx^2, \text{ etc.}$$

so that $z_n z_{n-1} \cdots z_2 z_1 = w_p$ is equivalent to

$$p = 1 + \sum_{z_j = x} 2^{j-1} + \sum_{z_k = y} 2^k.$$

Note that $n(w_p) = [\log_2(p)]$ holds and that $w_p w_q = w_r$ is equivalent to

$$r = 2^n(p-1) + q = 2^n p + (q - 2^n)$$

where $n = n(w_q)$ is the largest integer satisfying $2^n \leq q$. In particular $w_p w_q$ comes before $w_r w_s$ in the sequence if the indices satisfy $p \leq r$ and $q \leq s$ (unless $p = r$ and $q = s$).

Now let S be any semigroup. Let $\mathcal{A}_0(S)$ be the set of functions $f : S \to \mathbb{C}$ which are zero except at a finite number of elements of S and let $\mathcal{A}(S) = \ell^1(S)$ be the set of functions $f : S \to \mathbb{C}$ satisfying

$$\|f\|_1 = \sum_{s \in S} |f(s)| < \infty.$$

Then $\mathcal{A}_0(S)$ and $\mathcal{A}(S)$ are both algebras under the convolution product

$$f * g(s) = \sum_{\substack{uv = s \\ u,\, v \in S}} f(u)g(v).$$

Furthermore $(\mathcal{A}(S), \| \cdot \|_1)$ is a Banach algebra and $\mathcal{A}_0(S)$ is a dense subalgebra so that $\mathcal{A}(S)$ can be identified with the completion of the normed algebra $(\mathcal{A}_0(S), \| \cdot \|_1)$. We call $\mathcal{A}_0(S)$ the *algebraic semigroup algebra* of S and $\mathcal{A}(S)$ the ℓ^1-*semigroup algebra* of S. Note that the convolution product is obtained by identifying a function f with the formal sum $\sum_{s \in S} f(s)s$. We will represent the elements of $\mathcal{A}(S)$ by such formal sums whenever it is convenient to do so.

We denote $\mathcal{A}_0(FS(2))$ and $\mathcal{A}(FS(2))$ by $\mathcal{A}_0$ and $\mathcal{A}$, respectively, in the rest of this example (except the last Proposition). The following result gives the most interesting property of the Banach algebra $\mathcal{A}$. The author learned this result from William L. Paschke in 1972.

Proposition 1. *The Banach algebra $\mathcal{A}$ has no non-zero divisors of zero and no non-zero topologically nilpotent elements.*

Proof Suppose f and g are non-zero elements of $\mathcal{A}$ and p and q are the smallest indices satisfying $f(w_p) \neq 0$ and $g(w_q) \neq 0$, respectively. Then $f * g(w_p w_q) = f(w_p)g(w_q) \neq 0$ shows that $f * g$ is not zero. Furthermore for any $n \in \mathbb{N}$,

$$\|f^n\|^{1/n} \geq |f^n(w_p^n)|^{1/n} = |f(w_p)|$$

implies $\rho(f) \geq |f(w_p)| > 0$. Hence f is not topologically nilpotent. $\qquad\square$

Clearly the center of $\mathcal{A}$ is the set of multiples of the identity element. Although $\mathcal{A}$ is highly noncommutative it does have a large set $\Gamma_{\mathcal{A}}$ of algebra homomorphisms onto $\mathbb{C}$. For any word $w \in FS(2)$, let $n_x(w)$ and $n_y(w)$ be the number of occurrences of x and y, respectively, in w. For any $(\lambda, \mu) \in D \times D \setminus \{(0,0)\}$, the function $\gamma_{\lambda,\mu} : \mathcal{A} \to \mathbb{C}$ defined by

$$\gamma_{\lambda,\mu}\left(\sum_{n=1}^{\infty} \alpha_n w_n\right) = \sum_{n=1}^{\infty} \alpha_n \lambda^{n_x(w_n)} \mu^{n_y(w_n)}$$

belongs to Γ_A and it is easy to see that these are the only elements of Γ_A.

Bruce A. Barnes and John Duncan [1975] show that every element of $\mathcal{A}$ which is not in the center has connected spectrum with interior. A stronger result was proved for elements of $\mathcal{A}_0$ by Abdullah H. Al-Moajil [1975]. See also Duncan and Williamson [1982]. The first reference cited also contains the following result.

Proposition 2. *Let $\mathcal{A}$ be the ℓ^1-semigroup algebra of the free group on $\{x_n : n \in \mathbb{N}\}$. There is an unbounded linear functional on $\mathcal{A}$ which is bounded on each commutative subalgebra.*

Proof Let $\mathcal{B}$ be the closed subalgebra generated by all words of length at least two. By choosing a Hamel basis we may define a linear functional ω to satisfy

$$\omega(x_n) = n$$
$$\omega(\mathcal{B}) = \{0\}.$$

Clearly ω is unbounded on $\mathcal{A}$. Let $\mathcal{C}$ be a commutative subalgebra of $\mathcal{A}$ and choose an element $c = \sum \gamma_n x_n + d$ of $\mathcal{C}$ with $\gamma_n \in \mathbb{C}, d \in \mathcal{B}$ and c not equal to d. (If no such element exists there is nothing to prove.) If $a = \sum \alpha_n x_n + b$ with $\alpha_n \in \mathbb{C}$ and $b \in \mathcal{B}$ commuting with c, then it is easy to see that $\sum \alpha_n x_n$ and $\sum \gamma_n x_n$ commute so that $\alpha_n \gamma_m = \gamma_n \alpha_m$ holds for all $n, m \in \mathbb{N}$. Hence there is some $\beta \in \mathbb{C}$ satisfying $\sum \alpha_n x_n = \beta \sum \gamma_n x_n$. Hence the estimate

$$
\begin{aligned}
|\omega(a)| &= |\omega(\sum \alpha_n x_n)| = |\beta \omega(\sum \gamma_n x_n)| \\
&= \|\beta \sum \gamma_n x_n \|(|\omega(\sum \gamma_n x_n)|/\|\sum \gamma_n x_n\|) \\
&= \|\sum \alpha_n x_n\|(|\omega(c)|/\sum |\gamma_n|) \\
&= (|\omega(c)|/\sum |\gamma_n|)\|a\|
\end{aligned}
$$

shows that ω is bounded on $\mathcal{C}$. $\qquad\square$

4.8.6 Noncommutative Semiprime but Jacobson-radical Banach Algebras The following examples which are weighted cross products are taken from Julian M. Cusack [1976]. They seem to have been suggested by Rudi A. Hirschfeld and Stefan Rolewicz [1969] and by weighted semigroup algebras as described on p. 8 of Bonsall and Duncan [1973b].

Let $\mathcal{A}$ be a Banach algebra, let S be a semigroup and let W be a real valued weight function on S satisfying:

(a) $W(s) > 0 \quad s \in S$
(b) $W(st) \leq W(s)W(t) \qquad \forall\, s, t \in S.$

We use $\ell^1(S, \mathcal{A}, W)$ to denote the Banach space of all functions from S into $\mathcal{A}$ which are finite with respect to the norm:

$$\|f\| = \sum_{s \in S} \|f(s)\| W(s) < \infty.$$

This is a Banach algebra under the convolution product

$$f * g(s) = \sum_{rt=s} f(r)g(t) \qquad \forall \, f, \, g \in \ell^1(S, \mathcal{A}, W); s \in S.$$

Let V be a semigroup homomorphism of S into the group of isometric automorphisms of $\mathcal{A}$. Define a twisted convolution product on $\ell^1(S, \mathcal{A}, W)$ by

$$f *_V g(s) = \sum_{rt=s} f(r)V_r g(t).$$

Since each V_r is an isometry, the norm is still submultiplicative for this product, and since it is an automorphism the usual calculation establishes the associative law. We denote $\ell^1(S, \mathcal{A}, W)$ with this new product by $\ell^1(S, \mathcal{A}, W, V)$.

For any $t \in S$ and $a \in \mathcal{A}$, denote the function $s \mapsto \delta_{st}a$ by $k(t, a)$. These elements satisfy

$$k(t, a) *_V k(r, b) = k(tr, aV_t(b)) \qquad \forall \, t, \, r \in S; a, \, b \in \mathcal{A}$$

and hence their linear span is a dense subalgebra of $\ell^1(S, \mathcal{A}, W, V)$.

Proposition *With the notation as given, $\ell^1(S, \mathcal{A}, W)$ and $\ell^1(S, \mathcal{A}, W, V)$ are Banach algebras. If W satisfies the two additional conditions:*
  (c) $W(st) = W(ts)$ $\forall \, s, \, t \in S$;
  (d) $\lim W(s^n)^{1/n} = 0$ $\forall \, s \in S$;
in addition to (a) and (b), then $\ell^1(S, \mathcal{A}, W, V)$ is Jacobson-radical.

Proof It is enough to show that each $k(t, a)$, $t \in S, a \in \mathcal{A}$, belongs to the Jacobson radical. Hence we will show that for any $f \in \ell^1(S, \mathcal{A}, W, V)$, $t \in S$, and $a \in \mathcal{A}$, the element $g = f *_V k(t, a)$ is topologically nilpotent. Denoting the nth convolution power by g^n we find

$$g(s) = \sum_{rt=s} f(r)V_r(a) \qquad \forall \, s \in S$$

(where t occurs only in the summation index) and

$$g^n(s) = \sum_{s_1 s_2 \cdots s_n = s} g(s_1)V_{s_1}g(s_2) \cdots V_{s_1 s_2 \cdots s_{n-1}}g(s_n) \qquad \forall \, s \in S; \, n \geq 2.$$

From this we obtain

$$\|g^n\| = \sum_{s \in S} \|g^n(s)\| W(s)$$

$$\leq \sum_{s \in S} (\sum_{s_1 s_2 \cdots s_n = s} \prod_{j=1}^{n} (\sum_{r_j t = s_j} \|f(r_j)\| \, \|a\|)) W(s)$$

By conditions (b) and (c) on W we have $W(s) \leq W(t^n) \prod_{j=1}^{n} W(r_j)$ which gives:

$$\|g^n\| \leq \|a\|^n W(t^n) \sum_{s \in S} \sum_{r_1 t r_2 t \cdots r_n t = s} \prod_{j=1}^{n} \|f(r_j)\| W(r_j)$$

$$\leq \|a\|^n W(t^n) \|f\|^n.$$

Hence assumption (d) shows that g is topologically nilpotent so that $k(t, a)$ belongs to the Jacobson radical. □

Clearly this algebra is noncommutative if S or $\mathcal{A}$ is noncommutative or V is nontrivial.

Now let $FS(2)$ be the free semigroup on two generators described in the last example. For each word $s \in FS(2)$, define $W(s)$ by $W(s) = (n(s)!)^{-1}$ where $n(s)$ is the length of s and define V_s to be the identity map on $\mathbb{C}$. Then $\ell^1(FS(2), \mathbb{C}, W, V)$ is a noncommutative radical Banach algebra by the results we have just proved. Arguments similar to those given in the last example show that $\ell^1(FS(2), \mathbb{C}, W, V)$ has no non-zero divisors of zero. Hence it is prime (and thus *a fortiori* semiprime).

Another class of examples starts with a unital, prime commutative Banach algebra $\mathcal{A}$ and a nontrivial isometric automorphism V of $\mathcal{A}$ (e.g., take $\mathcal{A}$ to be the disc algebra $A(D)$ discussed in §3.2.13 and define V by $V(f)(\lambda) = f(\xi\lambda)$ for some fixed $\xi \in \mathbb{T} \setminus \{1\}$ and all $\lambda \in D$). Let S be the semigroup $\mathbb{N}$ of natural numbers and define W and V by $W(n) = (n!)^{-1}$ and $V_n = V^n$. Consider the Banach algebra $\ell^1(\mathbb{N}, \mathcal{A}, W, V)$. If f and g are two non-zero elements in this algebra with m and n, respectively, the smallest integers in $\mathbb{N}$ at which each is non-zero, then $f *_V g(n+m) = f(n) V^n f(m)$ is non-zero. Hence $\ell^1(\mathbb{N}, \mathcal{A}, W, V)$ is another noncommutative radical Banach algebra with no non-zero divisors of zero.

The last example constructed by this method has non-zero nilpotent elements although it is still a semiprime, Jacobson-radical Banach algebra. Let S be $\mathbb{N}$ and let $\mathcal{A}$ be the algebra $C([-1, 1])$. Define W and V by $W(n) = (n!)^{-1}$ and $V_n f(t) = f((-1)^n t)$ for all $n \in \mathbb{N}, f \in \mathcal{A}$, and $t \in [-1, 1]$.

In order to show that the algebra $\ell^1(\mathbb{N}, \mathcal{A}, W, V)$ is semiprime, we will show that if $f *_V g *_V f = 0$ holds for all g then f is zero. Suppose f is non-zero and let n be the least integer at which it is non-zero. Using the functions $k(s, a)$ introduced earlier, we get $f *_V k(m, 1) *_V f(2n + m) =$

$f(n)V^{n+m}f(n) = 0$. By choosing m to be 1 or 2, we get the contradiction $f(n) = 0$ since $f(n)^2 = 0$ holds. Hence the algebra is semiprime. However, we note that $k(1, f)$ has convolution square zero if f satisfies $f(t)f(-t) = 0$ for all $t \in [-1, 1]$.

4.8.7 A Jacobson Semisimple Banach Algebra with a Dense Nil Subalgebra
We use the terminology introduced in Example 4.8.5. Let S be the free unital semigroup with generators $\{x_n : n \in \mathbb{N}\}$ and let $\mathcal{A} = \mathcal{A}(S)$ be its ℓ^1-semigroup algebra. Let $\mathcal{I}$ be the closed ideal of $\mathcal{A}$ generated by the set of words $x_{i_1} x_{i_2} \cdots x_{i_n}$ in which x_p occurs more than p times, where p is defined by $p = \max\{i_1, i_2, \ldots, i_n\}$. Let $\mathcal{B}$ be $\mathcal{A}/\mathcal{I}$ and let $\mathcal{B}_0$ be the dense subalgebra of $\mathcal{B}$ which is the image of $\mathcal{A}_0 = \mathcal{A}_0(S)$ under the natural map of $\mathcal{A}$ onto $\mathcal{B}$.

If $a \in \mathcal{A}_0$ equals $\sum_{k=1}^q \alpha_k w_k$ and p is the largest index of any x_j occurring in any of the words w_k, then a^{qp+1} belongs to $\mathcal{I}$. Hence $\mathcal{B}_0$ is a nil algebra.

To show that $\mathcal{B}$ is semisimple we show that for each non-zero element $b \in \mathcal{B}$ there is some product bd which is not topologically nilpotent. Let $b = \sum_{q=1}^\infty \alpha_q w_q + \mathcal{I}$ be an expansion of b as a series of words w_q in S and assume $\alpha_1 w_1 \notin \mathcal{I}$. For simplicity we call such an expression an infinite linear combination of words. Let p be larger than the index of any x_j occurring in w_1. Define d to be $\sum_{k=0}^\infty 2^{-k} x_{p+k} + \mathcal{I}$. The element $(bd)^n$ will be an infinite linear combination of words of the form

$$w_{j_1} x_{p+k_1} w_{j_2} x_{p+k_2} \cdots w_{j_n} x_{p+k_n}.$$

A word of the above form with each j_k equal to one will be called *special*. We wish to show that there is no cancellation between special words and other words of the above general form, and we consider two cases. First, suppose one of the words w_{j_i} contains an x_r with $r \geq p$. Then the total number of letters x_r with $r \geq p$ occurring in the general form would be at least $n + 1$, and since each special word contains exactly n such letters no cancellation could occur. Second, suppose no w_{j_i} in the general form contains an x_r with $r \geq p$. Then equality forces each w_{j_i} to be w_1. Since the special words are obviously distinct for different choices of $p + k_1, \ldots, p + k_n$ we see that there is no cancellation. Now consider the special word with the k_i chosen by

$$k_i = \max\{q : p^q \text{ divides } i\}$$

so that this special word does not belong to $\mathcal{I}$. For $n = n_q p^q + \cdots + n_1 p + n_0$ with $0 \leq n_i < p$, k_i equals an s satisfying $0 \leq s \leq q$ exactly $n_s + n_{s+1}(p - 1) + n_{s+2}(p - 1)p + \cdots + n_q(p - 1)p^{q-s-1}$ times. Hence the coefficients of the special word with this choice of k_i is $2^{-r}|\alpha_1|^n$ with $r = (p - 1)^{-1}[n_q(p^q - 1) + \cdots + n_1(p - 1)] \leq (p - 1)^{-1}n$. From this we obtain the estimate

$$\|(bd)^n\|^{1/n} \geq 2^{-1/(p-1)}|\alpha_1|$$

which proves that bd is not topologically nilpotent. Hence the perfectly arbitrary non-zero element b does not belong to $\mathcal{A}_J$ so $\mathcal{A}$ is semisimple.

4.8.8 Topologically Nilpotent Banach Algebras These were introduced by J. K. Miziolek, T. Müldner and A. Rek [1972] as a class of topological algebras. Recently their study was revived by Peter G. Dixon [1991] and continued by Dixon and Vladimir Müller [1992], Dixon and George A. Willis [1992]. We briefly introduce the subject and refer the reader to the original papers for more detail. In particular, we omit most statements about actual rates of decrease contained in the original papers.

Definition A Banach algebra $\mathcal{A}$ is said to be *topologically nilpotent* if

$$\lim_{n\to\infty} \sup\{\|a_1 a_2 \cdots a_n\|^{1/n} : a_j \in \mathcal{A}_1 \text{ for } j = 1, 2, \ldots, n\} = 0.$$

It is called *uniformly topologically nil* if

$$\lim_{n\to\infty} \sup\{\|a^n\|^{1/n} : a \in \mathcal{A}_1\} = 0.$$

The following implications are obvious

$$\text{topologically nilpotent} \quad \Rightarrow \quad \text{uniformly topologically nil} \qquad (1)$$
$$\Rightarrow \quad \text{Jacobson radical.}$$

The next theorem combines results from Dixon [1991] and Dixon and Müller [1992]. Note that (f) is simply the assertion that $\mathcal{A}$ is uniformly topologically nil with a particular rate of decrease.

Theorem *The following are equivalent for a Banach algebra $\mathcal{A}$.*

(a) $\mathcal{A}$ *is topologically nilpotent.*

(b) *Every sequence* $\{a_n\}_{n\in\mathbb{N}} \subseteq \mathcal{A}_1$ *satisfies* $\lim_{n\to\infty} \|a_1 a_2 \cdots a_n\|^{1/n} = 0.$

(c) *For every closed ideal $\mathcal{I}$ of A, $\mathcal{I}$ and $A/\mathcal{I}$ are topologically nilpotent.*

(d) *For some closed ideal $\mathcal{I}$ of A, $\mathcal{I}$ and $A/\mathcal{I}$ are topologically nilpotent.*

(e) *For every element $a \in \mathcal{A}$, there is some finite constant $C = C(a)$ satisfying* $\|a^n\|^{1/n} \leq Cn^{-3\cdot 2^n/n}$ *for all $n \in \mathbb{N}$.*

(f) *There is some finite constant C satisfying*

$$\sup\{\|a^n\|^{1/n} : a \in \mathcal{A}_1\} \leq Cn^{-3\cdot 2^n/n} \qquad \forall \, n \in \mathbb{N}.$$

If $\mathcal{A}$ is commutative, these are equivalent to $\mathcal{A}$ being uniformly topologically nil, but there are noncommutative Banach algebras for which this fails.

The implication (e) $\Rightarrow$ (f) depends on a Baire category argument, but the rest of the proof is a series of careful estimates. Perhaps the most interesting is the proof that in a commutative algebra uniform topologically nil implies topologically nilpotent. The proof depends on writing $a_1 a_2 \cdots a_n$ as an n-fold contour integral of the polynomial $(\lambda_1 a_1 + \lambda_2 a_2 + \cdots + \lambda_n a_n)^n$ divided by $(\lambda_1 \lambda_2 \cdots \lambda_n)^2$. The rate of decrease in (e) and (f) comes from keeping track of the constants in a proof of the Nagata–Higman theorem. The given rate can presumably be greatly weakened, but the negative result at the end of the theorem shows that not every rate will give the result. Condition (e) makes it easy to see that in (c) $\mathcal{A}/\mathcal{I}$ is always topologically nilpotent.

Here are some explicit examples. Consider the Banach space $C([0,1])$ of continuous complex valued functions on the unit interval with the supremum norm as a Banach algebra $\mathcal{C}$ under cut-off convolution:

$$f * g(t) = \int_0^t f(s)g(t-s)ds \qquad \forall\, f,g \in \mathcal{C}. \tag{2}$$

A slight variation of the proof given above in Example 4.8.3 shows that any functions $f_1, f_2, \ldots, f_n$ which are bounded in absolute value by 1, satisfy $\|f_1 * f_2 * \cdots * f_n\|_\infty \le 1/(n-1)!$. Hence $\mathcal{C}$ is topologically nilpotent.

For the second example, simply alter the Banach space to be $L^1([0,1])$ but continue to use cut-off convolution defined by (2) as multiplication, and let f_k be 2^k times the characteristic function of $[0, 2_n]$, then $\|f_k = 1\|_1 = 1$ for all k, but $\|f_1 * f_2 * \cdots * f_n\|_1 = 1$ for all n. Thus this algebra is not topologically nilpotent.

Next, consider the ℓ^1 semigroup algebra $\mathcal{A}$ defined for the semigroup S generated by $\{u_n : n \in \mathbb{N}\}$ with the relations $u_i u_j = 0$ unless $j = i + 1$. The relationship $\|u_1 u_2 \cdots u_n\| = 1$ shows that $\mathcal{A}$ is far from topologically nilpotent. Dixon and Müller [1992] show that any element $a \in \mathcal{A}_1$ satisfies $\|a^n\| \le 1/n!$, so that $\mathcal{A}$ is uniformly topologically nil (with a rather slow rate of decrease).

As one would expect, it is hard to factor elements in topologically nilpotent Banach algebras. Because of the Cohen factorization theorem (Theorem 5.2.2), this is related to the nonexistence of bounded approximate identities. We will give a complete self-contained proof of a striking non-factorization result from Dixon [1991] and derive a corollary from Dixon and Willis [1992]. Corollary 8.1.3 below inspired this result.

4.8.9 Theorem *Let $\mathcal{A}$ be a topologically nilpotent Banach algebra and let $T: \mathcal{A} \to \mathcal{B}(\mathcal{X})$ be any continuous representation of $\mathcal{A}$ on a non-zero Banach space $\mathcal{X}$. Then $T_\mathcal{A}(\mathcal{X}) \ne \mathcal{X}$. In particular, $\mathcal{A}^2 \ne \mathcal{A}$.*

Proof Assume to the contrary that $T_\mathcal{A}(\mathcal{X}) = \mathcal{X}$. Then the map $\tilde{T}$ from the projective tensor product $\mathcal{A} \hat{\otimes} \mathcal{X}$ to $\mathcal{X}$ defined by $\tilde{T}(\sum_{j=1}^\infty a_j \otimes x_j) =$

$\sum_{j=1}^{\infty} T_{a_j}(x_j)$ for $a_j \in \mathcal{A}$, $x_j \in \mathcal{X}$ and $\sum_{j=1}^{\infty} \|a_j\| \|x_j\| < \infty$ is a continuous surjective linear map between Banach spaces. The open mapping theorem and Theorem 1.10.8 give a constant K such that any $x \in \mathcal{X}$ can be written as $x = \sum_{j=1}^{\infty} T_{a_j}(x_j)$ for $\sum_{j=1}^{\infty} \|a_j\| \|x_j\| \leq K \|x\|$. Iterating this construction by applying it to each $T_{a_j}(x_j)$ instead of x, and applying induction, allows us to write for any $n \in \mathbb{N}$

$$ x = \sum_{j=1}^{\infty} \left(\prod_{k=1}^{n} T_{a_{k,j}} \right) (x_j) \quad \text{where} \quad \sum_{j=1}^{\infty} \left(\prod_{k=1}^{n} \|a_{k,j}\| \right) \|x_j\| \leq K^n \|x\|. $$

Denote the nth supremum in the definition of topological nilpotence by $S(n)$ so that $\lim S(n) = 0$. In the display above we write $\prod_{k=1}^{n} T_{a_{k,j}}$ as $T_{a_{1,j} a_{2,j} \cdots a_{n,j}}$ which gives the estimate

$$
\begin{aligned}
\|x\| &\leq \sum_{j=1}^{\infty} \|T\| \|a_{1,j} a_{2,j} \cdots a_{n,j}\| \|x_j\| \\
&\leq \sum_{j=1}^{\infty} \|T\| S(n)^n \|a_{1,j}\| \|a_{2,j}\| \cdots \|a_{n,j}\| \|x_j\| \\
&\leq \|T\| (S(n)K)^n \|x\|.
\end{aligned}
$$

For any non-zero x, this is a contradiction, since $S(n) \to 0$. $\square$

4.8.10 Corollary *Let $\mathcal{A}$ be a commutative Banach algebra. If its Jacobson radical is both modular and topologically nilpotent, then $\mathcal{A}$ is unital if and only if $\mathcal{A}^2 = \mathcal{A}$.*

In particular, if $\mathcal{A}$ has a topologically nilpotent ideal of finite codimension, then $\mathcal{A}$ is unital if and only if $\mathcal{A}^2 = \mathcal{A}$.

Proof Any unital algebra $\mathcal{A}$ obviously satisfies $\mathcal{A}^2 = \mathcal{A}$. Suppose $\mathcal{A}_J$ is modular. Proposition 4.3.12(d) shows that an idempotent e may be chosen for its relative identity. Consider the algebra direct sum decomposition: $\mathcal{A} = \mathcal{A}e \oplus \mathcal{A}(1-e)$. If the second ideal is zero, e is a multiplicative identity for $\mathcal{A}$. Otherwise $\mathcal{A}(1-e)$ is included in $\mathcal{A}_J$ so it is topologically nilpotent. The theorem shows $(\mathcal{A}(1-e))^2 \neq \mathcal{A}(1-e)$ so we conclude that $\mathcal{A}^2 = (\mathcal{A}e)^2 \oplus (\mathcal{A}(1-e))^2$ is a proper subset of $\mathcal{A}$.

If $\mathcal{I}$ is any ideal of finite codimension, the Wedderburn theorem (Theorem 8.1.1) shows that there is an ideal $\mathcal{J}$ including $\mathcal{I}$ so that $\mathcal{A}/\mathcal{J}$ is semisimple and hence unital. Furthermore, $\mathcal{J}/\mathcal{I}$ is finite-dimensional and radical so that it is nilpotent. Hence, it is easy to see that $\mathcal{J} = \mathcal{A}_J$ is topologically nilpotent. Hence we are in the situation already considered.$\square$

4.8.11 Differences between the Topological and Geometric Categories of Banach Algebras In the geometric category if

$$ 0 \longrightarrow \mathcal{A} \overset{\varphi}{\longrightarrow} \mathcal{B} \overset{\psi}{\longrightarrow} \mathcal{C} \longrightarrow 0 $$

is an exact sequence, φ must be an isometry and C must carry the quotient norm.

The following result is called the *short five lemma*. Let

$$
\begin{array}{ccccccccc}
0 & \longrightarrow & \mathcal{A} & \overset{\varphi_1}{\longrightarrow} & \mathcal{B} & \overset{\varphi_2}{\longrightarrow} & \mathcal{C} & \longrightarrow & 0 \\
& & \downarrow{\scriptstyle\theta_1} & & \downarrow{\scriptstyle\theta_2} & & \downarrow{\scriptstyle\theta_3} & & \\
0 & \longrightarrow & \mathcal{A}^{\sim} & \overset{\psi_1}{\longrightarrow} & \mathcal{B}^{\sim} & \overset{\psi_2}{\longrightarrow} & \mathcal{C}^{\sim} & \longrightarrow & 0
\end{array}
$$

be a commutative diagram with exact rows. If θ_1 and θ_3 are isomorphisms, then θ_2 is also.

In the topological category of Banach algebras this was proved informally just before Theorem 1.1.16. However, it fails in the geometric category of Banach algebras as shown by the following easy example due to Robert C. Busby [1971a]. In the above diagram let $\mathcal{A} = \mathcal{A}^{\sim} = \mathcal{C} = \mathcal{C}^{\sim} = \mathbb{C}$ and $\mathcal{B} = \mathcal{B}^{\sim} = \mathbb{C} \times \mathbb{C}$. Let the vertical maps be the identity maps and $\varphi_1(\lambda) = \psi_1(\lambda) = (\lambda, 0)$ and $\varphi_2(\lambda, \mu) = \psi_2(\lambda, \mu) = \mu$. The norm on $\mathbb{C}$ is of course the absolute value, and the norms on $\mathcal{B}$ and $\mathcal{B}^{\sim}$ are $\|(\lambda, \mu)\|_1 = |\lambda| + |\mu|$ and $\|(\lambda, \mu)\|_{\infty} = \max\{|\lambda|, |\mu|\}$, respectively. In the geometric category, it is clear that θ_2 is not an isomorphism although all the hypotheses are fulfilled.

5

Approximate Identities and Factorization

Many Banach algebras which occur naturally are not unital. However, many of them do contain a sequence or net of elements which behaves like a multiplicative identity in the limit. Perhaps the most elementary example is the algebra $C_0(\mathbb{R})$ of continuous functions vanishing at infinity on the real line. Choose a sequence of functions $\{f_n\}_{n \in \mathbb{N}} \in C_0(\mathbb{R})$ each with values in the unit interval $[0, 1]$ and with f_n having the value 1 on all of $[-n, n]$. A moment's thought shows that any function $g \in C_0(\mathbb{R})$ satisfies

$$\lim_{n \to \infty} f_n g = g$$

where the limit is in the norm $\| \cdot \|_\infty$ of $C_0(\mathbb{R})$. The Fejér kernel, introduced in §1.8.15, provides another interesting example of the same phenomenon in L^1, C, AC or C^1, as defined there.

This chapter treats such approximate identities. The first section primarily shows the relationship between various possible definitions. The main result may be informally summarized as saying that the weakest definition implies the strongest. Section 5.1 also provides additional examples.

In a unital algebra every element a has the trivial factorization $a = 1a = a1$. Paul J. Cohen showed in [1959a] that every element in an algebra with a suitable approximate identity can also be factored. Allan M. Sinclair ([1978], [1979] and [1982]) showed the existence of a highly structured analytic semigroup of divisors. Sections 5.2 and 5.3 give versions of this theory. Section 5.3 also considers factorization results which depend only on separability of the algebra rather than the existence of an approximate identity.

5.1 Approximate Identities and Examples

In this Section we will discuss various definitions of approximate identities. We also introduce the strict topology and show that a nonunital Banach algebra $\mathcal{A}$ which has an approximate identity is dense in its double centralizer algebra in the strict topology. As a consequence, the double centralizer algebra is the completion of $\mathcal{A}$ in the strict topology. We begin by defining an approximate identity. For a more complete discussion of the material in this section, see Robert S.Doran and Josef Wichmann [1979].

519

In the literature, the terminology for approximate identities is some-what variable. We will use the strongest common condition as a definition and then show that various, apparently much weaker, conditions imply the existence of a bounded approximate identity in this strong sense. Recall that the weak or $\mathcal{A}^*$-topology on a normed linear space $\mathcal{A}$ is the weakest topology in which each element of $\mathcal{A}^*$ is continuous.

5.1.1 Definition Let $\mathcal{A}$ be a normed algebra. A $\langle$ left / weak left $\rangle$ approximate identity for $\mathcal{A}$ is a net $\{e_\alpha\}_{\alpha \in A} \subseteq \mathcal{A}$ such that, for each $a \in \mathcal{A}$, $\{e_\alpha a\}_{\alpha \in A}$ converges in the $\langle$ norm / weak topology $\rangle$ to a. A $\langle$ right / weak right $\rangle$ approximate identity is defined similarly and a $\langle$ two-sided / weak two-sided $\rangle$ approximate identity is a net which is both $\langle$ a left and a right / a weak left and a weak right $\rangle$ approximate identity. An approximate identity $\{e_\alpha\}_{\alpha \in A} \subseteq \mathcal{A}$ (of one of these types) is said to be bounded by $M \in \mathbb{R}_+$ if the set $\{\|e_\alpha\| : \alpha \in A\}$ is bounded by M and is said to be bounded if it is bounded by M for some $M \in \mathbb{R}_+$. Similarly, an approximate identity is said to be countable or commutative if $\{e_\alpha : \alpha \in A\}$ is countable or commutative, respectively. If the net is a sequence, the approximate identity is said to be sequential.

A normed algebra is said to be approximately unital if it has a bounded two-sided approximate identity.

It is obvious that an approximate identity of any type is a weak ap-proximate identity of the same type. Approximate identities which are not bounded are not particularly useful. See Dixon [1992] for some of the pathology in incomplete normed algebras. If any type of approximate iden-tity for a non-zero algebra is bounded by M, then it is clear that M is greater than or equal to 1. We will use this observation without comment in this section.

Approximate identities were apparently first considered explicitly by Irving E. Segal [1947a], who proved that any norm-closed self adjoint subal-gebra of the algebra of bounded linear operators on a Hilbert space contains an approximate identity. The group algebra $L^1(G)$ for any locally compact group also contains an approximate identity; special cases of this fact were used informally long before abstract approximate identities were defined. See, for instance, Hermann Weyl and F. Peter [1927], John von Neumann [1934], and Andre Weil [1940].

An approximately unital algebra $\mathcal{A}$ obviously shares some of the proper-ties of a unital algebra. For instance, the regular homomorphism of $\mathcal{A}$ into $\mathcal{D}(\mathcal{A})$ and both the left and right regular representations are homeomorphic isomorphisms. If $\mathcal{A}$ has an approximate identity bounded by M, we see

$$\frac{\|a\|}{M} \le \|L_a\| \le \|a\| \qquad \forall\, a \in \mathcal{A}$$

and similarly for R_a. Hence if $M = 1$, the three maps just mentioned are isometries. Also, any element belongs to the closed principal ⟨right / left / two-sided ⟩ ideal it generates. Similarly, we will show that any maximal left ideal in an approximately unital Banach algebra is closed. This is true in a unital algebra because the ideal would then be modular.

In a unital algebra, every element a is trivially a product: $a = 1a = a1$. From the beginning, approximate identities have been used to extend this useful result which is no longer trivial if the algebra is only approximately unital. We will pursue this in the next two sections after showing some other properties of approximate identities. We begin by showing how weaker conditions imply the existence of approximate identities.

Equivalent Conditions

In the next theorem, part (b) is due to Peter G. Dixon who has made a penetrating study of related results and counterexamples [1973a]. The proof given for (b) shows that a Banach algebra with a bounded left approximate identity and a right approximate identity has a two-sided approximate identity. Such an algebra need not have a bounded right or two-sided approximate identity, but a neat uniform boundedness argument shows that all sequential right approximate identities must be bounded. There is also a Banach algebra with a left and a right approximate identity (even with other desirable properties) but with no two-sided approximate identity. In the same paper, norms of approximate identities are studied. Part (c) was discovered independently by Mieczysław Altman [1972] and Joseph Wichmann [1973], and part (f) is essentially due to Teng Sun Liu, Arnoud van Rooij and Ju Kwei Wang [1973], Lemma 12. Barry E. Johnson [1972a], Proposition 1.6, gave a proof for (e) which was actually strong enough to establish (c).

5.1.2 Theorem *Let A be a normed algebra.*

(a) *There is a ⟨ left / right ⟩ approximate identity for A if and only if for every $\varepsilon > 0$ and every finite subset $\{a_1, a_2, \ldots, a_n\}$ of A there is an element $e \in A$ satisfying ⟨ $\|ea_j - a_j\| < \varepsilon$ / $\|a_j e - a_j\| < \varepsilon$ ⟩ for $j = 1, 2, \ldots, n$.*

(b) *If A has both a left and a right bounded approximate identity with bounds M and N, respectively, then it has a two-sided bounded approximate identity with bound $MN + M + N$.*

(c) *The following condition implies that A has a ⟨ left / right ⟩ approximate identity bounded by M:*

There is a number $M \in \mathbb{R}_+$ such that, for every $a \in A$ and every $\varepsilon > 0$, there is an $e \in A$ satisfying $\|e\| \leq M$ and ⟨ $\|ea - a\| < \varepsilon$ / $\|ae - a\| < \varepsilon$ ⟩.

(d) *The following condition implies that A has a bounded two-sided approximate identity:*

There is a number $M \in \mathbb{R}_+$ such that, for every $a \in A$, there are elements $e_1, e_2 \in A$ satisfying:

$$\|e_1\| \leq M, \ \|e_2\| \leq M, \ \|e_1 a - a\| < \varepsilon, \text{ and } \|a e_2 - a\| < \varepsilon.$$

(e) *If $\mathcal{A}$ has a bounded weak $\langle$ left / right / two-sided $\rangle$ approximate identity, then it has a bounded approximate identity of the same type.*

(f) *A closed commutative subalgebra $\mathcal{C}$ of $\mathcal{A}$ includes a bounded left approximate identity for $\mathcal{A}$ if for each $a \in \mathcal{A}$ there is a bounded net $\{e_\alpha\}_{\alpha \in A} \subseteq \mathcal{C}$ (with both the net and the bound possibly depending on a) such that $\{e_\alpha a\}_{\alpha \in A}$ converges to a in the weak topology. Similar statements hold for right and two-sided approximate identities for $\mathcal{A}$ included in $\mathcal{C}$.*

Proof (a): Suppose the "left" hypothesis holds. Let A be the set of finite subsets of $\mathcal{A}$ ordered by inclusion. For each $\alpha \in A$, choose $e_\alpha \in \mathcal{A}$ so that $\|e_\alpha a - a\| < n^{-1}$ holds for all $a \in \alpha$ where n is the number of elements in α. Now it is clear that $\{e_\alpha\}_{\alpha \in A}$ is a left approximate identity for $\mathcal{A}$. The converse is obvious. By symmetry, the same construction is possible if we replace "left" by "right".

(b): Let $\{e_\alpha\}_{\alpha \in A}$ and $\{f_\beta\}_{\beta \in B}$ be, respectively, a left and a right bounded approximate identity for $\mathcal{A}$ with bounds M and N. Let Δ be $A \times B$ ordered by setting $(\alpha, \beta) > (\alpha', \beta')$ if $\alpha > \alpha'$ and $\beta > \beta'$. Define a net $\{g_\delta\}_{\delta \in \Delta}$ by setting $g_{(\alpha, \beta)}$ equal to $f_\beta \circ e_\alpha$. An elementary calculation shows that $\{g_\delta\}_{\delta \in \Delta}$ is a two-sided bounded approximate identity with bound $MN + M + N$.

(c): We have assumed that for any $a \in \mathcal{A}$ and any $\varepsilon > 0$ there is an $e \in \mathcal{A}$ satisfying $\|e\| \leq M$ and $\|(1-e)a\| < \varepsilon$. Let $\{a_1, a_2, \ldots, a_n\}$ be a finite set in $\mathcal{A}$ and let ε be a positive number. Successively choose $e_1, e_2, \ldots, e_n$ in $\mathcal{A}$ satisfying $\|e_j\| \leq M$ and

$$\|(1 - e_j)(1 - e_{j-1}) \cdots (1 - e_1) a_j\| < \varepsilon (2(1 + M)^{n-j+1})^{-1}$$

for $j = 1, 2, \ldots, n$. Define $f \in \mathcal{A}$ by $1 - f = (1 - e_n)(1 - e_{n-1}) \cdots (1 - e_1)$. Then for any j between 1 and n, we have

$$
\begin{aligned}
\|f a_j - a_j\| &= \|(1 - f) a_j\| \\
&= \|[(1 - e_n)(1 - e_{n-1}) \cdots (1 - e_{j+1})][(1 - e_j) \cdots (1 - e_1)] a_j\| \\
&\leq (1 + M)^{n-j} \|(1 - e_j) \cdots (1 - e_1) a_j\| < \varepsilon (2(1 + M))^{-1}.
\end{aligned}
$$

Choose $e \in \mathcal{A}$ satisfying

$$\|e\| \leq M \quad \text{and} \quad \|ef - f\| < \varepsilon (2 \max\{\|a_j\| : j = 1, 2, \ldots, n\} + 1)^{-1}.$$

Then for each j between 1 and n, these inequalities give

$$
\begin{aligned}
\|e a_j - a_j\| &\leq \|e(f a_j - a_j)\| + \|(ef - f) a_j\| + \|f a_j - a_j\| \\
&\leq M \|f a_j - a_j\| + \|ef - f\| \, \|a_j\| + \|f a_j - a_j\| \\
&< M \varepsilon (2(1 + M))^{-1} + 2^{-1} \varepsilon + \varepsilon (2(1 + M))^{-1} = \varepsilon.
\end{aligned}
$$

The proof of (a) now establishes (c) since e_α can be chosen with norm bounded above by M.

(d): Results (b) and (c) together imply (d).

(e): Let $\{e_\alpha\}_{\alpha \in A}$ be a bounded weak left approximate identity. Since the norm closure of the convex hull $C = \mathrm{co}\{e_\alpha a : \alpha \in A\}$ is closed in the weak topology (Dunford and Schwartz [1958], Theorem V.3.13), it contains a. Since $C = \tilde{C}a$ where $\tilde{C} = \mathrm{co}\{e_\alpha : \alpha \in A\}$ is bounded by the bound for $\{e_\alpha\}_{\alpha \in A}$, (e) follows from (c). The other cases are similar.

(f): As in the proof of (e), we may assume that for each a there is a bounded net $\{e_\alpha\}_{\alpha \in A} \in C$ such that $e_\alpha a$ converges in norm to a. For each $n \in \mathbb{N}$, let $\mathcal{A}(n)$ be the closed set $\mathcal{A}(n) = \{a \in \mathcal{A} :$ there exists a net in C bounded by n which converges to a in norm$\}$. Since $\mathcal{A}$ equals $\cup_{n=1}^{\infty} \mathcal{A}(n)$ and each $\mathcal{A}(n)$ is norm closed, the Baire category theorem asserts that some $\mathcal{A}(N)$ includes an open ball. The set $\{a_1 - a_2 : a_1, a_2 \in \mathcal{A}(N)\}$ includes a neighborhood of zero and is closed under scalar multiplication, so it must equal $\mathcal{A}$. For any $a_1 - a_2 \in \mathcal{A}$ with $a_1, a_2 \in \mathcal{A}(N)$, let $\{e_{\alpha j}\}_{\alpha \in A}$ satisfy $\lim \|e_{\alpha j} a_j - a_j\| = 0$ for $j = 1, 2$. Then $\{e_{\alpha 1} \circ e_{\alpha 2}\}_{\alpha \in A} \subseteq C$ is a net bounded by $N^2 + 2N$ which satisfies $\lim \|(e_{\alpha 1} \circ e_{\alpha 2})a - a\| = 0$. Hence, (c) shows that $\mathcal{A}$ has a bounded approximate identity and the proof of (c) shows that it may be chosen in C. The proof for right approximate identities is similar. Use (b) for two-sided approximate identities. □

If $\mathcal{D}$ is a dense set in an algebra $\mathcal{A}$ with a $\langle$ left / right / two-sided $\rangle$ approximate identity or bounded approximate identity, then each element in this approximate identity can be approximated by an element in $\mathcal{D}$ giving an approximate identity in $\mathcal{D}$. Hence any separable approximately unital normed algebra has a countable bounded two sided approximate identity. This is an important result since Sinclair's theorem (Theorem 5.3.2) shows that much more is true in this case. Dixon [1987] shows that a left, right or two-sided approximate identity in a Banach algebra can be approximated by another of the same type in which all elements have their spectrum as close as desired to the unit interval $[0, 1]$ in $\mathbb{C}$.

The following result is due to Hans Reiter [1971].

5.1.3 Proposition *Let $\mathcal{I}$ be a closed ideal of a normed algebra $\mathcal{A}$.*

(a) *If $\mathcal{A}$ has $\langle$ left / right / two-sided $\rangle$ bounded approximate identity, then $\mathcal{A}/\mathcal{I}$ has an approximate identity of the same kind.*

(b) *If $\mathcal{I}$ has a $\langle$ left / right / two-sided $\rangle$ bounded approximate identity, then $\mathcal{A}$ has a $\langle$ left / right / two-sided $\rangle$ bounded approximate identity if and only if $\mathcal{A}/\mathcal{I}$ has a bounded approximate identity of the same kind.*

Proof (a): If $\{e_\alpha\}_{\alpha \in A}$ is any kind of approximate identity in $\mathcal{A}$, it is obvious that $\{e_\alpha + \mathcal{I}\}_{\alpha \in A}$ is an approximate identity of the same type in $\mathcal{A}/\mathcal{I}$ with the quotient norm.

(b): For the converse we will use (c) or (d) of Theorem 5.1.2. We consider the typical case of left approximate identities. Let M be the bound for the approximate identity in $\mathcal{I}$. Let $\varepsilon > 0$ and $a \in \mathcal{A}$ be arbitrary. Using the approximate identity in $\mathcal{A}/\mathcal{I}$ with bound N, we can find an e in $\mathcal{A}$ and a b in $\mathcal{I}$ satisfying $\|e\| \leq N$ and $\|ea - a + b\| < \varepsilon/(2M + 2)$. Using the approximate identity in $\mathcal{I}$, we can find $f \in \mathcal{A}$ satisfying $\|f\| \leq M$ and $\|fb - b\| < \varepsilon/2$. This gives

$$\|a - (f \circ e)a\| \leq \|(f - 1)(ea - a + b)\| + \varepsilon/2 < \varepsilon.$$

Since $\|f \circ e\|$ is bounded by $MN + M + N$, this verifies the hypotheses of Theorem 5.1.2(c). The case of right approximate identities is similar and then Theorem 5.1.2(d) handles the two-sided case. □

Dixon and George A. Willis [1992] prove an interesting result on approximately unital extensions which is quite unrelated to the last proposition. See Example 4.8.8 for the definition of topological nilpotence. A topologically nilpotent ideal is never approximately unital.

Proposition *Let $\mathcal{A}$ be a commutative Banach algebra satisfying $\mathcal{A}^2 = \mathcal{A}$. Then $\mathcal{A}$ is approximately unital if and only if it has some topologically nilpotent ideal $\mathcal{I}$ with $\mathcal{A}/\mathcal{I}$ approximately unital.*

It is easy to see that a direct product of finitely many Banach algebras is approximately unital if and only if each factor algebra is. The same result holds for tensor products. Richard J. Loy first established a similar result in [1970b] and the following proposition is due to James R. Holub [1972].

5.1.4 Proposition *Let $\mathcal{A}$ and $\mathcal{B}$ be Banach algebras and let $\mathcal{A} \otimes_\sigma \mathcal{B}$ be their complete algebra tensor product with respect to a reasonable norm σ. Then $\mathcal{A} \otimes_\sigma \mathcal{B}$ is approximately unital if and only if $\mathcal{A}$ and $\mathcal{B}$ are approximately unital.*

Proof The result actually holds for bounded one-sided approximate identities. We will give the proof for left identities.

First suppose $\{e_\alpha\}_{\alpha \in A}$ and $\{f_\beta\}_{\beta \in B}$ are bounded left approximate identities in $\mathcal{A}$ and $\mathcal{B}$ with bounds M and N, respectively. Let Δ be $A \times B$ with the usual product order. We will show that $\{e_\alpha \otimes f_\beta\}_{(\alpha,\beta) \in \Delta}$ is a bounded left approximate identity for $\mathcal{A} \otimes_\sigma \mathcal{B}$. For any $t = \sum_{j=1}^n a_j \otimes b_j$ in the algebraic tensor product, we get

$$t - (e_\alpha \otimes f_\beta)t = \sum_{j=1}^n (a_j - e_\alpha a_j) \otimes b_j + \sum_{j=1}^n a_j \otimes (b_j - f_\beta b_j)$$
$$- \sum_{j=1}^n (a_j - e_\alpha a_j) \otimes (b_j - f_\beta b_j),$$

and hence

$$\sigma(t - (e_\alpha \otimes f_\beta)t) \leq \sum_{j=1}^{n} \sigma((a_j - e_\alpha a_j) \otimes b_j) + \sum_{j=1}^{n} \sigma(a_j \otimes (b_j - f_\beta b_j))$$

$$+ \sum_{j=1}^{n} \sigma((a_j - e_\alpha a_j) \otimes (b_j - f_\beta b_j))$$

$$\leq \sum_{j=1}^{n} \|a_j - e_\alpha a_j\| \, \|b_j\| + \sum_{j=1}^{n} \|a_j\| \, \|b_j - f_\beta b_j\|$$

$$+ \sum_{j=1}^{n} \|a_j - e_\alpha a_j\| \, \|b_j - f_\beta b_j\|.$$

This estimate shows that $\{e_\alpha \otimes f_\beta\}_{(\alpha,\beta)\in\Delta}$ is an approximate identity bounded by MN in the algebraic tensor product. However, $\mathcal{A} \otimes_\sigma \mathcal{B}$ is just the completion of this algebraic tensor product in the norm σ, so that we have a bounded approximate identity on the complete tensor product.

Conversely, suppose $\{e_\alpha\}_{\alpha\in A}$ is an approximate identity for $\mathcal{A} \otimes_\sigma \mathcal{B}$. Let ω be an arbitrary element of $\mathcal{A}^*$ and consider the function $\omega \otimes I$ of the algebraic tensor product $\mathcal{A} \otimes \mathcal{B}$ into $\mathcal{B}$. This linear map is continuous in the injective tensor norm and hence in σ so we extend it to the complete tensor product. For any $a, c \in \mathcal{A}$, recall the notation $_c\omega(a) = \omega(ac)$ and note

$$((_c\omega \otimes I)(a \otimes b))(d) = (\omega \otimes I)((a \otimes b)(c \otimes d)) \qquad \forall \, a, c \in \mathcal{A}; b, d \in \mathcal{B}.$$

Linearity and continuity allow us to replace $a \otimes b$ by any tensor $t \in \mathcal{A} \otimes_\sigma \mathcal{B}$. Choose ω and c so that $\omega(c) = 1 = \|\omega\| \, \|c\|$ so that we get

$$(_c\omega \otimes I)(e_\alpha)(d) = (\omega \otimes I)(e_\alpha(c \otimes d))$$

for all $d \in \mathcal{B}$ and e_α in the approximate identity. This shows that $\{(_c\omega \otimes I)(e_\alpha)\}_{\alpha\in A}$ is a bounded approximate identity for $\mathcal{B}$. Obviously, $\mathcal{A}$ and right approximate identities can be handled similarly. □

The Strict Topology and Double Centralizers

We will now introduce the strict topology. It was first defined by R. Creighton Buck [1958] for the special case of $C_o(\Omega) \subseteq C(\Omega)$, where Ω is a locally compact space. The general definition given here is due to Robert C. Busby [1968], who also proved Propositions 5.1.6 and 5.1.7. See also Donald Curtis Taylor [1970].

5.1.5 Definition Let $\mathcal{A}$ be a normed algebra and let $\mathcal{I}$ be an ideal in $\mathcal{A}$. The *strict topology of $\mathcal{A}$ defined by $\mathcal{I}$* is the topology defined by the family of semi-norms $\{\lambda_b, \rho_b : b \in \mathcal{I}\}$ defined by

$$\begin{aligned} \lambda_b(a) &= \|ba\|, \\ \rho_b(a) &= \|ab\| \end{aligned} \quad \forall \, a \in \mathcal{A}.$$

Thus, a net $\{a_\alpha\}_{\alpha \in A}$ in $\mathcal{A}$ converges to $a \in \mathcal{A}$ in the strict topology defined by $\mathcal{I}$ if and only if $\lim \|ba_\alpha - ba\| = \lim \|a_\alpha b - ab\| = 0$ holds for all $b \in \mathcal{I}$. Clearly, the strict topology is locally convex. It is Hausdorff unless there is a non-zero element $a \in \mathcal{A}$ such that $ab = ba = 0$ for all $b \in \mathcal{I}$. It is also clear that the multiplication of $\mathcal{A}$ is continuous in each variable separately in the strict topology.

5.1.6 Proposition *Let $\mathcal{A}$ be a Banach algebra with a two-sided approximate identity. Then all double centralizers are continuous so $\mathcal{D}_B(\mathcal{A})$ is a Banach algebra which equals $\mathcal{D}(\mathcal{A})$ as an algebra. Furthermore, $\mathcal{A}$ is dense in $\mathcal{D}(\mathcal{A})$ in the strict topology of $\mathcal{D}(\mathcal{A})$ defined by $\mathcal{A}$.*

Proof The first result follows from Theorem 1.2.4. Let $\{e_\alpha\}_{\alpha \in A}$ be a two-sided approximate identity for $\mathcal{A}$. If $d \in \mathcal{D}(\mathcal{A})$ and $a \in \mathcal{A}$, then de_α belongs to $\mathcal{A}$ for each $\alpha \in A$. Also both $\lambda_a(d - de_\alpha) = \|(ad) - (ad)e_\alpha\|$ and $\rho_a(d - de_\alpha) = \|da - de_\alpha a\| \le \|d\| \, \|a - e_\alpha a\|$ converge to zero. $\quad\square$

A net $\{a_\alpha\}_{\alpha \in A}$ in a topological vector space $\mathcal{A}$ is called a *Cauchy net* if, for every neighborhood $\mathcal{U}$ of zero in $\mathcal{A}$, there is an $\alpha_0 \in A$ such that $a_\alpha \in a_{\alpha_0} + \mathcal{U}$ for all $\alpha \ge \alpha_0$. A net with a limit is always a Cauchy net. A topological linear space is said to be *complete* if each of its Cauchy nets has a limit. Given any Hausdorff locally convex topological linear space $\mathcal{A}$, it is possible to construct a complete Hausdorff locally convex topological linear space $\overline{\mathcal{A}}$ in which $\mathcal{A}$ is a dense linear subspace. Then $\overline{\mathcal{A}}$ is unique up to linear homeomorphism leaving $\mathcal{A}$ invariant and is called the *completion* of $\mathcal{A}$. It turns out that if $\mathcal{A}$ is a Banach algebra with a two-sided approximate identity, then $\mathcal{D}(\mathcal{A})$ is the completion of $\mathcal{A}$ in the strict topology.

5.1.7 Proposition *Let $\mathcal{A}$ be a Banach algebra with a two-sided approximate identity. The strict topology on $\mathcal{A}$ defined by $\mathcal{A}$ is inherited from the strict topology $\mathcal{A}$ defines on $\mathcal{D}(\mathcal{A})$. Furthermore, $\mathcal{D}(\mathcal{A})$ is the completion of $\mathcal{A}$ in this strict topology. If the two-sided approximate identity has norm one, then $\mathcal{A}_1$ is dense in $\mathcal{D}_B(\mathcal{A})_1$ so that $\mathcal{D}_B(\mathcal{A})_1$ is the completion of $\mathcal{A}_1$.*

Proof The first statement is obvious. In order to prove the second statement, Proposition 5.1.6 establishes that it is enough to show that $\mathcal{D}(\mathcal{A})$ is complete in the strict topology defined by $\mathcal{A}$. Let $\{(L^\alpha, R^\alpha)\}_{\alpha \in A} \subseteq \mathcal{D}(\mathcal{A})$ be a Cauchy net in the strict topology. Then for any $a \in \mathcal{A}, (L^\alpha, R^\alpha)a = L^\alpha(a)$ and $a(L^\alpha, R^\alpha) = R^\alpha(a)$ form Cauchy nets in the norm topology of $\mathcal{A}$. Denote their limits by $L(a)$ and $R(a)$, respectively. Then for any $a, b \in \mathcal{A}$

we have $aL(b) = \lim aL^\alpha(b) = \lim R^\alpha(a)b = R(a)b$. Hence, Theorem 1.2.4 shows that (L, R) is a double centralizer. Clearly, $\{(L^\alpha, R^\alpha)\}_{\alpha \in A}$ converges to (L, R) in the strict topology. Hence, $\mathcal{D}(\mathcal{A})$ is complete in the strict topology defined by $\mathcal{A}$. The final remark follows from the proof of Proposition 5.1.6.   □

Approximate Identities and Arens Multiplication

Recall the definition of the two Arens multiplications given in Section 1.4. Part of the following result was first noted by Paul Civin and Bertram Yood [1961], Lemma 3.8. Also recall that a mixed identity for $\mathcal{A}^{**}$ is a right identity for the first Arens product which is also a left identity for the second Arens product.

5.1.8 Proposition *Let $\mathcal{A}$ be a Banach algebra.*

(a) $\mathcal{A}$ has a bounded ⟨ left / right / two-sided ⟩ approximate identity if and only if $\mathcal{A}^{**}$ has a ⟨ left / right / mixed ⟩ identity for the ⟨ second / first / both ⟩ Arens multiplication(s).

(b) If $e \in \mathcal{A}^{**}$ is an identity for the ⟨ first / second ⟩ Arens multiplication, then e is also a ⟨ left / right ⟩ identity for the ⟨ second / first ⟩ Arens multiplication, and the canonical image in $\mathcal{A}^{**}$ of any bounded weak ⟨ left / right ⟩ approximate identity for $\mathcal{A}$ converges to e in the weak topology.

Proof Theorem 5.1.2 shows that a normed algebra with a bounded ⟨ left / right / two-sided ⟩ weak approximate identity has a bounded approximate identity of the same kind. Suppose $\{e_\alpha\}_{\alpha \in A}$ is an arbitrary subnet of a bounded weak left approximate identity for $\mathcal{A}$. Alaoglu's theorem shows that a subnet $\{\kappa(e_\gamma)\}_{\gamma \in \Gamma}$ of $\{\kappa(e_\alpha)\}_{\alpha \in A}$ converges to some $e \in \mathcal{A}^{**}$ in the weak topology where κ is the natural map. Any $a \in \mathcal{A}$ and $\omega \in \mathcal{A}^*$ satisfy $\omega(a) = \lim \omega(e_\gamma a) = \lim {}_a\omega(e_\gamma) = e({}_a\omega) = \omega_e(a)$. Therefore, any $f \in \mathcal{A}^{**}$ and $\omega \in \mathcal{A}^*$ satisfy $e \cdot f(\omega) = f(\omega_e) = f(\omega)$. Hence, e is a left identity for the second Arens multiplication. If $\mathcal{A}^{**}$ has an identity for the second Arens multiplication, e must be this identity. In this case, the canonical image of any bounded weak left approximate identity converges to e in the weak topology, since we have shown that some subnet of an arbitrary subnet converges to e. An entirely analogous argument shows that the canonical image in $\mathcal{A}^{**}$ of a bounded weak right approximate identity for $\mathcal{A}$ has a subnet converging to some right identity for the first Arens multiplication. Similarly, if $\mathcal{A}^{**}$ has an identity e for the first Arens product, the net itself converges to e.

Conversely, suppose $\mathcal{A}^{**}$ has a left identity element e for either Arens multiplication. Since $\kappa(\mathcal{A}_1)$ is dense in $(\mathcal{A}^{**})_1$ in the $\mathcal{A}^*$-topology, there is a bounded net $\{e_\alpha\}_{\alpha \in A}$ such that $\{\kappa(e_\alpha)\}_{\alpha \in A}$ converges to e. All $\omega \in \mathcal{A}^*$ and $a \in \mathcal{A}$ satisfy $\lim \omega(e_\alpha a) = \lim \kappa(e_\alpha a)(\omega) = \lim \kappa(e_\alpha)({}_a\omega) = e({}_a\omega) = e({}_{\kappa(a)}\omega) = e\kappa(a)(\omega) = e \cdot \kappa(a)(\omega) = \kappa(a)(\omega) = \omega(a)$. Hence $\{e_\alpha\}_{\alpha \in A}$ is a bounded weak left approximate identity. Similarly, if e is a right identity element for either Arens multiplication, then it is the limit of the canonical image in $\mathcal{A}^{**}$ of some bounded weak right approximate identity for $\mathcal{A}$. □

5.1.9 Approximate Identities in Group Algebras We now discuss approximate identities in $L^1(G)$ where G is a locally compact group. Certain concrete approximate identities were considered long before the abstract definition was introduced. In 1.8.15 we showed that the Fejér kernel, for example, is an approximate identity for $L^1(\mathbb{T})$.

We will now construct a simple two-sided approximate identity bounded by one in $L^1(G)$. For terminology, see Section 1.9. Let $\mathcal{U}$ be the collection of all compact neighborhoods of the identity in G ordered by reverse inclusion. For each $U \in \mathcal{U}$, let e_U be $\lambda(U)^{-1}$ times the characteristic function of U. Then each e_U belongs to $L^1(G)$ and satisfies $\|e_U\|_1 = 1$. Hence, $\{e_U\}_{U \in \mathcal{U}}$ is a net in $L^1(G)$ bounded by one. Theorem 20.4 in Hewitt and Ross [1963] asserts that, for any $\varepsilon > 0$ and any $f \in L^1(G)$, there is a neighborhood $V \in \mathcal{U}$ satisfying $\|_v f - f\|_1 < \varepsilon$ for all $v \in V$. Hence if $U \in \mathcal{U}$ satisfies $U \subseteq V$, then Fubini's theorem gives

$$
\begin{aligned}
\|e_U * f - f\|_1 &= \int_G |\int_U \lambda(U)^{-1} f(v^{-1}u)dv - f(u)|du \\
&\leq \int_G \lambda(U)^{-1} \int_U |_v f(u) - f(u)|dv\,du \\
&= \lambda(U)^{-1} \int_U \int_G |_v f(u) - f(u)|du\,dv \\
&= \lambda(U)^{-1} \int_U \|_v f - f\|_1 dv < \varepsilon.
\end{aligned}
$$

For any $\varepsilon > 0$ and any non-zero $f \in L^1(G)$, the same theorem guarantees the existence of a $V \in \mathcal{U}$ satisfying both $\|f_v - f\|_1 < \varepsilon/2$ and $|\Delta(v)^{-1} - 1| \leq \varepsilon/(2\|f\|_1 + 1)$ for all $v \in V$. Then for any $U \in \mathcal{U}$ satisfying $U \subseteq V$, Fubini's theorem gives

$$
\begin{aligned}
\|f * e_U - f\|_1 &= \int_G |\int_U \lambda(U)^{-1}\Delta(v)^{-1} f(uv^{-1})dv - f(u)|du \\
&\leq \int_G \lambda(U)^{-1} \int_U |\Delta(v)^{-1} f_v(u) - f(u)|dv\,du \\
&= \lambda(U)^{-1} \int_U \int_G |\Delta(v)^{-1} f_v(u) - f(u)|du\,dv
\end{aligned}
$$

$$\leq \quad \lambda(U)^{-1} \int\limits_{U} \left(\int\limits_{G} |\Delta(v)^{-1} f_v(u) - f_v(u)| \, du \right.$$

$$\left. + \int\limits_{G} |f_v(u) - f(u)| \, du \right) dv$$

$$= \quad \lambda(U)^{-1} \int\limits_{U} \left(|\Delta(v)^{-1} - 1| \, \|f_v\|_1 + \|f_v - f\|_1 \right) dv < \varepsilon.$$

Hence, $\{e_U\}_{U \in \mathcal{U}}$ is a two-sided approximate identity bounded by one.

We have constructed the simplest approximate identity bounded by one for $L^1(G)$ when G is an arbitrary locally compact group G. However, certain special approximate identities may prove useful in various circumstances. First, we mention another one which can be defined for any locally compact group G. Let $\mathcal{V}$ be the collection of all compact symmetric (*i.e.*, $u \in V \leftrightarrow u^{-1} \in V$) neighborhoods of the identity in G. For each $V \in \mathcal{V}$, let f_V be $\Delta^{-1/2}$ times the characteristic function of V. Then define e_V to be $\|f_V\|_1^{-1} f_V$ so that $\{e_V\}_{V \in \mathcal{V}}$ is a net of elements of norm one in $L^1(G)$. Arguments entirely similar to those given above show that this net is again a two-sided approximate identity bounded by one. It has the advantage that its elements are invariant under the natural involution on $L^1(G)$, which will be introduced in Volume II.

Some groups G have approximate identities in the center of $L^1(G)$. These are called *central approximate identities*.

Definition   Let G be a locally compact group. A subset of G is said to be *invariant* if it is invariant under all inner automorphisms. The group G is said to have ⟨ *an invariant neighborhood / small invariant neighborhoods* ⟩ (denoted by ⟨ $G \in$ [IN] / $G \in$ [SIN] ⟩) if ⟨ it has / included in every neighborhood of the identity it has ⟩ a compact neighborhood U of the identity which is invariant.

Proposition   *Denote the class of locally compact abelian groups by* [A], *the class of compact groups by* [K], *the class of discrete groups by* [D] *and the class of locally compact unimodular groups by* [Um]. *Then these classes satisfy*

$$[A] \cup [K] \cup [D] \subseteq [SIN] \subseteq [IN] \subseteq [Um].$$

The algebra $L^1(G)$ has a central approximate identity (which may be chosen to be of norm 1) if and only if $G \in$ [SIN].

Proof   If G is abelian, every neighborhood of the identity is invariant. If G is compact, the remark after Theorem 4.9 in Hewitt and Ross [1963] shows that G belongs to [SIN]. If G is discrete, $\{e\}$ is a compact invariant neighborhood of e. Thus the first inclusion holds. The second inclusion is obvious. Suppose U is a compact invariant neighborhood of e in G. Then $\lambda(U)$ is finite, non-zero and satisfies

$$\lambda(U) = \int{}_u f d\lambda = \int f_u d\lambda = \Delta(u)\lambda(U) \qquad \forall\, u \in G$$

where f is the characteristic function of U. This gives the final inclusion.

We shall only establish that $L^1(G)$ has a central approximate identity of norm 1 when $G \in [\mathrm{SIN}]$. (The opposite implication is proved in Robert D. Mosak [1971].) Let $\mathcal{U}$ be the set of all compact invariant neighborhoods of e ordered by reverse inclusion. For each $U \in \mathcal{U}$, let e_U be $\lambda(U)^{-1}$ times the characteristic function of U. Since this choice of $\mathcal{U}$ is cofinal in our earlier choice, $\{e_U\}_{U \in \mathcal{U}}$ is an approximate identity. To see that it is in the center, let $f \in L^1(G)$ and $u \in G$ be arbitrary and consider

$$
\begin{aligned}
e_U * f(u) &= \int e_U(uv)f(v^{-1})dv = \int {}_u e_U(v)f(v^{-1})dv \\
&= \int (e_U)_u(v)f(v^{-1})dv = \int \Delta(v)f(v^{-1})e_U(vu)dv = f * e_U(u).
\end{aligned}
$$

$\square$

Many other relations are known between these and other classes of locally compact groups. A locally compact group G is said to belong to $[\mathrm{FC}]^-$ if each of its conjugacy classes ($C_v = \{u^{-1}vu : u \in G\}$ for $v \in G$) has compact closure, and to belong to $[\mathrm{FIA}]^-$ if the closure of the group of inner automorphisms of G is compact in the topological group $\mathrm{Aut}(G)$ of all homeomorphic isomorphisms of G. (The natural *Braconnier topology* of $\mathrm{Aut}(G)$ is described in Hewitt and Ross [1963], 26.3. It is a slight variation on the compact open topology.) Since the closure of a conjugacy class is included in the continuous image of the closure of the group of inner automorphisms, the inclusion $[\mathrm{FIA}]^- \subseteq [\mathrm{FC}]^-$ is clear. An easy argument (which we omit) also shows $[\mathrm{FIA}]^- \subseteq [\mathrm{SIN}]$. The following important theorem was proved by Siegfried Grosser and Martin A. Moskowitz in [1967].

Theorem *A locally compact group belongs to* $[\mathrm{FIA}]^-$ *if and only if it belongs to both* $[\mathrm{FC}]^-$ *and* $[\mathrm{SIN}]$.

The Freudenthal–Weil theorem (*cf. e.g.*, Grosser and Moskowitz [1967], Theorem 4.3) asserts that a connected group lives in $[\mathrm{SIN}]$ if and only if it belongs to the class $[\mathrm{MAP}]$ introduced in §3.2.17 above, and that such a group is necessarily homeomorphically isomorphic to the direct product of $\mathbb{R}^n$ and a compact group for some n. (It is immediately obvious that such a direct product belongs to both $[\mathrm{MAP}]$ and $[\mathrm{SIN}]$.) We mention also the inclusion $[\mathrm{FC}]^- \subseteq [\mathrm{IN}]$ and refer the reader to Grosser and Moskowitz [1971a] Palmer [1978] and Volume II of the present work for detailed references, examples and many related results.

5.1.10   The semigroup algebra $L^1(\mathbb{R}_+)$ This algebra plays a special role relative to the theory of Section 5.3, so we will also discuss it briefly. If $\mathcal{A}$ is a Banach algebra with a countable approximate identity bounded by 1, then there is contractive homomorphism of $L^1(\mathbb{R}_+)$ into $\mathcal{A}$ and the concrete analytic semigroup of $L^1(\mathbb{R}_+)$ is mapped onto the more abstract analytic semigroup of $\mathcal{A}$ constructed in Section 5.3. From another vantage point, $L^1(\mathbb{R}_+)$ has the same relationship to the classical Laplace transform as $L^1(\mathbb{R})$ has to the Fourier integral transform and $L^1(\mathbb{T})$ has to Fourier series. More specifically, in each of these three cases the Gelfand transform can be identified with the classical transform mentioned. This example is related to Examples 3.4.10, 4.8.3 and 4.8.4.

As a Banach space, $L^1(\mathbb{R}_+)$ is the usual space of equivalence classes of absolutely integrable functions on $\mathbb{R}_+ = [0, \infty[$ with respect to Lebesgue measure. The norm of $f \in L^1(\mathbb{R}_+)$ is given by $\|f\|_1 = \int_0^\infty |f(t)|dt$, and the *convolution product* of $f, g \in L^1(\mathbb{R}_+)$ by

$$f * g(t) = \int_0^t f(t - s)g(s)ds \qquad \forall\, t \in \mathbb{R}_+.$$

A change of variables shows that $L^1(\mathbb{R}_+)$ is commutative.

Let $\mathbf{H}$ represent the open right half plane $\{\lambda \in \mathbb{C} : \mathrm{Re}(\lambda) > 0\}$. By identifying the closed right half plane $\overline{\mathbf{H}}$ with the Gelfand space of $L^1(\mathbb{R}_+)$ under the map $\lambda \mapsto \gamma_\lambda$ defined by

$$\gamma_\lambda(f) = \int_0^\infty f(t)e^{-\lambda t}dt \qquad \forall\, f \in L^1(\mathbb{R}_+)$$

we see that $L^1(\mathbb{R}_+)$ is semisimple but nonunital. We also see that under this identification the Gelfand transform $\hat{f}$ of $f \in L^1(\mathbb{R}_+)$ is just the Laplace transform: $\mathcal{L}(f)(\lambda) = \hat{f}(\gamma_\lambda)$, as already claimed. The Laplace transform $\mathcal{L}(f)$ is analytic on $\mathbf{H}$ as well as continuous on all of $\overline{\mathbf{H}}$.

The proof given in Example 3.4.10 shows that the sequence of functions $\{e_n\}_{n \in \mathbb{N}}$ (where $e_n(s)$ equals n for $s \leq 1/n$ and equals zero for $s > 1/n$) is a sequential approximate identity of norm 1 for $L^1(\mathbb{R}_+)$. The Titchmarsh convolution theorem [1926] shows that $L^1(\mathbb{R}_+)$ has no non-zero divisors of zero. Johnson [1964a] extended Wendel's theorem to identify the double centralizer algebra of $L^1(\mathbb{R}_+)$ as the convolution algebra $M(\mathbb{R}_+)$. As a Banach space this is the linear space of complex regular Borel measures on $\mathbb{R}_+$ with the total variation norm. The convolution product is given by

$$\mu * \nu(E) = \int \int \chi_E(s + t)d\mu(s)d\nu(t) \qquad \forall\ \text{Borel sets } E.$$

The basic analytic semigroup of $L^1(\mathbb{R}_+)$ is closely related to the *gamma function*:

$$\Gamma(\lambda) = \int_0^\infty t^{\lambda-1}e^{-t}dt \qquad \forall\, \lambda \in \mathbf{H}$$

(which is most famous for extending the factorial: $\Gamma(n+1) = n!$). For each $\lambda \in \mathbf{H}$, define $e(\lambda) \in L^1(\mathbb{R}_+)$ by

$$e(\lambda)(t) = t^{\lambda-1}e^{-t}/\Gamma(\lambda) \qquad \forall\, t \in \mathbb{R}_+.$$

Sinclair [1982] proves that this is an analytic function which is also a semi-group with interesting growth conditions. If $\lambda \in \mathbf{H}$ converges to 0 in any reasonable way, then $e(\lambda)$ becomes an approximate identity. For related results see Grabiner [1981], Dales [1983] and Dales and McClure [1987].

5.1.11 C*-algebras are Approximately Unital   C*-algebras were defined in §1.7.17. Proofs of all the following remarks will be found in Volume II. The functional calculus introduced in Example 3.4.17 is the chief tool in the proofs.

In a C*-algebra $\mathcal{A}$, the set $\mathcal{A}_H$ of hermitian elements (*i.e.*, those elements $h \in \mathcal{A}$ satisfying $h^* = h$) plays the role of the set of real numbers in the complex numbers. Each hermitian element has its spectrum in $\mathbb{R}$. (The converse is only true in the commutative case.) The set $\mathcal{A}_+$ of hermitian (or, equivalently, normal) elements with spectrum in $\mathbb{R}_+$ then plays the role of non-negative real numbers. These elements are actually called *positive elements*. In particular, the set of positive elements defines a partial order on $\mathcal{A}_H$:

$$h \leq k \quad \Leftrightarrow \quad k - h \in \mathcal{A}_+.$$

In a nonunital C*-algebra $\mathcal{A}$, it turns out that $\mathcal{A}_+ \cap \mathcal{A}_1$ is an approximate identity (obviously bounded by 1) in its own order. This approximate identity consisting of positive elements is attractive because of its natural order.

In the case of separable C*-algebras, it is easy to choose a sequential approximate identity. Johan F. Aarnes and Richard V. Kadison [1969] proved that any separable C*-algebra has a commutative approximate identity. We will use Sinclair's analytic semigroup construction in Section 5.3 to show that any separable approximately unital Banach algebra also has a commutative sequential approximate identity.

5.1.12 Approximate Identities in $\mathcal{B}_K(\mathcal{X})$ Let $\mathcal{X}$ be a Banach space which is not finite-dimensional. Then the ideals $\mathcal{B}_F(\mathcal{X})$, $\mathcal{B}_A(\mathcal{X})$ and $\mathcal{B}_K(\mathcal{X})$ of finite-rank, approximable and compact operators, respectively, are all nonunital. The following results are from Dixon [1986] and Niels Grønbaek and George A. Willis [1993]. In the latter paper, versions of some of the results are stated for any subalgebra of $\mathcal{B}_K(\mathcal{X})$ which includes $\mathcal{B}_F(\mathcal{X})$. See Definition 1.1.17 and §1.7.7 and compare Doran and Wichmann [1979], §30. We say that a normed algebra $\mathcal{A}$ has a ⟨ *left / right* ⟩ *approximate unit* if for every $a \in \mathcal{A}$ and every $\varepsilon > 0$ there is an element u satisfying ⟨ $\|ue - e\| < \varepsilon$ / $\|eu - e\| < \varepsilon$ ⟩. Theorem 5.1.2(c) shows that if there is a uniform upper bound for the norms of the elements u in a left or right approximate unit, then the algebra has a bounded approximate identity of the same type, but this fails if there is no such upper bound.

Theorem   *Let $\mathcal{X}$ be a Banach space.*

(a) $\mathcal{B}_F(\mathcal{X})$ *always has a left and right approximate unit.*

(b) *There is an example of a Banach space $\mathcal{X}$ for which neither $\mathcal{B}_A(\mathcal{X})$ nor $\mathcal{B}_K(\mathcal{X})$ has a left approximate unit.*

(c) $\langle$ $\mathcal{B}_F(\mathcal{X})$ *and* $\mathcal{B}_A(\mathcal{X})$ *have* / $\mathcal{B}_K(\mathcal{X})$ *has* $\rangle$ *a left approximate identity if $\mathcal{X}$ has the $\langle$ approximation property / compact approximation property $\rangle$.*

(d) $\langle$ $\mathcal{B}_F(\mathcal{X})$ *and* $\mathcal{B}_A(\mathcal{X})$ *have* / $\mathcal{B}_K(\mathcal{X})$ *has* $\rangle$ *a bounded left approximate identity if and only if $\mathcal{X}$ has the $\langle$ bounded approximation property / bounded compact approximation property $\rangle$.*

(e) $\langle$ $\mathcal{B}_F(\mathcal{X})$ / $\mathcal{B}_A(\mathcal{X})$ / $\mathcal{B}_K(\mathcal{X})$ $\rangle$ *has a bounded right approximate identity if and only if it has a bounded two-sided approximate identity and this occurs if and only if the identity operator in $\mathcal{B}(\mathcal{X}^*)$ can be approximated in the Grothendieck topology by the adjoints of operators in some bounded subset of $\langle$ $\mathcal{B}_F(\mathcal{X})$ / $\mathcal{B}_A(\mathcal{X})$ / $\mathcal{B}_K(\mathcal{X})$ $\rangle$. If $\mathcal{X}$ has a shrinking basis, then the associated sequence $\{P_n\}$ of basis projections in $\mathcal{B}_F(\mathcal{X})$ is a bounded two-sided approximate identity for each of these algebras.*

Proof (a): The collection of all finite-dimensional projections can serve as the approximate unit.

(b): The example is Szankowski's subspace of ℓ^1 from [1978]. Details are given in Dixon [1986], §3.

(c): We use the criterion of Theorem 5.1.2(a). Let $K_1, K_2, \ldots, K_n$ be an arbitrary finite set of operators in $\langle$ $\mathcal{B}_F(\mathcal{X})$ / $\mathcal{B}_A(\mathcal{X})$ / $\mathcal{B}_K(\mathcal{X})$ $\rangle$. Then $\bigcup_{j=1}^{n} T_j(\mathcal{X}_1)$ is compact so under our hypotheses, for any $\varepsilon > 0$, we can find an operator $E \in \langle$ $\mathcal{B}_F(\mathcal{X})$ / $\mathcal{B}_A(\mathcal{X})$ / $\mathcal{B}_K(\mathcal{X})$ $\rangle$ satisfying $\|EK_j - K_j\| < \varepsilon$ as we wished to show.

(d): The existence of the bounded approximate identity follows from our last argument and Theorem 5.1.2(b) since the operator E can be chosen of bounded norm.

The converse is an easy consequence of the proposition in §1.7.10. The existence of a bounded approximate identity with bound B in one of these algebras immediately implies that for any $\varepsilon > 0$ and any finite subset $\{x_1, x_2, \ldots, x_n\}$ of $\mathcal{X}$ there is an operator T in the algebra, satisfying $\|T\| \leq B$ and $\|Tx_j - x_j\| < \varepsilon$ for $j = 1, 2, \ldots, n$.

(e): The Banach space adjoint operation provides an injective isometric anti-homomorphism of $\mathcal{B}(\mathcal{X})$ into $\mathcal{B}(\mathcal{X}^*)$ which sends $\mathcal{B}_F(\mathcal{X})$ into $\mathcal{B}_F(\mathcal{X}^*)$ and $\mathcal{B}_K(\mathcal{X})$ into $\mathcal{B}_K(\mathcal{X}^*)$. Hence each of the algebras in (e) has a bounded right approximate identity if and only if the algebra of all its Banach space adjoints in $\mathcal{B}_K(\mathcal{X}^*)$ has a bounded left approximate identity. The arguments given in the proof of (d) establish the equivalence of the existence of these bounded left approximate identities and the corresponding approximation properties on $\mathcal{X}^*$. However, these approximation properties on $\mathcal{X}^*$ imply the corresponding approximation properties on $\mathcal{X}$ and hence the existence of bounded left approximate identities for the original algebras. Now apply Theorem 5.1.2(b).   □

5.2   General Factorization Theorems

We use the next simple proposition in both Theorem 5.2.2 and 5.3.2. It has been stated in a very general way to cover any possible application. We denote the closed linear span of a subset $\mathcal{S}$ in a normed linear space by $\overline{\operatorname{span}}(\mathcal{S})$.

5.2.1 Proposition   *Let $\mathcal{A}$ be a normed algebra with a bounded $\langle$ left / right / two-sided $\rangle$ approximate identity $\{e_\alpha\}_{\alpha \in A}$. Let $T^{(1)}, T^{(2)}, \ldots, T^{(n)}$ be continuous $\langle$ representations / anti-representations / representations and anti-representations $\rangle$ on normed linear spaces $\mathcal{X}^1, \mathcal{X}^2, \ldots, \mathcal{X}^n$, respectively. Consider $\{a_1, a_2, \ldots, a_p\} \subseteq \mathcal{A}$ and for each $j = 1, 2, \ldots, n$, arbitrary finite sets $\{x_{j1}, x_{j2}, \ldots, x_{jq_j}\} \subseteq \overline{\operatorname{span}}(T_A \mathcal{X}^j)$. Then for any positive number ε, there is an $\alpha_0 \in A$ such that*

$$\langle \; \|e_\alpha a_j - a_j\| < \varepsilon \quad \forall j = 1, 2, \ldots, p \; / \;\; \|a_j e_\alpha - a_j\| < \varepsilon \quad \forall j = 1, 2, \ldots, p \; /$$

$$\|e_\alpha a_j - a_j\| < \varepsilon \; \text{and} \; \|a_j e_\alpha - a_j\| < \varepsilon \quad \forall j = 1, 2, \ldots, p \; \rangle \qquad and$$

$$\|T^{(j)}_{e_\alpha} x_{jk} - x_{jk}\| < \varepsilon \qquad \forall \; j = 1, 2, \ldots, n; \; k = 1, 2, \ldots, q_j$$

hold for all $\alpha \geq \alpha_0$.

Proof Suppose the approximate identity is a left one bounded by M. First, we choose a representation $T \colon \mathcal{A} \to \mathcal{B}(\mathcal{X})$ and $x \in \overline{\operatorname{span}}(T_A \mathcal{X})$ from among all the given data. Since x is in $\overline{\operatorname{span}}(T_A \mathcal{X})$, we can choose $b_j \in \mathcal{A}$ and $x_j \in \mathcal{X}$ so that $\|x - \sum_{j=1}^m T_{b_j} x_j\| < \varepsilon (2 + 2M\|T\|)^{-1}$. To simplify notation, denote the difference $x - \sum_{j=1}^m T_{b_j} x_j$ by z. Choose $\alpha_0 \in A$ so that $\alpha \geq \alpha_0$ implies $\|e_\alpha b_j - b_j\| < \varepsilon (2\|T\| \sum_{j=1}^m \|x_j\| + 1)^{-1}$ for $j = 1, 2, \ldots, m$. Then, for $\alpha \geq \alpha_0$ we have

$$
\begin{aligned}
\|T_{e_\alpha} x - x\| &\leq \|T_{e_\alpha}(z)\| + \sum_{j=1}^m \|T_{e_\alpha} T_{b_j} x_j - T_{b_j} x_j\| + \| - z\| \\
&\leq \|T_{e_\alpha}\| \, \|z\| + \sum_{j=1}^n \|T_{e_\alpha b_j - b_j}\| \, \|x_j\| + \|z\| \\
&< (\|T\|M + 1)\varepsilon (2 + 2M\|T\|)^{-1} \\
&\quad + \varepsilon \|T\| (2\|T\| \sum_{j=1}^m \|x_j\| + 1)^{-1} \sum_{j=1}^m \|x_j\| = \varepsilon.
\end{aligned}
$$

If the approximate identity is right or two-sided, the argument for an anti-representation would be similar.

The approximation was achieved simply by choosing a large enough α_0 to approximate a finite list of elements from $\mathcal{A}$ sufficiently well. Obviously, given a whole list of representations and vectors, we will still only have to approximate a (larger) finite list of elements from $\mathcal{A}$. Since $\{e_\alpha\}_{\alpha \in A}$ is a net, this can be done. $\square$

If a normed algebra $\mathcal{A}$ has either a left or a right approximate identity, then each element $a \in \mathcal{A}$ is the limit of a net (actually a sequence) of products of elements in $\mathcal{A}$. Paul J. Cohen [1959a] proved that every element in a Banach algebra with a bounded approximate identity is actually the product of two elements (rather than just the limit of products) and that these elements can be chosen to satisfy useful additional conditions. We shall prove a version of this result discovered independently by Edwin Hewitt [1964] and by Philip C. Curtis, Jr. and Alessandro Figà-Talamanca [1966]. This version of the theorem is usually stated for Banach modules. Since we usually use the language of representations rather than the equivalent language of modules, we use it here also. A number of simpler proofs have been offered for theorems similar to Theorem 5.2.2. However, none of them establishes a sharp bound for $\|a\|$. (See Miecsław Altman [1971], Vlastimil Pták [1974] and Sidney L. Gulick, Teng Sun Liu and Arnoud van Rooij [1967].) A simpler proof of the full results has been given by Paul J. Koosis [1964] when the approximate identity is two-sided and bounded by 1. See also Section 5.3 below, Doran and Wichmann [1979], Pták [1979a] and Florian A. Potra and Pták [1984], Chapter 8.

5.2.2 Theorem *Let $\mathcal{A}$ be a Banach algebra with a $\langle$ left / right $\rangle$ approximate identity bounded by M. Let T be a continuous $\langle$ representation / anti-representation $\rangle$ of $\mathcal{A}$ on a Banach space $\mathcal{X}$. Then for each $y \in \overline{\mathrm{span}}(T_\mathcal{A}\mathcal{X})$ and each $\varepsilon > 0$, there are elements $a \in \mathcal{A}$ and $z \in \mathcal{X}$ which satisfy:*
(a) $y = T_a z$;
(b) $\|a\| \leq M$;
(c) $\|y - z\| < \varepsilon$;
(d) $z \in (T_\mathcal{A} y)^-$.
Hence, the set $\{T_a z : a \in \mathcal{A}, z \in \mathcal{X}\}$ is precisely the closed linear subspace $\overline{\mathrm{span}}(T_\mathcal{A}\mathcal{X})$ of $\mathcal{X}$.

Proof It is enough to prove the first case, since it gives the second when applied to the opposite algebra.

We make $\mathcal{A}^1$ into a Banach algebra by defining the norm as follows:

$$\|\lambda + a\| = |\lambda| + \|a\| \forall \lambda + a \in \mathcal{A}^1.$$

We define a representation of $\mathcal{A}^1$ on $\mathcal{X}$ which extends T, and which we will denote by T again, by

$$T_{\lambda+a}x = \lambda x + T_a x \qquad \forall \, \lambda + a \in \mathcal{A}^1; x \in \mathcal{X}.$$

Note that the norm $\|T\|$ of the bounded linear map $T: \mathcal{A}^1 \to \mathcal{X}$ is the maximum of 1 and the norm of the original map $T: \mathcal{A} \to \mathcal{X}$.

It is convenient to use P to denote the number $2M(1 + 2M)^{-1}$. Note $P < 1$, $P = 2M(1 - P)$ and $2M = P(1 - P)^{-1} = \sum_{n=1}^{\infty} P^n$.

The idea of the proof is to construct a convergent sequence of invertible elements $b_n = P^n + a_n$ in $\mathcal{A}^1$. Then we set $a = \lim b_n = \lim a_n \in \mathcal{A}$ and $z = \lim T_{d_n} y$ where $d_n = b_n^{-1}$. Thus $\{T_{d_n} y\}$ converges even though $\{\|T_{d_n}\|\}$ approaches infinity.

Let $\{e_\alpha\}_{\alpha \in A}$ be a left approximate identity for $\mathcal{A}$ bounded by M. We use $\mathcal{E}$ to denote the set $\{e_\alpha : \alpha \in A\}$. We will always denote elements of $\mathcal{E}$ by "e", often with some subscript. For any $e \in \mathcal{E}$, Proposition 2.2.9(e) shows that $P + (1 - P)e$ is invertible in $\mathcal{A}^1$. We will denote its inverse by c, i.e.,

$$c = (P + (1 - P)e)^{-1} \qquad \forall \, e \in \mathcal{E}. \tag{1}$$

Proposition 2.2.9(f) also shows

$$c - P^{-1} \in \mathcal{A} \text{ and } \|c\| \leq 2P^{-1} \qquad \forall \, e \in \mathcal{E}. \tag{2}$$

Hence any $e \in \mathcal{E}$ and any $a \in \mathcal{A}$ satisfy:

$$\begin{aligned}
\|ca - a\| &\leq \|c\| \, \|a - (P + (1 - P)e)a\| \\
&\leq 2P^{-1}(1 - P)\|a - ea\|.
\end{aligned}$$

Thus, the net $\{c_\alpha\}_{\alpha \in A} \subseteq \mathcal{A}^1$ satisfies the defining properties of a bounded left approximate identity in $\mathcal{A}$ except that the elements c_α are in $\mathcal{A}^1$ instead of in $\mathcal{A}$. It is easy to check that the proof of Proposition 5.2.1 remains valid, so that for any finite sets $\{a_1, a_2, \ldots, a_n\} \subset \mathcal{A}$ and $\{y_1, y_2, \ldots, y_m\} \subseteq \overline{\text{span}}(T_{\mathcal{A}} \mathcal{X})$ and any $\delta > 0$, there is an $e \in \mathcal{E}$ such that

$$\begin{aligned}
\|ca_j - a_j\| &< \delta && j = 1, 2, \ldots, n \text{ and} \\
\|T_c y_j - y_j\| &< \delta && j = 1, 2, \ldots, m
\end{aligned} \tag{3}$$

hold simultaneously.

Let $y \in \overline{\text{span}}(T_{\mathcal{A}} \mathcal{X})$ and $\varepsilon > 0$ be arbitrary. These will remain fixed throughout the rest of this proof. We will inductively choose elements $e_n \in \mathcal{E}$ so that $b_n = P^n + (2M)^{-1} \sum_{k=1}^{n} P^k e_k$ satisfies

$$b_n \in (\mathcal{A}^1)_G \tag{4}$$

and so that the inverse d_n of b_n satisfies

$$\|T_{d_n} y - T_{d_{n-1}} y\| < 2^{-n} \varepsilon \tag{5}$$

for all $n \geq 1$, where $d_0 = 1$.

We choose e_1 so that

$$\|T_{c_1}y - y\| < 2^{-1}\varepsilon.$$

Then, $b_1 = P + (2M)^{-1}Pe_1 = P + (1 - P)e_1$ is invertible with $c_1 = d_1$ as its inverse. Thus, e_1 satisfies (4) and (5). Now suppose $e_1, e_2, \ldots, e_m$ have been chosen satisfying (4) and (5). We will use (3) to find $e_{m+1} \in \mathcal{E}$ which satisfies (4) and (5). Let δ and δ' be the positive numbers $\delta = (2^{m+1}\|T\|(2P^{-1}\|y\| + \|d_m\|) + 1)^{-1}\varepsilon$ and $\delta' = \min\{\|d_m\|^{-1}, \delta(\|d_m\|^2 + \delta)^{-1}\}$ so $\|d_m\|^2\delta'(1 - \delta')^{-1} < \delta$. The inequalities of (3) show that we can find $e_{m+1} \in \mathcal{E}$ so that the following inequalities hold:

$$\begin{aligned} \|c_{m+1}e_k - e_k\| &< \delta' & k = 1, 2, \ldots, m \\ \|T_{c_{m+1}}y - y\| &< \delta. \end{aligned} \tag{6}$$

We show first that b_{m+1} is invertible. We have

$$\begin{aligned} c_{m+1}b_{m+1} - b_m &= c_{m+1}(P^{m+1} + (2M)^{-1}P^{m+1}e_{m+1}) \\ &\quad + (2M)^{-1}\sum_{k=1}^{m} P^k c_{m+1}e_k - P^m - (2M)^{-1}\sum_{k=1}^{m} P^k e_k \\ &= P^m c_{m+1}(P + (1 - P)e_{m+1}) - P^m \\ &\quad + (2M)^{-1}\sum_{k=1}^{m} P^k(c_{m+1}e_k - e_k) \\ &= (2M)^{-1}\sum_{k=1}^{m} P^k(c_{m+1}e_k - e_k) \end{aligned} \tag{7}$$

which gives

$$\|c_{m+1}b_{m+1} - b_m\| < (2M)^{-1}\sum_{k=1}^{m} P^k\delta' \leq \delta' < \|d_m\|^{-1}.$$

The induction hypothesis shows that b_m is invertible with inverse d_m. Hence, Proposition 2.2.9(b) shows that $c_{m+1}b_{m+1}$ is invertible and its inverse satisfies

$$\|(c_{m+1}b_{m+1})^{-1} - d_m\| < \|d_m\|^2\delta'(1 - \delta')^{-1} \leq \delta. \tag{8}$$

Thus e_{m+1} satisfies (4). We now check (5). Using (8), (6) and the definition of δ we get

$$\begin{aligned} \|T_{d_{m+1}}y - T_{d_m}y\| &= \|T_{(c_{m+1}b_{m+1})^{-1}c_{m+1}}y - T_{d_m}y\| \\ &\leq \|T_{[(c_{m+1}b_{m+1})^{-1} - d_m]c_{m+1}}y\| + \|T_{d_m}(T_{c_{m+1}}y - y)\| \\ &\leq \|T\|\,\|(c_{m+1}b_{m+1})^{-1} - d_m\|\,\|c_{m+1}\|\,\|y\| \\ &\quad + \|T\|\,\|d_m\|\,\|T_{c_{m+1}}y - y\| \\ &< \|T\|(\delta 2P^{-1}\|y\| + \|d_m\|\delta) = \varepsilon 2^{-(m+1)}. \end{aligned}$$

This completes the proof of (5) and hence the inductive construction of the $e_n \in \mathcal{E}$.

We now construct the elements $a \in \mathcal{A}$ and $z \in \overline{\mathrm{span}}(T_\mathcal{A}\mathcal{X})$ which satisfy (a), (b), (c), and (d) of the theorem relative to our choice of $y \in \overline{\mathrm{span}}(T_\mathcal{A}\mathcal{X})$ and $\varepsilon > 0$. For $m > n$, (5) gives

$$\|T_{d_m}y - T_{d_n}y\| \leq \sum_{k=n+1}^{m} \|T_{d_k}y - T_{d_{k-1}}y\| < \sum_{k=n+1}^{m} 2^{-k}\varepsilon < 2^{-n}\varepsilon.$$

Hence, $\{T_{d_n}y\}_{n\in\mathbb{N}}$ is a Cauchy sequence. Denote its limit by z. Then using the case $n = 0$ in the above inequality, we get $\|y - z\| < \varepsilon$. Since each $e_n \in \mathcal{E}$ has norm satisfying $\|e_n\| \leq M$, the series

$$(2M)^{-1} \sum_{n=1}^{\infty} P^n e_n$$

converges. Denote its sum by a. Then a is also given by $a = \lim_{n\to\infty} b_n$. Since T is continuous, we see $y = \lim_{n\to\infty} T_{b_n}T_{d_n}y = T_a z$. Thus, a and z satisfy conditions (a) and (c) of the theorem. We also have

$$\|a\| \leq (2M)^{-1} \sum_{n=1}^{\infty} P^n \|e_n\| \leq (2M)^{-1}P(1 - P)^{-1}M = M,$$

so a satisfies (b).

Finally, $\lim T_{e_a}y = y$ shows that y belongs to $(T_\mathcal{A}y)^-$. Since each d_n belongs to $\mathcal{A}^1, T_{d_n}y$ also belong to $(T_\mathcal{A}y)^-$. Therefore $z = \lim_{n\to\infty} T_{d_n}y$ belongs to $(T_\mathcal{A}y)^-$. This completes the proof of (d).

The last sentence follows immediately since we have shown that any $y \in \overline{\mathrm{span}}(T_\mathcal{A}\mathcal{X})$ has the form $T_a z$ for some $a \in \mathcal{A}$ and $z \in \mathcal{X}$.   □

We have proved Theorem 5.2.2 for factoring single elements. However, in some applications it is necessary to deal with the simultaneous factorization of more than one element. The first extensions of this sort were given independently by Nicholas Th. Varopoulos [1964a] and Johnson [1966], both of whom used their extensions to prove important theorems which we will present later. They proved the "algebra" (rather than the "representation") form of Corollary 5.2.3(c) below. The form we give was first established by Marc A. Rieffel [1969a], whose proof we use. The "algebra" form of Corollary 5.2.3(b) was established by Ian G. Craw [1969] for Fréchet algebras. Corollary 5.2.3(a) is a slight extension of his result in the Banach algebra case. Similar results have also been obtained by Heron S. Collins and William H. Summers [1969].

5.2.3 Corollary *Let $\mathcal{A}$ be a Banach algebra with a $\langle$ left / right $\rangle$ approximate identity bounded by M. Let T be a continuous $\langle$ representation / anti-representation $\rangle$ of $\mathcal{A}$ on a Banach space $\mathcal{X}$.*

(a) *If S is a countable union of compact subsets of $\overline{\mathrm{span}}(T_A\mathcal{X})$, then there is an $a \in \mathcal{A}$ so that $S \subseteq T_a\mathcal{X}$.*

(b) *If S is a compact subset of $\overline{\mathrm{span}}(T_A\mathcal{X})$ and ε is a positive number, then there is an element $a \in \mathcal{A}$ and a continuous function $z\colon S \to \mathcal{X}$ satisfying:*

(b$_1$) $y = T_a z(y) \qquad \forall\, y \in S;$

(b$_2$) $\|a\| \le M;$

(b$_3$) $\|y - z(y)\| < \varepsilon \qquad \forall\, y \in S;$

(b$_4$) $z(y) \in (T_A y)^- \qquad \forall\, y \in S.$

(c) *If $\{y_n\}_{n\in\mathbb{N}}$ is a sequence in $\overline{\mathrm{span}}(T_A\mathcal{X})$ converging to zero, and ε is a positive number, then there is an element $a \in \mathcal{A}$ and a sequence $\{z_n\}_{n\in\mathbb{N}}$ converging to zero and satisfying:*

(c$_1$) $y_n = T_a z_n \qquad \forall\, n \in \mathbb{N};$

(c$_2$) $\|a\| \le M;$

(c$_3$) $\|y_n - z_n\| < \varepsilon \qquad \forall\, n \in \mathbb{N};$

(c$_4$) $z_n \in (T_A y_n)^- \qquad \forall\, n \in \mathbb{N}.$

Proof We will prove (b) and then use this to prove (c) and (a).

(b): Let S be a compact subset of $\overline{\mathrm{span}}(T_A\mathcal{X})$ and let $\mathcal{Y}$ be the set of all continuous functions $f\colon S \to \mathcal{X}$. We make $\mathcal{Y}$ into a Banach space by using pointwise linear operations and the norm

$$\|f\| = \sup\{\|f(s)\| : s \in S\}.$$

Define a representation $\overline{T}$ of $\mathcal{A}$ on $\mathcal{Y}$ by setting

$$(\overline{T}_a f)(s) = T_a(f(s)) \qquad \forall\, a \in \mathcal{A}; f \in \mathcal{Y}; s \in S.$$

It is clear that $\overline{T}$ is bounded with $\|\overline{T}\| = \|T\|$.

We now apply Theorem 5.2.2 to $\mathcal{A}, \mathcal{Y}$, and $\overline{T}$. We choose $y \in \mathcal{Y}$ to be the identity function $y(s) = s$ for all $s \in S$. We must show $y \in \overline{\mathrm{span}}(\overline{T}_A\mathcal{Y})$. For any $\varepsilon > 0$, choose a finite cover $B(s_1, \varepsilon/2), B(s_2, \varepsilon/2), \ldots, B(s_K, \varepsilon/2)$ of S by $\varepsilon/2$-balls and a partition of unity $g_1, g_2 \ldots, g_K$ subordinate to this cover. (That is, for $k = 1, 2, \ldots, K$, $g_k\colon S \to [0,1]$ is continuous, and satisfies $g_k(s) = 0$ for $s \notin B(s_k, \varepsilon/2)$ and $\sum_{k=1}^{K} g_k \equiv 1$. *Cf.* Rudin [1974].) Then from the approximate identity, choose e_k satisfying $\|T_{e_k} s_k - s_k\| < \varepsilon/2$ for $k = 1, 2, \ldots, K$. Thus, $f_k(s) = g_k(s) s_k$ belongs to $\mathcal{Y}$ and

$$\left\| y - \sum_{k=1}^{K} \overline{T}_{e_k} f_k \right\| = \sup_{s\in S}\left\{ \left\| s - \sum_{k=1}^{K} g_k(s) T_{e_k} s_k \right\| \right\}$$

$$\le \sup_{s\in S}\left\{ \sum_{k=1}^{K} g_k(s)\big(\|s - s_k\| + \|s_k - T_{e_k} s_k\|\big) \right\}$$

$$< \sup_{s\in S}\left\{ \sum_{\|s-s_k\|<\varepsilon/2} g_k(s)\varepsilon + \sum_{\|s-s_k\|\ge\varepsilon/2} 0 \right\} \le \varepsilon.$$

Hence $y \in \overline{\text{span}}(\overline{T}_\mathcal{A}\mathcal{Y})$.

Now let $\varepsilon > 0$ be arbitrary. Then we find $a \in \mathcal{A}$ and $z \in \mathcal{Y}$ so that $\overline{T}_a z = y$, $\|a\| \le M$, $\|y - z\| < \varepsilon$ and $z \in (\overline{T}_\mathcal{A}y)^-$. We get conclusion (b) by evaluating y and z at each point $s \in \mathcal{S}$ in the expressions $\overline{T}_a z = y$, $\|y - z\| < \varepsilon$ and $z \in \overline{T}_\mathcal{A}(y)$.

(c): We specialize to the case where the compact set $\mathcal{S}$ is $\{y_n : n \in \mathbb{N}\} \cup \{0\}$, where $\{y_n\}_{n\in\mathbb{N}}$ is a sequence converging to $y_0 = 0$. We consider the subspace $\mathcal{Z}$ of $\mathcal{Y}$ consisting of all $f \in \mathcal{Y}$ such that $f(0) = 0$. Clearly, this is a closed $\overline{T}$-invariant subspace. Let $\tilde{T}$ be the restriction of the representation $\overline{T}$ to $\mathcal{Z}$. Apply Theorem 5.2.2 to $\mathcal{A}$, $\mathcal{Z}$ and $\tilde{T}$. Again, we choose $y \in \mathcal{Z}$ to be the identity function $y(s) = s$ for each $s \in \mathcal{S}$. Let $\varepsilon > 0$ be arbitrary. The theorem guarantees a $z \in \mathcal{Y}$ such that $\tilde{T}_a z = y$, $\|a\| \le M$, $\|y - z\| < \varepsilon$ and $z \in (\tilde{T}_\mathcal{A}y)^-$. Set z_n equal to $z(y_n)$ for each n. Clearly, z_n satisfies the desired properties. This proves (c).

(a): Suppose $\mathcal{S}$ is the union $\cup_{n=1}^\infty \mathcal{S}_n$ where each $\mathcal{S}_n$ is compact. Since the norm is a continuous function, $B_n = \sup\{\|s\| : s \in \mathcal{S}_n\}$ is finite for each n. Consider the set $\tilde{\mathcal{S}} = \cup_{n=1}^\infty (nB_n)^{-1}\mathcal{S}_n \cup \{0\}$ where $(nB_n)^{-1}\mathcal{S}_n$ is just $\{(nB_n)^{-1}s : s \in \mathcal{S}_n\}$. Then $\tilde{\mathcal{S}}$ is compact. (To see this, it is enough to check that each sequence in $\tilde{\mathcal{S}}$ has a convergent subsequence. If it contains a subsequence moving through the sequence of $\mathcal{S}_n$, we are done. If it does not, then there is a subsequence included in some $(nB_n)^{-1}\mathcal{S}_n$. Since this is a compact set, there is a convergent subsequence.) Apply (b) to $\tilde{\mathcal{S}}$.   □

5.2.4 Corollary *Let $\mathcal{A}$ be a Banach algebra with a $\langle$ left / right $\rangle$ approximate identity bounded by M. If $\{a_n\}_{n\in\mathbb{N}}$ is a sequence in $\mathcal{A}$ which converges to zero and ε is a positive number, then there is an element $e \in \mathcal{A}$ and a sequence $\{b_n\}_{n\in\mathbb{N}}$ in $\mathcal{A}$ converging to zero satisfying*

    (a) $\langle a_n = eb_n$ / $a_n = b_n e \rangle$ $\forall\, n \in \mathbb{N}$

    (b) $\|e\| \le M$

    (c) $\|a_n - b_n\| < \varepsilon$ $\forall\, n \in \mathbb{N}$

    (d) $\langle b_n \in (\mathcal{A}a_n)^-$ / $b_n \in (a_n\mathcal{A})^- \rangle$ $\forall\, n \in \mathbb{N}$.

Proof Apply Corollary 5.2.3(c) to the left or right regular representation. □

We now state Cohen's original [1959a] factorization theorem together with its converse, which is due to Altman [1975]. Note that not all of (b) is needed to establish (a).

5.2.5 Corollary *The following two conditions are equivalent for a Banach algebra $\mathcal{A}$.*

    (a) *$\mathcal{A}$ has a $\langle$ left / right $\rangle$ approximate identity bounded by M.*

    (b) *There is a constant M such that for each $a \in \mathcal{A}$ and $\varepsilon > 0$ there are elements $e \in \mathcal{A}$ and $\langle b \in (\mathcal{A}a)^-$ / $b \in (a\mathcal{A})^- \rangle$ satisfying $\langle a = eb$ / $a = be \rangle$, $\|e\| \le M$ and $\|a - b\| < \varepsilon$.*

In particular, when these conditions hold, each element of $\mathcal{A}$ can be written as a product.

Proof (a) $\Rightarrow$ (b) Immediate from Theorem 5.2.2 or Corollary 5.2.4.

(b) $\Rightarrow$ (a) Given $a \in \mathcal{A}$ and $\varepsilon > 0$, choose $e, b \in \mathcal{A}$ satisfying $\langle a = eb$ / $a = be \rangle$, $\|e\| \leq M$ and $\|b - a\| < \varepsilon/M$. Then $\|a - ea\| = \|eb - ea\| \leq \|e\| \, \|b - a\| < \varepsilon$. Hence (a) follows by Theorem 5.1.2(c). $\square$

Converses of various forms of this theorem have been sought by a number of workers. Donald C. Taylor [1968], F. Dennis Sentilles and Taylor [1969] and Liu, van Rooij, and Wang [1973] found partial converses. William L. Paschke [1973] showed by example that the converse of the final (weaker) statement in Corollary 5.2.5 is false. More recently, Dixon [1990] gave a four-dimensional algebra $\mathcal{A}$ in which every element may be factored since $\mathcal{A}$ is the direct sum of two algebras, one with a left identity and the other with a right identity, but which obviously lacks an approximate identity. He also noted the existence of (non-separable) commutative semisimple Banach algebras with factorization but without even unbounded approximate identities. However, he showed that every separable commutative Banach algebra with $\langle$ left / right $\rangle$ factorization of sequences converging to zero has a (possibly unbounded) approximate identity. By $\langle$ *left / right* $\rangle$ *factorization of sequences converging to zero*, we simply mean that for any sequence $\{a_n\}_{n \in \mathbb{N}} \subseteq \mathcal{A}$ converging to zero there is an element $e \in \mathcal{A}$ and a sequence $\{b_n\}_{n \in \mathbb{N}} \subseteq \mathcal{A}$ converging to zero satisfying $\langle a_n = eb_n$ / $a_n = b_n e \rangle$. Additional results and examples on converses to the factorization theorem, particularly for commutative algebras, are given in Willis [1992b].

Note that if a normed algebra $\mathcal{A}$ has a right approximate identity which is $\langle$ bounded by 1 / bounded / arbitrary $\rangle$, then the left regular representation of $\mathcal{A}$ is $\langle$ isometric / homeomorphic / faithful $\rangle$.

The factorization theorem gives new information on double centralizers. Theorem 1.2.4 shows that, if $\mathcal{A}$ is a Banach algebra and $\mathcal{A}_{LA} = \mathcal{A}_{RA} = \{0\}$ (*e.g.*, if $\mathcal{A}$ is a semiprime Banach algebra), then any pair of maps $L \colon \mathcal{A} \to \mathcal{A}$ and $R \colon \mathcal{A} \to \mathcal{A}$ satisfying

$$aL(b) = R(a)b \qquad a, b \in \mathcal{A}$$

will have both L and R linear and continuous. Of course if $\mathcal{A}$ has a right approximate identity, then $\mathcal{A}_{LA} = \{0\}$. We now establish an improvement of Theorem 1.2.4 which is due to Johnson [1966].

5.2.6 Proposition *Let $\mathcal{A}$ be a Banach algebra with a bounded $\langle$ left / right $\rangle$ approximate identity. Let $\langle$ L / R $\rangle \colon \mathcal{A} \to \mathcal{A}$ be a map which satisfies*

$$\langle L(ab) = L(a)b \text{ / } R(ab) = aR(b) \rangle \qquad \forall \, a, b \in \mathcal{A}.$$

Then $\langle L / R \rangle$ is linear and continuous. Thus it is a $\langle$ left / right $\rangle$ centralizer.

Proof We prove only the first case since the second follows by considering the reverse algebra. Let a_1, $a_2 \in \mathcal{A}$ and λ, $\mu \in \mathbb{C}$ be arbitrary. By Corollary 5.2.4, we can find elements e, b_1, $b_2 \in \mathcal{A}$ such that $a_1 = eb_1$ and $a_2 = eb_2$. Therefore, we get

$$
\begin{aligned}
L(\lambda_1 a_1 + \lambda_2 a_2) &= L(e(\lambda_1 b_1 + \lambda_2 b_2)) = L(e)(\lambda_1 b_1 + \lambda_2 b_2) \\
&= \lambda_1 L(e) b_1 + \lambda_2 L(e) b_2 = \lambda_1 L(eb_1) + \lambda_1 L(eb_2) \\
&= \lambda_1 L(a_1) + \lambda_2 L(a_2).
\end{aligned}
$$

Hence L is linear. Thus it is a left centralizer.

We will now prove that L is continuous. Suppose $\{a_n\}_{n \in \mathbb{N}}$ is a sequence in $\mathcal{A}$ which converges to zero. Corollary 5.2.4 shows that there is a sequence $\{b_n\}_{n \in \mathbb{N}}$ converging to zero and an element $e \in \mathcal{A}$ such that $a_n = eb_n$ for all n. Thus, $\lim_{n \to \infty} L(a_n) = \lim_{n \to \infty} L(eb_n) = \lim_{n \to \infty} L(e)b_n = 0$. Hence L is continuous.   □

Rieffel proved a module version in [1969a], but we shall not need his result.

Maximal One-sided Ideals in Approximately Unital Banach Algebras

In a unital Banach algebra every one- or two-sided ideal is modular. Hence, Theorem 2.4.7 shows that every maximal one- or two-sided ideal is closed. Michael D. Green [1976] has shown that the same conclusion holds true in an approximately unital algebra, at least for one-sided ideals. This is an example of ways in which approximately unital algebras resemble unital algebras.

5.2.7 Theorem *Let $\mathcal{A}$ be a Banach algebra with a bounded $\langle$ right / left $\rangle$ approximate identity. Then every maximal $\langle$ left / right $\rangle$ ideal is closed.*

Proof Consider the case in which $\mathcal{M}$ is a maximal left ideal. Let $\{b_n\}_{n \in \mathbb{N}}$ be a sequence in $\mathcal{M}$ converging to $b \in \mathcal{A}$. We wish to show $b \in \mathcal{M}$.

Let $b_0 = 0$ and consider the sequence $\{b - b_n\}_{n \in \mathbb{N}^0}$ which converges to zero. Corollary 5.2.4 gives a sequence $\{\tilde{c}_n\}_{n \in \mathbb{N}^0}$ converging to zero and an element d satisfying $b - b_n = \tilde{c}_n d$ for all $n \in \mathbb{N}^0$. To simplify notation write $c = \tilde{c}_0$ and for positive n $c_n = c - \tilde{c}_n$, so

$$
b_n = c_n d, \quad c_n \to c \text{ and } b = cd.
$$

Consider the left ideal $\mathcal{L} = \{a \in \mathcal{A} : ad \in \mathcal{M}\}$. Since it contains each c_n, it is enough to show that $\mathcal{L}$ is closed. We may assume that $\mathcal{L}$ is not equal to $\mathcal{A}$. The maximality of $\mathcal{M}$ shows that $\mathcal{A}d + \mathcal{M}$ equals $\mathcal{A}$. Hence, we may write $d = ed + m$ for some $e \in \mathcal{A}$ and $m \in \mathcal{M}$. Then any $a \in \mathcal{A}$ satisfies $(a - ae)d = am \in \mathcal{M}$. This shows that $\mathcal{L}$ is a modular left ideal. We will now show that $\mathcal{L}$ is also maximal. If $\mathcal{K}$ is a left ideal which properly includes $\mathcal{L}$, then the maximality of $\mathcal{M}$ implies $\mathcal{A} = \mathcal{K}d + \mathcal{M}$. As before, we may choose $e \in \mathcal{K}$ satisfying $e = ed + m$ for some $m \in \mathcal{M}$. Hence, any $a \in \mathcal{A}$ satisfies $a = ae + (a - ae) \in \mathcal{K} + \mathcal{L} = \mathcal{K}$. Since any strictly larger left ideal equals $\mathcal{A}$, we conclude that $\mathcal{M}$ is a maximal modular left ideal and therefore closed by Theorem 2.4.7. $\square$

5.3   Countable Factorization Theorems

In this section, we prove an important theorem, due to Sinclair, about Banach algebras with countable approximate identities. This theorem holds in any separable approximately unital algebra. We also prove some weaker factorization-like theorems which hold in any separable Banach algebra, whether or not it has an approximate identity.

The main theorem (Theorem 5.3.2) is a simplified version of the main theorem in Sinclair's book [1982] which was based on his papers [1978] and [1979]. By omitting a number of deeper properties, we are able to give a simpler induction argument as noted by Sinclair [1982]. Both the properties we prove and those we omit are useful, but Corollary 5.3.4 below is certainly the most important consequence of this theorem. For interesting applications of analytic semigroups in $L^1(G)$ for a locally compact group see José E. Galé [1991].

5.3.1 Definition   Let $\mathcal{A}$ be a Banach algebra. Let $\mathbf{H}$ be the open right half plane $\{\lambda \in \mathbb{C} : \text{Re}(\lambda) > 0\}$ and, for each α satisfying $0 < \alpha < \pi/2$, let $S(\alpha)$ be the sector

$$S(\alpha) = \{\lambda : |\arg(\lambda)| \leq \alpha; \ 0 < |\lambda| \leq 1\} \subseteq \mathbf{H}.$$

A function $e \colon \mathbf{H} \to \mathcal{A}$ is called an *analytic semigroup in $\mathcal{A}$* if it is an analytic function satisfying

$$e(\lambda + \mu) = e(\lambda)e(\mu) \qquad \forall \ \lambda, \mu \in \mathbf{H}.$$

An analytic semigroup is said to be *bounded* if for each α satisfying $0 < \alpha < \pi/2$, the set $\{\|e(\lambda)\| : \lambda \in S(\alpha)\}$ is bounded.

Note that if $e \colon \mathbf{H} \to \mathcal{A}$ is an analytic semigroup, then $\{e(1/n)\}_{n \in \mathbb{N}}$ is a sequential commutative approximate identity. If the semigroup is bounded, then this approximate identity is bounded by the bound for any sector $S(\alpha)$.

5.3.2 Theorem   *Let $\mathcal{A}$ be a Banach algebra with a countable two-sided bounded approximate identity. Let $\langle \ T \colon \mathcal{A} \to \mathcal{B}(\mathcal{X}) \ / \ S \colon \mathcal{A} \to \mathcal{B}(\mathcal{Y}) \ \rangle$ be a continuous $\langle$ representation / anti-representation $\rangle$. Then for any $x \in (T_{\mathcal{A}}\mathcal{X})^-$ and $y \in (S_{\mathcal{A}}(\mathcal{Y}))^-$, there are:*
     a bounded analytic semigroup $e \colon \mathbf{H} \to \mathcal{A}$,

an entire function $x\colon \mathbb{C} \to \mathcal{X}$ *with* $x(0) = x$, *and*
an entire function $y\colon \mathbb{C} \to \mathcal{Y}$ *with* $y(0) = y$
satisfying:

(a) $T_{e(\lambda)}(x(\mu + \lambda)) = x(\mu)$ $\forall\, \lambda \in \mathbf{H}; \mu \in \mathbb{C}$
    $S_{e(\lambda)}(y(\mu + \lambda)) = y(\mu)$ $\forall\, \lambda \in \mathbf{H}; \mu \in \mathbb{C}$.

(b) $(e(\lambda)\mathcal{A})^- = (\mathcal{A}e(\lambda))^- = \mathcal{A}$ $\forall\, \lambda \in \mathbf{H}$.

(c) $x(\lambda) \in (T_{\mathcal{A}}(x))^-$ *and* $y(\lambda) \in (S_{\mathcal{A}}(y))^-$ $\forall\, \lambda \in \mathbb{C}$.

(d) *For each* α *satisfying* $0 < \alpha < \pi/2$ *and* $a \in \mathcal{A}$,

$$a = \lim_{\lambda \to 0} e(\lambda)a = \lim_{\lambda \to 0} ae(\lambda)$$

when λ *is restricted to the sector* $S(\alpha)$.

(e) *If the original approximate identity has bound* 1, *then all positive real* t *satisfy* $\|e(t)\| \leq 1$.

Proof We will inductively choose a sequence $\{f_n\}_{n \in \mathbb{N}}$ from an approximate identity for $\mathcal{A}$ with bound M. For each $\lambda \in \mathbf{H}$, we will then construct an exponential element $e_n(\lambda) \in \mathcal{A}^1$ by

$$e_n(\lambda) = \exp\Big(\lambda \sum_{j=1}^{n}(f_j - 1)\Big) \tag{1}$$

and define $e(\lambda)$, $x(\lambda)$ and $y(\lambda)$ by

$$e(\lambda) = \lim_{n \to \infty} e_n(\lambda), \quad x(\lambda) = \lim_{n \to \infty} T_{e_n(-\lambda)}(x) \text{ and } y(\lambda) = \lim_{n \to \infty} T_{e_n(-\lambda)}(y).$$

The convergence of these limits depends on the careful choice of f_n so that for each n the "factor" $\exp(\lambda(f_n - 1))$ makes the elements which it multiplies very much smaller. (When the algebra is commutative, this really is a factor in the expression for e_n, but in the noncommutative case we must estimate how far it is from being an actual factor.) The limit of the sequence of exponentials is in $\mathcal{A}$ rather than $\mathcal{A}^1$ since each $e_n(\lambda)$ obviously has the form $e_n(\lambda) = e^{-n\lambda} + c_n$ where c_n is in $\mathcal{A}$. The group properties of the exponential function yield the semigroup properties of $e\colon \mathbf{H} \to \mathcal{A}$.

Although the calculations involved are not really difficult, the notation is sufficiently complicated that we will state a purely computational lemma.

5.3.3 Lemma *Let* $\mathcal{A}$ *be a Banach algebra and let* $T\colon \mathcal{A} \to \mathcal{B}(\mathcal{X})$ *be a continuous representation or anti-representation. Let* $\mu + a \in \mathcal{A}^1$, $M \in \mathbb{R}_+$, $f \in \mathcal{A}_M$ *and* $x \in \mathcal{X}$ *be arbitrary. To simplify notation, write* $N = M + \|a\| + 1$ *and* $P = \|(f-1)a\| + \|a(f-1)\|$.

(a) *All* $k \in \mathbb{N}$ *satisfy*

$$\|(a + f - 1)^k - a^k - (f-1)^k\| \leq N^k P.$$

(b) *All* $k \in \mathbb{N}$ *satisfy*

$$\|T_{(a+f-1)^k}(x) - T_{a^k}(x)\| \leq N^k \|T\|\, (\|x\|P + \|T_f(x) - x\|).$$

(c) *All $\lambda \in \mathbb{C}$ satisfy*

$$\|e^{\lambda(a+f-1)} - e^{\lambda a} - e^{\lambda(f-1)} + 1\| \le (e^{N|\lambda|} - 1)P.$$

(d) *For any positive $B \in \mathbb{R}$, there is a constant C, depending on $\|a\|$, B and M, such that any $\lambda \in B\mathbb{C}_1$ satisfies*

$$\|e^{\lambda(\mu+a+f-1)} - e^{\lambda(\mu+a)}\| \le e^{\mathrm{Re}(\lambda\mu)}(e^{|\lambda|(M+1)} - 1 + CP).$$

(e) *For any positive $B \in \mathbb{R}$, there is a constant D depending on $\|a\|$, $|\mu|$, B and M such that any $\lambda \in B\mathbb{C}_1$ satisfies*

$$\|T_{\exp(\lambda(\mu+a+f-1))}(x) - T_{\exp(\lambda(\mu+a))}(x)\| \le D(\|x\|P + \|T_f x - x\|).$$

Proof (a): Expanding $(a + (f - 1))^k$ without using the commutative law gives

$$
\begin{aligned}
\|(a + f - 1)^k - a^k - (f - 1)^k\| &\le \sum_{j=1}^{k-1} \|a\|^{k-1-j}\|f-1\|^{j-1}\left(\binom{k}{j} - 1\right)P \\
&\le \sum_{j=1}^{k-1} \|a\|^{k-1-j}(M+1)^{j-1}\binom{k}{j}P \\
&\le (\|a\| + M + 1)^k P = N^k P.
\end{aligned}
$$

(b): Since N is greater than both 1 and $\|f - 1\|$, (a) gives

$$
\begin{aligned}
\|T_{(a+f-1)^k}(x) - T_{a^k}(x)\| &\le \|T_{(f-1)^k}(x)\| + \|T\|\,\|x\|N^k P \\
&\le \|T\|N^{k-1}\|T_f(x) - x\| + \|T\|\,\|x\|N^k P \\
&\le N^k\|T\|(\|x\|P + \|T_f(x) - x\|).
\end{aligned}
$$

(c): Expanding the three exponentials and using (a) gives

$$
\begin{aligned}
\|e^{\lambda(a+f-1)} - e^{\lambda a} &- e^{\lambda(f-1)} + 1\| \\
&\le \sum_{k=1}^{\infty} \frac{|\lambda|^k}{k!}\|(a + f - 1)^k - a^k - (f - 1)^k\| \\
&\le \sum_{k=1}^{\infty} \frac{|\lambda|^k}{k!}N^k P = (e^{N|\lambda|} - 1)P.
\end{aligned}
$$

(d): By using (c) at the first inequality, we get

$$
\begin{aligned}
\|e^{\lambda(\mu+a+f-1)} - e^{\lambda(\mu+a)}\| &= e^{\mathrm{Re}(\lambda\mu)}\|e^{\lambda(a+f-1)} - e^{\lambda a}\| \\
&\le e^{\mathrm{Re}(\lambda\mu)}\left(\|e^{\lambda(f-1)} - 1\| + (e^{N|\lambda|} - 1)P\right).
\end{aligned}
$$

Hence, we may choose $C = e^{NB} - 1$.

(e): Expand the exponentials in series to get

$$\|T_{\exp(\lambda(\mu+a+f-1))}(x) \quad - \quad T_{\exp(\lambda(\mu+a))}(x)\|$$

$$\leq \quad e^{|\lambda\mu|} \sum_{k=1}^{\infty} \frac{|\lambda|^k}{k!} \|T_{(a+f-1)^k}(x) - T_{a^k}(x)\|$$

$$\leq \quad e^{B|\mu|} \sum_{k=1}^{\infty} \frac{B^k N^k}{k!} \|T\| \left(\|x\|P + \|T_f(x) - x\|\right)$$

$$\leq \quad D(\|x\|P + \|T_f(x) - x\|).$$

where $D = e^{B(N+|\mu|)}\|T\|$. □

We now resume the proof of the main theorem. Note that in (c), (d), and (e) of the lemma, the right hand side of the inequalities have factors of the form $P = \|(f-1)a\| + \|a(f-1)\|$ or $\|T_f(x) - x\|$ which can simultaneously be made as small as we please by suitable choice of f from an approximate identity. This is the key to the construction.

Let $\{g_n : n \in \mathbb{N}\}$ be a complete list of the elements in a countable bounded approximate identity. (We are not claiming that the approximate identity is sequential, so the order of this list is in no way related to the structure of the approximate identity as a net.)

We will now inductively choose the sequence $\{f_n\}_{n\in\mathbb{N}}$ from this or another countable approximate identity bounded by M to satisfy the following five conditions for all $\lambda \in n\mathbb{C}_1$:

$$\|e_{n-1}(\lambda) - e_n(\lambda)\| \quad \leq \quad e^{-(n-1)\text{Re}(\lambda)}\left(e^{|\lambda|(M+1)} - 1 + 2^{-n-1}\right) \quad (2)$$

$$\|T_{e_{n-1}(\lambda)}(x) - T_{e_n(\lambda)}(x)\| \quad \leq \quad 2^{-n} \quad\quad\quad\quad\quad\quad\quad (3)$$

$$\|S_{e_{n-1}(\lambda)}(y) - S_{e_n(\lambda)}(y)\| \quad \leq \quad 2^{-n} \quad\quad\quad\quad\quad\quad\quad (4)$$

$$\|e_{n-1}(\lambda)g_k - e_n(\lambda)g_k\| \quad \leq \quad 2^{-n} \quad \forall\, k \leq n \quad\quad\quad (5)$$

$$\|g_k\, e_{n-1}(\lambda) - g_k\, e_n(\lambda)\| \quad \leq \quad 2^{-n} \quad \forall\, k \leq n. \quad\quad (6)$$

Consider the case $n = 1$. Define $e_0(\lambda)$ to be $1 \in \mathcal{A}^1$. We use (d) and (e) in the lemma with μ, a, and hence P all equal to zero. Then (d) shows that (2) holds no matter how we choose f_1. Lemma 5.3.3(e) (with the indicated choices) and Proposition 5.2.1 applied to the representations T, S, L and R (where the latter two are the regular representations) show that we can choose f_1 from the approximate identity to simultaneously achieve (3), (4), (5) and (6).

Now suppose we have already chosen $f_1, f_2, \ldots, f_n$. We will use (d) and (e) of Lemma 5.3.3 again, but this time with $\mu = -n$ and $a = f_1 + f_2 + \cdots + f_n$. Then, as before, (d) gives (2) and we apply Proposition 5.2.1 to T,

S, L and R to achieve (3), (4), (5) and (6) simultaneously with (2). This completes the construction of the sequence of exponential functions e_n. Inequality (2) shows that for any $\lambda \in \mathbf{H}$

$$e(\lambda) = 1 + \lim_{n \to \infty} \sum_{k=1}^{n} \Big(e_k(\lambda) - e_{k-1}(\lambda) \Big) \tag{7}$$

converges. For $|\lambda| < 1$, it also gives the estimate

$$\|e(\lambda)\| \leq 1 + \frac{e^{|\lambda|(M+1)} - 1}{1 - e^{-\mathrm{Re}(\lambda)}} + \frac{2^{-1}}{1 - 2^{-1}e^{-\mathrm{Re}(\lambda)}}.$$

For λ in the sector $S(\alpha)$, this estimate becomes

$$\|e(\lambda)\| \leq 2 + \frac{e^{|\lambda|(M+1)} - 1}{1 - e^{-\cos(\alpha)|\lambda|}}$$

which has the limit $2 + (M+1)/\cos(\alpha)$ as $|\lambda|$ approaches 0. Hence, $e(\lambda)$ is bounded in $S(\alpha)$ for each $0 < \alpha < \pi/2$. As noted earlier, $e(\lambda)$ is in $\mathcal{A}$ not just $\mathcal{A}^1$ and the semigroup property follows from that of the exponential function. Moreover, our proof shows that convergence is uniform on any sector of the form $\{\lambda : 0 < |\lambda| \leq n; \arg(\lambda) \leq \alpha\}$ for any $n \in \mathbb{N}$ and $0 < \alpha < \pi/2$. Since every element of $\mathbf{H}$ is included in the interior of a sector of this type, e is analytic on $\mathbf{H}$ since it is the limit of the sequence of analytic functions e_n.

(a): Equations (3) and (4) show that we may define

$$x(\lambda) = \lim_{n \to \infty} T_{e_n(-\lambda)}(x) \quad \text{and} \quad y(\lambda) = \lim_{n \to \infty} S_{e_n(-\lambda)}(y)$$

since these limits exist. Then property (a) of the theorem follows directly from the group property of the exponentials $\lambda \mapsto e_n(\lambda)$.

(b): Similarly, inequalities (5) and (6) show that for all $m \in \mathbb{N}$ the limits

$$\lim_{n \to \infty} e_n(-\lambda)g_m \quad \text{and} \quad \lim_{n \to \infty} g_m e_n(-\lambda)$$

exist. Let us temporarily call them $h_m(\lambda)$ and $k_m(\lambda)$, respectively. These functions satisfy the analogue of condition (a) of the theorem. Hence for each $\lambda \in \mathbf{H}$ and each $m \in \mathbb{N}$, we have $g_m \in e(\lambda)\mathcal{A}$ and $g_m \in \mathcal{A}e(\lambda)$. Let $a \in \mathcal{A}$ be arbitrary. In the proper order of the approximate identity we have $a = \lim g_m a = \lim a\, g_m$, establishing (b).

(c): Proposition 5.2.1 shows that $\langle\ x\ /\ y\ \rangle$ belongs to $\langle\ (T_{\mathcal{A}}(x))^- /\ (S_{\mathcal{A}}(y))^- \rangle$ since by hypothesis it belongs to $\langle\ (T_{\mathcal{A}}(\mathcal{X}))^- /\ (S_{\mathcal{A}}(\mathcal{Y}))^- \rangle$. Hence, $T_{e_n(\lambda)}(x) \in (T_{\mathcal{A}}(x))^-$ and $S_{e_n(\lambda)}(y) \in (S_{\mathcal{A}}(y))^-$ are both obvious and imply (c).

(d): Let $a \in \mathcal{A}$, $0 < \alpha < \pi/2$ and $\varepsilon > 0$ be arbitrary. Define B to be $\sup\{\|e(\lambda)\| : \lambda \in S(\alpha)\}$. Property (b) allows us to choose an element c in $\mathcal{A}$ satisfying $\|a - e(1)c\| < \varepsilon/(2 + 2B)$. By the continuity of the function e on $\mathbf{H}$, any $\lambda \in S(\alpha)$ sufficiently close to 0 satisfies

$$\|e(\lambda)a - a\| \leq \|(e(\lambda) - 1)(a - e(1)c)\| + \|e(\lambda + 1)c - e(1)c\|$$
$$\leq (\|e(\lambda)\| + 1)\varepsilon/(2 + 2B) + \varepsilon/2 < \varepsilon.$$

Since $\varepsilon > 0$ was arbitrary, this proves (d).

(e): If the bound M of the approximate identity from which we chose the f_n is 1, then for positive real t we get

$$\|e_n(t)\| \leq e^{-nt}\|\exp(t\sum_{k=1}^{n} f_k)\| \leq e^{-nt}\exp(t\sum_{k=1}^{n} \|f_k\|) \leq e^{-nt}e^{nt} = 1. \quad \square$$

The following corollary summarizes the most important consequence of this theorem. See Sinclair [1982] for other consequences and additional properties which the analytic semigroup possesses if it is constructed by a still more careful induction.

5.3.4 Corollary *Any separable approximately unital Banach algebra or any Banach algebra with a countable bounded two-sided approximate identity has a commutative sequential two-sided approximate identity bounded by 1 in an equivalent algebra norm. If the original approximate identity was bounded by 1, no renorming is necessary.*

Proof We have already noted that if a normed algebra is separable, then any bounded two-sided approximate identity can be approximated by elements in the countable dense set, thus giving a countable approximate identity. Now suppose $\mathcal{A}$ satisfies the hypotheses of Theorem 5.3.2. Let $e: \mathbf{H} \to \mathcal{A}$ be the resulting analytic semigroup, and let S be the restriction of this semigroup to the sector $S(\alpha)$ for some $0 < \alpha < \pi/2$. Apply Proposition 1.1.9 to get an equivalent algebra norm in which S is in the unit ball. Then $\{e(1/n)\}_{n \in \mathbb{N}}$ is a commutative sequential approximate identity bounded by 1. If the original countable approximate identity was bounded by 1, result (e) of the theorem shows that no renorming is needed. $\square$

Dixon [1973a] shows that even in a commutative algebra with a bounded (uncountable) approximate identity of bound M there may be no approximate identity of norm less than M in any equivalent norm. He also exhibits an approximately unital Banach algebra which lacks any commutative approximate identity and an incomplete normed algebra with a sequential two-sided approximate identity but no commutative (even one-sided) approximate identity.

Factorization in General Separable Banach Algebras

We now wish to discuss some results of J. P. Reus Christensen [1976] and Richard J. Loy [1976] which give (among other consequences) a weak factorization theorem for representations of separable Banach algebras on separable Banach spaces. The proofs of the next two theorems depend on some simple results from the theory of analytic spaces. A topological space is called an *analytic space* if it is Hausdorff and is the continuous image of a separable complete metrizable space. The collection of analytic spaces is closed under forming continuous images, closed subsets, countable products and countable unions. Hence, the linear span of an analytic subset $\mathcal{G}$ of a topological linear space is analytic since it equals $\cup_{n=1}^{\infty}\{\sum_{j=1}^{n} \lambda_j x_j : \lambda_j \in \mathbb{C}; x_j \in \mathcal{G}\}$. We need two fundamental results about analytic spaces.

Pettis lemma: If $\mathcal{B}$ is an analytic second category subset of a topological linear space, then $\mathcal{B} - \mathcal{B}$ includes a neighborhood of zero.

Open mapping theorem: A continuous linear surjection of an analytic space onto a Banach space is an open map.

For proofs of these results, see either Christensen [1974], the appendix to Treves [1967] or Dales [1993].

The next result is due to Loy [1976]. We will give some less abstract consequences after we prove the theorem.

5.3.5 Theorem *Let $\mathcal{X}_1, \mathcal{X}_2, \ldots, \mathcal{X}_n$ and $\mathcal{Y}$ be Banach spaces with all but the last separable. Let $T: \mathcal{X}_1 \times \mathcal{X}_2 \times \cdots \times \mathcal{X}_n \to \mathcal{Y}$ be a continuous multilinear map. Let $\mathcal{Z}$ be a closed subspace of $\mathcal{Y}$ which is included in the linear span of the range of T. Then there exists an integer m such that each $z \in \mathcal{Z}$ can be written in the form*

$$z = \sum_{j=1}^{m} T(x_{1j}, x_{2j}, \ldots, x_{nj})$$

for suitable $x_{ij} \in \mathcal{X}_i$. Furthermore, the expression $\|z\|_T$ defined by

$$\inf\left\{\sum_{j=1}^{m} \|x_{1j}\| \, \|x_{2j}\| \cdots \|x_{nj}\| : z = \sum_{j=1}^{m} T(x_{1j}, x_{2j}, \ldots, x_{nj}); \ x_{ij} \in \mathcal{X}_i\right\}$$

gives a complete norm on $\mathcal{Z}$ equivalent to its original norm as a closed subspace of $\mathcal{Y}$.

Proof The continuity of T shows that there is a constant M' satisfying

$$\|z\| \leq M' \|z\|_T \qquad \forall \, z \in \mathcal{Z}.$$

The product space $\mathcal{X}_1 \times \mathcal{X}_2 \times \cdots \times \mathcal{X}_n$ is analytic, and its injection into the algebraic tensor product $\otimes_{i=1}^n \mathcal{X}_i$ is continuous when the latter carries its projective norm

$$\|x\| \;=\; \inf \left\{ \sum_{j=1}^m \|x_{1j}\| \, \|x_{2j}\| \cdots \|x_{nj}\| \; : \right.$$
$$\left. x = \sum_{j=1}^m x_{ij} \otimes x_{2j} \otimes \cdots \otimes x_{nj}; m \in \mathbb{N}; x_{ij} \in \mathcal{X}_i \right\}.$$

Hence, the range $\{x_1 \otimes x_2 \otimes \cdots \otimes x_n : x_i \in \mathcal{X}_i\}$ of this map is an analytic set. Therefore, its linear span $\otimes_{i=1}^n \mathcal{X}_i$ is analytic. Now T can be uniquely extended to a continuous map $\overline{T} : \otimes_{i=1}^n \mathcal{X}_i \to \mathcal{Y}$. Since $\mathcal{Z}$ is closed in $\mathcal{Y}$, $T^\leftarrow(\mathcal{Z})$ is analytic and hence the open mapping theorem shows the existence of a constant M'' satisfying

$$\|z\|_T \le M'' \|z\| \qquad \forall \, z \in \mathcal{Z}.$$

Let $M = \max\{M', 1 + M''\}$. Now for each positive integer p, define a set

$$\mathcal{B}_p \;=\; \{z \in \mathcal{Z} : z = \sum_{j=1}^p T(x_{1j}, x_{2j}, \ldots, x_{nj});$$
$$x_{ij} \in \mathcal{X}_i; \; \sum_{j=1}^p \|x_{1j}\| \, \|x_{2j}\| \ldots \|x_{nj}\| \le M \|z\|\}.$$

Since $\mathcal{B}_p$ is the continuous image of a closed set in $\prod_{j=1}^p (\prod_{i=1}^n \mathcal{X}_i)$, it is analytic. However, $\mathcal{Z}$ equals $\cup_{p=1}^\infty \mathcal{B}_p$ so the Baire category theorem shows that some $\mathcal{B}_q$ is second category. Hence, the Pettis lemma shows that $\mathcal{B}_q - \mathcal{B}_q$ contains a neighborhood of zero in $\mathcal{Z}$. However, $\mathcal{B}_{2q}$ includes $\mathcal{B}_q - \mathcal{B}_q$ and (like $\mathcal{B}_q - \mathcal{B}_q$) is closed under scalar multiples so we conclude $\mathcal{Z} = \mathcal{B}_{2q}$. Hence, we have proved the theorem with $m = 2q$. □

Next we interpret this as a factorization theorem.

5.3.6 Corollary *Let T be a continuous representation or anti-representation of a separable Banach algebra $\mathcal{A}$ on a separable Banach space $\mathcal{X}$, and let $\mathcal{Z}$ be a closed subspace of $\overline{\mathrm{span}}\{T_a x : a \in \mathcal{A}; x \in \mathcal{X}\}$. Then a positive integer m and a real number M exist such that each $z \in \mathcal{Z}$ can be written in the form*

$$z = \sum_{j=1}^m T_{a_j} x_j$$

for $a_j \in \mathcal{A}$ and $x_j \in \mathcal{X}$ satisfying

$$\sum_{j=1}^{m} \|a_j\| \, \|x_j\| \leq M \|z\|.$$

Proof Apply the theorem with $\mathcal{X}_1 = \mathcal{A}$, $\mathcal{X}_2 = \mathcal{X} = \mathcal{Y}$, $\mathcal{Z} = \mathcal{Z}$ and $T(a, x) = T_a x$. $\square$

The following remarks will be useful for our next application. Let $\mathcal{A}$ be an algebra. Recall that

$$\mathcal{A}^n = \operatorname{span}\{\prod_{j=1}^{n} a_j : a_j \in \mathcal{A}\}$$

is an ideal in $\mathcal{A}$ for each $n = 2, 3, \ldots$ and that these ideals satisfy

$$\cdots \subseteq \mathcal{A}^n \subseteq \cdots \subseteq \mathcal{A}^3 \subseteq \mathcal{A}^2 \subseteq \mathcal{A}.$$

We will be interested in the case in which some $\mathcal{A}^n$ has finite codimension in $\mathcal{A}$. In this case, $\mathcal{A}^2$ has finite codimension in $\mathcal{A}$. Assume that $\mathcal{A}^2$ has codimension p in $\mathcal{A}$, and choose $e_1, e_2, \ldots, e_p \in \mathcal{A}$ so that each $a \in \mathcal{A}$ may be written in the form $a = \sum_{k=1}^{q} a_k a'_k + \sum_{j=1}^{p} \lambda_j e_j$, with $q \in \mathbb{N}$, $\lambda_j \in \mathbb{C}$ and $a_k, a'_k \in \mathcal{A}$. Expand each a_k and each a'_k in the above expression into a similar expression. By collecting terms, we can write

$$a = \sum_{k=1}^{m} b_k b'_k b''_k + \sum_{i=1}^{p} \sum_{j=1}^{p} \lambda_{ij} e_i e_j + \sum_{j=1}^{p} \lambda_j e_j$$

with λ_j, $\lambda_{ij} \in \mathbb{C}$ and b_k, b'_k, $b''_k \in \mathcal{A}$. Proceeding in a similar fashion, we see that for any n, $\mathcal{A}^n$ has codimension at most $p + p^2 + \cdots + p^{n-1}$. Similarly, if $\mathcal{A}^2$ has at most countable codimension, then $\mathcal{A}^n$ has at most countable codimension for $n = 3, 4, \ldots$ also. Hence $\mathcal{A}^n$ has at most countable codimension for one integer $n \geq 2$ if and only if $\mathcal{A}^n$ has at most countable codimension for all integers $n \geq 2$.

Now we specialize the above considerations to separable Banach algebras and give a simple but fundamental result which combines work of Christensen [1976] and Loy [1976].

5.3.7 Theorem *Let $\mathcal{A}$ be a separable Banach algebra in which $\mathcal{A}^p$ has at most countable codimension for some integer $p \geq 2$. Then for each integer $n \geq 2$, $\mathcal{A}^n$ is closed and has finite codimension in $\mathcal{A}$. Furthermore, for each integer $n \geq 2$, there is a positive integer $m = m(n)$ and a real number $M = M(n)$ such that each element of $\mathcal{A}^n$ can be written in the form*

$$\sum_{j=1}^{m} \prod_{i=1}^{n} a_{ij} \text{ with } a_{ij} \in \mathcal{A} \text{ satisfying } \sum_{j=1}^{m} \prod_{i=1}^{n} \|a_{ij}\| \leq M \|a\|.$$

Proof By the remarks preceding this theorem, $\mathcal{A}^n$ has at most countable codimension for each integer $n \geq 2$. The subspace $\mathcal{A}^n$ in its relative topology is analytic since it is the countable union of the analytic spaces $\mathcal{P}_p = \{\sum_{k=1}^p \prod_{j=1}^n a_{jk} : a_{jk} \in \mathcal{A}\}$. Let $\mathcal{X}$ be an algebraic complementary space for $\mathcal{A}^n$ in $\mathcal{A}$. Then $\mathcal{X}$ with its relative topology is analytic, since it is at most of countable dimension. Hence, the map $\theta\colon \mathcal{A}^n \times \mathcal{X} \to \mathcal{A}$ defined by $\theta(a, x) = a + x$ is a continuous linear bijection from the analytic space $\mathcal{A}^n \times \mathcal{X}$ with its product topology onto a Banach space. The open mapping theorem shows that θ is a homeomorphism. Hence, if a sequence $\{b_k\}_{k \in \mathbb{N}} \subseteq \mathcal{A}^n$ converges to $a \in \mathcal{A}$ satisfying $a = b + x$ with $b \in \mathcal{A}^n$ and $x \in \mathcal{X}$, then x is zero and the sequence converges to b. Hence $\mathcal{A}^n$, and similarly $\mathcal{X}$, are closed in $\mathcal{A}$. However, the Baire category theorem shows that a closed subspace of at most countable dimension of a Banach space is finite-dimensional. Hence, each $\mathcal{A}^n$ has finite codimension and is closed in $\mathcal{A}$.

To obtain the remaining results in this theorem, we apply Theorem 5.3.5 with $\mathcal{X}_i = \mathcal{A}, \mathcal{Y} = \mathcal{Z} = \mathcal{A}^n$ and $T(a_1, a_2, \ldots, a_n) = a_1 a_2 \cdots a_n$.   □

In Volume II we will actually need a slight extension of the above theorem which we will now state. Its proof is entirely similar and will therefore be omitted.

5.3.8 Corollary *Let $\mathcal{A}$ be a separable Banach algebra with center $\mathcal{A}_Z$. If $\mathcal{A}_Z$ has countable codimension in $\mathcal{A}$, then $\mathcal{A}_Z \mathcal{A}^n$ is closed and has finite codimension in $\mathcal{A}$ for all positive integers n.*

6

Automatic Continuity

Introduction

This chapter discusses various linear maps between Banach algebras which preserve some aspect of the multiplicative structure. It investigates when algebraic restrictions make these maps automatically continuous and when they can be described more precisely or explicitly than assumed *a priori*. The first two sections deal with automatic continuity of homomorphisms. Jordan homomorphisms between associative algebras can be thought of as algebra homomorphisms which simply ignore noncommutativity. Explicitly, a Jordan derivation is a linear map $\varphi \colon \mathcal{A} \to \mathcal{B}$ between algebras which satisfies

$$\varphi(ab + ba) = \varphi(a)\varphi(b) + \varphi(b)\varphi(a) \qquad \forall\, a, b \in \mathcal{A}.$$

They are studied in Section 6.3. A derivation is a linear map δ defined on an algebra $\mathcal{A}$ which satisfies the derivation equation

$$\delta(ab) = \delta(a)b + a\delta(b) \qquad \forall\, a, b \in \mathcal{A}.$$

Section 6.4 is devoted to the study of derivations and shows their several intimate connections with homomorphisms. Finally, Section 6.5 deals briefly with Jordan derivations.

Banach algebra theory is a hybrid between algebra and analysis. In the most favorable situations, analytic proofs can be used to establish that certain desirable topological or geometric conditions follow from purely algebraic hypotheses. A Banach algebra $\mathcal{A}$ is said to have a *unique Banach algebra topology* if any two complete algebra norms on $\mathcal{A}$ are equivalent, so that the norm topology given by a Banach algebra norm is unique. An even more stringent condition is that $\mathcal{A}$ have a *unique normed algebra topology* in the sense that any two (not necessarily complete) algebra norms must be equivalent. Finite-dimensional Banach algebras have this property trivially. Meier Eidelheit [1940a] showed that for any Banach space $\mathcal{X}$, $\mathcal{B}(\mathcal{X})$ has a unique Banach algebra topology (*cf.* §1.7.15). Following Barry E. Johnson [1967c], we will show in Corollary 6.2.8 that $\mathcal{B}(\mathcal{X})$ has a unique normed algebra topology if $\mathcal{X}$ is linearly homeomorphic to $\mathcal{X} \otimes \mathcal{X}$. Israel Moiseevič Gelfand [1941a] showed that any commutative semisimple Banach algebra has a unique Banach algebra topology (our Corollary 3.1.11).

Charles E. Rickart made a penetrating study of the uniqueness of norm question [1950] and his results were completed by Johnson [1967a]. The resulting Johnson–Rickart theorem (Theorem 6.1.1) asserts that any semi-simple Banach algebra has a unique Banach algebra topology. This theorem includes the two earlier results mentioned above as well as many other partial results that were obtained before 1967. It is a cornerstone of general Banach algebra theory and will be used repeatedly in the sequel. We have already derived this result as Corollary 2.3.10 of the fundamental theorem of spectral norms, but because of its importance, we will also derive it, as Johnson did originally, from his theorem on irreducible representations (Theorem 4.2.15). This uniqueness of norm theorem shows that if a semi-simple algebra $\mathcal{A}$ has any complete algebra norm, then the Banach algebra topology can be described purely in terms of the algebraic structure of $\mathcal{A}$. An actual description would be most interesting.

Not all Banach algebras have a unique Banach algebra topology. A simple example can be obtained by the following rather silly construction. Let $(\mathcal{X}, \| \cdot \|_1)$ and $(\mathcal{Y}, \| \cdot \|_2)$ be two Banach spaces which have the same algebraic dimension but are not linearly homeomorphic (e.g., ℓ^1 and ℓ^2). Use Hamel bases for $\mathcal{X}$ and $\mathcal{Y}$ to define a linear isomorphism φ of $\mathcal{X}$ onto $\mathcal{Y}$. Make $\mathcal{X}$ into a Banach algebra by defining all products to be zero. Then $\| \cdot \|_1$ and $\|\varphi(\cdot)\|_2$ are two inequivalent complete algebra norms on $\mathcal{X}$. A much more interesting example, which is due to the combined work of Chester Feldman [1951] and William G. Bade and Phillip C. Curtis, Jr. [1960b] is given in Example 8.1.5. This example is commutative and has a one-dimensional nilpotent Jacobson radical. Theorem 6.4.23 gives other algebras with inequivalent norms due to Richard J. Loy [1974a]. None of these counterexample algebras are semiprime. (Of course the Johnson–Rickart theorem shows that they are not semisimple.) Thus a number of authors have asked whether all semiprime Banach algebras have a unique Banach algebra topology. The answer is still unknown, but Julian M. Cusack [1976], [1977] has shown that the existence of a semiprime Banach algebra which does not have a unique Banach algebra topology implies the existence of other weird mathematical structures. (See Example 6.4.24 for a discussion of these results.)

Let $\mathcal{A}$ be a Banach algebra under two norms $\| \cdot \|_1$ and $\| \cdot \|_2$. If the identity map from $(\mathcal{A}, \| \cdot \|_1)$ to $(\mathcal{A}, \| \cdot \|_2)$ is continuous, then the open mapping theorem shows that the two norms are equivalent. Hence if a Banach algebra fails to have a unique Banach algebra topology, then it is both the range and the domain of a discontinuous surjective isomorphism between Banach algebras. Conversely, if a Banach algebra is either the range or domain of a discontinuous surjective isomorphism between Banach algebras, then it can obviously be given two non-equivalent complete algebra norms (its own and the one induced by the isomorphism). Hence the question of whether a

Banach algebra $\mathcal{A}$ has a unique Banach algebra topology can be generalized by asking about the continuity of homomorphisms from $\mathcal{A}$ into, or onto, another Banach algebra or from another Banach algebra into, or onto, $\mathcal{A}$. Positive results of this type are said to deal with the *automatic continuity* of the maps involved. Theorem 6.1.5 collects a large number of similar relationships between various uniqueness of norm and automatic continuity conditions and shows their logical relationships in a diagram. This shows that the fundamental Johnson–Rickart result is implied by many conditions which have also been studied and shown to hold under various hypotheses. Following the statement of Theorem 6.1.5, we gather references to some of these results.

6.1 Automatic Continuity of Homomorphisms into $\mathcal{A}$

The next two sections contain several major results on automatic continuity of a homomorphism $\varphi \colon \mathcal{A} \to \mathcal{B}$ between Banach algebras. In the present section we consider some general techniques and the results putting restrictions on the target algebra $\mathcal{B}$. Theorem 6.1.3 states that any surjective homomorphism of a Banach algebra onto a semisimple Banach algebra is continuous. Corollary 6.1.14, which is due to Rickart [1950], states that any homomorphism of a Banach algebra onto a dense subalgebra of a strongly semisimple Banach algebra is continuous. Corollary 6.1.15, due to Georgi Evgenyevič Šilov [1947a], asserts that any homomorphism of a Banach algebra into a commutative semisimple Banach algebra is continuous. Theorem 6.1.18, taken from Nicholas P. Jewell and Allan M. Sinclair [1976], is too complicated to state here, but shows that all surjective homomorphisms of a Banach algebra onto certain radical Banach algebras are continuous. Example 6.1.19 gives algebras satisfying the hypotheses of this theorem and related results.

A simple example shows that at least density of the range is necessary in the first, second, and fourth results mentioned above. Let $\mathcal{A}$ be any infinite-dimensional Banach space made into a Banach algebra by defining all products to be zero. Let ω be a discontinuous linear functional on $\mathcal{A}$. Then the map $\varphi \colon \mathcal{A} \to M_2$ defined by

$$\varphi(a) = \begin{pmatrix} 0 & \omega(a) \\ 0 & 0 \end{pmatrix} \qquad \forall \, a \in \mathcal{A}$$

is a discontinuous homomorphism into the finite-dimensional strongly semisimple algebra M_2. Note that this construction depends on nothing except the fact that the matrix unit $e_{1,2}$ (with a 1 in the first row and second column) satisfies $(e_{1,2})^2 = 0$. Hence there is a discontinuous homomorphism into any algebra with a non-zero nilpotent element. Recall that §3.4.20 contains a discussion of the existence of point derivations. For certain interesting and nontrivial commutative Banach algebras, discontinuous point derivations are easily constructed. We show there how this gives rise to discontinuous homomorphisms into any unital normed algebra with a non-zero nilpotent element.

It is unknown whether the hypothesis in Theorem 6.1.3 that the target algebra be semisimple can be relaxed to substitute semiprime for semisimple. In fact we do not even know whether prime could replace semisimple in the hypotheses. Recall that Theorem 4.4.6 shows that a commutative algebra is semiprime if and only if it has no non-zero nilpotent elements. We have already raised the question of whether semiprime Banach algebras have a unique Banach algebra topology. See Cusack [1977] for further discussion of these questions. That paper shows that if any of the above questions have negative answers there is a topologically simple radical Banach algebra. No such Banach algebra is known.

There are a number of known results on uniqueness of norm and on automatic continuity of homomorphisms which we do not present in this section. Some relate to types of algebras which we have not yet studied. (Theorem 8.7.15 gives such a result.) Among results that assert that a homomorphism $\varphi \colon \mathcal{A} \to \mathcal{B}$ between Banach algebras is continuous or "almost continuous" if $\mathcal{A}$ satisfies certain conditions, we will only give a few of the easiest. Also results on *-homomorphisms will be discussed in Volume II. However, many other important results will not be given anywhere in the work. We feel this omission is justified because H. Garth Dales, a preeminent expert in the subject of automatic continuity, is currently completing a monograph [1994] on the subject.

In his seminal paper [1950], Rickart introduced the fundamental notion of the separating ideal which is expounded in this section. He proved that strongly semisimple Banach algebras have a unique Banach algebra topology and conjectured that any semisimple Banach algebra has this property. He showed how to reduce this question to the question of whether all primitive Banach algebras have a unique norm topology. (It was this special case which Johnson [1967a] settled.) Rickart [1950] also showed that a primitive Banach algebra with minimal ideals has a unique Banach algebra topology. He remarked that the last two results had been noted independently by Irving Kaplansky [unpublished]. Finally Rickart noted the continuity-of-homomorphism results which corresponded to his uniqueness of norm results. In the interim between Rickart's conjecture [1950] and Johnson's proof [1967a] that every semisimple Banach algebra has a unique Banach algebra topology, a number of partial results were obtained which have now been subsumed under the general theorem. We will not bother to refer to these results. For some related results, see the excellent short book by Sinclair [1976].

Uniqueness of Norm

We begin this section with the most important automatic continuity result—Johnson's uniqueness of norm theorem. The next three results, originally due to the combined work of Rickart [1950] and Johnson [1967a], were already stated as Corollaries 2.3.9 and 2.3.10 where they were derived from the fundamental theorem of spectral semi-norms. We shall now derive them, as Johnson did originally, from his result on irreducible normed representations (Theorem 4.2.15).

6.1.1 Theorem *A semisimple Banach algebra has a unique Banach algebra topology.*

Proof This simply means that any two Banach algebra norms are equivalent so that the topologies they induce are equal. It is an immediate consequence of the next theorem.   □

6.1.2 Theorem *Let $(\mathcal{A}, \|\cdot\|_1)$ and $(\mathcal{A}, \|\cdot\|_2)$ be Banach algebras. Then the quotient norms induced by $\|\cdot\|_1$ and $\|\cdot\|_2$ on $\mathcal{A}/\mathcal{A}_J$ are equivalent.*

Proof Again this follows from the next theorem. First set $\mathcal{A}$ equal to $(\mathcal{A}, \|\cdot\|_1)$ and $\mathcal{B}$ equal to $\mathcal{A}/\mathcal{A}_J$ with the quotient norm defined by $\|\cdot\|_2$. Then reverse the roles.   □

6.1.3 Theorem *Let $\mathcal{A}$ and $\mathcal{B}$ be Banach algebras with $\mathcal{B}$ semisimple. Any surjective homomorphism $\varphi\colon \mathcal{A} \to \mathcal{B}$ is continuous.*

Proof Suppose $\varphi\colon \mathcal{A} \to \mathcal{B}$ satisfies the hypotheses but is not continuous. Then it is not closed so that there is a sequence $\{a_n\}_{n\in\mathbb{N}}$ of elements in $\mathcal{A}$ which converges to zero but for which the sequence $\{\varphi(a_n)\}_{n\in\mathbb{N}}$ converges to a non-zero element $b \in \mathcal{B}$. Since $\mathcal{B}$ is semisimple and b is not zero, there is some irreducible representation T of $\mathcal{B}$ satisfying $T_b \neq 0$. Theorem 4.2.8 shows that $\mathcal{X}^T$ may be given a norm relative to which T is continuous. Since φ is surjective, $T \circ \varphi$ is an irreducible normed representative of $\mathcal{A}$ so that it is continuous by Theorem 4.2.15. Thus the continuity of T shows that the sequence $\{T_{\varphi(a_n)}\}_{n\in\mathbb{N}}$ converges to the non-zero operator T_b, but the continuity of $T \circ \varphi$ shows that $T_{\varphi(a_n)} = (T \circ \varphi)a_n$ converges to zero. This contradiction establishes the theorem.   □

The following is an elementary result related to Theorem 6.1.3. The Banach algebra version was noted by Rickart [1950].

6.1.4 Theorem *Let $\mathcal{A}$ be a spectral normed algebra, and let $\mathcal{B}$ be an algebra. Let $\varphi\colon \mathcal{A} \to \mathcal{B}$ be a surjective homomorphism. Then φ maps the closure of the kernel of φ into the Jacobson radical of $\mathcal{B}$. In particular, if $\mathcal{B}$ is semisimple, then the kernel of φ is closed.*

Proof Let $\mathcal{L}$ be a maximal modular left ideal of $\mathcal{B}$. The preimage of $\mathcal{L}$ is a maximal modular left ideal of $\mathcal{A}$ and hence is closed in $\mathcal{A}$ by Theorem 2.4.7. Thus, φ maps the closure of the kernel of φ into $\mathcal{L}$. Since $\mathcal{L}$ was arbitrary, φ maps the closure of its kernel into the Jacobson radical of $\mathcal{B}$. □

Each of the following conditions has been studied and shown to hold for some significant class of Banach algebras. Recall that a norm $||| \cdot |||$ dominates a norm $|| \cdot ||$ if there is some finite constant M satisfying $\|a\| \le M\||a\||$ for all $a \in \mathcal{A}$ and that two norms on the same linear space are consistent if the identity map between them has a closed graph. Conditions (d) and (e) are stated in the geometrical category of normed algebras while all the rest are stated in the (more fundamental) topological category. We give most conditions a brief name.

6.1.5 Theorem *The arrows in the diagram below show the implications known to hold among the following conditions on a Banach algebra $(\mathcal{A}, \| \cdot \|)$.*

(a) *$\mathcal{A}$ has a unique algebra norm topology: The $\| \cdot \|$-norm topology is the only algebra norm topology on $\mathcal{A}$ (i.e., any (not necessarily complete) algebra norm on $\mathcal{A}$ is equivalent to the norm $\| \cdot \|$ and hence is actually complete). $\Longleftrightarrow$ Any injective homomorphism of $\mathcal{A}$ into a normed algebra is a homeomorphism onto its image (which is therefore closed). $\Longleftrightarrow$ Any isomorphism from a normed algebra $\mathcal{B}$ onto $\mathcal{A}$ is a homeomorphism (so, in fact, $\mathcal{B}$ is a Banach algebra).*

(b) *The norm topology of $\mathcal{A}$ is minimum: The $\| \cdot \|$-norm topology is the smallest algebra norm topology (i.e., any (not necessarily complete) algebra norm $||| \cdot |||$ dominates the norm $\| \cdot \|$). $\Longleftarrow$ Any injective homomorphism of $\mathcal{A}$ into a normed algebra is open onto its image. $\Longleftarrow$ Any isomorphism from a normed algebra onto $\mathcal{A}$ is continuous.*

(c) *$\mathcal{A}$ has a unique Banach algebra norm topology: The $\| \cdot \|$-topology is the only Banach algebra topology on $\mathcal{A}$ (i.e., any Banach algebra norm on $\mathcal{A}$ is equivalent to (or consistent with) the norm $\| \cdot \|$). $\Longleftarrow$ Any isomorphism of $\mathcal{A}$ onto a Banach algebra is continuous. $\Longleftrightarrow$ Any isomorphism from a Banach algebra onto $\mathcal{A}$ is continuous.*

(d) *The norm of $\mathcal{A}$ is minimal: The norm $\| \cdot \|$ is minimal among algebra norms (i.e. any (not necessarily complete) algebra norm $||| \cdot |||$ satisfying $\||a\|| \le \|a\|$ for all $a \in \mathcal{A}$, actually equals $\| \cdot \|$). $\Longleftrightarrow$ Any contractive injective homomorphism of $\mathcal{A}$ into a normed algebra is actually an isometry.*

(e) *The norm of $\mathcal{A}$ is minimal spectral: The norm $\| \cdot \|$ is minimal among spectral norms (i.e. any (not necessarily complete) spectral norm $||| \cdot |||$ satisfying $\||a\|| \le \|a\|$ for all $a \in \mathcal{A}$ actually equals $\| \cdot \|$). $\Longleftrightarrow$ Any contractive isomorphism of $\mathcal{A}$ onto a spectral normed algebra is actually an isometry.*

(f) *The norm topology of $\mathcal{A}$ is minimal: The $\| \cdot \|$-topology is minimal among algebra norm topologies on $\mathcal{A}$ (i.e., any (not necessarily complete)*

algebra norm $|||\cdot|||$ which is dominated by $||\cdot||$ is actually equivalent to $||\cdot||$).
$\Longleftrightarrow$ Any continuous injective homomorphism of $\mathcal{A}$ into a normed algebra is actually a homeomorphism onto its image (which is therefore closed).

(g) $\mathcal{A}$ is permanently spectral: Any (not necessarily complete) algebra norm is spectral. $\Longleftrightarrow$ The range of any injective homomorphism into a Banach algebra is a spectral subalgebra.

(h) Any algebra semi-norm on $\mathcal{A}$ is continuous. $\Longleftrightarrow$ Any homomorphism of $\mathcal{A}$ into a normed algebra is continuous.

(i) Any homomorphism from a Banach algebra into $\mathcal{A}$ is continuous.

(j) Any homomorphism from a Banach algebra onto a dense subalgebra of $\mathcal{A}$ is continuous.

(k) Any surjective homomorphism from a Banach algebra onto $\mathcal{A}$ is continuous.

(l) Any surjective homomorphism from a Banach algebra onto $\mathcal{A}$ has a closed kernel and thus defines an isomorphism from the quotient Banach algebra onto $\mathcal{A}$.

(m) Any algebra norm on $\mathcal{A}$ is continuous. $\Longleftrightarrow$ Any injective homomorphism of $\mathcal{A}$ into a normed algebra is continuous.

(n) The norm $||\cdot||$ dominates every Banach algebra norm on $\mathcal{A}$. $\Longleftrightarrow$ Every surjective homomorphism from $\mathcal{A}$ onto a Banach algebra is continuous.

(o) Any two algebra norms on $\mathcal{A}$ are consistent with each other.

(p) Any algebra norm is consistent with the complete norm $||\cdot||$.

If $\mathcal{A}$ is also semisimple, then (c) and (k) hold, (g) implies (o) and hence (d)$\wedge$(g) is also equivalent to (f)$\wedge$(p), (f)$\wedge$(o), (d)$\wedge$(p) and (d)$\wedge$(o).

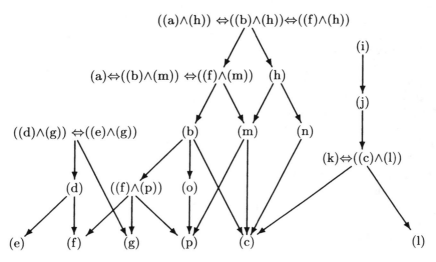

Before giving the easy proof, we will gather references to examples satisfying these conditions. Johnson [1967c] and [1969b] showed that condition (h) holds for any unital C*-algebra without proper closed cofinite ideals, for $\mathcal{B}(\mathcal{X})$ and for $\mathcal{B}(\mathcal{X})/\mathcal{B}_K(\mathcal{X})$ if the Banach space $\mathcal{X}$ has a continued bisection, and for $\mathcal{B}_K(\mathcal{X})$ if in addition $\mathcal{B}_K(\mathcal{X})$ has an approximate identity (cf. Theorems 6.2.3 and 6.2.6 and Corollary 6.2.7). Theorem 3.4.18 (due to Bonsall [1954a]) and §1.7.15 (due to Yood [1958]) show, respectively, that any C*-algebra satisfies (d) and both $\mathcal{B}(\mathcal{X})$ and $\mathcal{B}_K(\mathcal{X})$ (for any Banach space $\mathcal{X}$) satisfy (b). Hence C*-algebras without proper closed cofinite ideals, $\mathcal{B}(\mathcal{X})$ (for any Banach space $\mathcal{X}$ with a continued bisection) and $\mathcal{B}_K(\mathcal{X})$ (for any Banach space $\mathcal{X}$ with a continued bisection and the compact approximation property (§5.1.11)) satisfy (a)∧(h). Meyer [1992c] shows that the same is true for the generalized Calkin algebra $\mathcal{B}(\mathcal{X})/\mathcal{B}_K(\mathcal{X})$ for some classical sequence Banach spaces $\mathcal{X}$. Sandra B. Cleveland [1963] used the main boundedness lemma of Bade and Curtis to show that any C*-algebra actually satisfies condition (b). Dales [1989] and Ángel Rodríguez-Palacios [1990] give distinct but closely related proofs which use the above relatively elementary property of C*-algebras and Corollary 6.1.12 to obtain Cleveland's theorem much more easily. We give a version of this argument in Theorem 6.1.16. Cleveland's result extends Kaplansky's [1949a] discovery that commutative C*-algebras satisfy the stronger condition that any algebra norm actually majorizes rather than just dominates the C*-norm (our Theorem 2.4.15). In fact, Kaplansky's result is used in proving (d) for arbitrary C*-algebras.

Albrecht and Dales [1983] studied condition (n) and showed that (if one assumes the continuum hypothesis), it holds for a wide class of C*-algebras including the commutative ones for which (h) fails. Conditions (o) and (p) and results about them are due to Bohdan J. Tomiuk and Bertram Yood [1989] and Michael J. Meyer [1992a].

All those Banach algebras which we know to satisfy condition (c) also satisfy (k). These include all semisimple Banach algebras (Johnson [1967a] and Theorem 6.1.1 above) and various radical algebras of power series. Condition (e) was proved for approximate $B^{\#}$-norms (which need not be complete or spectral themselves a priori) by Smiley [1955] following work of Bonsall [1954a]. See Proposition 2.3.16. Since the complete norm on any C*-algebra and on any subalgebra of $\mathcal{B}(\mathcal{X})$ which includes $\mathcal{B}_F(\mathcal{X})$ (for any Banach space $\mathcal{X}$) satisfies the approximate $B^{\#}$-condition and condition (b), it also satisfies (d)∧(g).

Condition (g) was proved for completely regular semisimple commutative Banach algebras by Rickart [1953] (our Theorem 3.5.17) and for semisimple modular annihilator normed algebras by Yood [1958]. Theorem 8.4.12 removes the condition of semisimplicity. Condition (g) holds for radical Banach algebras trivially as noted in Corollary 4.3.9.

Corollary 6.1.15 shows that semisimple commutative Banach algebras have property (i), as first noted by Šilov [1947a]. Corollary 6.1.14 is Rickart's result that strongly semisimple Banach algebras satisfy (j). In [1967b] Barnes showed that semisimple modular annihilator Banach algebras also satisfy (j). In Theorem 8.7.15 we obtain the same result for the wider class of semisimple Duncan modular annihilator algebras. Theorem 6.1.4 has just given the elementary proof that semisimple algebras satisfy condition (l), but this follows from (k) which they also satisfy. Theorem 6.1.18 gives the result of Jewell and Sinclair [1976] that a wide class of (often Jacobson-radical) Banach algebras satisfy (k). Example 6.1.19 gives specific cases.

Proof First we establish the equivalence of the various alternate descriptions of the given conditions.

(a): Let $\varphi\colon \mathcal{A} \to \mathcal{B}$ be an injective homomorphism into a normed algebra $(\mathcal{B}, \|\cdot\|_{\mathcal{B}})$. Then $\|\|a\|\| = \|\varphi(a)\|_{\mathcal{B}}$ is an algebra norm on $\mathcal{A}$ and every algebra norm arises in this way since we may take $\mathcal{B}$ to be $\mathcal{A}$ with a different norm. Let $\psi\colon \mathcal{B} \to \mathcal{A}$ be an isomorphism of a normed algebra $(\mathcal{B}, \|\cdot\|_{\mathcal{B}})$ onto $\mathcal{A}$. Then $\|\|\psi(b)\|\| = \|b\|_{\mathcal{B}}$ defines an algebra norm on $\mathcal{A}$ and every algebra norm arises in this way since we may again take $\mathcal{B}$ to be $\mathcal{A}$ with a different norm. Thus the three conditions agree.

(b), (d), (e) and (f): The equivalence of the alternate conditions given in all these cases follows from the considerations introduced for (a).

(c): In this situation, the open mapping theorem shows that the continuous isomorphisms must be homeomorphisms. Note that we could get two more equivalent conditions by replacing "continuous" by "open" in the second and third conditions.

(g): Theorem 2.5.15 gives this equivalence.

(h): If $\varphi\colon \mathcal{A} \to \mathcal{B}$ is an arbitrary homomorphism into a normed algebra $(\mathcal{B}, \|\cdot\|_{\mathcal{B}})$, then $\|\|a\|\| = \|\varphi(a)\|_{\mathcal{B}}$ is an algebra semi-norm on $\mathcal{A}$ and every algebra semi-norm arises in this way.

(f)∧(h) $\Rightarrow$ (a): Condition (h) and the alternate condition under (f) immediately imply the first alternate condition for (a).

(f)∧(m) $\Leftrightarrow$ (a): In one direction this is the same as the previous case. The first alternate condition for (a) immediately implies (m) and the alternate condition for (f).

(b) $\Rightarrow$ (c): The second alternate condition should be used for both (b) and (c).

(b) $\Rightarrow$ (f): An injective homomorphism which is both continuous and open is certainly a homeomorphism.

(f)∧(p) $\Rightarrow$ (g): This implication requires Proposition 6.1.9(b). Let $|\cdot|$ be an algebra norm on $\mathcal{A}$. Let $\mathcal{B}$ be the completion of $\mathcal{A}$ in this norm and let $\varphi : \mathcal{A} \to \mathcal{B}$ be the natural embedding. Let $\tilde{\varphi} : \mathcal{A} \to \mathcal{B}/\mathcal{B}_\varphi$ be the induced map described and shown to be continuous in Proposition 6.1.9(b). The consistency of $|\cdot|$ and $\|\cdot\|$ shows that $\tilde{\varphi}$ is injective. Hence (f) shows that

$\tilde{\varphi}$ is a homeomorphism, so its dense range is closed, and thus equals $\mathcal{B}/\mathcal{B}_\varphi$. Since $|\cdot|$ majorizes the norm on $\mathcal{B}/\mathcal{B}_\varphi$, it is a spectral norm.

(d) $\Rightarrow$ (f): This implication was first noted by Rodríguez-Palacios [1990]. Suppose $(\mathcal{A}, \|\cdot\|)$ satisfies (d) and $\||\cdot\||$ is an algebra norm satisfying $\||a\|| \leq M\|a\|$ for all $a \in \mathcal{A}$. Then the closed unit ball $\mathcal{A}_1$ of $\mathcal{A}$ relative to $\||\cdot\||$ is a semigroup which is bounded relative to $\|\cdot\|$. Proposition 1.1.9 gives an algebra norm σ equivalent to $\|\cdot\|$ in which every element of $\mathcal{A}_1$ has norm at most 1. Thus $\sigma(a) \leq \||a\||$ for all $a \in \mathcal{A}$ together with (d) gives $\sigma(a) = \||a\||$ for all $a \in \mathcal{A}$. Hence $\|\cdot\|$ and $\||\cdot\||$ are equivalent.

(m) $\Rightarrow$ (c): Use the first alternate condition for (c).

(c)$\wedge$(l) $\Leftrightarrow$ (k): Let ψ be a surjective homomorphism of a Banach algebra $\mathcal{B}$ onto $\mathcal{A}$. Let $\mathcal{I}$ be the kernel of ψ which is closed by (l). If θ is the natural map of $\mathcal{B}$ onto $\mathcal{B}/\mathcal{I}$, then we can write $\psi = \tilde{\psi} \circ \theta$ where $\tilde{\psi}$ is an isomorphism of $\mathcal{B}/\mathcal{I}$ onto $\mathcal{A}$. Then ψ is continuous since $\tilde{\psi}$ and θ are. The implication in the other direction is immediate.

The other implications ((a) $\Rightarrow$ (b); (h) $\Rightarrow$ (m); (h) $\Rightarrow$ (n); (i) $\Rightarrow$ (j) $\Rightarrow$ (k); (n) $\Rightarrow$ (c); (m) $\Rightarrow$ (p); (b) $\Rightarrow$ (o) $\Rightarrow$ (p) and (e)$\wedge$(g) $\Rightarrow$ (d) $\Rightarrow$ (e)) are even more trivial.

If $\mathcal{A}$ is semisimple, Theorems 6.1.1 and 6.1.3 and Corollary 2.3.8 show that (c) and (k) hold and (g) implies (o). The rest of the claims are now immediate.   □

Here is another useful but almost trivial observation similar in spirit to the results of Theorem 6.1.5. It was noted by Albrecht and Dales [1983].

6.1.6 Proposition *If $\mathcal{I}$ is a closed ideal of a Banach algebra $\mathcal{A}$ and there is a discontinuous homomorphism from $\mathcal{A}/\mathcal{I}$, then there is a discontinuous homomorphism from $\mathcal{A}$.*

Proof Since the quotient map $\theta : \mathcal{A} \to \mathcal{A}/\mathcal{I}$ is open, $\varphi : \mathcal{A}/\mathcal{I} \to \mathcal{B}$ is continuous if and only if $\varphi \circ \theta$ is continuous.   □

Separating Spaces

Having restated, rederived, and investigated the relationship among these most fundamental results, we begin a more systematic study of automatic continuity. We introduce a notion (due to Rickart [1950]) which will be our major device in this study. The normed linear spaces of the definition will usually be Banach algebras in our applications. However, we will also use the case in which they are Banach modules, *i.e.*, Banach spaces together with a fixed continuous representation of a Banach algebra, and perhaps, a commuting continuous anti-representation also.

6.1.7 Definition Let $\mathcal{X}$ and $\mathcal{Y}$ be normed linear spaces and let $\varphi : \mathcal{X} \to \mathcal{Y}$ be a linear map. Define $\sigma_\varphi : \mathcal{Y} \to \mathbb{R}_+$ by

$$\sigma_\varphi(y) = \inf\{\|x\| + \|y - \varphi(x)\| : x \in \mathcal{X}\} \qquad \forall\, y \in \mathcal{Y}.$$

Then σ_φ is called the *separating function of φ*. The sets

$$\mathcal{X}^\varphi = \{x \in \mathcal{X} : \sigma_\varphi(\varphi(x)) = 0\} \quad \text{and}$$
$$\mathcal{Y}_\varphi = \{y \in \mathcal{Y} : \sigma_\varphi(y) = 0\}$$

are called the *separating spaces of φ in $\mathcal{X}$ and $\mathcal{Y}$*, respectively.

If there is no possibility of confusion, we write σ instead of σ_φ. The following characterizations of the separating spaces are easy to grasp

$$\mathcal{Y}_\varphi = \{y \in \mathcal{Y} : \text{ there is a sequence } \{x_n\}_{n \in \mathbb{N}} \subseteq \mathcal{X}$$
$$\text{satisfying } \lim_{n \to \infty} x_n = 0 \text{ and } \lim_{n \to \infty} \varphi(x_n) = y\} \quad (1)$$
$$\mathcal{X}^\varphi = \varphi^\leftarrow(\mathcal{Y}_\varphi).$$

6.1.8 Proposition *Let $\mathcal{X}$ and $\mathcal{Y}$ be normed linear spaces, and let $\varphi \colon \mathcal{X} \to \mathcal{Y}$ be a linear map.*

(a) *Then σ is a semi-norm on $\mathcal{Y}$ satisfying:*

$$\sigma(y) \leq \|y\| \quad \text{and} \quad \sigma(\varphi(x)) \leq \|x\| \qquad \forall\, x \in \mathcal{X};\, y \in \mathcal{Y}.$$

Hence $\mathcal{X}^\varphi$ and $\mathcal{Y}_\varphi$ are closed linear subspaces of $\mathcal{X}$ and $\mathcal{Y}$, respectively. Also σ is a norm if and only if $\mathcal{Y}_\varphi = \{0\}$.

(b) *If $S \in \mathcal{B}(\mathcal{X})$ and $T \in \mathcal{B}(\mathcal{Y})$ are bounded linear maps satisfying $\varphi \circ S = T \circ \varphi$, then σ satisfies*

$$\sigma \circ T(y) \leq \max\{\|S\|, \|T\|\}\sigma(y) \qquad \forall\, y \in \mathcal{Y}$$

which implies $T(\mathcal{Y}_\varphi) \subseteq \mathcal{Y}_\varphi$ and $S(\mathcal{X}^\varphi) \subseteq \mathcal{X}^\varphi$.

(c) *If $\mathcal{W}$ is another normed linear space and $\psi \colon \mathcal{W} \to \mathcal{X}$ is a bounded linear map, then the separating spaces satisfy $\mathcal{Y}_{\varphi \circ \psi} \subseteq \mathcal{Y}_\varphi$.*

(d) *If $\mathcal{X} = \mathcal{A}$ and $\mathcal{Y} = \mathcal{B}$ are normed algebras and φ is a homomorphism or anti-homomorphism, then they satisfy*

$$\sigma(\varphi(a)b) \leq \sigma(b) \max\{\|a\|, \|\varphi(a)\|\} \qquad \forall\, a \in \mathcal{A};\, b \in \mathcal{B}$$
$$\sigma(b\varphi(a)) \leq \sigma(b) \max\{\|a\|, \|\varphi(a)\|\} \qquad \forall\, a \in \mathcal{A};\, b \in \mathcal{B}.$$

Thus, $\mathcal{A}^\varphi$ is a closed ideal of $\mathcal{A}$ and $\mathcal{B}_\varphi$ is a closed ideal of $\varphi(\mathcal{A})^-$. If $\varphi(\mathcal{A})$ is dense in $\mathcal{B}$, $\mathcal{B}_\varphi$ is a closed ideal in $\mathcal{B}$.

Proof (a): The statements about the separating function are easily verified and imply the statements about the separating spaces.

(b): Any $y \in \mathcal{Y}$ satisfies

$$\sigma \circ T(y) \leq \inf\{\|Sx\| + \|y - \varphi(Sx)\| : x \in \mathcal{X}\}$$
$$\leq \inf\{\|S\|\,\|x\| + \|T\|\,\|y - \varphi(x)\| : x \in \mathcal{X}\}$$
$$\leq \max\{\|S\|, \|T\|\}\sigma(y).$$

This immediately implies $\sigma \circ \varphi \circ S(x) \leq \max\{\|S\|, \|T\|\}\sigma \circ \varphi(x)$ for any $x \in \mathcal{X}$. These two inequalities give the statements about the separating spaces.

(c): We use the characterization displayed above in (1). If y belongs to $\mathcal{Y}_{\varphi \circ \psi}$, then there is a sequence $\{w_n\}_{n \in \mathbb{N}} \subseteq \mathcal{W}$ satisfying $\lim_{n \to \infty} w_n = 0$ and $\lim_{n \to \infty} \varphi \circ \psi(w_n) = y$. Then $\lim_{n \to \infty} \psi(w_n) = 0$ shows that y belongs to $\mathcal{Y}_\varphi$.

(d): The two inequalities are special cases of the inequality in (b) when S and T are chosen from the regular representation operators L_a, $L_{\varphi(a)}$, R_a and $R_{\varphi(a)}$. The results about the separating spaces follow easily. $\quad\square$

The following properties show the usefulness of the separating spaces.

6.1.9 Proposition *Let $\mathcal{X}$ and $\mathcal{Y}$ be $\langle$ normed linear spaces / Banach spaces $\rangle$, and let $\varphi: \mathcal{X} \to \mathcal{Y}$ be a linear map.*

(a) *Then, $\varphi \langle$ has a closed graph / is continuous $\rangle$ if and only if $\mathcal{Y}_\varphi$ is $\{0\}$ (or, alternatively, if and only if σ is a norm).*

(b) *The linear maps $\tilde{\varphi}: \mathcal{X} \to \mathcal{Y}/\mathcal{Y}_\varphi$ and $\tilde{\tilde{\varphi}}: \mathcal{X}/\mathcal{X}^\varphi \to \mathcal{Y}/\mathcal{Y}_\varphi$ induced by φ always $\langle$ have a closed graph / are continuous $\rangle$.*

(c) *Let $\mathcal{Z}$ be a $\langle$ normed linear space / Banach space $\rangle$, and let $\psi: \mathcal{Y} \to \mathcal{Z}$ be a linear map. Then $\mathcal{Z}_{\psi \circ \varphi}$ is the closure of $\psi(\mathcal{Y}_\varphi)$. Furthermore, $\psi \circ \varphi$ $\langle$ has a closed graph / is continuous $\rangle$ if and only if $\psi(\mathcal{Y}_\varphi)$ is $\{0\}$. If $\psi \circ \varphi$ is continuous, then there is a constant M independent of $\mathcal{Z}$ and ψ satisfying $\|\psi \circ \varphi\| \leq M\|\psi\|$.*

Proof (a): Display (1), after the definition, shows that φ is closed if and only if the separating space vanishes. The closed graph theorem gives continuity if the spaces are Banach spaces.

(b): Let $\{x_n\}_{n \in \mathbb{N}} \subseteq \mathcal{X}$ be a sequence satisfying both $\lim x_n = 0$ and $\lim \tilde{\varphi}(x_n) = y + \mathcal{Y}_\varphi$. Then there is a sequence $\{y_n\}_{n \in \mathbb{N}} \subseteq \mathcal{Y}_\varphi$ satisfying $\lim(\varphi(x_n) - y - y_n) = 0$. Choose a sequence $\{z_n\}_{n \in \mathbb{N}} \subseteq \mathcal{X}$ with $\|z_n\| + \|\varphi(z_n) - y_n\|$ converging to 0. Then the sequence $\{x_n - z_n\}_{n \in \mathbb{N}}$ shows that y belongs to $\mathcal{Y}_\varphi$. Hence $\tilde{\varphi}$ is closed. Thus it is continuous if the spaces involved are complete.

(c): The inclusion $\psi(\mathcal{Y}_\varphi) \subseteq \mathcal{Z}_{\psi \circ \varphi}$ is immediate from the definitions of $\mathcal{Y}_\varphi$ and $\mathcal{Z}_{\psi \circ \varphi}$. Since the latter space is closed, we may take the closure of $\psi(\mathcal{Y}_\varphi)$ without altering the inclusion. We postpone the opposite inclusion until after we have proved the other results.

If $\psi \circ \varphi$ has a closed graph, then $\mathcal{Z}_{\psi \circ \varphi} = \{0\}$ implies $\psi(\mathcal{Y}_\varphi) = \{0\}$ by the inclusion we just noted. Suppose, conversely, that $\psi(\mathcal{Y}_\varphi)$ vanishes. Then $\psi \circ \varphi$ can be factored through the continuous map $\tilde{\varphi}$ of conclusion (b): $\psi \circ \varphi = \tilde{\psi} \circ \tilde{\varphi}$ where $\tilde{\psi}(y + \mathcal{Y}_\varphi) = \psi(y)$ for all $y \in \mathcal{Y}$. Thus $\psi \circ \varphi$ is continuous since both $\tilde{\psi}$ and $\tilde{\varphi}$ are. Furthermore, M may be chosen as $\|\tilde{\varphi}\|$.

It only remains to show the inclusion $\mathcal{Z}_{\psi \circ \varphi} \subseteq (\psi(\mathcal{Y}_\varphi))^-$. Let θ be the quotient map $\theta: \mathcal{Z} \to \mathcal{Z}/\mathcal{Z}_{(\psi(\mathcal{Y}_\varphi))^-}$. Then $\theta \circ \psi(\mathcal{Y}_\varphi) = \{0\}$ implies that

$\theta \circ \psi \circ \varphi$ has a closed graph. This in turn implies $\theta(\mathcal{Z}_{\psi \circ \varphi}) = \{0\}$. We conclude $\mathcal{Z}_{\psi \circ \varphi} \subseteq (\psi(\mathcal{Y}_\varphi))^-$.   □

We now give other technical results which will be used later. The first is due to Barnes [1967b].

6.1.10 Proposition *Let $\mathcal{A}$ and $\mathcal{B}$ be Banach algebras and let $\varphi: \mathcal{A} \to \mathcal{B}$ be a homomorphism. Every element of $\mathcal{B}_\varphi$ has connected spectrum containing zero. In particular $\mathcal{B}_\varphi$ contains no non-zero idempotent.*

Proof Suppose there is some $a \in \mathcal{B}_\varphi$ such that $S \subseteq Sp(a)$ is a nonempty open and closed subset of $Sp(a)$ which does not contain zero. Then we can choose an open set $V \subset \mathbb{C}$ satisfying $\text{dist}(V, 0) = \delta > 0$ and $V \cap Sp(a) = \bar{V} \cap Sp(a) = S$. Hence Lemma 3.4.6 asserts that there is a neighborhood $\mathcal{D}$ of a such that each $b \in \mathcal{D}$ satisfies $\delta < \rho(b)$. However, since a belongs to $\mathcal{B}_\varphi$ there is a sequence $\{a_n\}_{n \in \mathbb{N}} \subseteq \mathcal{A}$ satisfying $\lim a_n = 0$ and $\lim \varphi(a_n) = a$. This contradicts the inequality $0 < \delta < \rho(\varphi(a_n)) \leq \rho(a_n) \leq \|a_n\|$ which holds for all sufficiently large n. (The penultimate inequality is just Proposition 2.2.1(c).)   □

The following theorem and corollary were first derived (but not explicitly stated) by Bernard Aupetit in his important paper [1982a]. For Banach algebras they were explicitly stated by Garth Dales [1989] and Rodríguez-Palacios [1990], respectively, with proofs which are essentially due to Ransford [1989]. Both the theorem and its proof are small variations on the fundamental theorem of spectral norms (Theorem 2.3.6). Recall that $\mathcal{B}_{tN}$ is the set of topologically nilpotent elements in $\mathcal{B}$.

6.1.11 Theorem *Let $\mathcal{A}$ and $\mathcal{B}$ be spectral normed algebras and let $\varphi: \mathcal{A} \to \mathcal{B}$ be a linear map which satisfies $\rho_\mathcal{B}(\varphi(a)) \leq \rho_\mathcal{A}(a)$ for all $a \in \mathcal{A}$ (e.g. φ might be a homomorphism or anti-homomorphism). Then φ satisfies*

$$\rho_\mathcal{B}(\varphi(a))^2 \leq \rho_\mathcal{A}(a)\rho_\mathcal{B}(\varphi(a) - b) \qquad \forall\, a \in \mathcal{A}: b \in \mathcal{B}_\varphi.$$

6.1.12 Corollary *Under the hypotheses of the theorem,*

$$\varphi(\mathcal{A}^\varphi) = \varphi(\mathcal{A}) \cap \mathcal{B}_\varphi \subseteq \mathcal{B}_{tN}.$$

Proof The corollary follows by taking $b = \varphi(a)$ in the theorem. Proposition 2.2.1(c) gives the parenthetical remark in the statement of the theorem.

The theorem is proved by a simple application of Ransford's three circle theorem (Lemma 2.3.5). As noted after Theorem 2.2.14, for any $\varepsilon > 0$ Theorem 1.1.10(i) allows us to choose equivalent spectral norms satisfying $\|a\| < \rho_\mathcal{A}(a) + \varepsilon$ and $\|\varphi(a) - b\| < \rho_\mathcal{B}(\varphi(a) - b) + \varepsilon$. Let $\{a_n\}$ be a sequence converging to 0 in the norm of $\mathcal{A}$ and also satisfying $\varphi(a_n) \to b$ in the norm of $\mathcal{B}$. For each $n \in \mathbb{N}$, define a polynomial in $\mathcal{B}$ by $p_n(\lambda) =$

$\lambda\varphi(a_n) + \varphi(a - a_n)$. If we estimate the spectral radius of $p_n(\lambda)$ in the norm of $\mathcal{B}$, we get

$$\rho_\mathcal{B}(p_n(\lambda)) \leq |\lambda| \, \|\varphi(a_n)\| + \|\varphi(a - a_n)\|.$$

Using the norm in $\mathcal{A}$, we get

$$\rho_\mathcal{B}(p_n(\lambda)) \leq \rho_\mathcal{A}(\lambda a_n + (a - a_n)) \leq (|\lambda| + 1) \, \|a_n\| + \|a\|.$$

Substituting these estimates into Lemma 2.3.5 with arbitrary positive R gives

$$\rho_\mathcal{B}(\varphi(a))^2 \leq ((R + 1)\|a_n\| + \|a\|) \, (R^{-1}\|\varphi(a_n)\| + \|\varphi(a - a_n)\|).$$

When we first let n and then R approach ∞ we conclude

$$\rho_\mathcal{B}(\varphi(a))^2 \leq \|a\| \, \|\varphi(a) - b\| \leq (\rho_\mathcal{A}(a) + \varepsilon) \, (\rho_\mathcal{B}(\varphi(a)) - b) + \varepsilon).$$

Since $\varepsilon > 0$ was arbitrary, the proof is complete.   □

It is unknown whether Corollary 6.1.12 can be improved to state that $\mathcal{B}_\varphi \subseteq \mathcal{B}_{tN}$ which would imply $\mathcal{B}_\varphi \subseteq \mathcal{B}_J$ by Corollary 4.3.9. In fact this improvement is logically equivalent to the conjecture that every homomorphism with dense range of a Banach algebra into a semisimple Banach algebra is continuous. This conjecture combines the good features of Theorem 6.1.3 and Corollary 6.1.14. To see the equivalence, first suppose that $\mathcal{B}_\varphi \subseteq \mathcal{B}_{tN}$ always holds. Then $\mathcal{B}_\varphi$ would be a topologically nilpotent ideal and hence included in $\mathcal{B}_J = \{0\}$ whenever φ had dense range in a semisimple Banach algebra. Conversely, suppose the conjecture holds and $\varphi: \mathcal{A} \to \mathcal{B}$ is arbitrary. Then the map of $\mathcal{A}$ into $\overline{\varphi(\mathcal{A})}/\overline{\varphi(\mathcal{A})}_J$ would be continuous so $\mathcal{B}_\varphi$ would be included in $\overline{\varphi(\mathcal{A})}_J$ and hence in $\mathcal{B}_{tN}$. Of course the improved result does hold in the case where φ is surjective. This gives another proof of Theorem 6.1.3. We now derive additional results.

6.1.13 Proposition *Let $\mathcal{A}$ and $\mathcal{B}$ be spectral normed algebras and let $\varphi: \mathcal{A} \to \mathcal{B}$ be a homomorphism or anti-homomorphism with dense range.*
   (a) *$\rho(c) \leq \sigma(c)$ for all c in the center of $\mathcal{B}$.*
   (b) *$\mathcal{B}_\varphi \subseteq \mathcal{B}_M$.*
   (c) *If $\mathcal{B}$ is strongly semisimple, then φ has a closed graph.*

Proof (a): Let c belong to the center of $\mathcal{B}$. Theorem 2.4.3 implies: $\rho(c) \leq \rho(\varphi(a)) + \rho(c - \varphi(a)) \leq \rho(a) + \|c - \varphi(a)\| \leq \|a\| + \|c - \varphi(a)\|$ for all $a \in \mathcal{A}$. Hence (a) holds.

(b): Of course $\mathcal{B}_M$ is the Brown–McCoy or strong radical of $\mathcal{B}$. Let $\mathcal{M}$ be a maximal modular ideal of $\mathcal{B}$. Then, $\mathcal{B}/\mathcal{M}$ has an identity 1 which is in the center and satisfies $\rho_{\mathcal{B}/\mathcal{M}}(1) = 1$. Theorem 2.2.14 shows that $\mathcal{B}/\mathcal{M}$ is a spectral semi-normed algebra. Consider the map $\varphi': \mathcal{A} \to \mathcal{B}/\mathcal{M}$

which is the composition of φ and the natural map of $\mathcal{B}$ onto $\mathcal{B}/\mathcal{M}$. Let σ' be the separating semi-norm of this homomorphism. By part (a), σ' is not identically zero. Since $\mathcal{B}/\mathcal{M}$ is simple and $(\mathcal{B}/\mathcal{M})_{\varphi'}$ is an ideal by Proposition 6.1.8(d), σ' is a norm on $\mathcal{B}/\mathcal{M}$. Let $b \in \mathcal{B}_\varphi$ and $\varepsilon > 0$ be arbitrary. Then there is an $a \in \mathcal{A}$ satisfying $\|a\| + \|(b - \varphi(a)) + \mathcal{M}\| \leq \|a\| + \|b - \varphi(a)\| < \varepsilon$. Since $\varepsilon > 0$ was arbitrary, we conclude $\sigma'(b + \mathcal{M}) = 0$ and hence $b \in \mathcal{M}$. Thus, $\mathcal{B}_\varphi$ is included in $\mathcal{M}$. Since $\mathcal{M}$ was an arbitrary maximal modular ideal, we conclude $\mathcal{B}_\varphi \subseteq \mathcal{B}_M$.

(c): If $\mathcal{B}$ is strongly semisimple, then $\mathcal{B}_\varphi = \{0\}$ by (b). But this is exactly the assertion that φ has a closed graph.   □

Conditions on the Target Algebra

The surjectivity of the homomorphism in Theorem 6.1.3 severely limits the application of that theorem. There are several cases where the hypothesis of surjectivity can be relaxed. We give two.

6.1.14 Corollary *Let $\mathcal{A}$ and $\mathcal{B}$ be Banach algebras with $\mathcal{B}$ strongly semisimple. Then, any homomorphism $\varphi: \mathcal{A} \to \mathcal{B}$ with dense range is continuous.*

Proof Proposition 6.1.13(c) and the closed graph theorem prove this.   □

6.1.15 Corollary *Let $\mathcal{A}$ be a Banach algebra and let $\mathcal{B}$ be a commutative semisimple Banach algebra. Then any homomorphism $\varphi: \mathcal{A} \to \mathcal{B}$ is continuous.*

Proof We have already proved this result as Theorem 3.1.11, but we will give a different proof here. Any closed subalgebra of a commutative semisimple Banach algebra is strongly semisimple by Theorem 4.5.8. Thus we apply Corollary 6.1.14 to $\varphi: \mathcal{A} \to \varphi(\mathcal{A})^-$.   □

The next result was first established by Cleveland [1963] with a much more difficult proof. Proofs similar to the present elementary proof were discovered by Dales [1989] and Rodríguez-Palacios [1990].

6.1.16 Theorem *The algebra norm topology on a C^*-algebra $(\mathcal{A}, \|\cdot\|)$ is minimum, that is, for any (possibly incomplete) algebra norm $\|\|\cdot\|\|$ on $\mathcal{A}$ there is a finite constant M satisfying*

$$\|a\| \leq M\|\|a\|\| \forall\, a \in \mathcal{A}.$$

Proof Let $\mathcal{B}$ be the completion of $(\mathcal{A}, \|\|\cdot\|\|)$ and let $\varphi: \mathcal{A} \to \mathcal{B}$ be the embedding homomorphism which is injective and maps onto a dense subalgebra. Note that $\varphi(\mathcal{A})$ is a spectral subalgebra of $\mathcal{B}$ by Theorem 3.4.18 and Proposition 2.5.15. Corollary 6.1.12 shows that $\mathcal{A}^\varphi$ is an ideal of $\mathcal{A}$ included in $\mathcal{A}_{tN}$. Hence $\mathcal{A}^\varphi$ is included in the Jacobson radical of $\mathcal{A}$ which is just $\{0\}$ by Proposition 2.3.18. Hence the continuous homomorphism $\tilde{\varphi} \colon \mathcal{A} \to \mathcal{B}/\mathcal{B}_\varphi$ of Proposition 6.1.9(b) is actually injective. The algebra norm defined by $a \mapsto |||\tilde{\varphi}(a)|||$ satisfies $|||\tilde{\varphi}(a)||| \leq |||\tilde{\varphi}||| \, \|a\|$ for all $a \in \mathcal{A}$ and hence is equivalent to $\|\cdot\|$ by Theorem 3.4.18. However, this norm on $\mathcal{A}$ is certainly less than or equal to $|||\cdot|||$, so the conclusion follows. $\Box$

We conclude this section by proving a rather technical criterion, due to Jewell and Sinclair [1976], for a Banach algebra to have a unique norm topology. Example 6.1.19 will show that this result applies to several interesting Banach algebras including the Banach algebra $L^1(0,1)$ which was shown to have a discontinuous functional calculus in Example 3.4.10. The next lemma is a slight improvement of a lemma of Sinclair [1975]. Actually the original form of this lemma would suffice in the proof of Theorem 6.1.18, but we will use the present form of Lemma 6.1.17 again in Theorem 6.3.17 to prove the continuity of all derivations of an algebra satisfying the hypotheses of Theorem 6.1.18. In the reference already cited, Lemma 6.1.17 is also used to give additional proofs of Theorem 4.2.15 and Corollary 6.4.16.

If $\mathcal{X}$ and $\mathcal{Y}$ are normed linear spaces, recall that $\mathcal{L}(\mathcal{X}, \mathcal{Y})$ and $\mathcal{B}(\mathcal{X}, \mathcal{Y})$ denote the linear space of all linear maps of $\mathcal{X}$ into $\mathcal{Y}$ and the normed linear space of all continuous linear maps of $\mathcal{X}$ into $\mathcal{Y}$, respectively.

6.1.17 Lemma *Let $\mathcal{X}$ and $\mathcal{Y}$ be Banach spaces. Let $\{S_n\}_{n \in \mathbb{N}} \subseteq \mathcal{B}(\mathcal{X})$ and $\{T_n\}_{n \in \mathbb{N}} \subseteq \mathcal{B}(\mathcal{Y})$ and $\varphi \in \mathcal{L}(\mathcal{X}, \mathcal{Y})$ satisfy $T_n \varphi - \varphi S_n \in \mathcal{B}(\mathcal{X}, \mathcal{Y})$ for all $n \in \mathbb{N}$. Then there is an integer N satisfying*

$$(T_1 T_2 \cdots T_n \mathcal{Y}_\varphi)^- = (T_1 T_2 \cdots T_N \mathcal{Y}_\varphi)^- \qquad \forall\, n \geq N.$$

Proof First, we note that the identity

$$
\begin{aligned}
T_n T_{n+1} \cdots T_m \varphi \quad &- \quad \varphi S_n S_{n+1} \cdots S_m \\
&= \sum_{k=n}^{m} T_n T_{n+1} \cdots T_{k-1} (T_k \varphi - \varphi S_k) S_{k+1} S_{k+2} \cdots S_m
\end{aligned}
$$

proves that the operator on the left is continuous for each m.

Proposition 6.1.8(b) shows $T_{m+1} \mathcal{Y}_\varphi \subseteq \mathcal{Y}_\varphi$ for all $m \in \mathbb{N}$. Hence

$$(T_1 T_2 \cdots T_{m+1} \mathcal{Y}_\varphi)^- \subseteq (T_1 T_2 \cdots T_m \mathcal{Y}_\varphi)^-$$

for all $m \in \mathbb{N}$. Suppose this inclusion is strict for infinitely many m. By our first remark, we may group the T_j's and S_j's into finite products corresponding to the intervals of constancy of $(T_1 T_2 \cdots T_m \mathcal{Y}_\varphi)^-$ and hence assume that the inclusion is strict for all $m \in \mathbb{N}$. By scaling, we may also assume $\|S_m\| \leq 1$ for all $m \in \mathbb{N}$.

For each $m \in \mathbb{N}$, let $Q_m \colon \mathcal{Y} \to \mathcal{Y}/(T_1 T_2 \cdots T_m \mathcal{Y}_\varphi)^-$ denote the natural map. Proposition 6.1.9(c) shows that $Q_m T_1 T_2 \cdots T_m \varphi$ is continuous but $Q_m T_1 T_2 \cdots T_{m-1} \varphi$ is discontinuous for each $m \in \mathbb{N}$.

Inductively, choose a sequence $\{x_n\}_{n\in\mathbb{N}}$ satisfying $\|x_n\| \le 2^{-n}$ and

$$\|Q_nT_1T_2\cdots T_{n-1}\varphi x_n\| - \|Q_n(T_1T_2\cdots T_{n-1}\varphi - \varphi S_1S_2\cdots S_{n-1})\|$$

$$\ge n + \|Q_n\varphi \sum_{k=1}^{n-1} S_1S_2\cdots S_{k-1}x_k\|$$

$$+ \big(\|Q_n(T_1T_2\cdots T_n\varphi - \varphi S_1S_2\cdots S_n)\| + \|Q_nT_1T_2\cdots T_n\varphi\|\big)$$

for all $n \in \mathbb{N}$. Let z denote $\sum_{k=1}^{\infty} S_1S_2\cdots S_{k-1}x_k$. Since each tail of this series has norm at most 1, each $n \in \mathbb{N}$ satisfies

$$\|\varphi z\| \ge \|Q_n\varphi z\|$$

$$\ge \|Q_n\varphi S_1S_2\cdots S_{n-1}x_n\| - \|Q_n\varphi \sum_{k=1}^{n-1} S_1S_2\cdots S_{k-1}x_k\|$$

$$- \|Q_n\varphi S_1S_2\cdots S_n\left(\sum_{k=n+1}^{\infty} S_{n+1}S_{n+2}\cdots S_{k-1}x_k\right)\| \ge n.$$

This contradiction proves the lemma. $\qquad\square$

6.1.18 Theorem   *Let $\mathcal{B}$ be a Banach algebra satisfying:*

(a) *$\mathcal{B}$ has no non-zero finite-dimensional nilpotent ideals;*

(b) *For each infinite-dimensional closed ideal $\mathcal{I}$ of $\mathcal{B}$ there is a sequence $\{b_n\}_{n\in\mathbb{N}} \subseteq \mathcal{B}$ such that the sequence $\{(b_1b_2\cdots b_n\mathcal{I})^-\}_{n\in\mathbb{N}}$ of closed right ideals is constantly decreasing.*

Then any surjective homomorphism of a Banach algebra onto $\mathcal{B}$ is continuous, so $\mathcal{B}$ has a unique norm topology.

Proof The second portion of the conclusion follows from the first by Theorem 6.1.5. Let $\mathcal{A}$ be a Banach algebra and let $\varphi: \mathcal{A} \to \mathcal{B}$ be a surjective homomorphism. Proposition 6.1.8(d) shows that $\mathcal{B}_\varphi$ is a closed ideal. It is enough to prove $\mathcal{B}_\varphi = \{0\}$ by Proposition 6.1.9(a).

Suppose $\mathcal{B}_\varphi$ is infinite-dimensional. Choose a corresponding sequence $\{b_n\}_{n\in\mathbb{N}}$ as in condition (b). Choose $\{a_n\}_{n\in\mathbb{N}} \subseteq \mathcal{A}$ satisfying $\varphi(a_n) = b_n$ for each $n \in \mathbb{N}$. We obtain a contradiction by applying Lemma 6.1.17 to $\mathcal{X} = \mathcal{A}$, $\mathcal{Y} = \mathcal{B}$, $\varphi = \varphi$, $T_n(b) = b_n b$ for all $b \in \mathcal{B}$ and $S_n(a) = a_n a$ for all $a \in \mathcal{A}$. Hence, $\mathcal{B}_\varphi$ must be finite dimensional. Then the spectrum of each of its elements is a finite set. Proposition 6.1.9 then shows that each element has spectrum equal to $\{0\}$. Since we are assuming that $\mathcal{B}_\varphi$ is finite-dimensional, this shows that $\mathcal{B}_\varphi$ is nilpotent. Hence, (a) gives $\mathcal{B}_\varphi = \{0\}$ as desired. (Section 1.6 contains the well known arguments that each element in a finite dimensional algebra has a finite spectrum and that such an element is nilpotent if its spectrum is $\{0\}$.) $\qquad\square$

6.1.19 Examples Satisfying Theorem 6.1.18   We note that Theorems 6.1.18 and 6.4.20 have the same hypotheses, so that our examples satisfy both.

Let B be the Banach algebra $L^1([0,1])$ introduced in Example 3.4.10. Recall that multiplication is convolution

$$f * g(t) = \int_0^t f(s)g(t-s)ds \qquad \forall\, f,\, g \in B;\ t \in [0,1]$$

and that this algebra is commutative and neither semisimple nor semiprime. Indeed its Baer radical B_B is dense. Furthermore, Theorem 3.4.9 shows that B does not have a unique functional calculus and does contain a discontinuously embedded image of the disc algebra $A(\mathbf{D})$. Hence B is a very badly behaved Banach algebra. Nevertheless it does satisfy Theorem 6.1.18.

Example 3.4.10 shows that the closed ideals in B have the form

$$\mathcal{I}_a = \{f \in B : f = 0 \text{ a.e. on } [0,a]\} \qquad \forall\, a \in [0,1].$$

All these ideals (except $\mathcal{I}_0 = B$) are nilpotent, but none (except $\mathcal{I}_1 = \{0\}$) are finite-dimensional, so B has no non-zero finite dimensional nilpotent ideals. Now let $\mathcal{I}_a$ be an infinite-dimensional closed ideal in B so that a is strictly less than 1. For each $n \in \mathbb{N}$, let b_n be the characteristic function of $[1 - 2^{-n}(1-a), 1]$. Then it is easy to see that

$$(b_1 b_2 \cdots b_n \mathcal{I}_a)^- = \mathcal{I}_{1-2^{-n}(1-a)}$$

is a strictly decreasing sequence of closed ideals. Hence B satisfies hypothesis (b) of Theorem 6.1.18. We conclude that all homomorphisms of a Banach algebra onto B and all derivations of B (see Theorem 6.4.20) are continuous. Hence Kamowitz and Scheinberg [1969], whose results were originally stated for the continuous case only, give all the derivations of B and all the automorphisms in the connected component.

John C. Tripp [1983] provides two attractively stated theorems more special than Theorem 6.1.18, which still cover most known examples. We omit the proofs.

6.1.20   Theorem   *Let $\varphi : A \to B$ be a homomorphism between Banach algebras. Then φ is continuous if there is an element $b \in \varphi(A)$ which is not a left divisor of zero but satisfies $\bigcap_{n=1}^{\infty} \overline{b^n B} = \{0\}$.*

6.1.21   Theorem   *Let B be a Jacobson-radical, semiprime, commutative Banach algebra. Every homomorphism $\varphi : A \to B$ from another Banach algebra A is continuous if and only if no element $b \in B$ satisfies $\bigcap_{n=1}^{\infty} b^n B = \{0\}$.*

Theorem 6.1.3 can be derived from Theorem 6.1.18. Obviously a semisimple Banach algebra has no non-zero nilpotent ideals—finite-dimensional or otherwise. Furthermore, by applying some of the considerations used in the proof of Theorem 4.2.15 to a closed infinite-dimensional ideal of a semisimple Banach algebra $\mathcal{A}$, first under the hypothesis that it is not annihilated by some infinite-dimensional irreducible representation of $\mathcal{A}$ and then, under the hypothesis that it is annihilated by all such representation, $\mathcal{A}$ can be shown to satisfy the hypothesis of Theorem 6.1.18. For details see Jewell and Sinclair [1976].

Let $\mathcal{A} \subseteq \mathbb{C}[[t]]$ be a generalized Banach algebra of power series as described in §3.4.18. If $\mathcal{I}$ is any non-zero ideal and a is a non-zero element of $\mathcal{I}$, then $\{at^n : n \in \mathbb{N}\}$ is a linearly independent set in $\mathcal{I}$. Hence $\mathcal{A}$ has no non-zero finite-dimensional ideals—nilpotent or not. Suppose $\mathcal{I}$ is any closed non-zero ideal. Then $\{t^n \mathcal{I}^-\}_{n \in \mathbb{N}}$ is a constantly decreasing sequence of closed ideals. Hence any such algebra satisfies the hypotheses of Theorems 6.1.18 and 6.4.20. The conclusions of these theorems are known to hold on somewhat more general algebras of power series including some which are Fréchet algebras rather than Banach algebras (Loy [1969], [1970a]).

We remark here that Arens–Hoffman extensions (Example 3.4.13) are known to have a unique norm topology (David T. Brown [1966] and John A. Lindberg, Jr. [1971]). Other related results will be found in various examples, particularly Theorem 6.4.23 and Example 6.4.24 and Loy [1983] (*cf.* Lawrence Stedman [1983]).

Many interesting results on continuity and compactness of module maps are known. We mention Gerhard Racher [1981a], Johnson [1985] and Willis [1989a].

6.2   Automatic Continuity of Homomorphisms from $\mathcal{A}$

For a homomorphism $\varphi: \mathcal{A} \to \mathcal{B}$, we now investigate conditions on the initial algebra $\mathcal{A}$ which make φ automatically continuous. This theory is considerably more complicated than the corresponding theory (discussed above) of conditions on the final algebra $\mathcal{B}$. Since a comprehensive account of this theory will appear in H. Garth Dales [1994] about the same time that the present work is published, we will refer the reader to that account, Sinclair [1976] or other sources for a number of the more complicated facts. Here we will prove some basic results and outline the more recondite portions of the theory. The following definition and proposition are essentially due to Johnson [1969b].

6.2.1   Definition   Let $\varphi: \mathcal{A} \to \mathcal{B}$ be a homomorphism between normed algebras. The $\langle$ *left / right* $\rangle$ *continuity ideal* $\langle \mathcal{I}_L(\varphi) / \mathcal{I}_R(\varphi) \rangle$ of φ is defined by

$$\langle\ \mathcal{I}_L(\varphi) = \{a \in \mathcal{A} : \varphi(a)\mathcal{B}_\varphi = \{0\}\ \}\ /\ \mathcal{I}_R(\varphi) = \{a \in \mathcal{A} : \mathcal{B}_\varphi\varphi(a) = \{0\}\ \}\ \rangle.$$

The *two-sided continuity ideal* $\mathcal{I}(\varphi)$ of φ is $\mathcal{I}_L(\varphi) \cap \mathcal{I}_R(\varphi)$.

6.2.2 Proposition *Let* $\varphi\colon \mathcal{A} \to \mathcal{B}$ *be a homomorphism between normed algebras. The continuity ideals are ideals of* $\mathcal{A}$. *If* $\mathcal{A}$ *and* $\mathcal{B}$ *are Banach algebras, they satisfy*

$$
\begin{aligned}
\mathcal{I}_L(\varphi) &= \{a \in \mathcal{A}\colon b \mapsto \varphi(ab) \text{ is continuous from } \mathcal{A} \text{ to } \mathcal{B}\ \}; \\
\mathcal{I}_R(\varphi) &= \{a \in \mathcal{A}\colon b \mapsto \varphi(ba) \text{ is continuous from } \mathcal{A} \text{ to } \mathcal{B}\ \}; \text{ and} \\
\mathcal{I}(\varphi) &= \{a \in \mathcal{A}\colon b \mapsto \varphi(ab) \text{ and } b \mapsto \varphi(ba) \\
&\qquad\qquad\qquad\qquad \text{are both continuous from } \mathcal{A} \text{ to } \mathcal{B}\ \}.
\end{aligned}
$$

Proof Each continuity ideal is an ideal since $\mathcal{B}_\varphi$ is an ideal of $\overline{\varphi(\mathcal{A})}$. The alternate descriptions of the continuity ideals follow by applying Proposition 6.1.9(c) to the regular representation maps $L_{\varphi(a)}, R_{\varphi(a)}\colon \mathcal{B} \to \mathcal{B}$.   □

We give two of Johnson's results ([1969b], [1967c]) using these ideas.

6.2.3 Theorem *Let* $\mathcal{A}$ *be a* C^*-*algebra. If* $\mathcal{A}$ *is unital and has no proper closed cofinite ideals, then any homomorphism* $\varphi\colon \mathcal{A} \to \mathcal{B}$ *into a normed algebra is continuous. If* $\mathcal{A}$ *is not unital, the same result holds if the double centralizer algebra* $\mathcal{D}(\mathcal{A})$ *of* $\mathcal{A}$ *has no proper closed cofinite ideals.*

The proof of this depends on the following more technical theorem. It is enough to consider the case of double centralizer algebras since $\mathcal{D}(\mathcal{A})$ is always unital and it equals $\mathcal{A}$ if and only if $\mathcal{A}$ is unital. We use a variant of the left continuity ideal, introduced above, for this argument.

6.2.4 Theorem *Let* $\mathcal{A}$ *be a* C^*-*algebra and let* $\varphi\colon \mathcal{A} \to \mathcal{B}$ *be a homomorphism of* $\mathcal{A}$ *into a normed algebra* $\mathcal{B}$. *Then the ideal*

$$\mathcal{I}'_L(\varphi) = \{d \in \mathcal{D}(\mathcal{A}) : \varphi(ad)\mathcal{B}_\varphi = \{0\} \text{ for all } a \in \mathcal{A}\}$$

has cofinite closure in $\mathcal{D}(\mathcal{A})$.

The principal device in the proof of Theorem 6.2.4 is the following lemma. If the target algebra $\mathcal{B}$ of the homomorphism $\varphi\colon \mathcal{A} \to \mathcal{B}$ is just a normed algebra, we may replace it with the closure of $\varphi(\mathcal{A})$ and then complete it. This does not affect any of our results, so we may always assume that $\mathcal{B}$ is a Banach algebra in which $\varphi(\mathcal{A})$ is dense.

6.2.5 Lemma *Let* $\varphi\colon \mathcal{A} \to \mathcal{B}$ *be a homomorphism between Banach algebras and let* $\mathcal{I}'_L(\varphi)$ *be the ideal defined in Theorem 6.2.4. If* $\{f_n\}_{n\in\mathbb{N}}$, $\{g_n\}_{n\in\mathbb{N}}$

are sequences in $\mathcal{D}(\mathcal{A})$ satisfying

$$g_m g_n = 0 \quad \forall \, m \neq n \in \mathbb{N} \quad and \quad f_n g_n = f_n \quad \forall \, n \in \mathbb{N},$$

then $f_n \in \mathcal{I}'_L(\varphi)$ for all but finitely many $n \in \mathbb{N}$.

Proof If the result were false, we could choose a subsequence $\{f_n\}_{n \in \mathbb{N}}$ which was never in $\mathcal{I}'_L(\varphi)$, so we assume this for the original sequence. Define

$$\mathcal{W}_n = \{b \in \mathcal{B} : \varphi(a f_n) b = 0 \text{ for all } a \in \mathcal{A}\}$$

and denote the quotient map $\mathcal{B} \to \mathcal{B}/\mathcal{W}_n$ by ψ_n. Proposition 6.1.9(c) shows that $\psi_n \circ \varphi$ is discontinuous for each n. Hence we can find $h_n \in \mathcal{A}$ satisfying $\|h_n\| < 2^{-n} \|g_n\|^{-1}$ and $\|\psi_n \circ \varphi(h_n)\| > n$. Define h to be the convergent series $\sum_{n=1}^{\infty} g_n h_n$. The calculation $f_n(h - h_n) = f_n g_n h_n - f_n g_n = 0$ shows that $\varphi(h - h_n)$ belongs to $\mathcal{W}_n$. But this implies the contradiction $\|\varphi(h)\| \geq \|\psi_n \circ \varphi(h)\| = \|\psi_n \circ \varphi(h_n)\| > n$, establishing the lemma. $\square$

Proof of Theorem 6.2.4 At the end of §1.7.17, we showed that the double centralizer algebra of a C*-algebra is again a C*-algebra. In order to simplify notation we will henceforth denote the ideal $\mathcal{I}'_L(\varphi)$ by $\mathcal{I}$ and $\mathcal{D}(\mathcal{A})$ by $\mathcal{D}$. Any norm closed ideal in a C*-algebra (such as $\overline{\mathcal{I}}$ in $\mathcal{D}$) is invariant under the involution so that the quotient algebra has an involution. In fact this quotient algebra is a C*-algebra again. (These standard facts can be found in any book on C*-algebras, and will be proved in the second volume of this work.)

Let $h = h^*$ be an element of $\mathcal{D}$. We showed in §3.4.17 that there is a functional calculus for h under which any continuous function in $C(Sp(h))$ acts on h. Call a point $\lambda \in Sp(h)$ a *discontinuity point* if $f(\lambda) = 0$ for all $f \in C(Sp(h))$ for which $f(h)$ belongs to $\mathcal{I}$. Suppose there were infinitely many discontinuity points. Then we could find a sequence $\{\lambda_n\}$ of distinct discontinuity points with a discontinuity point λ_0 not in the sequence as its only limit point. By induction we could choose closed neighborhoods U_n and V_n of λ_n satisfying $V_n \subseteq U_n^o$ and $U_n \cap U_m = \emptyset$. Choose $f_n, g_n \in C(Sp(h))$ satisfying $f_n(\lambda_n) \neq 0$, $\mathrm{supp}(f_n) \subseteq V_n$, $g_n|_{V_n} \equiv 1$ and $\mathrm{supp}(g_n) \subseteq U_n$. The sequences $\{f_n\}$ and $\{g_n\}$ contradict the lemma, so we conclude that the set F of discontinuity points is finite.

Since $\mathcal{I}$ is an ideal in $\mathcal{D}$, $\mathcal{I}_h = \{f \in C(Sp(h)) : f(h) \in \mathcal{I}\}$ is an ideal in $C(Sp(h))$ which contains a function which is non-zero at each $\lambda \in Sp(h) \backslash F$. Thus Theorem 7.3.2 shows that $\mathcal{I}_h$ contains every function in $C(Sp(h))$ with support disjoint from F. (The ideal of such functions is denoted by $\mathcal{J}(H) = \mathcal{J}(H, \infty)$ in that theorem.)

Next we show that if $k = k^*$ belongs to $\mathcal{D}/\overline{\mathcal{I}}$, then its spectrum is finite. Choose $h = h^* \in \mathcal{D}$ so that $k = h + \overline{\mathcal{I}}$. (This is possible since if $k = a + \overline{\mathcal{I}}$ then we may take $h = (a + a^*)/2$.) Let F be the finite set of continuity points for h. Now we apply the functional calculus for continuous functions to h. For any $\lambda_0 \in \mathbb{C} \setminus F$ we can find a continuous function f with $f(\lambda) = (\lambda - \lambda_0)^{-1}$ in a neighborhood of F. Then $g(\lambda) = (\lambda - \lambda_0)f(\lambda) - 1$ is zero in a neighborhood of F. Hence $g(k)$ belongs to $\overline{\mathcal{I}}$ so $f(k) + \overline{\mathcal{I}}$ is an inverse for $k - \lambda_0$. Thus we have shown $Sp(k) \subseteq F$.

The proof of the theorem is now completed by noting that a C*-algebra (such as $\mathcal{D}/\overline{\mathcal{I}}$) is finite-dimensional if each of its elements has finite spectrum. This result can be found in many books on C*-algebras and a proof is included in the second volume of this work.   □

Proof of Theorem 6.2.3   We may assume that $\mathcal{B}$ is the closure of $\varphi(\mathcal{A})$. By hypothesis, the closure of $\mathcal{I}'_L(\varphi)$ is either $\{0\}$ (in which case $\mathcal{D}(\mathcal{A})$ is finite-dimensional so that the theorem is trivially true) or it is all of $\mathcal{A}$. But this implies $\mathcal{I}'_L(\varphi) = \mathcal{D}(\mathcal{A})$, so the identity element 1 of $\mathcal{D}(\mathcal{A})$ belongs to the ideal. But this implies $\mathcal{B}_\varphi$ is $\{0\}$ which gives the desired result by Proposition 6.1.9(a).   □

The above result applies to $\mathcal{B}(\mathcal{H})$ for any Hilbert space $\mathcal{H}$. It also applies to $\mathcal{B}_K(\mathcal{H})$ since we showed in §1.7.14 that $\mathcal{D}(\mathcal{B}_K(\mathcal{H})) = \mathcal{B}(\mathcal{H})$. Proposition 6.1.6 shows that it also applies to the Calkin algebra $\mathcal{B}(\mathcal{H})/\mathcal{B}_K(\mathcal{H})$. We wish to extend these results to Banach spaces more general than Hilbert spaces. Stefan Banach [1932] noted on p. 245 that many classical Banach spaces $\mathcal{X}$ are linearly homeomorphically (or even isometrically) isomorphic to $\mathcal{X} \oplus \mathcal{X}$. Whenever this is true, we can iterate the process to write $\mathcal{X}$ as the direct sum of 2^n subspaces each isomorphic to $\mathcal{X}$ and hence to each other. Fix this continued bisection for the rest of the next proof. From it we can construct many sequences $\{E_n\}_{n \in \mathbb{N}^0}$ of projections in $\mathcal{B}(\mathcal{X})$ satisfying: (1) $E_0 = I$ and (2) for each $n \in \mathbb{N}$, $E_n = E_{n-1}E_n = E_nE_{n-1}$ is an equivalent projection to $E_{n-1} - E_n$ by an element $U_n \in \mathcal{B}(\mathcal{X})$ which is an isometric map of $\mathcal{X}$ onto itself. (We require $E_n = U_n^{-1}(E_{n-1} - E_n)U_n$.) Any such sequence will be called a *complete chain* in the following proof.

6.2.6 Theorem   *Let $\mathcal{X}$ be a Banach space satisfying $\mathcal{X} = \mathcal{X} \oplus \mathcal{X}$. Then any homomorphism $\varphi \colon \mathcal{B}(\mathcal{X}) \to \mathcal{B}$ of $\mathcal{B}(\mathcal{X})$ into a normed algebra $\mathcal{B}$ is continuous.*

Proof   As noted above, we may as well take $\mathcal{B}$ to be the closure of $\varphi(\mathcal{B}(\mathcal{X}))$ in the completion of the original $\mathcal{B}$. Lemma 6.2.5 can be modified to show that only finitely many complete chains can miss the left continuity ideal $\mathcal{I}_L(\varphi)$. We will call these complete chains *exceptional*. Note that when E_n belongs to an ideal then the rest of any complete chain containing E_n must belong to the ideal. Conversely, if E_n does not belong to the ideal then at least one of E_{n+1} and $E_n - E_{n+1}$ must not belong to the ideal. Hence any projection from the continued bisection which is not in $\mathcal{I}_L(\varphi)$ belongs to an exceptional complete chain.

Since there are only finitely many exceptional complete chains we can choose p so large that each exceptional chain has a distinct projection as its pth entry. Hence if E_p comes from any complete chain at least one of E_{p+1} and $E_p - E_{p+1}$ belongs to $\mathcal{I}_L(\varphi)$. However, if one does the other does also, so they both do. Now there are only 2^p projections which can serve the role of E_{p+1} in any chain, so the sum of all of them is the identity operator I. This implies $I \in \mathcal{I}_L(\varphi)$ which gives $\mathcal{B}_\varphi = \{0\}$ and φ continuous.   □

6.2.7 Corollary *Let $\mathcal{X}$ be a Banach space satisfying $\mathcal{X} = \mathcal{X} \oplus \mathcal{X}$. Then any homomorphism of the generalized Calkin algebra $\mathcal{B}(\mathcal{X})/\mathcal{B}_K(\mathcal{X})$ into a normed algebra is continuous. If $\mathcal{B}_K(\mathcal{X})$ has a bounded left or right approximate identity, then any homomorphism of $\mathcal{B}_K(\mathcal{X})$ into a normed algebra is continuous.*

Proof The first result follows from Proposition 6.1.6.

The proof of the last theorem can be modified as in the proof of Theorem 6.2.4 to show $\varphi(\mathcal{B}_K(\mathcal{X}))\mathcal{B}_\varphi = \{0\}$. Now for $b \in \mathcal{B}_\varphi$, let $\{K_n\}$ be a sequence in $\mathcal{B}_K(\mathcal{X})$ which converges to zero and satisfies $\lim \varphi(K_n) = b$. The factorization theorem (Corollary 5.2.4) shows that (assuming a bounded left approximate identity) we can find T and $\{L_n\}$ in $\mathcal{B}_K(\mathcal{X})$ satisfying $\lim L_n = 0$ and $K_n = TL_n$ for all n. Proposition 6.1.9(c) shows that $\lim \varphi(L_n) + \mathcal{B}_\varphi = 0$ so there are elements b_n in $\mathcal{B}_\varphi$ satisfying $\lim(\varphi(L_n) + b_n) = 0$. This gives $b = \lim \varphi(K_n) = \lim \varphi(TL_n) = \lim \varphi(T)(\varphi(L_n) + b_n) = 0$ which establishes our claim.   □

6.2.8 Corollary *Let $\mathcal{X}$ be a Banach space satisfying $\mathcal{X} = \mathcal{X} \oplus \mathcal{X}$. Then any (not necessarily complete) algebra norm on $\mathcal{B}(\mathcal{X})$ is equivalent to the operator norm and hence is actually complete. If $\mathcal{B}_K(\mathcal{X})$ has a bounded left or right approximate identity, then the same result holds for $\mathcal{B}_K(\mathcal{X})$.*

Proof This is just property (a) of Theorem 6.1.5. Thus the corollary follows from the above results and the last theorem of §1.7.15.   □

Michael J. Meyer [1992c] extends the above corollary to certain Calkin algebras of classical Banach spaces. In [1989] Charles J. Read produced the first example of a Banach space $\mathcal{X}$ for which a discontinuous module derivation and hence (Proposition 6.4.4) a discontinuous homomorphism from $\mathcal{B}(\mathcal{X})$ into a Banach algebra could be constructed. His space was similar to the James space $\mathcal{J}$, so it is of interest that George A. Willis [1993] has shown that all homomorphisms from $\mathcal{J}$ are automatically continuous even though $\mathcal{J}$ does not satisfy the hypotheses of Theorem 6.1.18. If the continuum hypothesis holds, Dales, Loy and Willis [1992] produce a more fundamental example of a Banach space $\mathcal{X}$ for which there is a discontinuous homomorphisms on $\mathcal{B}(\mathcal{X})$ even though all derivations defined on it are continuous. We will discuss this unusual hypothesis below.

Homomorphisms from $C(\Omega)$

Not surprisingly, the first attempts to prove the continuity of all homomorphisms from a Banach algebra $\mathcal{A}$ into another normed algebra were directed towards commutative algebras and, in particular, towards the algebra $C(\Omega)$ of all continuous complex-valued functions on a compact Hausdorff space Ω. (The important paper by Gelfand and Naĭmark [1943] has a misstatement in Lemma 2 which, if accepted, would imply the continuity of such homomorphisms.) To avoid triviality, in the following remarks we will always assume that Ω is not a finite set. In [1949a] Kaplansky proved (Theorem 2.4.15 above) that any algebra norm on $C(\Omega)$ is necessarily larger than the uniform (or supremum) norm $\|\cdot\|_\infty$. This result, together with the open mapping theorem, shows that any complete algebra norm is equivalent to the uniform norm. Hence $C(\Omega)$ has an incomplete algebra norm if and only if it has an algebra norm inequivalent to $\|\cdot\|_\infty$. In 1948 Kaplansky asked whether all (not necessarily complete) algebra norms might be equivalent to $\|\cdot\|_\infty$. In [1960] William G. Bade and Phillip C. Curtis, Jr. published a fundamental paper directed towards proving this. The basis of their technique was a "main boundedness theorem", similar to Lemma 6.2.5, which applied in any commutative Banach algebra. It shows *inter alia* that if $\varphi \colon \mathcal{A} \to \mathcal{B}$ is any homomorphism from a commutative Banach algebra to a normed algebra, then for any sequence $\{e_n\}_{n \in \mathbb{N}}$ of orthogonal idempotents (*i.e.*, those satisfying $e_n e_m = \delta_{nm} e_n$) there is a constant M satisfying

$$\|\varphi(e_n)\| \le M \|e_n\|^2 \qquad \forall\, n \in \mathbb{N}.$$

For semisimple completely regular commutative Banach algebras $\mathcal{A}$, they showed that the discontinuity of any homomorphism from $\mathcal{A}$ was concentrated at a finite set F (called the *singularity set of* φ) in the Gelfand space of $\mathcal{A}$. In particular, φ is continuous on the ideal of functions vanishing on F. They were able to construct discontinuous homomorphisms (using point derivations as in §3.4.16) on certain semisimple completely regular commutative Banach algebras, but remarked that they knew no discontinuous homomorphisms when $\mathcal{A} = C(\Omega)$.

They showed that all homomorphisms from $C(\Omega)$ are "almost" continuous in a rather technical sense which we will state after a bit of preparation. For any $\omega \in \Omega$ we define the largest and smallest ideal corresponding to ω. The traditional notation is

$$\begin{aligned}
\mathcal{M}_\omega &= \{f \in C(\Omega) : f(\omega) = 0\} \quad \text{and} \\
\mathcal{J}_\omega &= \{f \in C(\Omega) : f \equiv 0 \text{ on some neighborhood of } \omega\}.
\end{aligned}$$

Our results in Example 3.2.15 show that each $\mathcal{M}_\omega$ is a maximal ideal of $C(\Omega)$ and that every maximal ideal arises in this way from some ω. Corollary 7.3.3 shows that $\mathcal{J}_\omega$ is the smallest ideal which is included in $\mathcal{M}_\omega$ but in no other maximal ideal. If F is a finite subset of Ω, let

$$\mathcal{M}(F) \;=\; \{f \in C(\Omega) : f(\omega) = 0 \;\;\forall\, \omega \in F\} \quad \text{and}$$
$$\mathcal{D}(F) \;=\; \{f \in C(\Omega) : f \text{ is constant in some}$$
$$\text{(variable) neighborhood of each } \omega \in F\}.$$

Then $\mathcal{M}(F)$ is an ideal and $\mathcal{D}(F)$ is a dense subalgebra of $C(\Omega)$. We now state (without proof) the main result of Bade and Curtis for a homomorphism onto a dense subalgebra of a Banach algebra essentially as given in Sinclair [1976], Theorem 10.3. If $\varphi \colon C(\Omega) \to \tilde{\mathcal{B}}$ is any homomorphism into a normed algebra $\tilde{\mathcal{B}}$, then $\mathcal{B}$ may be taken as the completion of the closure of $\varphi(C(\Omega))$ in $\tilde{\mathcal{B}}$, so this theorem makes no essential restriction on the range of the homomorphism. For hulls see Definition 3.2.5.

6.2.9 Theorem *Let Ω be a compact Hausdorff space and let $\varphi \colon C(\Omega) \to \mathcal{B}$ be a discontinuous homomorphism onto a dense subalgebra of a Banach algebra $\mathcal{B}$ with Jacobson radical $\mathcal{B}_J$. Then:*

(a) The hull $h(\mathcal{I}(\varphi))$ of the continuity ideal $\mathcal{I}(\varphi)$ of φ is a nonempty finite subset $F = \{\omega_1, \omega_2, \ldots, \omega_n\}$ of Ω.

(b) The restriction of φ to the dense subalgebra $\mathcal{D}(F)$ equals the restriction to $\mathcal{D}(F)$ of a continuous homomorphism $\psi \colon C(\Omega) \to \mathcal{B}$.

(c) The range $\psi(C(\Omega))$ of ψ is closed, $\mathcal{B}_J = \mathcal{B}_\varphi$, $\mathcal{B}_J \psi(\mathcal{M}_F) = \{0\}$ and $\mathcal{B}$ is the Banach space direct sum $\psi(C(\Omega)) \oplus \mathcal{B}_J$.

(d) The difference $\theta = \varphi - \psi \colon C(\Omega) \to \mathcal{B}$ when restricted to $\mathcal{M}_F$ is a homomorphism onto a dense subalgebra of $\mathcal{B}_J$.

(e) There are non-zero linear maps $\theta_1, \theta_2, \ldots, \theta_n \colon C(\Omega) \to \mathcal{B}$ satisfying $\theta = \theta_1 + \theta_2 + \cdots + \theta_n$ such that for $j = 1, 2, \ldots, n$ the restriction of θ_j to $\mathcal{M}_{\omega_j}$ is a homomorphism into $\mathcal{B}$.

(f) The Jacobson radical $\mathcal{B}_J$ is the Banach algebra direct sum

$$\mathcal{B}_J = \oplus_{j=1}^n \overline{\theta_j(C(\Omega))} \quad \text{where} \quad \overline{\theta_j(C(\Omega))}\,\overline{\theta_k(C(\Omega))} = \{0\} \text{ for } j \neq k.$$

For $j = 1, 2, \ldots, n$, the direct summands satisfy $\overline{\theta_j(C(\Omega))}\psi(\mathcal{M}_{\omega_j}) = \{0\}$.

Note that each of the non-zero homomorphisms $\theta_j|_{\mathcal{M}_{\omega_j}}$ is zero on the dense subalgebra $\mathcal{J}_{\omega_j}$ and hence is discontinuous. Thus this theorem shows that if there is a discontinuous homomorphism from $C(\Omega)$, then there is a discontinuous homomorphism from some $\mathcal{M}_\omega$ vanishing on $\mathcal{J}_\omega$ and with range dense in a radical Banach algebra.

Johnson [1976] extended the results of this theorem by considering the locally compact space $\Omega' = \Omega \backslash F$ and the algebra $C_0(\Omega')$ which is just the set of restrictions of functions in $\mathcal{M}_F$. Each function in $C_0(\Omega')$ has a unique extension to a continuous function on the Stone-Čech compactification $\beta(\Omega')$ of Ω'. Since the discontinuous portion θ of the above homomorphism is a homomorphism on $\mathcal{M}_F$, it can be re-analyzed as a homomorphism from

$C(\beta(\Omega'))$. For each point ω' in $\beta(\Omega') \setminus \Omega'$, let $\mathcal{J}'_{\omega'}$ be the intersection with $C_0(\Omega')$ of the ideal $\mathcal{J}_{\omega'}$ defined in $C(\beta(\Omega'))$. He proved that the discontinuity of the original homomorphism can be confined to a finite set of ideals $\mathcal{J}_{\omega'}$ for $\omega' \in \beta(\Omega')$.

Discontinuous Homomorphisms and the Continuum Hypothesis

Until the beginning of 1976, it seemed that a slightly more subtle approach might prove that all homomorphisms from any $C(\Omega)$ were continuous. During 1976 Jean Esterle ([1977], [1978b], [1978c]) and H. Garth Dales ([1979a], [1979b]) (also see their joint announcement [1977]) independently constructed discontinuous homomorphisms from $C(\Omega)$ for any infinite Ω. Both their constructions, which took somewhat different approaches, used the continuum hypothesis. This appeared briefly to be an artifact of their methods of proof. However in 1976, while their papers were being completed or awaiting publication, Robert M. Solovay used a construction due to an undergraduate at California Institute of Technology, W. Hugh Woodin, to show that there are models of set theory in which no discontinuous homomorphisms from any $C(\Omega)$ exist. In these models, every algebra norm on $C(\Omega)$ is equivalent to the uniform norm. Woodin soon simplified and amplified the result. He wrote them out for his doctoral thesis at the University of California, Berkeley. Unfortunately, none of the original work was ever published, so the book by Dales and Woodin [1987] is the only accessible source. We refer the reader to that book for details of the information we are about to briefly summarize.

First we need to address the seeming contradiction in the results we have just quoted. Gödel's incompleteness theorem shows that no system of axioms strong enough to define the positive integers can prove its own consistency. The standard axioms for set theory are named for Zermelo and Frankel and are denoted by ZFC. Despite Gödel's result, it seems reasonable to hope that no inconsistent results will be derived from these axioms. (Like most mathematicians, I believe that even if such an inconsistency were discovered, logicians would quickly find a new fundamental axiom system.) Because of Gödel's result, the best that can be proved for axiom systems which include ZFC is *relative consistency*. This simply means that if the new system leads to inconsistent results, then ZFC must already be internally inconsistent. These proofs are accomplished by a technique called *"forcing"* which was initiated by Paul J. Cohen. It essentially uses ZFC to construct a model that satisfies the axiom system in question. Thus if ZFC is consistent, so is the new system.

Here is the example which led us into this discussion of logic. The countably infinite cardinal is denoted by $\aleph_0$ and it is elementary to show that the cardinal number of the set $\mathbb{R}$ of real numbers is $2^{\aleph_0}$. Thus this cardinal is called *"the cardinal of the continuum"*. It is easy to show that the cardinal of the continuum is strictly larger than $\aleph_0$. The continuum hypothesis (denoted by CH) is the statement that the cardinal of the continuum is the next cardinal number after $\aleph_0$. This next cardinal is denoted by $\aleph_1$. The symbol ¬CH stands for the statement "the continuum hypothesis is false" or equivalently "$2^{\aleph_0} \neq \aleph_1$". It is known that both of the axiom systems ZFC + CH (the Zermelo-Frankel axioms together with the continuum hypothesis) and ZFC + ¬CH (the Zermelo-Frankel axioms together with the denial of the continuum hypothesis) are relatively consistent. Thus a mathematician who "believes" (or at least accepts) the Zermelo–Frankel axioms is free to either believe the continuum hypothesis or believe its denial. In the first of these axiom systems, for every infinite compact set Ω there is a discontinuous homomorphism from $C(\Omega)$. However, one can construct a model which satisfies the second axiom system and in which no $C(\Omega)$ has a discontinuous homomorphism. We again refer the reader to the book by Dales and Woodin [1987] for details and to the forthcoming book by Dales [1994] for related results. We will leave the subject after making the one additional remark that, if AC stands for the axiom of choice, then all of the four axiom systems ZFC + AC + CH, ZFC + AC + ¬CH, ZFC + ¬AC + CH and ZFC + ¬AC + ¬CH are relatively consistent.

One unsettling conclusion of this discussion is that the fundamental question of the automatic continuity of homomorphisms from $C(\Omega)$ is undecidable in ZFC. Curiously, Theorem 6.2.3 settled this question positively for certain noncommutative C*-algebras, but it cannot be decided for any infinite-dimensional commutative C*-algebra.

6.3   Jordan Homomorphisms

Jordan homomorphisms (under another name) were considered by Germán Ancochea [1942] and [1947]. Other early papers were by Kaplansky [1947b] and Loo-Keng Hua [1949]. In two fundamental papers [1950] and [1952] Jacobson and Rickart introduced the term "Jordan homomorphism" and made the first studies similar in spirit to the approach taken here. The name "Jordan homomorphism" comes from the generalization of quantum mechanical formalism introduced by Pascual Jordan [1932] and later studied in a classic paper by Jordan, John von Neumann and Eugene P. Wigner [1934]. (See also Harald Hanche-Olsen and Erling Størmer [1984].) This generalization is based on the "symmetrized" *Jordan product*

$$a \circ b = ab + ba$$

where a and b are elements of an associative algebra. A Jordan homomorphism is simply a linear map which preserves this product.

Our most definitive results on Jordan homomorphisms are in Volume II and deal with Jordan *-homomorphisms between C*-algebras and other special *-algebras. Here we give a fairly detailed analysis of Jordan homomorphisms between algebras and between Banach algebras.

We list a few of the main results. If $\mathcal{A}$ and $\mathcal{B}$ are both commutative algebras, it is obvious that a Jordan homomorphism $\varphi\colon \mathcal{A} \to \mathcal{B}$ is a homomorphism. Under suitable additional hypotheses, we show in Theorem 6.3.4 that the same result holds if either $\mathcal{A}$ or $\mathcal{B}$ is commutative. The proof depends on results of Jacobson and Rickart [1950] and on a theorem of Želazko [1968] which was given above as Theorem 2.4.13. Theorem 6.3.7 follows Herstein [1956] in showing that a Jordan homomorphism $\varphi\colon \mathcal{A} \to \mathcal{B}$ such that $\varphi(\mathcal{A})$ is dense in a prime topological algebra $\mathcal{B}$ is either a homomorphism or an anti-homomorphism. See also Smiley [1957]. As a consequence, Corollary 6.3.8 asserts that the kernel of a Jordan homomorphism $\varphi\colon \mathcal{A} \to \mathcal{B}$ is a closed ideal if $\mathcal{A}$ and $\mathcal{B}$ are spectral normed, $\mathcal{B}$ is semisimple and $\varphi(\mathcal{A})$ is dense in $\mathcal{B}$. Sinclair [1970b] showed that a surjective Jordan homomorphism of a Banach algebra onto a semisimple Banach algebra is continuous. Civin and Yood [1965] obtained the same result when the image is merely dense in a strongly semisimple Banach algebra. Theorem 6.3.10 gives both results. Finally, we give two results showing that under different hypotheses a Jordan homomorphism is the sum of a homomorphism and an anti-homomorphism. Theorem 6.3.11, due to Sinclair [1970b], applies to surjective Jordan homomorphisms onto special semisimple algebras. Theorem 6.3.12, due to Jacobson and Rickart [1950], applies to Jordan homomorphisms of any full $n \times n$ matrix algebra for $n > 1$ over any unital algebra. This second result will be of fundamental importance in our later discussion of Jordan *-homomorphisms between Banach *-algebras.

6.3.1 Definition Let $\mathcal{A}$ and $\mathcal{B}$ be algebras. A map $\varphi\colon \mathcal{A} \to \mathcal{B}$ is called a *Jordan homomorphism* if it is linear and satisfies

$$\varphi(ab + ba) = \varphi(a)\varphi(b) + \varphi(b)\varphi(a) \qquad \forall\, a, b \in \mathcal{A}. \tag{1}$$

For certain calculations with Jordan homomorphisms, it is convenient to use the *Lie bracket* notation:

$$[a, b] = ab - ba \qquad \forall\, a, b \in \mathcal{A}. \tag{2}$$

We begin with some trivial algebraic identities which we need later.

6.3.2 Lemma *Let $\mathcal{A}$ and $\mathcal{B}$ be algebras and let $\varphi\colon \mathcal{A} \to \mathcal{B}$ be a linear map. All of the following conditions are necessary and the first two are sufficient for φ to be a Jordan homomorphism.*
  (a) $\varphi(a^2) = \varphi(a)^2 \qquad \forall\, a \in \mathcal{A}.$
  (b) $\varphi(a^n) = \varphi(a)^n \qquad \forall\, a \in \mathcal{A};\ n \in \mathbb{N}.$
  (c) $\varphi(aba) = \varphi(a)\varphi(b)\varphi(a) \qquad \forall\, a, b \in \mathcal{A}.$
  (d) $\varphi(abc + cba) = \varphi(a)\varphi(b)\varphi(c) + \varphi(c)\varphi(b)\varphi(a) \qquad \forall\, a, b, c \in \mathcal{A}.$

(e) $\varphi([a,b]^2) = [\varphi(a), \varphi(b)]^2 \quad \forall\, a, b \in \mathcal{A}.$
(f) $\varphi([[a,b],c]) = [[\varphi(a), \varphi(b)], \varphi(c)] \quad \forall\, a, b, c \in \mathcal{A}.$
(g) $(\varphi(ab) - \varphi(a)\varphi(b))(\varphi(ab) - \varphi(b)\varphi(a)) = 0 \quad \forall\, a, b \in \mathcal{A}.$

Proof (a): If φ is a Jordan homomorphism (a) is obvious. Conversely, (a) implies

$$\varphi(ab + ba) = \varphi((a + b)^2) - \varphi(a^2) - \varphi(b^2) = \varphi(a)\varphi(b) + \varphi(b)\varphi(a).$$

(b): The sufficiency of (b) is now obvious. If $\varphi(a^m) = \varphi(a)^m$ and $\varphi(a^n) = \varphi(a)^n$ both hold then we get

$$\varphi(a^{m+n}) = (1/2)\varphi((a^m + a^n)^2 - a^{2m} - a^{2n}) = \varphi(a)^m\varphi(a)^n = \varphi(a)^{m+n}.$$

By induction (b) holds.

(c): Note $2\varphi(aba) = \varphi((ab + ba)a + a(ab + ba)) - \varphi(ba^2 + a^2b) = 2\varphi(a)\varphi(b)\varphi(a).$

(d): This follows from (c) and $abc + cba = (a + c)b(a + c) - aba - cbc.$

(e): This follows from (d) and $[[a,b],c] = abc + cba - (bac + cab).$

(f): This follows from (c), the definition of a Jordan homomorphism and the identity $[a,b]^2 = a(bab) + (bab)a - ab^2a - ba^2b.$

(g): Using (a), (d), (a) and (c), we get:

$$(\varphi(ab) - \varphi(a)\varphi(b))(\varphi(ab) - \varphi(b)\varphi(a))$$
$$= \varphi(abab) - \varphi(ab)\varphi(b)\varphi(a) - \varphi(a)\varphi(b)\varphi(ab) + \varphi(a)\varphi(b)^2\varphi(a)$$
$$= \varphi(abab) - \varphi(ab^2a + abab) + \varphi(ab^2a) = 0. \qquad \square$$

6.3.3 Lemma *Let $\mathcal{A}$ and $\mathcal{B}$ be algebras and let $\varphi \colon \mathcal{A} \to \mathcal{B}$ be a Jordan homomorphism. If e is an idempotent in $\mathcal{A}$ and a is an element of $\mathcal{A}$, then $\varphi(e)$ is an idempotent and*
(a) $ea = ae \Rightarrow \varphi(a)\varphi(e) = \varphi(e)\varphi(a) = \varphi(ea).$
(b) $ea = ae = a \Rightarrow \varphi(a)\varphi(e) = \varphi(e)\varphi(a) = \varphi(a).$
(c) $ea = ae = 0 \Rightarrow \varphi(a)\varphi(e) = \varphi(e)\varphi(a) = 0.$
In particular if $\mathcal{A}$ has an identity element 1, then $\varphi(1)$ is an identity element for the algebra generated by $\varphi(\mathcal{A})$.

Proof By Lemma 6.3.2(a), $\varphi(e)$ is an idempotent.

(a): Lemma 6.3.2(e) asserts $0 = [[\varphi(a), \varphi(e)], \varphi(e)]$ and this implies $\varphi(a)\varphi(e) - \varphi(e)\varphi(a)\varphi(e) = \varphi(e)\varphi(a)\varphi(e) - \varphi(e)\varphi(a).$ Multiplying on the left by $\varphi(e)$ gives $0 = \varphi(e)\varphi(a)\varphi(e) - \varphi(e)\varphi(a).$ Therefore, $\varphi(a)\varphi(e) - \varphi(e)\varphi(a)\varphi(e) = 0$ so $\varphi(a)\varphi(e) = \varphi(e)\varphi(a) = \varphi(eae) = \varphi(e^2a) = \varphi(ea).$ Thus (a) is proved. The rest of the statements in the proposition are now immediate. $\qquad \square$

In the next result, (a) and (b$_1$) are due to Nathan Jacobson and Rickart [1950], while (b$_2$) is due to Wiesław Żelazko [1968].

6.3.4 Theorem *Let A and B be algebras with B semisimple. Let $\varphi: A \to B$ be a Jordan homomorphism.*

(a) *For $a, b \in A$, $ab = ba$ implies $\varphi(a)\varphi(b) = \varphi(b)\varphi(a)$ if either:*

    (a_1) *$\varphi(A)$ generates B as an algebra, or*

    (a_2) *B is a topological algebra which is the closure of the subalgebra generated by $\varphi(A)$.*

(b) *Furthermore, φ is a homomorphism if either:*

    (b_1) *A is commutative and (a_1) or (a_2) holds, or*

    (b_2) *A is unital and B is a commutative spectral algebra.*

Proof (a): When a and b commute, Lemma 6.3.2(e) shows $[\varphi(a), \varphi(b)]^2 = 0$ and lemma 6.3.2(f) shows $[[\varphi(a), \varphi(b)], \varphi(c)] = 0$ for any $c \in A$. Thus $[\varphi(a), \varphi(b)]$ lies in the center of B, so it is enough to show that the center B_Z of the semisimple algebra B has no non-zero nilpotent elements. Consider

$$\mathcal{I} = \text{span}\{a + bc : a, c \in B_Z \text{ are nilpotent and } b \in B\}.$$

It is easy to check that this is a nil ideal and hence a quasi-regular ideal. Therefore $\mathcal{I} = \{0\}$ by Theorem 4.3.6, so there are no nilpotent elements in B_Z except zero.

(b): If (b_1) holds, the conclusion follows immediately from (a) and the definition of a Jordan homomorphism. Suppose (b_2) holds. Let γ be an arbitrary element in Γ_B. Theorem 2.4.13 shows that $\gamma \circ \varphi$ is a homomorphism. Since the elements of φ_B separate points in B, this implies that φ is a homomorphism. $\qquad \square$

6.3.5 Proposition *Let A and B be algebras and let $\varphi: A \to B$ be a Jordan homomorphism. If $a \in A$ is quasi-invertible, then $\varphi(a)$ is quasi-invertible and its quasi-inverse satisfies $\varphi(a^q) = \varphi(a)^q$.*

Proof We have

$$2\varphi(a) + 2\varphi(a^q) = \varphi(aa^q + a^qa) = \varphi(a)\varphi(a^q) + \varphi(a^q)\varphi(a). \tag{3}$$

Hence the proof will be complete if $\varphi(a)$ and $\varphi(a^q)$ commute. Multiplying (3) on the left by $\varphi(a)$ then multiplying it on the right by $\varphi(a)$ and subtracting the results gives

$$(2\varphi(a) - \varphi(a)^2)\varphi(a^q) = \varphi(a^q)(2\varphi(a) - \varphi(a)^2). \tag{4}$$

Note also

$$\begin{aligned}
2\varphi(a)^2 &+ \varphi(a)\varphi(a^q) + \varphi(a^q)\varphi(a) - \varphi(a)^2\varphi(a^q) - \varphi(a^q)\varphi(a)^2 \tag{5}\\
&= \varphi(2a^2 + aa^q + a^qa - (a^2a^q + a^qa^2)) = 0.
\end{aligned}$$

Adding the right side of (3) to the left side of (5) gives

$$2\varphi(a)^2 \;+\; (2\varphi(a) - \varphi(a)^2)\varphi(a^q) + \varphi(a^q)(2\varphi(a) - \varphi(a)^2) \qquad (6)$$
$$= \; 2\varphi(a) + 2\varphi(a^q).$$

Substituting (4) in (6) and dividing by 2 gives

$$\varphi(a)^2 \;+\; (2\varphi(a) - \varphi(a)^2)\varphi(a^q) = \varphi(a) + \varphi(a^q)$$
$$= \; \varphi(a)^2 + \varphi(a^q)(2\varphi(a) - \varphi(a)^2).$$

However this can be rewritten as

$$\varphi(a) \circ \varphi(a) \circ \varphi(a^q) = \varphi(a) = \varphi(a^q) \circ \varphi(a) \circ \varphi(a).$$

Hence we get

$$\varphi(a^q) \circ \varphi(a) = \varphi(a^q) \circ \varphi(a) \circ \varphi(a) \circ \varphi(a^q) = \varphi(a) \circ \varphi(a^q)$$

which is equivalent to the commutativity we wished to establish.   □

Recall that an algebra $\mathcal{A}$ is called prime if $\{0\}$ is a prime ideal. From Theorem 4.1.10 it follows that $\mathcal{A}$ is prime if and only if it satisfies one (hence all) of the following implications

$$a\mathcal{A}b = \{0\} \;\;\Rightarrow\;\; (a = 0 \text{ or } b = 0) \qquad \forall\, a,\, b \in \mathcal{A}; \qquad (7)$$
$$\mathcal{I}_1\mathcal{I}_2 = \{0\} \;\;\Rightarrow\;\; (\mathcal{I}_1 = \{0\} \text{ or } \mathcal{I}_2 = \{0\}) \qquad \forall\; \text{ideals } \mathcal{I}_1, \mathcal{I}_2 \text{ of } \mathcal{A}; \quad (8)$$
$$\mathcal{L}_1\mathcal{L}_2 = \{0\} \;\;\Rightarrow\;\; (\mathcal{L}_1 = \{0\} \text{ or } \mathcal{L}_2 = \{0\}) \;\; \forall\; \text{left ideals } \mathcal{L}_1, \mathcal{L}_2 \text{ of } \mathcal{A}. (9)$$

A primitive algebra (*i.e.*, an algebra in which $\{0\}$ is a primitive ideal) is prime by Theorem 4.1.8. Condition (8) shows that a simple algebra which is not the one-dimensional zero algebra is prime. We need two other simple properties of prime algebras which we include in a lemma.

6.3.6 Lemma *Let $\mathcal{A}$ be a prime algebra.*

(a) *If elements u, $v \in \mathcal{A}$ satisfy $ucv + vcu = 0$ for all $c \in \mathcal{A}$, then either $u = 0$ or $v = 0$.*

(b) *If $\mathcal{A}$ is a topological algebra, every dense subalgebra is prime.*

Proof Replace c by due. Then $uduev + vdueu = 0$. However $uev = -veu$ and $vdu = -udv$ so that $-udveu - udveu = 0$. Since d and e are arbitrary, $u\mathcal{A}v\mathcal{A}u = \{0\}$. Thus either $u = 0$ or $u\mathcal{A}v = \{0\}$. In the latter case, either $u = 0$ or $v = 0$ so that (a) is proved.

Statement (b) follows immediately from Condition (7) above.   □

The next result is a fundamental result of Israel N. Herstein [1956]. We will use it later to obtain results on Jordan homomorphisms into more general algebras.

6.3.7 Theorem *Let A and B be algebras with B prime. Let $\varphi: A \to B$ be a Jordan homomorphism such that either:*

(a) $\varphi(A) = B$, or

(b) *B is a topological algebra and $\varphi(A)$ is dense in B.*

Then φ is either a homomorphism or an anti-homomorphism.

Proof Let $u = \varphi(ab) - \varphi(a)\varphi(b)$ and $v = \varphi(ab) - \varphi(b)\varphi(a)$. We will use various formulas from Lemma 6.3.2 to show that u and v satisfy the criterion of Condition (a) in the last lemma. For any $c = \varphi(d)$ in $\varphi(A)$ we have

$$
\begin{aligned}
ucv &= \varphi(ab)\varphi(d)\varphi(ab) + \varphi(a)\varphi(b)\varphi(d)\varphi(b)\varphi(a) \\
&\quad -\varphi(a)\varphi(b)\varphi(d)\varphi(ab) - \varphi(ab)\varphi(d)\varphi(b)\varphi(a) \\
&= \varphi(abdab) + \varphi(abdba) - \varphi(a)\varphi(b)\varphi(d)\varphi(ab) - \varphi(ab)\varphi(d)\varphi(b)\varphi(a).
\end{aligned}
$$

Similarly we have

$$
vcu = \varphi(abdab) + \varphi(badab) - \varphi(b)\varphi(a)\varphi(d)\varphi(ab) - \varphi(ab)\varphi(d)\varphi(a)\varphi(b).
$$

Therefore we conclude

$$
\begin{aligned}
ucv + vcu &= \varphi(2abdab + abdba + badab) \\
&\quad -[(\varphi(a)\varphi(b) + \varphi(b)\varphi(a))\varphi(d)\varphi(ab) \\
&\quad +\varphi(ab)\varphi(d)(\varphi(a)\varphi(b) + \varphi(b)\varphi(a))] \\
&= \varphi(2abdab + abdba + badab) \\
&\quad -\varphi[(ab + ba)d(ab) + (ab)d(ab + ba)] = 0.
\end{aligned}
$$

If $\varphi(A)$ is dense in the topological algebra B, the result above holds for all $c \in B$. Hence by Lemma 6.3.6(a), either v or u is zero.

For $a \in A$, define $\mathcal{H}_a = \{b \in A : \varphi(ab) = \varphi(a)\varphi(b)\}$ and $\mathcal{K}_a = \{b \in A : \varphi(ab) = \varphi(b)\varphi(a)\}$. Then A is the union of the two linear subspaces $\mathcal{H}_a$ and $\mathcal{K}_a$. Therefore either $A = \mathcal{H}_a$ or $A = \mathcal{K}_a$. Now let $\mathcal{H} = \{a \in A : \varphi(ab) = \varphi(a)\varphi(b)$ for all $b \in A\}$ and let $\mathcal{K} = \{a \in A : \varphi(ab) = \varphi(b)\varphi(a)$ for all $b \in A\}$. Again A is the union of the two linear subspaces $\mathcal{H}$ and $\mathcal{K}$, so either $A = \mathcal{H}$ or $A = \mathcal{K}$. This proves the theorem. $\square$

6.3.8 Corollary *Let A and B be algebras and let $\varphi: A \to B$ be a Jordan homomorphism. Then $\ker(\varphi)$ is an ideal if either*

(a) *B is semiprime and $\varphi(A) = B$, or*

(b) *B is a spectral normed semisimple algebra and $\varphi(A)$ is dense in B.*

Under either hypothesis, if A is spectral normed and B is semisimple, then $\ker(\varphi)$ is closed.

Proof Let $\{\mathcal{P}_\alpha : \alpha \in A\}$ be the set of prime ideals of B. For each $\alpha \in A$, let ψ_α be the natural map $\psi_\alpha: B \to B/\mathcal{P}_\alpha$ and let $\varphi_\alpha = \psi_\alpha \circ \varphi$. Then by the theorem, $\ker(\varphi_\alpha)$ is an ideal for each α. Hence $\ker(\varphi) = \cap_{\alpha \in A} \ker(\varphi_\alpha)$ is an ideal.

If B is a spectral normed semisimple algebra, we make the same argument replacing prime ideals by primitive ideals. By Proposition 4.2.6 each primitive ideal is closed. Hence each B/P_α is spectral normed and $\varphi_\alpha(A)$ is dense. Thus the argument is completed in the same way using the alternative topological hypothesis of Theorem 6.3.7.

If A is spectral normed and B is semisimple, each φ_α (with P_α primitive) has a primitive kernel in A (if it is a homomorphism) or in the opposite algebra of A (if it is an anti-homomorphism). Since the opposite algebra is also spectral normed, each $\ker(\varphi_\alpha)$ is closed. Hence $\ker(\varphi) = \cap_{\alpha \in A} \ker(\varphi_\alpha)$ is closed. □

6.3.9 Theorem *Let A and B be algebras and let $\varphi \colon A \to B$ be a Jordan homomorphism with $\varphi(A) = B$.*
  (a) *If $\ker(\varphi) \subseteq A_J$, then $\varphi(A_J) = B_J$.*
  (b) *If $\ker(\varphi) \subseteq A_B$, then $\varphi(A_B) = B_B$.*
  (c) *If $\ker(\varphi) \subseteq A_M$, then $\varphi(A_M) = B_M$.*

Proof (a): Let $\psi \colon B \to B/B_J$ be the natural map. Then $\ker(\psi \circ \varphi) = \varphi^{-1}(B_J)$ is an ideal by Corollary 6.3.8. Let $\theta \colon B \to A/A_J$ be the map induced by φ^{-1} (*i.e.*, if $b = \varphi(a)$, set $\theta(b) = a + A_J$). It is well defined since $\ker(\varphi) \subseteq A_J$. It is easy to check that θ is a Jordan homomorphism. Hence Proposition 6.3.5 and Theorem 4.3.6(b) show that $\theta(B_J) = \varphi^{-1}(B_J)/A_J$ is a quasi-regular ideal. Thus each element $a \in \varphi^{-1}(A_J)$ satisfies $Sp_{A/A_J}(a + A_J) \subseteq \{0\}$. By Theorem 4.3.6(d), this implies that $Sp_A(a) \subseteq \{0\}$ for each $a \in \varphi^{-1}(B_J)$. Hence $\varphi^{-1}(B_J) \subseteq A_J$ by Theorem 4.3.6(b), so $B_J \subseteq \varphi(A_J)$. However, Corollary 6.3.8 also shows that $\ker(\theta) = \varphi(A_J)$ is an ideal, and it is quasi-regular by Proposition 6.3.5 so that $\varphi(A_J) \subseteq B_J$.

  (b): Let $\psi \colon B \to B/B_B$ be the natural map. Then $\ker(\psi \circ \varphi) = \varphi^{-1}(B_B)$ is an ideal by Corollary 6.3.8. Suppose $a \in A$ satisfies $aAa \subseteq \varphi^{-1}(B_B)$. Then Lemma 6.3.2(c) shows $\varphi(a)B\varphi(a) \subseteq B_B$ so that Theorem 4.4.3(b) gives $\varphi(a) \in B_B$. Hence by the same result $A/\varphi^{-1}(B_B)$ is semiprime. Thus Theorem 4.4.6 shows $\varphi^{-1}(B_B) \supseteq A_B$ so $\varphi(A_B) \subseteq B_B$. Now consider the Jordan homomorphism $\theta \colon B \to A/A_B$ induced by φ^{-1}. Then $\ker(\theta) = \varphi(A_B)$ is an ideal by Corollary 6.3.8 and it satisfies the condition of Theorem 4.4.3(b) as before so that $B_B \subseteq \varphi(A_B)$.

  (c): Let $\{M_\alpha : \alpha \in A\}$ be the set of maximal modular ideals of B. For each $\alpha \in A$, let $\psi_\alpha \colon B \to B/M_\alpha$ be the natural map and let $\varphi_\alpha = \psi_\alpha \circ \varphi$. Then each φ_α is a Jordan homomorphism of A onto a simple unital algebra. Hence each φ_α is a homomorphism or an anti-homomorphism by Theorem 6.3.7. However, whenever φ_α is an anti-homomorphism, if we merely replace B/M_α by its opposite algebra (which is also simple and unital) then each φ_α may be considered as a homomorphism. Thus $A_M = \cap\{\ker(\varphi) : \varphi$

is a homomorphism onto a simple unital algebra $\} \subseteq \cap\{\ker(\varphi_\alpha) : \alpha \in A\} = \cap\{\varphi^{-1}(\mathcal{M}_\alpha) : \alpha \in A\} = \varphi^{-1}(\cap\{\mathcal{M}_\alpha : \alpha \in A\}) = \varphi^{-1}(\mathcal{B}_M)$. Hence $\varphi(\mathcal{A}_M) \subseteq \mathcal{B}_M$.

Now let $\theta: \mathcal{B} \to \mathcal{A}/\mathcal{A}_M$ be the Jordan homomorphism induced by φ^{-1}. The same argument shows that $\mathcal{B}_M \subseteq \ker(\theta) = \varphi(\mathcal{A}_M)$.   □

Part (a) of the next theorem is due to Sinclair [1970b]. Part (b) is due to Civin and Yood [1965].

6.3.10 Theorem   *Let $\mathcal{A}$ and $\mathcal{B}$ be Banach algebras and let $\varphi: \mathcal{A} \to \mathcal{B}$ be a Jordan homomorphism. Then φ is continuous if either:*
(a) *$\mathcal{B}$ is semisimple and $\varphi(\mathcal{A}) = \mathcal{B}$, or*
(b) *$\mathcal{B}$ is strongly semisimple and $\varphi(\mathcal{A})$ is dense in $\mathcal{B}$.*

Proof (a): Let $\mathcal{P}$ be a primitive ideal of $\mathcal{B}$ and let $\psi: \mathcal{B} \to \mathcal{B}/\mathcal{P}$ be the natural map. Theorem 6.3.7 shows that $\psi \circ \varphi$ is either a homomorphism or an anti-homomorphism. If it is an anti-homomorphism, replace $\mathcal{A}$ by its reverse Banach algebra. Then $\psi \circ \varphi$ is a homomorphism of $\mathcal{A}$ onto $\mathcal{B}/\mathcal{P}$ so it is continuous by Theorem 6.1.3.

We now show φ is continuous by use of the closed graph theorem. Suppose $\{a_n\}_{n \in \mathbb{N}} \subseteq \mathcal{A}$ is a sequence converging to zero and $\{\varphi(a_n)\}_{n \in \mathbb{N}}$ converges to $b \in \mathcal{B}$. Then $\{\psi(\varphi(a_n))\}_{n \in \mathbb{N}}$ converges to zero by the continuity of $\psi \circ \varphi$ and converges to $\psi(b)$ by the continuity of the quotient map ψ. Since ψ was the quotient map modulo an arbitrary primitive ideal, b is zero. Hence φ is continuous since it has a closed graph.

(b): Let $\mathcal{M}$ be a maximal modular ideal of $\mathcal{B}$, and let $\psi: \mathcal{B} \to \mathcal{B}/\mathcal{M}$ be the natural map. Then $\psi \circ \varphi$ is a homomorphism or an anti-homomorphism as before. If it is an anti-homomorphism, we replace $\mathcal{A}$ by its reverse Banach algebra. Thus $\psi \circ \varphi$ can always be considered as a homomorphism. Its range is dense in $\mathcal{B}/\mathcal{M}$. Hence Corollary 6.1.14 shows it is continuous. The proof is finished by the same closed graph argument as before.   □

Part (a) of this theorem shows that the set of Jordan automorphisms of a semisimple Banach algebra $\mathcal{A}$ is a subgroup of $\mathcal{B}(\mathcal{A})$. Sinclair [1970b] shows that the connected component (in the norm topology) of this group consists of automorphisms. To obtain this, he uses a generalization to Jordan automorphisms of a theorem of Georges Zeller-Meier [1967] which we prove in the next section (Proposition 6.4.21). The generalization shows that each Jordan automorphism which is sufficiently close to the identity is the exponential of a Jordan derivation. Then Proposition 6.4.21 shows it lies on a norm continuous one-parameter group of automorphisms. Hence the subgroup generated by such one-parameter groups is a connected open (and hence closed) subgroup which must be the connected component of the identity.

The next theorem is a slight generalization of a theorem of Sinclair

[1970a]. Note that in this theorem conditions are placed on $\mathcal{B}$ in order to show that a Jordan homomorphism $\varphi\colon \mathcal{A} \to \mathcal{B}$ is the direct sum of a homomorphism and an anti-homomorphism. In Theorem 6.3.12 we will reach the same conclusion from restrictions on $\mathcal{A}$.

6.3.11 Theorem *Let $\mathcal{A}$ and $\mathcal{B}$ be algebras. Let $\mathcal{B}$ be unital and satisfy one of the following hypotheses:*

(a) $\mathcal{B}$ is semiprime and has no non-zero commutative prime algebra as a homomorphic image.

(b) $\mathcal{B}$ is semisimple and has no irreducible representations with commutative image.

(c) $\mathcal{B}$ is strongly semisimple and has no non-zero commutative simple homomorphic image.

If $\varphi\colon \mathcal{A} \to \mathcal{B}$ is a surjective Jordan homomorphism, then there is a direct sum decomposition $\mathcal{B} = \mathcal{I}_1 \oplus \mathcal{I}_2$ of $\mathcal{B}$ and surjective linear maps $\varphi_j\colon \mathcal{A} \to \mathcal{I}_j$ for $j = 1, 2$, such that φ_1 is a homomorphism, φ_2 is an anti-homomorphism, and $\varphi = \varphi_1 + \varphi_2$.

Proof Let $\{\mathcal{P}_\alpha : \alpha \in A\}$ be the set of prime, primitive or maximal modular ideals of $\mathcal{B}$, depending on whether $\mathcal{B}$ satisfies (a), (b), or (c), respectively. For each $\alpha \in A$, let ψ_α be the natural map $\mathcal{B} \to \mathcal{B}/\mathcal{P}_\alpha$ and let $\varphi_\alpha = \psi_\alpha \circ \varphi$. Theorem 6.3.7 shows that each φ_α is either a homomorphism or an anti-homomorphism. Condition (a), (b), or (c) shows that there is no φ_α which is both a homomorphism and an anti-homomorphism. Let A_1 be the set of $\alpha \in A$ such that φ_α is an anti-homomorphism and let A_2 be the set of $\alpha \in A$ such that φ_α is a homomorphism. We have seen that A is the disjoint union of A_1 and A_2. Let $\mathcal{I}_j = \cap\{\mathcal{P}_\alpha : \alpha \in A_j\}$ for $j = 1, 2$. Condition (a), (b), or (c) shows that $\mathcal{I}_1 \cap \mathcal{I}_2 = \{0\}$. On the other hand, $\mathcal{I}_1 + \mathcal{I}_2$ is modular since $\mathcal{B}$ is unital. If $\mathcal{I}_1 + \mathcal{I}_2$ is a proper ideal it is contained in a maximal modular ideal $\mathcal{M}$ by Theorem 2.4.6. Since a maximal modular ideal belongs to $\{\mathcal{P}_\alpha : \alpha \in A\}$ no matter which of the three meanings this set has, this proves (by contradiction) that $\mathcal{I}_1 + \mathcal{I}_2 = \mathcal{B}$ so that we do have a direct sum decomposition. The direct sum decomposition of the identity element in $\mathcal{B}$ gives central idempotents p_1 and p_2 such that $\mathcal{I}_j = p_j\mathcal{B}$ for $j = 1, 2$. The maps φ_1 and φ_2 defined by $\varphi_1(a) = p_1\varphi(a)$ and $\varphi_2(a) = p_2\varphi(a)$ for all $a \in \mathcal{A}$ obviously satisfy the conditions of the theorem. $\square$

One can easily prove (as in Sinclair [1970a]) that the decomposition given in the last theorem is unique. If the second half of conditions (a), (b), and (c) are omitted, this uniqueness will obviously fail. However, Sinclair gives an example to show that even the theorem fails without this restriction. We reproduce this interesting example.

Let $\mathcal{C} = C([-1, 1])$ and let $\mathcal{D} = \{f \in \mathcal{C} : f(0) = 0\}$. Let $\mathcal{A}$ be the algebra of 2×2 matrices with entries from $\mathcal{C}$ on the principal diagonal and entries from $\mathcal{D}$ in the other positions. This is obviously an algebra under the usual matrix operations. Let $\varphi: \mathcal{A} \to \mathcal{A}$ leave the principal diagonal entries fixed, but replace each of the other two entries by a function which agrees with it on $[-1, 0]$ but agrees with the opposite entry on $[0, 1]$. It is clear that φ is a Jordan isomorphism and a little calculation shows the lack of ideals with the properties of $\mathcal{I}_1$ and $\mathcal{I}_2$.

We now turn to a theorem which derives essentially the same conclusion as Theorem 6.3.11 from very different hypotheses. We begin by recalling some facts about matrix algebras.

Given any algebra $\mathcal{D}$, the family of $n \times n$ matrices with elements from $\mathcal{D}$ forms an algebra $\mathcal{D} \otimes M_n$ under the customary matrix algebraic operations. It is natural to identify each $d \in \mathcal{D}$ with the diagonal matrix with d in each position on the diagonal. We will assume from now on that $\mathcal{D}$ is unital. Denote the matrix with 1 in the ith row and jth column and zero in all other entries by $E^{ij} \in \mathcal{D} \otimes M_n$. These elements satisfy:

$$
\begin{aligned}
E^{ij} E^{k\ell} &= \delta_{jk} E^{i\ell} \qquad \forall\, 1 \le i, j, k, \ell \le n \\
\sum_{i=1}^{n} E^{ii} &= 1 \qquad\qquad\qquad\qquad\qquad (10) \\
d E^{ij} &= E^{ij} d \qquad \forall\, d \in \mathcal{D};\ 1 \le i, j \le n
\end{aligned}
$$

under the above embedding of $\mathcal{D}$ into $\mathcal{D} \otimes M_n$. In fact it is easy to see that an algebra $\mathcal{A}$ is isomorphic to $\mathcal{B} \otimes M_n$ for some algebra $\mathcal{B}$ if and only if it contains a subalgebra $\mathcal{D}$ isomorphic to $\mathcal{B}$ and elements E^{ij} for $1 \le i, j \le n$ such that these identities hold. (Theorem 1.1.20 contains an easier characterization.) In this case every element of $\mathcal{A}$ can be written uniquely as

$$
\sum_{i=1}^{n} \sum_{j=1}^{n} d_{ij}\, E^{ij} \tag{11}
$$

for suitable $d_{ij} \in \mathcal{D}$. It is easy to check that such an algebra is isomorphic to the algebraic tensor product $\mathcal{D} \otimes M_n$ as a linear space with multiplication given by

$$
(d_1 \otimes b_1)(d_2 \otimes b_2) = d_1 d_2 \otimes b_1 b_2 \qquad \forall\, d_1, d_2 \in \mathcal{D};\ b_1, b_2 \in M_n.
$$

We say that $\mathcal{A}$ is an $n \times n$ matrix algebra over $\mathcal{D}$ when it is isomorphic to $\mathcal{D} \otimes M_n$. In this case, we usually consider $\mathcal{D}$ as embedded in $\mathcal{A}$ as indicated above.

The following important theorem and its proof are due to Jacobson and Rickart [1950].

6.3.12 Theorem *Let $\mathcal{A}$ and $\mathcal{B}$ be algebras, and let $\mathcal{A}$ be a full $n \times n$ matrix algebra for $n > 1$ over a unital algebra $\mathcal{D}$. Let $\varphi: \mathcal{A} \to \mathcal{B}$ be a Jordan homomorphism and assume that $\mathcal{B}$ is the algebra generated by $\varphi(\mathcal{A})$.*

Then there are orthogonal central idempotents g_1 and g_2 in $\mathcal{B}$ such that $g_1\mathcal{B}$ and $g_2\mathcal{B}$ are full matrix algebras over $\varphi(\mathcal{D})$ and the maps $\varphi_j\colon\mathcal{A}\to g_j\mathcal{B}$ defined by $\varphi_j(a)=g_j\varphi(a)$ are surjective for $j=1,2$. Furthermore φ_1 is a homomorphism, φ_2 is an anti-homomorphism and $\varphi=\varphi_1+\varphi_2$.

Proof Assume that $\mathcal{D}$ is embedded in $\mathcal{A}$ and let $\{E^{ij}:1\le i,j\le n\}$ be a system of matrix units for $\mathcal{A}$ satisfying (10). Throughout this proof we will use Lemma 6.3.2 and Lemma 6.3.3 without always citing them specifically. For $i\ne j$, let $E^{ij\prime}=E^{ii}+E^{ij}$. Then $(E^{ij\prime})^2=E^{ij\prime}$ and for $d\in\mathcal{D}$, $[d,E^{ij\prime}]=0$. Hence Lemma 6.3.3 shows that $\varphi(E^{ij\prime})$ is idempotent and $[\varphi(E^{ij\prime}),\varphi(d)]=0$ for all $d\in\mathcal{D}$. Similarly $\varphi(E^{ij})$ is idempotent and $[\varphi(E^{ii}),\varphi(d)]=0$. Hence all i and j satisfy $[\varphi(E^{ij}),\varphi(d)]=0$ and $\varphi(dE^{ii})=\varphi(d)\varphi(E^{ii})$. For $i\ne j$, we have

$$
\begin{aligned}
\varphi(dE^{ij}) &= \varphi(dE^{ij}E^{jj}+E^{jj}E^{ij}d)\\
&= \varphi(d)\varphi(E^{ij})\varphi(E^{jj})+\varphi(E^{jj})\varphi(E^{ij})\varphi(d)\\
&= \varphi(d)(\varphi(E^{ij})\varphi(E^{jj})+\varphi(E^{jj})\varphi(E^{ij}))\\
&= \varphi(d)\varphi(E^{ij}E^{jj}+E^{jj}E^{ij})=\varphi(d)\varphi(E^{ij}).
\end{aligned}
$$

Since we have proved this for $i=j$ already, we have

$$
\varphi(dE^{ij})=\varphi(d)\varphi(E^{ij})=\varphi(E^{ij})\varphi(d)\qquad\forall\,d\in\mathcal{D};\ 1\le i,j\le n.
$$

Hence, any element of $\mathcal{A}$ written as in (11) satisfies

$$
\varphi\Big(\sum_{i=1}^{n}\sum_{j=1}^{n}d_{ij}E^{ij}\Big)=\sum_{i=1}^{n}\sum_{j=1}^{n}\varphi(d_{ij})\varphi(E^{ij}).\tag{12}
$$

We now introduce two orthogonal sets of matrix units $\{g_{ij}:1\le i,j\le n\}$ and $\{h_{ij}:1\le i,j\le n\}$ in $\mathcal{B}$. For $i\ne j$ define

$$
g_{ij}=\varphi(E^{ii})\varphi(E^{ij})\varphi(E^{jj})\quad\text{and}\quad h_{ij}=\varphi(E^{ii})\varphi(E^{ji})\varphi(E^{jj}).
$$

(We define g_{ii} and h_{ii} a little later.) Note that $E^{ij}=E^{ii}E^{ij}E^{jj}+E^{jj}E^{ij}E^{ii}$ implies

$$
\varphi(E^{ij})=g_{ij}+h_{ji}.\tag{13}
$$

Lemma 6.3.3 gives $g_{ij}\varphi(E^{jj})=\varphi(E^{ii})g_{ij}=g_{ij}$ and $h_{ij}\varphi(E^{jj})=\varphi(E^{ii})h_{ij}=h_{ij}, g_{ij}\varphi(E^{ii})=\varphi(E^{jj})g_{ij}=h_{ij}\varphi(E^{ii})=\varphi(E^{jj})h_{ij}=0$. We conclude

$$
\begin{aligned}
g_{ij} &= \varphi(E^{ii})\varphi(E^{ij})=\varphi(E^{ij})\varphi(E^{jj})\\
h_{ij} &= \varphi(E^{jj})\varphi(E^{ji})=\varphi(E^{ji})\varphi(E^{ii}).
\end{aligned}
$$

For $j\ne k$, this shows $g_{ij}g_{k\ell}=0$. For $i\ne j\ne k\ne i$, it also shows

$$
\begin{aligned}
g_{ij}g_{jk} &= \varphi(E^{ij})\varphi(E^{jj})g_{jk}=\varphi(E^{ij})g_{jk}=\varphi(E^{ij})\varphi(E^{jk})\varphi(E^{kk})\\
&= (\varphi(E^{ij})\varphi(E^{jk})+\varphi(E^{jk})\varphi(E^{ij}))\varphi(E^{kk})\\
&= \varphi(E^{ij}E^{jk}+E^{jk}E^{ij})\varphi(E^{kk})\\
&= \varphi(E^{ik})\varphi(E^{kk})=g_{ik},
\end{aligned}
$$

where we used Lemma 6.3.3 and $E^{kk}E^{ij} = E^{ij}E^{kk} = 0$ for the fourth equality. Thus, any i, j, k, ℓ satisfy

$$g_{ij}g_{k\ell} = \delta_{jk}g_{i\ell}. \tag{14}$$

Similarly, any i, j, k, ℓ satisfy

$$h_{ij}h_{k\ell} = \delta_{jk}h_{i\ell}. \tag{15}$$

Now for each i choose $j \neq i$ and define

$$g_{ii} = g_{ij}g_{ji} \quad h_{ii} = h_{ij}h_{ji}.$$

Then g_{ii} is independent of the choice of $j \neq i$ since $g_{ik}g_{ki} = g_{ij}g_{jk}g_{ki} = g_{ij}g_{ji}$ when $i \neq j \neq k \neq i$. Similarly h_{ii} is well defined. It is easy to check that (14) and (15) now hold for all choices of indices. Furthermore for $j \neq i$, $\varphi(E^{ii}) = \varphi(E^{ij}E^{jj}E^{jj}E^{ji} + E^{ij}E^{ii}E^{ii}E^{ji}) = g_{ij}g_{ji} + h_{ij}h_{ji} = g_{ii} + h_{ii}$ so that (13) hold for all choices of indices.

We now prove that $g_{ij}h_{k\ell} = 0$ for all choices of indices. Because of the definition of g_{ii} and h_{ii}, we may assume $i \neq j$ and $k \neq \ell$. Thus $g_{ij}h_{k\ell} = \varphi(E^{ij})\varphi(E^{jj})\varphi(E^{kk})\varphi(E^{\ell k})$. If $j \neq k$, the middle product is zero so we may assume $j = k$. Then $g_{ij}h_{j\ell} = g_{ij}h_{ji}h_{i\ell} = \varphi(E^{ij})\varphi(E^{jj})\varphi(E^{ij})h_{i\ell} = \varphi(E^{ij}E^{jj}E^{ij})h_{i\ell} = 0$.

Let $g_1 = \sum_{i=1}^{n} g_{ii}$ and let $g_2 = \sum_{i=1}^{n} h_{ii}$. Since $1 = \sum_{i=1}^{n} E^{ii}$, equation (13) gives $\varphi(1) = g_1 + g_2$. Clearly g_1 and g_2 are orthogonal idempotents which satisfy $g_1\varphi(E^{ij}) = \varphi(E^{ij})g_1 = g_{ij}, g_2\varphi(E^{ij}) = \varphi(E^{ij})g_2 = h_{ji}$, and $[g_1, d] = [g_2, d] = 0$ for each $d \in \mathcal{D}$. Hence g_1 and g_2 are central. Using formula (12), we have

$$\varphi_1\Big(\sum_{i=1}^{n}\sum_{j=1}^{n} d_{ij}E^{ij}\Big) = \sum_{i=1}^{n}\sum_{j=1}^{n}\varphi(d_{ij})g_{ij} \tag{16}$$

$$\varphi_2\Big(\sum_{i=1}^{n}\sum_{j=1}^{n} d_{ij}E^{ij}\Big) = \sum_{i=1}^{n}\sum_{j=1}^{n}\varphi(d_{ij})h_{ji}.$$

For $a, b \in \mathcal{D}$ and $i \neq j$, we have $\varphi(ab)\varphi(E^{ii}) + \varphi(ba)\varphi(E^{jj}) = \varphi(abE^{ii} + baE^{jj}) = \varphi((aE^{ij} + bE^{ji})^2) = (\varphi(a)\varphi(E^{ij}) + \varphi(b)\varphi(E^{ji}))^2 = (\varphi(a)(g_{ij} + h_{ji}) + \varphi(b)(g_{ji} + h_{ij}))^2 = \varphi(a)\varphi(b)(g_{ii} + h_{jj}) + \varphi(b)\varphi(a)(g_{jj} + h_{ii})$. Multiplying by g_{ij} and h_{ij} on the right gives $\varphi(ab)g_{ij} = \varphi(a)\varphi(b)g_{ij}$ and $\varphi(ab)h_{ij} = \varphi(b)\varphi(a)h_{ij}$. These formulas together with (16) show that φ_1 and φ_2 satisfy the asserted conditions. $\square$

Lie Homomorphisms

In this section, we have studied homomorphisms preserving the Jordan product $a \circ b = ab + ba$ for a, b in an associative algebra $\mathcal{A}$. Along with this symmetrized product, it is natural to consider the anti-symmetrized *Lie product*

$$[a, b] = ab - ba \qquad \forall\, a, b \in \mathcal{A}$$

and the *Lie homomorphisms* which preserve it. However, the most significant results on this product require some sort of left–right symmetry in the algebra, such as that given by an involution. Since we are postponing the study of algebras with an involution until Volume II, we will postpone the discussion of Lie homomorphisms. The reader may wish to consult papers by Wallace S. Martindale, III [1969], [1977] and Richard A. Howland [1969]. Here is a typical result from the first reference.

6.3.13 Theorem *Let $\varphi\colon \mathcal{A} \to \mathcal{B}$ be a Lie isomorphism where $\mathcal{A}$ is unital, prime and has a nontrivial idempotent and $\mathcal{B}$ is unital and simple. Then φ is the sum $\varphi = \psi + \tau$ where ψ is either an isomorphism or the negative of an anti-isomorphism and τ is a linear map into the center of $\mathcal{B}$ which sends all commutators to zero.*

See Civin and Yood [1965] for other interesting results on the Jordan and Lie structure of Banach algebras. See also Rodríguez-Palacios [1979].

6.4 Derivations

In our discussion of derivations we will use the language of bimodules, normed bimodules and Banach bimodules introduced in Definition 4.1.17. We always regard any algebra $\mathcal{A}$ as a bimodule under the *regular module action* under which ab has the same meaning whether considered as a product in $\mathcal{A}$, the left module action or the right module action. If $\mathcal{A}$ is a $\langle$ normed algebra / Banach algebra $\rangle$, this regular bimodule is a $\langle$ normed bimodule / Banach bimodule $\rangle$.

6.4.1 Definition Let $\mathcal{A}$ be an algebra and let $\mathcal{X}$ be an $\mathcal{A}$-bimodule. A *bimodule derivation of $\mathcal{A}$ into $\mathcal{X}$* is a linear map $\delta\colon \mathcal{A} \to \mathcal{X}$ which satisfies

$$\delta(ab) = \delta(a)b + a\delta(b) \qquad a,\, b \in \mathcal{A}.$$

The set of derivations of $\mathcal{A}$ into $\mathcal{X}$ is denoted by $\Delta(\mathcal{A}, \mathcal{X})$. If $\mathcal{A}$ is a normed algebra and $\mathcal{X}$ is a normed $\mathcal{A}$-bimodule, then the set of bounded derivations, $\Delta(\mathcal{A}, \mathcal{X}) \cap \mathcal{B}(\mathcal{A}, \mathcal{X})$, is denoted by $\Delta_B(\mathcal{A}, \mathcal{X})$.

When $\mathcal{X}$ is just the regular $\mathcal{A}$-bimodule $\mathcal{A}$ itself, we speak of an *algebra derivation of $\mathcal{A}$* and denote the sets by $\Delta(\mathcal{A})$ and $\Delta_B(\mathcal{A})$.

Bimodule derivations arise naturally in a number of situations (*cf.* Ringrose [1972a], Johnson [1972a]). In particular they are at the center of the cohomology theory of Banach algebras given in Volume II. When no confusion is likely, we will usually shorten "bimodule derivation" and "algebra derivation" to just "derivation".

It is clear that $\Delta(\mathcal{A}, \mathcal{X})$ and $\Delta_B(\mathcal{A}, \mathcal{X})$ are linear spaces under pointwise operations with the latter normed and complete under the norm of $\mathcal{B}(\mathcal{A}, \mathcal{X})$. It is also trivial to check that both $\Delta(\mathcal{A})$ and $\Delta_B(\mathcal{A})$ are closed under the Lie bracket

$$[\delta_1, \delta_2] = \delta_1\delta_2 - \delta_2\delta_1 \qquad \forall\ \delta_1, \delta_2 \in \Delta. \tag{1}$$

Readers acquainted with Lie algebras will recognize these spaces as (often infinite-dimensional) Lie algebras.

The following example explains the term "derivation". Let $\mathcal{A}$ be the set, $C^\infty(\mathbb{R})$, of infinitely differentiable complex-valued functions on $\mathbb{R}$ made into an algebra by pointwise operations. Define $\delta\colon \mathcal{A} \to \mathcal{A}$ by setting δf equal to the derivative of f for any $f \in C^\infty(\mathbb{R})$. It was, of course, in this context that Gottfried Wilhelm Leibniz discovered the formula

$$\delta^n(ab) = \sum_{k=0}^{n} \binom{n}{k} \delta^{n-k}(a)\delta^k(b) \tag{2}$$

for repeated differentiation. Since the usual inductive proof depends only on δ being a derivation, we will use Leibniz' formula for derivations without further comment.

The formula for the derivative of a polynomial allows derivations to be defined on algebras over any field, if the elements of the algebra can be interpreted as polynomials. A similar remark holds for algebras of formal power series over any field. We have little to say about these purely algebraic types of derivations although they have important applications. Our next example, which is fundamental to our subsequent discussion, can be defined in any algebra.

6.4.2 Definition Let $\mathcal{A}$ be an algebra and let $\mathcal{X}$ be an $\mathcal{A}$-bimodule. For any element $x \in \mathcal{X}$, the map $\delta_x\colon \mathcal{A} \to \mathcal{X}$ defined by

$$\delta_x(a) = xa - ax \quad a \in \mathcal{A}$$

is called the *inner derivation determined by* x. A derivation δ of $\mathcal{A}$ into $\mathcal{X}$ is called an *inner derivation* if it is equal to δ_x for some $x \in \mathcal{X}$. The set of inner derivations is denoted by $\Delta_I(\mathcal{A}, \mathcal{X})$. If $\mathcal{X} = \mathcal{A}$, we write $\Delta_I(\mathcal{A})$.

It is trivial to check that an inner derivation is a derivation. In fact, the set $\Delta_I(\mathcal{A})$ of inner derivations is a *Lie ideal* of the Lie algebra $\Delta(\mathcal{A})$. This means that $[\delta, \delta_a]$ belongs to $\Delta_I(\mathcal{A})$ for all $\delta \in \Delta(\mathcal{A})$ and $a \in \mathcal{A}$ and it follows from the useful formula

$$[\delta, \delta_a] = \delta\delta_a - \delta_a\delta = \delta_{\delta(a)} \quad a \in \mathcal{A}, \delta \in \Delta(\mathcal{A}). \tag{3}$$

If $\mathcal{A}$ is a normed algebra and $\mathcal{X}$ a normed bimodule, then inner derivations are clearly bounded. Hence for a normed algebra $\mathcal{A}$, $\Delta_I(\mathcal{A})$ is a Lie ideal of

$\Delta_B(\mathcal{A})$. Note also that the map $x \mapsto \delta_x$ is a linear map of $\mathcal{X}$ into $\Delta_I(\mathcal{A}, \mathcal{X})$. The kernel of this map is the center $\mathcal{X}_Z$ of $\mathcal{X}$ where the center of an $\mathcal{A}$-bimodule is the set $\{x \in \mathcal{X} : ax = xa \text{ for all } a \in \mathcal{A}\}$, so that the center of the regular $\mathcal{A}$-bimodule is just the usual center of the algebra $\mathcal{A}$.

We will show that all derivations on certain algebras are inner. For a commutative algebra the only inner derivation is the zero map. This special case will also prove important.

We need to consider the separating space of a derivation. Proposition 6.1.8(d) shows that the separating space $\mathcal{B}_\varphi$ of a homomorphism $\varphi : \mathcal{A} \to \mathcal{B}$ is a closed ideal of $\overline{\varphi(\mathcal{A})} \subseteq \mathcal{B}$. More is true for derivations.

6.4.3 Proposition *Let $\delta : \mathcal{A} \to \mathcal{X}$ be a bimodule derivation of a normed algebra $\mathcal{A}$ into a normed $\mathcal{A}$-bimodule $\mathcal{X}$. Then $\mathcal{X}_\delta$ is a closed $\mathcal{A}$-sub-bimodule. If $\mathcal{X}$ is the regular $\mathcal{A}$-bimodule $\mathcal{A}$, then $\mathcal{A}_\delta$ is a closed ideal of $\mathcal{A}$.*

Proof   Suppose $\{a_n\}_{n \in \mathbb{N}} \subseteq \mathcal{A}$ converges to 0 and $\{\delta(a_n)\}$ converges to $x \in \mathcal{X}_\delta$. Then the computation

$$\delta(ba_nc) = \delta(b)a_nc + b\delta(a_n)c + ba_n\delta(c) \to bxc \qquad \forall\, b, c \in \mathcal{A}$$

gives the desired result. $\square$

Homomorphisms constructed from Derivations

There are two major connections between derivations and homomorphisms. We discuss the more elementary case first. Let $\mathcal{A}$ be an algebra and let $\mathcal{X}$ be an $\mathcal{A}$-bimodule. Then $\mathcal{A} \times \mathcal{X}$ can be made into an algebra by pointwise linear operations and the multiplication

$$(a, x)(b, y) = (ab, ay + xb) \qquad \forall\, a, b \in \mathcal{A};\ x, y \in \mathcal{X}. \tag{4}$$

We denote this algebra by $\mathcal{A} \oplus \mathcal{X}$. If $\mathcal{A}$ is a $\langle$ normed algebra / Banach algebra $\rangle$ and $\mathcal{X}$ is a $\langle$ normed / Banach $\rangle$ $\mathcal{A}$-bimodule, then it is clear that these operations make $\mathcal{A} \oplus \mathcal{X}$ into a $\langle$ normed algebra / Banach algebra $\rangle$ under the algebra norm

$$\|(a, x)\| = \|a\| + |||x||| \qquad \forall\, a \in \mathcal{A};\ x \in \mathcal{X}$$

where $||| \cdot |||$ is the equivalent norm on $\mathcal{X}$ defined by

$$|||x||| = \sup\{||x||, ||ax||, ||xb||, ||axb|| : a, b \in \mathcal{A}_1\} \qquad \forall\, x \in \mathcal{X}.$$

Note that $\mathcal{A} \oplus \mathcal{X}$ is just the normed algebra semidirect product of the subalgebra $\{(a, 0) : a \in \mathcal{A}\}$ which is naturally isometrically isomorphic to $\mathcal{A}$ and the ideal $\{(0, x) : x \in \mathcal{X}\}$ which is naturally homeomorphically linearly isomorphic to $\mathcal{X}$ considered as an algebra with all products zero.

6.4.4 Proposition *Let $\mathcal{A}$ be an algebra, $\mathcal{X}$ an $\mathcal{A}$-bimodule and $\mathcal{A} \oplus \mathcal{X}$ the algebra defined above. If $\delta : \mathcal{A} \to \mathcal{X}$ is a derivation, then $\varphi_\delta : \mathcal{A} \to \mathcal{A} \oplus \mathcal{X}$ defined by $\varphi_\delta(a) = (a, \delta(a))$ is an injective homomorphism. If $\mathcal{A}$ and $\mathcal{X}$ are normed, then the separating spaces satisfy $(\mathcal{A} \oplus \mathcal{X})_{\varphi_\delta} = \{0\} \oplus \mathcal{X}_\delta$ so that φ_δ is continuous if and only if δ is continuous. Furthermore,*

$$||a||_\delta = ||a|| + |||\delta(a)||| \qquad \forall\, a \in \mathcal{A}$$

is an algebra norm on $\mathcal{A}$ which is equivalent to the original norm (or, equivalently, continuous) if and only if δ is continuous.

Hence if every injective homomorphism of a Banach algebra $\mathcal{A}$ into a Banach algebra is continuous, then every derivation of $\mathcal{A}$ into a Banach $\mathcal{A}$-bimodule is continuous.

Proof Straightforward.   □

The last statement of the above proposition extends several of the results in Section 6.2 about automatic continuity of homomorphisms to cover derivations. For instance, Theorem 6.2.6 shows that all homomorphisms from $\mathcal{B}(\mathcal{X})$ are continuous when $\mathcal{X}$ is a Banach space linearly homeomorphically isomorphic to $\mathcal{X} \times \mathcal{X}$. Hence, all derivations from $\mathcal{B}(\mathcal{X})$ into Banach $\mathcal{B}(\mathcal{X})$-bimodules are continuous when $\mathcal{X}$ has this property. Dales, Loy and Willis [1992] give the first example of a Banach space $\mathcal{X}$ for which $\mathcal{B}(\mathcal{X})$ is known to have discontinuous derivations but no discontinuous homomorphisms. (See §1.7.8 above, and compare Loy and Willis [1989], Read [1989] and Willis [1993].)

Here is a very simple concrete construction of an inequivalent algebra norm using the process described in the proposition. Let $\mathcal{A}$ be the disc algebra of §1.5.2 with its usual supremum norm $|| \cdot ||_\mathbb{D}$ on $\mathbb{D}$. Then the second norm is just

$$||f||_\delta = ||f||_\mathbb{D} + |f'(1)| \qquad \forall\, f \in \mathcal{A}.$$

This norm arises by letting $\mathbb{C}$ be a Banach $\mathcal{A}$-bimodule under the action $f\lambda = \lambda f = f(1)\lambda$ and considering the bimodule derivation $\delta(f) = f'(1)$. Later we will extend these remarks about inequivalent algebra norms to obtain two inequivalent complete norms on a commutative Banach algebra.

One Parameter Groups of Homomorphisms and Derivations

We now turn to the second connection between derivations and homomorphisms. It only makes sense in the case of derivations with range in the original algebra which must be normed. The interesting case is that of a continuous derivation on a Banach algebra.

Let $\mathcal{A}$ be a Banach algebra and let δ be a derivation of $\mathcal{A}$. If the series $\sum_{n=0}^\infty (n!)^{-1} \delta^n(a)$ converges for all $a \in \mathcal{A}$ (which is certainly true if δ is bounded), then we will denote the sum by $e^\delta(a)$. For any $a, b \in \mathcal{A}$ and any $\lambda \in \mathbb{C}$, Leibniz' formula gives

$$
\begin{aligned}
e^{\lambda\delta}(ab) &= \sum_{n=0}^{\infty} \frac{\lambda^n \delta^n(ab)}{n!} = \sum_{n=0}^{\infty} \frac{\lambda^n}{n!} \sum_{k=0}^{n} \binom{n}{k} \delta^{n-k}(a)\delta^k(b) \qquad (5) \\
&= \sum_{n=0}^{\infty} \sum_{k=0}^{n} \frac{\lambda^{n-k}\delta^{n-k}(a)}{n-k!} \frac{\lambda^k \delta^k(b)}{k!} = \sum_{m=0}^{\infty} \frac{\lambda^m \delta^m(a)}{m!} \sum_{k=0}^{\infty} \frac{\lambda^k \delta^k(b)}{k!} \\
&= e^{\lambda\delta}(a) e^{\lambda\delta}(b).
\end{aligned}
$$

Thus for each $\lambda \in \mathbb{C}$, $e^{\lambda\delta}$ is an automorphism which is continuous if δ is.

We will further investigate the case in which δ is continuous. It is clear that $t \mapsto e^{t\delta}$ is a continuous group homomorphism from the additive group of $\mathbb{R}$ into the group of automorphisms of $\mathcal{A}$ in $\mathcal{B}(\mathcal{A})$. In order to fully describe this interesting situation we need more terminology. For related concepts see Jerome A. Goldstein [1985].

6.4.5 Definition Let $\mathcal{B}$ be an algebra. A *one-parameter group* (of invertible elements) in $\mathcal{B}$ is a group homomorphism α from the additive group of $\mathbb{R}$ into the multiplicative group of $\mathcal{B}$. If $\mathcal{B}$ is a topological or normed algebra, the terms *continuous one-parameter group* and *bounded one-parameter group* have the obvious meaning.

Let α be a one-parameter group in a topological algebra $\mathcal{B}$. When the limit $\lim_{t\to 0}(\alpha_t - 1)/t$ exists, it is called the *generator of α*.

Of course, $\mathcal{B}$ is unital if there is a one-parameter group in $\mathcal{B}$. We have checked that if δ is a continuous derivation of a Banach algebra $\mathcal{A}$, then $t \mapsto e^{t\delta}$ is a norm continuous one-parameter group of automorphisms of $\mathcal{A}$ in $\mathcal{B}(\mathcal{A})$. As arguments from elementary calculus should suggest, the generator of this group is δ. The case in which $\mathcal{B}$ is $\mathcal{B}(\mathcal{X})$ with its strong operator topology for some Banach space $\mathcal{X}$ is important. For our present purposes, it is enough to consider the case in which $\mathcal{B}$ (usually $\mathcal{B}(\mathcal{A})$) is a Banach algebra.

6.4.6 Proposition *Let $\mathcal{B}$ be a Banach algebra and let α be a norm continuous one-parameter group in $\mathcal{B}$.*
  (a) *Then the generator of α exists.*
  (b) *If the generator is denoted by $\delta \in \mathcal{B}$, then it satisfies $\alpha_t = e^{t\delta}$ for all $t \in \mathbb{R}$.*
  (c) *Any $\delta \in \mathcal{B}$ satisfies $\delta = \lim_{t\to 0}(e^{t\delta} - 1)/t$.*
  Thus $\delta \mapsto (t \mapsto e^{t\delta})$ is a bijection of $\mathcal{B}$ onto the set of norm continuous one-parameter groups in $\mathcal{B}$.

Proof Since α is norm continuous, there is some $\varepsilon > 0$ such that $|t| < \varepsilon$ implies $\|1 - \alpha_t\| < 1$. For any such t the series $-\sum_{n=1}^{\infty} n^{-1}(1 - \alpha_t)^n$ converges to an element that we denote by $\delta(t) = \log(1 - (1 - \alpha_t)) = \log(\alpha_t)$.

We note that any $t \in \mathbb{R}$ and $n \in \mathbb{N}$ satisfying $|nt| < \varepsilon$ must also satisfy $\delta(nt) = \log(\alpha_{nt}) = \log((\alpha_t)^n) = n \log(\alpha_t) = n\delta(t)$. Hence any $m \in \mathbb{N}$ and any s satisfying $|s| < \varepsilon$ must also satisfy $m^{-1}\delta(s) = \delta(m^{-1}s)$. This implies $\delta(rt) = r\delta(t)$ for any rational number r satisfying $0 < r < 1$ and any t satisfying $|t| < \varepsilon$. However $\delta(\cdot)$ is a continuous function of t for $|t| < \varepsilon$ since $t \mapsto \alpha_t$ is continuous and the series defining $\delta(\cdot)$ converges uniformly. Hence $\delta(st) = s\delta(t)$ holds for all $s \in [0, 1]$ and t satisfying $|t| < \varepsilon$. Define δ by $\delta = s^{-1}\delta(s)$ for some (arbitrary) $s \in]0, \varepsilon[$. This implies

$$\alpha_t = e^{t\delta} \tag{6}$$

for all $t \in]0, \varepsilon[$. Now for any $t \in \mathbb{R}, t/n$ belongs to $]0, \varepsilon[$ for some $n \in \mathbb{N}$. Hence (6) holds for any $t \in \mathbb{R}$. Thus $\delta = \lim_{t \to 0} t^{-1}(e^{t\delta} - 1) = \lim_{t \to 0} t^{-1}(\alpha_t - 1)$ holds.

For any $\delta \in \mathcal{B}$, we have

$$\lim_{t \to 0} \frac{e^{t\delta} - 1}{t} = \lim_{t \to 0} \left(\delta + t \sum_{n=2}^{\infty} \frac{t^{n-2}\delta^n}{n!} \right) = \delta.$$

This, together with (6), proves that $\delta \mapsto (t \mapsto e^{t\delta})$ is a bijection of $\mathcal{B}$ onto the set of norm continuous one-parameter groups in $\mathcal{B}$. $\square$

We mention a suggestive example for which continuity and convergence to the generator are not in the norm topology. Let $\mathcal{A}$ be $C^{\infty}(\mathbb{R})$ and let $\alpha \colon \mathbb{R} \to \mathcal{L}(C^{\infty}(\mathbb{R}))_G$ be the translation map defined by

$$(\alpha_t f)(s) = f(t + s) \qquad \forall\, f \in C^{\infty}(\mathbb{R}); s, t \in \mathbb{R}.$$

Then the generator (in a generalized sense) of α is the map $\delta \in \mathcal{L}(C^{\infty}(\mathbb{R}))$ defined by

$$(\delta f)(s) = \lim_{t \to 0} \frac{f(t + s) - f(s)}{t} = f'(s) \qquad \forall\, s \in \mathbb{R}.$$

Thus δ is the ordinary derivative map given as our first example of a derivation. This example points up the fact that Proposition 6.4.6 covers a very special case of a more general situation. In fact, the unbounded generators of groups and semigroups of operators continuous in the strong operator topology have received more attention than the norm continuous ones studied here. We will not explore the more general situation here.

We turn now to the analogue of Proposition 6.4.6 in which we are chiefly interested.

6.4.7 Theorem *Let $\mathcal{A}$ be a Banach algebra. If δ is a bounded derivation of $\mathcal{A}$, then $t \mapsto e^{t\delta}$ and $t \mapsto e^{it\delta}$ are norm continuous one-parameter groups of automorphisms of $\mathcal{A}$ in $\mathcal{B}(\mathcal{A})$. If α is a norm continuous one-parameter* group of automorphisms of $\mathcal{A}$ in $\mathcal{B}(\mathcal{A})$, then its generator is a bounded derivation of $\mathcal{A}$. The maps

$$\delta \mapsto (t \mapsto e^{t\delta}) \qquad and \qquad \alpha \mapsto (generator\ of\ \alpha) \qquad (7)$$

are mutually inverse. They establish a bijection between $\Delta_B(\mathcal{A})$ and the set of norm continuous one-parameter groups of automorphisms of $\mathcal{A}$.

Proof Calculation (5) shows that $t \mapsto e^{t\delta}$ and $t \mapsto e^{it\delta}$ are norm continuous one-parameter groups of automorphisms of $\mathcal{A}$ in $\mathcal{B}(\mathcal{A})$. Suppose, conversely, that α is a norm continuous one-parameter group of automorphisms of $\mathcal{A}$ in $\mathcal{B}(\mathcal{A})$ and that δ is its generator. Then δ is a bounded linear map and for any $a, b \in \mathcal{A}$, we get

$$\begin{aligned}
\delta(ab) &= \lim_{t \to 0} \frac{\alpha_t(ab) - ab}{t} \\
&= \lim_{t \to 0} \alpha_t(a) \lim_{t \to 0} \frac{\alpha_t(b) - b}{t} + \lim_{t \to 0} \frac{\alpha_t(a) - a}{t} b \\
&= a\delta(b) + \delta(a)b.
\end{aligned}$$

Hence δ is a bounded derivation. Proposition 6.4.6 shows that the maps of (7) are mutually inverse. Thus they obviously establish a bijection between $\Delta_B(\mathcal{A})$ and the set of norm continuous one-parameter groups of automorphisms of $\mathcal{A}$. $\qquad \square$

6.4.8 Proposition Let $\mathcal{A}$ be a Banach algebra. Let a be an element of $\mathcal{A}$. Any $t \in \mathbb{R}$ and any $b \in \mathcal{A}$ satisfy

$$e^{t\delta_a}(b) = e^{ta}be^{-ta}. \qquad (8)$$

Conversely, if $t \mapsto u_t$ is a norm continuous one-parameter group in $\mathcal{A}$ and $a \in \mathcal{A}$ is its generator, then δ_a is the generator of the norm continuous one-parameter group

$$t \mapsto (b \mapsto u_t b u_{-t}).$$

Furthermore, the inclusions

$$Sp_{\mathcal{B}(\mathcal{A})}(\delta_a) \subseteq \{\lambda - \mu : \lambda, \mu \in Sp(a)\} \qquad and$$
$$Sp_{\mathcal{B}(\mathcal{A})}(\alpha_a) \subseteq \{e^{\lambda - \mu} : \lambda, \mu \in Sp(a)\}$$

hold where $\alpha_a \in \mathcal{B}(\mathcal{A})$ is the map defined by

$$\alpha_a(b) = e^a be^{-a} \qquad \forall\, b \in \mathcal{A}.$$

Proof By induction one shows

$$(\delta_a)^n(b) = \sum_{k=0}^n (-1)^k \binom{n}{k} a^{n-k} b a^k \qquad \forall\, b \in \mathcal{A}.$$

A calculation parallel to that given in (5) above gives (8), establishing the first statement. The second statement follows from the first, equation (6) and the uniqueness statement in Theorem 6.4.7.

Finally, we turn to calculating the spectra. Note that the second inclusion follows from the first together with the spectral mapping theorem, Theorem 3.3.7(b). Let $a \in \mathcal{A}$ be arbitrary but fixed. Note that L_a and R_a commute where L and R are the left and right regular representations of $\mathcal{A}$. Then Corollary 3.1.6 and Propositions 2.5.3 and 2.5.2 give

$$
\begin{aligned}
Sp_{\mathcal{B}(\mathcal{A})}(\delta_a) &= Sp_{\mathcal{B}(\mathcal{A})}(L_a - R_a) \subseteq Sp_{\mathcal{B}(\mathcal{A})}(L_a) - Sp_{\mathcal{B}(\mathcal{A})}(R_a) \\
&= Sp_{\mathcal{A}}(a) - Sp_{\mathcal{A}}(a) \\
&= \{\lambda - \mu : \lambda, \mu \in Sp_{\mathcal{A}}(a)\}. \qquad \square
\end{aligned}
$$

Having established the connection between bounded derivations and norm continuous one-parameter groups of automorphisms, we turn now to some simple properties of derivations before showing that any derivation of a semisimple Banach algebra is continuous.

6.4.9 Proposition *Let δ be a derivation of an algebra $\mathcal{A}$. Then*

(a) $\delta(e) = 0$ *for any central idempotent e. In particular $\delta(1) = 0$ if 1 is an identity element in $\mathcal{A}$.*

(b) δ *maps the center of $\mathcal{A}$ into itself.*

(c) *If $\mathcal{I}$ is any ideal satisfying $\mathcal{I}^2 = \mathcal{I}$, then δ maps $\mathcal{I}$ into itself.*

Proof (a): Any central idempotent e satisfies $\delta(e) = \delta(e^2) = \delta(e)e + e\delta(e) = 2e\delta(e)$. Multiplying by e and subtracting gives $e\delta(e) = 0$. Hence $\delta(e) = 2e\delta(e)$ is zero.

(b): Any central element z and any element a satisfy

$$
\delta(z)a = \delta(za) - z\delta(a) = \delta(az) - \delta(a)z = a\delta(z).
$$

(c): Finally if a belongs to $\mathcal{I} = \mathcal{I}^2$, then there are elements $a_j, b_j \in \mathcal{I}$ satisfying $a = \sum_{j=1}^{n} a_j b_j$. Thus $\delta(a) = \sum_{j=1}^{n} (\delta(a_j)b_j + a_j\delta(b_j))$ belongs to $\mathcal{I}$ again. $\square$

We mention an interesting connection between derivations and the functional calculus. Let δ be a bounded derivation of a unital Banach algebra $\mathcal{A}$. Let a be an element of $\mathcal{A}$ commuting with $\delta(a)$ and let f belong to $\mathcal{F}_a$, so that $f(a)$ is defined by the functional calculus. Then we get

$$
\delta(f(a)) = f'(a)\delta(a).
$$

To prove this, note first that any $\lambda \in \mathbb{C} \setminus Sp(a)$ satisfies

$$
\begin{aligned}
0 &= \delta(1) = \delta((\lambda - a)(\lambda - a)^{-1}) \\
&= -\delta(a)(\lambda - a)^{-1} + (\lambda - a)\delta((\lambda - a)^{-1}).
\end{aligned}
$$

This implies $\delta((\lambda - a)^{-1}) = (\lambda - a)^{-2}\delta(a)$. Hence for any $\omega \in \mathcal{A}^*$, we get

$$\omega(\delta(f(a))) = \frac{1}{2\pi i} R^*_{\delta(a)}(\omega)(\int_\Gamma f(\lambda)(\lambda - a)^{-2}d\lambda),$$

where $R^*_{\delta(a)}$ is the adjoint of the right regular representation $R_{\delta(a)}$ of $\delta(a)$. Expanding the integrand in the neighborhood of each $\lambda \in \Gamma$ and applying the residue theorem and the Hahn–Banach theorem, we obtain the desired result.

Let $\mathcal{A}$ be an algebra and let δ be a derivation of $\mathcal{A}$. Then δ belongs to the algebra $\mathcal{L}(\mathcal{A})$ of linear maps on $\mathcal{A}$. We may consider the inner derivation δ_δ of $\mathcal{L}(\mathcal{A})$ determined by δ. For any $a \in \mathcal{A}$, L_a and R_a belong to $\mathcal{L}(\mathcal{A})$. The obvious computation shows

$$\delta_\delta(L_a) = \mathcal{L}_{\delta(a)} \quad \text{and} \quad \delta_\delta(R_a) = R_{\delta(a)} \quad a \in \mathcal{A}. \tag{9}$$

This computation is the main part of the proof of the next simple result which shows that any derivation of a "good" algebra $\mathcal{A}$ is "almost" an inner derivation when extended to $\mathcal{D}(\mathcal{A})$. For some algebras which have outer derivations (e.g., simple nonunital C*-algebras) the "almost" can be dropped. In this case $\mathcal{D}(\mathcal{A})$ is an algebra with only inner derivations.

6.4.10 Proposition *Let $\mathcal{A}$ be an algebra and let δ be a derivation of $\mathcal{A}$. Let $\theta: \mathcal{A} \to \mathcal{D}(\mathcal{A})$ be the regular homomorphism and let δ_δ be the derivation on $\mathcal{D}(\mathcal{A})$ defined by*

$$\delta_\delta(L, R) = (\delta_\delta L, \delta_\delta R) = (\delta L - L\delta, \delta R - R\delta) \qquad \forall \, (L, R) \in \mathcal{D}(\mathcal{A}).$$

Then δ satisfies

(a) *$\theta \circ \delta = \delta_\delta \circ \theta$.*

(b) *δ maps the annihilator ideal $\mathcal{A}_A$ into itself.*

(c) *When $\mathcal{A}_A$ vanishes and $\mathcal{A}$ is regarded as embedded in $\mathcal{D}(\mathcal{A})$, then δ_δ is a derivation of $\mathcal{D}(\mathcal{A})$ which extends the derivation δ of $\mathcal{A}$.*

(d) *If $\mathcal{A}_A = \{0\}$ and $\mathcal{A}^2 = \mathcal{A}$ both hold, then the map $\delta \mapsto \delta_\delta$ is a linear bijection of $\Delta(\mathcal{A})$ onto $\Delta(\mathcal{D}(\mathcal{A}))$.*

Proof Result (a) is a restatement of (3). Since $\ker(\theta) = \mathcal{A}_A$, (b) follows from (a). (The calculations $0 = \delta(ab) = \delta(a)b + a\delta(b) = a\delta(b)$ and $0 = \delta(ba) = \delta(b)a + b\delta(a) = \delta(b)a$ for any $a \in \mathcal{A}$ and $b \in \mathcal{A}_A$ prove (b) directly.) For any $a, b \in \mathcal{A}$ and $(L, R) \in \mathcal{D}(\mathcal{A})$, we get

$$\begin{aligned}
\delta_\delta(L)(ab) &= \delta(L(ab)) - L(\delta(ab)) \\
&= \delta(L(a)b) - L(\delta(a)b + a\delta(b)) \\
&= \delta(L(a))b + L(a)\delta(b) - L(\delta(a))b - L(a)\delta(b) \\
&= (\delta L - L\delta)(a)b = \delta_\delta(L)(a)b.
\end{aligned}$$

Hence $\delta_\delta(L)$ is a left centralizer. Similarly $\delta_\delta(R)$ is a right centralizer. With the same notation we also find

$$
\begin{aligned}
a\delta_\delta(L)(b) &= a((\delta L - L\delta)(b)) = a\delta(L(b)) - aL(\delta(b)) \\
&= a\delta(L(b)) - R(a)\delta(b) + \delta(R(a)b) - \delta(aL(b)) \\
&= \delta(a)L(b) - \delta(R(a))b = R(\delta(a))b - \delta(R(a))b \\
&= \delta_\delta(R)(a)b.
\end{aligned}
$$

Hence δ_δ maps $\mathcal{D}(\mathcal{A})$ into $\mathcal{D}(\mathcal{A})$. It is a derivation since it is the restriction of the inner derivation determined by δ. These facts together with (a) prove (c).

Suppose $\mathcal{A}_A = \{0\}$ and $\mathcal{A}^2 = \mathcal{A}$ both hold. Regard $\mathcal{A}$ as embedded in $\mathcal{D}(\mathcal{A})$. Then any derivation of $\mathcal{D}(\mathcal{A})$ maps $\mathcal{A}$ into itself by Proposition 6.4.9(c). Thus restriction to $\mathcal{A}$ maps $\Delta(\mathcal{D}(\mathcal{A}))$ into $\Delta(\mathcal{A})$ and this map is the inverse of $\delta \mapsto \delta_\delta$. Hence the latter linear map is a bijection of $\Delta(\mathcal{A})$ onto $\Delta(\mathcal{D}(\mathcal{A}))$. □

The next proposition is essentially due to Ivan Vidav [1955], Feliks Vladimirovič Širokov [1956] and David C. Kleinecke [1957], independently.

6.4.11 Proposition *Let $(\mathcal{A}, \|\cdot\|)$ be a normed algebra and δ a continuous derivation on $\mathcal{A}$. If $\delta(a)$ commutes with a for some $a \in \mathcal{A}$, then $\|\delta(a)\|^\infty = 0$.*

Proof For $a, b \in \mathcal{A}$ suppose $\delta_b(a) = -\delta_a(b)$ commutes with a. Then $(\delta_a)^2(b) = 0$ holds and thus Leibniz' formula (2) shows $\delta_a^2(b^2) = 2(\delta_a(b))^2$. By induction, we will prove $\delta_a^n(b^n) = n!(\delta_a(b))^n$ for all $n \in \mathbb{N}$. Leibniz' formula and the induction hypothesis give

$$
\begin{aligned}
\delta_a^n(b^n) &= \delta_a^n(b^{n-1})b + n\delta_a^{n-1}(b^{n-1})\delta_a(b) + 0 \\
&= \delta_a((n-1)!(\delta_a(b))^{n-1})b + n((n-1)!)(\delta_a(b))^{n-1}\delta_a(b) \\
&= 0 + n!(\delta_a(b))^n \quad \text{and} \\
\|\delta_b(a)\|^\infty &= \|\delta_a(b)\|^\infty = \lim_{n\to\infty} \|\delta_a(b)^n\|^{1/n} \\
&\leq \lim_{n\to\infty} (n!)^{-1/n}\|\delta_a(b)\| = 0.
\end{aligned}
$$

Thus we have proved the proposition for any inner derivation δ_b.

However if δ is an arbitrary derivation of $\mathcal{A}$ and $\delta(a)$ commutes with a for some $a \in \mathcal{A}$, then (3) shows that $\delta_\delta(L_a) = L_{\delta(a)}$ commutes with L_a. Hence $\|L_{\delta(a)}\|^\infty = 0$ holds by the first paragraph of this proof. This implies $\|\delta(a)\|^\infty = 0$, since $\|(\delta(a))^{n+1}\|^{n^{-1}} \leq \|(L_{\delta(a)})\|^{n^{-1}}\|\delta(a)\|^{n^{-1}}$ holds. □

The next important result is due to Isadore M. Singer and John Wermer [1955]. See also Kaplansky [1958a], [1958b]. As they noted, it implies that there is no complete algebra norm on $C^\infty(\mathbb{R})$. For an extension of this result see Chang Pao Chen [1978]. Marc P. Thomas [1988] verified their conjecture that the word "continuous" could be removed from the hypotheses. We will outline his quite technical proof below.

6.4.12 Corollary *The range of any continuous derivation of a commutative spectral normed algebra is included in its Jacobson radical.*

Proof For a commutative spectral normed algebra, $\|\delta(a)\|^\infty = 0$ implies that $\delta(a)$ belongs to $\mathcal{A}_J$ by Theorem 4.5.8.   □

The original proof for Banach algebras given by Singer and Wermer [1955] is so attractive that we will give it here although it is less elementary than the proof above. After Proposition 6.4.16, we will indicate still another attractive proof which is due to Sinclair [1969].

Let δ be a continuous derivation of the commutative Banach algebra $\mathcal{A}$. Then $e^{\lambda\delta}$ is an automorphism for each $\lambda \in \mathbb{C}$ by the calculation (5). Hence, if γ is an element of $\Gamma_{\mathcal{A}}$ (*i.e.*, any homomorphism of $\mathcal{A}$ onto $\mathbb{C}$), then $\gamma \circ e^{\lambda\delta}$ is also an element of $\Gamma_{\mathcal{A}}$. Theorem 3.1.5 shows that $|\gamma(e^{\lambda\delta}(a))| \le \|a\|$ holds for each $a \in \mathcal{A}$. Thus $\lambda \mapsto \gamma(e^{\lambda\delta}(a))$ is a bounded analytic function, so it is a constant. This implies $0 = \lim_{\lambda\to 0} \lambda^{-1}[\gamma(e^{\lambda\delta}(a)) - \gamma(a)] = \gamma(\delta(a))$. Since $\gamma \in \Gamma_{\mathcal{A}}$ was arbitrary, Theorem 4.5.8 implies that a belongs to $\mathcal{A}_J$. Therefore γ maps $\mathcal{A}$ into $\mathcal{A}_J$.

We remark that this proof for commutative Banach algebras can easily be extended to commutative spectral normed algebras. Suppose δ is a bounded derivation on a commutative spectral normed algebra $\mathcal{A}$. Let $\overline{\mathcal{A}}$ be the completion of $\mathcal{A}$ and let $\overline{\delta}: \overline{\mathcal{A}} \to \overline{\mathcal{A}}$ be the extension of δ by continuity. Clearly $\overline{\delta}$ is a derivation of $\overline{\mathcal{A}}$ so that $\overline{\delta}(\overline{\mathcal{A}}) \subseteq \overline{\mathcal{A}}_J$ by the proof just given. Hence $\delta(\mathcal{A})$ is included in $\mathcal{A} \cap \overline{\mathcal{A}}_J$. However $\mathcal{A} \cap \overline{\mathcal{A}}_J$ is an ideal in $\mathcal{A}$ which consists of topologically nilpotent elements. Hence $\mathcal{A} \cap \overline{\mathcal{A}}_J \subseteq \mathcal{A}_J$ holds by Theorem 4.3.6(b).

Thomas [1988] settled the more than thirty year old conjecture of Singer and Wermer [1955] that the last corollary remains valid even for not necessarily continuous derivations. Martin Mathieu and Volker Runde [1992] show that if $\delta: \mathcal{A} \to \mathcal{A}$ is a derivation on an arbitrary Banach algebra satisfying $[a, \delta(a)] \in \mathcal{A}_Z$ for all $a \in \mathcal{A}$, a *centralizing* derivation, then its range is in the Jacobson radical. See also Mathieu and Murphy [1992] and Yood [1984a].

6.4.13 Theorem *Let $\mathcal{A}$ be a commutative Banach algebra. The range of any derivation $\delta: \mathcal{A} \to \mathcal{A}$ is included in the radical of $\mathcal{A}$.*

Outline of Proof Since a derivation always sends idempotents to 0, we may assume that $\mathcal{A}$ is unital. In [1969a] Johnson showed that if this result fails for a commutative unital Banach algebra $\mathcal{A}$, then there is a finite nonempty set of orthogonal idempotents $\{e_1, e_2, \ldots, e_n\}$ such that the restriction of δ

to $\mathcal{B} = (1 - e_1 - e_2 - \cdots - e_n)\mathcal{A}$ has range in $\mathcal{B}_J$. Moreover δ sends each $e_k\mathcal{A}$ into itself and each of these algebras has the form $\mathcal{R}^1$ where $\mathcal{R}$ is a radical algebra. Hence, if the theorem fails, there is a derivation $\tilde{\delta}\colon \mathcal{R}^1 \to \mathcal{R}^1$ and an element $b \in \mathcal{R}$ with $\tilde{\delta}(b) \notin \mathcal{R}$. Thus $\tilde{\delta}(b)$ is invertible. Let its inverse be c. Then $\delta = c\tilde{\delta}$ is a derivation on $\mathcal{R}^1$ with $\delta(b) = 1$ for some radical Banach algebra $\mathcal{R}$ and some $b \in \mathcal{R}$. It is enough to show that this cannot happen.

In the notation of 3.4.10, let $\mathcal{D} = \mathcal{D}(b, \mathcal{R}^1)$ be the largest b-divisible subspace of $\mathcal{R}^1$. The proof shows that b is not nilpotent and that for some $m \in \mathbb{N}$ the closure $\overline{\mathcal{D}}$ of the ideal $\mathcal{D}$ coincides with the closure of $b^m\mathcal{R}^1 \neq \{0\}$. The quotient algebra $\mathcal{R}^1/\mathcal{D}$ is isomorphic to the algebra $\mathcal{C}[[z]]$ of formal power series over a certain unital commutative algebra $\mathcal{C}$. The isomorphism sends b to z and sends the derivation δ to the ordinary derivative of the formal series. Thomas discovers intricate structure in this algebra which he calls *recalcitrant systems*. They preclude the possibility that $\delta(b) = 1$, thus concluding the proof. See the original paper, Volker Runde [1990] and [1991] or Dales [1994] for full details. □

The following important corollary is due to Singer and Wermer [1955] for bounded derivations. We prove the result for arbitrary derivations. It was first proved in this form by Johnson [1969a]. See also Curtis [1961] where it was established for completely regular algebras.

6.4.14 Corollary *Any derivation of a commutative semisimple Banach algebra is zero.*

Proof This follows immediately from Theorem 6.4.13 or from Corollary 6.4.12 and Theorem 6.4.18. □

6.4.15 Corollary *Let $\mathcal{A}$ be a Banach algebra with a semisimple center. Then any derivation is zero on the center. Hence any norm continuous one-parameter group of automorphisms leaves each element of the center fixed.*

Proof Proposition 6.4.9(b) shows that a derivation of $\mathcal{A}$ maps the center $\mathcal{A}_Z$ of $\mathcal{A}$ into itself. Hence the restriction of the derivation to $\mathcal{A}_Z$ is zero by Corollary 6.4.14. This proves the first statement. The second then follows from Theorem 6.4.7. □

We turn now to a result of Sinclair [1969]. The proof is similar to the proof of Kleinecke [1957] given in the proof of Proposition 6.4.11. We also obtain a third proof of Corollary 6.4.12 from it.

6.4.16 Proposition *Any continuous derivation of a spectral normed algebra $\mathcal{A}$ maps each primitive ideal of $\mathcal{A}$ into itself. If $\mathcal{A}$ is a semisimple Banach algebra, this is true of any derivation.*

Proof Let $\delta: \mathcal{A} \to \mathcal{A}$ be a derivation and let $\mathcal{P}$ be an ideal. First we show by induction on n that any $b \in \mathcal{P}$ satisfies $\delta^n(b^n) + \mathcal{P} = n!(\delta(b))^n + \mathcal{P}$. The case $n = 1$ is trivial. For any $0 \le j < k$ repeated application of Leibniz' formula shows $\delta^j(b^k)$ can be expressed as a sum of products $\delta^{j_1}(b)\delta^{j_2}(b) \cdots \delta^{j_k}(b)$, where each j_k is a non-negative integer. Since some j_i must be zero in each product, $\delta^j(b^k)$ belongs to $\mathcal{P}$. Now Leibniz' formula and the induction hypothesis show

$$
\begin{aligned}
\delta^{n+1}(b^{n+1}) + \mathcal{P} &= \delta(\delta^n(b^n b)) + \mathcal{P} = \delta(\sum_{j=0}^{n} \binom{n}{j} \delta^j(b^n)\delta^{n-j}(b)) + \mathcal{P} \\
&= \sum_{j=0}^{n} \binom{n}{j} [\delta^{j+1}(b^n)\delta^{n-j}(b) + \delta^j(b^n)\delta^{n-j+1}(b)] + \mathcal{P} \\
&= n\delta^n(b^n)\delta(b) + \delta^n(b^n)\delta(b) + \mathcal{P} \\
&= (n+1)\delta^n(b^n)\delta(b) + \mathcal{P} = (n+1)!(\delta(b))^{n+1} + \mathcal{P}.
\end{aligned}
$$

If δ is continuous, Stirling's formula gives for each $b \in \mathcal{P}$

$$
\|(\delta(b) + \mathcal{P})^n\|^{n^{-1}} = [(n!)^{-1}\|\delta^n(b^n) + \mathcal{P}\|]^{n^{-1}} < (e/n)\|\delta\| \, \|b\|.
$$

Hence each element of $(\delta(\mathcal{P}) + \mathcal{P})/\mathcal{P}$ is quasi-nilpotent. The definition of a derivation shows that $(\delta(\mathcal{P}) + \mathcal{P})/\mathcal{P}$ is an ideal. Hence $(\delta(\mathcal{P}) + \mathcal{P})/\mathcal{P}$ is a quasi-regular ideal and thus is contained in the Jacobson radical of $\mathcal{A}/\mathcal{P}$ by Theorem 4.3.6. Finally, if $\mathcal{P}$ is primitive $(\delta(\mathcal{P}) + \mathcal{P})\mathcal{P} = \{0\}$ so δ maps $\mathcal{P}$ into itself. The second statement follows from Corollary 6.4.19. $\quad\square$

Here is the third proof of Corollary 6.4.12 which was promised above. Let δ be a continuous derivation on a commutative spectral normed algebra. We replace $\mathcal{A}$ by $\mathcal{A}^1$, defining $\delta^1: \mathcal{A}^1 \to \mathcal{A}^1$ by $\delta^1(\lambda + a) = \delta(a)$ for all $\lambda + a \in \mathcal{A}^1$. (If $\mathcal{A}$ is already unital, this is consistent by Proposition 6.4.9(a).) Now Theorems 4.1.9 and 2.4.12 show that $\mathcal{A}/\mathcal{P}$ is isomorphic to $\mathbb{C}$ for each primitive ideal $\mathcal{P}$. Hence any element $a \in \mathcal{A}$ can be written as $\lambda + b$ with $\lambda \in \mathbb{C}$ and $b \in \mathcal{P}$. Thus $\delta(a) = \delta(b)$ belongs to $\mathcal{P}$ by Proposition 6.4.16. Since $\mathcal{P}$ was an arbitrary primitive ideal, $\delta(a)$ belongs to $\mathcal{A}_J$. Since a was arbitrary, δ maps $\mathcal{A}$ into $\mathcal{A}_J$.

We now turn to Johnson and Sinclair's proof [1968] that every derivation of a semisimple Banach algebra is continuous. They actually prove more. Since we see only a small simplification in proving Corollary 6.4.19 directly instead of deriving it from Theorem 6.4.18, we follow Johnson and Sinclair's original procedure. Thus for the next few results we will deal with a map $\delta: \mathcal{A} \to \mathcal{A}$ which satisfies $\delta(a + b) = \delta(a) + \delta(b)$ and $\delta(ab) = \delta(a)b + a\delta(b)$ for all $a, b \in \mathcal{A}$, but which is not necessarily linear. We call such a map an *additive derivation*. Of course, an additive derivation is linear over the rational numbers, and it is therefore bounded if and only if it is continuous. Continuity also implies real linearity.

Theorem 6.4.18 and Corollary 6.4.19 were conjectured by Kaplansky [1958b]. He also conjectured Corollary 6.4.19 for the special case of C*-algebras [1958a]. The latter conjecture was verified by Shoichiro Sakai [1960]. We will prove this special case directly in Volume II.

6.4.17   Lemma   *Let $\mathcal{A}$ be a Banach algebra and let δ be an additive derivation on $\mathcal{A}$. Let T be a bounded representation of $\mathcal{A}$ on a Banach space $\mathcal{X}$.*

(a) If T is irreducible and $\mathcal{X}$ is infinite-dimensional, then the map

$$A \mapsto \delta(A)x \quad A \in \mathcal{A}$$

is continuous on $\mathcal{A}$ for each $x \in \mathcal{X}$.

(b) If $T = \oplus_{n=1}^{\infty} T^n$ is the direct topological sum of infinitely many irreducible subrepresentations on finite-dimensional T-invariant subspaces $\{\mathcal{X}^n : n \in \mathbb{N}\}$ and no pair of the T^n are equivalent, then the above map is continuous for each x in each $\mathcal{X}^n$ except for a finite set of $n \in \mathbb{N}$.

Proof   In order to avoid large numbers of double subscripts, we suppress the symbol T and denote elements of $\mathcal{A}$ by capital letters as if they were operators on $\mathcal{X}$. (This notation is not logically defensible since T may have a nontrivial kernel $\mathcal{I}$, and it is not clear *a priori* that $\mathcal{I}$ is δ-invariant. (After Corollary 6.4.19 has been proved, Proposition 6.4.16 will show that $\mathcal{I}$ is δ-invariant.) The reader can check that no confusion arises.)

The proof is by contradiction.

In case (a), suppose $x_0 \in \mathcal{X}$ is a non-zero element such that

$$A \mapsto \delta(A)x_0$$

is continuous on $\mathcal{A}$. Let x be any other element of $\mathcal{X}$. Since T is irreducible, there is an element $B \in \mathcal{A}$ such that $Bx_0 = x$. Hence $A \mapsto \delta(A)x = \delta(A)Bx_0 = \delta(AB)x_0 - A\delta(B)x_0$ is continuous on $\mathcal{A}$. Thus if $A \mapsto \delta(A)x$ is discontinuous for any $x \in \mathcal{X}$, it is discontinuous for all non-zero $x \in \mathcal{X}$. We will derive a contradiction from assuming $A \mapsto \delta(A)x$ is discontinuous for all non-zero x. This will prove the lemma in case (a).

In case (b), the above argument shows that $A \to \delta(A)x$ is discontinuous for all non-zero $x \in \mathcal{X}^n$ if it is discontinuous for any $x_0 \in \mathcal{X}^n$. For the sake of proof by contradiction, we may therefore assume that $A \mapsto \delta(A)x$ is discontinuous for each non-zero x in each $\mathcal{X}^n$. (Any $\mathcal{X}^n$ for which this does not hold is simply omitted from the list.) Since the $\mathcal{X}^n$ are all closed and $\mathcal{A}$-invariant, it follows that $A \mapsto \delta(A)x$ is discontinuous for all non-zero $x \in \mathcal{X}$.

In case (a) we choose a sequence $\{x_n\}_{n \in \mathbb{N}}$ of linearly independent unit vectors. Corollary 4.2.14 shows that we can choose a sequence of elements

$\{C_n\}_{n\in\mathbb{N}}$ in $\mathcal{A}$ satisfying

$$C_n C_{n-1} \cdots C_1 x_{n-1} = 0$$

$$\{C_n C_{n-1} \cdots C_1 x_j : j \geq n\} \text{ is linearly independent} \tag{10}$$

$$\|C_n\| \leq 2^{-n}.$$

(The main point of this choice is simply to make $C_n C_{n-1} \cdots C_1 x_n$ non-zero for each $n \in \mathbb{N}$.)

In case (b) we will find $\{x_n\}_{n\in\mathbb{N}}$ in $\mathcal{X}$ and $\{C_n\}_{n\in\mathbb{N}}$ in $\mathcal{A}$ satisfying these same conditions. Choose each x_n to be a unit vector in $\mathcal{X}^n$. For each $n \in \mathbb{N}$, consider the image of $\mathcal{A}$ under $T^1 \oplus T^2 \oplus \cdots \oplus T^n$. This is a subalgebra of $\mathcal{B}(\mathcal{X}^1) \oplus \mathcal{B}(\mathcal{X}^2) \oplus \cdots \oplus \mathcal{B}(\mathcal{X}^n)$ which includes each direct summand. Hence by Wedderburn's Theorem (Theorem 8.1.1), it is isomorphic to a direct sum of the algebras $\mathcal{B}(\mathcal{X}^k)$. Since the T^k are all inequivalent, it must be the direct sum of all of them. Hence we may choose an element C_n in $\mathcal{A}$ such that $C_n x_j = 0$ for $j < n$ and the restriction of C_n to $\mathcal{X}^n$ is invertible. Clearly these C_n can be chosen to satisfy $\|C_n\| < 2^{-n}$. Then they satisfy all the conditions of (10).

By assumption $A \to \delta(A)x$ is discontinuous for each non-zero $x \in \mathcal{X}$, so we can choose elements $A_1, A_2, \ldots$ in $\mathcal{A}$ satisfying

$$\|A_n\| \leq \min\{1 + \|\delta(C_j C_{j-1} \cdots C_1)\|^{-1} : j = 1, 2, \ldots, n\} \tag{11}$$

$$\|\delta(A_n) C_n C_{n-1} \cdots C_1 x_n\| \geq n + \sum_{j=1}^{n-1} \|\delta(A_j C_j C_{j-1} \cdots C_1)\|.$$

The restrictions on the norms of the C_n and A_n allow us to define elements D and D_n in $\mathcal{A}$ by

$$D = \sum_{j=1}^{\infty} A_j C_j C_{j-1} \cdots C_1 \tag{12}$$

$$D_n = A_{n+1} + \sum_{j=n+2}^{\infty} A_j C_j C_{j-1} \cdots C_{n+2}.$$

These elements satisfy

$$\|D_n\| \, \|\delta(C_{n+1} C_n \cdots C_1)\| \leq 2 \tag{13}$$

since

$$\|D_n\| \leq \|A_{n+1}\| + \sum_{j=n+2}^{\infty} \|A_j\| \, \|C_j C_{j-1} \cdots C_{n+2}\|$$

$$\leq (1 + \|\delta(C_{n+1} C_n \cdots C_1)\|)^{-1} (1 + \sum_{j=n+2}^{\infty} \|C_j C_{j-1} \cdots C_{n+2}\|)$$

$$< \quad (1 + \|\delta(C_{n+1}C_n \cdots C_1)\|)^{-1}(1 + \sum_{j=n+2}^{\infty} 2^{-j})$$

holds. For any $n \geq 2$ we have

$$\begin{aligned}
D &= (\sum_{j=1}^{n-1} A_j C_j C_{j-1} \cdots C_2 C_1) + A_n C_n C_{n-1} \cdots C_2 C_1 \\
&\quad + D_n C_{n+1} C_n \cdots C_2 C_1
\end{aligned}$$

which implies

$$\begin{aligned}
\delta(D)x_n &= \delta(\sum_{j=1}^{n-1} A_j C_j C_{j-1} \cdots C_1)x_n + \delta(A_n)C_n C_{n-1} \cdots C_2 C_1 x_n \\
&\quad + \delta(D_n)C_{n+1}C_n \cdots C_2 C_1 x_n + A_n\delta(C_n C_{n-1} \cdots C_2 C_1)x_n \\
&\quad + D_n\delta(C_{n+1}C_n \cdots C_2 C_1)x_n.
\end{aligned}$$

However, equations (11), (10) and (13) give $\|A_n\delta(C_n \cdots C_2 C_1)x_n\| \leq 1$, $\delta(D_N)C_{n+1} \cdots C_2 C_1 x_n = 0$ and $\|D_n\delta(C_{n+1} \cdots C_2 C_1)x_n\| \leq 2$. These inequalities together with (11) give

$$\|\delta(D)\| \geq \|\delta(D)x_n\| \geq n - 3 \qquad \forall\, n > 0.$$

This contradiction proves the lemma.   □

In the next theorem it is not assumed that $\mathcal{A}$ is unital. The notation $(1 - e)\mathcal{A}$ simply denotes the set $\{a - ea : a \in \mathcal{A}\}$.

6.4.18 Theorem *Let $\mathcal{A}$ be a semisimple Banach algebra, and let δ be an additive derivation of $\mathcal{A}$. Then $\mathcal{A}$ contains a central idempotent e satisfying:*
   (a) *$e\mathcal{A}$ and $(1 - e)\mathcal{A}$ are invariant under δ.*
   (b) *δ restricted to $(1 - e)\mathcal{A}$ is continuous.*
   (c) *$e\mathcal{A}$ is finite-dimensional.*

Proof For each primitive ideal $\mathcal{P}$ of $\mathcal{A}$, let $\psi_{\mathcal{P}} \colon \mathcal{A} \to \mathcal{A}/\mathcal{P}$ be the natural map, and choose an irreducible representation $T^{\mathcal{P}}$ of $\mathcal{A}$ on a linear space $\mathcal{X}_{\mathcal{P}}$ with kernel $\mathcal{P}$. Theorem 4.2.8 shows that $\mathcal{X}_{\mathcal{P}}$ can be given a Banach space norm so that $T^{\mathcal{P}}$ is continuous. Assume for a moment that

$$a \mapsto T^{\mathcal{P}}_{\delta(a)}x \quad a \in \mathcal{A}$$

is continuous for each $x \in \mathcal{X}$. Then this rational linear map is real linear. Thus $(tT^{\mathcal{P}}_{\delta(a)} - T^{\mathcal{P}}_{\delta(ta)})x = 0$ holds for all $t \in \mathbb{R}, a \in \mathcal{A}$ and $x \in \mathcal{X}$. This implies $tT^{\mathcal{P}}_{\delta(a)} - T^{\mathcal{P}}_{\delta(ta)} = 0$, so $t\delta(a) - \delta(ta)$ belongs to $\mathcal{P}$ for each $t \in \mathbb{R}$ and $a \in \mathcal{A}$. Hence $\psi_{\mathcal{P}} \circ \delta$ is real linear. We will now prove that it is continuous by the closed graph theorem. Suppose $\{a_n\}_{n\in\mathbb{N}} \subseteq \mathcal{A}$ converges to zero and $\{\psi_{\mathcal{P}}(\delta(a_n)) : n \in \mathbb{N}\}$ converges to $b + \mathcal{P}$ in $\mathcal{A}/\mathcal{P}$. Then all $x \in \mathcal{X}^{\mathcal{P}}$ satisfy $T_b^{\mathcal{P}} x = \lim T_{\delta(a_n)}^{\mathcal{P}} x = 0$, so $T_b^{\mathcal{P}} = 0$. Thus $b + \mathcal{P} = \mathcal{P}$ holds. Hence $\psi_{\mathcal{P}} \circ \delta$ has a closed graph and is therefore continuous.

If $\mathcal{A}/\mathcal{P}$ is infinite-dimensional, then $\mathcal{X}_{\mathcal{P}}$ is infinite dimensional (since $\mathcal{A}/\mathcal{P}$ is isomorphic to a subalgebra of $\mathcal{B}(\mathcal{X}_{\mathcal{P}})$). Hence Lemma 6.4.17(a) shows that our assumption above (on the continuity of the map $a \mapsto T_{\delta(a)}^{\mathcal{P}} x$) is correct in this case, so $\varphi_{\mathcal{P}} \circ \delta$ is continuous.

Next we consider the case where $\mathcal{A}/\mathcal{P}$ is finite-dimensional. If our assumption (on the continuity of the map $a \mapsto T_{\delta(a)}^{\mathcal{P}} x$) were false for infinitely many choices of $\mathcal{P}$, we could construct $T = \oplus_{n=1}^{\infty} T^{\mathcal{P}_n}$ as a direct topological sum contradicting Lemma 6.4.17(b). Thus our assumption above is correct except for finitely many such $\mathcal{P}$. Let $\mathcal{P}_1, \mathcal{P}_2, \ldots, \mathcal{P}_n$ be these exceptional cofinite primitive ideals.

Let $\mathcal{P}_0$ be the intersection of all primitive ideals of $\mathcal{A}$ such that $\psi_{\mathcal{P}} \circ \delta$ is continuous and hence real linear. Let $\psi_0 \colon \mathcal{A} \to \mathcal{A}/\mathcal{P}_0$ be the natural map. Then $\psi_0 \circ \delta$ is real linear, and a now familiar closed graph argument shows that it is continuous. If $\mathcal{P}_0$ is $\{0\}$, then the theorem is proved with $e = 0$. Hence we assume $\mathcal{P}_0$ is not zero.

Theorem 4.3.2 and the semisimplicity of $\mathcal{A}$ show that the ideal $\mathcal{P}_0$ is semisimple. Also $\{0\} = \cap_{j=0}^n \mathcal{P}_j = \cap_{j=1}^n (\mathcal{P}_0 \cap \mathcal{P}_j)$ holds. Thus the map $b \mapsto (b + \mathcal{P}_0 \cap \mathcal{P}_1, b + \mathcal{P}_0 \cap \mathcal{P}_2, \ldots, b + \mathcal{P}_0 \cap \mathcal{P}_n)$ is an injective homomorphism of $\mathcal{P}_0$ into $\mathcal{P}_0/(\mathcal{P}_0 \cap \mathcal{P}_1) \times \mathcal{P}_0/(\mathcal{P}_0 \cap \mathcal{P}_2) \times \cdots \times \mathcal{P}_0/(\mathcal{P}_0 \cap \mathcal{P}_n)$ which is isomorphic to $(\mathcal{P}_0 + \mathcal{P}_1)/\mathcal{P}_1 \times (\mathcal{P}_0 + \mathcal{P}_2)/\mathcal{P}_2 \times \cdots \times (\mathcal{P}_0 + \mathcal{P}_n)/\mathcal{P}_n$ which, in turn, is a subalgebra of the finite-dimensional algebra $\mathcal{A}/\mathcal{P}_1 \times \mathcal{A}/\mathcal{P}_2 \times \cdots \times \mathcal{A}/\mathcal{P}_n$. Hence $\mathcal{P}_0$ is a finite dimensional semisimple algebra. Theorem 8.1.1 shows that $\mathcal{P}_0$ contains an identity element e. For any $a \in \mathcal{A}$, ea and ae belong to $\mathcal{P}_0$ and thus satisfy $ea - eae = ae$. Thus e is a central idempotent so $\delta(e) = 0$ follows from Proposition 6.4.9 (which obviously does not depend on linearity). Therefore $\delta(ea) = \delta(e)a + e\delta(a) = e\delta(a)$ holds, so $e\mathcal{A} = \mathcal{P}_0$ and $(1 - e)\mathcal{A}$ are δ-invariant. The natural map $\psi_0 \colon \mathcal{A} \to \mathcal{A}/\mathcal{P}_0$ induces a continuous isomorphism θ of $(1 - e)\mathcal{A}$ onto $\mathcal{A}/\mathcal{P}_0$. The open mapping theorem shows that θ^{-1} is also continuous. The restriction of δ to $(1 - e)\mathcal{A}$ is continuous since it can be expressed as the composition $\theta^{-1} \circ (\psi_0 \circ \delta)$ of the two continuous maps θ^{-1} and $\psi_0 \circ \delta$. $\qquad\square$

An additional argument proves that δ is actually complex linear on $(1 - e)\mathcal{A}$. We refer the reader to the original paper for this (Johnson and Sinclair [1968]).

6.4.19 Corollary *A derivation of a semisimple Banach algebra is continuous.*

Proof Since $e\mathcal{A}$ is finite-dimensional and a derivation is linear, its restriction to $e\mathcal{A}$ is continuous.   □

Small modifications of the proofs of Lemma 6.4.17 and Theorem 6.4.18 show that these results hold for additive left centralizers on semisimple Banach algebras. In fact such a map L on $\mathcal{A}$ is continuous on all of $\mathcal{A}$ since $L(a) = L(ea) = L(e)a$ for any $a \in e\mathcal{A}$. We have already noted that such maps are linear.

We will now prove the theorem of Jewell and Sinclair [1976] on the continuity of derivations on certain not necessarily semisimple Banach algebras. The hypotheses on the algebra are exactly the same as in Theorem 6.1.18. See Example 6.1.19 for a description of some Banach algebras to which this theorem applies.

6.4.20 Theorem *Every derivation of a Banach algebra $\mathcal{A}$ is continuous if $\mathcal{A}$ satisfies:*

(a) *$\mathcal{A}$ has no non-zero finite-dimensional nilpotent ideals.*

(b) *For each closed infinite-dimensional ideal $\mathcal{I}$ of $\mathcal{A}$, there is a sequence $\{a_n\}_{n\in\mathbb{N}} \subseteq \mathcal{A}$ such that the sequence $\{(\Pi_{j=1}^{n} a_j)\mathcal{I}^{-}\}_{n\in\mathbb{N}}$ of closed right ideals is constantly decreasing.*

Proof Let δ be a derivation of $\mathcal{A}$ and let $\mathcal{A}_\delta$ be the separating space of Definition 6.1.7. Propositions 6.1.8 and 6.1.9 show that $\mathcal{A}_\delta$ is a closed linear space which equals $\{0\}$ if and only if δ is continuous. If $a \in \mathcal{A}_\delta$ and $\{a_n\}_{n\in\mathbb{N}}$ are as in the definition and $b \in \mathcal{A}$ is arbitrary, then the limits $ab = \lim(\delta(a_n)b + a_n\delta(b)) = \lim \delta(a_n b)$ and $ba = \lim(\delta(b)a_n + b\delta(a_n)) = \lim \delta(ba_n)$ show that $\mathcal{A}_\delta$ is an ideal.

Suppose $\mathcal{A}_\delta$ is infinite-dimensional. Choose a sequence $\{a_n\}_{n\in\mathbb{N}}$ satisfying condition (b) in the statement of the theorem. Lemma 6.1.17 applied to $\mathcal{X} = \mathcal{Y} = \mathcal{A}$, $\varphi = \delta$, $S_n(a) = T_n(a) = a_n a$ for all $a \in \mathcal{A}$ gives a contradiction.

Hence $\mathcal{A}_\delta$ must be finite-dimensional. Therefore, the linear map δ is continuous when restricted to $\mathcal{A}_\delta$. We will show that $\mathcal{A}_\delta^2$ is zero. Let $a, b \in \mathcal{A}_\delta$ be arbitrary and let $\{a_n\}_{n\in\mathbb{N}}$ be a sequence converging to zero such that $\{\delta(a_n)\}_{n\in\mathbb{N}}$ converges to a. Then we have $ab = \lim \delta(a_n)b = \lim(\delta(a_n b) - a_n\delta(b)) = 0$. Since the finite dimensional ideal $\mathcal{A}_\delta$ is nilpotent, it is zero. Hence δ is continuous by Proposition 6.1.9.   □

We remark that all derivations are known to be continuous on certain other non-semisimple Banach algebras. Loy [1970a] proves this for certain Banach algebras of formal power series. Loy [1970a], John A. Lindberg, Jr. [1971], and Julian M. Cusack [1976] prove the continuity of every derivation on an Arens–Hoffman extension of a commutative Banach algebra with the same property, and in particular on an Arens–Hoffman extension of a commutative semisimple Banach algebra (*cf.* Example 3.4.13). Cusack [1976] shows that if there is a discontinuous derivation on a semiprime

Banach algebra, then there is a topologically simple radical Banach algebra. It is not known whether such a Banach algebra exists. See Examples 6.4.24, 6.4.25 and 6.4.26.

The next result is due to Georges Zeller-Meier [1967]. The proof uses an argument due to L. Terrell Gardner [1965]. Jacques Dixmier [1957], p. 314, gives a similar result, where α is assumed to be an isometry and the constant $\sqrt{3}$ replaces 1. He attributes the proof to Jean-Pierre Serre. A similar result for *-automorphisms of C*-algebras is due to Richard V. Kadison and John R. Ringrose [1967]. Johnson [1975], p. 88 shows by a three-dimensional example that Dixmier's result [1957], which states that an automorphism α satisfying $Sp(\alpha) \subseteq \{\lambda \in \mathbb{C} : \arg(\lambda) < 2\pi/3\}$ is the exponential of a derivation, is the best possible result of its type.

6.4.21 Proposition   *Let α be a continuous automorphism of a Banach algebra $\mathcal{A}$. If α has spectrum in the open right half plane or satisfies $\|\alpha - I\| < 1$, then α belongs to a norm continuous one-parameter group of automorphisms.*

Proof If $\|\alpha - I\| < 1$, then the spectrum of α lies in the open right half plane, so we need only consider the latter hypothesis. If we extend α to $\mathcal{A}^1$ in the natural way (*i.e.*, $\alpha(\lambda + a) = \lambda + \alpha(a)$), then the spectrum of α still satisfies the hypothesized restriction. Thus we may assume that $\mathcal{A}$ is unital.

Let $\delta = \text{Log}(\alpha)$, where Log is the principal branch of the logarithm function. We will show that δ is a derivation so that $\alpha = e^\delta$ lies on the norm continuous one-parameter group $(t \mapsto e^{t\delta})$. Let $\beta : \mathcal{B}(\mathcal{A}) \to \mathcal{B}(\mathcal{A})$ be the automorphism defined by

$$\beta(T) = \alpha T \alpha^{-1} \qquad \forall \; T \in \mathcal{B}(\mathcal{A}).$$

For any $a, b \in \mathcal{A}$, we get $\beta(L_a)b = \alpha L_a \alpha^{-1} b = \alpha(a\alpha^{-1}(b)) = \alpha(a)b = L_{\alpha(a)}b$. Hence $L_{\mathcal{A}}$ is a closed invariant subspace of β. Note that β is $\exp(\delta_\delta)$, where $\delta_\delta \in \mathcal{B}(\mathcal{B}(\mathcal{A}))$ is the inner derivation defined by $\delta \in \mathcal{B}(\mathcal{A})$. The spectral restriction on α shows $Sp(\delta) \subseteq \{\lambda \in \mathbb{C} : |\text{Im}(\lambda)| < \pi/2\}$. Proposition 6.4.8 shows

$$Sp(\delta_\delta) \subseteq Sp(L_\delta - R_\delta) \subseteq \{\lambda \in \mathbb{C} : |\text{Im}(\lambda)| < \pi\}.$$

This implies $\text{Log}(\beta) = \text{Log}(\exp(\delta_\delta)) = \delta_\delta$ so $L_{\mathcal{A}}$ is invariant under $\text{Log}(\beta) = \delta_\delta$. Hence for each $a \in \mathcal{A}$, there is an element $a' \in \mathcal{A}$ satisfying $\delta_\delta L_a = L_{a'}$. For any $b \in \mathcal{A}$, we get $a'b = L_{a'}b = \delta_\delta L_a b = \delta(ab) - a\delta(b)$, so in particular for $b = 1$ we have $a' = \delta(a) - a\delta(1)$. Thus if we show $\delta(1) = 0$, then δ satisfies the derivation equation. Use Runge's theorem (Theorem 3.3.3) to approximate Log by polynomials $\{p_n\}_{n \in \mathbb{N}}$ on $\text{co}(Sp(\alpha))$. Then $\delta(1) = \text{Log}(\alpha)(1) = \lim p_n(\alpha)1 = \lim p_n(1) = \text{Log}(1) = 0$. Thus δ is a derivation, so $\alpha = e^\delta$ belongs to $\{e^{t\delta} : t \in \mathbb{R}\}$. $\square$

In the above proposition, if α satisfies $\|\alpha - I\| < 1$, then $\mathrm{Log}(\alpha)$ can be defined by the series expansion in powers of $\alpha - I$ as in the proof of Theorem 3.4.4. Then Runge's Theorem would no longer be needed. Next we give an analogue of Theorem 3.4.2.

6.4.22 Corollary *Let $\mathcal{A}$ be a Banach algebra and let G be the topological group of continuous automorphisms of $\mathcal{A}$ in the norm topology of $\mathcal{B}(\mathcal{A})$. Then the principal component G_0 of G is the subgroup generated by $\{e^\delta : \delta \in \Delta_B(\mathcal{A})\}$ and G/G_0 is discrete.*

Proof By the proposition the group generated by $\{e^\delta : \delta \in \Delta_B(\mathcal{A})\}$ is an open subgroup. It is obviously connected since $t \mapsto e^{t\delta}$ is an arc connecting e^δ to the identity automorphism. Hence it is the component.   □

Examples of Discontinuous Derivations and Homomorphisms

Let $\mathcal{A}$ be a commutative Banach algebra and let $\mathcal{X}$ be a Banach $\mathcal{A}$-bimodule. Then $\mathcal{X}$ is said to be a *commutative bimodule* if it satisfies

$$ax = xa \qquad \forall\, x \in \mathcal{X}; a \in \mathcal{A}.$$

The algebra $\mathcal{A} \oplus \mathcal{X}$ defined before Proposition 6.4.3 is commutative if and only if both $\mathcal{A}$ and $\mathcal{X}$ are commutative. Let $\delta \colon \mathcal{A} \to \mathcal{X}$ be a discontinuous derivation. We will consider the inequivalent algebra norm $\|\cdot\|_\delta$ on $\mathcal{A} \oplus \mathcal{X}$ introduced in Proposition 6.4.4. Note that $\|a\|_\delta$ is just the norm of $\mathcal{A} \oplus \mathcal{X}$ pulled back through φ_δ. Since the graph of δ is just the range of φ_δ, it is also the graph norm of δ in $\mathcal{A} \times \mathcal{X}$ where $\mathcal{X}$ has the norm $\|\|\cdot\|\|$. We will denote the completion of $\mathcal{A}$ under the norm $\|\cdot\|_\delta$ by $\overline{\mathcal{A}}^\delta$. Rather than considering the abstract completion, we will take this Banach algebra to be the closure of the graph of δ in the Banach algebra $\mathcal{A} \oplus \mathcal{X}$. The next theorem is due to Loy [1974a].

6.4.23 Theorem *Let $\mathcal{A}$ be a commutative Banach algebra and let $\mathcal{X}$ be a commutative Banach $\mathcal{A}$-bimodule. Let $\delta \colon \mathcal{A} \to \mathcal{X}$ be a non-zero derivation which is zero on a dense subset of $\mathcal{A}$. Then the Banach algebra $\overline{\mathcal{A}}^\delta$ has two inequivalent complete norms.*

Proof The definition of the separating space $\mathcal{X}_\delta$ shows that it is included in the closure $\delta(\mathcal{A})^-$ of the range of δ. We now show that the two spaces are equal. An element $x \in \mathcal{X}$ is in $\delta(\mathcal{A})^-$ if there is a sequence $\{b_n\}_{n \in \mathbb{N}} \subseteq \mathcal{A}$ with $\delta(b_n)$ converging to x. By the density of $\ker(\delta)$ we can choose a sequence $\{c_n\}_{n \in \mathbb{N}} \subseteq \ker(\delta)$ with $a_n = b_n - c_n$ converging to zero. But the sequence $\{a_n\}_{n \in \mathbb{N}}$ shows that $x \in \mathcal{X}_\delta$.

Next we show that the equality of $\mathcal{X}_\delta$ and $\delta(\mathcal{A})^-$ implies the equality of $\mathcal{A} \oplus (\mathcal{X}_\delta)$ and $\overline{\mathcal{A}}^\delta$. The inclusion $\overline{\mathcal{A}}^\delta = (\text{graph of } \delta)^- \subseteq \mathcal{A} \oplus \delta(\mathcal{A})^- = \mathcal{A} \oplus \mathcal{X}_\delta$ is obvious. However, any $a \oplus x \in \mathcal{A} \oplus \mathcal{X}_\delta$ can be written as $a \oplus x =$

$a \oplus \delta(a) + 0 \oplus (x - \delta(a))$ and both terms belong to the linear space (Graph of δ)$^-$ since $x - \delta(a)$ belongs to $\delta(\mathcal{A})^- = \mathcal{X}_\delta$.

The equality of $\delta(\mathcal{A})^-$ and $\mathcal{A} \oplus \mathcal{X}_\delta$ shows that $\delta(\mathcal{A})^-$ is a Banach algebra under both of the norms $\|\cdot\|_\delta$ and $\|\cdot\|$ (restricted from $\mathcal{A} \oplus \mathcal{X}$). Since δ is non-zero with dense kernel, it is discontinuous and hence these norms are not equivalent. $\square$

6.4.24 A Specific Example   This example, due to Loy [1974a], satisfies the hypotheses of the above theorem. Let $\mathcal{A}$ be the disc algebra introduced in §1.5.2 and discussed further in §3.2.13. Let $\mathcal{X}$ be a Banach space and let $T \in \mathcal{B}(\mathcal{X})$ have norm at most one. If $a \in \mathcal{A}$ has the power series expansion $\sum_{n=0}^{\infty} \alpha_n z^n$, then the module action is given by

$$ax = xa = \sum_{n=0}^{\infty} \alpha_n 2^{-n} T^n(x) \qquad \forall\, x \in \mathcal{X}.$$

This makes $\mathcal{X}$ into a commutative Banach $\mathcal{A}$-bimodule since all $a \in \mathcal{A}$ and $x \in \mathcal{X}$ satisfy

$$\begin{aligned}
\|ax\| &= \sum_{n=0}^{\infty} |\alpha_n| 2^{-n} \|x\| \le \sum_{n=0}^{\infty} |\alpha_n| 2^{-n/2} 2^{-n/2} \|x\| \\
&\le \left(\sum_{n=0}^{\infty} |\alpha_n|^2 2^{-n}\right)^{1/2} \left(\sum_{n=0}^{\infty} 2^{-n}\right)^{1/2} \|x\| \\
&\le \frac{1}{2\pi} \int_0^{2\pi} |a(2^{-1/2} e^{i\theta})|^2 d\theta \sqrt{2} \|x\| \le \sqrt{2} \|a\| \|x\|.
\end{aligned}$$

Note that if we denote the function $\lambda \mapsto a(\lambda/2)$ for $\lambda \in 2\mathbb{D}$ by $\tilde{a}$, then our module action is

$$ax = \tilde{a}(T)x \qquad \forall\, a \in \mathcal{A}; x \in \mathcal{X}$$

in terms of the analytic functional calculus.

To complete this construction, we use results of Dales [1973] discussed briefly in example 3.4.11. Let $\mathcal{X}$ be $C([0,1])$ and let T be the indefinite integral operator or Volterra operator:

$$Tf(t) = \int_0^t f(s)ds \qquad \forall\, f \in \mathcal{X}; t \in [0,1].$$

In the reference just cited, a non-zero linear map β is defined from the space $\mathcal{F}_\mathbb{D}$ of functions analytic in a neighborhood of $\mathbb{D}$ into $C([0,1])$ which vanishes on the set of polynomials and satisfies

$$\beta(fg) = f(T)\beta(g) + g(T)\beta(f) \qquad \forall\, f, g \in \mathcal{F}_\mathbb{D}.$$

Define δ by $\delta(a) = \beta(\tilde{a})$ for all $a \in \mathcal{A}$. Then $\delta \colon \mathcal{A} \to \mathcal{X}$ is a derivation which vanishes on the dense subset of polynomials. Thus we have satisfied all the hypotheses of the theorem and hence have constructed a commutative Banach algebra with two distinct norms. The construction given has other interesting properties which are discussed in Loy [1974a]. It is shown that for any particular non-algebraic function f in $\mathcal{F}_D$ (such as the exponential function), β and $a \in \mathcal{A}$ may be chosen so as to make $f(a)$ have different meanings in the functional calculi belonging to the two norms $\| \cdot \|$ and $\| \cdot \|_\delta$ on $\overline{\mathcal{A}}^\delta$. This implies that the radical of $\overline{\mathcal{A}}^\delta$ is infinite-dimensional by Theorem 3.4.12. The paper cited above contains further related results.

All bimodule derivations on certain algebras are shown to be continuous in Ringrose [1972] and Bade and Curtis, Jr. [1974], [1978] and [1984]. Because of point derivations, it is obvious that this can only happen on a commutative unital Banach algebra when $\mathcal{M}^2$ has finite codimension for each maximal ideal $\mathcal{M}$. The [1978] paper shows that in a separable algebra this condition and the absence of prime ideals of infinite codimension are necessary and sufficient. Bade and Dales [1989a], [1989b] give interesting partial results in algebras of power series and weighted convolution algebras. See also Loy [1973], Jewell [1977], Fereidoun Ghahramani [1980], Dales and McClure [1981], Willis [1983a], [1986], Ramesh V. Garimella [1987], [1991], Viet Ngo [1988], J. Vukman [1988], Niels Grønbaek [1982a] [1989a], Kil Woung Jun, Young-Whan Lee and Dal-Won Park [1990] and Brian Forrest [1988],[1992a].

6.4.25 Cusack's Results on Automatic Continuity The following discussion is taken from Chapter 2 of Cusack's thesis [1976]. The basic method used is an extension of the argument used to prove Theorems 6.1.18 and 6.4.20 in the present chapter. In the course of these proofs it is shown that certain separating ideals in a Banach algebra $\mathcal{A}$ satisfy the following conditions.

If $\{a_n\}_{n \in \mathbb{N}} \subseteq \mathcal{A}$ is any sequence, then the decreasing sequence of right ideals $(a_1 a_2 \cdots a_n \mathcal{A})^-$ eventually becomes constant.

We shall call a closed ideal with this property an *asymptotic ideal*. (Cusack's name "separating ideal" can lead to confusion.) Non-nilpotent asymptotic ideals are studied in depth in Cusack's thesis with the objective of showing that many (possibly all) Banach algebras fail to have non-nilpotent, radical asymptotic ideals. In particular, it is shown that if $\mathcal{I}$ is an asymptotic ideal of a Banach algebra $\mathcal{A}$, then the Baer radical of the algebra $\mathcal{I}$ is a closed, and therefore, nilpotent ideal of $\mathcal{A}$.

This detailed study shows that any of the following eight conditions imply the existence of a topologically simple radical Banach algebra:

(1) There is an asymptotic ideal with a non-nilpotent Jacobson radical.

(2) There is a homomorphism of one Banach algebra onto another which

has a non-nilpotent separating ideal.

(3) There is a discontinuous homomorphism of a Banach algebra onto a semiprime Banach algebra.

(4) There is a semiprime Banach algebra which fails to have unique norm topology.

(5) There is a derivation of some Banach algebra with a non-nilpotent separating ideal.

(6) There is a discontinuous derivation of a semiprime Banach algebra.

(7) There is a derivation of a commutative Banach algebra with range not contained in the Jacobson radical.

(8) There is a derivation of some Banach algebra which fails to map each primitive ideal into itself.

Whether such an algebra exists remains unknown. However, it is shown that (2) and (3) are equivalent and that (6), (7) or (8) implies (5).

6.4.26 Derivations on radical commutative Banach algebras We give some interesting examples of continuous non-zero derivations on certain radical commutative Banach algebras.

Let $\mathcal{A}$ be the disc algebra with convolution multiplication which was described in Example 4.8.3. Define $\delta\colon \mathcal{A} \to \mathcal{A}$ by

$$(\delta f)(z) = zf(z) \qquad \forall \, f \in \mathcal{A}.$$

Any $f,\, g \in \mathcal{A}$ and $z \in D$ satisfy

$$
\begin{aligned}
\delta(f * g)(z) &= zf * g(z) = z \int_0^z f(w)g(z-w)dw \\
&= \int_0^z wf(w)g(z-w)dw + \int_0^z f(w)(z-w)f(z-w)dw \\
&= (\delta(f) * g)(z) + (f * \delta(g))(z)
\end{aligned}
$$

so that δ is a derivation on $\mathcal{A}$ which is obviously bounded with norm one.

Note that the same construction can be applied to the convolution algebra $L^1([0,1])$ introduced in Example 3.4.10. In fact, each of these examples is a special case of the general results on derivations of Banach algebras of power series presented in Sandy Grabiner [1974]. This paper determines all the derivations on each Banach algebra in a rather large class of radical Banach algebras of power series. See Example 3.4.14 above for notation. If h is a formal power series with zero constant term, then $g \mapsto g \circ h$ (formal composition of power series) defines an endomorphism of $\mathbb{C}[[z]]$ and if k is any formal power series then $g \mapsto kg'$ (formal differentiation followed by formal power series multiplication) is a derivation of $\mathbb{C}[[z]]$. Suppose $\mathcal{A}$ is a Banach algebra of power series invariant under these operations. Then it is fairly easy to see that the restrictions of these maps are continuous endomorphisms and derivations of $\mathcal{A}$ and that all continuous endomorphisms and derivations of $\mathcal{A}$ have this form. These simple results are extended in various ways in Grabiner [1974]. Since all the automorphisms and derivations of various Banach algebras of power series are known to be continuous (Example 6.1.19 and Loy [1969]), many of the results in the cited reference do not need continuity in their hypotheses.

6.5   Jordan Derivations

We briefly investigate Jordan derivations. **Warning:** In this section we use the notation $a \circ b$ to denote the *Jordan product* $a \circ b = ab + ba$ rather than the quasi-product.

6.5.1   Definition   Let $\mathcal{A}$ be an algebra. A map $\delta \colon \mathcal{A} \to \mathcal{A}$ is called a *Jordan derivation* if it is linear and satisfies

$$\delta(a \circ b) = \delta(a) \circ b + a \circ \delta(b) \qquad \forall \, a, \, b \in \mathcal{A}.$$

Obviously a derivation is a Jordan derivation. In fact there are not many examples of Jordan derivations which are not derivations. A number of results help to explain this scarcity. We will give the elementary algebraic proof due to Cusack [1975], [1976] that every Jordan derivation on a semiprime algebra is a derivation. Although this proof is straightforward, it is long and highly computational. Hence before giving it we will prove the much weaker result of Sinclair [1970b] that a continuous Jordan derivation on a semisimple Banach algebra is a derivation. Unlike the proof given here, Sinclair's original proof depends on Herstein's theorem [1957] that any Jordan derivation on a prime algebra is a derivation.

6.5.2   Proposition   *A bounded Jordan derivation of a semisimple normed algebra is a derivation.*

Proof Let δ' be a bounded Jordan derivation on a normed algebra $\mathcal{A}'$. Let $\mathcal{A}$ be the completion of $\mathcal{A}'$ and let δ be the closure of δ' in $\mathcal{A} \times \mathcal{A}$. Then δ is a bounded Jordan derivation on $\mathcal{A}$. As in the proof of Proposition 6.4.16, we find $\delta^n(a^n) + \mathcal{P} = n!\delta(a)^n + \mathcal{P}$ for any $n \in \mathbb{N}$, any $a \in \mathcal{A}$ and any ideal $\mathcal{P}$ of $\mathcal{A}$. Hence, as before, $(\delta(b) + \mathcal{P})/\mathcal{P}$ is topologically nilpotent in $\mathcal{A}/\mathcal{P}$.

Next we show that each primitive ideal of $\mathcal{A}$ is invariant under δ. Suppose $\mathcal{P}$ is a primitive ideal of $\mathcal{A}$, b is an element of $\mathcal{P}$ and $\delta(b)$ does not belong to $\mathcal{P}$. Let T be an irreducible representation of $\mathcal{A}$ on a linear space $\mathcal{X}$ with kernel $\mathcal{P}$. Choose $x \in \mathcal{X}$ so that $y = T_{\delta(b)}x$ is not zero. Since $\delta(b) + \mathcal{P}$ and $T_{\delta(b)}$ have $\{0\}$ as spectrum, x and y are linearly independent.

Thus Theorem 4.2.13 shows that there is an element $a \in \mathcal{A}$ so that $T_a x = x$ and $T_a y = x - y$. Then $T_{\delta(b)a + a\delta(b)} x = T_{\delta(b)} x + T_a y = y + x - y = x$, so that $T_{\delta(b)a + a\delta(b)}$ is not topologically nilpotent. On the other hand, $\delta(b)a + a\delta(b) + \mathcal{P} = \delta(ba + ab) + \mathcal{P}$ shows that $T_{\delta(b)a + a\delta(b)}$ is topologically nilpotent. This contradiction proves that $\delta(b)$ belongs to $\mathcal{P}$. Hence each primitive ideal is δ-invariant. (This paragraph of the proof is valid for any spectral normed algebra $\mathcal{A}$.)

A Jordan derivation satisfies Leibniz' formula relative to Jordan multiplication, i.e., $\delta^n(a \circ b) = \sum_{k=0}^{n} \binom{n}{k} \delta^k(a) \circ \delta^{n-k}(b)$, where $a \circ b = ab + ba$. Hence for any $t \in \mathbb{R}$, $e^{t\delta}(a \circ b) = e^{t\delta}(a) \circ e^{t\delta}(b)$ holds for all $a, b \in \mathcal{A}$. Each primitive ideal $\mathcal{P}$ of $\mathcal{A}$ is invariant under $e^{t\delta}$ so we define a norm continuous one-parameter group of Jordan automorphisms $\{\alpha_t : t \in \mathbb{R}\}$ of $\mathcal{A}/\mathcal{P}$ by

$$\alpha_t(a + \mathcal{P}) = e^{t\delta}(a) + \mathcal{P}.$$

Theorem 6.3.7 shows that each α_t is either an automorphism or an anti-automorphism of $\mathcal{A}/\mathcal{P}$. By continuity, all α_t for t sufficiently close to zero must be automorphisms. Now the group property shows that $\{\alpha_t : t \in \mathbb{R}\}$ is a group of automorphisms. Hence its generator is a derivation. Thus

$$\delta(ab) - \delta(a)b - a\delta(b)$$

belongs to each primitive ideal of $\mathcal{A}$. By semisimplicity, δ is a derivation. Hence δ' is also a derivation. $\qquad\square$

Note that the second paragraph of the above proof shows that any continuous Jordan derivation on a spectral normed algebra leaves the primitive ideals invariant.

In order to organize the computations for Cusack's theorem (which are taken directly from [1976]) we will adopt two notational conventions:

$$[a, b, c] = abc + cba \qquad \forall\, a, b, c \in \mathcal{A}$$

$$\delta(a, b) = \delta(ab) - [\delta(a)b + a\delta(b)] \qquad \forall\, a, b, c \in \mathcal{A}$$

where δ is a Jordan derivation. Note that δ is a derivation exactly when $\delta(a, b)$ vanishes for all $a, b \in \mathcal{A}$. Our first lemma does not involve Jordan derivations. From Theorem 4.1.10, recall that if $\mathcal{P}$ is a prime ideal, then $a\mathcal{A}b \subseteq \mathcal{P}$ implies either $a \in \mathcal{P}$ or $b \in \mathcal{P}$. As a slight extension we note that $a\mathcal{A}b\mathcal{A} \subseteq \mathcal{P}$ implies either $a \in \mathcal{P}$ or $b \in \mathcal{P}$. The proof is the same.

6.5.3 Lemma Let $\mathcal{P}$ be a prime ideal in an algebra $\mathcal{A}$.
(a) If $a, b \in \mathcal{A}$ satisfy

$$[a, c]b \in \mathcal{P} \qquad \forall\, c \in \mathcal{A},$$

then either b belongs to $\mathcal{P}$ or $[a, c]$ belongs to $\mathcal{P}$ for all $c \in \mathcal{A}$.

(b) *If a, $b \in \mathcal{A}$ satisfy*

$$[a, c, b] \in \mathcal{P} \qquad \forall\, c \in \mathcal{A},$$

then either $a \in \mathcal{P}$ or $b \in \mathcal{P}$.

Proof (a): Suppose $[a, c] \notin \mathcal{P}$ for some $c \in \mathcal{A}$. Any $d \in \mathcal{A}$ satisfies $[a, c]db = [acd - cda]b + c(da - ad)b = [a, cd]b - c[a, d]b \in \mathcal{P}$. Hence $[a, c]\mathcal{A} \not\subseteq \mathcal{P}$ implies $b \in \mathcal{P}$.

(b): For arbitrary d, $e \in \mathcal{A}$, we get $adaeb + bdaea \in \mathcal{P}$. Then $adb + bda \in \mathcal{P}$ and $bea + aeb \in \mathcal{P}$ imply $2adbea \in \mathcal{P}$. This gives $a\mathcal{A}b\mathcal{A}a \subseteq \mathcal{P}$ from which we conclude either $a \in \mathcal{P}$ or $b \in \mathcal{P}$. □

The second lemma, due to Herstein [1957], is also computational.

6.5.4 Lemma *If δ is a Jordan derivation on an algebra $\mathcal{A}$, all a, b, $c \in \mathcal{A}$ satisfy:*

(a) $\delta(a^2) = a\delta(a) + \delta(a)a$;
(b) $\delta(aba) = \delta(a)ba + a\delta(b)a + ab\delta(a)$;
(c) $\delta([a, b, c]) = [\delta(a), b, c] + [a, \delta(b), c] + [a, b, \delta(c)]$;
(d) $\delta(a, b)[a, b] = 0$;
(e) $[a, b]\delta(a, b) = 0$;
(f) $[c, b]\delta(a, b) + [a, b]\delta(c, b) = 0$;
(g) $[c, a]\delta(a, b) + [a, b]\delta(c, a) = 0$.

Proof (a): All $a \in \mathcal{A}$ satisfy

$$2\delta(a^2) = \delta(a \circ a) = a \circ \delta(a) + \delta(a) \circ a = 2(a\delta(a) + \delta(a)a).$$

(b): Note the identity

$$d \circ (d \circ e) = d \circ (de + ed) = d^2 \circ e + 2ded \qquad \forall\, d, e \in \mathcal{A}.$$

From this we obtain

$$\begin{aligned}
2\delta(aba) &= \delta(a \circ (a \circ b)) - \delta(a^2 \circ b) \\
&= a \circ \delta(a \circ b) + \delta(a) \circ (a \circ b) - a^2 \circ \delta(b) - \delta(a^2) \circ b \\
&= 2\delta(a)ba + 2a\delta(b)a + 2ab\delta(a).
\end{aligned}$$

(c): Replace a by $a + c$ in (b). This gives

$$\delta((a + c)b(a + c)) = \delta(aba) + [\delta(a), b, c] + [a, \delta(b), c] + [a, b, \delta(c)] + \delta(cbc).$$

Then (c) follows by subtracting $\delta(aba + cbc)$ from each side.

(d): From (c) we get $\delta([a, b, ab]) = [\delta(a), b, ab] + [a, \delta(b), ab] + [a, b, \delta(ab)]$. The identity $[a, b, ab] = (ab)^2 + ab^2a$, together with (a) and (b) implies $\delta([a, b, ab]) = ab\delta(ab) + \delta(ab)ab + \delta(a)b^2a + a\delta(b^2)a + ab^2\delta(a) = ab\delta(ab) +$

$\delta(ab)ab + \delta(a)b^2a + ab\delta(b)a + a\delta(b)ba + ab^2\delta(a)$. By comparing the two expressions for $\delta([a, b, ab])$, we get $0 = (\delta(ab) - \delta(a)b - a\delta(b))ab - (\delta(ab) - a\delta(b) - \delta(a)b)ba = \delta(a, b)[a, b]$.

(e): Since δ is obviously a Jordan derivation on the reverse algebra, this follows from (d) by reversing the product.

(f): From (e) we see $[a + c, b]\delta(a + c, b) = 0$. Combining this with $[a + c, b] = [a, b] + [c, b]$, $\delta(a + c, b) = \delta(a, b) + \delta(c, b)$, and $[a, b]\delta(a, b) = 0 = [c, b]\delta(c, b)$, we have $[c, b]\delta(a, b) + [a, b]\delta(c, b) = 0$, as required.

(g): Interchange a and b in (f). This gives $[c, a]\delta(b, a) + [b, a]\delta(c, a) = 0$. But $\delta(a, b) + \delta(b, a) = \delta(a \circ b) - a \circ \delta(b) - \delta(a) \circ b = 0$ implies $\delta(b, a) = -\delta(a, b)$. Thus $[b, a] = [-a, b]$ gives the desired conclusion. $\qquad\square$

6.5.5 Lemma   *Let $\mathcal{P}$ be a prime ideal of the algebra $\mathcal{A}$, and let δ be a Jordan derivation on $\mathcal{A}$. If $[a, b]$ is in $\mathcal{P}$, then $\delta(a, b)$ is in $\mathcal{P}$.*

Proof We first consider the case when $\mathcal{A}/\mathcal{P}$ is noncommutative. We assume $[a, b] \in \mathcal{P}$. Lemma 6.5.4(f) and (g) give $[c, b]\delta(a, b) \in \mathcal{P}$ and $[c, a]\delta(a, b) \in \mathcal{P}$ for all $c \in \mathcal{A}$. It follows from Lemma 6.5.3(a) that if there is an element $c \in \mathcal{A}$ satisfying either $[c, a] \notin \mathcal{P}$ or $[c, b] \notin \mathcal{P}$, then $\delta(a, b) \in \mathcal{P}$ holds. Suppose that $[c, a]$ and $[c, b]$ are in $\mathcal{P}$ for all c in A. By noncommutativity, let c and e be any elements of $\mathcal{A}$ satisfying $[c, e] \notin \mathcal{P}$. Then, $[e, b]$ and $[a + e, b]$ are in $\mathcal{P}$ and $[c, e]$ and $[c, a + e]$ are not in $\mathcal{P}$. It follows that $\delta(e, b)$ and $\delta(a + e, b)$ are in $\mathcal{P}$, and that $\delta(a, b) = \delta(a + e, b) - \delta(e, b)$ is therefore in $\mathcal{P}$.

Now suppose that $\mathcal{A}/\mathcal{P}$ is commutative. Lemma 6.5.4(d) and (e) give $[a, b] \circ \delta(a, b) = 0$ and therefore, since δ is a Jordan derivation, $0 = \delta([a, b]) \circ \delta(a, b) + [a, b] \circ \delta(\delta(a, b))$. Since $\mathcal{A}/\mathcal{P}$ is commutative $\delta([a, b]) \circ \delta(a, b)$, and therefore $2\delta([a, b])\delta(a, b)$, are in $\mathcal{P}$. Since $\mathcal{A}/\mathcal{P}$ has no divisors of zero, either $\delta(a, b)$ or $\delta([a, b])$ is in $\mathcal{P}$. If $\delta([a, b])$ is in $\mathcal{P}$, we get $2\delta(a, b) = \delta(ab + ba) + \delta(ab - ba) - 2(a\delta(b) + \delta(a)b) = \delta([a, b]) + (a \circ \delta(b) - 2\delta(a)b) + (\delta(a) \circ b - 2\delta(a)b) \in \mathcal{P}$. Thus in either case, $\delta(a, b)$ belongs to $\mathcal{P}$. $\qquad\square$

The final lemma is concerned with the case in which $[a, b]$ does not belong to a prime ideal $\mathcal{P}$.

6.5.6 Lemma   *Let δ be a Jordan derivation on an algebra $\mathcal{A}$. All a, b, $c \in \mathcal{A}$ satisfy:*

(a) $\delta([a, b]c) = (\delta(ab) - b\delta(a) - \delta(b)a)c + [a, b]\delta(c) + c\delta(a, b)$;

(b) $\delta(c[a, b]) = c(\delta(ab) - b\delta(a) - \delta(b)a) + \delta(c)[a, b] + \delta(a, b)c$;

(c) $[[a, b], c, \delta(a, b)] = 0$.

Proof (a): The identity $[a, b]c = [a, b, c] - c \circ (ba)$ and Lemma 6.5.4(c) give $\delta([a, b]c) = \delta([a, b, c]) - \delta(c \circ (ba)) = [\delta(a), b, c] + [a, \delta(b), c] + [a, b, \delta(c)] - c\delta(ba) - \delta(ba)c - \delta(c)ba - ba\delta(c) = c(b\delta(a) + \delta(b)a - \delta(ba)) + (\delta(a)b + a\delta(b) - \delta(ba))c - \delta(c)ba - ba\delta(c) + ab\delta(c) + \delta(c)ba$.

Since δ is a Jordan derivation, we have $\delta(a)b + a\delta(b) - \delta(ba) = \delta(ab) - \delta(b)a - b\delta(a)$ and $b\delta(a) + \delta(b)a - \delta(ba) = \delta(a,b)$, and therefore $\delta([a,b]c) = c\delta(a,b) + (\delta(ab) - b\delta(a) - \delta(b)a)c + [a,b]\delta(c)$ as required.

(b): This may be proved by a similar argument, or by reversing the product.

(c): The idea of the proof is to evaluate $\delta([a,b]c[a,b])$ in two different ways and then obtain the required result from Lemma 6.5.4(d) and (e).

Replacing c by $[a,b]c$ in (b), we get $\delta([a,b]c[a,b]) = \delta(a,b)[a,b]c + \delta([a,b]c)[a,b] + [a,b]c(\delta(ab) - b\delta(a) - \delta(b)a) = [a,b]\delta(c)[a,b] + [\delta(ab) - b\delta(a) - \delta(b)a, c, [a,b]]$, by (a) and Lemma 6.5.4(d) and (e). We now observe $\delta([a,b]) = (\delta(ab) - b\delta(a) - \delta(b)a) + \delta(a,b)$. This implies

$$\delta([a,b]c[a,b]) = \delta([a,b]c[a,b]) - [\delta(a,b), c, [a,b]],$$

by Lemma 6.5.4(b). The result now follows by subtracting $\delta([a,b]c[a,b])$ from both sides.   □

Cusack's theorem is actually a little stronger than so far noted.

6.5.7 Theorem *A Jordan derivation δ on any algebra A satisfies*

$$\delta(ab) - [\delta(a)b + a\delta(b)] \in A_B \qquad \forall\, a,\, b \in A$$

and hence induces a derivation on A/A_B. In particular every Jordan derivation on a semiprime algebra is a derivation.

Proof Let a and b be any elements of A and let P be any prime ideal of A. Lemma 6.5.6 shows $[[a,b], c, \delta(a,b)] = 0$, for all c in A. According to Lemma 6.5.3(b), either $[a,b]$ or $\delta(a,b)$ belong to P. But Lemma 6.5.5 shows that $[a,b] \in P$ implies $\delta(a,b) \in P$. Since A_B is the intersection of all the prime ideals of A, this completes the proof.   □

7

Structure Spaces

Let $\mathcal{A}$ be an algebra. In this chapter we will study the space $\langle\, P_{\mathcal{A}}\,/\,\Pi_{\mathcal{A}}\,/\,\Xi_{\mathcal{A}}\,\rangle$ of $\langle\,$ prime / primitive / maximal modular $\rangle$ ideals of $\mathcal{A}$ as a topological space under the hull–kernel or Jacobson topology. For certain classes of algebras $\mathcal{A}$, e.g., completely regular algebras (Section 7.2) and strongly harmonic algebras (Section 7.4), we will show that the subdirect product representation relative to $\Xi_{\mathcal{A}}$ (introduced in Definition 1.3.3 and Section 4.6) yields significant information about $\mathcal{A}$. Section 7.3 deals with more detailed questions in ideal theory revolving around primary ideals. We also consider central and weakly central algebras and show that they are completely regular under fairly weak additional hypotheses.

7.1   The Hull–Kernel Topology

In Section 3.2 we introduced the hull–kernel topology on the Gelfand space $\Gamma_{\mathcal{A}}$ of a commutative Banach algebra $\mathcal{A}$. It is comparatively little used except in the case of completely regular commutative spectral algebras where it is Hausdorff and coincides with the Gelfand topology. In the commutative case, Proposition 3.1.3 shows that the Gelfand space of $\mathcal{A}$ can be identified with the set $\Xi_{\mathcal{A}}$ of maximal modular ideals of $\mathcal{A}$, and Theorem 4.1.9 shows that the latter set coincides with the set $\Pi_{\mathcal{A}}$ of primitive ideals.

In a noncommutative algebra $\mathcal{A}$, the set $P_{\mathcal{A}}$ of prime ideals and its subsets, $\Pi_{\mathcal{A}}$ and $\Xi_{\mathcal{A}}$, can each be given the hull–kernel topology. Again, this topology seems to be of comparatively little use unless further restrictions are placed on the algebra. Generally, these restrictions need to be sufficient to force the topology to be Hausdorff. The most important such restriction is the notion of complete regularity for noncommutative Banach algebras, which was introduced by Alfred B. Willcox in his thesis published as [1956a] and further elaborated by Charles E. Rickart in his book [1960]. Another is the notion of strongly harmonic algebras introduced by Kwangil Koh [1972] building on ideas of Silviu Teleman [1971]. We develop both of these theories in this chapter.

Given any collection $\tilde{P}$ of ideals in an algebra $\mathcal{A}$, there is a corresponding subdirect product decomposition. This can be thought of as the function $a \mapsto \hat{a}$ given by

$$\hat{a}(\mathcal{I}) = a + \mathcal{I} \in \mathcal{A}/\mathcal{I} \qquad \forall\ \mathcal{I} \in \tilde{P}$$

619

which maps $\mathcal{A}$ into the full direct product $\prod_{\mathcal{I} \in \tilde{P}}(\mathcal{A}/\mathcal{I})$. Such subdirect product representations ordinarily give little useful information about the algebra even when they are faithful. One main purpose of this chapter is to investigate situations in which the representation in $\prod_{\mathcal{M} \in \Xi_{\mathcal{A}}}(\mathcal{A}/\mathcal{M})$, and a closely related representation, do become useful.

The hull–kernel topology on the space of primitive ideals of a noncommutative ring was first systematically studied by Nathan Jacobson [1945b]. However, Marshall Harvey Stone [1937] had used a very similar idea in his theory of structure spaces of Boolean rings, and Gelfand and Georgi E. Šilov [1941] had explicitly considered the hull–kernel topology on the carrier space of a commutative Banach algebra. Most of the ideas of this chapter were first investigated in the commutative case, particularly in Šilov [1947a].

For a general algebra $\mathcal{A}$, the set $P_{\mathcal{A}}$ of prime ideals of $\mathcal{A}$ seems to be the natural setting for the hull–kernel topology. However since prime ideals are not necessarily closed in a Banach algebra, they play a comparatively small role in the theory of this work. It seems likely that useful classes of closed prime ideals larger than the class of primitive ideals will eventually be found. Partly for this reason, we will introduce the hull–kernel topology for arbitrary subsets of the set $P_{\mathcal{A}}$ of prime ideals of an algebra $\mathcal{A}$. The subsets of chief interest are $P_{\mathcal{A}}$ itself, $\Pi_{\mathcal{A}} = \{$ primitive ideals of $\mathcal{A}\}$ and $\Xi_{\mathcal{A}} = \{$ maximal modular ideals of $\mathcal{A}\}$. Theorems 4.1.8 and 4.1.9 show the inclusions $\Xi_{\mathcal{A}} \subseteq \Pi_{\mathcal{A}} \subseteq P_{\mathcal{A}}$. Some of the following notation was introduced in a preliminary way in Chapter 4.

7.1.1 Definition   Let $\mathcal{A}$ be an algebra and let $\tilde{P}_{\mathcal{A}}$ be a nonempty subset of the set $P_{\mathcal{A}}$ of prime ideals of $\mathcal{A}$. For each subset $\mathcal{S}$ of $\mathcal{A}$, the *hull of $\mathcal{S}$ relative to $\tilde{P}_{\mathcal{A}}$* is the set

$$\tilde{h}(\mathcal{S}) = \{\mathcal{P} \in \tilde{P}_{\mathcal{A}} : \mathcal{S} \subseteq \mathcal{P}\}.$$

A subset of $\tilde{P}_{\mathcal{A}}$ which has the form $\tilde{h}(\mathcal{S})$ for some $\mathcal{S} \subseteq \mathcal{A}$ is called a *hull* in $\tilde{P}_{\mathcal{A}}$. When $\tilde{P}_{\mathcal{A}}$ is $\langle\, P_{\mathcal{A}} \,/\, \Pi_{\mathcal{A}} \,/\, \Xi_{\mathcal{A}} \,\rangle$, we write $\langle\, h^P \,/\, h^\Pi \,/\, h^\Xi \,\rangle$ instead of $\tilde{h}$. For each subset B of $\tilde{P}_{\mathcal{A}}$, the *kernel of B* is the set

$$k(B) = \cap\{\mathcal{P} : \mathcal{P} \in B\}.$$

The next proposition shows that $\tilde{h}$ and k set up a Galois correspondence between the subsets of $\tilde{P}_{\mathcal{A}}$ and the subsets of $\mathcal{A}$.

7.1.2 Proposition   *Let $\mathcal{A}$ be an algebra and let $\tilde{P}_{\mathcal{A}}$ be a nonempty subset of $P_{\mathcal{A}}$. Consider $\mathcal{S}_1 \subseteq \mathcal{S}_2 \subseteq \mathcal{A}$ and $B_1 \subseteq B_2 \subseteq \tilde{P}_{\mathcal{A}}$. Then $k(B_1)$ is an ideal of $\mathcal{A}$ and these subsets satisfy*

$$\tilde{h}(\mathcal{S}_2) \;\subseteq\; \tilde{h}(\mathcal{S}_1) \qquad k(B_2) \subseteq k(B_1) \tag{1}$$

$$\mathcal{S}_1 \subseteq \tilde{h}k(\mathcal{S}_1) \qquad B_1 \subseteq \tilde{h}k(B_1) \qquad (2)$$

$$\tilde{h}k\tilde{h}(\mathcal{S}_1) = \tilde{h}(\mathcal{S}_1) \qquad k\tilde{h}k(B_1) = k(B_1). \qquad (3)$$

Hence, a subset B of $\tilde{P}_A$ is a hull if and only if it satisfies $B = \tilde{h}k(B)$. Any family $\{\mathcal{S}_\alpha : \alpha \in A\}$ of subsets of $\mathcal{A}$ and any family $\{B_\alpha : \alpha \in A\}$ of subsets of $\tilde{P}_A$ satisfy:

$$\tilde{h}(\bigcup_{\alpha \in A} \mathcal{S}_\alpha) = \tilde{h}(\sum_{\alpha \in A} \mathcal{S}_\alpha) = \tilde{h}(\mathrm{span}(\bigcup_{\alpha \in A} \mathcal{S}_\alpha)) = \bigcap_{\alpha \in A} \tilde{h}(\mathcal{S}_\alpha) \qquad (4)$$

$$k(\bigcup_{\alpha \in A} B_\alpha) = \bigcap_{\alpha \in A} k(B_\alpha). \qquad (5)$$

Any finite set $\{\mathcal{I}_1, \mathcal{I}_2, \ldots, \mathcal{I}_n\}$ of ideals of $\mathcal{A}$ satisfies

$$\tilde{h}(\bigcap_{j=1}^{n} \mathcal{I}_j) = \tilde{h}(\mathcal{I}_1 \mathcal{I}_2 \cdots \mathcal{I}_n) = \bigcup_{j=1}^{n} \tilde{h}(\mathcal{I}_j). \qquad (6)$$

Proof These results are either obvious or are established by the same easy arguments used for Proposition 3.2.6. For (6), we note that the last equation is immediate since each ideal in $\tilde{P}$ is prime. Also, the first set is included in the second by (1) while it includes the third set from the definition of a hull. □

7.1.3 Theorem *Let $\mathcal{A}$ be an algebra and let $\tilde{P}_A$ be a nonempty subset of P_A. Then the family*

$$\{\tilde{P}_A \setminus H : H \text{ is a hull in } \tilde{P}_A\}$$

of complements of hulls is a topology for $\tilde{P}_A$ which satisfies the T_0 separation axiom.

Proof Let $\mathcal{T}$ be the family which is asserted to be a topology. The equations $\tilde{h}(\mathcal{A}) = \emptyset$ and $\tilde{h}(\{0\}) = \tilde{P}_A$ show that $\tilde{P}_A$ and $\emptyset$ belong to $\mathcal{T}$. Proposition 7.1.2 shows that $\mathcal{T}$ is closed under arbitrary unions and under (finite) intersections. The T_0 separation axiom is the statement that if two points $\mathcal{P}_1$ and $\mathcal{P}_2$ are distinct, then at least one is not in the closure of the other. This is obvious since $\mathcal{P}_1 \in \tilde{h}k(\{\mathcal{P}_2\})$ is equivalent to $\mathcal{P}_2 \subseteq \mathcal{P}_1$. □

7.1.4 Definition Let $\mathcal{A}$ be an algebra and let $\tilde{P}_A$ be a nonempty subset of P_A. The topology described above is called the *hull–kernel* (or *Jacobson*) *topology* on $\tilde{P}_A$. Let $\langle\, P_A \,/\, \Pi_A \,/\, \Xi_A \,\rangle$ be the topological spaces of all $\langle$ prime ideals / primitive ideals / maximal modular ideals $\rangle$ under the hull–kernel topology. This topological space will be called the $\langle$ *prime structure space / structure space / strong structure space* $\rangle$ of $\mathcal{A}$.

We have already noted that Theorems 4.1.8 and 4.1.9 show the inclusions $\Xi_A \subseteq \Pi_A \subseteq P_A$ as sets. The definition of the hull–kernel topology shows immediately that the topologies of Π_A and Ξ_A agree with their relativized topologies as subsets of P_A or Π_A.

7.1.5 Proposition *For any algebra A, Ξ_A satisfies the T_1 separation axiom. Furthermore, if each $\langle$ primitive / prime $\rangle$ ideal of A is modular or is included in a maximal modular ideal (e.g., if A is unital) and if $\langle$ Π_A / P_A $\rangle$ satisfies the T_1 separation axiom, then $\langle$ Π_A / P_A $\rangle$ equals Ξ_A.*

Proof Since $\mathcal{P}_1 \in \{\mathcal{P}_2\}^-$ is equivalent to $\mathcal{P}_1 \supseteq \mathcal{P}_2$, this is obvious. □

7.1.6 Proposition *Let A be an algebra. Let $\tilde{P}_A$ be a nonempty subset of P_A which includes Ξ_A. If $\mathcal{I}$ is a modular ideal in A, then $\tilde{h}(\mathcal{I})$ is compact. In particular, $\tilde{P}_A$ itself is compact if A is unital.*

Proof We establish the finite intersection property for $\tilde{h}(\mathcal{I})$. Let $\{H_\alpha : \alpha \in A\}$ be a family of closed subsets of $\tilde{h}(\mathcal{I})$ satisfying $\cap_{\alpha \in A} H_\alpha = \emptyset$. Each $\alpha \in A$ satisfies $\mathcal{I} \subseteq k\tilde{h}(\mathcal{I}) \subseteq k(H_\alpha)$. Hence $\mathcal{R} = \sum_{\alpha \in A} k(H_\alpha)$ includes the modular ideal $\mathcal{I}$ and is therefore modular. Therefore, Theorem 2.4.6(b) shows that $\mathcal{R}$ is included in some $\mathcal{M} \in \Xi_A \subseteq \tilde{P}_A$ unless $\mathcal{R}$ equals A. Equation (4) of Proposition 7.1.2 shows $\tilde{h}(\sum_\alpha k(H_\alpha)) = \cap_{\alpha \in A} \tilde{h}k(H_\alpha) = \cap_{\alpha \in A} H_\alpha = \emptyset$. We therefore conclude $\mathcal{R} = A$. Hence the relative identity e for $\mathcal{I}$ can be written in the form $e = a_1 + \cdots + a_n$ for some $a_j \in k(H_{\alpha_j})$. Since any $a \in A$ can be written as $a = (1-e)a + a_1 a + \cdots + a_n a \in k(H_{\alpha_1}) + k(H_{\alpha_2}) + \cdots + k(H_{\alpha_n})$, we get $\cap_{j=1}^n H_{\alpha_j} = \tilde{h}(\sum_{j=1}^n k(H_{\alpha_j})) = \tilde{h}(A) = \emptyset$ as we wished to show.

The last statement of the proposition is an immediate consequence of the first conclusion. □

7.1.7 Theorem *Let A be an algebra, and let $\mathcal{I}$ be an ideal in A. The hull $\langle$ $h^P(\mathcal{I})$ / $h^\Pi(\mathcal{I})$ / $h^\Xi(\mathcal{I})$ $\rangle$ of $\mathcal{I}$ in $\langle$ P_A / Π_A / Ξ_A $\rangle$ is homeomorphic to $\langle$ $P_{A/\mathcal{I}}$ / $\Pi_{A/\mathcal{I}}$ / $\Xi_{A/\mathcal{I}}$ $\rangle$ under the map $\mathcal{P} \mapsto \mathcal{P}/\mathcal{I}$. Also, $\langle$ $P_A \backslash h^P(\mathcal{I})$ / $\Pi_A \backslash h^\Pi(\mathcal{I})$ / $\Xi_A \backslash h^\Xi(\mathcal{I})$ $\rangle$ is homeomorphic to $\langle$ $P_\mathcal{I}$ / $\Pi_\mathcal{I}$ / $\Xi_\mathcal{I}$ $\rangle$ under the map $\mathcal{P} \mapsto \mathcal{P} \cap \mathcal{I}$.*

Proof It is easy to see that $\mathcal{P} \mapsto \mathcal{P}/\mathcal{I}$ is a bijection of $\langle$ $h^P(\mathcal{I})$ / $h^\Pi(\mathcal{I})$ / $h^\Xi(\mathcal{I})$ $\rangle$ onto $\langle$ $P_{A/\mathcal{I}}$ / $\Pi_{A/\mathcal{I}}$ / $\Xi_{A/\mathcal{I}}$ $\rangle$, and this fact is included in Theorem $\langle$ 4.1.8 / 4.1.10 / 4.5.2 $\rangle$. If B is any subset of $\langle$ $h^P(\mathcal{I})$ / $h^\Pi(\mathcal{I})$ / $h^\Xi(\mathcal{I})$ $\rangle$ and B' is its image under this map, then $k(B')$ equals $k(B)/\mathcal{I}$. Hence any $\mathcal{P} \in \langle$ P_A / Π_A / Ξ_A $\rangle$ satisfies $\mathcal{P} \supseteq k(B)$ if and only if it satisfies $\mathcal{P}/\mathcal{I} \supseteq k(B')$. Therefore the map $\mathcal{P} \mapsto \mathcal{P}/\mathcal{I}$ is a homeomorphism of $\tilde{h}(\mathcal{I})$ in $\langle$ P_A / Π_A / Ξ_A $\rangle$ onto $\langle$ $P_{A/\mathcal{I}}$ / $\Pi_{A/\mathcal{I}}$ / $\Xi_{A/\mathcal{I}}$ $\rangle$.

The last sentence of Theorem 4.1.10 asserts that $\mathcal{P} \mapsto \mathcal{P} \cap \mathcal{I}$ is a bijection of $P_A \backslash h^P(\mathcal{I})$ onto $P_\mathcal{I}$. In order to prove the second half of the theorem for

$P_\mathcal{A}$, it only remains to show that this bijection $\mathcal{P} \mapsto \mathcal{P} \cap \mathcal{I}$ is a homeomorphism of $P_\mathcal{A} \backslash h^P(\mathcal{I})$ onto $P_\mathcal{I}$. We must show that for any $\mathcal{P} \in P_\mathcal{A} \backslash h^P(\mathcal{I})$ and any $B \subseteq P_\mathcal{A} \backslash h(\mathcal{I}), \mathcal{P} \supseteq k(B)$ is equivalent to $\mathcal{P} \cap \mathcal{I} \supseteq k\{\mathcal{P}' \cap \mathcal{I} : \mathcal{P}' \in B\}$. The implication $\mathcal{P} \supseteq k(B) \Rightarrow \mathcal{P} \cap \mathcal{I} \supseteq k\{\mathcal{P}' \cap \mathcal{I} : \mathcal{P}' \in B\}$ is obvious and the opposite implication follows since $(\cap_{\mathcal{P}' \in B}\mathcal{P}') \cap \mathcal{I} \subseteq \mathcal{P}$ and $\mathcal{I} \not\subseteq \mathcal{P}$ imply $\cap_{\mathcal{P}' \in B}\mathcal{P}' \subseteq \mathcal{P}$ by the prime nature of $\mathcal{P}$. Since $\Pi_\mathcal{A}$ and $\Xi_\mathcal{A}$ are homeomorphically embedded in $P_\mathcal{A}$, the map $\mathcal{P} \mapsto \mathcal{P} \cap \mathcal{I}$ restricted to $\langle \Pi_\mathcal{A} \backslash h^\Pi(\mathcal{I}) / \Xi_\mathcal{A} \backslash h^\Xi(\mathcal{I}) \rangle$ must be a homeomorphism onto its range. It thus remains only to prove that each $\langle$ primitive / maximal modular $\rangle$ ideal of $\mathcal{I}$ has the form $\mathcal{P} \cap \mathcal{I}$ for some $\langle$ primitive / maximal modular $\rangle$ ideal of $\mathcal{A}$. However, this follows immediately from Theorem 4.1.8 in the case of primitive ideals and from Theorem 4.5.2 in the case of maximal modular ideals. $\qquad\square$

7.1.8 Corollary *Let $\mathcal{A}$ be an algebra and $\mathcal{I}$ the ideal $\langle \mathcal{A}_B / \mathcal{A}_J / \mathcal{A}_M \rangle$. Then the map $\mathcal{P} \mapsto \mathcal{P}/\mathcal{I}$ is a homeomorphism of $\langle P_\mathcal{A} / \Pi_\mathcal{A} / \Xi_\mathcal{A} \rangle$ onto $\langle P_{\mathcal{A}/\mathcal{I}} / \Pi_{\mathcal{A}/\mathcal{I}} / \Xi_{\mathcal{A}/\mathcal{I}} \rangle$.*

Proof The choice of $\mathcal{I}$ gives $\langle h^P(\mathcal{I}) = P_\mathcal{A} / h^\Pi(\mathcal{I}) = \Pi_\mathcal{A} \ h^\Xi(\mathcal{I}) = \Xi_\mathcal{A} \rangle . \square$

7.1.9 Proposition *Let $\mathcal{A}$ and $\tilde{P}_\mathcal{A}$ be as described in* Definition 7.1.1 *and also assume $\Xi_\mathcal{A} \subseteq \tilde{P}_\mathcal{A}$. Any two ideals $\mathcal{I}_1$ and $\mathcal{I}_2$ of $\mathcal{A}$ satisfy:*

(a) *The sum $\mathcal{I}_1 + \mathcal{I}_2$ equals $\mathcal{A}$ if and only if $\mathcal{I}_1 + \mathcal{I}_2$ is modular and $\tilde{h}(\mathcal{I}_1)$ and $\tilde{h}(\mathcal{I}_2)$ are disjoint.*

(b) *If H is a hull in $\tilde{P}_\mathcal{A}$ with $k(H)$ modular and $\tilde{h}(\mathcal{I}_1) \cap H$ empty, then $\mathcal{I}_1$ contains an identity for $\mathcal{A}$ modulo $k(H)$.*

Proof (a): If $\mathcal{A} = \mathcal{I}_1 + \mathcal{I}_2$ holds, then $\mathcal{I}_1 + \mathcal{I}_2$ is trivially modular, and equation (4) of Proposition 7.1.2 shows that $\tilde{h}(\mathcal{I}_1)$ and $\tilde{h}(\mathcal{I}_2)$ are disjoint. Conversely, if $\tilde{h}(\mathcal{I}_1)$ and $\tilde{h}(\mathcal{I}_2)$ are disjoint, the same result shows that $\tilde{h}(\mathcal{I}_1 + \mathcal{I}_2)$ is empty. If $\mathcal{I}_1 + \mathcal{I}_2$ is modular, Theorem 2.4.6 then shows $\mathcal{I}_1 + \mathcal{I}_2 = \mathcal{A}$.

(b): Since $k(H) + \mathcal{I}_1$ includes $k(H)$, it is modular. Since $\tilde{h}k(H) \cap \tilde{h}(\mathcal{I}_1) = H \cap \tilde{h}(\mathcal{I}_1)$ is empty, (a) shows $k(H) + \mathcal{I}_1 = \mathcal{A}$. Let e be a relative identity for $k(H)$. Write $e = c + d$ with $c \in k(H)$ and $d \in \mathcal{I}_1$. Then d is a relative identity for $k(H)$ since $d - e = c$ belongs to $k(H)$. $\qquad\square$

The next result has important consequences in the next chapter. Note that $\Pi_\mathcal{A}$ could be replaced with any set $\tilde{P}_\mathcal{A}$ of prime ideals (each one σ-closed for part (b)) satisfying $\tilde{P}_\mathcal{A} \supseteq \Pi_\mathcal{A}$.

7.1.10 Proposition *Let $\mathcal{A}$ be an algebra, let $\mathcal{I}$ be an ideal of $\mathcal{A}$ and let a be an element of $\mathcal{A}$. In the following statements, the word "quasi-invertible" can be replaced by "invertible" if $\mathcal{A}$ is unital.*

(a) *The element $a + \mathcal{I}$ is quasi-invertible in $\mathcal{A}/\mathcal{I}$ if and only if $a + k\,h^\Pi(\mathcal{A})$ is quasi-invertible in $\mathcal{A}/k\,h^\Pi(\mathcal{I})$.*

(b) *If σ is a spectral semi-norm on $\mathcal{A}$, then $a + \mathcal{I}$ is quasi-invertible in $\mathcal{A}/\mathcal{I}$ if and only if $a + \overline{\mathcal{I}}$ is quasi-invertible in $\mathcal{A}/\overline{\mathcal{I}}$ where $\overline{\mathcal{I}}$ is the σ-closure of $\mathcal{I}$.*

Proof Note that $kh^{\Pi}(\mathcal{I})/\mathcal{I}$ is naturally isomorphic to $(\mathcal{A}/\mathcal{I})_J$ by the definition of the Jacobson radical. Hence (a) follows from Theorem 4.3.6. If σ is a spectral semi-norm, its quotient norm $\tilde{\sigma}$ on $\mathcal{A}/\mathcal{I}$ is also spectral, so Corollary 4.3.10 implies $\overline{\mathcal{I}}/\mathcal{I} \subseteq (\mathcal{A}/\mathcal{I})_J$. Hence (b) follows from (a).   □

Theorem 7.1.7 shows that if $\mathcal{I}$ is an ideal in an algebra $\mathcal{A}$, then the map $\mathcal{B} \mapsto \mathcal{B} \cap \mathcal{I}$ is a homeomorphism of $\langle\, P_{\mathcal{A}} \setminus h^P(\mathcal{I}) \,/\, \Pi_{\mathcal{A}} \setminus h^{\Pi}(\mathcal{I}) \,/\, \Xi_{\mathcal{A}} \setminus h^{\Xi}(\mathcal{I}) \,\rangle$ onto $\langle\, P_{\mathcal{I}} \,/\, \Pi_{\mathcal{I}} \,/\, \Xi_{\mathcal{I}} \,\rangle$. The same results hold for certain subsets other than ideals. The next instance is essentially due to Irving Kaplansky [1950]. Proposition 7.2.14 gives another instance due to Kaplansky [1949a]. We do not know whether the corresponding results hold for prime ideals.

7.1.11 Proposition *Let $\mathcal{A}$ be an algebra, and let e be an idempotent in $\mathcal{A}$. The map $\mathcal{P} \mapsto \mathcal{P} \cap e\mathcal{A}e$ is a homeomorphism of $\langle\, \Pi_{\mathcal{A}} \setminus h^{\Pi}(e\mathcal{A}e) \,/\, \Xi_{\mathcal{A}} \setminus h^{\Xi}(e\mathcal{A}e) \,\rangle$ onto $\langle\, \Pi_{e\mathcal{A}e} \,/\, \Xi_{e\mathcal{A}e} \,\rangle$.*

Proof If $\mathcal{P}$ is a primitive ideal in $\Pi_{\mathcal{A}} \setminus h^{\Pi}(e\mathcal{A}e)$, let T be an algebraically irreducible representation of $\mathcal{A}$ on $\mathcal{X}$ satisfying $\ker(T) = \mathcal{P}$. Then T_e is nonzero and each non-zero $y \in T_e\mathcal{X}$ satisfies $T_{e\mathcal{A}e}y = T_eT_{\mathcal{A}}y = T_e\mathcal{X}$. Hence $b \mapsto T_b|_{T_e\mathcal{X}}$ is an irreducible representation of $\mathcal{A}$ on $T_e\mathcal{X}$. The kernel of this representation is obviously $\mathcal{P} \cap e\mathcal{A}e$. Hence the map $\mathcal{P} \mapsto \mathcal{P} \cap e\mathcal{A}e$ maps $\Pi_{\mathcal{A}} \setminus h^{\Pi}(e\mathcal{A}e)$ into $\Pi_{e\mathcal{A}e}$. Let $\mathcal{P}$ be a primitive ideal in $\Pi_{e\mathcal{A}e}$ and let $\mathcal{M}$ be a maximal modular left ideal of $e\mathcal{A}e$ satisfying $\mathcal{P} = \mathcal{M} : e\mathcal{A}e$ (Definition 4.1.7 and Theorem 4.1.8). Define $\overline{\mathcal{M}}$ to be the modular left ideal $\overline{\mathcal{M}} = \mathcal{A}\mathcal{M} + \mathcal{A}(1-e)$ of $\mathcal{A}$ which is proper since it satisfies $e\overline{\mathcal{M}}e \subseteq \mathcal{M} \neq e\mathcal{A}e$. Therefore, there is a maximal modular left ideal $\mathcal{L}$ of $\mathcal{A}$ which includes $\overline{\mathcal{M}}$. Let $\overline{\mathcal{P}}$ be the primitive ideal $\mathcal{L} : \mathcal{A}$ of $\mathcal{A}$. We wish to show $\overline{\mathcal{P}} \cap e\mathcal{A}e = \mathcal{P}$. Any $b \in \mathcal{P}$ satisfies $b\mathcal{A}e = be\mathcal{A}e \subseteq \mathcal{M} \subseteq \mathcal{L}$ and $b\mathcal{A}(1 - e) \subseteq \mathcal{A}(1 - e) \subseteq \mathcal{L}$, so we conclude $\mathcal{P} \subseteq \overline{\mathcal{P}} \cap e\mathcal{A}e$. Since e (the identity of $e\mathcal{A}e$) does not belong to $\mathcal{L}$, it does not belong to $\mathcal{M} + (\overline{\mathcal{P}} \cap e\mathcal{A}e) \subseteq \mathcal{M} + \overline{\mathcal{P}} \subseteq \mathcal{L}$. Since $\mathcal{M}$ is maximal, this shows $\mathcal{M} = \mathcal{L} \cap e\mathcal{A}e, \overline{\mathcal{P}} \cap e\mathcal{A}e \subseteq \mathcal{M}$ and therefore $\overline{\mathcal{P}} \cap e\mathcal{A}e \subseteq \mathcal{P}$. This concludes the proof of the identity $\overline{\mathcal{P}} \cap e\mathcal{A}e = \mathcal{P}$ and shows that $\mathcal{P} \mapsto \mathcal{P} \cap e\mathcal{A}e$ maps $\Pi_{\mathcal{A}} \setminus h^{\Pi}(e\mathcal{A}e)$ onto $\Pi_{e\mathcal{A}e}$.

Next we show that the map $\mathcal{P} \mapsto \mathcal{P} \cap e\mathcal{A}e$ is injective. If $\mathcal{P}_1, \mathcal{P}_2 \in \Pi_{\mathcal{A}} \setminus h^{\Pi}(e\mathcal{A}e)$ satisfy $\mathcal{P}_1 \cap e\mathcal{A}e = \mathcal{P}_2 \cap e\mathcal{A}e$, then we have $(\mathcal{A}e)\mathcal{P}_1(\mathcal{A}e) \subseteq \mathcal{A}(e\mathcal{P}_1 e) \subseteq (\mathcal{P}_1 \cap e\mathcal{A}e) \subseteq \mathcal{A}\mathcal{P}_2 \subseteq \mathcal{P}_2$. Since $\mathcal{P}_2$ is prime and $\mathcal{A}e \supseteq e\mathcal{A}e$ is not included in $\mathcal{P}_2$, we conclude $\mathcal{P}_1 \subseteq \mathcal{P}_2$. Similarly, $\mathcal{P}_2 \subseteq \mathcal{P}_1$ holds. This shows $\mathcal{P}_1 = \mathcal{P}_2$ as we wished.

If $\mathcal{P}$ is a maximal modular ideal in $\Xi_{\mathcal{A}} \setminus h^{\Xi}(e\mathcal{A}e)$ with relative identity d, then $\mathcal{P} \cap e\mathcal{A}e$ is a modular ideal in $e\mathcal{A}e$ with relative identity ede. Furthermore, this ideal is maximal in $e\mathcal{A}e$ since $b \in e\mathcal{A}e \setminus (\mathcal{P} \cap e\mathcal{A}e)$ implies

$e\mathcal{A}e = e(\mathcal{A}^1 b\mathcal{A}^1 + \mathcal{P})e = (e\mathcal{A}e)b(e\mathcal{A}e) + \mathcal{P}$. Conversely, if $\mathcal{P} \subseteq e\mathcal{A}e$ is a maximal modular ideal of $e\mathcal{A}e$ with relative identity $d \in e\mathcal{A}e$, then the ideal $\mathcal{P} + (1-e)\mathcal{A} + \mathcal{A}(1-e) + \mathcal{A}(1-e)\mathcal{A}$ is a modular ideal of $\mathcal{A}$ with relative identity d (since $ad - a = (eaed - eae) + e(ad - a)(1-e) + (1-e)(ad - a)$ holds for all $a \in \mathcal{A}$) and any maximal modular ideal $\mathcal{M}$ of $\mathcal{A}$ including this ideal with relative identity d must satisfy $\mathcal{M} \cap e\mathcal{A}e = \mathcal{P}$. We have shown that the map $\mathcal{P} \mapsto \mathcal{P} \cap e\mathcal{A}e$ sends $\Xi_{\mathcal{A}} \setminus h^{\Xi}(e\mathcal{A}e)$ bijectively onto $\Xi_{e\mathcal{A}e}$.

Finally, it remains to show that the map is a homeomorphism. It is enough to consider the case of primitive ideals. Suppose that F is a closed subset of $\Pi_{e\mathcal{A}e}$, and let $B \subseteq \Pi_{\mathcal{A}} \setminus h^{\Pi}(e\mathcal{A}e)$ be defined by $B = \{\mathcal{P} \in \Pi_{\mathcal{A}} \setminus h^{\Pi}(e\mathcal{A}e) : \mathcal{P} \cap e\mathcal{A}e \in F\}$. In order to show that B is closed in $\Pi_{\mathcal{A}} \setminus h^{\Pi}(e\mathcal{A}e)$, we must show that any $\mathcal{P}_0 \in \Pi_{\mathcal{A}} \setminus h^{\Pi}(e\mathcal{A}e)$ belongs to B if it includes $k(B)$. However, this is obvious from the definition of B and the fact that F is closed in $\Pi_{e\mathcal{A}e}$. Conversely, suppose that B is closed in $\Pi_{\mathcal{A}} \setminus h^{\Pi}(e\mathcal{A}e)$ and define $F \subseteq \Pi_{e\mathcal{A}e}$ by $F = \{\mathcal{P} \cap e\mathcal{A}e : \mathcal{P} \in B\}$. In order to show that F is closed, we need to show that $\mathcal{P}_0 \in \Pi_{e\mathcal{A}e}$ belongs to F if it includes $k(F)$. We have shown that there is some $\mathcal{P}_1 \in \Pi_{\mathcal{A}} \setminus h^{\Pi}(e\mathcal{A}e)$ satisfying $\mathcal{P}_0 = \mathcal{P}_1 \cap e\mathcal{A}e$. Then $\mathcal{P}_1$ satisfies

$$e\, k(B)e \subseteq k(B) \cap e\mathcal{A}e = \cap\{\mathcal{P} \cap e\mathcal{A}e : \mathcal{P} \in B\} = k(F) \subseteq \mathcal{P}_0 \subseteq \mathcal{P}_1.$$

Since $\mathcal{P}_1$ is a prime ideal satisfying $(\mathcal{A}e)k(B)(\mathcal{A}e) \subseteq \mathcal{P}_1$ and since $\mathcal{P}_1$ does not include $\mathcal{A}e$, it must include $k(B)$. Since B is closed, this implies $\mathcal{P}_1 \in B$ and hence $\mathcal{P}_0 \in F$. $\qquad\square$

Subdirect Product Representations

We will briefly review the construction of a subdirect product representation associated with a collection of ideals. (See Definition 1.3.3 and the discussion surrounding Theorem 4.6.1.) Let $\mathcal{A}$ be an algebra and let B be a nonempty collection of ideals of $\mathcal{A}$. (The cases of particular interest are, of course, $B = P_{\mathcal{A}}, B = \Pi_{\mathcal{A}}$ or $B = \Xi_{\mathcal{A}}$.) For each $a \in \mathcal{A}$, let $\hat{a}$ be the cross section in the direct product $\Pi_{\mathcal{P} \in B}(\mathcal{A}/\mathcal{P})$ defined by

$$\hat{a}(\mathcal{P}) = a + \mathcal{P} \qquad \forall\ \mathcal{P} \in B.$$

Let $\hat{\mathcal{A}}$ be the subalgebra $\{\hat{a} : a \in \mathcal{A}\} \subseteq \Pi_{\mathcal{P} \in B}(\mathcal{A}/\mathcal{P})$. Clearly, $\hat{\mathcal{A}}$ satisfies $\{\hat{a}(\mathcal{P}) : a \in \mathcal{A}\} = \mathcal{A}/\mathcal{P}$ for each $\mathcal{P} \in B$ so that $\hat{\mathcal{A}}$ is a subdirect product. The map $a \mapsto \hat{a}$ is a homomorphism of $\mathcal{A}$ onto $\hat{\mathcal{A}}$ with kernel $\cap_{\mathcal{P} \in B}\mathcal{P}$.

For a special case in which B equals $\langle\ P_{\mathcal{A}}\ /\ \Pi_{\mathcal{A}}\ /\ \Xi_{\mathcal{A}}\ \rangle$, the kernel of the representation $a \mapsto \hat{a}$ described above is $\langle\ \mathcal{A}_B\ /\ \mathcal{A}_J\ /\ \mathcal{A}_M\ \rangle$. In this case, the quotients $\mathcal{A}/\mathcal{P}$ are $\langle$ prime / primitive / unital simple $\rangle$ algebras and hence are reasonably tractable so that the structure of the direct product $\Pi_{\mathcal{P} \in B}(\mathcal{A}/\mathcal{P})$ is comparatively well understood. However, the subdirect product representation will still be of comparatively little use unless useful information can be found describing which elements of the direct product belong to $\hat{\mathcal{A}}$. Situations in which this type of information is available are a main theme of the present chapter.

A change in viewpoint may be instructive. Instead of the direct product $\Pi_{\mathcal{P} \in B}(\mathcal{A}/\mathcal{P})$, consider the disjoint union $S = S_B(\mathcal{A}) = \cup_{\mathcal{P} \in B}(\mathcal{A}/\mathcal{P})$. Define a function $\pi \colon S \to B$ by $\pi(a + \mathcal{P}) = \mathcal{P}$ for all $a \in \mathcal{A}$ and $\mathcal{P} \in B$. Then for each $\mathcal{P} \in B$, the subset $\pi^{-1}(\mathcal{P})$ of S is called the *fiber* (or *stalk*) over $\mathcal{P}$. Each fiber has the structure of an algebra. The sets B and S are called the *base space* and *total space*, respectively, and the map $\pi \colon S \to B$ is called the *fiber map*. A function $f \colon B \to S$ is called a *section* (or *global section*) if $\pi \circ f$ is the identity map on B. The set $\Gamma(B, S)$ of all sections is obviously an algebra under pointwise operations. For each $a \in \mathcal{A}$, the function $\hat{a} \colon B \to S$ defined as before by $\hat{a}(\mathcal{P}) = a + \mathcal{P}$ is a section and $a \mapsto \hat{a}$ is a homomorphism of $\mathcal{A}$ into the algebra $\Gamma(B, S)$. In terms of this language, the subdirect product representation will become useful if an effective description can be given of which sections belong to $\hat{\mathcal{A}}$. The hull–kernel topology would seem to offer some hope for finding such a description. However, in general, it is difficult to find meaningful relations between elements of neighboring fibers, so that the topology of the structure space has only a very weak interaction with the subdirect product. This is in part due to the poor separation properties of the hull–kernel topology. Except on $\Xi_{\mathcal{A}}$, it is not often even T_1, and it is even less often Hausdorff. There are two ways to avoid this difficulty. Any topological space can be made T_1, Hausdorff or completely regular by collapsing certain sets of points into points and giving the quotient space its quotient topology. If this is done to $P_{\mathcal{A}}, \Pi_{\mathcal{A}}$ or $\Xi_{\mathcal{A}}$ (with the hull–kernel topology), then the new space may even be identified with a certain space of semiprime ideals. (The point in the quotient space obtained by collapsing $B \subseteq P_{\mathcal{A}}$ to a point can be identified with the ideal $k(B)$ of the original algebra.) These Hausdorff or completely regular quotients of structure spaces have proved valuable and are exhaustively discussed in John Dauns and Karl Heinrich Hofmann [1968], [1969]. In the first reference, several classes of topological algebras are exhibited as the algebra of continuous sections of a type of bundle or field of algebras over a suitable quotient structure space. This important theory is too technical to be included here. See James Michael G. Fell and Robert S. Doran [1988a], [1988b], Joseph W. Kitchen, Jr. and David A. Robbins [1982], [1983] and [1985], Maurice J. Dupré and Richard M. Gillette [1983], Stephen Allan Selesnick [1979], Anthony Karel Seda [1981] and Gerhard Gierz [1982].

We will pursue the second approach. We place more severe restrictions on the algebra $\mathcal{A}$ which force its structure space to be well behaved. Proposition 7.1.5 suggests that the best theory (*i.e.*, the most results with the least restrictive hypotheses) will be obtained by concentrating attention on the strong structure space. This leads to the theory developed in the next section.

Algebras Satisfying Polynomial Identities

This specialized subject will illustrate some of the ideas just introduced. We begin our discussion with some simple examples. An algebra $\mathcal{A}$ is commutative if and only if any two of its elements are roots of the noncommuting polynomial

$$S(x, y) = xy - yx.$$

Here is a second example. Recall that M_n is the algebra of all $n \times n$ complex matrices. If $C \in M_2$ has trace zero, Theorem 2.7.15 (the Jordan decomposition) shows that it can be written as $G^{-1}DG$ where G is an invertible matrix and D is either a diagonal matrix with diagonal entries λ and $-\lambda$ for some $\lambda \in \mathbb{C}$ or a strictly lower triangular matrix. In either case, C^2 is a multiple of the identity matrix and is therefore in the center of M_2. However, the commutator $AB - BA$ of any two matrices A and B in M_2 has trace zero. This shows that any three matrices in M_2 are roots of the noncommuting polynomial

$$W(x, y, z) = (xy - yx)^2 z - z(xy - yx)^2$$

known as Wagner's polynomial. We can easily define much more complicated algebras which satisfy this same polynomial identity. For instance, let Ω be a compact Hausdorff space and consider the algebra $C(\Omega, M_2)$ of all continuous functions from Ω into M_2. (We have already noted the identification of this algebra with the algebra tensor product $C(\Omega) \otimes M_2$.) Clearly any three elements of $C(\Omega, M_2)$ are roots of Wagner's polynomial W. Any subalgebra satisfies the same condition.

These examples should make clear what we mean by a noncommuting polynomial over the complex field. In order to keep the discussion as elementary as possible, we will consider a *noncommuting polynomial* to be a finite linear combination of monomials where each monomial is just a finite product of symbols. The *degree* of a noncommuting polynomial is the length of its longest monomial. A noncommuting polynomial P of n variables may be evaluated at an ordered n-tuple $(a_1, a_2, \ldots, a_n)$ of elements from any algebra $\mathcal{A}$ simply by replacing each symbol by its corresponding element in $\mathcal{A}$. Obviously, this gives an element in $\mathcal{A}$. We say that $(a_1, a_2, \ldots, a_n)$ is a *root* of P if it satisfies $P(a_1, a_2, \ldots, a_n) = 0$.

7.1.12 Definition Let $\mathcal{A}$ be an algebra and let $P = P(x_1, x_2, \ldots, x_n)$ be a noncommuting polynomial. Then $\mathcal{A}$ is said to *satisfy the polynomial identity* P if any ordered n-tuple $(a_1, a_2, \ldots, a_n)$ of elements from $\mathcal{A}$ is a root of P. It is said to *satisfy a polynomial identity of degree d* if it satisfies some noncommuting polynomial of degree d.

The *standard polynomial* in n noncommuting variables is defined by

$$S_n(x_1, x_2, \ldots, x_n) = \sum_{\sigma \in \Sigma_n} \mathrm{sgn}(\sigma) x_{\sigma(1)} x_{\sigma(2)} \cdots x_{\sigma(n)} \qquad \forall \; x_1, x_2, \ldots, x_n$$

where Σ_n is the group of all permutations of $\{1, 2, \ldots, n\}$ and $\mathrm{sgn}(\sigma)$ is ± 1 depending on whether σ is an even or an odd permutation.

Obviously the standard polynomial S_n is of degree n and *multilinear*. (The latter means that for any $k = 1, 2, \ldots, n$, if $a_1, a_2, \ldots, a_k, \ldots, a_n$ and $a_1, a_2, \ldots, b_k, \ldots, a_n$ are roots of S_n, then so is $a_1, a_2, \ldots, \lambda a_k + \mu b_k, \ldots, a_n$ for any λ, μ in $\mathbb{C}$.) It is also *alternating*. (This means that interchanging the order of two elements in an n-tuple substituted into it changes the sign. Hence if the same element occurs twice in an n-tuple, then S_n is zero on the n-tuple.) Multilinearity and the alternating property are useful since they show that an algebra $\mathcal{A}$ satisfies the polynomial identity S_n if and only if every set of n distinct elements chosen from a linear space basis for $\mathcal{A}$ is a root when arranged in some (hence in any) order.

In this subject, the degree of a polynomial is more important than the number of noncommuting variables or the number of terms. Hence the next theorem shows that Wagner's fifth degree polynomial identity for M_2 with six terms in three noncommuting variables is eclipsed by the 24-term, fourth degree polynomial S_4 in four noncommuting variables just introduced. We shall omit the proof of the following result of Shimshon Amitsur and Jacob Levitzki [1950] since full proofs can be found in any of the standard references listed at the end of this discussion.

7.1.13 Theorem *The algebra M_n satisfies the standard polynomial S_{2n} of degree $2n$ and does not satisfy any polynomial identity of degree less than $2n$.*

Perhaps the most fundamental result in the theory is due to Irving Kaplansky [1948e]. We will give the special case for Banach algebras and indicate how that follows from the strictly algebraic statement.

7.1.14 Theorem *A primitive Banach algebra which satisfies a polynomial identity of degree d is isomorphic to a full matrix algebra M_n with $n \leq d/2$.*

Proof The version given on page 17 of Nathan Jacobson's book [1975] states that $\mathcal{A}$ is a simple algebra of dimension n^2 (where $n \leq d/2$) over the center $\mathcal{A}_Z$ of $\mathcal{A}$ and that $\mathcal{A}_Z$ is a field. However, the Gelfand–Mazur theorem (Corollary 2.2.3) shows that this field $\mathcal{A}_Z$ must be just the complex field $\mathbb{C}$. Theorem 8.1.1 shows that any finite-dimensional simple algebra over $\mathbb{C}$ is a full matrix algebra.   □

7.1.15 Example In order to have a simple but reasonably representative example, the reader may wish to consider the algebra $\mathcal{A}$ defined by

$$\{f \colon [0, 2] \to M_2 : f(0)_{11} = f(0)_{22}; f(s)_{12} = f(s)_{21} = 0 \text{ for } s \leq 1; f(2)_{12} = 0\}$$

so the matrices $f(t)$ are multiples of the identity for $t = 0$, diagonal for $t \leq 1$ and lower triangular for $t = 2$. The set of primitive ideals is:

$$\Pi_{\mathcal{A}} = \{\mathcal{P}_t, \ \mathcal{P}_{s,k} : t \in \{0\} \cup]1, 2[; \ s \in]0, 1] \cup \{2\}; \ k \in \{1, 2\} \}$$

where $\mathcal{P}_t = \{f \in \mathcal{A} : f(t) = 0\}$ and $\mathcal{P}_{s,k} = \{f \in \mathcal{A} : f(s)_{kk} = f(s)_{21} = 0\}$. Note that the quotient algebra of $\mathcal{A}$ modulo $\mathcal{P}_0$ or $\mathcal{P}_{s,k}$ for $s \in]0,1] \cup \{2\}$ and $k \in \{1,2\}$ is $\mathbb{C}$ while the quotient modulo $\mathcal{P}_t$ for $t \in]1,2[$ is M_2. The hull–kernel topology on $\Pi_{\mathcal{A}}$ is T_1 since all primitive ideals are maximal. It assigns the topology of $]1,2[$ to $\{\mathcal{P}_t : t \in]1,2[\}$, the topology of $]0,1]$ to $\{\mathcal{P}_{s,k} : s \in]0,1] \}$ for $k \in \{1,2\}$, and for $s \in \{1,2\}$ it puts both $\mathcal{P}_{s,1}$ and $\mathcal{P}_{s,2}$ in the closure of any subset of $\{\mathcal{P}_t, : t \in]1,2[\}$ which has s in the closure of its set of indices. Finally for $k \in \{1,2\}$, $\mathcal{P}_0$ is in the closure of a subset of $\{\mathcal{P}_{s,k} : s \in]0,1]\}$ if 0 is in the closure of the set of indices. The center of $\mathcal{A}$ is $\mathcal{A}_Z = \{f \in \mathcal{A} : f(t)$ is a multiple of the identity for $t \in \{0\} \cup [1,2]$ and $f(t)$ is a diagonal matrix for $t \in]0,1[\}$.

We derive some results about semisimple unital algebras satisfying a polynomial identity which are suggested by the main results of Naum Ya. Krupnik [1980a], [1980b], [1981] and [1987] and Krupnik and E. M. Shpigel' [1982]. These results show that the last example is somewhat typical.

7.1.16 Theorem   *Let $\mathcal{A}$ be a semisimple unital Banach algebra which satisfies a polynomial identity of degree d.*

(a) *Each primitive ideal $\mathcal{P}$ of $\mathcal{A}$ is maximal with quotient algebra isomorphic to $M_{n_{\mathcal{P}}}$ for some $n_{\mathcal{P}} \le d/2$.*

(b) *The center $\mathcal{A}_Z$ of $\mathcal{A}$ is the set of elements c such that for each $\mathcal{P} \in \Pi_{\mathcal{A}}$ there is a (necessarily unique) complex number $\gamma_{\mathcal{P}}(c)$ satisfying $c - \gamma_{\mathcal{P}}(c)1 \in \mathcal{P}$.*

(c) *The map $\mathcal{P} \mapsto \gamma_{\mathcal{P}}$ is an open surjection of $\Pi_{\mathcal{A}}$ onto $\Gamma_{\mathcal{A}_Z}$ with $\ker(\gamma_{\mathcal{P}}) = \mathcal{P} \cap \mathcal{A}_Z$. Hence $\mathcal{A}_Z$ is semisimple.*

(d) *There are continuous functions τ and δ from $\mathcal{A}$ to $B(\Pi_{\mathcal{A}})$ with the former linear satisfying*

$$\tau(ab) = \tau(ba) \quad \delta(ab) = \delta(ba) \qquad \forall \; a,b \in \mathcal{A} \tag{7}$$

$$\tau(c)(\mathcal{P}) = n_{\mathcal{P}}\gamma_{\mathcal{P}}(c) \quad \delta(c)(\mathcal{P}) = \gamma_{\mathcal{P}}(c)^{n_{\mathcal{P}}} \qquad \forall \; c \in \mathcal{A}_Z; \mathcal{P} \in \Gamma_{\mathcal{A}} \tag{8}$$

$$a \in \mathcal{A}_G \quad \text{if and only if} \quad \delta(a) \text{ is never zero} \tag{9}$$

$$Sp(a) = \{\lambda \in \mathbb{C} : \delta(\lambda - a) \text{ is zero at some } \mathcal{P} \in \Pi_{\mathcal{A}}\} \qquad \forall \; a \in \mathcal{A} \tag{10}$$

$$\delta : \mathcal{A}_G \to B(\Pi_{\mathcal{A}})_G \text{ is a group homomorphism.} \tag{11}$$

Proof (a): If $\mathcal{A}$ satisfies a polynomial identity it is clear that $\mathcal{A}/\mathcal{P}$ satisfies the same identity. Hence the last theorem shows that this quotient is isomorphic to a full matrix algebra satisfying the indicated bound on its dimension. Since matrix algebras are simple, $\mathcal{P}$ is maximal.

(b): If c satisfies the condition, then $ac - ca$ belongs to each primitive ideal. Hence c is in the center of $\mathcal{A}$ because $\mathcal{A}$ is semisimple. Conversely, if c belongs to the center of $\mathcal{A}$ then $c + \mathcal{P}$ must be in the center of $\mathcal{A}/\mathcal{P}$ for each $\mathcal{P}$. Since this quotient is isomorphic to a full matrix algebra, $c + \mathcal{P}$ is a multiple of the identity $1 + \mathcal{P}$.

(c): It is easy to see that $\gamma_\mathcal{P}$ is an element of the Gelfand space of $\mathcal{A}_Z$ with the indicated kernel. The proof of Proposition 7.2.14(c) and (d), which begins by assuming the algebra $\mathcal{A}$ is unital as we have supposed here, shows that this is a surjection. Since hulls are always Gelfand closed, it is an open map.

(d): For each $\mathcal{P}$, let $\varphi_\mathcal{P}$ be the natural map of $\mathcal{A}$ onto $\mathcal{A}/\mathcal{P}$ and choose an isomorphism $\theta_\mathcal{P}: \mathcal{A}/\mathcal{P} \to M_{n_\mathcal{P}}$. Define $\tau(a)(\mathcal{P})$ and $\delta(a)(\mathcal{P})$ to be $\mathrm{Tr}(\theta_\mathcal{P} \circ \varphi_\mathcal{P}(a))$ and $\mathrm{Det}(\theta_\mathcal{P} \circ \varphi_\mathcal{P}(a))$, respectively, for all $a \in \mathcal{A}$. Since all automorphisms of M_n are inner, the definitions of τ and δ are independent of the choice of $\theta_\mathcal{P}$. Equation (10) (which we are about to prove) shows that for each $\mathcal{P}$ the absolute values of the characteristic values of $\theta_\mathcal{P} \circ \varphi_\mathcal{P}(a)$ are all bounded above by $||a||$ which implies $||\tau(a)||_\infty \leq n_\mathcal{P}||a|| \leq (d/2)||a||$ and $||\delta(a)||_\infty \leq ||a||^{n_\mathcal{P}} \leq ||a||^{d/2}$ so that these functions are bounded. The continuity of τ follows immediately from the above estimate and its linearity. To see the continuity of δ, we note that $\varphi_\mathcal{P}$ is continuous for each $\mathcal{P}$ and that, since $\mathcal{A}/\mathcal{P}$ is finite-dimensional it has a unique linear space norm topology. Hence $\mathrm{Det} \circ \theta_\mathcal{P}$ is also continuous. Formulas (7), (8) and (11) are immediate consequences of the definition. Theorem 4.1.8(d), which states that an element is invertible if and only if it is invertible modulo each primitive ideal, makes (9) and (10) clear.   □

Krupnik [1987] considers topologies on $\Pi_\mathcal{A}$ which make the functions above and the coordinate functions of the various matrix representations continuous. We believe that more work along this line would be desirable. The following definition and results are also based on Krupnik [1987].

7.1.17 Definition   A unital Banach algebra $\mathcal{A}$ is said to have *a sufficient family of n-dimensional representations* if there is a set Φ of homomorphisms $\varphi: \mathcal{A} \to M_{n_\varphi}$ satisfying $n_\varphi \leq n$ for all $\varphi \in \Phi$ and

$$(a \in \mathcal{A}_G) \iff (\mathrm{Det}(\varphi(a)) \neq 0 \quad \forall \ \varphi \in \Phi).$$

7.1.18 Theorem   *The following are equivalent for a unital Banach algebra $\mathcal{A}$.*

(a) $\mathcal{A}$ *has a sufficient family of n-dimensional representations.*

(b) $\mathcal{A}/\mathcal{A}_J$ *satisfies a polynomial identity of degree $2n$.*

(c) $\mathcal{A}/\mathcal{A}_J$ *satisfies the standard polynomial identity of degree $2n$.*

If $\mathcal{A}/\mathcal{A}_J$ has a sufficient family of n-dimensional representations, then $\mathcal{A}$ enjoys the above properties.

A Banach algebra $\mathcal{A}$ has a sufficient family of n-dimensional representations for some n if and only if $\mathcal{A}/\mathcal{A}_J$ satisfies some polynomial identity.

Proof (a) $\Rightarrow$ (c): Let Φ be a sufficient family of n-dimensional representations. Note that $b \in \cap\{\ker(\varphi) : \varphi \in \Phi\}$ implies $\mathrm{Det}(\varphi(\lambda 1 - b)) \neq 0$ for all non-zero λ and all $\varphi \in \Phi$. Hence $Sp(b) = \{0\}$ for all b in $\cap\{\ker(\varphi) : \varphi \in \Phi\}$. Therefore this ideal is included in the Jacobson radical. Since each $2n$-tuple $a_1, a_2, \ldots, a_{2n}$ chosen from $\mathcal{A}$ and each $\varphi \in \Phi$ satisfy $S_{2n}(a_1, a_2, \ldots, a_{2n}) \in \ker(\varphi)$, $S_{2n}(a_1, a_2, \ldots, a_{2n})$ is in $\mathcal{A}_J$. Therefore $\mathcal{A}/\mathcal{A}_J$ satisfies the polynomial identity S_{2n}.

(c) $\Rightarrow$ (b): Obvious.

(b) $\Rightarrow$ (a): Theorem 7.1.16, particularly (10), (11) and the definition of δ, establishes this.   □

Here is a very special case. Theorem 1.1.20 describes the case in which a Banach algebra $\mathcal{B}$ is a full matrix algebra over some subalgebra $\mathcal{A}$ which is itself a unital algebra. If $\mathcal{A}$ is commutative we easily show that $\mathcal{B}$ has a sufficient family of n-dimensional representations. In fact we only need to assume that $\mathcal{B}$ is a unital spectral subalgebra of $M_n(\mathcal{A})$. Let $\varphi: \mathcal{B} \to M_n(\mathcal{A})$ be an isomorphism onto a spectral subalgebra. Let $\Phi = \{\varphi_\gamma : \gamma \in \Gamma_\mathcal{A}\}$ where $\varphi_\gamma: \mathcal{B} \to M_n$ is defined by $(\varphi_\gamma(b))_{ij} = \gamma((\varphi(a))_{ij})$. If $\mathrm{Det}(\gamma \circ \varphi(b)) = \gamma(\mathrm{Det}(\varphi(b)))$ never vanishes, then $\mathrm{Det}(\varphi(b))$ is invertible in $\mathcal{A}$. Thus Cramer's rule (see the remarks following Theorem 1.1.20) allows us to write the inverse for $\varphi(a)$ in $M_n(\mathcal{A})$. Since φ is an isomorphism onto a unital spectral subalgebra, this is enough.

The modern theory of algebras satisfying polynomial identities (often called *PI-algebras*) started with Kaplansky's paper [1948e]. There is a vast literature which concentrates mainly on rings and non-normed algebras over an arbitrary field. We mention Israel N. Herstein [1968], Nathan Jacobson [1975], Aleksandr Robertovič Kemer [1991], Naum Ya. Krupnik [1987], Krupnik and Bernd Silbermann [1989], Denis Luminet [1986], Vladimir Müller [1990], Claudio Procesi [1973] and Louis Halle Rowen [1980]. In the last two decades, there has been considerable emphasis on *central polynomials* which, like $(xy - yx)^2$ on M_2, send an algebra onto its center. See particularly the last reference listed above.

7.1.19 Structure Spaces of Tensor Products   The results given in Example 3.2.18 suggest that each structure space of the projective tensor product of two Banach algebras ought to be the Cartesian product of the corresponding structure spaces of the factors. Neither completely general theorems of this type nor any relevant counterexamples seem to be known. Bernard R. Gelbaum [1959], [1961], [1970] (the last reference corrects errors in the previous two) obtains a continuous map of $\Xi_\mathcal{A} \times \Xi_\mathcal{B}$ onto $\Xi_{\mathcal{A} \hat{\otimes} \mathcal{B}}$ and shows that it is a surjective homeomorphism in certain cases. In particular, he gives a natural homeomorphism of $\Xi_\mathcal{A} \times \Xi_\mathcal{B}$ onto a subset of $\Xi_{\mathcal{A} \hat{\otimes} \mathcal{B}}$. No case seems to be known in which this subset is proper. Kjeld B. Laursen [1970], using a method due to Arnold Lebow [1968], shows that if $\mathcal{A}$ and $\mathcal{B}$ are Banach algebras with $\mathcal{A}$ commutative, then $\Gamma_\mathcal{A} \times \Xi_\mathcal{B}$ is in bijective correspondence with $\Xi_{\mathcal{A} \hat{\otimes} \mathcal{B}}$ under the map which sends the pair $(\gamma, \mathcal{M})$ to the ideal

$$\{\sum_{k=1}^{\infty} a_k \otimes b_k \in \mathcal{A}\hat{\otimes}\mathcal{B} : \sum_{k=1}^{\infty} \|a_k\|\, \|b_k\| < \infty \text{ and } \sum_{k=1}^{\infty} \varphi(a_k)b_k \in \mathcal{M}\}.$$

For C*-algebras, closely related but more definitive results will be given in Volume II, including those of Laursen [1969a].

7.2 Completely Regular Algebras

The following concept was introduced by Willcox [1956a] and extended by Rickart [1960]. On first sight, the second condition in the next definition appears somewhat artificial. Proposition 7.2.2 explains it: it is not enough to have $\Xi_{\mathcal{A}}$ Hausdorff, $\Xi_{\mathcal{A}^1}$ must also be Hausdorff.

7.2.1 Definition An algebra $\mathcal{A}$ is said to be *completely regular* if it satisfies:

(a) $\Xi_{\mathcal{A}}$ is Hausdorff; and

(b) Each $\mathcal{M} \in \Xi_{\mathcal{A}}$ has a neighborhood V in $\Xi_{\mathcal{A}}$ such that $k(V)$ is modular.

If $\mathcal{A}$ is unital, the condition (b) is automatically fulfilled by choosing $V = \Xi_{\mathcal{A}}$. Hence, a unital algebra is completely regular if and only if its strong structure space is Hausdorff. Note also that Proposition 7.1.6 applied to (b) shows that $\Xi_{\mathcal{A}}$ is a locally compact topological space and hence completely regular. The next proposition extends this observation.

7.2.2 Proposition *Let $\mathcal{A}$ be a nonunital algebra. Define $\langle\, \Pi_{\mathcal{A}}^1\, /\, \Xi_{\mathcal{A}}^1\, \rangle$ to be the set $\langle\, \Pi_{\mathcal{A}} \cup \{\infty\}\, /\, \Xi_{\mathcal{A}} \cup \{\infty\}\, \rangle$ with the topology consisting of the open subsets of $\langle\, \Pi_{\mathcal{A}}\, /\, \Xi_{\mathcal{A}}\, \rangle$ and all the sets of the form $\langle\, (\Pi_{\mathcal{A}} \setminus H) \cup \{\infty\}\, /\, (\Xi_{\mathcal{A}} \setminus H) \cup \{\infty\}\, \rangle$, where H is a hull (possibly empty) in $\langle\, \Pi_{\mathcal{A}}\, /\, \Xi_{\mathcal{A}}\, \rangle$ with $k(H)$ modular. Then $\langle\, \Pi_{\mathcal{A}}^1\, /\, \Xi_{\mathcal{A}}^1\, \rangle$ is a compact $\langle\, T_0\, /\, T_1\, \rangle$ space which is homeomorphic to $\langle\, \Pi_{\mathcal{A}^1}\, /\, \Xi_{\mathcal{A}^1}\, \rangle$ in a natural way.*

Furthermore, $\Xi_{\mathcal{A}}^1$ is Hausdorff (equivalently $\mathcal{A}^1$ is completely regular) if and only if $\mathcal{A}$ is completely regular. In this case, $\Xi_{\mathcal{A}}^1$ is the one-point compactification of $\Xi_{\mathcal{A}}$. Hence, $\Xi_{\mathcal{A}}$ is a completely regular topological space when $\mathcal{A}$ is completely regular.

Proof As usual we regard $\mathcal{A}$ as embedded in $\mathcal{A}^1$. Note that the second paragraph is an immediate consequence of the first. (The remarks following Definition 3.2.9 show that $\Xi_{\mathcal{A}}$ is completely regular.)

Theorem 7.1.7 shows that the map $\mathcal{P} \mapsto \mathcal{P} \cap \mathcal{A}$ is a homeomorphism of $\langle\, \Pi_{\mathcal{A}^1} \setminus h^{\Pi}(\mathcal{A})\, /\, \Xi_{\mathcal{A}^1} \setminus h^{\Xi}(\mathcal{A})\, \rangle$ onto $\langle\, \Pi_{\mathcal{A}}\, /\, \Xi_{\mathcal{A}}\, \rangle$. Clearly, $\mathcal{A}$ itself is the unique element of $\langle\, h^{\Pi}(\mathcal{A})\, /\, h^{\Xi}(\mathcal{A})\, \rangle$. Thus the bijection of $\langle\, \Pi_{\mathcal{A}^1}\, /\, \Xi_{\mathcal{A}^1}\, \rangle$

onto $\langle\ \Pi^1_{\mathcal{A}}\ /\ \Xi^1_{\mathcal{A}}\ \rangle$ is given by

$$\begin{cases} \mathcal{P}\mapsto\mathcal{P}\cap\mathcal{A}, & \text{if } \mathcal{P}\neq\mathcal{A} \\ \mathcal{A}\mapsto\infty. \end{cases}$$

We need to identify the inverse of this map on $\langle\ \Pi_{\mathcal{A}}\ /\ \Xi_{\mathcal{A}}\ \rangle$. If $\mathcal{M}$ is a maximal modular $\langle$ left ideal / ideal $\rangle$ in $\mathcal{A}$ with $\langle$ right relative / relative $\rangle$ identity e, define $\overline{\mathcal{M}}\subset\mathcal{A}^1$ by $\overline{\mathcal{M}}=\{\lambda+a\in\mathcal{A}^1:\lambda e+a\in\mathcal{M}\}$. Then $\overline{\mathcal{M}}\cap\mathcal{A}=\mathcal{M}$ holds and the calculation $\mathcal{A}^1/\overline{\mathcal{M}}=(\mathcal{A}+\overline{\mathcal{M}})/\overline{\mathcal{M}}\simeq\mathcal{A}/(\overline{\mathcal{M}}\cap\mathcal{A})=\mathcal{A}/\mathcal{M}$ shows that $\overline{\mathcal{M}}$ is also a maximal modular $\langle$ left ideal / ideal $\rangle$ of $\mathcal{A}^1$. Theorem 4.1.8 shows that we may write any $\mathcal{P}\in\Pi_{\mathcal{A}}$ as $\mathcal{M}:\mathcal{A}$ for some maximal modular left ideal $\mathcal{M}$ and then define $\overline{\mathcal{P}}$ to be the primitive ideal $\overline{\mathcal{M}}:\mathcal{A}^1$ of $\mathcal{A}^1$ which satisfies $\overline{\mathcal{P}}\cap\mathcal{A}=\{a\in\mathcal{A}:\lambda a+ab\in\mathcal{M}$ for all $\lambda+b\in\mathcal{A}^1\}=\mathcal{P}$. (If $\mathcal{P}$ is maximal modular, we have now given two definitions of $\overline{\mathcal{P}}$ and these definitions even appear to depend on the choice of the relative identities involved, but since $\overline{\mathcal{P}}\cap\mathcal{A}=\mathcal{P}$ holds, Theorem 7.1.7 shows that all these definitions of $\overline{\mathcal{P}}$ must agree.) Thus $\mathcal{P}\mapsto\overline{\mathcal{P}}$ from $\langle\ \Pi_{\mathcal{A}}\ /\ \Xi_{\mathcal{A}}\ \rangle$ to $\langle\ \Pi_{\mathcal{A}^1}\setminus h^{\Pi}(\mathcal{A})\ /\ \Xi_{\mathcal{A}^1}\setminus h^{\Xi}(\mathcal{A})\ \rangle$ is the inverse of the above map.

To complete the proof, we need only show that the neighborhood system for ∞ in $\langle\ \Pi^1_{\mathcal{A}}\ /\ \Xi^1_{\mathcal{A}}\ \rangle$ agrees with the neighborhood system for $\mathcal{A}$ in $\langle\ \Pi_{\mathcal{A}^1}\ /\ \Xi_{\mathcal{A}^1}\ \rangle$. If the nonempty hull H in $\langle\ \Pi_{\mathcal{A}^1}\ /\ \Xi_{\mathcal{A}^1}\ \rangle$ does not contain $\mathcal{A}$, then the kernel of its image in $\langle\ \Pi^1_{\mathcal{A}}\ /\ \Xi^1_{\mathcal{A}}\ \rangle$ is $k(H)\cap\mathcal{A}$ which is modular since it satisfies $\mathcal{A}^1/k(H)=(\mathcal{A}+k(H))/\mathcal{A}\simeq\mathcal{A}/(k(H)\cap\mathcal{A})$. Conversely, if H is a hull in $\langle\ \Pi_{\mathcal{A}}\ /\ \Xi_{\mathcal{A}}\ \rangle$ with $k(H)$ modular and e is a relative identity for $k(H)$, then the kernel of the image of H in $\langle\ \Pi_{\mathcal{A}^1}\ /\ \Xi_{\mathcal{A}^1}\ \rangle$ contains $1-e$ and hence is not included in $\mathcal{A}$. Since this is equivalent to saying that the complement of the image of H is a neighborhood of $\mathcal{A}$ in $\langle\ \Pi_{\mathcal{A}^1}\ /\ \Xi_{\mathcal{A}^1}\ \rangle$, this completes the proof.   □

7.2.3 Proposition *Let $\mathcal{A}$ be completely regular and let $\mathcal{I}$ be an ideal of $\mathcal{A}$. Then $\mathcal{I}$ and $\mathcal{A}/\mathcal{I}$ are completely regular.*

Proof Theorem 7.1.7 shows that $\Xi_{\mathcal{I}}$ and $\Xi_{\mathcal{A}/\mathcal{I}}$ are Hausdorff, so it only remains to show that they satisfy condition (b) of the definition. In order to show this for $\mathcal{I}$, choose $\tilde{\mathcal{M}}\in\Xi_{\mathcal{I}}$. Theorem 7.1.7 shows that $\tilde{\mathcal{M}}=\mathcal{M}\cap\mathcal{I}$ for some $\mathcal{M}\in\Xi_{\mathcal{A}}\setminus h^{\Xi}(\mathcal{I})$. Choose a closed neighborhood H of $\mathcal{M}$ in $\Xi_{\mathcal{A}}$ disjoint from $h^{\Xi}(\mathcal{I})$ and with $k(H)$ modular. (This is possible since $\Xi_{\mathcal{A}}$ is completely regular and $h^{\Xi}(\mathcal{I})$ is closed.) Let $\tilde{H}$ be the image of H in $\Xi_{\mathcal{I}}$ under the map described in Theorem 7.1.7 so that $\tilde{H}$ is a neighborhood of $\tilde{\mathcal{M}}$ and equals $k(H)\cap\mathcal{I}$. Proposition 7.1.9(b) shows that $\mathcal{I}$ contains a relative identity e for $k(H)$ as an ideal of $\mathcal{A}$. Thus, e is a relative identity for $k(\tilde{H})=k(H)\cap\mathcal{I}$ as an ideal of $\mathcal{I}$. Therefore (b) holds in $\mathcal{I}$.

In order to establish condition (b) in $\mathcal{A}/\mathcal{I}$, let $\tilde{\mathcal{M}}\in\Xi_{\mathcal{A}/\mathcal{I}}$ be arbitrary. Theorem 7.1.7 shows $\tilde{\mathcal{M}}=\mathcal{M}/\mathcal{I}$ for some $\mathcal{M}\in h(\mathcal{I})$. Let V be a neighborhood of $\mathcal{M}$ in $\Xi_{\mathcal{A}}$ with $k(V)$ modular. Then $k(V \cap h(\mathcal{I}))$ is also modular since it includes $k(V)$. Let $\tilde{V}$ be the image of $V \cap h(\mathcal{I})$ in $\Xi_{\mathcal{A}/\mathcal{I}}$ under the map described in Theorem 7.1.7 so that $\tilde{V}$ is a neighborhood of $\tilde{\mathcal{M}}$ and $k(V)$ is $k(V \cap h(\mathcal{I}))/\mathcal{I}$. Then $k(\tilde{V})$ is modular in $\mathcal{A}/\mathcal{I}$ with relative identity $e + \mathcal{I}$, where e is a relative identity for $k(V \cap h(\mathcal{I}))$ in $\mathcal{A}$. Hence $\mathcal{A}/\mathcal{I}$ satisfies condition (b) of the definition. □

7.2.4 Corollary *An algebra $\mathcal{A}$ is completely regular if and only if $\mathcal{A}/\mathcal{A}_M$ is completely regular.*

Proof The last proposition shows that $\mathcal{A}/\mathcal{A}_M$ is completely regular if $\mathcal{A}$ is. Corollary 7.1.8 shows that $\Xi_{\mathcal{A}}$ and $\Xi_{\mathcal{A}/\mathcal{A}_M}$ are homeomorphic under the map $\mathcal{M} \mapsto \mathcal{M}/\mathcal{A}_M$. Hence we need only check that condition (b) holds in $\mathcal{A}$ when $\mathcal{A}/\mathcal{A}_M$ is completely regular. Let $\mathcal{M} \in \Xi_{\mathcal{A}}$ be arbitrary and let V be a neighborhood of $\mathcal{M}/\mathcal{A}_M$ in $\Xi_{\mathcal{A}/\mathcal{A}_M}$ with $k(V)$ modular. Let $\overline{V}$ be the corresponding neighborhood of $\mathcal{M}$ in $\Xi_{\mathcal{A}}$ so that $k(V)$ equals $k(\overline{V})/\mathcal{A}_M$. Since $\mathcal{A}/k(\overline{V})$ is isomorphic to $(\mathcal{A}/\mathcal{A}_M)/(k(\overline{V})/\mathcal{A}_M) = (\mathcal{A}/\mathcal{A}_M)/k(V)$, $k(\overline{V})$ is modular. Hence (b) holds in $\mathcal{A}/\mathcal{A}_M$. □

The following result gives a converse to Proposition 7.1.6 in a special case.

7.2.5 Proposition *Let $\mathcal{A}$ be completely regular. An ideal $\mathcal{I}$ of $\mathcal{A}$ is modular if and only if $h^{\Xi}(\mathcal{I})$ is compact.*

Proof Proposition 7.1.6 shows that $h^{\Xi}(\mathcal{I})$ is compact if $\mathcal{I}$ is modular. Assume that $h^{\Xi}(\mathcal{I})$ is compact. For each $\mathcal{M} \in h^{\Xi}(\mathcal{I})$, choose a neighborhood $V_{\mathcal{M}}$ of $\mathcal{M}$ with $k(V_{\mathcal{M}})$ modular. Choose a finite subcover $V_1, V_2, \ldots, V_n$. The second inclusion of (1) and (5) in Proposition 7.1.2 show $k(V) \supseteq k(\cup_{j=1}^n V_j) = \cap_{j=1}^n k(V_j)$. Hence $k(V)$ is modular, since $e_1 \circ e_2 \circ \ldots \circ e_n$ is a relative identity for $k(V)$, where e_j is a relative identity for $k(V_j)$, $j = 1, 2, \ldots, n$. □

We will now obtain the detailed structure of the subdirect product representation $\hat{\mathcal{A}} \subseteq \Pi_{\mathcal{M} \in \Xi_{\mathcal{A}}}(\mathcal{A}/\mathcal{M})$ for a completely regular $\mathcal{A}$. If $\mathcal{A}$ is strongly semisimple as well as completely regular, this will give structural details for $\mathcal{A}$. We begin by obtaining a *partition of the identity* for the relative identity of a modular ideal.

7.2.6 Proposition *Let $\mathcal{A}$ be a completely regular algebra. Let K be a compact subset of $\Xi_{\mathcal{A}}$ and let $V_1, V_2, \ldots, V_n$ be a finite open cover of K. Then there exist elements $e_j \in k(\Xi_{\mathcal{A}} \setminus V_j)$ for $j = 1, 2, \ldots, n$ such that $e_1 + e_2 + \cdots + e_n$ is a relative identity for $k(K)$.*

Proof For $j = 1, 2, \ldots, n$, let $\mathcal{I}_j$ be the ideal $k(\Xi_{\mathcal{A}} \setminus V_j)$ and let $\mathcal{I}$ be $\mathcal{I}_1 + \mathcal{I}_2 + \cdots + \mathcal{I}_n$. Proposition 7.1.2 shows $h(\mathcal{I}) = \cap_{j=1}^n h(\mathcal{I}_j) = \cap_{j=1}^n (\Xi_{\mathcal{A}} \setminus$

$V_j) = \Xi_A \setminus \cup_{j=1}^n V_j$. Hence K and $h(\mathcal{I})$ are disjoint. Since Proposition 7.2.5 shows that $k(K)$ is modular, Proposition 7.1.9 gives a relative identity $e \in \mathcal{I}$ for $k(K)$. Since $\mathcal{I}$ is the sum $\mathcal{I}_1 + \mathcal{I}_2 + \cdots + \mathcal{I}_n$, the existence of $e_j \in \mathcal{I}_j$ satisfying $e = e_1 + e_2 + \cdots + e_n$ is immediate. $\qquad\square$

7.2.7 Definition   Let A be an algebra and let $\mathcal{L}$ be a subset of A. An element $f \in \Pi_{M \in \Xi_A}(A/M)$ is said to *belong to $\mathcal{L}$ near an ideal* $M \in \Xi_A$ if there is some neighborhood V of M in Ξ_A and some $a \in \mathcal{L}$ satisfying $\hat{a} = f$ on V. An element $f \in \Pi_{M \in \Xi_A}(A/M)$ is said to *belong to $\mathcal{L}$ at infinity* if there is some open subset V of Ξ_A with $\Xi_A \setminus V$ compact and some $a \in \mathcal{L}$ satisfying $\hat{a} = f$ on V. If f belongs to $\mathcal{L}$ at each point of Ξ_A and at infinity, then f is said to *belong to $\mathcal{L}$ locally*.

The following lemma shows that we can "paste together" elements of the algebra if their representations in the subdirect product agree on the overlapping parts of certain open sets.

7.2.8 Lemma *Let A be a completely regular algebra. Let $\mathcal{L}$ be a one-sided ideal of A. Let H be a closed subset of Ξ_A which is covered by open sets $V_0, V_1, \ldots, V_n$ such that $V_1, \ldots, V_n$ have compact closures. If $a_0, a_1, \ldots, a_n$ are elements of $\mathcal{L}$ satisfying $\hat{a}_j = \hat{a}_k$ on $V_j \cap V_k$ for all $j, k = 0, 1, \ldots, n$, then there is an element $a \in \mathcal{L}$ satisfying*

$$\hat{a} = \hat{a}_j \text{ on } V_j \cap H \qquad \forall \, j = 0, 1, \ldots, n.$$

Proof   Let $\mathcal{L}$ be a left ideal. Proposition 7.2.6 gives elements $e_1, e_2, \ldots, e_n \in A$ satisfying $e_j \in k(\Xi_A \setminus V_j)$ (*i.e.*, $\hat{e}_j$ vanishes off V_j) for $j = 1, 2, \ldots, n$ and such that $e = \sum_{j=1}^n e_j$ is a relative identity for $k(H \setminus V_0)$. For each $M \in \Xi_A$, let $j(M)$ be the set $j(M) = \{j : M \in V_j\}$, so that $\hat{e}(M) = \sum_{j \in j(M)} \hat{e}_j(M)$ holds. Define $a \in \mathcal{L}$ by $a = (1 - e)a_0 + e_1 a_1 + e_2 a_2 + \cdots + e_n a_n$. For $M \in H \cap V_k \cap V_0$, we get

$$
\begin{aligned}
\hat{a}(M) &= (1 - \hat{e}(M))\hat{a}_0(M) + \sum_{j=1}^n \hat{e}_j(M)\hat{a}_j(M) \\
&= [(1 - \hat{e}(M)) + \sum_{j \in j(M)} \hat{e}_j(M)]a_k(M) = \hat{a}_k(M),
\end{aligned}
$$

and for $M \in (H \cap V_k) \setminus V_0$ we get

$$\hat{a}(M) = ((\widehat{1-e})\hat{a}_0)(M) + \sum_{j \in j(M)} \widehat{e_j}(M)\widehat{a_j}(M) = \widehat{ea_k}(M) = \widehat{a_k}(M). \square$$

7.2.9 Lemma *Let A be a completely regular algebra and let $\mathcal{L}$ be a one-sided ideal of A. If $f \in \prod_{M \in \Xi_A} (A/M)$ belongs to $\mathcal{L}$ near each point of a* *compact set $K \subseteq \Xi_{\mathcal{A}}$, then there is some $b \in \mathcal{L}$ satisfying $\hat{b} = f$ on an open neighborhood of K.*

Proof For each $\mathcal{M} \in K$, choose an open neighborhood $V_{\mathcal{M}}$ of $\mathcal{M}$ in $\Xi_{\mathcal{A}}$ and an element $a \in \mathcal{L}$ satisfying $\hat{a}_{\mathcal{M}} = f$ on $V_{\mathcal{M}}$. Since $\Xi_{\mathcal{A}}$ is locally compact, we may assume each $V_{\mathcal{M}}$ has compact closure. By compactness, we can choose a finite subcover $V_1, V_2, \ldots, V_n$ with corresponding elements $a_1, a_2, \ldots, a_n$ in $\mathcal{L}$ satisfying $\hat{a}_j = f$ on V_j for $j = 1, 2, \ldots, n$. Lemma 7.2.8 gives the desired result.   □

We can now obtain a useful description of which elements of the full direct product $\Pi_{\mathcal{M} \in \Xi_{\mathcal{A}}}(\mathcal{A}/\mathcal{M})$ belong to the subdirect product $\hat{\mathcal{A}}$ or to the image in $\hat{\mathcal{A}}$ of a left ideal $\mathcal{L}$ of $\mathcal{A}$.

7.2.10 Theorem *Let $\mathcal{A}$ be a completely regular algebra and let $\mathcal{L}$ be a (not necessarily proper) left ideal of $\mathcal{A}$. The image $\hat{\mathcal{L}}$ of $\mathcal{L}$ in the subdirect product representation $\hat{\mathcal{A}} \subseteq \Pi_{\mathcal{M} \in \Xi_{\mathcal{A}}}(\mathcal{A}/\mathcal{M})$ is the set of $f \in \Pi_{\mathcal{M} \in \Xi_{\mathcal{A}}}(\mathcal{A}/\mathcal{M})$ which belong to $\mathcal{L}$ locally. If $\mathcal{A}$ is unital, $\hat{\mathcal{A}}$ is the set of $f \in \Pi_{\mathcal{M} \in \Xi_{\mathcal{A}}}(\mathcal{A}/\mathcal{M})$ which belong to $\mathcal{A}$ near each point of $\Xi_{\mathcal{A}}$.*

Proof If f belongs to $\mathcal{L}$ locally, then it belongs to $\mathcal{L}$ at infinity so there is some compact set $K \subseteq \Xi_{\mathcal{A}}$ and some $a \in \mathcal{L}$ satisfying $\hat{a} = f$ on the complement of K. Furthermore, f belongs to $\mathcal{L}$ near each point of $\Xi_{\mathcal{A}}$ and *a fortiori* near each point of K so there is some $b \in \mathcal{L}$ satisfying $\hat{b} = f$ on an open neighborhood of K. Apply Lemma 7.2.8. If $\mathcal{A}$ is unital, every $f \in \Pi_{\mathcal{M} \in \Xi_{\mathcal{A}}}(\mathcal{A}/\mathcal{M})$ belongs to $\hat{\mathcal{A}}$ at infinity.   □

We can use the above description to obtain some remarkably detailed information about the structure of a completely regular algebra. We give an example.

7.2.11 Proposition *Let $\mathcal{A}$ be a completely regular algebra. For any $n \in \mathbb{N}$ let $H_0, H_1, \ldots, H_n$ be disjoint closed subsets of $\Xi_{\mathcal{A}}$ with $H_1, H_2, \ldots, H_n$ compact and let $a_0, a_1, \ldots, a_n \in \mathcal{A}$ be arbitrary. Then there is some $a \in \mathcal{A}$ satisfying*

$$\hat{a} = \hat{a}_j \text{ on } H_j, \text{ for } j = 0, 1, \ldots, n.$$

Proof Define H to be the set $\cup_{j=0}^{n} H_j$ and $\mathcal{B}$ to be the algebra $\mathcal{A}/k(H)$. Then $\mathcal{B}$ is completely regular and $\Xi_{\mathcal{B}}$ may be identified with $H = \cup_{j=0}^{n} H_j$. Regard each H_j as a subset of $\Xi_{\mathcal{B}}$ and define $f \in \Pi_{\mathcal{M} \in \Xi_{\mathcal{B}}}(\mathcal{B}/\mathcal{M})$ by $f(\mathcal{M}) = (a_j + k(H)) + \mathcal{M}$ for $\mathcal{M} \in H_j, j = 0, 1, 2, \ldots, n$. Then f belongs locally to $\hat{\mathcal{B}}$, so Theorem 7.2.10 shows that some $b = a + k(H) \in \mathcal{B}$ satisfies $\hat{b} = f$ on $\Xi_{\mathcal{B}} = H$. Thus $a \in \mathcal{A}$ has the desired property.   □

Theorem 7.2.10 describes which elements $f \in \Pi_{\mathcal{M} \in \Xi_{\mathcal{A}}}(\mathcal{A}/\mathcal{M})$ have the form $\hat{b}$ for some element b in a left ideal $\mathcal{L}$. If the algebra is assumed to be strongly semisimple as well as completely regular, so that the subdirect sum representation is faithful, we obtain information on which elements $a \in \mathcal{A}$ belong to $\mathcal{L}$. We will now apply this result to the case of an ideal, where it can be sharpened by the following lemma.

7.2.12 Lemma *Let $\mathcal{A}$ be completely regular and let $\mathcal{I}$ be an ideal of $\mathcal{A}$. For each $a \in \mathcal{A}$, $\hat{a}$ belongs to $\mathcal{I}$ near each point in $h(\{a\})^\circ$ or in the complement of $h(\mathcal{I})$.*

Proof Since zero belongs to $\mathcal{I}$ and $\hat{a} = \hat{0}$ holds on $h(\{a\})^\circ$, a belongs to $\mathcal{I}$ near each point of $h(\{a\})^\circ$. If $\mathcal{M}$ belongs to $\Xi_\mathcal{A} \setminus h(\mathcal{I})$, we can choose a neighborhood V of $\mathcal{M}$ with compact closure disjoint from $h(\mathcal{I})$. Proposition 7.2.5 shows that $k(V)$ is modular so Proposition 7.1.9 shows that $\mathcal{I}$ contains a relative identity e for $k(V)$. Hence $(ea)\hat{} = \hat{a}$ holds on V. Since ea belongs to $\mathcal{I}$, this shows that a belongs to $\mathcal{I}$ near each point of $\Xi_\mathcal{A} \setminus h(\mathcal{I})$. □

7.2.13 Corollary *Let $\mathcal{A}$ be a strongly semisimple completely regular algebra, and let $\mathcal{I}$ be an ideal of $\mathcal{A}$.*

(a) An element $a \in \mathcal{A}$ belongs to $\mathcal{I}$ if and only if $\hat{a}$ belongs to $\mathcal{I}$ near each point of $h(\mathcal{I}) \setminus h(\{a\})^\circ$ and at infinity.

(b) If $V \subseteq \Xi_\mathcal{A}$ is an open set containing $h(\mathcal{I})$ and possessing a compact complement, then $k(V)$ is included in $\mathcal{I}$.

Proof Theorem 7.2.10 and Lemma 7.2.12 establish (a). Conclusion (b) is an immediate consequence. □.

We now investigate the connection between the structure spaces of an algebra and the Gelfand space of its center. The next result is related to Proposition 7.1.11. For a commutative spectral algebra $\mathcal{C}$, there is a close connection between its Gelfand space $\Gamma_\mathcal{C}$ and its strong structure space $\Xi_\mathcal{C}$. The map which takes $\gamma \in \Gamma_\mathcal{C}$ into its kernel is a bijection of $\Gamma_\mathcal{C}$ onto $\Xi_\mathcal{C} = \Pi_\mathcal{C}$. However, we endow the Gelfand space with its Gelfand topology and the structure spaces with their hull–kernel topologies. Hence this map is a homeomorphism if and only if $\mathcal{C}$ is completely regular.

7.2.14 Proposition *Let $\mathcal{A}$ be a spectral algebra with center $\mathcal{A}_Z$.*

(a) Each primitive ideal in $\Pi_\mathcal{A} \setminus h^\Pi(\mathcal{A}_Z)$ is modular with a relative identity in $\mathcal{A}_Z$.

(b) The map $\mathcal{P} \mapsto \mathcal{P} \cap \mathcal{A}_Z$ sends $\Pi_\mathcal{A} \setminus h^\Pi(\mathcal{A}_Z)$ into $\Pi_{\mathcal{A}_Z} = \Xi_{\mathcal{A}_Z}$.

(c) If $\mathcal{A}_Z$ is completely regular, this map is continuous and the images of both $\Pi_\mathcal{A} \setminus h^\Pi(\mathcal{A}_Z)$ and $\Xi_\mathcal{A} \setminus h^\Xi(\mathcal{A}_Z)$ are closed.

(d) If (c) holds and $\mathcal{A}$ is $\langle$ semisimple / strongly semisimple $\rangle$ then the image of $\langle \Pi_\mathcal{A} \setminus h^\Pi(\mathcal{A}_Z) / \Xi_\mathcal{A} \setminus h^\Xi(\mathcal{A}_Z) \rangle$ is all of $\Pi_{\mathcal{A}_Z} = \Xi_{\mathcal{A}_Z}$.

Proof (a), (b): Theorem 4.2.11 shows that for each primitive ideal $\mathcal{P}$ of $\mathcal{A}$ the center of $\mathcal{A}/\mathcal{P}$ is isomorphic to $\mathbb{C}$ when it is non-zero. However, the map $c + (\mathcal{P} \cap \mathcal{A}_Z) \mapsto c + \mathcal{P}$ is obviously an injective homomorphism of $\mathcal{A}_Z/(\mathcal{P} \cap \mathcal{A}_Z)$ into the center of $\mathcal{A}/\mathcal{P}$ which has non-zero range when $\mathcal{P}$ belongs to $\Pi_{\mathcal{A}} \setminus h^{\Pi}(\mathcal{A}_Z)$. Thus, the preimage of $1 \in \mathbb{C} \simeq (\mathcal{A}/\mathcal{P})_Z$ contains an element $e \in \mathcal{A}_Z$ which is a relative identity for $\mathcal{P}$. Furthermore, the isomorphism $\mathcal{A}_Z/(\mathcal{P} \cap \mathcal{A}_Z) \simeq \mathbb{C}$ shows that $\mathcal{P} \cap \mathcal{A}_Z$ belongs to $\Xi_{\mathcal{A}_Z}$. Theorem 4.1.9 shows $\Xi_{\mathcal{A}_Z} = \Pi_{\mathcal{A}_Z}$.

(c), (d): At first, we assume that $\mathcal{A}$ is unital (so that $h^{\Pi}(\mathcal{A}_Z)$ is empty) and we work exclusively with $\Pi_{\mathcal{A}} \setminus h^{\Pi}(\mathcal{A}_Z)$.

If B is any closed set in $\Pi_{\mathcal{A}_Z}$ then $\mathcal{P}_0 \in \Pi_{\mathcal{A}}$ belongs to the closure of its preimage in $\Pi_{\mathcal{A}}$ if and only if it satisfies $\mathcal{P}_0 \supseteq \cap\{\mathcal{P} \in \Pi_{\mathcal{A}} : \mathcal{P} \cap \mathcal{A}_Z \in B\}$. But this is equivalent to $\mathcal{P}_0 \cap \mathcal{A}_Z \supseteq \cap_{\mathcal{P} \in B} \mathcal{P}$, which implies $\mathcal{P}_0 \cap \mathcal{A}_Z \in B$ since B is closed. Since the preimage in $\Pi_{\mathcal{A}}$ of each closed set is closed, the map $\mathcal{P} \mapsto \mathcal{P} \cap \mathcal{A}_Z$ is continuous from $\Pi_{\mathcal{A}}$ to $\Pi_{\mathcal{A}_Z}$. Since $\Pi_{\mathcal{A}}$ is compact and $\Pi_{\mathcal{A}_Z}$ is Hausdorff, this shows that the image of $\Pi_{\mathcal{A}}$ is closed. Finally if $\mathcal{A}$ is semisimple, then $\cap_{\mathcal{P} \in \Pi_{\mathcal{A}}}(\mathcal{P} \cap \mathcal{A}_Z) = (\cap_{\mathcal{P} \in \Pi_{\mathcal{A}}} \mathcal{P}) \cap \mathcal{A}_Z = \{0\}$ implies that the image of $\Pi_{\mathcal{A}}$ is dense. Since this image is also closed, it is all of $\Pi_{\mathcal{A}_Z}$. This establishes the proposition for $\Pi_{\mathcal{A}}$ when $\mathcal{A}$ is unital.

Suppose $\mathcal{A}$ is nonunital. Then $(\mathcal{A}^1)_Z$ is naturally isomorphic to $(\mathcal{A}_Z)^1$. Proposition 7.2.2 identifies $\Pi_{\mathcal{A}^1}$ and $\Pi_{\mathcal{A}^1_Z}$ with $\Pi_{\mathcal{A}} \cup \{\infty\}$ and $\Pi_{\mathcal{A}_Z} \cup \{\infty\}$, respectively, under suitable topologies. Since $\mathcal{P} \mapsto \mathcal{P} \cap \mathcal{A}_Z$ carries the point ∞ $(= \mathcal{A}$ in $\Pi_{\mathcal{A}^1})$ onto ∞ $(= \mathcal{A}_Z$ in $\Pi_{\mathcal{A}^1_Z})$, the nonunital case follows from the unital case. The proof for the strong structure space is similar. $\square$

Note that the above theorem does not assert that the map $\mathcal{P} \mapsto \mathcal{P} \cap \mathcal{A}_Z$ is injective on either $\Pi_{\mathcal{A}} \setminus h^{\Pi}(\mathcal{A}_Z)$ or $\Xi_{\mathcal{A}} \setminus h(\mathcal{A}_Z)$. This is not true in general, but the cases in which it is true are of sufficient interest to have been studied. The two parts of the following definition are due to Kaplansky [1949a] and Yosinao Misonou [1952], respectively.

7.2.15 Definition An algebra $\mathcal{A}$ is said to be $\langle$ *central / weakly central* $\rangle$ if $\langle h^{\Pi}(\mathcal{A}_Z) \, / \, h^{\Xi}(\mathcal{A}_Z) \rangle$ is empty and the map $\mathcal{P} \mapsto \mathcal{P} \cap \mathcal{A}_Z$ is injective on $\langle \Pi_{\mathcal{A}} \, / \, \Xi_{\mathcal{A}} \rangle$. A central algebra is obviously weakly central. An algebra which satisfies only $h^{\Pi}(\mathcal{A}_Z) = \emptyset$ is called *quasi-central*.

7.2.16 Theorem *A central spectral algebra $\mathcal{A}$ satisfies $\Pi_{\mathcal{A}} = \Xi_{\mathcal{A}}$.*

Proof Proposition 7.2.14(a) shows that each primitive ideal $\mathcal{P}$ of $\mathcal{A}$ is modular and hence is included in a maximal modular ideal $\mathcal{M}$ (Theorem 2.4.6). This implies $\mathcal{P} \cap \mathcal{A}_Z \subseteq \mathcal{M} \cap \mathcal{A}_Z$. Proposition 7.2.14(b) shows that $\mathcal{P} \cap \mathcal{A}_Z$ is maximal and hence equal to $\mathcal{M} \cap \mathcal{A}_Z$. Since $\mathcal{P} \mapsto \mathcal{P} \cap \mathcal{A}_Z$ is injective on $\Pi_{\mathcal{A}}$, we conclude $\mathcal{P} = \mathcal{M}$. $\square$

Finally we use Proposition 7.2.14 to show that certain weakly central spectral algebras are completely regular. This result is due to Willcox [1956a].

7.2.17   **Theorem**   *Let A be a weakly central spectral algebra with a completely regular center. Then A is completely regular and $M \mapsto M \cap A_Z$ is a homeomorphism of Ξ_A onto a closed subset of Ξ_{A_Z}. If A is either strongly semisimple or central and semisimple, then $M \mapsto M \cap A_Z$ is a homeomorphism of Ξ_A onto Ξ_{A_Z}.*

Proof Proposition 7.2.14(c) shows that $M \mapsto M \cap A_Z$ is a continuous map of Ξ_A onto a closed subset of Ξ_{A_Z}. Since A is weakly central, this map is a bijection onto its range, and since A_Z is completely regular, this range is Hausdorff. Hence Ξ_A is Hausdorff.

Let M_0 be an arbitrary point in Ξ_A. Use the complete regularity of A_Z to choose a neighborhood $\tilde{V}$ of $M_0 \cap A_Z$ in Ξ_{A_Z} such that $k(\tilde{V})$ is a modular ideal in A_Z with relative identity e. Let V be the preimage of $\tilde{V}$ in Ξ_A (*i.e.*, $V = \{M \in \Xi_A : M \cap A_Z \in \tilde{V}\}$). Proposition 7.2.14(a) shows that each $M \in V$ has a relative identity e' in A_Z. Since $e + (M \cap A_Z)$ and $e' + (M \cap A_Z)$ both equal the identity element of $A_Z/(M \cap A_Z)$, we conclude $e - e' \in M \cap A_Z$. Hence, e is a relative identity for each M in V. Thus, e is a relative identity for $k(V)$. Therefore, V is a neighborhood of M_0 with modular kernel. This completes the proof that A is completely regular.

Let M_0 be an arbitrary point in Ξ_A and let K be a compact neighborhood of M_0. Then the restriction of the map $M \mapsto M \cap A_Z$ to K is a continuous bijection of a compact space onto a Hausdorff space and is therefore a homeomorphism. Thus, the map $M \mapsto M \cap A_Z$ is a homeomorphism of Ξ_A onto its range. If A is strongly semisimple, Proposition 7.2.14 shows that the range of this map is Ξ_{A_Z}. If A is central, Theorem 7.2.16 shows $\Pi_A = \Xi_A$. Therefore, semisimplicity implies strong semisimplicity. $\square$

See also John Francis Rennison [1982], [1987], [1988], [1990] and Sin-ei Takahasi [1981], [1983a], [1983b], [1986].

7.3   Primary Ideals and Spectral Synthesis

Identifying the ideals of an algebra A is always one of the most basic and most important tasks in understanding the structure of the algebra. The maximal modular ideals and the primitive ideals are usually much easier to find than the general ideals. (In many Banach algebras the prime ideals are not easily found.) The ideals which are intersections of maximal modular ideals or of primitive ideals are therefore tractable. Another way to express this property of an ideal is to say that it is the kernel of its hull, taken with respect to Ξ_A or Π_A, respectively. Thus, one is led to ask which ideals in an algebra are of this type. When A is a spectral normed algebra, any intersection of maximal modular or of primitive ideals is closed so that one should ask which closed ideals are the intersection of maximal modular or of primitive ideals. This is often called the problem of *spectral synthesis*.

In order to use the theory of completely regular algebras, we concentrate on the question of which closed ideals are the intersection of maximal modular ideals. The following definitions are essentially due to Georgi Evgenyevič Šilov [1947a] and Willcox [1956a].

7.3.1 Definition Let $\mathcal{A}$ be an ⟨ algebra / spectral normed algebra ⟩. Then $\mathcal{A}$ is called an ⟨ *N-algebra* / *topological N-algebra* ⟩ if each of its ⟨ ideals / closed ideals ⟩ with nonempty hull in $\Xi_{\mathcal{A}}$ is equal to the kernel of its hull. It is called ⟨ *Tauberian* / *topologically Tauberian* ⟩ if each of its ⟨ ideals / closed ideals ⟩ is included in some maximal modular ideal.

An ideal which is included in exactly one maximal modular ideal is said to be *primary*. If a primary ideal $\mathcal{P}$ satisfies $h^{\Xi}(\mathcal{P}) = \{\mathcal{M}\}$, then $\mathcal{P}$ is said to be $\mathcal{M}$-*primary*. We call $\mathcal{A}$ a ⟨ *weak N-algebra* / *weak topological N-algebra* ⟩ if each of its ⟨ primary ideals / closed primary ideals ⟩ is maximal modular.

Let $\mathcal{A}$ be an algebra and let $H \subseteq \Xi_{\mathcal{A}}$ be a hull. Let $\mathcal{J}(H)$ be the sum of all ideals of the form $k(V)$ where V is an open set in $\Xi_{\mathcal{A}}$ which includes H, and let $\mathcal{J}(H, \infty)$ be the sum of all ideals of the form $k(V)$, where V is an open subset of $\Xi_{\mathcal{A}}$ including H with compact complement. If $H = \{\mathcal{M}\}$ is a singleton, we write $\mathcal{J}(H) = \mathcal{J}(\mathcal{M})$ and $\mathcal{J}(H, \infty) = \mathcal{J}(\mathcal{M}, \infty)$ and if H is empty, we write $\mathcal{J}(\emptyset, \infty) = \mathcal{J}(\infty)$.

Note that an ⟨ algebra / spectral normed algebra ⟩ is an ⟨ N-algebra / topological N-algebra ⟩ if and only if each ⟨ homomorphic image / continuous homomorphic image ⟩ of $\mathcal{A}$ is either Brown–McCoy radical or strongly semisimple. (The term N*-algebra used by Šilov [1947a], Willcox [1956a], Rickart [1960] and others must be changed since an asterisk in the name of a class of algebras is now universally used to denote a class of algebras with involution.)

Our definition of Tauberian algebras differs from that given by Willcox [1956a] and Rickart [1960]. We will show that a strongly semisimple completely regular Banach algebra is topologically Tauberian in our terminology exactly when it is Tauberian in their terminology. For other algebras, our definition seems more useful. The terminology arises from a close connection with the Tauberian theorems of harmonic analysis. Note that an ⟨ algebra / spectral normed algebra ⟩ is a ⟨ Tauberian N-algebra / topologically Tauberian topological N-algebra ⟩ if and only if each of its ⟨ ideals / closed ideals ⟩ is the kernel of its hull. This happens if and only if each of its ⟨ homomorphic / continuous homomorphic ⟩ images is strongly semisimple.

Our definition of primary ideals is a considerable generalization of the classical notion of primary ideals used in commutative algebra (*cf., e.g.,* Oscar Zariski and Pierre Samuel [1958]). However, the primary ideals defined here play approximately the same role in the present theory as the classical ones do in commutative algebra. Note that in a spectral normed algebra, the closure of any $\mathcal{M}$-primary ideal is $\mathcal{M}$-primary. A maximal modular ideal $\mathcal{M}$ is always $\mathcal{M}$-primary. An $\mathcal{M}$-primary ideal which is not equal to $\mathcal{M}$ is obviously not the kernel of its hull. Hence, any ⟨ N-algebra / topological N-algebra ⟩ is a ⟨ weak N-algebra / weak topological N-algebra ⟩. If an algebra is not a weak N-algebra, we may hope to write some ideals as an intersection of primary ideals even if they are not an intersection of maximal modular ideals.

Corollary 7.2.13 suggested the definition of $\mathcal{J}(H)$ and $\mathcal{J}(H, \infty)$. If V_1 and V_2 are two open sets in $\Xi_\mathcal{A}$ both including H, then $V_1 \cap V_2$ is another such set. Hence, the equation $k(V_1 \cap V_2) = k(V_1) + k(V_2)$ shows that $\mathcal{J}(H)$ can also be described as the union of all ideals of the form $k(V)$ for V an open set including H. The same reasoning shows that $\mathcal{J}(H, \infty)$ is the union of all ideals of the form $k(V)$, where V is an open subset of $\Xi_\mathcal{A}$ including H with compact complement. Of course if $\mathcal{A}$ is unital, $\mathcal{J}(H)$ and $\mathcal{J}(H, \infty)$ coincide.

If $\mathcal{A}$ is strongly semisimple and we consider it as identified with its subdirect product representation in $\Pi_{\mathcal{M} \in \Xi_\mathcal{A}} \mathcal{A}/\mathcal{M}$ or in $S_{\Xi_\mathcal{A}}(\mathcal{A})$, then the ideal $\mathcal{J}(H)$ is the set of elements which vanish near H, and $\mathcal{J}(H, \infty)$ is the set of elements which vanish both near H and at infinity. Thus for $\mathcal{M} \in \Xi_\mathcal{A}$, $\mathcal{A}/\mathcal{J}(\mathcal{M})$ can be identified with the algebra of germs of functions vanishing near $\mathcal{M}$. This is of course a standard construction in sheaf theory (*cf.* Roger Godement [1958]).

7.3.2 Proposition *Let $\mathcal{A}$ be a strongly semisimple completely regular algebra. For any hull $H \subseteq \Xi_\mathcal{A}$, $\mathcal{J}(H, \infty)$ is minimal among ideals with hull H. (That is, $\mathcal{J}(H, \infty)$ satisfies $h(\mathcal{J}(H, \infty)) = H$ and is included in any ideal $\mathcal{I}$ of $\mathcal{A}$ which satisfies $h(\mathcal{I}) = H$.) In particular, for each $\mathcal{M} \in \Xi_\mathcal{A}$, $\mathcal{J}(\mathcal{M}, \infty)$ is minimal among all $\mathcal{M}$-primary ideals.*

Proof First we note that $h(\mathcal{J}(H, \infty))$ equals H. The inclusion $H \subseteq h(\mathcal{J}(H, \infty))$ is obvious. Suppose $\mathcal{M} \in \Xi_\mathcal{A}$ does not belong to H. Since $\Xi_\mathcal{A}$ is locally compact, we can choose an open neighborhood V of $\mathcal{M}$ in $\Xi_\mathcal{A}$ with compact closure disjoint from H. Then $U = \Xi_\mathcal{A} \setminus V^-$ satisfies $k(U) \subseteq \mathcal{J}(H, \infty)$. Hence, $h(\mathcal{J}(H, \infty)) \subseteq hk(U) \subseteq \Xi_\mathcal{A} \setminus V$ does not include $\mathcal{M}$. Since $\mathcal{M} \notin H$ was arbitrary, this shows $H = h(\mathcal{J}(H, \infty))$.

If $\mathcal{I}$ is an ideal satisfying $h(\mathcal{I}) = H$, and V is an open set including H with compact complement, then Corollary 7.2.13(b) shows $k(V) \subseteq \mathcal{I}$. Hence we conclude $\mathcal{J}(H, \infty) \subseteq \mathcal{I}$. □

7.3.3 Corollary *If $\mathcal{A}$ is a strongly semisimple completely regular spectral normed algebra, then $\mathcal{J}(H, \infty)^-$ is minimal among closed ideals with hull H. (That is, $\mathcal{J}(H, \infty)^-$ is a closed ideal with hull H and is included in* *any other closed ideal with hull H.) In particular, for each* $\mathcal{M} \in \Xi_{\mathcal{A}}$, $\mathcal{J}(\mathcal{M},\infty)^-$ *is minimal among all closed* $\mathcal{M}$-*primary ideals.*

Proof As noted above, the closure of an $\mathcal{M}$-primary ideal in a spectral normed algebra is $\mathcal{M}$-primary.   □

7.3.4 Theorem *Let* $\mathcal{A}$ *be a strongly semisimple completely regular* ⟨ *algebra / spectral normed algebra* ⟩.

(a) $\mathcal{A}$ *is an* ⟨ *N-algebra / topological N-algebra* ⟩ *if and only if* ⟨ $\mathcal{J}(H,\infty)$ / $\mathcal{J}(H,\infty)^-$ ⟩ *equals* $k(H)$ *for each nonempty hull H in* $\Xi_{\mathcal{A}}$.

(b) $\mathcal{A}$ *is a* ⟨ *weak / weak topological* ⟩ *N-algebra if and only if* ⟨ $\mathcal{J}(\mathcal{M},\infty)$ / $\mathcal{J}(\mathcal{M},\infty)^-$ ⟩ *equals* $\mathcal{M}$ *for each maximal modular ideal* $\mathcal{M}$ *of* $\mathcal{A}$.

(c) $\mathcal{A}$ *is* ⟨ *Tauberian / topologically Tauberian* ⟩ *if and only if* ⟨ $\mathcal{J}(\infty)$ / $\mathcal{J}(\infty)^-$ ⟩ *equals* $\mathcal{A}$.

Proof These results are immediate consequences of ⟨ Proposition 7.3.2 / Corollary 7.3.3 ⟩ which shows that any ⟨ ideal / closed ideal ⟩ of $\mathcal{A}$ satisfies ⟨ $\mathcal{J}(h(\mathcal{I}),\infty) \subseteq \mathcal{I} \subseteq kh(\mathcal{I})$ / $\mathcal{J}(h(\mathcal{I}),\infty)^- \subseteq \mathcal{I} \subseteq kh(\mathcal{I})$ ⟩ whether $h(\mathcal{I})$ is arbitrary nonempty (a), a singleton (b) or empty (c).   □

Specializing to the case of a strongly semisimple completely regular Banach algebra $\mathcal{A}$, we wish to find conditions on the faithful subdirect product representation $\mathcal{A} \to \hat{\mathcal{A}} \subseteq \Pi_{\mathcal{M}\in\Xi_{\mathcal{A}}}(\mathcal{A}/\mathcal{M})$ related to the properties of Theorem 7.3.4. Note first that $\hat{\mathcal{A}}$ is a normed subdirect product in the sense that

$$\|\hat{a}\|_\infty = \sup\{\|\hat{a}(\mathcal{M})\| : \mathcal{M} \in \Xi_{\mathcal{A}}\}$$
$$= \sup\{\|a + \mathcal{M}\| : \mathcal{M} \in \Xi_{\mathcal{A}}\}$$

is finite and hence defines an algebra norm $\|\cdot\|_\wedge$ on $\mathcal{A}$ (where, of course, $\|\hat{a}(\mathcal{M})\| = \|a+\mathcal{M}\| = \inf\{\|a+b\| : b \in \mathcal{M}\}$ is the quotient norm on $\mathcal{A}/\mathcal{M}$). Clearly this norm satisfies

$$\|a\|_\wedge \le \|a\| \quad \forall \ a \in \mathcal{A}.$$

It is not usually complete; in fact, the inverse boundedness theorem shows that it is complete if and only if it is equivalent to $\|\cdot\|$.

We say that $\hat{\mathcal{A}} \subseteq \Pi_{\mathcal{M}\in\Xi_{\mathcal{A}}}(\mathcal{A}/\mathcal{M})$ is *closed under multiplication by* $C(\Xi_{\mathcal{A}})$ if for each $f \in C(\Xi_{\mathcal{A}})$ and $a \in \mathcal{A}$, there is a $b \in \mathcal{A}$ satisfying

$$\hat{b}(\mathcal{M}) = f(\mathcal{M})\hat{a}(\mathcal{M}) \quad \forall \ \mathcal{M} \in \Xi_{\mathcal{A}}.$$

The element b is then denoted by fa.

7.3.5 Theorem *Let* $\mathcal{A}$ *be a strongly semisimple completely regular Banach algebra. Let* $\|\cdot\|_\wedge$ *be equivalent to* $\|\cdot\|$ *and let* $\hat{\mathcal{A}}$ *be closed under* *multiplication by* $C(\Xi_{\mathcal{A}})$. *Then* $\mathcal{A}$ *is a topological N-algebra if it is a weak topological N-algebra.*

Proof The last theorem shows that we may assume $\mathcal{J}(\mathcal{M}, \infty)^- = \mathcal{M}$ for all $\mathcal{M} \in \Xi_{\mathcal{A}}$. Hence, we need only show $\mathcal{J}(H, \infty)^- \supseteq k(H)$ for all nonempty hulls H in $\Xi_{\mathcal{A}}$. Since $\mathcal{J}(\infty)^-$ includes $\mathcal{J}(\mathcal{M}, \infty)^- = \mathcal{M}$ for each $\mathcal{M} \in \Xi_{\mathcal{A}}$, it includes $k(H)$. Hence, for any $b \in k(H)$ and any $\varepsilon > 0$, there is some open set U_0 in $\Xi_{\mathcal{A}}$ with compact complement and an element $c_0 \in k(U_0)$ satisfying $\|b - c_0\| < \varepsilon$. Let K_0 be the compact set $H \setminus U_0$. The equation $\mathcal{J}(\mathcal{M}, \infty)^- = \mathcal{M}$ shows that for each $\mathcal{M}$ in K_0 there is an open neighborhood $U_{\mathcal{M}}$ of $\mathcal{M}$ and an element $c_{\mathcal{M}} \in k(U_{\mathcal{M}})$ satisfying $\|b - c_{\mathcal{M}}\| < \varepsilon$. Since K_0 is compact, it is covered by a finite family $U_1, U_2, \dots, U_n$ of these neighborhoods $U_{\mathcal{M}}$. For $j = 1, 2, \dots, n$, let $c_j \in k(U_j)$ be an element satisfying $\|b - c_j\| < \varepsilon$. Since $\Xi_{\mathcal{A}}$ is locally compact and $U_0 \cup U_1 \cup \cdots \cup U_n$ is an open set including H with compact complement, there is an open set U_{n+1} which includes the complement of $U_0 \cup U_1 \cup \cdots \cup U_n$ but has compact closure disjoint from H. Since $U_0, U_1, \dots, U_{n+1}$ is a finite cover of $\Xi_{\mathcal{A}}$, there exists a partition of unity $f_0, f_1, \dots, f_{n+1}$ subordinate to this cover. (That is, for $j = 0, 1, 2, \dots, n+1$ there exists a function $f_j \in C(\Xi_{\mathcal{A}})$ with range in $[0,1]$ satisfying $f_j(\mathcal{M}) = 0$ for $\mathcal{M} \notin U_j$ and $\sum_{j=0}^{n+1} f_j(\mathcal{M}) = 1$ for all $\mathcal{M} \in \Xi_{\mathcal{A}}$. See Bourbaki [1966] IX.4.3 for a proof.) Define $c \in \mathcal{A}$ by

$$c = f_0 c_0 + f_1 c_1 + \cdots + f_n c_n + f_{n+1} b$$

and $U \subseteq \Xi_{\mathcal{A}}$ by $U = \Xi_{\mathcal{A}} \setminus U_{n+1}^-$. Since each c_j belongs to $k(U_j)$ and each f_j vanishes off U_j for $j = 1, 2, \dots, n$, the element c belongs to $k(U)$. Since U is an open neighborhood of H with compact complement, the element c belongs to $\mathcal{J}(H, \infty)$. For each $\mathcal{M} \in \Xi_{\mathcal{A}}$, we get

$$
\begin{aligned}
\|\hat{c}(\mathcal{M}) - \hat{b}(\mathcal{M})\| &= \left\| \sum_{j=0}^{n} f_j(\mathcal{M}) \hat{c}_j(\mathcal{M}) - \sum_{j=0}^{n} f_j(\mathcal{M}) \hat{b}(\mathcal{M}) \right\| \\
&\leq \sum_{j=0}^{n} f_j(\mathcal{M}) \|\hat{c}_j(\mathcal{M}) - \hat{b}(\mathcal{M})\| \\
&\leq \sum_{j=0}^{n} f_j(\mathcal{M}) \|c_j - b\| < \varepsilon.
\end{aligned}
$$

Hence $\|c - b\|_{\wedge} \leq \varepsilon$ holds. Since $\| \cdot \|_{\wedge}$ and $\| \cdot \|$ are equivalent and $\varepsilon > 0$ was arbitrary, we conclude that b belongs to $\mathcal{J}(H, \infty)^-$ as we wished to show. $\square$

7.3.6 Theorem *Let $\mathcal{A}$ be a strongly semisimple completely regular Banach algebra. Let $\| \cdot \|_{\wedge}$ be equivalent to $\| \cdot \|$ and let $\hat{\mathcal{A}}$ be closed under multiplication by $C(\Xi_{\mathcal{A}})$. For each $a \in \mathcal{A}$, let the function $\mathcal{M} \mapsto \|\hat{a}(\mathcal{M})\|$ from $\Xi_{\mathcal{A}}$ to*

$\mathbb{R}_+$ be continuous at zero and vanish at infinity. Then every closed ideal $\mathcal{I}$ of $\mathcal{A}$ satisfies $\mathcal{I} = kh(\mathcal{I})$, so that $\mathcal{A}$ is a topologically Tauberian topological N-algebra.

Proof The previous two theorems show that it is enough to establish $\mathcal{J}(\infty)^- = \mathcal{A}$ and $\mathcal{J}(\mathcal{M}, \infty)^- \supseteq \mathcal{M}$ for each $\mathcal{M} \in \Xi_{\mathcal{A}}$. We verify both simultaneously. Let $\mathcal{M}_0$ be $\langle$ a maximal modular ideal / all of $\mathcal{A}$ $\rangle$ and let a be an element of $\mathcal{M}_0$. For any $\varepsilon > 0$ there is an open set V with compact complement satisfying $\|\hat{a}(\mathcal{M})\| < \varepsilon$ for each $\mathcal{M} \in V$ and $\langle$ $\mathcal{M}_0 \in V$ / no additional condition $\rangle$. Let U be an open set satisfying $\overline{U} \subseteq V$ and $\langle$ $\mathcal{M}_0 \in U$ / no additional condition $\rangle$. Urysohn's lemma (*cf. e.g.* Rudin [1974]) shows that there is an $f \in C(\Xi_{\mathcal{A}})$ with range in [0,1] satisfying $f(\mathcal{M}) = 1$ for $\mathcal{M} \notin V$ and $f(\mathcal{M}) = 0$ for $\mathcal{M} \in U$. Define b to be fa. Then $\|a - b\|_{\wedge}$ is the supremum of the numbers

$$\|\hat{a}(\mathcal{M}) - \hat{b}(\mathcal{M})\| = \begin{cases} 0 & \text{if } \mathcal{M} \notin V \\ f(\mathcal{M})\|\hat{a}(\mathcal{M})\| < \varepsilon & \text{if } \mathcal{M} \in V. \end{cases}$$

The equivalence of $\| \cdot \|_{\wedge}$ and the complete norm shows $\langle$ $a \in \mathcal{J}(\mathcal{M}_0, \infty)^-$ / $a \in \mathcal{J}(\infty)^-$ $\rangle$. □

The abstract commutative Banach algebra formulation of Ditkin's theorem in Šilov [1947a] is generalized to the setting of semisimple completely regular weak topological Banach N-algebras by Willcox [1956a], Corollary 1.6.1. (In Definition 2.3 of this paper, change "$I(x) \subset \mathcal{I}(M_0)$" to "$I(x) \cap \mathcal{I}(M_0)$".)

We now explore the theory of primary ideals in more detail beginning with a beautiful result of Willcox [1956a].

7.3.7 Theorem *Let $\mathcal{A}$ be a strongly semisimple completely regular algebra. Then each ideal $\mathcal{I}$ of $\mathcal{A}$ with nonempty hull is the intersection of the set*

$$\{\mathcal{I} + \mathcal{J}(\mathcal{M}, \infty) : \mathcal{M} \in h(\mathcal{I})\}$$

of primary ideals.

Proof Proposition 7.3.2 shows that for each $\mathcal{M} \in h(\mathcal{I})$, the hull of $\mathcal{I} + \mathcal{J}(\mathcal{M}, \infty)$ must be $\{\mathcal{M}\}$ so that this ideal is $\mathcal{M}$-primary. Obviously, the intersection $\cap\{\mathcal{I} + \mathcal{J}(\mathcal{M}, \infty) : \mathcal{M} \in h(\mathcal{I})\}$ includes $\mathcal{I}$. On the other hand, for each element $a \in \mathcal{I} + \mathcal{J}(\mathcal{M}, \infty)$, it is clear that $\hat{a}$ belongs to $\mathcal{I}$ near $\mathcal{M}$ and at infinity. Hence, each element of the intersection satisfies the criterion of Corollary 7.2.13(a) and thus belongs to $\mathcal{I}$. □

It does not appear to be known whether the analogous topological version of this theorem holds. We would want to show that if a spectral normed algebra or Banach algebra satisfies the hypotheses of the above theorem, then each of its closed ideals $\mathcal{I}$ with nonempty hull satisfies

$$\mathcal{I} = \cap\{(\mathcal{I} + h(\mathcal{M}, \infty))^- : \mathcal{M} \in h(\mathcal{I})\}.$$

The set of ideals which are intersected in this equation is obviously the set of closed primary ideals which are minimal among closed primary ideals including $\mathcal{I}$, so that this equation holds if $\mathcal{I}$ can be written as the intersection of closed primary ideals. Unfortunately, no proof seems to be known to show that the above intersection is included in $\mathcal{I} = (\cap\{\mathcal{I} + h(\mathcal{M}, \infty) : \mathcal{M} \in h(\mathcal{I})\})^-$. Some partial results are given in Willcox [1956a].

If $\mathcal{A}$ is a strongly semisimple completely regular $\langle$ algebra / spectral normed algebra $\rangle$, then $\mathcal{A}$ has another subdirect product representation besides the representation $\hat{\mathcal{A}} \subseteq \Pi_{\mathcal{M} \in \Xi_\mathcal{A}}(\mathcal{A}/\mathcal{M})$. For each $\mathcal{M} \in \Xi_\mathcal{A}$, $\langle$ Proposition 7.3.2 / Corollary 7.3.3 $\rangle$ asserts that $\langle \mathcal{J}(\mathcal{M}, \infty) / \mathcal{J}(\mathcal{M}, \infty)^- \rangle$ is an $\mathcal{M}$-primary ideal. Hence the intersection of these primary ideals is the zero ideal and for each $\mathcal{M} \in \Xi_\mathcal{A}$, $\langle \mathcal{A}/\mathcal{J}(\mathcal{M}, \infty) / \mathcal{A}/\mathcal{J}(\mathcal{M}, \infty)^- \rangle$ is an $\langle$ algebra / spectral normed algebra $\rangle$ which is *local* in the sense that it has exactly one maximal modular ideal. For each $a \in \mathcal{A}$ and $\mathcal{M} \in \Xi_\mathcal{A}$, define $\tilde{a}(\mathcal{M})$ to be $\langle a + \mathcal{J}(\mathcal{M}, \infty) / a + \mathcal{J}(\mathcal{M}, \infty)^-) \rangle$ in this quotient algebra. Then the following result is immediate.

7.3.8 Proposition *Let $\mathcal{A}$ be a strongly semisimple completely regular $\langle$ algebra / spectral normed algebra $\rangle$. Then the function $a \mapsto \tilde{a}$ defined above is a faithful $\langle$ representation / continuous representation $\rangle$ of $\mathcal{A}$ as a $\langle$ subdirect product / normed subdirect product $\rangle$ of local $\langle$ algebras / spectral normed algebras $\rangle$.*

7.4 Strongly Harmonic Algebras

The hypothesis of strong semisimplicity (*i.e.*, $\cap_{\mathcal{M} \in \Xi_\mathcal{A}} \mathcal{M} = \{0\}$) seems unnecessarily strong for the above result since the conclusions may hold if $\langle \cap_{\mathcal{M} \in \Xi_\mathcal{A}} \mathcal{J}(\mathcal{M}, \infty) / \cap_{\mathcal{M} \in \Xi_\mathcal{A}} \mathcal{J}(\mathcal{M}, \infty)^- \rangle$ is zero and each $\mathcal{J}(\mathcal{M}, \infty)$ is a primary ideal. We will now introduce a class of algebras which are not all strongly semisimple, but have a very satisfactory subdirect product representation in terms of primary direct factors. As a matter of fact, for these algebras this subdirect product representation can be replaced by a representation as an algebra of sections of a suitable sheaf over $\Xi_\mathcal{A}$ as base space in which the fibers are local algebras.

This class of algebras was introduced by Kwangil Koh [1972]. His work is strongly influenced by a series of papers by Silviu Teleman [1969a], [1969b] and [1969c] which are presented in a connected way in Teleman's survey [1971]. These results are also related to the works by John Dauns and Karl Heinrich Hofmann [1966] and [1968].

7.4.1 Definition An algebra $\mathcal{A}$ is said to be *strongly harmonic* if for each pair $\mathcal{M}_1$, $\mathcal{M}_2$ of distinct maximal modular ideals of $\mathcal{A}$, there are ideals $\mathcal{I}_1$ and $\mathcal{I}_2$ satisfying $\mathcal{M}_j \notin h(\mathcal{I}_j)$ for $j = 1, 2$ and $\mathcal{I}_1\mathcal{I}_2 = \{0\}$.

These algebras are closely related to completely regular algebras as shown in the next proposition.

7.4.2 Proposition *The strong structure space of a strongly harmonic algebra is Hausdorff. Hence, a unital strongly harmonic algebra is completely regular. Conversely, a strongly semisimple algebra with a Hausdorff strong structure space is strongly harmonic.*

Proof Let $\mathcal{M}_1$ and $\mathcal{M}_2$ be distinct elements of $\Xi_{\mathcal{A}}$ and let $\mathcal{I}_1$ and $\mathcal{I}_2$ be the corresponding ideals guaranteed by Definition 7.4.1 satisfying $\mathcal{I}_1\mathcal{I}_2 = \{0\}$ and $\mathcal{M}_j \notin h(\mathcal{I}_j)$ for $j = 1, 2$. Then $h(\mathcal{I}_1) \cup h(\mathcal{I}_2) \supseteq h(\mathcal{I}_1\mathcal{I}_2) = \Xi_{\mathcal{A}}$ shows that $\Xi_{\mathcal{A}} \setminus h(\mathcal{I}_2)$ and $\Xi_{\mathcal{A}} \setminus h(\mathcal{I}_1)$ are disjoint neighborhoods of $\mathcal{I}_1$ and $\mathcal{I}_2$, respectively. Hence, $\Xi_{\mathcal{A}}$ is Hausdorff, so if $\mathcal{A}$ is unital it is completely regular.

Suppose $\mathcal{A}$ is strongly semisimple and its strong structure space is Hausdorff. Then for any two distinct ideals $\mathcal{M}_1$ and $\mathcal{M}_2$ in $\Xi_{\mathcal{A}}$, choose disjoint open neighborhoods V_1 and V_2 of $\mathcal{M}_1$ and $\mathcal{M}_2$, respectively. Then the ideals $\mathcal{I}_1$ and $\mathcal{I}_2$ of Definition 7.4.1 may be chosen as $\mathcal{I}_1 = k(\Xi_{\mathcal{A}} \setminus V_1)$ and $\mathcal{I}_2 = k(\Xi_{\mathcal{A}} \setminus V_2)$ since these ideals satisfy $\mathcal{M}_1 \notin \Xi_{\mathcal{A}} \setminus V_1 = h(\mathcal{I}_1)$, $\mathcal{M}_2 \notin \Xi_{\mathcal{A}}\setminus V_2 = h(\mathcal{I}_2)$ and $\mathcal{I}_1\mathcal{I}_2 \subseteq \mathcal{I}_1 \cap \mathcal{I}_2 = k(\Xi_{\mathcal{A}}\setminus V_1) \cap k(\Xi_{\mathcal{A}}\setminus V_2) = k(\Xi_{\mathcal{A}}) = \{0\}$. □

For an algebra $\mathcal{A}$ which is strongly harmonic rather than strongly semisimple and completely regular, it is not clear that $\mathcal{J}(\mathcal{M}, \infty)$ will be a primary ideal when $\mathcal{M}$ is a maximal modular ideal of $\mathcal{A}$. The next definition gives a suitable substitute for $\mathcal{J}(\mathcal{M})$.

7.4.3 Definition For each maximal modular ideal $\mathcal{M}$ in $\mathcal{A}$, the $\mathcal{M}$-*component of the zero ideal* is the ideal $\mathcal{I}(\mathcal{M})$ defined by

$$\mathcal{I}(\mathcal{M}) = \{a \in \mathcal{A} : a\mathcal{A}^1 b = \{0\} \text{ for some } b \in \mathcal{A} \setminus \mathcal{M}\}.$$

Since a maximal modular ideal is prime, it is clear that $\mathcal{I}(\mathcal{M})$ is included in $\mathcal{M}$. (To see this, note that $a \in \mathcal{I}(\mathcal{M})$ implies $\mathcal{A}^1 a\mathcal{A}^1 b = \{0\} \subseteq \mathcal{M}$ for some b in $\mathcal{A} \setminus \mathcal{M}$. Since $\mathcal{M}$ is prime, this implies $a \in \mathcal{A}^1 a \subseteq \mathcal{M}$.) We can describe $\mathcal{I}(\mathcal{M})$ in terms of the left and right annihilators introduced in Definition 1.2.1.

$$\mathcal{I}(\mathcal{M}) = \cup\{(\mathcal{A}^1 b)_{LA} : b \in \mathcal{A} \setminus \mathcal{M}\} = \{a \in \mathcal{A} : (a\mathcal{A}^1)_{RA} \not\subseteq \mathcal{M}\}.$$

Hence, if a belongs to $\mathcal{I}(\mathcal{M})$ for some $\mathcal{M} \in \Xi_{\mathcal{A}}$, then $U = \Xi_{\mathcal{A}} \setminus h((a\mathcal{A}^1)_{RA})$ is a neighborhood of $\mathcal{M}$ in $\Xi_{\mathcal{A}}$ and a belongs to each $\mathcal{M}'$ in U.

We will now show that $\mathcal{I}(\mathcal{M})$ is well-behaved in certain strongly harmonic algebras including unital ones. A satisfactory theory depends on making some hypothesis similar to assuming that the algebra is unital, since $\mathcal{I}(\mathcal{M})$ is an analogue of $\mathcal{J}(\mathcal{M})$ rather than $\mathcal{J}(\mathcal{M}, \infty)$.

7.4.4 Proposition *The following two conditions are equivalent for an algebra $\mathcal{A}$ in which each primary ideal is modular.*

(a) $\mathcal{A}$ *is strongly harmonic.*

(b) *For each $\mathcal{M} \in \Xi_\mathcal{A}$, $\mathcal{I}(\mathcal{M})$ is the smallest $\mathcal{M}$-primary ideal (i.e., $\mathcal{I}(\mathcal{M})$ satisfies $h^\Xi(\mathcal{I}(\mathcal{M})) = \{\mathcal{M}\}$ and $\mathcal{I}(\mathcal{M})$ is included in any $\mathcal{P}$ satisfying $h^\Xi(\mathcal{P}) = \{\mathcal{M}\}$).*

Proof (a)$\Rightarrow$(b): Suppose $\mathcal{M}$ is a maximal modular ideal of $\mathcal{A}$ and $\mathcal{P}$ is an $\mathcal{M}$-primary ideal. We first show that any $a \in \mathcal{I}(\mathcal{M})$ belongs to $\mathcal{P}$. We have noted that $(a\mathcal{A}^1)_{RA}$ is not included in $\mathcal{M}$. If $\mathcal{P} + (a\mathcal{A}^1)_{RA}$ were not equal to $\mathcal{A}$, it would be included in some maximal modular ideal $\mathcal{M}' \neq \mathcal{M}$ by Theorem 2.4.6. Hence $\mathcal{A} = \mathcal{P} + (a\mathcal{A}^1)_{RA}$ holds so that the relative identity e for $\mathcal{P}$ can be written as $e = p + b$ for $p \in \mathcal{P}$ and $b \in (a\mathcal{A}^1)_{RA}$. Therefore, $a = a(p + b) - (ae - a) = ap + (ae - a)$ belongs to $\mathcal{P}$ as we wished to show. This again shows $\mathcal{I}(\mathcal{M}) \subseteq \mathcal{M}$ (i.e., $\mathcal{M} \in h(\mathcal{I}(\mathcal{M}))$) and that any $\mathcal{M}$-primary ideal of $\mathcal{A}$ includes $\mathcal{I}(\mathcal{M})$.

Next we show that no ideal $\mathcal{M}' \in \Xi_\mathcal{A}$ different from $\mathcal{M}$ belongs to $h(\mathcal{I}(\mathcal{M}))$. Given such an $\mathcal{M}'$, we have ideals $\mathcal{I}$ and $\mathcal{I}'$ of $\mathcal{A}$ satisfying $\mathcal{I} \not\subseteq \mathcal{M}, \mathcal{I}' \not\subseteq \mathcal{M}'$ and $\mathcal{I}'\mathcal{I} = \{0\}$. Hence we may choose $b \in \mathcal{I} \setminus \mathcal{M}$ satisfying $\mathcal{I}'\mathcal{A}^1 b = \{0\}$. This implies $\mathcal{I}' \subseteq \mathcal{I}(\mathcal{M})$. Thus $\mathcal{I}' \not\subseteq \mathcal{M}'$ implies $\mathcal{I}(\mathcal{M}) \not\subseteq \mathcal{M}'$ as we wished to show.

(b)$\Rightarrow$(a): Let $\mathcal{M}_1$ and $\mathcal{M}_2$ be distinct ideals in $\Xi_\mathcal{A}$. By condition (b), we may choose $a \in \mathcal{I}(\mathcal{M}_2) \setminus \mathcal{M}_1$. Hence, $\mathcal{I}_1 = \mathcal{A}^1 a \mathcal{A}^1 \not\subseteq \mathcal{M}_1$ and $\mathcal{I}_2 = (a\mathcal{A}^1)_{RA} \not\subseteq \mathcal{M}_2$ satisfy $\mathcal{I}_1 \mathcal{I}_2 = \{0\}$.   □

For a strongly harmonic algebra in which each primary ideal is modular, the above proposition and Theorem 7.1.7 show that $\mathcal{M}/\mathcal{I}(\mathcal{M})$ is the unique maximal modular ideal of $\mathcal{A}/\mathcal{I}(\mathcal{M})$ and that if $\mathcal{I}$ is any ideal of $\mathcal{A}$ properly included in $\mathcal{I}(\mathcal{M})$, then $\mathcal{A}/\mathcal{I}$ fails to have a unique maximal modular ideal.

7.4.5 Proposition *An algebra $\mathcal{A}$ satisfies $\cap_{\mathcal{M} \in \Xi_\mathcal{A}} \mathcal{I}(\mathcal{M}) = \{0\}$ if any of the following conditions hold:*

(a) $\mathcal{A}$ *is unital;*

(b) $\mathcal{A}_{LA} = \{0\}$ *and $\mathcal{A}$ is Tauberian;*

(c) $\mathcal{A}_{LA} = \{0\}$ *and $\mathcal{A}$ is a topologically Tauberian spectral normed algebra.*

Proof (a) If a is non-zero, then $a\mathcal{A} = a\mathcal{A}^1$ is non-zero so that $(a\mathcal{A})_{RA}$ is a proper ideal. Theorem 2.4.6 shows that there is some maximal (necessarily modular) ideal $\mathcal{M}$ which includes $(a\mathcal{A})_{RA}$. This implies $a \notin \mathcal{I}(\mathcal{M})$. The conclusion follows.

(b) and (c): If a is non-zero, either hypothesis shows that there is a maximal modular ideal $\mathcal{M}$ which includes $(a\mathcal{A}^1)_{RA}$. Hence, a does not belong to $\mathcal{I}(\mathcal{M})$. Since a was an arbitrary non-zero element, $\cap_{\mathcal{M} \in \Xi_A} \mathcal{I}(\mathcal{M})$ is zero.   □

We will now give the representation theory for a unital strongly harmonic algebra as the algebra of all continuous sections of a sheaf of algebras. The last result shows that the representation is faithful.

7.4.6 Definition A *sheaf of algebras* is a triple (S, π, B) consisting of two topological spaces S and B, a continuous surjection $\pi \colon S \to B$, and, for each $b \in B$, an algebra structure for $\pi^{-1}(b)$, satisfying:

(a) Each $s \in S$ has an open neighborhood U in S such that the restriction of π to U is a homeomorphism of U onto $\pi(U)$ which is an open set in B; and

(b) For each $\lambda \in \mathbb{C}$, the function $s \to \lambda s$ is continuous from S to S. The functions $(s, t) \to s + t$ and $(s, t) \to st$ are continuous from $\{(s, t) \in S \times S : \pi(s) = \pi(t)\} \subseteq S \times S$ in its relativized product topology to S.

For each subset V of B, a function $f \colon V \to S$ is called a *continuous section over* V if it is continuous and $\pi \circ f$ is the identity map on V. The algebra (under pointwise operations) of continuous sections over V is denoted by $\Gamma(V, S)$. The algebra $\Gamma(B, S)$ is denoted by $\Gamma(S)$ and is called the *algebra of global continuous sections of* S. Finally the sheaf is said to be *soft* if each continuous section over a compact subset of B is the restriction of a global continuous section, and is said to be *unital* if each fiber is a unital algebra and the section which takes the value 1 at each point of the base space is continuous.

For a more detailed general discussion of sheaves see Godement [1958]. Other more specialized papers are cited below.

7.4.7 Proposition *Let $\mathcal{A}$ be an algebra and let $S(\mathcal{A})$ be the disjoint union $\cup\{\mathcal{A}/\mathcal{I}(\mathcal{M}) : \mathcal{M} \in \Xi_A\}$. Let $\pi \colon S(\mathcal{A}) \to \Xi_A$ be defined by $\pi(a + \mathcal{I}(\mathcal{M})) = \mathcal{M}$. For each $a \in \mathcal{A}$, let $\tilde{a} \colon \Xi_A \to S(\mathcal{A})$ be the global section of π defined by $\tilde{a}(\mathcal{M}) = a + \mathcal{I}(\mathcal{M})$. Then $\{\tilde{a}(U) : a \in \mathcal{A}; U \text{ open in } \Xi_A\}$ is the base for a topology on $S(\mathcal{A})$. Under this topology, $(S(\mathcal{A}), \pi, \Xi_A)$ is a sheaf of algebras and for each $a \in \mathcal{A}$, $\tilde{a} \colon \Xi_A \to S(\mathcal{A})$ is a continuous section. Moreover, $f \colon \Xi_A \to S(\mathcal{A})$ is a continuous section if and only if each point in Ξ_A has a neighborhood on which f agrees with some $\tilde{a}$.*

Proof First we show that the set described in the proposition is the base for a topology. If $b + \mathcal{I}(\mathcal{M}) \in S(\mathcal{A})$ is a point in $\tilde{a}(U) \cap \tilde{c}(V)$ for some $a, c \in \mathcal{A}$ and open subsets U, V of Ξ_A, then $a - b, c - b$ and hence $a - c$ belong to $\mathcal{I}(\mathcal{M})$. By definition of $\mathcal{I}(\mathcal{M})$, this implies $((a - c)\mathcal{A}^1)_{RA} \nsubseteq \mathcal{M}$. Let W be the open subset $U \cap V \cap (\Xi_A \setminus h(((a - c)\mathcal{A}^1)_{RA}))$ of Ξ_A which includes $\mathcal{M}$.

Then $b + \mathcal{I}(\mathcal{M})$ is contained in the basic open set $\tilde{a}(W) = \tilde{c}(W)$ which is included in $\tilde{a}(U) \cap \tilde{c}(V)$. Hence we have a base for a topology.

Clearly for each $a + \mathcal{I}(\mathcal{M}) \in S(A)$ and each open neighborhood of $\mathcal{M}$ in $\Xi_\mathcal{A}$, $\tilde{a}(V)$ is mapped by π bijectively onto $V = \pi(\tilde{a}(V))$. However, $\{\tilde{a}(U) : U \subseteq V$ is open in $\Xi_\mathcal{A}\}$ is a base for the topology of $\tilde{a}(V)$ since $\tilde{c}(W) \cap \tilde{a}(V) = \tilde{a}(W \cap V)$ holds. Thus π is also a homeomorphism when restricted to V so that (a) of Definition 7.4.6 is satisfied. Condition (b) of this definition is trivial in this case so that $(S(\mathcal{A}), \pi, \Xi_\mathcal{A})$ is a sheaf of algebras. From the definition of the topology, it is obvious that $\tilde{a}$ is a continuous section for each $a \in \mathcal{A}$. In order to prove the last sentence, let $f \colon \Xi_\mathcal{A} \to S(\mathcal{A})$ be an arbitrary section, and let $\mathcal{M}_0 \in \Xi_\mathcal{A}$ be arbitrary. Let $\tilde{a}(V)$ be a basic open neighborhood of $f(\mathcal{M}_0)$. Then $f^\leftarrow(\tilde{a}(V)) = \{\mathcal{M} \in V : \tilde{a}(\mathcal{M}) = f(\mathcal{M})\}$ is an open set if and only if the condition is satisfied.□

The above proposition shows that any algebra $\mathcal{A}$ has a homomorphic image $\tilde{\mathcal{A}}$ which is a subalgebra of $\Gamma(S(\mathcal{A}))$. Clearly, the kernel of this representation is $\cap_{\mathcal{M} \in \Xi_\mathcal{A}} \mathcal{I}(\mathcal{M})$ so that the representation is faithful if this intersection is zero and *a fortiori* if $\mathcal{A}$ is unital or strongly semisimple. However, this is just a reformulation of the subdirect product representation of $\mathcal{A}$ in $\Pi_{\mathcal{M} \in \Xi_\mathcal{A}} \mathcal{A}/\mathcal{I}(\mathcal{M})$ and suffers from all the weaknesses of such a representation unless it can be shown that $\tilde{\mathcal{A}}$ is actually all of $\Gamma(S(\mathcal{A}))$. A number of hypotheses have been explored which force this equality. We mention the following papers: Alexander Grothendieck [1960], Dauns and Hoffmann [1966], [1968], Richard S. Pierce [1967], Teleman [1969a], [1969b], [1971], Koh [1971], [1972]. (Several of these papers speak of the base space as $P_\mathcal{A}$, but if $P_\mathcal{A}$ is Hausdorff in a unital algebra $\mathcal{A}$, then $P_\mathcal{A}$ equals $\Xi_\mathcal{A}$.) We now give the theorem from the last of the above references starting with a result which we would call a lemma except for its independent interest.

7.4.8 Proposition *Let $\mathcal{A}$ be a strongly harmonic algebra and let $K \subseteq \Xi_\mathcal{A}$ be compact.*

(a) *For any $\mathcal{M}_0 \notin K$, there are ideals $\mathcal{I}_1$ and $\mathcal{I}_2$ of $\mathcal{A}$ satisfying $\mathcal{M}_0 \notin h(\mathcal{I}_1), K \cap h(\mathcal{I}_2) = \emptyset$ and $\mathcal{I}_1\mathcal{I}_2 = \{0\}$.*

(b) *Furthermore, K equals the hull $h(\cap\{\mathcal{I}(\mathcal{M}) : \mathcal{M} \in K\})$.*

Proof (a): For each $\mathcal{M} \in K$, choose ideals $\mathcal{I}_1^\mathcal{M}$ and $\mathcal{I}_2^\mathcal{M}$ satisfying $\mathcal{M}_0 \notin h(\mathcal{I}_1^\mathcal{M}), \mathcal{M} \notin h(\mathcal{I}_2^\mathcal{M})$ and $\mathcal{I}_1^\mathcal{M}\mathcal{I}_2^\mathcal{M} = \{0\}$. Since K is compact, it is covered by a finite collection of the open sets $\Xi_\mathcal{A} \setminus h(\mathcal{I}_2^\mathcal{M})$. Let the corresponding pairs of ideals be $(\mathcal{I}_1^1, \mathcal{I}_2^1), (\mathcal{I}_1^2, \mathcal{I}_2^2), \ldots, (\mathcal{I}_1^n, \mathcal{I}_2^n)$ and define $\mathcal{I}_1$ and $\mathcal{I}_2$ by $\mathcal{I}_1 = \mathcal{I}_1^1\mathcal{I}_1^2 \cdots \mathcal{I}_1^n$ and $\mathcal{I}_2 = \sum_{k=1}^n \mathcal{I}_2^k$. Then K is covered by $\cup_{k=1}^n (\Xi_\mathcal{A} \setminus h(\mathcal{I}_2^k)) = \Xi_\mathcal{A} \setminus \cap_{k=1}^n h(\mathcal{I}_2^k) = \Xi_\mathcal{A} \setminus h(\mathcal{I}_2)$ and $\mathcal{M}_0$ is not an element of $h(\mathcal{I}_1)$. Also $\mathcal{I}_1\mathcal{I}_2 = \{0\}$ is obvious.

(b): The obvious inclusion $\cap_{\mathcal{M} \in K} \mathcal{I}(\mathcal{M}) \subseteq \cap_{\mathcal{M} \in K} \mathcal{M} = k(K)$ implies $K \subseteq h(\cap_{\mathcal{M} \in K} \mathcal{I}(\mathcal{M}))$. Suppose some $\mathcal{M}_0 \in h(\cap_{\mathcal{M} \in K} \mathcal{I}(\mathcal{M}))$ is not in K. Then, (a) gives ideals $\mathcal{I}_1$ and $\mathcal{I}_2$ satisfying $\mathcal{M}_0 \notin h(\mathcal{I}_1), K \cap h(\mathcal{I}_2) = \emptyset$ and

$\mathcal{I}_1\mathcal{I}_2 = \{0\}$. Hence any $\mathcal{M} \in K$ satisfies $\mathcal{I}_2 \not\subseteq \mathcal{M}$ which implies $\mathcal{I}_1 \subseteq \mathcal{I}(\mathcal{M})$. This gives $\mathcal{I}_1 \subseteq \cap_{\mathcal{M}\in K}\mathcal{I}(\mathcal{M})$. Since we assumed $\mathcal{M}_0 \in h(\cap_{\mathcal{M}\in K}\mathcal{I}(\mathcal{M}))$, this leads to the contradiction $\mathcal{I}_1 \subseteq \mathcal{M}_0$.   □

7.4.9 Theorem *Let $\mathcal{A}$ be a unital strongly harmonic algebra. Then the sheaf $(S(\mathcal{A}), \pi, \Xi_{\mathcal{A}})$ is a unital, soft sheaf of unital local algebras. Furthermore, $a \mapsto \tilde{a}$ is an isomorphism of $\mathcal{A}$ onto the algebra $\Gamma(S(\mathcal{A}))$ of global continuous sections of this sheaf. Conversely if (S, π, B) is any unital, soft sheaf of unital local algebras over a compact Hausdorff space B, then $\Gamma(S)$ is a unital strongly harmonic algebra and B is naturally homeomorphic to $\Xi_{\Gamma(S)}$.*

Proof Proposition 7.4.5 shows that $a \mapsto \tilde{a}$ is an isomorphism. Next we show that $S(\mathcal{A})$ is a soft sheaf and that $\tilde{\mathcal{A}}$ includes (and hence equals) $\Gamma(S(\mathcal{A}))$ by showing that, for any compact subset K_0 of $\Xi_{\mathcal{A}}$ ($K_0 = \Xi_{\mathcal{A}}$ is possible since $\mathcal{A}$ is unital) and for any $f \in \Gamma(K_0, S(\mathcal{A}))$, there is an $a \in \mathcal{A}$ satisfying $\tilde{a}|_{K_0} = f$. Since $\Xi_{\mathcal{A}}$ is Hausdorff (Proposition 7.4.2), K_0 is closed so $U_0 = \Xi_{\mathcal{A}} \setminus K_0$ is open. Since f is continuous, Proposition 7.4.7 shows that for each $\mathcal{M} \in K_0$ there is an open set $U_{\mathcal{M}}$ in $\Xi_{\mathcal{A}}$ and an element $a_{\mathcal{M}} \in \mathcal{A}$ satisfying $\tilde{a}_{\mathcal{M}} = f$ on $U_{\mathcal{M}} \cap K_0$. Relying on the compactness of K_0, choose a finite subcover $U_1, U_2, \ldots, U_n$ of the cover $\{U_{\mathcal{M}} : \mathcal{M} \in K_0\}$ and let $a_1, a_2, \ldots, a_n$ be the corresponding elements of $\mathcal{A}$. Then $\Xi_{\mathcal{A}}$ equals $\cup_{k=0}^n U_k$. Define K_k and $\mathcal{I}_k$ by $K_k = \Xi_{\mathcal{A}} \setminus U_k$ and $\mathcal{I}_k = \cap\{\mathcal{I}(\mathcal{M}) : \mathcal{M} \in K_k\}$ for $k = 0, 1, \ldots, n$. Each K_k is compact since it is closed in Ξ which is compact by Proposition 7.1.6. Proposition 7.4.8(b) now shows $K_k = h(\mathcal{I}_k)$ for $k = 0, 1, \ldots, n$. But this gives $\emptyset = \cap_{k=0}^n K_k = \cap_{k=0}^n h(\mathcal{I}_k) = h(\sum_{k=0}^n \mathcal{I}_k)$, which implies $\mathcal{A} = \sum_{k=0}^n \mathcal{I}_k$. Hence, there are elements $e_k \in \mathcal{I}_k$ satisfying $1 = \sum_{k=0}^n e_k$. For each $\mathcal{M} \in K_0 \cap K_k$, we have $\tilde{a}_k(\mathcal{M})\tilde{e}_k(\mathcal{M}) = 0 = f(\mathcal{M})\tilde{e}_k(\mathcal{M})$ and for each $\mathcal{M} \in K_0 \cap U_k$, we have $\tilde{a}_k(\mathcal{M})\tilde{e}_k(\mathcal{M}) = f(\mathcal{M})\tilde{e}_k(\mathcal{M})$, so that we get $\tilde{a}_k\tilde{e}_k = f\tilde{e}_k$ on K_0 for all $k = 1, 2, \ldots, n$. Define $a \in \mathcal{A}$ by $a = e_0 + \sum_{k=1}^n a_k e_k$. Then for each $\mathcal{M} \in K_0$, we find

$$\tilde{a}(\mathcal{M}) = \tilde{e}_0(\mathcal{M}) + \sum_{k=1}^n \tilde{a}_k(\mathcal{M})\tilde{e}_k(\mathcal{M})$$

$$= f(\mathcal{M})\tilde{e}_0(\mathcal{M}) + \sum_{k=1}^n f(\mathcal{M})\tilde{e}_k(\mathcal{M})$$

$$= f(\mathcal{M})(\sum_{k=0}^n \widetilde{e_k})(\mathcal{M}) = f(\mathcal{M})$$

since $\tilde{e}_0(\mathcal{M})$ is zero. Thus we have established that $(S(\mathcal{A}), \pi, \Xi_{\mathcal{A}})$ is a soft sheaf and (taking $K_0 = \Xi_{\mathcal{A}}$) that $\tilde{\mathcal{A}}$ equals $\Gamma(S(\mathcal{A}))$.

Let (S, π, B) be a unital soft sheaf of unital local algebras over a compact Hausdorff space. The constant section 1 is an identity element for $\Gamma(S)$. For each $b \in B$, let $\tilde{\mathcal{M}}_b$ be the unique maximal ideal of $\pi^{-1}(b)$. Since the sheaf is soft, for any $b \in B$ the set $\mathcal{M}_b = \{f \in \Gamma(S) : f(b) \in \tilde{\mathcal{M}}_b\}$ is a proper subset of $\Gamma(S)$ which is obviously a (necessarily modular) ideal of $\Gamma(S)$. Since $f \mapsto f(b)$ is a homomorphism of $\Gamma(S)$ onto $\pi^{-1}(b)$ and $\tilde{\mathcal{M}}_b$ is maximal in $\pi^{-1}(b)$, $\mathcal{M}_b$ is maximal in $\Gamma(S)$. Furthermore, since the sheaf is soft, $\mathcal{M}_b$ and $\mathcal{M}_{b'}$ are distinct when b and b' are distinct elements of B. Hence, $\{\mathcal{M}_b : b \in B\}$ is a subset of $\Xi_{\Gamma(S)}$. Next we show that this subset equals $\Xi_{\Gamma(S)}$.

If $\mathcal{I}$ is any proper ideal of $\Gamma(S)$, we show that there is some $b \in B$ satisfying $\mathcal{I} \subseteq \mathcal{M}_b$. Suppose not. Then for each $b \in B$ the ideal $\{f(b) : f \in \mathcal{I}\}$ of $\pi^{-1}(b)$ must be $\pi^{-1}(b)$ itself. Hence for each b we can find $f_b \in \mathcal{I}$ satisfying $f_b(b) = 1$. For each $b \in B$, there is an open neighborhood U_b of b satisfying $f_b(d) = 1$ for all $d \in U_b$. Choose a finite subcover $\{U_{b_k} : k = 1, 2, \ldots, n\}$ and apply Lemma 7.2.8 to this subcover and the corresponding family of functions $\{f_{b_k} : k = 1, 2, \ldots, n\}$. This gives the contradiction: $1 \in \mathcal{I}$. We conclude that $b \mapsto \mathcal{M}_b$ gives a bijection of B onto $\Xi_{\Gamma(S)}$.

Next we must show that the map $b \mapsto \mathcal{M}_b$ is a homeomorphism. Since both B and $\Xi_{\Gamma(S)}$ are compact and Hausdorff, this will be true if the inverse of this map is continuous. Hence we show that if U is an open set in B, then $\{\mathcal{M}_b : b \in U\}$ is open in $\Xi_{\Gamma(S)}$. Let $b_0 \in U$ be arbitrary. Since the sheaf is soft and B is compact, we can find $f \in \Gamma(S)$ satisfying $f(b_0) = 1$ and $f(b) = 0$ for all $b \in B \setminus U$. Hence, $\Xi_{\Gamma(S)} \setminus \{\mathcal{M}_b : b \in U\} = \{\mathcal{M}_b : b \in B \setminus U\}$ is included in $h(\{f\})$. But then, $\mathcal{M}_{b_0} \in \Xi_{\Gamma(S)} \setminus h(\{f\}) \subseteq \{\mathcal{M}_b : b \in U\}$ shows that the arbitrary point $\mathcal{M}_{b_0}$ in the image (under $b \mapsto \mathcal{M}_b$) of U is in the interior of this image so that the image of U is open and $b \mapsto \mathcal{M}_b$ is a homeomorphism. $\qquad\square$

The representation described in the last theorem applies to a number of classes of Banach algebras which are important in analysis. However, since the fibers in a sheaf are necessarily discrete, this representation ignores the topological structure of these algebras. This is at the very least unaesthetic, and suggests the possibility of a deeper representation theory of topological algebras by continuous sections of some structure more general than a sheaf. Such a theory has been worked out in detail by Dauns and Hofmann [1968]. This beautiful theory is so technically complicated that we will not present even a survey of its results. However, we remark that this theory gives a detailed faithful representation theory for arbitrary C*-algebras and in this respect far transcends the theory described here. One reason for the wider applicability of this representation theory is the fact that it is based on the refinement of the subdirect product representation of an algebra $\mathcal{A}$ over $\Pi_{\mathcal{A}}$ (rather than $\Xi_{\mathcal{A}}$) and hence is faithful for semisimple rather than strongly semisimple algebras. (All C*-algebras are semisimple.)

7.4.10 Example There are many examples of completely regular and strongly harmonic algebras. Theorem 3.2.10 shows that a completely regular commutative Banach algebra (in the sense of Definition 3.2.9) is a completely regular algebra in the sense of Definition 7.2.1. Examples of commutative completely regular algebras include: (1) $C_0(\Omega)$ for any locally compact Hausdorff space Ω (equivalently any commutative Banach algebra satisfying the hypotheses of Theorem 3.2.12, *i.e.*, any commutative C*-algebra); (2) for each $n \in \mathbb{N}$, the algebra $C^{(n)}([0,1])$ of n-times continuously differentiable functions on $[0,1]$ with the norm

$$\|f\| = \sum_{k=0}^{n} \frac{\|f^{(k)}\|_\infty}{k!} \qquad \forall \ f \in C^{(n)}([0,1]);$$

(3) the algebra $CBV([0,1])$ of continuous functions of bounded variation on $[0,1]$ with norm

$$\|f\| = \|f\|_\infty + \|f\|_{BV} \qquad \forall \ f \in CBV([0,1]);$$

and (4) the classical group algebra $L^1(G)$ for any locally compact abelian group. Most of these commutative examples have suitable noncommutative generalizations.

Any algebra $\mathcal{A}$ with a discrete strong structure space $\Xi_\mathcal{A}$ is obviously completely regular since any $\mathcal{M} \in \Xi_\mathcal{A}$ has the neighborhood $\{\mathcal{M}\}$ which has $\mathcal{M}$ as its modular kernel. A *fortiori*, any algebra with a discrete structure space $\Pi_\mathcal{A}$ is completely regular. Theorem 8.4.5 shows that all modular annihilator algebras have discrete structure spaces and hence are completely regular. The next Chapter has many examples in Section 8.8, the diagram of Theorem 8.7.11 and Example 8.1.2. In particular, $L^1(K)$ is completely regular when K is a compact group.

John R. Liukkonen and Richard D. Mosak ([1974b], Corollary 2.5) and Eberhardt Kaniuth and Detlef Steiner ([1973], p. 324) independently and by different methods extend this last observation (and a result of Willcox [1956b]) to show that $L^1(G)$ is completely regular for any group G in $[\text{FIA}]^-$. A group is said to belong to $[\text{FIA}]^-$ if it is locally compact and the closure of the group of inner automorphisms in the group of all homeomorphic automorphisms (with its natural topology) is compact. (The abbreviation stands for "finite inner automorphism" and the superscript bar indicates that "finite" should be changed to its topological equivalent "compact".) Note that this is a common generalization of the result for compact and for abelian groups. Kaniuth and Günter Schlichting ([1970] Satz 2) and Mosak [1972] showed that the Banach algebra $L^1(G)$ is also strongly semisimple for groups G in $[\text{FIA}]^-$. Siegfried Grosser and Martin A. Moskowitz [1971b] obtained the same result for [MAP] groups. The class

[MAP] was introduced in §3.2.17. The abbreviation stands for "maximally almost periodic" and a group G belongs to the class [MAP] if it is locally compact and has enough finite-dimensional irreducible representations to separate points or, equivalently, if there is some continuous injective homomorphism of it into a compact group. For many related results see the present author's survey [1978].

In Volume II, we will show that a C*-algebra $\mathcal{A}$ has a Hausdorff structure space $\Pi_{\mathcal{A}}$ if and only if the function $\mathcal{P} \mapsto \|a + \mathcal{P}\|$ is continuous on $\Xi_{\mathcal{A}}$ for each $a \in \mathcal{A}$. It then follows easily that $\mathcal{A}$ is completely regular. In addition, Yosinao Misonou [1952] shows that W*-algebras are weakly central. Fred B. Wright [1954] obtains the same result for AW*-algebras. Theorem 7.2.17 then shows that these algebras are completely regular.

Willcox [1956a] notes that the algebra of matrices over $C^n([0,1])$ is completely regular and that in this case $\mathcal{J}(t_0)^-$ is the set of matrix functions satisfying $f'_{ij}(t_0) = f_{ij}(t_0) = 0$ for all i, j. The results cited in §1.10 shows that certain other tensor products are completely regular.

Proposition 7.4.2 shows that strongly semisimple algebras with Hausdorff structure spaces are strongly harmonic. A semisimple commutative algebra with Hausdorff Gelfand space is therefore strongly harmonic. This includes all the commutative examples mentioned above. Similarly we conclude that $L^1(G)$ is strongly harmonic if G is an [FIA]$^-$ group.

If $\mathcal{A}$ is any AW*-algebra and $\mathcal{M}_1$ and $\mathcal{M}_2$ are distinct ideals in $\Xi_{\mathcal{A}}$, then there are central idempotents $e_j \notin \mathcal{M}_j$ for $j = 1, 2$, satisfying $e_1 e_2 = 0$. Hence AW*-algebras (and *a fortiori* W*-algebras) are strongly harmonic. These algebras are not strongly semisimple unless they are finite (in the sense defined below).

7.4.11   Centroids, Double Centralizer Algebras and Extensions

Proposition 7.2.14 shows that the structure space of a spectral algebra $\mathcal{A}$ and of its center $\mathcal{A}_Z$ may be identified if, for instance, $\mathcal{A}$ is strongly semisimple and central and $\mathcal{A}_Z$ is completely regular. Theorem 7.2.17 explicitly states the analogous conclusion for strong structure spaces. Quite generally there is a connection between the structure spaces of an algebra and its center. However, the center of a nonunital algebra may well reduce to $\{0\}$ making it useless. The center of $\mathcal{A}^1$ is not a useful substitute in the present context. For instance, if $\mathcal{A}$ is nonunital with a locally compact Hausdorff structure space, then $\mathcal{A}^1$ may have a structure space which fails to be Hausdorff or even T_1. These considerations lead us to consider the centroid which is very well behaved.

Definition   The *centroid* of an algebra $\mathcal{A}$ is the center of the double centralizer algebra $\mathcal{D}(\mathcal{A})$ of $\mathcal{A}$.

Proposition 1.2.3 shows that the centroid of $\mathcal{A}$ can also be described as the subalgebra of $\mathcal{L}(\mathcal{A})$ consisting of maps T satisfying

$$T(a)b = T(ab) = aT(b) \qquad \forall \ a, b \in \mathcal{A}.$$

For a Banach algebra with trivial left and right annihilator ideals, Theorem 1.2.4(b) shows that all the maps in the centroid are continuous.

The idea of the centroid was independently introduced by Robert C. Busby [1968] and Jacques Dixmier [1968], both working in the context of C*-algebras. A systematic investigation was begun by Dauns and Hofmann [1968], III.6 and [1969]. Under various hypotheses, they determine the structure space of the centroid. This work has been continued by Dauns [1969], [1974] and closely related work is found in Busby [1967], [1971b]. In general, these references deal with various special cases of the problem of finding the structure space of an algebra $\mathcal{A}$ in terms of the structure space of an ideal $\mathcal{I}$ of $\mathcal{A}$ and the structure space of $\mathcal{A}/\mathcal{I}$. The best results are obtained in the case of C*-algebras, and some of them will be discussed further in Volume II.

Another result on C*-algebras is closely enough related to the present considerations to warrant discussion here. This is the Dauns–Hofmann theorem which asserts that the centroid of a C*-algebra $\mathcal{A}$ can be identified with $C(\Pi_{\mathcal{A}})$, so that for each T in the centroid there is an $f \in C(\Pi_{\mathcal{A}})$ satisfying

$$T(a) + \mathcal{P} = f(\mathcal{P})a + \mathcal{P} \qquad \forall \ \mathcal{P} \in \Pi_{\mathcal{A}}; \ a \in \mathcal{A}.$$

This was first proved in Dauns and Hofmann [1968]. Other more transparent proofs have been given in Dixmier [1968], Theorem 5, Gert K. Pedersen [1972] and George A. Elliott and Dorte Olesen [1974]. The last cited proof is also the easiest and establishes a Banach space generalization.

8

Algebras with Minimal Ideals

Introduction

In this chapter we investigate the role which minimal left ideals play in those algebras which possess them. (*Minimal* here simply means minimal under inclusion among all non-zero left ideals.) Significant results usually depend on assuming that the algebra is semiprime. However, the stronger assumption of semisimplicity, which is common in other areas of our theory, is comparatively seldom needed here. Furthermore, norms behave particularly well on minimal ideals so that a number of results can be proved for normed algebras without assuming completeness. In particular, any norm is spectral on many of the algebras considered in this chapter.

We will naturally be concerned mainly with algebras which not only have some minimal left ideals, but "enough" minimal left ideals in some sense. Such algebras are quite special, but they have a correspondingly rich theory. Many classes of algebras with "enough" minimal left ideals have been defined, and towards the end of the section we will prove all the inclusions which hold between these various classes. (These results are summarized in a diagram of implications in Theorem 8.8.11.) However, we will concentrate attention on the class of modular annihilator algebras which was introduced by Bertram Yood [1958] and has been intensively studied by Yood [1964] and by Bruce A. Barnes [1964], [1966], [1968a], [1968b], and [1971a]. This class seems to give a particularly good compromise between axioms strong enough to give a significant theory and axioms weak enough to be satisfied by most important examples. Furthermore, modular annihilator algebras are defined purely algebraically while many of the other classes are defined in terms of mixed algebraic and topological criteria. This provides a particularly elegant and straightforward theory.

In a semiprime algebra $\mathcal{A}$, every minimal left ideal has the form $\mathcal{A}e$ for an idempotent $e \in \mathcal{A}$ such that $e\mathcal{A}e$ is a division algebra. Conversely, $\mathcal{A}e$ is a minimal left ideal for any such idempotent. These idempotents are called *minimal idempotents*. If $\mathcal{A}$ is a normed or spectral algebra and e and f are minimal idempotents, then $e\mathcal{A}e$ and $e\mathcal{A}f$ are both one-dimensional (unless the latter is $\{0\}$).

The sum of all minimal left ideals in a semiprime algebra $\mathcal{A}$ is the ideal (not just left ideal) generated by the set $\mathcal{A}_{MI}$ of all minimal idempotents. This ideal, which was introduced by Jean Dieudonné [1942], is called the *socle* of $\mathcal{A}$ and is denoted by $\mathcal{A}_F$. It is also the sum of all minimal right ideals. In a semiprime algebra $\mathcal{A}$, the socle and Jacobson radical satisfy $\mathcal{A}_F\mathcal{A}_J = \mathcal{A}_J\mathcal{A}_F = \{0\}$ and this makes the Jacobson radical tractable.

If $\mathcal{L}$ is a minimal left ideal of $\mathcal{A}$, then the restriction of the left regular representation of $\mathcal{A}$ to act on $\mathcal{L}$ is an (algebraically) irreducible representation of $\mathcal{A}$. Conversely, any irreducible restriction of the left regular representation has this form. These irreducible representations have pleasant properties which are described in Proposition 8.2.6 and in Theorem 8.3.6 for semiprime normed or spectral algebras. In particular, these representations are uniquely determined by their kernels.

Any faithful irreducible representation T of an algebra $\mathcal{A}$ with a nonzero socle maps the socle onto the algebra of those finite-rank operators on the representation space which belong to $T_{\mathcal{A}}$. This allows a very concrete description of any such representation, which we give in Proposition 8.3.3 and Theorems 8.3.4 and 8.3.6.

Theorem 8.4.5 contains eight equivalent conditions on a semiprime algebra which we use to define modular annihilator algebras. The irreducible representations of these algebras are restrictions of the left regular representation and hence are uniquely determined by their kernels. Furthermore if $\mathcal{A}$ is such an algebra, then the structure space $\Pi_{\mathcal{A}}$ of $\mathcal{A}$ is discrete, the hull of $\mathcal{A}_F$ is empty, every left ideal of $\mathcal{A}$ which is not a subset of $\mathcal{A}_J$ includes a minimal left ideal, and $\mathcal{A}_J$ is the left and right annihilator of $\mathcal{A}_F$. Theorem 8.4.12 shows that all norms on modular annihilator algebras are spectral.

Proposition 8.3.3 shows that ideals and semiprime quotients of modular annihilator algebras are again of the same type. Furthermore, in any semiprime algebra $\mathcal{A}$, any ideal of $k(h(\mathcal{A}_F))$ is a modular annihilator algebra. This allows the introduction in Theorem 8.5.5 of a very detailed Fredholm theory (due to Barnes [1968a]) for elements in $k(h(\mathcal{A}_F))$ for any semiprime algebra $\mathcal{A}$. Since any modular annihilator algebra $\mathcal{A}$ satisfies $\mathcal{A} = k(h(\mathcal{A}_F))$, this Fredholm theory holds for all elements in these algebras. This in turn gives a purely spectral characterization of semisimple modular annihilator algebras among semisimple algebras as those for which no element has a non-zero limit point for its spectrum. This result of Barnes [1968b] is given in Theorem 8.6.4.

Section 8.7 introduces many types of algebras which have been shown to have a large socle in some sense. These include dual, annihilator, left annihilator, completely continuous, left completely continuous, compact, modular annihilator, and Duncan modular annihilator algebras. After defining all these terms, we establish their logical relationships. Theorem 8.8.11 presents all these results in the form of a diagram of implications. Enough examples are included to show that no implication beyond those in the theorem are possible between these concepts for semisimple Banach algebras. A number of results are proved for left annihilator and annihilator algebras.

Finally, we prove the automatic continuity result of Barnes [1967b] which asserts that any homomorphism of a Banach algebra onto a dense subalgebra of a semisimple Duncan modular annihilator is necessarily continuous.

Since minimal ideals play so important a role throughout this section it may be well to review the simplest and most concrete representative case before beginning to consider the more abstract results. Let $\mathcal{X}$ be a normed linear space and let $\mathcal{B}(\mathcal{X})$ be the normed algebra of all continuous linear operators on $\mathcal{X}$. For any $x \in \mathcal{X}$ and $\omega \in \mathcal{X}^*$ the symbol $x \otimes \omega$ denotes the element of $\mathcal{B}(\mathcal{X})$ defined by

$$x \otimes \omega(y) = \omega(y)x \qquad \forall\, y \in \mathcal{X}.$$

For each fixed non-zero $\omega \in \mathcal{X}^*$ the set

$$\mathcal{X} \otimes \omega = \{x \otimes \omega : x \in \mathcal{X}\}$$

is obviously a non-zero left ideal of $\mathcal{B}(\mathcal{X})$ which contains no smaller non-zero left ideal of $\mathcal{B}(\mathcal{X})$. Conversely, if $\mathcal{L}$ is any minimal left ideal of $\mathcal{B}(\mathcal{X})$, choose a non-zero $T \in \mathcal{L}$ and $\omega' \in \mathcal{X}^*$ to satisfy $\omega = T^*(\omega') \neq 0$. Then for any $x \in \mathcal{X}, x \otimes \omega = (x \otimes \omega')T$ belongs to $\mathcal{L}$, so that $\mathcal{L}$ includes (and therefore equals) $\mathcal{X} \otimes \omega$. Hence we have described the family of all minimal left ideals of $\mathcal{B}(\mathcal{X})$.

If we normalize ω to have norm one, then the map $x \mapsto x \otimes \omega$ is a linear isometry of $\mathcal{X}$ onto $\mathcal{X} \otimes \omega$ which is also an equivalence between the usual representation of $\mathcal{B}(\mathcal{X})$ on $\mathcal{X}$ and the representation of $\mathcal{B}(\mathcal{X})$ on $\mathcal{X} \otimes \omega$ obtained by restricting the operators in the left regular representations of $\mathcal{B}(\mathcal{X})$ to $\mathcal{X} \otimes \omega \subseteq \mathcal{B}(\mathcal{X})$. If we choose $z \in \mathcal{X}$ to satisfy $\omega(z) = 1$, then we can write $\mathcal{X} \otimes \omega = \mathcal{B}(\mathcal{X})E$, where E is the one-dimensional projection $z \otimes \omega \in \mathcal{B}(\mathcal{X})$. Note that E satisfies $E\mathcal{B}(\mathcal{X})E = \{ETE : T \in \mathcal{B}(\mathcal{X})\} = \{z \otimes \omega(Tz \otimes \omega) : T \in \mathcal{B}(\mathcal{X})\} = \{\omega(Tz)z \otimes \omega : T \in \mathcal{B}(\mathcal{X})\} = \mathbb{C}E$.

Similarly each minimal right ideal of $\mathcal{B}(\mathcal{X})$ has the form $z \otimes \mathcal{X}^* = \{z \otimes \omega : \omega \in \mathcal{X}^*\}$ for some fixed non-zero $z \in \mathcal{X}$. If z is normalized to have norm one, the map $\omega \mapsto z \otimes \omega$ is an isometric linear isomorphism of $\mathcal{X}^*$ onto $z \otimes \mathcal{X}^*$ which is an equivalence between the usual anti-representation of $\mathcal{B}(\mathcal{X})$ on $\mathcal{X}^*$ (i.e., $T \mapsto T^* \in \mathcal{B}(\mathcal{X}^*)$) and the anti-representation of $\mathcal{B}(\mathcal{X})$ on $z \otimes \mathcal{X}^*$ obtained by restricting the operators of the right regular anti-representation of $\mathcal{B}(\mathcal{X})$ to $z \otimes \mathcal{X}^* \subseteq \mathcal{B}(\mathcal{X})$. As before we can write $z \otimes \mathcal{X}^* = E\mathcal{B}(\mathcal{X})$, where $E = z \otimes \omega$ is a one-dimensional projection obtained by choosing any $\omega \in \mathcal{X}^*$ satisfying $\omega(z) = 1$.

8.1   Finite-Dimensional Algebras

We begin this chapter by devoting a section to consideration of finite-dimensional algebras. A simple dimension argument shows that any non-zero left ideal in a finite dimensional algebra contains a minimal left ideal.

Starting from this observation, we could prove the next two theorems by arguments similar to those which we use in the rest of this chapter. However, we prefer to give simple proofs based on Theorem 4.2.13.

Wedderburn Theorem

Our first theorem is due to Elie Cartan [1898]. The theorem was generalized to finite-dimensional algebras over arbitrary fields by Joseph H. Maclagan Wedderburn [1907] and to rings satisfying the ascending and descending chain conditions on left ideals by Emil Artin [1926]. Hence this theorem is often called the Artin–Wedderburn theorem. Charles Hopkins [1939] showed that only the descending chain condition on left ideals is needed for its proof. We could easily prove the theorem with this hypothesis, but we will not do so. (Of course the generalized form of the theorem without further hypotheses does not give full matrix rings over the complex numbers, but rather matrix rings over some division ring.) This brief review of history shows that the theorem deserves to be called the Cartan theorem, but, in part to avoid confusion with other theorems by Cartan, we will follow common usage and name it after Wedderburn.

We mention first that every finite-dimensional algebra is a Banach algebra under some norm and that any two such norms are equivalent. In fact any two linear space norms on a finite-dimensional space are easily seen to be equivalent by a simple basis argument. The same argument shows that any linear operator on a finite-dimensional space is continuous with respect to any norm on the space. To see that any finite-dimensional algebra $\mathcal{A}$ can be made into a Banach algebra one can proceed as follows. Choose a basis for $\mathcal{A}^1$. This establishes a linear isomorphism between $\mathcal{A}^1$ and $\mathbb{C}^n$ for some n. Transfer the Euclidean linear space norm (or any other norm) from $\mathbb{C}^n$ to $\mathcal{A}^1$. Now the extended left regular representation is an isomorphism of $\mathcal{A}$ into $\mathcal{B}(\mathcal{A}^1)$. Hence the algebra norm of $\mathcal{B}(\mathcal{A}^1)$ can be transferred to $\mathcal{A}$. Since $\mathcal{A}$ is finite-dimensional it is complete in this norm.

8.1.1 Wedderburn Theorem *A finite-dimensional prime algebra is isomorphic to a full matrix algebra and hence is simple and unital. A finite-dimensional semiprime algebra is isomorphic to a direct sum of finitely many full matrix algebras and hence is strongly semisimple and unital.*

Proof Let $\mathcal{A}$ be a finite-dimensional prime algebra. Finite dimensionality implies the existence of some minimal left ideal $\mathcal{L}$. (Simply look for a left ideal of smallest dimension.) Let $T: \mathcal{A} \to \mathcal{L}(\mathcal{L})$ be the restriction of the left regular representation to $\mathcal{L}$, i.e.,

$$T_a b = ab \qquad \forall\, a \in \mathcal{A}; b \in \mathcal{L}.$$

Then T is irreducible since $\mathcal{L}$ is minimal. The kernel of T satisfies $\ker(T)\mathcal{L} = \{0\}$. Since $\mathcal{A}$ is a prime algebra (i.e., $\{0\}$ is a prime ideal) and $\mathcal{L}$ is non-zero, this implies $\ker(T) = \{0\}$. Hence T is both faithful and irreducible. By the remarks preceding this theorem $\mathcal{A}$ is a Banach algebra. Hence Theorem 4.2.13 applies and shows that T is strictly dense. However, it is an immediate consequence of the definition of strict density that a strictly dense representation on the finite-dimensional linear space $\mathcal{L}$ must have all of $\mathcal{L}(\mathcal{L})$ as range. Hence T induces an isomorphism of $\mathcal{A}$ onto the full matrix algebra M_n where n is the dimension of $\mathcal{L}$. Since M_n is simple and unital this concludes the proof of the first assertion of the theorem.

Now let $\mathcal{A}$ be a finite-dimensional semiprime algebra. The last paragraph shows that each prime ideal $\mathcal{P}$ of $\mathcal{A}$ is maximal modular since $\mathcal{A}/\mathcal{P}$ is a full matrix algebra which is simple and unital. A dimension argument shows that there is a finite set $\mathcal{P}_1, \mathcal{P}_2, \ldots, \mathcal{P}_n$ of prime ideals in $\mathcal{A}$ satisfying $\cap_{j=1}^n \mathcal{P}_j = \{0\}$. We may assume that no proper subset has intersection $\{0\}$. For each $a \in \mathcal{A}$ define $\theta(a)$ by

$$\theta(a) = (a + \mathcal{P}_1, a + \mathcal{P}_2, \ldots, a + \mathcal{P}_n).$$

Then θ is an injective homomorphism of $\mathcal{A}$ into the algebraic direct sum $\oplus_{j=1}^n \mathcal{A}/\mathcal{P}_j$. By what we have proved above, each quotient algebra $\mathcal{A}/\mathcal{P}_j$ is isomorphic to a full matrix algebra. Thus it only remains to show that θ is surjective. This will follow if we show that for any k satisfying $1 \le k \le n$ and any $b \in \mathcal{A}/\mathcal{P}_k$ there is an $a \in \mathcal{A}$ satisfying $\theta(a) = (0, 0, \ldots, 0, b, 0, \ldots, 0)$, where the non-zero entry is in the kth position. This is equivalent to showing

$$\mathcal{A} = (\bigcap_{\substack{1 \le j \le n \\ j \ne k}} \mathcal{P}_j) + \mathcal{P}_k$$

for any k satisfying $1 \le k \le n$. However this is an immediate consequence of the fact that $\mathcal{P}_k$ is a maximal ideal.   □

8.1.2 Example   We give a simple, completely explicit example of what this decomposition into full matrix algebras looks like in a concrete case. Indeed we have chosen the simplest example which is interesting—namely the six-dimensional algebraic group algebra of the symmetric group on three objects. Since this algebra is not commutative it must contain a full matrix algebra of dimension strictly greater than one, and hence one knows *a priori* that it is the direct sum of M_2 and two copies of $\mathbb{C}$. Here are the details.

Denote the symmetric group of all permutations of the numbers $\{1, 2, 3\}$ by Σ_3. As usual let (12) and (123) denote, respectively, the permutation interchanging 1 and 2 and the permutation which replace 1 by 2, 2 by 3, and 3 by 1. Using this notation, we denote the six group elements by: e = identity transformation, $a = (12)$, $b = (23)$, $c = (13)$, $d = (123)$ and $f = (132)$. The group multiplication table follows.

	e	a	b	c	d	f
e	e	a	b	c	d	f
a	a	e	d	f	b	c
b	b	f	e	d	c	a
c	c	d	f	e	a	b
d	d	c	a	b	f	e
f	f	b	c	a	e	d

Let $\mathcal{A}$ be the algebraic group algebra (Section 1.10) of Σ_3 which in the case of a finite group such as Σ_3 is just the underlying algebra of the normed ℓ^1- or L^1-group algebra. It is natural to identify each group element with the delta function concentrated at it. Then e is an identity for $\mathcal{A}$. Consider the following elements of $\mathcal{A}$:

$$h = \frac{1}{6}(a + b + c + d + e + f), \quad k = \frac{1}{6}(-a - b - c + d + e + f)$$

$$j = \frac{1}{3}(2e - d - f).$$

These are orthogonal central idempotents with sum $e = 1$ so that $\mathcal{A}$ has the algebra direct sum decomposition

$$\mathcal{A} = h\mathcal{A} \oplus k\mathcal{A} \oplus j\mathcal{A}$$

with $h\mathcal{A}$ and $k\mathcal{A}$ one-dimensional and $j\mathcal{A}$ four-dimensional. If we define $\chi_0, \chi_1 : \Sigma_3 \to \mathbb{C}$ by

$$\chi_0(g) \quad = \quad 1 \qquad g \in \Sigma_3$$
$$\chi_1(g) \quad = \quad \begin{cases} +1 & g = d, e \text{ or } f \\ -1 & g = a, b \text{ or } c \end{cases}$$

we see:

$$gh = hg = \chi_0(g)h, \quad gk = kg = \chi_1(g)k \qquad \forall \ g \in \Sigma_3.$$

This gives the multiplicative structure of the one-dimensional minimal ideals $h\mathcal{A} = \mathbb{C}h$ and $k\mathcal{A} = \mathbb{C}k$ in $\mathcal{A}$. In order to obtain the structure of the four-dimensional minimal ideal $j\mathcal{A}$ we let ζ be a complex cube root of 1 (e.g., $\zeta = (-1 \pm \sqrt{3}i)/2$ and consider the two orthogonal noncentral projections $m = \frac{1}{3}(e + \zeta d + \zeta^2 f)$ and $n = \frac{1}{3}(e + \zeta^2 d + \zeta f)$.) It is easy to check $j = m + n$ and that $\pi : j\mathcal{A} \to M_2$ is an isomorphism when we define:

$$\pi(m) = \begin{pmatrix} 1 & 0 \\ 0 & 0 \end{pmatrix}, \qquad \pi(n) = \begin{pmatrix} 0 & 0 \\ 0 & 1 \end{pmatrix},$$

$$\pi(am) = \begin{pmatrix} 0 & 0 \\ 1 & 0 \end{pmatrix} \quad \text{and} \quad \pi(an) = \begin{pmatrix} 0 & 1 \\ 0 & 0 \end{pmatrix}.$$

Thus we have written $\mathcal{A}$ as a direct sum of minimal ideals isomorphic to $\mathbb{C}, \mathbb{C}$ and M_2. The restriction of the left regular representation of $\mathcal{A}$ to the two dimensional subspace spanned by m and am yields the group (or algebra) representations $\pi: \Sigma_3 \to M_2$ (or $\pi: \mathcal{A} \to M_2$) given by:

$$\pi(e) = \begin{pmatrix} 1 & 0 \\ 0 & 1 \end{pmatrix} \qquad \pi(a) = \begin{pmatrix} 0 & 1 \\ 1 & 0 \end{pmatrix} \qquad \pi(b) = \begin{pmatrix} 0 & \zeta \\ \zeta^2 & 0 \end{pmatrix}$$

$$\pi(c) = \begin{pmatrix} 0 & \zeta^2 \\ \zeta & 0 \end{pmatrix} \qquad \pi(d) = \begin{pmatrix} \zeta^2 & 0 \\ 0 & \zeta \end{pmatrix} \qquad \pi(f) = \begin{pmatrix} \zeta & 0 \\ 0 & \zeta^2 \end{pmatrix}.$$

Similar explicit calculations have been carried out for many small finite groups. The difficulty increases rapidly with the size of the group. We will show later that a similar analysis is possible for compact groups. Again the details are complicated in all but the simplest cases.

The following corollary is taken from Peter G. Dixon and George A. Willis [1993]. The second statement depends on Theorem 8.1.1. Theorem 4.8.9 and Corollary 4.8.10 are closely related.

8.1.3 Corollary *Let $\mathcal{A}$ be a commutative Banach algebra. Then $\mathcal{A}$ is unital if and only if it satisfies $\mathcal{A}^2 = \mathcal{A}$ and it also has a modular closed nilpotent ideal.*

In particular, a finite-dimensional commutative algebra $\mathcal{A}$ is unital if and only if it satisfies $\mathcal{A}^2 = \mathcal{A}$.

Proof If $\mathcal{A}$ is unital, $\mathcal{A}^2 = \mathcal{A}$ and we may choose the ideal $\{0\}$. Suppose, conversely, that $\mathcal{I}$ is a modular closed ideal and n is a positive integer such that all $b_1, b_2, \ldots, b_n \in \mathcal{I}$ satisfy $b_1 b_2 \cdots b_n = 0$. Let e be a relative identity for $\mathcal{I}$. Then any $a_1, a_2, \ldots, a_n \in \mathcal{A}$ satisfy

$$(1 - e)^n a_1 a_2 \cdots a_n = (1 - e)a_1(1 - e)a_2 \cdots (1 - e)a_n = 0.$$

Hence the n-fold quasi-product power of e, which we denote by u, satisfies $u a_1 a_2 \cdots a_n = a_1 a_2 \cdots a_n$ for any n-fold product $a_1 a_2 \cdots a_n$ in $\mathcal{A}$. However, by induction, $\mathcal{A}^2 = \mathcal{A}$ implies that every element of $\mathcal{A}$ is a sum of n-fold products, so that u is an identity for $\mathcal{A}$.

If $\mathcal{A}$ is finite-dimensional, Theorem 8.1.1 shows that $\mathcal{A}/\mathcal{A}_J$ is unital, so $\mathcal{A}_J$ is modular. Any finite-dimensional algebra can be given a norm in which it is a Banach algebra. Furthermore $\mathcal{A}_J$ is a closed nilpotent ideal, so that the last paragraph shows that $\mathcal{A}$ is unital if $\mathcal{A}^2 = \mathcal{A}$. □

Feldman Theorem

The next theorem is due to Chester Feldman [1951]. It is an improvement of another theorem due to Wedderburn [1907] which deals with finite-dimensional algebras. The theorem shows a case in which one can be sure that a nonsemisimple algebra can be suitably decomposed into its Jacobson radical and a subalgebra isomorphic to its semisimple quotient. After the theorem we will give an example, also due to Feldman, of a Banach algebra $\mathcal{A}$ in which there is no subalgebra $\mathcal{B}$ such that $\mathcal{A}$ is the linear space direct sum of $\mathcal{A}_R$ and $\mathcal{B}$. We will also introduce some standard terminology and additional results.

8.1.4 Feldman Theorem *Let $\mathcal{A}$ be a Banach algebra such that $\mathcal{A}/\mathcal{A}_J$ is finite-dimensional. Let $\Phi\colon \mathcal{A} \to \mathcal{A}/\mathcal{A}_J$ be the natural map. Then there is a closed subalgebra $\mathcal{B} \subseteq \mathcal{A}$ satisfying:*
(a) $\mathcal{B} + \mathcal{A}_J = \mathcal{A}$.
(b) $\mathcal{B} \cap \mathcal{A}_J = \{0\}$.
(c) Φ *induces a homeomorphic isomorphism of $\mathcal{B}$ onto $\mathcal{A}/\mathcal{A}_J$.*

Proof The last theorem shows that $\mathcal{A}/\mathcal{A}_J$ is isomorphic to $\sum_{k=1}^{n} M_{n_k}$. Choose an isomorphism of $\sum_{k=1}^{n} M_{n_k}$ onto $\mathcal{A}/\mathcal{A}_J$. Let $\{a_{ij}^k : 1 \le i,j \le n_k\}$ be the image in $\mathcal{A}/\mathcal{A}_J$ of the matrix units in the kth matrix under the chosen isomorphism. Proposition 4.3.12(e) shows that we may choose orthogonal idempotents $\{e_{ii}^k : 1 \le k \le n; 1 \le i \le n_k\}$ in $\mathcal{A}$ such that $\Phi(e_{ii}^k) = a_{ii}^k$ for all i and k. The identities $a_{ii}^k a_{i1}^k a_{11}^k = a_{i1}^k$ and $a_{11}^k a_{1j}^k a_{jj}^k = a_{1j}^k$ show that for all k, i, j satisfying $1 \le k \le n, 1 \le i \le n_k$ and $1 \le j \le n_k$ we can choose $b_{i1}^k \in e_{ii}^k \mathcal{A} e_{11}^k$ and $b_{1j}^k \in e_{11}^k \mathcal{A} e_{jj}^k$ satisfying $\Phi(b_{i1}^k) = a_{i1}^k$ and $\Phi(b_{1j}^k) = a_{1j}^k$. Then the equation $\Phi(b_{1j}^k b_{j1}^k) = a_{1j}^k a_{j1}^k = a_{11}^k$ shows that there is an element $c_j^k \in \mathcal{A}_J \cap e_{11}^k \mathcal{A} e_{11}^k$ satisfying $b_{1j}^k b_{j1}^k = e_{11}^k - c_j^k$. Proposition 2.5.3 shows that $\mathcal{A}_J \cap e_{11}^k \mathcal{A} e_{11}^k$ is a spectral subalgebra of $\mathcal{A}$ and Theorem 4.3.6(b) shows that c_j^k has a quasi-inverse, which we will denote by d_j^k, in $\mathcal{A}$ and hence in $\mathcal{A}_J \cap e_{11}^k \mathcal{A} e_{11}^k$. Define e_{i1}^k and e_{1j}^k by $e_{i1}^k = b_{i1}^k(e_{11}^k - d_i^k)$ and $e_{1j}^k = b_{1j}^k$. Then we have

$$
\begin{aligned}
e_{1j}^k e_{j1}^k &= b_{1j}^k b_{j1}^k (e_{11}^k - d_j^k) = (e_{11}^k - c_j^k)(e_{11}^k - d_j^k) \\
&= e_{11}^k - c_j^k - d_j^k + c_j^k d_j^k = e_{11}^k.
\end{aligned}
$$

Now we define e_{ij}^k by $e_{ij}^k = e_{i1}^k e_{1j}^k$. Clearly these elements satisfy

$$
\Phi(e_{ij}^k) = a_{ij}^k \quad \text{and} \quad e_{ij}^k e_{pq}^m = \delta_{km}\delta_{jp}e_{iq}^k
$$

for all $1 \le k, m \le n, 1 \le i,j \le n_k, 1 \le p,q \le n_m$ (where δ_{km} is the Kronecker delta and the second equation just reflects the definition of matrix units in the direct sum decomposition $\sum_{k=1}^{n} M_{n_k}$). Hence if we denote the linear span of these elements by $\mathcal{B}$, then Φ induces an isomorphism of $\mathcal{B}$ onto $\mathcal{A}/\mathcal{A}_J$. Since $\mathcal{B}$ is finite-dimensional, it is complete and hence closed in $\mathcal{A}$ and the isomorphism from $\mathcal{B}$ onto $\mathcal{A}/\mathcal{A}_J$ induced by Φ is a homeomorphism. Since Φ maps $\mathcal{B}$ onto $\mathcal{A}/\mathcal{A}_J$ we conclude, $\mathcal{B} + \mathcal{A}_J = \mathcal{A}$. Finally, since $\mathcal{B}$ has the form $e\mathcal{A}e$, where $e = \sum_{k=1}^{n} \sum_{i=1}^{n_k} e_{ii}^k$ is idempotent, and since $\mathcal{B}$ is semisimple, Proposition 4.3.12(b) implies $\mathcal{B} \cap \mathcal{A}_J = \{0\}$. $\square$

8.1.5 Feldman's Example The following example is due to Feldman [1951]. However some of the results given here were first obtained by William G. Bade and Philip C. Curtis, Jr. [1960a]. Consider the commutative algebra $\mathcal{A}^0$ of all finite sums

$$\sum_{j=1}^{n} \lambda_j e_j + \mu r \quad \forall \, \lambda_j, \mu \in \mathbb{C}$$

where $\{e_j : j \in \mathbb{N}\}$ is a family of mutually orthogonal idempotents and r is an element satisfying $r^2 = re_j = e_j r = 0$ for all $j \in \mathbb{N}$. Norm these elements by

$$\| \sum_{j=1}^{n} \lambda_j e_j + \mu r \| = \max\{(\sum |\lambda_j|^2)^{1/2}, |\mu - \sum_{j=1}^{n} \lambda_j|\}.$$

It is straightforward to check that this is an algebra norm (submultiplicativity is checked in the first reference above). Let $\mathcal{A}$ be the completion of $\mathcal{A}^0$ in this norm. It is easy to see that the Jacobson radical of $\mathcal{A}$ is the one-dimensional space $\mathbb{C}r$. Furthermore, in the quotient norm we have

$$\| \sum_{j=1}^{n} \lambda_j e_j + \mu r + \mathcal{A}_J \| = (\sum_{j=1}^{n} |\lambda_j|^2)^{1/2}$$

so that the map

$$\lambda \mapsto \lim_{n \to \infty} \sum_{j=1}^{n} \lambda_j (e_j + \mathcal{A}_J) \quad \forall \, \lambda \in \ell^2(\mathbb{N})$$

is an isometric isomorphism of $\ell^2(\mathbb{N})$ with pointwise multiplication onto $\mathcal{A}/\mathcal{A}_J$. Hence $\mathcal{A}/\mathcal{A}_J$ contains the element $\sum_{j=1}^{\infty} j^{-1}(e_j + \mathcal{A}_J)$ (convergence is in the quotient norm since $\sum_{j=1}^{\infty} j^{-2} = \pi^2/6$) while $\mathcal{A}$ does not contain any element of the form $\sum_{j=1}^{\infty} j^{-1} e_j$ because this series diverges in $\mathcal{A}$ since $\sum_{j=1}^{\infty} j^{-1} = \infty$.

     The above remarks imply that there is no closed subalgebra $\mathcal{B}$ of $\mathcal{A}$ satisfying $\mathcal{A} = \mathcal{B} \oplus \mathcal{A}_J$ (see Theorem 6.1(b) of our second reference for a proof). Hence Feldman's Theorem (8.1.4, above) fails for this Banach algebra. Since the radical of $\mathcal{A}$ is one dimensional and nilpotent this shows that no simple hypothesis on the radical itself will be strong enough to imply the conclusion of Theorem 8.1.4.

We follow Theorem 8.6.1(c) of our second reference in constructing a nonclosed subalgebra $\mathcal{B}$ of $\mathcal{A}$ which satisfies $\mathcal{A} = \mathcal{B} \oplus \mathcal{A}_J$. The map of ℓ^2 onto $\mathcal{B}$ which we are about to define is a discontinuous isomorphism. Since $\ell^1(\mathbb{N})$ is a linear subspace of $\ell^2(\mathbb{N}) \simeq \mathcal{A}/\mathcal{A}_J$ we may use a Hamel basis or Zorn's lemma argument to construct a linear subspace $\mathcal{Y}$ of $\ell^2(\mathbb{N})$ satisfying $\ell^2(\mathbb{N}) = \ell^1(\mathbb{N}) \oplus \mathcal{Y}$. For $\lambda \in \ell^1(\mathbb{N})$ define $\varphi(\lambda)$ by

$$\varphi(\lambda) = \sum_{j=1}^{\infty} \lambda_j e_j$$

which converges absolutely in the norm of $\mathcal{A}$. For any $\lambda \in \mathcal{Y}$ (which must satisfy $\sum_{j=1}^{\infty} |\lambda_j| = \infty$) we define $\varphi(\lambda)$ as the limit of the Cauchy sequence $\varphi(\lambda)_n = \sum_{j=1}^{n} \lambda_j e_j + (\sum_{j=1}^{n} \lambda_j)r$ in $\mathcal{A}$. It is easy to show that the natural map of $\mathcal{A}$ onto $\mathcal{A}/\mathcal{A}_J$ (or equivalently the Gelfand transform) maps $\varphi(\ell^2(\mathbb{N}))$ onto $\ell^2(\mathbb{N})$ (with suitable interpretation of $\mathcal{A}/\mathcal{A}_J$ or of $\Phi_{\mathcal{A}}$, respectively). It remains to show that φ is a homomorphism and that $\varphi(\ell^2(\mathbb{N}))$ is a subalgebra of $\mathcal{A}$. Clearly the restriction of φ to $\ell^1(\mathbb{N})$ is a homomorphism and $\varphi(\ell^1(\mathbb{N}))$ is a subalgebra. If λ, μ are two elements of $\mathcal{Y}$, then we have

$$\varphi(\lambda)_n \varphi(\mu)_n = \Big(\sum_{j=1}^{n} \lambda_j e_j + (\sum_{j=1}^{n} \lambda_j)r\Big)\Big(\sum_{j=1}^{n} \mu_j e_j + (\sum_{j=1}^{n} \mu_j)r\Big) = \sum_{j=1}^{n} \lambda_j \mu_j e_j$$

and this sequence converges to $\varphi(\lambda\mu)$ in $\varphi(\ell^1(\mathbb{N}))$ since the pointwise product $\lambda\mu$ belongs to $\ell^1(\mathbb{N})$ by the Cauchy–Schwarz inequality. Hence, the product of any two elements $\varphi(\lambda)$ and $\varphi(\mu)$ in $\varphi(\mathcal{Y})$ equals $\varphi(\lambda\mu)$ in $\varphi(\ell^1(\mathbb{N}))$. Similarly, the product of an element $\varphi(\lambda)$ in $\varphi(\ell^1(\mathbb{N}))$ and an element $\varphi(\mu)$ in $\varphi(\mathcal{Y})$ equals $\varphi(\lambda\mu)$ in $\varphi(\ell^1(\mathbb{N}))$. Therefore, $\varphi(\ell^2(\mathbb{N}))$ is a subalgebra and φ is a homomorphism.

Since the above remarks show that $\mathcal{B}$ is not closed, φ must be discontinuous. To see this explicitly, note that the idempotent element $f_n = (1, 1, \ldots, 1, 0, 0, \ldots)$ in $\ell^2(\mathbb{N})$ with 1 in its first n entries satisfies $\|f_n\| = n^{1/2}$ and $\|\varphi(f_n)\| = n$.

Another very important property of this example is the fact that $\mathcal{A}$ is a Banach algebra under two inequivalent norms. A second complete algebra norm is given by

$$|||\varphi(\lambda) + \mu r||| = \|\lambda\|_2 + |\mu|$$

where we have split each element of $\mathcal{A}$ according to the direct sum decomposition $\mathcal{A} = \varphi(\ell^2(\mathbb{N})) \oplus \mathcal{A}_J$.

Using the same direct sum decomposition again, note that we can define a map $\delta \colon \mathcal{A} \to \mathcal{A}$ by

$$\delta(\varphi(\lambda) + \mu r) = \mu r.$$

This is a derivation since each term in the derivation equation $\delta(ab) = \delta(a)b + a\delta(b)$ is zero. Similarly δ is also an endomorphism. However, in the original norm of $\mathcal{A}$ this map is discontinuous. In particular, for any $n \in \mathbb{N}$ we have $\| \sum_{j=1}^{n} e_j + nr \| = n^{1/2}$ but $\|\delta(\sum_{j=1}^{n} e_j + nr)\| = \|\delta(\varphi(f_n) + nr)\| = \|nr\| = n$.

Note that the algebra $\mathcal{A}$ has a dense socle in either of the two norms given here. Of course, the socle of any commutative algebra is defined. It is important to note that $\mathcal{A}$ is not semiprime since $\mathcal{A}_J^2 = \mathcal{A}_J\mathcal{A} = \{0\}$ holds. However it is obvious from the definition that $\mathbb{C}r$ and $\mathbb{C}e_i$ for any $i \in \mathbb{N}$ are minimal (two-sided) ideals of $\mathcal{A}$ so that the socle contains any element of the form $\sum_{j=1}^{n} \lambda_j e_j + \mu r$. By definition, this set of elements is dense in $\mathcal{A}$ in its original norm and it is also obviously dense in the norm $\||\cdot\||$.

We introduce some terminology and other examples related to this one.

8.1.6 Definition Let $\mathcal{A}$ be a Banach algebra and let $\mathcal{A}_J$ be its Jacobson radical. We say that $\mathcal{A}$ has an ⟨ *algebraic / topological* ⟩ Wedderburn splitting if there is a ⟨ subalgebra / closed subalgebra ⟩ $\mathcal{B}$ satisfying

$$\mathcal{A} = \mathcal{A}_J + \mathcal{B} \quad \text{and} \quad \mathcal{A}_J \cap \mathcal{B} = \{0\}.$$

Of course, in either case $\Phi : \mathcal{A} \to \mathcal{A}/\mathcal{A}_J$ induces a continuous algebra isomorphism of $\mathcal{B}$ onto $\mathcal{A}/\mathcal{A}_J$ which is homeomorphic in the case of a topological splitting by the open mapping theorem. In this terminology, Feldman's theorem (Theorem 8.1.4) assures a topological Wedderburn splitting and his example (Example 8.1.5) has an algebraic but no topological Wedderburn splitting.

The following elegant theorem of Curtis and Richard J. Loy [1989] improves a similar result of Gregory F. Bachelis and Sadahiro Saeki [1987]. The corollary indicates how this relates to the last example.

8.1.7 Theorem *Let $\mathcal{A}$ and $\mathcal{B}$ be commutative Banach algebras with $\mathcal{A}$ unital. Let $\varphi, \psi : \mathcal{A} \to \mathcal{B}$ be two continuous homomorphisms satisfying $\varphi(a) - \psi(a) \in \mathcal{B}_J$ for all $a \in \mathcal{A}$.*

(a) For $a \in \mathcal{A}_G$, $\lim \|a^n\| \, \|a^{-n}\|/n = 0$ implies $\varphi(a) = \psi(a)$.

(b) For $a \in \mathcal{A}$, $\lim \|e^{na}\| \, \|e^{-na}\|/n = 0$ implies $\varphi(a) = \psi(a)$.

(c) If $\mathcal{A}$ is the closed linear span of elements satisfying (a) or (b), then φ and ψ are equal.

Proof Define ⟨ e / f ⟩ to be ⟨ $\varphi(1)$ / $\psi(1)$ ⟩. Then $e - ef$ and $f - ef$ are idempotents in $\mathcal{B}_J$, so $e = ef = f$ by Proposition 4.3.12. Hence we may replace $\mathcal{B}$ by $e\mathcal{B}$ and thus assume that $\mathcal{B}$, φ and ψ are all unital.

(a): For any $a \in \mathcal{A}_G$, define $b = \varphi(a)^{-1}(\psi(a) - \varphi(a))$ so $1 + b = \varphi(a^{-1})\psi(a)$. Then $b \in \mathcal{B}_J$ satisfies $\lim \|(1 + b)^n\| = \lim \|\varphi(a^{-n})\psi(a^n)\| \leq \|\varphi\| \, \|\psi\| \lim \|a^n\| \, \|a^{-n}\|/n = 0$. Theorem 2.6.12(a) shows $b = 0$ and hence $\varphi(a) = \psi(a)$.

(b): For any $a \in \mathcal{A}$, part (a) shows $\varphi(e^a) = \psi(e^a)$. Hence we get $e^\varphi(a) = \varphi(e^a) = \psi(e^a) = e^{\varphi(a)+r}$ where $r = \psi(a) - \varphi(a)$. This implies $0 = r(1 + \sum_{n+1}^\infty r^{n-1}/n!)$ where the second factor is invertible since the series is in the Jacobson radical. We conclude $r = 0$.
(c) Obvious.   □

8.1.8 Corollary *Let $\mathcal{A}$ be a non-semisimple unital commutative Banach algebra which is the closed linear span of either of the following two sets:*

$$\{a \in \mathcal{A}_G : \lim \|a^n\| \, \|a^{-n}\|/n = 0\} \quad or \quad \{a \in \mathcal{A} : \lim \|e^{na}\| \, \|e^{-na}\|/n = 0.\}$$

Then $\mathcal{A}$ does not have a topological Wedderburn splitting.

Proof If $\mathcal{A}$ had a topological Wedderburn splitting, we could choose $\varphi: \mathcal{A} \to \mathcal{A}$ to be the identity map and $\psi: \mathcal{A} \to \mathcal{A}$ to be the composition of the natural map $\mathcal{A} \to \mathcal{A}/\mathcal{A}_J$ and the inverse of the continuous (hence homeomorphic) isomorphism of $\mathcal{B}$ onto $\mathcal{A}/\mathcal{A}_J$. The theorem would imply $\mathcal{A}_J = \{0\}$, contrary to hypothesis.   □

8.1.9 Non-splitting and Sets of Non-synthesis The following interesting examples are due to the combined work of Bachelis and Saeki [1987], Bade [1989], Curtis and Loy [1989] and Curtis [1989], based on the above theorem or variations of it. Sets of non-synthesis were introduced in §3.6.16 where the ideal $\mathcal{J}(S)$, which is minimal among all closed ideals with hull S, is also described (cf. Definition 7.3.1). (Our notation is essentially dual to that used in some of these references.)

Theorem *Let G be a noncompact locally compact abelian group and let S be a compact set of non-synthesis in $\hat{G}$. Then $L^1(G)/\mathcal{J}(S)$ has no algebraic Wedderburn splitting.*

Theorem *Let S be a closed set of non-synthesis in $\mathbb{R}$. Then $L^1(G)/\mathcal{J}(S)$ has no topological Wedderburn splitting.*

A number of hypotheses are also known which force the existence of algebraic or topological Wedderburn splittings. The question of the uniqueness of the subalgebra or closed subalgebra $\mathcal{B}$ has also been discussed. See our chapter on cohomology in Volume II and Feldman [1951], Bade and Curtis [1960a], [1960b], Barry E. Johnson [1968a], [1972b], Nikolai Valeryevich Yakovlev [1989] and the references above.

Here are some references to the numerous papers giving criteria for a Banach algebra to be finite-dimensional: Kaplansky [1954], Dixon [1974], Abdullah H. Al-Moajil [1981], [1982], Leoni Dalla, S. Giotopoulos and Nelli Katseli [1989]

8.2   Minimal Ideals and the Socle

Idempotents and Ideals

In our discussion of algebras with minimal ideals certain idempotents will play an important role. For this reason we recall some simple identities which we have used before. For any idempotent e in any algebra $\mathcal{A}$ we have:

$$\mathcal{A}e = \{a \in \mathcal{A} : a = ae\} = \{a \in \mathcal{A} : a(1 - e) = 0\} \tag{1}$$

$$e\mathcal{A} = \{a \in \mathcal{A} : a = ea\} = \{a \in \mathcal{A} : (1 - e)a = 0\} \tag{2}$$

$$e\mathcal{A}e = \{a \in \mathcal{A} : a = ea = ae\} = \{a \in \mathcal{A} : (1 - e)a = a(1 - e) = 0\}. \tag{3}$$

All three of these sets are spectral subalgebras and are closed if $\mathcal{A}$ is a topological algebra. Furthermore, Proposition 4.3.12(b) shows that the last set satisfies $(e\mathcal{A}e)_J = e\mathcal{A}e \cap \mathcal{A}_J$. Recall also that two idempotents e and f are said to be orthogonal if they satisfy $ef = fe = 0$.

Many of our preliminary results will deal with one-sided ideals. We will sometimes state these results, and even definitions, merely for left ideals. In most cases the result for right ideals follows immediately by considering the reverse algebra. It is easy to see when this is the case. However, the reader should bear in mind that a primitive ideal is not necessarily primitive in the reverse algebra.

Minimal Ideals

8.2.1 Definition   Let $\mathcal{A}$ be an algebra. An idempotent $e \in \mathcal{A}$ is called *minimal* if $e\mathcal{A}e$ is a division algebra. The set of minimal idempotents in $\mathcal{A}$ is denoted by $\mathcal{A}_{MI}$. A left ideal of $\mathcal{A}$ is called *minimal* if it is minimal among the set of non-zero left ideals ordered by inclusion. If $\mathcal{A}$ is a topological algebra, a left ideal of $\mathcal{A}$ is called a *minimal closed left ideal* if it is closed and is minimal among the set of non-zero closed left ideals of $\mathcal{A}$. Minimal ideals, minimal closed ideals, minimal right ideals and minimal closed right ideals are defined similarly.

The term "minimal idempotent" as used in the above definition seems to have been introduced by Charles E. Rickart [1960] although closely related concepts have a long history. (The related idea of primitive idempotents even occurs in Wedderburn [1907].) It is justified by the next proposition.

Partial justification can also be given in terms of the order structure of idempotents. If $\mathcal{A}$ is an algebra and e and f are idempotents in $\mathcal{A}$, we write $e \leq f$ if $fe = e$ holds. This is equivalent to $e\mathcal{A} \subseteq f\mathcal{A}$. If $\mathcal{A}$ is an algebra of linear operators on a linear space $\mathcal{X}$, it is equivalent to $e\mathcal{X} \subseteq f\mathcal{X}$. (This is the reason we are using principal right ideals rather than principal left ideals here. There is a different, but completely similar, partial order among idempotents based upon left ideals, and it enjoys the

same relationship to minimal idempotents that we are about to establish.)
Now suppose f is a minimal idempotent and e is an idempotent satisfying
$e \leq f$. Then we have $efef = eef = ef$ and $fef = ef$. Hence $ef = fef$ is
an idempotent in the division algebra $f\mathcal{A}f$ so that either $ef = f$ or $ef = 0$
holds. Therefore f is minimal in the ordering just introduced if it is minimal
according to Definition 8.2.1. The converse is not true. For instance, for
any compact connected topological space Ω, 1 is an idempotent in the
commutative semisimple algebra $C(\Omega)$ which is minimal in this ordering
but does not satisfy Definition 8.2.1 unless Ω is a singleton.

8.2.2 Proposition *Let $\mathcal{L}$ be a minimal left ideal of an algebra $\mathcal{A}$. If $\mathcal{L}^2$ is
non-zero, then there is an idempotent $e \in \mathcal{A}$ which satisfies $\mathcal{L} = \mathcal{A}e$. Any
such idempotent is minimal. Conversely, if $\mathcal{A}$ is semiprime and $e \in \mathcal{A}$ is
a minimal idempotent in $\mathcal{A}$, then $\mathcal{A}e$ is a minimal left ideal. Analogous
results hold for right ideals.*

Proof Let $\mathcal{L}$ be a minimal left ideal satisfying $\mathcal{L}^2 \neq \{0\}$. Then for some
(necessarily non-zero) $a \in \mathcal{L}, \mathcal{L}a$ is a non-zero subset of $\mathcal{A}$ which is a left
ideal included in $\mathcal{L}$. The minimality of $\mathcal{L}$ implies $\mathcal{L} = \mathcal{L}a$. This implies
the existence of a (necessarily non-zero) element $e \in \mathcal{L}$ satisfying $ea = a$.
Similarly $\mathcal{L}(e - 1)$ is a left ideal included in $\mathcal{L}$. Minimality implies either
$\mathcal{L}(e - 1) = \mathcal{L}$ or $\mathcal{L}(e - 1) = \{0\}$. However, the first equation is impossible
since it would imply the existence of an element $b \in \mathcal{L}$ satisfying $b(e-1) = e$
which in turn implies the contradiction $0 = ba - ba = bea - ba = b(e-1)a =
ea = a$. Hence $\mathcal{L}(e - 1) = \{0\}$ holds and this implies $e(e - 1) = 0$ so that
e is idempotent. Moreover, the inclusions $\mathcal{L} \supseteq \mathcal{A}e \supseteq \{e\}$ and minimality
imply $\mathcal{L} = \mathcal{A}e$.

Now suppose e is an idempotent satisfying $\mathcal{L} = \mathcal{A}e$. Then $ee = e$
belongs to $\mathcal{L}$ and the subalgebra $e\mathcal{A}e$ has e as an identity. Any element
$eae \in e\mathcal{A}e \setminus \{0\}$ satisfies $\mathcal{L} \supseteq \mathcal{A}eae \supseteq \{eeae\}$. Hence minimality implies
$\mathcal{L} = \mathcal{A}eae$. Thus there is an element $b \in \mathcal{A}$ satisfying $beae = e$ and hence
satisfying $(ebe)(eae) = e$. Thus each non-zero element of $e\mathcal{A}e$ has a right
inverse. The proof of Corollary 2.2.3 shows that $e\mathcal{A}e$ is a division algebra.
Hence e is a minimal idempotent.

Finally, in order to establish the converse, we assume that $\mathcal{A}$ is semiprime
and that e is a minimal idempotent. Let $\mathcal{L}$ be a non-zero ideal included
in $\mathcal{A}e$. Since $\mathcal{L}^2 \subseteq \mathcal{A}e\mathcal{L}$ is not zero, $e\mathcal{L}$ is not zero. Hence there is some
element $ae \in \mathcal{L}$ satisfying $eae \neq 0$. Since $e\mathcal{A}e$ is a division algebra there is
an element $b \in \mathcal{A}$ satisfying $ebeae = e$. Hence e belongs to the left ideal $\mathcal{L}$
so that $\mathcal{L}$ equals $\mathcal{A}e$. Thus $\mathcal{A}e$ is a minimal left ideal.   □

8.2.3 Corollary *Let $\mathcal{A}$ be a semiprime algebra. Every minimal left ideal
of $\mathcal{A}$ has the form $\mathcal{A}e$ for some minimal idempotent e. Conversely, if e is
a minimal idempotent, then $\mathcal{A}e$ is a minimal left ideal.*

If $\mathcal{A}$ is also a topological algebra, then every minimal left ideal is closed, so that a left ideal of $\mathcal{A}$ is a minimal closed left ideal if and only if it is a minimal left ideal.

Consideration of the simple case of a full matrix algebra shows that the idempotent e of the last corollary is not usually uniquely defined.

8.2.4 Proposition *The following two conditions are equivalent for a pair e and f of idempotents in an algebra $\mathcal{A}$.*
(a) $ef = e$ and $fe = f$.
(b) $\mathcal{A}e = \mathcal{A}f$.
These conditions establish an equivalence relation both on the set $\mathcal{A}_I$ of all idempotents in $\mathcal{A}$ and on the set $\mathcal{A}_{MI}$ of all minimal idempotents in $\mathcal{A}$.

If $\mathcal{A}$ is semiprime the map $e \mapsto \mathcal{A}e$ establishes a bijective correspondence between the set of equivalence classes of minimal projections and the set of minimal left ideals of $\mathcal{A}$.

Proof Condition (b) is equivalent to assuming both $e \in \mathcal{A}f$ and $f \in \mathcal{A}e$. Thus it is equivalent to (a) by equation (1) above. The conditions obviously define an equivalence relation on either set mentioned. The last proposition shows that the map mentioned establishes the stated bijective correspondence. □

We now mention another useful equivalence relation on $\mathcal{A}_{MI}$ with larger equivalence classes. Let $\mathcal{A}$ be a semiprime algebra and let e be a minimal idempotent of $\mathcal{A}$ so that $\mathcal{A}e$ is a minimal left ideal and $e\mathcal{A}$ is a minimal right ideal. Then it is obvious that the restriction of the left regular representation to $\mathcal{A}e$ is an irreducible representation. This representation is so useful that we introduce the special notation L^e for it. Thus we have

$$L^e_a(be) = abe \qquad \forall\, a,\, b \in \mathcal{A}. \tag{4}$$

The right regular representation is an anti-representation of $\mathcal{A}$. An anti-representation can be thought of as a representation of the reverse algebra $\mathcal{A}^R$. Using this viewpoint, we carry over all the terminology concerning representations to anti-representations. Then the restriction of the right regular representation to $e\mathcal{A}$ is an irreducible anti-representation. Again we give this anti-representation a special notation, R^e. Hence we have

$$R^e_a(eb) = eba \qquad \forall\, a,\, b \in \mathcal{A}. \tag{5}$$

Now we introduce an equivalence relation on $\mathcal{A}_{MI}$ by considering when L^e and L^f are equivalent for $e,\, f \in \mathcal{A}_{MI}$. The notation $Equiv(L^e, L^f)$ and some remarks about equivalences were introduced following Definition 4.1.1.

8.2.5 Proposition *Let $\mathcal{A}$ be a semiprime algebra and let e and f be minimal idempotents of $\mathcal{A}$. For each $b \in e\mathcal{A}f$, let $V_b: \mathcal{A}e \to \mathcal{A}f$ be defined* by $V_b(ae) = aeb$ for all $ae \in \mathcal{A}e$. Then the map $b \mapsto V_b$ establishes a linear bijection between $e\mathcal{A}f \setminus \{0\}$ and $Equiv(L^e, L^f)$. The following are equivalent:

(a) L^e is equivalent to L^f.
(b) R^e is equivalent to R^f.
(c) $e\mathcal{A}f$ is non-zero.
(d) $f\mathcal{A}e$ is non-zero.

Proof Clearly V_b belongs to $Equiv(L^e, L^f)$ for each non-zero $b \in e\mathcal{A}f$. Suppose V belongs to $Equiv(L^e, L^f)$. Let $b = V(e)$. Then $b \in \mathcal{A}f$ and $eb = eV(e) = L^f_e V(e) = V(L^e_e d) = V(e) = b$ imply $b \in e\mathcal{A}f$. Furthermore $V(ae) = V(L^e_{ae}e) = L^f_{ae}V(e) = aeb = b$ for all $a \in \mathcal{A}$ implies $V = V_b$. Hence $b \mapsto V_b$ is a linear bijection. This proves the equivalence of (a) and (c). The same argument in $\mathcal{A}^R$ proves the equivalence of (b) and (d). Since e and f play symmetrical roles in (a) and (b), (c) and (d) are equivalent. □

Properties (a) and (b) in the last proposition are easily seen to be equivalent when e and f are arbitrary idempotents in an arbitrary algebra. In this more general case they are each equivalent to the existence of elements $c \in e\mathcal{A}f$ and $d \in f\mathcal{A}e$ satisfying $cd = e$ and $dc = f$. However, in our more special setting the easier criteria (c) and (d) will be useful. Note that the conditions just mentioned for pairs of elements of $\mathcal{A}_I$ define an equivalence relation on $\mathcal{A}_I$ which generalizes the equivalence relation on $\mathcal{A}_{MI}$ defined by the conditions of the last proposition.

The next proposition provides an alternative proof of the equivalence of the four conditions in the last proposition.

8.2.6 Proposition *Let $\mathcal{A}$ be a semiprime algebra and let e be a minimal idempotent of $\mathcal{A}$. Then L^e is an irreducible representation of $\mathcal{A}$ and any irreducible representation of $\mathcal{A}$ with the same kernel as L^e is equivalent to L^e. Furthermore, if $\mathcal{A}$ is a prime algebra, then L^e is faithful so that $\mathcal{A}$ is in fact primitive.*

Proof Since Proposition 8.2.2 shows that $\mathcal{A}e$ is a minimal left ideal, L^e is irreducible. Suppose T is another irreducible representation of $\mathcal{A}$ with the same kernel as L^e. Then T_e is non-zero so we may choose a non-zero vector z in $T_e \mathcal{X}^T$. Define a map $V : \mathcal{A}e \to \mathcal{X}^T$ by $V(ae) = T_{ae}z = T_a z$. Then we get $T_a V(be) = T_a T_{be} z = T_{abe} z = V(abe) = V L^e_a(be)$ for all $b \in \mathcal{A}$. Hence $V(\mathcal{A}e)$ is a non-zero T-invariant subspace of $\mathcal{X}_T$. By irreducibility V is surjective. Similarly $\ker(V)$ is a left ideal of $\mathcal{A}e$ which does not contain e so that V is injective. Hence V is an equivalence.

If $\mathcal{A}$ is a prime algebra then $\ker(L^e)\mathcal{A}e = \{0\}$ and $\mathcal{A}e \neq \{0\}$ imply that L^e is faithful so $\mathcal{A}$ is primitive. □

The following definition is due to Dieudonné [1942].

8.2.7 Definition   Let $\mathcal{A}$ be an algebra. The *left socle* of $\mathcal{A}$ is the sum of all the minimal left ideals of $\mathcal{A}$. The *right socle* of $\mathcal{A}$ is the sum of all minimal right ideals of $\mathcal{A}$. If the left and right socle of $\mathcal{A}$ coincide then their common value is called the *socle of $\mathcal{A}$* and is denoted by $\mathcal{A}_F$.

Here, as usual, we follow the common convention that the sum of an empty family of one-sided ideals or ideals of $\mathcal{A}$ is $\{0\}$. Curiously, this common convention is not followed by many authors in dealing with left socles, right socles and socles. Hence they say that a socle exists if the left and right socles exist, are non-zero, and coincide. In the same case we would say that $\mathcal{A}$ has a non-zero socle. When we say that $\mathcal{A}$ has a socle we do not imply that the socle is non-zero.

Obviously the left socle is a left ideal and the right socle is a right ideal. We will now prove that in fact they are both ideals.

8.2.8 Proposition *Let $\mathcal{A}$ be an algebra. The left socle and the right socle of $\mathcal{A}$ are both ideals. If $\mathcal{A}$ is semiprime, these ideals coincide so $\mathcal{A}$ has a socle. In this case the socle equals the left ideal, the right ideal and the ideal generated by the set of minimal idempotents of $\mathcal{A}$, and therefore $\mathcal{A}_F \cap \mathcal{A}_J$ equals $\{0\}$.*

Proof   Let $\mathcal{L}$ be a minimal left ideal and let a be an element of $\mathcal{A}$. Then $\mathcal{L}a$ is a left ideal. If $\mathcal{R}$ is a left ideal included in $\mathcal{L}a$, then $\{b \in \mathcal{L} : ba \in \mathcal{R}\}$ is a left ideal included in $\mathcal{L}$ and hence is $\{0\}$ or $\mathcal{L}$. Therefore $\mathcal{L}a$ is either a minimal left ideal or zero. Hence the left socle $\oplus\{\mathcal{L} : \mathcal{L}$ is a minimal left ideal$\}$ is closed under multiplication from the right. Thus it is an ideal. Similarly, or by considering the reverse algebra, the right socle is an ideal.

If $\mathcal{A}$ is semiprime, Corollary 8.2.3 shows that the left socle is the left ideal generated by $\mathcal{A}_{MI}$ and that the right socle is the right ideal generated by $\mathcal{A}_{MI}$. The result just proved shows that in fact both the left and the right socle are the ideal generated by $\mathcal{A}_{MI}$. Hence they agree, so $\mathcal{A}$ has a socle.

Suppose a non-zero a belongs to $\mathcal{A}_F \cap \mathcal{A}_J$. Then since a belongs to $\sum_{e \in \mathcal{A}_{MI}} \mathcal{A}e$ there is some $e \in \mathcal{A}_{MI}$ satisfying $ae \neq 0$. By minimality $\mathcal{A}e$ is included in $\mathcal{A}_F \cap \mathcal{A}_J$. This implies $e \in \mathcal{A}_J$ which is impossible by Proposition 4.3.12(a). Hence $\mathcal{A}_F \cap \mathcal{A}_J$ is $\{0\}$. $\square$

The first argument in the above proof is useful. It shows that if $\mathcal{L}$ is a minimal left ideal of $\mathcal{A}$ and a is an element of $\mathcal{A}$, then $\mathcal{L}a$ is either $\{0\}$ or a minimal left ideal. See also Abdullah H. Al-Moajil [1979] and Nelli Katseli [1984] for a new characterization of the socle.

The Presocle

Most of the theory in this chapter is developed for semiprime algebras and even semisimplicity is sometimes required in other treatments. In order to extend some of the results to a wider context, M. R. F. Smyth [1976] introduced the concept of the presocle in an arbitrary algebra.

8.2.9 Definition The presocle $\mathcal{A}_{PF}$ of an algebra $\mathcal{A}$ is the preimage in $\mathcal{A}$ of the socle in the semisimple quotient $\mathcal{A}/\mathcal{A}_J$ of $\mathcal{A}$, *i.e.*,

$$\mathcal{A}_{PF} = \{a \in \mathcal{A} : a + \mathcal{A}_J \in (\mathcal{A}/\mathcal{A}_J)_F\}.$$

8.2.10 Proposition *In any algebra, the presocle is an ideal which includes both the left and the right socle.*

Proof The presocle is obviously an ideal. Suppose $\mathcal{L}$ is a minimal left ideal in $\mathcal{A}$. If $\mathcal{L}^2$ is $\{0\}$ then it is included in the Jacobson radical. Otherwise, Proposition 8.2.2 shows that it has the form $\mathcal{L} = \mathcal{A}e$ for some minimal idempotenet e. Since $e + \mathcal{A}_J$ is a minimal idempotent in $\mathcal{A}/\mathcal{A}_J$, $\mathcal{L}$ is again included in the presocle. The argument for the right socle is similar.   □

In a semiprime algebra, $\mathcal{A}_F + \mathcal{A}_J$ is a subset of the presocle. In fact, this sum is a direct sum by the last proposition. When are the two sets equal? Proposition 4.3.12(d) shows that the preimage of a minimal idemoptent in $\mathcal{A}/\mathcal{A}_J$ can be chosen to be an idempotent in $\mathcal{A}$. When is it minimal? For additional results we refer the reader to the paper already cited and Smyth [1982], Barnes, Murphy, Smyth and West [1982] and Pietro Aiena [1987].

8.3   Algebras of Operators with Minimal Ideals

We will now investigate the socle of a strictly dense algebra of operators on a linear space. Note that such an algebra is primitive, hence semisimple, hence semiprime. Therefore the socle exists. We introduce some concepts and notation which will allow a rather concrete description of the socle in this case.

8.3.1 Definition   A *pairing* between two linear spaces $\mathcal{V}$ and $\mathcal{W}$ is a bilinear map $\langle \cdot, \cdot \rangle : \mathcal{V} \times \mathcal{W} \to \mathbb{C}$ which satisfies

$$\big(\langle v, w \rangle = 0 \text{ for all } v \in \mathcal{V} \big) \quad \Rightarrow \quad w = 0 \tag{1}$$
$$\big(\langle v, w \rangle = 0 \text{ for all } w \in \mathcal{W} \big) \quad \Rightarrow \quad v = 0.$$

If T is an operator in $\mathcal{L}(\mathcal{V})$, then an operator $T^{\ddagger} \in \mathcal{L}(\mathcal{W})$ is said to be an *adjoint of T relative to* $\langle \cdot, \cdot \rangle$ if it satisfies

$$\langle Tv, w \rangle = \langle v, T^{\ddagger}w \rangle \qquad \forall\, v \in \mathcal{V}; w \in \mathcal{W}. \tag{2}$$

The set of operators in $\mathcal{L}(\mathcal{V})$ with adjoints in $\mathcal{L}(\mathcal{W})$ relative to a pairing $\langle \cdot, \cdot \rangle$ is denoted by $\mathcal{L}_{\langle\ \rangle}(\mathcal{V})$.

A pairing between $\mathcal{V}$ and $\mathcal{W}$ is called a *normed pairing* if $\mathcal{V}$ and $\mathcal{W}$ are normed linear spaces and the bilinear map is continuous. The set of operators in $\mathcal{B}(\mathcal{V})$ with adjoints in $\mathcal{B}(\mathcal{W})$ relative to a normed pairing $\langle \cdot, \cdot \rangle$ is denoted by $\mathcal{B}_{\langle\ \rangle}(\mathcal{V})$.

It is easy to check that an adjoint is unique when it exists, and that $\mathcal{L}_{\langle\ \rangle}(\mathcal{V})$ and $\mathcal{B}_{\langle\ \rangle}(\mathcal{V})$ are subalgebras of $\mathcal{L}(\mathcal{V})$ and $\mathcal{B}(\mathcal{V})$, respectively. We will denote the adjoint of $T \in \mathcal{L}_{\langle\ \rangle}(\mathcal{V})$ by $T^{\ddagger}$. Then the map $T \mapsto T^{\ddagger}$ is an anti-isomorphism of $\mathcal{L}_{\langle\ \rangle}(\mathcal{V})$ into $\mathcal{L}(\mathcal{W})$ and of $\mathcal{B}_{\langle\ \rangle}(\mathcal{V})$ into $\mathcal{B}(\mathcal{W})$. For any subset $\mathcal{B} \subseteq \mathcal{L}_{\langle\ \rangle}(\mathcal{V})$ we will use $\mathcal{B}^{\ddagger}$ to denote the subset $\mathcal{B}^{\ddagger} = \{T^{\ddagger} : T \in \mathcal{B}\} \subseteq \mathcal{L}(\mathcal{W})$. If $\mathcal{V}$ is a normed linear space and $\mathcal{W}$ is a Banach space, the closed graph theorem shows $\mathcal{B}_{\langle\ \rangle}(\mathcal{V}) = \mathcal{B}(\mathcal{V}) \cap \mathcal{L}_{\langle\ \rangle}(\mathcal{V})$. (A special case of this situation plays a prominent role in Volume II.)

Next we will investigate the finite-rank operators in $\mathcal{L}_{\langle\ \rangle}(\mathcal{V})$ and $\mathcal{B}_{\langle\ \rangle}(\mathcal{V})$. (The *rank* of an operator is the dimension of its range.) If $\mathcal{V}$ is a linear space or normed linear space, we will denote the set of finite-rank operators in $\mathcal{L}(\mathcal{V})$ and in $\mathcal{B}(\mathcal{V})$ by $\mathcal{L}_F(\mathcal{V})$ and $\mathcal{B}_F(\mathcal{V})$, respectively. The results given here are due to Nathan Jacobson [1947]. We need an elementary lemma.

8.3.2 Lemma *Let $\mathcal{V}$ and $\mathcal{W}$ be linear spaces and let $\langle \cdot, \cdot \rangle$ be a pairing for $\mathcal{V}$ and $\mathcal{W}$. For any finite linearly independent set $\{v_1, v_2, \ldots, v_n\}$ in $\mathcal{V}$ there exists a (necessarily linearly independent) set $\{w_1, w_2, \ldots, w_n\}$ in $\mathcal{W}$ satisfying*

$$\langle v_i, w_j \rangle = \delta_{ij} \quad 1 \leq i, j \leq n.$$

Proof Let $\mathcal{X}$ be the linear subspace of $\mathcal{V}$ spanned by $\{v_1, v_2, \ldots, v_n\}$ and let $\mathcal{Y}$ be the subspace of linear functionals on $\mathcal{X}$ induced by the pairing with elements of $\mathcal{W}$. Since $\mathcal{X}$ is finite-dimensional, if $\mathcal{Y}$ is not the entire space $\mathcal{X}^{\dagger}$ of linear functionals on $\mathcal{X}$ there must be an element $v \in \mathcal{X}$ satisfying $\omega(v) = 0$ for all $\omega \in \mathcal{Y}$. This is impossible since it would contradict the second implication in (1). Hence $\mathcal{Y}$ is just $\mathcal{V}^{\dagger}$ and in particular there are elements of $\mathcal{W}$ implementing each of the linear functionals $v_i \mapsto \delta_{ij}$.   □.

Let $\langle \cdot, \cdot \rangle$ be a pairing between the linear spaces $\mathcal{V}$ and $\mathcal{W}$. For any $v \in \mathcal{V}$ and $w \in \mathcal{W}$ define maps $v \otimes w : \mathcal{V} \to \mathcal{V}$ and $w \otimes^{\ddagger} v : \mathcal{W} \to \mathcal{W}$ by

$$v \otimes w(x) = \langle x, w \rangle v \quad \forall\, x \in \mathcal{V} \tag{3}$$

$$w \otimes^{\ddagger} v(y) = \langle v, y \rangle w \quad \forall\, y \in \mathcal{W}.$$

It is clear that $v \otimes w$ belongs to $\mathcal{L}(\mathcal{V})$, $w \otimes^{\ddagger} v$ belongs to $\mathcal{L}(\mathcal{W})$ and that $w \otimes^{\ddagger} v$ is the adjoint for $v \otimes w$. Furthermore, these operators have rank one when they are non-zero. Hence any operator of the form

$$\sum_{j=1}^{n} v_j \otimes w_j \tag{4}$$

belongs to $\mathcal{L}_F(\mathcal{V}) \cap \mathcal{L}_{\langle\ \rangle}(\mathcal{V})$ and has the operator

$$\sum_{j=1}^{n} w_j \otimes^{\ddagger} v_j \qquad (5)$$

as its adjoint. If $\langle\cdot,\cdot\rangle$ is a normed pairing, the operator (4) belongs to $\mathcal{B}_F(\mathcal{V}) \cap \mathcal{B}_{\langle\ \rangle}(\mathcal{V})$ since both (4) and (5) are clearly bounded operators.

Next we will show that, conversely, any operator in $\mathcal{L}_F(\mathcal{V}) \cap \mathcal{L}_{\langle\ \rangle}(\mathcal{V})$ (or in $\mathcal{B}_F(\mathcal{V}) \cap \mathcal{B}_{\langle\ \rangle}(\mathcal{V})$ if the pairing is normed) has the form (4).

Suppose T is a finite-rank operator in $\mathcal{L}_{\langle\ \rangle}(\mathcal{V})$. Then we may find a basis $\{v_1, v_2, \ldots, v_n\}$ for its range and we have

$$T x = \sum_{j=1}^{n} \omega_j(x) v_j \qquad \forall\ x \in \mathcal{V}$$

where each ω_j is a linear functional on $\mathcal{V}$. Use Lemma 8.3.2 to find elements $y_1, y_2, \ldots, y_n$ in $\mathcal{W}$ satisfying $\langle v_j, y_i \rangle = \delta_{ij}$ for $1 \le i, j \le n$. Then we have

$$\langle T x, y_i \rangle = \sum_{j=1}^{n} \omega_j(x) \langle v_j, y_i \rangle = \omega_i(x)$$

for $1 \le i \le n$. Hence if we define $w_i \in \mathcal{W}$ by $w_i = T^{\ddagger} y_i$ for $1 \le i \le n$ we have $T = \sum_{j=1}^{n} v_j \otimes w_j$ as desired.

Hence we have established that $\mathcal{L}_F(\mathcal{V}) \cap \mathcal{L}_{\langle\ \rangle}(\mathcal{W})$ or in the normed case $\mathcal{B}_F(\mathcal{V}) \cap \mathcal{B}_{\langle\ \rangle}(\mathcal{V})$ is the set of all operators of the form (4). We will now discuss the most important case in which such operators arise. We separate the nonnormed case and the normed case. Recall that strict density was introduced in Definition 4.2.12.

8.3.3 Proposition *Let $\mathcal{B}$ be a strictly dense algebra of operators on a linear space $\mathcal{V}$ and let $\mathcal{B}$ have at least one minimal left (or right) ideal. Then there is a linear space $\mathcal{W}$ and a pairing $\langle\cdot,\cdot\rangle$ of $\mathcal{V}$ and $\mathcal{W}$ satisfying the following three conditions:*

$$\mathcal{B} \ \subseteq\ \mathcal{L}_{\langle\ \rangle}(\mathcal{V}); \qquad (6)$$

$$\mathcal{B}_F \ =\ \mathcal{L}_F(\mathcal{V}) \cap \mathcal{B} = \mathcal{L}_F(\mathcal{V}) \cap \mathcal{L}_{\langle\ \rangle}(\mathcal{V})$$

$$=\ \{\sum_{j=1}^{n} v_j \otimes w_j\ :\ n \in \mathbb{N}; v_j \in \mathcal{V}\ ;\ w_j \in \mathcal{W}\ for\ j = 1, 2, \ldots, n\}$$

$$=\ \cap\{\mathcal{I} : \mathcal{I}\ is\ a\ non\text{-}zero\ ideal\ of\ \mathcal{B}\}; \qquad (7)$$

$$\mathcal{B}^{\ddagger}\ is\ strictly\ dense\ on\ \mathcal{W}. \qquad (8)$$

Proof Since $\mathcal{B}$ is semiprime (it is even primitive), Corollary 8.2.3 shows that $\mathcal{B}$ has a minimal left ideal if and only if it has a minimal idempotent. First we will show that each minimal idempotent E of $\mathcal{B}$ has rank one. If E has rank greater than one then there are independent vectors v_1, v_2 in $E\mathcal{V}$. Since $\mathcal{B}$ is strictly dense there is an operator $T \in \mathcal{B}$ satisfying $Tv_1 = 0, Tv_2 = v_2$. Hence ETE is non-zero so there is an operator $S \in E\mathcal{B}E$ satisfying $SETE = E$. However, this contradicts $Ev_1 = v_1$. Hence E has rank one.

Since each minimal projection of $\mathcal{B}$ has rank one, Proposition 8.2.8 shows $\mathcal{B}_F = \mathcal{B}\mathcal{B}_{M I}\mathcal{B} \subseteq \mathcal{B}\cap\mathcal{L}_F(\mathcal{V})$. In order to establish the opposite inclusion and hence prove the first equality of (7), we choose an arbitrary $T \in \mathcal{B}\cap\mathcal{L}_F(\mathcal{V})$. Consider the natural pairing

$$\langle v, w\rangle = w(v) \qquad \forall \; v \in \mathcal{V}; \; w \in \mathcal{V}^\dagger$$

between $\mathcal{V}$ and $\mathcal{V}^\dagger$. The remarks preceding the statement of this proposition show that we may write

$$T = \sum_{j=1}^{n} v_j \otimes w_j,$$

where $v_j \in \mathcal{V}, w_j \in \mathcal{V}^\dagger$ and $\{v_1, v_2, \ldots, v_n\}$ is linearly independent. The strict density of $\mathcal{B}$ shows that for any $k = 1, 2, \ldots, n$ there is an operator $T_k \in \mathcal{B}$ satisfying $T_k v_j = \delta_{kj} v_j$. Then

$$\mathcal{B}T_k T = \mathcal{B}(v_k \otimes w_k) = \{v \otimes w_k : v \in \mathcal{V}\}$$

is obviously a minimal left ideal of $\mathcal{B}$. Hence $T = \sum_{j=1}^{n} T_k T$ belongs to $\mathcal{B}_F$ since $T_k T = T_k T_k T$ belongs to $\mathcal{B}T_k T$ for each $k = 1, 2, \ldots, n$. Hence we have established

$$\mathcal{B}_F = \mathcal{L}_F(\mathcal{V}) \cap \mathcal{B}.$$

Now define $\mathcal{W}$ by

$$\mathcal{W} = \{w \in \mathcal{V}^\dagger : v \otimes w \in \mathcal{B} \text{ for some non-zero } v \in \mathcal{V}\}.$$

Furthermore, for any $T \in \mathcal{B}$ we have $v \otimes wT = v \otimes T^\dagger w$ so that $T^\dagger$ maps $\mathcal{W}$ into itself.

We continue to use the pairing $\langle v, w\rangle = w(v)$ for any $v \in \mathcal{V}$ and $w \in \mathcal{W}$. This is a bilinear mapping which satisfies the first implication in (1). To prove the second implication in (1), let $v \in \mathcal{V} \setminus \{0\}$ be arbitrary. We must find $w \in \mathcal{W}$ satisfying $\langle v, w\rangle \neq 0$. Let $y \in \mathcal{W} \setminus \{0\}$ be arbitrary and choose $x \in \mathcal{V}$ to satisfy $\langle x, y\rangle \neq 0$. The irreducibility of $\mathcal{B}$ guarantees a $T \in \mathcal{B}$ satisfying $Tv = x$. Choosing w to be $w = T^\dagger y$ we have $\langle v, w\rangle = \langle v, T^\dagger y\rangle = \langle Tv, y\rangle = \langle x, y\rangle \neq 0$. Hence we have shown that $\langle \cdot, \cdot\rangle$ is a pairing of $\mathcal{V}$ and $\mathcal{W}$.

For any $T \in \mathcal{B}$ the restriction of $T^\dagger$ to $\mathcal{W}$ is the adjoint of T relative to this pairing. This proves (6) which implies $\mathcal{B}_F = \mathcal{L}_F(\mathcal{V}) \cap \mathcal{B} \subseteq \mathcal{L}_F(\mathcal{V}) \cap \mathcal{L}_{\langle\ \rangle}(\mathcal{V})$. However we proved the penultimate equality of (7) before stating this proposition, and our remarks on the set $\mathcal{W}$ show that any operator of the form (4) belongs to $\mathcal{L}_F(\mathcal{V}) \cap \mathcal{B}$. This completes the proof of (7) except for the last equality which we will prove after establishing (8).

In order to prove (8), let $\{w_1, w_2, \ldots, w_n\}$ and $\{y_1, y_2, \ldots, y_n\}$ be arbitrary subsets of $\mathcal{W}$ with the first linearly independent. Lemma 8.3.2 guarantees a set $\{x_1, x_2, \ldots, x_n\}$ in $\mathcal{V}$ satisfying $\langle x_j, w_k \rangle = \delta_{jk}$. Define $T \in \mathcal{B}$ by $T = \sum_{j=1}^n x_j \otimes y_j$. Then $T^\ddagger(w_k) = \sum_{j=1}^n y_j \otimes^\ddagger x_j(w_k) = y_k$ holds. Therefore $\mathcal{B}^\ddagger$ is strictly dense, proving (8).

To establish the last equality of (7), it is enough to show that any non-zero ideal $\mathcal{I}$ of $\mathcal{B}$ includes $\mathcal{B}_F$. In fact it is enough to show that $\mathcal{I}$ contains some rank one operator $v_0 \otimes w_0$ since the irreducibility of $\mathcal{B}$ and $\mathcal{B}^\ddagger$ then implies that $\mathcal{I}$ contains every rank one operator $v \otimes w$ and hence includes $\mathcal{B}_F$. If $T \in \mathcal{I}$ is non-zero, choose $x \in \mathcal{V}$ satisfying $Tx \neq 0$ and $w \in \mathcal{W} \setminus \{0\}$. Then $Tx \otimes w = T(x \otimes w)$ is a rank one operator in $\mathcal{B}$. Thus (7) is established. □

We now give a variant of the last proposition which takes into account possible norms on $\mathcal{B}$. For Banach algebras it is due to Rickart [1950].

8.3.4 Theorem *Let $\mathcal{B}$ be an irreducible spectral algebra of operators on a linear space $\mathcal{V}$ and let $\mathcal{B}$ have at least one minimal left (or right) ideal. Then there is a norm on $\mathcal{V}$, a normed linear space $\mathcal{W}$ and a normed pairing $\langle \cdot, \cdot \rangle$ between $\mathcal{V}$ and $\mathcal{W}$ satisfying*

$$\mathcal{B} \subseteq \mathcal{B}_{\langle\ \rangle}(\mathcal{V}); \tag{9}$$

$$\mathcal{B}_F = \mathcal{B}_F(\mathcal{V}) \cap \mathcal{B} = \mathcal{B}_F(\mathcal{V}) \cap \mathcal{B}_{\langle\ \rangle}(\mathcal{V}); \tag{10}$$

$$\mathcal{B}^\ddagger \text{ is strictly dense on } \mathcal{W}. \tag{11}$$

If $\mathcal{B}$ is a Banach algebra under any norm $\|\cdot\|$, then $\mathcal{V}$ and $\mathcal{W}$ may be taken as Banach spaces and any operator T in $\mathcal{B}$ satisfies $\|T^\ddagger\|_{op} \leq \|T\|_{op} \leq \|T\|$, where $\|\cdot\|_{op}$ denotes the operator norm on $\mathcal{W}$ and $\mathcal{V}$.

Proof Theorem 4.2.13 shows that $\mathcal{B}$ is strictly dense. Choose a spectral semi-norm σ on $\mathcal{B}$. Theorem 4.2.8 shows that the choice of any non-zero $z \in \mathcal{V}$ gives rise to a norm on $\mathcal{V}$ by the construction of Theorem 4.2.7. Furthermore, relative to this norm, each operator in $\mathcal{B}$ is bounded. We obtain the linear space $\mathcal{W}$ as in the proof of the last proposition except that this time the elements of $\mathcal{W}$ are all continuous linear functionals in $\mathcal{V}^*$. Give $\mathcal{W}$ the norm which is the restriction of the norm of $\mathcal{V}^*$. Then the pairing between $\mathcal{V}$ and $\mathcal{W}$ is a normed pairing. For each $T \in \mathcal{B}$, $T^\ddagger$ will be the restriction to $\mathcal{W}$ of $T^\dagger$. This proves (9). Hence (10) and (11) follow directly from the corresponding results in the last proposition. When $\mathcal{B}$ is a Banach algebra we use its complete norm in the above construction. Then Theorem 4.2.7 gives all the results stated here. ⃞

Next we introduce some topological considerations on the algebra.

8.3.5 Proposition *Let $\mathcal{A}$ be either a normed or a spectral algebra. Every minimal idempotent e of $\mathcal{A}$ satisfies $e\mathcal{A}e = \mathbb{C}e$. If σ is either an algebra norm or a spectral semi-norm for $\mathcal{A}$, then there is a continuous linear functional ω on $(\mathcal{A}, \sigma)$ satisfying*

$$\omega(a)e = eae. \tag{12}$$

Proof Let σ be a spectral semi-norm on $\mathcal{A}$. Then $\sigma(e)$ is non-zero since no idempotent is quasi-invertible. Hence under either hypothesis on σ, the restriction of σ to $e\mathcal{A}e$ is a nontrivial algebra norm. The Gelfand–Mazur theorem (Corollary 2.2.3) shows that $e\mathcal{A}e$ consists of complex multiples of its identity element, e. Define $\omega : \mathcal{A} \to \mathbb{C}$ by $eae = \omega(a)e$. Clearly ω is a linear functional. It is also continuous since it satisfies $|\omega(a)| = \sigma(e)^{-1}\sigma(\omega(a)e) = \sigma(e)^{-1}\sigma(eae) \le \sigma(a)\sigma(e)$. ⃞

When $\mathcal{A}$ is a semiprime normed or spectral algebra and e is a minimal idempotent of $\mathcal{A}$, then the irreducible representations L^e and R^e have some special properties which we wish to describe.

8.3.6 Theorem *Let $\mathcal{A}$ be a semiprime normed or spectral algebra. Let e be a minimal idempotent of $\mathcal{A}$. Then L^e and R^e are strictly dense, and for any spectral semi-norm or algebra norm on $\mathcal{A}$ there exist norms on $\mathcal{A}e$ and $e\mathcal{A}$ and a normed pairing $\langle \cdot, \cdot \rangle$ for $\mathcal{A}e$ and $e\mathcal{A}$ satisfying:*

$$\langle a, b \rangle e = ba \qquad \forall\ a \in \mathcal{A}e; b \in e\mathcal{A}; \tag{13}$$

$$L^e_{\mathcal{A}} \subseteq \mathcal{B}_{\langle\ \rangle}(\mathcal{A}e); \tag{14}$$

$$(L^e_a)^{\ddagger} = R^e_a \in \mathcal{B}(e\mathcal{A}) \qquad \forall\ a \in \mathcal{A}; \tag{15}$$

$$L^e_{\mathcal{A}_F} = \mathcal{B}_F(\mathcal{A}e) \cap \mathcal{B}_{\langle\ \rangle}(\mathcal{A}e); \tag{16}$$

$$L^e \text{ and } R^e \text{ are contractive.} \tag{17}$$

If $\mathcal{A}$ is a Banach algebra, then the norms on $e\mathcal{A}$ and $\mathcal{A}e$ may be taken as the complete norms which are the restrictions of the complete norm of $\mathcal{A}$.

Finally, if e and f are two minimal idempotents then either $e\mathcal{A}f = \{0\}$ or $e\mathcal{A}f = \mathbb{C}c \ne \{0\}$, where c is any non-zero element in $e\mathcal{A}f$. In the latter case any equivalence between L^e and L^f is a complex multiple of V_c and hence is a topological equivalence.

Proof Proposition 8.2.2 shows that $\mathcal{A}e$ and $e\mathcal{A}$ are minimal left and right ideals, respectively, so that L^e and R^e are irreducible. Hence Theorem 4.2.13 shows that L^e and R^e are strictly dense. Let σ be a norm or a spectral semi-norm on $\mathcal{A}$ which is the complete norm of $\mathcal{A}$ if $\mathcal{A}$ is a Banach algebra. Then $\sigma(e)$ is non-zero since e is not quasi-invertible and hence $\sigma(e) = \sigma(e^2) \leq \sigma(e)^2$ implies $\sigma(e) \geq 1$. Furthermore, the set $\{b \in \mathcal{A}e : \sigma(b) = 0\}$ is a left ideal of $\mathcal{A}$ which is a proper subset of the minimal ideal $\mathcal{A}e$ and hence is zero. Thus the restriction of σ to $\mathcal{A}e$ is a norm. Similarly σ induces a norm on $e\mathcal{A}$. Obviously L^e and R^e are contractive relative to these norms. Using the notation of Proposition 8.3.5 we define the pairing between $\mathcal{A}e$ and $e\mathcal{A}$ by

$$\langle a, b \rangle = \omega(ba) \qquad \forall \, a \in \mathcal{A}e; b \in e\mathcal{A}.$$

Equation (13) is an immediate consequence. It in turn implies

$$\{a \in \mathcal{A}e : \langle a, b \rangle = 0 \text{ for } b \in e\mathcal{A}\} = \{a \in \mathcal{A}e : ba = 0 \text{ for } b \in e\mathcal{A}\}.$$

The minimality of $\mathcal{A}e$ implies that this last set is $\{0\}$. Similarly $\{b \in e\mathcal{A} : \langle a, b \rangle = 0 \text{ for } a \in \mathcal{A}e\}$ equals $\{0\}$. Hence $\langle \cdot, \cdot \rangle$ is a pairing. Equation (13) also gives $|\langle a, b \rangle| \leq \sigma(\langle a, b \rangle e) = \sigma(ba) \leq \sigma(b)\sigma(a)$ which implies that $\langle \cdot, \cdot \rangle$ is a normed pairing.

Using (13) yet again, we get $\langle L_c^e a, b \rangle e = bca = \langle a, R_c^e b \rangle$ for any elements $a \in \mathcal{A}e, b \in e\mathcal{A}$ and $c \in \mathcal{A}$. This establishes (15) and hence (14). Equation (16) is now a mere restatement of (10).

Proposition 8.2.5 shows that the last paragraph of the present theorem will follow if $e\mathcal{A}f$ is at most one-dimensional. Suppose $a, b \in e\mathcal{A}f$ are linearly independent. Since L^e and R^e are strictly dense, there are elements $c, d \in \mathcal{A}$ satisfying $ca = b, cb = b, ad = b$ and $bd = 0$. This implies $b = cb = cad = bd = 0$ which contradicts the independence of a and b. This establishes the last result of the theorem.   □

8.3.7 Proposition *Let $\mathcal{A}$ be a normed algebra.*

    (a) *If e is a minimal idempotent in $\mathcal{A}$, then $\kappa(e)$ is a minimal idempotent in $\mathcal{A}^{**}$ with respect to either Arens product.*

    (b) *If $\mathcal{A}$ and $\mathcal{A}^{**}$ are also semiprime and $\mathcal{A}$ has a minimal $\langle$ left / right $\rangle$ ideal, then $\mathcal{A}^{**}$ also has a minimal $\langle$ left / right $\rangle$ ideal.*

Proof By Proposition 8.3.5 $e\mathcal{A}e$ is just $\mathbb{C}e$. Hence $\kappa(e)\mathcal{A}^{**}\kappa(e)$ is just $\mathbb{C}\kappa(e)$ since $\kappa(e\mathcal{A}e)$ is weak* dense in it by (15) of Theorem 1.4.2.

Suppose $\mathcal{A}$ has a minimal left ideal. Corollary 8.2.3 shows that the minimal left ideal has the form $\mathcal{A}e$ for some minimal idempotent e. Thus $\mathcal{A}^{**}e$ is a minimal ideal in $\mathcal{A}^{**}$ by the last portion of Proposition 8.2.2.   □

8.4   Modular Annihilator Algebras

Up to the present point in this chapter we have either assumed directly that the algebras under consideration have minimal one-sided ideals or minimal idempotents or assumed that they were finite-dimensional. Many large and important classes of infinite-dimensional algebras have minimal one-sided ideals. In most cases proof depends either on the spectral properties of compact operators or on Proposition 8.4.3 below. This proposition shows that if certain left annihilator ideals are non-zero they are minimal left ideals. Before stating this important proposition we need to summarize some elementary facts about annihilator ideals.

8.4.1 Definition Let S be a subset of an algebra $\mathcal{A}$. The *left annihilator* of S in $\mathcal{A}$ will be denoted by

$$\mathcal{A}_{LA}S = \{a \in \mathcal{A} : ab = 0 \text{ for } b \in S\}.$$

The *right annihilator* of S in $\mathcal{A}$ will be denoted by $\mathcal{A}_{RA}S$. If the algebra $\mathcal{A}$ is clearly understood, we simply write $_{LA}S$ and $_{RA}S$. For the left annihilator in $\mathcal{A}$ of the right annihilator of S in $\mathcal{A}$ we write $\mathcal{A}_{LRA}S$ or $_{LRA}S$. Similar notation has the analogous meaning. A ⟨ left / right ⟩ ideal of $\mathcal{A}$ which is the ⟨ left / right ⟩ annihilator of some subset of $\mathcal{A}$ is called a ⟨ *left / right* ⟩ *annihilator ideal*.

8.4.2 Proposition *Let S be a subset of an algebra $\mathcal{A}$. Then $_{LA}S$ is a left ideal and $_{RA}S$ is a right ideal. Similarly the left annihilator of a left ideal or the right annihilator of a right ideal is an ideal. Furthermore $S \subseteq \mathcal{I} \subseteq \mathcal{A}$ implies*

$$_{LA}S \supseteq {}_{LA}\mathcal{I}, \quad _{RA}S \supseteq {}_{RA}\mathcal{I}, \tag{1}$$

$$S \subseteq {}_{RLA}S, \quad S \subseteq {}_{LRA}S. \tag{2}$$

From these results we get easily

$$_{LRLA}S = {}_{LA}S, \quad _{RLRA}S = {}_{RA}S. \tag{3}$$

An ideal $\mathcal{I}$ is a ⟨ left / right ⟩ annihilator ideal if and only if it satisfies ⟨ $\mathcal{I} = {}_{LRA}\mathcal{I}$ / $\mathcal{I} = {}_{RLA}\mathcal{I}$ ⟩. If $\mathcal{A}$ is semiprime and $\mathcal{I}$ is an ideal, they satisfy:

$$\mathcal{A}_J \subseteq {}_{LA}(\mathcal{A}_F) = {}_{RA}(\mathcal{A}_F); \tag{4}$$

$$\mathcal{A}_F \subseteq {}_{LA}(\mathcal{A}_J) = {}_{RA}(\mathcal{A}_J);$$

$$_{LA}\mathcal{I} = {}_{RA}\mathcal{I} \quad and \quad \mathcal{I} \cap {}_{LA}\mathcal{I} = \{0\}. \tag{5}$$

Finally, if $\mathcal{A}$ is a topological algebra it is obvious that both $_{LA}S$ and $_{RA}S$ are closed.

Proof Most of these remarks are obvious. The results on semiprime algebras follow from the last remark in Proposition 8.2.8 and the obvious equations $(\mathcal{I}_{LA}\mathcal{I})^2 = \{0\}$ and $(_{RA}\mathcal{I}\mathcal{I})^2 = \{0\}$. □

We will often use these easy results without comment. The next two results are preparatory for Theorem 8.4.5 which plays a fundamental role throughout the rest of this chapter. The first is due to Yood ([1964], Lemma 3.3, p. 38.)

8.4.3 Proposition *Let A be a semiprime algebra. The following conditions are equivalent for a maximal modular left ideal M of A.*

(a) $_{RA}M \neq \{0\}$.

(b) *There is a minimal idempotent e of A satisfying:* $M = {}_{LRA}M = A(1-e)$ *and* $_{RA}M = eA$.

(c) $A_F \not\subseteq M$.

The corresponding results hold for right ideals.

Proof (a) $\Rightarrow$ (b): If $_{RA}M$ is a quasi-regular right ideal, then Theorem 4.3.6 shows $_{RA}M \subseteq A_J \subseteq M$. However this implies $(_{RA}M)^2 \subseteq M_{RA}M = \{0\}$ which contradicts the semiprime character of A. Lemma 4.3.5 gives a right quasi-singular element $e \in {}_{RA}M$. Then $A(1-e)$ is a proper left ideal which includes $M(1-e) = M$. The maximality of M implies $_{LRA}M = M = A(1-e)$. Since e belongs to $_{RA}M$, we conclude $(1-e)e \in {}_{RA}A = \{0\}$. Hence e is an idempotent and thus $_{RA}M = {}_{RA}(A(1-e)) = eA$. In order to prove that e is a minimal idempotent, Proposition 8.2.2 shows that it is enough to establish the minimality of Ae. If $\mathcal{L}$ is a non-zero left ideal included in Ae, the maximality of $A(1-e)$ shows the existence of an element $be \in \mathcal{L} \subseteq Ae$ satisfying $be + a(1-e) = e$. Multiplying on the right by e we get $e = be \in \mathcal{L}$ and hence $\mathcal{L} = Ae$. Therefore e is minimal.

(b) $\Rightarrow$ (c): Obviously $e \in A_F$ does not belong to M.

(c) $\Rightarrow$ (a): Corollary 8.2.3 shows that there is a minimal idempotent e satisfying $Ae \not\subseteq M$ and hence $Ae \cap M = \{0\}$ and $Ae \oplus M = A$. Let d be a right relative identity for M and let $d = b \oplus c$ be the corresponding direct sum decomposition for d so that $b = be \in Ae$ and $c \in M$ hold. Then we have $a(b-1) = a(d-1) - ac \in M$. Hence $ae(b-1) \in Ae \cap M = \{0\}$ for all $a \in A$, implies $b^2 = b$ and $Ab = Ae$. So finally, $A = Ae \oplus M = Ab \oplus M = Ab \oplus A(1-b)$ and $A(1-b) \subseteq M$ imply $M = A(1-b)$ and $_{RA}M = bA \neq \{0\}$. $\square$

The following result is due to Barnes ([1966], Proposition 3.1, p. 567.)

8.4.4 Proposition *Let A be a semiprime algebra and let e be a minimal idempotent of A. Then $A(1-e)$ and $(1-e)A$ are maximal modular left and right ideals of A, respectively. Moreover, the set*

$$\begin{aligned} \mathcal{P}^e &= {}_{LA}(Ae) = {}_{RA}(eA) \\ &= A(1-e) : A = \{a \in A : Aa \subseteq (1-e)A\} \end{aligned}$$

is a primitive ideal in both A and A^R which includes no other primitive ideal of A or A^R. Every irreducible representation of A with kernel $\mathcal{P}^e$ is equivalent to both L^e and $L^{A/A(1-e)}$. Any prime ideal which does not contain e includes $_{LA}(Ae)$.

Proof Obviously $\mathcal{A}(1-e)$ is a modular left ideal. Suppose $\mathcal{L}$ is a left ideal which properly includes $\mathcal{A}(1-e)$. The Peirce decomposition, $\mathcal{A} = \mathcal{A}e \oplus \mathcal{A}(1-e)$, implies that there is some $b \in \mathcal{L}$ satisfying $b = ae + c(1-e)$ with $ae \neq 0$. Since $\mathcal{A}e$ is a minimal left ideal by Proposition 8.2.2, $ae \in \mathcal{L} \cap \mathcal{A}e$ implies $\mathcal{L} \supseteq \mathcal{A}e$ and hence $\mathcal{L} = \mathcal{A}$. Thus $\mathcal{A}(1-e)$ is a maximal modular left ideal of $\mathcal{A}$. The same argument applied to $\mathcal{A}^R$ shows that $(1-e)\mathcal{A}$ is a maximal modular right ideal of $\mathcal{A}$. Theorem 4.1.8(a) now shows that $\mathcal{A}(1-e) : \mathcal{A}$ is primitive in $\mathcal{A}$ and $\{a \in \mathcal{A} : \mathcal{A}a \subseteq (1-e)\mathcal{A}\}$ is primitive in $\mathcal{A}^R$.

The equations

$$_{LA}(\mathcal{A}e) = \mathcal{A}(1-e) : \mathcal{A} \quad \text{and} \quad _{RA}(e\mathcal{A}) = \{a \in \mathcal{A} : \mathcal{A}a \subseteq (1-e)\mathcal{A}\}$$

are immediate. Let $\mathcal{P}^e$ denote the primitive ideal $\mathcal{A}(1-e) : \mathcal{A} = {}_{LA}(\mathcal{A}e)$. If a belongs to $_{RA}(e\mathcal{A})$ then ea is zero so $\mathcal{A}e(_{RA}(e\mathcal{A})) = \{0\} \subseteq \mathcal{P}^e$ holds. Theorem 4.1.8(b) shows that $\mathcal{P}^e$ is prime so that either $\mathcal{A}e \subseteq \mathcal{P}^e$ or $_{RA}(e\mathcal{A}) \subseteq \mathcal{P}^e$ holds. It must be the latter since $e \in \mathcal{A}e$ does not belong to $\mathcal{P}^e$. The same argument in the reverse algebra shows $\mathcal{P}^e = {}_{LA}(\mathcal{A}e) \subseteq \{a : \mathcal{A}a \subseteq (1-e)\mathcal{A}\} = {}_{RA}(e\mathcal{A})$. Hence all four sets are equal as asserted.

The ideal $\mathcal{P}^e = {}_{LA}(\mathcal{A}e)$ is the kernel of L^e and the ideal $\mathcal{P}^e = \mathcal{A}(1-e) : \mathcal{A}$ is the kernel of $L^{\mathcal{A}/\mathcal{A}(1-e)}$ so that the last sentence of the proposition follows from Proposition 8.2.6.

Suppose $\mathcal{P}$ is primitive and satisfies $\mathcal{P} \subseteq \mathcal{P}^e$ for some $e \in \mathcal{A}_{MI}$. Then $\mathcal{P}^e {}_{RA}\mathcal{P}^e = \{0\} \subseteq \mathcal{P}$ implies $\mathcal{P}^e \subseteq \mathcal{P} \subseteq \mathcal{P}^e$ or $_{RA}\mathcal{P}^e \subseteq \mathcal{P} \subseteq \mathcal{P}^e$ by Theorem 4.1.8(b). However $_{RA}\mathcal{P}^e = {}_{RLA}(\mathcal{A}e) \supseteq \mathcal{A}e \neq \{0\}$ and the semiprime nature of $\mathcal{A}$ contradicts $_{RA}\mathcal{P}^e \subseteq \mathcal{P}^e$. If $\mathcal{P}$ is a prime ideal, then $e \notin \mathcal{P}$ and $_{LA}(\mathcal{A}e)\mathcal{A}e = \{0\} \subseteq \mathcal{P}$ imply $\mathcal{P}^e = {}_{LA}(\mathcal{A}e) \subseteq \mathcal{P}$. $\qquad\square$

The next theorem combines results of Yood ([1964], Theorem 3.4, p. 38) and Barnes [1966].

8.4.5 Theorem *The following conditions are equivalent for a semiprime algebra $\mathcal{A}$.*

(a) *Each maximal modular left ideal $\mathcal{M}$ satisfies $_{RA}\mathcal{M} \neq \{0\}$.*

(b) *Each maximal modular right ideal $\mathcal{M}$ satisfies $_{LA}\mathcal{M} \neq \{0\}$.*

(c) $\mathcal{A}/\mathcal{A}_F$ *is a radical algebra.*

(d) *Every primitive ideal of $\mathcal{A}$ has the form $\mathcal{P}^e$ (defined in Proposition 8.4.4) for some $e \in \mathcal{A}_{MI}$.*

(e) *Every irreducible representation of $\mathcal{A}$ is equivalent to L^e for some $e \in \mathcal{A}_{MI}$.*

(f) *The hull of $\mathcal{A}_F$ is empty.*

(g) $\Pi_\mathcal{A}$ *is discrete and $\mathcal{A}$ satisfies $_{LA}(\mathcal{A}_F) = {}_{RA}(\mathcal{A}_F) = \mathcal{A}_J$.*

(h) $\Pi_\mathcal{A}$ *is discrete and every left ideal which is not a subset of $\mathcal{A}_J$ includes a minimal left ideal.*

Proof (a)⟺(c): The map $\mathcal{M} \to \mathcal{M}/\mathcal{A}_F$ is easily seen to send the set of maximal modular left ideal of $\mathcal{A}$ including $\mathcal{A}_F$ onto the set of maximal modular left ideals of $\mathcal{A}/\mathcal{A}_F$. Hence $\mathcal{A}/\mathcal{A}_F$ has no maximal modular left ideals (*i.e.*, it is a radical algebra) if and only if no maximal modular left ideal of $\mathcal{A}$ includes $\mathcal{A}_F$. Proposition 8.4.3 shows that this is equivalent to (a).

(b)⟺(c): Since (c) holds in $\mathcal{A}$ if and only if it holds in $\mathcal{A}^R$, and since (b) holds in $\mathcal{A}$ if and only if (a) holds in $\mathcal{A}^R$, these two conditions are equivalent by what we just proved.

(c)⟺(f): Theorem 4.1.6(b) shows that $\mathcal{A}/\mathcal{A}_F$ has an irreducible representation if and only if there is an irreducible representation of $\mathcal{A}$ with kernel containing $\mathcal{A}_F$. Hence $\mathcal{A}/\mathcal{A}_F$ is radical if and only if $h(\mathcal{A}_F)$ is empty.

(d)⟹(f): This result, which is not needed to close the circuit of implications, follows easily from $e \notin \mathcal{P}^e$.

(f)⟹(d): Let $\mathcal{P}$ be a primitive ideal of $\mathcal{A}$. Proposition 8.2.8 shows there is some $e \in \mathcal{A}_{MI}$ satisfying $e \notin \mathcal{P}$, and hence $e \notin e\mathcal{P} \subseteq \mathcal{P}$. Thus the inclusion $e\mathcal{P} \subseteq e\mathcal{A}$ and the minimality of $e\mathcal{A}$ imply $e\mathcal{P} = \{0\}$. Hence $\mathcal{P}$ is included in $_{RA}e\mathcal{A} = \mathcal{P}^e$. We conclude that $\mathcal{P} = \mathcal{P}^e$ since $\mathcal{P}^e$ is minimal among primitive ideals by Proposition 8.4.4.

(d)⟺(e): Proposition 8.4.4 shows this.

(d)⟹(g): Proposition 8.4.4 and (d) imply

$$\mathcal{A}_J = \cap_{e \in \mathcal{A}_{MI}} \mathcal{P}^e = \cap_{e \in \mathcal{A}_{MI}} {}_{LA}(\mathcal{A}e) = {}_{RA}(\mathcal{A}_F).$$

In order to show that $\Pi_{\mathcal{A}}$ is discrete we will show that the complement $\triangle_e$ of each singleton subset $\{\mathcal{P}^e\}$ of $\Pi_{\mathcal{A}}$ is closed. Any $\mathcal{P}^f$ in $\triangle_e$ is a prime ideal and hence satisfies either $\mathcal{P}^e \subseteq \mathcal{P}^f$ or $_{LA}\mathcal{P}^e \subseteq \mathcal{P}^f$. Since $\mathcal{P}^f$ is a minimal primitive ideal by Proposition 8.4.4 we have $_{LRA}(e\mathcal{A}) = {}_{LA}\mathcal{P}^e \subseteq \mathcal{P}^f$. Since this holds for all $\mathcal{P}^f \in \triangle_e$ we have $_{LRA}(e\mathcal{A}) \subseteq k(\triangle_e)$. However $e \in {}_{LRA}(e\mathcal{A})$ implies $\mathcal{P}^e \notin h(_{LRA}e\mathcal{A})$. We conclude $\mathcal{P}^e \notin hk(\triangle_e)$ as we wished to show.

(g) ⟹ (h): Let $\mathcal{L}$ be a left ideal which is not included in $\mathcal{A}_J = {}_{RA}\mathcal{A}_F = \cap\{_{RA}\mathcal{A}e : e \in \mathcal{A}_{MI}\}$, where the last equality follows from Proposition 8.2.8. Let $e \in \mathcal{A}_{MI}$ satisfy $\mathcal{L} \not\subseteq {}_{RA}\mathcal{A}e$ so that there is some $b \in \mathcal{L}$ with $\mathcal{A}eb \neq \{0\}$. The remark following Proposition 8.2.8 shows that $\mathcal{A}eb$ is a minimal left ideal included in $\mathcal{L}$.

(h) ⟹ (f): Let $\mathcal{P}$ be a primitive ideal and let $\mathcal{I}$ be the kernel $k(\Pi_{\mathcal{A}} \backslash \{\mathcal{P}\})$ so that $\mathcal{A}_J = \mathcal{I} \cap \mathcal{P}$ holds. Since $\Pi_{\mathcal{A}}$ is discrete we conclude $\mathcal{I} \not\subseteq \mathcal{P}$ and hence $\mathcal{I} \not\subseteq \mathcal{A}_J$. Hence there is a minimal left ideal $\mathcal{L}$ included in $\mathcal{I}$. However $\mathcal{L}$ is not included in $\mathcal{A}_J = \mathcal{I} \cap \mathcal{P}$ since $\mathcal{L}$ contains a non-zero idempotent by Proposition 8.2.2 and $\mathcal{A}_J$ contains none by Proposition 4.3.12. Hence $\mathcal{L}$ is not included in $\mathcal{P}$ and thus $\mathcal{P}$ is not in the hull of $\mathcal{A}_F$. Since $\mathcal{P} \in \Pi_{\mathcal{A}}$ was arbitrary, $h(\mathcal{A}_F)$ is empty. $\quad\square$

8.4.6 Definition   A *modular annihilator algebra* is a semiprime algebra satisfying the equivalent conditions of Theorem 8.4.5.

Yood ([1958], p. 376) defined modular annihilator algebras and their theory has been considerably advanced by Yood [1964] and by several papers of Barnes, particularly [1964], [1966], [1968a], [1968b], and [1971a]. Barnes does not require in his definition that modular annihilator algebras be semiprime, although Yood did in [1964]. Most results require that the algebra be semiprime. John Duncan [1967] used the term modular annihilator algebra for a much weaker concept introduced in Definition 8.6.1 below. The idea of modular annihilator algebras evolved from the much more restrictive idea of annihilator algebras introduced by Alfred W. Goldie [1954] and in Definition 8.7.1 below. We will develop the theory of modular annihilator algebras very fully here.

Theorem 8.4.5 gives a great deal of information on modular annihilator algebras. If $\mathcal{A}$ is a modular annihilator algebra then $\Pi_{\mathcal{A}}$ is discrete, $h(\mathcal{A}_F)$ is empty, $\mathcal{A}/\mathcal{A}_F$ is a radical algebra, the Jacobson radical satisfies $\mathcal{A}_J = {}_{LA}(\mathcal{A}_F) = {}_{RA}(\mathcal{A}_F)$, every irreducible representation is equivalent to L^e for some $e \in \mathcal{A}_{MI}$, and every left ideal which is not a subset of $\mathcal{A}_J$ includes a minimal left ideal. Furthermore, we note that $\mathcal{A}$ satisfies

$$_{LA}(\mathcal{A}_F \oplus \mathcal{A}_J) = {}_{RA}(\mathcal{A}_F \oplus \mathcal{A}_J) = \{0\} \tag{6}$$

so that $\mathcal{A}_F \oplus \mathcal{A}_J$ is "most" of $\mathcal{A}$ in a certain sense. To see this, note that the semiprime character of $\mathcal{A}$ gives $\mathcal{A}_F \cap \mathcal{A}_J = \mathcal{A}_F \cap {}_{LA}\mathcal{A}_F = \{0\}$, and $_{RA}(\mathcal{A}_F \oplus \mathcal{A}_J) = {}_{LA}(\mathcal{A}_F \oplus \mathcal{A}_J) = {}_{LA}(\mathcal{A}_F) \cap {}_{LA}(\mathcal{A}_J) = \mathcal{A}_J \cap {}_{LA}(\mathcal{A}_J) = \{0\}$. For other algebras with discrete structure spaces see the references above.

Note that any semiprime Jacobson radical algebra is a modular annihilator algebra by default since it has no maximal modular left ideals at all. This limits the theory somewhat. Characterizations of algebras with $\mathcal{A}/\mathcal{A}_J$ modular annihilator are given in Theorem 8.6.4. In particular, one of these characterizations is purely in terms of the spectrum of each element. We now state a result which gives a reasonably concrete representation of $\Pi_{\mathcal{A}}$.

8.4.7 Proposition   *Let $\mathcal{A}$ be a modular annihilator algebra. For each irreducible representation T of $\mathcal{A}$, let $[T]$ be the equivalence class of T relative to equivalence of representations, and let $\hat{\mathcal{A}}$ be the set of these equivalence classes. Let $\sim$ be the equivalence relation on $\mathcal{A}_{MI}$ defined by*

$$[L^e] = [L^f] \iff [R^e] = [R^f] \iff e\mathcal{A}f \neq \{0\} \iff f\mathcal{A}e \neq \{0\}. \tag{7}$$

Then the maps

$$[e] \mapsto \mathcal{P}^e; \quad [e] \mapsto [L^e]; \quad [L^e] \mapsto \ker(L^e) = \mathcal{P}^e \tag{8}$$

establish (compatible) bijections of $(\mathcal{A}_{MI}/\sim)$ onto $\Pi_{\mathcal{A}}$; of $(\mathcal{A}_{MI}/\sim)$ onto $\hat{\mathcal{A}}$ and of $\hat{\mathcal{A}}$ onto $\Pi_{\mathcal{A}}$.

Proof Proposition 8.2.5 establishes (7). Theorem 8.4.5 and Proposition 8.2.6 give the other results.   □

The next result shows that if $\mathcal{I}$ is an ideal in a semiprime algebra $\mathcal{A}$ then $\mathcal{A}_{MI}$ is the disjoint union of $\mathcal{I}_{MI} = \mathcal{A}_{MI} \cap \mathcal{I}$ and $\mathcal{A}_{MI} \cap {}_{LA}\mathcal{I} = \mathcal{A}_{MI} \cap {}_{RA}\mathcal{I}$. If e and f belong to $\mathcal{A}_{MI}$ and satisfy $e \sim f$ (so that $e\mathcal{A}f$ and $f\mathcal{A}e$ are non-zero), it is clear that e and f either both belong to $\mathcal{I}_{MI}$ or both belong to $\mathcal{A}_{MI} \cap {}_{LA}\mathcal{I}$. Proposition 8.4.9 shows that if $\mathcal{I}$ is an ideal in a modular annihilator algebra then $\mathcal{I}$ and $\mathcal{A}/\mathcal{I}$ are modular annihilator algebras, provided that $\mathcal{I}$ is semiprime in the second case. Furthermore, it is easy to see that $e, f \in \mathcal{I}_{MI}$ satisfy $e \sim f$ relative to $\mathcal{I}$ if and only if they satisfy the same relation relative to $\mathcal{A}$. Similarly e, f in $\mathcal{A}_{MI} \cap {}_{LA}\mathcal{I}$ satisfy $e \sim f$ in $\mathcal{A}$ if and only if they satisfy $e + \mathcal{I} \sim f + \mathcal{I}$ in $\mathcal{A}/\mathcal{I}$. Hence, when $\mathcal{I}$ is a semiprime ideal in a modular annihilator algebra, the construction of the last proposition gives bijections among $h(\mathcal{I}), \Pi_{\mathcal{A}/\mathcal{I}}, (\mathcal{A}_{MI} \cap {}_{LA}\mathcal{I}/ \sim)$, and $((\mathcal{A}/\mathcal{I})_{MI}/ \sim)$ and among $(\Pi_{\mathcal{A}} \setminus h(\mathcal{I}))$, $\Pi_{\mathcal{I}}$, and $(\mathcal{A}_{MI} \cap \mathcal{I}/ \sim)$ and $(\mathcal{I}_{MI}/ \sim)$.

8.4.8 Proposition *An ideal $\mathcal{I}$ of semiprime algebra $\mathcal{A}$ satisfies:*
(a) $\mathcal{I}_{MI} = \mathcal{A}_{MI} \cap \mathcal{I}$.
(b) $\mathcal{A}_{MI} \setminus \mathcal{I}_{MI} = \mathcal{A}_{MI} \cap {}_{LA}\mathcal{I} = \mathcal{A}_{MI} \cap {}_{RA}\mathcal{I}$.
(c) $\mathcal{I}_F = \mathcal{A}_F \cap \mathcal{I} = \mathcal{A}_F\mathcal{I}$.
(d) $\mathcal{A}_F \cap {}_{LA}\mathcal{I} = \mathcal{A}_F \cap {}_{RA}\mathcal{I}$.
(e) $\mathcal{A}_F = (\mathcal{A}_F \cap \mathcal{I}) \oplus (\mathcal{A}_F \cap {}_{LA}\mathcal{I})$.

Proof (a): Let e be an idempotent in $\mathcal{I}$. Then we have $\mathcal{I}e \subseteq \mathcal{A}e = (\mathcal{A}e)e \subseteq \mathcal{I}e$ which implies $\mathcal{I}e = \mathcal{A}e$ and $e\mathcal{I}e = e\mathcal{A}e$. This implies (a).

(b): The inclusion $\mathcal{A}_{MI} \cap {}_{RA}\mathcal{I} \subseteq \mathcal{A}_{MI} \setminus \mathcal{I}_{MI}$ is clear. To prove the opposite inclusion, let e be an element of $\mathcal{A}_{MI} \setminus \mathcal{I}_{MI}$. Then $\mathcal{I}e \subseteq \mathcal{A}e$ is either $\{0\}$ or $\mathcal{A}e$. Since the latter case implies $e \in \mathcal{I}$, it does not hold. Hence e belongs to ${}_{RA}\mathcal{I}$. The second equality follows from (5).

(c), (d), (e): Theorem 8.4.5(a) shows that $\mathcal{I}$ is semiprime. Hence Proposition 8.2.8 and the proof of part (a) above show $\mathcal{I}_F = \sum_{e \in \mathcal{I}_{MI}} \mathcal{I}e = \sum_{e \in \mathcal{I}_{MI}} \mathcal{A}e$ and

$$\mathcal{A}_F = \sum_{e \in \mathcal{A}_{MI}} \mathcal{A}e = \sum_{e \in \mathcal{I}_{MI}} \mathcal{A}e + \sum_{e \in \mathcal{A}_{MI} \setminus \mathcal{I}_{MI}} \mathcal{A}e.$$

Since (b) shows that the second summand is included in ${}_{LA}\mathcal{I}$, the last expression is a direct sum decomposition.   □

The first statement in the next proposition is due to Yood [1964] and its proof is due to Barnes [1971a]. The last statement is due to Barnes [1968a]. Remarks following Theorem 8.6.4 will show certain cases when subalgebras of modular annihilator algebras are again of the same type.

8.4.9 Proposition *Let $\mathcal{A}$ be a modular annihilator algebra and let $\mathcal{I}$ be an ideal of $\mathcal{A}$. Then $\mathcal{I}$ is a modular annihilator algebra. If $\mathcal{I}$ is a semiprime ideal, then $\mathcal{A}/\mathcal{I}$ is also a modular annihilator algebra. Further if $\mathcal{A}$ is any semiprime algebra and $\mathcal{I}$ is an ideal of $k(h(\mathcal{A}_F))$, then $\mathcal{I}$ is a modular annihilator algebra.*

Proof In order to show that $\mathcal{I}$ is a modular annihilator algebra it is enough (by Theorem 8.4.5(f)) to show that $h(\mathcal{I}_F) \subseteq \Pi_{\mathcal{I}}$ is empty. Suppose $\mathcal{P}$ is a primitive ideal of $\mathcal{I}$ which includes $\mathcal{I}_F$. Theorem 4.1.8(e) asserts that there is some primitive ideal $\overline{\mathcal{P}}$ of $\mathcal{A}$ satisfying $\overline{\mathcal{P}} \cap \mathcal{I} = \mathcal{P}$. However $\mathcal{I}\mathcal{A}_F \subseteq \mathcal{A}_F \subseteq \mathcal{P}$ implies $\mathcal{A}_F \subseteq \overline{\mathcal{P}}$ which contradicts Theorem 8.4.5(f). Hence $h(\mathcal{I}_F)$ is empty so $\mathcal{I}$ is a modular annihilator algebra.

Theorem 8.4.5(e) shows that $\mathcal{A}/\mathcal{I}$ will be a modular annihilator algebra if and only if every irreducible representation of $\mathcal{A}/\mathcal{I}$ is equivalent to $L^{\tilde{e}}$ for some minimal idempotent $\tilde{e} \in (\mathcal{A}/\mathcal{I})_{MI}$. Let T be an irreducible representation of $\mathcal{A}/\mathcal{I}$ on a linear space $\mathcal{X}$. For each $a \in \mathcal{A}$ denote $a + \mathcal{I}$ by $\tilde{a}$. Define $\overline{T}: \mathcal{A} \to \mathcal{L}(\mathcal{X})$ by $\overline{T}_a = T_{\tilde{a}}$ for each $a \in \mathcal{A}$. Then $\overline{T}$ is an irreducible representation of $\mathcal{A}$ so that there is an element $e \in \mathcal{A}_{MI}$ and a linear bijection $U: \mathcal{A}e \to \mathcal{X}$ satisfying $U(bae) = U(L_b^e ae) = \overline{T}_b U(ae)$ for all $a, b \in \mathcal{A}$. Suppose ae belongs to $\mathcal{I}$. Then we get $U(ae) = U((ae)ee) = \overline{T}_{ae}U(ee) = T_{\widetilde{ae}}U(e) = 0$. Hence we can define a linear bijection $\tilde{U}: (\mathcal{A}/\mathcal{I})\tilde{e} \to \mathcal{X}$ by $\tilde{U}(\tilde{a}\tilde{e}) = U(ae)$. Furthermore $\tilde{U}(\tilde{b}\tilde{a}\tilde{e}) = U(bae) = \overline{T}_b U(ae) = T_{\tilde{b}}U(ae) = T_{\tilde{b}}\tilde{U}(\tilde{a}\tilde{e})$ holds for all $\tilde{a}, \tilde{b} \in \mathcal{A}/\mathcal{I}$. Therefore $\tilde{U}$ is an equivalence between $L^{\tilde{e}}$ and T. This shows that $(\mathcal{A}/\mathcal{I})\tilde{e}$ is a minimal ideal so that $\tilde{e}$ belongs to $(\mathcal{A}/\mathcal{I})_{MI}$ by Proposition 8.2.2. Thus we have verified condition 8.4.5(e) for $\mathcal{A}/\mathcal{I}$.

Suppose $\mathcal{A}$ is semiprime and $\mathcal{I} = k(h(\mathcal{A}_F))$. Then $\mathcal{I}$ is semiprime by Theorem 4.4.8(a) and satisfies $\mathcal{I}_F = \mathcal{A}_F$ by Proposition 8.4.8(c). Hence the ideal $\mathcal{I}/\mathcal{I}_F = \mathcal{I}/\mathcal{A}_F$ in $\mathcal{A}/\mathcal{A}_F$ is radical by Theorem 4.3.2(a). Thus $\mathcal{I}$ is a modular annihilator algebra by Theorem 8.4.5(c). The first statement of this proposition now yields the last statement.   □

The last statement in Proposition 8.4.9 shows that any semiprime algebra $\mathcal{A}$ contains a largest ideal $k(h(\mathcal{A}_F))$ which is a modular annihilator algebra. Of course it may happen that $\mathcal{A}/k(h(\mathcal{A}_F))$ has a non-zero socle so that $kh(((\mathcal{A}/k(h(\mathcal{A}_F)))_F)$ is non-zero. Proceeding in this way, transfinitely if necessary, and then looking back at preimages in $\mathcal{A}$ we would eventually arrive at a largest ideal of $\mathcal{A}$ which could be analyzed in terms of modular annihilator algebras. In this way we could define a radical property which would be something of a curiosity since the radical part would be susceptible of analysis and the radical free part would be intractable.

The following result is due to Barnes.

8.4.10 Proposition   *Let A be a modular annihilator algebra.*

(a) *If $\mathcal{P}$ is a prime ideal of A then either: (1) $\mathcal{P}$ does not include A_F and both $A/\mathcal{P}$ and $\mathcal{P}$ are primitive so $\mathcal{P}$ has the form $\mathcal{P}^e$; or (2) $\mathcal{P}$ includes A_F and $A/\mathcal{P}$ is radical.*

(b) *Let T be a topologically irreducible normed representation of A on a normed linear space $\mathcal{X}$. Then either $\ker(T)$ does not include A_F and there is a dense T-invariant subspace $\mathcal{Y}$ of $\mathcal{X}$ such that the restriction of T to $\mathcal{Y}$ is algebraically irreducible and has the same kernel as T, or $\ker(T)$ includes A_F and $A/\ker(T)$ is radical.*

Proof   (a): Proposition 8.4.9 shows that $A/\mathcal{P}$ is a modular annihilator algebra. If $A/\mathcal{P}$ is not radical then it has a primitive ideal which Theorem 8.4.5(d) shows must be of the form $_{LA}(A/\mathcal{P})e$ for some $e \in (A/\mathcal{P})_{MI}$. Since $(A/\mathcal{P})e$ is non-zero and $A/\mathcal{P}$ is prime, the primitive ideal $_{LA}(A/\mathcal{P})e$ must be zero. Hence $\mathcal{P}$ is primitive. If $\mathcal{P}$ does not include A_F then Proposition 8.2.8 shows that there is some $e \in A_{MI}$ which does not belong to $\mathcal{P}$. The minimality of Ae implies that $Ae \cap \mathcal{P} = \{0\}$. Hence the restriction of the left regular representation of $A/\mathcal{P}$ to $(Ae + \mathcal{P})/\mathcal{P}$ is irreducible. Hence if $A/\mathcal{P}$ is radical $\mathcal{P}$ includes A_F. If $\mathcal{P}$ includes A_F then $A/\mathcal{P}$ is radical by Theorems 8.4.5(c) and 4.3.2(a).

(b): Proposition 4.2.5 asserts that $\ker(T)$ is prime. Assume that $\ker(T)$ does not include A_F. Then Theorem 8.2.8 shows the existence of an $e \in A_{MI}$ which does not belong to $\ker(T)$. Choose $z \in \mathcal{X}$ satisfying $T_e z \neq 0$. Since T is topologically irreducible the T-invariant subspace $\mathcal{Y} = \{T_{ae}z : a \in A\}$ is dense in $\mathcal{X}$. The surjection $V: Ae \to \mathcal{Y}$ defined by $V(ae) = T_{ae}z$ for all $a \in A$ certainly satisfies $VL_b^e = T_v V$ for all $b \in A$. Hence its kernel is a left ideal properly included in the minimal left ideal Ae and therefore equal to zero. Therefore V is an equivalence between L^e and the restriction of T to $\mathcal{Y}$ which must therefore be algebraically irreducible. The argument given in the proof of (a) shows that $\ker(L^e) = {}_{LA}Ae$ equals $\ker(T)$.   □

8.4.11 Example   The following simple example shows that a faithful topologically irreducible representation T of a semisimple modular annihilator algebra need not be algebraically irreducible. Let $\mathcal{X}_0$ be an incomplete normed linear space with completion $\mathcal{X}$ and let A be $\{x \otimes \omega \in \mathcal{B}_F(\mathcal{X}) : x \in \mathcal{X}_0, \omega \in \mathcal{X}^*\}$. Then the identity representation of A on $\mathcal{X}$ is topologically irreducible but not algebraically irreducible. Its restriction to $\mathcal{X}_0$ is algebraically irreducible.

The first statement in the next result (for the semisimple modular annihilator algebras) is due to Yood ([1958], Lemma 2.8, p. 376). The original proof, which we give, evolved from Frank F. Bonsall [1954a].

8.4.12 Theorem *Any algebra norm on a modular annihilator algebra is a spectral norm. An algebra semi-norm τ on a modular annihilator algebra is a spectral semi-norm if and only if the Jacobson radical is τ-closed. This occurs if $\mathcal{A}_\tau$ is a semiprime ideal included in $\mathcal{A}_J$.*

Proof Let $\mathcal{A}$ be a modular annihilator algebra and let $\|\cdot\|$ be an algebra norm on $\mathcal{A}$. Let $\overline{\mathcal{A}}$ be the completion of $(\mathcal{A}, \|\cdot\|)$. Suppose a is an element of $\mathcal{A}$ and λ is a non-zero element of the spectrum of a. Then $\lambda^{-1}a$ is not quasi-invertible in $\mathcal{A}$. Suppose it lacks a left quasi-inverse. Then $\mathcal{A}(\lambda^{-1}a - 1)$ is a proper modular left ideal since it does not include $\lambda^{-1}a$. Theorem 2.4.6(b) shows that there is a maximal modular left ideal $\mathcal{L}$ which includes $\mathcal{A}(\lambda^{-1}a - 1)$. Since $\mathcal{A}$ is a modular annihilator algebra we may choose a non-zero element b in $_{RA}\mathcal{L} \subseteq {}_{RA}\mathcal{A}(\lambda^{-1}a - 1)$. Since $\lambda^{-1}ab - b$ belongs to $_{RA}\mathcal{A}$ and $\mathcal{A}$ is semiprime, we conclude $\lambda^{-1}ab = b$. Now if $c \in \overline{\mathcal{A}}$ were a left quasi-inverse for $\lambda^{-1}a$ in $\overline{\mathcal{A}}$ we would have $0 \neq b = \lambda^{-1}ab = (c \circ \lambda^{-1}a)b + c\lambda^{-1}ab - cb = 0 + cb - cb = 0$. Hence $\lambda^{-1}a$ is not quasi-invertible in $\overline{\mathcal{A}}$. If $\lambda^{-1}a$ lacks a right quasi-inverse the argument is similar. Hence $\rho_\mathcal{A}(a) \leq \rho_{\overline{\mathcal{A}}}(a) \leq \|a\|$ holds. Therefore $\|\cdot\|$ is a spectral norm.

Let τ be an algebra semi-norm on $\mathcal{A}$. If τ is a spectral semi-norm, Corollary 4.3.11 shows that $\mathcal{A}_J$ is τ-closed and $\mathcal{A}_\tau \subseteq \mathcal{A}_J$. Conversely, if $\mathcal{A}_J$ (which is always semiprime since it is the intersection of prime ideals) is τ-closed or if $\mathcal{A}_\tau = \{a \in \mathcal{A} : \tau(a) = 0\} \subseteq \mathcal{A}_J$ is semiprime, then τ induces an algebra norm $\tilde{\tau}$ on $\tilde{\mathcal{A}} = \mathcal{A}/\mathcal{A}_J$ or on $\tilde{\mathcal{A}} = \mathcal{A}/\mathcal{A}_\tau$, respectively. Since $\tilde{\mathcal{A}}$ is a modular annihilator algebra by Proposition 8.4.9, $\tilde{\tau}$ is a spectral norm by the argument above. Hence τ is a spectral semi-norm since it satisfies $\tau(a) \geq \tilde{\tau}(\tilde{a}) \geq \rho(a + \mathcal{A}_J) = \rho(a)$, where $\tilde{a}$ represents the residue class of a in $\tilde{\mathcal{A}}$ and we have used Theorem 4.3.6(d). $\qquad\square$

Theorem 8.1.1 asserts that a finite-dimensional semiprime algebra is the direct sum of finitely many full matrix algebras. Hence it is obviously a semisimple unital algebra and Theorem 8.4.5(c) shows that it is a modular annihilator algebra since $\mathcal{A} = \mathcal{A}_F$. The next result, which is essentially due to Barnes ([1966], Proposition 6.3, p. 573), extends this observation and provides a converse. The proof is new. First, we provide examples of semiprime algebras $\mathcal{A}$ which are not modular annihilator algebras even though $\mathcal{A}/\mathcal{A}_J$ is a modular annihilator algebra or even finite-dimensional.

8.4.13 Example Suppose $\mathcal{B}$ is one of the semiprime radical Banach algebras described in Example 8.8.3, 8.8.4 or 8.8.6 and let $\mathcal{A}$ be the unitization $\mathcal{B}^1$ of $\mathcal{B}$. Then $\mathcal{A}_J = \mathcal{B}$ is the unique maximal modular left ideal of $\mathcal{A}$. (To see this note that $\lambda + a$ has spectrum $\{\lambda\}$ in $\mathcal{A}$.) Hence $\mathcal{A}/\mathcal{A}_J$ is isomorphic to $\mathbb{C}$ and is therefore a modular annihilator algebra. However, the equation $_{RA}(\mathcal{A}_J) = \{0\}$ shows that $\mathcal{A}$ is not a modular annihilator algebra although it is semiprime. (*Cf.* Theorem 8.6.4 below.)

8.4.14 Proposition *Let $\mathcal{A}$ be a non-radical modular annihilator algebra. Then $\mathcal{A}$ is a unital spectral algebra if and only if $\mathcal{A}/\mathcal{A}_J$ is finite-dimensional. In particular a unital, semisimple, modular annihilator, spectral (or normed) algebra is finite-dimensional.*

Proof Theorem 8.1.1 and the remarks preceding it show that $\mathcal{A}/\mathcal{A}_J$ is unital and normable if it is finite-dimensional (without even assuming that $\mathcal{A}$ is a modular annihilator algebra). Now suppose $\mathcal{A}$ is a modular annihilator spectral algebra and $\mathcal{A}/\mathcal{A}_J$ is unital. Then Proposition 8.4.9 shows that we may as well replace $\mathcal{A}$ by $\mathcal{A}/\mathcal{A}_J$ and hence assume that $\mathcal{A}$ is a unital semisimple modular annihilator algebra. Hence if $\mathcal{A}_F$ is proper it is included in some maximal modular left ideal and therefore has a non-zero right annihilator. Since Theorem 8.4.5(g) shows $_{RA}\mathcal{A}_F = \mathcal{A}_J = \{0\}$ we conclude that $\mathcal{A} = \mathcal{A}_F$. Hence we can write $1 = \sum_{j=1}^{n} a_j e_j$ for some $a_j \in \mathcal{A}$ and $e_j \in \mathcal{A}_{MI}$. Theorem 8.3.6 therefore shows that any $b \in \mathcal{A}$ can be written as $b = 1b1 = \sum_{j,k=1}^{n} a_j e_j b a_k e_k = \sum_{j,k=1}^{n} \lambda_{j,k} a_j c_{jk}$ where we have used the final remark in Theorem 8.3.6 to write $e_j \mathcal{A} e_k = \mathbb{C} c_{jk}$. Hence $\{a_j c_{jk} : 1 \leq j, k \leq n\}$ is a finite set which spans $\mathcal{A}$. □

8.5   Fredholm Theory

We wish to introduce the Fredholm theory of certain elements in semi-prime algebras which is due to Barnes [1968a]. This requires a number of preliminaries. The major results are stated in Theorem 8.5.5.

Let T be a linear operator on a linear space $\mathcal{X}$. For any $n \in \mathbb{N}^0$ we clearly have the inclusions

$$\ker(T^n) \subseteq \ker(T^{n+1}); \qquad (1)$$
$$\text{and } T^n \mathcal{X} \supseteq T^{n+1} \mathcal{X}.$$

Furthermore, it is easy to see that if there is some $m \in \mathbb{N}^0$ satisfying $(\ker(T^m) = \ker(T^{m+1}); \ T^m \mathcal{X} = T^{m+1} \mathcal{X})$ then for any $n \in \mathbb{N}$ we will have $(\ker(T^m) = \ker(T^{m+n}); \ T^m \mathcal{X} = T^{m+n} \mathcal{X})$. We are ready for a definition.

8.5.1 Definition The *ascent* of T is the smallest integer in $\mathbb{N}^0$ satisfying $\ker(T^m) = \ker(T^{m+1})$ and the *descent* of T is the smallest integer m in $\mathbb{N}^0$ satisfying $T^m \mathcal{X} = T^{m+1} \mathcal{X}$. If there is no such integer we will say the ascent or descent is ∞.

Let $\mathcal{A}$ be an algebra and let a be an element of $\mathcal{A}^1$. Let $\langle \tilde{L}_a / \tilde{R}_a \rangle$ denote the restriction to $\mathcal{A}$ of the $\langle$ left / right $\rangle$ regular representation of a on $\mathcal{A}^1$, respectively, *i.e.*,

$$\tilde{L}_a b = ab \qquad \tilde{R}_a b = ba \qquad a \in \mathcal{A}^1 \quad b \in \mathcal{A}.$$

We will denote the *ascent of* $\langle \tilde{L}_a / \tilde{R}_a \rangle$ by $\langle \alpha_L(a) / \alpha_R(a) \rangle$, and the *descent of* $\langle \tilde{L}_a / \tilde{R}_a \rangle$ by $\langle \delta_L(a) / \delta_R(a) \rangle$.

Note that the ascent of T is zero if and only if T is injective and the descent of T is zero if and only if T is surjective. If $\mathcal{A}$ is semiprime, then we have the formulas

$$
\begin{aligned}
\alpha_L(a) &= \min\{m \in \mathbb{N}^0 : {}_{RA}\mathcal{A}a^m = {}_{RA}\mathcal{A}a^{m+1}\}; \\
\delta_L(a) &= \min\{m \in \mathbb{N}^0 : a^m\mathcal{A} = a^{m+1}\mathcal{A}\}; \\
\alpha_R(a) &= \min\{m \in \mathbb{N}^0 : {}_{LA}a^m\mathcal{A} = {}_{LA}a^{m+1}\mathcal{A}\}; \\
\delta_R(a) &= \min\{m \in \mathbb{N}^0 : \mathcal{A}a^m = \mathcal{A}a^{m+1}\}.
\end{aligned}
\tag{2}
$$

8.5.2 Lemma *If a linear operator $T \in \mathcal{L}(\mathcal{X})$ has finite ascent and descent, then they are equal and*

$$
\mathcal{X} = T^m\mathcal{X} \oplus \ker(T^m)
$$

holds where m is their common value.

Proof Denote the ascent and descent of T by $\alpha(T)$ and $\delta(T)$, respectively. First assume $\delta(T) = 0$ and suppose $\alpha(T) > 0$. Then T is surjective but not injective. Choose $x_1 \in \ker(T) \setminus \{0\}$. Inductively choose $x_2, x_3, \ldots$ satisfying $Tx_{n+1} = x_n$. Then for each n, x_n belongs to $\ker(T^n)$ but not to $\ker(T^{n-1})$. This contradicts the finite character of $\alpha(T)$. Hence $\delta(T) = 0$ implies $\alpha(T) = 0$.

Next we will show $\alpha(T) \leq \delta(T)$. Denote $\delta(T)$ by m. Then consider the restriction T' of the operator T to the space $T^m(\mathcal{X})$. Since this operator is surjective $\delta(T') = 0$ and hence $\alpha(T')$ (which is clearly finite and less than or equal to $\alpha(T)$) must also be zero. Thus T' is invertible. Suppose x belongs to $\ker(T^{m+1})$. Then $T'(T^m x) = T^{m+1}x = 0$ implies $T^m x = 0$. Hence $\alpha(T)$ satisfies $\alpha(T) \leq m = \delta(T)$.

Next we will show $\delta(T) \leq \alpha(T)$. In the first paragraph we have covered the case $\delta(T) = 0$ so we may assume $m = \delta(T) > 0$. Let $y = T^{m-1}x$ belong to $T^{m-1}\mathcal{X} \setminus T^m\mathcal{X}$. By the choice of m, the restriction of T^m to $T^m\mathcal{X}$ is surjective so that there is some $z \in T^m\mathcal{X}$ satisfying $T^m z = T^m x$. Hence we have $T^m(z - x) = 0$ and $T^{m-1}(z - x) = T^{m-1}z - y \neq 0$. Thus we conclude $\alpha(T) \geq \delta(T)$ and by our previous result $\alpha(T) = \delta(T)$.

Finally, let y be any element of $T^m\mathcal{X} \cap \ker(T^m)$ where m satisfies $m = \alpha(T) = \delta(T)$. Then some $x \in \mathcal{X}$ satisfies $y = T^m x$ so $T^{2m}x = T^m y = 0$. However, this implies $y = T^m x = 0$ since $m = \alpha(T)$. Hence we have $\ker(T^m) \cap T^m\mathcal{X} = \{0\}$. Now suppose x is any element of $\mathcal{X}$. Since restriction of T^m to $T^m\mathcal{X}$ is surjective we may write $T^m x = T^m y$ for some $y \in T^m\mathcal{X}$. Hence we have $x = y + (x - y)$ where $y \in T^m\mathcal{X}$ and $T^m(x - y) = T^m x - T^m y = 0$ hold. Hence $\mathcal{X}$ is the direct sum of $T^m\mathcal{X}$ and $\ker(T^m)$. □

There is a natural concept of dimension or rank in the socle of a semiprime algebra. We introduce this notion and then prove a simple lemma which explains the concept.

8.5.3 Definition A left ideal has *finite-rank* if it is included in the sum of finitely many minimal left ideals. In this case we will call the *rank* of the left ideal the smallest number of minimal left ideals the sum of which contains it. The ideal $\{0\}$ will be said to have rank zero. Clearly a similar concept can be defined for right ideals.

The next lemma shows that there is no ambiguity in speaking of the rank of a (two-sided) ideal. Note that any principal left ideal $\mathcal{A}a$ generated by an element a in the socle has finite-rank since we may write $a = \sum_{j=1}^{n} a_j$ with each a_j in a minimal left ideal $\mathcal{L}_j$. Hence $\mathcal{A}a$ is included in $\sum_{j=1}^{n} \mathcal{L}_j$. In essence the following lemma develops the theory of finite-rank left and right ideals.

8.5.4 Lemma *Let $\mathcal{A}$ be a semiprime algebra. A left ideal $\mathcal{L}$ in $\mathcal{A}$ has finite-rank if and only if there is an idempotent e in $\mathcal{A}_F$ satisfying $\mathcal{L} = \mathcal{A}e$. The rank of a finite-rank left ideal $\mathcal{L}$ is the number of orthogonal minimal idempotents in any maximal family included in $\mathcal{L}$. It is also the number of terms in any direct sum decomposition of $\mathcal{L}$ as the sum of minimal left ideals. Any left ideal properly included in a finite-rank left ideal $\mathcal{L}$ has finite-rank strictly less than the rank of $\mathcal{L}$. The analogous results hold for right ideals also.*

Proof By symmetry we only need to deal with left ideals. We use Corollary 8.2.3 repeatedly and without further comment. Let $\mathcal{L}$ be a non-zero left ideal with rank n so that there is a set $\{e_j : j = 1, 2, \ldots, n\}$ of minimal idempotents satisfying $\mathcal{L} \subseteq \sum_{j=1}^{n} \mathcal{A}e_j$. Theorem 8.4.5(h) applied to $\mathcal{L}$ considered as a left ideal of $\mathcal{A}_F$ shows that $\mathcal{L}$ includes some minimal ideal and hence contains some minimal idempotent. Zorn's Lemma guarantees a maximal family $\{f_\alpha : \alpha \in A\}$ of orthogonal minimal idempotents in $\mathcal{L}$. Then $\sum_{\alpha \in A} \mathcal{A}f_\alpha \subseteq \mathcal{L}$ is a direct sum. For each $\alpha_1 \in A, f_{\alpha_1} = \sum_{j=1}^{n} a_j e_j$ holds for certain $a_j \in \mathcal{A}$. There must be some k with $a_k e_k \neq 0$ which implies in turn $\mathcal{A}a_k e_k \neq \{0\}, \mathcal{A}e_k = \mathcal{A}a_k e_k \subseteq \mathcal{A}f_{\alpha_1} + \sum_{j=1}^{n} \mathcal{A}e_j$, and hence $\mathcal{L} \subseteq \mathcal{A}f_{\alpha_1} + \sum_{j=1}^{n} \mathcal{A}e_j$. Hence if A has more than one element, then for any $\alpha_2 \in A \setminus \{\alpha_1\}$ we have $f_{\alpha_2} = a_{\alpha_1} f_{\alpha_1} + \sum_{\substack{j=1 \\ j \neq k}}^{n} b_j e_j$. Since the sum $\sum_{\alpha \in A} \mathcal{A}f_\alpha$ is direct, there must be some $i \neq k$ with $b_i e_i \neq 0$, so we may write $\mathcal{L} \subseteq \mathcal{A}f_{\alpha_1} + \mathcal{A}f_{\alpha_2} + \sum_{\substack{j=1 \\ j \neq i,k}}^{n} \mathcal{A}e_j$. Proceeding in this way at each step we replace an ideal $\mathcal{A}e_j$ by an ideal $\mathcal{A}f_{\alpha_k}$. Hence eventually we exhaust A (since $\{f_\alpha : \alpha \in A\}$ is an orthogonal family). We want to show $\mathcal{L} \subseteq \sum_{\alpha \in A} \mathcal{A}f_\alpha \subseteq \mathcal{L}$.

Let e be the sum of the finite set $\{f_\alpha : \alpha \in A\}$ so that e belongs to $\mathcal{L} \cap \mathcal{A}_F$. If $\mathcal{L}(1 - e)$ is a non-zero left ideal included in $\mathcal{L}$, it must contain a minimal idempotent f'_β by the argument given above. Define f_β by $f_\beta = (1-e)f'_\beta \in \mathcal{L}(1-e) \subseteq \mathcal{L}$. Then $f_\beta = (1-e)f'_\beta, f'_\beta = (1-e)f'_\beta(1-e)f'_\beta = f_\beta^2$ is an idempotent which is non-zero since it satisfies $f'_\beta = (f'_\beta)^2 = f'_\beta(1-e)f'_\beta =$

$f'_\beta f_\beta$. It is minimal by Proposition 8.2.2 since $\mathcal{A}f_\beta = \mathcal{A}f'_\beta$ is minimal. Then $\{f_\alpha : \alpha \in A\} \cup \{f_\beta\}$ contradicts the maximality of $\{f_\alpha : \alpha \in A\}$. Hence $\mathcal{L}(1 - e) = 0$ implies $\mathcal{L} = \mathcal{L}e \subseteq \mathcal{A}e \subseteq \sum_{\alpha \in A} \mathcal{A}f_\alpha \subseteq \mathcal{L}$. The definition of rank shows that A has at least n elements, and the construction shows that it has at most n elements. Hence the rank of a finite-rank ideal is the number of orthogonal minimal idempotents contained in any maximal family. The equation $\mathcal{L} = \mathcal{A}e$ shows that a finite-rank ideal has the form $\mathcal{A}e$ for some $e \in \mathcal{A}_F$. The converse is a special case of the remark that any principal left ideal $\mathcal{A}a$ generated by an element $a \in \mathcal{A}_F$ is finite-rank.

If we apply the replacement procedure described at the beginning of this proof to a direct sum $\mathcal{L} = \oplus_{j=1}^n \mathcal{A}e_j$ then linear independence shows that the process cannot terminate with less than n terms and the proof already given shows that it does terminate with not more than n terms. Hence the number of elements in a maximal family of orthogonal idempotents is exactly n, so any direct sum decomposition contains exactly n terms.

Finally, let $\mathcal{M}$ be a left ideal included in a finite-rank left ideal $\mathcal{L}$. Clearly $\mathcal{M}$ has finite-rank. Let $\{f_j : j = 1, 2, \ldots, m\}$ and $\{e_j : j = 1, 2, \ldots, n\}$ be maximal families of orthogonal minimal idempotents in $\mathcal{M}$ and in $\mathcal{L}$, respectively. The argument given earlier in this proof proves $m \leq n$. If $m = n$, the replacement procedure described earlier shows $\mathcal{M} = \mathcal{L}$. This proves the inequality of ranks.   □

The next theorem contains Barnes' [1968a] major results on the Fredholm theory of certain elements in semiprime algebras. Note that for any ideal $\mathcal{I}$ in an algebra $\mathcal{A}$ the preimage in $\mathcal{A}$ of the Jacobson radical in $\mathcal{A}/\mathcal{I}$ is the set $k(h(\mathcal{I}))$. Hence this set is the same in $\mathcal{A}$ and in the reverse algebra $\mathcal{A}^R$ even though it is defined in terms of primitive ideals. Since a semiprime algebra $\mathcal{A}$ is a modular annihilator algebra if and only if $\mathcal{A}/\mathcal{A}_F$ is radical, we see that $\mathcal{A}$ is a modular annihilator algebra if and only if $\mathcal{A} = k(h(\mathcal{A}_F))$. Furthermore $\mathcal{I} = k(h(\mathcal{A}_F))$ is always a modular annihilator algebra by Proposition 8.4.9. Thus any modular left ideal in $\mathcal{I}$ has a non-zero right annihilator in $\mathcal{I}$ and hence in $\mathcal{A}$. We will use these facts in the next theorem.

8.5.5 Theorem *Any element a in the socle $\mathcal{A}_F$ of a semiprime algebra $\mathcal{A}$ satisfies:*

(a) *$Sp(a)$ is finite.*

(b) *For any non-zero $\lambda \in Sp(a)$ and any $n \in \mathbb{N}$, $_{LA}((\lambda - a)^n \mathcal{A})$ and $_{RA}(\mathcal{A}(\lambda - a)^n)$ have finite-rank.*

(c) *For any non-zero $\lambda \in Sp(a)$,*

$$\alpha_L(\lambda - a) = \delta_L(\lambda - a) = \alpha_R(\lambda - a) = \delta_R(\lambda - a)$$

are all finite.

(d) *For any non-zero $\lambda \in Sp(a)$ and any $n \in \mathbb{N}$, there is an idempotent $e \in \mathcal{A}_F$ satisfying $_{RA}(\mathcal{A}(\lambda - a)^n) = e\mathcal{A}$ and $\mathcal{A}(\lambda - a)^n = \mathcal{A}(1 - e)$.*

(e) *Any non-zero $\lambda \in Sp(a)$ satisfies*

$$\mathcal{A} = {}_{RA}\mathcal{A}(\lambda - a)^m \oplus (\lambda - a)^m \mathcal{A} = {}_{LA}(\lambda - a)^m \mathcal{A} \oplus \mathcal{A}(\lambda - a)^m$$

where $m = \alpha_L(\lambda - a)$.

Any element a in $k(h(\mathcal{A}_F))$ satisfies (b) and if $\mathcal{A}$ is a normed or spectral algebra then a also satisfies (c), (d), (e) and:

(a') *$Sp(a)$ is either a finite set, or a countably infinite set containing zero which is its only limit point.*

In particular any element a in a modular annihilator algebra $\mathcal{A}$ satisfies (b) and if $\mathcal{A}$ is also normed or spectral, then it also satisfies (c), (d), (e), and (a').

Proof (b): We discuss only $_{RA}(\mathcal{A}(\lambda-a)^n)$ since this result in $\mathcal{A}^R$ establishes the other half of (b). Let a belong to $\mathcal{A}_F$ and let λ and n be as stated. The we have $(\lambda - a)^n = \lambda^n - ab$ where b is a polynomial in a. Hence any $c \in {}_{RA}(\mathcal{A}(\lambda-a)^n)$ satisfies $(\lambda^n - ab)c \in {}_{RA}\mathcal{A} = \{0\}$ so $c = \lambda^{-n}abc$ belongs to $a\mathcal{A}$. Lemma 8.5.4 now shows that $_{RA}(\mathcal{A}(\lambda - a)^n) \subseteq a\mathcal{A}$ has finite-rank.

Suppose next that a belongs to $k(h(\mathcal{A}_F))$ and that λ, n, and b are as before. Since $k(h(\mathcal{A}_F))/\mathcal{A}_F$ is radical, Theorem 4.3.6(b) shows that there is some $c \in k(h(\mathcal{A}_F))$ and some $d \in \mathcal{A}_F$ satisfying $c \circ \lambda^{-n}ba = d$. The equation in $\mathcal{A}^1$, $(1-c)(\lambda-a)^n = \lambda^n(1-d)$ shows $\mathcal{A}(\lambda-a)^n \supseteq \mathcal{A}(1-d)$ and hence $_{RA}(\mathcal{A}(\lambda - a)^n) \subseteq {}_{RA}(\mathcal{A}(1 - d))$. Since the last ideal has finite-rank by the first part of the proof above, we have established (b).

(c): Again the symmetry of the situation allows us to prove only one half of the relations. We will first show that $\alpha_L(\lambda - a)$ is finite when $a \in \mathcal{A}_F$. Let m be the rank of $a\mathcal{A}$. If $\alpha_L(\lambda - a) > m$ we can find an orthogonal set $e_0, e_1, e_2, \ldots, e_m$ of idempotents with $e_0 \in {}_{RA}(\mathcal{A}(\lambda-a))$ and $e_k \in {}_{RA}(\mathcal{A}(\lambda - a)^{k+1}) \setminus {}_{RA}(\mathcal{A}(\lambda - a)^k)$ for $k = 1, 2, \ldots, m$. However, as in the proof of (b), we conclude that each e_k belongs to $a\mathcal{A}$. This contradicts the rank of $a\mathcal{A}$ and shows that $\alpha_L(\lambda - a) \leq m$.

Next we assume that a belongs to $k(h(\mathcal{A}_F))$ and that $\mathcal{A}$ is a normed or spectral algebra. Again we will show that $\alpha_L(\lambda - a)$ is finite. For each $n \in \mathbb{N}$ define a right ideal $\mathcal{R}_n$ by $\mathcal{R}_n = {}_{RA}(\mathcal{A}(\lambda - a)) \cap (\lambda - a)^n\mathcal{A}$. If $\alpha_L(\lambda - a)$ is not finite each $\mathcal{R}_n$ is non-zero. We will derive a contradiction from assuming this. Since $_{RA}(\mathcal{A}(\lambda - a))$ has finite-rank by (b) and the sequence $\{\mathcal{R}_n\}$ of these right ideals is clearly decreasing, this sequence must eventually become constant (and by assumption, unequal to $\{0\}$). Hence there is some $e \in \mathcal{A}_{MI}$ in $\cap_{n=1}^{\infty} \mathcal{R}_n$. Therefore, for all $n \geq 0$, we have

$$e \in {}_{RA}(\mathcal{A}(\lambda - a)) \cap (\lambda - a)^n\mathcal{A}. \tag{3}$$

We consider the representation L^e described in Theorem 8.3.6. Choose an algebra norm or spectral semi-norm σ on $\mathcal{A}$ and a corresponding norm on $\mathcal{A}e$. Proposition 8.4.10 shows that σ is a spectral semi-norm in any case. Since $k(h(\mathcal{A}_F))/\mathcal{A}_F$ is radical, Theorem 2.2.5 and Corollary 4.3.9 show $\lim_{n\to\infty}(\sigma'((a+\mathcal{A}_F)^n))^{1/n} = 0$, where σ' is the quotient semi-norm. Hence we can find an integer $p \in \mathbb{N}$ and an element $c \in \mathcal{A}_F$ satisfying $\sigma(a^p - c) < |\lambda|^p/6$. To simplify notation we put $L_a^e = K, L_c^e = J$ and $K^p - J = D$. Then Theorem 8.3.6 shows $\|D\| < |\lambda|^p/6$.

Now we will use this inequality to show that $\lambda - K$ has finite ascent. Suppose not. The Riesz Lemma (in §1.7.7) shows that we can find $y_n \in \ker((\lambda - K)^{n+1})$ with $\|y_n\| = 1$ and $\mathrm{dist}(y_n, \ker((\lambda - K)^n)) \geq \frac{1}{2}$. Let $z_n = \lambda^{-p}Jy_n$. Then we have

$$
\begin{aligned}
z_n &= \lambda^{-p}(K^p y_n - D y_n) \\
&= y_n - \lambda^{-1}(\lambda - K)\sum_{k=0}^{p-1}(\lambda^{-1}K)^k y_n - \lambda^{-p}D y_n.
\end{aligned}
$$

Since the middle term belongs to $\ker((\lambda - K)^n)$, we have for any $n > m$

$$z_n - z_m = y_n - y - \lambda^{-p}D y_n + \lambda^{-p}D y_m$$

with $y \in \ker((\lambda - K)^n)$. Hence we have

$$
\begin{aligned}
\|z_n - z_m\| &\geq \|y_n - y\| - \|\lambda^{-p}D y_n\| - \|\lambda^{-p}D y_m\| \\
&\geq \frac{1}{2} - \frac{1}{6} - \frac{1}{6} = \frac{1}{6}.
\end{aligned}
$$

However, this is impossible since Theorem 8.3.6 shows that $J = L_c^e$ has finite-dimensional range so that each z_n belongs to the compact ball in $J\mathcal{A}e$ of radius $\|J\|\,|\lambda|^{-p}$. Therefore $\lambda - L_a^e$ has finite ascent. This contradicts (3). Therefore the original assumption that $\alpha_L(\lambda - a)$ was infinite must be rejected. Hence $\alpha_L(\lambda - a)$ is finite under either hypothesis.

Next we will show that $\delta_R(\lambda - a)$ is finite. Let a satisfy either hypothesis. Denote $\alpha_L(\lambda - a)$ by m. We may suppose $m > 0$, so part (b) shows that there is an idempotent $e \in \mathcal{A}_F$ satisfying $_{RA}(\mathcal{A}(\lambda - a)^m) = e\mathcal{A}$. Consider the modular left ideal $\mathcal{L} = \mathcal{A}((\lambda - a)^m - e)$. We will prove $_{RA}\mathcal{L} = \{0\}$, and then $\mathcal{L} = \mathcal{A}$. Suppose $b \in {}_{RA}\mathcal{L}$ so that $(\lambda - a)^m b = eb$ and $(\lambda - a)^{2m}b = (\lambda - a)^m eb = 0$. Since $m = \alpha_L(\lambda - a)$ we have $(\lambda - a)^m b = 0$. However this implies $b \in {}_{RA}(\mathcal{A}(\lambda - a)^m) = e\mathcal{A}$, so $b = eb = (\lambda - a)^m b = 0$ as we wished to show. Now suppose $\mathcal{L} \neq \mathcal{A}$. Theorem 2.4.6(e) shows that $f = \lambda^{-m}((\lambda - a)^m + e) \in kh(\mathcal{A}_F)$ is the right relative identity for some maximal modular left ideal $\mathcal{M} \supseteq \mathcal{L} \cap kh(\mathcal{A}_F) \supseteq kh(\mathcal{A}_F)(1 - f)$. But Proposition 8.4.9 asserts that $kh(\mathcal{A}_F)$ is a modular annihilator algebra, so Theorem 8.4.5(a) gives a non-zero element b in the right annihilator of $\mathcal{M}$ in $kh(\mathcal{A}_F)$. Since $kh(\mathcal{A}_F)$ is semiprime, we conclude $(1 - f)b = 0$. This contradicts $_{RA}(\mathcal{A}(1 - f)) = \{0\}$.

Let b be any element of $\mathcal{A}(\lambda - a)^m$. Then we may write $b = c(\lambda - a)^m$ and $c = d((\lambda - a)^m - e)$. Combining these and recalling the definition of e, gives $b = d(\lambda - a)^{2m}$. Hence $\delta_R(\lambda - a)$ is less than or equal to $\alpha_L(\lambda - a)$. By symmetry $\alpha_L(\lambda - a)$, $\alpha_R(\lambda - a)$, $\delta_L(\lambda - a)$ and $\delta_R(\lambda - a)$ are all finite. Lemma 8.5.2 shows that $\alpha_L = \delta_L$, $\alpha_R = \delta_R$ and the inequality just obtained together with its translation into the reverse algebra give the rest of the equations in (c).

(e): This is an immediate consequence of (c) and Lemma 8.5.2.

(d): Lemma 8.5.4 and conclusion (b) above show the existence of an idempotent e satisfying $_{RA}\mathcal{A}(\lambda - a)^n = e\mathcal{A}$. Clearly we have $\mathcal{A}(\lambda - a)^n \subseteq$ $_{LRA}\mathcal{A}(\lambda-a)^n = _{LA}e\mathcal{A} = \mathcal{A}(1-e)$. On the other hand, $_{RA}(\mathcal{A}(\lambda-a)^n + \mathcal{A}e) = _{RA}\mathcal{A}(\lambda - a)^n \cap _{RA}(\mathcal{A}e) = e\mathcal{A} \cap (1 - e)\mathcal{A} = \{0\}$. Hence $\mathcal{A}(\lambda - a)^n + \mathcal{A}e = \mathcal{A}$ holds and any $b \in \mathcal{A}(1 - e)$ may be written as $b = (c(\lambda - a)^n + de)(1 - e) = c(\lambda - a)^n$. This gives the opposite inclusion.

(a): We have $a \in \mathcal{A}_F$. Suppose $\{\lambda_n\}_{n \in \mathbb{N}}$ is an infinite sequence of distinct non-zero points in $Sp(a)$. For each $n \in \mathbb{N}$, $\lambda_n^{-1}a$ is either left or right quasi-singular. Hence by working in either $\mathcal{A}$ or $\mathcal{A}^R$, and choosing the subsequence where $\lambda_n^{-1}a$ is left quasi-singular, we may assume that $\mathcal{A}(\lambda_n - a)$ is a proper left ideal for each n. Hence, for each $n \in \mathbb{N}$, conclusion (b) shows that there is some $e_n \in \mathcal{A}_{MI}$ satisfying $\lambda_n e_n = ae_n$. Therefore each e_n belongs to the finite-rank ideal $a\mathcal{A}$. If for some $n > 1$ there exist $a_j \in \mathcal{A}$ satisfying $e_1 a_1 + e_2 a_2 + \cdots + e_n a_n = 0$ with $e_n a_n$ non-zero, we obtain the contradiction $0 = (\lambda_1 - a)(\lambda_2 - a) \cdots (\lambda_n - a)(e_1 a_1 + e_2 a_2 + \cdots + e_{n-1}a_{n-1}) = (\lambda_1 - a)(\lambda_2 - a) \cdots (\lambda_n - a)(-e_n a_n) = -(\lambda_1 - \lambda_n)(\lambda_2 - \lambda_n) \cdots (\lambda_{n-1} - \lambda_n)e_n a_n$. Hence $e_1 \mathcal{A}_1 + e_2 \mathcal{A}_2 + \cdots + e_n \mathcal{A}_n$ is a direct sum for any $n \in \mathbb{N}$. This contradicts the finite-rank of $a\mathcal{A}$.

(a'): Theorem 8.4.12 shows that $k(h(\mathcal{A}_F))$ is a spectral algebra under either hypothesis. If $Sp(a)$ has no non-zero limit points then each compact annulus $\{\lambda : n^{-1} \le |\lambda| \le \rho(a)\}$ must contain only finitely many points so $Sp(a)$ is countable. Hence it is enough to derive a contradiction from the assumption that $\lambda \ne 0$ is the limit of a sequence $\{\lambda_n\}_{n \in \mathbb{N}}$ in $Sp(a)$. The same argument used in the proof of (a) shows that for each $n \in \mathbb{N}$ there is an $e_n \in \mathcal{A}_{MI}$ satisfying $(\lambda_n - a)e_n = 0$. Conclusion (e) shows that we can write

$$\mathcal{A} = \mathcal{A}(\lambda - a)^m \oplus _{LA}((\lambda - a)^m \mathcal{A}),$$

where $m = \alpha_R(\lambda - a)$. Let $\mathcal{L}$ be the left ideal

$$\mathcal{L} = \{b \in \mathcal{A} : \lim_{n \to \infty} \|be_n\|/\|e_n\| = 0\}.$$

Since $(\lambda - a)^m e_n = (\lambda - \lambda_n)^m e_n$ holds for all $n \ge 0$, we have $\mathcal{A}(\lambda - a)^m \subseteq \mathcal{L}$ and $e_n \in (\lambda - a)^m \mathcal{A}$ for all $n \in \mathbb{N}$. Therefore $_{LA}((\lambda - a)^m \mathcal{A})e_n$ vanishes for all $n \in \mathbb{N}$, so that $_{LA}((\lambda - a)^m \mathcal{A})$ is also included in $\mathcal{L}$. The direct sum decomposition above shows that $\mathcal{L} = \mathcal{A}$. However $\lim_{n \to \infty} \|ae_n\|/\|e_n\| = $

$\lim_{n\to\infty} \|\lambda_n e_n\|/\|e_n\| = |\lambda|$ shows that this contradicts the assumption that λ was non-zero. Hence (a') holds.   □

8.6   Algebras with Countable Spectrum for Elements

In this section we will characterize two kinds of algebras by assuming, in part, that the spectrum of every element is countable. We begin by introducing a class of algebras larger than the class of modular annihilator algebras. Duncan [1967] defined a class of Banach algebras which he called modular annihilator algebras but his definition differs from that given in this work which is due to Yood ([1958], p. 376). His modular annihilator algebras were Banach algebras with a family $\{\mathcal{L}_\alpha : \alpha \in A\}$ of maximal modular left ideals satisfying

$$\cap_{\alpha \in A}\mathcal{L}_\alpha = \{0\} \quad \text{and} \quad _{RA}\mathcal{L}_\alpha \neq 0 \quad \forall\,\alpha \in A.$$

The first equation implies that the algebra is semisimple, hence semiprime. Proposition 8.4.3 shows that the second condition implies the existence of a minimal idempotent e_α for each $\alpha \in A$ satisfying $\mathcal{L}_\alpha = \mathcal{A}(1 - e_\alpha)$. Hence we have $\{0\} = \cap_{\alpha \in A}\mathcal{L}_\alpha = \cap_{\alpha \in A}\mathcal{A}(1 - e_\alpha) = \cap_{\alpha \in A}\,_{LA}(e_\alpha\mathcal{A}) \supseteq _{LA}(\mathcal{A}_F)$. Therefore Duncan's definition is equivalent to requiring that the Banach algebra $\mathcal{A}$ be semisimple and satisfy $_{LA}(\mathcal{A}_F) = \{0\}$. In keeping with the spirit of the development given here we offer the following definition.

8.6.1 Definition   An algebra $\mathcal{A}$ is called a *Duncan modular annihilator algebra* if it is semiprime and satisfies

$$_{LA}\mathcal{A}_F = _{RA}\mathcal{A}_F = \mathcal{A}_J.$$

Theorem 8.4.5(g) shows that modular annihilator algebras are Duncan modular annihilator algebras. However, if $\mathcal{X}$ is any infinite-dimensional Banach space, then $\mathcal{A} = \mathcal{B}(\mathcal{X})$ is a semisimple (in fact primitive) Duncan modular annihilator algebra which is not a modular annihilator algebra. To see that $\mathcal{B}(\mathcal{X})$ is a Duncan modular annihilator algebra note that $T \in _{LA}(\mathcal{A}_F)$ implies $0 = T(x \otimes \omega) = Tx \otimes \omega$ for all $x \in \mathcal{X}$ and $\omega \in \mathcal{X}^*$. Hence $Tx = 0$ for all $x \in \mathcal{X}$. The algebra $\mathcal{B}(\mathcal{X})$ is not a modular annihilator algebra since it is infinite-dimensional and unital contradicting Proposition 8.4.14.

The following theorem is due to Barnes [1968a], Theorem 2.2.

8.6.2 Theorem   *If every element of a Banach algebra $\mathcal{A}$ has finite or countable spectrum, then $\mathcal{A}/\mathcal{A}_J$ is a Duncan modular annihilator algebra.*

Proof First we will show that a semisimple commutative Banach algebra $\mathcal{A}$ without minimal ideals must contain a point with uncountable spectrum. If the Gelfand space $\Gamma_{\mathcal{A}}$ contains a connected subset which is not a singleton, then there is some $a \in \mathcal{A}$ for which $Sp(a) = \hat{a}(\Gamma_{\mathcal{A}})$ is a connected subset of $\mathbb{C}$ that is not a singleton. Since such sets are uncountable we may assume that $\Gamma_{\mathcal{A}}$ is totally disconnected. If $\Gamma_{\mathcal{A}}$ contained an isolated point then the Šilov idempotent theorem (Corollary 3.5.13) would give a minimal idempotent. Hence $\Gamma_{\mathcal{A}}$ is totally disconnected without isolated points and each open set is infinite.

Choose a nonempty, open, compact subset N in $\Gamma_{\mathcal{A}}$. Then choose two disjoint, nonempty, open, compact subsets N_0 and N_1 satisfying $N_0 \cup N_1 = N$. For $k = 0, 1$ choose disjoint, nonempty, open, compact subsets N_{k0} and N_{k1} satisfying $N_{k0} \cup N_{k1} = N_k$. Proceeding inductively in this fashion, we obtain nonempty, open, compact subsets $N_{k_1 k_2 \ldots k_n}$, where $k_1, k_2, \ldots, k_n$ is any list of n 0's and 1's. The Šilov idempotent theorem (Corollary 3.5.13) gives corresponding idempotents $e_{k_1 k_2 \ldots k_n}$ such that $\hat{e}_{k_1 k_2 \ldots k_n}$ is the characteristic function of $N_{k_1 k_2 \ldots k_n}$. For each $n \in \mathbb{N}$ define $a_n \in \mathcal{A}$ by

$$a_n = \sum_{k_n = 1} e_{k_1 k_2 \ldots k_n}.$$

Choose a strictly increasing sequence $\{m_n\}_{n \in \mathbb{N}}$ so that $a = \sum_{n=1}^{\infty} 2^{-m_n} a_n$ converges in $\mathcal{A}$.

Given any sequence $\{k_n\}_{n \in \mathbb{N}}$ of 0's and 1's, the finite intersection property shows that there is a point φ in $\cap_{n=1}^{\infty} N_{k_1 k_2 \ldots k_n}$ and we have

$$\sum_{n=1}^{\infty} k_n 2^{-m_n} = \hat{a}(\varphi) \in Sp(a).$$

Therefore $Sp(a)$ is uncountable. This completes the proof of our first assertion.

Now suppose $\mathcal{A}$ satisfies the hypotheses of this theorem. Theorem 4.3.6(d) shows that we may assume that $\mathcal{A}$ is semisimple by replacing it by $\mathcal{A}/\mathcal{A}_J$. Denote the closed ideal $_{LA}\mathcal{A}_F$ by $\mathcal{I}$. If no element of $\mathcal{I}$ has spectrum different from $\{0\}$ then Theorem 4.3.6(b) shows $\mathcal{I} \subseteq \mathcal{A}_J = \{0\}$. Thus this case is complete.

Now suppose that $\mathcal{I}$ contains an element b with spectrum different from $\{0\}$. We will show that this implies the existence of a minimal idempotent $e \in \mathcal{I}$. This is a contradiction since Proposition 8.4.8(a) shows that e also belongs to $\mathcal{A}_F$ but the semisimplicity of $\mathcal{A}$ gives $\mathcal{A}_F \cap {}_{LA}\mathcal{A}_F = \{0\}$. Since $\mathcal{I}$ is an ideal we have $Sp_{\mathcal{I}}(b) = Sp_{\mathcal{A}}(b)$. Note that there is a non-zero isolated point in the spectrum of b. Hence Proposition 3.4.1 gives a non-zero idempotent in $\mathcal{I}$. Let E be a maximal family of commuting idempotents in $\mathcal{I}$ and let $\mathcal{C}$ be a maximal commutative subalgebra of $\mathcal{I}$ including E. Since $\mathcal{C}$ is a closed spectral subalgebra of $\mathcal{A}$ we may apply the first portion of this proof to $\mathcal{C}$ and thus obtain a minimal idempotent e in $\mathcal{C}$. We wish to show that e is minimal in $\mathcal{I}$, or equivalently (by Proposition 8.3.5) that $e\mathcal{I}e$ equals $\mathbb{C}e$. If f is an idempotent in $e\mathcal{I}e$ and g is an idempotent in E, then Proposition 8.3.5 gives $ge = ege = \lambda e$ for some $\lambda \in \mathbb{C}$ and hence $gf = gef = \lambda ef = f\lambda e = fg$. Therefore f belongs to E by the maximality of E. The last sentence now shows $f = fe = \lambda e$ for some $\lambda \in \mathbb{C}$. However, $\lambda e = f = f^2 = \lambda^2 e$ shows that λ is zero or one. Therefore the unital algebra $e\mathcal{I}e$ contains no non-zero idempotent except its identity e. Hence Proposition 3.4.1 shows that the spectrum of each $a \in e\mathcal{I}e$ is connected. Since $e\mathcal{I}e$ is a spectral subalgebra of $\mathcal{A}$, each $a \in e\mathcal{I}e$ has countable spectrum. Thus the spectrum of each $a \in e\mathcal{I}e$ is a singleton.

Theorem 4.3.2(a) and Proposition 4.3.12(b) show that $e\mathcal{I}e$ is semisimple. We apply an argument of Irving Kaplansky ([1954], Lemma 4, p. 375) to show that the set $\mathcal{N}$ of elements with spectrum $\{0\}$ is the Jacobson radical of $e\mathcal{I}e$ and hence each element $a \in e\mathcal{I}e$ satisfies $a = \lambda e$ for $\lambda \in Sp(a)$. The remark following Theorem 4.3.6 shows that it is enough to show that $\mathcal{N}$ is a left ideal. If $a \in \mathcal{N}$, $b \in \mathcal{A}$ and $ba \notin \mathcal{N}$ then Proposition 2.1.8 shows $ab \notin \mathcal{N}$. Since the spectrum of these elements is a singleton but not $\{0\}$, $(ba)^{-1}b$ and $b(ab)^{-1}$ are right and left inverses for a. Hence $a \in \mathcal{N}$ implies $ba \in \mathcal{N}$. If $a, b \in \mathcal{N}$ and $a + b \notin \mathcal{N}$ then $a(a+b)^{-1} = 1 - b(a+b)^{-1}$ contradicts our last conclusion that $a(a+b)^{-1}$ and $b(a+b)^{-1}$ both belong to $\mathcal{N}$. Hence $a + b$ belongs to $\mathcal{N}$ as we wished to show.   □

8.6.3 Example   The converse of this result is far from true. Consider for instance $\mathcal{A} = \ell^\infty(I)$ (under pointwise algebra operations) where I is the unit interval (or any other uncountable subset of $\mathbb{C}$). Then $\mathcal{A}_F$ is the set of functions in $\mathcal{A}$ with only finitely many non-zero values so that $\mathcal{A}$ is clearly a commutative semisimple, Duncan modular annihilator algebra. However, the function $f(t) = t$ for all $t \in I$ has uncountable spectrum.

Consider the set $\mathcal{A} = c(I)$ of all $f: I \to \mathbb{C}$ for which there is some $\lambda_f \in \mathbb{C}$ such that the set $\{t \in I : |f(t) - \lambda_f| > \varepsilon\}$ is finite for all $\varepsilon > 0$. Under pointwise multiplication this is a unital commutative semisimple Banach algebra where each element has countable spectrum but $\mathcal{A}$ is not a modular annihilator algebra by Proposition 8.4.14 or Theorem 8.6.4.

If we consider $\mathcal{A} = \ell^p(I)$ for $1 \le p \le \infty$ or $\mathcal{A} = c_0(I) = \{f: I \to \mathbb{C} :$ the set $\{t \in I : |f(t)| > \varepsilon\}$ is finite for all $\varepsilon > 0\}$, each element in these algebras has countable spectrum and they are modular annihilator algebras by Theorem 8.6.4.

Example 8.4.13 shows that we cannot replace $\mathcal{A}/\mathcal{A}_J$ in the conclusion of Theorem 8.6.2 by $\mathcal{A}$.

Property (b) of the next theorem shows that a major difference between modular annihilator algebras and Duncan modular annihilator algebras is that the latter may contain idempotents not in their socle. The equivalence of (a) and (c) below is due to Barnes ([1968b], Theorem 4.2, p. 516) but our proof is much simpler than his. A closely related result is given in Proposition 8.7.8. Example 8.4.13 also shows that this characterization cannot be extended to non-semisimple modular annihilator algebras. Since elements in $\mathcal{A}$ and their images in $\mathcal{A}/\mathcal{A}_J$ have exactly the same spectrum, no spectral condition can characterize non-semisimple modular annihilator algebras.

8.6.4 Theorem *The following are equivalent for a Banach algebra* $\mathcal{A}$.
(a) $\mathcal{A}/\mathcal{A}_J$ *is a modular annihilator algebra.*
(b) *Each element of* $\mathcal{A}$ *has finite or countable spectrum and each idempotent of* $\mathcal{A}/\mathcal{A}_J$ *belongs to the socle of* $\mathcal{A}/\mathcal{A}_J$.
(c) *No element of* $\mathcal{A}$ *has a non-zero limit point for its spectrum.*

In fact any modular annihilator algebra has all its idempotents in $\mathcal{A}_F$ *and if it is normed or spectral it satisfies (c).*

Proof Theorem 8.5.5 shows that if $\mathcal{A}$ or $\mathcal{A}/\mathcal{A}_J$ is a modular annihilator algebra then (c) holds. In particular (a) implies (c). If e is any idempotent in a modular annihilator algebra then $e + \mathcal{A}_F$ is an idempotent in the radical algebra $\mathcal{A}/\mathcal{A}_F$, hence Proposition 4.3.12(a) implies $e \in \mathcal{A}_F$.

For the rest of the argument, Theorem 4.3.6(d) allows us to replace $\mathcal{A}$ by $\mathcal{A}/\mathcal{A}_J$ and thus assume that $\mathcal{A}$ is semisimple.

(c) $\Rightarrow$ (b): Since any uncountable set in $\mathbb{C}$ must have more than one limit point we need only prove the statement about idempotents. Suppose e is a non-zero idempotent in $\mathcal{A}$. Then $e\mathcal{A}e$ is semisimple by Proposition 2.5.3(d). Hence Theorem 8.4.5(e) shows that $e\mathcal{A}e$ contains minimal idempotents which are obviously minimal idempotents in $\mathcal{A}$ since $fe\mathcal{A}ef = f\mathcal{A}f$ holds for any $f \in e\mathcal{A}e$. Let E be a maximal family of orthogonal minimal idempotents in $e\mathcal{A}e$. If E is infinite we may choose a sequence $\{f_n\}_{n\in\mathbb{N}}$ of distinct elements in E. Choose a sequence $\{\lambda_n\}_{n\in\mathbb{N}}$ of distinct numbers converging to zero and satisfying $\sum_{n=1}^{\infty} |\lambda_n| \, \|f_n\| < \infty$. Then the element $e + \sum_{n=1}^{\infty} \lambda_n f_n$ has f_n as a characteristic vector with characteristic value $1 + \lambda_n$ for each $n \in \mathbb{N}$. Since this contradicts (c), we conclude that E is finite, say $E = \{f_j : j = 1, 2, \ldots, n\}$. If $f = e - \sum_{j=1}^{n} f_j$ is non-zero then the argument at the beginning of this paragraph shows that $f\mathcal{A}f$ contains minimal idempotents. Since this would contradict the maximality of E we conclude that $e = \sum_{j=1}^{n} f_j$ belongs to the socle of $\mathcal{A}$.

(b) $\Rightarrow$ (a). (Recall that we are assuming that $\mathcal{A}$ is semisimple.) Let $\mathcal{M}$ be a maximal modular left ideal of $\mathcal{A}$. Let c be a right relative identity for $\mathcal{M}$ so that 1 belongs to $Sp(c)$. The spectral condition in (b) ensures that we may apply Proposition 3.4.1 to obtain an idempotent e commuting with c and satisfying $c \circ d = d \circ c = e$ for some $d \in \mathcal{A}$. Hence we have $\mathcal{A}(1 - e) \subseteq \mathcal{A}(1 - c) \subseteq \mathcal{M}$. Condition (b) shows that e belongs to $\mathcal{A}_F$. Hence we have $\mathcal{A}_F \not\subseteq \mathcal{M}$. Thus $\mathcal{A}$ is a modular annihilator algebra by Proposition 8.4.3 and Theorem 8.4.5(a).   □

Barnes ([1969b], Theorem 3.1, p. 7) deduced the following result as a corollary of the above theorem and of Theorem 8.4.12. If $\mathcal{A}$ is a normed, semisimple, modular annihilator algebra, then $\mathcal{B}\mathcal{B}_J$ is a modular annihilator algebra (where $\mathcal{B}$ is a subalgebra of $\mathcal{A}$ which is a Banach algebra in some norm) if and only if the norm of $\mathcal{A}$ is a spectral norm on $\mathcal{B}$. In particular, a closed subalgebra of a modular annihilator algebra satisfies this condition. The general result follows since the spectral properties proved in this theorem show that the spectrum and the boundary of the spectrum are equal for any $a \in \mathcal{A}$, and hence by Proposition 2.5.10, $\mathcal{B}$ is a spectral subalgebra if and only if the norm of $\mathcal{A}$ is a spectral norm on $\mathcal{B}$. In the same paper Barnes also shows that if $\mathcal{A}$ is a normed, semisimple, modular annihilator algebra and $\mathcal{B}$ is a semisimple subalgebra then $\mathcal{B}_F = \mathcal{A}_F \cap \mathcal{B}$. (A mistake in the proof can be corrected.) We remark that the inclusions $\mathcal{B}_F \subseteq \mathcal{A}_F \cap \mathcal{B}$ and $\mathcal{A}_{MI} \cap \mathcal{B} \subseteq \mathcal{B}_{MI}$ are easy. If $e \in \mathcal{B}$ is idempotent, then it must belong to $\mathcal{A}_F$, since $e + \mathcal{A}_F$ is an idempotent in the radical algebra $\mathcal{A}/\mathcal{A}_F$. The first inclusion follows. On the other hand if $e \in \mathcal{A}_{MI} \cap \mathcal{B}$ holds, then the inclusion $e\mathcal{B}e \subseteq e\mathcal{A}e = \mathbb{C}e$ proves its minimality. See also Aupetit [1986].

8.7 Classes of Algebras with Large Socle

We will now discuss the relationship between various conditions on algebras or topological algebras which ensure that they have a large socle in some sense. All these relations will be summarized in a diagram in Theorem 8.7.11. Counterexamples will be provided at the end of the chapter to show that we have exhausted all of the possible implications. Along the way we will show that many such algebras can be written as topological direct sums of topologically simple ideals of the same type and that each topologically simple algebra has a particularly satisfactory faithful representation as compact operators.

We begin by defining the remaining classes of algebras which have large socles in some sense and which have been studied in detail in the literature.

8.7.1 Definition A topological algebra $\mathcal{A}$ is:

(a) A *dual algebra* if it is semiprime and each closed left ideal $\mathcal{L}$ satisfies $\mathcal{L} = {}_{LRA}\mathcal{L}$ and each closed right ideal $\mathcal{R}$ satisfies $\mathcal{R} = {}_{RLA}\mathcal{R}$.

(b) A *left annihilator algebra* if it is semiprime and the left annihilator of each proper closed right ideal is non-zero.

(c) An *annihilator algebra* if both $\mathcal{A}$ and $\mathcal{A}^R$ are left annihilator algebras.

(d) A *left completely continuous algebra* if it is normed and L_a is a compact operator for each $a \in \mathcal{A}$.

(e) A *completely continuous algebra* if both $\mathcal{A}$ and $\mathcal{A}^R$ are left completely continuous.

(f) A *compact algebra* if it is normed and $L_a R_a$ is a compact operator for each $a \in \mathcal{A}$.

Dual algebras, annihilator algebras and left annihilator algebras were defined by Irving Kaplansky [1948a], Frank F. Bonsall and Alfred W. Goldie [1954], and Malcolm F. Smiley [1955], respectively. In their definitions, they did not require that the algebras be semiprime, but the majority of their results were for semisimple algebras. On the other hand, the definitions of the last two classes referred only to Banach algebras. The concept of dual algebras seems a little too strong since it leaves out important examples without providing much better results. (However, Kaplansky [1948a] shows in Theorem 15 that for any comnpact group G, $L^p(G)$ is a dual algebra under convolution.) The existence of an annihilator Banach algebra which is not dual was first established by Johnson [1967b]. His example was semisimple and commutative.

The definitions of completely continuous, left completely continuous and compact algebras are due to Irving E. Segal [1940] and [1947a], Definition 1.9, p. 80 (cf. Kaplansky [1948b], p. 698), to A. Olubummo [1957] (whose attribution of the definition to Kaplansky we have been unable to verify), and to Klaus Vala [1967] and James C. Alexander [1968], independently. The last definition grew out of an observation of Vala [1964] (*cf.* Example 8.8.2). See also Marianne Freundlich [1949], Bonsall [1969b], Alexander [1969], Kari Ylinen [1968], Yood [1982] and Abdullah H. Al-Moajil [1984].

Note that a normed algebra $\mathcal{A}$ is completely continuous if and only if both L_a and R_a are compact operators for each $a \in \mathcal{A}$. The implications:

dual $\Rightarrow$ annihilator $\Rightarrow$ left annihilator $\Rightarrow$ modular annihilator   and

completely continuous $\Rightarrow$ left completely continuous $\Rightarrow$ compact

are obvious. For any reflexive Banach space $\mathcal{X}$, $\mathcal{B}_F(\mathcal{X})^-$ is a topologically simple semisimple annihilator Banach algebra, but a reflexive Banach space $\mathcal{X}$ was constructed by Alexander M. Davie [1973a] such that $\mathcal{B}_F(\mathcal{X})^-$ is not dual. We will show in Proposition 8.7.4 below that $\mathcal{B}_F(\mathcal{X})^-$ is a topologically simple, semisimple, left annihilator Banach algebra for any Banach space $\mathcal{X}$. Bonsall and Goldie ([1954], Theorem 16, p. 164) showed that $\mathcal{B}_F(\mathcal{X})^-$ is never an annihilator algebra, unless $\mathcal{X}$ is reflexive.

Example 8.8.8, which is due to M. R. F. Smyth [1980], shows that a semisimple left completely continuous Banach algebra need not be completely continuous.

Among the classes of algebras introduced here, annihilator algebras have probably received the most attention. A full presentation of their theory is given in Bonsall and Duncan ([1973b], Section 32). Nevertheless, we believe that modular annihilator algebras and left annihilator algebras (particularly the former) provide more convenient compromises between strong axioms (to give a rich theory) and weak axioms (to give a multitude of examples). Note that, unlike the classes defined here, modular annihilator algebras are defined purely algebraically without reference to any topology.

We begin investigating the relationship among these classes of algebras and the size of their socle by considering certain topological algebras $\mathcal{A}$ in which the socle or the ideal $\mathcal{A}_F + \mathcal{A}_J$ is dense. Result (b) is due to Smiley [1955]. Proposition 8.7.3 contains a partial converse of (b) as well as a structure theorem for topological algebras with dense socle.

8.7.2 Proposition *Let $\mathcal{A}$ be a spectral normed algebra.*
  (a) *If $\mathcal{A}$ is a left annihilator algebra, then $\mathcal{A}_F + \mathcal{A}_J$ is dense.*
  (b) *If $\mathcal{A}$ is a semisimple left annihilator algebra, then $\mathcal{A}_F$ is dense.*
  (c) *If $\mathcal{A}_F + \mathcal{A}_J$ is dense in $\mathcal{A}$ and $\mathcal{A}$ is semiprime, then $\mathcal{A}$ is a modular annihilator algebra.*

Proof (a) Theorems 2.4.7 and 8.4.5(b) show that $\mathcal{A}$ is a modular annihilator algebra. Hence equation (6) of Section 8.4 gives $_{LA}(\mathcal{A}_F + \mathcal{A}_J) = \{0\}$. Since $(\mathcal{A}_F + \mathcal{A}_J)^-$ is a closed (right) ideal with zero left annihilator, it must equal $\mathcal{A}$.

  (b) This follows immediately.

  (c) Let $\mathcal{L}$ be a maximal modular left ideal in $\mathcal{A}$. Then $\mathcal{L}$ is closed (by Theorem 2.4.7) and proper, so it does not include $\mathcal{A}_F + \mathcal{A}_J$. However, it does include $\mathcal{A}_J$ by Theorem 8.3.6(a). Hence it does not include $\mathcal{A}_F$. Thus Proposition 8.4.3 shows $_{RA}\mathcal{L} \neq \{0\}$. Hence $\mathcal{A}$ is a modular annihilator algebra by Theorem 8.4.5(a).   □

The main idea of the next proposition is due to Kaplansky ([1948b], Theorem 5, p. 692).

8.7.3 Proposition *Let $\mathcal{A}$ be a semiprime topological algebra with dense socle. Then $\mathcal{A}$ is semisimple and equal to the normed direct sum of its closed topologically simple ideals. Furthermore, the closed topologically simple ideals are exactly those of the form $\mathcal{A}e\mathcal{A}^-$ for some $e \in \mathcal{A}_{MI}$.*

Proof Inclusion (4) of Section 8.4 shows $\mathcal{A}_J \subseteq {_{LA}}(\mathcal{A}_F)$. Clearly anything which annihilates $\mathcal{A}_F$ annihilates its closure, so this inclusion becomes $\mathcal{A}_J \subseteq {_{LA}}\mathcal{A} = \{0\}$. Hence $\mathcal{A}$ is semisimple.

  Let e be any minimal idempotent in $\mathcal{A}$ and consider the closed ideal $\mathcal{K} = \mathcal{A}e\mathcal{A}^-$ generated by e. First we will show that $\mathcal{K}$ is topologically simple. Theorem 4.3.2(a) shows that $\mathcal{K}$ is semisimple. Let $\mathcal{I}$ be a non-zero closed ideal of $\mathcal{K}$. If $\mathcal{I}e$ is $\{0\}$ and a and b are arbitrary elements of $\mathcal{A}$ then $ae \in \mathcal{K}$ implies $\mathcal{I}aeb = (\mathcal{I}ae)eb \subseteq \mathcal{I}eb = \{0\}$. It follows that $\mathcal{A}e\mathcal{A} = \{\sum_{j=1}^{n} a_j eb_j : n \in \mathbb{N}; a_j, b_j \in \mathcal{A}\}$ satisfies $\mathcal{I}\mathcal{A}e\mathcal{A} = \{0\}$. Since $\mathcal{K}$ is the closure of $\mathcal{A}e\mathcal{A}$ we conclude $\mathcal{I}^2 \subseteq \mathcal{I}\mathcal{K} = \{0\}$, which is a contradiction since $\mathcal{K}$ is semiprime by Theorem 8.4.5(a). Hence $\mathcal{I}e$ is not zero so there is some $ae \in \mathcal{I}e \setminus \{0\}$. Since $\mathcal{K}ae$ is not zero and $\mathcal{K}ae$ is included in the minimal left ideal $\mathcal{A}e$, we have $\mathcal{I} \supseteq \mathcal{K}\mathcal{I} = \mathcal{A}e$. Similarly we find $\mathcal{I} \supseteq e\mathcal{A}$. Hence the closed ideal $\mathcal{I}$ includes $\mathcal{A}e\mathcal{A}$ and hence $\mathcal{K}$. Thus $\mathcal{K}$ is topologically simple.

Conversely, suppose $\mathcal{K}$ is any topologically simple ideal in $\mathcal{A}$. Then $\mathcal{K}\mathcal{A}_F$ is not $\{0\}$ since this would imply $\mathcal{K}^2 \subseteq \mathcal{K}\mathcal{A} = \mathcal{K}(\mathcal{A}_F)^- = \{0\}$. Hence $\mathcal{K} \cap \mathcal{A}_F$ is a non-zero ideal in the modular annihilator algebra $\mathcal{A}_F$ and thus contains a minimal idempotent e by Theorem 8.4.5(h). Thus the inclusion $\mathcal{A}e\mathcal{A}^- \subseteq \mathcal{K}$ together with the topologically simple nature of $\mathcal{K}$ gives $\mathcal{K} = \mathcal{A}e\mathcal{A}^-$.

Let A be a subset of $\mathcal{A}_{MI}$ so that each ideal of the form $\mathcal{A}e\mathcal{A}^-$ with $e \in \mathcal{A}_{MI}$ can be expressed in the same form for a unique $e \in A$. Clearly $(\sum_{e \in A}(\mathcal{A}e\mathcal{A})^-)^-$ includes and hence equals $\mathcal{A}_F = \mathcal{A}$. Hence $\mathcal{A}$ equals the topological sum (*i.e.*, the closure of the sum) of its topologically simple closed ideals. However, this sum must be topologically direct. This can be seen as follows. Suppose $\{\mathcal{I}_\alpha : \alpha \in A\}$ is the set of all distinct topologically simple ideals in $\mathcal{A}$. Then $\mathcal{I}_\alpha \mathcal{I}_\beta = \mathcal{I}_\alpha \cap \mathcal{I}_\beta$ is $\{0\}$ for each $\alpha \neq \beta$. Suppose there is a subset $B \subseteq A$ and an element $\alpha \in A \setminus B$ such that $\mathcal{I}_\alpha \cap (\sum_{\beta \in B} \mathcal{I}_\beta)^-$ is not $\{0\}$. Then this intersection must equal $\mathcal{I}_\alpha$, and since $\mathcal{I}_\alpha \sum_{\beta \in B} \mathcal{I}_\beta = \{0\}$ holds by the previous remark, we have $\mathcal{I}_\alpha^2 = \{0\}$, contrary to assumption. (Note that we have shown that any sum of distinct topologically simple ideals in a semiprime topological algebra is a topological direct sum.) □

In applying the last proposition to semisimple dual, annihilator, or left annihilator algebras, it is useful to know that the topologically simple direct summands are again of the same type. Any closed ideal in a dual topological algebra is again a dual algebra as shown by Kaplansky ([1948b], Theorem 2, p. 690). It is shown in Bonsall and Duncan ([1973b], Theorem 9, p. 163) that a closed ideal $\mathcal{I}$ in a semiprime annihilator Banach algebra $\mathcal{A}$ is an annihilator algebra if it satisfies $(\mathcal{I}\mathcal{A})^- = \mathcal{A}\mathcal{I}^- = \mathcal{I}$. Since this is obviously true of topologically simple ideals in a semiprime algebra, the result holds here also. For left annihilator algebras the result is included in Smiley [1955]. We remark that Bonsall and Duncan ([1973b], Proposition 15, p. 165) also contains a condition under which the quotient of an annihilator Banach algebra is again of the same type.

In Proposition 8.7.4 we will use the approximate $B^\#$-condition introduced in Definition 2.3.9 above to characterize those topologically simple direct summands which occur in Proposition 8.7.3.

8.7.4 Proposition *The following are equivalent for a left annihilator Banach algebra $\mathcal{A}$.*

(a) *$\mathcal{A}$ is primitive.*

(b) *$\mathcal{A}$ is topologically simple and semisimple.*

(c) *There is a faithful, contractive, strictly dense representation T of $\mathcal{A}$ on some Banach space $\mathcal{X}$ satisfying $\mathcal{B}_F(\mathcal{X}) \subseteq T_\mathcal{A} \subseteq \mathcal{B}_K(\mathcal{X})$ and such that the preimage of $\mathcal{B}_F(\mathcal{X})$ under T is dense in $\mathcal{A}$.*

Moreover the norm of $\mathcal{A}$ satisfies the approximate $B^\#$ condition if and only if T is an isometry.

Proof (b) $\Rightarrow$ (a). Since $\mathcal{A}$ is semisimple, it has at least one primitive ideal, which must be $\{0\}$ by topological simplicity.

(c) $\Rightarrow$ (a). This follows immediately since we have already noted that $\mathcal{B}_F(\mathcal{X})$ is irreducible on $\mathcal{X}$.

As noted above $\mathcal{A}$ is a modular annihilator algebra.

(a) $\Rightarrow$ (b). The socle of $\mathcal{A}$ is dense, since otherwise $\mathcal{A}_J = {}_{RA}\mathcal{A}_F$ is non-zero. Theorem 8.4.5(d) shows that there is some minimal idempotent e satisfying $\{0\} = {}_{LA}\mathcal{A}e$. Let $\mathcal{I}$ be a non-zero closed ideal of $\mathcal{A}$. Then for each non-zero $b \in \mathcal{I}$ there is some $ae \in \mathcal{A}e$ with $bae \neq 0$. Hence $\mathcal{I}e$ is a non-zero ideal included in the minimal ideal $\mathcal{A}e$. Therefore $\mathcal{A}e = \mathcal{I}e \subseteq \mathcal{I}$ implies $e \in \mathcal{I}$. Therefore $\mathcal{I}$ includes $\mathcal{A}e\mathcal{A}^-$ which is topologically simple by Proposition 8.7.3. If $\mathcal{A}e\mathcal{A}^-$ includes $\mathcal{A}_{MI}$ it includes $\mathcal{A}_F$ and hence includes $\mathcal{A}_F^- = \mathcal{A}$ and we are done. Otherwise there is some $f \in \mathcal{A}_{MI}$ satisfying $f \notin \mathcal{A}e\mathcal{A}^-$ and hence $\mathcal{A}f\mathcal{A}^- \cap \mathcal{A}e\mathcal{A}^- = \{0\}$. However, taking $\mathcal{I} = \mathcal{A}f\mathcal{A}^-$ in the previous argument we have the contradiction $e \in \mathcal{A}f\mathcal{A}^-$.

(a) $\Rightarrow$ (c) Theorem 8.4.5(e) shows that $\mathcal{A}$ has a faithful representation L^e for some $e \in \mathcal{A}_{MI}$. Theorem 8.3.6 shows the properties of L^e. Proposition 8.7.2(b) shows that $\mathcal{A}_F$ is dense in $\mathcal{A}$. Since L^e is contractive, $L^e_{\mathcal{A}_F} \subseteq \mathcal{B}_F(\mathcal{A}e)$ is dense in $L^e_\mathcal{A}$. This implies $L^e_\mathcal{A} \subseteq \mathcal{B}_F(\mathcal{A}e)^- \subseteq \mathcal{B}_K(\mathcal{A}e)$.

Next we show that all of $\mathcal{B}_F(\mathcal{A}e)$ is included in $L^e_\mathcal{A}$. Suppose S is a rank one operator in $\mathcal{B}(\mathcal{A}e)$ and suppose ${}_{LA}\ker(S)$ is $\{0\}$. This implies successively, ${}_{LA}(\ker(S)\mathcal{A}) = \{0\}$, $(\ker(S)\mathcal{A})^- = \mathcal{A}$, and $\mathcal{A}e = (\ker(S)\mathcal{A})^-e = \ker(S)e\mathcal{A}e^- = \ker(S)e^- = \ker(S)$. Since this is impossible, we may choose $b \neq 0$ which satisfies $b\ker(S) = \{0\}$. Let $ce \in \mathcal{A}e$ satisfy $S(ce) \neq 0$. Then we have $\mathcal{A}e = \ker(S) \oplus \mathbb{C}ce$ and hence $bce \neq 0$ since otherwise we would have $L^e_b = 0$. Now since L^e is irreducible we can find $d \in \mathcal{A}$ satisfying $L^e_d bce = S(ce)$. Hence S equals L^e_{db}, so $L^e_\mathcal{A}$ includes all rank one operators. Hence it includes all finite-rank operators.

If T is an isometry then $\mathcal{A}$ satisfies the approximate $B^\#$-condition, since we have already noted that $T_\mathcal{A}$ does. Now suppose $\mathcal{A}$ satisfies the approximate $B^\#$-condition. Since $a \mapsto \|T_a\|$ is an algebra norm on $\mathcal{A}$ satisfying $\|T_a\| \leq \|a\|$ for all $a \in \mathcal{A}$ and a spectral norm by Proposition 8.4.10, Proposition 8.7.5 shows

$$\|T_a\| = \|a\| \qquad \forall\, a \in \mathcal{A}. \qquad \square$$

The following easy result is essentially due to Kaplansky ([1948b], p. 698).

8.7.5 Proposition *Let $\mathcal{A}$ be a normed algebra satisfying one of the last three properties of* Definition 8.7.1. *Let $\mathcal{B}$ be a closed subalgebra and let $\mathcal{I}$ be a closed ideal of $\mathcal{A}$. Then $\mathcal{B}$ and $\mathcal{A}/\mathcal{I}$ satisfy the same property as $\mathcal{A}$.*

Proof The case of closed subalgebras is obvious. So is the case of quotient algebras after noting the equality $\varphi \circ L_a = L_{a+\mathcal{I}} \circ \varphi$, where $\varphi \colon \mathcal{A} \to \mathcal{A}/\mathcal{I}$ is the natural map.   □

The next result is due to Alexander [1968] (Theorem 7.3, p. 15) and Barnes for Banach algebras.

8.7.6 Proposition *Let $\mathcal{A}$ be a semiprime spectral normed algebra with dense socle. Then $\mathcal{A}$ is a semisimple modular annihilator algebra and if T is any normed topologically irreducible representation of $\mathcal{A}$ on a normed linear space $\mathcal{X}$, then there is a dense T-invariant linear subspace $\mathcal{Y}$ of $\mathcal{X}$ such that the restriction of T to $\mathcal{Y}$ is algebraically invariant and has the same kernel as T.*

If $\mathcal{A}$ is a Banach algebra, then it is compact.

Proof Since the socle is dense, its annihilator is zero and hence $\mathcal{A}$ is semisimple by inclusion (4) of Section 8.4. (Note that the norm need not be spectral in this argument.) Since primitive ideals are closed in a spectral norm, none can contain the dense socle of $\mathcal{A}$. Theorem 8.4.5(f) implies that $\mathcal{A}$ is a modular annihilator algebra and Proposition 8.4.10 implies the condition on T since the primitive ideal $\ker(T)$ does not include $\mathcal{A}_F$.

Next we will show that if a belongs to the socle of a semiprime Banach algebra $\mathcal{A}$, then $L_a R_a$ has finite-dimensional range. (The converse is proved by Alexander in [1968], Theorem 7.2, p. 14.) We can write $a = \sum_{j=1}^n a_j e_j$, where $a_j \in \mathcal{A}$ and $e_j \in \mathcal{A}_{MI}$. Hence Theorem 8.3.6 shows that $a\mathcal{A}a \subseteq \sum_{j,k=1}^n a_j e_j \mathcal{A} a_k e_k \subseteq \sum_{j,k=1}^n \mathbb{C}a_j c_{jk}$ for any non-zero $c_{jk} \in e_j \mathcal{A} e_k$. Thus $L_a R_a$ has finite-dimensional range so it is compact.

The map $a \mapsto L_a R_a$ (although nonlinear) is obviously continuous. Hence if $\mathcal{A}_F$ is dense in $\mathcal{A}$ and $L_a R_a$ is compact for each $a \in \mathcal{A}_F$, then $L_a R_a$ is compact for each $a \in \mathcal{A}$.   □

8.7.7 Corollary *If $\mathcal{A}$ is a semiprime Banach algebra with $\mathcal{A}_F \oplus \mathcal{A}_J$ dense and $\mathcal{A}_J$ compact then $\mathcal{A}$ is compact.*

Proof If a belongs to $\mathcal{A}_F + \mathcal{A}_J$, we can write $a = c + \sum_{j=1}^n a_j e_j$, with $c \in \mathcal{A}_J$, $a_j \in \mathcal{A}$, and $e_j \in \mathcal{A}_{MI}$. Then for any $b \in \mathcal{A}$, we have $L_a R_a(b) = cbc + \sum_{j=1}^n cba_j e_j + \sum_{j=1}^n a_j e_j bc + \sum_{j,k=1}^n a_j e_j ba_k e_k$. Inclusion (4) of Section 8.4 shows that the second and third sums are zero. Hence if $\mathcal{A}_J$ is compact, the argument in the previous proof shows that $\mathcal{A}$ is compact.   □

Recall that a bounded operator T on a normed linear space $\mathcal{X}$ is called a *Riesz operator* if its image in $\mathcal{B}(\mathcal{X})/\mathcal{B}_K(\mathcal{X})$ is topologically nilpotent.

(These operators were briefly discussed in Section 2.8.) The next result is due to Barnes [1971a]. Note that both the necessity and the sufficiency of the first condition are proved under (different) weaker hypotheses.

8.7.8 Proposition *A semisimple Banach algebra is a modular annihilator algebra if and only if $L_a R_a$ is a Riesz operator for each $a \in \mathcal{A}$. Hence for any compact Banach algebra $\mathcal{A}$, $\mathcal{A}/\mathcal{A}_J$ is a modular annihilator algebra.*

Proof If $\mathcal{A}$ is a compact Banach algebra, Proposition 8.7.5 shows that $\mathcal{A}/\mathcal{A}_J$ is a compact semisimple Banach algebra. Hence the last statement is an immediate consequence of the first.

Suppose $L_a R_a$ is a Riesz operator and $\mathcal{C}$ is the commutant of a in $\mathcal{A}$. If $\lambda \neq 0$ belongs to the boundary of $Sp_{\mathcal{C}}(a^2)$, then Theorem 2.5.7 shows that there is a sequence $\{b_n\}_{n \in \mathbb{N}}$ with $\|b_n\| = 1$ for all n and $\{\lambda^{-1} a^2 b_n - b_n\}_{n \in \mathbb{N}}$ converging to zero. Hence $Sp(L_a R_a)$ contains λ since $\{(\lambda - L_a R_a) b_n\}_{n \in \mathbb{N}}$ converges to zero. Also $\partial Sp_{\mathcal{A}}(a^2)$ equals $\partial Sp_{\mathcal{C}}(a^2)$ by Proposition 2.5.3. So $Sp(a^2)$ is included in $Sp(L_a R_a)$ which has no limit points except zero as shown in Section 2.8. Thus the same is true of $Sp(a)$ since $Sp(a^2) = Sp(a)^2$ holds. Theorem 8.6.4 shows that $\mathcal{A}$ is a modular annihilator algebra.

Conversely, suppose $\mathcal{A}$ is a semisimple modular annihilator Banach algebra. Then $\mathcal{A}/\mathcal{A}_F^-$ is a radical algebra by Theorems 8.4.5(c) and 4.3.2(a). Hence Corollary 4.3.8 shows that for any $a \in \mathcal{A}$ and any $n \in \mathbb{N}$ there is an element $b_n \in \mathcal{A}_F$ and an $\varepsilon_n > 0$ satisfying

$$\|a^n - b_n\|^{1/n} = \varepsilon_n \to 0.$$

Any $c \in \mathcal{A}$ satisfies

$$
\begin{aligned}
\|((L_a R_a)^n - L_{b_n} R_{b_n})c\| \ &= \ \|a^n c a^n - b_n c b_n\| \\
&\leq \ \|a^n c a^n - b_n c a^n\| + \|b_n c a^n - b_n c b_n\| \\
&\leq \ \|a^n - b_n\|(\|a^n\| + \|b_n\|)\|c\| \qquad \text{and hence} \\
\|(L_a R_a)^n - L_{b_n} R_{b_n}\|^{1/n} \ &\leq \ \varepsilon_n(\|a^n\| + \|b_n\|)^{1/n} \leq \varepsilon_n(2\|a^n\| + \varepsilon_n^n)^{1/n}.
\end{aligned}
$$

The last sequence converges to zero. The proof of the last propositon shows that $L_{b_n} R_{b_n}$ belongs to $\mathcal{B}_F(\mathcal{A}) \subseteq \mathcal{B}_K(\mathcal{A})$, so $L_a R_a$ is a Riesz operator. □

The next proposition is due to Barnes ([1966], Theorem 7.2) for semisimple algebras. The last remark is due to Kaplansky ([1948b], p. 698).

8.7.9 Proposition *Let $\mathcal{A}$ be a semiprime left completely continuous normed algebra. Then $\mathcal{A}$ is a modular annihilator algebra and all its irreducible representations are finite-dimensional.*

Proof Since $\mathcal{A}$ is semiprime, $L \colon \mathcal{A} \to \mathcal{B}_K(\mathcal{A}) \subseteq \mathcal{B}(\mathcal{A})$ is faithful and Proposition 2.5.3 shows $Sp_{\mathcal{A}}(a) = Sp_{\mathcal{L}(\mathcal{A})}(L_a)$. In order to prove that $\mathcal{A}$ is a modular annihilator algebra, Theorem 8.4.5(c) and 4.3.2(a) show

that it is enough to prove that there is no maximal modular left ideal including $\mathcal{A}_F$. Suppose $\mathcal{M}$ is an exception. Choose $c \in \mathcal{A}$ satisfying $\mathcal{A}(1 - c) \subseteq \mathcal{M}$. Then 1 belongs to $Sp_\mathcal{A}(c) = Sp(L_c)$. Since L_c is compact, $\ker(I - L_c) = {}_{RA}\mathcal{A}(1 - c)$ is finite-dimensional and the ascent of $I - L_c$ is finite by the results of Section 2.8. If m denotes this ascent, then
$$\mathcal{R} = {}_{RA}(\mathcal{A}(1 - c)^m) = \ker((I - L_c)^m) = \ker((I - L_c)^{2m}) = {}_{RA}(\mathcal{A}(1 - c)^{2m})$$
is a finite-dimensional right ideal.

We wish to find an idempotent $e \in \mathcal{A}_F$ satisfying $e\mathcal{A} = \mathcal{R}$. Since $\mathcal{R}$ is finite-dimensional, it includes a minimal right ideal of $\mathcal{A}$ which contains a minimal idempotent by Lemma 8.5.4. Let e be the sum of a maximal (necessarily finite) family of orthogonal minimal idempotents in $\mathcal{R}$. Obviously e belongs to the socle of $\mathcal{A}$. Suppose $e\mathcal{A} \subseteq \mathcal{R}$ does not equal $\mathcal{R}$. Then we may choose $a \in \mathcal{R} \setminus e\mathcal{A}$ and consider $b = a - ea \neq 0$. Since $b\mathcal{A} \subseteq \mathcal{R}$ is non-zero and finite-dimensional it includes a minimal idempotent f, by the above argument. The minimality of f and the equation $ef = 0$ (which follows from $f \in b\mathcal{A}$), show that $f \in \mathcal{R}$ is orthogonal to all the idempotents chosen previously with sum e. This contradiction establishes $\mathcal{R} = e\mathcal{A}$.

Since we are assuming $\mathcal{A}_F \subseteq \mathcal{M}$, we have $\mathcal{A}(1 - (d+e)) \subseteq \mathcal{M}$, where $d \in \mathcal{A}$ is defined by $1 - d = (1-c)^m$. As before, the fact that L^1_{d+e} is compact and that 1 belongs to its spectrum implies that ${}_{RA}\mathcal{A}(1 - (d+e)) = \ker(I - L^1_{d+e})$ is non-zero. Let b be a non-zero element in this ideal so that it satisfies $(1 - d)b = eb \in {}_{RA}(\mathcal{A}(1 - c)^m)$. Then we have $\mathcal{A}(1 - d)^2 b = \mathcal{A}(1 - d)eb = 0$ which implies $b \in {}_{RA}(\mathcal{A}(1 - d)^2) = {}_{RA}(\mathcal{A}(1 - c)^{2m}) = {}_{RA}(\mathcal{A}(1 - d))$. This implies $eb = (1 - d)b = 0$. Since b was chosen from $e\mathcal{A}$ this implies $b = 0$, contrary to its choice. Hence we conclude that no maximal modular left ideal includes $\mathcal{A}_F$ so that $\mathcal{A}$ is a modular annihilator algebra by Theorem 8.4.5(a) and Proposition 8.4.3.

Theorem 8.4.5(d) now shows that any irreducible representation has a kernel of the form $\mathcal{P}^e$ for some $e \in \mathcal{A}_{MI}$. Proposition 8.4.4 shows that $\mathcal{P}^e = \{a \in \mathcal{A} : \mathcal{A}a \subseteq (1 - e)\mathcal{A}\}$ is the kernel of the anti-representation R^e.

Since L_e is compact by hypothesis, and its restriction to $e\mathcal{A}$ is the identity operator on $e\mathcal{A}$, $e\mathcal{A}$ must be finite dimensional. Hence $\mathcal{A}/\mathcal{P}^e$ is also. Thus any irreducible representation is finite-dimensional. $\square$

8.7.10 Theorem *Let $\mathcal{A}$ be a compact Banach algebra. Every irreducible representation of $\mathcal{A}$ is equivalent to a representation on a Banach space in which each representing operator is compact.*

Proof Since we are only interested in irreducible representations we may replace $\mathcal{A}$ by $\mathcal{A}/\mathcal{A}_J$ and hence assume that $\mathcal{A}$ is semisimple. Proposition 8.7.10 asserts that $\mathcal{A}$ is a modular annihilator algebra. Theorem 8.4.5(e) shows that each irreducible representation is equivalent to L^e for some $e \in \mathcal{A}_{MI}$. Theorem 8.3.6 gives the structure of these representations and we will use the notation defined there.

Choose and fix an arbitrary $e \in \mathcal{A}_{MI}$ and an arbitrary $a \in \mathcal{A}$ for which L_a^e is non-zero. We will show that L_a^e is compact.

Any b, c, $d \in \mathcal{A}$ satisfy

$$\|bec\| \, \|de\| \geq \|becde\| = \|b\langle de, ec\rangle e\| = |\langle de, ec\rangle| \, \|be\|.$$

Hence any b, $c \in \mathcal{A}$ satisfy

$$\|bec\| \geq \|be\| sup\{|\langle de, ec\rangle| : de \in (\mathcal{A}e)_1\}.$$

Since L_a^e is non-zero, we may choose $d \in \mathcal{A}$ satisfying $ade \neq 0$. Since $\langle \cdot, \cdot \rangle$ is a pairing we may choose $c \in \mathcal{A}$ satisfying $0 \neq \langle ade, ec\rangle e = ecade = \langle de, eca\rangle e$. Denote $|\langle de, eca\rangle|/\|de\|$ by δ. The inequality above shows that any $b \in \mathcal{A}$ satisfies

$$\|beca\| \geq \delta\|be\|.$$

Now let $\{b_n e\}_{n \in \mathbb{N}} \subseteq (\mathcal{A}e)_1$ be arbitrary. Since $\mathcal{A}$ is compact there is a subsequence $\{ab_{n_k}eca\}_{k \in \mathbb{N}}$ of $L_a R_a(b_n ec) = ab_n eca$ which converges. The last inequality shows

$$\|L_a^e b_{n_k} e - L_a^e b_{n_j} e\| \leq \delta^{-1}\|ab_{n_k}eca - ab_{n_j}eca\|$$

which in turn shows that a subsequence of $\{L_a^e b_n e\}_{n \in \mathbb{N}}$ converges. Hence L_a^e is compact as we wished to show. □

We now summarize the major results of this chapter.

8.7.11 Theorem *The diagram on the next page summarizes known implications between the various classes of algebras studied in this chapter. Arrows indicate implications and letters beside the arrows indicate extra conditions which are required in the proof. Some of these extra conditions may be unnecessary but no other implications, beyond those shown here, can hold between these classes of algebras for semisimple Banach algebras.*

Proof Results not proved in Propositions 8.7.2, 8.7.6, 8.7.8, 8.7.9, or Theorem 8.7.10 or in the discussions surrounding the various definitions are all trivial. Example 8.8.8 or Barnes ([1966], Theorem 7.1) show that even for Banach algebras with dense socle the implication LCC⇒CC fails.

When $\mathcal{A}$ is a Banach algebra, there are two ways to define the condition KIR given in the diagram. First, any irreducible representation of the algebra $\mathcal{A}$ is equivalent to a continuous representation on a Banach space such that $T_{\mathcal{A}}$ consists of compact operators. Second, if T is a strongly continuous representation of the algebra $\mathcal{A}$ on a Banach space then $T_{\mathcal{A}}$ consists of compact operators. Theorem 4.2.8 shows the equivalence of these two definitions. □

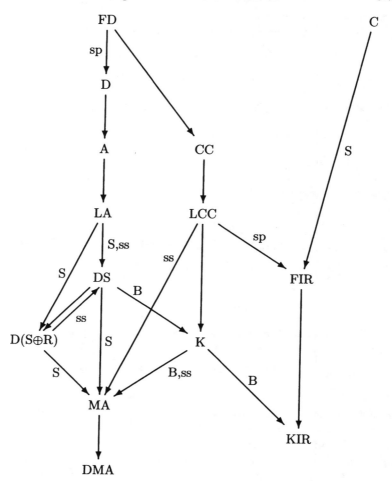

FD = finite-dimensional
D = dual
A = annihilator
LA = left annihilator
DS = semiprime with dense socle
MA = modular annihilator

C = commutative
CC = completely continuous
LCC = left completely continuous
K = compact
DMA = Duncan modular annihilator

D(S⊕R) = semiprime with (socle ⊕ radical) dense
FIR = all irreducible representations are finite-dimensional
KIR = all irreducible representations are equivalent to representations on
      Banach spaces in which each representing operator is compact.
Abbreviations for the conditions are: sp = semiprime; ss = semisimple;
S = spectral normed; B = complete normed.

Annihilator Algebras

We wish to include a few results on the much studied class of annihilator Banach algebras. Let $\mathcal{A}$ be such an algebra. Since it is a left annihilator algebra, Proposition 8.7.2 shows that $\mathcal{A}_F + \mathcal{A}_J$ is dense. Hence the socle $\mathcal{A}_F$ is dense if $\mathcal{A}$ is semisimple. In that case Proposition 8.7.3 shows that $\mathcal{A}$ is the normed direct sum of its closed topologically simple ideals, all of which are of the form $(\mathcal{A}e\mathcal{A})^-$ for some minimal idempotent e. Proposition 8.7.4 gives more information on these ideals. None of the above results require more than that the Banach algebra be a left annihilator algebra. The next theorem, which improves on Bonsall and Duncan ([1973b] 32.20), captures the difference between annihilator algebras and left annihilator algebras. We have already noted that $\mathcal{B}_A(\mathcal{X})$ is a left annihilator algebra for any Banach space $\mathcal{X}$, but this theorem shows that it is an annihilator algebra if and only if $\mathcal{X}$ is reflexive. For another approach to this theory using *single elements* (an abstraction of finite-rank operators), see John A. Erdos [1971] and S. Giotopoulos, S. [1985], *cf.* Joachim Puhl [1978].

8.7.12 Theorem *Let $\mathcal{A}$ be a semiprime annihilator Banach algebra and let e be a minimal idempotent. Then the pairing of Theorem 8.3.6 establishes homeomorphic linear isomorphisms of $e\mathcal{A}$ onto $(\mathcal{A}e)^*$ and of $\mathcal{A}e$ onto $(e\mathcal{A})^*$ when $e\mathcal{A}$ and $\mathcal{A}e$ have the complete norms they inherit as closed linear subspaces of $\mathcal{A}$. Hence $e\mathcal{A}$ and $\mathcal{A}e$ are both reflexive Banach spaces and the linear functional of Proposition 8.3.5 is autoperiodic.*

Proof By symmetry it is enough to establish the homeomorphic linear isomorphism of $e\mathcal{A}$ onto $(\mathcal{A}e)^*$. Recall from Proposition 8.3.5 and Theorem 8.3.6 that there is a linear functional $\omega \in \mathcal{A}^*$ satisfying $\omega(a)e = eae$ for all $a \in \mathcal{A}$ and a pairing between $\mathcal{A}e$ and $e\mathcal{A}$ satisfying $\langle a, b \rangle e = ba = \omega(ba)e$ for all $a \in \mathcal{A}e$ and all $b \in e\mathcal{A}$. Hence the map $b \in e\mathcal{A} \mapsto \omega_b|_{\mathcal{A}e}$ is a continuous linear injection of $e\mathcal{A}$ into $(\mathcal{A}e)^*$. (As usual, ω_b is defined by $\omega_b(a) = \omega(ba)$ for all $a \in \mathcal{A}$.) It is enough to show that this map is surjective, since the inverse boundedness theorem will then establish that it is a homeomorphism. Let τ be an arbitrary non-zero element of $(\mathcal{A}e)^*$. Define $\overline{\tau} \in \mathcal{A}^*$ by $\overline{\tau}(a) = \tau(ae)$ for all $a \in \mathcal{A}$. In the notation of Definition 4.2.20 we have ${}_{\overline{\tau}}\mathcal{A} = \{b \in \mathcal{A} : \overline{\tau}(ba) = 0 \ \forall \ a \in \mathcal{A}\}$. We claim $\ker(\tau) = \mathcal{A}e \cap {}_{\overline{\tau}}\mathcal{A}$. (If $b = be$ belongs to the kernel of τ then for all $a \in \mathcal{A}$, we have $\overline{\tau}(ba) = \tau(beae) = \tau(b\omega(a)e) = \omega(a)\tau(b) = 0$. For the opposite inclusion, note that $b \in \mathcal{A}e$ implies $\tau(b) = \overline{\tau}(b) = \overline{\tau}(be)$ which gives $\mathcal{A}e \cap {}_{\overline{\tau}}\mathcal{A} \subseteq \ker(\tau)$.) Since τ is non-zero, we may find $c \in \mathcal{A}e$ satisfying $\tau(c) = 1$.

We now switch attention to the topologically simple Banach algebra $\mathcal{B} = (\mathcal{A}e\mathcal{A})^-$ and its right ideal $\mathcal{R} = {}_{\overline{\tau}}\mathcal{A} \cap \mathcal{B}$. We use Theorem 32.9 of Bonsall and Duncan [1973b] to establish that $\mathcal{B}$ is an annihilator algebra. Note that $\mathcal{R}$ is a proper ideal of $\mathcal{B}$, since it does not contain $c \in \mathcal{A}e \subseteq \mathcal{B}$. Hence the annihilator algebra property shows that there is a non-zero element d

of $\mathcal{B}$ in $_{LA}\mathcal{R}$. However, the proper two sided ideal $_{LA}\mathcal{A}e$ of $\mathcal{B}$ is $\{0\}$ by topological simplicity. For any $a \in \mathcal{A}e$, $a - \tau(a)c$ belongs to $\ker(\tau) \subseteq \mathcal{R}$ so we conclude $d(a - \tau(a)c) = 0$. Since $d\mathcal{A}e$ is not $\{0\}$, $dc \in \mathcal{A}e$ is not zero. Using the fact that we have a pairing again, we find $b \in e\mathcal{A}$ satisfying $bdc = e$. Hence any $a \in \mathcal{A}e$ satisfies $\tau(a)e = \tau(a)bdc$ so $\tau = \omega_{bd}|_{e\mathcal{A}}$ as we wished to show. Finally, ω is autoperiodic since Theorem 8.3.6 shows that $\mathcal{A}e$ and $e\mathcal{A}$ are homeomorphically isomorphic to the left and right representation spaces defined by ω.   □

8.7.13 Corollary *The representation space of any irreducible representation of a semiprime annihilator Banach algebra can be given the sturcture of a reflexive Banach space, relative to which the representation is continuous. If the representation is already strongly continuous on a Banach space, the Banach space is reflexive.*

Proof Theorem 8.4.5 shows that every irreducible representation is equivalent to the representation L^e on the Banach space $\mathcal{A}e$ for some minimal idempotent e. Theorem 8.7.12 and Corollary 4.2.4 complete the proof.   □

8.7.14 Corollary *A semisimple annihilator Banach algebra is an ideal in its double dual with respect to either Arens product.*

Proof By Proposition 1.4.13 we must show that the left and right regular representations L_a and R_a are weakly compact for each a in a semisimple annihilator algebra $\mathcal{A}$. The theorem shows this for any minimal idempotent $a = e$ since $\mathcal{A}e$ and $e\mathcal{A}$ are reflexive. However, $\mathcal{A}$ is the closed ideal generated by its minimal idempotents. Since L and R are continuous and the weakly compact operators form a closed ideal in $\mathcal{B}(\mathcal{A})$, we get the desired result.   □

Automatic Continuity

We close this section with an automatic continuity result too specialized to be given in Chapter 6. The result is due to Barnes [1967b]. The reader may wish to compare this result with Proposition 6.1.13 and Corollary 6.1.14. The set $\mathcal{B}_\varphi$ was introduced in Definition 6.1.7.

8.7.15 Theorem *Let $\mathcal{A}$ and $\mathcal{B}$ be Banach algebras and let $\varphi: \mathcal{A} \to \mathcal{B}$ be a homomorphism with dense range. Then $_{LA}(\mathcal{B}_F)$ includes $\mathcal{B}_\varphi$. Hence if $_{LA}(\mathcal{B}_F)$ vanishes (i.e., if $\mathcal{B}$ is a semisimple Duncan modular annihilator algebra) then φ is continuous. In particular a normed representation T of $\mathcal{A}$ on a normed linear space $\mathcal{X}$ is continuous if $T_\mathcal{A} \cap \mathcal{B}_F(\mathcal{X})$ is dense in $T_\mathcal{A}$.*

Proof Proposition 6.1.8(d) shows that $\mathcal{B}_\varphi$ is a closed ideal of $\mathcal{B}$. For any minimal left ideal $\mathcal{L}$ of $\mathcal{B}$, $\mathcal{B}_\varphi\mathcal{L} \subseteq \mathcal{L}$ equals either $\mathcal{L}$ or $\{0\}$. The former case is impossible since $\mathcal{L} = \mathcal{B}_\varphi\mathcal{L} \subseteq \mathcal{B}_\varphi$ contains a non-zero idempotent by

Corollary 8.2.3, but $\mathcal{B}_\varphi$ contains none by Proposition 6.1.10. Hence $\mathcal{B}_\varphi$ is included in $_{LA}(\mathcal{B}_F)$.

If $_{LA}(\mathcal{B}_F)$ equals $\{0\}$ then φ is continuous by Proposition 6.1.9(a). The last remark follows from the previous one by noting that $\mathcal{X}$ may be completed without altering the hypothesis.　　　　　　　　　　　　　　　　□

We mention two other results on automatic continuity of homomorphisms which involve minimal ideals. The first result is contained in Bachelis [1972]. Let $\mathcal{A}$ and $\mathcal{B}$ be Banach algebras with $\mathcal{A}$ semisimple and let $\varphi\colon \mathcal{A} \to \mathcal{B}$ be a homomorphism. Then φ is continuous on each minimal ideal of $\mathcal{A}$. Furthermore there is a constant K satisfying

$$\|\varphi(ab)\| \leq K\|a\|\,\|b\| \qquad \forall\ a \in \mathcal{A}_F; \quad b \in \mathcal{A}_F^-.$$

From this it is shown that if $\mathcal{A}_F^-$ has a bounded approximate identity, then φ is continuous on $\mathcal{A}_F$. Note that this result puts no restrictions on $\mathcal{B}$ or on $\varphi(\mathcal{A})$.

The last result we mention is from Barnes [1967b]. Let $\mathcal{A}$ be a Banach algebra such that any closed ideal $\mathcal{I}$ of $\mathcal{A}$ with $\mathcal{A}/\mathcal{I}$ finite-dimensional satisfies $\mathcal{I}^2 = \mathcal{I}$. (He shows that C*-algebras and dual algebras with bounded approximate identities (such as $L^1(G)$ for G a compact group) satisfy this condition.) Let $\mathcal{B}$ be a strongly semisimple modular annihilator algebra (e. g., a semisimple completely continuous algebra). Then any homomorphism $\varphi\colon \mathcal{A} \to \mathcal{B}$ is continuous without any restriction on the range of φ.

8.8　Examples

8.8.1 Example The argument given before Theorem 8.6.2 shows that $\mathcal{B}(\mathcal{X})$ is a semisimple (indeed primitive) Duncan modular annihilator normed algebra or Banach algebra, respectively, whenever $\mathcal{X}$ is a normed linear space or Banach space. It was also shown there that when $\mathcal{X}$ is an infinite-dimensional Banach space, then $\mathcal{B}(\mathcal{X})$ is not a modular annihilator algebra.

8.8.2 Example Vala [1964] (cf. Alexander [1968], Theorem 3.2) shows that for any compact linear operator T on a Banach space $\mathcal{X}$ the map from $\mathcal{B}(\mathcal{X})$ to $\mathcal{B}(\mathcal{X})$ defined by

$$S \mapsto TST \quad S \in \mathcal{B}(\mathcal{X})$$

is a compact linear operator on $\mathcal{B}(\mathcal{X})$. This proves that $\mathcal{B}_K(\mathcal{X})$ is a compact Banach algebra for any Banach space $\mathcal{X}$. Hence $\mathcal{B}_K(\mathcal{X})$ is a modular annihilator algebra and thus Theorem 8.4.5(e) shows easily that every irreducible representation of $\mathcal{B}_K(\mathcal{X})$ is topologically equivalent to its given representation on $\mathcal{X}$. Thus every irreducible representation of $\mathcal{B}_K(\mathcal{X})$ consists of compact operators and none of them are finite-dimensional if $\mathcal{X}$ is finite-dimensional. Actually $\mathcal{B}_K(\mathcal{X})$ is the largest irreducible subalgebra of $\mathcal{B}(\mathcal{X})$ which is a compact Banach algebra with respect to a complete algebra norm majorizing the operator norm (Bonsall and Duncan [1973b], p. 179). Furthermore, $\mathcal{B}_K(\mathcal{X})$ is topologically simple when $\mathcal{B}_K(\mathcal{X}) = \mathcal{B}_F(\mathcal{X})^-$ holds, as it does for many Banach spaces $\mathcal{X}$. That this identity fails for some Banach spaces was first shown by Per Enflo [1973]. For such a Banach space, $\mathcal{B}_K(\mathcal{X})$ is a compact Banach algebra in which its socle $\mathcal{B}_F(\mathcal{X})$ is not dense.

8.8.3 Example The algebra $\mathcal{B}_A(\mathcal{X}) = \mathcal{B}_F(\mathcal{X})^-$ of approximable operators obviously has $\mathcal{B}_F(\mathcal{X})$ as its dense socle. Since it is a semisimple (in fact, primitive) Banach algebra, it is compact and a modular annihilator algebra by results in this section. It is also easy to see that it is topologically simple. More is true. $\mathcal{B}_A(\mathcal{X})$ is an approximate $B^\#$-left annihilator algebra. This is proved in Smiley ([1955], Theorem 2). Furthermore, $\mathcal{B}_A(\mathcal{X})$ is an annihilator algebra if and only if $\mathcal{X}$ is reflexive. See Antonio Fernández López and Ángel Rodríguez-Palacios [1985]. Hence $\mathcal{B}_A(\mathcal{X})$, for any nonreflexive Banach space $\mathcal{X}$, is a topologically simple, semisimple, left annihilator Banach algebra which is not an annihilator algebra. This too can be generalized. Let $\mathcal{B}$ be any algebra of operators which includes $\mathcal{B}_F(\mathcal{X})$ and is complete in some norm which dominates the operator norm and relative to which $\mathcal{B}_F(\mathcal{X})$ is dense in $\mathcal{B}$. Then $\mathcal{B}$ is an annihilator algebra if and only if $\mathcal{X}$ is reflexive (Rickart [1960], Theorem 2.8.23), and $\mathcal{B}$ is a dual algebra if and only if each $T \in \mathcal{B}$ belongs to $\overline{T\mathcal{B}} \cap \overline{\mathcal{B}T}$ (*ibid.*, Theorem 2.8.27). The C^p algebras of operators on a Hilbert space and the algebras $\mathcal{B}_{HS}(\mathcal{H})$ and $\mathcal{B}_T(\mathcal{H})$ discussed in §1.7.17 satisfy all the above hypotheses and hence are infinite-dimensional, topologically simple, dual algebras.

We remark here that all these examples of semisimple (in fact primitive) dual Banach algebras have natural infinite-dimensional (faithful) irreducible representations and thus are not left completely continuous or completely continuous algebras by Proposition 8.7.9.

8.8.4 Example Davie [1973a] uses Enflo's example [1973] of a reflexive Banach space which does not have the approximation property to provide a topologically simple annihilator Banach algebra $\mathcal{B}_A(\mathcal{X})$ which is not dual, thus settling a question which remained open since Bonsall and Goldie's paper [1954].

8.8.5 Example The algebra of inessential operators on a Banach space $\mathcal{X}$ introduced by Kleinecke [1963] is a modular annihilator algebra by Proposition 8.4.10 above. Similarly one sees that the algebra $\mathcal{B}_{WK}(\mathcal{X})$ of weakly compact operators on $\mathcal{X} = C(\Omega)$ (Ω, a compact Hausdorff space) and on

$\mathcal{X} = L^1(M)$ (M, a measure space) are modular annihilator algebras since Dunford and Schwartz [1958], VI.7.5 and VI.8.13, show that the square of these algebras is included in $\mathcal{B}_K(\mathcal{X})$. By previous remarks these algebras $\mathcal{B}_{WK}(\mathcal{X})$ provide examples of modular annihilator algebras which are not compact algebras since they properly contain $\mathcal{B}_K(\mathcal{X})$. If $\mathcal{X}$ is a locally convex topological linear space then $\mathcal{B}_K(\mathcal{X})$ is shown to be a modular annihilator algebra by Barnes ([1971a], 4.2).

8.8.6   A Normed Algebra of Operators with a Dense Socle which is not a Modular Annihilator Algebra.

This example is due to Yood [1964]. The assertions which we do not check here are all verified in the cited reference. Let $\mathcal{X}$ be an incomplete normed linear space and let $\overline{\mathcal{X}}$ be its completion. Let $\mathcal{A} = \mathcal{B}_F(\mathcal{X})$ be the algebra of bounded linear operators on $\mathcal{X}$ which have finite-rank. Let z belong to $\overline{\mathcal{X}} \setminus \mathcal{X}$ and let U belong to the closure of $\mathcal{A}$ in $\mathcal{B}(\mathcal{X})$ but satisfy $\overline{U}z = z$ where $\overline{U}$ is the closure of the graph of U in $\overline{\mathcal{X}} \times \overline{\mathcal{X}}$. Let $\mathcal{B}$ be the subalgebra of $\mathcal{B}(\mathcal{X})$ generated by $\mathcal{A}$ and U. Clearly $\mathcal{B}$ is a normed algebra with dense socle $\mathcal{A}$. However, the left ideal $\mathcal{M} = \{T \in \mathcal{B} : Tz = 0\}$ is maximal among closed modular left ideals but is not a maximal modular left ideal. Hence $\mathcal{B}$ is not a modular annihilator algebra since it is normed but contains nonclosed maximal left ideals contrary to Theorem 8.4.12.

8.8.7   Example

The algebra of all precompact operators on a normed linear space need not be a modular annihilator algebra or even a Duncan modular annihilator algebra. As pointed out by Barnes [1971a], the example on p. 590 of John Ringrose [1957] contains a precompact operator with the whole plane as spectrum. Hence Theorems 8.6.2 and 8.6.4 show this.

8.8.8   A Left Completely Continuous Semisimple Banach Algebra which is not Completely Continuous

Smyth [1980] gave the first such example settling a question which had been long open. Let $\mathcal{A}$ be the algebra (under pointwise operations) of complex sequences. Consider the following four norms:

$$\|a\|_1 = \sum_{n=1}^{\infty} |a_n| \qquad \|a\|_2 = \sup\{|a_n| : n \in \mathbb{N}\}$$

$$\|a\|_3 = \sum_{n=1}^{\infty} n^2 |a_n| \qquad \|a\|_4 = \sum_{n=1}^{\infty} n |a_n|$$

Let $\mathcal{A}(k)$ be the subalgebra of elements in $\mathcal{A}$ with finite $\|\cdot\|_k$-norm. Define $\mathcal{L}$ to be the following algebra under matrix multiplication:

$$\mathcal{L} = \{\begin{pmatrix} a_1 & a_2 \\ a_3 & a_4 \end{pmatrix} : a_k \in \mathcal{A}(k); k = 1, 2, 3, 4\}$$

with norm

$$\left\| \begin{pmatrix} a_1 & a_2 \\ a_3 & a_4 \end{pmatrix} \right\| = \sum_{k=1}^{4} \|a_k\|_k.$$

It is tedious but straightforward to show that $\mathcal{L}$ is a semisimple Banach algebra. (The ideals

$$\mathcal{M}_n = \{ \begin{pmatrix} a_1 & a_2 \\ a_3 & a_4 \end{pmatrix} : a_{kn} = 0; k = 1, 2, 3, 4 \}$$

are maximal modular since $\mathcal{L}/\mathcal{M}_n$ is isomorphic to M_2. Hence $\mathcal{L}$ is even strongly semisimple.

In order to show that $\mathcal{L}$ is left completely continuous, we choose an arbitrary element $b = \begin{pmatrix} b_1 & b_2 \\ b_3 & b_4 \end{pmatrix} \in \mathcal{L}_1$ and show that $b\mathcal{L}_1$ is totally bounded. Let $\varepsilon > 0$ be arbitrary and choose $N \in \mathbb{N}$ satisfying

$$\sum_{n=N}^{\infty} |b_1 n| < \frac{\varepsilon}{9} \qquad \sum_{n=N}^{\infty} n^{-2} < \frac{\varepsilon}{9}$$

$$N^{-1} < \frac{\varepsilon}{9} \qquad \sum_{n=N}^{\infty} n^2 |b_{3n}| < \frac{\varepsilon}{9}$$

Let $\mathcal{T}$ be the finite-dimensional ideal:

$$\mathcal{T} = \{ a \in \mathcal{L} : a_{kn} = 0 \quad n > N; \quad k = 1, 2, 3, 4 \}.$$

It is easy to verify that the distance from any $c \in b\mathcal{L}_1$ to $\mathcal{L}_1$ is less than $8\varepsilon/9$. Since $\mathcal{T}$ is finite dimensional, $\mathcal{T}_1$ can be covered by a finite number of $(\varepsilon/9)$ - balls. Hence $b\mathcal{L}_1$ is totally bounded as we wished to show.

To see that $\mathcal{L}$ is not completely continuous, we define $\{a_m\}_{m\in\mathbb{N}} \in \mathcal{A}_1$ by $a_{mn} = \delta_{mn}$ (the Kronecker delta). Clearly $\begin{pmatrix} a_m & 0 \\ 0 & 0 \end{pmatrix}$ belongs to $\mathcal{L}_1$ but $\begin{pmatrix} a_m & 0 \\ 0 & 0 \end{pmatrix} \begin{pmatrix} 0 & 1 \\ 0 & 0 \end{pmatrix} = \begin{pmatrix} 0 & a_m \\ 0 & 0 \end{pmatrix}$ has no convergent sequence, so that right multiplication by $\begin{pmatrix} 0 & 1 \\ 0 & 0 \end{pmatrix}$ is not a compact operator.

In the cited reference an easy proof is given that a left completely continuous semiprime algebra with a dense socle is completely continuous.

8.8.9 Example Smyth [1980] also gave an example of a semisimple compact Banach algebra $\mathcal{A}$ and elements $b, c \in \mathcal{A}$ so that

$$a \mapsto bac \quad a \in \mathcal{A}$$

is not compact. This contradicts a conjecture of Alexander [1968] who had proved that all such maps are compact if $\mathcal{A}$ is a C*-algebra.

BIBLIOGRAPHY[1]

Aarnes, Johan F.; Kadison Richard V.

[1969] Pure states and approximate identities, *Proc. Amer. Math. Soc.* **21** (1969), 749–752. **MR 39 #1980.** 5.1

Ackermans, S. T. M.

[1982] On the local spectrum at the identity of a Banach algebra, *Nederl. Akad. Wetensch. Indag. Math.* **44** (1982), no. 3, 261–263. **MR 84c: 46050b.** 2.1

Ackermans, S. T. M.; Gerards, A. M. H.

[1982] Spectral localization in Banach algebras, *Nederl. Akad. Wetensch. Indag. Math.* **44** (1982), no. 3, 251–260. **MR 84c: 46050a.** 2.1

Aiena, Pietro

[1983] Relative regularity and Fredholm theory in semiprime algebras with identity, (Italian summary), *Rend. Circ. Mat. Palermo (2)* **32** (1983), no. 1, 100–109. **MR 85e: 46034.** 2.1

[1987] Relatively regular ideals in semi-prime Banach algebras, *Rend. Circ. Mat. Palermo (2)* **36** (1987), no. 2, 194–204. **MR 90f: 46077.** 8.2

[1990] Riesz multipliers on commutative semisimple Banach algebras, *Arch. Math. (Basel)* **54** (1990), no. 3, 293–303. **MR 91e: 46064.** 3.1

Akkar, Mohamed

[1978] Théorie spectrale et calcul fonctionnel holomorphe d'un élément non régulier d'une algèbre *m*-complète, *Afrika Mat.* **1** (1978) no. 1, 19–33. **MR 81m: 46066.** 3.3

Albert, A. Adrian

[1939] Structure of Algebras, *American Mathematics Society Colloquium Publication*, Vol. 24, *American Mathematical Society New York*, 1939. **MR 1-99.** 4.0, 4.1

Albrecht, Ernst

[1982a] Decomposable systems of operators in harmonic analysis, *Toeplitz centennial (Tel Aviv, 1981)*, pp., 19–35, *Operator theory: Adv. Appl.* 4, *Birkhäuser, Basel–Boston, Mass.*, 1982. **MR 83m: 47030.** 3.2

[1982b] Several variable spectral theory in the noncommutative case, *Spectral theory (Warsaw, 1977)*, 9–30, *Banach Center Publ.*, 8, *PWN, Warsaw*, 1982. **MR 85h: 47015.** 3.5

Albrecht, Ernst; Dales, H. Garth

[1983] Continuity of homomorphisms from C*-algebras and other Banach algebras, *Radical Banach algebras and automatic continuity (Long Beach, Calif. 1981)*, 375–396, *Lecture Notes in Mathematics*, 975, *Springer, Berlin–New York*, 1983. **MR 84f: 46063.** 6.1

---

[1]The numbers on the right margin show the sections to which the reference is related. All reference for Volume II have been removed except those for Chapter 9 (now 13) Cohomology and Chapter 10 (now 14) K-theory. These chapters were originally part of Volume I but were moved to Volume II because of space limitations.

Albrecht, Ernst; Eschmeier, Jórge

[1993]   Analytic functional models and local spectral theory, preprint.   2.8

Albrecht, Ernst; Eschmeier, Jórge; Neumann, Michael M.

[1986]   Some topics in the theory of decomposable operators, *Advances in invariant subspaces and other results of operator theory* (*Timişoara and Herculane*, 1984), 15–34, *Oper. Theory: Adv. Appl.*, 17, *Birkhäuser, Basel–Boston, MA* 1986. **MR 89c: 47046.**   2.8

Alexander, James C.

[1968]   Compact Banach algebras, *Proc. London Math. Soc.* (3) **18** (1968), 1–18. **MR 37 #4618.**   8.7, 8.8

Alexander, James C.

[1969]   On Riesz operators, *Proc. Edinburgh Math. Soc.* (2) **16** (1968/69), 227–232. **MR 40 #783.**   8.7

Alfsen, Erik M.

[1963]   A simplified constructive proof of the existence and uniqueness of Haar measure, *Math. Scand.* **12** (1963), 106–116. **MR 28 #1250.**   1.9

Allan, Graham R.

[1968]   A form of local characterization of Gelfand transforms, *J. London Math. Soc.* **43** (1968), 623–625. **MR 37 #6756.**   3.1, 3.5

[1970]   On lifting analytic relations in commutative Banach algebras, *J. Funct. Anal.* **5** (1970), 37–43. **MR 41 #7437.**   3.5

[1971]   Some aspects of the theory of commutative Banach algebras and holomorphic functions of several variables, *Bull. London Math. Soc.* **43** (1971), 1–17. **MR 44 #804.**   3.5

[1972]   Embedding the algebra of formal power series in a Banach algebra, *Proc. London Math. Soc.* (3) **25** (1972), 329–340. **MR 46 #4201.**   2.9, 3.3, 3.4, 4.8

[1989]   Power-bounded elements in a Banach algebra and a theorem of Gelfand, *Conference on automatic continuity and Banach algebras* (*Canberra*, 1989), 1–12, *Proc. Centre Math. Anal. Austral. Nat. Univ.*, 21, *Austral. Nat. Univ., Canberra*, 1989. **MR 91a: 46069.**   2.6

[1989]   Locally inner derivations, *Conference on automatic continuity and Banach algebras* (Canberra, 1989), 13–24, *Proc. Centre Math. Anal. Austral. Nat. Univ.*, 21, *Austral. Nat. Univ., Canberra*, 1989. **MR 91c: 46066.**   9

Allan, Graham R.; Dales, H. Garth; McClure, J. Peter

[1971]   Pseudo-Banach algebras, *Studia Math.* **40** (1971), 55–69. **MR 48 #12061.** 3.3

Allan, Graham R.; O'Farrell, Anthony G.; Ransford, Thomas J.

[1987]   A Tauberian theorem arising in operator theory, *Bull. London Math. Soc.* **19** (1987), no. 6, 537–545. **MR 89c: 47003.**   2.6

Allan, Graham R.; Ransford, Thomas J.

[1989]   Power dominated elements in Banach algebras, *Studia Math.* **94** (1989), no. 6, 63–79. **MR 91a: 46050.**   2.6

Al-Moajil, Abdullah H.

[1975]   The spectrum of some special elements in the free Banach algebra, *Proc. Amer. Math. Soc.* **50** (1975), 218–222. **MR 51 #6420.**   4.8

[1979]   Elements of the socle of a semisimple Banach algebra, *Studia Sci. Math. Hungar.* **14** (1979), no. 4, 427–430. **MR 84c: 46047.**   8.2

[1981]   Characterization of finite dimensionality for semisimple Banach algebras, *Manuscripta Math.* **33** (1980/81), no. 3–4, 315–325. **MR 83f: 46054.**   8.1

[1982]   The compactum and finite dimensionality in Banach algebras, *Internat. J. Math. Math. Sci.* **5** (1982), no. 2, 275–280. **MR 83g: 46043.**   8.1

[1984]   The compactum of a semisimple commutative Banach algebra, *Internat. J. Math. Math. Sci.* **7** (1984), no. 4, 821–822. **MR 86f: 46054.**   8.7

Altman, Mieczysław

[1971]   Factorisation dans les algèbres de Banach, *C. R. Acad. Sci. Paris Sér A-B* **272** (1971), A1388–A1389. **MR 44 #2038.**   5.2

Bibliography

[1972]   Contracteurs dans les algèbres de Banach, *C. R. Acad. Sci. Paris Sér A-B* **274**(1972), A399–A400. **MR 45 #2473.**   5.1

[1975]   A generalization and the converse of Cohen's factorization theorem, *Duke Math. J.* **42** (1975), 105–110. **MR 52 #15010.**   5.2

Ambrose, Warren

[1945]   Structure theorems for a special class of Banach algebras, *Trans. Amer. Math. Soc.* **57** (1945), 364–386. **MR 7-126.**   1.1

Amitsur, Shimshon

[1952]   A general theory of radicals. I. Radicals in complete lattices, *Amer. J. Math.* **74** (1952), 774–786. **MR 14-347.**   4.7

Amitsur, Shimshon; Levitzki, Jacob

[1950]   Minimal identities for algebras, *Proc. Amer. Math. Soc.* **1** (1950), 449–463. **MR 12-155.**   7.1

Ancochea, Germán

[1942]   Le Théorème de von Staudt en géométrie projective quaternionienne, *J. Reine Angew. Math.* **184** (1942), 193–198. **MR 5-72.**   6.3

[1947]   On semi-automorphisms of division algebras, *Ann. of Math.* (2) **48** (1947), 147–153. **MR 8-310.**   6.3

Andrew, Alfred; Green, William L.

[1980]   On James' quasireflexive Banach space as a Banach algebra, *Canad. J. Math.* **32** (1980), no. 5, 1080–1101. **MR 82d: 46077.**   1.7

Andrunakievič, Vladimir A.

[1961]   Radicals in associative rings. II. Examples of special radicals. (Russian), *Mat. Sb.* **55** (1961), 329–346. **MR 25 #1186.**   4.7

Aniszczyk, Bohdan; Frankiewicz, Ryszard; Ryll-Ardzewski, Czesław

[1991]   Continuity of a homomorphism on commutative subalgebras is not sufficient for continuity, *Studia Math.* **98** (1991), no. 3, 247–248. **MR 92i: 46058.**   6.1

Antonyan, S. A.

[1983]   Banach *G*-algebras, (Russian), *Funktsional Anal. i Prilozhen* **17** (1983), no. 2, 62–63. (English translation; Functional Anal. Appl. **17**(1983), no. 2, 129–130.) **MR 84j: 46080.**   1.9

Ara, Pere; Mathieu, Martin

[1991]   On ultraprime Banach algebras with a non-zero socle, *Proc. Roy. Irish Acad., Sect. A* **91** (1991), no. 1, 89–98. **MR 93h: 46061.**   4.1

Arens, Richard F.

[1946]   The space L^ω and convex topological rings, *Bull. Amer. Math. Soc.* **52** (1946), 931–935. **MR 8-165.**   2.4

[1947a]   Linear topological division algebras, *Bull. Amer. Math. Soc.* **53** (1947), 623–630. **MR 9-6.**   2.2, 2.9

[1947b]   Representations of *-algebras, *Duke Math. J.* **14** (1947), 269–282. **MR 9-44.**

[1951a]   Operations induced in function classes, *Monatsh. Math.* **55** (1951), 1–19. **MR 13-372.**   1.4

[1951b]   The adjoint of a bilinear operation, *Proc. Amer. Math. Soc.* **2** (1951), 839–848. **MR 13-659.**   1.4

[1958]   Inverse-producing extensions of normed algebras, *Trans. Amer. Math. Soc.* **88** (1958), 536–548. **MR 20 #1921.**   2.4, 3.4

[1960]   Extensions of Banach algebras, *Pacific J. Math.* **10** (1960), 1–16. **MR 23#A4031.**   2.4, 3.4

[1961]   The analytic-functional calculus in commutative topological algebras, *Pacific J. Math.* **11** (1961), 405–429. **MR 25 #4373.**   1.1, 3.2, 3.5

[1963]   The group of invertible elements of a commutative Banach algebra, *Studia Math.* (Ser. Specjalna) Zeszyt **1** (1963), 21–23. **MR 26 #4198.**   3.0, 3.5

[1966]   To what extent does the space of maximal ideals determine the algebra?, *Function Algebras, (Proc. Internat. Sympos. on Function Algebras, Tulane Univ.),* 1965, Scott–Foresman, Chicago, Ill. 1966, 164–168. **MR 33 #3126.**   3.5

[1987]   On a theorem of Gleason, Kahane and Żelazko, *Studia Math.* **87** (1987), no. 2, 193–196. **MR 89c: 46069.**   2.4

Arens, Richard F.; Calderon, Alberto P.

[1955] Analytic functions of several Banach algebra elements, *Ann. of Math.* (2) **62** (1955), 204–216. **MR 17-177.**
3.0, 3.5

Arens, Richard F.; Hoffman, Kenneth M.

[1956] Algebraic extension of normed algebras, *Proc. Amer. Math. Soc.* **7** (1956), 203– 210. **MR 17-1113.**
3.4

Arikan, Nilgün

[1981] Arens regularity and reflexivity, *Quart. J. Math. Oxford Ser.* (2) **32** (1981), no. 128, 383–388. **MR 83a: 46055.**
1.4

[1982] A simple condition ensuring the Arens regularity of bilinear mappings, *Proc. Amer. Math. Soc.* **84** (1982), no. 4, 525–531. **MR 83i: 46053.**
1.4

[1983] Arens regularity and retractions, *Glasgow Math. J.* **24** (1983), no. 1, 17–21. **MR 84e: 46046.**
1.4

Aronszajn, Nachman; Smith, Kennan T.

[1954] Invariant subspaces of completely continuous operators, *Ann. of Math.* (2) **60** (1954), 345–350. **MR 16-488.**
2.8

Arosio, Alberto; Ferreira, Artur V.

[1978a] Une caractérisation des fonctionnelles linéaires multiplicatives, *C. R. Acad. Sci. Paris Sér A-B* **287** (1978), no. 5, A327–A330. **MR 80b: 46054.**
2.4

[1978b] Caractérisation spectrale des fonctionnelles linéaires multiplicatives, *C. R. Acad. Sci. Paris Sér A-B* **287** (1978), no. 16, A1085–A1088. **MR 80b: 46059.**
2.4

Artin, Emil

[1926] Zur theorie der hyperkomplexen zahlen, *Abh. Math. Sem. Hamburgischen Univ.* **5** (1926), 251–260.
1.6, 4.8, 8.1

Atiyah, Michael F.

[1967] *K*-Theory. Lecture Notes by D. W. Anderson, *W. A. Benjamin, New York– Amsterdam*, 1967. **MR 36 #7130.**
10

Aupetit, Bernard H.

[1974] Caractérisation des éléments quasi-nilpotents dans les algèbres de Banach, *Atti Accad. Naz. Lincei Rend. Cl. Sci. Fis. Mat. Natur.* (8) **56** (1974), no. 5, 672– 674. **MR 52 #8939.**
2.3

[1975] Continuité du spectre dans les algèbres de Banach avec involution, (English summary), *Pacific J. Math.* **56** (1975), no. 2, 321–324. **MR 51 #11117.**
2.2, 2.4

[1979a] Propriétés spectrales des algèbres de Banach, (French) [Spectral properties of Banach algebras], Lecture Notes in Mathematics, 735, *Springer, Berlin*, 1979. **MR 81i: 46055.**
2.2, 2.3, 2.4

[1979b] Une généralisation du théorème de Gleason-Kahane-Żelazko pour les algèbres de Banach, (English summary), *Pacific J. Math.* **85** (1979), no. 1, 11–17. **MR 82a: 46044.**
2.4

[1982a] The uniqueness of the complete norm topology in Banach algebras and Banach Jordan algebras, *J. Funct. Anal.* **47** (1982), no. 1, 1–6. **MR 83g: 46044.**
2.3, 6.1

[1982b] Analytic multivalued functions in Banach algebras and uniform algebras, *Adv. in Math.* **44** (1982), no. 1, 18–60. **MR 84b: 46059.**
2.3

[1983] Automatic continuity conditions for a linear mapping from a Banach algebra onto a semisimple Banach algebra, *Radical Banach algebras and automatic continuity (Long Beach, Calif., 1981)*, 313–316, *Lecture Notes in Mathematics, 975, Springer, Berlin–New York*, 1983. **MR 84h: 46060.**
2.3

[1986] Inessential elements of Banach algebras, *Bull. London Math. Soc.* **18** (1986), no. 5, 493–497. **MR 87k: 46100.**
8.6

[1991] A Primer on Spectral Theory, *Universitext. Springer-Verlag, N.Y.*, 1991. 193 pp. **MR 92c: 46001.**
2.3, 2.4

Aupetit, Bernard H.; Gardner, L. Terrell

[1981] On a theorem of S. Berberian and I. Halperin, *C. R. Math. Rep. Acad. Sci. Canada 3* (1981), no. 4, 229–233. **MR 82i: 46070.**
2.1

Aupetit, Bernard H.; Laffey, Thomas J.; Zemánek, Jaroslav

[1981]   Spectral classification of projections, *Linear Algebra Appl.* **41** (1981), 131–135. MR **83d: 46057.** 10

Aupetit, Bernard H.; Wermer, John

[1978]   Capacity and uniform algebras, *J. Funct. Anal.* **28** (1978), *no.* 3, 386–400. MR **80a: 31001.** 2.3

Aupetit, Bernard; Zemánek, Jaroslav

[1978]   On the spectral radius in real Banach algebras, (Russian summary), *Bull. Acad. Polon. Sci. Sér. Sci. Math. Astronom. Phys.* **26** (1978), *no.* 12, 969–973, (1979). MR **80e: 46034.** 2.4

[1981]   Local behaviour of the spectral radius in Banach algebras, *J. London Math. Soc.* (2) **23** (1981), *no.* 1, 171–178. MR **83d: 46056.** 2.4

[1983a]   Local characterizations of the radical in Banach algebras, *Bull. London Math. Soc.* **15** (1983), *no.* 1, 25–30. MR **84g: 46068.** 2.4

[1983b]   Local behavior of the spectrum near algebraic elements, *Linear Algebra Appl.* **52/53** (1983), 39–44. MR **84h: 46063.** 2.3, 2.4

Bachelis, Gregory F.

[1972]   Homomorphisms of Banach algebras with minimal ideals, *Pacific J. Math.* **41** (1972), 307–311. MR **47 #5597.** 8.7

[1983]   Some radical quotients in harmonic analysis, *Radical Banach algebras and automatic continuity (Long Beach, Calif.*, 1981), 301–308, *Lecture Notes in Mathematics, 975, Springer, Berlin–New York*, 1983. MR **84k: 43001.** 4.8

Bachelis, Gregory F.; Saeki, Sadahiro

[1987]   Banach algebras with uncomplemented radicals, *Proc. Amer. Math. Soc.* **100** (1987), *no.* 2, 271–274. MR **88f: 46104.** 8.1

Bade, William G.

[1983]   Multipliers of weighted ℓ^1-algebras, *Radical Banach algebras and automatic continuity (Long Beach, Calif.* 1981), 227–247, *Lecture Notes in Mathematics, 975, Springer, Berlin–New York*, 1983. MR **84e: 46055.** 3.4, 4.8

[1989]   The Wedderburn decomposition for quotient algebras arising from sets of non-synthesis, *Conference on automatic continuity and Banach algebras (Canberra*, 1989), 25–31, *Proc. Centre Math. Anal. Austral. Nat. Univ.*, 21, *Austral. Nat. Univ., Canberra*, 1989. MR **91d: 46067.** 8.1

Bade, William G.; Curtis, Philip C., Jr.

[1960a]   Homomorphisms of commutative Banach algebras, *Amer. J. Math.* **82** (1960), 589–608. MR **22 #8354.** 8.1

[1960b]   The Wedderburn decomposition of commutative Banach algebras, *Amer. J. Math.* **82** (1960), 851–866. MR **23 #A529.** 6.0, 6.2, 8.1

[1974]   The continuity of derivations of Banach algebras, *J. Funct. Anal.* **16** (1974), 372–387. MR **50 #10820.** 6.4

[1978]   Prime ideals and automatic continuity problems for Banach algebras, *J. Funct. Anal.* **29** (1978), *no.* 1, 88–103. MR **80j: 46077.** 6.4

[1984]   Module derivations from commutative Banach algebras, *Proceedings of the conference on Banach algebras and several complex variables (New Haven, Conn.*, 1983), 71–81, *Contemp. Math.*, 32, *Amer. Math. Soc. Providence, R. I.*, 1984. MR **86b: 46083.** 6.4

Bade, William G.; Curtis, Philip C. Jr.; Dales, H. Garth

[1987]   Amenability and weak amenability for Beurling and Lipschitz algebras, *Proc. London Math. Soc.* (3) **55** (1987), *no.* 2, 359–377. MR **88f: 46098.** 1.9, 9

Bade, William G.; Curtis, Philip C., Jr.; Laursen, Kjeld B.

[1980]   Divisible subspaces and problems of automatic continuity, *Studia Math.* **68** (1980), *no.* 2, 159–186. MR **82b: 46059.** 3.4, 6.2

Bade, William G.; Dales, H. Garth

[1989a]   Discontinuous derivations from algebras of power series, *Proc. London Math. Soc.* (3) **59** (1989), *no.* 1, 133–152. MR **90j: 46048.** 6.4

720   Bibliography

[1989b]   Continuity of derivations from radical convolution algebras, *Studia Math.* **95**
          (1989), no. 1, 59–91. **MR 90k: 46115**.   6.4
[1992]   The Wedderburn decomposability of some commutative Banach algebras, *J.*
          Funct. Anal. **107** (1992), no. 1, 105–112. **MR 93d: 46090**.   8.1

Bade, William G.; Dales, H. Garth; Laursen, Kjeld B.

[1984]   Multipliers of radical Banach algebras of power series, *Mem. Amer. Math. Soc.*
          49 (1984), no. 303. **MR 85j: 46094**.   3.4

Badea, Cătălin

[1991]   The Gleason–Kahane–Żelazko theorem and Ransford's generalized spectra, *C.*
          R. Acad. Sci. Paris Sér. I Math. **313** (1991), no. 10., 679–683. **MR 93a:**
          46086.   2.1, 2.4

Baer, Reinhold

[1943]   Radical ideals, *Amer. J. Math.* **65** (1943), 537–568. **MR 5-88**.   4.3, 4.4, 4.8

Baker, John Walter; Pym, John S.

[1971]   A remark on continuous bilinear mappings, *Proc. Edinburgh Math. Soc.* (2) **17**
          (1970/71), 245–248. **MR 46 #2429**.   3.1

Baker, John Walter; Rejali, Ali

[1989]   On the Arens regularity of weighted convolution algebras, *J. London Math.*
          Soc. (2) **40** (1989), no. 3, 535–546. **MR 91i 46054**.   1.4, 1.9

Balsam, Zofia

[1980]   On topological divisors of zero and permanently singular elements in two-
          norm algebras, *Funct. Approx. Comment. Math.* **9** (1980), 137–144. **MR 84m:**
          46060a.   2.5

Banach, Stefan

[1932]   Théorie des operations lineaires, Warsaw, 1932.   6.2
[1948]   Remarques sur les groupes et les corps métriques, *Studia Math.* **10** (1948),
          178–181. **MR 10-590**.   2.9

Barnes, Bruce A.

[1964]   Modular annihilator algebras, *Dissertation, Cornell University, Ithaca, New*
          York, 1964.   8.0, 8.4
[1966]   Modular annihilator algebras, *Canad. J. Math.* **18** (1966), 566–578. **MR 33**
          #2681.   8.0, 8.4, 8.7
[1967a]   Algebras with the spectral expansion property, *Ill. J. Math.* **11** (1967), 284–290.
          MR 35 #747.   8.0
[1967b]   Some theorems concerning the continuity of algebra homomorphisms, *Proc.*
          Amer. Math. Soc. **18** (1967), 1035–1038. **MR 36 #3125**.   6.1, 8.0, 8.7
[1968a]   A generalized Fredholm theory for certain maps in the regular representations
          of an algebra, *Canad. J. Math.* **20** (1968), 495–504. **MR 38 #534**. 8.4, 8.5, 8.6
[1968b]   On the existence of minimal ideals in a Banach algebra, *Trans. Amer. Math.*
          Soc. **133** (1968), 511–517. **MR 37 #2008**.   8.0, 8.4, 8.6
[1969a]   The Fredholm elements of a ring, *Canad. J. Math.* **21** (1969), 84–95. **MR 38**
          #5823.   8.5
[1969b]   Subalgebras of modular annihilator algebras, *Proc. Cambridge Philos. Soc.* **66**
          (1969), 5–12. **MR 41 #4236**.   8.4
[1971a]   Examples of modular annihilator algebras, *Rocky Mountain J. Math.* **1** (1971),
          no. 4, 657–665. **MR 44 #5777**.   8.0, 8.4,8.7, 8.8
[1971b]   Banach algebras which are ideals in Banach algebras, *Pacific J. Math.* **38**
          (1971), 1–7. **MR 46 #9738**.   1.9
[1976]   Representations of the ℓ^1-algebra of an inverse semigroup, *Trans. Amer. Math.*
          Soc. **218** (1976), 361–396. **MR 53 #1169**.
[1979]   Properties of representations of Banach algebras with approximate identities,
          Proc. Amer. Math. Soc. **74** (1979), no. 2, 235–241. **MR 80g: 46045**. 5.2
[1981]   Ideal and representation theory of the ℓ^1-algebra of a group with polynomial
          growth, *Colloq. Math.* **45** (1981), no. 2, 301–315. **MR 84h: 43008**.   1.9
[1983]   Inverse closed subalgebras and Fredholm theory, *Proc. Roy. Irish Acad. Sect.*
          A **83** (1983), no. 2, 217–224. **MR 85c: 46042**.   2.5, 8.5
[1985]   Algebraic elements of a Banach algebra modulo an ideal, *Pacific J. Math.* **117**
          (1985), no. 2, 219–231. **MR 86i: 46051**.   2.3

Barnes, Bruce A.; Duncan, John

[1975]   The Banach algebra $\ell^1(S)$, *J. Funct. Anal.* **18** (1975), 96–113. **MR 51 #13587**.
           4.8

Barnes, Bruce A.; Murphy, Gerard J.; Smyth, M. R. F.; West, Trevor T.

[1982]   Riesz and Fredholm Theory in Banach algebras, Research Notes in Mathematics 67, *Pitman (Advanced Publishing Program), Boston, Mass.–London* 1982, **MR 84a: 46108**.   1.7, 8.2

Bass, Hyman

[1968]   Algebraic K-theory, *W. A. Benjamin, Inc., New York–Amsterdam*, 1968. **MR 40 #2736**.   10

Batty, Charles J. K.

[1978]   Tensor products of compact convex sets and Banach algebras, *Math. Proc. Cambridge Philos. Soc.* **83** (1978), no. 3, 419–427. **MR 80c: 46060**.   1.10

Bauer, F. L.

[1962]   On the field of values subordinate to a norm, *Numer. Math.* **4** (1962), 103–113. **MR 26 #2860**.   2.6

Beauzamy, Bernard

[1982]   Introduction to Banach spaces and their geometry, North-Holland Mathematics Studies, 68. Notas de Matemática. [Mathematical Notes], 86. *North-Holland Publishing Co., Amsterdam–New York*, 1982. **MR 84g: 46017**.   1.7

[1985]   Un opérateur sans sous-espace invariant: simplification de l'example de P. Enflo, *Integral Equations Operator Theory* **8** (1985), no. 3, 314–384. **MR 88b: 47011**.   2.8

Beddaa, Abdelouahhab; Oudadess, Mohamed

[1989]   On a question of A. Wilansky in normed algebras, *Studia Math.* **95** (1989), no. 2, 175–177. **MR 91d: 46057**.   2.2

Bekka, Mohammed El Bacher

[1986]   Primitive ideals with bounded approximate units in L^1-algebras of exponential Lie groups, *J. Austral. Math. Soc. Ser. A* **41** (1986), no. 3, 411–420. **MR 87k: 43006**.   5.1

[1990]   Complemented subspaces of $L^\infty(G)$, ideals of $L^1(G)$ and amenability, *Monatsh. Math.* **109** (1990), no. 3, 195–203. **MR 91f: 43006**.   9

Belfi, Victor A.; Doran, Robert S.

[1980]   Norm and spectral characterizations in Banach algebras, *Enseign. Math.* (2) **26** (1980), no. 1-2, 103–130. **MR 81j: 46071**.   2.4

Berglund, John F.; Junghenn, Hugo D.; Milnes, Paul

[1978]   Compact right topological semigroups and generalizations of almost periodicity, Lecture Notes in Mathematics, 663, *Springer-Verlag, Berlin–New York*, 1978. **MR 80c: 22003**.   1.9

[1989]   Analysis on semigroups: Function spaces, compactifications, representations, Canadian Mathematical Society Series of Monographs and Advanced Texts, A Wiley-Interscience Publication, *John Wiley & Sons, Inc., New York*, 1989, 334 pp. **MR 91b: 43001**.   1.9

Berkani, M.; Esterle, Jean

[1987]   Banach algebras with left sequential approximate identities close to their square, *Operators in indefinite metric spaces, scattering theory and other topics (Bucharest, 1985)*, 29–40, Oper. Theory: Adv. Appl., 24, *Birkhäuser, Basel–Boston, Mass.*, 1987. **MR 88i: 46063**.   5.1

Berkson, Earl

[1963]   A characterization of scalar type operators on reflexive Banach spaces, *Pacific J. Math.* **13** (1963), 365–373. **MR 27 #5131**.   2.6

Beurling, Arne

[1938]   Sur les intégrales de Fourier absolument convergentes, *Congrès des Math. Scand. Helsingfors*, 1938, pp. 345–366.   1.1, 1.8, 1.9, 2.2, 3.3

Bhatt, Subhash J.

[1983]   On spectra and numerical ranges in locally m-convex algebras, *Indian J. Pure Appl. Math.* **14** (1983), no. 5, 596–603. MR **85g: 46057**.   2.6

Bhatt, Subhash J.; Karia, D. J.

[1992]   Uniqueness of the uniform norm with an application to topological algebras, *Proc. Amer. Math. Soc.* **116** (1992), no. 2, 499–503.   1.1, 3.1

Bishop, Errett

[1959]   A duality theorem for an arbitrary operator, *Pacific J. Math.* **9** (1959), 379–397. MR **22 #8339**.   3.2

[1960]   Boundary measures of analytic differentials, *Duke Math. J.* **27** (1960), 331–340. MR **22 #9621**.   2.8

Blackadar, Bruce E.

[1986]   K-theory for operator algebras, *Mathematical Sciences Research Institute Publications* 5, *Springer-Verlag, New York–Berlin*, 1986. 338 pp. MR **88g: 46082**.

                                                                              3.3, 10
Blum, Edward K.

[1953]   The fundamental group of the principal component of a commutative Banach algebra, *Proc. Amer. Math. Soc.* **4** (1953), 397–400. MR **14-1096**.   3.5

Bochner, Salomon

[1927]   Beiträge zur Theorie der fastperiodischen, *Functionen Math. Ann.* **96** (1927), 119–147.   3.2

[1933]   Integration von Funktionen deren Werte die Elemente eines Vektorraumes sind, *Fund. Math.* **20** (1933), 262–276.   1.9

[1934]   A theorem on Fourier–Stieltjes integrals, *Bull. Am. Math. Soc.* **40** (1934), 271–276.   3.1

Bochner, Salomon; Phillips, Ralph S.

[1942]   Fourier expansions for non-commutative normed rings, *Ann. of Math.*(2) **43** (1942), 409–418. MR **4-218**.   1.1

Bohnenblust, H. Frederic; Karlin, Samuel

[1955]   Geometrical properties of the unit sphere of Banach algebras, *Ann. of Math.*(2) **62** (1955), 217–229. MR **17-177**.   2.6

Bohr, Harald

[1932]   Fastperiodische Funktionen, *Ergebnisse der Math.* I, 5, *Springer, Berlin*, 1932. (English translation), *Chelsea Publishing Co., New York*, 1947.   3.2

Bonnard, Michel

[1969]   Sur le calcul fonctionnel holomorphe multiforme dans les algèbres topologiques, *Ann. Sci. École Norm. Sup.* (4) **2** (1969), 397–422. MR **41 #2396**.   3.5

Bonsall, Frank F.

[1954a]   A minimal property of the norm in some Banach algebras, *J. London Math. Soc.* **29** (1954), 156–164. MR **15-803**.   2.0, 2.3, 3.4, 6.1, 8.4
          MR **16-936**.

[1969a]   The numerical range of an element of a normed algebra, *Glasgow Math. J.* **10** (1969), 68–72. MR **39 #4675**.   2.6

[1969b]   Operators that act compactly on an algebra of operators, *Bull. London Math. Soc.* **1** (1969), 163–170. MR **40 #750**.   8.7

Bonsall, Frank F.; Crabb, Michael J.

[1970]   The spectral radius of a Hermitian element of a Banach algebra, *Bull. London Math. Soc.* **2** (1970), 178–180. MR **42 #853**; erratum MR **43**, p.**1696**. 2.6

Bonsall, Frank F.; Duncan, John

[1967]   Dual representations of Banach algebras, *Acta. Math*, **117** (1967), 79–102. MR **34 #4929**.   2.3, 4.2

[1971]   Numerical ranges of operators on normed spaces and of elements of normed algebras, *London Mathematics Society Lecture Note Series*, 2, *Cambridge University Press, London–New York*, 1971. MR **44 #5779**.   2.6

[1973a] Numerical ranges II. *London Math. Soc. Lecture Note Series*, No. 10, *Cambridge University Press, New York–London*, 1973. **MR 56 #1063.** 2.4, 2.6

[1973b] Complete normed algebras, *Ergebnisse der Mathematik und ihrer Grenzgebeite, Band* 80, *Springer-Verlag, New York-Heidelberg*, 1973. **MR 54 #11013.** general

[1980] Numerical ranges, *Studies in functional analysis*, pp. 1–49, *MAA Stud. Math.*, 21, *Math. Assoc. America, Washington, D. C.*, 1980. **MR 81m: 47011.** 2.6

Bonsall, Frank F.; Goldie, Alfred W.

[1954] Annihilator algebras, *Proc. London Math. Soc.* (3) **4** (1954), 154–167. **MR 15-881.** 8.4, 8.7, 8.8

Bourbaki, Nicolas

[1966] Elements of Mathematics: General Topology, Part I and Part 2, *Addison-Wesley Publishing Company, Reading, Mass.–London–Don Mills, Ont.* 1966. **MR 34 #5044a,b.** erratum, 40, p. 1704. 1.10, 7.3

[1967] Éléments de mathématique. Fasc. XXXII. *Théories spectrales.* Chapitre I: Algèbres normés. Chapitre II: Groupes localement compacts commutatifs. *Actualités Scientifiques et Industrielles*, No. 1332. *Herman, Paris*, 1967. **MR 35 #4725.** 2.0, 3.5

[1974] Elements of Mathematics, *Algebra*, Part I, Chapters 1-3. Translated from French. *Hermann, Paris; Addison-Wesley Publishing Co., Reading, Mass.*, 1974. **MR 50 #6689.** 1.10

Bredon, Glen E.

[1963] A new treatment of the Haar integral, *Michigan Math. J.* **10** (1963), 365–373. **MR 29 #4836.** 1.9

Browder, Andrew

[1969] Introduction to function algebras, *W. A. Benjamin, Inc., New York–Amsterdam*, 1969. **MR 39 #7431.** 3.2, 3.5

[1971] On Bernstein's inequality and the norm of Hermitian operators, *Amer. Math. Monthly* **78** (1971), 871–873. **MR 45 #896.** 2.6

Brown, Bailey; McCoy, Neal H.

[1947] Radicals and subdirect sums, *Amer. J. Math.* **69** (1947), 46–58. **MR 8-433.** 4.5, 4.8

[1948] The radical of a ring, *Duke Math. J.* **15** (1948), 495–499. **MR 10-6.** 4.8

Brown, David T.

[1966] A class of Banach algebras with a unique norm topology, *Proc. Amer. Math. Soc.* **17** (1966), 1429–1434. **MR 34 #4938.** 3.4, 6.1

Brown, Herbert I.; Cass, F. P.; Robinson, Irvine J. W.

[1978] On isomorphisms between certain subalgebras of $B(X)$, *Studia Math.* **63** (1978), no. 2, 189–197. **MR 80a: 47073.** 1.7

de Bruijn, Nicolaas Govert

[1958] Function theory in Banach algebras, *Ann. Acad. Sci. Fenn. Ser. A I no.* 250/5 (1958), 13 pp. **MR 20 #3463.** 3.3, 3.4

Buck, R. Creighton

[1958] Bounded continuous functions on a locally compact space, *Mich. Math. J.* **5** (1958), 95–104. **MR 21 #4350.** 1.9, 5.1

Buoni, John J.; Klein, Albert J.

[1979] On the generalized Calkin algebra, *Pacific J. Math.* **80** (1979), no. 1, 9–12. **MR 80e: 47039.** 1.7

Burckel, Robert B.

[1970] Weakly almost periodic functions on semigroups, *Gordon and Breach Science Publishers, New York–London–Paris*, 1970. **MR 41 #8562.** 1.4

[1972] Characterizations of $C(X)$ among its subalgebras. Lecture Notes in Pure and Applied Mathematics, Vol. 6, *Marcel-Dekker, Inc., New York*, 1972. **MR 56 #1068.** 1.5

[1984] Bishop's Stone Weierstrass theorem, *Amer. Math. Monthly* **91** (1984), no. 1, 22–32. **MR 85i: 46071.** 1.5

724 Bibliography

[1994]   An introduction to classical complex analysis, Vol. II, Pure and Applied Mathematics, *Academic Press, New York*, to appear.   3.3

Burlando, Laura

[1986]   On two subsets of a Banach algebra that are related to the continuity of spectrum and spectral radius, Proceedings of the symposium on operator theory (Athens, 1985) *Linear Algebra Appl.* **84** (1986), 251–269. **MR 87m: 46098.**
                                                                                    2.2
[1990]   On continuity of the spectral radius function in Banach algebras, *Ann. Mat. Pure Appl.* (4) **156** (1990), 357–380. **MR 92e: 46103.**   2.2
[1991]   On continuity of the spectrum functor in Banach algebras with good ideal structure, *Riv. Mat. Pura Appl.* **9** (1991), 7–21. **MR 93a: 46088.**   2.2

Burnside, William

[1911]   Theory of groups of finite order, 2nd Edition, *Cambridge University Press, Cambridge*, 1911.   4.0

Busby, Robert C.

[1967]   On structure spaces and extensions of C*-algebras, *J. Funct. Anal.* **1** (1967), 370–377. **MR 37 #771.**   1.2, 7.4
[1968]   Double centralizers and extensions of C*-algebras, *Trans. Amer. Math. Soc.* **132** (1968), 79–99. **MR 37 #770.**   1.2, 5.1, 7.4
[1971a] Extensions in certain topological algebraic categories, *Trans. Amer. Soc.* **159** (1971), 41–56. **MR 43# 7937.**   1.2, 4.8
[1971b] On a theorem of Fell, *Proc. Amer. Math. Soc.* **30** (1971), 133–140. **MR 44 #814.**
                                                                                    7.4

Calkin, John Williams

[1941]   Two-sided ideals and congruences in the ring of bounded operators in Hilbert space, *Ann. of Math.* (2) **42** (1941), 839–873. **MR 3-208.**   1.1, 1.7

Caradus, Selwyn R.

[1978]   Generalized inverses and operator theory. Queen's Papers in Pure and Applied Mathematics, 50, *Queen's University, Kingston, Ont.*, 1978. iv + 206 pp. **MR 81m: 47003.**
                                                                                    2.1

Caradus, Selwyn R.; Pfaffenberger, William E.; Yood, Bertram

[1974]   Calkin algebras and algebras of operators on Banach spaces, Lecture Notes in Pure and Applied Mathematics, Vol. 9, *Marcel Dekker, Inc., New York*, 1974. **MR 54 #3434.**
                                                                                    1.7

Carne, T. Keith

[1978]   Tensor products of Banach algebras, *J. London Math. Soc.* (2) **17** (1978), *no.* 3, 480–488. **MR 58 #12417.**   1.10
[1979]   Not all H'-algebras are operator algebras, *Math. Proc. Cambridge Philos. Soc.* **86** (1979), *no.* 2, 243–249. **MR 81m: 46063.**   1.10
[1981]   Representation theory for tensor products of Banach algebras, *Math. Proc. Cambridge. Philos. Soc.* **90** (1981), 445–463. **MR 83c: 46067.**   1.10, 4.3

Cartan, Élie

[1894]   Sur la structure des groupes de transformations finis et continus, *These, Paris, Nancy*, (1894). Also in Oeuvres Complètes, Vol I, *Paris, Gauthier–Villars*, 1952.
                                                                                    1.1
[1898]   Les groupes bilineaires et les systemes de nombres complexes, *Ann. Fac. Sci. Toulouse* **12** (1898), 1–99. Also in Oeuvres Complètes, Vol II, *Paris, Gauthier–Villars*, 1952.   1.1, 1.6, 4.4, 4.5, 4.8, 8.1

Cartan, Henri; Godement, Roger

[1947]   Théorie de la dualité et analyse harmonique dans les groupes abéliens localement compacts, *Ann. Sci. École Norm. Sup.* (3) **64** (1947), 79–99. **MR 9-326.**   3.6

Cayley, Arthur

[1855]   Sept differents mémoires d'analyse, *J. Reine Angew. Math.* **50** (1855), 277–317.
                                                                                    1.6
[1858]   A memoir on the theory of matrices, *Phil. Trans. Royal Soc. London* **148** (1858), 17–38.   1.6

Cedilnik, A.

[1983] On Banach algebras without zero divisors, (Slovenian summary), *Glas. Mat. Ser. III* **18(38)** (1983), *no.* 1, 67–69. **MR 84j: 46078.** 2.2

Cheikh, O. H.; Oudadess, Mohamed

[1988] On a commutativity question in Banach algebras. (Arabic summary), *Arab Gulf J. Sci. Res. A. Math. Phys. Sci.* **6** (1988), *no.* 2, 173–179. **MR 89i: 46053.** 2.3

Chen, Chang Pao

[1978] On the intersections and the unions of Banach algebras, *Tamkang J. Math.* **9** (1978), *no.* 1, 21–27. **MR 80k: 46082.** 6.4

[1983] A generalization of the Gleason–Kahane–Żelazko theorem, *Pacific J. Math.* **107** (1983), *no.* 1, 81–87. **MR 85d: 46070.** 2.4

Chen, Gong Ning

[1987] The Schwarz lemma for the spectral radius in Banach algebras, (Chinese. English summary), *Beijing Shifan Daxue Xuebao* (1987), *no.* 1, 13–17. **MR 88i: 46065.** 2.3

Chernoff, Paul R.

[1973] Representations, automorphisms, and derivations of some operator algebras, *J. Funct. Anal.* **12** (1973), 275–289. **MR 50 #2934.** 6.4, 9

Cho, Tae Geun

[1982] A power factorization theorem in Banach modules, *J. Korean Math. Soc.* **18** (1981/82), *no.* 2, 175–180. **MR 84a: 46105.** 5.3

Cho, Tae Geun; Rho, Jae Chul

[1989] Continuity of certain homomorphisms of Banach algebras, *J. Korean Math. Soc.* **26** (1989), *no.* 1, 105–110. **MR 90h: 46081.** 6.2

Christensen, Jens Peter Reus

[1974] Topology and Borel structure. Descriptive topology and set theory with applications to functional analysis and measure theory, North Holland Math. Studies Vol. 10; (Notas de Matemática *no.* 51), *North-Holland Publishing Co., Amsterdam–London; American Elsevier Publishing Co., New York*, 1974. **MR 50 #1221.** 5.3

[1976] Codimension of some subspaces in a Fréchet algebra, *Proc. Amer. Math. Soc.* **57** (1976), *no.* 2, 276–278. **MR 53 #8902.** 5.3

Chryssakis, Thanassis

[1986] Numerical ranges in locally *M*-convex algebras. I., *J. Austral. Math. Soc. Ser. A* **41** (1986), *no.* 3, 304–316. **MR 87k: 46110.** 2.6

Cigler, Johann; Losert, Viktor; Michor, Peter W.

[1979] Banach modules and functors on categories of Banach spaces. Lecture Notes in Pure and Applied Mathematics, 46, *Marcel Dekker, Inc. New York*, 1979. **MR 80j: 46112.** 1.10, 4.7

Civin, Paul

[1961] Extensions of homomorphisms, *Pacific J. Math.* **11** (1961), 1223–1233. **MR 26 #6811.** 1.4

[1962a] Ideals in the second conjugate algebra of a group algebra, *Math. Scand.* **11** (1962), 161–174. **MR 27 #5139.** 1.4

[1962b] Annihilators in the second conjugate algebra of a group algebra, *Pacific J. Math.* **12** (1962), 855–862. **MR 26 #2894.** 1.4

Civin, Paul; Yood, Bertram

[1959] Involutions on Banach algebras, *Pacific J. Math.* **9** (1959), 415–436. **MR 21 #4365.** 1.1

[1961] The second conjugate space of a Banach algebra as an algebra, *Pacific J. Math.* **11** (1961), 847–870. **MR 26 #622.** 1.4, 3.1, 4.1, 5.1

[1965] Lie and Jordan structures in Banach algebras, *Pacific J. Math.* **15** (1965), 775–797. **MR 32 #6245.** 6.3

Clarkson, James A.

[1936]   Uniformly convex spaces, *Trans. Amer. Math. Soc.* **40** (1936), 396–414.   1.7

Cleveland, Sandra Barkdull

[1963]   Homomorphisms of non-commutative *-algebras, *Pacific J. Math.* **13** (1963), 1097–1109. **MR 28 #1500.**   2.0, 6.1

Cohen, Paul J.

[1959a]   Factorization in group algebras, *Duke Math. J.* **26** (1959), 199–205. **MR 21 #3729.**
                                                                        5.0, 5.2
[1959b]   Homomorphisms and idempotents of group algebras, *Bull. Amer. Math. Soc.* **65** (1959), 120–122. **MR 27 #1533.**
                                                                        3.5, 3.6
[1960a]   On a conjecture of Littlewood and idempotent measures, *Amer. J. Math.* **82** (1960), 191–212. **MR 24 #A3231.**   3.6
[1960b]   On homomorphisms in group algebras, *Amer. J. Math.* **82** (1960), 213–226. **MR 24 #A3232.**
                                                                        3.5, 3.6

Collins, Heron S.; Summers, William H.

[1969]   Some applications of Hewitt's factorization theorem, *Proc. Amer. Math. Soc.* **21** (1969), 727–733. **MR 39 #1982.**   5.2

Colojoară, Ion; Foiaş, Ciprian

[1968]   Theory of generalized spectral operators, Mathematics and its Applications, Vol. 9. *Gordon and Breach, Scientific Publishers, New York,* 1968. xvi + 232 pp. **MR 52 #15085.**   2.8

Combes, Francois

[1969]   Sur les faces d'une C*-algèbre, *Bull. Sci. Math.* (2) **93** (1969), 37–62. **MR 42 #856.**   1.1

Connes, Alain

[1988]   Entire cyclic cohomology of Banach algebras and characters of θ-summable Fredholm modules, *K-Theory* **1** (1988), no. 6, 519–548. **MR 90c: 46094.**   9, 10

Conway, John B.

[1985]   A course in functional analysis. Graduate Tests in Mathematics, 96. *Springer-Verlag, New York–Berlin,* 1985. **MR 86h: 46001.**   9

Conway, John B.; Herrero, Domingo A.; Morrel, Bernard B.

[1989]   Completing the Riesz–Dunford functional calculus, *Mem. Amer. Math. Soc.,* **82** (1989), no. 417, 104 pp. **MR 90m: 47023.**   3.3

Corach, Gustavo

[1982]   Algebraic elements in topological algebras, (Spanish. English summary) *Rev. Un. Mat. Argentina* **30** (1981/82), no. 2, 118–127. **MR 84i: 46056.**   2.3

Corach, Gustavo; Larotonda, Angel R.

[1984]   Stable range in Banach algebras, *J. Pure Appl. Algebra* **32** (1984), no. 3, 289–300. **MR 86g: 46070.**   9, 10
[1986]   A stabilization theorem for Banach algebras, *J. Algebra* **101** (1986), no. 2, 433–449. **MR 87h:46103.**   10

Corach, Gustavo; Porta, Horacio A.; Recht, Lázaro

[1990]   Differential geometry of systems of projections in Banach algebras, *Pacific J. Math.* **143** (1990), no. 2, 209–228. **MR 91g: 46056.**   10

Corach, Gustavo; Suárez, Fernando Daniel

[1987a]   Extension of characters in commutative Banach algebras, *Studia Math.* **85** (1987), no. 2, 199–202. **MR 88d: 46098.**   3.5
[1987b]   Dense morphisms in commutative Banach algebras, *Trans. Amer. Math. Soc.* **304** (1987), no. 2, 537–547. **MR 88k: 46064.**   9
[1988]   Thin spectra and stable range conditions, *J. Funct. Anal.* **81** (1988), no. 2, 432–442. **MR 89k: 46061.**   10

Cousin, P.

[1895]   Sur les fonctions de n-variables complexes, *Acta Math.* **19** (1895), 1–62.   3.5

Cowling, Michael; Fendler, Gero

[1984]   On representations in Banach spaces, *Math. Ann.* **266** (1984), no. 3, 307–315.
MR 85j: 46083.   1.3

Crabb, Michael J.; Duncan, John

[1979]   Some inequalities for norm unitaries in Banach algebras, *Proc. Edinburgh Math. Soc.* (2) **21** (1978/1979), no. 1, 17–23. MR 58 #12368.   2.6

Crabb, Michael J.; McGregor, Colin M.

[1988]   Polynomials in a Hermitian element, *Glasgow Math. J.* **30** (1988), no. 2, 171–176. MR 90b: 46105.   2.6

Craw, Ian G.

[1969]   Factorisation in Fréchet algebras, *J. London Math. Soc.* **44** (1969), 607–611. MR 39 #3311.   5.2

[1971]   A condition equivalent to the continuity of characters on a Fréchet algebra, *Proc. London Math. Soc.* (3) **22** (1971), 452–464. MR 44 #4524.   3.5

[1978]   Commutative abelian Galois extensions of a Banach algebra, *J. Funct. Anal.* **27** (1978), no. 2, 170–178. MR 80g: 46059.   3.4

Craw, Ian, Raeburn, Iain; Taylor, Joseph L.

[1983]   Automorphisms of Azumaya algebras over a commutative Banach algebra, *Amer. J. Math.* **105** (1983), no. 4, 815–841. MR 85c: 46055.   9

Craw, Ian G.; Ross, Susan

[1983]   Separable algebras over a commutative Banach algebra, *Pacific J. Math.* **104** (1983), no. 2, 317–336. MR 85c: 46054.   3.4, 10

Craw, Ian G.; Young, Nicholas J.

[1974]   Regularity of multiplication in weighted group and semigroup algebras, *Quart. J. Math. Oxford Ser.* (2) **25** (1974), 351–358. MR 51 #1282.   1.4, 1.9

Cuntz, Joachim

[1983a] Generalized homomorphisms between C*-algebras and KK-theory, *Dynamic and processes (Bielefeld, 1981)*, 31–45, Lecture Notes in Mathematics, 1031, Springer, Berlin–New York, 1983. MR 85j: 46126.   10

[1983b] K-theoretic amenability for discrete groups, *J. Reine Angew. Math.* **344** (1983), 180–195. MR 86e: 46064.   9, 10

[1987]   A new look at KK-Theory, *K-Theory* **1** (1987), no. 1, 31–51. MR 89a: 46142.   9, 10

Curtis, Charles W.; Reiner, Irving

[1981]   Methods of representation theory, Vol. I, With applications to finite groups and orders, Pure and Applied Mathematics, A Wiley-Interscience Publication, *John Wiley & Sons, Inc., New York*, 1981. MR 82i: 20001.   1.9

[1987]   Methods of representation theory, Vol. II, With applications to finite groups and orders, Pure and Applied Mathematics, A Wiley-Interscience Publication, *John Wiley & Sons, Inc., New York*, 1987. MR 88f: 20002.   1.9

Curtis, Philip C., Jr.

[1961]   Derivations of commutative Banach algebras, *Bull. Amer. Math. Soc.* **67** (1961), 271–273. MR 23 #A3475.   6.4

[1989]   Complementation problems concerning the radical of a commutative amenable Banach algebra, *Conference on automatic continuity and Banach algebras (Canberra 1989)*, 56–60, Proc. Centre Math. Analysis Austral. Nat. Univ., **21**, Austral. Nat. Univ., Canberra, 1989. MR 91b: 46048.   8.1, 9

Curtis, Philip C., Jr.; Figà-Talamanca, Alessandro

[1966]   Factorization theorems for Banach algebras, *Function Algebras (Proc. Internat. Sympos. on Function Algebras, Tulane University, 1965)*, 169–185. Scott–Foresman, Chicago, Ill., 1966. MR 34 #3350.   5.2

Curtis, Philip C., Jr.; Loy, Richard J.

[1989]   The structure of amenable Banach algebras, *J. London Math. Soc.* (2) **40** (1989), no. 1, 89–104. MR 90k: 46114.   8.1, 9

728   Bibliography

Curto, Raul E.

[1982]   Spectral permanence for joint spectra, *Trans. Amer. Math. Soc.* **270** (1982) no. 2, 659–665. **MR 83i: 46061.**   3.5

Cusack, Julian M.

[1975]   Jordan derivations on rings, *Proc. Amer. Math. Soc.* **53** (1975), no. 2, 321–324. **MR 53 #3033.**   6.5

[1976]   Homomorphisms and derivations on Banach algebras, *Dissertation, Edinburgh,* 1976.   3.4, 4.8, 6.0, 6.4, 6.5

[1977]   Automatic continuity and topologically simple radical Banach algebras, *J. London Math. Soc.* (2) **16** (1977), no. 3, 493–500. **MR 57 #1121.**   6.0, 6.1

Cutland, Nigel J. (editor)

[1988]   Nonstandard analysis and its applications, Papers from a conference held at the University of Hull, 1986. London Mathematical Society Student Texts, 10., *Cambridge University Press, Cambridge–New York,* 1988. 346 pp. **MR 89m: 03060.**   1.3

Dales, H. Garth

[1973]   The uniqueness of the functional calculus, *Proc. London Math. Soc.* (3) **27** (1973), 638–648. **MR 48 #12062.**   3.4, 6.4

[1978]   Automatic continuity: a survey, *Bull. London Math. Soc.* **10** (1978), no. 2, 129–183. **MR 80c: 46053.**   6.1, 6.2

[1979a]   Discontinuous homomorphisms from topological algebras, *Amer. J. Math.* **101** (1979), no. 3, 635–646. **MR 81i: 46059.**   6.2

[1979b]   A discontinuous homomorphism from $C(X)$, *Amer. J. Math.* **101** (1979), no. 3, 647–734. **MR 81g: 46066.**   6.2

[1981b]   Norming nil algebras, *Proc. Amer. Math. Soc.* **83** (1981), no. 1, 71–74. **MR 82j: 46069.**   1.1, 2.5

[1983a]   Convolution algebras on the real line, *Radical Banach algebras and automatic continuity (Long Beach, Calif.,* 1981), 180–209, *Lecture Notes in Mathematics,* 975, *Springer, Berlin–New York,* 1983. **MR 84m: 46052.**   5.1

[1983b]   The continuity of traces, *Radical Banach algebras and automatic continuity (Long Beach, Calif.,* 1981), 451–458, *Lecture Notes in Mathematics,* 975, *Springer, Berlin–New York,* 1983. **MR 84m: 46065.**

[1984]   Automatic continuity of homomorphisms from C*-algebras, *Functional analysis: surveys and recent results, III (Paderborn,* 1983), 197–218, *North-Holland Math. Stud.* 90, *North-Holland, Amsterdam–New York,* 1984. **MR 86g: 46084a.**   6.2

[1989]   On norms on algebras, *Conference on Automatic Continuity and Banach Algebras (Canberra,* 1989), 61–96, *Proc. Centre Math. Anal. Austral. Nat. Univ.,* 21, *Austral. Nat. Univ., Canberra,* 1989. **MR 90k: 46110.**   6.1

[1994]   Banach algebras and automatic continuity, London Mathematical Society Monographs, *Oxford University Press,* to appear.   3.4, 5.3, 6.1, 6.2, 6.4

Dales, H. Garth; Davie, Alexander M.

[1973]   Quasianalytic Banach function algebras, *J. Funct. Anal.* **13** (1973), 28–50. **MR 49 #7782.**   3.4

Dales, H. Garth; Esterle Jean

[1977]   Discontinuous homomorphisms from $C(X)$, *Bull. Amer. Math. Soc.* **83** (1977), no. 2, 257–259. **MR 55 #3791.**   6.2

Dales, H. Garth; Loy, Richard J.; Willis, George A.

[1992]   Homomorphisms and derivations from $B(E)$, *J. London. Math. Soc.,* to appear.   1.7, 6.2, 6.4

Dales, H. Garth; McClure, J. Peter

[1981]   Higher point derivations on commutative Banach algebras. III, *Proc. Edinburgh Math. Soc.* (2) **24** (1981), no. 1, 31–40. **MR 82j: 46062.**   6.4

[1987]   Nonstandard ideals in radical convolution algebras on a half-line, *Canad. J. Math.* **39** (1987), no. 2, 309–321. **MR 88m: 46067.**   5.1

Dales, H. Garth; Willis, George A.

[1983]   Cofinite ideals in Banach algebras, and finite-dimensional representations of group algebras, *Radical Banach algebras and automatic continuity (Long Beach, Calif.,* 1981), 397–407, *Lecture Notes in Mathematics,* 975, *Springer, Berlin–New York,* 1983. **MR 84h: 46061.**   3.4

Dales, H. Garth; Woodin, W. Hugh

[1987] An introduction to independence for analysts, *London Mathematical Society Lecture Note Series*, 115, *Cambridge University Press, Cambridge–New York*, 1987. 241 pp. **MR 90d: 03101.**   6.2

Dalla, Leoni; Giotopoulos, S.; Katseli, Nelli

[1989] The socle and finite-dimensionality of a semiprime Banach algebra, *Studia Math.* **92** (1989), *no.* 2, 201–204. **MR 90f: 46079.**   8.1, 8.2

Daniel, James W.; Palmer, Theodore W.

[1969] On $\sigma(T)$, $||T||$ and $||T^{-1}||$, *Linear Algebra and Appl.* **2** (1969), 381–386. **MR 40 #4788.**   2.6

Dauns, John

[1969] Multiplier rings and primitive ideals, *Trans. Amer. Math. Soc.* **145** (1969), 125–158. **MR 41 #1805.**   1.2, 7.4

[1974] The primitive ideal space of a C*-algebra, *Canad. J. Math.* **26** (1974), 42–49. **MR 49 #1131.**   7.4

Dauns, John; Hofmann, Karl Heinrich

[1966] The representation of biregular rings by sheaves, *Math. Z.* **91** (1966), 103–123. **MR 32 #4151.**   7.1, 7.4

[1968] Representation of rings by sections, *Memoirs American Mathematical Society no.* 83, *American Mathematical Society, Providence, R. I.*, 1968. **MR 40 #752.**   1.2, 7.1, 7.4

[1969] Spectral theory of algebras and adjunction of identity, *Math. Ann.* **179** (1969), 175–202. **MR 40 #734.**   1.2, 7.1, 7.4

Davie, Alexander M.

[1973a] A counterexample on dual Banach algebras, *Bull. London Math. Soc.* **5** (1973), 79–80. **MR 49 #3498.**   8.7, 8.8

[1973b] The approximation problem for Banach spaces, *Bull. London Math. Soc.* **5** (1973), 261–266. **MR 49 #3499.**   1.7

[1973c] Quotient algebras of uniform algebras, *J. London Math. Soc.* (2) **7** (1973), 31–40. **MR 48 #2779.**   3.1

Davis, William J.; Figiel, Tadeusz; Johnson, William B.; Pełczyński, Aleksander

[1974] Factoring weakly compact operators, *J. Funct. Anal.* **17** (1974), 311–327. **MR 50 #8010.**   1.7

Day, Mahlon M.

[1957] Amenable semigroups, *Ill. J. Math.* **1** (1957), 509–544. **MR 19-1067.**   9

[1973] Normed linear spaces, Third edition. Ergebnisse der Mathematik und ihrer Grenzgebiete, Band 21, *Springer-Verlag, New York–Heidelberg*, 1973. **MR 49 #9588.**   1.7, 9

[1976] Lumpy subsets in left-amenable locally compact semigroups, *Pacific J. Math.* **62** (1976), *no.* 1, 87–92. **MR 53 #14008.**   9

De Cannière, Jean

[1979] An illustration of the noncommutative duality theory for locally compact groups, *Bull. Soc. Math. Belg. Sér. B* **31** (1979), *no.* 1, 57–66. **MR 82k: 22006.**   1.9

De Cannière, Jean; Enock, Michel; Schwartz, Jean-Marie

[1979] Algèbres de Fourier associées à une algèbre de Kac, *Math. Ann.* **245** (1979), *no.* 1, 1–22. **MR 80k: 43009.**   1.9

De Cannière, Jean; Rousseau, Ronny J. E.

[1984] The Fourier algebra as an order ideal of the Fourier–Stieltjes algebra, *Math. Z.* **186** (1984), *no.* 4, 501–507. **MR 86b: 43016.**   1.9

Denk, Hans; Riederle, Max

[1982] A generalization of a theorem of Pringsheim, *J. Approx. Theory* **35** (1982), *no.* 4, 355–363. **MR 83j: 46061.**   2.0

Derighetti, Antoine

[1978]   Some remarks on $L^1(G)$, *Math. Z.* **164** (1978), *no.* 2, 189–194. **MR 80f: 43009.**
                                                                                    1.9
Dhar, R. K.; Vasudeva, H. L.

[1981] $L^1(I, X)$ with order convolution, *Proc. Amer. Math. Soc.* **83** (1981), *no.* 3, 499–505. **MR 82j: 46070.**
                                                                                    3.1
[1982] $M(R, X)$ with order convolution, *Math. Nachr.* **105** (1982), 271–279. **MR 84b: 43003.**
                                                                                    3.1
Di Bucchianico, Alessandro

[1991]   Banach algebras, logarithms, and polynomials of convolution type, *J. Math. Anal. Appl.* **156** (1991), *no.* 1, 253–273. **MR 92d: 46123.**
                                                                                    2.1
Dickson, Leonard Eugene

[1927]   Algebren und ihne Zahlentheorie, *Orell Füssiverlag, Zurich and Leipzig*, 1927.
                                                                          1.1, 4.0, 4.8
Diestel, Joe

[1984]   Sequences and series in Banach spaces. Graduate Texts in Mathematics, 92, *Springer-Verlag, New York–Berlin*, 1984. **MR 85i: 46020.**
                                                                                    1.7
Diestel, Joe; Uhl, John Jerry, Jr.

[1977]   Vector Measures. With a forward by B. J. Pettis. Mathematical Surveys No. 15, *American Mathematical Society, Providence, R. I.*, 1977. **MR 56 #12216.**
                                                              1.7, 1.9, 1.10, 4.2
[1983]   Progress in vector measures-1977–83, *Measure theory and its applications* (Sherbrooke, Que., 1982), 144–192, *Lecture Notes in Mathematics*, 1033, *Springer, Berlin–New York*, 1983. **MR 85g: 28008.**
                                                                                    1.7
Dieudonné, Jean

[1942]   Sur le socle d'un anneau et les anneaux simples infinis, *Bull. Soc. Math. France* **70** (1942), 46–75. **MR 6-144.**
                                                                              8.0, 8.2
Dirichlet, Peter Lejeune

[1829]   Sur la convergence des séries trigonométriques, *Journal für die reine und ange-wandte Mathematik. Herausgegaben von A. L. Crelle* **4** (1829), 157–169.   1.8
Ditkin, Vitalli Arsenievič

[1939]   On the structure of ideals in certain normed rings, *Uchenye Zapiski Moskov. Gos. Univ. Matematika* **30** (1939), 83–130. (Russian. English summary). **MR 1-336.**
                                                                                    1.1
Divinsky, Nathan J.

[1965]   Rings and radicals, Mathematical Expositions No. 14, *University of Toronto Press, Toronto, Ont.* 1965. **MR 33 #5654.**
                                                                          4.4, 4.7, 4.8
Dixmier, Jacques

[1957]   Les algèbres d'operateurs dans l'espace hilbertien, (Algèbres de von Neumann). *Cahiers scientifiques, Fascicule* XXV, *Gauthier-Villars, Paris*, 1957. **MR 20 #1234.**
                                                                                    6.4
[1968]   Ideal center of a C*-algebra, *Duke Math. J.* **35** (1968), 375–382. **MR 37 #5703.**
                                                                                    7.4
Dixon, Peter G.

[1972]   Generalized B*-algebras, II, *J. London Math. Soc.* (2) **5** (1972), 159–165. **MR 46 #4214.**
[1973a]   Approximate identities in normed algebras, *Proc. London Math. Soc.* (3) **26** (1973), 485–496. **MR 47 #9286.**
                                                                              5.1, 5.3
[1973b]   Semiprime Banach algebras, *J. London Math. Soc.* (2) **6** (1973), 676–678. **MR 47 #7432.**
                                                                              4.4, 4.8
[1974]   Locally finite Banach algebras, *J. London Math. Soc.* (2) **8** (1974), 325–328. **MR 50 #996.**
                                                                                    8.1
[1977]   Nonseparable Banach algebras whose squares are pathological, *J. Funct. Anal.* **26** (1977), *no.* 2, 190–200. **MR 56 #12888.**
                                                                                    5.3
[1978]   Approximate identities in normed algebras. II, *J. London Math. Soc.* (2) **17** (1978), *no.* 1, 141–151. **MR 80b: 46055.**
                                                                                    5.1
[1979b]   Spectra of approximate identities in Banach algebras, *Math. Proc. Cambridge Philos. Soc.* **86** (1979), *no.* 2, 271–278. **MR 80h: 46070.**
                                                                                    5.1

[1981]   Automatic continuity of positive functionals on topological involution algebras,
          Bull. Austral. Math. Soc. **23** (1981), *no.* 2, 265–281. **MR 82f: 46061**.
[1982]   An example for factorization theory in Banach algebras, *Proc. Amer. Math.
          Soc.* **86** (1982), *no.* 1, 65–66. **MR 83h: 46063**.   5.3
[1983]   On the intersection of the principal ideals generated by powers in a Banach alge-
          bra, *Radical Banach algebras and automatic continuity (Long Beach, Calif.*,
          1981), 340–341, *Lecture Notes in Mathematics*, 975, *Springer, Berlin–New
          York*, 1983. **MR 85c: 46043b**.   6.2
[1986]   Left approximate identities in algebras of compact operators on Banach spaces,
          Proc. Roy. Soc. Edinburgh Sect. A **104** (1986), *no.* 1-2, 169–175. **MR 88a:
          47044**.   5.1
[1987]   Spectra of left approximate identities in Banach algebras, *Bull. London Math.
          Soc.* **19** (1987), *no.* 2, 169–173. **MR 88b: 46074**.   5.1
[1990]   Factorization and unbounded approximate identities in Banach algebras, *Math.
          Proc. Cambridge Philos. Soc.* **107** (1990), *no.* 3, 557–572. **MR 91e: 46057**.
                                                                                             5.2
[1991]   Topologically nilpotent Banach algebras and factorization, *Proc. Roy. Soc. Ed-
          inburgh Sect.* A **119** (1991), *no.* 3-4, 329–341.**MR 93d: 46073**.   4.8
[1992]   Unbounded approximate identities in normed algebras, *Glasgow Math. J.* **34**
          (1992), *no.* 2, 189–192. **MR 93c: 46081**.   5.1

Dixon, Peter G.; Müller, Vladimir
[1992]   A note on topologically nilpotent Banach algebras, *Studia Math.*, **102** (1992),
          no. 3, 269–275. **MR 93d: 46074**.   4.8

Dixon, Peter G.; Willis, George A.
[1992]   Approximate identities in extensions of topologically nilpotent Banach algebras,
          Proc. Roy. Soc. Edinburgh, **122A** (1992), *no.* 1-2, 45–52.   4.8, 5.1, 8.1

Domar, Yngve
[1956]   Harmonic analysis based on certain commutative Banach algebras, *Acta Math.*
          96 (1956), 1–66. **MR 17-1228**.   1.9
[1983]   Extensions of the Titchmarsh convolution theorem with applications in the the-
          ory of invariant subspaces, *Proc. London Math. Soc.* (3) **46** (1983), *no.* 2,
          288–300. **MR 85b: 46062**.   3.4
[1987]   On the unicellularity of $\ell^p(\omega)$, *Monatsh. Math.* **103** (1987), *no.* 2, 103–113.
          MR 88c: 47050.   3.4

Donoghue, William F., Jr.
[1957]   The lattice of invariant subspaces of a completely continuous quasi-nilpotent
          transformation, *Pacific J. Math.* **7** (1957), 1031–1035. **MR 19-1066**.   3.4

Doran, Robert S.
[1980]   An application of idempotents in the classification of complex algebras, *Elem.
          Math.* **35** (1980), *no.* 1, 16–17. **MR 81b: 46065**.   2.1

Doran, Robert S.; Tiller, Wayne
[1983]   P-commutativity of the Banach algebra $L_1^{**}(G)$, *Manuscripta Math.* **43** (1983),
          no. 1, 85–86. **MR 85a: 46030**.   1.4

Doran, Robert S.; Wichmann, Josef
[1979]   Approximate identities and factorization in Banach modules, Lecture Notes in
          Mathematics, 768, *Springer-Verlag, Berlin–New York*, 1979. **MR 83e: 46044**.
                                                                                             5.1, 5.2

Douglas, Ronald G.
[1980]   C*-algebra extensions and K-homology. Annals of Mathematics Studies, 95,
          *Princeton University Press, Princeton, N. J.: University of Tokyo Press,
          Tokyo*, 1980. **MR 82c: 46082**.   10

Dowson, Henry R.
[1980]   Spectral theory of linear operators. London Mathematical Society Mono-
          graphs, 12, *Academic Press, Inc. [Harcort Brace Jovanovich, Publishers] Lon-
          don/New York*, 1978. xii+422 pp. **MR 80c: 47022**.   2.8

Dugundji, James; Granas, Andrzej
[1982]   Fixed point theory. I, Monografie Martematyczne [Mathematical Monographs]
          Państwowe Wydawnictwo Naukowe (PWN), Warsaw, 1982. 209 pp. **MR 83j:
          54038**.   general

van Dulst, Dick

[1978]   Reflexive and superreflexive Banach spaces. *Mathematical Centre Tracts*, 102, *Mathematisch Centrum, Amsterdam*, 1978. **MR 80d: 46019.**   1.7

van Dulst, Dick; Singer, Ivan

[1976]   On Kadec–Klee norms on Banach spaces, *Studia Math.* **54** (1975/76), *no.* 3, 205–211. **MR 52 #14937.**   1.7

Duncan, John

[1967]   B$^{\#}$ modular annihilator algebras, *Proc. Edinburgh Math. Soc.* (2) **15** (1966/67), 89–102. **MR 36 #6945.**   8.4, 8.6

[1981]   Banach algebra. Report on the 70's, prospects for the 80's, *Southeast Asian Bull. Math.* **5** (1981), *no.* 1-2, 4–15. **MR 83f: 46053.**   1.1

Duncan, John; Hosseiniun, Seyed Ali Reza

[1979]   The second dual of a Banach algebra, *Proc. Roy. Soc. Edinburgh Sect. A* **84** (1979), *no.* 3-4, 309–325. **MR 81f: 46057.**   1.4

Duncan, John; Paterson, Alan L. T.

[1990]   Amenability for discrete convolution semigroup algebras, *Math. Scand.* **66** (1990), 141–146. **MR 91m: 43001.**   1.9, 9

Duncan, John; Ülger, Ali

[1992]   Almost periodic functionals on Banach algebras, *Rocky Mountain J. Math.* **22** (1992),*no.* 3, 837–848. **MR 93h: 46057.**   1.4

Duncan, John; Williamson, John Hunter

[1982]   Spectra of elements in the measure algebra of a free group, *Proc. Roy. Irish Acad. Sect. A* **82** (1982), *no.* 1, 109–120. **MR 84a: 43005.**   4.8

Dunford, Nelson

[1938]   Uniformity in linear spaces, *Trans. Amer. Math. Soc.* **44** (1938), 305–356. 1.9

[1943]   Spectral theory, I. Convergence to projections, *Trans. Amer. Math. Soc.* **54** (1943), 185–217. **MR 5-39.**   3.3

[1954]   Spectral operators, *Pacific J. Math.* **4** (1954), 321–354. **MR 16-142.**   2.7

Dunford, Nelson; Schwartz, Jacob T.

[1958]   Linear operators Part I: General theory. With the assistance of W. G. Bade and R. G. Bartle. Pure and Applied Mathematics, Vol. 7, *Interscience Publishers, Inc., New York; Interscience Publishers, Ltd., London;* 1958. **MR 22 #8302.**   general

[1963]   Linear operators. Part II: Spectral theory. Self adjoint operators in Hilbert space. With the assistance of William G. Bade and Robert G. Bartle. *Interscience Publishers [John Wiley & Sons], New York–London,* 1963. **MR 32 #6181.**   general

[1971]   Linear operators. Part III: Spectral operators. With the assistance of William G. Bade and Robert G. Bartle. Pure and Applied Mathematics, Vol. VII. *Interscience Publishers [John Wiley & Sons], New York–London,* 1971. **MR 54 #1009.**   2.8

Dunkl, Charles F.; Ramirez, Donald E.

[1972]   Existence and norm uniqueness of invariant means on $L^{\infty}(\hat{G})$, *Proc. Amer. Math. Soc.* **32** (1972), 525–530. **MR 45 #5668.**   9

Dupré, Maurice J.; Gillette, Richard M.

[1983]   Banach bundles, Banach modules and automorphisms of C*-algebras. Research Notes in Mathematics, 92, *Pitman (Advanced Publishing Program), Boston, Mass.–London,* 1983. **MR 85j: 46127.**   7.1

Dutta, Turini Kumar

[1990]   Hermitian elements in projective tensor products, *Indian J. Pure Appl. Math.* **21** (1990), *no.* 7, 605–616. **MR 91k: 46047.**   2.6

Eberlein, William F.

[1955]   A note on Fourier–Stieltjes transforms, *Proc. Amer. Math. Soc.* **6** (1955), 310–312. **MR 16-817.**   3.1

Edwards, Robert E.

[1967]   Fourier series: A modern introduction. Vol I, *Holt, Rinehart and Winston, Inc., New York–Montreal, Quebec–London,* 1967. **MR 35 #7062.**   1.7

Effros, Edward G.

[1981]   Dimensions and C*-algebras. CBMS Regional Conference Series in Mathematics, 46, *Conference Board of the Mathematical Sciences, Washington, D. C.,* 1981, 74 pp. **MR 84K: 46042.**   10

[1988]   Amenability and virtual diagonals for von Neumann algebras, *J. Funct. Anal.* **78** (1988), no. 1, 137–153. **MR 89e: 46072.**   9

Eidelheit, Meier

[1940a]   On isomorphisms of rings of linear operators, *Studia Math.* **9** (1940), 97–105. (English. Ukrainian summary). **MR 2-224; 7-620; 3-51.**   1.1, 1.7, 6.0

[1940b]   Concerning rings of continuous functions, *Ann. of Math.* **41** (1940), 391–393. **MR 1-240.**   1.1

Eilenberg, Samuel; Moore, John C.

[1965]   Foundations of relative homological algebra, *Mem. Amer. Math. Soc. No.* 55 (1965), 39 pp. **MR 31 #2294.**   9

Elliott, George A.; Olesen, Dorte

[1974]   A simple proof of the Dauns–Hofmann theorem, *Math. Scand.* **34** (1974), 231–234. **MR 50 #8091.**   7.4

Enflo, Per

[1972]   Banach spaces which can be given an equivalent uniformly convex norm, Proceedings of the International Symposium on Partial Differential Equations and the Geometry of Normed Linear Spaces (Jerusalem, 1972) *Israel J. Math.* **13** (1972), 281–288. **MR 49 #1073.**   1.7

[1973]   A counterexample to the approximation problem in Banach spaces, *Acta. Math.* **130** (1973), 309–317. **MR 53 #6288.**   1.7, 8.8

[1976]   On the invariant subspace problem in Banach spaces, Séminaire Maurey–Schwartz (1975-1976), Espaces L^p, applications radonifiantes et géométrie des espaces de Banach, Exp. Nos. 14-15, 7 pp. *Centre Math. École Polytech. Palaiseau,* 1976. **MR 57 #13530.**   2.8

[1987]   On the invariant subspace problem in Banach spaces, *Acta. Math.* **158** (1987), no. 3-4, 213–313. **MR 88j: 47006.**   2.8

Enock, Michel; Schwartz, Jean-Marie

[1979]   Une dualité dans les algèbre de von Neumann, Bull. Soc. Math. France. Mém., No. 44. Supplément au Bull. Soc. Math. France. Tome 103, no. 4, *Société Mathematique de France, Paris,* 1975. 144 pp. **MR 56 #1091.**   1.9

Erdos, John A.

[1971]   On certain elements in C*-algebras, *Illinois J. Math.* **15** (1971), 682–693. **MR 44 #7305.**   8.7

Ernest, John

[1967]   Hopf–von Neumann algebras. *Functional Analysis (Proc. Conf., Irvine, Calif.,* 1966), pp. 195–215. *Academic Press, London; Thompson Book Co., Washington, D. C.,* 1967. **MR 36 #6956.**   1.9, 9

Eschmeier, Jörg; Putinar, Mihai

[1984]   Spectral theory and sheaf theory III. *J. Reine Angew. Math.* **354** (1984), 150–163. **MR 88b: 47041.**   3.5

Esterle, Jean

[1977]   Solution d'un problèm d'Erdös, Gillman et Henriksen et application à létude des homorphismes de $C(K)$, *Acta Math. Acad. Sci. Hungar.* **30** (1977), no. 1-2, 113–127. **MR 58 #2298.**   6.2

[1978a]   Semi-norms sur $C(K)$, *Proc. London Math. Soc.* (3) **36** (1978), no. 1, 27–45. **MR 58 #2297.**   6.2

[1978b]   Sur l'existence d'un homomorphisme discontinu de $C(K)$, *Proc. London Math. Soc.* (3) **36** (1978), no. 1, 46–58. **MR 58 #2299.**   6.2

[1978c]   Injection de semi-groupes divisibles dans des algèbres de convolution et construction d'homomorphismes discontinus de $C(K)$, *Proc. London Math. Soc.* (3) **36** (1978), no. 1, 59–85. **MR 58 #2300.**   6.2

[1979a]   A very discontinuous homomorphism from $A(D)(\ast)$, *Amer. J. Math.* **101** (1979), no. 5, 969–977. **MR 81g: 46071.**   6.2

[1979b]   Homomorphismes discontinus des algèbres de Banach commutatives séparables, (English summary), *Studia Math.* **66** (1979), no. 2, 119–141. **MR 81m: 46067.**   6.2

[1980]   Theorems of Gel'fand–Mazur type and continuity of epimorphisms from $C(K)$, *J. Funct. Anal.* **36** (1980), no. 3, 273–286. **MR 81m: 46068.**   6.2

[1981a]   Universal properties of some commutative radical Banach algebras, *J. Reine Angew. Math.* **321** (1981), 1–24. **MR 82i: 46078.**   4.8

[1981b]   Elements for a classification of commutative radical Banach algebras, *Radical Banach algebras and automatic continuity* (*Long Beach, Calif.*, 1981), 4–65, *Lecture Notes in Mathematics*, 975, Springer, Berlin–New York, 1983. **MR 84h: 46064.**   4.8

[1982]   Discontinuous homomorphisms from $C(K)$. *Spectral Theory* (*Warsaw*, 1977), 251–262, *Banach Center Publ.* 8 PWN, Warsaw, 1982. **MR 85g: 46062.**   6.2

[1983]   Polynomial connections between projections in Banach algebras, *Bull. London Math. Soc.* **15** (1983), no. 3, 253–254. **MR 84g: 46069.**   10

[1984]   Mittag-Leffler methods in the theory of Banach algebras and a new approach to Michael's problem, *Proceedings of the conference on Banach algebras and several complex variables* (*New Haven, Conn.*, 1983), 107–129, *Contemp. Math.*, 32, Amer. Math. Soc., Providence, R. I., 1984. **MR 86a: 46056.**   3.1

Esterle, Jean; Galé, José E.

[1982]   Regularity of Banach algebras generated by analytic semigroups satisfying some growth conditions, *Proc. Amer. Math. Soc.* **92** (1982), no. 3, 377–380. **MR 86a: 46055.**   3.4, 3.6

Esterle, Jean; Strouse, E.; Zouakia, F.

[1990]   Theorems of Katznelson–Tzafriri type for contractions, *J. Funct. Anal.* **94** (1990), 273–287. **MR 92c: 47016.**   2.6

Eymard, Pierre

[1964]   L'algèbre de Fourier d'un groupe localement compact, *Bull. Soc. Math. France* **92** (1964), 181–236. **MR 37 #4208.**   1.9, 9

Fack, Thierry Laurent

[1983]   K-théorie bivariante de Kasparov. [Kasparov's bivariant K-theory] *Bourbaki seminar, Vol. 1982/83*, 149–166, *Astérisque*, 105–106, Soc. Math. France, Paris, 1983. **MR 86f: 46075.**   10

Fack, Thierry Laurent; Maréchal, Odile

[1979]   Application de la K-théorie algébrique aux C*-algèbres, *Algèbres d'opérateurs* (*Sém., Les Plans-sur-Bex*, 1978), pp. 144–169, *Lecture Notes in Mathematics*, 725, Springer, Berlin, 1979. **MR 81i: 46093.**   10

Fack, Thierry Laurent; Skandalis, Georges

[1981]   Connes' analogue of the Thom isomorphism for the Kasparov groups, *Invent. Math.* **64** (1981), no. 1, 7–14. **MR 82g: 46113.**   10

Falcón Rodríguez, C. M.

[1988]   The denseness of the group of invertible elements of a uniform algebra, (Spanish. English summary), *Cienc. Mat.* (Havana) **9** (1988), no. 2, 11–17. **MR 91c: 46072.**   2.1

Falconer, K. J.

[1982]   Growth conditions on powers of Hermitian elements, *Math. Proc. Cambridge Philos. Soc.* **92** (1982), no. 1, 115–119. **MR 83j: 46058.** Corrigenda : *Math. Proc. Cambridge Philos. Soc.* **95** (1984), no. 3, 513–515. **MR 85g: 46058.** 2.6

Feichtinger, Hans G.

[1987]   Individual factorization in Banach modules, *Colloq. Math.* **51** (1987), 107–117. **MR 88k: 46052.**   5.2

Feichtinger, Hans G.; Graham, Colin C.; Lakien, Eric H.

[1979]   Nonfactorization in commutative, weakly selfadjoint Banach algebras, *Pacific J. Math.* **80** (1979), *no.* 1, 117–125. **MR 81f: 46060**.   5.2

Feldman, Chester

[1951]   The Wedderburn principal theorem in Banach algebras, *Proc. Amer. Math. Soc.* **2** (1951), 771–777. **MR 13-361**.   4.3, 6.0, 8.1

Fel'dman, I. A.; Gohberg, Israel Tzutikovič; Markus, A. S.

[1960]   Normally solvable operators and ideals associated with them (Russian. Moldavian summary), *Bull. Akad. Štiince RSS Moldoven, no.* 10, **76** (1960), 51–70. **MR 36 #2004**.   1.7

Fell, James Michael G.; Doran, Robert S.

[1988a]   Representations of *-algebras, locally compact groups, and Banach *-algebraic bundles. Vol. 1. Basic representation theory of groups and algebras, Pure and Applied Mathematics, 125, *Academic Press, Inc., Boston, MA*, 1988. 746 pp. **MR 90c: 46001**.   7.1

[1988b]   Representations of *-algebras, locally compact groups, and Banach *-algebraic bundles. Vol. 2. Banach *-algebraic bundles, induced representations, and the generalized Mackey analysis, Pure and Applied Mathematics, 126, *Academic Press, Inc., Boston, MA*, 1988. pp. 747–1486. **MR 90c: 46002**.   7.1

Fernández López, Antonio; Florencio, Miguel; Paúl, Pedro J.; Müller, Vladimir

[1990]   Extensions of topological algebras, *Collect. Math.* **40** (1989), *no.* 1, 55–65. **MR 91h: 46100**.   2.5

Fernández López, Antonio; Rodríguez–Palacios, Ángel

[1985]   A note on annihilator algebras, *Quart. J. Math. Oxford Ser.* (2) **36** (1985), *no.* 143, 279–281. **MR 86m: 46049**.   2.3, 8.8

[1986a]   A Wedderburn theorem for nonassociative complete normed algebras, *J. London Math. Soc.* (2) **33** (1986), *no.* 2, 328–338. **MR 88d: 46096**.   8.1

[1986b]   On the socle of a noncommutative Jordan algebra, *Manuscripta Math.* **56** (1986), *no.* 3, 269–278. **MR 88f: 17003**.   8.2

Figà-Talamanca, Alessandro

[1979a]   A remark on multipliers of the Fourier algebra of the free group, (Italian summary), *Boll. Un. Mat. Ital.* A(5) **16** (1979), *no.* 3, 577–581. **MR 81c: 43003**.   1.9

Filali, M.

[1992]   The ideal structure of some Banach algebras, *Math. Proc. Cambridge Philos. Soc.* **111** (1992), *no.* 3, 567–576. **MR 93e: 22005**.   1.4

Fitting, Hans

[1935]   Primärkomponentenzerlegung in nichtkommutativen Ringen, *Math. Ann.* **111** (1935), 19–41.   4.8

Flinn, P. H.; Smith, Roger R.

[1984] M-structure in the Banach algebra of operators on $C_0(\Omega)$, *Trans. Amer. Math. Soc.* **281** (1984), *no.* 1, 233–242. **MR 86j: 46049**.   2.6

Foiaş, Ciprian

[1963]   Spectral maximal spaces and decomposable operators in Banach spaces, *Arch. Math.*, (1963), 341–349. **MR 27 #2865**.   2.8

Ford, James W. M.

[1966]   Subalgebras of Banach algebras generated by semigroups, *Thesis, University of Newcastle on Tyne*, 1966.   3.4

[1967]   A square root lemma for Banach (*)-algebras, *J. London Math. Soc.* **42** (1967), 521–522. **MR 35 #5950**.   3.4

Forrest, Brian

[1988]   Amenability and derivations of Fourier algebras, *Proc. Amer. Math. Soc.* **104** (1988), *no.* 2, 437–442. **MR 89c: 43001**.   6.4, 9

[1991]   Arens regularity and discrete groups, *Pacific J. Math.* **151** (1991), *no.* 2, 217–227. **MR 93c: 43001**.   1.4

[1992a]   Some Banach algebras without discontinuous derivations, *Proc. Amer. Math. Soc.* **114** (1992), *no.* 4, 965–970. **MR 92g: 43001.**   6.4

[1992b]   Complemented ideals in Fourier algebras and the Radon Nikodým property, *Trans. Amer. Math. Soc.* **333** (1992), *no.* 2, 689–700. **MR 92m: 43003.**   1.9

Fourier, Jean Baptiste Joseph

[1822]   Théorie analytique de la chaleur, *F. Didot, Paris,* 1822; English Translation: Analytic theory of heat, *Dover, New York,* 1955.   1.8

Fredholm, Eric Ivar

[1900]   Sur une nouvelle mëthode pour la rësolution du problème de Dirichlet, *Förh. Svenska Tek. Vetensk Akad. Finl.* **54** (1900), 34–46.   1.8, 2.0

Freundlich, Marianne

[1949]   Completely continuous elements of a normed ring, *Duke Math. J.* **16** (1949), 273–283. **MR 10-612.**   8.7

Friedberg, Stephen H.

[1979]   Compact multipliers on Banach algebras, *Proc. Amer. Math. Soc.* **77** (1979), *no.* 2, 210. **MR 80f: 43011.**   9

Frobenius, George

[1878]   Über lineare substitutionen und bilineare Formen, *J. Reine. Angew. Math.* **84** (1878), 1–63.   2.2

Fuster i Capilla, Robert; Marquina, Aguiló

[1984]   Geometric series in incomplete normed algebras, *Amer. Math. Monthly* **91** (1984), *no.* 1, 49–51. **MR 85g: 46059.**   2.2

Gaal, Steven A.

[1973]   Linear analysis and representation theory, Die Grundlehren der Mathematischen Wissenschaften, Band 198, *Springer-Verlag, New York-Heidelberg,* 1973. **MR 56 #5777.**   1.9

Galé, José E.

[1991]   Compactness properties of a locally compact group and analytic semigroups in the group algebra, Conference on Mathematical Analysis (El Escorial, 1989). *Publ. Mat.* **35** (1991), *no.* 1, 131–140. **MR 92i: 46055.**   5.3

Galé, José E.; Ransford, Thomas J.; White, M. C.

[1992]   Weakly compact homomorphisms, *Trans. Amer. Math. Soc.* **331** (1992),*no.* 2, 815–824. **MR 92h: 46074.**   9

Gamelin, Theodore W.

[1969]   Uniform algebras, *Prentice Hall, Inc., Englewood Cliffs, N. J.,* 1969. **MR 53 #14137.**   3.2, 3.5

Garcia Cruz, Marìa Antonia

[1985]   On the subsets of a Banach algebra with localized exponential spectrum, (Spanish. English summary), *Cienc. Mat.* (Havana) **6** (1985), *no.* 2, 70–73. **MR 87i: 46105.**   2.1

Gardner, Barry James

[1989]   Radical theory, Pitman Research Notes in Mathematics Series, 198, *Longman Scientific & Technical, Harlow; John Wiley and Sons, Inc., New York,* 1989. x + 199 pp. **MR 90h: 16019.**   4.7

Gardner, L. Terrell

[1965]   An invariance theorem for representations of Banach algebras, *Proc. Amer. Math. Soc.* **16** (1965), 983–986. **MR 32 #372.**   6.4

[1966]   Square roots in Banach algebras, *Proc. Amer. Math. Soc.* **17** (1966), 132–134. **MR 32 #6247.**   3.4

Garimella, Ramesh V.

[1987]   Continuity of derivations on some semiprime Banach algebra, *Proc. Amer. Math. Soc.* **99** (1987), *no.* 2, 289–292. **MR 88b: 46075.**   6.4

[1991]   Prime ideals and continuity of derivations on integral domains, *Indian J. Pure Appl. Math.* **22** (1991), *no.*12, 1013–1017. **MR 92k: 46088.**   6.4

Gaur, A. K.; Husain, Taqdir
[1989]   Spatial numerical ranges of elements of Banach algebras, *Internat. J. Math. Math. Sci.* **12** (1989), *no.* 4, 633–640. **MR 91a: 46052.**   2.6

Gelbaum, Bernard R.
[1959]   Tensor products of Banach algebras, *Canad. J. Math.* **11** (1959), 297–310. **MR 21 #2922.**   1.10, 3.2, 7.1
[1961]   Note on the tensor product of Banach algebras, *Proc. Amer. Math. Soc.* **12** (1961), 750–757. **MR 24 #A1039.**   1.10, 3.2, 7.1
[1962]   Tensor products and related questions, *Trans. Amer. Math. Soc.* **103** (1962), 525–548. **MR 25 #2406.**   1.10, 3.2, 4.3
[1965]   Tensor products over Banach algebras, *Trans. Amer. Math. Soc.* **118** (1965), 131–149. **MR 31 #2629.**   1.10
[1970]   Tensor products of Banach algebras. II, *Proc. Amer. Math. Soc.* **25** (1970), 470–474. **MR 41 #4238.**   1.10, 3.2, 7.1
[1980]   The Cayley–Hamilton theorem in Banach algebras, *Bull. Inst. Math. Acad. Sinica* **8** (1980), *no.* 4, 615–616. (Author's summary). **MR 82e: 46067.**   2.3

Gelfand, Israel Moiseevič
[1938]   Abstrakte funktionen und lineare operatoren, *Rec. Math. N. S.* **4** (1938), 235–284.   1.9
[1939a]   On normed rings, *Dokl. Akad. Nauk SSSR* **23** (1939), 430–432.   1.1
[1939b]   To the theory of normed rings II. On absolutely convergent trigonometric series and integrals, *Dokl. Akad. Nauk SSSR* **25** (1939), 570–572. **MR 1-330.**   1.1
[1939c]   To the theory of normed rings III. On the ring of almost periodic functions, *Dokl. Akad. Nauk SSSR* **25** (1939), 573–574. **MR 1-331.**   1.1
[1941a]   Normierte Ringe, *Rec. Math.* (Mat. Sbornik) N. S **51** (1941), 3–24. (German. Russian summary). **MR 3-51.**   1.1, 2.0, 2.2, 3.0, 3.1, 3.3, 6.0, 6.1
[1941b]   Ideale und primäre Ideale in normierten Ringen, *Rec. Math.* (Mat. Sbornik) N. S. 9 **51** (1941), 41–48. (German. Russian summary). **MR 3-52.**   1.1
[1941c]   Zur Theorie der Charaktere der Abelschen topologischen Gruppen, *Rec. Math.* (Mat. Sbornik) N. S. 9 **51** (1941), 49–50. (German. Russian summary). **MR 3-36.**   1.1, 2.6
[1941d]   Über absolut konvergente trigonometrische Reihen und Integrale, *Rec. Math.* (Mat. Sbornik) N. S. 9 **51** (1941), 51–66. (German. Russian summary). **MR 3-51.**   1.1
[1987]   Collected papers, Vol. I, With remarks by V. W. Guillemin and S. Sternberg, *Springer-Verlag, Berlin-New York,* 1987. 883 pp. **MR 89k: 01042a.**   1.1

Gelfand, Israel Moiseevič; Kolmogoroff, Andrei Nikolaevič
[1939]   On rings of continuous functions in topological spaces, *Dokl. Akad. Nauk SSSR* **22** (1939), 1–15.   1.1

Gelfand, Israel Moiseevič; Naĭmark, Mark Aronovič
[1943]   On the imbedding of normed rings into the ring of operators in Hilbert space, *Rec. Math.* (Mat. Sbornik) N.S. 12 **54** (1943), 197–213. (English. Russian summary). **MR 5-147.**   3.2, 6.2
[1948]   Normed rings with involutions and their representations, *Izvestiya Akad. Nauk SSSR Ser. Mat.* **12** (1948), 445–480. **MR 10-199.**   1.1

Gelfand, Israel Moiseevič; Raĭkov, Dmitriĭ Abramovič; Šilov, Georgi Evgenyevič
[1946]   Commutative normed rings. *Uspehi Mat. Nauk* (N.S.) **1** (1946), *no.* 2(12), 48–146. (Russian). **MR 10-258.** (English translation) *Amer. Math. Soc. Transl.* (2) 5 (1957), 115–220. **MR 18-714.**   1.1, 3.4
[1960]   Commutative normed rings. (Russian) *Sovremennye Problemy Matematiki. Gosudarstv. Izdat. Fiz.-Mat. Lit., Moscow,* 1960. 316 pp. **MR 23A #1242** (English translation) *Chelsea Publishing Co., New York,* (1964), 306 pp. **MR 34 #4940** cf. **MR 29 #6327** .   1.1

Gelfand, Israel Moiseevič; Šilov, Georgi Evgenyevič
[1941]   Über verschiedene Methoden der Einführung der Topologie in die Menge der maximalen Ideale eines normierten Ringes, *Rec. Math.* (Mat. Sbornik) N. S. 9 **51** (1941), 25–39. (German. Russian summary). **MR 3-52.**   1.1, 3.2, 7.1

Ghahramani, Fereidoun

[1980]   Homomorphisms and derivations on weighted convolution algebras, *J. London Math. Soc.* (2) **21** (1980), *no.* 1, 149–161. **MR 81j: 43009.**   1.9, 6.4

[1982]   Endomorphisms of $L^1(\mathbf{R}^+)$, *J. Math. Anal. Appl.* **85** (1982), *no.* 2, 308–314. **MR 83g: 46047.**   1.9

[1983]   Isomorphisms between radical weighted convolution algebras, *Proc. Edinburgh Math. Soc.* (2) **26** (1983), *no.* 3, 343–351. **MR 85h: 43002.**   1.9

[1984]   Weighted group algebra as an ideal in its second dual space, *Proc. Amer. Math. Soc.* **90** (1984), *no.* 1, 71–76. **MR 85i: 43007.**   1.4

[1989]   Automorphisms of weighted measure algebras, *Conference on Automatic Continuity and Banach Algebras, Canberra, 1989*, pp. 144–154, Proc. Centre Math. Anal. Austral. Nat. Univ., 21, *Austral. Nat. Univ., Canberra*, 1989. **MR 91c: 43001.**   1.9

Ghahramani, Fereidoun; Lau, Anthony To Ming

[1988]   Isometric isomorphisms between the second conjugate algebras of group algebras, *Bull. London Math. Soc.* **20** (1988), *no.* 4, 342–344. **MR 89e: 43008.**   1.4

Ghahramani, Fereidoun; Lau, Anthony To Ming; Losert, Viktor

[1990]   Isometric isomorphisms between Banach algebras related to locally compact groups, *Trans. Amer. Math. Soc.* **321** (1990), *no.* 1, 273–283. **MR 90m: 43010.**   1.4

Ghahramani, Fereidoun; McClure, J. Peter

[1990]   Automorphisms of radical weighted convolution algebras, *J. London Math. Soc.* 2 **41** (1990), *no.* 1, 122–132. **MR 91k: 46056.**   1.9, 3.4

[1992]   Module homomorphisms of the dual modules of convolution Banach algebras, *Canad. Math. Bull.* **35** (1992), *no.* 2, 180–185. **MR 93f: 43004.**   1.4, 1.9, 3.4

Ghahramani, Fereidoun; McClure, J. Peter; Grabiner, Sandy

[1990]   Standard homomorphisms and regulated weights on weighted convolution algebras, *J. Funct. Anal.* **91** (1990), *no.* 2, 278–286. **MR 91k: 43007.**   3.4

Gierz, Gerhard

[1982]   Bundles of topological vector spaces and their duality, Lecture notes in Mathematics, 955, *Springer, Berlin–New York*, 1982. iv+296 pp. **MR 84c: 46076.**   7.1

Gil de Lamadrid, Jesús

[1963]   Uniform cross norms and tensor products of Banach algebras, *Bull. Amer. Math. Soc.* **69** (1963), 797–803. **MR 28 #5353.**   1.10

[1965a] Uniform cross norms and tensor products of Banach algebras, *Duke Math. J.* **32** (1965), 359–368. **MR 32 #8125.**   1.10

[1965b] Measures and tensor products, *Trans. Amer. Math. Soc.* **114** (1965), 98–121. **MR 31 #3851.**   1.10

[1967]   Topological modules. Banach algebras, tensor products, algebras of kernels, *Trans. Amer. Math. Soc.* **126** (1967), 361–419. **MR 34 #4939.**   1.10, 2.4

[1984]   Extending positive definite linear forms, *Proc. Amer. Math. Soc.* **91** (1984), *no.* 4, 593–594. **MR 85g: 46068.**   9.4

Gil de Lamadrid, Jesús; Argabright, Loren N.

[1990]   Almost periodic measures, *Mem. Amer. Math. Soc.* **85** (1990), *no.* 428. 219 pp. **MR 90k: 43001.**   1.4

Gillman, Leonard; Jerison, Meyer

[1960]   Rings of continuous functions, The University Series in Higher Mathematics, *D. Van Nostrand, Princeton–Toronto–London–New York*, 1960. **MR 22 #6994.**   1.5

Giotopoulos, S.

[1985]   A note on annihilator Banach algebras, *Math. Proc. Cambridge Philos. Soc.* **97** (1985), *no.* 1, 101–106. **MR 86b: 46072.**   8.7

Gleason, Andrew M.

[1967]   A characterization of maximal ideals, *J. Analyse Math.* **19** (1967), 171–172. **MR 35 #4732.**   2.4

Glicksberg, Irving

[1964] Some uncomplemented function algebras, *Trans. Amer. Math. Soc.* **111** (1964), 121–137. **MR 28 #4383**.   3.1

[1973] Some remarks on absolute continuity on groups, *Proc. Amer. Math. Soc.* **40** (1973), 135–139. **MR 47 #9166**.   1.9

Godefroy, Gilles; Iochum, Bruno

[1988] Arens regularity of Banach algebras and the geometry of Banach spaces, *J. Funct. Anal.* **80** (1988), no. 1, 47–59. **MR 89j: 46051**.   1.4

Godement, Roger

[1958] Topologie algébrique et théorie des faisceaux, *Actualitiés Sci. Ind. no. 1252 Publ. Math. Univ. Strasbourg, no. 13, Hermann, Paris*, 1958. **MR 21 #1583**.
                                                                              3.5, 7.3, 7.4

Goldberg, Seymour

[1966] Unbounded linear operators: Theory and applications, *McGraw–Hill, New York–Toronto, Ont.–London*, 1966. **MR 34 #580**.   1.7, 2.8

Goldie, Alfred W.

[1960] Semi-prime rings with maximum condition, *Proc. London Math. Soc.* (3) **10** (1960), 201–220. **MR 22 #2627**.   4.8

Goldmann, Helmut

[1990] Uniform Fréchet algebras. North-Holland Mathematics Studies, 162, *North-Holland Publishing Co., Amsterdam*, 1990. 355 pp. **MR 91f: 46073**.   2.9

Goldstein, Jerome A.

[1985] Semigroups of linear operators and applications. Oxford Mathematical Monographs, *The Clarendon Press, Oxford University Press, New York*, 1985. 245 pp. **MR 87c: 47056**.   6.4

Gourdeau, Frédéric

[1989] Amenability of Banach algebras, *Math. Proc. Cambridge Philos. Soc.* **105** (1989), no. 2, 351–355. **MR 90a: 46125**.   9

Grabiner, Sandy

[1969] The nilpotency of Banach nil algebras, *Proc. Amer. Math. Soc.* **21** (1969), 510. **MR 38 #4995**.   4.4

[1971] A formal power series operational calculus for quasinilpotent operators, *Duke Math. J.* **38** (1971), 641–658. **MR 45 #2513**.   4.8

[1973] A formal power series operational calculus for quasi-nilpotent operators. II., *J. Math. Anal. Appl.* **43** (1973), 170–192. **MR 50 #10876**.   4.8

[1974] Derivations and automorphisms of Banach algebras of power series, *Memoirs American Mathematical Society No. 146, American Mathematical Society, Providence, R. I.*, 1974. **MR 54 #3410**.   3.4, 4.8, 6.4

[1975] Weighted shifts and Banach algebras of power series, *Amer. J. Math.* **97** (1975), 16–42. **MR 52 #8982**.   3.4, 4.8

[1976] Nilpotents in Banach algebras, *J. London Math. Soc.* (2) **14** (1976), no. 1, 7–12. **MR 56 #1064**.   4.4

[1981] Weighted convolution algebras on the half line, *J. Math. Anal. Appl.* **83** (1981), no. 2, 531–553. **MR 83c: 46045**.   4.8, 5.1

[1983] Weighted convolution algebras as analogues of Banach algebras of power series, *Radical Banach algebras and automatic continuity (Long Beach, Calif., 1981)*, 282–289, *Lecture Notes in Mathematics, 975, Springer, Berlin–New York*, 1983. **MR 84g: 46073**.   4.8, 5.1

[1984a] Compact endomorphisms and closed ideals in Banach algebras, *Proc. Amer. Math. Soc.* **92** (1984), no. 4, 547–548. **MR 85j: 46079**.   9

[1984b] The spectral diameter in Banach algebras, *Proc. Amer. Math. Soc.* **91** (1984), no. 1, 59–63. **MR 85k: 46057**.   2.2

[1984c] Extremely nonstandard ideals and noninjective operational calculi, *J. London Math. Soc.* (2) **30** (1984), no. 1, 129–135. **MR 86b: 46081**.   3.4

[1988] Homomorphisms and semigroups in weighted convolution algebras, *Indiana Univ. Math. J.* **37** (1988), no. 3, 589–615. **MR 90f: 43007**.   3.4

Grabiner, Sandy; Thomas, Marc P.

[1985]   Nonunicellular strictly cyclic quasinilpotent shifts on Banach spaces, *J. Operator Theory* **13** (1985), no. 1, 163–170. **MR 86j: 47040.**   3.4, 4.8

Granirer, Edmond E.

[1973]   Criteria for compactness and for discreteness of locally compact amenable groups, *Proc. Amer. Math. Soc.* **40** (1973), 615–624. **MR 49 #5712.**   9

[1974a]   Properties of the set of topological invariant means on P. Eymard's W*-algebra VN(G), *Nederl. Akad. Wetensch. Proc. Ser. A* **77** = *Indag. Math.* **36** (1974), 116–121. **MR 50 #10682.**   9

[1974b]   Weakly almost periodic and uniformly continuous functionals on the Fourier algebra of any locally compact group, *Trans. Amer. Math. Soc.* **189** (1974), 371–382. **MR 49 #1017.**   9

Graven, A. W. M.

[1979]   Injective and projective Banach modules, *Nederl. Akad. Wetensch. Indag. Math.* **41** (1979), no. 3, 253–272. **MR 81e: 46063.**   4.7

Gray, Mary W.

[1967]   Radical subcategories, *Pacific J. Math.* **23** (1967), 79–89. **MR 35 #5490.** 4.7

[1970]   A radical approach to algebra, *Addison–Wesley Publishing Co. Reading, Mass.–London–Dan Mills, Ont.*, 1970. **MR 42 #306.**   4.7, 4.8

Green, Michael D.

[1976]   Maximal one-sided ideals in Banach algebras, *Math. Proc. Cambridge Philos. Soc.* **80** (1976), no. 1, 109–111. **MR 53 #14129.**   5.2

Greenleaf, Frederick P.

[1965]   Norm decreasing homomorphisms of group algebras, *Pacific J. Math.* **15** (1965), 1187–1219. **MR 33 #3117.**   1.9

[1966]   Characterization of group algebras in terms of their translation operators, *Pacific J. Math.* **18** (1966), 243–276. **MR 34 #626.**   1.9

[1969]   Invariant means on topological groups and their applications, *Van Nostrand Mathematical Studies*, No. 16, *Van Nostrand; Reinhold Co., New York–Toronto, Ont.–London*, 1969. **MR 40 #4776.**   9

Grigoryan, Souren Arshakovič

[1984]   Polynomial extensions of commutative Banach algebras. (Russian), *Uspekhi Mat. Nauk* **39** (1984), no. 1 (235), 129–130. **MR 85j: 46086.**   3.4

[1989]   Generalized analytic functions in the sense of Arens and Singer. (Russian. English and Armenian summaries), *Izv. Akad. Nauk Armyan. SSR Ser. Mat.* **24** (1989), no. 3, 226–247, 306. **MR 91e: 46070.**   3.5

Grobler, J. Jacobus; Raubenheimer, H.

[1991]   Spectral properties of elements in different Banach algebras, *Glasgow Math. J.* **33** (1991), no. 1, 11–20. **MR 92a: 46054.**   2.5

Groenewald, Lenore; Raubenheimer, Heinrich

[1988]   A note on the singular and exponential spectrum in Banach algebras, *Quaestiones Math.* **11** (1988), no. 4, 399–408. **MR 89k: 46064.**   2.1, 2.5

Grønbaek, Niels

[1982a]   Power factorization in Banach modules over commutative radical Banach algebras, *Math. Scand.* **50** (1982), no. 1, 123–134. **MR 83i: 46055.**   4.8, 6.4

[1982b]   Discontinuous homomorphisms and the separating space, *Proc. Edinburgh Math. Soc.* (2) **25** (1982), no. 1, 35–39. **MR 83g: 46045.**   6.1

[1983]   Weighted discrete convolution algebras, *Radical Banach algebras and automatic continuity (Long Beach, Calif., 1981)*, 295–300, *Lecture Notes in Mathematics*, 975, *Springer, Berlin–New York*, 1983. **MR 84h: 46068.**   1.9, 4.8

[1988]   Amenability of weighted discrete convolution algebras on cancellative semigroups, *Proc. Roy. Soc. Edinburgh Sect. A.* **110** (1988), no. 3-4, 351–360. **MR 89k: 43004.**   1.9, 9

[1989a]   Commutative Banach algebras, module derivations, and semigroups, *J. London Math. Soc.* (2) **40** (1989), no. 1, 137–157. **MR 91b: 46045.**   1.9, 6.4

[1989b]   A characterization of weakly amenable Banach algebras, *Studia Math* **94** (1989), no. 2, 149–162. **MR 92a: 46055.**   9

[1989c] Constructions preserving weak amenability, *Conference on Automatic Continuity and Banach Algebras*, Canberra, 1989, 186–202, Proc. Centre Math. Anal. Austral. Nat. Univ. 21, Austral. Nat. Univ., Canberra, 1989. **MR 92a: 46056.** 9

[1990a] Amenability of discrete convolution algebras, the commutative case, Pacific J. Math. **143** (1990), no. 2, 243–249. **MR 91d: 43008.** 9

[1990b] Amenability of weighted convolution algebras on locally compact groups, Trans. Amer. Math. Soc. **319** (1990), no. 2, 765–775. **MR 90j: 43003.** 9

[1992] Weak and cyclic amenability for noncommutative Banach algebras, Proc. Edinburgh Math. Soc. (2) **35** (1992), no. 2, 315–328. **MR 93d: 46082.** 9

[1993] Morita equivalence for Banach algebras. A transcription, 53 page preprint, Københavens Universitet Matematisk Institut. 10

Grønbaek, Niels; Johnson, Barry E. Willis, George A.

[1993] Ammenability of Banach algebras of compact operators, preprint. 9

Grønbaek, Niels; Willis, George A.

[1993] Approximate identities in Banach algebras of compact operators, Canad. Math. Bull. **36** (1993), no. 1, 45–53. 5.1

Grone, Robert; Johnson, Peter D., Jr.

[1982] Spectral radius and seminorms in finite-dimensional algebras. II, Colloq. Math. **46** (1982), no. 1, 85–88. **MR 84j: 46079.** 2.2

Grosser, Michael

[1979a] Bidualräume und Vervollständigungen von Banach-moduln. (German) [Bidual spaces and completions of Banach modules] Lecture Notes in Mathematics, 717, Springer, Berlin, 1979. **MR 82i: 46075.** 1.4, 4.8

[1979b] $L^1(G)$ as an ideal in its second dual space, Proc. Amer. Math. Soc. **73** (1979), no. 3, 363–364. (Author's summary). **MR 82i: 43005.** 1.4

[1984] Arens semiregular Banach algebras, Monatsh. Math. **98** (1984), no. 1, 41–52. **MR 86d: 46042.** 1.4

[1987] Arens semiregularity of the algebra of compact operators, Ill. J. Math. **31** (1987), 544–573. **MR 90e: 46038.** 1.4

Grosser, Michael; Losert, Viktor

[1984] The norm-strict bidual of a Banach algebra and the dual of $C_u(G)$, Manuscripta Math. **45** (1984), no. 2, 127–146. (Summary). **MR 86b: 46073.** 1.4

Grosser, Siegfried; Moskowitz, Martin A.

[1967] On central topological groups, Trans. Amer. Math. Soc. **127** (1967), 317–340. **MR 35 #292.** 5.1

[1971a] Compactness conditions in topological groups, J. Reine Angew. Math. **246** (1971), 1–40. **MR 44 #1766.** 5.1

[1971b] Harmonic analysis on central topological groups, Trans. Amer. Math. Soc. **156** (1971), 419–454. **MR 43 #2165.** 7.4

Grothendieck, Alexander

[1954] Résumé des resultats essentiels dans la théorie des produits tensoriels topologiques et des espaces nucléaires, Ann. Inst. Fourier Grenoble **4** (1952), 73–112 (1954). **MR 15-879; 1140; MR 16-1336.** 1.4, 1.10

[1955] Produits tensoriels topologiques et espaces nucléaires, Mem. Amer. Math. Soc. **16** (1955). **MR 17-763.** 1.7, 1.10

[1956] Résumé de la théorie métrique des produits tensoriels topologiques, Bol. Soc. Mat. São Paulo **8** (1953), 1–79, (1956). **MR 20 #1194.** 1.7

[1960] Éléments de géométrie algébrique. I. Le langage des schémas. Rédigés avec la collaboration de J. Dieudonné, Inst. Hautes Études Sci. Publ. Math. No. 4 (1960), 228 pp. **MR 36 #177a.** 7.4

Guichardet, Alain

[1966] Sur l'homologie et la cohomologie des algèbres de Banach, C. R. Acad. Sci. Paris Sér. A-B **262** (1966), A38–A41. **MR 32 #8199.** 9

[1973] Cohomology of topological groups and positive definite functions, J. Multivariate Analysis **3** (1973), 249–261. **MR 49 #460.** 9

[1980]   Cohomologie des groupes topologiques et des algèbres de Lie. (French) [Cohomology of topological groups and Lie algebras] Textes Mathématiques, [Mathematical Texts], 2. *CEDIC, Paris*, 1980. **MR 83f: 22004.**   9

Gulick, Sidney L., 3d.

[1966a]   Commutativity and ideals in the biduals of topological algebras, *Pacific J. Math.* **18** (1966), 121–137. **MR 33 #3118.**   1.4

[1966b]   The bidual of a locally multiplicatively-convex algebra, *Pacific J. Math.* **17** (1966), 71–96. **MR 34 #627.**   1.4

Gulick, Sidney L., 3d; Liu, Teng Sun; van Rooij, Arnoud C. M.

[1967]   Group algebra modules. II, *Canad. J. Math.* **19** (1967), 151–173. **MR 36 #5713.**   5.2

Gunning, Robert C.; Rossi, Hugo

[1965]   Analytic functions of several complex variables, *Prentice–Hall, Inc., Englewood Cliffs, N. J.*, 1965. **MR 31 #4927.**   3.5

Gurarie, David

[1984]   Representations of compact groups on Banach algebras, *Trans. Amer. Math. Soc.* **285** (1984), no. 1, 1–55. **MR 86h: 22007.**   1.9

Gustafson, Karl

[1968]   Compact-like operators and the Eberlein theorem, *Amer. Math. Monthly* **75** (1968), 958–964. **MR 38 #4962.**   1.7

Haar, Alfred

[1933]   Der Massbegriff in der Theorie der Kontinuierlichen Gruppen, *Ann. of Math.* (2) **34** (1933), 147–169.   1.9

Halmos, Paul R.

[1950]   Measure theory, *D. Van Nostrand Co. Inc., New York, N. Y.*, 1950. **MR 11-504.**   1.9

[1971]   Capacity in Banach algebras, *Indiana Univ. Math. J.* **20** (1970/71), no. 9, 855–863. **MR 42 #3569.**   2.3

Hanche-Olsen, Harald

[1980]   A note on the bidual of a JB-algebra, *Math. Z.* **175** (1980), no. 1, 29–31. **MR 81m: 46079.**   1.4

Hanche-Olsen, Harald; Størmer, Erling

[1984]   Jordan operator algebras. Monographs and Studies in Mathematics, 21, *Pitman (Advanced Publishing Program), Boston, Mass,–London*, 1984. **MR 86a: 46092.**   6.3

de la Harpe, Pierre; Skandalis, G.

[1984]   Déterminant associé à une trace sur une algébre de Banach, *Ann. Inst. Fourier* (Grenoble) **34** (1984), no. 1, 241–260. **MR 87i: 46146a.**   10

Harris, Lawrence A.

[1981]   A generalization of C*-algebras, *Proc. London Math. Soc.* (3) **42** (1981), no. 2, 331–361. **MR 82e: 46089.**

Harte, Robin

[1972]   Spectral mapping theorems, *Proc. Roy. Irish Acad. Sect. A* **72** (1972), 89–107. **MR 48 #4738.**   3.5

[1973]   The spectral mapping theorem for quasicommuting systems, *Proc. Roy. Irish Acad. Sect. A* **73** (1973), 7–18. **MR 47 #7426.**   3.5

[1976]   The exponential spectrum in Banach algebras, *Proc. Amer. Math. Soc.* **58** (1976), 114–118. **MR 53 #11375.**   2.1

[1977]   A spectral mapping theorem for holomorphic functions, *Math. Zeit.* **154** (1977), no. 1, 67–69. **MR 56 #6391.**   3.5

[1981]   Invertibility, singularity and Joseph L. Taylor, *Proc. Roy. Irish Acad. Sect. A* **81** (1981), no. 1, 71–79. **MR 82m: 47003.**   3.5

[1987]   Regular boundary elements, *Proc. Amer. Math. Soc.* **99** (1987), no. 2, 328–330. **MR 88d: 46088.**   2.1

[1988]   Invertibility and singularity for bounded linear operators, Monographs and Textbooks in Pure and Applied Mathematics, 109, *Marcel Dekker, Inc., New York*, 1988. 590 pp. **MR 89d: 47001.**   2.8, 3.5

Hartmann, Klaus

[1981]   Pedersen ideal and group algebras, *Proc. Amer. Math. Soc.* **83** (1981), *no.* 1,
         183–188. **MR 82k: 46089.**   1.9

Hatori, Osamu

[1992]   Range transformations on a Banach function algebra, *Proc. Amer. Math. Soc.*
         116 (1992), *no.* 1, 149–156. **MR 92k: 46082.**   3.3

Hauenschild, Wilfried; Kaniuth, Eberhard; Kumar, Ajay

[1983]   Ideal structure of Beurling algebras on [FC]⁻ groups, *J. Funct. Anal.* **51** (1983),
         no. 2, 213–228. **MR 84m: 22007.**   1.9

Hawkins, Thomas

[1972]   Hypercomplex numbers, Lie groups, and the creation of group representation
         theory, *Arch. Hist. Exact Sci.* **8** (1972), 243–287.   1.2, 1.6, 4.0

Hayman, Walter Kurt; Kennedy, P. B.

[1976]   Subharmonic functions, Volume I, London Mathematical Society Monographs,
         No. 9,*Accademic Press [Harcourt Brace Javonovich, Publishers], London,
         New York,* 1976, xvii+284 pp.**MR 57 #665.**   2.3

Hebroni, Pessach

[1938]   Über linear Differentialgleichungen in Ringen, *Comp. Math.* **5** (1938), 403–429.
                                                                                1.1

Heinrich, Stefan

[1980]   Ultraproducts in Banach space theory, *J. Reine Angew. Math.* **313** (1980),
         72–104. **MR 82b: 46013.**   1.3

Helemskiĭ, Alexandr Yakovlevič

[1970a] The homological dimension of normed modules over Banach algebras, (Russian)
         Mat. Sb. (N. S.) **81(123)** (1970), 430–444. (Translated in *Math. USSR-Sb.* **10**
         (1970), 399–411.)MR 41 #7436.   9

[1970b] The homological dimension of the Banach algebras of analytic functions, (Rus-
         sian) *Mat. Sb.* (N. S.) **83(125)** (1970), 222–233. (Translated in *Math. USSR-Sb.*
         12 (1970), 221–233.) **MR 42 #6617.**   9

[1972a] A certain method of computing and estimating the global homology dimen-
         sion of Banach algebras, (Russian) *Mat. Sb.* (N. S.) **87(129)** (1972), 122–135.
         (Translated in *Math. USSR-Sb.* **16** (1972), 125–138.) **MR 45 #9135.**   9

[1972b] The global dimension of a functional Banach algebra is different from one, (Rus-
         sian), *Funkcional Anal. i Priložen* **6** (1972), *no.* 2, 95–96. (Translated in *Func-
         tional Anal. Appl.* **6** (1972), 166–168.) **MR 46 #4202.**   9

[1974]   The global dimension of a Banach function algebra is different from one, (Rus-
         sian) *Trudy Kaf. Teorii Funkcii i Funkcional. Anal. Moskov. Gos. Univ. No.*
         1 (1974), 39–58. **MR 53 #1273.**   9

[1981]   Homological methods in the holomorphic calculus of several operators in Banach
         spaces, after Taylor (Russian) *Uspekhi Mat. Nauk.* **36** (1981), *no.* 1 (217), 127–
         172, 248. (Translated in *Russian Math. Surveys,* **36**, *no.* 1 (1981).) **MR 82e:
         47020.**   3.5, 9

[1984]   Flat Banach modules and amenable algebras, (Russian) *Trudy Moskov. Mat.
         Obshch.* **47** (1984), 179–218, 247. (Translated in *Trans. Moskow Math.
         Soc.(1984),* (1985), pp. 229–250.) **MR 86g: 46108.**   9

[1989a] Some remarks about ideas and results of topological homology, *Conference on
         Automatic Continuity and Banach Algebras (Canberra,* 1989), 203–238, *Proc.
         Centre Math. Anal. Austral. Nat. Univer.,* 21, *Austral. Nat. Univ. Canberra,*
         1989. **MR 91c: 46099.**   3.5, 9

[1989b] The homology of Banach and topological algebras. Translated from Russian
         by Alan West. Mathematics and its applications (Soviet Series), 41, *Kluwer,
         Academic Publishers Group, Dordrecht,* 1989. xx+334 pp. **MR 92d: 46178.**
                                                                       1.10, 3.5, 4.7, 9

[1993]   Banach algebras and locally convex algebras. Translated from Russian by Alan
         West. Clarendon Press, *Oxford University Press, Oxford,* 1993. xii+462 pp.
                                                                          1.10, 2.9, 9

Helemskiĭ, Alexandr Yakovlevič; Šeinberg, Mark Viktovič

[1979]　Amenable Banach algebras, *Funktional Anal. i Priložen* **13** (1979), *no.* 1, 42–48, 96, *Trans. Functional Anal. Appl.* **13** (1979), 32–37. **MR 81g: 46061.**
9

Helgason, Sigurdur

[1956]　Multipliers of Banach algebras, *Ann. of Math.* (2) **64** (1956), 240–254. **MR 18-494.**
1.2

Helson, Henry

[1983]　Harmonic analysis, *Addison Wesley Publishing Co., Reading Mass.*, 1983. 190 pp. **MR 85e: 43001.**
1.9

Hennefeld, Julien O.

[1968]　A note on the Arens products, *Pacific J. Math.* **26** (1968), 115–119. **MR 37 #6755.**
1.4

[1979]　$L(X)$ as a subalgebra of $K(X)^{**}$, *Illinois J. Math.* **23** (1979), *no.* 4, 681–686. **MR 80j: 47057.**
1.4

[1980]　A note on M-ideals in $B(X)$, *Proc. Amer. Math. Soc.* **78** (1980), *no.* 1, 89–92. **MR 81b: 46021.**
1.4, 1.7

[1983]　M-ideals, related spaces and some approximation properties, *Banach space theory and its applications (Bucharest, 1981)*, 96–102, *Lecture Notes in Mathematics*, 991, *Springer, Berlin–New York*, 1983. **MR 85a: 46015.**
1.4, 1.7

Herstein, Israel N.

[1956]　Jordan homomorphisms, *Trans. Amer. Math. Soc.* **81** (1956), 331–341. **MR 17-938.**
6.3

[1957]　Jordan derivations of prime rings, *Proc. Amer. Math. Soc.* **8** (1957), 1104–1110. **MR 20 #2362.**
6.5

[1968]　Noncommutative rings, The Carus Mathematical Monographs, No. **15**, *Mathematical Association of America, Menascha, Wis., Distributed by John Wiley & Sons, Inc.*, 1968. **MR 37 #2790.**
7.1

Heuser, Harro G.

[1982]　Functional Analysis, Translated from the German by John Horváth, A Wiley-Interscience Publication. *John Wiley & Sons, Ltd., Chichester*, 1982. **MR 83m: 46001.**
2.1

Hewitt, Edwin

[1964]　The ranges of certain convolution operators, *Math. Scand.* **15** (1964), 147–155. **MR 32 #4471.**
5.2

Hewitt, Edwin; Ross, Kenneth A.

[1963]　Abstract harmonic analysis. Vol. I: Structure of topological groups. Integration theory, group representations, Die Grundlehren der Mathematischen Wissenschaften, Bd. 115, *Academic Press, Inc., Publishers, New York; Springer-Verlag, Berlin–Göttingen–Heidelberg*, 1963. **MR 28 #158.**
general

[1970]　Abstract harmonic analysis. Vol. II: Structure and analysis for compact groups. Analysis on locally compact Abelian groups, Die Grundlehren der Mathematischen Wissenschaften, Band **152**, *Springer-Verlag, New York–Berlin*, 1970. **MR 41 #7378**; erratum, *42*, p. 1825.
general

Hewitt, Edwin; Stromberg, Karl R.

[1965]　Real and abstract analysis. A modern treatment of the theory of functions of a real variable, *Springer-Verlag, New York*, 1965. **MR 32 #5826.**
1.8

Higman, Graham

[1956]　On a conjecture of Nagata, *Proc. Camb. Philos. Soc.* **52** (1956), 1–4. **MR 17-453.**
4.4

Higson, Nigel

[1988]　Algebraic K-theory of stable C*-algebras, *Adv. in Math.* **67** (1988), *no.* 1, 140 pp. **MR 89g: 46110.**
10

Hilbert, David

[1904]　Über das Dirichletsche Prinzip, *Math. Annalen* **59** (1904), 63–67.
1.8, 2.0

[1906]　Zur Variationsrechnung, *Math. Annalen* **62** (1906), 351–370.
1.7, 1.8, 2.0

[1912]   Über den Begriff der Klasse von Differentialgleichungen, *Math. Annalen* **73**
          (1912), 95–108.   2.0
[1970]   Gesammelte Abhandlungen, Band III: Analysis, Grundlagen der Mathematik,
          Physik, Verschiedenes Lebensgeschichte, *Springer-Verlag, Berlin–New York*,
          1970. **MR 41 #8201c**.   general

Hile, Gerald N.; Pfaffenberger, William E.

[1985]   Generalized spectral theory in complex Banach algebras, *Canad. J. Math.* **37**
          (1985), *no.* 6, 1211–1236. **MR 88b: 46072**.   10
[1987]   Idempotents in complex Banach algebras, *Canad. J. Math.* **39** (1987), *no.* 3,
          625–630. **MR 89a: 46107**.   10

Hille, Einar

[1948]   Functional analysis and semi-groups, American Mathematical Society Collo-
          quium Publications, vol. 31. *Amer. Math. Soc., New York*, 1948. **MR 9-594**.
                                                  1.1, 2.1, 4.1, 4.2, 4.3, 4.8
[1958]   On roots and logarithms of elements of a complex Banach algebras, *Math. Ann.*
          136 (1958), 46–57. **MR 20 #2632**.   3.4
[1959]   On the inverse function theorem in Banach algebras, *Calcutta Math. Soc.*
          Golden Jubilee Commemoration Vol. (1958/59), Part I, pp 65–69. Calcutta
          Math. Soc., Calcutta, 1963. **MR 27 #5141**.   3.5

Hille, Einar; Phillips, Ralph S.

[1957]   Functional analysis and semi-groups, American Mathematical Society Collo-
          quium Publications, vol. 31, rev. ed. *American Mathematical Society, Provi-
          dence, R. I.*, 1957. **MR 19-664**.   1.1, 1.9, 2.3

Hirschfeld, Rudi A.; Rolewicz, Stefan

[1969]   A class of non-commutative Banach algebras without divisors of zero, *Bull.
          Acad. Polon. Sci. Sér. Sci. Math. Astronom. Phys.* **17** (1969), 751–753. **MR
          40 #7802**.   4.8

Hirschfeld, Rudolf A.; Żelazko, Wiesław

[1968]   On spectral norm Banach algebras, *Bull. Acad. Polon. Sci. Sér. Sci. Math.
          Astronom. Phys.* **16** (1968), 195–199. **MR 37 #4621**.   3.1

Hochschild, Gerhard P.

[1942]   Semisimple algebras and generalized derivations, *Amer. J. Math.* **64** (1942),
          677–694. **MR 4-71**.   9
[1945]   On the cohomology groups of an associative algebra, *Ann. of Math.* (2) **46**
          (1945), 58–67. **MR 6-114**.   9
[1946]   On the cohomology theory for associative algebras, *Ann. of Math.* (2) **47** (1946),
          568–579. **MR 8-64**.   9
[1947]   Cohomology and representations of associative algebras, *Duke Math. J.*, **14**
          (1947), 921–948. **MR 9-267**.   1.2, 9

Hochwald, Scott H.; Morrel, Bernard

[1987]   Some consequences of left invertibility, *Proc. Amer. Math. Soc.* **100** (1987),
          no. 1, 109–110. **MR 89a: 46105**.   2.1

Holub, James R.

[1972]   Bounded approximate identities, *Bull. AUstral. Math. Soc.* **7** (1972), 443–445.
          MR 48: #6964.   5.1

Hopkins, Charles

[1939]   Rings with minimal condition for left ideals, *Ann. of Math.* **40** (1939), 712–730.
          MR 1-2.   4.8, 8.1

Hormander, Lars V.

[1966]   An introduction to complex analysis in several variables, *D. Van Nostrand Co.,
          Inc., Princeton, N. J.–Toronto, Ont.–London* 1966. **MR 34 #2933**. 3.2, 3.5

Howland, Richard A.

[1969]   Lie isomorphisms of derived rings of simple rings, *Trans. Amer. Math. Soc.*
          145 (1969), 383–396. **MR 40 #5661**.   6.3

Hu, Chuanpu

[1988]   A generalization of K-theory for complex Banach algebras, *Quart. J. Math. Oxford Ser* (2) **39** (1988), *no.* 155, 349–359. **MR 90b: 46135.**   10

Hu, Sze-Tsen

[1959]   Homotopy theory. Pure and Applied Mathematics. Vol. VIII, *Academic Press, New York–London*, 1959. **MR 21 #5186.**   3.5

Hua, Loo-Keng

[1949]   On the automorphisms of a field, *Proc. Nat. Acad. Sci. U.S.A.*, **35** (1949), 386–389. **MR 10-675.**   6.2

Huang, Dan Run

[1992]   Generalized inverses over Banach algebras, *Integr. Equat. Oper. Th.* **15** (1992), *no.* 3, 454–469. **MR 93a: 46092.**   1.6, 2.1

Husain, Taqdir

[1983]   Multiplicative functionals on topological algebras, Research Notes in Mathematics, 85. *Pitman (Advanced Publishing Program), Boston, Mass.–London,* 1983. 147 pp. **MR 85a: 46025.**   3.1

Husain, Taqdir; Srinivasan, Anitha

[1979]   Numerical range theory for pseudo-Banach algebras, *Kyungpook Math. J.* **19** (1979), *no.* 1, 75–84. **MR 80m: 46047.**   2.6

Igari, Satoru; Kanjin, Yuichi

[1979]   Homomorphisms of measure algebras on the unit circle, *J. Math. Soc. Japan* **31** (1979), *no.* 3, 503–512. **MR 80i: 43002.**   1.9

Illoussamen, E.; Oudadess, Mohamed

[1992]   Sur des criteres de commutativite dans les algebres de Banach, *C. R. Math. Rep. Acad. Sci. Canada* **14** (1992), *no.* 5, 183–188. **MR 86g: 46078.**   2.4, 3.1, 4.3

Inasaridze, Hverdi Nikolaecič

[1985]   K-theory of special normed algebras, *Uspekhi Mat. Nauk* **40** (1985), *no.* 4 (244), 169–170. **MR 87f: 46135.**   10

Inoue, Jyunji

[1971]   A note on the largest subalgebra of a Banach algebra, *J. Math. Soc. Japan* **23**, (1971), *no.* 2, 278–294. **MR 44 #7223.**   1.9, 3.6

Inoue, Jyunji; Takahasi, Sin-ei

[1992]   A note on the largest regular subalgebra of a Banach algebra, *Proc. Amer. Math. Soc.* **116** (1992), *no.* 4, 961–962. **MR 93b: 46099.**   3.2

Iochum, Bruno; Loupias, G.

[1989]   Arens regularity and local reflexivity principle for Banach algebras, *Math. Ann.* **284** (1989), *no.* 1, 23–40. **MR 90c: 46061.**   1.4

Isik, Nilgun; Pym, John S.; Ülger, Ali

[1987]   The second dual of the group algebra of a compact group, *J. London Math.* (2) **35** (1987), *no.* 1, 135–148. **MR 88f: 43012.**   1.4

Istrǎţescu, Vasile I.

[1969]   On maximum theorems for operator functions, *Rev. Roumaine Math. Pures Appl.* **14** (1969), 1025–1029. **MR 40: #4761.**   2.3

[1981]   Introduction to linear operator theory, Monographs and Textbooks in Pure and Applied Math., 65.*Marcel Dekker, Inc., New York* (2), 1981, xii + 579 pp. **MR 83d: 47002.**   2.6

Jacobson, Nathan

[1945a]   The radical and semi-simplicity for arbitrary rings, *Amer. J. Math.* **67** (1945), 300–320. **MR 7-2.**   2.1, 3.1, 4.1, 4.3, 4.8

[1945b]   A topology for the set of primitive ideals in an arbitrary ring, *Proc. Nat. Acad. Sci. U. S. A.* **31** (1945), 333–338. **MR 7-110.**   2.0, 3.2, 4.1, 4.8, 7.1

[1945c] Structure theory of simple rings without finiteness assumptions, *Trans. Amer. Math. Soc.* **57** (1945), 228–245. **MR 6-200.**   4.0

[1945d] Structure theory for algebraic algebras of bounded degree, *Ann. of Math.* (2) **46** (1945), 695–707. **MR 7-238.**   1.1, 2.3

[1947] On the theory of primitive rings, *Ann. of Math.* (2) **48** (1947), 8–21. **MR 8-433.**   8.3

[1956] Structure of rings, *American Mathematical Society Colloquium Publications*, vol. 37, *American Mathematical Society, Providence, R. I.* 1956. **MR 18-373.**   2.4, 4.1, 4.8

[1964] Structure of rings, *American Mathematical Society Colloquium Publications*, vol. 37, Revised edition, *American Mathematical Society, Providence, R. I.*, 1964. **MR 36 #5158.**   4.4

[1975] PI-algebras. An introduction. Lecture Notes in Mathematics 441, *Springer-Verlag, Berlin*, 1975. iv + 115 pp. **MR 51 #5654**   7.1

Jacobson, Nathan; Rickart, Charles E.

[1950] Jordan homomorphisms of rings, *Trans. Amer. Math. Soc.* **69** (1950), 479–502. **MR 12-387.**   6.3

[1952] Homomorphisms of Jordan rings of self-adjoint elements, *Trans. Amer. Math. Soc.* **72** (1952), 310–322. **MR 13-719.**   6.3

Jakobczak, Piotr

[1985] On a new integral formula in the symbolic calculus, *Comment. Math. Prace Mat.* **25** (1985), no. 2, 215–225. **MR 87j: 46094.**   3.5

James, Robert C.

[1951] A non-reflexive Banach space isometric with its second conjugate space, *Proc. Nat. Acad. Sci. U. S. A.* **37** (1951), 174–177. **MR 13-356.**   1.7

[1972] Super-reflexive Banach spaces, *Canad. J. Math.* **24** (1972), 896–904. **MR 47 #9248.**   1.7

Jameson, Graham James Oscar

[1987] Summing and nuclear norms in Banach space theory. London Mathematical Society Student Texts, 8, *Cambridge University Press, Cambridge–New York*, 1987. 174 pp. **MR 89c: 46020.**   1.7

Jarosz, Krzysztof

[1985a] Perturbations of Banach algebras. Lecture Notes in Mathematics, 1120, *Springer-Verlag, Berlin–New York*, 1985. **MR 86k: 46074.**   9

[1985b] Isometries in semisimple, commutative Banach algebras, *Proc. Amer. Math. Soc.* **94** (1985), no. 1, 65–71. **MR 86d: 46044.**   2.4

[1991] Generalizations of the Gleason–Kahane–Żelazko, *Rocky Mountain J. Math.* **21** (1991), 915–921.   5.2, 7.4, 9

Jewell, Nicholas P.

[1977] The existence of discontinuous module derivations, *Pacific J. Math.* **71** (1977), no. 2, 465–475. **MR 56 #9268.**   6.4

Jewell, Nicholas P.; Sinclair, Allan M.

[1976] Epimorphisms and derivations on $L^1(0, 1)$ are continuous, *Bull. London Math. Soc.* **8** (1976), no. 2, 135–139. **MR 53 #6326.**   6.1, 6.4

Johnson, Barry E.

[1964a] An introduction to the theory of centralizers, *Proc. London Math. Soc.* (3) **14** (1964), 299–320. **MR 28 #2450.**   1.2, 1.7, 4.5, 5.1

[1964b] Centralizers on certain topological algebras, *J. London Math. Soc.* **39** (1964), 603–614. **MR 29 #5115.**   1.2

[1966] Continuity of centralizers on Banach algebras, *J. London Math. Soc.* **41** (1966), 639–640. **MR 34 #629.**   1.2, 5.1, 5.2

[1967a] The uniqueness of the (complete) norm topology, *Bull. Amer. Math. Soc.* **73** (1967), 537–539. **MR 35 #2142.**   2.0, 2.3, 4.2, 6.0, 6.1

[1967b] A commutative semisimple annihilator Banach algebra which is not dual, *Bull. Amer. Math. Soc.* **73** (1967), 407–409. **MR 34 #6556.**   8.7

[1967c] Continuity of homomorphisms of algebras of operators, *J. London Math. Soc.* **42** (1967), 537–541. **MR 35 #5953.**   6.0, 6.1, 6.2

[1968a]   The Wedderburn decomposition of Banach algebras with finite dimensional rad-
          ical, *Amer. J. Math.* **90** (1968), 866–876. **MR 38 #1529**.   8.1, 9
[1968b]   Continuity of derivations and a problem of Kaplansky, *Amer. J. Math.* **90**
          (1968), 1067–1073. **MR 39 #776**.   6.4
[1969a]   Continuity of derivations on commutative algebras, *Amer. J. Math.* **91** (1969),
          1–10. **MR 39 #7433**.   6.4
[1969b]   Continuity of homomorphisms of algebras of operators. II, *J. London Math.
          Soc.* (2) **1** (1969), 81–84. **MR 40 #753**.   6.1, 6.2
[1972a]   Cohomology in Banach algebras, *Memoirs of the American Mathematical So-
          ciety* no. **127**, *American Mathematical Society, Providence, R. I.* (1972), 96
          pp. **MR 51 #11130**.   5.1, 6.4, 9
[1972b]   Approximate diagonals and cohomology of certain annihilator Banach algebras,
          Amer. J. Math. **94** (1972), 685–698. **MR 47 #5598**.   8.1, 9
[1975]   Introduction to cohomology in Banach algebras, *Algebras in analysis* (Prov.
          Instructional Conf. and NATO Advanced Study Inst., Birmington, 1973) pp.
          84–100. *Academic Press, London*, 1975. **MR 54 #5835**.   6.4, 9
[1976]   Norming $C(U)$ and related algebras, *Trans. Amer. Math. Soc.* **220** (1976),
          37–58. **MR 54 #3415**.   6.2
[1977]   Perturbations of Banach algebras, *Proc. London Math. Soc.* (3) **34** (1977), *no.*
          3, 439–458. **MR 56 #1094**.   9
[1980]   Low-dimensional cohomology of Banach algebras, *Operator algebras and appli-
          cation, Part 2 (Kingston, Ont.,* 1980), pp. 253–259, *Proc. Sympos. Pure Math.*
          38, *Amer. Math. Soc., Providence, R. I.*, 1982. **MR 84k: 46054**.   9
[1985]   Continuity of homomorphisms of Banach G-modules, *Pacific J. Math.* **120**
          (1985), *no.* 1, 111–121. **MR 87c: 46057**.   6.1
[1986]   Approximately multiplicative functionals, *J. London Math. Soc.* (2) **34** (1986),
          no. 3, 489–510. **MR 87k: 46105**.   9
[1987]   Continuity of generalized homomorphisms, *Bull. London Math. Soc.* **19** (1987),
          no. 1, 67–71. **MR 88i: 46066**.   9
[1988a]   Approximately multiplicative maps between Banach algebras, *J. London Math.
          Soc.* (2) **37** (1988), *no.* 2, 294–316. **MR 89h: 46072**.   9
[1988b]   Derivation from $L^1(G)$ into $L^1(G)$ and $L^\infty(G)$, *Harmonic Analysis,* (*Lux-
          embourg, 1987), 191–198 Lecture Notes Math.*, 1359, *Springer, Berlin–New
          York*, 1988. **MR 90a: 46122**.   6.4, 9
[1988c]   Perturbations of multiplication and homomorphisms, *Deformation theory of
          algebras and structures and applications (Il Ciocco,* 1986) 565–579. *NATO
          Adv. Sci. Inst. Sec. C: Math. Phys. Sci.* **247** *Kluwer Acad. Publ. Dordrecht*,
          1988. **MR 90c: 46062**.   9
[1991a]   Weak amenability of group algebras of connected complex semisimple Lie
          groups, *Proc. Amer. Math. Soc.* **111** (1991), *no.* 1, 177–185. **MR 91k: 43009**.
                                                                                           9
[1991b]   Weak amenability of group algebras, *Bull. London Math. Soc.* **23** (1991), *no.*
          3, 281–284. **MR 92k: 43004**.   9
[1992]   Weakly compact homomorphisms between Banach algebras, *Math. Proc. Cam-
          bridge Philos. Soc.* **112** (1992), *no.* 1, 157–163. **MR 93d: 460578**.   9

Johnson, Barry E.; Ringrose, John R.
[1969]   Derivations of operator algebras and discrete group algebras, *Bull. London
          Math. Soc.* **1** (1969), 70–74. **MR 39 #6091**.   9

Johnson, Barry E.; Sinclair, Allan M.
[1968]   Continuity of derivations and a problem of Kaplansky, *Amer. J. Math.* **90**
          (1968), 1067–1073. **MR 39 #776**.   6.4

Johnson, Peter D., Jr.
[1978]   Spectral radius and seminorms in finite-dimensional algebras. *Colloq. Math.* **39**
          (1978), *no.* 2, 331–341. **MR 80e: 46032**.   2.2

Jordan, Pascual
[1932]   Über eine Klasse nichtassociativer hypercomplexer algebren, Nachr. Ges. Wiss.
          Götingen (1932), 569–575.   6.3

Jordan, Pascual; von Neumann, John; Wigner, Eugene P.
[1934]   On an algebraic generalization of the quantum mechanical formalism, *Ann. of
          Math.* (2) **36** (1934), 29–64.   6.3

Julg, Pierre; Valette, Alain

[1985] Group actions on trees and K-amenability, *Operator algebras and their connections with topology and ergodic theory (Buşteni, 1983)*, 289–296, *Lecture Notes in Mathematics*, 1132, *Springer, Berlin–New York*, 1985. **MR 86m: 46064.** 9

Jun, Kil Woung

[1986] Homomorphisms on Banach algebras, *J. Korean Math. Soc.* **23** (1986), *no.* 1, 35–41. **MR 87g: 46081.** 6.1

Jun, Kil Woung; Lee, Young-Whan; Park, Dal-Won

[1990] The continuity of derivations and module homomorphisms, *Bull. Korean Math. Soc.* **27** (1990), *no.* 2, 197–206. **MR 92b: 46076.** 6.4

Kadison, Richard V.; Ringrose, John R.

[1967] Derivations and automorphisms of operator algebras, *Comm. Math. Phys.* **4** (1967), 32–63. **MR 34 #6552.** 6.4

[1986] Fundamentals of the theory of operator algebras, Vol. II, Advanced theory, Pure and Applied Mathematics, **100**, *Academic Press, Inc., Orlando, Fla.*, 1986. pp. 399–1074. **MR 88d: 46106.** 1.10

Kahane, Jean-Pierre

[1970] Series de Fourier absolument convergentes, Ergebnisse der Mathematik und ihrer Grenzgebiete, Band 50, *Springer-Verlag, Berlin–New York*, (1970). viii+169 pp. **MR 43 #801.** 1.10

Kahane, Jean-Pierre; Żelazko, Wiesław

[1968] A characterization of maximal ideals in commutative Banach algebras, *Studia Math.* **29** (1968), 339–343. **MR 37 #1998.** 2.4

Kaijser, Sten

[1976] Some remarks on injective Banach algebras, Spaces of analytic functions (Sem. Funct. Anal. and Function theory, Kristiansand, 1975) 84–95. *Lecture Notes in Mathematics*, 512, *Springer, Berlin*, 1976. **MR 58 #17879.** 1.5, 1.7, 4.2, 4.7

[1981] On Banach modules. I, *Math. Proc. Cambridge Philos. Soc.* **90** (1981), *no.* 3, 423–444. **MR 83f: 46057.** 4.2

[1983] A simple-minded proof of the Pisier–Grothendieck inequality, *Banach spaces, harmonic analysis, and probability theory (Storrs, Conn.)*, 1980/1981, 33–44, *Lecture Notes in Mathematics*, 995, *Springer, Berlin–New York*, 1983. **MR 85d: 46097.** 1.10

Kajetanowicz, Przemysław

[1984] Extreme spectral states and multiplicative extensions in Banach algebras, *Colloq. Math.* **48** (1984), *no.* 1, 111–116. **MR 86a: 46052.** 2.4

Kakutani, Shizuo

[1938] Iteration of linear operations in complex Banach spaces, *Proc. Imp. Acad. Tokyo* **14** (1938), 295–300. 1.7

Kallin, Eva

[1963] A nonlocal function algebra, *Proc. Nat. Acad. Sci. U. S. A.* **49** (1963), 821–824. **MR 27 #2878.** 1.1, 3.2, 3.5

[1965] Polynomial convexity: The three sphere problem, Proceedings of the Conference on Complex Analysis, Minneapolis, 1964, 301–304, *Springer, Berlin*, 1965. **MR 31 #3631.** 3.5

Kalton, Nigel J.

[1988] Sums of idempotents in Banach algebras, *Canad. Math. Bull.* **31** (1988), *no.* 4, 448–451. **MR 89i: 46052.** 10

Kalton, Nigel J.; Wood, Geoffrey V.

[1976] Homomorphisms of group algebras with norm less than $\sqrt{2}$, *Pacific J. Math.* **62** (1976), *no.* 2, 439–460. **MR 54 #3296.** 1.9

Kametani, Shunji

[1952] An elementary proof of the fundamental theorem of normed fields, *J. Math. Soc. Japan* **4** (1952), 96–99. **MR 14-240.** 2.2

Kamowitz, Herbert

[1962]   Cohomomology groups of commutative Banach algebras, *Trans. Amer. Math. Soc.* **102** (1962), 352–372. **MR 30 #458.**   9

[1980]   Compact endomorphisms of Banach algebras, *Pacific J. Math.* **89** (1980), no. 2, 313–325. **MR 82c: 46063.**   9

[1981a] On compact multipliers of Banach algebras, *Proc. Amer. Math. Soc.* **81** (1981), no. 1, 79–80. **MR 81j: 47020.**   9

[1981b] Compact weighted endomorphisms of $C(X)$, *Proc. Amer. Math. Soc.* **83** (1981), no. 3, 517–521. **MR 82m: 47020.**   9

Kamowitz, Herbert; Scheinberg, Stephen

[1969]   Derivations and automorphisms of $L^1(0,1)$, *Trans. Amer. Math. Soc.* **135** (1969), 415–427. **MR 38 #1533.**   3.4, 6.1

Kandelaki, Tamaz Konstantinovič

[1990]   K-theory of Z_2-graded Banach categories,I. *K-theory and homological algebra* (*Tbilisi*, 1987-88), 180–221, Lecture Notes in Math., 1437, Springer, Berlin, 19890. **MR 92c:46078.**   10

Kaniuth, Eberhardt; Schlichting, Günter

[1970]   Zur harmonischen analyse klassenkompakter gruppen, II, *Invent. Math.* **10** (1970), 332–345. **MR 47 #5521.**   7.4

Kaniuth, Eberhardt; Steiner, Detlef

[1973]   On complete regularity of group algebras, *Math. Ann.* **204** (1973), 305–329. **MR 57 #10361.**   7.4

Kaplan, Samuel

[1985]   The bidual of $C(X)$, I. North-Holland Mathematics Studies, 101, *North-Holland Publishing Co., Amsterdam–New York*, 1985. xv+423 pp. **MR 86k: 46001.**   1.4

Kaplansky, Irving

[1947a] Topological rings, *Amer. J. Math.* **69** (1947), 153–183. **MR 8-434.** 2.1, 2.2, 4.2

[1947b] Semi-automorphisms of rings, *Duke Math. J.* **14** (1947), 521–525. **MR 9-172.**   6.2, 6.3

[1948a] Dual rings, *Ann. of Math.* (2) **49** (1948), 689–701. **MR 10-7.**   8.7

[1948b] Locally compact rings, *Amer. J. Math.* **70** (1948), 447–459. **MR 9-562.**   2.2, 8.7

[1948c] Topological rings, *Bull. Amer. Math. Soc.* **54** (1948), 809–826. **MR 10-179.**   2.4

[1948d] Regular Banach algebras, *J. Indian Math. Soc.* **12** (1948), 57–62. **MR 10-549.**   2.1

[1948e] Rings with a polynomial identity, *Bull. Amer. Math. Soc.* **54** (1948), 575–580. **MR 10-7.**   7.1

[1949a] Normed algebras, *Duke Math. J.* **16** (1949), 399–418. **MR 11-115.**   2.4, 2.5, 3.5, 4.3, 6.1, 6.2, 7.1, 7.2

[1949b] Primary ideals in group algbras, *Proc. Nat. Acad. Sci. U. S. A.* **35** (1949), 133–136. **MR 10-428.**   1.9, 7.3

[1950]   Topological representations of algebras. II, *Trans. Amer. Math. Soc.* **68** (1950), 62–75. **MR 11-317.**   7.1

[1952]   Symmetry of Banach algebras *Proc. Amer. Math. Soc.* **3** (1952), 396–399. **MR 14-58.**   6.4

[1953]   Modules over operator algebras, *Amer. J. Math.* **75** (1953), 839–858. **MR 15-327.**   9

[1954]   Ring isomorphisms of Banach algebras, *Canadian J. Math.* **6** (1954), 374–381. **MR 16-49.**   8.1, 8.6

[1958a] Functional analysis. Some aspects of analysis and probability, 1–34, Surveys in Applied Mathematics. Vol 4. *John Wiley & Sons, Inc., New York: Chapman & Hall, Ltd, London*, 1958. **MR 21 #286.**   6.4

[1958b] Derivations of Banach algebras, Seminar on Analytic Functions. Vol II. *Institute for Advanced Study, Princeton, N. J.*, 1958.   6.4

Karakhanyan, M. I.

[1984]   Asymptotic properties of commutators of elements of Banach algebras. (Russian. English and Armenian summaries), *Izv. Akad. Nauk Armyan. SSR Ser. Mat.* **19** (1984), no. 3, 242–247, 258. (English translation: *Soviet J. Contemp. Math. Anal.* **19** (1984), /em no. 3, 56–61.) **MR 86f: 46049.**   2.4

Kato, Tosio

[1958]   Perturbation theory for nullity, deficiency, and other quantities of linear opera-
         tors, *J. Analyse Math.* **6** (1958), 261–322. **MR 21 #6541**.   1.7
[1966]   Perturbation theory for linear operators. Die Grundlehren der mathematischen
         Wissenschaften Band *132*, *Springer-Verlag, New York*, 1966. **MR 34 #3324**.
                                                                                         1.7

Katseli, Nelli

[1985]   Compact elements of Banach algebras, *Bull. Soc. Math. Grèce* (N.S.) **26** (1985),
         65–72. **MR 87h: 46106**.   8.2

Katznelson, Yitzhak; Tzafriri, Lior

[1986]   On power bounded operators, *J. Funct. Anal.* **68** (1986), *no.* 3, 313–328. **MR
         88e: 47006**.   2.6

Kelley, John L.

[1955]   General topology, *D. Van Nostrand Co., Inc., Toronto–New York–London*,
         1955. **MR 16-1136**.   1.1

Kelley, John L.; Namioka, Isaac *et al.*

[1963]   Linear topological spaces. University Series in Higher Math., *D. Van Nostrand
         Co., Princeton, N. J.*, 1963. **MR 29 #3851**.   1.1

Kemer, Aleksandr Robertovič

[1991]   Ideals of identities of associative algebras, Translated from the Russian by C. W.
         Kohls. Translations of Mathematical Monographs 87, *American Mathematical
         Society, Providence, R. I.*, 1991. vi.+81 pp.   7.1

Kesler, S. Š.; Krupnik, Naum Ya.

[1967]   The invertibility of matrices over a ring (Russian), *Kišinev Gos. Univ. Učen.
         Zap.*, **91** 1967, 51–54. **MR 41 #5386**.   1.6

Khourguine, Iakov Isaevič; Tschetinine, N.

[1940]   Sur les sous-anneaux fermés de l'anneau des fonctions à n derivées continues.
         Dokl. Akad. Nauk SSSR **29** (1940), 288–291. **MR 2-223**.   1.1

Killing, Wilhelm

[1888]   Die Zusammensetzung der endlichen stetigen transformations gruppppen, Erster
         Theil *Math. Ann.* **31** (1888), 252–290; Zweiter Theil *Math. Ann.* **33** (1889),
         1–48; Dritter Theil *Math. Ann.*34 (1889), 57–122; Vierten Theil *Math. Ann.*
         36 (1890), 181–189.   4.8

Kim, Byung–Do; Jun, Kil–Woung

[1987]   Hermitian elements of a Banach algebra, *J. Korean Math. Soc.* **24** (1987), *no.*
         1, 21–23. **MR 88e: 46039**.   2.6

Kiran, Shashi; Singh, Ajit Iqbal

[1988]   The *J*-sum of Banach algebras and some applications, *Yokohama Math. J.* **36**
         (1988), *no.* 1, 1–20. **MR 90a: 46115**.   1.7

Kislyakov, Sergei Vitalyevič

[1989]   Proper uniform algebras are uncomplemented, (Russian) *Dokl. Akad. Nauk
         (SSSR)* **309** (1989), *no.* 4, 795–798. **MR 91e: 46065**.   3.1

Kitchen, Joseph W. Jr.; Robbins, David A.

[1982]   Gel'fand representation of Banach modules, *Dissertationes Math.* (*Rozprawy
         Mat.*) **203** (1982), 47 pp. **MR 85g: 46060**.   7.1
[1983]   Sectional representations of Banach modules, *Pacific J. Math.* **109** (1983), *no.*
         1, 135–156. **MR 85a: 46026**.   7.1
[1985]   Two notions of spectral synthesis for Banach modules, *Tamkang J. Math.* **16**
         (1985), *no.* 2, 59–65. **MR 87h: 46109**.   7.1

Kleinecke, David C.

[1957]   On operator commutators, *Proc. Amer. Math. Soc.* **8** (1957), 535–536. **MR
         19-435**.   6.4
[1963]   Almost-finite, compact and inessential operators, *Proc. Amer. Math. Soc.* **14**
         (1963), 863–868. **MR 27 #5136**.   1.7, 8.8

Koh, Kwangil

[1971]   On functional representations of a ring without nilpotent elements, *Canad. Math. Bull.* **14** (1971), 349–352. **MR 51 #5673**.   7.4

[1972]   On a representation of a strongly harmonic ring by sheaves, *Pacific J. Math.* **41** (1972), 459–468. **MR 47 #8626**.   7.1, 7.4

Kokk, Arne

[1989]   A joint spectrum and extension of homomorphisms of topological algebras, *Tartu Riikl Ül Toimetised No*, 836 (1989), 95–110. **MR 90i: 46087**.   3.5

Koosis, Paul J.

[1964]   Sur un théorème de Paul Cohen, *C. R. Acad. Sci. Paris* **259** (1964), 1380–1382. **MR 30 #2295**.   5.2

Köthe, Gottfried

[1930]   Die struktor der Ringe, deren Restklassenring nach dem Radikal vollständig reduzibel ist, *Math. Zeit.* **32** (1930), 161–186.   1.3, 4.4, 4.8

[1971]   Topological vector spaces. I. Translated from the German by D. J. H. Garling, Die Grundlehren der mathematischen Wissenschaften, Band 159, *Springer–Verlag, New York*, 1969. **MR 40 #1750**.   1.1

Kowalski, Sergiusz; Słodkowski, Zbigniew

[1980]   A characterization of multiplicative linear functionals in Banach algebras, *Studia Math.* **67** (1980), *no.* 3, 215–223. **MR 82d: 46070**.   2.4

Kraljević, Hrvoje

[1982]   Rank and index in Banach algebras, *Functional analysis* (*Dubrovnik*), 1981, pp'. 98–117, *Lecture Notes in Mathematics*, 948, *Springer, Berlin–New York*, 1982. **MR 84f: 46065**.   8.5

Kraljević, Hrvoje; Suljagić, Salih; Veselić, K.

[1982]   Index in semisimple Banach algebras. (Serbo-Croatian summary), *Glas. Mat. Ser. III* **17(37)** (1982), *no.* 1, 73–95. **MR 84c: 46048**.   8.5

Kreĭn, Mark Grigorevič

[1940a] A ring of functions on a topological group, *C. R.* (*Doklady*) *Acad. Sci.* URSS (N.S.) **29** (1940), 275–280. **MR 2-316**.   1.1

[1940b] On a special ring of functions, *C. R.* (*Doklady*) *Acad. Sci.* URSS (N.S.) **29** (1940), 355–359. **MR 2-316**.   1.1

[1941a] On almost periodic functions on a topological group, *C. R.* (*Doklady*) *Acad. Sci.* URSS (N.S.) **30** (1941), 5–8. **MR 2-316**.   1.1

[1941b] On positive functionals on almost periodic functions, *C. R.* (*Doklady*) *Acad. Sci.* URSS (N.S.) **30** (1941), 9–12. **MR 2-316**.   1.1

[1941c] Sur une généralisation du théorème de Plancherel au cas des intégrales de Fourier sur les groupes topologiques commutatifs, *Dokl. Akad. Nauk SSSR* **30** (1941), 484–488. **MR 2-316**.   1.1

Krull, Wolfgang

[1928]   Zur Theorie der zweiseitigen Ideale in nichtkommutativen Bereichen, *Math. Zeit.* **28** (1928), 481–503.   4.1

Krupnik, Naum Ya.

[1980a] A sufficient set of n-dimensional representations of a Banach algebra and the n-symbols. (Russian) *Funktsional. Anal. i Prilozhen.* **14** (1980), *no.* 1, 63–64. **MR 82c: 46060**.   7.1

[1980b] Conditions for the existence of an n-symbol and a sufficient set of n-dimensional representations of a Banach algebra. (Russian) Linear operators, *Mat. Issled. No.* 54 (1980), 84–97, 167. **MR 81j: 46076**.   7.1

[1981]   Banach PI-algebras. (Russian), *Investigations in functional analysis and differential equations*, pp. 54–59, 140, "Shtiintsa", Kishinev, 1981. **MR 83a: 46054**.   7.1

[1987]   Banach algebras with symbol and singular integral operators, Translated from the Russian by A. Iacob, Operator Theory: Advances and Appoications, 26. *Birkhäuser-Verlag, Basel*, 1987, x + 205 pp. **MR 90g: 47092**   1.6, 7.1

Krupnik, Naum Ya.; Markus, M. A.

[1988]   Inverse-closedness of some Banach subalgebras. (Russian) *Investigat. in differential equations and Math. Anal.* (Russian), 98–99, 174 *"Shtiintsa"* Kishinev, 1988. **MR 89m: 46087.**   2.5

Krupnik, Naum Ya.; Shpigel', E. M.

[1982]   On the boundedness of a family of homomorphisms. (Russian) Differential equations and operator theory, *Mat. Issled. No.* 67 (1982), 69–71, 172. **MR 84d: 46076.**   7.1

Krupnik, Naum Ya.; Silbermann, Bernd

[1989]   The structure of some Banach algebras fulfilling a standard identity, *Math. Nachr.* 142 (1989), 175–180. **MR 91b: 46043.**   7.1

Kulkarni, S. H.

[1984]   Gleason–Kahane–Želazko theorem for real Banach algebras, *J. Math. Phys. Sci.* 18 (1983/84), *Special Issue*, S19–S28. **MR 85j: 46080.**   2.4

Kuroš, Aleksandr Gennadievič

[1953]   Radicals of rings and algebras, *Mat. Sbornik* N.S. 33 **75** (1953), 13–26. **MR 15-194.**   4.7

Larotonda, Angel; Zalduendo, Ignacio M.

[1983]   Continuity of characters in products of algebras, *Comment. Math. Univ. Carolin.* **24** (1983), *no.* 3, 453–459. **MR 85b: 46056.**   3.1

[1984]   The holomorphic functional calculus, *Rev. Un. Mat. Argentina* 31 (1984), *no.* 3, 139–148. **MR 88h: 46092.**   3.5

Larsen, Ronald

[1971]   An introduction to the theory of multipliers, Die Grundlehren der Mathematischen Wissenschaften, Band 175, *Springer-Verlag, New York–Heidelberg*, 1971. **MR 55 #8695.**   1.2

[1973]   Banach algebras. An introduction. Pure and Applied Mathematics, No. 24, *Marcel Dekker, Inc., New York*, 1973. **MR 58 #7010.**   1.1

Lau, Anthony To Ming

[1978]   Characterizations of amenable Banach algebras, *Proc. Amer. Math. Soc.* **70** (1978), *no.* 2, 156–160. **MR 80a: 46036.**   9

[1979]   Uniformly continuous functionals on the Fourier algebra of any locally compact group, *Trans. Amer. Math. Soc.* **251** (1979), 39–54. **MR 80m: 43009.**   1.4

[1981]   The second conjugate algebra of the Fourier algebra of a locally compact group, *Trans. Amer. Math. Soc.* **267** (1981), *no.* 1, 53–63. **MR 83e: 43009.**   1.4

[1983]   Analysis on a class of Banach algebras with applications to harmonic analysis on locally compact groups and semigroups, *Fund. Math.* **118** (1983), *no.* 3, 161–175. **85k: 43007.**   1.4, 1.9, 9

[1986]   Continuity of Arens multiplication on the dual space of bounded uniformly continuous functions on locally compact groups and topological semigroups, *Math. Proc. Cambridge Philos. Soc.* **99** (1986), *no.* 2. 273–283. **MR 87i: 43001.** 1.4

[1987]   Uniformly continuous functionals on Banach algebras, *Colloq. Math.* **51** (1987), 195–205. **MR 88f: 43006.**   1.4

Lau, Anthony To Ming; Losert, Viktor

[1988]   On the second conjugate algebra of $L_1(G)$ of a locally compact group, *J. London Math. Soc.* (2) **37** (1988), *no.* 3, 464–470. **MR 89e: 43007.**   1.4

Lau, Anthony To Ming; Pym, John

[1990]   Concerning the second dual of the group algebra of a locally compact group, *J. London Math. Soc.* (2) **41** (1990), *no.* 3, 445–460. **MR 91h: 22009.**   1.4

Lau, Anthony To Ming; Wong, James C. S.

[1988]   Invariant subspaces for algebras of linear operators and amenable locally compact groups, *Proc. Amer. Math. Soc.* **102** (1988), *no.* 3, 581–586. **MR 89c: 43007.**   9

754   Bibliography

Laursen, Kjeld B.

[1969a] Tensor products of Banach algebras with involution, *Trans. Amer. Math. Soc.* **136** (1969), 467–487. **MR 38 #2610.**   1.10, 7.1

[1969b] Ideal structure in generalized group algebras, *Pacific J. Math.* **30** (1969), 155–174. **MR 40 #1792.**   1.9, 1.10

[1970] Maximal two-sided ideals in tensor products of Banach algebras, *Proc. Amer. Math. Soc.* **25** (1970), 475–480. **MR 44 #802.**   1.10, 3.2, 4.3, 7.1

[1981] Automatic continuity of generalized intertwining operators, *Dissertationes Math.* (Rozprawy Mat.) **189** (1981), **MR 82i: 46076.**   6.1

[1983a] Epimorphisms of C*-algebras, *Functional analysis: surveys and recent results*, III (Paderborn, 1983), 219–232, North-Holland Math. Stud., 90, *North-Holland, Amsterdam–New York*, 1984. **MR 86g: 46084b.**   6.2

[1983b] Ideal structure in radical sequence algebras, *Radical Banach algebras and automatic continuity* (*Long Beach, Calif.*, 1981), 248–257, Lecture Notes in Mathematics, 975, *Springer, Berlin–New York*, 1983. **MR 84d: 46085.**   4.8

[1988] Algebraic spectral subspaces and automatic continuity, *Czechoslovak Math. J.* **38** (113) (1988), *no.* 1, 157–172. **MR 89c: 47048.**   6.1

Laursen, Kjeld B.; Neumann, Michael M.

[1992] Decomposable multipliers and applications to harmonic analysis, *Studia Math.* **101** (1992), *no.* 2, 193–214. **MR 93a: 46102.**   2.8

Laursen, Kjeld B.; Stein, James D. Jr.

[1973] Automatic continuity in Banach spaces and algebras, *Amer. J. Math.* **95** (1973), *no.* 2, 495–506. **MR 49 #7791.**   6.2

Lebow, Arnold

[1968] Maximal ideals in tensor products of Banach algebras, *Bull. Amer. Math. Soc.* **74** (1968), 1020–1022. **MR 38 #539.**   1.10, 3.2, 7.1

Lebow, Arnold; Schecter, Martin

[1971] Semigroups of operators and measures of noncompactness, *J. Funct. Anal.* **7** (1971), 1–26. **MR 42 #8301.**   4.3

Le Page, Claude

[1967] Sur quelques conditions entraînant la commutativité dans les algèbres de Banach, *C. R. Acad. Sci. Paris Sér.* A-B **265** (1967), A235–237. **MR 37 #1999.**   2.4, 3.1

Leptin, Horst

[1968] Darstellungen verallgemeinerter L^1-Algebren, *Invent. Math.* **5** (1968), 192–215. **MR 38 #5022.**   1.9

Levitzki, Jacob

[1951] Prime ideals and the lower radical, *Amer. J. Math.* **73** (1951), 25–29. **MR 12-474.**   4.4, 4.8

Li, Bing Ren; Yang, Yi Min

[1989] Solutions of equations involving analytic functions in a Banach algebra. A Chinese summary appears in Acta Math. Sinica **33** (1990), *no.* 4, 576. *Acta. Math. Sinica* (*N.S.*) **5** (1989), *no.* 3, 271–280. **MR 91a: 46056.**   3.3

Lin, V. Ja

[1973] Holomorphic fiberings and multivalued functions of elements of a Banach algebra (Russian), *Funkcional. Anal. i Prilozen.* **7** (1973), *no.* 2, 43–51. **MR 47 #7444.**   3.5

Lindberg, John A., Jr.

[1964] Algebraic extensions of commutative Banach algebras, *Pacific J. Math.* **14** (1964), 559–583. **MR 30 #3380.**   3.4

[1971] A class of commutative Banach algebras with unique complete norm topology and continuous derivations, *Proc. Amer. Math. Soc.* **29** (1971), 516–520. **MR 43 #3803.**   3.4, 6.1, 6.4

[1979] Renorming a normed algebra having a semigroup of near isometries, *Rev. Roumaine Math. Pures Appl.* **24** (1979), *no.* 7, 1065–1074. **MR 81a: 46056.**   1.1

Lindenstrauss, Joram; Tzafriri, Lior

[1977]   Classical Banach spaces. I. Sequence spaces, Ergebnisse der Mathematik und ihrer Grenzgebiete, [Results in Mathematics and Related Areas], Vol. **92**, Springer-Verlag, Berlin–New York, 1977. **MR 58 #17766**.   1.7

[1979]   Classical Banach spaces. II. Function spaces. Ergebnisse der Mathematik und ihrer Grenzgebiete, 97, Springer-Verlag, Berlin–New York, 1979. **MR 81c: 46001**.   1.7

Liu, Teng Sun, van Rooij, Arnoud C. M.; Wang, Ju Kwei

[1973]   Projections and approximate identities for ideals in group algebras, Trans. Amer. Math. Soc. **175** (1973), 469–482. **MR 47 #7327**.   5.1, 5.2

Liukkonen, John R.; Mosak, Richard D.

[1974a]   The primitive dual space of [FC]⁻ groups, J. Funct. Anal. **15** (1974), 279–296. **MR 49 #10814**.   7.4

[1974b]   Harmonic analysis and the centers of group algebras, Trans. Amer. Math. Soc. **195** (1974), 147–163. **MR 50 #2815**.   7.4

[1977]   Harmonic analysis and centers of Beurling algebras, Comment. Math. Helv. **52** (1977), no. 3, 297–315. **MR 57 #10559**.   1.9

Lomonosov, Victor I.

[1974]   Invariant subspaces of the family of operators that commute with a completely continuous operator, Funkcional Anal. i Prilozen, 7, (1973), no. 3, 55–56. Translated in Funct. Anal. and its Appl. **7** (1974), 213–214. **MR 54 #8319**.   2.8

Loomis, Lynn H.

[1953]   An introduction to abstract harmonic analysis, D. Van Nostrand Company, Inc., Toronto–New York–London, 1953. **MR 14-883**.   1.1, 3.2

Lorch, Edgar Raymond

[1942]   The spectrum of linear transformations, Trans. Amer. Math. Soc. **52** (1942), 238–248. **MR 4-247**.   3.3, 3.4

[1943]   The theory of analytic functions in normed Abelian vector rings, Trans. Amer. Math. Soc. **54** (1943), 414–425. **MR 5-100**.   2.1, 3.3, 3.4

[1944]   The structure of normed Abelian rings, Bull. Amer. Math. Soc. **50** (1944), 447–463. **MR 6-69**.   3.4

[1962]   Spectral Theory, University Texts in the Mathematical Sciences. Oxford University Press, New York, 1962. **MR 25 #427**.   1.1, 2.1

Losert, Viktor

[1984]   Properties of the Fourier algebra that are equivalent to amenability, Proc. Amer. Math. Soc. **92** (1984), no. 3, 347–354. **MR 86b: 43010**.   9

Losert, Viktor; Rindler, Harald

[1984]   Asymptotically central functions and invariant extensions of Dirac measure. Probability measures on groups, VII (Oberwolfach, 1983), 368–378, Lecture Notes in Mathematics, 1064, Springer, Berlin–New York, 1984. **MR 86e: 43007**.   1.4

Loy, Richard J.

[1969]   Continuity of derivations on topological algebras of power series, Bull. Austral. Math. Soc. **1** (1969), 419–424. **MR 54 #3423**.   6.1, 6.4

[1970a]   Uniqueness of the complete norm topology and continuity of derivations on Banach algebras, Tôhoku Math. J. (2) **22** (1970), 371–378. **MR 43 #2508**.   3.4, 6.1, 6.4

[1970b]   Identities in tensor products of Banach algebras, Bull. Austral. Math. Soc. **2** (1970), 253–260. **MR 41 #5970**.   1.10, 5.1

[1973]   Continuity of higher derivations, Proc. Amer. Math. Soc. **37** (1973), 505–510. **MR 47 #827**.   6.4

[1974a]   Commutative Banach algebras with non-unique complete norm topology, Bull. Austral. Math. Soc. **10** (1974), 409–420. **MR 50 #2918**.   3.4, 6.0, 6.1, 6.4

[1974b]   Banach algebras of power series. Collection of articles dedicated to the memory of Hanna Neumann VII, J. Austral. Math. Soc. **17** (1974), 263–273. **MR 50 #10824**.   3.4

[1976]   Multilinear mappings and Banach algebras, *J. London Math. Soc.* (2) **14** (1976), *no.* 3, 423–429. **MR 56 #12890.**   5.3
[1983]   The uniqueness of norm problem in Banach algebras with finite-dimensional radical. *Radical Banach algebras and automatic continuity (Long Beach, Calif.,* 1981), 317–327, *Lecture Notes in Mathematics,* 975, *Springer, Berlin–New York,* 1983. **MR 85h: 46071.**   6.1
[1989]   Subalgebras of amenable algebras, *Conference on Automatic Continuity and Banach Algebras (Canberra,* 1989), 288–296, *Proc. Centre Math. Anal. Austral. Nat. Univer.,* 21, *Austral. Nat. Univ. Canberra,* 1989. **MR 91e: 46066.**   9

Loy, Richard J.; Willis, G. A.

[1989]   Continuity of derivations on $B(E)$ for certain Banach spaces E, *J. London Math. Soc.* (2) **40** (1989), *no.* 2, 327–346. **MR 91f: 46069.**   6.4

Lumer, Günter

[1961]   Semi-inner-product spaces, *Trans. Amer. Math. Soc.* **100** (1961), 29–43. **MR 24 #A2860.**   2.6

Lumer, Günter; Rosenblum, Marvin

[1959]   Linear operator equations, *Proc. Amer. Math. Soc.* **10** (1959), 32–41. **MR 21 #2927.**   4.1

Luminet, Denis

[1986]   A functional calculus for Banach PI-algebras, *Pacific J. Math.* **125** (1986), *no.* 1, 127–160. **MR 88c: 46060.**   3.5, 7.1

Luong, N. C.; Száz, Árpád

[1979]   Remarks on a functional calculus based on Fourier series. *Publ. Math. Debrecen* **26** (1979), *no.* 3-4, 167–172. **MR 82a: 43004.**   3.5

Malliavin, Paul

[1959]   Impossibilité de la synthèse spectrale sur les groupes abéliens non compacts, *Inst. Hautes Études Sci. Publ. Math.* (1959), 85–92. **MR 21 #5854c.**   3.6

Maltese, George

[1979]   Integral representation theorem via Banach algebras, *Enseign. Math.* (2) **25** (1979), *no.* 3-4, 273–284. **MR 82j: 46067.**   2.4

Mantero, Anna Maria; Tonge, Andrew

[1979a]   Some remarks on the injective tensor product and Banach algebras, (Italian summary), *Boll. Un. Mat. Ital. A* (5) **16** (1979), *no.* 1, 109–114. **MR 80d: 46123.**   1.10, 4.2
[1979b]   Banach algebras and von Neumann's inequality, *Proc. London Math. Soc.* (3) **38** (1979), *no.* 2, 309–334. **MR 80j: 46081.**
[1980]   The Schur multiplication in tensor algebras, *Studia Math.* **68** (1980), *no.* 1, 1–24. **MR 81k: 46074.**   1.10

Martindale, Wallace S. III

[1969]   Lie Isomorphisms of prime rings, *Trans. Amer. Math. Soc.* **142** (1969), 437–455. **MR 40 #4308.**   6.3
[1977]   Lie and Jordan mappings in associative rings, *Ring theory (Proc. Conf., Ohio Univ., Ohio,* 1976), pp. 71–84. *Lecture Notes in Pure and Appl. Math., Vol.* 25, *Dekker, New York,* 1977. **MR 56 #5660.**   6.3

Martínez-Moreno, J.; Mojtar-Kaïdi, Amin; Rodríguez-Palacios, Ángel

[1981]   On a nonassociative Vidav–Palmer theorem, *Quart. J. Math. Oxford Ser.* (2) **32** (1981), *no.* 128, 435–442. **MR83a: 46059.**   2.6

Martínez-Moreno, J.; Rodríguez-Palacios, Ángel

[1985]   Imbedding elements whose numerical range has a vertex at zero in holomorphic semigroups, *Proc. Edinburgh Math. Soc.* (2) **28** (1985), *no.* 1, 91–95. **MR 86f: 46052.**   2.6

Mascioni, Vania

[1987]   Some characterizations of complex normed Q-algebras, *Elem. Math.* **42** (1987), 10–14. **MR 88k: 46053.**   2.5

Máté, László
[1965] On the factor theory of commutative Banach algebras, (Russian summary) *Magyar Tud. Akad. Mat. Kutató Int. Közl.* **9**, (1964), 359–364 (1965). **MR 33 #6444.** 1.4
[1967] The Arens product and multiplier operators, *Studia Math.* **28** (1966/67), 227–234. **MR 35 # 5938.** 1.4
[1984] Generalized Arens product for the multiplier problem, *Operator algebras and group representations*, Vol. II (*Neptun*, 1980), 65–68, *Monographs Stud. Math.*, 18, Pitman, Boston, Mass.–London, 1984. **MR 85b: 46059.** 1.4

Mathieu, Martin
[1989] Rings of quotients of ultraprime Banach algebras with applications to elementary operators, *Conference on Automatic Continuity and Banach Algebras* (Canberra, 1989), 297–317, *Proc. Centre Math. Anal. Austral. Nat. Univ.*, 21, *Austral. Nat. Univ., Canberra*, 1989. **MR 91a: 46054.** 4.1
[1991] The symmetric algebra of quotients of an ultraprime Banach algebra, *J. Austral. Math. Soc. (Series A)* **50** (1991), 75–87. **MR 92g: 46061.** 4.1

Mathieu, Martin; Murphy, Gerard J.
[1991] Derivations mapping into the radical, *Arch. Math. (Basel)* **57** (1991), no. 5, 469–474. **MR 92j: 46085.** 6.4

Mathieu, Martin; Runde, Volker
[1992] Derivations mapping into the radical, II, *Bull. London Math. Soc.*, **24** (1992), no. 5, 485–487. **MR 93f: 46073.** 6.4

Mazur, Stanisław
[1938] Sur les anneaux linéaires, *C. R. Acad. Sci. Paris* **207** (1938), 1025–1027. 1.1, 2.2

McCoy, Neal H.
[1947] Subdirect sums of rings, *Bull. Amer. Math. Soc.* **53** (1947), 856–877. **MR 9-77.** 1.3, 4.6
[1949] Prime ideals in general rings, *Amer. J. Math.* **71** (1949), 823–833. **MR 11-311.** 4.1, 4.4, 4.8

McKilligan, Sheila A.; White, Alan J.
[1972] Representations of L-algebras, *Proc. London Math. Soc.* (3) **25** (1972), 655–674. **MR 47 #2368.** 1.4, 9

McLean, Roderick G.; Kummer, Hans
[1988] Representations of the Banach algebra $\ell^1(S)$, *Semigroup Forum* **37** (1988), no. 1, 119–122. **MR 89e: 46062.** 1.9

Mehta, R. D.; Vasavada, Mahavirendra
[1985] A note on automorphisms of tensor products of Banach algebras, *Indian J. Pure Appl. Math.* **16** (1985), no.1, 13–16. **MR 86i: 46074.** 1.10

Mergelyan, Sergei Nikitovič
[1951] On the representation of functions by series of polynomials of closed sets, *Dokl. Akad. Nauk SSSR* **78** (1951), 405–408. [Amer. Math. Soc. Transl. No. 85.] **MR 13-23.** 3.3

Meyer, Michael J.
[1991] The spectral extension property and extension of multiplicative linear functionals, *Proc. Amer. Math. Soc.* **112** (1991), no. 3, 855–861. **MR 91j: 46059.** 2.5, 3.2, 3.4
[1992a] Minimal incomplete norms on Banach algebras, *Studia Math.* **102** (1992), no. 1, 77–85. **MR 93c: 46085.** 2.3, 6.1
[1992b] Continuous dense embeddings of strong Moore algebras, *Proc. Amer. Math. Soc.*, **116** (1992), no. 3, 727–735. **MR 93a: 46095.** 2.5, 3.4
[1992c] On a topological property of certain Calkin algebras, *Bull. London Math. Soc.*, **24** (1992), no. 6, 591–598. **MR 93i: 46081.** 1.7, 6.1, 6.2
[1992d] Weak invertibility and strong spectrum, *Studia Math.* **105** (1992), no. 3, 255–269. 2.1, 2.2, 4.5

Michael, Ernest A.

[1952]   Locally multiplicatively-convex topological algebras, *Mem. Amer. Math. Soc.* No. 11, (1952). **MR 14-482**.   2.2, 2.9, 3.1

Michal, Aristotle D.; Martin, Robert S.

[1934]   Some expansions in vector spaces, *Jour. de Mathematiques Pures et Appliques*, New Series **13** (1934), 69–71.   1.1

Michor, Peter W.

[1978]   Functors and categories of Banach spaces. Tensor products, operator ideals and functors on categories of Banach spaces, Lecture Notes in Mathematics, 651. *Springer, Berlin–New York*, 1978. iii+99 pp. **MR 80h: 46116**.   4.7

Mijangos, Ma. José

[1982]   Banach spaces with multiplication, (Spanish), *Gaceta Mat.* (1) (*Madrid*) **34** (1982), no. 1-3, 29–46. **MR 86k: 46079**.   1.1

Miller, Vivien G.

[1993]   Restrictions and quotients of decomposable operators, *Dissertation, Mississippi State University, 1993.*   2.8

Milman, D. P.

[1938]   On some criteria for the regularity of spaces of type (B), *Doklady Akad. Nauk SSSR* **20** (1938), 234.   1.7

Milman, Vitali D.; Schechtman, Gideon

[1986]   Asymptotic theory of finite-dimensional normed spaces. With an appendix by M. Gromov. Lecture Notes in Mathematics, 1200, *Springer-Verlag, Berlin–New York*, 1986. 156 pp. **MR 87m: 46038**.   1.7

Misonou, Yosinao

[1952]   On a weakly central operator algebra, *Tôhoku Math. J.* (2) **4** (1952), 194–202. **MR 14-566**.   7.2, 7.4

Mitjagin, Boris Samuilovič; Švarc, Albert Solomonovič

[1964]   Functors in categories of Banach spaces, (Russian) *Uspehi Mat. Nauk* **19** (1964), no. 2 (116), 65–130. 194–202. English translation: *Russian Math. Surveys* **19** (1964), no. 2, 65–127.**MR 29 #3866**.   4.7

Miziolek, J. K.; Müldner, T.; Rek, A.

[1972]   On topologically nilpotent algebras, *Studia Math.* **43** (1972), 41–50. **MR 46 #6030**.   4.8

Mocanu, Gh.

[1967]   Sur quelques critères de commutativité dans les algebres de Banach, *C. R. Acad. Sci. Paris* **265** (1967), 235–237.   3.1

[1971]   Sur quelques critères de commutativité pour une algébre de Banach, *An. Univ. Bucuresti Mat.-Mec.* **20** (1971), no. 2, 127–129. **MR 46 #9739**.   3.1

Molien, Theodor

[1893]   Über systeme höherer complexer Zahlen, *Math. Ann. Bd.* **41** (1893), 83–156.   1.1, 4.0, 4.8

[1897]   Eine Bemerkung zur Theorie der homogenen substitutionsgruppen *Sitzber. Naturfouscher Ges. Univ. Jurjeff (Dorpat)* **11** (1897), 259–274.   4.0

Monna, Antonie F.

[1973]   Functional analysis in historical perspective, *John Wiley & Sons, New York–Toronto, Ont.*, 1973. **MR 58 #2112**.   1.1

Moore, Calvin C.

[1976a]   Group extensions and cohomology for locally compact groups. III, *Trans. Amer. Math. Soc.* **221** (1976), no. 1, 1–33. **MR 54 #2867**.   9

[1976b]   Group extensions and cohomology for locally compact groups. IV, *Trans. Amer. Math. Soc.* **221** (1976), no. 1, 35–58. **MR 54 #2868**.   9

Mosak, Richard D.

[1971]   Central functions in group algebras, *Proc. Amer. Math. Soc.* **163** (1972), 277–310. **MR 43 #5323**.   5.1

[1972]   The L^1-and C*-algebras of $[FIA]_B^-$ groups and their representations, *Trans. Amer. Math. Soc.* **163** (1972), 277–310. **MR 45 #2096.**   7.4

[1975]   Banach algebras, Chicago Lectures in Mathematics, *The University of Chicago Press, Chicago,* 1975. **MR 54 #3406.**   1.1

Muhly, Paul S.

[1983]   Radicals, crossed products and flows, *Ann. Polon. Math.* **43** (1983), no. 1, 35–42. **MR 85i: 46093.**   4.3

Müller, Vladimir

[1979]   On domination problem in Banach algebras, *Comment. Math. Univ. Carolin.* **20** (1979), no. 3, 475–481. **MR 81c: 46043.**   3.4

[1982]   On domination and extensions of Banach algebras, *Studia Math.* **73** (1982), no. 1, 75–80. **MR 83j: 46062.**   3.4

[1983]   The inverse spectral radius formula and removability of spectrum, (Russian summary), *Časopis Pěst. Mat.* **108** (1983), no. 4, 412–415. **MR 85f: 46090.**   3.4

[1988]   Adjoining inverses to noncommutative Banach algebras and extensions of operators, *Studia Math.* **91** (1988), no. 1, 73–77. **MR 90a: 46117.**   3.4

[1990]   Kaplansky's theorem and Banach PI-algebras, *Pacific J. Math.* **141** (1990), no. 2, 355–361. **MR 91k: 46048.**   7.1

Müller, Vladimir; Sołtysiak, Andrzej

[1988]   On the largest generalized joint spectrum, *Comment. Math. Univ. Carolin.* **29** (1988), no. 2, 255–259. **MR 90a: 46116.**   3.5

[1989]   Spectrum of generators of a noncommutative Banach algebra, *Studia Math.* **93** (1989), no. 1, 87–95. **MR 90c: 46063.**   3.5

Murphy, Gerard J.

[1981]   Continuity of the spectrum and spectral radius, *Proc. Amer. Math. Soc.* **82** (1981), no. 4, 619–621. **MR 82h: 46066.**   2.2

Murphy, Gerard J.; West, Trevor T.

[1979]   Spectral radius formulae, *Proc. Edinburgh Math. Soc.* (2) **22** (1979), no. 3, 271–275. **MR 81a: 46054.**   2.2

[1980a] Removing the interior of the spectrum, *Comment. Math. Univ. Carolin.* **21** (1980), no. 3, 421–431. **MR 81m: 46069.**   2.5, 3.4

[1980b] Decomposition algebras of Riesz operators, *Glasgow Math. J.* **21** (1980), no. 1, 75–79. **MR 81a: 47023.**   2.8

[1992]   The index group, the exponential spectrum, and some spectral containment theorems, *Proc. Roy. Irish Acad. Sect. A* **92A** (1992), no. 2, 229–238.   2.1

Murray, Francis J.; von Neumann, John

[1935]   On rings of operators, *Ann. of Math.* **37** (1935), 116–229.   1.1

[1937]   On rings of operators II, *Amer. Math. Soc.* **41** (1937), 208–248.   1.1

[1943]   On rings of operators IV, *Ann. of Math.* (2) **44** (1943), 716–808. **MR 5-101.** 1.1

Nachbin, Leopoldo

[1965]   The Haar integral, *D. Van Nostrand, Co. Inc. Princeton, N. J.–Toronto, Ont.–London,* 1965. **MR 31 #271.**   1.9

Nagasawa, Masao

[1959]   Isomorphisms between commutative Banach algebras with an application to rings of analytic functions, *Kōdai. Math. Sem. Rep.* **11** (1959), 182–188. **MR 22 #12379.** 2.4

Nagata, Masayoshi

[1951]   On the theory of radicals in a ring, *J. Math. Soc. Japan* **3** (1951), 330–344. **MR 13-902.**   4.4, 4.8

[1952]   On the nilpotency of nil-algebras, *J. Math. Soc. Japan* **4** (1952), 296–301. **MR 14-719.**   4.4

Nagumo, Mitio

[1936]   Einige analytische Untersuchungen in linearen metrischen Ringen, *Japan J. Math.* **13** (1936),61–80.   1.1, 2.1

Naĭmark, Mark Aronovič

[1948]   Rings with involutions, *Uspehi Matem. Nauk (N.S.)* 3 no. 5(27)(1948), 52–145.
         MR 10-308.   1.1
[1956]   Normed rings (in Russian), *Gosudarstv. Izdat. Tehn-Teor. Lit., Moscow,* 1956.
         487 pp. **MR 19-870.** (Most recent edition is: Normed algebras, *Wolters-Noordhoff Publishing Co., Groningen,* 1972. **MR 55 #11042.**)   1.1

Neumann, Michael M.

[1992a] Commutative Banach algebras and decomposable operators *Monatsh. Math.*
         113 (1992), no. 3, 227–243. **MR 93e: 46056.**   2.8, 6.1
[1992b] Banach algebras, decomposable convolution operators, and a spectral mapping
         property, *Function spaces (Edwardsville, IL,* 1990), 307–323, *Lecture Notes in
         Pure and Appl. Math.,* 136, Dekker, New York, 1992. **MR 93e: 46057.**   2.8

von Neumann, John

[1927]   Mathematische Begründung der Quantenmechanik, *Gött. Nach.* 1 (1927), 1–57.
                                                                                   2.0
[1929]   Zur Algebra der Funktionaloperatoren und Theorie der normalen Operatoren,
         Math. Annalen **102** (1929), 370–427.   1.1
[1934]   Almost periodic functions in a group, I, *Trans. Amer. Math. Soc.* **36** (1934),
         445–492.   3.2, 5.1
[1936a] On a certain topology for rings of operators, *Ann. of Math.* **37** (1936), 111–115.
                                                                                   1.1
[1936b] On regular rings, *Proc. Nat. Acad. Sci. U. S. A.* **22** (1936), 707–713.   2.1
[1940]   On rings of operators. III, *Ann. of Math.* **41** (1940), 94–161. **MR 1-146.** 1.1
[1943]   On some algebraical properties of operator rings, *Ann. of Math.* (2) **44** (1943),
         709–715. **MR 5-100.**   1.1

Newburgh, John D.

[1951]   The variation of spectra, *Duke Math. J.* **18** (1951), 165–176. **MR 14-481.**
                                                                              2.2, 3.4

Ngo, Viet

[1988]   A structure theorem for discontinuous derivations of Banach algebras of differ-
         ential functions, *Proc. Amer. Math. Soc.* **102** (1988), no. 3, 507–513. **MR 89d:
         46057.**   6.4

Nickerson, Helen K.; Spencer, Donald C.; Steenrod, Norman E.

[1959]   Advanced Calculus, *D. Van Nostrand Co., Inc., Toronto–Princeton, N. J.*
         1959. **MR 23 #A976.**   1.6, 3.5

Niestegge, Gerd

[1983]   Extreme spectral states of commutative Banach algebras, *Math. Ann.* **264**
         (1983), no. 2, 179–188. **MR 84k: 46040.**   2.4
[1984]   A note on criteria of Le Page and Hirschfeld–Żelazko for the commutativity of
         Banach algebras, *Studia Math.* **79** (1984), no. 1, 87–90. **MR 86g: 46071.** 2.4

Niknam, Assadollah

[1986]   A remark on closability of linear mappings in normed algebras, *Arch. Math.*
         (Basel) **47** (1986), no. 2, 131–134. **MR 88a: 46050.**   6.4

Noether, Emmy

[1929]   Hyperkomplexe Grössen und Darstellungstheorie, *Math. Zeit.* **30** (1929), 641–
         692.   4.0

Nylen, Peter; Leiba, Rodman

[1990]   Approximation numbers and Yamamoto's theorem in Banach algebras, *Integral
         Equations Operator Theory* **13** (1990), no. 5, 728–749. **MR 91m: 46071.** 8.5

Oka, K.

[1936]   Sur les fonctions rationnelles de plusieurs variables, I. Domaines convexes pur
         rapport aux fonctions rationnelles, *J. Sci. Hiroshima Univ.* **6** (1936), 245–256.
                                                                              3.3, 3.5

Okada, Nolio

[1983]   Some norm conditions for a Banach algebra to be either the reals, complexes or
         quaternions, *TRU Math.* **19** (1983), no. 1, 67–74. **MR 85d: 46063.**   2.2

Okniński, Jan

[1982]   Spectrally finite algebras and tensor products, *Comm. Algebra* **10** (1982), *no.*
          18, 1939–1950. **MR 84j: 16014.**   1.10

Olubummo, Adegoke

[1957]   Left completely continuous B#-algebras, *J. London Math. Soc.* **32** (1957), 270–
          276. **MR 19-665.**   8.7
[1979]   Unbounded multiplier operators, *J. Math. Anal. Appl.* **71** (1979), *no.* 2, 359–
          365. **MR 80m: 43004.**   1.9

Oshobi, E. O.

[1986]   A class of Banach algebras whose duals are the multiplier algebras, *Simon
          Stevin* **60** (1986), *no.* 4, 339–345. **MR 88d: 46092.**   1.4

Oshobi, E. O.; Pym, John S.

[1981]   Banach algebras whose duals are multiplier algebras, *Bull. London Math. Soc.*
          13 (1981), *no.* 1, 66–68. **MR 82i: 46071.**   1.4
[1984]   Banach algebras whose duals are multiplier algebras. II, *Ann. Soc. Sci. Brux-
          elles Sér. I* **98** (1984), *no.* 2-3, 65–68. **MR 86e: 46041.**   1.4
[1987]   Banach algebras whose duals consist of multipliers, *Math. Proc. Cambridge
          Philos. Soc.* **102** (1987), *no.* 3, 481–505. **MR 88j: 46044.**   1.4

Oudadess, Mohamed

[1983a]   Continuité des caractères dans les algèbres uniformément A-convexes. (English
          summary), *Ann. Sci. Math. Québec* **7** (1983), *no.* 2, 193–201. **MR 85a: 46029.**
                                                                                    3.1
[1983b]   Commutativité de certaines algèbres de Banach. (English summary), *Bol. Soc.
          Mat. Mexicana* (2) **28** (1983), *no.* 1, 9–14. **MR 86c: 46052.**   2.4
[1985]   Sur le radical de Jacobson dans les algèbres localement A-convexes. (English
          summary), *C. R. Math. Rep. Acad. Sci. Canada* **7** (1985), *no.* 1, 21–25. **MR
          86g: 46078.**   2.4, 4.3
[1987]   Commutativity of some A-convex algebras, *J. Univ. Kuwait Sci.* **14** (1987), *no.*
          2, 205–211. **MR 89i: 46054.**   2.4
[1990b]   Discontinuity of the product in multiplier algebras, *Publ. Mat.* **34** (1990), *no.*
          2, 397–401. **MR 92a: 46053.**   3.3

Palmer, Theodore W.

[1965]   Unbounded normal operators on Banach spaces, Dissertation, Harvard Univer-
          sity, 1965.   2.6
[1968a]   Characterizations of C*-algebras, *Bull. Amer. Math. Soc.* **74** (1968), 538–540.
          MR 36 #5709.   2.6
[1968b]   Unbounded normal operators on Banach spaces, *Trans. Amer. Math. Soc.* **133**
          (1968), 385–414. **MR 37 #6768.**   2.6
[1970]   Characterizations of C*-algebras II, *Trans. Amer. Math. Soc.* **148** (1970), 577–
          588. **MR 41 #7447.**   2.6
[1972]   Hermitian Banach *-algebras, *Bull. Amer. Math. Soc.* **78** (1972), 522–524. **MR
          45 #7481.**   2.0
[1973]   The Gelfand-Raikov theorem: an elementary proof, mimeographed notes, 118
          pages.   1.9
[1974]   Characterizations of W*-homomorphisms and expectations, *Proc. Amer. Math.
          Soc.* **46** (1974), 265–272. **MR 50 #14249.**
[1978]   Classes of nonabelian, noncompact, locally compact groups, *Rocky Mountain
          J. Math.* **8** (1978), 683–741. **MR 81j: 22003.**   5.1, 5.2, 7.4, 9
[1985]   The bidual of the compact operators, *Trans. Amer. Math. Soc.* **288** (1985),
          no. 2, 827–839. **MR 86f: 47027.**   1.4, 1.7
[1989]   A simple proof of the fundamental theorem of spectral pseudo-norms, privately
          circulated notes.   2.3, 2.4
[1992]   Spectral algebras, *Rocky Mountain J. Math.* **22** (1992), 293–328. **MR 93d:
          46079.**   2.3, 2.4

Parshall, Karen H.

[1985]   Joseph H. M. Wedderburn and the structure theory of algebras, *Arch. Hist.
          Exact Sci.* **32** (1985), *no.* 3-4, 223–349. **MR 86h: 01050.**   1.1, 1.6

Parsons, D. J.

[1984]   The centre of the second dual of a commutative semigroup algebra, *Math. Proc.
          Cambridge Philos. Soc.* **95** (1984), *no.* 1, 71–92. **MR 85c: 43004.** 1.4

Paschke, William L.

[1973b]   A factorable Banach algebra without bounded approximate unit, *Pacific J. Math.* **46** (1973), 249–251. **MR 48 #2765.**   5.2

Paterson, Alan L. T.

[1988]   Amenability, Mathematical Surveys and Monographs 29, *American Mathematical Society, Providence, R. I.*, 1988, 452 pp. **MR 90e: 43001.**   9

Paulsen, Vern I.

[1982]   The group of invertible elements in a Banach algebra, *Colloq. Math.* **47** (1982), no. 1, 97–100. **MR 83m: 46073.**   2.1

[1986]   Completely bounded maps and dilations. Pitman Research Notes in Mathematics Series, 146, *Longman Scientific and Technical, Harlow; John Wiley and Sons, Inc. New York*, 1986, 187 pp. **MR 88h: 46111.**

Pedersen, Gert K.

[1966]   Measure theory for C*-algebras, *Math. Scand.* **19** (1966), 131–145. **MR 34 #3453.**   1.1

[1972]   Applications of weak* semicontinuity in C*-algebra theory, *Duke Math. J.* **39** (1972), 431–450. **MR 47 #4012.**   7.4

[1976]   Spectral formulas in quotient C*-algebras, *Math. Z.*, **148** (1976), no. 3, 299–300. **MR 53 #14151.**   2.2

Peirce, Benjamin

[1870]   Linear associative algebra, (Distributed privately in 1870) Published posthumously in *Amer. J. Math.* **4** (1881), 97–229.   1.1, 1.6

Peirce, Charles Saunders

[1873]   Description of a notation for the logic of relatives resulting from an amplification of the conceptions of Boole's calculus of logic, *Mem. Amer. Acad. Arts and Sci.* **9** (1873), 317–389.   1.6

[1881]   [Notes and addenda to B. Peirce [1870]], *Amer. J. Math.* **4** (1881), 97–229.   1.2, 4.0

Perlis, Sam

[1942]   A characterization of the radical of an algebra, *Bull. Amer. Math. Soc.* **48** (1942), 128–132. **MR 3-264.**   2.1, 4.3, 4.8

Pettis, Billy J.

[1938]   On integration in vector spaces, *Trans. Amer. Math. Soc.* **44** (1938), 277–304.   1.9

Petz, Dénes

[1983]   On spectral and central states of Banach algebras, *Acta Math. Hungar.* **42** (1983), no. 1-2, 19–24. **MR 85m: 46043.**   2.4

Pfaffenberger, William E.

[1970]   On the ideals of strictly singular and inessential operators, *Proc. Amer. Math. Soc.* **25** (1970), 603–607. **MR 41 #9036.**   1.7

Phillips, N. Christopher

[1987]   Equivariant K-theory and freeness of group actions on C*-algebras, Lecture Notes in Mathematics, 1274, *Springer-Verlag, Berlin–New York*, 1987. 371 pp. **MR 89k: 46086.**   10

[1988]   Inverse limits of C*-algebras and applications, *Operator algebras and applications, Vol.* 1, 127–185, London Math. Soc. Lecture Note Ser., 135, *Cambridge Univ. Press, Cambridge–New York*, 1988. **MR 90j: 46052.**   10

[1991]   K-theory for Fréchet algebras, *Internat. J. Math.* **2** (1991), no. 1, 77–129. **MR 92e: 46143.**   10

Phillips, John; Raeburn, Iain

[1983]   Central cohomology of C*-algebras, *J. London Math. Soc.* (2) **28** (1983), no. 2, 363–375. **MR 85d: 46095.**   9

Phillips, John; Raeburn, Iain; Taylor, Joseph L.

[1982]   Automorphisms of certain C*-algebras and torsion in second Čech cohomology, *Bull. London Math. Soc.* **14** (1982), no. 1, 33–38. **MR 83a: 46080.**   9

Pier, Jean-Paul

[1984]   Amenable locally compact groups, Pure and Applied Mathematics, *John Wiley and Sons, New York*, 1984. 418 pp. **MR 86a: 43001.**   9

[1988] Amenable Banach algebras, Pitman Research Notes in Mathematics Series 172. *Longman Scientific & Technical, Harlow, John Wiley & Sons, Inc., New York*, 1988. 161 pp. **MR 89g: 46093.** 9

[1990] L'Analyse harmonique son développement historique, *Masson, Paris*, 1990. 330 pp. **MR 91f: 43001.** 1.9

Pierce, Richard S.

[1967] Modules over commutative regular rings, *Mem. Amer. Math. Soc.* No. 70, *American Mathematical Society, Providence, R. I.*, (1967). **MR 36 #151.** 7.4

Pietsch, Albrecht

[1980] Operator ideals. Translated from the German by the author. Mathematische Monographien [Mathematical Monographs], 16, *North-Holland Publishing Company, Amsterdam–New York–Oxford*, 1980, (English translation). **MR 81a: 47002.** 1.7

Pimsner, Mihai; Popa, Sorin

[1978] The Ext-groups of some C*-algebras considered by J. Cuntz, *Rev. Roumaine Math. Pures Appl.* **23** (1978), no. 7, 1069–1076. **MR 81j: 46094.** 10

Pisier, Gilles

[1979] A remarkable homogeneous Banach algebra, *Israel J. Math.* **34** (1979), no. 1-2, 38–44. **MR 81k: 43009.** 3.4

[1981] Contre-exemple à une conjecture de Grothendieck. (English summary), *C. R. Acad. Sci. Paris Sér. I Math.* **293** (1981), no. 15, 681–683. **MR 83b: 46023.** 1.10

[1983] Counterexamples to a conjecture of Grothendieck. *Acta. Math.* **151** (1983), no. 3-4, 181–208. **MR 85m: 46017.** 1.10

[1986] Factorization of linear operators and geometry of Banach spaces. CBMS Regional Conference Series in Mathematics, 60, *Published for the Conference Board of the Mathematical Sciences, Washington, D. C.: by the American Mathematical Society, Providence, R. I.*, 1986. 154 pp. **MR 88a:47020.** 1.7

Potra, Florian A.; Pták, Vlastimil

[1984] Nondiscrete induction and iterative processes. *Research Notes in Mathematics*, 103, *Pitman (Advanced Publishing Program), Boston, Mass.–London*, 1984. 207 pp. **MR 86i: 65003.** 5.2

Procesi, Claudio

[1973] Rings with polynomial identities. Pure and Applied Mathematics, 17, *Marcel Dekker, Inc. New York*, 1973. 190 pp. **MR 51 #3214.** 7.1

Prosser, Reese T.

[1963] On the ideal structure of operator algebras, *Mem. Amer. Math. Soc.* **45**, *American Mathematical Society, Providence, R. I.*, 1963. 28 pp. **MR 27 #1846.** 1.1

Prüfer, Heinz

[1925] Neue Begründung der algebraischen Zahlentheorie, *Math. Ann.* **94**, (1925), 198–143. 1.3

Pryde, Alan James

[1988] A noncommutative joint spectral theory, *Workshop/Miniconference on Functional Analysis and Optimization* (Canberra, 1988), 153–161, *Proc. Centre Math. Anal. Austral. Nat. Univ.*, 20, *Austral. Nat. Univ., Canberra*, 1988. **MR 90k: 47008.** 3.5

Pták, Vlastimil

[1968] A uniform boundedness theorem and mappings into spaces of operators, *Studia Math.* **31** (1968), 425–431. **MR 38 #4967.** 6.1

[1970] On the spectral radius in Banach algebras with involution, *Bull. London Math. Soc.* **2** (1970), 327–334. **MR 43 #932.** 2.0

[1972] Banach algebras with involution, *Manuscripta Math.* **6** (1972), 245–290. **MR 45 #5764.** 2.0

[1974] Deux théorèmes de factorisation, *C. R. Acad. Sci. Paris Ser. A* **278** (1974), 1091–1094. **MR 49 #5846.** 5.2

[1977]   Perturbations and continuity of the spectrum, *Proceedings of the International Conference on Operator Algebras, Ideals, and their Applications in Theoretical Physics (Leipzig, 1977)*, 177–186, Leipzig, 1978. **MR 80g: 47016.**   2.4

[1979a]   Factorization in Banach algebras, *Studia Math.* **65** (1979), no 3, 279–285. **MR 81i: 46057.**   5.2

[1979b]   Commutators in Banach algebras, *Proc. Edinburgh Math. Soc.* (2) **22** (1979), no. 3, 207–211. **MR 81a: 46055.**   4.1

Pták, Vlastimil; Zemánek, Jaroslav

[1977]   On uniform continuity of the spectral radius in Banach algebras, *Manuscripta Math.* **20** (1977), *no.* 2, 177–189. **MR 56 #1065.**   2.4

Puhl, Joachim

[1978]   The trace of finite and nuclear elements in Banach algebras, *Czechoslovak Math. J.* **28** (**103**) (1978), *no.* 4, 656–676. **MR 81a: 47024.**   8.7

Putinar, Mihai

[1980]   Functional calculus with sections of an analytic space, *J. Operator Theory* **4** (1980), *no.* 2, 297–306. **MR 82i: 32033.**   3.5

[1982]   The superposition property for Taylor's functional calculus, *J. Operator Theory* **7** (1982), *no.* 1, 149–155. **MR 83f: 47015.**   3.5

[1983a]   Spectral theory and sheaf theory. I, *Dilation theory, Toeplitz operators and other topics (Timişoara/Herculane, 1982)*, 283–297, *Oper. Theory: Adv. Appl.*, 11, Birkhäuser, Basel–Boston, Mass., 1983. **MR 86h: 46105.**   3.5

[1983b]   Uniqueness of Taylor's functional calculus, *Proc. Amer. Math. Soc.* **89** (1983), *no.* 4, 647–650. **MR 86i: 47023.**   3.5

[1984a]   Functional calculus and the Gel'fand transformation, *Studia Math.* **79** (1984), *no.* 1, 83–86. **MR 86c: 46056.**   3.5

[1984b]   Spectral theory of representations of Stein algebras. I. (Romanian. English summary) *Stud. Cerc. Mat.* **36** (1984), *no.* 3, 193–220. **MR 86c: 46055a.**   3.5

[1984c]   Spectral theory of representations of Stein algebras. II. (Romanian. English summary) *Stud. Cerc. Mat.* **36** (1984), *no.* 5, 387–408. **MR 87a: 46080.**   3.5

[1984d]   Spectral theory of representations of Stein algebras. III. (Romanian. English summary) *Stud. Cerc. Mat.* **36** (1984), *no.* 4, 293–310. **MR 86c: 46055b.** 3.5

Pym, John S.

[1965]   The convolution of functionals on spaces of bounded functions, *Proc. London Math. Soc.* (3) **15** (1965), 84–104. **MR 30 #3367.**   1.4

[1983]   Remarks on the second duals of Banach algebras, *J. Nigerian Math. Soc.* **2** (1983), 31–33. **MR 86g: 46072.**   1.4, 1.7

Pym, John S.; Ülger, Ali

[1989]   On the Arens regularity of inductive limit algebras and related matters, *Quart. J. Math. Oxford Ser.* (2) **40** (1989), *no.* 157, 101–109. **MR 90e: 46039.**   1.4

Racher, Gerhard

[1981a]   On the continuity of convolution in $L^p(G)$, (Italian summary), *Boll. Un. Math. Ital.* A(5) **18** (1981), *no.* 1, 122–126. **MR 82g: 43006.**   1.10, 6.1

[1981b]   Remarks on a paper of G. F. Bachelis and J. E. Gilbert: "Banach spaces of compact multipliers and their dual spaces" [Math. Z. **125** (1972), 285–297; MR 49 #3457], *Monatsh. Math.* **92** (1981), *no.* 1, 47–60. **MR 83i: 43008.**   6.1

[1983]   A proposition of A. Grothendieck revisited, *Banach space theory and its applications* (Bucharest, 1981), 215–227, *Lecture Notes in Mathematics*, 991, Springer, Berlin–New York, 1983. **MR 85e: 43008.**   1.10, 9

Raeburn, Iain and Taylor, Joseph L.

[1977]   Hochschild cohomology and perturbation of Banach algebras, *J. Functional Analysis* **25** (1977), *no.* 3, 258–266. **MR 58 #30334.**   3.5, 9

Raǐkov, Dmitriǐ Abramovič

[1945]   Harmonic analysis on commutative groups with the Haar measure and the theory of characters, *Trav. Inst. Math. Stekloff* **14** (1945), 86 pp. **MR 8-133.** (German translation: **MR 15-932.**)   1.1, 3.6

[1946]   To the theory of normed rings with involution, *C. R. (Doklady) Acad. Sci. URSS* (N.S.) **54** (1946), 387–390. **MR 8-469.**   2.0

Rakočević, Vladimir

[1984]   Polynomially compact elements of Banach algebras. (Serbo-Croatian summary) *Mat. Vesnik* **36** (1984), no. 2, 167–172. **MR 86c: 46053.**   8.7

[1988]   Moore-Penrose inverse in Banach algebras, *Proc. Roy. Irish Acad. Sect. A* **88** (1988), no. 1, 57–60. **MR 90a: 46118.**   2.1

Ransford, Thomas J.

[1984a] Generalised spectra and analytic multivalued functions, *J. London Math. Soc.* (2) **29** (1984), no. 2, 306–322. **MR 85f: 46091.**   2.1

[1984b] A short elementary proof of the Bishop–Stone–Weierstrass theorem, *Math. Proc. Cambridge Philos. Soc.* **96** (1984), no. 2, 309–311. **MR 86c: 46023.**
1.5

[1989]   A short proof of Johnson's uniqueness-of-norm theorem, *Bull. London Math. Soc.* **21** (1989), no. 5, 487–488. **MR 90g: 46069.**   2.0, 2.3, 6.1

Read, Charles J.

[1984]   A solution to the invariant subspace problem, *Bull. London Math. Soc.* **16** (1984), no. 4, 337–401. **MR 86f: 47005.**   2.8

[1985]   A solution to the invariant subspace problem on the space ℓ_1, *Bull. London Math. Soc.* **17** (1985), no. 4, 305–317. **MR 87e: 47013.**   2.8

[1989]   Discontinuous derivations on the algebra of bounded operators on a Banach space, *J. London Math. Soc.* (2) **40** (1989), no. 2, 305–326. **MR 91f: 46068.**
6.2, 6.4

Reid, Constance

[1970]   Hilbert. With an appreciation of Hilbert's mathematical work by Hermann Weyl, *Springer-Verlag, New York–Berlin*, 1970. **MR 42 #5767.**   2.0

Reiter, Hans

[1968]   Classical harmonic analysis and locally compact groups, *Clarendon Press, Oxford*, 1968. **MR 46 #5933.**   1.9

[1971] L^1-algebras and Segal algebras, Lecture Notes in Mathematics, Vol. 231, *Springer-Verlag, Berlin–New York*, 1971. **MR 55 #13158.**   5.1

Renaud, Peter

[1969a] Invariant means on a class of von Neumann algebras, *Trans. Amer. Math. Soc.* **170** (1972), 285–291. **MR 46 #3688.**   9

Rennison, John Francis

[1982]   Conditions related to centrality in a Banach algebra, *J. London Math. Soc.* (2) **26** (1982), no. 1, 155–168. **MR 84a: 46104.**   7.2

[1987]   Conditions related to centrality in a Banach algebra, II, *J. London Math. Soc.* (2) **35** (1987), no. 3, 499–513. **MR 88h: 46090.**   7.2

[1988]   The quasicentre of a Banach algebra, *Math. Proc. Cambridge Philos. Soc.* **103** (1988), no. 2, 333–337. **MR 89f: 46104.**   7.2

[1990]   Closure properties of the quasi-centre of a Banach algebra, *Math. Proc. Camb. Philos. Soc.* **108** (1990), no. 2, 355–364. **MR 92f: 46059.**   7.2

Rickart, Charles E.

[1947]   The singular elements of a Banach algebra, *Duke Math. J.* **14** (1947), 1063–1077. **MR 9-358.**   2.1

[1950]   The uniqueness of norm problem in Banach algebras, *Ann. of Math.* (2) **51** (1950), 615–628. **MR 11-670.**   2.0, 2.3, 2.5, 4.0, 4.2, 6.0, 6.1, 8.3

[1953]   On spectral permanence for certain Banach algebras, *Proc. Amer. Math. Soc.* **4** (1953), 191–196. **MR 14-660.**   2.0, 3.5, 6.1

[1958]   An elementary proof of a fundamental theorem in the theory of Banach algebras, *Michigan Math. J.* **5** (1958), 75–78. **MR 20 #4786.**   2.2

[1960]   General theory of Banach algebras, The University Series in Higher Mathematics, *D. Van Nostrand, Co., Princeton, N. J.–Toronto–London–New York*, 1960. **MR 22 #5903.**   general

[1969]   Analytic functions of an infinite number of complex variables, *Duke Math. J.* **36** (1969), 581–597. **MR 40 #7819.**   3.5

Rieffel, Marc A.

[1965]   A characterization of commutative group algebras and measure algebras, *Trans. Amer. Math. Soc.* **116** (1965), 32–65. **MR 33 #6300.**   1.9, 1.10

766   Bibliography

[1966]   A characterization of the group algebras of the finite groups, *Pacific J. Math.*
          16 (1966), 347–363. **MR 32 #2493**.   1.9
[1969a]   On the continuity of certain intertwining operators, centralizers, and positive
          linear functionals, *Proc. Amer. Math. Soc.* **20** (1969), 455–457. **MR 38 #5000**.
                                                                                       5.2
[1969b]   Square-integrable representations of Hilbert algebras, *J. Funct. Analysis* **3**
          (1969), 265–300. **MR 39 #6094**.   2.0
[1979a]   Normal subrings and induced representations, *J. Algebra* **59** (1979), no. 2, 364–
          386. **MR 80h: 16029**.   1.1
[1979b]   Unitary representations of group extensions; an algebraic approach to the theory
          of Mackey and Blattner, *Studies in analysis*, pp. 43–82, *Adv. in Math. Suppl.*
          Stud., 4, *Academic Press, New York–London*, 1979. **MR 81h: 22004**.   1.1
[1982]   Morita equivalence for operator algebras. pp. 285–298, *Proc. Sympos. Pure*
          Math., 38, *Amer. Math. Soc., Providence, R. I.*, 1982. **MR 84k: 46045**.   10
[1983]   Dimension and stable rank in the K-theory of C*-algebras, *Proc. London Math.*
          Soc. (3) **46** (1983), no. 2, 301–333. **MR 84g: 46085**.   10

Riesz, Frigyes

[1911]   Sur certains sytèms singuliers d'equations integrales, *Ann. École Norm. Sup.*
          (3) **28** *(1911)*, *33–62*.   3.3
[1918]   Über lineare Funktionalgleichungen, *Acta. Math.* **41** (1918), 71–98.   1.7

Rigelhof, Roger P.

[1969a]   Norm decreasing homomorphisms of measure algebras, *Trans. Amer. Math.*
          Soc. **136** (1969), 361–371. **MR 39 #364**.   1.9
[1969b]   A characterization of $M(G)$, *Trans. Amer. Math. Soc.* **136** (1969), 373–379.
          MR 38 #2536.   1.9

Ringrose, John R.

[1957]   Precompact linear operators in locally convex spaces, *Proc. Cambridge Philos.*
          Soc. **53** (1957), 581–591. **MR 19-869**.   8.8
[1972]   Automatic continuity of derivations of operator algebras, *J. London Math. Soc.*
          (2) **5** (1972), 432–438. **MR 51 #11123**.   6.4

Robertson, Alexander Provan; Robertson, Wendy Jean

[1964]   Topological vector spaces, Cambridge Tracts in Mathematics and Mathematical
          Physics No. 53, *Cambridge University Press, New York*, 1964. **MR 28 #5318**.
                                                                                        1.1
Rodinò, Nicola

[1983]   On the Gel'fand-Mazur theorem. (Italian. English summary) *Atti Accad. Naz.*
          Lincei Rend. Cl. Sci. Fis. Mat. Natur. (8) **72** (1982), no. 1, 1–5 (1983). **MR**
          85i: 46068.   2.2

Rodríguez Macías, Raúl

[1983]   Some characteristics of the algebraic subset of a Banach algebra. (Spanish. En-
          glish summary) *Cienc. Mat. (Havana)* **4** (1983), no. 3, 45–50. **MR 85j: 46082**.
                                                                                        2.3
Rodríguez-Palacios, Ángel

[1979]   Continuity of the Jordan product implies ordinary continuity in the com-
          plete semiprime case. (Spanish) *Contributions in probability and mathemat-*
          ical statistics, teaching of mathematics and analysis (Spanish), pp. 280–288,
          Grindley, Granada, 1979. **MR 83b: 46063**.   6.3
[1985]   The uniqueness of the complete norm topology in complete normed nonasso-
          ciative algebras, *J. Funct. Anal.* **60** (1985), no. 1, 1–15. **MR 86f: 46053**.
                                                                                        2.3
[1987]   A note on Arens regularity, *Quart. J. Math. Oxford Ser.* (2) **38** (1987), no.
          149, 91–93. **MR 88c: 46059**.   1.4
[1990]   Automatic continuity with application to C*-algebras, *Math. Proc. Cambridge*
          Philos. Soc. **107** (1990), no. 2, 345–347. **MR 90k: 46111**.   6.1

Rogozin, B. A.; Sgibnev, M. S.

[1980]   Banach algebras of measures on the line. (Russian) *Sibirsk. Mat. Zh.* **21** (1980),
          no 2, 160–169, 239. **MR 81e: 43003**.   1.9

Roitman, Moche; Sternfield, Yitzak

[1981]   When is a linear functional multiplicative?, *Trans. Amer. Math. Soc.* **267** (1981), *no.* 1, 111–124. **MR 82j: 46061.**   2.4

Rolewicz, Stefan

[1957]   On a certain class of linear metric spaces, *Bull. Acad. Polon. Sci. Cl. III* **5** (1957), 471–473. **MR 19-562.**   2.9

Romo, John G.

[1980a] Spectral synthesis in Banach modules. I, *Tamkang J. Math.* **11** (1980), *no.* 1, 91–109. **MR 82g: 46118.**   1.9

[1980b] Spectral synthesis in Banach modules. II, *Tamkang J. Math.* **11** (1980), *no.* 2, 191–201. **MR 84g: 46072.**   1.9

Rosenberg, Jonathan

[1982a] Homological invariants of extensions of C*-algebras, *Operator algebras and applications, Part 1 (Kingston, Ont.,* 1980), pp. 35–75, *Proc. Sympos. Pure Math.*, 38, *Amer. Math. Soc., Providence, R. I.,* 1982. **MR 85h: 46099.**
                                                                                              9, 10
[1982b] The role of K-theory in noncommutative algebraic topology, *Operator algebras and K-theory (San Francisco, Calif.),* 1981, pp. 155–182, *Contemp. Math.,* 10, *Amer. Math. Soc. Providence, R. I.,* 1982. **MR 84h: 46097.**   9, 10

Rosenberg, Jonathan; Schochet, Claude

[1981]   Comparing functors classifying extensions of C*-algebras, *J. Operator Theory* **5** (1981), *no.* 2, 267–282. **MR 84j: 46101.**   10

Rosenblatt, Joseph Max

[1976a] Invariant means and invariant ideals in $L_\infty(G)$ for a locally compact group *G*, *J. Functional Analysis* **21** (1976), *no.* 1, 31–51. **MR 53 #1163.**   9

[1976b] Invariant means for the bounded measurable functions on a non-discrete locally compact group, *Math. Ann.* **220** (1976), *no.* 3, 219–228. **MR 53 #1164.**   9

Rosenblum, Marvin

[1958]   On a theorem of Fuglede and Putnam, *J. London Math. Soc.* **33** (1958), 376–377. **MR 20 #6037.**   2.4

Rowell, James W.

[1984]   Unilateral Fredholm theory and unilateral spectra, *Proc. Roy. Irish Acad. Sect. A* **84** (1984), *no.* 1, 69–85. **MR 86g: 46076.**   2.1

Rowen, Louis Halle

[1980]   Polynomial identities in ring theory, Pure and Applied Mathematics, 84. *Academic Press [Harcourt Brace Jovanovich, Publishers,], New York–London–Toronto, Ont.,* 1980. 365 pp. **MR 82a: 16021.**   7.1

Royden, Halsey L.

[1963]   Function algebras, *Bull. Amer. Math. Soc.* **65** (1959), 227–247. **MR 26 #6817.**
                                                                                       3.0, 3.5

Rudin, Walter

[1958]   Measure algebras on abelian groups, *Bull. Amer. Math. Soc.* **69** (1963), 281–298. **MR 21 #7404.**   1.9

[1962]   Fourier analysis on groups, Interscience Tracts in Pure and Applied Mathematics No. 12, *Interscience Publishers, J. Wiley, New York–London,* 1962. **MR 27 #2808.**   1.8, 3.6

[1972]   Invariant means on L^∞, Collection of articles honoring the completion by Antoni Zygmund of 50 years of scientific activity, III, *Studia Math.* **44** (1972), 219–227. **MR 46 #4105.**   9

[1973]   Functional analysis. McGraw–Hill Series in Higher Mathematics, *McGraw Hill Book Co., New York–Düsseldorf–Johannesburg,* 1973. 397 pp. **MR 51 #1315.** Second edition **MR 92k: 46001.**   3.3

[1974]   Real and complex analysis. Second edition, McGraw–Hill Series in Higher Mathematics, *McGraw Hill Book Co., New York–Düsseldorf–Johannesburg,* 1974. 452 pp. **MR 49 #8783.**   1.9, 3.3, 5.2, 7.3

Rudol, Krzysztof

[1983]   On spectral mapping theorems, *J. Math. Anal. Appl.* **97** (1983), *no.* 1, 131–139. **MR 85c: 46045.**   3.5

Runde, Volker

[1990]   Derivationen auf kommutativen Banachalgebren, Schriftenreihe des Mathematischen Instituts der Universität Münster, 3, Serie, 1 *Universität Münster, Mathematisches Institut, Münster*, 1990. 75 pp.MR **91d: 46062.**   6.4

[1991]   Automatic continuity of derivations and epimorphisms, *Pacific J. Math.* **147** (1991), 365–374. MR **92k: 46080.**   6.4

[1992]   A functorial approach to weak amenability for commutative Banach algebras, *Glasgow Math. J.* **34** (1992), *no.* 2, 241–251. MR **93d: 46084.**   9

Runge, Carl David Tolmé

[1885]   Zur Theorie der eindeutigen analytischen Functionen, *Acta. Math.* **6** (1885), 245–248.   3.3

Rupp, Rudolf

[1990a]   Stable rank of holomorphic function algebras, *Studia Math.* **97**, 1990, *no.* 2, 85–90. MR **92c: 46058.**   10

[1990b]   Stable rank of finitely generated algebras, *Arch. Math.* (Basel) **55** (1990), *no.* 5, 438–444. MR **92c: 46055.**   10

[1991]   Stable rank and boundary principle, *Topology Appl.* **40**, 1991, *no.* 3, 307–316. MR **92i: 46063.**   10

Russo, Bernard; Dye, Henry A.

[1966]   A note on unitary operators in C*-algebras, *Duke Math. J.* **33** (1966), 413–416. MR **33 #1750.**

Ruston, Anthony F.

[1954]   Operators with a Fredholm theory, *J. London Math. Soc.* **29** (1954), 318–326. MR **15-965.**   1.7

Sakai, Shoichiro

[1956]   A characterization of W*-algebras, *Pacific J. Math.* **6** (1956), 763–773. MR **18-811.**   1.10

[1960]   On a conjecture of Kaplansky, *Tôhoku Math. J.* (2) **12** (1960), 31–33. MR **22 #2913.**   6.4

Sasvári, Zoltán

[1991]   On a refinement of the Gelfand–Raĭkov theorem, *Math. Nachr.* **150** (1991), 185–187. MR **92f: 43008.**   1.9

Schaefer, Helmut H.

[1971]   Topological vector spaces, Third printing corrected. Graduate Texts in Math. Vol. 3. *Springer-Verlag, New York*, 1971. MR **33 #1689.**   1.1

Schanuel, Stephen H.; Zame, William R.

[1982]   Naturality of the functional calculus, *Bull. London Math. Soc.* **14** 1982, *no.* 3, 218–220. MR **83i: 46084.**   3.5

Schatten, Robert

[1943a]   On the direct product of Banach spaces, *Trans. Amer. Math. Soc.* **53** (1943), 195–217. MR **4-161.**   1.10

[1943b]   On reflexive norms for the direct product of Banach spaces, *Trans. Amer. Math. Soc.* **54** (1943), 498–506. MR **5-99.**   1.10

[1946]   The cross-space of linear transformations, *Ann. of Math.* 2 **47** (1946), 73–84. MR **7-455.**   1.10

[1950]   A theory of cross-spaces, *Ann. Math. Studies no.* 26, *Princeton University Press, Princeton, N. J.*, 1950. MR **12-186.**   1.10

[1960]   Norm ideals of completely continuous operators. Ergebnisse der Mathematik und ihrer Grenzgebiete. N. F., Heft 27. *Springer-Verlag, Berlin-Göttingen-Heidelberg*, 1960. MR **22 #9878.**   1.10

Schatten, Robert; von Neumann, John

[1946]   The cross-space of linear transformations. II, *Ann. Math.* (2) **47** (1946), 608–630. MR **8-31.**   1.10

[1948]   The cross-space of transformations. III, *Ann. Math.* (2) **49** (1948), 557–582. MR **10-256.**   1.10

Schatz, Joseph A.

[1953]   Review of "On a Theorem of Gelfand and Naimark and the B*-algebra" by M. Fukamiya, *Math. Reviews* **14** (1953), 884.   2.6

Schauder, Juliusz

[1930]   Der Fixpunkt satz in Funktionraümen, *Studia Math.* **2** (1930), 170–179.   2.8

Schechter, Martin

[1971]   Principles of functional analysis, *Academic Press, New York–London*, 1971. MR **56** #3607.   1.7

Scheffold, Egon

[1988]   Über Banachverbandsalgebren mit multiplikativer Zerlegungseigenschaft, *Acta Math. Hungar.* **52** (1988), no. 3-4, 273–289. MR **90i: 46090**.   3.1

Schmidt, Bernd

[1970]   Spektrum, numerische Wertebereich und ihre Maximumprinzipien Banachalgebren (English summary) *Manuscripta Math.* **2** (1970), 191–202. MR **41** #5971.

2.3

Schmidt, Erhard

[1907]   Auflösung der allgemeinen linearen Gleichungen, *Math. Ann.* **64** (1907), 161–174.   1.8

[1908]   Über die Auflösung linearen Gleichungen mit unendlich vielen Unbekannten, *Rend. Circ. Mat. Palermo* **25** (1908), 53–77.   1.8

Schmitt, Lothar M.

[1991]   Quotients of local Banach algebras are local Banach algebras, *Publ. Res. Inst. Math. Sci.* **27** (1991), no. 6, 837–843. MR **93c: 46133**.   3.3, 10

Schoenberg, Isaac J.

[1934]   A remark on the preceding note by Bochner, *Bull. Am. Math. Soc.* **40**, 277–278.

3.1

Schulz, Jürgen

[1985]   Störungsideale in Banachalgebren—ein Beitrag zur Spektraltheorie, *Wiss. Z. Tech. Hochsch. Karl-Marx-Stadt* **27** (1985), no. 1, 139–144. MR **87e: 46062**.

4.3

Schur, Issai

[1905]   Neue Begründung der Theorie der Gruppencharaktere, Sitzungsbrichte der Preussischen Akademie der Wissenschaften 1905 Physik-Math. Klasse, 406–432 (in Vol I pp. 143–169 Gesammelte Abhandlungen, *Springer, Berlin*, 1973). 4.1

Schweitzer, Larry B.

[1991]   Representations of dense subalgebras of C*-algebra with applications to spectral invariance, doctoral thesis, University of California at Berkeley   3.3

[1992]   A short proof that $M_n(A)$ is local if A is local and Fréchet, *Internat. J. Math.*, **3** (1992), no. 4, 581–589. MR **93i: 46082**.   3.3, 4.2, 10

[1993]   A non-spectral dense Banach subalgebra of the irrational rotation algebra, *Proc. Amer. Math. Soc.*, to appear.   2.5, 3.3

Segal, Irving E.

[1940]   Ring properties of certain classes of functions, Dissertation, Yale, 1940. 1.1, 8.7

[1941]   The group ring of a locally compact group. I, *Proc. Nat. Acad. Sci. U. S. A.* **27** (1941), 348–352. MR **3-36**.   1.1, 4.5, 4.8

[1947a]   Irreducible representations of operator algebras, *Bull. Amer. Math. Soc.* **53** (1947), 73–88. MR **8-520**.   1.1, 2.4, 4.1, 4.2, 4.3, 4.5, 4.8, 5.1, 8.7

[1947b]   Postulates for general quantum mechanics, *Ann. of Math.* (2) **48** (1947), 930–948. MR **9-241**.   1.1

[1947c]   The group algebra of a locally compact group, *Trans. Amer. Math. Soc.* **61** (1947), 69–105. MR **8-438**.   1.9

Šeinberg, Mark Viktovitch

[1971]   Relatively injective modules over Banach algebras, (Russian. English summary) *Vestnik Moskov. Univ. Ser. I Mat. Meh.* **26** (1971), no. 3, 53–58. MR **43** #6729.   9

[1972]   Homological properties of closed ideals that have a bounded approximative unit, (Russian. English summary) *Vestnik Moskov. Univ. Ser. I Mat. Meh.* **27** (1972), no. 4, 39–45. MR **46** #9765.   9

[1973]   The relative homological dimension of group algebras of locally compact groups,
          (Russian) *Izv. Akad. Nauk SSSR Ser. Mat.* **37** (1973), 308–318. **MR 48**
          #6965.   9

Selesnick, Stephen Allan

[1979]   Rank one projective modules over certain Fourier algebras, *Applications of*
          sheaves (Proc. Res. Sympos. Appl. Sheaf Theory to Logic, Algebra and Anal.
          Univ. Durham, Durham, 1977 pp. 702–713, *Lecture notes in Mathematics,*
          753, *Springer, Berlin–New York,* 1979. **MR 83j: 46083.**   7.1

Selivanov, Yurii Vasilyevič

[1978]   Projectivity of certain Banach modules and the structure of Banach algebras,
          (Russian) *Izv. Vyssh. Uchebn. Zaved. Mat.* (1978), *no.* 1(188), 110–116. **MR**
          80g: 46046.   4.7
[1979]   Biprojective Banach algebras, (Russian) *Izv. Akad. Nauk SSSR Ser. Mat.* **43**
          (1979), *no.* 5, 1159–1174, 1198. **MR 81h: 46050.**   8.6
[1992]   Homological characterizations of the approximation property for Banach spaces,
          Glasgow Math. J. **34** (1992), *no.* 2, 229–239.   1.7, 9

Sentilles, F. Dennis; Taylor, Donald C.

[1969]   Factorization in Banach algebras and the general strict topology, *Trans. Amer.*
          Math. Soc. **142** (1969), 141–152. **MR 40 #703.**   5.2

Šilov, Georgi Evgenyevič

[1939]   Ideals and subrings of the ring of continuous functions, *Dokl. Akad. Nauk* SSSR
          22 (1939), 7–10.   1.1
[1940a]   Sur la théorie des idéaux dans les anneaux normés de fonctions, *C. R. (Doklady)*
          Acad. Sci. URSS (N.S.) **27** (1940), 900–903. **MR 2-224.**   1.1, 2.2, 3.2
[1940b]   On the extension of maximal ideals, *C. R. (Doklady) Acad. Sci.* URSS (N.S.)
          29 (1940), 83–84. **MR 2-314.**   1.1, 2.5, 3.2
[1947a]   On regular normed rings, *Trav. Inst. Math. Stekloff* **21** (1947), 118 pp. (Rus-
          sian. English summary). **MR 9-596.**   1.1, 3.0, 3.1, 3.2, 3.4, 4.8, 6.1, 7.1, 7.3
[1947b]   On normed rings possessing one generator, *Rec. Math. [Mat. Sbornik]* N.S. 21
          63 (1947), 25–47. (Russian. English summary). **MR 9-445.**   2.5, 3.4
[1950]   On a theorem of I. M. Gel'fand and its generalizations, *Doklady Akad. Nauk*
          SSSR (N. S.) **72** (1950), 641–644 (Russian). **MR 12-111.**   2.5
[1953]   On decomposition of a commutative normed ring in a direct sums of ideals, *Mat.*
          Sbornik N.S. 32 **74** (1953), 353–364. **MR 14-884, 1278.** also *Amer. Math. Soc.*
          Transl. (2) **1** (1955), 37–48. **MR 17-512.**   1.5, 3.0, 3.5

Sims, Brailey

[1982]   "Ultra"-techniques in Banach space theory. Queen's Papers in Pure and Applied
          Mathematics, 60, *Queen's University, Kingston, Ont.,* 1982. 117 pp. **MR 86h:**
          46032.   1.3

Sinclair, Allan M.

[1969]   Continuous derivations on Banach algebras, *Proc. Amer. Math. Soc.* **20** (1969),
          166–170. **MR 38 #1530.**   6.4
[1970a]   Jordan homomorphisms and derivations on semisimple Banach algebras, *Proc.*
          Amer. Math. Soc. **24** (1970), 209–214. **MR 40 #3310.**   6.3
[1970b]   Jordan automorphisms on a semisimple Banach algebra, *Proc. Amer. Math.*
          Soc. **25** (1970), 526–528. **MR 41 #4241.**   6.3, 6.5
[1971]   The norm of a hermitian element in a Banach algebra, *Proc. Amer. Math. Soc.*
          28 (1971), 446–450. **MR 43 #921.**   2.6
[1974]   Homomorphisms from C*-algebras, *Proc. London Math. Soc.* (3) **29** (1974),
          435–452. **MR 50 #10834.** *cf.* Corrigendum *ibid.* **32** (1976), *no.* 2, 322. **MR**
          52 #11612.   6.2
[1975]   Homomorphisms from $C_0(R)$, *J. London Math. Soc.,* (2) **11** (1975), *no.* 2,
          165–174. **MR 51 #13689.**
[1976]   Automatic continuity of linear operators, *London Mathematical Society Lec-*
          ture Note Series, No. 21, *Cambridge University Press, Cambridge–New York–*
          Melbourne, 1976. 92 pp. **MR 58 #7011.**   4.2, 6.1, 6.2
[1978]   Bounded approximate identities, factorization, and a convolution algebra, *J.*
          Funct. Anal. **29** (1978), *no.* 3, 308–318. **MR 80c: 46054.**   5.0, 5.3
[1979]   Cohen's factorization method using an algebra of analytic functions, *Proc. Lon-*
          don Math. Soc. (3) **39** (1979), *no.* 3, 451–468. **MR 80m: 46048.**   5.0, 5.3

[1982] Continuous semigroups in Banach algebras, *London Mathematical Society Lecture Note Series*, 63, *Cambridge University Press, Cambridge–New York*, 1982. 145 pp. **MR 84b: 46053.** 5.0, 5.1, 5.3

Singer, Ivan

[1970] Bases in Banach spaces. I. Die Grundlehren der mathematischen Wissenschaften, Band 154, *Springer-Verlag, New York, Berlin*, 1979. **MR 45 #7451.** 1.7

[1981] Bases in Banach spaces. II, *Editura Academiei Republicii Socialiste România, Bucharest: Springer-Verlag, Berlin–New York*, 1981. 880 pp. **MR 82k: 46024.** 1.7

Singer, Isadore M.; Wermer, John

[1955] Derivations on commutative normed algebras, *Math. Ann.* **129** (1955), 260–264. **MR 16-1125.** 2.4, 6.4

Širokov, Feliks Vladimirovič

[1956] Proof of a conjecture of Kaplansky, *Uspehi Mat. Nauk* (N.S.) **11** (1956), *no.* 4 (70), 167–168. (Russian). **MR 19-435.** 2.4, 6.4

Słodkowski, Zbigniew

[1977] An infinite family of joint spectra, *Studia Math.* **61** (1977), *no.* 3 239–255. **MR 57 #1157.** 3.5

Słodkowski, Zbigniew; Żelazko, Wiesław

[1974] On joint spectra of commuting families of operators, *Studia Math.* **50** (1974), 127–148. **MR 49 #11280.** 3.5

Smiley, Malcolm F.

[1955] Right annihilator algebras, *Proc. Amer. Math. Soc.* **6** (1955), 698–701. **MR 17-386.** 2.3, 6.1, 8.7, 8.8

[1957] Jordan homomorphisms onto prime rings, *Trans. Amer. Math. Soc.* **84** (1957), 426–429. **MR 18-715.** 6.2

Smith, Roger R.

[1978] An addendum to "*M*-ideal structure in Banach algebras" [J. Funct. Anal. **27** (1978), *no.* 3, 337–349; MR 57 #7175] by Smith and J. D. Ward, *J. Funct. Anal.* **32** (1979), *no.* 3, 269–271. **MR 80j: 46087.** 1.4

[1979] On Banach algebra elements of thin numerical range, *Math. Proc. Cambridge Philos. Soc.* **86** (1979), *no.* 1, 71–83. **MR 80f: 46051.** 2.6

[1981] The numerical range in the second dual of a Banach algebra, *Math. Proc. Cambridge Philos. Soc.* **89** (1981), *no.* 2, 301–307. **MR 82e: 46065.** 2.6

Smith, Roger R.; Ward, Joseph D.

[1979] Applications of convexity and *M*-ideal theory to quotient Banach algebras, *Quart. J. Math. Oxford Ser.* (2) **30** (1979), *no.* 119, 365–384. **MR 80h: 46071.** 1.4, 2.6

Šmulian, Vitold L.

[1940] On multiplicative linear functionals in certain special normed rings, *C. R. (Doklady) Acad. Sci. URSS* (N.S.) **26** (1940), 13–16. **MR 2-222.** 1.1

Smyth, M. R. F.

[1976] Riesz algebras, *Proc. Roy. Irish Acad. Sect. A*, **76** (1976), *no.* 31, 327–333. **MR 55 #6189.** 8.2

[1980] On problems of Olubummo and Alexander, *Proc. Roy. Irish Acad. Sect. A*, **80** (1980), *no.* 1, 69–74. **MR 82a: 46046.** 8.7, 8.8

[1982] Fredholm theory in Banach algebras, *Spectral theory (Warsaw, 1977)*, 403–414, *Banach Center Publ.*, 8, *PWN, Warsaw*, 1982. **MR 85i: 46063.** 8.2, 8.5

Smyth, M. R. F.; West, Trevor

[1975] The spectral radius formula in quotient algebras, *Math. Z.*, **145** (1975), *no.* 2, 157–161. **MR 52 #6429.** 2.2

Sołtysiak, Andrzej

[1978] On Banach algebras with closed set of algebraic elements, *Comment. Math. Prace Mat.* **20** (1977/78), *no.* 2, 479–484. **MR 80c: 46064.** 2.3

[1982]   Algebraic elements in Banach algebras without nonzero quasi-nilpotents, *Funct. Approx. Comment. Math.* **12** (1982), 45–47. **MR 87b: 46050.**   2.3
[1984]   On quasi-algebraic elements in Banach algebra, *Funct. Approx. Comment. Math.* **14** (1984), 15–16. **MR 87b: 46051.**   2.3
[1987]   Approximate point joint spectra and multiplicative functionals, *Studia Math.* **86** (1987), no. 3, 277–286. **MR 89a: 46106.**   3.5
[1988]   Joint spectra and multiplicative functionals, *Colloq. Math.* **56** (1988), no. 2, 357–366. **MR 90f: 46080.**   3.5

Srinivasan, V. K.

[1979]   On some Gel'fand-Mazur like theorems in Banach algebras, *Bull. Austral. Math. Soc.* **20** (1979), no. 2, 211–215. **MR 81e: 46029a.** Corrigenda: *Bull. Austra. Math. Soc.* **23** (1981), no. 3, 479–480. **MR 82h: 46067.**   2.2

Stedman, Lawrence

[1983]   Banach algebras with one-dimensional radical, *Bull. Austral. Math. Soc.* **27** (1983), no. 1, 115–119. **MR 84m: 46053.**   6.1

Steen, Lynn Arthur

[1973]   Highlights in the history of spectral theory, *Amer. Math. Monthly* **80** (1973), 359–381. **MR 47 #5643.**   2.0

Steen, Stourton William Peile

[1936]   An introduction to the theory of operators, *Proc. London Math. Soc.* **41** (1936), 361–392.   1.1
[1937]   An introduction to the theory of operators, *Proc. London Math. Soc.* **43** (1937), 529–543.   1.1
[1938]   An introduction to the theory of operators, *Proc. London Math. Soc.* **44** (1938), 398–411.   1.1
[1939]   Introduction to the theory of operators. IV. Linear functionals, *Proc. Cambridge Philos. Soc.* **35** (1939), 562–578. **MR 1-147.**   1.1
[1940]   Introduction to the theory of operators. V. Metric rings, *Proc. Cambridge Philos. Soc.* **36** (1940), 139–149. **MR 2-104.**   1.1

Stein, James D., Jr.

[1969]   Homomorphisms of B*-algbras, *Pacific J. Math.* **28** (1969), 431–439. **MR 39 #4687.**   6.2

Stone, Marshall Harvey

[1932]   Linear transformations in Hilbert space and their applications to analysis, *Amer. Math. Soc. Colloquium Pub.*, No. 15, New York, (1932).   1.1, 3.3
[1937]   Application of the theory of Boolean rings to general topology, *Trans. Amer. Math. Soc.* **41** (1937), 375–481.   3.2, 7.1
[1953]   On the theorem of Gelfand–Mazur, *Ann. Soc. Polon. Math.* **25** (1952), 238–240 (1953). **MR 15-132.**   2.2

Størmer, Erling

[1980]   Regular abelian Banach algebras of linear maps of operator algebras, *J. Funct. Anal.* **37** (1980), no. 3, 331–373. **MR 81k: 46057.**   4.1

Stout, Edgar Lee

[1971]   The theory of uniform algebras, *Bogden and Quigley, Inc., Tarrytown-on-Hudson, N. Y.*, 1971. **MR 54 #11066.**   3.1, 3.2, 3.3, 3.5

Stromberg, Karl R.

[1981]   Introduction to classical real analysis. Wadsworth International Mathematics Series, *Wadsworth International, Belmont, CA.*, 1981. 575 pp. **MR 82c: 26002.**   2.3

Suliński, Adam

[1958]   Some questions in the general theory of radicals, *Mat. Sb. N.S.* 44 **86** (1958), 273–286. **MR 20 #4581.**   4.7

Suliński, A.; Anderson, Robert F. V.; Divinsky, Nathan

[1966]   Lower radical properties for associative and alternative rings, *J. London Math. Soc.* **41** (1966), 417–424. **MR 33 #4095.**   4.4, 4.8

Swan, Richard G.

[1977]   Topological examples of projective modules, *Trans. Amer. Math. Soc.* **230** (1977), 201–234. **MR 56 #6657.**   2.2

Sylvester, James Joseph

[1884]   Lectures on the principles of universal algebra, *Amer. J. Math.* **6** (1884), 270–286.   1.6

Szankowski, Andrzej

[1978]   Subspaces without the approximation property, *Israel J. Math.* **30** (1978), *no.* 1-2, 123–129. **MR 80b: 46032.**   1.7, 5.1

[1981]   B(H) does not have the approximation property, *Acta Math.* **147** (1981), *no.* 1-2, 89–108. **MR 83a: 46033.**   1.7

Takahasi, Sin-ei

[1981]   On the center of quasicentral Banach algebras with bounded approximate identity, *Canad. J. Math.* **33** (1981), *no.* 1, 68–90. **MR 83a: 46053.**   7.2

[1983a]   Quasicentrality of Banach algebras and approximate identity, *Bull. Fac. Sci. Ibaraki Univ. Ser. A No.* 15 (1983), 17–18. **MR 84i: 46059.**   7.2

[1983b]   Central double centralizers on quasicentral Banach algebras with bounded approximate identity, *Canad. J. Math.* **35** (1983), *no.* 2, 373–384. **MR 85m: 46048.**   7.2

[1984]   Finite dimensionality in socle of Banach algebras, *Internat. J. Math. Math. Sci.* **7** (1984), *no.* 3, 519–522. **MR 86c: 46054.**   8.7

[1984c]   Remarks on quasicentral approximate identities, *Manuscripta Math.* **47** (1984), *no.* 1-3, 229–232. **MR 86a: 46051.**   1.4, 7.2

[1986]   Remarks on Tauberian theorem for quasicentral Banach algebras, *Bull. Fac. Sci. Ibaraki Univ. Ser. A, No.* 18, (1986), 55–56. **MR 87i: 46107.**   7.2

Takahasi, Sin-ei; Hatori, Osamu

[1990]   Commutative Banach algebras which satisfy a Bochner–Schoenberg–Eberlein type theorem, *Proc. Amer. Math. Soc.* **110** (1990), *no.* 1, 149–158. **MR 90m: 46086.**   3.1

[1992]   Commutative Banach algebras and BSE-inequlities, *Math. Japonica* **37** (1992), *no.* 4, 607–614. **MR 93h: 46069.**   3.1

Takesaki, Masamichi

[1969]   A characterization of group algebras as a converse of Tannaka–Stinespring–Tatsuuma duality theorem, *Amer. J. Math.* **91** (1969), 529–564. **MR 39 #5752.**   1.9, 9

[1972]   Duality and von Neumann algebras, *Lectures on operator algebras; Tulane Univ. Ring and Operator Theory Year* 1970-1971, *Vol. II; (dedicated to the memory of David M. Topping)*, pp. 665–786. *Lecture Notes in Math., Vol. 247, Springer, Berlin,* **91** 1972. **MR 53 #704.**   1.9, 9

Taylor, Angus E.

[1938a]   The resolvent of a closed transformation, *Bull. Amer. Math. Soc.* **44** (1938), 70–74.   2.2

[1938b]   Linear operators which depend analytically on a parameter, *Ann. Math.* (2) **39** (1938), 574–593.   2.2, 3.3

Taylor, Donald Curtis

[1968]   A characterization of Banach algebras with approximate unit, *Bull. Amer. Math. Soc.* **74** (1968), 761–766. **MR 37 #2003.**   5.2

[1970]   The strict topology for double centralizer algebras, *Trans. Amer. Math. Soc.* **150** (1970), 633–643. **MR 44 #7302.**   5.1

Taylor, Joseph L.

[1965]   The structure of convolution measure algebras, *Trans. Amer. Math. Soc.* **119** (1965), 150–166. **MR 32 #2932.**   3.6, 9

[1969]   Noncommutative convolution measure algebras, *Pacific J. Math.* **31** (1969), 809–826. **MR 41 #844.**   3.6, 9

[1970a]   A joint spectrum for several commuting operators, *J. Functional Analysis* **6** (1970), 172–191. **MR 42 #3603.**   3.5

774   Bibliography

[1970b]   Several variable spectral theory, *Functional Analysis (Proc. Sympos., Monterey, Calif.,1969)*, 1–10. *Academic Press, New York*, 1970. **MR 42 #6623**.
     3.5

[1970c]   The analytic functional calculus for several commuting operators, *Acta. Math.* **125** (1970), 1–38. **MR 42 #6622**.
     3.5

[1971]   The cohomology of the spectrum of a measure algebras, *Acta. Math.* **126** (1971), 195–225. **MR 53 #14013**.
     3.5, 3.6, 9, 10

[1972a]   Homology and cohomology for topological algebras, *Advances in Math.* **9** (1972), 137–182. **MR 48 #6966**.
     3.5, 9, 10

[1972b]   A general framework for multi-operator functional calculus, *Advances in Math.* **9** (1972), 183–252. **MR 48 #6967**.
     3.5, 9

[1973a]   Measure algebras, Expository lectures from the CBMS Regional Conference held at the University of Montana, Missoula, Mont., June 1972. CBMS Conference Board of the Mathematical Sciences Regional Conference Series in Mathematics, No. 16. *Published for the Conference Board of the Mathematical Sciences by the American Mathematical Society, Providence, R. I.*, 1973. **MR 55 #979**.
     1.10, 3.5, 3.6, 9

[1973b]   Functions of several noncommutative variables, *Bull. Amer. Math. Soc.* **79** (1973), 1–34. **MR 47 3995**.
     3.5

[1975]   Banach algebras and topology, *Algebras in Analysis (Proc. Instructional Conf. and NATO Advanced Study Inst., Birmingham, 1973)*, pp. 118–186. *Academic Press, London*, 1975. **MR 54 #5837**.
     3.5, 9, 10

[1976]   Topological invariants of the maximal ideal space of a Banach algebra, *Advances in Math.* **19** (1976), 149–206. **MR 53 #14134**.
     3.5, 9, 10

[1982]   A bigger Brauer group, *Pacific J. Math.* **103** (1982), *no.* 1, 163–203. **MR 84g: 13007**.
     9

Teleman, Silviu

[1969a]   La representation des anneaux tauberiens discrets par des faisceaux, *Rev. Roumaine Math. Pures Appl.* **14** (1969), 249–264. **MR 40 #737a**.
     7.4

[1969b]   La representation des anneaux réguliers par les faisceaux, *Rev. Roumaine Math. Pures Appl.* **14** (1969), 703–717. **MR 40 #737b**.
     7.4

[1969c]   Representation par faisceaux des modules sur les anneaux harmoniques, *C. R. Acad. Sc. Paris Sér. A-B* **269** (1969), A753–A756. **MR 41 #276**.
     7.4

[1971]   Theory of harmonic algebras with applications to von Neumann algebras and cohomology of locally compact spaces (de Rham's theorem), *Lectures on the applications of sheaves to ring theory*, 99–315, *Lecture Notes in Mathematics*, 248, *Springer, Berlin*, 1971. **MR 50 #13165**.
     7.1, 7.4

Tewari, U. B.; Dutta, M.; Madan, Shobha

[1982]   Tensor products of commutative Banach algebras, *Internat. J. Math. Math. Sci.* **5** (1982), *no.* 3, 503–512. **MR 84e: 46082**.
     1.10

Thomas, Marc P.

[1978]   Algebra homomorphisms and the functional calculus, *Pacific J. Math.* **79** (1978), *no.* 1, 251–269. **MR 81g: 46063**.
     3.4

[1983]   A nonstandard ideal of a radical Banach algebra of power series, *Bull. Amer. Math. Soc. (N.S)* **9** (1983), *no.* 3, 331–333. **MR 85i: 46070a**.
     3.4

[1984a]   A nonstandard closed subalgebra of a radical Banach algebra of power series, *J. London Math. Soc.* (2) **29** (1984), *no.* 1, 153–163. **MR 85i: 46069**.
     3.4

[1984b]   A nonstandard ideal of a radical Banach algebra of power series, *Acta Math.* **152** (1984), *no.* 3-4, 199–217. **MR 85i: 46070b**.
     3.4

[1988]   The image of a derivation is contained in the radical, *Ann. of Math.* **128** (2) (1988), *no.* 3, 435–460. **MR 90d: 46075**.
     6.4

[1991]   Principal ideals and semi-direct products in commutative Banach algebras, *J. Funct. Anal.* **101** (1991), *no.* 2, 312–328. **MR 92h: 46078**.
     6.4

Titchmarsh, Edward Charles

[1926]   The zeros of certain integral functions, *Proc. London Math. Soc.* (2) **25** (1926), 283–302.
     3.4, 4.8, 5.1

Tits, Jacques

[1972]   Free subgroups in linear groups, *J. Algebra* **20** (1972), 250–270. **MR 44 #4105**.
     9

Toeplitz, Otto

[1918]   Pas algebraische Analogen zu einem Satze von Fejer, *Math. Zeit.* **2** (1918), 187–197.   2.6

Tomczak-Jaegermann, Nicole

[1989]   Banach–Mazur distances and finite-dimensional operator ideals. Pitman Monographs and Surveys in Pure and Applied Mathematics, 38, *Longman Scientific and Technical, Harlow; copublished in the United States with John Wiley and Sons, Inc. New York*, 1989. 395 pp. **MR 90k: 46039.**   1.7

Tomiuk, Bohdan J.

[1979]   On some properties of Segal algebras and their multipliers, *Manuscripta Math.* **27** (1979), *no.* 1, 1–18. **MR 80h: 46072.**   8.7

[1982]   Arens regularity of Banach algebras which are conjugate spaces, *Bull. Inst. Math. Acad. Sinica* **10** (1982), *no.* 2, 155–163. **MR 83m: 46074.**   1.4

[1983]   Arens regularity of conjugate Banach algebras with dense socle, *Rocky Mountain J. Math.* **13** (1983), *no.* 1, 117–124. **MR 84d: 46071.** (Correction: *Proc. Amer. Math. Soc.* **91** (1984), *no.* 1, 171. **MR 85b: 46058.**)   1.4

[1986]   Isomorphisms of multiplier algebras, *Glasgow Math. J.* **28** (1986), *no.* 1, 73–77. (Introduction). **MR 88a: 46051.**   8.7

[1987]   The strong radical and the left regular representation, *J. Austral. Math. Soc. Ser. A* **43** (1987), *no.* 1, 1–9. **MR 88e: 46041.**   4.5

[1988a]   Dual complementors on Banach algebras, *Proc. Amer. Math. Soc.* **103** (1988), *no.* 3, 815–822. **MR 89h: 46073.**   8.7

[1988b]   Biduals of Banach algebras which are ideals in a Banach algebra, *Acta Math. Hungar.* **52** (1988), *no.* 3-4, 255–263. **MR 90b: 46101.**   1.4, 8.7

Tomiuk, Bohdan J.; Yood, Bertram

[1978]   Topological algebras with dense socle, *J. Funct. Anal.* **28** (1978), *no.* 2, 254–277. **MR 80f: 46052.**   8.7

[1989]   Incomplete normed algebra norms on Banach algebras, *Studia Math.* **95** (1989), *no.* 2, 119–132. **MR 91e: 46063.**   2.3, 6.1

Tomiyama, Jun

[1960]   Tensor products of commutative Banach algebras, *Tôhoku Math. J.* (2) **12** (1960), 147–154. **MR 22 #5910.**   1.10, 3.2

[1972]   Primitive ideals in tensor products of Banach algebras, *Math. Scand.* **30** (1972), 257–262. **MR 48 #12079.**   1.10, 4.3

Tonge, Andrew

[1980]   La presque-périodicité et les coalgèbres injectives, (English summary), *Studia Math.* **67** (1980), *no.* 2, 103–118. **MR 81j: 46075.**   1.4, 1.10, 4.2

Tornheim, Leonard

[1952]   Normed fields over the real and complex fields, *Michigan Math. J.* **1** (1952), 61–68. **MR 14-131.**   2.2

Torrance, Ellen

[1970]   Maximal C*-subalgebras of a Banach algebra, *Proc. Amer. Math. Soc.* **25** (1970), 622–624. **MR 41 #4265.**   2.6

Tremon, Michel

[1985]   Polynômes de degré minimum connectant deux projections dans une algèbre de Banach, *Linear Algebra Appl.* **64** (1985), 115–132. **MR 86g: 46074.**   10

Treves, Francois

[1967]   Topological vector spaces, distributions and kernels, *Academic Press, New York–London*, 1967. **MR 37 #726.**   5.3

Tripp, John C.

[1983]   Automatic continuity of homomorphisms into Banach algebras. *Radical Banach algebras and automatic continuity (Long Beach, Calif., 1981)*, 334–339, *Lecture Notes in Mathematics*, 975, *Springer, Berlin–New York*, 1983. **MR 85c: 46043a.**   6.1

Tsertos, Yiannis

[1986] On the circle-exponent function, *Bull. Soc. Math. Grèce (N. S.)* **27** (1986), 137–147. **MR 89e: 46052.**   2.4

Turovskiĭ, Yu. V.

[1984a] The property of mapping the Harte spectrum by polynomials for *n*-commutative families of elements of a Banach algebra, (Russian), *Spectral theory of operators and its applications, No.* 5, 152–177, *"Ĕlm", Baku,* 1984. **MR 86k: 46076.**   3.5

[1984b] Spectral properties of elements of normed algebras, and invariant subspaces, (Russian) *Funktsional. Anal. i Prilozhen.* **18** (1984), *no.* 2, 77–78. **MR 85f: 46092.**   3.5

[1985] Spectral properties of certain Lie subalgebras and the spectral radius of subsets of a Banach algebra, (Russian) *Spectral theory of operators and its applications, No.* 6, 144–181, *"Ĕlm", Baku,* 1985. **MR 87k: 46102.**   2.4

Ülger, Ali

[1987] Weakly compact bilinear forms and Arens regularity, *Proc. Amer. Math. Soc.* **101** (1987), *no.* 4, 697–704. **MR 88k: 46056.**   1.4

[1988a] Arens regularity of the algebra $A\hat{\otimes}B$, *Trans. Amer. Math. Soc.* **305** (1988), *no.* 2, 623–639. **MR 89c: 46064.**   1.4, 1.10

[1988b] Arens regularity of $K(X)$, *Monatsh. Math.* **105** (1988), *no.* 4, 313–318. **MR 90b: 47081.**   1.4, 1.7

[1990a] Arens regularity sometimes implies the RNP, *Pacific J. Math.* **143** (1990), *no.* 2, 377–399. **MR 91f: 46067.**   1.4

[1990b] Multiplicativity factors for seminorms, *J. Math. Anal. Appl.* **146** (1990), *no.* 2 469–481. **MR 91f: 46066.**   1.1

[1990c] Arens regularity of the algebra $C(K, A)$, *J. London Math. Soc.* (2) **42** (1990), 354–364. **MR 92d: 46127.**   1.4

Vaĭnerman, Leonid I.; Kac, Grigorii Isaakovič

[1974] Nonunimodular ring groups and Hopf–von Neumann algebras, (Russian), *Mat. Sb. (N. S.)* **94(136)** (1974), 194–225, 335. **MR 50 #536.**   1.9

Vaĭs, I.

[1988] Algebras that are generated by two idempotents, (Russian), *Seminar Analysis (Berlin, 1987/1988),* 139–145, *Akademie-Verlag, Berlin,* 1988. **MR 90h: 46083.**   10

Vala, Klaus

[1964] On compact sets of compact operators, *Ann. Acad. Sci. Fenn. Ser. A I No.* 351 (1964). **MR 29 #6333.**   8.7, 8.8

[1967] Sur les éléments compacts d'une algèbre normée, *Ann. Acad. Sci. Fenn. Ser. A I No.* 407 (1967). **MR 36 #5692.**   8.7

Van Daele, Alfons

[1988a] K-theory for graded Banach algebras. I, *Quart. J. Math. Oxford Ser.* (2) **39** (1988), *no.* 154, 185–199. **MR 89g: 46113.**   10

[1988b] K-theory for graded Banach algebras. II, *Pacific J. Math.* **134** (1988), *no.* 2, 377–392. **MR 89k: 46089.**   10

[1988c] A note on the K-group of a graded Banach algebra, *Bull. Soc. Math. Belg.* Sér B **40** (1988), *no.* 3, 353–359. **MR 90d: 46102.**   10

Varopoulos, Nicholas Th.

[1964] Continuité des formes linéaires positives sur une algèbre de Banach avec involution, *C. R. Acad. Sci.* Paris 258 (1964), 1121–1124. **MR 28 #4387.**   5.2

[1965] Sur les ensembles parfaits et les séries trigonométriques, *C. R. Acad. Sci. Paris* 260 (1965), 4668–4670, 5165–5168, 5997–6000. **MR 31 #2567.**   1.10

[1967] Tensor algebras and harmonic analysis, *Acta math.* **119** (1967), 51–112. **MR 39 #1911.**   1.10

Vasilescu, Florian Horia

[1982] Analytic functional calculus and spectral decompositions. Translated from the Romanian. Mathematics and its Applications (East European Series), 1, *D. Reidel Publishing Co., Dordrecht-Boston, Mass.; Editura Academiei Republicii Socialiste România, Bucharest,* 1982. 378 pp. **MR 85b: 47016.**   3.3, 3.5

Vasudevan, R.; Goel, Satya

[1984]   Embedding of quasimultipliers of a Banach algebra into its second dual, *Math. Proc. Cambridge Philos. Soc.* **95** (1984), *no.* 3, 457–466. **MR 85m: 46045.**
1.4

[1985]   Quasimultipliers and normed full direct-sum of Banach algebras, *Ann. Soc. Sci. Bruxelles Sér. I* **9** (1985), no 2-3, 85–95. **MR 88k: 46054.**
1.4

Vesentini, Edoardo

[1968]   On the subharmonicity of the spectral radius (Italian summary) *Boll. Un. Mat. Ital. A* (4) **1** (1968), 427–429. **MR 39: #6080.**
2.3

[1970]   Maximum theorems for spectra *Essays on Topology and Related Topics* (*Mémoires dédiée à Georges de Rham*), pp. 111–117, Springer, New York, 1970. **MR 42 #6612.**
2.3

[1983]   Carathéodory distances and Banach algebras, *Adv. in Math.* **47** (1983), *no.* 1, 50–73. **MR 85h: 32041.**
2.3

[1984a]   Subharmonicity in Banach algebras, International conference on analytical methods in number theory and analysis (Moscow, 1981), *Trudy Mat. Inst. Steklov* **163** (1984), 28–31. **MR 86m: 46047.**
2.3

[1984b]   Some recent advances in the spectral theory of locally convex algebras, *Topology* (*Leningrad, 1982*), 367–377, *Lecture Notes in Math.*, 1060, *Springer, Berlin–New York*, 1984. **MR 86m: 46048.**
2.3

Vidav, Ivan

[1955]   Über eine Vermutung von Kaplansky, *Math. Z.* **62** (1955), 330. **MR 16-1125.**
2.4, 6.4

[1956]   Eine metrische Kennzeichnung der selbstadjungierten Operatoren, *Math. Z.* **66** (1956), 121–128. **MR 18-912.**
2.6

Vrbová, Paula

[1981]   A remark concerning commutativity modulo radical in Banach algebras, *Comment. Math. Univ. Carolin.* **22** (1981), *no.* 1, 145–148. **MR 82g: 46088.** 2.4

Vũ Quóc Phóng

[1992]   A short proof of the Y. Katznelson's and L. Tzafriri's theorem, *Proc. Amer. Math. Soc.* **115** (1992), *no.* 4, 1023–1024. **MR 92j: 47012.**
2.6

Vukman, J.

[1988]   Some remarks on derivations in Banach algebras and related results, *Aequationes Math.* **36** (1988), *no.* 2-3, 165–175. **MR 90a: 46120.**
6.4

Waelbroeck, Lucien

[1954]   Le calcul symbolique dans les algèbres commutatives, *J. Math. Pures Appl.* (9) **33** (1954), 147–186. **MR 17-513.**
3.0, 3.5

[1982a]   The holomorphic functional calculus as an operational calculus, *Spectral theory* (*Warsaw, 1977*), 513–552, *Banach Center Publ.*, 8, PWN, Warsaw, 1982. **MR 85g: 46061.**
3.5

[1982b]   The Taylor spectrum and quotient Banach spaces, *Spectral theory* (*Warsaw, 1977*), 573–578, *Banach Center Publ.*, 8, PWN, Warsaw, 1982. **MR 85m: 46072c.**
3.5

[1983a]   Holomorphic functional calculus and quotient Banach algebras, *Studia Math.* **75** (1983), *no.* 3, 273–286. **MR 85j: 46088a.**
3.5

[1983b]   Quasi-Banach algebras, ideals, and holomorphic functional calculus, *Studia Math.* **75** (1983), *no.* 3, 287–292. **MR 85j: 46088b.**
3.5

Walter, Martin E.

[1974]   A duality between locally compact groups and certain Banach algebras, *J. Functional Analysis* **17** (1974), 131–160. **MR 50 #14067.**
9

[1986]   On the norm of a Schur product, *Linear Algebra Appl.* **79** (1986), 209–213. **MR 87j: 15048.**
1.9

[1989]   On a new method for defining the norm of Fourier–Stieltjes algebras, *Pacific J. Math.* **137** (1989), *no.* 1, 209–223. **MR 90e: 43004.**
1.9

Wang, Ju Kwei

[1961]   Multipliers of commutative Banach algebras, *Pacific J. Math.* **11** (1961), 1131–1149. **MR 25 #1462.**
1.2, 3.1

778   Bibliography

Ward, Josephine A.

[1983]   Closed ideals of homogeneous algebras, *Monatsh. Math.* **96** (1983), *no.* 4, 317–324. **MR 85i: 43005.**   1.9

Warner, Seth

[1979]   A new approach to Gel'fand–Mazur theory and the extension theorem, *Michigan Math. J.* **26** (1979), *no.* 1, 13–17. **MR 81b: 16030.**   2.2

Watanabe, Seiji

[1974]   A Banach algebra which is an ideal in the second dual space, *Sci. Rep. Niigata Univ. Ser. A No. 11* (1974), 95–101. **MR 52 #3960.**   1.4

Wedderburn, Joseph H. Maclagan

[1907]   On hypercomplex numbers, *Proc. London Math. Soc.* (2) **6** (1907), 77–118.
          1.1, 4.3, 4.4, 8.1

Weil, André

[1940]   L'intégration dans les groupes topologiques et ses applications, *Actual. Sci. Ind.*, *no.* 869, *Hermann et Cie, Paris,* (1940) **MR 3-198.** 2nd ed. *Actual. Sci. Ind. no.* 1145, *Hermann, Paris,* 1951.   1.1, 1.9, 5.1

Wendel, James G.

[1952]   Left centralizers and isomorphisms of group algebras, *Pacific J. Math.* **2** (1952), 251–261. **MR 14-246.**   1.2, 1.9

Wermer, John

[1976]   Banach algebras and several complex variables, 2nd ed. Graduate Texts in Mathematics, No. 35, *Springer-Verlag, New York–Heidelberg,* 1976. **MR 52 #15021.**   3.2, 3.3, 3.5

West, Trevor T.

[1966]   The decomposition of Riesz operators, *Proc. London Math. Soc.* (3) **16** (1966), 737–752. **MR 33 #6417.**   1.7

Weston, Jeffrey D.

[1960]   Positive perfect operators, *Proc. London Math. Soc.* (3) **10** (1960), 545–565. **MR 22 #9822.**   1.2

Weyl, Hermann; Peter, F.

[1927]   Die Vollständigkeit der primitiven Darstellungen einer gescholssenen Kontinuierlichen Gruppe, *Math. Ann.* **97** (1927), 737–755.   5.1

Whitley, Robert J.

[1967]   An elementary proof of the Eberlein–Šmulian theorem, *Math. Ann.* **172** (1967), 116–118. **MR 35 #3419.**   1.7

Wichmann, Josef

[1973]   Bounded approximate units and bounded approximate identities, *Proc. Amer. Math. Soc.* **41** (1973), 547–550. **MR 48 #2767.**   5.1

Wiener, Norbert

[1923]   Notes on a paper of M. Banach, *Fund. Math.* **4** (1923), 136–143.   3.3
[1932]   Tauberian theorems, *Annals of Math.* **33** (1932), 1–100.   1.1, 1.8, 3.6

Willcox, Alfred B.

[1956a]   Some structure theorems for a class of Banach algebras, *Pacific J. Math.* **6** (1956), 177–192. **MR 18-53.**   7.1, 7.2, 7.3
[1956b]   Note on certain group algebras, *Proc. Amer. Math. Soc.* **7** (1956), 874–879. **MR 19-46.**   7.4

Wille, Regina

[1985]   The theorem of Gleason–Kahane–Żelazko in a commutative symmetric Banach algebra, *Math. Z.* **190** (1985), *no.* 2, 301–304. **MR 86j: 46060.**   2.4

Williams, James P.

[1967]   Spectra of products and numerical ranges, *J. Math. Anal. Appl.* **17** (1967), 214–220. **MR 34 #3341.**   2.6

Willis, George A.

[1982]   Approximate units in finite codimensional ideals of group algebras, *J. London Math. Soc.* (2) **26** (1982), *no.* 1, 143–154. **MR 84f: 43006.**   5.2

[1983a]   The continuity of derivations from group algebras and factorization in cofinite ideals, *Radical Banach algebras and automatic continuity (Long Beach, Calif.,* 1981), 408–421, *Lecture Notes in Mathematics,* 975, *Springer, Berlin–New York,* 1983. **MR 85g: 43002.**   6.4

[1983b]   Factorization in codimension two ideals of group algebras, *Proc. Amer. Soc.* **89** (1983), *no.* 1, 95–100. **MR 85c: 43005.**   5.2

[1986]   The continuity of derivations and module homomorphisms, *J. Austral. Math. Soc. Ser.* (*A*) **40** (1986), *no.* 3, 299–320. **MR 87j: 46093.**   6.4

[1989a]   Factorization in group algebras, *Conference on Automatic Continuity and Banach Algebras (Canberra.,* 1989), 334–344, *Proc. Centre Math. Anal. Austral. Nat. Univ.,* 21, *Austral. Nat. Univ., Canberra,* 1989. **MR 91c: 43006.**   6.1

[1989b]   Ultraprime group algebras, *Conference on Automatic Continuity and Banach Algebras (Canberra.,* 1989), 345–349, *Proc. Centre Math. Anal. Austral. Nat. Univ.,* 21, *Austral. Nat. Univ., Canberra,* 1989. **MR 90m: 43011.**   4.1

[1992a]   Homological properties of compact operators on a Banach space, (to appear). 9

[1992b]   Examples of factorization without bounded approximate units, *Proc. London Math. Soc.* (3) **64** (1992), *no.* 3, 602–624. **MR 93c: 46102.**   5.2

[1992c]   The continuity of derivations from group algebras: factorizable and connected groups, *J. Austral. Math. Soc. Ser.* (*A*) **52** (1992), *no.* 2, 185–204. **MR 93b: 46096.**   6.4

[1992d]   The compact approximation does not imply the approximation property, *Studia Math.* **103** (1992), *no.* 1, 99–108. **MR 93i: 46035.**   1.7

[1993]   Iterated factorization of operators on the Banach space *E* and continuity of Homomorphisms from *B*(*E*), (to appear).   6.2, 6.4

Wojtaszczyk, Przemysław

[1991]   Banach spaces for analysts, *Cambridge Studies in Advanced Mathematics,* 25, *Cambridge University Press, Cambridge,* 1991. xiii, 382 pp.   1.7

Wong, James C. S.

[1973]   An ergodic property of locally compact amenable semigroups, *Pacific J. Math.* **48** (1973), 615–619. **MR 52 #14845.**   9

Wong, Pak Ken

[1978]   On certain quasi-complemented and complemented Banach algebras, *Internat. J. Math. Math. Sci.* **1** (1978), *no.* 3, 307–317. **MR 80d: 46092.**   8.7

[1985]   Arens product and the algebra of double multipliers, *Proc. Amer. Math. Soc.* **94** (1985), *no.* 3, 441–444. **MR 86g: 46075.**   1.4

[1988]   A note on annihilator Banach algebras, *Bull. Austral. Math. Soc.* **38** (1988), *no.* 1, 77–81. **MR 90a: 46121.**   8.7

[1990]   On dual Banach algebras, *Proc. Amer. Math. Soc.* **108** (1990), *no.* 4, 899–904. **MR 90g: 46070.**   8.7

Wood, Geoffrey V.

[1983]   Isomorphisms of group algebras, *Bull. London Math. Soc.* **15** (1983), *no.* 3, 247–252. **MR 85b: 43002.**   1.9

[1991]   Small isomorphisms between group algebras, *Glasgow Math. J.* **33** (1991), *no.* 1, 21–28. **MR 92b: 43007.**   1.9

Wright, Fred B.

[1954]   A reduction for algebras of finite type, *Ann. of Math.* (2) **60** (1954), 560–570. **MR 16-375.**   7.4

Yakovlev, Nikolai Valeryevič

[1989]   Examples of Banach algebras with a radical that is noncomplementable as a Banach space, *Uspekhi Mat. Nauk.* **44** (1989), *no.* 5(269), 185–186; *translation in Russian Math. Surveys* **44** (1989), *no.* 5, 224–225. **MR 91h: 46089.**   8.1

Yang, Qing De

[1986]   The existence of the socle of a semiradial Banach algebra. (Chinese), *J. Math. Res. Exposition* **6** (1986), *no.* 3, 23–26. **MR 89h: 46074.**   8.2

780   Bibliography

Ylinen, Kari

[1968] Compact and finite-dimensional elements of normed algebras, *Ann. Acad. Sci. Fenn. Ser. A I No.* 428 (1968). **MR 38 #6365**.   8.7

Yood, Bertram

[1949] Additive groups and linear manifolds of transformations between Banach spaces, *Amer. J. Math.* **71** (1949), 663–677. **MR 11-114**.   4.2

[1958] Homomorphisms on normed algebras, *Pacific J. Math.* **8** (1958), 373–381. **MR 21 #2924**.   1.7, 2.0, 2.2, 2.5, 6.1, 8.0, 8.4, 8.6

[1963] On the extension of modular maximal ideals, *Proc. Amer. Math. Soc.* **14** (1963), 615–620. **MR 27 #1848**.   3.2

[1964] Ideals in topological rings, *Canad. J. Math.* **16** (1964), 28–45. **MR 28 #1505**.   2.4, 8.0, 8.4

[1972] Closed prime ideals in topological rings, *Proc. London Math. Math. Soc.* (3) **24** (1972), 307–323. **MR 45 #7475**.   4.4

[1979] On the strong radical of certain Banach algebras, *Proc. Edinburgh Math. Soc.* (2) **21** (1978/79), no. 1, 81–85. **MR 58 #12374**.   4.5

[1982] One-sided ideals in Banach algebras, *J. Nigerian Math. Soc.* **1** (1982), 25–30. **MR 84c: 46049**.   8.7

[1984a] Continuous homomorphisms and derivations on Banach algebras, *Proceedings of the conference on Banach algebras and several complex variables (New Haven, Conn., 1983)*, 279–284, *Contemp. Math.*, 32, Amer. Math. Soc. Providence, R. I., 1984. **MR 86b: 46074**.   6.4

[1984b] Structure spaces of rings and Banach algebras, *Michigan Math. J.* **31** (1984), no. 2, 181–189. **MR 85m: 46044**.   8.4

[1990] Commutativity theorems for Banach algebras, *Mich. Math. J.* **37** (1990), no. 2, 203–210. **MR 91d: 46055**.   2.4

Yosida, Kôsaku

[1936a] On the group embedded in the metrical complete ring, *Jap. J. Math.* **13** (1936), 7–26.   1.1, 1.7

[1936b] On the group embedded in the metrical complete ring, *Jap. J. Math.* **13** (1936), 459–472.   1.1

[1938] Mean ergodic theorem in Banach space, *Imp. Acad. Tokyo* **14** (1938), 292–294.   1.7

[1965] Functional analysis, Die Grundlehren der Mathematischen Wissenschaften Band **123**, *Academic Press, Inc., New York*, 1965. **MR 31 #5054**.   1.7, 1.9

Young, Nicholas J.

[1973a] The irregularity of multiplication in group algebras, *Quart. J. Math. Oxford Ser.* (2) **24** (1973), 59–62. **MR 47 #9290**.   1.4

[1973b] Semigroup algebras having regular multiplication, *Studia Math.* **47** (1973), 191–196. **MR 48 #9260**.   1.4, 1.9

[1976] Periodicity of functionals and representations of normed algebras on reflexive spaces, *Proc. Edinburgh Math. Soc.* (2) **20** (1976/77), no. 2, 99–120. **MR 55 #8800**.   1.4, 1.7, 4.2

[1980] Norm and spectral radius for algebraic elements of a Banach algebra, *Math. Proc. Cambridge Philos. Soc.* **88** (1980), no. 1, 129–133. **MR 82d: 46071**. 2.3

Yuan, Chuan Kuan

[1988] Inner invariant means and the regular conjugation representation of $L^1(G)$, *Harmonic analysis (Luxembourg, 1987)*, 283–287, *Lecture Notes in Math.*, 1359, Springer, Berlin–New York, 1988. **MR 90g: 43001**.   9

Zalduendo, Ignacio

[1989] A geometric condition equivalent to commutativity in Banach algebras, *Studia Math.* **94** (1989), no. 2, 187–192. **MR 91a: 46055**.   2.1, 2.4

Zame, William R.

[1979] Existence, uniqueness and continuity of functional calculus homomorphisms, *Proc. London Math. Soc.* (3) **39** (1979), no. 1, 73–92. **MR 81i: 46065**.   3.5

[1984] Covering spaces and the Galois theory of commutative Banach algebras, *J. Funct. Anal.* **55** (1984), no. 2, 151–175. **MR 86j: 46051**.   3.4

Zariski, Oscar; Samuel, Pierre

[1958] Commutative algebra, I and II, The University Series in Higher Mathematics, *D. Van Nostrand, Princeton, N. J.*, 1958 and 1960. **MR 19-833** and **MR 22 #11006.** 7.3

Żelazko, Wiesław

[1960] On the locally bounded and *m*-convex topological algebras, *Studia Math.* **19** (1960), 333–356. **MR 23 #A4033.** 2.9

[1965] Metric generalizations of Banach algebras, *Rozprawy Matematyczne* **47** (1965), 70 pp. **MR 33 #1752** 2.9

[1967] On generalized topological divisors of zero in real *m*-convex algebras, *Studia Math.* **28** (1966/67), 241–244. **MR 35 #7128.** 2.5

[1968] A characterization of mutliplicative linear functionals in complex Banach algebras, *Studia Math.* **30** (1968), 83–85. **MR 37 #4620.** 2.4, 6.3

[1973a] On multiplicative linear functionals, *Colloq. Math.* **28** (1973), 251–253. **MR 48 #6939.** 2.4

[1973b] Banach algebras, Translated from the Polish by Marcin E. Kuczuma. *Elsevier Publishing Co., Amsterdam–London–New York; PWN-Polish Scientific Publishers, Warsaw*, 1973. 182 pp. **MR 56 #6389.** 2.2, 2.4

[1979] An axiomatic approach to joint spectra. I, *Studia Math.* **64** (1979), no. 3, 249–261. **MR 80h: 46076.** 3.5

[1984] On ideal theory in Banach and topological algebras. Monografías del Instituto de Matemáticas, 15, *Universidad Nacional Autónoma de México, Mexico City*, 1984. 152 pp. **MR 86h: 46086.** 2.5

[1986] Topological divisors of zero, their applications and generalization, *Geometry seminars, 1985 (Italian) (Bologna, 1985)*, 175–191, *Univ. Stud. Bologna, Bologna*, 1986. **MR 88d: 46089.** 2.5

[1987] On generalized topological divisors of zero, *Studia Math.* **85** (1987), no. 2, 137–148. **MR 88f: 46096.** 2.5

[1991] On a problem of Fell and Doran, *Colloq. Math.* **62** (1991), no. 1, 31–37. **MR 92k: 46093.** 4.2

Zeller-Meier, Georges

[1967] Sur les automorphismes des algèbres de Banach, *C. R. Acad. Sci. Paris Ser. A-B* **264** (1967), A1131–A1132. **MR 35 #3443.** 6.3, 6.4

Zemánek, Jaroslav

[1977a] A survey of recent results on the spectral radius in Banach algebras, *General topology and its relations to modern analysis and algebra, IV (Proc. Fourth Prague Topological Sympos., Prague, 1976), Part B*, pp. 531–540. *Soc. Czechoslovak Mathematicians and Physicists, Prague*, 1977. **MR 58 #7088.** 2.4

[1977b] Spectral radius characterizations of commutativity in Banach algebras, *Studia Math.* **61** (1977), no. 3, 257–268. **MR 57 #1124.** 2.4

[1978] A simple proof of the Weierstrass–Stone theorem, *Comment. Math. Prace Mat.* **20** (1977/78), no. 2, 495–497 (loose errata). **MR 80c: 46062.** 1.5, 2.3

[1979] Idempotents in Banach algebras, *Bull. London Math. Soc.* **11** (1979), no. 2, 177–183. **MR 80h: 46073.** 10

[1980] Spectral characterization of two-sided ideals in Banach algebras, (Russian), *Studia Math.* **67** (1980), no. 1, 1–12. **MR 81h: 46051.** 2.4

[1982] Properties of the spectral radius in Banach algebras, *Spectral theory (Warsaw, 1977)*, 579–595, *Banach Center Publ.*, 8, *PWN, Warsaw*, 1982. **MR 85d: 46064.** 2.4, 3.0, 4.2

[1983] The essential spectral radius and the Riesz part of the spectrum, *Functions, series, operators, Vol. I, II (Budapest, 1980)*, 1275–1289, *Colloq. Math. Soc. János Bolyai, 35, North-Holland, Amsterdam–New York*, 1983. **MR 86b: 47003.** 2.8

INDEX

SYMBOL INDEX

791

Printed in the United States
By Bookmasters